EIGHTH EDITION, ENHANCED

Nutrition Now

Judith E. Brown

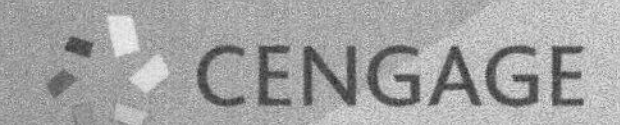

Australia • Brazil • Mexico • Singapore • United Kingdom • United States

***Nutrition Now,* Eighth Edition, Enhanced**

Judith E. Brown

Product Manager: Courtney Heilman

Production Manager: Julia White

Learning Designer: Miriam Myers

Content Manager: Oden Connolly

Art Director: Helen Bruno

Production Service: Lori Hazzard, MPS Limited

Marketing Manager: Julie Nusser

Photo Researcher: Lumina Datamatics Ltd.

Text Researcher: Lumina Datamatics Ltd.

Text Designer: Andrei Pasternak

Cover Designer: Michael Cook

Cover Image: Getty/Andy Roberts

Compositor: MPS Limited

Library of Congress Control Number: 2018947211

Student Edition:
ISBN: 978-0-357-02165-1

Loose-leaf Edition:
ISBN: 978-0-357-02166-8

Cengage
20 Channel Center Street
Boston, MA 02210
USA

Printed at CLDPC, USA, 04-19

About the Author

JUDITH E. BROWN is Professor Emerita of Nutrition at the School of Public Health, and of the Department of Obstetrics and Gynecology at the University of Minnesota. She received her Ph.D. in human nutrition from Florida State University and her M.P.H. in public health nutrition from the University of Michigan. Dr. Brown has received competitively funded research grants from the National Institutes of Health, the Centers for Disease Control and Prevention, and the Maternal and Child Health Bureau and has over 100 publications in the scientific literature including the *New England Journal of Medicine*, the *Journal of the American Medical Association*, and the *Journal of the American Dietetic Association*. A recipient of the Agnes Higgins Award in Maternal Nutrition from the March of Dimes, Dr. Brown is a registered dietitian and the successful author of *Everywoman's Guide to Nutrition*, *Nutrition for Your Pregnancy*, and *What to Eat Before, During, and After Pregnancy*.

Courtesy of the author

Contents in Brief

Contents

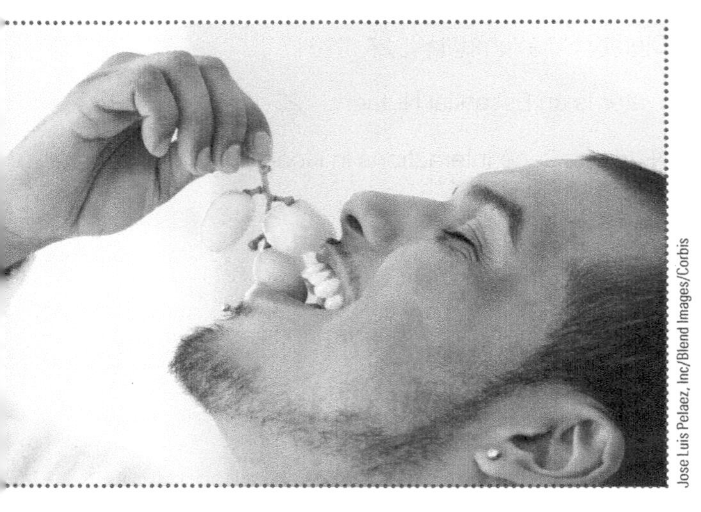
Jose Luis Pelaez, Inc/Blend Images/Corbis

nutrition timeline

1621
First Thanksgiving feast at Plymouth colony

PhotoDisc

1702
First coffeehouse in America opens in Philadelphia

PhotoDisc

1734
Scurvy recognized

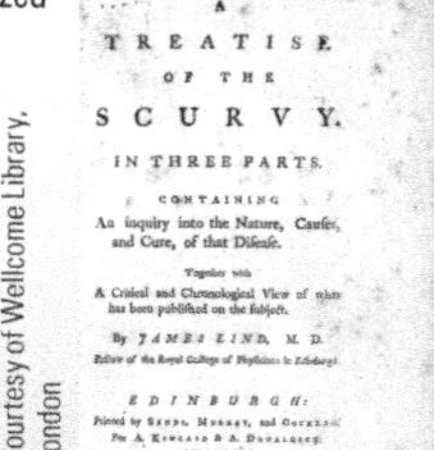
A TREATISE OF THE SCURVY. IN THREE PARTS. CONTAINING An inquiry into the Nature, Causes, and Cure, of that Disease. Together with A Critical and Chronological View of what has been published on the Subject. By JAMES LIND, M.D. EDINBURGH: MDCCLIII.

Courtesy of Wellcome Library, London

1744
First record of ice cream in America at Maryland colony

PhotoDisc

Alistair Berg/Digital Vision/Getty Images

1747
Lind publishes "Treatise on Scurvy," citrus identified as cure

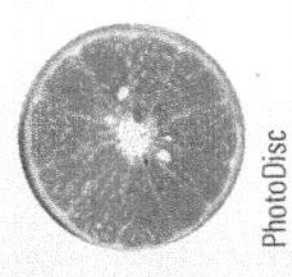
PhotoDisc

1750
Ojibway and Sioux war over control of wild rice stands

1762
Sandwich invented by the Earl of Sandwich

PhotoDisc

1771
Potato heralded as famine food

1774
Americans drink more coffee in protest over Britain's tea tax

Digital Vision/Alamy

nutrition timeline

1775

Lavoisier ("the father of the science of nutrition") discovers the energy-producing property of food

Stefano Bianchetti/CORBIS

1816

Protein and amino acids identified, followed by carbohydrates and fats in the mid-1800s

1833

Beaumont's experiments on a wounded man's stomach greatly expand knowledge about digestion

© Bettmann/Corbis

1862

U.S. Department of Agriculture founded by authorization of President Abraham Lincoln

1871

Proteins, carbohydrates, and fats determined to be insufficient to support life; there are other "essential" components

Scott Goodwin Photography

1895
First milk station providing children with uncontaminated milk opens in New York City

Bettman/CORBIS

1896
Atwater publishes *Proximate Composition of Food Materials*

1906
Pure Food and Drug Act signed by President Theodore Roosevelt to protect consumers against contaminated foods

Bettman/CORBIS

1910
Pasteurized milk introduced

PhotoDisc

1912
Funk suggests scurvy, beriberi, and pellagra caused by deficiency of "vitamines" in the diet

John E. Kelly/Photolibrary/Getty Images

nutrition timeline

1913
First vitamin discovered (vitamin A)

PhotoDisc

1914
Goldberger identifies the cause of pellagra (niacin deficiency) in poor children to be a missing component of the diet rather than a germ as others believed

1916
First dietary guidance material produced for the public released; title is "Food for Young Children"

1917
First food groups published—the Five Food Groups: Milk and Meat; Vegetables and Fruits; Cereals; Fats and Fat Foods; Sugars and Sugary Foods

1921
First fortified food produced: iodized salt, needed to prevent widespread iodine-deficiency goiter in many parts of the United States

LunaseeStudios/Shutterstock.com

ROSENFELD/AGE Fotostock

1928
American Society for Nutritional Sciences and the *Journal of Nutrition* founded

1929
Essential fatty acids identified

PhotoDisc

1930s
Vitamin C identified in 1932, followed by pantothenic acid and riboflavin in 1933 and vitamin K in 1934

1937
Pellagra found to be due to a deficiency of niacin

PhotoDisc

1938
Health Canada issues nutrient intake standards

1941
First refined grain enrichment standards developed

Creatas/Fotosearch

nutrition timeline

1941
First Recommended Dietary Allowances (RDAs) announced by President Franklin Roosevelt on radio

Franklin D. Roosevelt Presidential Library and Museum

1946
National School Lunch Act passed

PhotoDisc

1947
Vitamin B_{12} identified

1953
Double helix structure of DNA discovered

PhotoDisc

1956
Basic Four Food Groups released by the U.S. Department of Agriculture

©JIL Photo/Shutterstock.com

1965
Food Stamp Act passed, Food Stamp program established

1966
Child Nutrition Act adds school breakfast to the National School Lunch Program

PhotoDisc

1968
First national nutrition survey in U.S. launched (Ten State Nutrition Survey)

1970
First Canadian national nutrition survey launched (Nutrition Canada National Survey)

1972
Special Supplemental Food and Nutrition Program for Women, Infants, and Children (WIC) established

Digital Vision/Getty Images

nutrition timeline

1977
Dietary Goals for the United States issued

1978
First Health Objectives for the Nation released

1989
First national scientific consensus report on diet and chronic disease published

1992
The Food Guide pyramid is released by the USDA

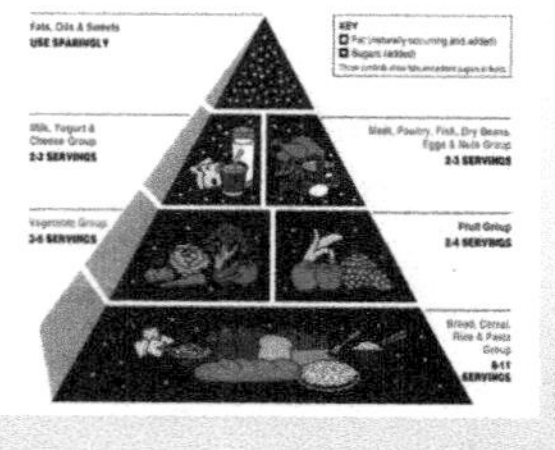
USDA

1997
RDAs expanded to Dietary Reference Intakes (DRIs)

Bill Milne/StockFood Creative/Getty Images

1998
Folic acid fortification of refined grain products begins

Richard Anderson

2003
Sequencing of DNA in the human genome completed; marks beginning of new era of research in nutrient–gene interactions

2015
Increasing rates of obesity and type 2 diabetes become global epidemics

2018
Advances in knowledge of the effects of gene variants on nutritional status and health begin the "personalized approach" to dietary recommendations.

Kablonk! RF/Golden Pixels LLC/Alamy

©Stephane Bidouze/Shutterstock.com

Preface

"Everything should be made as simple as possible. But not simpler."
—ALBERT EINSTEIN

Welcome to the enhanced 8th edition of *Nutrition Now*, an introductory, science-based, and application-oriented textbook for your nutrition courses. The text contains multiple critical thinking and decision-making activities for students, organized around key nutrition concepts and specific learning objectives. This updated release of *Nutrition Now* reflects a move toward increased use of a broad variety of digital resources intended to enhance student learning, and to lighten the weight and cost of this text book.

This edition of *Nutrition Now* catches up with advances in the field of nutrition:

- MyPlate resources
- healthy dietary patterns and health
- prevention of peanut allergy
- the prevalence of food allergies and likely reasons for it
- microbiota, diet, and inflammation
- nutrition label components
- health effects of refined carbohydrates and highly processed foods
- "relative energy deficiency in sports" (RED-S)
- wheat intolerance syndrome
- effects of total fat and saturated fat on health
- functions and health effects of brown fat and white fat
- "best sources" of fish and recommendations for intake
- origins of gene variants, their interaction with dietary intake, and health effects related to gene variants and diet
- the lack of benefit of vitamin and mineral supplements, and other dietary supplements
- salt intake, salt sensitivity, and blood pressure
- physical activity recommendations
- dietary and supplement recommendations for pregnancy
- diagnostic standards for fetal alcohol syndrome
- health effects of artificial sweeteners
- infant feeding recommendations and the weight status group of "severe obesity" in children
- dietary and lifestyle factors related to recent decline in longevity
- nutrition and menopause in women and andropause in men

- health effects of worldwide, increased availability of inexpensive sugary, fast, and highly processed foods that replace foods traditionally consumed
- the effects of climate changes on availability of croplands, water, food supplies, and malnutrition.

In addition, "Nutrition Up Close," "Reality Check," "Take Action," and review questions have been modified to reflect content changes.

Nutrition Now continues to be oriented toward enhancing instructors' teaching experiences and helping students build a firm foundation of scientific knowledge and understanding about nutrition that will serve them well throughout their careers and life.

Pedagogical Features

There are 33 units in *Nutrition Now*, and all but the first unit can be used in any order. Each unit begins with learning objectives, and content and review questions at the end of the units are organized around the learning objectives. Student group and individual activities based on real-life situations are presented online in MindTap, along with a variety of videos, review questions, and interactive learning activities. Activities include taste testing to identify genetically determined sensitivity to bitterness, developing a dietary behavioral change plan, anthropometry lab, designing fraudulent nutrition products, a physical activity assessment, and an assessment of three days of dietary intake.

Resources for the Instructor

- **Nutrition MindTap for *Nutrition Now*.** Instant Access Code, ISBN-13: 9781305868304. MindTap is well beyond an e-book, a homework solution or digital supplement, a resource center website, a course delivery platform, or a learning management system. More than 70 percent of students surveyed said it was unlike anything they have seen before. MindTap is a new personal learning experience that combines all your digital assets—readings, multimedia, activities, and assessments—into a singular learning path to improve student outcomes.
- **Diet & Wellness Plus** The Diet & Wellness Plus app in MindTap helps you gain a better understanding of how nutrition relates to your personal health goals. It enables you to track your diet and activity, generate reports, and analyze the nutritional value of the food you eat! It includes more than 55,000 foods in the database, custom food and recipe features, the latest dietary references, as well as your goal and actual percentages of essential nutrients, vitamins, and minerals. It also helps you identify a problem behavior and make a positive change. After completing a wellness profile questionnaire, Diet & Wellness Plus will rate the level of concern for eight different areas of wellness, helping you determine the areas where you are most at risk. It then helps you put together a plan for positive change by helping you select a goal to work toward—complete with a reward for all your hard work.

  The Diet & Wellness Plus app is accessed from the app dock in MindTap and can be used throughout the course for students to track their diet and activity and behavior change. There are activities and labs in the course that have students access the app to further extend learning and integrate course content.
- **Instructor Companion Site** Everything you need for your course in one place! This collection of book-specific lecture and class tools is available online via **www.cengage.com/login**. Access and download PowerPoint presentations, images, instructor's manual, videos, and more.

- **Cengage Learning Testing Powered by Cognero** Cengage Learning Testing Powered by Cognero is a flexible online system that allows you to:
    - author, edit, and manage test bank content from multiple Cengage Learning solutions
    - create multiple test versions in an instant
    - deliver tests from your LMS, your classroom, or wherever you want
- **Global Nutrition Watch** Bring currency to the classroom with Global Nutrition Watch from Cengage Learning! This user-friendly website provides convenient access to thousands of trusted sources, including academic journals, newspapers, videos, and podcasts, for you to use for research projects or classroom discussion. Global Nutrition Watch is updated daily to offer the most current news about topics related to nutrition.

Acknowledgments

My thanks and appreciation go out to Oden Connolly, Content Manager for Sciences at Cengage. Oden effectively managed to build the complex web that now incorporates updates to the *Nutrition Now* text and its expanding digital components.

It is said that instructors adopt a specific textbook but that students play a major role in instructors' decision to keep it. I am honored that you chose to adopt *Nutrition Now* and deeply pleased with the thought that students are helping you decide to keep it.

Reviewers' feedback is the lifeline of text writing, and the reviewers of the eighth edition conveyed very useful advice that was incorporated into the eighth edition. The advice led me to some very interesting places on specific topics that changed my thinking and writing. Thank you for the helpful information and please keep your comments coming.

Key Nutrition Concepts and Terms

NUTRITION SCOREBOARD

1 Calories are a component of food. **True/False**

2 Nutrients are substances in food that are used by the body for growth and health. **True/False**

3 Inadequate intakes of vitamins and minerals can harm health, but high intakes do not. **True/False**

4 "Dietary Reference Intakes" (DRIs) provide science-based standards for nutrient intake. **True/False**

Answers can be found at the end of the unit.

Jose Luis Pelaez, Inc/Blend Images/Corbis

LEARNING OBJECTIVES

After completing Unit 1 you will be able to:

- Explain the scope of nutrition as an area of study.
- Demonstrate a working knowledge of the meaning of the 10 nutrition concepts.

The Meaning of Nutrition

- **Explain the scope of nutrition as an area of study.**

What is nutrition? It can be explained by situations captured in photographs as well as by words. This introduction presents a photographic tour of real-life situations that depict aspects of the study of nutrition.

Before the tour begins, take a moment to make yourself comfortable and clear your mind of clutter. Take a careful look at the photographs shown below and on the next two pages, pausing to mentally describe in two or three sentences what each photograph shows.

Not everyone who thinks about the photographs will describe them in the same way. Reactions will vary somewhat due to personal experiences, interests, beliefs, and cultural background. An individual trying to gain weight will probably react differently to the photograph of the person on the scale than someone who is trying to lose weight. The photo of a dad measuring his son's growth progress may bring back memories of the "measuring wall" you knew as a child, and how you were encouraged to eat your vegetables to grow up strong and tall. Depending on your experience and background, you may recognize the photo of ham hocks, greens, and beans as your favorite holiday meal. The final photo, showing a crowd of children and adults at a soup kitchen, may have reminded you that food is essential for life.

Although knowledge about nutrition is generated by impersonal and objective methods, it can be a very personal subject.

Nutrition Defined

nutrition The study of foods, their nutrients and other chemical constituents, and the effects that foods and food constituents have on health.

In a nutshell, **nutrition** is the study of foods and health. It is a science that focuses on foods, their nutrient and other chemical constituents, and the effects of food and food constituents on body processes and health. The scope of nutrition extends from food choices to the effects of diet and specific food components on biological processes and health.

Gary Conner/PhotoEdit

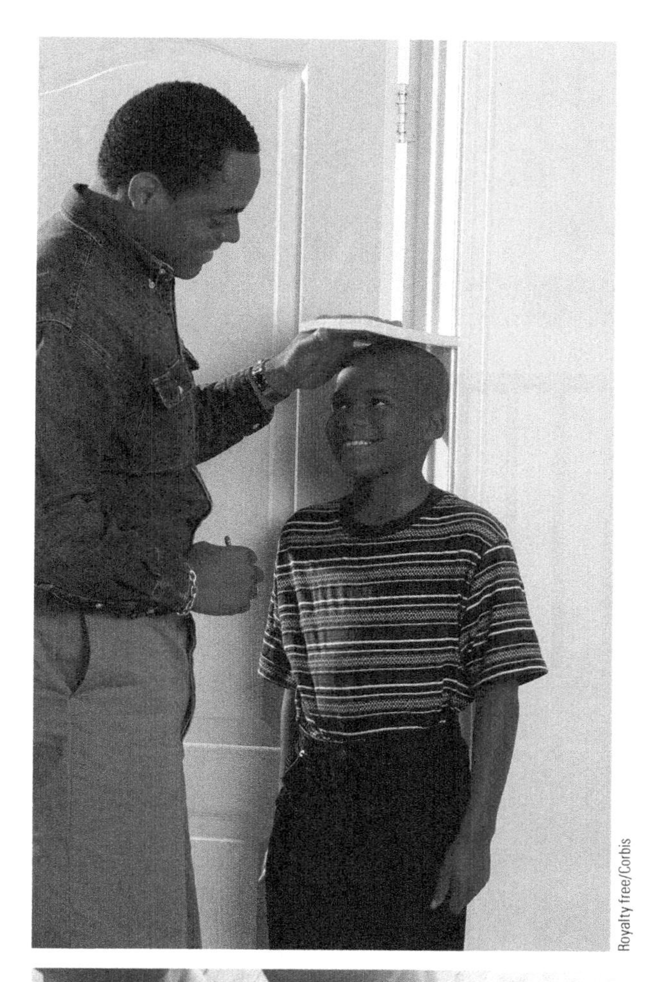

Royalty free/Corbis

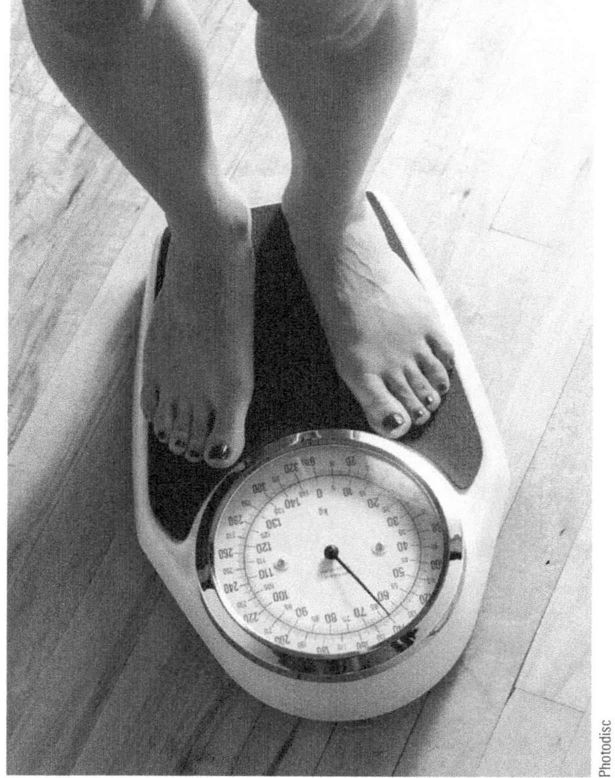

Photodisc

Jupiterimages/Photolibrary/Getty Images

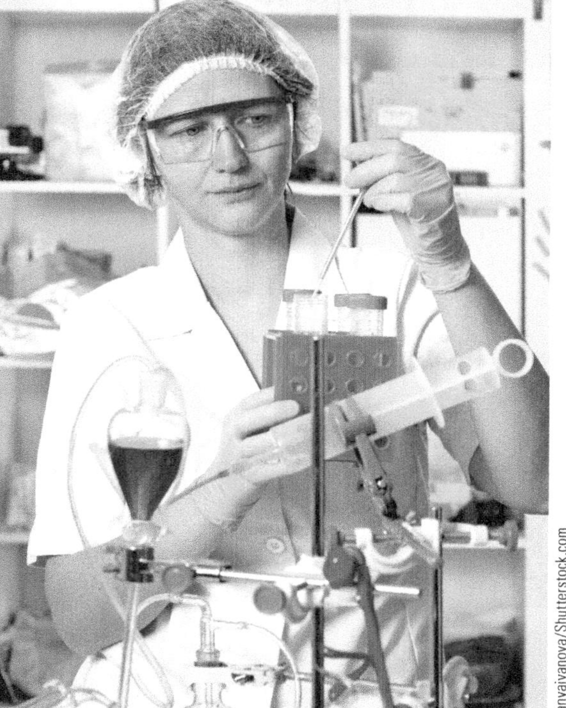

anyaivanova/Shutterstock.com

Uschi Gerschner/Newscom

Photodisc

Pat Shearman/Alamy

Budimir Jevtic/Shutterstock.com

Christian Science Monitor/Getty Images

Nutrition Is a "Melting Pot" Science The broad scope of nutrition makes it an interdisciplinary science. Knowledge provided by the behavioral and social sciences, for example, is needed in studies that examine how food preferences develop and how they may be changed. Information generated by the biological, chemical, physical, and food sciences is required to propose and explain diet and disease relationships. The knowledge and skills of mathematicians and statisticians are needed to develop and implement appropriate research designs and analysis strategies that produce objective, reliable research results. The study of nutrition will bring you into contact with information from a variety of disciplines.

Nutrition Knowledge Is Applicable As you study the science of nutrition, you will discover answers to a number of questions about your own diet, health, and eating behaviors. Is obesity primarily due to eating habits, physical inactivity, the food environment, or your genes? How do you know whether new information you hear about nutrition is true? Can sugar harm more than your teeth? Can the right diet or supplement give you a competitive edge? What is a healthful diet and how do you know if you have one? If improvements seem warranted, what's the best way to go about changing your diet for the better? These are just a few of the questions that will be addressed during the course of your study of nutrition. You will take from this learning experience not only knowledge about nutrition and health, but also skills that will keep the information and insights working to your advantage for a long time to come.

Foundation Knowledge for Thinking about Nutrition

- **Demonstrate a working knowledge of the meaning of the 10 nutrition concepts.**

You don't have to be a bona fide nutritionist to think like one. What you need is a grasp of the language and basic concepts of the science. The purpose of this unit is to give you that background. The essential topics covered here are explored in greater depth in units to come, and they build on this foundation of knowledge. With a working knowledge of nutrition terms and concepts, you will have an uncommonly good sense of nutrition.

food security Access at all times to a sufficient supply of safe, nutritious foods.

food insecurity Limited or uncertain availability of safe, nutritious foods—or the ability to acquire them in socially acceptable ways.

NUTRITION CONCEPT #1

Food is a basic need of humans.

Humans need enough food to live, and they need the right amount and assortment of foods for optimal health. In the best of all worlds, the need for food is combined with the condition of **food security**. People who experience food security have access at all times to a sufficient supply of the safe, nutritious foods that are needed for an active and healthy life. They are able to acquire acceptable foods in socially acceptable ways; they do not have to scavenge or steal food in order to survive or to feed their families. It is important to note that food security emphasizes access to an adequate supply of *nutritious* food rather than simply sufficient food. **Food insecurity** exists whenever the availability of safe, nutritious foods—or the ability to acquire them in socially acceptable ways—is limited or uncertain (Illustration 1.1).[1]

Adults who live in food-insecure households are more likely to have poor-quality diets, to be obese, and to have hypertension, heart disease, or diabetes than adults who are poor but food secure.[2] The ready availability of inexpensive, high-calorie foods; poverty; the absence of local supermarkets; limited opportunities

David Bacon/Alamy

Illustration 1.1 "It is possible to go an entire lifetime without knowing about people's experiences with hunger."
—*Meghan LeCates, Capital Area Food Bank*

for exercise; and lack of cooking facilities may be partly responsible for the higher rates of obesity and chronic disease among food-insecure adults.[3] Although many children living in food-insecure households are nourished adequately, successful in school, and develop high levels of social skills, as a group they are at higher risk of poor school performance and social and behavioral problems.[4]

Food insecurity exists in 12.3% of U.S. and 7.7% of Canadian households.[5,6] The rate in the United States is over twice as high as the target of 6% established as a national health goal.[7]

Who Are the Food Insecure?

Although unemployment, disabilities, and homelessness contribute to food insecurity, most food-insecure individuals live in households with one or more full-time worker.[8]

One in four of such households include an adult serving in the military.[9] College students are being increasingly recognized as a risk group for food insecurity. Over 120 U.S. colleges have established food pantries for students, and more are being developed as you read this.[10] Low wages and high food costs can limit the ability of households and college students to afford sufficient, nutritious food.

Food Security and Stainable Diets Climate change is affecting worldwide food security. Floods and drought precipitated by global warming can displace fertile farmlands and water supplies, and reduce crop yields and fish and seafood availability.[11] Food production, in turn, affects climate change by contributing greenhouse gases that trap heat in the atmosphere.[12] Fossil fuels, which are widely used in food production, processing, and transport, are the largest source of greenhouse gas emissions in the United States.[13]

The extent of production of greenhouse gases is generally assessed by measuring the amount of carbon dioxide released into the atmosphere as a result of food production and other processes. The result is referred to as the "**carbon footprint**." In general, plant foods such as grains and vegetables leave a lower carbon footprint than dairy and meats.

The world's history of famine due to crop failures, floods, drought, and epidemics of crop and livestock diseases has taught us the importance of using farming and environmental practices that support healthy crops, livestock, and people. Farming and environmental methods that support nutritious food production and farm profitability, and protect the environment are part and parcel of what is now called "**sustainable agriculture**."[14,15]

carbon footprint Generally defined as the amount of the greenhouse gas carbon dioxide emitted by farming, food production, a person's activities, or a product's manufacture and transport.

sustainable agriculture Broadly defined as the use of farming methods, environmental practices, and economic policies that meet society's present food needs without compromising the ability of future generations to meet their own needs.

calorie A unit of measure of the amount of energy supplied by food. (Also known as a kilocalorie, abbreviated kcal, or the *large Calorie* with a capital C.)

nutrients Chemical substances in food that are used by the body for growth and health. The six categories of nutrients are carbohydrates, proteins, fats, vitamins, minerals, and water.

NUTRITION CONCEPT #2

Foods provide energy (calories), nutrients, and other substances needed for growth and health.

People eat foods for many different reasons. The most compelling reason is that we need the calories, nutrients, and other substances supplied by foods for growth and health.

Brand X Pictures/Getty Images

Illustration 1.2 Foods provide nutrients. "Please pass the complex carbohydrates, thiamin, and niacin . . . I mean, the bread!"

Calories

A **calorie** is a unit of measure of the amount of energy in a food—and of how much energy will be transferred to the person who eats it. Although we often refer to the number of calories in this food or that one, calories are not a substance present in food. And, because calories are a unit of measure, they do not qualify as a nutrient.

Nutrients

Nutrients are chemical substances present in food that are used by the body to sustain growth and health (Illustration 1.2). Essentially everything that's in our body was once a component of the food we consumed.

There are six categories of nutrients (Table 1.1). Each category (except water) consists of a number of different substances that are used by the body for growth and health. The carbohydrate category includes simple sugars and complex carbohydrates (starches and dietary fiber). The protein category includes 20 amino acids, the chemical units that serve as the "building blocks" for proteins. Several different types of fat are included in the fat category. Of primary concern are the saturated fats, unsaturated fats, essential fatty acids, *trans* fats, and cholesterol. The vitamin category consists of 14 vitamins, and the mineral category includes 15 minerals. Water makes up a nutrient category by itself.

Carbohydrates, proteins, and fats supply calories and are called the *energy nutrients.* Although each of these three types of nutrients performs a variety of functions, they share the property of being the body's only sources of fuel. Vitamins, minerals, and water are chemicals that the body needs for converting carbohydrates, proteins, and fats into energy and for building and maintaining muscles, blood components, bones, and other tissues in the body.

Table 1.1 The six categories of nutrients

1. Carbohydrate
2. Protein
3. Fat
4. Vitamins
5. Minerals
6. Water

Other Substances in Food

Food also contains many other substances, some of which are biologically active in the body. One major type of such substances is the **phytochemicals**. There are thousands of them in plants. Illustration 1.3 presents examples of plant foods that are particularly rich sources of phytochemicals. Phytochemicals provide plants with color, give them flavor, foster their growth, and protect them from insects and diseases. In humans, consumption of certain phytochemicals in diets is strongly related to a reduced risk of developing certain types of cancer, heart disease, infections, and other disorders.[16]

phytochemicals Chemical substances in plants (phyto = plant). Some phytochemicals perform important functions in the human body. They give plants color and flavor, participate in processes that enable plants to grow, and protect plants against insects and diseases. Also called phytonutrients.

Specific phytochemicals have names that may be hard to pronounce and difficult to remember. Nevertheless, here are a few examples. Plant pigments, such as lycopene (like-o-peen), which help make tomatoes red, anthocyanins (an-tho-sigh-an-ins), which give blueberries their characteristic blue color, and beta-carotene (bay-tah-kar-o-teen), which imparts an orange color to carrots, are phytochemicals that act as **antioxidants.**

They protect plant cells—and in some cases, human cells, too—from damage that can make them susceptible to disease. Various types of sulfur-containing phytochemicals are present in cabbage, broccoli, cauliflower, brussels sprouts, and other vegetables of the same family. These substances help prevent a number of different types of cancer in people with specific gene types.[17]

antioxidants Chemical substances that prevent or repair damage to cells caused by exposure to oxidizing agents such as environmental pollutants, smoke, ozone, and oxygen. Oxidation reactions are a normal part of cellular processes.

Some Nutrients Must Be Provided by the Diet Many nutrients are required for growth and health. The body can manufacture some of these from raw materials supplied by food, but others must come assembled. Nutrients that the body cannot generally produce, or cannot produce in sufficient quantity, are referred to as **essential nutrients.** Here *essential* means "required in the diet." Vitamin A, iron, and calcium are examples of essential nutrients. Table 1.2 lists all the known essential nutrients.

essential nutrients Nutrients required for normal growth and health that the body can generally not produce, or produce in sufficient amounts. Essential nutrients must be obtained in the diet.

Nutrients used for growth and health that can be manufactured by the body from components of food in our diet are considered nonessential. Cholesterol, creatine, and

Illustration 1.3 Examples of good food sources of beneficial phytochemicals.

Table 1.2 **Essential nutrients for humans: A reference table**

Energy Nutrients	Vitamins	Minerals	Water
Carbohydrates	Biotin	Calcium	Water
Fats[a]	Folate	Chloride	
Proteins[b]	Niacin (B_3)	Chromium	
	Pantothenic acid	Copper	
	Riboflavin (B_2)	Fluoride	
	Thiamin (B_1)	Iodine	
	Vitamin A	Iron	
	Vitamin B_6 (pyroxidine)	Magnesium	
	Vitamin B_{12}	Manganese	
	Vitamin C (ascorbic acid)	Molybdenum	
	Vitamin D	Phosphorus	
	Vitamin E	Potassium	
	Vitamin K	Selenium	
	Choline[c]	Sodium	
		Zinc	

[a]Fats supply the essential nutrients linoleic and alpha-linolenic acids.
[b]Proteins are the source of nine "essential amino acids": histidine, isoleucine, leucine, lysine, methionine, phenylalanine, threonine, tryptophan, and valine. The other 11 amino acids are not a required part of our diet; they are considered "nonessential."
[c]A dietary source of choline, a vitamin-like nutrient, may not be required during all stages of the life cycle.

nonessential nutrients Nutrients required for normal growth and health that the body can manufacture in sufficient quantities from other components of the diet. We do not require a dietary source of nonessential nutrients.

glucose are examples of **nonessential nutrients**. Nonessential nutrients are present in food and used by the body, but they are not required parts of our diet because we can produce them ourselves.

Both essential and nonessential nutrients are required for growth and health. The difference between them is whether or not we need to obtain the nutrient from a dietary source. A dietary deficiency of an essential nutrient will cause a specific deficiency disease, but a dietary lack of a nonessential nutrient will not. People develop scurvy (the vitamin C–deficiency disease), for example, if they do not consume enough vitamin C. But you could have zero cholesterol in your diet and not become "cholesterol deficient," because your liver produces cholesterol.

Our Requirements for Essential Nutrients The amount of essential nutrients humans need each day varies a great deal, from amounts measured in cups to micrograms. (See Table 1.3 to get a notion of the amount represented by a gram, milligram, and other measure.) Generally speaking, adults need 11 to 15 cups of water from fluids and foods, 9 tablespoons of protein, one-fourth teaspoon of calcium, and only one-thousandth teaspoon (a 30-microgram speck) of vitamin B_{12} each day.

We all need the same nutrients, but not always in the same amounts. The amounts needed vary among people based on:

- Age
- Sex
- Growth status
- Body size
- Genetic traits

and the presence of conditions such as:

- Pregnancy
- Breastfeeding

Table 1.3 **Units of measure commonly employed in nutrition**

Measure	Abbreviation	Equivalents
Kilogram	kg	1 kg = 2.2 lb = 1,000 grams
Pound	lb	1 lb = 16 oz = 454 grams = 2 cups (liquid)
Ounce	oz	1 oz = 28 grams = 2 tablespoons (liquid)
Gram	g	1 g = 1/28 oz = 1,000 milligrams
Milligram	mg	1 mg = 1/28,000 oz = 1,000 micrograms
Microgram	mcg, μg	1 mcg = 1/28,000,000 oz

1 egg = 50 grams or $1^3/_4$ oz; 212 milligrams (0.2 grams) of cholesterol in yolk

1 slice of bread = 1 oz = 28 grams

1 nickel = 5 grams

Photodisc

1 teaspoon of sugar = 4 grams, 1 grain of sugar = 200 micrograms

- Illnesses
- Drug/medication use
- Exposure to environmental contaminants

Each of these factors, and others, can influence nutrient requirements. General diet recommendations usually make allowances for major factors that influence the level of nutrient need, but they cannot allow for all of the factors.

Nutrient Intake Standards Recommendations for daily levels of nutrient intake were first developed in the United States in 1943 and have been updated periodically since then. Called the *Recommended Dietary Allowances* (RDAs), these standards were established in response to the high rejection rate of World War II recruits, many of whom were underweight and had nutrient deficiencies. The recommended levels of nutrient intake provided are based on age, gender, and condition (pregnant or breastfeeding). Because the science underlying nutrient intake and health advances with time, these standards are periodically revised and expanded (Illustration 1.4).

Judith Brown

Illustration 1.4 The latest DRI report (2010) on calcium and vitamin D.

Dietary Reference Intakes Nutrient intake standards now in place are referred to as *Dietary Reference Intakes* (DRIs); they include categories of nutrient intake in addition to the RDAs. The current RDAs referenced in the DRIs reflect nutrient intake levels that protect almost all healthy individuals from developing deficiency disease and that also reduce the risk of common chronic diseases. Table 1.4 provides examples of endpoints aimed at chronic disease prevention used by the DRI Committee to determine the RDAs. A category labeled *Adequate Intakes* (AIs) has been added to indicate "tentative RDAs" for a few nutrients such as vitamin K and fluoride, for which too few reliable scientific studies have been done to establish an RDA.

The DRI standards include a category called *Estimated Average Requirement* (EAR). This category represents nutrient intake levels that are estimated to meet the nutrient intake requirements of 50% of individuals within an age, sex, and condition (pregnant or breastfeeding) group.

DRI standards consider the effects of excessively high intake of nutrients, primarily from supplements and fortified foods, on health. These standards are labeled *Tolerable Upper Intake Levels*, abbreviated *ULs* for "upper levels." Table 1.5 graphically displays the

Table 1.4 **Examples of primary endpoints used to estimate DRIs**

Carbohydrate: Amount needed to supply optimal levels of energy to the brain.
Total Fiber: Amount shown to provide the greatest protection against heart disease.
Folate: Amount that maintains normal red blood cell folate concentration.
Iodine: Amount that corresponds to optimal functioning of the thyroid gland.
Selenium: Amount that maximizes its function in protecting cells from damage.

Table 1.5 Terms and abbreviations used in the DRIs and a graphic representation of their meaning

Estimated Average Requirement (EAR)
Recommended Dietary Allowance (RDA)
Adequate Intake (AI)
Tolerable Upper Intake Level (UL)
Risk of dietary deficiency
Risk of overdose reactions
100%
50%
Nutrient intake
Low
High

- **Dietary Reference Intakes (DRIs).** This is the general term used for nutrient intake standards for healthy people.
- **Recommended Dietary Allowances (RDAs).** These are levels of essential nutrient intake judged to be adequate to meet the known nutrient needs of practically all healthy persons while decreasing the risk of certain chronic diseases.
- **Adequate Intakes (AIs).** These are "tentative" RDAs. AIs are based on less conclusive scientific information than are the RDAs.
- **Estimated Average Requirements (EARs).** These are nutrient intake values that are estimated to meet the requirements of half the healthy individuals in a group. The EARs are used to assess adequacy of intakes of population groups.
- **Tolerable Upper Intake Levels (ULs).** These are upper limits of nutrient intake compatible with health. The ULs do not reflect desired levels of intake. Rather, they represent total daily levels of nutrient intake from food, fortified foods, and supplements that should not normally be exceeded.

relationships between nutrient intake level and the various categories of the DRI standards now in use, and presents definitions of the DRI nutrient intake categories.

Developed by nutrition scientists from the United States and Canada, the RDAs apply to 97 to 98% of all healthy people in both countries. The fundamental premise of the first RDAs—that nutrient intake should come primarily from foods—is maintained in the current nutrient intake standards.

DRI tables are located at the back of this text. Check out these tables. They can be used to identify recommended daily levels of essential nutrient intake and levels of intake that should not normally be exceeded.

Illustration 1.5 Schematic representation of the structure of a human cell.

NUTRITION CONCEPT #3

Health problems related to nutrition originate within cells.

Cells are the main employers of nutrients (Illustration 1.5). All body processes required for growth and health take place within cells and the fluid that surrounds them. Humans contain over 100 trillion cells in body tissues (and around 100 times more than that number if bacteria and other microorganisms in the large intestine are included).[18] The functions of each cell are maintained by the nutrients it receives. Problems begin when a cell's need for nutrients differs from the available supply.[19]

Nutrient Functions at the Cellular Level Cells are the building blocks of tissues (such as muscles and bones), organs (such as the kidneys, heart, and liver), and systems (such as the respiratory, reproductive, circulatory, and nervous systems). Normal cell health and functioning are maintained when a nutritional and environmental utopia exists within and around the cells. Such circumstances allow **metabolism**—the chemical changes that take place within and outside of cells—to proceed flawlessly. Disruptions in the availability of nutrients—or the presence of harmful substances in the cell's environment—initiate diseases and disorders that eventually affect tissues, organs, and systems. Here are two examples of how cell functions can be disrupted by the presence of low or high concentrations of nutrients:

metabolism The chemical changes that take place in the body. The formation of energy from carbohydrates is an example of a metabolic process.

1. Folate, a B vitamin, is required for protein synthesis within cells. When too little folate is available, cells produce proteins with abnormal shapes and functions. Abnormalities in the shape of red blood cell proteins, for example, lead to functional changes that produce loss of appetite, weakness, and irritability.[20]
2. When too much iron is present in cells, the excess reacts with and damages cell components. If cellular levels of iron remain high, the damage spreads, impairing the functions of organs such as the liver, pancreas, and heart.[21]

Health problems in general begin with disruptions in the normal activity of cells. Humans are as healthy as their cells.

NUTRITION CONCEPT #4

Poor nutrition can result from both inadequate and excessive levels of nutrient intake.

For each nutrient, every individual has a range of optimal intake that produces the best level for cell and body functions. On either side of the optimal range are levels of intake associated with impaired body functions.[22] This concept is presented in Illustration 1.6. Inadequate essential nutrient intake, if prolonged, results in obvious deficiency diseases. Marginally deficient nutrient intake generally produces subtle changes in behavior or physical condition. If the optimal intake range is exceeded, mild to severe changes in mental and physical functions occur, depending on the amount of the excess and the nutrient. Severe zinc deficiency, for example, is related to diarrhea, respiratory infection, and stunted growth. Mild zinc deficiency causes disturbances in the sense of taste and smell, and reduces appetite and food intake. Excessive intake of zinc is associated with vomiting and a decline in the body's ability to fight infections.[23] Nearly all cases of vitamin and mineral overdose result from excessive use of supplements or errors made in the level of nutrient fortification of food products. They are almost never caused by foods. For nutrients, "enough is as good as a feast."

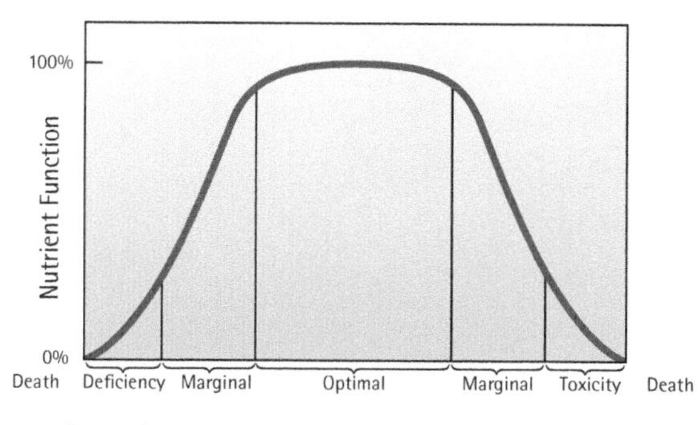

Illustration 1.6 For every nutrient, there is a range of optimal intake that corresponds to the optimal functioning of that nutrient in the body.

Steps in the Development of Nutrient Deficiencies and Toxicities Poor nutrition due to inadequate diet generally develops in the stages outlined in Illustration 1.7. To help explain the stages, this illustration includes an example of how vitamin A deficiency develops.

After a period of deficient intake of an essential nutrient, the body's tissue reserves of the nutrient become depleted. Blood levels of the nutrient then decrease because there are no reserves left to replenish the blood supply. Without an adequate supply of the nutrient in the blood, cells get shortchanged. They no longer have the supply of nutrients needed to maintain normal function. If the dietary deficiency is prolonged, the malfunctioning cells cause

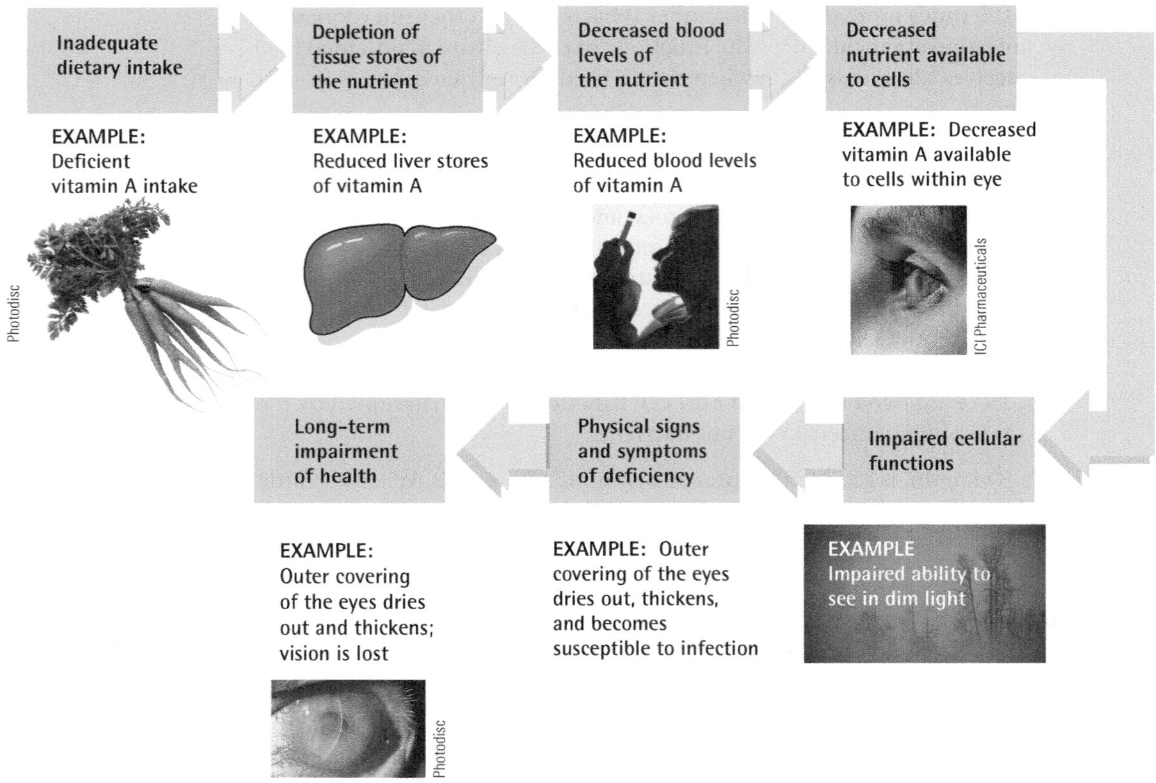

Illustration 1.7 The usual sequence of events in the development of a nutrient deficiency and an example of how vitamin A deficiency develops.[24,25]

Inadequate dietary intake

EXAMPLE: Deficient vitamin A intake

Photodisc

Depletion of tissue stores of the nutrient

EXAMPLE: Reduced liver stores of vitamin A

Decreased blood levels of the nutrient

EXAMPLE: Reduced blood levels of vitamin A

Photodisc

Decreased nutrient available to cells

EXAMPLE: Decreased vitamin A available to cells within eye

ICI Pharmaceuticals

Impaired cellular functions

EXAMPLE Impaired ability to see in dim light

Physical signs and symptoms of deficiency

EXAMPLE: Outer covering of the eyes dries out, thickens, and becomes susceptible to infection

Long-term impairment of health

EXAMPLE: Outer covering of the eyes dries out and thickens; vision is lost

Photodisc

sufficient impairment to produce physically obvious signs of a deficiency disease. Eventually, some of the problems produced by the deficiency may no longer be repairable, and permanent changes in health and function may occur. In most cases, the problems resulting from the deficiency can be reversed if the nutrient is supplied before this final stage occurs.

Excessively high intake of many nutrients such as vitamin A and selenium produces toxicity diseases. The vitamin A toxicity disease is called *hypervitaminosis A,* and the disease for selenium toxicity is called *selenosis.* Signs of the toxicity disease stem from increased levels of the nutrient in the blood and subsequent oversupply of the nutrient to the cells. The high nutrient load upsets the balance needed for normal cell function. The changes in cell functions lead to the signs and symptoms of the toxicity disease.

For both deficiency and toxicity diseases, the best time to correct the problem is usually at the level of dietary intake, before tissue stores are adversely affected. In that case, no harmful effects on health and cell function occur—they are prevented.[26]

Chad Johnston/Masterfile

Illustration 1.8 This woman has iron deficiency anemia. Chances are good that she has poor status of other nutrients in addition to iron.

Nutrient Deficiencies Are Often Multiple Most foods contain many nutrients, so poor diets will affect the intake level of more than one nutrient (Illustration 1.8). Inadequate diets generally produce a spectrum of signs and symptoms related to multiple nutrient deficiencies. For example, protein, vitamin B_{12}, iron, and zinc are packaged together in many high-protein foods. The protein-deficient, starving children you may see in news reports are rarely deficient just in protein. They may also be deficient in iron, zinc, and vitamin B_{12}.

The "Ripple Effect" Dietary changes affect the level of intake of many nutrients. Switching from a high-fat to a low-fat diet, for instance, may result in a higher intake of protein, carbohydrate, or both; a higher intake of cholesterol and vitamin E; and increased intake of vitamin A, vitamin C, and iron. So dietary changes introduced for the purpose of improving the intake level of a particular nutrient produce a ripple effect on the intake of other nutrients.

NUTRITION CONCEPT #5

Humans have adaptive mechanisms for managing fluctuations in nutrient intake.

Healthy humans are equipped with a number of adaptive mechanisms that partially protect the body from poor health due to fluctuations in dietary intake. In the context of nutrition, adaptive mechanisms act to conserve nutrients when dietary supply is low and to eliminate them when they are present in excessively high amounts. Dietary surpluses of energy and some nutrients—such as vitamin A and vitamin B_{12}—can be stored within tissues for later use. In the case of iron, copper, and calcium, the body regulates the amounts absorbed in response to its need for them. The body is unable to store excess amino acids for very long. It uses excess amino acids consumed primarily as a source of energy.[27]

Here are some other examples of how the body adapts to changes in dietary intake:

- When calorie intake is reduced by fasting, starvation, or dieting, the body adapts to the decreased supply by lowering energy expenditure. Declines in body temperature and the capacity to do physical work also act to decrease the body's need for calories. When caloric intake exceeds the body's need for energy, the excess is stored as fat for energy needs in the future.
- The ability of the gastrointestinal tract to absorb dietary iron increases when the body's stores of iron are low. To help protect the body from iron overdose, the mechanisms that facilitate iron absorption in times of need shut down when enough iron has been stored.
- The body can protect itself from excessively high levels of intake of vitamin C from supplements by excreting the excess in the urine.

Although these built-in mechanisms do not protect humans from all the consequences of poor diets, they do provide an important buffer against the development of nutrient-related health problems.

NUTRITION CONCEPT #6

Malnutrition can result from poor diets and from disease states, genetic factors, or combinations of these factors.

Malnutrition means poor nutrition; it results from both inadequate and excessive availability of calories and nutrients in the body. Vitamin A toxicity, obesity, vitamin C deficiency (scurvy), and underweight are examples of malnutrition.

malnutrition Poor nutrition resulting from an excess or lack of calories or nutrients.

Malnutrition can result from poor diets and also from diseases that interfere with the body's ability to use the nutrients consumed. Diarrhea, alcoholism, cancer, bleeding ulcers, and HIV/AIDS, for example, may be primarily responsible for the development of malnutrition in people with these disorders.

In addition, a percentage of the population is susceptible to malnutrition and increased disease risk due to genetic factors. For example, people may be born with a genetic tendency to produce excessive amounts of cholesterol, absorb high levels of iron, or use folate poorly. Some cases of obesity, diabetes, heart disease, and cancer are related to a combination of genetic and dietary factors.[28]

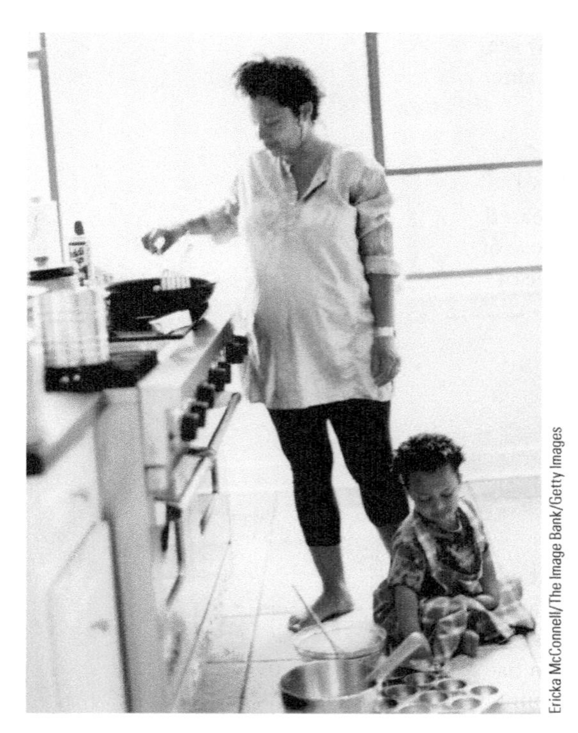
Ericka McConnell/The Image Bank/Getty Images

Illustration 1.9 Women who are pregnant or breastfeeding and infants are among the people who are at a higher risk of becoming inadequately nourished.

NUTRITION CONCEPT #7

Some groups of people are at higher risk of becoming inadequately nourished than others.

Women who are pregnant or breastfeeding, infants, growing children, the frail elderly, the ill, and those recovering from illness have a greater need for nutrients than other people. As a result, they are at higher risk than other people of becoming inadequately nourished (Illustration 1.9). The fetus during pregnancy and infants are developing rapidly and are particularly vulnerable to the adverse affects of poor nutrition. Poor nutrition experienced early in life can induce changes in gene function that adversely affect health status for a lifetime.[29] In cases of widespread food shortages, such as those induced by natural disasters or war, the health of these nutritionally vulnerable groups is compromised the soonest and the most.

Within the nutritionally vulnerable groups, certain people and families are at particularly high risk of malnutrition. These are people and families who are poor and least able to secure food, shelter, and high-quality medical services. The risk of malnutrition is not shared equally among all persons within a population.

NUTRITION CONCEPT #8

Poor nutrition can influence the development of certain chronic and other diseases.

chronic diseases Slow-developing, long-lasting diseases that are not contagious (e.g., heart disease, diabetes, and cancer). They can be treated but not always cured.

dietary pattern The quantities, proportions, variety, or combination of different foods, drinks, and nutrients in diets, and the frequency with which they are habitually consumed.

Poor nutrition does not result only in nutrient deficiency or toxicity diseases. Faulty diets play important roles in the development of **chronic diseases** such as hypertension, heart disease, cancer, and osteoporosis. Diets high in salt, for example, are related to the development of hypertension; those low in vegetables and fruit to cancer; low-calcium diets and poor vitamin D status to osteoporosis; and high-sugar diets to tooth decay. The harmful effects of poor dietary practices on chronic disease development generally accumulate over the course of years.[30]

NUTRITION CONCEPT #9

Adequacy, variety, and balance are key characteristics of healthy dietary patterns.

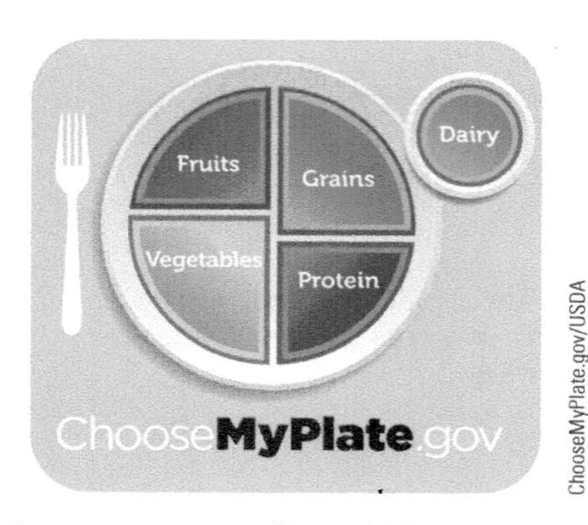

ChooseMyPlate.gov/USDA

Illustration 1.10 ChooseMyPlate is the icon for the USDA's food guidance system. It is intended to visually help people choose healthy meals.

Healthy diets correspond to a **dietary pattern** associated with normal growth and development, a healthy body weight, health maintenance, and disease prevention. One such pattern is represented by the USDA's ChooseMyPlate food group intake guide (Illustration 1.10). Several other types of dietary patterns, including the Mediterranean-style dietary pattern and the DASH Eating Plan, have also been found to promote health and foster disease prevention.[15] Healthy dietary patterns are characterized by regular consumption of moderate portions of a variety of foods from each of the basic food groups. No specific foods or food preparation techniques are excluded in a healthy dietary pattern.

Healthy dietary patterns are plant-food based and include the regular consumption of vegetables, fruits, dried beans, fish and seafood, low-fat dairy products, poultry and lean

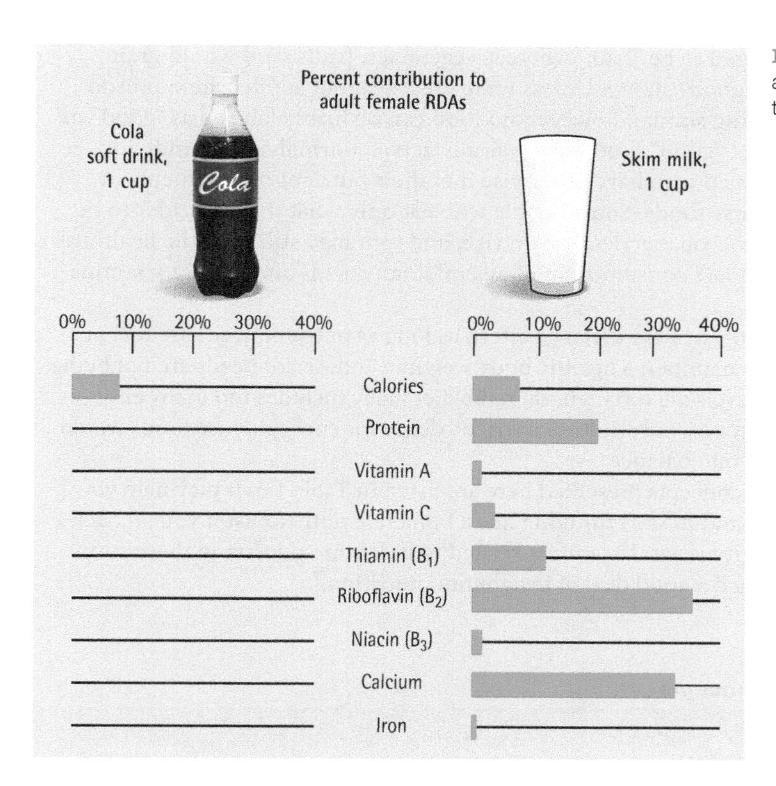

Illustration 1.11 Calorie and nutrient content of an empty-calorie and a nutrient-dense food. Percentages given represent percent contributions to adult female RDAs.

meats, nuts and seeds, and whole grains. Dietary patterns that include large or frequent servings of foods containing high amounts of *trans* fat, added sugars, salt, or alcohol miss the healthy dietary pattern mark. Healthy dietary patterns supply needed nutrients and beneficial phytochemicals through food rather than supplements or special dietary products.[31]

Energy and Nutrient Density Most Americans consume more calories than needed, become overweight as a result, *and* consume inadequate diets. This situation is partly due to over-consumption of **energy-dense foods** such as processed and high-fat meats, chips, candy, many desserts, and full-fat dairy products. Energy-dense foods have relatively high calorie values per unit weight of the food. Intake of energy-dense diets is related to the consumption of excess calories and to the development of overweight and diabetes.[32]

Many energy-dense foods are nutrient poor, meaning they contain low levels of nutrients given their caloric value. These foods are sometimes referred to as **empty-calorie foods**; these include products such as soft drinks, sherbet, hard candy, alcohol, and cheese twists. Excess intake of energy-dense and empty-calorie foods increases the likelihood that calorie needs will be met or exceeded before nutrients needs are met.[33] Diets most likely to meet nutrient requirements without exceeding calorie needs contain primarily **nutrient-dense foods**, foods with high levels of nutrients and relatively low calorie value. Nutrient-dense foods such as non-fat milk and yogurt, lean meat, dried beans, vegetables, and fruits provide relatively high amounts of nutrients compared to their calorie value.[32] Illustration 1.11 shows a comparison of the calorie and nutrient content of an empty-calorie and a nutrient-dense food.

energy-dense foods Foods that provide relatively high levels of calories per unit weight of the food. Fried chicken; cheeseburgers; a biscuit, egg, and sausage sandwich; and potato chips are energy-dense foods.

empty-calorie foods Foods that provide an excess of energy or calories in relation to nutrients. Soft drinks, candy, sugar, alcohol, and animal fats are considered empty-calorie foods.

nutrient-dense foods Foods that contain relatively high amounts of nutrients compared to their calorie value. Broccoli, collards, bread, cantaloupe, and lean meats are examples of nutrient-dense foods.

NUTRITION CONCEPT #10

There are no "good" or "bad" foods.

People tend to classify foods as being "good" or "bad," but such opinions over-simplify each food's potential contribution to a diet.[31] Typically hot dogs, ice cream, candy, bacon, and french fries are judged to be "bad," whereas vegetables, fruits, and whole-grain products are given the "good" stamp. Unless we're talking about spoiled stew, poison mushrooms, or something similar, however, no food can be firmly labeled as "good" or "bad." Ice cream can be a "good" food for physically active, normal-weight individuals with a high calorie need who have otherwise met their nutrient requirements by consuming nutrient-dense foods. Some people who eat only what they consider to be "good" foods such as broccoli, berries, brown rice, and tofu may still miss the healthful diet mark due to inadequate consumption of essential fatty acids and certain vitamins and minerals.

All foods can fit into a healthy dietary pattern as long as nutrient needs are met at calorie intake levels that maintain a healthy body weight.[31] If nutrient needs are not being met and calorie intake levels are too high, then the diet likely includes too many energy-dense or empty-calorie foods. Substituting nutrient-dense for energy-dense foods would help bring the diet back into balance.

The basic nutrition concepts presented here are listed in Table 1.6. It may help you remember the concepts and to start thinking like a bona fide nutritionist if you go back over each concept and give several examples related to it. If you understand these concepts, you will have gained a good deal of insight into nutrition.

Table 1.6 Nutrition concepts

1.	Food is a basic need of humans.
2.	Foods provide energy (calories), nutrients, and other substances needed for growth and health.
3.	Health problems related to nutrition originate within cells.
4.	Poor nutrition can result from both inadequate and excessive levels of nutrient intake.
5.	Humans have adaptive mechanisms for managing fluctuations in nutrient intake.
6.	Malnutrition can result from poor diets and from disease states, genetic factors, or combinations of these factors.
7.	Some groups of people are at higher risk of becoming inadequately nourished than others.
8.	Poor nutrition can influence the development of certain chronic and other diseases.
9.	Adequacy, variety, and balance are key characteristics of healthy dietary patterns.
10.	There are no "good" or "bad" foods.

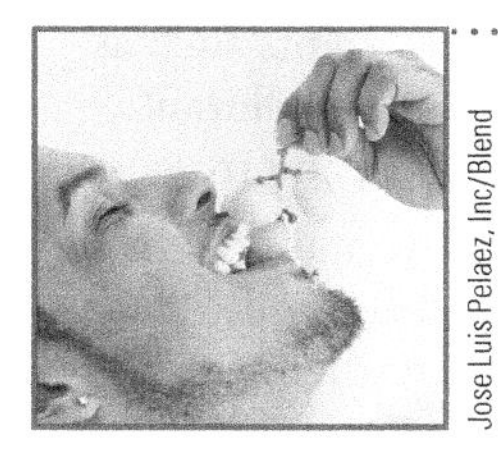
Jose Luis Pelaez, Inc/Blend Images/Corbis

NUTRITION up close

Nutrition Concepts Review

Focal Point: Nutrition concepts apply to diet and health relationships.

Write the number of the nutrition concept from Table 1.6 that most closely applies to the situation described. Use each concept and do not repeat concept numbers in your response.

Nutrition concept number	Situation
1. ____	The Irish potato famine caused thousands of deaths.
2. ____and ____	Otis mistakenly thought that as long as he consumed enough calories from food along with vitamin and mineral supplements, he would stay healthy no matter what he ate.
3. ____	I feel guilty every time I eat potato chips. I wish they weren't bad for me.
4. ____	Phyllis was relieved to learn that her chronic diarrhea was due to the high level of vitamin C supplements she had been taking.
5. ____	A low amount of iron in Tawana's red blood cells was the reason for her loss of appetite and low energy level.
6. ____	Far more young children than soldiers died as a result of the 10-year civil war in Sudan.
7. ____	For the past 20 years, Don's idea of dinner was a big steak and potatoes. His recent heart attack changed his view of what's for dinner.
8. ____	During the two weeks they were backpacking in the Netherlands, Tomás and Ozzie ate very few vegetables and fruits. Their health remained robust, however.
9. ____	Zhang wasn't aware that he had the inherited condition hemochromatosis until he began taking iron supplements and developed iron overload symptoms.

Feedback on Nutrition Up Close is located in Appendix F.

REVIEW QUESTIONS

- **Explain the scope of nutrition as an area of study.**

1. Nutrition is defined as "the study of foods, their nutrients and other chemical constituents, and the effects that food constituents have on health." **True/False**
2. ____ Cassandra is on her way to her first nutrition class and is thinking about what the course will cover. Listed below are her ideas. Three of the ideas correspond to the scope of the study of nutrition. Which idea is *not* a component of the scope of the study of nutrition?
   a. diet and disease relationships
   b. components of healthful diets
   c. magical powers of super foods for weight loss
   d. nutrient composition of foods

- **Demonstrate a working knowledge of the meaning of the 10 nutrition concepts.**

3. The word *nonessential* as in *nonessential nutrient* means that the nutrient is *not* required for growth and health. **True/False**
4. Food insecurity is a problem in developing countries, but it is *not* a problem in the United States or Canada. **True/False**
5. Nutrients are classified into five basic groups: carbohydrates, protein, fats, vitamins, and water. **True/False**
6. The development of standards for nutrient intake levels was prompted in part by the high rejection rate of World War II recruits due to underweight and nutrient deficiencies. **True/False**
7. Tissue stores of nutrients decline after blood levels of the nutrients decline. **True/False**
8. To maintain health, all essential nutrients must be consumed at the recommended level daily. **True/False**
9. An individual's genetic traits play a role in how nutrient intake affects disease risk. **True/False**

10. ____ If you wanted to know the Recommended Dietary Allowance for protein for a 7-month-old infant, you would refer to tables on the:

a. Dietary Guidelines for Americans
b. Infant Nutritional Recommendations for Americans
c. Recommended Daily Intakes
d. Recommended Dietary Allowances

11. ____ The Tolerable Upper Intake Level for iron for a 65-year-old male in milligrams per day (mg/d) is:

a. 1,100
b. 45
c. 350
d. 3.5

12. Certain phytochemicals, or________, such as lycopene and anthocyanins, act as ____________.

13. Groups of people at higher risk than others of becoming inadequately nourished include _______ and ________.

For the questions below, match the term in column A with its definition in column B.

Column A	Column B
____14. Essential nutrients	a. Chemical substances in food that are used by the body for growth and health.
____15. Nutrients	b. Chemical substances that prevent or repair damage to cells caused by exposure to oxidizing agents such as environmental pollutants, smoke, ozone, and oxygen.
____16. Food insecurity	c. Foods that contain relatively high amounts of nutrients in relation to their calorie value.
____17. Antioxidants	d. The chemical changes that take place in the body.
____18. Calorie	e. Limited or uncertain availability of safe, nutritious foods—or the ability to acquire them in socially acceptable ways.
____19. Metabolism	f. A unit of measure of the amount of energy supplied by food.
____20. Malnutrition	g. Nutrients required for normal growth and health that the body can generally not produce, or produce in sufficient amounts.
____21. Nutrient dense	h. Poor nutrition resulting from an excess or lack of calories or nutrients.

Answers to these questions can be found in Appendix F.

NUTRITION SCOREBOARD ANSWERS

1. Calories are a measure of the amount of energy supplied by food. They're a property of food, not a substance present in food. **False**
2. That's the definition of nutrients. **True**
3. Excessive as well as inadequate intake levels of vitamins and minerals can be harmful to health. **False**
4. The DRIs are located at the back of this text. **True**

UNIT 2

The Inside Story about Nutrition and Health

NUTRITION SCOREBOARD

1. How long people live and how healthy they are primarily depends on four factors: lifestyle behaviors, the environments to which they are exposed, genetic factors, and access to quality health care. **True/False**
2. Diet is related to the top two causes of death in the United States. **True/False**
3. Biological processes of modern humans were designed over 40,000 years ago. **True/False**
4. Most economically developed countries regularly monitor levels of various contaminants and nutrients in foods and diets, but the United States does not. **True/False**

Answers can be found at the end of the unit.

Ariel Skelley/Blend Images/Jupiter Images

LEARNING OBJECTIVES

After completing Unit 2 you will be able to:

- Identify characteristics of diets related to the development of specific diseases.
- Explain how differences in diets of early versus modern humans may promote the development of certain diseases.
- List the types of food that are core components of healthful diets.

Nutrition in the Context of Overall Health

- **Identify characteristics of diets related to the development of specific diseases.**

Think of your body as a machine. How well this machine performs depends on a number of related factors: the quality of its design and construction, the appropriateness of the materials used to produce it, and how well it is maintained.

A machine designed to produce 10,000 copies a day will break down sooner if it is used to make 20,000 copies a day. The repair call will, in all probability, come earlier if the machine is overused *and* poorly maintained or if it has a part that doesn't work well. On the other hand, chances are good the copy machine will function at full capacity if it is free from design flaws, skillfully constructed from appropriate materials, properly used, and kept in good shape through regular maintenance.

Although much more complex and sophisticated, the human body is like a machine in some important ways. How well the body works and how long it lasts depend on a variety of interrelated factors. The health and fitness of the human machine depend on genetic traits (the design part of the machine), the quality of the materials used in its construction (your diet), and regular maintenance (your diet, other lifestyle factors, and health care).

Lifestyles exert the strongest overall influence on health and longevity (Illustration 2.1).[2] Behaviors that constitute our lifestyle—such as diet, smoking habits, illicit drug use or excessive drinking, level of physical activity or psychological stress, and the amount of sleep we get—largely determine whether we are promoting health or disease. Of the lifestyle factors that affect health, our diet is one of the most important.[3] In a sense, it is fortunate that diet is related to disease development and prevention. Unlike age, gender, and genetic makeup, our diets are within our control.

People have an intimate relationship with food—each year we put over a thousand pounds of it into our bodies! Food supplies the raw materials the body needs for growth and health; these, in turn, are affected by the types of food we usually eat. The diet we feed the human machine can hasten, delay, or prevent the onset of an impressive group of today's most common health problems.

KEY NUTRITION CONCEPTS

Unit 1 presents 10 key nutrition concepts that are fundamental to the science of nutrition. The content on diet and health covered in this unit directly relates to three of them:

1. Nutrition Concept #3: Health problems related to nutrition originate within cells.

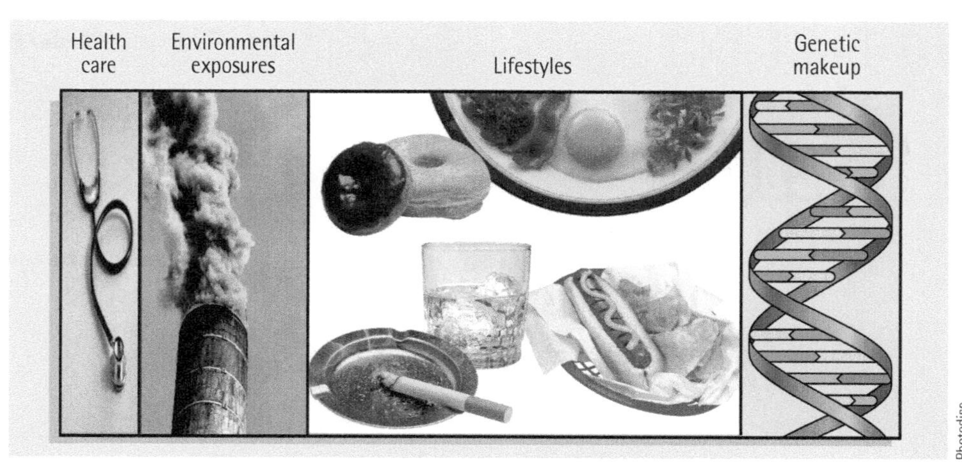

Illustration 2.1 Conditions that contribute to death among adults under the age of 75 in the United States. Health care refers to access to quality care; environmental exposures include the safety of one's surroundings and the presence of toxins and disease-causing organisms in the environment; lifestyle factors include diet, exercise, obesity, smoking, genetic traits, and alcohol and drug use.[3,4]

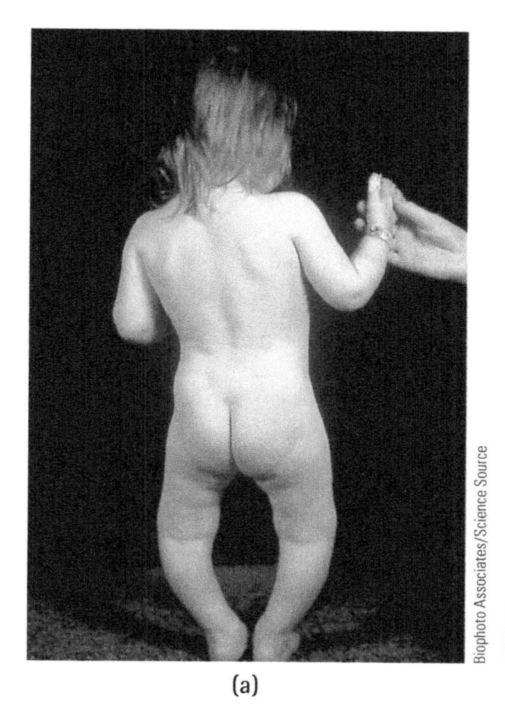

(a)

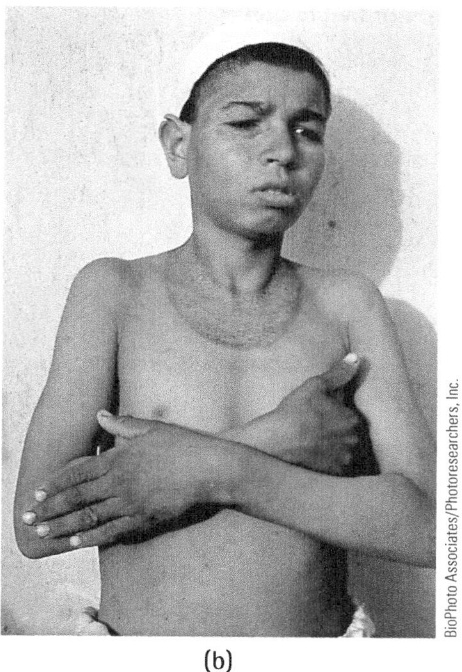

(b)

Illustration 2.2 Vitamin D deficiency (rickets shown on the left) and niacin deficiency (pellagra pictured on the right) were leading causes of hospitalization of children in the United States in the 1930s.

2. Nutrition Concept #6: Malnutrition can result from poor diets and from disease states, genetic factors, or combinations of these factors.
3. Nutrition Concept #8: Poor nutrition can influence the development of certain chronic diseases.

chronic diseases Slow-developing, long-lasting diseases that are not contagious (e.g., heart disease, cancer, diabetes). They can be treated but not always cured.

diabetes Short for the term *diabetes mellitus*, a disease characterized by abnormal utilization of glucose by the body and elevated blood glucose levels. There are three main types of diabetes: type 1, type 2, and gestational diabetes. The word *diabetes* in this text refers to type 2 diabetes, by far the most common.

hypertension High blood pressure. It is defined as blood pressure exerted inside of blood vessel walls that typically exceeds 140/90 mm Hg (millimeters of mercury).

The Nutritional State of the Nation

Since early in the twentieth century, researchers have known that what we eat is related to the development of vitamin and mineral deficiency diseases, to compromised growth and impaired mental development in children, and to the body's ability to fight infectious diseases. Seventy years ago in the United States, widespread vitamin deficiency diseases filled children's hospital wards and contributed to serious illness and death in adults (Illustration 2.2). Now, however, dietary excesses are filling hospital beds and reducing the quality of life for millions of Americans.

Today, the major causes of death among Americans and in other developed countries are slow-developing, lifestyle-related **chronic diseases**.[6] Based on government survey data, 44% of Americans have a chronic condition such as **diabetes**, heart disease, cancer, **hypertension**, or high cholesterol levels; 13% have three or more of these conditions.[7]

The leading causes of death among Americans are heart disease and cancer (see Illustration 2.3). Together they account for 47% of all deaths. Western-type dietary patterns that are generally low in vegetables, fruits, whole grains, dried beans, poultry, nuts, and fish and relatively high in meat, refined grains, sugars, calories, and salt are linked to the development of a number of chronic diseases.[8] People who consume Western-type diets are at higher risk of developing obesity, diabetes, cancer, heart disease, and hypertension.[3,9] Examples of diseases and disorders associated with dietary intake are shown in Table 2.1.

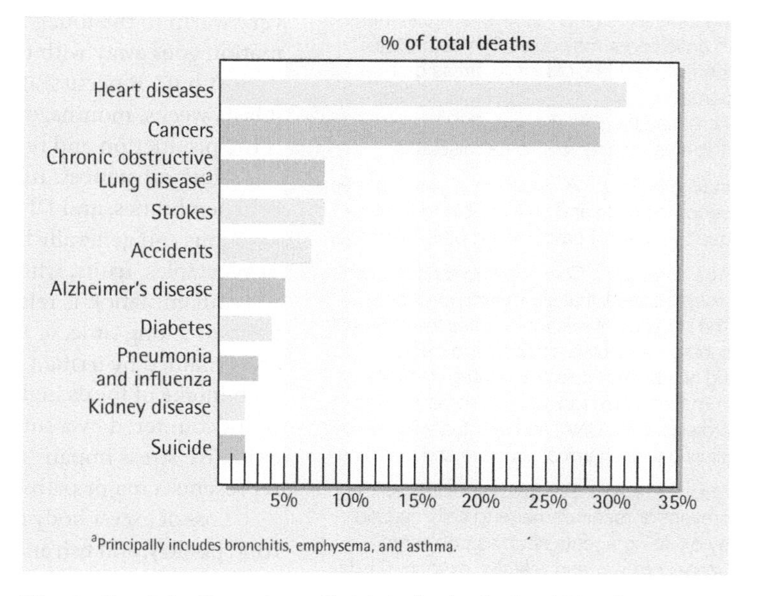

Illustration 2.3 Percentage of total deaths for the top 10 leading causes of death in the United States, 2015.[5]
Source: NCHS Data Brief No. 267, December 2016.

Table 2.1 Examples of diseases and disorders linked to diet[8–14]

Disease or disorder	Dietary connections
Heart disease	Excessive body fat, high intake of *trans* fat, added sugar, and salt; low vegetable, fruit, fish, nuts, and whole-grain intake
Cancer	Low vegetable and fruit intake; excessive body fat and alcohol intake; regular consumption of processed meats
Stroke	Low vegetable and fruit intake; excessive alcohol intake; high animal-fat diets
Diabetes (type 2)	Excessive body fat; low vegetable, whole grain, and fruit intake; high added sugar intake
Cirrhosis of the liver	Excessive alcohol consumption; poor overall diet
Hypertension	Excessive sodium (salt) and low potassium intake; excess alcohol intake; low vegetable and fruit intake; excessive levels of body fat
Iron-deficiency anemia	Low iron intake
Tooth decay and gum disease	Excessive and frequent sugar consumption; inadequate fluoride intake
Osteoporosis	Inadequate calcium and vitamin D; low intake of vegetables and fruits
Obesity	Excessive calorie intake; overconsumption of energy-dense, nutrient-poor foods
Chronic inflammation and oxidative stress	Excessive calorie intake; excessive body fat; high animal-fat diets; low intake of whole grains, vegetables, fruit, and fish
Alzheimer's disease	Regular intake of high animal-fat products; low intake of olive oil, vegetables, fruits, fish, wine, and whole grains

chronic inflammation Low-grade inflammation that lasts weeks, months, or years. Inflammation is the first response of the body's immune system to infectious agents, toxins, or irritants. It triggers the release of biologically active substances that promote oxidation and other reactions to counteract the infection, toxin, or irritant. A side effect of chronic inflammation is that it may also damage lipids, cells, tissues, and certain body processes. Also called low-grade inflammation.

oxidative stress A condition that occurs when cells are exposed to more oxidizing molecules (such as free radicals) than to antioxidant molecules that neutralize them. Over time, oxidative stress causes damage to lipids, DNA, cells and tissues. It increases the risk of heart disease, type 2 diabetes, cancer, and other diseases.

osteoporosis A condition in which bones become fragile and susceptible to fracture due to a loss of calcium and other minerals.

free radicals Chemical substances (often oxygen-based) that are missing electrons. The absence of electrons makes the chemical substance reactive and prone to oxidizing nearby molecules by stealing electrons from them. Free radicals can damage lipids, proteins, cells, DNA, and eventually tissues by altering their chemical structure and functions.

antioxidants Chemical substances that prevent or repair damage to cells caused by oxidizing agents such as pollutants, ozone, smoke, and reactive oxygen. Oxidation reactions are a normal part of cellular processes. Vitamins C and E and certain phytochemicals function as antioxidants.

Shared Dietary Risk Factors A number of the diseases and disorders listed in Table 2.1 share the common risk factors of low intake of vegetables, fruits, and whole grains; excess body fat and alcohol intake; and high animal-fat intake. These risk factors are associated with the development of **chronic inflammation** and **oxidative stress**, conditions that are strongly related to the development of heart disease, diabetes, **osteoporosis**, Alzheimer's disease, cancer, and other chronic diseases.[15]

Inflammation and Oxidative Stress Inflammation is an important part of the body's defense systems against cell and tissue damage due to the presence of infectious agents, chemical irritants, toxins, or physical injury. It can be classified as acute or chronic. Most of us are familiar with acute inflammation. That's a temporary reaction that occurs when you sprain your ankle or develop a fever. The ankle you injured or your head feels very warm to the touch and hurts, and your head may ache from the fever. Acute inflammation goes away with time as your body heals itself. Chronic, low-grade inflammation doesn't hurt; it occurs silently within the fluids, tissues, and cells inside your body; and it lasts weeks, months, or years. An important part of the body's inflammatory response is the production and release of oxidizing agents such as **free radicals** that destroy the offending substances. In the process, however, free radicals may also oxidize fats (lipids), cell membranes, and DNA inside of cells. In the short term, damage induced by oxidation reactions can generally be reduced by **antioxidants** produced by the body and consumed in vegetables, fruits, whole-grain products, and other plant foods.[16]

Inflammation is related to chronic disease development when it is present at a low level for a long time, or is *chronic*. Chronic inflammation and the resulting oxidative stress are sustained by irritants continually present in the body. Excess body fat and habitually high intake of foods high in saturated fats and added sugar are examples of such irritants. If not countered by a sufficient supply of antioxidants, chronic inflammatory processes and oxidative stress impair the normal functioning of cells and tissues. Chronic inflammation represents a major pathway by which diet influences the development of chronic disease.[39]

Loss of excess body fat and a dietary pattern high in whole grains, colorful vegetables, fruit, poultry, and fish and low in refined grains, added sugars, red and processed meat, high-fat dairy products, and sweetened beverages is associated with lower levels of inflammation compared to dietary patterns that do not.[17, 18] Table 2.2 summarizes the types of foods included in dietary patterns that tend to decrease or increase oxidative stress and inflammation.

Nutrient–Gene Interactions and Health Some diseases are promoted by interactions between nutrients and genes. One example of such an interaction involves gene types and the health effects of cruciferous vegetables such as broccoli, cauliflower, brussels sprouts, and cabbage. These vegetables contain isothiocyanates (pronounced ice-o-thee-o-sigh-ah-nates), which are involved in the prevention of cancer development. Individuals carrying certain types of genes that lead to the rapid breakdown of isothiocyanates and related compounds are more susceptible to cancer development than others who break down these substances slowly. Isothiocyanates appear to block mechanisms that promote tumor development.[19] Another example of nutrient–gene interactions that influence health relates to obesity. The causes of obesity are complex and include interactions between a number of gene types and environmental factors. One contributing genetic factor appears to be the form of the FTO gene present. This gene participates in processes that regulate appetite and food intake. People with the "high-risk" form of the FTO gene experience a higher lifetime risk of becoming overweight or obese compared to those who have the "low-risk" form of the gene.[20]

Knowledge of nutrient–gene interactions in health and disease is expanding rapidly and is greatly enhancing our understanding of the relationship between diet and health. This knowledge will contribute to the development of personalized nutritional interventions targeted at an individual's genetic characteristics.[21]

Table 2.2 Types of food associated with decreased or increased inflammation, oxidative stress, or both[11,14,15]

Decreased
Colorful fruits and vegetables
Dried beans
Whole grains
Fish and seafood, fish oils
Red wines
Dark chocolates
Olive oil
Nuts
Coffee, tea
Increased
Processed and high-fat meats
High-fat dairy products
Baked products, snack foods with *trans* fats
Soft drinks, other high-sugar beverages
Excess alcohol
Refined grain products (rice, white bread, pastries)

The Importance of Food Choices

People are not born with an internal compass that directs them to select a healthy diet—and it shows. If given access to a food supply like that available in the United States, people show a marked tendency to choose a diet that is high in energy-dense, nutrient-poor foods[22] (Illustration 2.4). Such a diet tends to include processed foods high in saturated fat, salt, or sugar and low in whole grains, vegetables, fruits, and other basic foods. This type of diet poses the greatest risks to the health of Americans.[22]

Diet and Diseases of Western Civilization

- **Explain how differences in diets of early versus modern humans may promote the development of certain diseases.**

Why is the U.S. diet—a "Western" style of eating—hazardous to our health in so many ways? What is it about a diet that is high in animal fat, salt, and sugar and low in vegetables, fruits, and whole grains that promotes certain chronic diseases? A good deal of evidence indicates that the chronic diseases now prevalent in the United States and other Westernized countries have roots in dietary changes that have taken place over centuries.

Our Bodies Haven't Changed

The biological processes that control what the human body does with food were developed over the course of our evolution tens of thousands of years ago. These processes exist today because they are linked to the genetic makeup of humans and continue to influence how diet affects health.[23,24]

Then . . . For the first 200 centuries of their existence, humans survived by hunting and gathering (Illustration 2.5). They were constantly on the move, pursuing wild game or following the seasonal maturation of fruits and vegetables. Meat, berries, and many other plant products obtained from successful hunting and gathering journeys spoiled quickly, so they had to be consumed in a short time. Feasts would be followed by famines that lasted until the next successful hunt or harvest.[23]

. . . and Now The bodies of modern humans, adapted to exist on a diet of wild game, fish, fruits, nuts, seeds, roots, vegetables, and grubs; to survive periods of famine; and to sustain a physically demanding lifestyle are now exposed to a different set of circumstances.

Illustration 2.4 Lopsided, all-American food choices.

The foods we eat bear little resemblance to the foods available to our early ancestors (Illustration 2.6). Sugar, salt, alcohol, food additives, oils, margarine, dairy products, refined grain products, and processed foods were not a part of their diets. These ingredients and foods came with Western civilization.[23] Furthermore, we do not have to engage in strenuous physical activity to obtain food, and our feasts are no longer followed by famines.

The human body developed other survival mechanisms that are not the assets they used to be. Mechanisms that stimulate hunger in the presence of excess body fat stores, conserve the body's supply of sodium, and confer an innate preference for sweet-tasting foods—as well as a digestive system that functions best on a high-fiber diet—were advantages for early humans. They are not advantageous for modern humans, however, because our diets and lifestyles are now vastly different.

(a)

(b)

Illustration 2.5 Hunter-gatherers still exist in the world, but their numbers are diminishing. It is estimated that hunter-gatherers consume approximately 3,000 calories daily due to their physically demanding way of life.

Irven DeVore/Anthro-Photo

Richard Anderson

Illustration 2.6 The disconnect between high animal fat, high salt, high sugar, and processed foods in Western-type diets (right) and wild plants and animal foods consumed by our early ancestors (left). Foods consumed by hunter-gatherers shown in the photograph include birds' eggs, wild cucumbers, roots, nuts, and berries. Not shown are grubs and other insects, which might be consumed as quickly as they are discovered.

Although the human body has a remarkable ability to adapt to changes in diet, the health problems of modern civilization such as heart disease, cancer, hypertension, and diabetes are thought to result, in part, from diets that are greatly different from those of our early ancestors. The human body was built to function best on a diet that is low in sugar and sodium, contains lean sources of protein, and is high in fiber, vegetables, and fruits.[25] Strong evidence for this conclusion is provided by studies that track how disease rates change as people adopt a Western style of eating.

Different Diets, Different Disease Rates

Many countries are adopting the Western diet and the pattern of disease that accompanies it. People in Japan, for example, live longer than almost anyone else in the world—until they move to the United States (Table 2.3). In Japan, the traditional diet consists mainly of rice, vegetables, fish, shellfish, broths, tofu, noodles, seaweed, eggs, tea, and small portions of meat

Table 2.3 Life expectancy at birth for countries with high life expectancies, 2015[33]

Country	Life expectancy (years)	Country	Life expectancy (years)
Japan	84	Greece	82
Switzerland	83	Austria	82
France	83	Finland	81
Italy	83	Germany	81
Spain	83	Belgium	81
Australia	82	New Zealand	81
United Kingdom	82	Denmark	81
Canada	82	Costa Rica	80
Ireland	82	Cuba	80
Portugal	82	United States	79

Illustration 2.7 Traditional Japanese meal. Compare these to the "all-American" food choices shown in Illustration 2.4.

JTB MEDIA CREATION, Inc./Alamy

(Illustration 2.7).[26] When Japanese people move to the United States, their diets often change to include, on average, more fat, sugar, and calories and less fish and vegetables.[27] Japanese living in the United States are much more likely to develop diabetes (Illustration 2.8), heart disease, breast cancer, and colon cancer than people who remain in Japan.[29,31]

Dietary habits in Japan are rapidly becoming similar to those in the United States. Hamburgers, fries, steak, ice cream, and other high-fat foods are gaining in popularity. Rates of diabetes, heart disease, and cancer of the breast and colon are on the rise in Japan.[27,30] Similarly, the "diseases of Western civilization" are occurring at increasing rates in Russia, Greece, Israel, and other countries adopting the Western diet.[32] Obesity, diabetes, and heart disease rates tend to increase among some population groups after they immigrate to the United States. Latinos who move to the United States, for example, tend to consume poorer quality diets, are more likely to become obese, and are more likely to develop diabetes than people in their home countries.[34]

Today's food supply makes it a bit challenging to eat more like our early ancestors did. What types of foods would you choose if you wanted to shape a diet that is closer to that of hunter-gathers? Join Beth and Shandra in making these dietary decisions in the Reality Check.

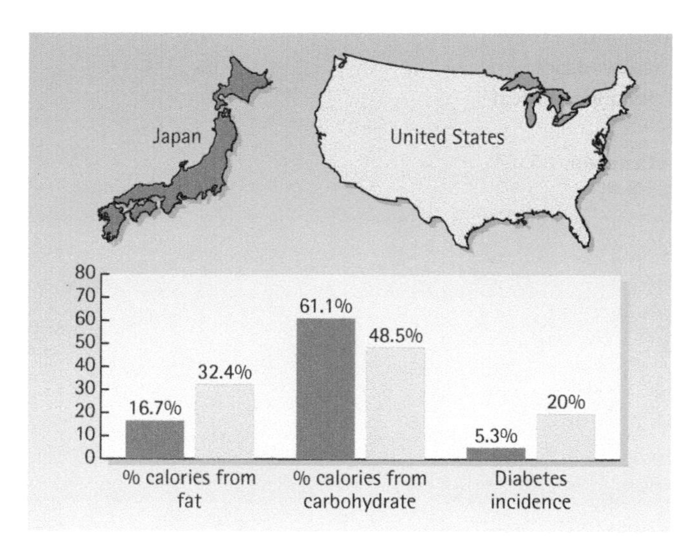

Illustration 2.8 An increased rate of diabetes in Japanese men immigrating to Seattle corresponds to dietary changes.
Source: Tsunehara CH, et al. Diet of second-generation Japanese-American men with and without non-insulin-dependent diabetes. *Am J Clin Nutr.* 1990;52:731–38.

The Power of Prevention

Heart disease, cancer, and other chronic diseases are not the inevitable consequence of Westernization. The types of diets that promote chronic disease can be avoided or changed. Although heart disease is still the leading cause of death in the United States, its rate has declined by over 50% in the last 30 years. About half of this decline is related to improvements in risk factors such as smoking, hypertension, and elevated blood cholesterol levels; the other half is primarily related to medical interventions for people with heart disease.[35] The American Heart Association concludes that future gains in heart health among Americans will stem primarily from improved dietary intakes, declines in rates of overweight and obesity, increased physical activity, and decreased smoking.[36]

Table 2.4 Examples of the *Healthy People 2020* nutrition objectives for the nation[37]

Healthier food access
• Increase the proportion of schools that offer nutritious foods and beverages outside of school meals.
• Increase the proportion of Americans who have access to a food retail outlet that sells a variety of foods that are encouraged by the Dietary Guidelines.
Weight status
• Increase the proportion of adults who are at a healthy weight.
• Reduce the proportion of adults who are obese.
• Reduce the proportion of children and adolescents who are considered obese.
• Prevent inappropriate weight gain in youths and adults.
Food insecurity
• Eliminate very low food security among children.
• Reduce household food insecurity and in doing so reduce hunger.
Food and nutrient consumption
• Increase the variety and contribution of vegetables to diets.
• Increase the contribution of whole grains to diets.
• Reduce the consumption of calories from solid fats and added sugars.
• Reduce the consumption of saturated fat.
• Reduce the consumption of sodium.
• Increase the consumption of calcium.
• Reduce the consumption of sodium.
Iron deficiency
• Reduce iron deficiency among young children and females of childbearing age.
• Reduce iron deficiency among pregnant females.
Health care and worksite settings
• Increase the proportion of physician office visits that include counseling or education related to nutrition and weight.

Improving the American Diet

- **List the types of food that are core components of healthful diets.**

Many efforts are under way to improve the diet and health status of Americans. Like other countries with high rates of "Western" diseases, the United States sets national health goals and implements programs aimed at improving health. Goals and objectives for changes in health status in the United States are presented in the report *Healthy People 2020*. Examples of the *Healthy People 2020* objectives for nutrition are shown in Table 2.4. These objectives highlight the national emphasis on improving weight status and dietary intake of the population by the year 2020.

REALITY CHECK

Getting Back to the Basics—But How?

Beth and Shandra have been roommates for a year and usually shop for groceries together. For the next trip to the grocery store, they decide that each of them will make up a shopping list that includes foods that resemble those their early ancestors might have eaten. Here are the results.

Which list do you think comes closest to matching the basic foods consumed by our early ancestors?

Answers on page 2-10.

Photodisc

Beth: rice, yogurt, pork, honey, olive oil

Photodisc

Shandra: carrots, nuts, asparagus, fish, blueberries

ANSWERS TO REALITY CHECK

Getting Back to the Basics—But How?

Shandra's list contains unprocessed plant foods and fish, which would have been available to our early ancestors.

Beth:

Shandra:

Table 2.5 ChooseMyPlate.gov food guidance

Balancing calories
• Enjoy your food, but eat less.
• Avoid oversized portions.
Foods to increase
• Make half your plate fruits and vegetables.
• Make at least half your grains whole grains.
• Switch to fat-free or low-fat (1%) milk.
Foods to reduce
• Compare sodium in foods like soup, bread, and frozen meals—and choose the foods with lower numbers.
• Drink water instead of sugary drinks.

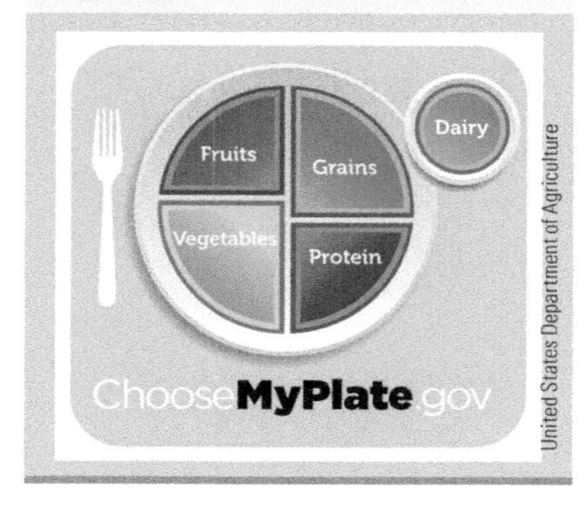

What Should We Eat?

Evidence-based conclusions about nutrition and health are translated into food choices by the U.S. Department of Agriculture's (USDA) food guidance materials. Intended for public consumption, the materials specify food choices that build healthy dietary patterns and contribute to the prevention of common diseases such as diabetes, hypertension, and heart disease.

USDA's advice on healthy food choices changes as knowledge about diet and health relationships advances, and so do the names it uses to label the new food guidance materials. In the past nine decades, USDA's food choice guidance has changed from the "Basic Five Food Groups," to the "Basic Four Food Groups," to "MyPyramid." Beginning in 2011, food guidance materials became labeled "MyPlate" and "ChooseMyPlate" and reflect current concerns about food choices, nutrition, and health (Table 2.5).

Current food intake recommendations focus on basic, nutrient-dense foods such as whole-grain products, vegetables, fruits, lean meats, low-fat dairy products, dried beans, and fish. They call for reduced consumption of soft drinks and other sweetened beverages, high-fat meat and dairy products, and foods high in salt. Sweets, desserts, and packaged snacks are not excluded from the recommendations. They can be included in a healthy dietary pattern as long as food group recommendations are met and overall calorie needs are not exceeded. ChooseMyPlate interactive tools for planning dietary intake are available at ChooseMyPlate.gov.

Nutrition Surveys: Tracking the American Diet

The food choices people make and the quality of the American diet and food supply are regularly evaluated by national surveys (Table 2.6). The first survey began in 1936. It was conducted in conjunction with the original national program aimed at reducing hunger, poor growth in children, and vitamin and mineral deficiency diseases. Results of nutrition surveys are used to identify problem areas within the food supply, characteristics of diets consumed by the public, and the prevalence of nutrition-related health disorders. The surveys provide information ranging from the amount of lead and pesticides in certain foods to the adequacy of diets of low-income families. Together with the results of studies conducted by university researchers and others, they provide the information needed to give direction to food and nutrition programs and to policies aimed at improving the availability and quality of the food supply. Many other countries, including Canada and Australia, have similar programs.

Table 2.6 Periodic national surveys of food, diet, and health in the United States

Survey	Purpose
1. National Health and Nutrition Examination Survey (NHANES)	Assesses dietary intake, health, and nutritional status in a sample of adults and children in the United States on a continual basis
2. Nationwide Food Consumption Survey (NFCS)	Performs regular surveys of food and nutrient intake and understanding of diet and health relationships among a national sample of individuals in the United States
3. Total Diet Study (sometimes called Market Basket Study)	Ongoing studies begun in 1961 that determine the levels of various contaminants and nutrients in foods and diets

Ariel Skelley/Blend Images/ Jupiter Images

NUTRITION up close

Food Types for Healthful Diets

Focal Point: Separating the types of foods that fit into the ChooseMyPlate basic food groups from those that characterize Western-type diets.

The ChooseMyPlate food groups consist of foods that make up a healthful, disease-preventing dietary pattern. The typical Western-type dietary pattern, on the other hand, is related to the development of a number of diseases and includes many types of food that are not part of the ChooseMyPlate basic food groups. To which dietary pattern do the following types of food most appropriately belong?

Food types	Dietary pattern	
	ChooseMyPlate	Western
Mixed vegetables		
Cold cuts (ham, bologna, salami)		
Fish and seafood		
Whole-grain breads		
Fruit jams and jellies		
Potato, tortilla, and other snack chips		
Dried beans		
Poultry (chicken, turkey)		
Fruit		
Vegetables		
Gravy		
Ice cream		
Soft drinks		
Salad dressing, mayonnaise		
Skim milk		
Cake		

Feedback to the Nutrition Up Close is located in Appendix F.

REVIEW QUESTIONS

- **Identify characteristics of diets related to the development of specific diseases.**

1. Vitamin deficiency diseases remain a major health problem for children in the United States. **True/False**
2. Diets are related to the top two causes of death. **True/False**
3. Low intake of vegetables and fruits is related to the development of heart disease, cancer, hypertension, and osteoporosis. **True/False**
4. The incidence of chronic diseases such as diabetes, cancer, and heart disease increases as countries adopt a Western style of eating. **True/False**

- **Explain how differences in diets of early versus modern humans may promote the development of certain diseases.**

5. Our genetic makeup changes as our diets change. **True/False**

6. List two major characteristics of the lifestyles of early humans cited in this unit that are rare in modern humans and don't involve types of foods available.
   a. ______________________________
   b. ______________________________
7. ___ Chronic inflammation and oxidative stress are sustained by irritants continually present in the body. Two examples of irritants that promote the presence of these conditions in the body are:
   a. physical inactivity and meal skipping
   b. weight loss and lack of sleep
   c. excess body fat and habitually high intake of added sugars
   d. high intake of water and excess dietary fiber
8. ___ Adverse effects of chronic inflammation and oxidative stress can be diminished by:
   a. loss of excess body fat
   b. increased water consumption
   c. decreased intake of high-fiber foods
   d. daily consumption of potatoes
9. Although individuals cannot change their genetic makeup, they can change their risk for chronic disease development by making healthful eating and lifestyle changes. **True/False**

- **List the types of food that are core components of healthful diets.**

10. National surveys in the United States assess food intake, nutritional health of the population, and the safety of the food supply. **True/False**
11. The *Basic Four Food Groups* is the most recently published guide to selection of a healthful diet. **True/False**
12. ___ Which of the following statements about food choices is true?
    a. Given the availability of a wide assortment of foods, people tend to select and consume a healthful diet.
    b. People are *not* born with an internal compass that directs them to select and consume a healthy diet.
    c. Food choices usually change very little as people move from one country to another.
    d. Recommendations for healthy diets include a narrow range of food choices.

Questions 13–15 refer to the following case scenario:

Assume you are attending a family picnic and are staring at a table full of food. Your choices from the table are: fried chicken, cold-cut platter (bologna, salami), whole-grain rolls, shrimp, jell-o, spinach salad, fruit salad, baked beans, potato chips, zucchini squash, cake, and low-fat milk.

13. ___ From the food options available, which three foods would you put on your plate if you wanted to consume the basic foods?
    a. fried chicken, fruit salad, and baked beans
    b. zucchini squash, jell-o, and baked beans
    c. spinach salad, cold cuts, and shrimp
    d. spinach salad, zucchini squash, and shrimp
14. ___ Which of the following sets of foods would be considered creations of modern humans?
    a. potato chips and cold cuts
    b. cake and spinach salad
    c. jell-o and fruit salad
    d. fried chicken and zucchini squash
15. ___ You decide to have a glass of milk along with shrimp, fruit salad, baked beans, and spinach salad. Which basic food group is missing from your plate?
    a. vegetables
    b. fruits
    c. grains
    d. protein foods

Answers to these questions can be found in Appendix F.

NUTRITION SCOREBOARD ANSWERS

1. There are no secrets to a long, healthy life. **True**
2. Diet is associated with the development of heart disease and cancer, which cause over half of all deaths in the United States.[1] **True**
3. Hairstyles may be different, but our bodies are the same as they were 40,000 years ago. **True**
4. The Total Diet Study has been ongoing in the United States since 1961.[38] False

Ways of Knowing about Nutrition

Mimagephotography/Shutterstock.com

NUTRITION SCOREBOARD

1 It is illegal to convey false or misleading information about nutrition in magazine and newspaper articles and on television. **True/False**

2 Knowledge about nutrition is gained by scientific studies. **True/False**

3 "Double-blind" studies are used to diminish the "placebo effect." **True/False**

4 Assume that studies have shown that a raw food diet increases life expectancy in animals. These results would indicate that raw food diets would increase life expectancy in adults. **True/False**

5 You read in an online science blog that fructose intake is associated with the development of type 2 diabetes in men. This means that fructose intake causes type 2 diabetes. **True/False**

Answers can be found at the end of the unit.

LEARNING OBJECTIVES

After completing Unit 3 you will be able to:

- Evaluate the reliability of advertisements and other information about nutrition and nutrition-related products and services.
- Identify sources of reliable nutrition information.
- Explain each component of the scientific methods used to identify reliable information about nutrition.

How Do I Know if What I Read or Hear about Nutrition Is True?

- **Evaluate the reliability of advertisements and other information you hear or read about nutrition and nutrition-related products and services.**

"Fresh fruits and vegetables are more nutritious than canned or frozen fruits and vegetables."

"Coconut oil helps you lose weight."

"Organic foods are nutritionally superior to conventionally grown foods."

In the words of Dr. Martin Luther King, Jr., "The function of education is to teach one to think intensively and to think critically." That's what this unit is about. How do you know if what you read or hear about nutrition is true? With only the information given in these examples, you couldn't know. In reality, however, this is how much of the information we receive about nutrition comes to us—in bits and pieces. The nutrition information offered to the public is a mix of truths, half-truths, and gossip. The information does not have to meet any standard of truth before it can be represented as true in books, magazines, newspapers, TV and radio reports and interviews, pamphlets, the Internet, and speeches.

Opinions expressed about nutrition are protected by the freedom of speech provisions of the U.S. Constitution. Although it is misleading and fraudulent for a tabloid article or an Internet site to announce "19 foods have negative calories," it is not illegal. The promotion of nutritional remedies that are not known to work, such as amino acid supplements for hair growth and coffee bean extract for weight loss, is likewise protected by freedom of speech. It is illegal, however, to put false or misleading information about a product or service on a product label, in a product insert, or in an advertisement. In addition, the U.S. and Canadian mail systems cannot be used to send or to receive payments for products that are fraudulent.

With so much misinformation available, it is difficult to know what to do when we hear or read something about nutrition that may benefit us personally or perhaps help a friend or relative. Why does such a mix of nutrition sense and nonsense exist? How can you separate the sound information from the highly questionable? Where does nutrition information you can trust come from? These questions are answered in this unit.

KEY NUTRITION CONCEPTS

Content in Unit 3 on how trustworthy information about nutrition is generated and how to identify it relates to these key nutrition concepts:

1. Nutrition Concept #3: Health problems related to nutrition originate within cells.
2. Nutrition Concept #6: Malnutrition can result from poor diets and from disease states, genetic factors, or combinations of these factors.

Profits, Prophets, and Proof

Consumers are bombarded with nutrition information, much of it about products and services that may be ineffective or scientifically untested. Likely reasons for this situation can be grouped into two major categories: profit and personal beliefs and convictions. The first and most important reason is the profit motive.

Motivation for Nutrition Misinformation #1: Profit

. . . [join] a special company with 22nd century breakthrough nutritional products and unequaled compensation plan!
—USA TODAY *BUSINESS ADVERTISEMENT*

Pump Up Sales With Heart Healthy Nutrients:...
people everywhere are turning to better nutrition.
Satisfy their needs with custom nutrient premixes....
—AN INGREDIENT INDUSTRY-SPONSORED NEWSLETTER

As long as consumers seek quick and easy ways to lose weight, build muscle, slow aging, and reduce stress—goals that cannot be achieved quickly or easily—there will continue to be a huge market for nutrition products and services that offer assistance (Illustration 3.1).[1,2]

Not everyone who is in the business of nutrition has the goal of maintaining or improving people's health. Many seek to make money from people who are willing to believe their advertisements and buy products or services because they sound like they will work.

People may believe highly questionable claims about nutrition-related products, and be willing to buy and try the products for many reasons:[2,3]

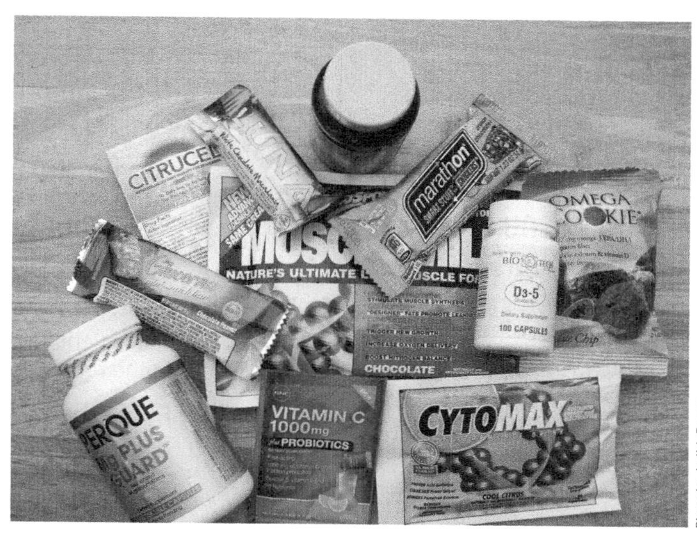

Photo by Judith Brown

Illustration 3.1 Some examples of the many nutrition products available.

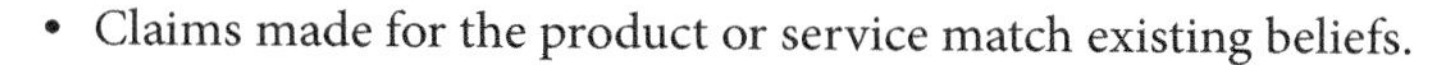

- Claims made for the product or service match existing beliefs.
- Friends or family members told you that the product or service worked for them.
- The product or service offers a money back guarantee.
- Promotional materials sound scientific and true.
- Individuals may lack the knowledge and skills needed to evaluate product claims.
- The products offer solutions to important problems that have few or no solutions in orthodox health care.
- Promotional materials may appeal to people who are dissatisfied with traditional medical care, fear the costs or side effects of medications, or want a "natural" remedy.
- Many people believe that vitamins, minerals, proteins, amino acids, herbs, and similar products are healthful and harmless.

Instead of evidence and facts, profit-oriented companies may use testimonials ("It worked for me, it will work for you!"), "medical experts," and Hollywood stars and sports heroes to promote their products. Their advertisements promote ideas like "a wonder to science" or "miraculous" that appeal to some people's inclinations toward the mysterious or the divine. Real remedies aren't advertised in such terms. Nor are they portrayed as being effective because they come from Europe, the Ecuadoran highlands, the ancient Orient, or organic algae ponds.

Illustration 3.2 offers a formula for developing and marketing a fraudulent nutrition product. Try following the steps and make up your own miraculous nutritional cure. Once you've devised your own product, you'll find it easier to detect fraudulent marketing.

Controlling Profit-Motivated Nutrition Frauds The Federal Trade Commission (FTC) has the authority to remove advertisements that make false claims from the airwaves and the Internet. In the past, the FTC has exerted its authority by removing blatantly false and misleading advertisements, including those for the "European Weight Loss Patch," colon detoxifiers, juices with herbal extracts, coral calcium, male enhancement dietary supplements, and bee pollen. Nevertheless, misleading and inaccurate advertisements still appear because enforcement efforts are weak and are concentrated on very dangerous products. The FTC and other federal agencies do respond when several consumers register complaints about a nutritional product or advertisement. You can notify the FTC of a complaint from the website ftccomplaintassistant.gov.

Science Interrupted Here's a true story:

Friday, 2:00 pm. A call came into the Nutrition Department from a man who wanted to talk to a nutrition expert. He had heard on last night's news that zinc lozenges were good for treating

of Tennesse... made $7,000 in commission. No Gimmicks – No competition – Big money every week! For more information call
(800) 555-1234

HAVE YOU EVER SEEN A FAT PLANT?

Scientists in our laboratory have discovered the secret. For a lifetime of being reed thin, send $29.95 to:

Obesity Research Center
P.O. Box 281752,
Miami, FL

HOT! HOT! HOT!

Banks & Financial Institutions made HUGE PROFITS trading foreign currencies & so ca...

1. Identify a common problem people really want fixed that cannot easily or quickly be fixed another way.

   EXAMPLES: Obesity, low energy, weak muscles.

2. Make up a nutritional remedy and connect it to a biological process in the body. Try to think of a remedy that probably won't harm anyone. While you are at it, create a catchy name for your product.

   EXAMPLES: An herb that burns fat by speeding up metabolism, vitamins that boost the body's supply of energy, protein supplements that go directly to your muscles.

3. Develop a scientific-sounding explanation for the effect the product has on the body. Refer to the results of scientific studies.

   EXAMPLES: "Research has shown that the combination of herbs in this product stimulates the production of chemicals that trigger the energy-production cycle in the body." "As nutritional scientists have known for a century, the body requires B vitamins to form energy. The more you consume of this special formulation of B vitamins, the more energy you can produce and the more energy you will have!" "The unique combination of amino acids in this product is the same as that found in the jaw muscles of African lions. It is well known that an African lion can lift a 600-pound animal in its teeth."

4. Dream up testimonials from bogus users of the product, before-and-after photos, or "expert" opinions to quote. Use terms like "magical," miraculous," "suppressed by traditional medicine," "secret," or "natural" as much as possible.

   EXAMPLES: Jane Fondu, Indianapolis: "I didn't think this product was going to work. The other products I tried didn't. My doctor couldn't help me lose weight. Thank God for [insert name of product]! I miraculously lost 20 pounds a week while eating everything I wanted." Dr. J. R. Whatsit, DM, NtD, Director of Nutritional Research: "We discovered this secret herb in our laboratories after years of looking for the substance in plants that keeps them from getting fat. Voilà! We found it. This discovery may win us a Nobel Prize." Or show photos of a skinny person and a beefed-up person. (The photos don't have to be of the same person; the people only need to look similar.)

5. Offer customer a money-back guarantee.

Illustration 3.2 Create your own fraudulent nutrition product.

cold symptoms. Because he had a cold and didn't think the zinc would hurt, he purchased and consumed a whole roll of lozenges. Now he had a horrible, metallic taste in his mouth that he couldn't get rid of, no matter how often he brushed his teeth or gargled with mouthwash. He was worried that the taste was going to stay in his mouth forever. The nutrition expert assured him that it wouldn't. He had overdosed on zinc, and the taste would go away slowly.

This man was not the only one who heard the newscast about zinc and consumed too many lozenges the next day. Sales of zinc lozenges skyrocketed after the newscast, and as it happened, the author of the research study profited handsomely from the increased sales. After completing the study, the author bought shares of stock in the company that made the lozenges.[4] Such a financial tie should have been reported in the article. It is possible that a financial incentive could have influenced how the author presented the study's results.

Some researchers have a vested interest in their research results. A study of 1,000 Massachusetts scientists who had published research articles found that one-third held a patent for the product tested, were paid industry consultants, or had another form of financial stake in the research.[4] Another study found that authors of research supportive of a new artificial fat were four times more likely to have financial affiliations with the company producing the product than were authors with no such company ties.[5,6]

Who Is Conducting the Research? Nutrition research is sometimes conducted by people or companies that have a stake in the results. Although this doesn't mean the research results are invalid, it does mean that the study design should be carefully scrutinized before it is published in a scientific journal and broadcast on the news. People and companies with vested interests in the results of studies may not report findings that reflect negatively on a product.[7] Promotional materials may neglect to mention studies that produced different results. For these reasons, it is important to consider whether the results of the study and other claims may be tainted by purely financial motives.

A Checklist for Identifying Nutrition Misinformation
Illustration 3.3 lists some common features of fraudulent information for nutrition products and services. If you find any of these characteristics in articles, advertisements, websites, or pamphlets, or hear them on infomercials or TV and radio interviews, beware of the product or service. People offering legitimate information are cautious and don't exaggerate nutritional benefits to health. They usually aren't selling anything but rather are trying to inform people about new findings so they can make better decisions about optimal nutrition.

The Business of Nutrition News Newspapers, TV stations, magazines, books, and other sources of information make money when the number of viewers or readers is high. To increase viewers or readers, the media attempts to present information that will pique people's interest. Nutrition and health news tends to do that, and the poor journalism confuses the public a lot.[1,8] The media have access to a large number of nutrition studies; in a recent 6-month period over 680,000 papers on nutrition or diet were published in peer-reviewed journals (Illustration 3.4).

Studies reporting new results related to obesity, cancer, diabetes, heart disease, vitamins, and food safety are particularly hot topics. To keep people interested, the media may sensationalize and oversimplify nutrition-related stories. They may report on one study one day and describe another study with opposite results the next. A classic example of headlines about nutrition that have confused the public is shown in Illustration 3.5. Such study-by-study coverage of nutrition news leaves consumers not knowing what to believe or what to do. A Nutrition Trends Survey by the American Dietetic Association uncovered a strong consumer preference (81% of those sampled) for learning about the latest nutrition breakthroughs only *after* they had been generally accepted by nutrition and health professionals.

Nutrition studies are complex, and the results of one study are almost never enough to prove a point. Decisions about personal nutrition should be based on accumulated evidence that is broadly supported by nutrition and other scientists.

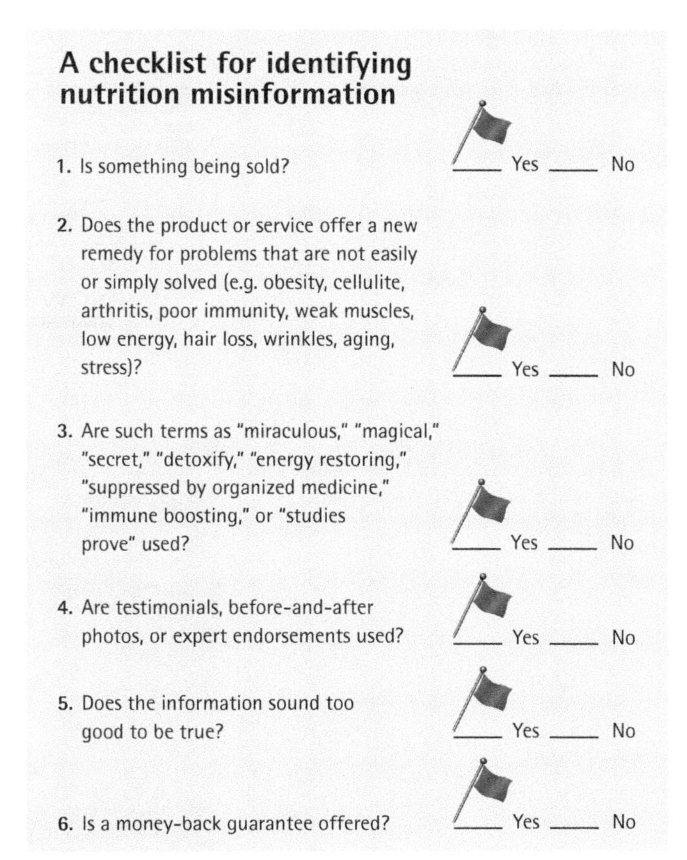

Illustration 3.3 If a red flag comes up as you read about a product or service, beware.

Motivation for Nutrition Misinformation #2: Personal Beliefs and Convictions Numerous alternative health practitioners (such as nutripaths, iridologists, electrotherapists, Scientologists, and faith healers) use nutrition remedies. Although their approaches to and philosophies about health may vary, these practitioners often have strong beliefs in the benefits of what they do.

There's no way to say whether most of the remedies used by alternative nutrition practitioners work. (Some of the unproven remedies employed are shown in Illustration 3.6.) The nutritional cures offered are often not based on evidence provided by scientific

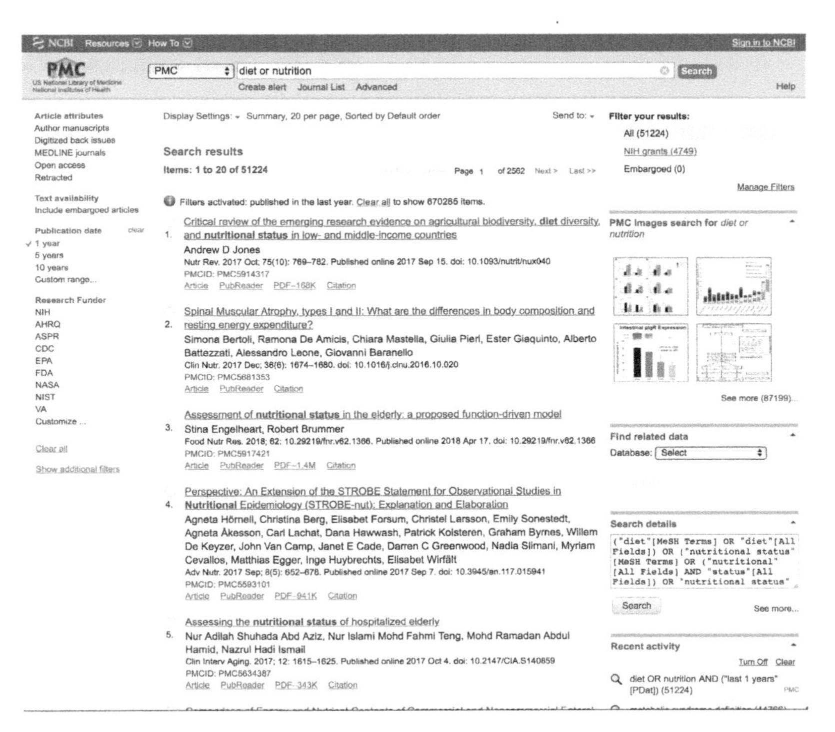

Illustration 3.4 Searching the term "nutrition OR diet" on the National Library of Medicine's PubMed site indicates that almost 39,000 research papers were published within a recent 12-month period.

Illustration 3.5 How to confuse the public: a lesson delivered by the headlines about vitamin E and heart disease.

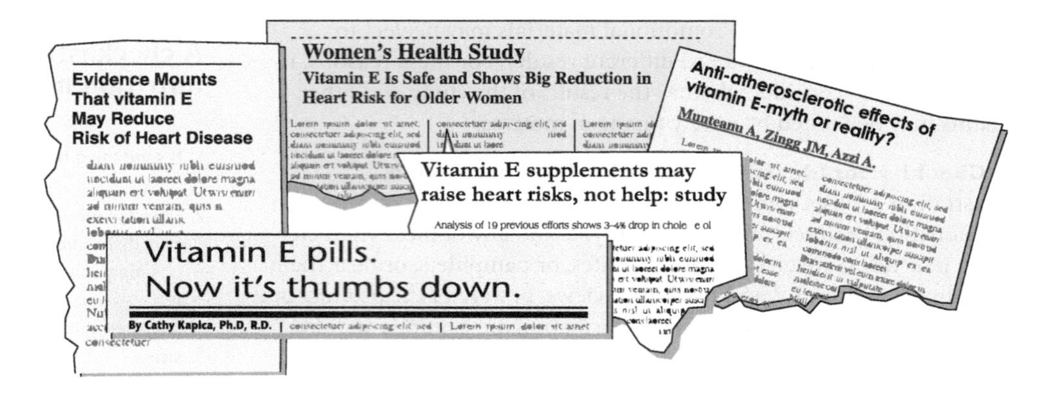

Illustration 3.6 Supplements that may be used to treat a variety of health problems. The safety and effectiveness of many of these products have yet to be proved.

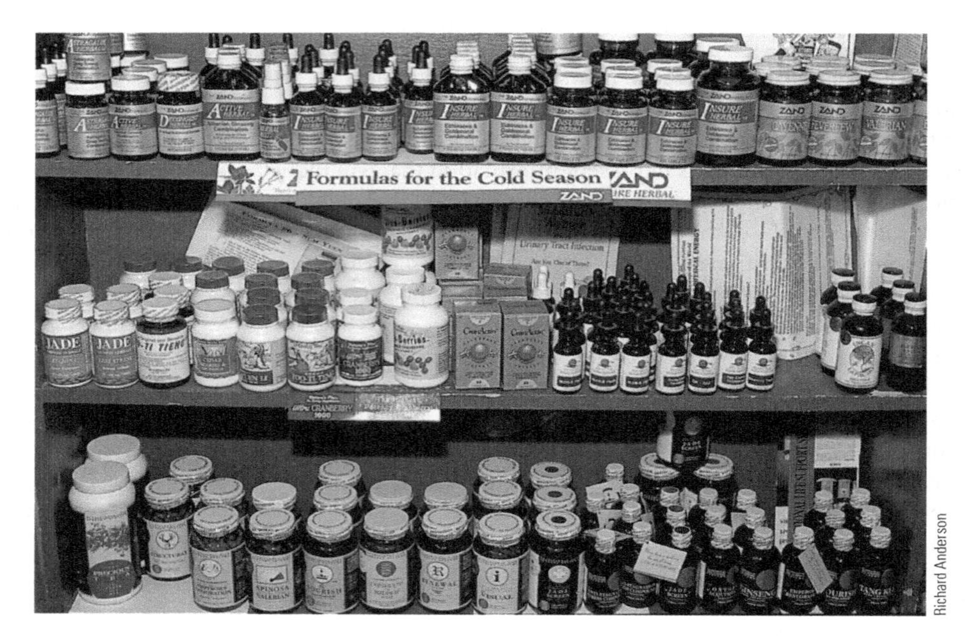

Richard Anderson

studies. Until they are evaluated scientifically, the logical conclusion is that they cannot be considered effective or safe.

Professionals with Embedded Beliefs Professionals who work in health care and research are not immune to the pull of deeply rooted convictions about diet and health relationships. They sometimes remain wedded to theories and beliefs even after they have been disproved. The nurse or doctor who holds onto the belief that salt intake should be restricted in pregnancy and the university professor who is convinced that pesticides on foods pose no risk to health are two examples of professional sources of misinformation. The failure of such professionals to give up erroneous convictions about diet and health adds to the flow of nutrition misinformation as well as to consumer confusion and increased health risk.[3]

How to Identify Nutrition Truths

- **Identify sources of reliable nutrition information.**

Where does accurate nutrition information come from? There is only one way to identify sound nutrition information: put it to the test and see if it survives the dispassionate, systematic examination dictated by science.

Science is a unique and powerful explanatory system that separates facts from assumptions and beliefs. It produces information based on facts and evidence. Our understanding of nutrition is based on scientifically determined facts and evidence obtained from laboratory, animal, and human studies. The studies provide information that qualifies for inclusion when

Illustration 3.7 The FDA's Food Safety Modernization Act of 2011 requires that science-based standards for the safe production and harvesting of fruits and vegetables be implemented by the produce industry.

developing public policies about nutrition and health (Illustration 3.7) and clinical nutrition practices. Science delivers information eligible for inclusion in textbooks about nutrition.

Sources of Reliable Nutrition Information

Reliable sources of information about nutrition meet the standards of proof required by science. They report decisions about nutrition and health relationships that are based on multiple studies and achieve scientific consensus. These decisions represent the majority opinion of scientists who are knowledgeable about a particular nutrition topic.

Nutrition recommendations made to the public, such as the Dietary Guidelines for Americans and the Dietary Reference Intakes, are based on the consensus of scientific opinion derived from available research. The information is made available to the public not to sell a product or to promote an ideology, but to inform consumers honestly about nutrition and to help people use the information to maintain or improve their health. Organizations and individuals offering reliable nutrition information are listed in Table 3.1.

Table 3.1 Reliable sources of nutrition information[a]

Source of nutrition information	Examples
Nonprofit, professional health organizations	American Heart Association American Cancer Society American Academy of Nutrition and Dietetics (formerly known as the American Dietetic Association)
Scientific organizations	National Academy of Sciences American Society for Nutrition
Government publications: nutrition, diet, and health reports	National Institutes of Health Surgeon General Food and Drug Administration Centers for Disease Control U.S. Department of Agriculture
Registered dietitians/nutritionists	Hospitals Public health departments Extension service Universities
Nutrition textbooks	College and university nutrition courses Nutrition faculty of accredited universities

[a]See Appendix B for more details about reliable sources of nutrition information.

Table 3.2 Who's who in nutrition and dietetics: Laws and regulations governing practice[14]

Title	Qualifications
Registered dietitian	Individual who has met accredited, baccalaureate academic requirements, and acquired the knowledge, skills, and experience necessary to pass a national registration examination and participates in continuing professional education. Qualification is conferred by the Commission on Dietetic Registration and expressed in the title RD (registered dietitian) or RDN (registered dietitian/nutritionist). RDs and RDNs can order therapeutic diets, and monitor and manage dietary plans for patients.[15]
Dietetic Technician, Registered	Individual who has completed an associate's or baccalaureate degree from an accredited institution and successfully completed the didactic program required in dietetics, a supervised, accredited practice program, and passes the Registration Examination for Dietetic Technician/ Registered (DTR).

Nutrition Information on the Internet Social media and Internet-based search engines are transforming the way we get information about foods, diets, supplements, and health. Eight in ten Internet users, including scientists and consumers, look for health and nutrition information online.[11,12] As is the case with many other media sources, the accuracy of nutrition information made available on the Web varies considerably. In general, the most reliable sites are those developed by government health agencies (Web addresses that end in or include .gov), educational institutions (.edu), and nonprofit professional health and science organizations (.org).[13]

Who Are Qualified Nutrition Professionals? Many people refer to themselves as nutritionists, but only some of them are qualified based on education and experience in **dietetics**. These individuals are registered, licensed, or certified dietitians or nutritionists who meet qualifications established by national and state regulations. They have demonstrated a mastery of knowledge about the science of nutrition and appropriate clinical practices. Table 3.2 describes the qualifications of those who legitimately use the title *dietitian* or *registered dietitian (RD)/nutritionist (RDN)*. The specific titles vary somewhat depending on state regulations and laws governing the practice of nutrition and dietetics.[14]

dietetics The integration, application, and communication of practice principles derived from food, nutrition, social, business, and basic sciences, to achieve and maintain optimal nutrition status of individuals and groups.

It is difficult to make sound decisions about every nutrition question or issue that arises. When trying to make sense out of the information you read or hear, don't hesitate to get help. Visit reliable websites, check out the index of this book for the relevant topic, or call your local health department or area Food and Drug Administration (FDA) office. The more solid your information, the better your decisions will be.

REALITY CHECK

Microwaves and Plastic

You've just gotten a tweet: DO NOT microwave foods in plastic! Hot plastic releases highly toxic dioxins that can cause cancer . . .

No source for the facts stated is given. Cyndi's and Scott's responses:

Who gets the thumbs up?

Answers on page 3-9

Photodisc

Cyndi: I'll check it out at the WebMD .com.

Photodisc

Scott: I was afraid of that. Now I know microwaving food in plastic containers is dangerous.

ANSWERS TO REALITY CHECK

Microwaves and Plastic

Good thinking, Cyndi. Information gained from the fda.gov website indicates that plastic containers may contain bisphenol A, a softening compound that makes plastic flexible. Heat increases its release from plastic. Exposure to bisphenol A at current levels in the United States does not appear to harm health. It is recommended, however, that people use glass or ceramic containers in microwave ovens. The use of bisphenol A is banned in baby bottles, sippy cups, and infant formula containers.[9,10]

Photodisc

Cyndi:

Photodisc

Scott:

The Methods of Science

- **Explain each component of the scientific methods used to identify reliable information about nutrition.**

Nikita G. Sidorov/Shutterstock.com

Illustration 3.8 The scientific method is not mysterious at all, but a carefully planned process for answering a specific question.

Does the term *scientific method* conjure up an image of beady-eyed, white-coated scientists working diligently in windowless basement laboratories? Although we tend to assume that scientists are busy advancing knowledge about nutrition and many other fields, for most people the methods of science are a mystery. They should not be (Illustration 3.8). Consumers use a lot of nutrition information. To make sound judgments about nutrition and health, people need to know how to distinguish results produced by scientific studies from those generated by personal opinion, product promotions, and bogus studies.

There are many established methods for conducting scientific studies. The specific methods employed vary from study to study depending on the type of research conducted. Nevertheless, all types of scientific studies have one feature in common: they are painstakingly planned. Planning is the most important, and often the most time-consuming, part of the entire research process.

Developing the Plan

The first part of the planning process entails clearly stating the question to be addressed—and, one hopes, answered—by the research. For purposes of illustration, let's say our research will address the effects of vitamin *X* supplements on hair loss (Table 3.3 and

Table 3.3 Overview of a human nutrition research study: Vitamin *X* supplements and hair loss[a]

A. **Pose a clear question:** "Do vitamin *X* supplements increase hair loss?"
B. **State the hypothesis to be tested:** "Vitamin *X* supplements of 25,000 IU per day taken for three months increase hair loss in healthy adults."
C. **Design the research:** 1. What type of research design should be used? "In this study, a clinical trial will be used. The supplement and placebo will be allocated by a double-blind procedure." 2. Who should the research subjects be? "The study will exclude subjects who may be losing or gaining hair due to balding, hair treatments, medications, or current use of vitamin *X* supplements." 3. How many subjects are needed in the study? "The required sample size is calculated to be 20 experimental and 20 control subjects." 4. What information needs to be collected? "Information on hair loss, conditions occurring that might affect hair loss, and the use of supplements and placebos will be collected." 5. What method of measuring hair loss should be used? "The study will use the reliable "60-second hair count" technique.[12] 6. What statistical tests should be used to analyze the results? "Appropriate tests identified."
D. **Obtain approval for the study from the committee on the use of humans in research:** "Approval obtained."
E. **Implement the study design:** "Implemented."
F. **Evaluate the findings:** "Subjects receiving the 25,000 IU of vitamin *X* for three months lost significantly more hair than subjects receiving the placebo. Hypothesized relationship found to be true."
G. **Submit paper on the research for publication in a scientific journal or other document.**

[a]This is a fictitious study used for illustrative purposes only.

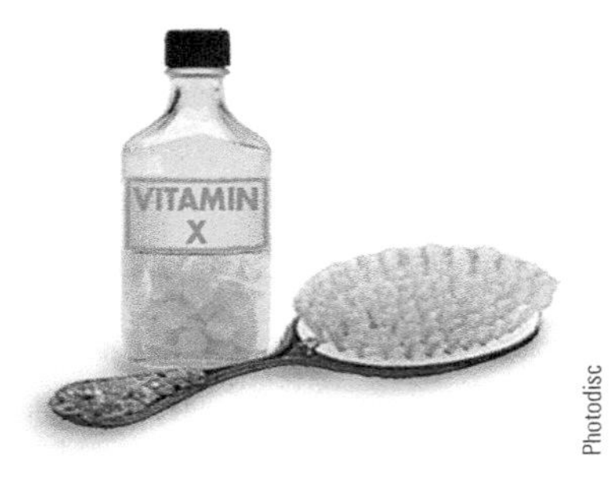

Photodisc

Illustration 3.9 Do high doses of vitamin *X* increase hair loss?

Illustration 3.9). The idea for the research came from a study that hinted that users of vitamin *X* supplements had increased hair loss. In that study, adults were given 25,000 international units (IU) of vitamin *X* for three months to test its safety. Although the study found that vitamin *X* had no adverse effects on health and offered some benefits, several subjects who had received the supplements complained of hair loss. Accordingly, our study poses the question: Do vitamin *X* supplements increase hair loss?

The Hypothesis: Making the Question Testable

The question is then transformed into an explicit hypothesis that can be proved or disproved by the research. (**Hypothesis** and other research terms used in this unit are defined in Table 3.4.) The results of the research must provide a true or false response to the hypothesis and not an explanatory sentence or paragraph. So, in this example, "Vitamin *X* increases hair loss" wouldn't do as a hypothesis because it leaves too many questions about the relationship unanswered. The effect of vitamin *X* on hair loss may depend on the amount of vitamin *X* given; how long it is taken; or on the research subjects' health, age, and other characteristics. The hypothesis "Vitamin *X* supplements of 25,000 IU per day taken for three months by healthy adults increase hair loss" is concrete enough to be addressed by research.

Table 3.4 A short glossary of research terms

Term	Definition
Association or associated	The finding that one condition is correlated with, or related to another condition, such as a disease or disorder. For example, diets low in vegetables are associated with breast cancer. Associations do not prove that one condition (such as a diet low in vegetables) causes an event (such as breast cancer). They indicate that a statistically significant relationship between a condition and an event exists.
Cause and effect	A finding that demonstrates that a condition causes a particular event. For example, vitamin C deficiency causes the deficiency disease scurvy.
Clinical trial	A study design in which one group of randomly assigned subjects (or subjects selected by the "luck of the draw") receives an active treatment and another group receives an inactive treatment, or "sugar pill," called a placebo.
Control group	Subjects in a study who do not receive the active treatment or who do not have the condition under investigation. Control periods, or times when subjects are not receiving the treatment, are sometimes used instead of a control group.
Double blind	A study in which neither the subjects participating in the research nor the scientists performing the research know which subjects are receiving the treatment and which are getting the placebo. Both subjects and investigators are "blind" to the treatment administered.
Epidemiological study	Research that seeks to identify conditions related to particular events within a population. This type of research does not identify cause-and-effect relationships. For example, much of the information known about diet and cancer is based on epidemiological studies that have found that diets low in vegetables and fruits are associated with the development of heart disease.
Experimental group	Subjects in a study who receive the treatment being tested or have the condition that is being investigated.
Hypothesis	A statement made prior to initiating a study of the relationship sought to be tested by the research.
Meta-analysis	An analysis of data from multiple studies. Results are based on larger samples than the individual studies and are therefore more reliable. Differences in methods and subjects among the studies may bias the results of meta-analyses.
Peer review	Evaluation of the scientific merit of research or scientific reports by experts in the area under review. Studies published in scientific journals have gone through peer review prior to being accepted for publication.
Placebo	A "sugar pill," an imitation treatment given to subjects in research.
Placebo effect	Changes in health or perceived health that result from expectations that a "treatment" will produce an effect on health.
Statistically significant	Research findings that likely represent a true or actual result and not one due to chance.

The Research Design: Gathering the Right Information

Poor research design, or the lack of a solid plan on how the research will be conducted, is often a weak link that renders studies useless. It's the "oops, we forgot to get this critical piece of information!" or the "how did all the measurements come out wrong?" at the end of a study that can ruin months or even years of work. Each step in the research process must be thoroughly planned. In research, there are no miracles. If something can go wrong because of incomplete planning, it probably will.

Research designs are often based on the answers to the following questions:

- What type of research design should be used?
- Who should the research subjects be?
- How many subjects are needed in the study?
- What information needs to be collected?
- What are accurate ways to collect the needed information?
- What statistical tests should be used to analyze the findings?

What Type of Research Design Should Be Used? Several different research designs can be used to test hypotheses. We could use an **epidemiological study** design to determine if hair loss is more common among people who take vitamin *X* supplements than among people who do not. Researchers commonly use this type of study to identify conditions that are related to specific health events in humans. To provide preliminary evidence, we could use animal studies that determine hair loss in supplemented and unsupplemented animals. Or we could use another design, such as a clinical trial. Since this design would work well for the proposed hypothesis about vitamin *X* supplements, we'll follow the rules for conducting a **clinical trial** in this example. (You can see a map of the research design used for this hypothetical study in Illustration 3.10.)

The purpose of clinical trials is to test the effects of a treatment or intervention on a specific biological event or other measureable outcome. In our example, we will test the effect of vitamin *X* supplements on hair loss.

The Experimental and Control Groups Clinical trials, as well as other research studies that address questions about nutrition, require an **experimental group** (the group of subjects who receive vitamin *X* in this example) and a **control group** (the comparison group that receives a **placebo** and not vitamin *X*). Never trust the results of a study that didn't employ both. Here's why. How do we know the effect of a certain treatment isn't due to something *other* than the treatment? What if we gave a group of adults vitamin *X* supplements and they lost hair? Does this mean the vitamin *X* caused the hair loss? Could it be that the subjects lost no more hair during the treatment period than they would have lost without the vitamin *X*? We can't know whether vitamin *X* supplements produce hair loss if we don't know how much hair is lost without vitamin *X*.

After measuring their usual hair loss, we will randomly assign (by the "luck of the draw") people in the study to serve in either the experimental or the control group. Individuals assigned to the control group will be given pills that look, taste, and feel like the vitamin *X* supplements but have no effect on hair loss or gain. Neither the research staff nor the people in the study will know who is getting vitamin *X* and who is getting the placebo (Illustration 3.11). Scientists use this **double-blind** procedure because knowing which group is which may affect people's expectations and thus change the results.

The "Placebo Effect" The placebo effect can cause a good deal of confusion in research. That's because people tend to have expectations about what a treatment will do, and those expectations can influence what happens.

A good example of the **placebo effect** occurred in a study that tested the effectiveness of a medication meant to reduce binge eating among people with bulimia.[16] After the usual number of binge-eating episodes was determined, 22 women with bulimia were given either the medication or a placebo in a double-blind fashion. The number of

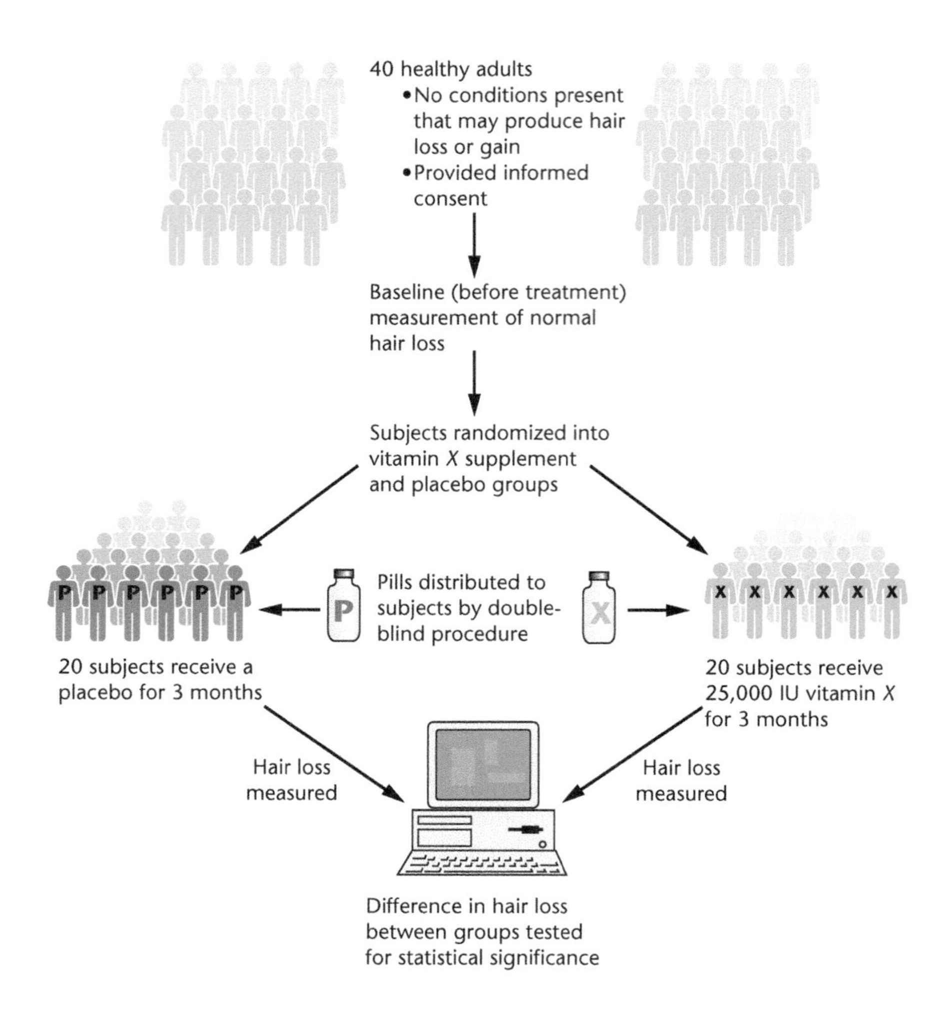

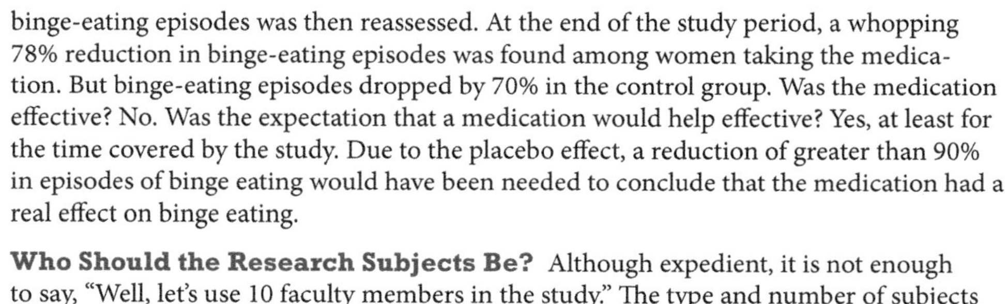

Illustration 3.10 Map of the design of the vitamin *X* supplement study.

binge-eating episodes was then reassessed. At the end of the study period, a whopping 78% reduction in binge-eating episodes was found among women taking the medication. But binge-eating episodes dropped by 70% in the control group. Was the medication effective? No. Was the expectation that a medication would help effective? Yes, at least for the time covered by the study. Due to the placebo effect, a reduction of greater than 90% in episodes of binge eating would have been needed to conclude that the medication had a real effect on binge eating.

Who Should the Research Subjects Be? Although expedient, it is not enough to say, "Well, let's use 10 faculty members in the study." The type and number of subjects employed by the research are important considerations.

The type of subjects involved in research is important because we need to exclude people who have conditions that might produce the problem the research is examining. For example, hair loss may result from periodic balding, a bad perm, the use of mousse or hair spray, or illnesses or medications. If we include people with these conditions and circumstances in the study, it will be difficult to determine whether hair loss is due to vitamin *X* or something else. In addition, it is important to exclude from the study people who already take vitamin *X* supplements or are bald. An inappropriate or biased selection of subjects could make the results useless or, in the worst case, make the study's results come out in a predetermined way.

Stan Maddock

Illustration 3.11 Which is the vitamin *X* supplement? Neither the subject nor the investigator should know which is vitamin *X* and which is the placebo.

How Many Subjects Are Needed in the Study? The number of subjects needed for a study is a mathematical question, and we won't go into the formulas here. The number of subjects is based on the number needed to separate true differences between the experimental and control groups from differences that are due to chance

Larry St. Pierre/Shutterstock.com

Photodisc

Illustration 3.12 How many subjects are needed in the study? It's important to get the number right.

(Illustration 3.12). Daily hair loss normally ranges from 50 to 150 strands per day.[17] We'll need to include enough subjects in the study to make sure differences in hair loss between the groups are not due to normal fluctuations in hair loss. This point is important because studies employing too few subjects lead to inconclusive results and a great deal of controversy about nutrition. One or a few subjects are never enough to prove a point about nutrition.

Let's assume that the mathematical formulas applied to the vitamin *X* study show that 20 people are needed in both the experimental group and the control group.

What Information Needs to Be Collected? Information recorded by researchers must represent an evenhanded approach to discovering the facts. We must identify not only the findings that may support the hypothesis, but also *those that may refute it.* To know if vitamin *X* supplements increase hair loss, we need to know how much hair is lost, whether the subjects took the supplement faithfully, and whether something came up during the study (such as a bad perm) that might alter hair loss. The presence of such conditions should be determined by methods known to provide accurate results.

What Are Accurate Ways to Collect the Needed Information? "Garbage in—garbage out." If the information obtained on and from subjects is inaccurate, then the study is worthless. To avoid this problem, good research employs methods of collecting information known to produce accurate results.

Let's consider the problem of measuring hair loss. How can we do that accurately? First, we look for a method that has already been demonstrated by research to be accurate. We find one called "the 60 second hair count"[18] and employ it.

What Statistical Tests Should Be Used to Analyze the Findings? Before collecting the first piece of information, we'll select appropriate tests for identifying **statistically significant** results of the research. Statistical tests are used to identify significant differences between the findings from the experimental and control groups. Without such tests, it is often very hard to decide what the findings mean. What if the study shows the group that took vitamin *X* lost an average of 5% more hair than the control group? Is that 5% difference due to the vitamin *X* or to something else, such as chance fluctuations in normal hair loss? Well-chosen statistical tests will tell us whether the differences between groups are in all probability real or due to chance occurrences or coincidence.

Obtaining Approval to Study Human Subjects

An important step in the planning process is applying for approval to conduct the proposed research on human subjects. Universities and other institutions that conduct research have formal committees, called institutional review boards, that scrutinize plans to make sure proposed studies follow the rules governing research on human subjects. For this study, we will have to show to the committee's satisfaction that the level of vitamin *X* employed is safe. As a part of this process, the committee usually requires that human subjects consent, in writing, to participate in the study.

Implementing the Study

With the design in place and the appropriate approvals obtained, it is time to implement the study. Assume that, due to the study's solid design and importance, we have been awarded a grant to fund the research. We can now recruit subjects for the study and enroll them if they are eligible and consent to participate. Measurements of hair loss are performed according to a schedule, use of the pills by subjects is monitored, results are checked for errors and entered into a computerized data file, and statistical tests are applied. This process usually ends with computer printouts that exhibit the findings.

Making Sense of the Results

Let's say subjects who received the supplement lost 30% more hair than the control subjects did. Having applied the appropriate statistical test, we find that 30% is a highly significant difference. Does that mean that vitamin *X* supplements *cause* hair loss? *Cause* is a strong word in research terms. It implies that a **cause-and-effect** relationship exists (that the vitamin *X* supplements caused the hair loss). But many factors can contribute to hair loss, and since many causes may be unknown or not measured by the study, it is difficult to conclude with absolute certainty that vitamin *X* by itself caused the hair loss. Assume that 4 of the 20 subjects in the experimental group who took the vitamin *X* supplement lost less hair than usual. The vitamin *X* supplements *didn't* cause them to lose hair.

We could conclude from this research (if, as a reminder, there was such a thing as vitamin *X*) that vitamin *X* supplements, given to healthy adults at a dose of 25,000 IU per day for three months, are strongly *associated* with hair loss. The term **associated** as used here means that the vitamin *X* supplements were strongly *related* to hair loss but may not have *caused* the hair loss. The hypothesized relationship between supplemental vitamin *X* and hair loss would be found to be true. Although the research strongly indicates a cause-and-effect relationship, additional studies would be needed to prove the relationship exists in other groups of people and at different doses of vitamin *X*.

After determining the results, a paper is written describing the research and its conclusions and is submitted for publication in a scientific journal. If judged to be acceptable after **peer review** (review by qualified experts), the paper is published.

Complete coverage of how nutrition questions are addressed through research would take a three-course sequence. Nevertheless, it is hoped that the information presented here makes it easier to judge the likely accuracy of reports about nutrition studies in the popular media. There is a lot more to identifying nutrition truths than meets the eye.

Science and Personal Decisions about Nutrition Science is based on facts and evidence. The grounding ethic of scientists is that facts and evidence are more sacred than any other consideration. These characteristics of science and scientists are strong assets for the job of identifying truths.

Although imperfect because they are undertaken by humans in an environment of multiple constraints, only scientific studies produce information about nutrition that you can count on. Evidence is the single best ingredient for decision making about nutrition and your health.

NUTRITION up close

Checking Out a Fat-Loss Product

Focal Point: How to make an informed decision about the truthfulness of an advertisement for a fat-loss product.

Read the accompanying advertisement for a fat-loss product. Then check out the information using the checklist for identifying nutrition misinformation.

P.O. Box 281752,
Miami, FL

Lose Over 2 Inches In 3 Weeks

A Scientifically Proven Fat Reduction Cream That Actually Works!

A scientific, double-blind, placebo-controlled research study of both sexes demonstrated that LIPID MELT fat reduction cream reduces inches from the thigh area. Subjects lost an average of 2 inches from each thigh in only 3 weeks!

Subjects from a preliminary pilot study reported that LIPID MELT cream reduced inches from their abdomen. Both studies reported no skin rashes, discomfort or sensitivity. Formulated with patented liposome technology, LIPID MELT contains no drugs, only natural active ingredients. Initially sold only in salons and spas.

Six-week supply: $25 + $4.95 s/h.
We guarantee that you will lose inches or your money will be refunded.

1. Is something being sold?
   Yes No
2. Is a new remedy for problems that are not easily or simply solved being offered?
   Yes No
3. Are terms such as *miraculous, magical, secret, detoxify, energy restoring, suppressed by organized medicine, immune boosting*, or *studies prove* used?
   Yes No
4. Are testimonials, before-and-after photos, or expert endorsements used?
   Yes No
5. Does the information sound too good to be true?
   Yes No
6. Is a money-back guarantee offered?
   Yes No

Optional: Repeat the activity using a nutrition-related advertisement from the Internet.

Feedback to the Nutrition Up Close is located in Appendix F.

REVIEW QUESTIONS

- **Evaluate the reliability of advertisements and other information about nutrition and nutrition-related products and services.**

1. By law, claims about nutrition that are presented on product labels and packaging must be scientifically accurate. **True/False**
2. The leading motivation underlying the presentation of nutrition misinformation is the profit motive. **True/False**
3. The Food and Drug Administration (FDA) has the authority to remove advertisements making false claims about nutrition from the airwaves and the Internet. **True/False**
4. Individuals can help combat the proliferation of bogus nutrition advertisements and products by notifying the Federal Trade Commission (FTC) of their existence. **True/False**
5. Nutrition products and services that offer a "money-back guarantee" are most likely to work as advertised. **True/False**
6. Nutrition information offered by trained sales staff at nutrition product stores can be counted on as being accurate. **True/False**

- **Identify sources of reliable nutrition information.**

7. The term *registered dietitian (RD) or registered dietitian/nutritionist (RDN)* means that individuals with the title have acquired the knowledge and skills necessary to pass a national examination on nutrition and participate in continuing education. **True/False**
8. ____ Recently, a talk show–induced scare about the safety of apple juice due to its arsenic content spread across the Internet. The news was hard to believe, but it might be true so you decide to look it up on the Web. From the options

listed below, select the most reliable source of information on the safety of apple juice.

a. www.doctoroz.com
b. www.cbsnews.com
c. www.fda.gov
d. www.usapple.org

- **Explain each component of the scientific methods used to identify reliable information about nutrition.**

9. ____ While talking with your brother on the phone he complains that he "feels a cold coming on." You tell him to take a vitamin D supplement because you have not gotten a cold since you started taking vitamin D. Your brother takes the pill and the next day he feels better. Which of the following statements is *least* likely to be true regarding the relationship between the vitamin D pill and prevention of your brother's cold?
   a. Your brother's experience proves vitamin D prevents colds.
   b. Although it is possible vitamin D helps prevent colds, your brother may have felt better the next day even if he hadn't taken the vitamin D.
   c. Reliable information about the effects of vitamin D on colds cannot be based solely on one person's experience.
   d. Believing that vitamin D prevents colds is not the same as scientifically demonstrating that it does.
10. ____ You hear on *The Nightly News* that scientists have discovered an association between high levels of soft drink consumption and violent behavior in teens. Which of the following statements represents the correct interpretation of this study's results?
   a. Violent behavior in teens is triggered by soft drink consumption.
   b. Teens who engage in violent behavior tend to consume more soft drinks than teens who do not engage in violent behavior.
   c. Violent behavior is only exhibited by teens who consume high amounts of soft drinks.
   d. Eliminating soft drink consumption would end violent behavior among teens.
11. An advantage of clinical trials over other study designs is that they consistently identify cause-and-effect relationships. **True/False**
12. Research results may be unreliable if too few subjects are used in studies. **True/False**
13. Research articles published in scientific journals undergo peer review, a process by which experts in the areas under investigation evaluate articles prior to their acceptance for publication. **True/False**

For the following items, match the term in Column A with its abbreviated definition shown in Column B.

Column A	Column B
____ 14. Meta-analysis	a. "Sugar pill" or imitation treatment used in clinical trials.
____ 15. Double blind	b. Subjects in a study who do not receive the active treatment or who not have the condition under investigation.
____ 16. Placebo	c. An analysis of data from multiple studies.
____ 17. Statistically	d. Research finding that is likely true significant and not significant due to chance.
____ 18. Control group	e. A study in which both research subjects and investigators do not know who is receiving an active or placebo treatment.

Answers to these questions can be found in Appendix F.

NUTRITION SCOREBOARD ANSWERS

1. Freedom of speech applies to information about nutrition in articles, speeches, pamphlets, and broadcasts. However, it is illegal to make false or misleading claims about nutrition in advertisements or on product labels and packaging. **False**
2. The "way of knowing" for the field of nutrition is science. **True**
3. In a double-blind study, neither the research staff nor the research subjects know which subjects are getting the real treatment and which are receiving the fake treatment. This reduces the placebo ("sugar pill") effect, or changes in health that are due to the expectation that a specific treatment will have a particular impact on health. **True**
4. Biologically, animals differ from humans in many ways. Results from animal studies cannot be assumed to apply to humans. **False**
5. Association does not equal causation. **False**

UNIT 4

Understanding Food and Nutrition Labels

NUTRITION SCOREBOARD

1 Nutrition labels are required on all foods and dietary supplements sold in the United States. **True/False**

2 Nutrition labeling rules allow health claims to be made on the packages of certain food products. **True/False**

3 A food product labeled "less sugar" means the product contains very little sugar. **True/False**

4 Nutrition labels contain all of the information people need to make healthy decisions about what to eat. **True/False**

Answers can be found at the end of the unit.

allesalltag/Alamy

LEARNING OBJECTIVES

After completing Unit 4 you will be able to:

- Apply knowledge about the four key elements of nutrition labeling to decisions about the nutritional value of foods.
- Evaluate nutrient content and health claims made on dietary supplement labels.
- Compare the characteristics of organically and conventionally produced food products.
- Identify strengths and weaknesses of various nutrition labeling systems on food packaging and calorie listings for food items.

Nutrition Labeling

- **Apply knowledge about the four key elements of nutrition labeling to decisions about the nutritional value of foods.**

Misleading messages, hazy health claims, and the slippery serving sizes that characterized food labels in the past led to a revolution in nutrition labeling. Consumers—especially those responsible for buying food for the family; people with weight concerns, food allergies, or diabetes; and the health-conscious—made it clear they wanted to end the mystery about what's in many foods. Passage of the 1990 Nutrition Labeling and Education Act by Congress indicated that their concerns had been heard. In 1993, the Food and Drug Administration (FDA) published rules for nutrition labeling; implementation and revision of the standards continue.[2]

KEY NUTRITION CONCEPTS

Content in this unit on nutrition labeling directly relates to three key nutrition concepts:

1. Nutrition Concept #2: Foods provide energy (calories), nutrients, and other substances needed for growth and health.
2. Nutrition Concept #4: Poor nutrition can result from both inadequate and excessive levels of nutrient intake.
3. Nutrition Concept #6: Malnutrition can result from poor diets and from disease states, genetic factors, or combinations of these factors.

Illustration 4.1 Nutrition information for produce, fish, and seafood can be presented on posters.[a]

Seafood Nutrition Facts

Cooked (by moist or dry heat with no added ingredients), edible weight portion. Percent Daily Values (%DV) are based on a 2,000 calorie diet.

Seafood Serving Size (84 g/3 oz)	Calories	Calories from Fat	Total Fat g	Total Fat %DV	Saturated Fat g	Saturated Fat %DV	Cholesterol mg	Cholesterol %DV	Sodium mg	Sodium %DV	Potassium mg	Potassium %DV	Total Carbohydrate g	Total Carbohydrate %DV	Protein g	Vitamin A %DV	Vitamin C %DV	Calcium %DV	Iron %DV
Blue Crab	100	10	1	2	0	0	95	32	330	14	300	9	0	0	20g	0%	4%	10%	4%
Catfish	130	60	6	9	2	10	50	17	40	2	230	7	0	0	17g	0%	0%	0%	0%
Clams, about 12 small	110	15	1.5	2	0	0	80	27	95	4	470	13	6	2	17g	10%	0%	8%	30%
Cod	90	5	1	2	0	0	50	17	65	3	460	13	0	0	20g	0%	2%	2%	2%
Flounder/Sole	100	15	1.5	2	0	0	55	18	100	4	390	11	0	0	19g	0%	0%	2%	0%
Haddock	100	10	1	2	0	0	70	23	85	4	340	10	0	0	21g	2%	0%	2%	6%
Halibut	120	15	2	3	0	0	40	13	60	3	500	14	0	0	23g	4%	0%	2%	6%
Lobster	80	0	0.5	1	0	0	60	20	320	13	300	9	1	0	17g	2%	0%	6%	2%
Ocean Perch	110	20	2	3	0.5	3	45	15	95	4	290	8	0	0	21g	0%	2%	10%	4%
Orange Roughy	80	5	1	2	0	0	20	7	70	3	340	10	0	0	16g	2%	0%	4%	2%
Oysters, about 12 medium	100	35	4	6	1	5	80	27	300	13	220	6	6	2	10g	0%	6%	6%	45%
Pollock	90	10	1	2	0	0	80	27	110	5	370	11	0	0	20g	2%	0%	0%	2%
Rainbow Trout	140	50	6	9	2	10	55	18	35	1	370	11	0	0	20g	4%	4%	8%	2%
Rockfish	110	15	2	3	0	0	40	13	70	3	440	13	0	0	21g	4%	0%	2%	2%
Salmon, Atlantic/Coho/Sockeye/Chinook	200	90	10	15	2	10	70	23	55	2	430	12	0	0	24g	4%	4%	2%	2%
Salmon, Chum/Pink	130	40	4	6	1	5	70	23	65	3	420	12	0	0	22g	2%	0%	2%	4%
Scallops, about 6 large or 14 small	140	10	1	2	0	0	65	22	310	13	430	12	5	2	27g	2%	0%	4%	14%
Shrimp	100	10	1.5	2	0	0	170	57	240	10	220	6	0	0	21g	4%	4%	6%	10%
Swordfish	120	50	6	9	1.5	8	40	13	100	4	310	9	0	0	16g	2%	2%	0%	6%
Tilapia	110	20	2.5	4	1	5	75	25	30	1	360	10	0	0	22g	0%	2%	0%	2%
Tuna	130	15	1.5	2	0	0	50	17	40	2	480	14	0	0	26g	2%	2%	2%	4%

Seafood provides negligible amounts of *trans* fat, dietary fiber, and sugars.

U.S. Food and Drug Administration (January 1, 2008)

[a]Download posters from the FDA that show nutrition information for the twenty most frequently consumed raw fruits, vegetables, and fish in the United States by searching the term: Nutrition Information for Raw Fruits, Vegetables, and Fish.

Key Elements of Nutrition Labeling Standards

Nutrition labeling regulations cover four major areas:

- The Nutrition Facts panel
- Nutrient content claims
- Health claims
- Structure/function claims

The Nutrition Facts panel is required on most foods sold in grocery stores. Rules established for the other three areas must be followed when a claim about the product is made on the packaging.

The Nutrition Facts Panel With the exception of foods sold in very small packages or in stores like local bakeries and of alcohol-containing beverages such beer, wine, and liquor, foods containing more than one ingredient must display a Nutrition Facts panel. Single fresh foods, such as pears, a head of cabbage, or fresh shrimp, do not have to be labeled, but grocery stores are encouraged to present nutrition information on posters (Illustration 4.1).

Nutrition Facts panels provide specific information about a food's serving size, the calorie value and nutrient content of a serving, and ingredients. Illustration 4.2 shows a Nutrition Facts panel and provides explanations of what various components of the panel mean. The panel highlights the fat, saturated fat, *trans* fat, cholesterol, sodium, carbohydrate, dietary fiber, sugars, vitamins A and C, and calcium and iron content of a serving of food.

All labeled foods must provide the information shown on the Nutrition Facts panel in Illustration 4.2. Additional information

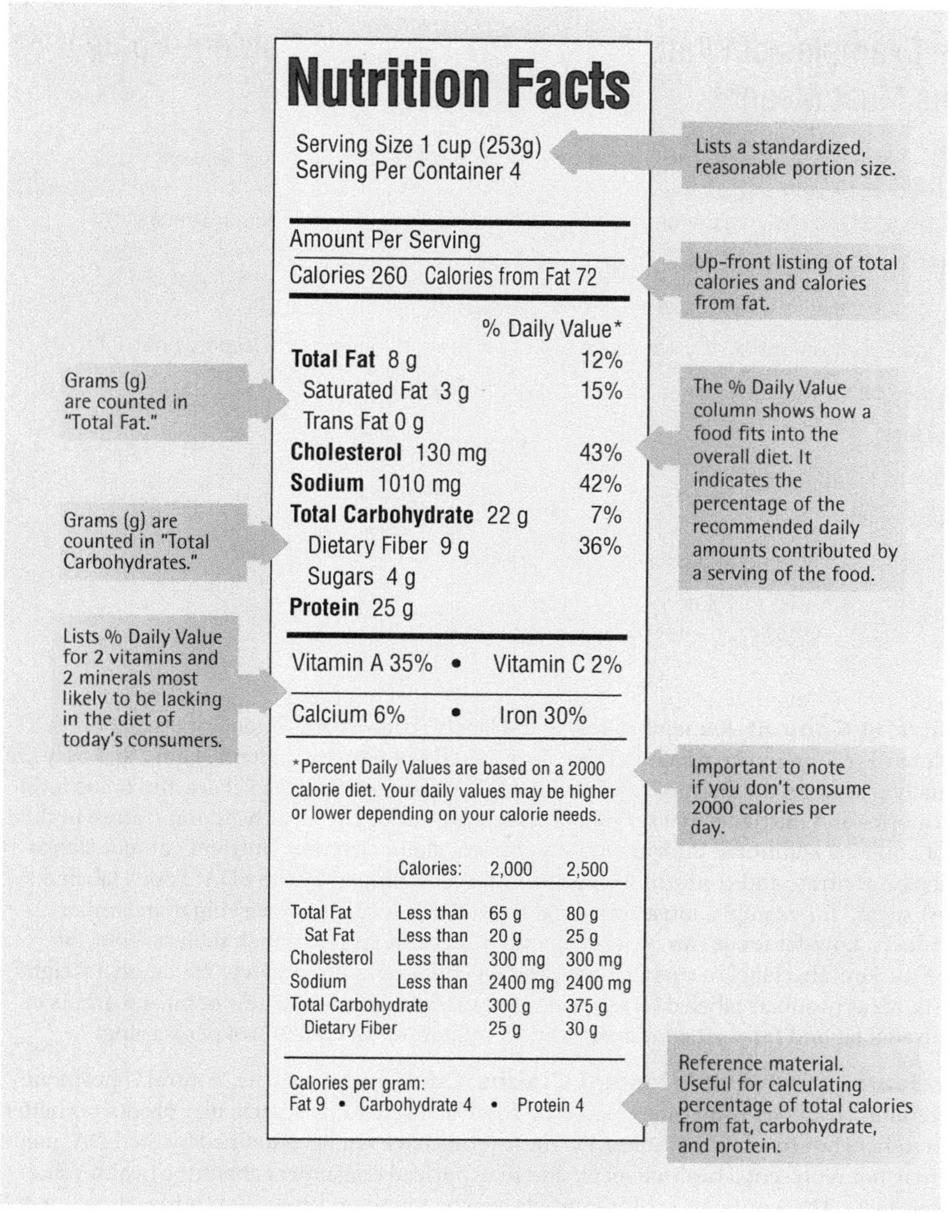

Illustration 4.2 Inside the Nutrition Facts panel.

Nutrition Facts

Serving Size 1 cup (253g)
Serving Per Container 4

Amount Per Serving
Calories 260 Calories from Fat 72

	% Daily Value*
Total Fat 8 g	12%
Saturated Fat 3 g	15%
Trans Fat 0 g	
Cholesterol 130 mg	43%
Sodium 1010 mg	42%
Total Carbohydrate 22 g	7%
Dietary Fiber 9 g	36%
Sugars 4 g	
Protein 25 g	

Vitamin A 35% • Vitamin C 2%

Calcium 6% • Iron 30%

*Percent Daily Values are based on a 2000 calorie diet. Your daily values may be higher or lower depending on your calorie needs.

	Calories:	2,000	2,500
Total Fat	Less than	65 g	80 g
Sat Fat	Less than	20 g	25 g
Cholesterol	Less than	300 mg	300 mg
Sodium	Less than	2400 mg	2400 mg
Total Carbohydrate		300 g	375 g
Dietary Fiber		25 g	30 g

Calories per gram:
Fat 9 • Carbohydrate 4 • Protein 4

Lists a standardized, reasonable portion size.

Up-front listing of total calories and calories from fat.

Grams (g) are counted in "Total Fat."

The % Daily Value column shows how a food fits into the overall diet. It indicates the percentage of the recommended daily amounts contributed by a serving of the food.

Grams (g) are counted in "Total Carbohydrates."

Lists % Daily Value for 2 vitamins and 2 minerals most likely to be lacking in the diet of today's consumers.

Important to note if you don't consume 2000 calories per day.

Reference material. Useful for calculating percentage of total calories from fat, carbohydrate, and protein.

on specific nutrients such as vitamin D, vitamin E, folate, and potassium can be added to the panel on a voluntary basis. However, if the package makes a claim about the food's content of a particular nutrient that is not on the "must have" list, then information about that nutrient has to be added to the Nutrition Facts panel. Nutrition Facts panels also contain a column headed **% Daily Value (%DV)**. Figures given in this column are intended to help consumers answer such questions as "Does a serving of this macaroni and cheese provide more protein than the other brand?" and "How much fiber does this cereal provide compared to my daily need for it?"

% Daily Value (%DV) Daily Values are scientifically agreed-upon standards of daily intake of nutrients from the diet developed for use on nutrition labels. The "% Daily Values" listed in nutrition labels represent the percentages of the standards obtained from one serving of the food product.

Daily Values (DVs) are standard amounts of nutrients developed specifically for use on nutrition labels. They are based on the 1968 edition of the Recommended Dietary Allowances (RDAs).[3] The %DV figures listed on labels represent the percentages of the standard nutrient amounts obtained from one serving of the food product. Standard values for total fat, saturated fat, and carbohydrates are based on a daily intake of 2,000 calories. In general, % Daily Values of 5 or less are considered "low," Daily Values of 10 to 19% are considered "good," and those listed as 20% or more "high" sources of the nutrient listed on the label.[2]

health action Some Examples of What "Front of the Package" Nutrient-Content Claims Must Mean[4,5]

Term	Means that a serving of the product contains:
More	At least 10% more of the Daily Value for a vitamin, mineral, protein, dietary fiber, or potassium
Good source	From 10 to 19% of the Daily Value for a particular nutrient
High	20% or more of the Daily Value for a particular nutrient
Less	At least 25% less of a nutrient or of calories than appropriate reference foods (similar products)
Less sugar	At least 25% less sugar than appropriate reference foods (similar products)
Low calorie	40 calories or less
Low sodium	140 grams or less sodium
Low fat	3 grams or less of fat
Free	No, or negligible amounts of, fat, sugars, *trans* fat, or sodium
Gluten free	Less than 20 parts per million of gluten

Table 4.1 Examples of undefined claims used on food labels

Natural
All natural
Pure
Antibiotic-free
Raised without antibiotics
Additive-free
Pesticide-free
Hormone-free
Nutritionally improved
No cholesterol (on plant foods)
Free-range
Eco-friendly
Pasture-fed
Contains whole grains
Made with real fruit
Dairy-free
Probiotic
Vegan
Free of fructose
Real cane sugar
Agave nectar

Nutrient Content Remember seeing the labels "High Fiber," "Good Source," or "Gluten-Free" on food packages? These are examples of nutrient content claims, and they are usually placed on the front of food packages. The claims are used to characterize the level of calories and nutrients in a serving of a food product. (The Health Action feature in this unit contains additional information on nutrient claim criteria.) Nutrient content claims must be accurate and conform to specific criteria developed by the FDA. Foods labeled "low sugar," for example, must contain at least 25% less sugar per serving than similar products. Low-fat foods can be labeled with a "percent fat free" label, such as "98% fat-free" turkey. This label means that the product contains approximately 2% fat on a weight basis. Meat products labeled "lean" must contain fewer than 10 grams of fat, 4.5 grams of saturated fat and *trans* fat combined, and 95 milligrams of cholesterol per serving.

Evaluating Undefined Content Claims Other claims, such as "natural," "hormone-free," and "clean" are sometimes placed on food labels to imply a particular product is better than others not so labeled (Table 4.1). These terms have not been defined by the FDA, may or may not represent a health benefit, and may mislead consumers about the health value of products. The word *natural*, for example, can be found on labels for foods such as potato chips, butter-flavored microwave popcorn, and ice cream (see Illustration 4.3). About 60% of consumers look for the "natural" label on food packages and often assume that the label means the food has no artificial ingredients, pesticides, or **genetically modified organisms (GMOs)**. It is also commonly assumed that animals raised for meat and poultry and sold as "natural" were not given growth hormones or antibiotics. This is not the case.[6]

A label may declare that a product is "low-sugar, "made with real fruit," or "made with whole grains." Products labeled as low-sugar may be high in fat and calories, and the fruit or whole-grain content of foods labeled "made with real fruit" or "made with whole grains" may be as little as a gram of fruit or whole grains per serving. A look at the ingredients label will help you estimate how much sugar, fruits, vegetables, or whole grains are in a product because ingredients are listed by weight. If fruit, for example, is listed as the first or second ingredient, the product likely contains a significant amount of fruit. If the fruit is listed further down, you know it probably doesn't have much fruit in it.

Health Claims In 1984, the Kellogg Company launched an ad campaign for All-Bran cereal that announced "eating the right foods may reduce your risk of some kinds of cancer." Sales of the high-fiber cereal increased 37% in one year, but then Kellogg had to withdraw the ads. The FDA ruled the All-Bran statement was equivalent to a claim for a drug. The campaign, however, started the nutrition and health claims revolution.[7,8]

On approval by the FDA, foods or food components with scientifically agreed-upon benefits for disease prevention can be labeled with a health claim (Table 4.2). However, the health claim must be based on the FDA's "model claim" statements. For instance, scientific consensus holds that diets high in fruits and vegetables may lower the risk of cancer, so a health claim to this effect is allowed. The FDA's model claim for labeling fruits and vegetables is "Low-fat diets rich in fruits and vegetables may reduce the risk of some types of cancer, a disease associated with many factors." The FDA approves health claims only for food products that are not high in fat, saturated fat, cholesterol, or sodium. Model health claims approved by the FDA are shown in Table 4.3.

Photo by Judith Brown

Illustration 4.3 Here's the ingredient list for the ice cream labeled "Made with natural ingredients": skim milk, cream, sugar, corn syrup, whey, molasses, acacia gum, guar gum, ground vanilla beans, vanilla extract, carob bean gum, carrageenan, xanthan gum.

Labeling Foods as Enriched or Fortified The vitamin and mineral content of foods can be increased by **enrichment** and **fortification**. Definitions for these terms were established more than 50 years ago. Enrichment pertains only to refined grain products, which lose vitamins and minerals when the germ and bran are removed during processing. Enrichment replaces the thiamin, riboflavin, niacin, and iron lost with the germ and bran. By law, producers of bread, cornmeal, pasta, crackers, white rice, and other products made from refined grains must use enriched flours. Beginning in 1998, federal regulations mandated that folate (a B vitamin) in the form of folic acid be added to refined grain products. This regulation was put into effect and is helping to reduce the incidence of a particular type of inborn, structural problem in children (neural tube defects) related to low blood levels of folate early in pregnancy.[9]

Any food product can be fortified with vitamins and minerals—and many are. One of the few regulations governing the fortification of foods is that the amount of vitamins and minerals added must be listed in the Nutrition Facts panel. Illustration 4.4 shows some examples of enriched and fortified foods.

Food enrichment and fortification began in the 1930s to help prevent deficiency diseases such as rickets (vitamin D deficiency), goiter (from iodine deficiency), pellagra

genetically modified organisms (GMOs) Food products that contain genetic material (usually from a bacteria) that has been transferred from the genetic material of another organism to give the food specific properties when produced.

enrichment The replacement of the thiamin, riboflavin, niacin, and iron lost when grains are refined.

fortification The addition of one or more vitamins and/or minerals to a food product.

Table 4.2 FDA-approved health claims[7]

1. Development of cancer depends on many factors. A diet low in fat may reduce the risk of some cancers.
2. Diets low in sodium may reduce the risk of high blood pressure, a disease associated with many factors.
3. Healthful diets with adequate folate may reduce a woman's risk of having a child with brain or spinal cord defect.
4. Frequent eating of foods high in sugars and starches as between-meal snacks can promote tooth decay.
5. Early introduction of peanuts may reduce the risk of peanut allergy in high risk infants.

Scott Goodwin Photography

Scott Goodwin Photography

Scott Goodwin Photography

Table 4.3 Examples of model health claims approved by the FDA for labels of foods that qualify based on nutrient content; model claims are often abbreviated on food labels[a]

Food and related health issues	Model health claim
Whole-grain foods and heart disease, certain cancers	Diets rich in whole-grain foods and other plant foods and low in fat, saturated fat, and cholesterol may reduce the risk of heart disease and certain cancers.
Sugar alcohols and tooth decay	Frequent between-meal consumption of food high in sugars and starches promotes tooth decay. Xylitol, the sugar alcohol in this food, may reduce the risk of tooth decay.
Saturated fat, cholesterol, and heart disease	Development of heart disease depends on many factors. Eating a diet low in saturated fat and cholesterol and high in fruits, vegetables, and grain products that contain fiber may lower blood cholesterol levels and reduce your risk of heart disease.
Calcium, vitamin D, and osteoporosis	Regular exercise and a healthy diet with enough calcium and vitamin D help maintain good bone health and may reduce the risk of osteoporosis later in life.
Fruits and vegetables and cancer	Low-fat diets rich in fruits and vegetables may reduce the risk of some types of cancer, a disease associated with many factors.
Folate and neural tube defects	Women who consume adequate amounts of folate daily throughout their childbearing years may reduce their risk of having a child with a brain or spinal cord defect.

[a]Nutrition labeling regulations exist in many countries. If you travel to Europe, you may notice that the health claim "helps maintain normal bowel function" is allowed on prune juice.[16]

(niacin deficiency), and iron-deficiency anemia.[10–12] Today, foods are increasingly being fortified for the purpose of reducing the risk of chronic diseases such as osteoporosis, cancer, and heart disease. Other foods, such as snack bars, fruit drinks, and sweetened ready-to-eat cereals, are being fortified with an array of vitamins and minerals. Regular consumption of fortified foods increases the risk that people will exceed Tolerable Upper Intake Levels (ULs) of nutrients designated in the RDAs. Regular use of multiple vitamin and mineral supplements, along with liberal intake of fortified foods, enhances the likelihood that excessive amounts of some nutrients will be consumed.[12]

The Ingredient Label Still more useful information about the composition of food products is listed on ingredient labels. The label of any food that contains more than one ingredient must list the ingredients in order of their contribution to the weight of the food (Table 4.4).

The FDA requires that ingredient labels include the presence of common food allergens in products. Potential food allergens include milk, eggs, fish, shellfish, tree nuts, wheat, peanuts,

Illustration 4.4 Examples of foods that are enriched (*left*) and fortified (*right*).

Table 4.4 Almost all foods with more than one ingredient must have an ingredient label. Here are the rules.

1. Ingredients must be listed in order of their contribution to the weight of the food, from highest to lowest.
2. Beverages that contain juice must list the percentage of juice on the ingredient label.
3. The terms *colors* and *color added* cannot be used. The name of the specific color ingredients must be given (for example, caramel color, turmeric).
4. Milk, eggs, fish, and five other foods to which some people are allergic must be listed on the label.

and soybeans. These foods, sometimes called the "Big Eight," account for 90% of food allergies. Precautionary statements such as "may contain," or "processed in a facility with" are allowed.[13]

Food Additives on the Label Specific information about **food additives** must be listed on the ingredient label. About 10,000 chemical substances may be added to food to enhance its flavor, color, texture, cooking properties, shelf life, or nutrient content. Food additives considered GRAS (*G*enerally *R*ecognized *A*s *S*afe) by the FDA can be used in food without preapproval. Food dyes, however, must be approved by the FDA prior to their use in food.[14] Specific ingredients allowed on the GRAS list change with time as new evidence becomes available. For example, *trans* fatty acids were considered to be generally recognized as safe until 2017, when research studies made it clear that small amounts may harm health. It is expected that food ingredients approved as GRAS will change with time as new information becomes available.[17]

food additives Any substances added to food that become part of the food or affect the characteristics of the food. The term applies to substances added both intentionally and unintentionally to food.

Table 4.5 provides examples of the functions of some additives used in food. The most common food additives are sugar and salt, but trace amounts of polysorbate, potassium benzoate, and many other additives that are not so familiar are also included in foods. Appendix D lists many of the most common additives and indicates their function in foods.

Trace amounts of substances such as pesticides; hormones and antibiotics given to livestock; fragments of packaging materials such as plastic, wax, aluminum, or tin; very small fragments of bone; and insects may end up in foods. These are considered "unintentional additives" and do not have to be included on the label.

Irradiated Foods Food irradiation is an odd example of a food additive. It is actually a process that doesn't add anything to foods. Irradiation uses X-rays, gamma rays, or electron beams to kill insects, bacteria, molds, and other microorganisms in food. Food irradiation

Table 4.5 Solving the mystery of ingredient label terms[a]

Cake mix ingredient label	Additive	Function
Ingredients: Sugar, enriched flour bleached (wheat flour, niacin), iron, thiamin mononitrate (vitamin B_1), riboflavin (vitamin B_2), vegetable shortening (contains partially hydrogenated soybean cottonseed oil), sodium aluminum phosphate, dextrose, leavening (baking soda, monocalcium phosphate, dicalcium phosphate, aluminum sulfate), wheat starch, propylene glycol monoesters, modified corn starch, salt, egg white, vanilla, dried corn syrup, polysorbate 60, nonfat milk, **mono-** and **diglycerides**, sodium citrate, xanthan gum, soy lecithin.	Sodium aluminum phosphate	Gives baked products a light texture
	Propylene glycol monoesters	Helps blend ingredients uniformly, enhances moisture content and texture
	Mono- and diglycerides	Maintains product softness after baking
	Xanthan gum	Thickening agent, helps hold product together after baking

Scott Goodwin Photography

Note: All foods with more than one ingredient must have an ingredient label.
[a]See Appendix D for more information on the definitions of ingredient terms.

Judith Brown

Illustration 4.5 The "radura," as the symbol is called, must be displayed on irradiated foods. The words "treated by irradiation, do not irradiate again" or "treated with radiation, do not irradiate again" must accompany the symbol.

enhances the shelf life of food products and decreases the risk of food-borne illness. Irradiation must be performed according to specific federal rules.[15]

Irradiated foods retain no radioactive particles. The process is like having your luggage X-rayed at the airport. Your luggage doesn't become radioactive, nor has anything in it changed. The process leaves no evidence of having occurred. Actually, this lack of change creates a challenge because inspection agencies can't determine whether a food has been irradiated or not. All irradiated foods—except spices that are added to processed foods—are required to display the international "radura" symbol and to indicate that the food has been irradiated (Illustration 4.5). Irradiation is approved for use on chicken, turkey, pork, beef, eggs, grains, fresh fruits and vegetables, and other foods in the United States.[15]

Genetically Modified Organisms (GMOs) GMOs contain selected, individual genes transferred from one organism to another or by editing existing genes. It is generally done to improve a crop's resistance to diseases, increase production, improve drought resistance, or enhance the nutrient content of the GMO product. GMO foods have been consumed by humans since the mid-1990s. They have passed risk assessment tests and have not been found to pose risks to human health. Consequently, the FDA determined that the labeling of GMO foods is not warranted and GMO foods do not need to be labeled in the United States or Canada.[24,25] Other countries do require GMO foods to be labeled, based primarily on concerns such as crop biodiversity and the control of seed crops by a few agricultural companies (Illustration 4.6). Some people are fearful about consuming genes, but they are a component of most of the food we eat, are digested, and do not function as genes in our bodies.[24]

dietary supplement Any product intended to supplement the diet, including vitamins, minerals, proteins, enzymes, herbs, hormones, and organ tissues. Such products must be labeled "Dietary Supplement."

structure/function claim Statement appearing primarily on dietary supplement labels that describes the effect a supplement may have on the structure or function of the body. Such statements cannot claim to diagnose, cure, mitigate, treat, or prevent disease.

Dietary Supplement Labeling

- **Evaluate nutrient content and health claims made on dietary supplement labels.**

A **dietary supplement** is a product taken by mouth that contains a "dietary ingredient" intended to supplement the diet. The dietary ingredients in these products include vitamins, minerals, proteins, enzymes, herbs, hormones, and organ tissues. Nutrition labeling regulations place dietary supplements in a special category under the general umbrella of "foods," not drugs. Dietary supplements differ from drugs in that they do not have to undergo rigorous testing and obtain FDA approval before they are sold. In return, dietary supplement labels cannot claim that the products treat, cure, or prevent disease.[18]

According to FDA regulations, dietary supplements must be labeled as such and include a "Supplement Facts" panel that lists serving size, ingredients, and percent Daily Value (%DV) of essential nutrient ingredients (Illustration 4.7). Like foods, qualifying dietary supplements can be labeled with nutrient claims (e.g., "high in calcium"), and health claims (e.g., "diets low in saturated fat and cholesterol that include sufficient soluble fiber may reduce the risk of heart disease"). Dietary supplement labels can also include **structure/function claims**.

Nutrition Facts
Per 125mL (87g)

Amount	% Daily Value
Calories 80	
Fat 0.5 g	1%
Saturated 0 g + Trans 0 g	0%
Cholesterol 0 mg	
Sodium 0 mg	0%
Carbohydrate 18 g	6%
Fibre 2 g	8%
Sugars 2 g	
Protein 3 g	
Vitamin A 2% Vitamin C	10%
Calcium 0% Iron	2%

Scott Goodwin Photography

Illustration 4.7 Dietary supplement label.

INGREDIENTS: SUGAR, WATER, MAIZE FLOUR (PRODUCED FROM GENETICALLY MODIFIED MAIZE), EGG,FLAVOURINGS.

Illustration 4.6 Example of GMO labeling in the United Kingdom for a corn (maize) product.[26]

Structure or Function Claims

Dietary supplements can be labeled with statements that describe effects the supplement may have on body structures or functions (shown in Illustration 4.8). Under this regulation, for example, certain supplements can be labeled with the following statements: "Promotes healthy heart and circulatory function," "Helps maintain mental health," or "Supports the immune system." If a structure/function claim is made on the label or package inserts, the label or insert must acknowledge that the FDA does not support the claim: "*This statement has not been evaluated by the FDA. This product is not intended to diagnose, treat, cure, or prevent any disease.*" Dietary supplements labeled with misleading or untruthful information and those that are not safe can be taken off the market and the manufacturers fined.

Structure and function claims do not have to be approved before they can appear on product labels. The FDA and FTC have taken action against companies responsible for fraudulent claims.[18] The FDA has issued new guidelines that define the quality of scientific evidence needed to substantiate structure/function claims.[18]

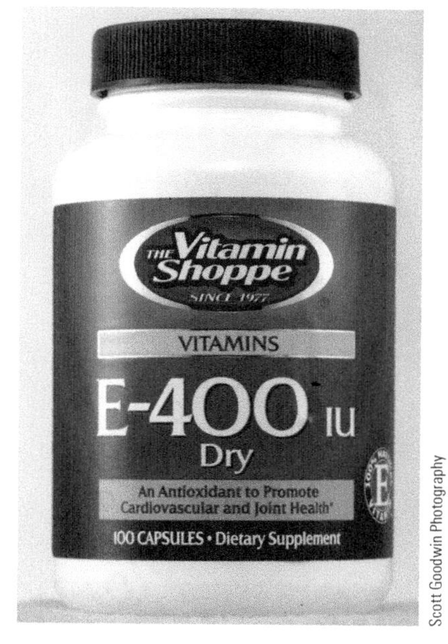

Illustration 4.8 Example of a structure/function claim.

The COOL Rule

The United States Department of Agriculture (USDA) requires retailers to display a country-of-origin label (COOL) on certain products (Illustration 4.9). The rule is meant to expand informed consumer choices and to help track down food-borne illness outbreaks. The rule applies to meats, fish, seafood, fruits, vegetables, many nuts, and some herbs.[19]

Illustration 4.9 Country-of-origin labels are now required on certain foods.

Organic Foods

- **Compare the characteristics of organically and conventionally produced food products.**

For many years, consumers and producers of organic foods urged Congress to set criteria for the use of the term *organic* on food labels. Consumers wanted to be assured that foods were really organic, and producers wanted to keep the business honest. Standards were needed because consumers cannot distinguish organically produced foods from others by looking at them, tasting them, or reading their nutrient values. Demand for and production of organic foods in the United States are growing. Organic foods comprised about 5% of total food sales in 2016.[21,34]

Labeling Organic Foods

Rules that qualify foods as organic are shown in Illustration 4.10. If organic growers and processors qualify according to USDA-approved certifying organizations, they can place the green-and-white USDA Organic seal on product labels (Illustration 4.11). The Canadian organic label, also shown in Illustration 4.11, is equivalent to the USDA

Illustration 4.10 USDA rules for qualifying foods as organic.

1. Plants
   - Must be grown in soils not treated with synthetic fertilizers, pesticides, and herbicides for at least three years
   - Cannot be fertilized with sewer sludge
   - Cannot be treated by irradiation
   - Cannot be grown from genetically modified seeds or contain genetically modified ingredients
2. Animals
   - Cannot be raised in "factory-like" confinement conditions
   - Cannot be given antibiotics or hormones to prevent disease or promote growth
   - Must be given feed products that are 100% organic

United States Department of Agriculture

Canada Organic Regime (COR), Canadian Food Inspection Agency (CFIA)

Illustration 4.11 Foods certified as organic by the USDA can display the "USDA Organic" seal on food packages. The approved seal for Canada is below it.

Organic label. The USDA and Canadian authorities can impose financial penalties on companies that use the seal inappropriately. Organically grown and produced foods can be labeled in four other ways:

1. "100% Organic" if they contain entirely organically produced ingredients.
2. "Organic" if they contain at least 95% organic ingredients.
3. "Made with organic ingredients" if they contain at least 70% organic ingredients.
4. "Some Organic Ingredients" if the product contains less than 70% organic ingredients.

Some people choose organic foods because they perceive them to be better for the environment and for health than conventionally grown foods.[22] Organic foods tend to have lower detectable levels of pesticide residues and heavy metals, higher concentrations of certain antioxidants, and lower presence of antibiotic-resistant bacteria than conventionally grown foods. Although more expensive to produce and purchase than conventionally produced foods, organic foods are "friendlier" to the environment and leave a low carbon footprint.[22,23]

Other Nutrition Labeling Systems

- **Identify strengths and weaknesses of various nutrition labeling systems on food packaging and calorie listings for food items.**

If you are a nutrition label reader, you have probably noticed that a variety of food company–initiated labeling systems appear on the front of food packages. Three of these, "Nutrition at a Glance," "Smart Choices Made Easy," and the "Smart Choices Program," are shown in Illustration 4.12. The Smart Choices Program features a symbol that identifies more nutritious choices within specific food product categories, and lists the calories per serving and number of servings per package. This system has been selected for use by a number of food companies in Europe and the United States.[27] Other nutrition labeling systems, including those from the American Heart Association and various grocery store chains, are being used on food product packaging.

Industry-initiated labeling systems are voluntary and generally employ science-based criteria for judging the nutrient qualities of food products.[28] The labels supplement the Nutrition Facts panel of the food products. The primary weaknesses of the food industry–backed labeling systems are that they may cause confusion among consumers, criteria used to label foods vary, they may favor the company's products, and the labels only appear on selected food products.[27,29]

Labeling systems in place may not adequately emphasize information consumers need to know to help curb high-calorie intake and obesity. This concern is being addressed through increased emphasis on the labeling of calories in portions of food sold in chain restaurants.

Scott Goodwin Photography

Illustration 4.12 A few examples of food industry–initiated nutrition labeling systems used on food packages.

Calories on Display

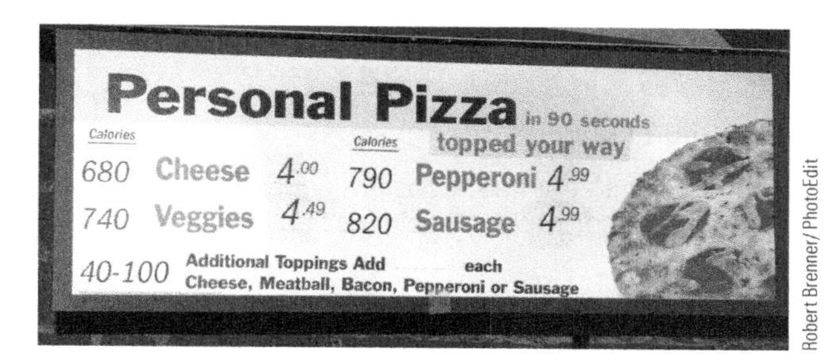

Robert Brenner/PhotoEdit

Illustration 4.13 An example of the required calorie labeling of chain restaurant menu items.

Since 2012, several large cities and states have implemented legislation that requires chain restaurants with more than 20 locations to post the calorie value of a serving of each standardized menu item offered (Illustration 4.13). Details regarding the item's content of 11 other nutrients must be available on request. This calorie labeling system for chain restaurants was adopted for nationwide implementation beginning in 2012. The labeling requirement also extends to food offered in vending machines for establishments with 20 or more locations nationwide.[30]

Study results indicate that calorie labeling of restaurant menu items is associated with little change or somewhat lower-calorie food choices among diners, and increased availability of lower-calorie menu options in some restaurants.[31,32]

Upcoming Nutrition Label Revisions

With the exception of the addition of *trans* fats to the label in 2006, few major changes have occurred in nutrition labeling in the United States. Science marches on, however, and knowledge about the behavioral science behind the design of effective communication tools has advanced. The new labels will reflect advances in nutrition science and national dietary intake recommendations, reality-based serving size standards, and an improved design.[3,4]

The major changes proposed for the new labels include:

1. Reality-based serving sizes. The serving size listed on the Nutrition Facts Panel would represent the average portion consumed by people over the age of 2 in the United States.
2. Calorie content of a serving of the food would be highlighted in large type on the Nutrition Facts Panel.
3. *Added sugars* would be placed on the Nutrition Facts Panel under *Sugars*. It is proposed that added sugars be defined as "sugar or sugar-containing ingredient that is added during processing."
4. The vitamin D and potassium content of a serving of the food would be required on the Nutrition Facts Panel because intake of these two nutrients tends to be low. Vitamins A and C would no longer be required because intakes are rarely low.
5. Updated Daily Values (DVs) for most vitamins and minerals based on updates to the RDAs since 1968 will be used on the new labels. A DV for added sugars of less than 10% of total calories and reporting %DV, grams, and teaspoons of added sugar will appear on the new labels. High-sugar foods will likely not qualify for health claims in the future.[3,4]
6. "Calories from fat" will not appear on the new label because it is now clear that the types of fat consumed are more important than the amount.

REALITY CHECK

Nutrition Labeling

Foods labeled as "fat free" have few or no calories.

Who gets the thumbs up?

Answers on page 4-12

PhotoDisc

Jared: Right. You take the fat out of food, and calories go away.

PhotoDisc

Ronald: Maybe, maybe not. "Fat-free" foods could still contain sugars, protein, and other ingredients that have calories.

ANSWERS TO REALITY CHECK

Nutrition Labeling

Ronald got it. "Fat free" does not equal "calorie free." Fat-free Caesar salad dressing, for example, provides 40 calories in a 2-tablespoon serving.

PhotoDisc

Jared:

PhotoDisc

Ronald:

The new standards will begin to be implemented in 2018.[3] A sample of the Nutrition Facts Panel highlighting the proposed changes appears in Illustration 4.14.

Beyond Nutrition Labels

Even with the new labels, consumers can't be stupid.
—Max Brown

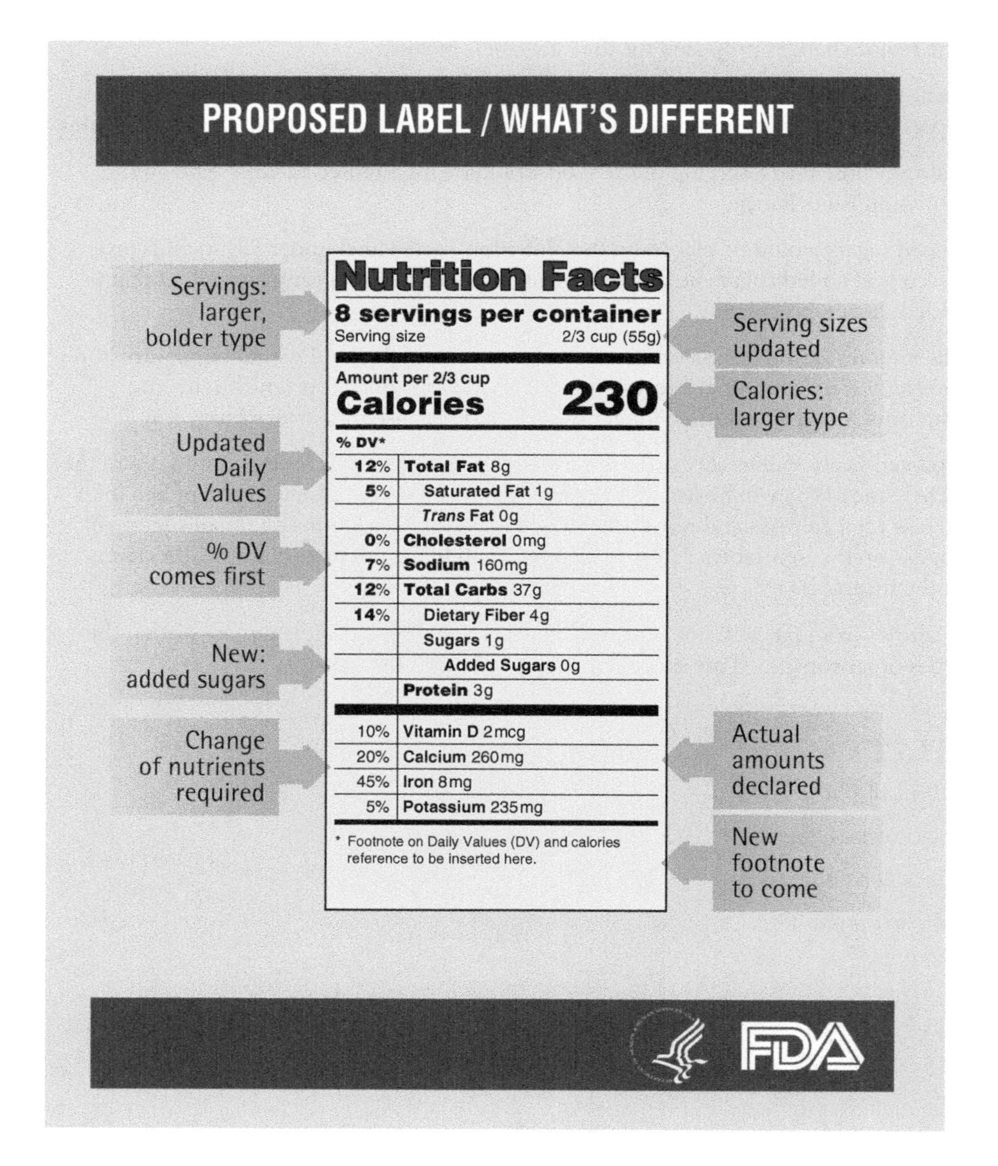

Illustration 4.14 The proposed Nutrition Facts Panel.

Understanding and applying the information on nutrition labels calls for more nutrition knowledge on the public's part, and labeled information that is more user-friendly, easier to understand, and monitored to ensure factual honesty. As highlighted in the Reality Check, people need to know about nutrition before they can understand nutrition labels and incorporate labeled foods appropriately into an overall diet. Good diets include more than foods with nutrition labels. The ice cream cone from the stand in the mall; the orange, potato, or fish we buy in the store; and the pizza delivered to the dorm are unlabeled parts of many diets. We need to know enough about the composition of unlabeled foods to fit them into a healthy diet. Nutrition labels often list a few vitamins and minerals, but many more are required for health. It's particularly important for people to know if their diet is varied enough to supply needed vitamins and minerals. In addition, nutrient and health claims made on food packaging do not address potentially negative aspects of food products, such as "low fiber," "high saturated fat," or "may promote tooth decay." Use of the MyPlate food group guide should go hand in hand with label reading for diet planning.

Not every food we eat has to have the "right" label profile. Serving mostly low-fat or low-calorie foods to children, for example, might have unintended unhealthy effects. Young children need calories and fat for growth and development. If diets are severely restricted, growth and development will be impaired.

Nutrition labels are an important tool for helping people make informed food-purchasing decisions. About 6 in 10 adults use them to guide their food purchasing decisions, and those who do use them tend to consume healthier diets than those who don't.[33] However, labels do not now—nor will they ever—provide all the information needed to make wise decisions about food. Only people who are well informed about nutrition can do that.

allesalltag/Alamy

NUTRITION up close

Understanding Food Labels

Focal Point: Understanding nutrition labels so you know what you are getting.

Are you ready to put your knowledge of nutrition labels to use? Examine the front-of-package, facts panel, and ingredient labels shown in the Illustrations below. Then answer the following questions:

Courtesy of Brown

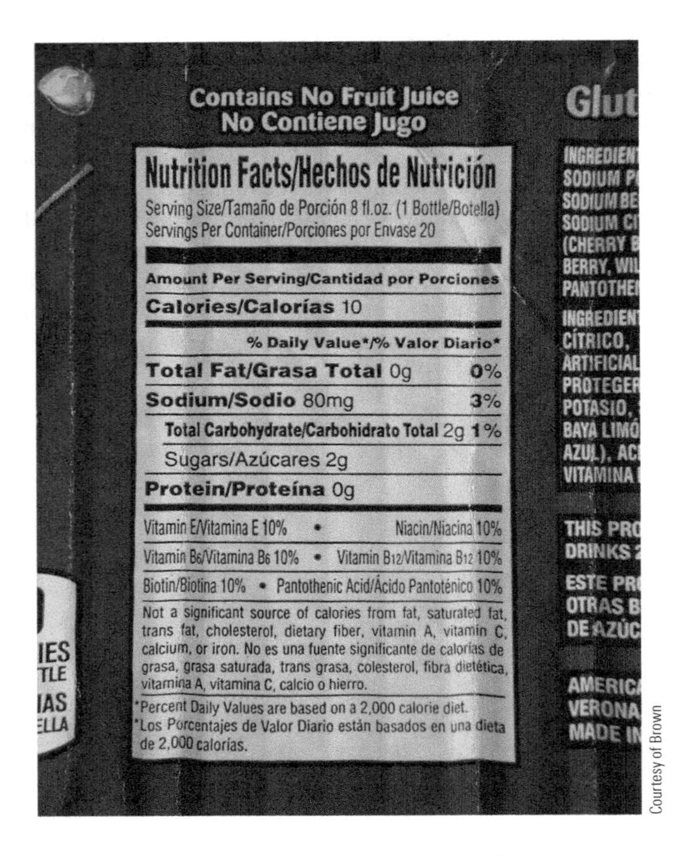

Courtesy of Brown

1. Fruit drinks are not part of a healthy dietary pattern. Does the lower sugar content and the addition of B vitamins and vitamin E to a fruit drink change that?
2. Does the product contain any fruit or fruit juice?
3. The product is fortified with biotin, pantothenic acid, niacin, vitamin B_6, and vitamin B_{12}. Which of these vitamins are included in the group of vitamins considered to be of public health significance?
4. The fruit drink is labeled gluten-free even though there is no reason to expect that a fruit drink would contain gluten. Why do you think this label was added?
5. If nutrition labels were required to label both the positive and negative characteristics of a product, which positive and negative characteristics would you attribute to this product?

Feedback to the Nutrition Up Close is located in Appendix F.

REVIEW QUESTIONS

- **Apply knowledge about the four key elements of nutrition labeling to decisions about the nutritional value of foods.**

1. Almost all multiple-ingredient foods must be labeled with nutrition information. **True/False**
2. Food manufacturers can list any serving size they want on Nutrition Facts panels. **True/False**
3. In general, %DV listed for nutrients in Nutrition Facts panels of 10% or more are considered "low," and those listed as 50% or more are considered "high." **True/False**

4. An overriding principle of nutrition labeling regulations is that nutrient content and health claims made about a food on the packaging must be truthful. **True/False**
5. The term *enriched* on a food label means that extra vitamins and minerals have been added to the food to bolster its nutritional value. **True/False**
6. The ingredient that makes up the greatest portion of a food product's weight must be listed first on ingredients labels. **True/False**
7. Food labeling regulations do *not* require food manufacturers to list the presence of major food allergens on ingredient labels. **True/False**
8. ___ A tea manufacturer labels a green tea product as "high in the antioxidant vitamin C." After brewing, a serving of the green tea was found to contain 10 mg of vitamin C. Which of the following statements about the green tea nutrient content claim is true?
   a. The green tea qualifies for the "high in" nutrient content claim made.
   b. The green tea qualifies for a "natural" content claim.
   c. The green tea qualifies for a "healthy" nutrient content claim.
   d. The green tea qualifies for a "good source of" nutrient content claim.
9. ___You see a yogurt product at the grocery store labeled "fat free." This nutrient content claim means that a standard serving of the yogurt contains:
   a. No, or negligible amounts of, fat.
   b. No, or negligible amounts of, fat, *trans* fat, and sodium.
   c. Fewer than 10 grams of fat and 4.5 grams of saturated fat.
   d. Three grams of fat or less.
10. ___John eats three cookies for a bedtime snack while reading the Nutrition Facts panel on the cookie package. He notices that a three-cookie serving provides 2 grams of saturated fat and 0 grams of *trans* fat. The cookie qualifies for a nutrient content claim of:
   a. Low saturated fat.
   b. *Trans* fat free.
   c. Extra lean.
   d. None of the above.

Questions 11 through 14 refer to the following case scenario.

You have just been informed by your health care provider that your cholesterol level is "borderline high" and that you need to lose 10 pounds. The health care provider recommends that you reduce your calorie intake by 400 calories a day, cut back on saturated fats, and come back to the clinic in three months.

Here's your plan on how to decease calories and saturated fat: You will cut back on your intake of snack food. You have been eating three 1-ounce servings of potato chips, and two 1-ounce servings of cheese crackers nightly. You notice from the Nutrition Facts panels on the chips and crackers that each ounce of potato chips (15 chips) provides 160 calories, and each ounce of cheese crackers (25 crackers) provides 140. Also, one serving of potato chips provides 1 gram (5% Daily Value) of saturated fat, and one serving of cheese crackers has 1.5 grams (8% Daily Value) of saturated fat.

11. How many calories were you consuming from the chips and crackers daily? ___ calories
12. If you reduced your intake of both potato chips and cheese crackers to one serving per day, you would reduce your intake from these snack foods by ___ calories.
13. The amount of saturated fat in terms of grams and % Daily Value provided by three servings of potato chips and two servings of cheese crackers is ___ grams and ___ % Daily Value.
14. If you decrease your intake of potato chips and cheese crackers to one ounce of each a day, you will reduce your saturated fat intake from these snacks by ___ grams and by ___ % Daily Value.

- **Evaluate nutrient content and health claims made on dietary supplement labels.**

15. Dietary supplements, such as those for herbs and vitamins, are considered drugs and are not regulated by FDA's Nutrition Labeling rules. **True/False**
16. Foods, but *not* dietary supplements, can be labeled with nutrient content and health claims. **True/False**
17. ___Assume a dietary supplement made from pomegranate is labeled with the claims "lowers plaque formation in the arteries" and "improves blood flow." It turns out that the FDA becomes aware of the claims and subsequently sues the maker of the dietary supplement. What would be the most likely reason for the FDA suit?
   a. There are no approved nutrient claims for "lowers plaque formation" and "improves blood flow."
   b. Dietary supplement labels cannot claim that a product treats a disease.
   c. There was insufficient research to support the claims.
   d. The FDA did not preapprove the claim before it was made for the juice.

- **Compare the characteristics of organically and conventionally produced food products.**

18. Animals providing meats labeled organic cannot be given antibiotics or hormones. **True/False**
19. Foods bearing the USDA Organic seal are certified as organic by the U.S. Department of Agriculture. **True/False**

- **Identify strengths and weaknesses of various nutrition labeling systems on food packaging and calorie listings for food items.**

20. Nutrition labels provide all the nutrition information we need to make healthful food choices. **True/False**

Answers to these questions can be found in Appendix F.

NUTRITION SCOREBOARD ANSWERS

1. Labeling is required for almost all processed foods and for dietary supplements, but remains largely voluntary for fresh fruits, vegetables, and fish.[1,2] **False**
2. Health claims for food products are allowed on many food packages. The claims must be truthful and adhere to FDA standards. **True**
3. Less sugar doesn't necessarily mean *low-sugar*. The "less sugar" label means the labeled product has at least 25% less sugar per serving than comparable products. It still may contain a good deal of sugar.[2] **False**
4. Nutrition labels are necessarily short and can't tell the whole story about food and health. They help people make several key decisions about a food's composition. **False**

Nutrition, Attitudes, and Behavior

NUTRITION SCOREBOARD

1 Food preferences are primarily determined by genetic factors. **True/False**

2 Food habits never change. **True/False**

3 Sweet, sour, salty, and bitter tastes can be sensed over all parts of the tongue. **True/False**

4 Children's acceptance of a wide variety of vegetables and fruits can be increased by frequently offering them an assortment of vegetables and fruits. **True/False**

Answers can be found at the end of the unit.

LEARNING OBJECTIVES

After completing Unit 5 you will be able to:

- Identify factors that influence an individual's food choices and preferences.
- Apply the process for making healthful changes in food choices to a specific change in food intake.
- Differentiate between scientifically supported and unsupported conclusions about diet and behavior relationships.

Origins of Food Choices

- **Identify factors that influence an individual's food choices and preferences.**

Horse meat is a favorite food in a large area of north-central Asia. Pork, which is widely consumed in North and South America, Europe, and other areas, is rigidly avoided by many people in Islamic countries. Bone-marrow soup and sautéed snails are delicacies in France, while kidney pie is traditional in England. Dog is a popular food in Borneo, New Guinea, the Philippines, and other countries, whereas snake is a delicacy in China. Some people enjoy insects (Illustration 5.1). And then there are steamed clams and raw oysters—food passions for some, but absolutely disgusting to others.[3]

When did you first think "yecck!"? The food choices just described would elicit that response among people from a variety of cultures, but they would not necessarily be responding to the same foods.

Why do people eat what they do? People learn from their family and the society in which they live what animals and plants are considered food and which are not.[4] Once items are identified as food, they develop a legacy of strong symbolic, emotional, and cultural meanings. Comfort foods, health foods, junk foods, fun foods, soul foods, fattening foods, mood foods, and pig-out foods, for example, have been identified in the United States. All cultures have their "super food": in Russia and Ireland, it's potatoes; in Central America, it's corn and yucca (a starchy root, also called *manioc*); in Somalia, it's rice. The designation refers to the cultural significance of the food and not to its nutritional value.[5]

In countries such as the United States where a wide variety of foods are available and people have the luxury of selecting which foods they will eat, food choices are influenced by a range of factors (Illustration 5.2). Of these factors, food preference has the largest impact. Food preferences vary a good deal among individuals and lead to a wide array of specific food choices. Rather than being inborn, food preferences are primarily learned.[7]

Gavriel Jecan/Photodisc/Getty Images

Illustration 5.1 Grasshoppers are Mexican delicacies, served at Girasoles Restaurant in Mexico City.

KEY NUTRITION CONCEPTS

Our examination of the factors involved in food choices relates to three of the key nutrition concepts introduced in Unit 1:

1. Nutrition Concept #1: Food is a basic need of humans.
2. Nutrition Concept #7: Some groups of people are at higher risk of becoming inadequately nourished than others.
3. Nutrition Concept #8: Poor nutrition can influence the development of certain chronic diseases.

We Don't Instinctively Know What to Eat

The food choices people make are not driven by a need for nutrients or guided by food selection genes. People deficient in iron, for example, do not seek out iron-rich foods. If we're overweight, no inner voice tells us to reject high-calorie foods. Women who are pregnant don't instinctively know what to eat to nourish their growing fetuses. No evidence indicates that young children, if offered a wide variety of foods, would select and ingest a well-balanced diet.[8]

Humans are born with mechanisms that help them decide when and how much to eat, however.[7] An inborn attraction to sweet-tasting foods, a dislike for the taste of bitter or sour, and the response of thirst when water is needed all influence food and fluid intake to an extent (Illustration 5.3).[6,9] There is evidence to suggest that people deficient in sodium experience an increased preference for salty foods.[7]

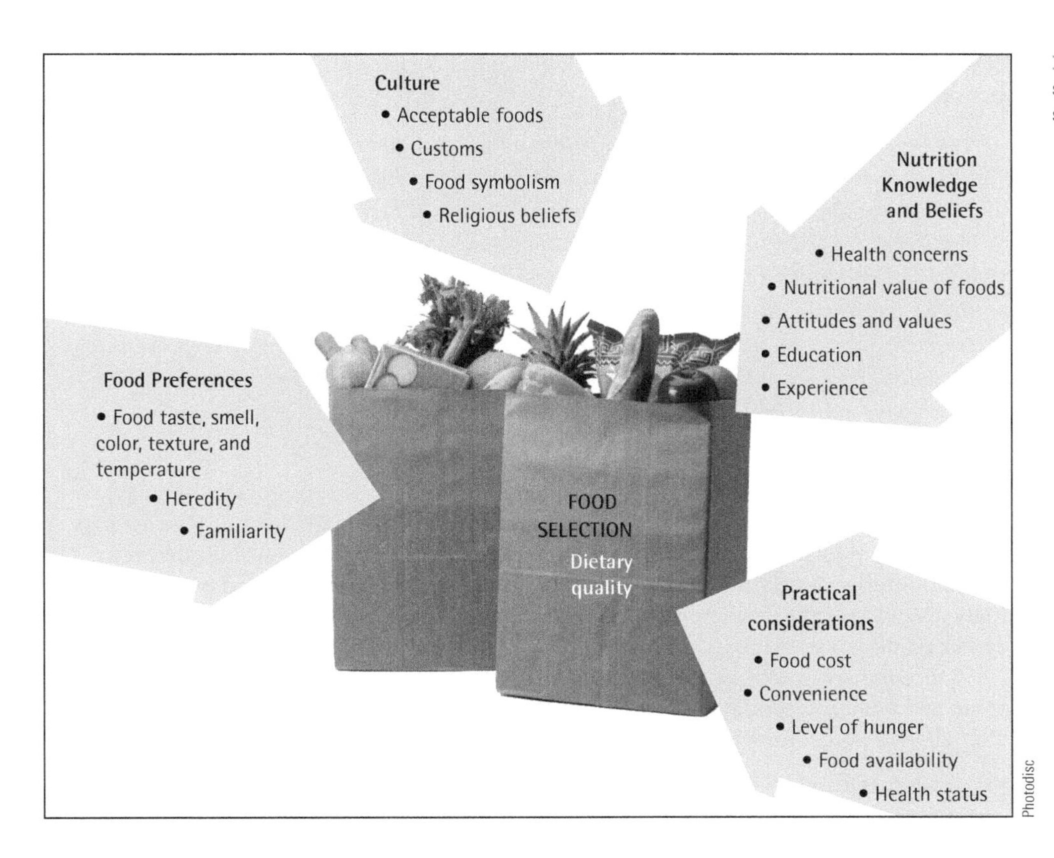

Illustration 5.2 Factors influencing food selection and dietary quality. Each of these sets of factors interacts with the others.[6]

Food Preferences

You're going to eat that!?! Do you have any idea what's in that hot dog?

Yeah, I do. There's barbecue in the backyard, ball games with my mom and dad, and the parties at my friend's house. The memories taste great!

The strong symbolic, emotional, and cultural meanings of food come to life in the form of food preferences. We choose foods that, based on our cultural background and other learning experiences, give us pleasure. Foods give us pleasure when they relieve our hunger pains, delight our taste buds, provide the right feeling of texture in our mouth, or give us comfort and a sense of security. We find foods pleasurable when they outwardly demonstrate our superior intelligence, our commitment to total fitness, or our pride in our ethnic heritage. We reject foods that bring us discomfort, guilt, and unpleasant memories, and those that run contrary to our values and beliefs.[10]

The Symbolic Meaning of Food Food symbolism, cultural influences, and emotional reasons for food choices are broad concepts that may become clearer with concrete examples. Here are a few examples to consider.

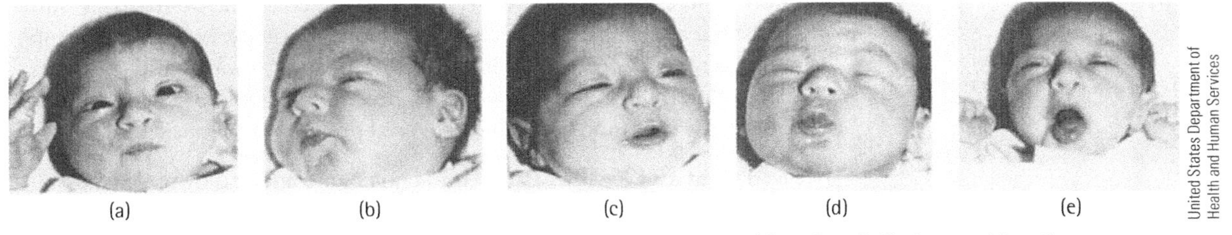

Illustration 5.3 Newborn infants respond to different tastes: (a) baby at rest, (b) tasting distilled water, (c) tasting sugar, (d) tasting something sour, and (e) tasting something bitter.
Source: Taste-induced facial expressions of neonate infants from the classic studies of J. E. Steiner, in Taste and Development: *The Genesis of Sweet Preference*, ed. J. M. Weiffenbach, HHS Publication no. NIH 77-1068 (Bethesda, MD: U.S. Department of Health and Human Services, 1977), pp. 173–89, with permission of the author.

Status Foods Vance Packard, in his book *The Status Seekers*, provides a memorable example of the symbolic value of food:

> *As a lad, this man had grown up in a poor family of Italian origin. He was raised on blood sausages, pizza, spaghetti, and red wine. After completing high school, he went to Minnesota and began working in logging camps, where—anxious to be accepted—he soon learned to prefer beef, beer, and beans, and he shunned "Italian" food. Later, he went to a Detroit industrial plant, and eventually became a promising young executive. . . . In his executive role he found himself cultivating the favorite foods and beverages of other executives: steak, whiskey, and seafood. Ultimately, he gained acceptance in the city's upper class. Now he began winning admiration from people in his elite social set by going back to his knowledge of Italian cooking, and serving them, with the aid of his manservant, authentic Italian treats such as blood sausage, spaghetti, and red wine!*[11]

Comfort Foods Ice cream, apple pie, chicken noodle soup, boxed chocolates, meat loaf and mashed potatoes: these are popular comfort foods in the United States. The feelings of security and love that came along with the tea and honey or chicken soup that your mother or father gave you when you had a cold, or with the ice cream and popsicles lovingly given to soothe a sore throat, are renewed with comfort foods. Some comfort foods can bring pleasure and reduce anxiety just by their image (Illustration 5.4).[12,14]

Once the symbolic value of a food is established as a comfort food, its nutritional value will remain secondary.[13] Food status is a strong determinant of food choices; and after all, as a noted nutritionist once said, "Life needs a little bit of cheesecake."[14]

"Discomfort Foods" Memories of bad experiences with food and expectations that certain foods will harm us in some way, each contribute to our learning about food and affect our food preferences. Eating a piece of blueberry pie right before an attack of the flu or overindulging in sweet pickles or olives, for example, may take these foods off your preferred list for a long time. Children who have had the experience of not being able to leave a table until they ate their green peas (or another food) may hold on to that "discomforting" memory for a lifetime.

Cultural Values Surrounding Food A team of scientists observed that the diet of certain groups in the Chin States of Upper Burma (now Myanmar) was seriously deficient in animal protein. After considerable study, a way was found to improve the situation by cross-breeding the small, local black pigs raised by the farmers with an improved strain to obtain progeny giving a greater yield of meat. The entire operation, however, completely failed to benefit the nutrition of the population because of one fact, which had been viewed as irrelevant. The cross-bred pigs were spotted. And it was firmly believed—as firmly as we believe that to eat, say, mice, would be disgusting—that spotted pigs were unfit to eat.[3]

Dietary change introduced into a culture for the purpose of improving health can be successful only if it is accepted by the culture. Cultural norms are not easily modified.[4]

Jon Edwards Photograp/Age Fotostock

Illustration 5.4 Do you find comfort in this photo?

Other Factors Influencing Food Choices and Preferences Food preferences and selections are also affected by the desire to consume foods that are considered healthy. Reducing fat intake, eating more fruits and vegetables, and cutting down on sweets bring rewards and pleasures such as weight loss and maintenance, an end to constipation, lower blood cholesterol level, and a newly discovered preference for basic foods.[15]

Food Cost and Availability Food choices are affected by the cost of food. Researchers found that college students eating in dining halls have better diets when they prepay for their meals for the entire term rather than paying at each meal.[16] The cost of vegetables and fruits in grocery stores affects intake; the higher the price, the lower the intake. Intake of vegetables and fruits tends to increase when they become cheaper to buy.[17]

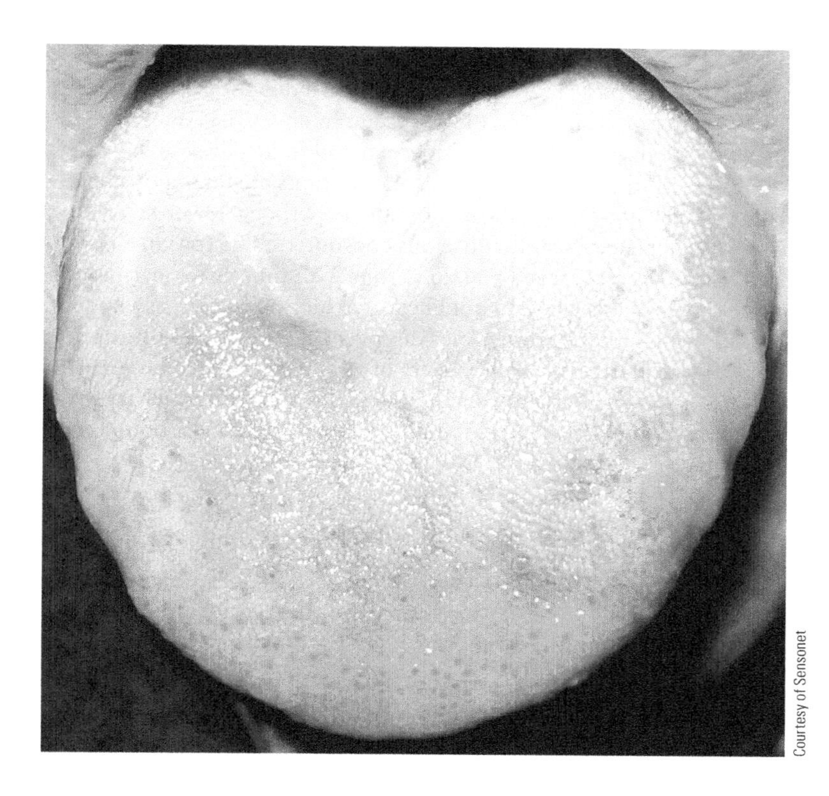
Courtesy of Sensonet

Illustration 5.5 This tongue is colored with blue food dye to show the small "bumps" on the tongue called papillae. Taste receptor cells are located in taste buds within the papillae.[1]

Taste Sensitivities and Food Preferences Taste and smell are major factors influencing food choices and health. How foods taste to individuals is largely determined by genetic factors, whereas recognizing smells is learned. The flavor of food as we perceive it is created in our brain by taste and smell signals generated by food consumption.[18]

Five basic tastes have been identified: sweet, salty, bitter, sour, and umami (savory or meaty).[18] Evidence that a sixth basic taste exists for fatty is growing.[18] When food is consumed, taste receptor cells located primarily in tiny taste buds all over the tongue transmit to the brain information on the sensation and intensity of the basic tastes present (Illustration 5.5). The perceived intensity of the tastes will vary among individuals. Some people, for example, are very sensitive to a bitter taste (supertasters), others will detect it to a limited extent (medium tasters), and still other people will not taste bitter at all (nontasters).[1] Supertasters tend not to eat vegetables such as brussels sprouts, cabbage, and broccoli, and to dislike bitter-tasting teas, wine, and tonic water.[19-22] About 25% of the U.S. population are supertasters, 50% are medium tasters, and 25% are nontasters.[23] Adults with above-average taste intensity scores, especially for sugar, appear to be more likely to gain weight over time than others with lower scores.[19]

Individual differences in smell perceptions also influence food preferences. A dislike of cilantro or arugula, for example, is related to a perceived adverse smell of the food. Most people immediately dislike and will decide not to eat a food that that tastes very bitter or sour, or that smells "bad."[18]

Food Preferences and Choices Do Change

- **Apply the process for making healthful changes in food choices to a specific change in food intake.**

Who says old dogs can't learn new tricks? Most Americans aren't eating the way they used to. Recent surveys indicate that food choices made in the United States are trending toward healthier dietary patterns. On average, people are consuming more fish, poultry, nuts, kale, and sweet potatoes, and lower amounts of red meat, white potatoes, whole milk, and ice cream than in the past.[26–28] Caloric intake is on a downward trend, as are portion sizes and the frequency of dining at fast food restaurants.[26–28] The American consumer is increasingly interested in flavorful ethnic and spicy foods, locally sourced vegetables and meats, sustainable seafood, and organic foods. Restaurants are taking this development seriously, with menus that offer less meat and more appetizing, lower-calorie, plant-centered dishes that appeal to health-conscious consumers (Illustration 5.6).[26-30]

Food choices are largely learned and do change with time as we learn more about foods and health. Individuals who did not like foods such as oysters, asparagus, beets, mushrooms, or sushi as children sometimes discover that they really like these foods later in life.[25] Children's acceptance of a wide variety of vegetables and fruits can be increased by frequently offering them an assortment of vegetables and fruits.[2] Perhaps your food choices have changed over time. How do the food choices you make now compare with the choices you made five years ago?

How Do Food Choices Change?

What are the ingredients for change in food choices? Why do some people succeed in improving their food choices while other people find that very hard to do? Nutrition knowledge, attitudes, and values have a lot to do with changing food choices for the better.

Nutrition Knowledge and Food Choices Sound knowledge about good nutrition necessarily precedes the selection of a healthful diet. But is knowledge enough to ensure that healthy changes in diet will be made? The answer is "yes" for some people and "no" for others.[6] Many people know far more about the components of a good diet than they put into practice, but between knowledge and practice lie multiple beliefs and experiences that act as barriers to change (Illustration 5.7). Change of any type is most likely to succeed when the benefits of making the change outweigh the disadvantages. This makes changes in food choices a very individual decision, with each person deciding whether a change is in his or her best interest.[6] But what sorts of circumstances, in addition to

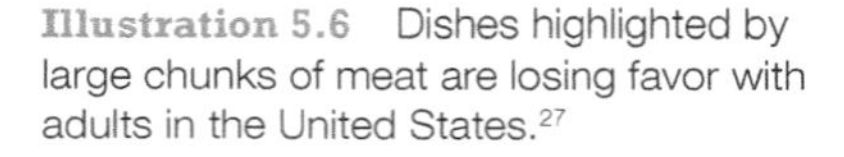

Illustration 5.6 Dishes highlighted by large chunks of meat are losing favor with adults in the United States.[27]

©stockcreations/Shutterstock.com

Photodisc

Illustration 5.7 Why knowledge about a good diet may not be enough to improve food choices.

increased knowledge, make changes in food choices worthwhile for individuals—and even highly desired?

Nutrition Attitudes, Beliefs, and Values The value individuals place on diet and health is reflected in the food choices they make. A survey of restaurant patrons found that food choices varied according to the consumer's perceptions of the importance of diet to health.[31]

- "Unconcerned" consumers—people who are not concerned about the connection between diet and health and who tend to describe themselves as "meat and potato eaters"—select foods for reasons other than health.
- "Committed" consumers believe that a good diet plays a role in the prevention of illness. They tend to consume a diet consistent with their commitment to good nutrition.
- "Vacillating" consumers—people who describe themselves as concerned about diet and health but who do not consistently base food choices on this concern—tend to vary their food choices depending on the occasion. These consumers are likely to abandon diet and health concerns when eating out or on special occasions, but they generally adhere to a healthy diet.

Avoiding illness and curing or diminishing current health problems are likewise strong incentives for changing food choices (Table 5.1). In almost all instances, the key to lasting improvement in diet is to make changes you feel confident you can make, and can do without a lot of "will power." Changes that bring you more pleasure than inconvenience or discomfort have staying power, while those that require a lot of will power to maintain rarely last.[6]

Successful Changes in Food Choices

The primary reason efforts to improve food choices fail is that the changes attempted are too drastic. Improvements that last tend to be the smallest acceptable changes needed to do the job.[15]

The Process of Changing Food Choices Assume you need to lose weight and want to keep it off by modifying your food choices. A promising plan to accomplish this goal would begin by identifying food choices you would like to change and lower-calorie food options you would be willing and able to eat (Table 5.2). Then you could make the plan more specific by identifying the changes that would be easiest to implement. For example, assume the low-calorie foods you like include frozen nonfat yogurt and oranges. You might decide to eat yogurt or an orange in place of your usual bedtime snack of ice cream. A specific dietary change such as this is much easier to implement than a broad notion,

Table 5.1 Factors that enhance food intake changes[6]

• Attitude that nutrition is important
• Belief that diet affects health
• Perceived susceptibility to diet-related health problems
• Perception that benefits of change outweigh barriers to change
• Confidence that the behavior change can be made

Table 5.2 Changing food choices for the better[6,15]

The process	An example
1. Identify a healthful change in your diet you'd like to make.	1. I'd like to lower my fat intake.
2. Identify two food choices you make that should change because they contribute to the need for the healthful change you identified.	2. I eat at fast food restaurants three times a week. I usually have a large order of fries and once a week I eat fried chicken.
3. Identify two or more specific, acceptable options for more healthful food choices than the ones identified under number 2.	3. Options identified: • Order tossed salad with low-calorie dressing instead of fries. • Eat a grilled chicken sandwich every other week instead of fried chicken. • Eat Mexican fast food more often.
4. Decide which option is easiest to accomplish and requires the smallest change to get the job done.	4. I love Mexican food. It would be easy to eat tacos instead of french fries or fried chicken.
5. Plan how to incorporate the change into your diet.	5. Mondays and Fridays, I'll eat tacos.
6. Implement the change. Be prepared for midcourse corrections.	6. Midcourse correction: On Fridays, when I'm with my friends, it's easier to eat at the restaurant they like. I'll order the grilled chicken sandwich and coleslaw.

such as "eat less." Although weight loss will take a while, such a small acceptable change has a much better chance of working than a drastic change in diet.[15]

Planning for Relapses When making a change in your diet, be prepared for relapses. Relapses happen for a number of reasons, and they don't mean the attempt has failed. People often return to old habits because the change they attempted was too drastic or because they tried to make too many changes at once. If the change undertaken doesn't work out, rethink your options and make a midcourse correction.

Strategies for Improving Diets in Groups of People Improvements in the diets of people in the United States are needed to reduce the risk of obesity and diet-related diseases. A number of behavioral changes that help people in general eat better are:

- Making healthy, delicious, affordable food available at home, in restaurants, schools, colleges, at work, and in military settings
- Decreasing fast food restaurant dining
- Increasing family meals and home cooking
- Learning to cook in school and at home
- Self-monitoring of calorie intake and weight
- Maintaining the healthful eating patterns of a home country after immigrating to the United States

Improvements in farm-to-table transfer of locally produced foods and reliable access to healthy, affordable foods also tend to improve dietary intakes.[15]

Does Diet Affect Behavior?

- **Differentiate between scientifically supported and unsupported conclusions about diet and behavior relationships.**

Food affects behavior in some rather striking ways. Irritable, hungry, crying infants rapidly change into cooing, sleepy angels after they are fed. Low-on-sleep employees perk up after their morning coffee. A high-calorie lunch makes some people feel calm and sleepy, and eating favorite foods with family and friends can make people feel happy.[32] Not only

Table 5.3 Examples of ways in which diet may affect behavior

Dietary characteristic	Behavioral outcomes
Malnutrition, growth stunting in early childhood	Lower intellectual functioning and school performance, increased antisocial behavior in childhood[33]
Nutritional supplementation of malnourished young children	Improved growth and intellectual functioning in adulthood[34]
Very low carbohydrate intake (less than 20 grams per day)	Reduced short-term memory, slower reaction times, increased attention span[35]
Ingestion of certain color dye additives	Moderate increases in hyperactivity and other behaviors in susceptible children[36,37]
Lead consumption	Higher risk of violent and aggressive behavior, hyperactivity, and mental and behavioral problems[38]
Iron deficiency in young children	Long-term deficits in learning ability and social skills[39]

do our behaviors affect our diets, but also our diets can affect our behaviors. Examples of associations between dietary characteristics and behavior are listed in Table 5.3. One common belief about food and behavior is the subject of this unit's Reality Check.

Malnutrition and Mental Performance

Like growth and health, mental development and intellectual capacity can be affected by diet. The effects range from mild and short term to serious and lasting, depending on when the malnutrition occurs, how long it lasts, and how severe it is. The effects are most severe when malnutrition occurs while the brain is growing and developing.

Severe deficiency of protein, calories, or both early in life leads to growth retardation, low intelligence, poor memory, short attention span, and social passivity. When the nutritional insult is early and severe, some or all of these effects may be lasting (Illustration 5.8).[40] In Barbados, for example, children who experienced protein-calorie malnutrition in the first year of life did not fully recover even with nutritional rehabilitation. Growth improved but academic performance did not. Compared to well-nourished children, those experiencing protein and calorie deficits during infancy were more likely to drop out of school and were four times more likely to have symptoms of attention deficit hyperactivity disorder (ADHD).[41]

Protein-calorie malnutrition that occurs later in childhood, after the brain has developed, produces behavioral effects that can be corrected with nutritional rehabilitation.

REALITY CHECK

Can Food Be a Love Potion?

Toward the end of a friend's birthday party, Glenda and Cassell got into a heated debate about the existence of food aphrodisiacs. Part of the conversation went like this:

Is it all in Glenda's head?

Answers on page 5–10.

Glenda: Chocolate and vanilla are natural love potions, there's no doubt about it. If I eat a chocolate truffle and spritz on vanilla flavoring like perfume before a date, the guy always goes nuts for me!

Cassell: Are you kidding? No way! I've heard about oysters and this tree bark that are supposed to work miracles, too. It's all in your head.

ANSWERS TO REALITY CHECK

Can Food Be a Love Potion?

The idea that food can act as an aphrodisiac has been around since ancient times. Although many have looked and others have tried, no one has found a food that acts like a love potion.[51]

Photodisc

Glenda:

Photodisc

Cassell:

REUTERS/Enrique Marcarian/Landov

Illustration 5.8 Malnutrition in early childhood has long-lasting effects. Some children never fully recover.

Correction of other deficits that often accompany malnutrition, such as lack of educational and emotional stimulation and harsh living conditions, hastens and enhances recovery.[40]

Protein-calorie malnutrition severe enough to cause permanent delays in mental development rarely occurs in the United States. When it does, malnutrition is usually due to neglect or inadequate care giving. More common dietary events that impair learning in U.S. children are skipping breakfast, fetal exposure to alcohol, iron deficiency, and lead toxicity.

Early Exposure to Alcohol Affects Mental Performance Mental development can be permanently delayed by exposure to alcohol during fetal growth. Although growth is also retarded, the most serious effects of fetal exposure to alcohol are permanent delays in mental development and behavioral problems associated with them. Women are advised not to drink if they are pregnant or may become pregnant.[42]

Iron Deficiency Impairs Learning Most cases of iron-deficiency anemia in children result from inadequate intake of dietary iron. Iron-deficiency anemia in children is a widespread problem in developed and developing countries and likely is the most common single nutrient deficiency.[43] The potential impact of iron-deficiency anemia on the functional capacity of humans represents staggering possibilities.

Until recently, it was thought that the effects of iron-deficiency anemia on intellectual performance were short term and could be reversed by treating the anemia. Although some of the effects can be lessened by treatment early in life, some are lasting. Five-year-old children in Costa Rica treated for iron-deficiency anemia during infancy scored lower on hand–eye coordination and other motor skill tests than similar children without a history of anemia. Studies in the United States have identified fearful behavior, shyness, lack of playfulness, and reduced problem-solving ability in iron-deficient children.[39,44]

Tony Freeman/PhotoEdit

Illustration 5.9 There are many opportunities for overexposure to lead. Young children are especially vulnerable.

Overexposure to Lead There are many opportunities for overexposure to lead. Approximately 84% of U.S. houses built before 1978 contain some lead-based paint.[45] Children living in or near these houses may eat the paint flakes (they taste sweet), or the old paint may contaminate the soil near the houses (Illustration 5.9). Lead also ends up in soil from industrial and agricultural chemicals, in water from lead-based pipes and solder, and in the air from the days when leaded gas was used.

Although the use of lead in cans, pipes, and gasoline has decreased dramatically, lead remains in the environment for long periods. Lead also stays in the body, stored principally in the bones, for a long time—20 years or more. It takes over a year of treatment to reduce blood lead levels. The effects of excessive exposure to lead in children include increased absenteeism from school, impaired reading skills, higher dropout rates, and increased aggressive behavior.[46] Occupational exposure to lead in adults is associated with poor kidney function and the risk of hypertension.[47] Blood lead levels in children have dropped substantially in recent decades. Despite this drop, 500,000 young children in the United States, especially in parts of the Midwest, still have elevated blood lead levels.[48]

Food Additives, Sugar, and Hyperactivity The notion that certain food additives are related to hyperactivity in children has been popular since the related Feingold Hypothesis was announced in the mid-1970s. Since then, multiple studies have been undertaken to test whether the hypothesis is true. In a study involving a large group of healthy 3-, 8-, and 9-year-old children, intake of a beverage containing four food colorants (types of yellow, orange, and red color additives) and a preservative (sodium benzoate) was related to development of hyperactivity. Signs of hyperactivity detected in the children consuming the beverage containing the food dyes included overactivity, impulsiveness, and short attention spans. Not all children consuming the beverage demonstrated hyperactive behavior, however, and the same effects were observed in children with or without attention deficit hyperactivity disorder.[36] It appears that some children may be genetically vulnerable to the effects of these food additives while others are not.[36,37]

Studies examining the effects of sugar intake on hyperactivity in children have not demonstrated that such a relationship exits.[49] The excitement that often accompanies high-sugar eating occasions such as Halloween and birthday parties—or the expectation that sugar causes hyperactivity—may be responsible for the reported effect.[50] (See Illustration 5.10.)

The Future of Diet and Behavior Research

Identifying the effects of nutrition on behavior is a tricky business. Many factors in addition to diet influence behavior, making it difficult to separate diet from social, economic, educational, and genetic factors. We still have much to learn, and many assumptions about diet and behavior must await confirmation through research.

Ruizluquepaz/iStockphoto.com

Illustration 5.10 Do food additives or lots of sugary foods make children hyperactive?

© Denis Rozhnovsky/Shutterstock.com

NUTRITION up close

Improving Food Choices

Focal Point: Developing a plan for healthier eating.

Identify a change in your diet that you would like to make. Then develop a plan for making the change by thinking through and responding to each element of the dietary change process listed. (Refer to Table 5.2 for examples of responses.)

Dietary Change Process	Your Response
1. Identify a healthful change in your diet you'd like to make.	1. ____________________
2. Identify two food choices you make that should change because they contribute to the need for the healthful change you identified.	2. ____________________
3. Identify two specific, acceptable options for food choices more healthful than those identified under number 2.	3. ____________________
4. Decide which option is easiest to accomplish and requires the smallest change to get the job done.	4. ____________________
5. Plan how to incorporate the change into your diet.	5. ____________________

Feedback to the Nutrition Up Close is located in Appendix F.

REVIEW QUESTIONS

- **Identify factors that influence an individual's food choices and preferences.**

1. Food preferences are universal—everyone likes the same foods. **True/False**
2. Foods will be rejected by a population, no matter how nutritious they may be, if the foods don't fit into a culture's definition of what foods are appropriate to eat. **True/False**
3. Food choices are driven largely by a person's need for nutrients. **True/False**
4. Potatoes, rice, corn, and yucca are examples of "super foods" in specific countries due to their cultural significance. **True/False**
5. Nutrition knowledge is an important prerequisite for making healthful food choices. **True/False**
6. ____ Emanuel is at a dinner party and the hostess asks, "Would you like to try the brussels sprouts? They're fresh from the garden and delicious!" Emanuel immediately responds, "No, thank you. I've tried brussels sprouts at least ten times before and they just don't taste good to me." Chances are good that Emanuel
   a. is being a picky eater.
   b. dislikes all vegetables.
   c. is upset about something.
   d. really does not like the taste of brussels sprouts.
7. ____ Last week Yu asked her neighbor where she should take her visiting parents to dinner. The neighbor recommends Al's Restaurant, saying it has "the best food in town." Later that week Yu spoke with her neighbor and the neighbor asked how her parents liked Al's Restaurant. Yu responds, "Well, actually, we thought it was awful. They used way too much garlic in everything." The neighbor is shocked. What's the most likely reason for this reaction?
   a. The neighbor assumed that because she likes the food at Al's, everyone will.
   b. Yu's family doesn't recognize good food.
   c. Yu's family members are genetically sensitive to the taste of garlic.
   d. The neighbor is unable to detect the presence of garlic in food.

- **Apply the process for making healthful changes in food choices to a specific change in food intake.**

8. Broad dietary changes, such as a decision to simply "eat less," are more likely to produce lasting behavioral changes than are small changes in diets, such as snacking on favorite fruits rather than candy. **True/False**

9. Changes in dietary intake that are acceptable to an individual and easy to implement are the types of changes that are most likely to last. **True/False**
10. Changing food choices for the better takes planning and includes individual decisions on specifically how the change in food choices will be implemented. **True/False**
11. Individuals need to plan for modifying their approach to improving food choices because even the best-planned changes in food choices sometimes fail. **True/False**

The next two questions refer to this scenario:

Assume your wife decides to increase her vegetable intake. She travels from place to place for work and eats out a lot. She loves the premade mixed greens salad and the single-serve yogurt dressing packet you can get at some gas station convenience stores. She decides she would happy to have that for lunch twice a week instead of her usual burger or pizza slices.

12. ____Does this plan have a chance of working?
    a. Yes, because she seems to be a person with great will power.
    b. Yes, she is planning a specific and acceptable change.
    c. No, the plan isn't likely to work because the change is too drastic.
    d. No, the plan won't work because she will grow tired of eating salads.
13. ____The salad-for-lunch plan worked so well that your wife decides she will lose 10 pounds in the next two months by only eating salads. Will this plan likely work?
    a. It has a good chance of working. She really loves salads.
    b. Yes, she won't be consuming very many calories.
    c. It's unlikely to work because it is too drastic a change in her diet.
    d. It probably won't work because she will start craving burgers and pizza.

- **Differentiate between scientifically supported and unsupported conclusions about diet and behavior relationships.**

14. Examples of ways in which dietary intake affects behavior include the relationship between sugar intake and hyperactivity in children. **True/False**
15. Excessive exposure to lead during childhood is related to impaired reading skills and higher school dropout rates. **True/False**

Answers to these questions can be found in Appendix F.

NUTRITION SCOREBOARD ANSWERS

1. Although genetics plays a role in food preferences, the predominant influences are environmentally determined. **False**
2. The idea that food habits don't change is a myth. **False**
3. Sweet, sour, salty, and bitter tastes are sensed by taste cells in all parts of the tongue, and not, as popularly thought, clustered by type of taste in different zones on the tongue. Taste receptor cells are also located in the back of the throat and nasal cavity.[18] **True**
4. **True**[2]

Healthy Dietary Patterns, Dietary Guidelines, MyPlate, and More

NUTRITION SCOREBOARD

1 Healthy dietary patterns include the regular consumption of legumes, nuts, poultry, and whole-grain products. **True/False**

2 Over half of U.S. adults fail to consume three or more servings of vegetables and two or more servings of fruits daily. **True/False**

3 The basic food groups include a "healthy snack" group. **True/False**

4 "Value" fast food meals may provide double the calories of regular-sized fast food meals. **True/False**

Answers can be found at the end of the unit.

Alistair Berg/Digital Vision/Getty Images

LEARNING OBJECTIVES

After completing Unit 6 you will be able to:

- Apply the characteristics of healthy dietary patterns to the design of one.
- Identify characteristics of dietary patterns that promote health and those that do not.
- Utilize MyPlate.gov guidance materials and tools for dietary planning and evaluation.

Healthy Eating: Achieving a Balance between Good Taste and Good for You

- **Apply the characteristics of healthy dietary patterns to the design of one.**

May I have your attention please? For a moment, think about the foods in Illustration 6.1. If your mouth is watering and you're ready to go out and buy some ripe peaches, you have found the balance between good taste and good for you. Who said foods that taste good aren't good for you?

KEY NUTRITION CONCEPTS

This unit explores what constitutes a healthy diet and ways of evaluating and assessing diets. The discussion directly relates to four key nutrition concepts:

1. Nutrition Concept #2: Foods provide energy (calories), nutrients, and other substances needed for growth and health.
2. Nutrition Concept #4: Poor nutrition can result from both inadequate and excessive levels of nutrient intake.
3. Nutrition Concept #8: Poor nutrition can influence the development of certain chronic and other diseases.
4. Nutrition Concept #9: Adequacy, variety, and balance are key characteristics of healthy dietary patterns.

Susie M. Eising/StockFood Creative/Getty Images

Photodisc

Photodisc

Wally Eberhart/Visuals Unlimited/Getty Images

Illustration 6.1 Can you smell it? Can you taste it? A plump, golden peach. It's so ripe that juice spurts from it and drips down your chin when you take a bite. A golden brown turkey just taken out of the oven. The wonderful smell fills the kitchen. A steaming loaf of homemade bread just set out to cool. A perfect ripe tomato just picked from the garden. It melts in your mouth.

Characteristics of Healthy Dietary Patterns

Healthy **dietary patterns** come in a variety of forms. They may include bread, olives, nuts, fruits, beans, vegetables, and lamb (as in Greece); rice, vegetables, fish, and seaweed (as in China); or black beans, rice, chicken, and tropical fruits (as in Cuba and Costa Rica). Illustration 6.2 gives examples of the diverse foods that can be part of a healthy dietary pattern. Although the types of foods that go into them can vary substantially, dietary guidelines support healthful dietary patterns that share four basic characteristics: adequacy, variety, balance, and health maintenance.[4,5]

Adequate diets include a wide variety of foods that together provide sufficient levels of calories and **essential nutrients**. What's considered sufficient in the United States and Canada? For calories, it's the number that maintains a healthy body weight. For essential nutrients, sufficiency corresponds to intakes that are in line with recommended intake levels represented by the *Recommended Dietary Allowances (RDAs)* and *Adequate Intakes (AIs)*. (Tables showing these levels appear on the inside covers of this book.) Recommended amounts of essential nutrients should be obtained from foods to reap the benefits offered by the variety of naturally occurring substances in foods that promote health.

Variety is a core characteristic of healthy diets because the nutrient content of food differs. Consumption of an assortment of foods from each of the basic food groups increases the probability that the diet will provide enough of them all. You could, for example, eat three servings of vegetables a day by eating only potatoes. But you would consume a much broader variety of vitamins, minerals, and beneficial substances in plants—such as antioxidants—if you consumed spinach and tomatoes along with the potatoes.

dietary patterns The quantities, proportions, variety, or combination of different foods, drinks, and nutrients in diets, and the frequency with which they are habitually consumed. Also called "eating patterns".

adequate diet A diet consisting of foods that together supply sufficient protein, vitamins, and minerals and enough calories to meet a person's need for energy.

essential nutrients Substances the body requires for normal growth and health but cannot manufacture in sufficient amounts; they must be obtained in the diet.

variety A diet consisting of many different foods from all of the food groups.

Illustration 6.2 Foods that contribute to healthy diets in different countries.
(a) Dinner in Italy might be linguine primavera (pasta with vegetables). (b) Pad Thai, rice noodles and vegetables, is a favorite dish in Thailand. (c) Tamales are a celebration food in many Latin cultures. (d) Dal, curry dishes, vegetables, and chicken are popular parts of the cuisine of India.

a.
Photodisc

b.
Photodisc

c.
Photodisc

d.
Photodisc

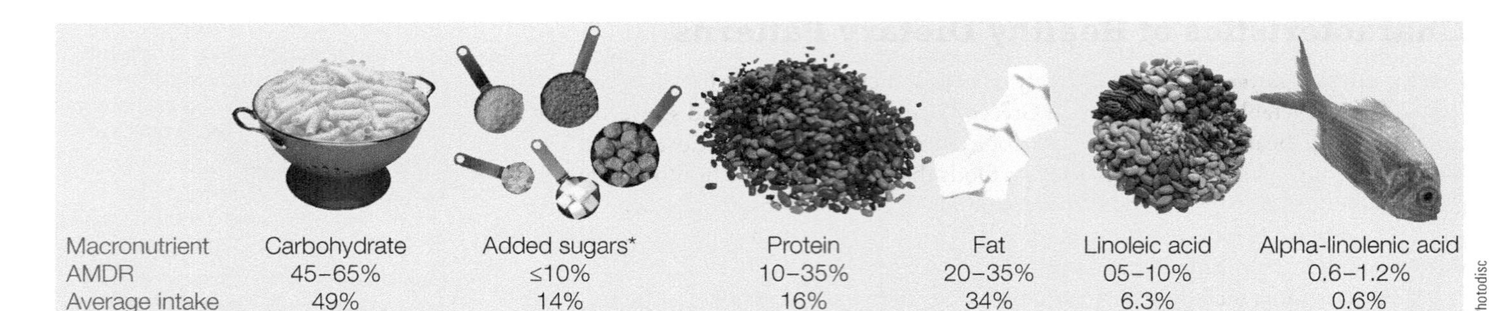

Macronutrient	Carbohydrate	Added sugars*	Protein	Fat	Linoleic acid	Alpha-linolenic acid
AMDR	45–65%	≤10%	10–35%	20–35%	05–10%	0.6–1.2%
Average intake	49%	14%	16%	34%	6.3%	0.6%

Photodisc

*The 2015 Dietary Guidelines Advisory Committee concluded that total added sugar intake should be10% or less of total calories.[1]

Illustration 6.3 The match between Acceptable Macronutrient Distribution Ranges (AMDRs) and average intake by adults in the United States.[1,6]

A **balanced diet** provides calories, nutrients, and other components of food in the right proportion—neither too much nor too little. Diets that contain too much sodium or too little fiber, for example, are out of balance. Diets that provide more calories than needed to maintain a healthy body weight are also out of balance.

People need relatively large amounts of carbohydrates, proteins, fats, and water. These nutrients are collectively classified as **macronutrients**. Guidelines indicating the percentages of total caloric intake that should consist of carbohydrate, protein, and fat are listed in a DRI table labeled *Acceptable Macronutrient Distribution Ranges*, or *AMDRs*. AMDRs have also been set for linoleic acid and alpha-linolenic acid (the two essential fatty acids). Illustration 6.3 shows the acceptable ranges for carbohydrate, protein, fat, **added sugars**, and essential fatty acid intake, as well as the average intake levels of U.S. adults.

Another key attribute of healthy dietary patterns is health maintenance. As methods used to assess dietary patterns have advanced, researchers have been able to examine the relationship between complex dietary patterns and health. Although the utility of dietary adequacy, moderation, and balance remains, it is important to consider the effects of overall long-term dietary pattern on health. People consume foods, not nutrients, and the nutrients and other components of food such as phytochemicals interact with each other in ways that can modify health outcomes over time. Several dietary patterns have been shown to reduce the risk of cardiovascular disease, obesity, hypertension, and type 2 diabetes compared to Western-style dietary patterns. Healthy dietary patterns provide levels of calories and nutrients needed for growth, health, and life. In particular, they provide adequate amounts of potassium, vitamins D, A, E, C, and folate, calcium, magnesium, and fiber—nutrients that are most likely to be found in low amounts in the diets of people in the United States.[1] Only 16% of adults consume three or more servings of vegetables and two or more servings of fruits daily, and about 10% consume at least one serving of dark-green or colorful vegetable daily.[6,7]

balanced diet A diet that provides neither too much nor too little of nutrients and other components of food such as fat and fiber.

macronutrients The group name for carbohydrate, protein, fat, and water. They are called macronutrients because we need relatively large amounts of them in our daily diet.

added sugars Sugars that are either added during the processing of foods, or are packaged as such and include sugars (free, mono-, and disaccharides), syrups, naturally occurring sugars that are isolated from a whole food and concentrated, and other caloric sweeteners.

whole grains Cereal grains that consist of the intact, ground, cracked, or flaked kernel, which includes the bran, the germ, and the innermost part of the kernel (the endosperm).

Healthy Dietary Patterns Identified for the United States

Also called healthy "eating patterns," healthy dietary patterns most consistently associated with positive health and nutrient intake outcomes are the:

- Healthy Mediterranean-Style Dietary Pattern
- Healthy U.S. Dietary Pattern (represented by USDA's MyPlate recommendations)
- Healthy Vegetarian-Style Dietary Pattern

The foods represented in these dietary patterns, and their proportionate representation in pattern, vary somewhat. However, they all are associated with reduced the risk of obesity, type 2 diabetes, heart disease, and other diseases and disorders, and with improved nutrient intake compared to our existing Western-style dietary pattern.[32]

Healthy dietary patterns are anchored by plant foods. They are represented by the regular consumption of fruits, vegetables, **whole grains**, nuts, legumes, oils, low-fat dairy products, poultry, lean meat, fish and seafood, and moderate alcohol consumption (by adults who choose to drink). There is room in these patterns for the occasional dessert or other treat, and individuals can decide which specific foods they will eat from the overall food types represented in the pattern. Table 6.1 summarizes the general characteristics of healthy dietary patterns. Two specific healthy dietary patterns, the Healthy U.S.-Style Dietary Pattern and Mediterranean-Style Dietary Pattern, are presented in Table 6.6. The table provides examples of these dietary patterns based on 2,000 calories. Recommendations based on 11 other calorie need levels are included in the 2015 Dietary Guidelines report.[30] The third healthy dietary pattern is the Vegetarian-Style Dietary Pattern. It is covered in Unit 16 on vegetarian diets.

National Guides for Healthful Diets

Due to the impact of food choices on the health of individuals and population groups, many countries establish recommendations for dietary intake. Such recommendations are periodically updated as new knowledge about diet and health emerges. Most of the guidelines include recommendations for physical activity and food safety as well.

National recommendations for diet and physical activity usually apply to children over the age of 2 and aren't appropriate for every individual in a population. General guidelines may not completely match the needs of individuals who, for example, are strict vegetarians, have food allergies, or have specific genetic traits that affect nutrient utilization. A basic premise of national dietary guidelines is that nutrient needs should be met through food consumption without reliance on dietary or nutrient supplements.[1]

Benefits of population-based dietary and physical activity recommendations are multiple. The information is science-based, free, and made widely available to the public on the Internet.

Table 6.1 Characteristics of healthy dietary patterns[1,4,5]

Healthy dietary patterns include the regular consumptions of:
• Fruits and vegetables
• Whole grains and whole-grain products, and other high-fiber foods
• Nuts, all types
• Oils (such as olive oil, seed and vegetable oils)
• Legumes (such as navy and pinto beans)
• Low-fat dairy products
• Poultry, lean meats
• Fish and seafood
• Alcohol in moderation (for adults who choose to drink).
Healthy dietary patterns are *not* characterized by the regular consumptions of:
• Foods and beverages high in added sugar such as sugar-sweetened soft drinks, fruit drinks, and sweetened teas
• Foods high in sodium such as some fast and **processed foods**
• Refined grain products such as white bread, rice cakes, pasta
• Processed meats such as salami and bologna
• Foods high in saturated fats, such as animal fat and tropical oils

processed foods Any food other than raw agricultural commodities (such as carrots and eggs) that have been modified from their original state through processing. They are classified based on the extent of processing from minimal (frozen, canned fruits, and vegetables) to ultra-processed (foods such as soft drinks, cheese twists, processed meats) that contain little or no whole foods.

Table 6.2 **Examples of dietary guidelines from around the world**[8]

Country	Example of dietary guidelines
Japan	Eat 30 or more different kinds of foods daily.
China	Eat clean and safe food.
Norway	FOOD + JOY = HEALTH.
United Kingdom	Encourage and support the production of lower saturated fat foods.
Mexico	Eat more dried beans and less food of animal origin.
South Africa	Enjoy a variety of foods. Be active.
Cuba	Fish and chicken are the healthiest meats.

Adherence to the information can help people stay healthy and lower their risk of developing disorders such as diabetes, heart disease, cancer, osteoporosis, and obesity.[5]

Some of the national guidelines give credit to the cultural and social importance of food. Guidelines for Japanese people, for example, include "Happy eating makes for happy family life; sit down and eat together and talk; treasure family taste and home cooking." Table 6.2 provides examples of dietary guidelines established for a variety of countries.

National guidance on healthful diets for some countries is accompanied by extensive information on how to select a healthful diet and achieve recommended levels of physical activity. In the United States, the national guidelines for diet are called the Dietary Guidelines for Americans, and the major how-to guide for implementing the guidelines is represented by MyPlate.

Dietary Guidelines for Americans

- **Identify characteristics of diets that promote health and those that do not.**

The Dietary Guidelines for Americans provide science-based recommendations to promote health and reduce the risk for major chronic diseases through diet and physical activity. Due to their credibility and focus on health promotion and disease prevention for the public, the Dietary Guidelines form the basis of federal food and nutrition programs and policies.[1]

By law, the Dietary Guidelines for Americans are updated every five years. The first edition of the Dietary Guidelines was published in 1980. The 2015 Dietary Guidelines highlight dietary patterns associated with decreased risk of obesity, heart disease, diabetes, and nutrient inadequacies. The Guidelines also encourage adequate physical activity. The 2015 Dietary Guidelines Advisory Committee, which consisted of scientific experts on nutrition and health, concluded that the health of the U.S. population could be improved, and common diseases and disorders prevented, if Americans consume a healthy dietary pattern (shown in Table 6.1) and exercise regularly. Key elements of the 2015 Dietary Guidelines related to food and nutrient intake, dietary pattern, and physical activity are listed in Table 6.3.

Most Americans do not currently consume diets that match the recommendations presented in the Dietary Guidelines.[1,10] There are many reasons for this, including access to affordable and nutritious foods, food preferences, opportunities for physical activity, and fast-paced lifestyles. Sometimes the reason can be that we simply don't think about the broad assortment of foods that can be included in a healthy dietary pattern. The Take

Table 6.3 2015 Dietary Guidelines for Americans: Key Recommendations[30,a] Guidelines that encourage a healthy eating pattern:

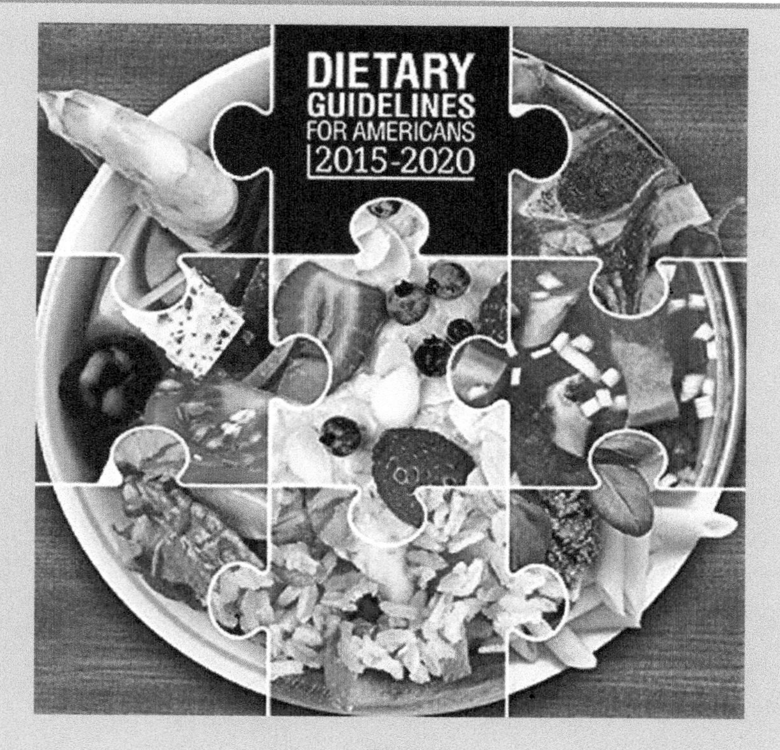

1. Follow a healthy eating pattern at an appropriate calorie level across the lifespan. Healthy eating patterns include:
   - Vegetables, including dark green, red, and orange colored vegetables
   - Fruits, especially whole fruits
   - Grains, at least half of which are whole grains
   - Fat-free or low-fat fairy products, including milk, yogurt, cheese, and/or fortified soy beverages
   - A variety of protein foods, including seafood, lean meats and poultry, eggs, legumes (beans and peas), and nuts and seeds.
   - Oils
   - Alcohol consumption by adults of drinking age who chose to consume it. Alcohol should be consumed in moderation—up to one drink a day for women and two for men.
   - The variety of foods needed to fully meet people's need for nutrients.

   Healthy eating patterns limit intake of:
   - Saturated fat and added sugars to less than 10% of total calories
   - Exclude *trans* fats, and limit sodium intake to less than 2,300 mg per day.
2. Shift to healthier food and beverage choices.
   - Choose nutrient-dense foods and beverages across and within all food groups in place of less healthy choices.
3. Support healthy eating patterns for all.
   - Everyone has a role in helping to create and support healthy eating patterns from home to school to work communities.
4. Physical Activity Recommendations
   In addition to healthy eating patterns, adults should meet the Physical Activity Guidelines for Americans (at least 150 minutes of moderate intensity activity per week if physically able) to help promote health and reduce the risk of chronic disease.

[a]The Dietary Guidelines apply to the United States population over the age of 2 years.

Action feature is intended to highlight a variety of specific foods that can both be part of a healthy diet and match our food preferences.

Application of the Dietary Guidelines to Public Programs The Dietary Guidelines for Americans form the basis of federal food and nutrition programs and policies (Table 6.4). Food options served in school lunch programs, nutrition labeling standards, USDA's food guidance materials, and foods served to military personnel, for example, are based on the Dietary Guideline recommendations. The most important and useful materials on how to implement the Dietary Guidelines are found at ChooseMyPlate.gov, USDA's site for the MyPlate food intake guidance materials.

Take Action

To bring variety to your diet by identifying the types of whole grain products, legumes, and fish and seafood you already like (but may not think about) or would be willing to try.

Please a √ in the appropriate column across from the food choice.

	Do not like	Never tried it	I like this food	I would try this one
Whole-Grain Foods				
popcorn	____	____	____	____
whole-grain crackers	____	____	____	____
cornmeal bread	____	____	____	____
brown rice	____	____	____	____
buckwheat pancakes	____	____	____	____
wild rice	____	____	____	____
barley	____	____	____	____
millet	____	____	____	____
hominy grits	____	____	____	____
polenta (whole grain type)	____	____	____	____
Legumes				
black beans	____	____	____	____
pinto beans	____	____	____	____
red beans	____	____	____	____
split peas	____	____	____	____
navy beans	____	____	____	____
black-eyed peas	____	____	____	____
chickpeas (garbanzos)	____	____	____	____
soy nuts	____	____	____	____
lentils	____	____	____	____
fava beans	____	____	____	____
lima beans	____	____	____	____
Fish and Seafood				
salmon	____	____	____	____
tuna	____	____	____	____
catfish	____	____	____	____
cod	____	____	____	____
tilapia	____	____	____	____
clams	____	____	____	____
shrimp	____	____	____	____
squid	____	____	____	____
anchovies	____	____	____	____
crab	____	____	____	____
scallops	____	____	____	____
lobster	____	____	____	____

Table 6.4 **Examples of federal food and nutrition programs that utilize the U.S. Dietary Guidelines recommendations[9]**

• Supplemental Food and Nutrition Assistance Program (SNAP, formerly known as Food Stamps)
• Head Start
• WIC (Special, Supplemental Nutrition Program for Women, Infants, and Children)
• National School Lunch Program
• Military food allowance program
• Nutrition labeling
• MyPlate educational materials
• Indian Health Service
• Healthy People 2020 (national objectives for improvements in weight status and diet)
• Older Americans Nutrition Program

ChooseMyPlate

- **Utilize ChooseMyPlate.gov guidance materials and tools for dietary planning and evaluation.**

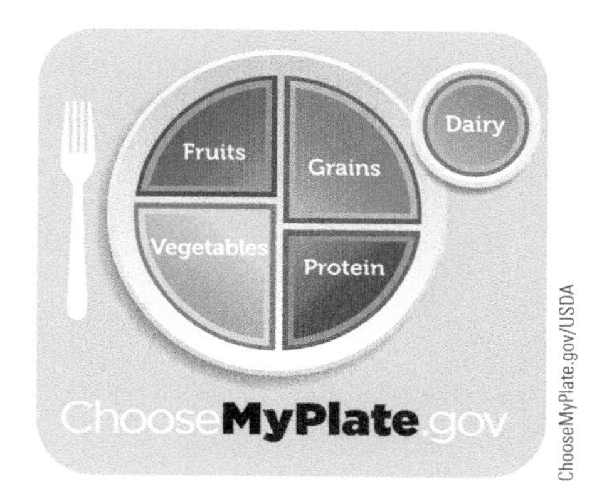

Food group guides from the USDA have been available in the United States since 1916. Known by such names as the Basic Four Food Groups, and the Food Guide Pyramid, these guides are periodically updated to reflect advances in knowledge about foods, diets, and health. The latest revision of USDA's food guidance materials is called MyPlate and is represented by the MyPlate logo. Illustration 6.4 shows this logo and a plate of foods set up to match its messages. The MyPlate logo is intended to give consumers a visual reminder of the types and proportions of food that make up healthy meals. The logo shows a plate with four sections in different colors representing the proportion of vegetables, fruits, grains, and protein foods that should be on your plate. Next to the plate is a circle that represents a dairy product such as low-fat milk or other low-fat dairy product. The types of foods included on the plate are basic and nutrient dense. Information offered by ChooseMyPlate is available in English, Spanish, Chinese, Arabic, and other languages at ChooseMyPlate.gov.

Illustration 6.4 Foods shown on the plate consist of stewed turkey, barley, Swiss chard, mandarin oranges, and 1% milk.

ChooseMyPlate.gov Healthy Eating Messages

ChooseMyPlate.gov supports the healthy eating messages in the Dietary Guidelines by offering the following key pieces of advice:

- Make at least half your plate fruits and vegetables.
- Enjoy your food but eat less.
- Make half your grains whole grains.
- Eat fewer foods that are high in saturated fat, added sugar, and sodium.
- Avoid oversized portions.
- Switch to fat-free or low-fat milk.
- Drink water instead of sugary drinks.

Fruits
Focus on fruits.
>> See Fruit Group

Vegetables
Vary your veggies.
>> See Vegetable Group

Grains
Make at least half your grains whole.
>> See Grains Group

Protein Foods
Go lean with protein.
>> See Protein Foods Group

Dairy
Get your calcium-rich foods.
>> See Dairy Group

ChooseMyPlate.gov/USDA

Illustration 6.5 USDA's ChooseMyPlate basic food groups and priority messages related to each group.

- Compare sodium in foods like soup, bread, and frozen meals and choose the foods with lower numbers.

The importance of consuming enough calories for growth and health while not eating extra calories and gaining weight is also stressed. Regular physical activity (60 minutes per day for children and adolescents and 2½ hours or more per week of moderate-intensity activity for adults weekly) is stressed because it contributes to weight control and provides many other health benefits.

Healthy U.S.-Style Dietary Pattern

This dietary pattern is based on the Dietary Approaches to Stop Hypertension Eating Plan ("The DASH Diet") and is used to formulate dietary recommendations for USDA's MyPlate educational materials (Illustration 6.5). Grains, vegetables, fruits, dairy, and protein foods are the designated groups. Interactive, educational material provided by MyPlate includes details about the types of foods that belong to each group.

Portion Sizes and Food Measure Equivalents Unlike previous food group guides, the current version does not recommend serving sizes or numbers of servings individuals in general should consume from the food groups. This information is provided if requested using a personalized interactive tool on menu planning such as the My Plan. The personalized information generated shows amounts of each food group to consume and food portion sizes that correspond to that amount.

Information on amounts of basic foods recommended for different levels of calorie need in the MyPlate materials are based on cup- and ounce equivalents. Table 6.5 provides a listing of cup- and ounce equivalents for the food groups. You can click on the food group name within the plan to access additional information on food choices within each group.

Sample Menus What does an eating pattern based on USDA's food group guidelines look like? Two weeks of menus based on USDA's food group recommendations and calorie and nutrient needs are available at ChooseMyPlate.gov (Illustration 6.6).

Table 6.5 How much food counts as a cup or an ounce?

Vegetables and Fruits: 1 cup (c-eq) = 1 cup raw or cooked vegetables or fruit, 1 cup vegetable or fruit juice, or 2 cups leafy salad greens, ½ cup dried vegetable or fruit
Dairy: 1 cup equivalent (c-eq) = 1 cup milk, yogurt, or fortified soy milk, or 1½ ounces natural or 2 ounces processed cheese
Grains: 1 ounce equivalent (oz-eq) = 1 slice of bread, ½ cup cooked rice, cereal, or pasta; or 1 ounce ready-to-eat cereal (about 1 cup flaked cereal)
Protein: 1 ounce equivalent (oz-eq) = 1 ounce lean meat, poultry, or seafood; 1 egg; 1 Tbsp peanut butter; ½ ounce nuts or seeds; ¼ cup cooked dried beans or peas

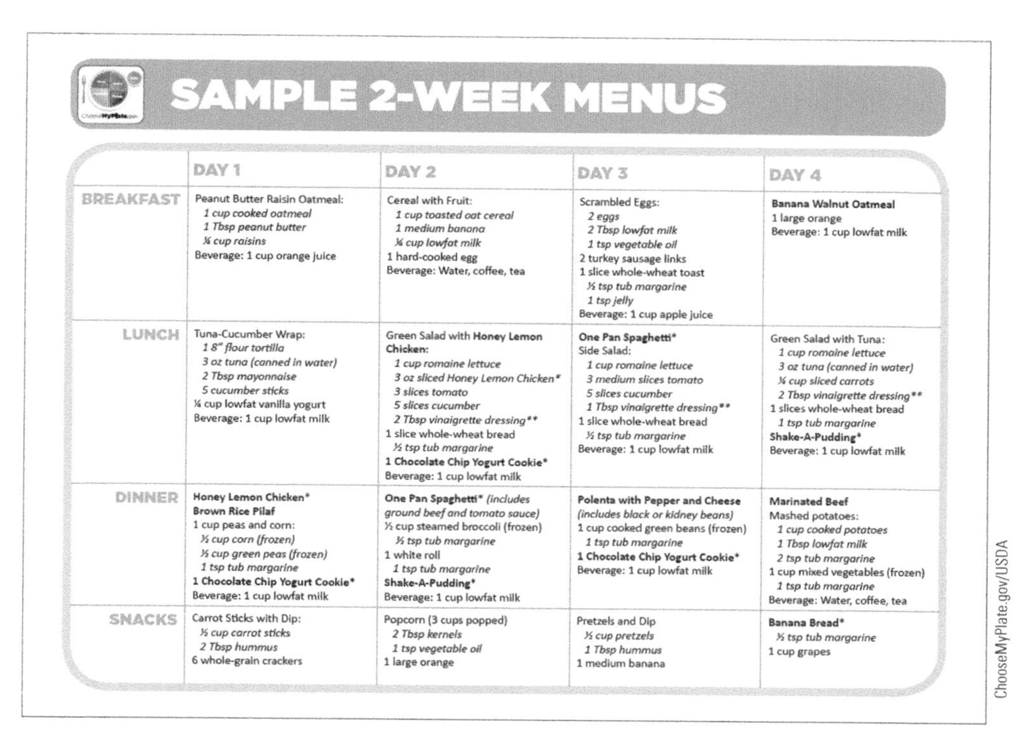
SAMPLE 2-WEEK MENUS

	DAY 1	DAY 2	DAY 3	DAY 4
BREAKFAST	Peanut Butter Raisin Oatmeal: *1 cup cooked oatmeal* *1 Tbsp peanut butter* *¼ cup raisins* Beverage: 1 cup orange juice	Cereal with Fruit: *1 cup toasted oat cereal* *1 medium banana* *¼ cup lowfat milk* 1 hard-cooked egg Beverage: Water, coffee, tea	Scrambled Eggs: *2 eggs* *2 Tbsp lowfat milk* *1 tsp vegetable oil* 2 turkey sausage links 1 slice whole-wheat toast *½ tsp tub margarine* *1 tsp jelly* Beverage: 1 cup apple juice	**Banana Walnut Oatmeal** 1 large orange Beverage: 1 cup lowfat milk
LUNCH	Tuna-Cucumber Wrap: *1 8" flour tortilla* *3 oz tuna (canned in water)* *2 Tbsp mayonnaise* *5 cucumber sticks* ¼ cup lowfat vanilla yogurt Beverage: 1 cup lowfat milk	Green Salad with **Honey Lemon Chicken:** *1 cup romaine lettuce* *3 oz sliced Honey Lemon Chicken** *3 slices tomato* *5 slices cucumber* *2 Tbsp vinaigrette dressing*** 1 slice whole-wheat bread *½ tsp tub margarine* **1 Chocolate Chip Yogurt Cookie*** Beverage: 1 cup lowfat milk	**One Pan Spaghetti*** Side Salad: *1 cup romaine lettuce* *3 medium slices tomato* *5 slices cucumber* *1 Tbsp vinaigrette dressing*** 1 slice whole-wheat bread *½ tsp tub margarine* Beverage: 1 cup lowfat milk	Green Salad with Tuna: *1 cup romaine lettuce* *3 oz tuna (canned in water)* *¼ cup sliced carrots* *2 Tbsp vinaigrette dressing*** 1 slices whole-wheat bread *1 tsp tub margarine* **Shake-A-Pudding*** Beverage: 1 cup lowfat milk
DINNER	**Honey Lemon Chicken*** **Brown Rice Pilaf** 1 cup peas and corn: *½ cup corn (frozen)* *½ cup green peas (frozen)* *1 tsp tub margarine* **1 Chocolate Chip Yogurt Cookie*** Beverage: 1 cup lowfat milk	**One Pan Spaghetti*** *(includes ground beef and tomato sauce)* ½ cup steamed broccoli (frozen) *½ tsp tub margarine* 1 white roll *1 tsp tub margarine* **Shake-A-Pudding*** Beverage: 1 cup lowfat milk	**Polenta with Pepper and Cheese** *(includes black or kidney beans)* 1 cup cooked green beans (frozen) *1 tsp tub margarine* **1 Chocolate Chip Yogurt Cookie*** Beverage: 1 cup lowfat milk	**Marinated Beef** Mashed potatoes: *1 cup cooked potatoes* *1 Tbsp lowfat milk* *2 tsp tub margarine* 1 cup mixed vegetables (frozen) *1 tsp tub margarine* Beverage: Water, coffee, tea
SNACKS	Carrot Sticks with Dip: *½ cup carrot sticks* *2 Tbsp hummus* 6 whole-grain crackers	Popcorn (3 cups popped) *2 Tbsp kernels* *1 tsp vegetable oil* 1 large orange	Pretzels and Dip *½ cup pretzels* *1 Tbsp hummus* 1 medium banana	**Banana Bread*** *½ tsp tub margarine* 1 cup grapes

ChooseMyPlate.gov/USDA

Illustration 6.6 Sample 2-Week Menus available from ChooseMyPlate.gov.

Resources Available from ChooseMyPlate.gov The ChooseMyPlate website provides access to educational materials on topics such as meal preparation and grocery shopping, food choices for preschoolers, pregnancy weight gain, and portion distortion. Examples of these materials are shown and briefly described in Illustrations 6.7 and 6.8.

MyPlate Plan

Find your Healthy Eating Style

Everything you eat and drink matters. Find your healthy eating style that reflects your preferences, culture, traditions, and budget—and maintain it for a lifetime! The right mix can help you be healthier now and into the future. The key is choosing a variety of foods and beverages from each food group—*and making sure that each choice is limited in saturated fat, sodium, and added sugars.* Start with small changes—**"MyWins"**—to make healthier choices you can enjoy.

Food Group Amounts for 1,600 Calories a Day

Fruits	Vegetables	Grains	Protein	Dairy
1 1/2 cups	**2 cups**	**5 ounces**	**5 ounces**	**3 cups**
Focus on whole fruits	Vary your veggies	Make half your grains whole grains	Vary your protein routine	Move to low-fat or fat-free milk or yogurt
Focus on whole fruits that are fresh, frozen, canned, or dried.	Choose a variety of colorful fresh, frozen, and canned vegetables—make sure to include dark green, red, and orange choices.	Find whole-grain foods by reading the Nutrition Facts label and ingredients list.	Mix up your protein foods to include seafood, beans and peas, unsalted nuts and seeds, soy products, eggs, and lean meats and poultry.	Choose fat-free milk, yogurt, and soy beverages (soy milk) to cut back on your saturated fat.

Limit

Drink and eat less sodium, saturated fat, and added sugars. Limit:
- Sodium to **2,300 milligrams** a day.
- Saturated fat to **18 grams** a day.
- Added sugars to **40 grams** a day.

Be active your way: Children 6 to 17 years old should move **60 minutes** every day. Adults should be physically active at least **2 1/2 hours** per week.

ChooseMyPlate.gov/USDA

Illustration 6.7 MyPlate Daily Checklist shows the amounts of each basic food group that should be included in a day's diet for an individual with a 1,600 calorie need. It also includes a check list that can be used to compare your intake with that recommended.

Source: ChooseMyPlate.gov, https://www.cnpp.usda.gov/sites/default/files/dietary_guidelines_for_americans/MyPlateDailyChecklist_1600cals_Age4-8.pdf].

Get Your MyPlate Plan

After you answer a few questions, the MyPlate Plan calculates the approximate number of calories you should consume daily to lose, gain, or maintain your weight.

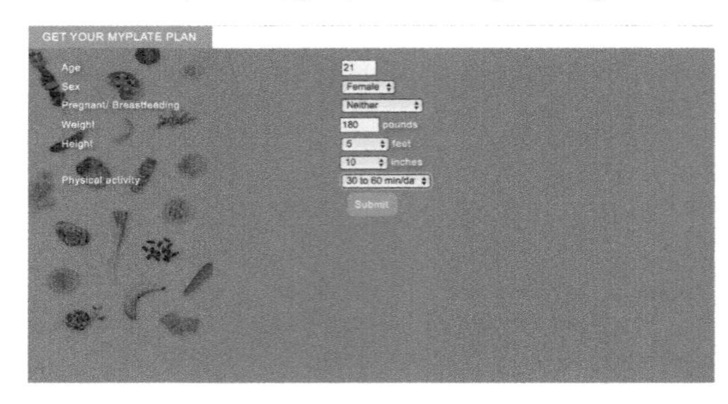

Tips for Vegetarians

Intended to help vegetarians plan their diets, this site lists tips for food sources of key nutrients and suggestions for selecting vegetarian foods.

MyPlate Moms/Moms-To-Be

This site provides information about health and nutrition for pregnant and breastfeeding women.

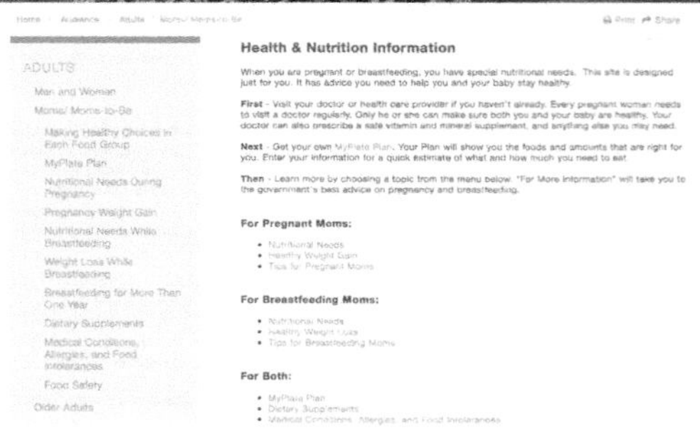

Food Fact Cards

Each card contains fun facts and tips about each produce item and how their local produce fits into MyPlate.

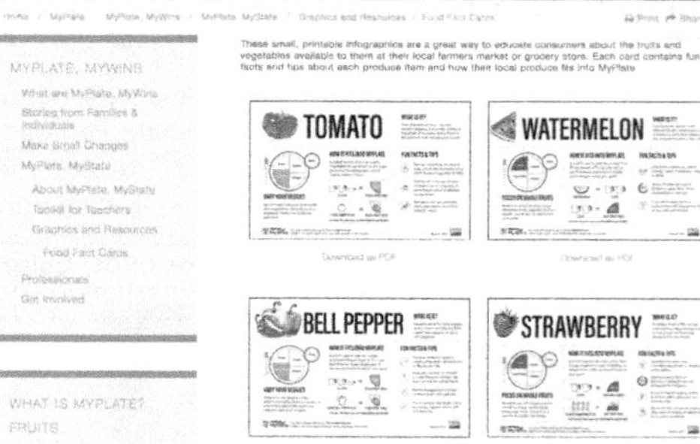

Illustration 6.8 Tools and tips available from ChooseMyPlate.gov.

Stay tuned to MyPlate.gov. Additional useful tools are periodically added to this site.

Limitations of MyPlate Materials available from MyPlate are almost entirely made available on the Internet, making the information inaccessible to people who do not use computers or have access to the Internet. MyPlate does not provide specific recommendations for infants, individuals on therapeutic diets, or vegans.

Menus suggested by MyPlate may not correspond to individual food preferences and contain relatively few ethnic foods. As with past food guides, planning and evaluating how mixed dishes (such as stews, soups, salads, and various types of pizza) fit into the food groups can still be perplexing.

In 2018, interactive features for dietary analysis, physical activity and body weight tracking available on ChooseMyPlate were reduced or removed and are being replaced by other, reliable interactive programs, including the Diet and Wellness Plus available on MindTap.

Other Healthy Dietary Patterns Other types of dietary patterns have been shown to promote health and prevent disease. Two such patterns are the Dietary Approaches to Stop Hypertension (DASH) and the Healthy Mediterranean Dietary Pattern.

The DASH Eating Plan

Originally identified as a diet that helps control mild and moderate **hypertension**, the DASH Eating Plan has also been found to reduce the risk of certain types of cancer, osteoporosis, and heart disease. Improvements in blood pressure are generally seen within two weeks of starting this dietary pattern.[11,12]

The DASH dietary pattern emphasizes fruits, vegetables, low-fat dairy foods, whole-grain products, poultry, fish, and nuts. Only small amounts of fats, red meats, sweets, and sugar-containing beverages are included. This dietary pattern provides ample amounts of potassium, magnesium, calcium, fiber, and protein, and limited amounts of saturated and ***trans* fats**.[13,14] Although two calorie levels are shown in the table, DASH Eating Plans are available for 12 levels (1,600 to 3,200 calories) online.[13]

hypertension High blood pressure. It is defined as blood pressure exerted inside blood vessel walls that typically exceeds 140/90 millimeters of mercury.

***trans* fats** A type of unsaturated fatty acid produced by the addition of hydrogen to liquid vegetable oils to make them more solid. Small amounts of naturally occurring *trans* fat are found in milk and meat.

The Healthy Mediterranean Dietary Pattern

The traditional Mediterranean diet ranks with the USDA's Food Guide and the DASH Eating Plan when it comes to health promotion and chronic disease prevention.[5] The Mediterranean diet was originally based on foods consumed by people in Greece, Crete, southern Italy, and other Mediterranean areas where rates of chronic disease were low and life expectancy long.[15] The DASH Eating Plan and the Mediterranean Diet were used to formulate the "Healthy-U.S.-Style Dietary Pattern" and the "Healthy Mediterranean-Style Dietary Pattern" recommended in the Dietary Guidelines (Table 6.6).

The Mediterranean countries consist of Italy, France, Monaco, Slovenia, Croatia, Bosnia and Herzegovina, Albania, Turkey, Syria, Lebanon, Israel, Greece, Portugal, Egypt, Libya, Tunisia, Algeria, and Spain. People in these countries consume a similar dietary pattern and also experience longer than average life expectancy. Lifestyles generally include ample physical activity and consumption of locally grown foods and fish from the sea. Farming methods in the Mediterranean region are largely based on sustainable agricultural practices that match usual climate and soil conditions, and seasonal water availability.[35]

Table 6.6 Composition of the Healthy U.S. Style-and the Healthy Mediterranean-Style Eating Patterns based on a need for 2,000 calories daily

Food group		Healthy U.S.-Style Eating Pattern		Healthy Mediterraanean-Style Eating Pattern
Grains[a]	©David Kay/ Shutterstock.com	6 oz-eq		6 oz-eq (half or more should be whole grains)
Vegetables (include dark green, red, and orange vegetables)	©Feng Yu/ Shutterstock.com	2½ c-eq	2½ c-eq	2½ c-eq
				½ cup raw or cooked vegetables
				½ cup vegetable juice
Fruits	©Tatiana Popova/ Shutterstock.com	2 c-eq	2 c-eq	2½ c-eq
				½ cup fresh, frozen, or canned fruit
				½ cup fruit juice
Dairy	©DenisNata/ Shutterstock.com	3 c-eq	3 c-eq	2 c-eq
Protein Foods	©PHB.cz (Richard Semik)/Shutterstock .com	5½ oz-eq	5½ oz-eq	6½ oz-eq
Oils		6 tsp	6 tsp	6 tsp
Calories, other[a]		270/day	270/day	260/day

[a]Includes calories from added sugars, solid fats, alcohol, or additional basic foods

The Mediterranean dietary pattern, represented by the Mediterranean Diet Pyramid shown in Illustration 6.9, emphasizes plant foods, such as fruits, vegetables, grains (mostly whole), beans, and nuts. Fish and seafood is represented in the diet at least twice a week, poultry and eggs twice weekly or less, and cheese and yogurt one to seven times a week. Meats and sweets form the smallest part of the pyramid and are consumed infrequently. Wine in moderation is a traditional part of the Mediterranean diet, and water intake is encouraged. A number of studies have shown that this dietary pattern is associated with lower risks of heart disease, stroke, diabetes, several forms of cancer, and overall mortality.[16,17]

Realities of the Food Environment

Decisions people make about which foods to eat and how much of them to consume are affected by nutrition knowledge; food affordability, availability, and preferences; time availability; and other factors. The selection and consumption of a healthy dietary pattern is related to all of these factors, and largely takes place in an environment that promotes the overconsumption of relatively inexpensive, convenient, and highly palatable foods. In this environment, the larger the portion of foods that can be served that meets consumers' preferences, time constraints, and budgets, the higher the profit for those selling the foods. This situation produces a food environment that encourages the over-consumption of foods, especially when eating out.[32]

Portion Distortion

Do you know how much food you ate yesterday? If you had meat, what was the size of the serving of meat you consumed? Did you drink something with that? How much of that beverage did you consume?

Few people are aware of how much food they eat. Some people think a serving of food is the same as the portion of food they are served or eat. Portion sizes or "servings" of food today tend to exceed standard serving sizes developed by the USDA for use in planning healthful diets. In this era of large portion and people sizes it's particularly important to become aware of how much food is eaten.[31]

Individual ideas of normal serving amounts are based on past experiences at family meals, the size of portions provided by restaurants, and packaged food and beverage sizes. Supersized meals at fast food restaurants, large portions served by other restaurants, large bakery products, and larger cups of soft drinks are contributing to the problem of portion distortion. Table 6.7 provides examples of how food portion sizes have changed with time. Table 6.8 provides a guide for estimating food portion sizes.

Mediterranean Diet Pyramid
A contemporary approach to delicious, healthy eating

Meats and Sweets — Less often
Wine — In moderation
Poultry and Eggs — Moderate portions, every two days or weekly
Cheese and Yogurt — Moderate portions, daily to weekly
Drink Water
Fish and Seafood — Often, at least two times per week
Fruits, Vegetables, Grains (mostly whole), Olive oil, Beans, Nuts, Legumes and Seeds, Herbs and Spices — Base every meal on these foods
Be Physically Active; Enjoy Meals with Others

Illustration by George Middleton
© 2009 Oldways Preservation and Exchange Trust www.oldwayspt.org

Illustration 6.9 The Mediterranean Diet Pyramid.

Are Supersized Portions Supersizing Americans? Supersized fast food meals or "value" meals can contain double the caloric content compared to their regular-sized counterparts. A single, supersized meal of a double quarter-pound cheeseburger, large fries, and thick shake provides more calories (about 2,200) than many people need in a day. Larger portions don't cost restaurants much more than smaller portions, they increase sales volume, and they encourage people to eat more. Many Americans are eating a good deal more food than they need, and it is appears that rising rates of obesity are partly related to large portion sizes.[3]

Table 6.7 Typical portion sizes and calorie content of foods in the marketplace versus calorie content and portion sizes in the past[18,19]

Food	Portion size Calories 20 years ago	Marketplace portion size Calories now
Bagel	3-inch diameter	6-inch diameter
	140 calories	350 calories
Cheeseburger	4.3 ounces	7.1 ounces
	343 calories	535 calories
French fries	2.4 ounces	6.9 ounces
	210 calories	610 calories
Soft drink	6.5 ounces	20 ounces
	85 calories	820 calories
Muffin	1.5 ounce	6.5 ounce
	167 calories	724 calories

Table 6.8 Portion size estimators

Portion	Photo credit
1 cup = baseball	Skizer/Shutterstock.com
½ cup = tennis ball	Vlad09/Shutterstock.com
¼ cup = golf ball or extra-large egg	Cameramannz/Shutterstock.com
2 Tablespoons = ping-pong ball	Tomas1111/Dreamstime.com
1 teaspoon = fingertip	RunPhoto/StockbyteGetty Images
3 ounces of meat = deck of cards, palm of hand	Dedyukhin Dmitry/Shutterstock.com
For single-serve foods, check the weight given on the food package label.	

nutrient-dense foods Foods that contain relatively high amounts of nutrients compared to their calorie value.

Children and adults tend to eat more when offered larger portions of foods than smaller portions. Children and adolescents have been reported to consume 5–12% more soft drinks, pizza, french fries, and salty snacks when offered large portions.[20] Among adults, a 50% increase in portion sizes of meals has been found to increase daily energy intake by 423 calories.[21] Frequent dining at fast food restaurants (three or more times per week) is associated with higher intakes of calories, sugar, and sodium and with a higher risk of overweight and obesity than less frequent dining at these restaurants.[22] On a positive note, some restaurants are offering smaller portions and healthier menu options than in the past.[23]

Can You Still Eat Right When Eating Out? The question about what to eat often boils down to choosing the right restaurant. According to USDA data, 43% of American adults eat out at least weekly.[24] In general, foods eaten away from home have lower nutrient content and are higher in fat than foods eaten at home (see Illustration 6.10).[25] In addition, children and teenagers who eat dinner with their families most days tend to have more healthful dietary patterns than those who never or only occasionally eat dinner with their family.[26]

Staying on Track While Eating Out You'll find it easier to stick to a healthy diet if you decide what to eat before you enter a restaurant and look over the menu (Illustration 6.11). You could make the decision to order soup and a salad, broiled meat, a half-portion of the entrée (or to split an entrée with someone else), or no dessert *before* entering the restaurant. "Impulse ordering" is a hazard that can throw diets out of balance. If you're going to a party or a business event where food will be served, decide before you go what types of food you will eat and what you will drink. If only high-calorie foods are offered, plan on taking a small portion and stopping there. Some people find it helps to have a healthy snack before going to a party or an event where high-calorie food will be served to avoid being really hungry when they get there.

Can Fast Foods Be Part of a Healthy Diet? As the information in Table 6.9 demonstrates, many of the foods served in fast food restaurants deserve their reputation of being high in calories, sodium, and sugar. In recognition of this fact, and in response to legislative requirements that chain restaurants label the caloric value of menu items, lower-calorie and more **nutrient-dense foods** are being added to menus. The addition of low-fat milk,

Illustration 6.10 Hamburgers, french fries, and pizza are top-selling food items in U.S. restaurants.[27]

oatmeal, fruited yogurt, coleslaw, multigrain breads, apple slices, a variety of salads, and bean burritos to fast food and fast causal food menus makes it easier to eat well when eating out.

Illustration 6.11 "No, no thank you. Extra cheese isn't part of what I planned to eat."

Photodisc

The Slow Food Movement An interesting trend in food preparation and consumption is making its way across the globe. The trend is away from fast and processed foods, and toward sustainable, eco-friendly agricultural practices and locally grown foods. The Slow Food and similar movements represent some aspects of this trend. Part of an international group, Slow Food USA is an educational organization that supports ecologically sound food production; the revival of the kitchen and the table as centers of pleasure, culture, and community; and living a slower and more harmonious rhythm of life.[28] The trend is placing the topic of healthy eating in a new light for many individuals and communities and may help bring people closer to family, friends, and the environment.

What If You Don't Know How to Cook? With so many convenience foods available and time at a premium, there is growing concern that we're becoming a nation of cooking illiterates. Cooking at home gives you control over what you eat and how it's prepared, and is associated with better diet quality than eating out.[33] Some people immensely enjoy cooking and get a thrill out of making their specialties for friends and family. It's becoming a popular leisure-time activity: 43% of U.S. adults have taken it up for enjoyment.[29]

If you don't know how to cook, there are several ways to learn. You could start on your own by using the recipes on food packages like pasta, tomato sauce, or dried beans. You could search for recipes online or buy a basic cookbook and make dishes like salads, tacos, shish kebab, and lentil soup. You could even take a community education course. Illustration 6.12 shows some examples of good starter cookbooks. Read the sections on basic cooking skills and learn what types of equipment and utensils you need to prepare basic dishes. Get the foods and other supplies you need. Select the recipes that look doable and be sure they pass your taste and nutritional standards tests. Voilà! You're cooking.

Illustration 6.12 Some good starter cookbooks and recipe sources.

Richard Anderson

Bon Appétit!

Dietary guidelines from some other countries contain one other rule of healthy eating that would serve Americans well: "Enjoy your meals." Eating a healthy diet should be enjoyable. If it's too much of a struggle, the healthy diet won't last. The best diets are those that keep us healthy and enhance our sense of enjoyment and wellbeing. The trick is to remember the broad array of nutritious foods we like that give us good taste and enjoyment when we eat them. And remember not to feel guilty when you occasionally eat a hamburger or some ice cream. Enjoy them to the utmost—as much as ripe oranges, papaya, homemade soups, roast turkey, hummus, and countless other nutritious delicacies.

REALITY CHECK

Portion Distortion

Mohammad and Kevin decided to eat at an Italian restaurant after soccer practice. Just for the fun of it, they agreed to guess how many cups of spaghetti they would consume if they ate the portion served to them.

Who has the better guess?

Answers on page 6-18

Photodisc

Mohammad: Let me see . . . I'd guess it would be 3 cups.

Photodisc

Kevin: I bet it's about a cup. It looks like a normal serving to me.

ANSWERS TO REALITY CHECK

Portion Distortion

The average portion size of pasta served by restaurants is nearly 3 cups.[18] For people on a 2,000 calorie per day diet, that's a day's allotment for the foods from the grain group.

Photodisc

Mohammad:

Photodisc

Kevin:

Table 6.9 Examples of the calorie, fiber, and sodium content of large and small portions of fast food meals. The numbers in parentheses represents the % of the RDA/AI for a 19-year old male[a]

Menu item	Calories	Fiber, g	Sodium, mg
Shake Shack Double Shackburger®	855 (28%)	3 (18%)	1,200 (80%)
Single Shack Burger® (Shake Shack)	550 (18%)	3 (18%)	824 (55%)
Jelly donut, (Dunkin' Donut),	270 (9%)	1 (3%)	330 (22%)
Plain Cake Donut (Dunkin')	320 (10%)	1 (3%)	300 (20%)
Chicken Naked Strips®, 3 (Popeye's)	170 (6%)	0 (0%)	550 (37%)
Chicken Tender, 3 pc (Popeye's)	310 (10%)	2 (5%)	1,240 (83%)
McChicken Sandwich® (McDonald's)	350 (11%)	2 (5%)	600 (40%)
Buttermilk Crispy Chicken Sandwich® (McDonalds)	570 (19%)	4 (11%)	1,050 (70%)
Whopper® Jr. Sandwich (Burger King)	310 (10%)	1 (3%)	390 (26%)
Whopper® Sandwich (Burger King)	660 (22%)	2 (5%)	980 (65%)

[a]Data obtained from internet fast food nutrition information sites, 11/17.

NUTRITION up close

The Pros and Cons of Fast Food Dining

Focal Point: Fast foods can be a major source of calories and sodium, and poor sources of fiber—or not, depending on how often and which fast foods are consumed.

Meet Benson, a person with a busy life who usually gets her meals at fast food or fast casual restaurants on most days during the work week. Benson has a daily calorie need of 2,200. An "adequate intake" of fiber for Benson is 28 g, and 2,300 mg for sodium.

1. Below is a list of the foods Benson consumed at fast food restaurants on one day. Using the information provided in Table 6.9, fill in the columns and complete the calculations indicated in sections a–d.

	Calories	Fiber, g	Sodium, mg
Breakfast:			
Jelly donuts, 2	______	______	_____
Lunch			
Buttermilk Crispy Chicken Sandwich®	______	______	_____
Dinner			
Shake Shack Double Shakeburger®	______	______	_____
a. Column Totals:	______	______	_____

b. Percent of Benson's daily calorie need obtained from fast foods: _______%

c. Difference in sodium intake from fast foods and her AI for sodium of 2,300 mg (add a + or a – in front of your number): _______mg

d. By how many grams was Benson's intake of fiber less than the recommended intake amount? _____g

Feedback to the Nutrition Up Close is located in Appendix F.

REVIEW QUESTIONS

- **Apply the characteristics of healthful diets to the design of one.**

1. Adequate diets are defined as those that provide sufficient calories to relieve hunger and maintain a person's body weight. **True/False**
2. The diets of Americans tend to be out of balance in a number of ways. **True/False**
3. ______ "Macronutrients" consist of:
   a. fiber, vitamins, carbohydrates, and minerals
   b. water, calories, fat, and fiber
   c. protein, minerals, vitamins, and fats
   d. carbohydrates, proteins, fat, and water
4. ______ Andre, a normal-weight man with little body fat, takes a multivitamin and mineral supplement each morning to make sure he gets all the vitamins and minerals needed daily. The rest of the day he eats whatever he wants, such as candy, donuts, pizza, burgers, and soft drinks. Is Andre consuming a healthful diet?
   a. Most likely yes, because he is getting all the vitamins and minerals he needs.
   b. Yes, because he is normal weight.
   c. Probably not because his diet lacks balance and variety.
   d. It depends on the composition of the multivitamin and mineral supplement.

- **Identify characteristics of diets that promote health and those that do not.**

5. The Healthy Mediterranean-Style Dietary Pattern is an example of a vegetarian diet plan that promotes weight loss. **True/False**

6. Healthy dietary patterns are represented by the Healthy U.S.-Style Dietary Pattern, Healthy Vegetarian-Style Dietary Pattern, and the Healthy Mediterranean-Style Dietary Pattern **True/False**
7. The DASH Eating Plan represents a dietary pattern that is intended for vegetarians. **True/False**
8. The 2015 Dietary Guidelines include a recommendation to lower total fat intake. **True/False**
9. The 2015 Dietary Guidelines include recommendations related to saturated fat intake and physical activity. **True/False**
10. ______ As the head of food service for Lincoln Elementary School you have been charged with making sure the new cafeteria lunch menus conform to the Dietary Guidelines recommendations. Which of the following sets of menu items would you *limit* offering in the lunch menu?
    a. tossed salads and salad dressing
    b. corn and tuna fish
    c. hot dogs and fruit drinks
    d. fruit salad and corn bread
11. Healthy dietary patterns share the basic characteristics of providing
    a. adequate levels of nutrients
    b. low amounts of fat
    c. low amounts of carbohydrate
    d. high amounts of protein

- **Utilize MyPlate.gov guidance materials for dietary planning and evaluation.**

12. ChooseMyPlate provides information used in the United States to help people implement the Dietary Guidelines. **True/False**
13. Which of the following foods does not belong to a basic food group?
    a. Popcorn
    b. Peanuts
    c. Cheddar cheese
    d. Energy bar
14. How many servings of grains would you consume if you ate 2 slices of whole grain bread and a cup of oatmeal?
    a. 3
    b. 4
    c. 5
    d. 6
15. ______ Maya lost 10 pounds and would like to keep the weight off. Tonight she's headed to a party where lots of delicious pastries will be served and she doesn't want to eat too many of them. What can she do to help with that?
    a. Try to convince herself that she really doesn't like pastries.
    b. Decide before she goes to the party to enjoy one pastry.
    c. Skip the party. It would be impossible to resist the pastries.
    d. Go to the party late because the pastries may be gone.
16. ______ Which of the following statements about large portion size is true?
    a. Adults and children tend to eat more when offered large portions of foods rather than small portions.
    b. People eat until they feel full and then stop eating, regardless of portion size.
    c. Large portions tend to make adults and children eat less than if given smaller portions.
    d. Children tend to eat more when offered more food but adults do not.

Answers to these questions can be found in Appendix F.

NUTRITION SCOREBOARD ANSWERS

1. Healthy dietary patterns also include fish, seafood, vegetables, fruits, seeds, low-fat dairy products, and a moderate amount of alcohol (for adults who choose to drink).[1] **True**
2. Eighty-four percent of U.S. adults fail to meet this recommendation.[2] **True**
3. The basic food groups do not include a "healthy snack" food group. **False**
4. "Value" (aka supersized) fast food meals can pile on calories.[3] **True**

How the Body Uses Food: Digestion and Absorption

NUTRITION SCOREBOARD

1 Almost all of the carbohydrate and fat you consume in foods is absorbed by the body, but only about half of the protein is absorbed. **True/False**

2 Disorders of the digestive system are a leading cause of hospitalizations in the United States and Canada. **True/False**

3 Most stomach ulcers are caused by overeating spicy foods. **True/False**

4 Periodic colon cleansing decreases the toxin load in the body and "reboots" the body's digestive processes. **True/False**

Answers can be found at the end of the unit.

Digital Vision/Alamy

LEARNING OBJECTIVES

After completing Unit 7 you will be able to:

- Outline specific mechanical and chemical processes involved in digestion of carbohydrates, proteins, and fats.
- Describe the ways in which diet is related to common types of digestive disorders.

My Body, My Food

- **Outline specific mechanical and chemical processes involved in digestion of carbohydrates, proteins, and fats.**

You are not the same person you were a month ago. Although your body looks the same and you don't notice the change, the substances that make up the organs and tissues of your body are constantly changing. Tissues we generally think of as solid and permanent, such as bones, the heart, blood vessels, and nerves, are continually renewing themselves. The raw materials used in the body's renewal processes are the nutrients you consume in foods.

Each day, about 5% of our body weight is replaced by new tissue. Existing components of cells are renewed, the substances in our blood are replaced, and body fluids are recycled. Taste cells, for example, are replaced about every seven days, and the cells lining the intestinal tract are replaced every one to three days. All of the cells of the skin are replaced every month. Red blood cells turn over every 120 days.[5] If you thought it was hard to maintain a car, an apartment, or a house, just imagine the difficulty of maintaining a body! Maintenance is just one of the body's ongoing functions that require nutrients as raw material.

KEY NUTRITION CONCEPTS

Two key nutrition concepts directly relate to the content on digestion, absorption, and digestive disorders covered in this unit:

1. Nutrition Concept #2: Foods provide energy (calories), nutrients, and other substances needed for growth and health.
2. Nutrition Concept #5: Humans have adaptive mechanisms for managing fluctuations in nutrient intake.

How Do Nutrients in Food Become Available for the Body's Use?

La vie est une fonction chimique. (Life is a chemical process.)
—Antoine Lavoisier, late eighteenth century

The components of food used to form and maintain body tissues are nutrients. Through the processes of **digestion** and **absorption**, nutrients are made available for use by every cell in the body.

digestion The mechanical and chemical processes whereby ingested food is converted into substances that can be absorbed by the intestinal tract and utilized by the body.

absorption The process by which nutrients and other substances are transferred from the digestive system into body fluids for transport throughout the body.

enzymes Protein substances that speed up chemical reactions. Enzymes are found throughout the body but are present in particularly large amounts in the digestive system.

The Internal Travels of Food: An Overview The "food processor" of the body is the digestive system, shown in Illustration 7.1. It consists of a 25- to 30-foot-long muscular tube and organs such as the liver and pancreas that secrete digestive juices. The digestive juices break foods down into their molecular components, which can be absorbed and utilized by the body.

Much of the work of digestion is accomplished by **enzymes** manufactured by components of the digestive system such as the salivary glands, stomach, and pancreas. Enzymes are complex protein substances that speed up the reactions that break down food. A remarkable feature of enzymes is that they are not changed by the chemical reactions they affect. This makes them reusable.

Carbohydrates, fats, and proteins all have their own sets of digestive enzymes. Altogether, over a hundred different enzymes participate in the digestion of carbohydrates, fats, and proteins. Table 7.1 presents information on some of the enzymes involved in digestion and highlights their specific roles. In Table 7.2, you will see how these enzymes are involved in the digestion of carbohydrate, fat, and protein.

Digestive Processes Digestive processes actually begin before the first bite of food enters the mouth. All a person needs to do to get digestive juices flowing is to think about food, smell food, or see it.[6] You can put this information to the test by clearing your mind

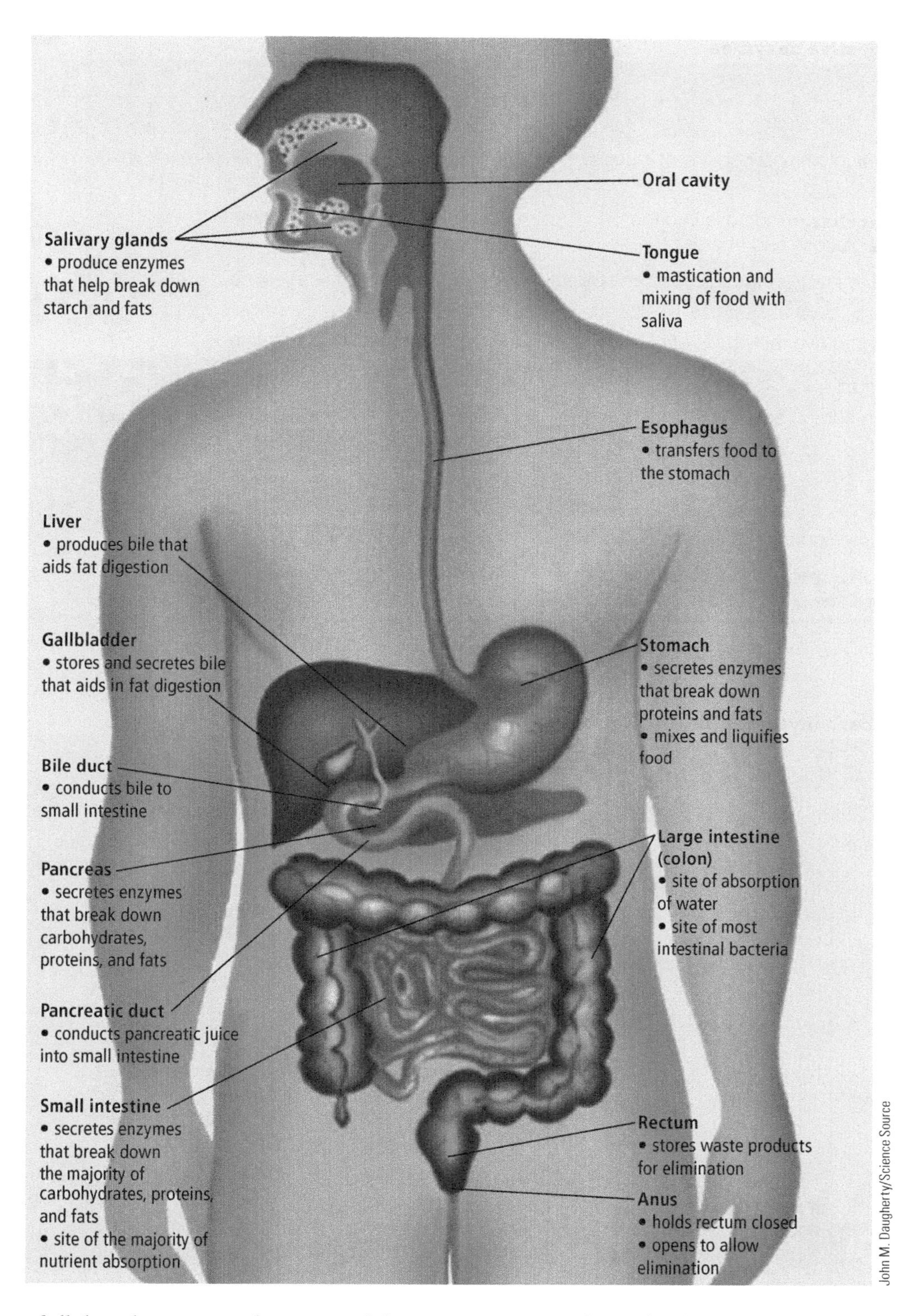

Illustration 7.1 The digestive system.

of all thoughts, turning the page, and then concentrating only on the photo shown in Illustration 7.2.

As you chew, glands under the tongue release saliva, which lubricates the food so that it can be swallowed and easily passed along the intestinal tract. Saliva gets food digestion started. It contains salivary amylase and lingual lipase, which begin to break down carbohydrates (amylase) and fats (lipase).[7]

The amount of salivary amylase produced by individuals varies a good deal based on genetic traits. Differences in salivary amylase production can alter the "feel," or texture, of food in the mouth and food preferences. People who produce high levels of amylase

Table 7.1 Primary function of some digestive enzymes

Enzyme	Enzyme function	Enzyme source
A. Carbohydrate digestion		
Amylase	Breaks down **starch** into smaller chains of glucose molecules	Produced in the salivary glands (salivary amylase) and the pancreas (pancreatic amylase)
Sucrase	Separates the **disaccharide** sucrose into the **monosaccharides** glucose and fructose	Produced in the small intestine
Lactase	Splits the disaccharide lactose into glucose and galactose	Produced in the small intestine
Maltase	Separates maltose into two molecules of glucose	Produced in the small intestine
B. Fat digestion		
Lipase	Breaks down fats into fragments of fatty acids and glycerol	Produced in salivary glands (lingual lipase) and the pancreas (pancreatic lipase). The action of lipase is enhanced by **bile**.
C. Protein digestion		
Pepsin	Separates protein into shorter chains of amino acids	Produced by the stomach
Trypsin	Splits short chains of amino acids into molecules containing one, two, or three amino acids	Produced by the pancreas

Table 7.2 Summary of the digestion of carbohydrates, fats, and proteins

	Mouth	Stomach	Small intestine, pancreas, liver, and gallbladder	Large intestine (colon)
Carbohydrates (excluding fiber)	The salivary glands secrete saliva to moisten and lubricate food; chewing crushes and mixes the food with salivary amylase, which initiates starch digestion.	Digestion of starch continues while food remains in the stomach. Some alcohol is absorbed here. Acid produced in the stomach aids digestion and destroys many bacteria in food.	Pancreatic amylase continues starch digestion. Sucrase, lactase, and maltase break down disaccharides into monosaccharides that are absorbed. Some alcohol is absorbed here.	Undigested carbohydrates reach the colon and can be partly broken down by intestinal bacteria.
Fiber	The teeth crush fiber and mix it with saliva to moisten it for swallowing.	No action.	Fiber binds cholesterol and some minerals.	Most fiber is excreted with feces; some fiber is digested by bacteria in the colon.
Fat	Fat-rich foods are mixed with saliva. Small amounts of lingual lipase accomplish some fat breakdown.	Fat tends to separate from the watery stomach fluid and foods and float on top of the mixture. About 10–30% of fat is broken down by lingual lipase. Fat is last to leave the stomach.	Bile readies fat for the ac tion of lipase from the pancreas. Lipase splits fats into fatty acids and glycerol, which are absorbed.	A small amount of fatty material escapes absorption and is carried out of the body with other wastes.
Protein	In the mouth, chewing crushes and softens protein-rich foods and mixes them with saliva.	Stomach acid works to uncoil protein strands and to activate the stomach's protein-digesting enzyme. Pepsin breaks the protein strands into smaller chains of amino acids.	Trypsin splits protein into molecules containing one, two, or three amino acids. These amino acids are absorbed.	The large intestine concentrates and carries undigested fiber and other residues out of the body.

Michael Newman/PhotoEdit

Illustration 7.2 Testing, testing. This is a test of your salivary secretions. Did the lemon speak directly to your salivary glands? If you want to turn the digestive processes off, quit thinking about food.

break down and liquefy starch in foods to a greater extent than do people who produce low amounts. This gives the foods a soft feel and enhances their flavor. The same foods consumed by low amylase producers may feel firm and have a less desirable flavor.[8,9]

After food is chewed, it is swallowed and passed down the esophagus to the stomach. Muscles that act as one-way valves at the entrance and exit of the stomach ensure that the food stays there until it's liquefied, mixed with digestive juices, and ready for the digestive processes of the small intestine. Solid foods tend to stay in the stomach for two to four hours, whereas most liquids pass through it in about 20 minutes.[9]

When the stomach has finished with its work, it ejects 1 to 2 teaspoons of its liquefied contents into the small intestine through the muscular valve at its end. Stomach contents continue to be ejected in this fashion until they are totally released into the small intestine. These small pulses of liquefied food stimulate muscles in the intestinal walls to contract and relax; these movements churn and mix the food as it is digested by enzymes. When the diet contains sufficient fiber, the bulge of digesting food in the intestine tends to be larger. Larger food bulges stimulate a higher level of intestinal muscle activity than do smaller food bulges. Thus, high-fiber meals pass through the digestive system somewhat faster than low-fiber meals.

Digestion, as well as the absorption of nutrients, is greatly enhanced by the structure of the intestines (Illustration 7.3). Fingerlike projections called "villi" line the inside of the

starch Complex carbohydrates made up of complex chains of glucose molecules. Starch is the primary storage form of carbohydrate in plants. The vast majority of carbohydrate in our diet consists of starch, monosaccharides, and disaccharides.

disaccharides Simple sugars consisting of two sugar molecules. Sucrose (table sugar) consists of a glucose and a fructose molecule, lactose (milk sugar) consists of glucose and galactose, and maltose (malt sugar) consists of two glucose molecules.

monosaccharides (*mono* = one, *saccharide* = sugar) Simple sugars consisting of one sugar molecule. Glucose, fructose, and galactose are monosaccharides.

bile A yellowish-brown or green fluid produced by the liver, stored in the gallbladder, and secreted into the small intestine. It acts like a detergent, breaking down globs of fat entering the small intestine to droplets, making the fats more accessible to the action of lipase.

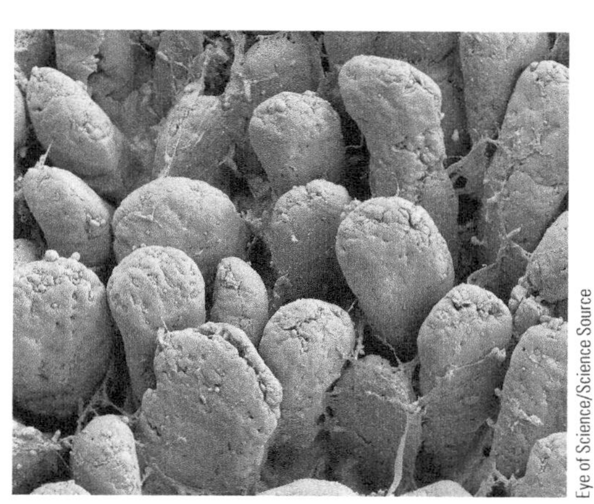

Eye of Science/Science Source

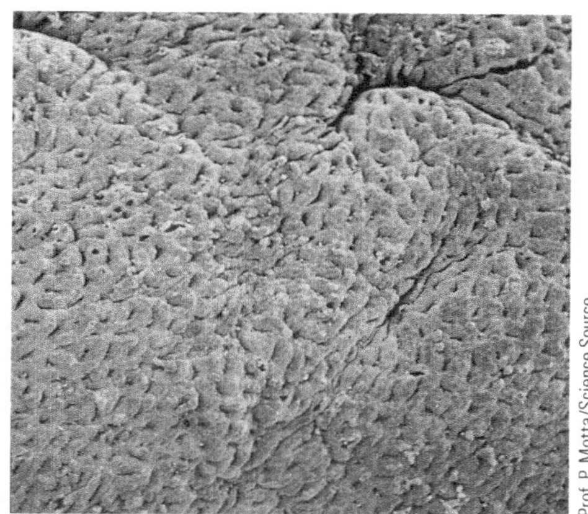

Prof. P. Motta/Science Source

Illustration 7.3 Scanning electron micrographs of cross-sections of the small intestine (*left*) and the large intestine (*right*). Note the high density of villi in the small intestine and the relative flatness of the lining of the large intestine.

intestinal wall and increase its surface area tremendously. If laid flat, the surface area of the small intestine would be about the size of a baseball infield, or approximately 675 square feet. This large mass of tissue requires a high level of nutrients for its own maintenance. Much of this need (50% in the small intestine and 80% in the large intestine) is met by foods that are being digested.[10]

Absorption Digestion is complete when carbohydrates, fats, and proteins are reduced to substances that can be absorbed, and when vitamins and minerals are released from food. The end products of the digestion of approximately 99% of the carbohydrate, 92% of the protein, and 95% of the fat in our diets are absorbed. The primary end products of carbohydrate, fat, and protein digestion that are absorbed are

- Carbohydrate: glucose
- Fat: fatty acids and glycerol
- Protein: amino acids

Some of the end products of digestion are absorbed in the stomach and large intestine, though nutrient absorption occurs primarily in the small intestine. About 30% of alcohol consumed with meals is absorbed in the stomach; the remainder is absorbed in the small intestine.[11] Water, sodium, and chloride are mainly absorbed by the large intestine. Substances in food that cannot be digested or absorbed, along with bacteria and fragments of cells from the intestinal lining that are being discarded, are concentrated by the large intestine and excreted as stools. The Reality Check feature addresses the digestion of raw vegetables and the question some people have about whether our bodies can absorb nutrients from them at all.

lymphatic system A network of vessels that absorb some of the products of digestion and transport them to the heart, where they are mixed with the substances contained in blood.

circulatory system The heart, arteries, capillaries, and veins responsible for circulating blood throughout the body.

The Lymphatic and Circulatory Systems The end products of digestion are taken up by the **lymphatic system** (Illustration 7.4) and the **circulatory system** (Illustration 7.5) for eventual distribution to all cells in the body. Lymph vessels and blood vessels infiltrate the villi that line the inside of the intestines (Illustration 7.6) and transport absorbed nutrients toward the major branches of the lymphatic and circulatory systems. The broken-down products of fat digestion are largely absorbed into the lymph vessels, whereas carbohydrate and protein broken down products enter the blood vessels.

The nutrient-rich contents of the lymphatic system are transferred to the bloodstream at a site near the heart where vessels from both systems merge into one vessel. From there the lymph and blood mixture is sent to the heart and subsequently throughout the body by way of the circulatory system. Nutrients delivered by the circulatory system reach every organ and tissue in the body, supplying cells with the nutrients obtained from food.

Beyond Absorption Cells can use nutrients directly for energy, body structures, or the regulation of body processes. For example, glucose can be used as is for energy formation or converted to glycogen and stored for later use. Fatty acids, an end product of fat digestion, can be incorporated into cell membranes, used in the synthesis of certain hormones, or used as fuel for energy formation. Vitamins and minerals freed from food by digestion can be used by cells to regulate enzyme activity or tissue maintenance. The body has a limited storage capacity for some vitamins and minerals. Consequently, excessive amounts of certain vitamins and minerals such as vitamin C, thiamin, and sodium are largely excreted in urine.

inflammation Reactions of the body to the presence of infectious agents, toxins, or irritants. Inflammation triggers the release of biologically active substances that promote oxidation and other reactions that counteract the infectious agent, toxin, or irritant.

umami (u-mam-e) A Japanese word meaning "pleasant savory taste." The taste is described as brothy or meaty and is recognized as the fifth basic taste. Foods containing glutamate, an amino acid derivative, elicit the umami taste; these include fish, meats, mushrooms, ripe tomatoes, and aged cheese.

Other Functions of the Gastrointestinal Tract The gastrointestinal tract is involved in body processes beyond digestion and absorption. These functions include regulation of digestion and absorption processes by taste sensors and the roles of gut bacteria in the prevention of infection and **inflammation**-related disorders, fiber digestion, and vitamin production.

Functions of Taste Sensors Our gastrointestinal tract plays an important role in identifying the five basic tastes of food (salty, sweet, bitter, sour, and **umami**). The taste of food helps humans identify which foods are safe and beneficial to eat and which may not be.

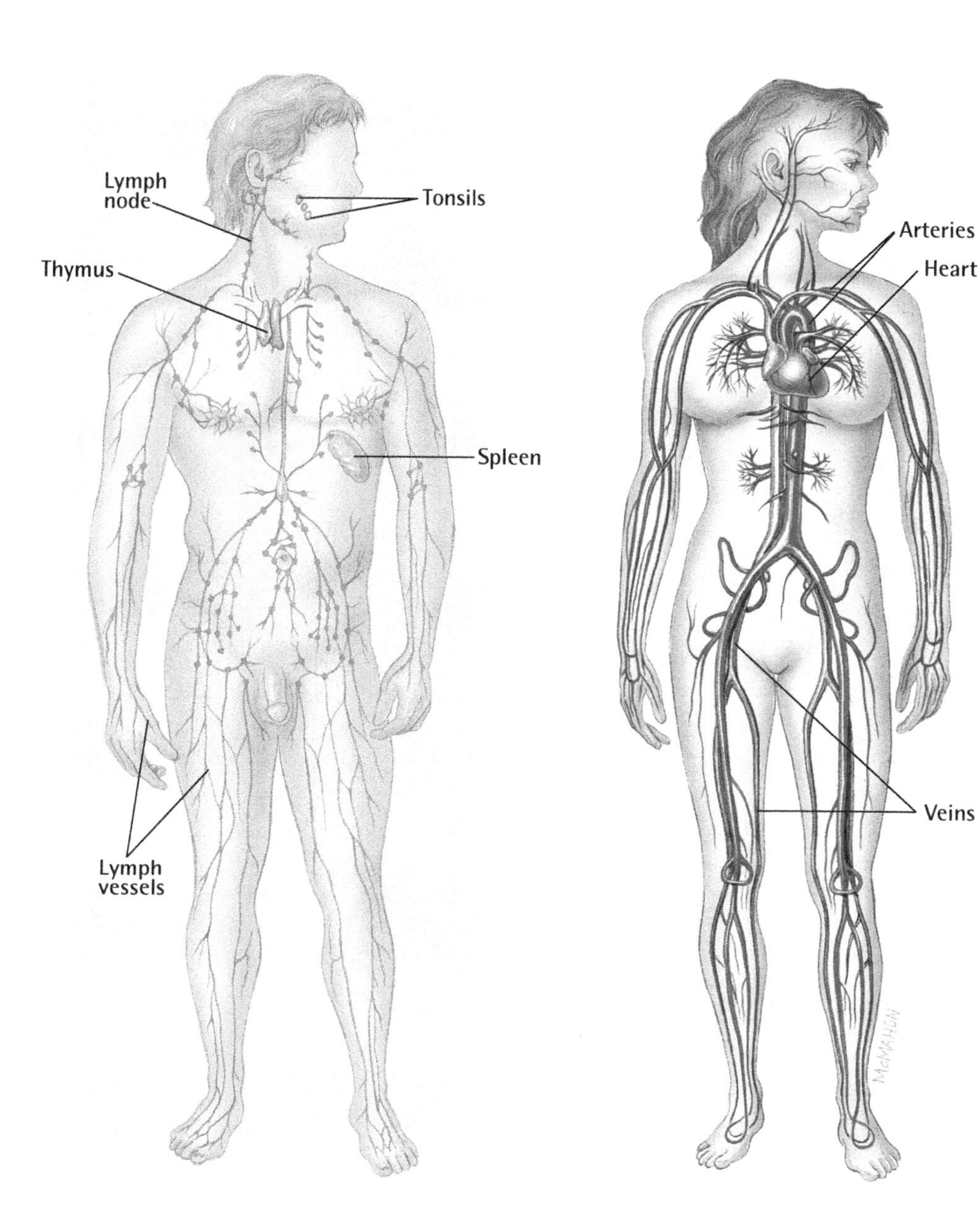

Illustration 7.4 (*left*) The lymphatic system.

Illustration 7.5 (*right*) The circulatory system includes the heart and blood vessels. This system serves as the nutrient transportation system of the body.

REALITY CHECK

Can You Digest Raw Vegetables?

Owen and Raja had just finished eating a shredded cabbage salad when Owen suddenly comes up with a statement about something he recently read on the Internet: "You can't digest raw cabbage. It's like trying to digest raw carrots, broccoli, or corn. They just go right through you."

Do they?

Answers on page 7-8.

AISPIX by Image Source/Shutterstock.com

Owen: Our digestive juices can't break down raw vegetables, so we don't get any nutrients from them.

iStockphoto.com/Juanmonino

Raja: Really? I eat raw vegetables and think they're nutritious.

ANSWERS TO REALITY CHECK

Can You Digest Raw Vegetables?

Nutrients are absorbed from raw vegetables, but the extent of nutrient availability depends largely on cell wall breakdown. Plant cell walls resist digestion because they are fibrous, and nutrients contained in vegetable cells will not be made available for absorption if cell walls stay intact. Disrupting cell walls by thoroughly chewing raw vegetables increases nutrient availability and absorption.[12]

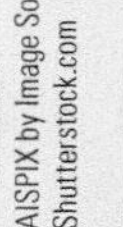

AISPIX by Image Source/ Shutterstock.com

iStockphoto.com/Juanmonino

Owen:

Raja:

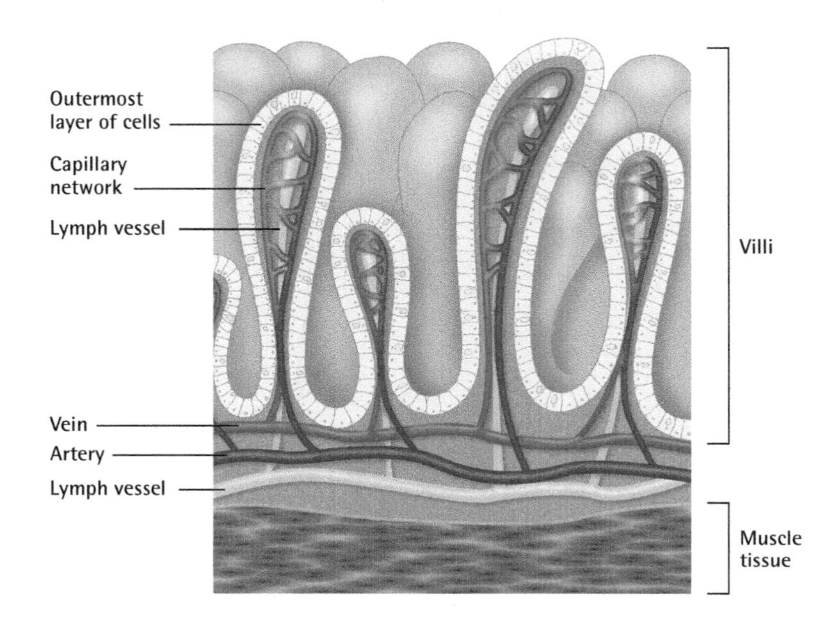

Illustration 7.6 Structure of villi, showing blood and lymph vessels.

Sweet, savory, and salty foods generally provide pleasurable sensations (Illustration 7.7), whereas the taste of bitter and sour may be disliked and identified as potentially toxic.[13]

Taste and other sensors that identify the composition of foods consumed are located throughout the gastrointestinal tract, and not just in the mouth. Messages about how a food tastes and its composition are relayed from these sensors to tissues involved in digestion and absorption, and in the utilization of nutrients by the body. The sweet taste message, for example, begins processes that prepare the body to absorb and utilize glucose. The detection of bitter substances can lead to mechanisms that promote the development of diarrhea that will flush potential toxins out of the body. Taste signals also activate processes that decrease appetite and food intake over the course of a meal.[14]

microbes Microscopic organisms, including bacteria and fungi. Some microbes are beneficial, some pathogenic (harmful), and some have little effect on the body. Also called *microorganisms*.

gut microbiota The microbial population living in our intestine. Formerly called gut flora.

probiotics Live bacteria that are taken orally to restore beneficial bacteria to the gastrointestinal tract. Strains of *Lactobacillus* (lac-toe-bah-sil-us) and *Bifidobacteria* (bif-id-dough-bacteria) are the best-known probiotics.

Gut Microbiota The human body is home to trillions of **microbes**, predominantly bacteria, introduced into the body through food, air, and bodily contact with the external environment. The collection of microorganisms in and on humans is referred to as the "human microbiota," The weight of the body's microbiota is estimated to be 3–4 pounds (1.5–2 kg). **Gut microbiota** is the name given to the population of microorganisms that reside in the intestines (illustration 7.8). It is estimated that our intestines contain around 1,000 bacterial species and 100 times more genes than are found in the human genome. Humans begin to acquire their gut microbiota in utero, and that changes over time largely through diet and environmental exposures. The general composition of gut microbiota is similar in most people, but the species composition is unique to individuals because diets and environmental exposures differ.[15,16]

Microorganisms in the gut perform specific functions that promote health. They:

- consume and break down fiber and other undigested food residues
- excrete fatty acids as an end product of fiber digestion; the fatty acids are absorbed and serve as a source of energy
- help fight infection and prevent inflammation by signaling the presence of harmful microbes which are then targeted by the body's infection-fighting immune system
- produce, and thereby contribute to, the body's supply of biotin, vitamin B_{12}, and vitamin K[17–20]

Research is uncovering a host of other functions of gut bacteria. Exposure to digestive products from ultra-processed foods or diets high in red meat or sugar, or low in fiber appear to promote overgrowth of microbes capable of triggering a chronic, low-grade inflammation that predisposes the body to store fat.[21,22] These types of diets appear to promote the development of obesity, type 2 diabetes, colon cancer, irritable bowel syndrome, and other disorders through changes induced by microbes in the gut. These effects are not observed when a high-fiber, high-plant food diet is consumed.[15,21,22] The population of beneficial microbes in the intestines can be disrupted by antibiotics and other medications, but can be renewed to some extent by **probiotics** and dietary changes.[16,23]

iStockphoto.com/ArthurHidden

Illustration 7.7 The taste of food activates many digestive tract processes.

Digestive Disorders

- **Describe the ways in which diet is related to common types of digestive disorders.**

Digestive disorders such as **heartburn**, **hemorrhoids**, **irritable bowel syndrome**, and **duodenal and stomach ulcers** are the leading cause of hospitalization among U.S. adults aged 45 to 64 and Canadian adults aged 35 to 54 years. Digestive disorders account for more than 104 million medical visits each year in the United States alone.[2,24] Diet and weight status are among the important factors that influence the development and treatment of many of the common disorders of the digestive tract.

heartburn A condition that results when acidic stomach contents are released into the esophagus, usually causing a burning sensation.

hemorrhoids (hem-or-oids) Swelling of veins in the anus or rectum.

irritable bowel syndrome (IBS) A disorder of bowel function characterized by chronic or episodic gas; abdominal pain; diarrhea, constipation, or both.

duodenal (do-odd-en-all) and stomach ulcers Open sores in the lining of the duodenum (the uppermost part of the small intestine) or the stomach.

Constipation

Constipation is medically defined to exist when an individual has fewer than three bowel movements per week.[25] Some people think they are constipated if they do not have a bowel movement every day. However, normal stool elimination may be three times a day or three times a week, depending on the person. Constipation is characterized by difficulty passing stools because they are hard and dry. People with constipation feel "blocked up" and lousy overall. They may see small amounts of bright red blood on the stool or toilet paper caused by bleeding hemorrhoids or a slight tearing of the anus. This generally disappears after constipation is controlled. Constipation is a symptom, not a disease. Almost everyone experiences constipation at some point in life, and a poor diet is typically the cause.[26]

There are a number of other causes of constipation: the presence of a disorder or disease in the intestinal tract, immobility, medication use, habitual use of laxatives, or a slow transit time of food in the digestive tract. Constipation caused by a slow transit time of food through the digestive tract is often due to the consumption of too little fiber.[27] This is a common cause of constipation and can generally be relieved, and subsequently

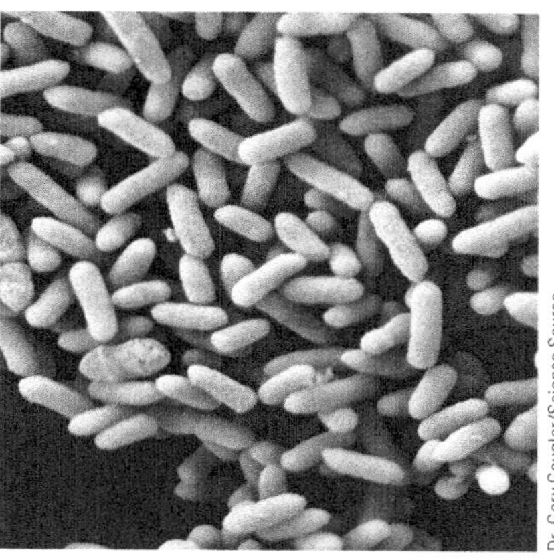
Dr. Gary Gaugler/Science Source

Illustration 7.8 An example of a health-promoting species of bacteria (*B. infantis*) found in the large intestine.

Scott Goodwin Photography

Illustration 7.9 Food sources of dietary fiber. Together, the foods shown provide 29 grams of dietary fiber, an amount that helps prevent constipation.

prevented, by including 25–30 grams of dietary fiber daily. Good sources of fiber are shown in Illustration 7.9. People with severe constipation may not benefit from high fiber intake, however.[26]

Fiber increases a person's fluid requirement, so people who increase their fiber intake should also increase their fluid intake. Fiber can increase the production of gas by bacteria that digest fiber. It should be introduced gradually to allow the bacteria in the colon to adjust to the new levels of fiber.[28]

Myths Related to Constipation Some common beliefs about constipation do not hold up to scientific scrutiny (see Table 7.3). For example, there is no evidence that stools contain toxins that can be absorbed and cause harm to the body, or that you can improve your health by periodically "cleansing" or "detoxifying" the colon.[29] In fact, colon cleansing alters the diversity of the gut microbiota.[31] Unless dehydration is a problem, consuming plenty of fluids will not treat constipation.[30] The presence of soft stools that are easily excreted is a strong sign that constipation does not exist.[32]

Table 7.3 Myths about constipation[25,38]

1. Poisonous substances are absorbed from stools and cause "autointoxication" diseases.
2. Colon cleansing "detoxifies" the body.
3. Extra-long colons cause constipation.
4. All cases of constipation are caused by inadequate fiber intake.
5. You can treat constipation by drinking plenty of fluids.
6. You can lose weight and stay healthy if you take laxatives regularly.
7. If you do not have a bowel movement every day there is something wrong with you.

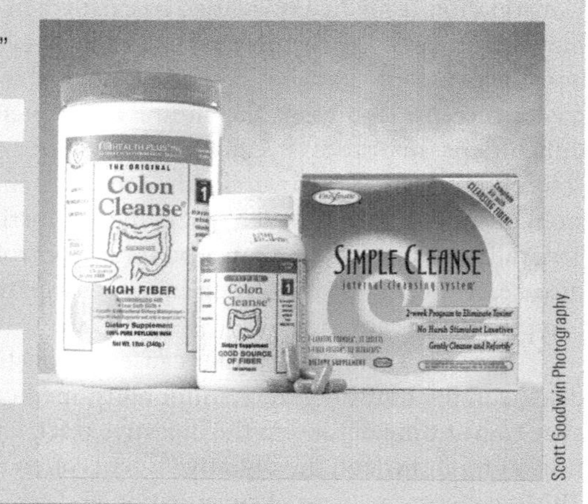

Scott Goodwin Photography

Ulcers

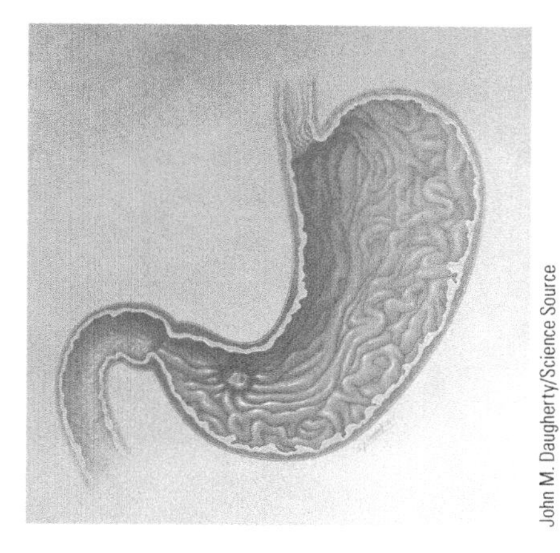

Illustration 7.10 A stomach ulcer.

Ulcers are sores that occur primarily in the lining of the stomach and duodenum (Illustration 7.10). They occur in millions of Americans each year.[33] Ulcers are caused by *Helicobacter pylori* bacterial infection or the overuse of aspirin, ibuprofen, and similar medications. *H. pylori* is acquired by the ingestion of foods and other substances contaminated with saliva, vomit, or feces from a person harboring the bacteria in the stomach. Rates of *H. pylori* infection are highest in countries with poor sanitary conditions.[34]

H. pylori infection and overuse of some types of pain medications can break down the protective mucus layer that coats the inside of the stomach and small intestine. When this happens, digestive juices and stomach acid are allowed to erode the stomach and intestinal lining. Symptoms of ulcers include abdominal pain, reduced appetite, weight loss, and feelings of being bloated or nauseated after eating.[35]

The development of ulcers has not been found to be related to the intake of specific foods or to high levels of stress. Ulcers can be treated in almost all cases by antibiotics that destroy *H. pylori* bacteria or drugs that reduce stomach acid.[36]

Heartburn

Heartburn occurs at least weekly in 20% of American adults.[37] The most common symptom of heartburn is a painful, burning feeling in the chest or throat near the area where the heart is located. It occurs when the valve at the top of the stomach, which normally keeps digesting food in the stomach, relaxes. The weakened valve allows the acidic contents of the stomach to back up into the esophagus.[38]

The causes of heartburn are not completely understood, but a variety of factors such as obesity, overeating, pregnancy, and the use of certain medications appear to be related to its development. Certain dietary factors, such as those listed in Table 7.4, have been reported to worsen the symptoms of heartburn in some people. Lifestyle changes, such as weight loss, dietary changes, elevation of the head during sleep, chewing sugar-free gum for a half hour after eating, and medications that reduce stomach acid are commonly used to relieve heartburn.[38,39]

Table 7.4 Diet-related factors and foods that may worsen the symptoms of heartburn[40,44,45]

- Obesity
- Overeating
- High-fat foods
- High-carbohydrate foods
- Alcohol
- Spicy foods
- Tomato sauce
- Inadequate fiber intake
- Coffee

Irritable Bowel Syndrome

Irritable bowel syndrome (IBS) has a name that matches its primary symptom: an irritated bowel. People with irritable bowel syndrome have trouble moving food along the intestines. Food tends to go though too quickly (which produces diarrhea), or too slowly (which leads to constipation). They experience pain and cramping. Overgrowth of bacteria in the large intestine and sensitivity to certain foods and stress appear to be parts of the problem in IBS. The diagnosis of IBS is made when people experience symptoms of abdominal pain, bloating, and diarrhea or constipation three times a week for three months or longer.[40] About one in seven American adults have symptoms of IBS. Although IBS has not been found to be associated with the development of serious disease, it has important, negative impacts on quality of life.[41]

Most people with IBS can control their symptoms with a healthy dietary pattern, stress management, and medications (if needed). Avoidance of large meals, adding **soluble fibers** to the diet, and the use of specific probiotics have been found to ease the symptoms of IBS.[42] It is recommended that people with IBS identify foods that seem to promote their IBS. Individuals with the syndrome tend to report that their symptoms improve when they consume fewer sugar foods, fatty foods, coffee, alcohol, and hot spices. Consultation with a health care provider such as a registered dietitian can help identify potentially offending foods.[40,43]

soluble fiber Types of dietary fiber that dissolve in water, forming a gel. Good sources of soluble fiber include oats, oatmeal, oat bran, apples, pears, dried beans, carrots, and psyllium.

Diarrhea

Diarrhea is a common problem in the United States, occurring an average of four times a year in adults.[46] It is a leading public health problem in many developing countries. Most cases of diarrhea are due to bacterial- or viral-contaminated food or water, lack of

diarrhea The presence of three or more liquid stools in a 24-hour period.

immunizations against infectious diseases, and interactions between malnutrition and infection. Children are particularly susceptible to diarrhea. It can deplete the body of fluids and nutrients and produce malnutrition. If it lasts more than two weeks or is severe, diarrhea can lead to dehydration, heart and kidney malfunction, and death. An estimated 3.5 million deaths from diarrheal diseases occur each year to the world's population of children 5 years of age or under.[47]

The vast majority of cases of diarrhea can be prevented through food and water sanitation programs, immunizations, and adequate diets. The early use of oral rehydration fluids (e.g., the formula provided by the World Health Organization and commercial formulas such as Pedialyte and Rehydralyte) shortens the duration of diarrhea. Rehydration generally takes four to six hours after the fluids are begun.[48]

Rather than "resting the gut" during diarrhea, children and adults, once rehydrated, should eat solid foods. Foods such as yogurt, lactose-free or regular milk, chicken, potatoes and other vegetables, dried beans, and rice and other cereals are generally well tolerated and provide nutrients needed for the repair of the intestinal tract. It is best to avoid sugary fluids such as soft drinks. High-sugar beverages tend to draw fluid into the intestinal tract rather than increase the absorption of fluid.[49]

Flatulence

flatulence (flat-u-lens) Presence of excess gas in the stomach and intestines.

Everyone experiences **flatulence**—it's normal. Gas can occur in the esophagus, stomach, small intestine, and large intestine due to swallowed air or bacterial breakdown of food in the large intestine. Eating and drinking while in a rush generally increases air ingestion. Bacterial production of gas in the large intestine can be related to the ingestion of dried beans, broccoli, cauliflower, brussels sprouts, onions, corn, and other vegetables containing "resistant starch" that bacteria digest. Fructose, which is used to sweeten a variety of food products and beverages, and sorbitol (used in some types of candy and gum) may lead to gas formation by bacteria that produce gas as a waste product of carbohydrate digestion. Heartburn and other gastrointestinal tract disorders and medications such as antibiotics are also associated with gas production.[46]

People often think they produce too much gas, even when they don't. The amount of gas swallowed and produced by gut bacteria varies a good deal among individuals and within the same individual. Gas production changes depending on what foods are eaten, the types of bacteria populating the large intestine, medications used, and the presence of gastrointestinal tract disorders. Severe and painful symptoms related to gas production may signal the presence of a digestive disorder.[46]

Stomach Growling Gas in the stomach can make your stomach growl. When your stomach growls, you know that gas and food or fluids are mixing in your stomach. The growling tends to be louder when your stomach is empty, when there's no food to muffle the noise. The thought, sight, or smell of food can also trigger stomach growling.[50]

microbiome The totality of microorganisms and their collective genetic material present in or on the human body.

Feeding Your Digestive System Right Research on digestive processes and the health effects of the gut **microbiome** has boomed since the discovery that laboratory animals raised in sterile environment and lacking gut microbiota are at high risk of particular types of infections.[16] Scientists are rapidly advancing knowledge of the effects of the gastrointestinal tract microbiomes as well as that of the oral cavity, respiratory system, and urinary tract on health. It appears that for the vast majority of people, consumption of a healthy dietary pattern that includes lots of plant foods and fiber, little sugar, red meat, and highly processed foods best serve the interest of our microbiomes and health.[15,16]

Digital Vision/Alamy

NUTRITION up close

Carbohydrate, Fat, and Protein Digestion

Focal Point: Digestion makes carbohydrates, proteins, and fats available for absorption and utilization by the body.

Nikki just finished lunch. She had a chicken sandwich with mayonnaise and a glass of skim milk. She's wondering what is happening to the food in her digestive tract. What is happening?

List the primary enzymes involved in the digestion of the carbohydrate, fat, and protein in Nikki's meal. Also list the primary end products of carbohydrate, fat, and protein digestion that are absorbed.

	Primary enzymes	Primary end products of digestion
Carbohydrate	1.	1.
	2.	2.
	3.	3.
	4.	
Fat	1.	1.
		2.
Protein	1.	1.
		2.

Feedback to the Nutrition Up Close is located in Appendix F.

REVIEW QUESTIONS

- **Outline specific mechanical and chemical processes involved in digestion of carbohydrates, proteins, and fats.**

1. Cells lining the intestinal tract are replaced every 30 days. **True/False**
2. Enzymes are protein substances that speed up chemical reactions and are reusable. **True/False**
3. The end products of digestion are taken up by the lymphatic and circulatory systems. **True/False**
4. The nutrient-rich contents of the lymphatic system do not mix with blood. These nutrients are distributed throughout the body by a separate system of lymph vessels. **True/False**
5. The body has a limited capacity to store some vitamins and minerals. **True/False**
6. About 75% of the protein, carbohydrate, and fat we consume in food is digested and absorbed. **True/False**
7. The amount of salivary lipase produced by individuals affects the "feel," or texture, of food in the mouth and food preferences. **True/False**
8. The digestion of fats in the small intestine is largely accomplished by the action of pancreatic lipase. **True/False**
9. Sweet, sour, bitter, salty, and umami represent the five basic tastes. **True/False**
10. Taste sensors are located only in the mouth. **True/False**
11. Bacteria present in our large intestine digest a portion of the fiber in our diets. This is a function human digestive enzymes *cannot* perform. **True/False**
12. Bacteria in the large intestine produce several vitamins, but they are not in a form humans can absorb. **True/False**
13. ____ Chris underwent surgery for weight loss that greatly reduced the amount of the small intestine that could be used for food digestion and nutrient absorption. Digestion and absorption of which of the following nutrients would be most adversely affected by the surgery?
    a. fat
    b. protein
    c. carbohydrate
    d. alcohol

The next three questions refer to the following situation:

Assume a pharmaceutical company makes a drug that must be swallowed whole. It comes in a pill coated with protein.

14. ______ In which part of the gastrointestinal tract would the drug release begin?
    a. mouth
    b. stomach
    c. small intestine
    d. large intestine
15. ______ This same company makes another drug that comes in a pill coated with starch that must be swallowed whole. In which part of the gastrointestinal tract would this drug start to be released?
    a. mouth
    b. stomach
    c. small intestine
    d. large intestine
16. ______ The company also manufacturers a fiber supplement. In which part of the gastrointestinal tract would the fiber be digested to some extent?
    a. mouth
    b. stomach
    c. small intestine
    d. large intestine

- **Describe the ways in which diet is related to common types of digestive disorders.**

17. Avoidance of high-protein meals, adding calcium to the diet, and the use of specific probiotics are recommended for the treatment of constipation. **True/False**
18. Overeating and consumption of high-fat foods and carbonated beverages worsen heartburn symptoms in some people. **True/False**
19. Foods containing resistant starch, such as dried beans and corn, may increase gas production by bacteria in the large intestine. **True/False**
20. Constipation is medically defined to exist when an individual has fewer than three bowel movements per week. **True/False**
21. Colon cleansing has been shown to prevent disease by removing toxic products from the large intestine. **True/False**
22. It is recommended that people with diarrhea who are not dehydrated consume solid foods. **True/False**
23. Although some types of bacteria in our intestines are beneficial to health, overall they are related to disease development. **True/False**

Answers to these questions can be found in Appendix F.

NUTRITION SCOREBOARD ANSWERS

1. Over 90% of all the carbohydrates, fats, *and* proteins consumed in food are absorbed and become part of the body. **False**
2. Digestive disorders are the leading cause of hospitalizations among 35–64 year old adults.[1,2] **True**
3. Stomach ulcers have not been shown to be caused by overeating spicy or other types of foods.[3] **False**
4. That doesn't appear to happen.[4] **False**

Calories! Food, Energy, and Energy Balance

Howard Kingsnorth/The Image Bank/Getty Images

NUTRITION SCOREBOARD

1 Carbohydrates provide more calories than do fats. **True/False**

2 A teaspoon of butter has a higher calorie value than a teaspoon of margarine. **True/False**

3 Energy can be neither created nor destroyed. It can, however, change from one form to another. **True/False**

4 In some instances, a person's appetite can override the feeling of satiety. **True/False**

Answers can be found at the end of the unit.

LEARNING OBJECTIVES

After completing Unit 8 you will be able to:

- Estimate total calorie need using the formulas presented.
- Describe how the first law of thermodynamics applies to energy balance in humans.
- Calculate the calorie value of a food from its content of carbohydrate, protein, fat, and alcohol (if present).

calorie (calor = heat) A unit of measure used to express the amount of energy produced by foods in the form of heat. The calorie used in nutrition is the large calorie, or the kilocalorie (kcal). It equals the amount of energy needed to raise the temperature of 1 kilogram of water (about 4 cups) from 15 to 16°C (59 to 61°F). The term *kilocalorie*, or *calorie* as used in this text, is gradually being replaced by the kilojoule (kJ) in the United States; 1 kcal = 4.2 kJ.

basal metabolism Energy used to support body processes such as growth, health, tissue repair and maintenance, and other functions. Assessed while at rest, basal metabolism includes energy the body expends for breathing, pumping of the heart, maintenance of body temperature, and other life-sustaining, ongoing functions.

Energy!

- **Estimate total calorie need using the formulas presented.**

When you think of calories, do you think of energy? That is the "scientifically correct" way to think about them. Energy is what calories are all about.

The **calorie** is like a centimeter or pound in that it is a unit of measure. Rather than serving as a measure of length or weight, the calorie is used as a measure of energy. Specifically, a calorie is the amount of energy needed to raise the temperature of 1 kilogram of water (about 4 cups) from 15°C to 16°C, or 59°F to 61°F (Illustration 8.1). This amount of energy is used as a standard for assigning caloric values to foods. Because calories are a unit of measure, they are not a component of food like vitamins or minerals. When we talk about the caloric content of a food, we're really talking about the caloric value of the food's energy content.

The caloric value of food is determined in a "bomb calorimeter" by burning it completely in a container surrounded by a specific amount of water (Illustration 8.2). The energy released by the food in the form of heat raises the temperature of the surrounding water. The rise in temperature indicates how many calories were released from the portion of food. Although the body doesn't literally burn food, the amount of heat released by food while burning is approximately the same as the amount of energy it supplies to the body.

KEY NUTRITION CONCEPTS

The following key nutrition concepts relate to this unit's content on calories, energy, and energy balance:

1. Nutrition Concept #2: Foods provide energy (calories), nutrients, and other substances needed for growth and health.
2. Nutrition Concept #4: Poor nutrition can result from both inadequate and excessive levels of nutrient intake.
3. Nutrition Concept #10: There are no good or bad foods.

Illustration 8.1 A calorie is the amount of energy needed to raise the temperature of 1 kilogram of water (about 4 cups) from 15°C to 16°C (59°F to 61°F).

The Body's Need for Energy

The body uses energy from foods to fuel muscular activity, growth, and tissue maintenance and repair; to chemically process nutrients; and to maintain body temperature (to name a few examples). These needs for energy are subdivided into three categories: **basal metabolism**, physical activity, and **dietary thermogenesis** (Table 8.1 on page 8-3 and Illustration 8.3 on page 8-4).

The largest single contributor to energy need is basal metabolism, or "resting metabolism" as it is also called. It accounts for 60 to 75% of the total need for calories in the vast majority of people.[1] Energy-requiring processes of basal metabolism include breathing, energy used by metabolically active tissues and organs, the beating of the heart, maintenance of body temperature, renewal of muscle and bone tissue, and other ongoing activities that sustain life and health. Growth is considered a component of basal metabolism. The proportion of total calories needed for basal metabolism is particularly high during the growing years.

Some organs and tissues in the body are more metabolically active—that is, they require more energy to sustain their functions—than others. Body fat is less metabolically active than lean tissues, accounting for less than 20% of basal metabolic calorie expenditure in most people. The brain, liver, kidneys, and muscles are metabolically active; together, they account for 80% or more of the energy used for basal metabolism.[2]

Only a small percentage of people have an unusually low or high **basal metabolic rate (BMR)**. In healthy individuals, slight differences in BMR rarely account for the ease with which weight is gained, maintained, or lost.[2] Uncommon metabolic disorders, such as underactive thyroid and Cushing's syndrome, can modify basal metabolic processes and can lead to changes in weight.[3] The Reality Check for this unit addresses the topic of slow metabolism and weight loss.

Energy-using activities of basal metabolic processes require no conscious effort on our part; they are continuous activities that the body must perform to sustain life. The energy needed to carry out basal metabolic functions is assessed when a person is in a state of complete physical and emotional rest and has not eaten for at least eight hours.

Basal metabolic rate, or resting energy expenditure as it is also called, is determined clinically and in research studies by **indirect calorimetry**. Illustration 8.4 shows an example of an indirect calorimeter and its use in a clinical setting. This method determines energy expenditure indirectly through a determination of the body's oxygen utilization for energy production in the body.[4]

Indirect calorimetry is a fairly expensive and time-consuming process. Equations for calculating basal metabolic rate are used when estimates suffice.[5,6]

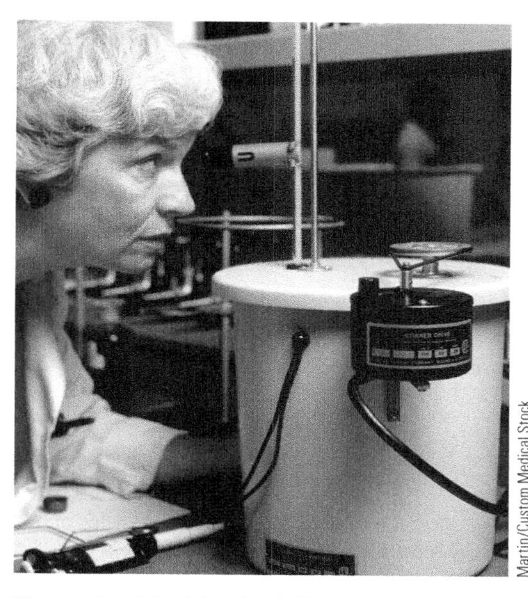

Martin/Custom Medical Stock

Illustration 8.2 A bomb calorimeter used to measure the calorie value of foods. A food's calorie value is determined by the amount of heat released and transferred to water when the food is completely burned.

How Much Energy Do I Expend for Basal Metabolism? You can quickly estimate the calories needed for basal metabolic processes as follows:

- For men: Multiply body weight in pounds by 11.
- For women: Multiply body weight in pounds by 10.

Thus, a man who weighs 170 pounds needs approximately 170×11, or 1,870 calories per day for basal metabolic processes. A 135-pound woman needs 135×10, or 1,350 calories.

This formula gives an estimate of calories used for basal metabolism in adults based on sex and weight. Physical activity level, muscle mass, height, health status, and genetic

Table 8.1 The three energy-requiring processes of the body

Process
1. Basal metabolism Energy use related to maintenance of normal body functions while at rest
2. Physical activity Energy use related to muscular work
3. Dietary thermogenesis Energy use related to food ingestion. (The process gives off heat.)

dietary thermogenesis Thermogenesis means the production of heat. Dietary thermogenesis is the energy expended during the digestion of food and the absorption, utilization, storage, and transport of nutrients. Some of the energy escapes as heat. It accounts for approximately 10% of the body's total energy need. Also called diet-induced thermogenesis and thermogenic effect of food or feeding.

basal metabolic rate (BMR) The rate at which energy is used by the body when it is at complete rest. BMR is expressed as calories used per unit of time, such as an hour, per unit of body weight in kilograms. Also called *resting metabolic rate* (RMR) or *resting energy expenditure* (REE).

indirect calorimetry A method of measuring energy expenditure from determination of the amount of oxygen utilized by the body during a specific unit of time.

REALITY CHECK

Is There Any Such Thing as a Slow Metabolism?

Answers on page 8-4

Photodisc

Mel: I've got a slow metabolism. It doesn't matter how little I eat, I still can't lose weight.

Photodisc

Juanita: Some of my friends who have trouble losing weight say that too, but it can't be right. "Slow metabolism" is a myth.

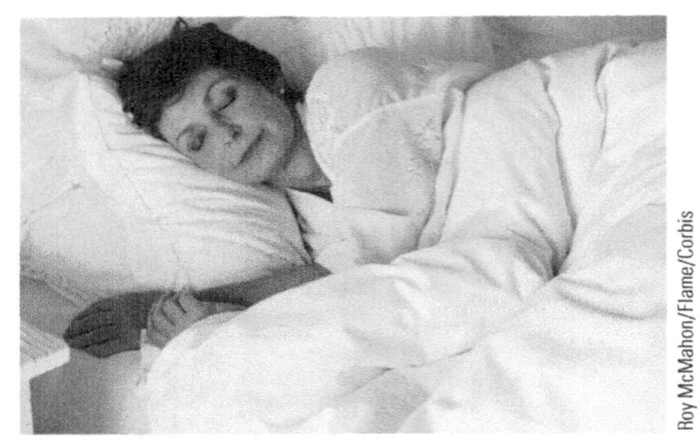

Roy McMahon/Flame/Corbis

basal metabolism

PhotoDisc

physical activity

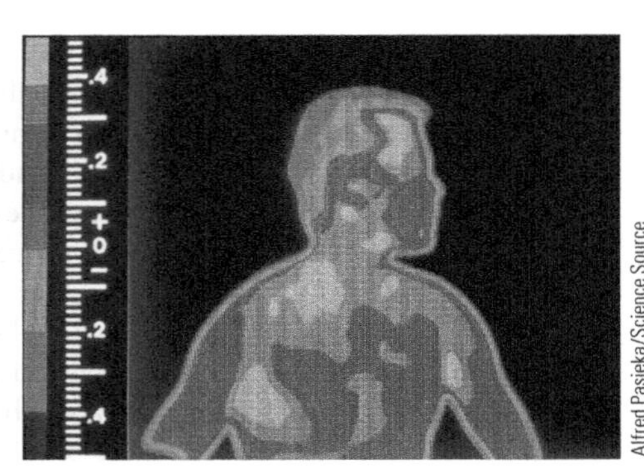

Alfred Pasieka/Science Source

dietary thermogenesis

Illustration 8.3 Examples of the three types of energy-requiring processes in the body.

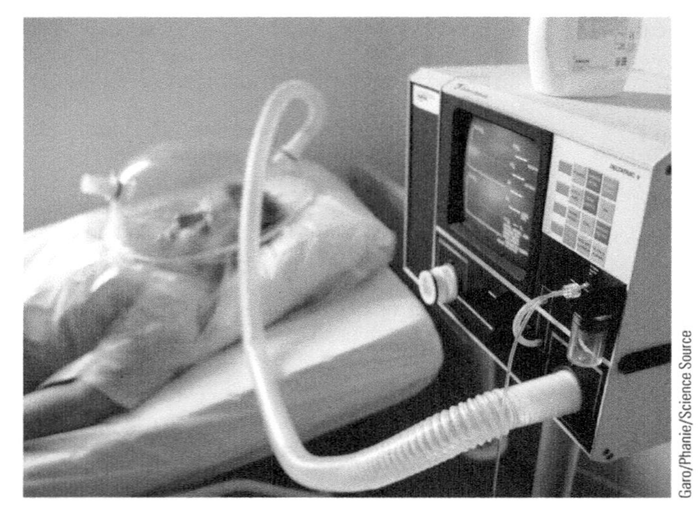

Garo/Phanie/Science Source

Illustration 8.4 An example of an indirect calorimeter used to assess energy expenditure in clinical and research settings.

traits also influence calorie expenditure for basal metabolism to some extent. Consequently, results obtained using this quick formula may be 10 to 20% lower or higher than the true number of calories required for basal metabolism.[6,7]

How Much Energy Do I Expend in Physical Activity? The caloric level needed for physical activity can vary a lot, depending on how active a person is. It usually accounts for the second highest amount of calories we expend. The energy cost of supporting a physically inactive lifestyle (Table 8.2) is about 30% of the number of calories needed for basal metabolism. An "average" activity level requires roughly 50% of the calories needed for basal metabolism, and an "active" level requires approximately 75%.[6] A physically inactive person needing 1,500 calories a day for basal metabolism, for example, would require about 450 calories (1,500 calories $\times$ 0.30 = 450 calories) for physical activity.

People have a tendency to overestimate time spent in physical activity. This in turn tends to overestimate calories needed for physical activity, and depending on the amount of overestimation, can lead to highly inaccurate estimates of total calorie need.[7] Physical activity level should be based on time spent actually engaged in physical activity and should not include time spent getting ready for the activity, standing around, or between activities.

ANSWERS TO REALITY CHECK

Is There Any Such Thing as a Slow Metabolism?

Mel's basal metabolism may be lower than average BMR for his weight, but his metabolism wouldn't be "slow." Calories needed for basal metabolism can be somewhat lower in people with relatively high amounts of body fat relative to muscle mass.[2]

Photodisc

Mel:

Photodisc

Juanita:

Table 8.2 Estimated energy expenditure by usual level of activity[16]

Activity level	Percentage of basal metabolism calories
Inactive. Sitting most of the day; less than two hours of moving about slowly or standing	30
Average. Sitting most of the day; walking or standing two to four hours, but no strenuous activity	50
Active. Physically active four or more hours each day; little sitting or standing; some physically strenuous activities	75

How Many Calories Does Dietary Thermogenesis Take? A portion of the body's energy expenditure is used for digesting foods; absorbing, utilizing, and storing nutrients; and transporting nutrients into cells. Some of the energy involved in such activities escapes as heat. These processes are referred to as dietary thermogenesis. Calories expended for dietary thermogenesis are estimated as 10% of the sum of basal metabolic and usual physical activity calories. For instance, say a person's basal metabolic need is 1,500 calories and 450 calories are required for usual activity: 1,500 calories + 450 calories = 1,950 calories. Calories expended for dietary thermogenesis would equal approximately 10% of the 1,950 calories, or 195 calories.

Adding It All Up Your estimated total daily need for calories is the sum of calories used for basal metabolism, physical activity, and dietary thermogenesis. In the preceding example, total energy needed would be 2,145 calories (1,500 + 450 + 195). A complete example of calculations involved in estimating a person's total calorie need is provided in Table 8.3. Although the caloric level calculated won't be exactly right, it should provide a reasonable estimate of your total caloric need.

Where's the Energy in Foods?

- **Calculate the calorie value of a food from its content of carbohydrate, protein, fat, and alcohol (if present).**

Any food that contains carbohydrates, proteins, or fats (the "energy nutrients") supplies the body with energy. (That *includes* the foods listed in Table 8.4!) Carbohydrates and proteins supply the body with four calories per gram, and fat provides nine calories per

Table 8.3 Summary of calculations for estimating total calorie need of a 130-pound, inactive woman

	Calories
1. Basal metabolism Multiply body weight in pounds by 10. (For men the figure is 11.)	130 × 10 = 1,300
2. Physical activity Multiply basal metabolism calories by 0.30 (for 30%) based on the usual energy expenditure level of "inactive" (see Table 8.2).	1,300 × 0.30 = 390
3. Dietary thermogenesis Add calories needed for basal metabolism and physical activity together: 1,300 + 390 = 1,690 Multiply the result by 0.10 (for 10%).	1,690 × 0.10 = 169
4. Total calorie need Add calories needed for basal metabolism, physical activity, and dietary thermogenesis together. 1,300 + 390 + 169 = 1,859	**Total calorie need = 1,859**

Table 8.4 Foods that *do* have calories

1. A candy bar eaten with a diet soda
2. Celery and grapefruit
3. Hot chocolate, cheesecake, or soft drinks consumed to make you feel better
4. Cookie pieces
5. Foods "taste-tested" during cooking
6. Foods you eat while on the run
7. Foods you eat straight from their original containers (like ice cream from the carton, milk from a jug, or peanut butter from the jar)

Table 8.5 Calorie values of the energy nutrients of alcohol

	Cals/gm
Carbohydrate	4
Protein	4
Fat	9
Alcohol	7

gram. Alcohol also serves as a source of energy. There are seven calories in each gram of alcohol (Table 8.5).[2]

If you enjoy grilling foods on an outdoor barbecue, you have probably observed firsthand the high level of stored energy in fats (Illustration 8.5). Unlike the drippings of low-fat foods such as shrimp or vegetables, drips from high-fat foods cause bursts of flame to shoot up from the grill. The high-energy content of alcohol can be seen in the alcohol-fueled flames that adorn cherries jubilee and bananas Foster.

If you know the carbohydrate, protein, fat, and (if present) alcohol content of a food or beverage, you can calculate how many calories it contains. For example, say a cup of soup contains 15 grams of carbohydrate, 10 grams of protein, and 5 grams of fat. To calculate the caloric value of the soup, multiply the number of grams of carbohydrate and protein by 4 and the number of grams of fat by 9. Then add the results together:

$$\begin{aligned} \text{15 grams carbohydrate} \times \text{4 calories/gram} &= \text{60 calories} \\ \text{10 grams protein} \times \text{4 calories/gram} &= \text{40 calories} \\ \text{5 grams fat} \times \text{9 calories/gram} &= \underline{\text{45 calories}} \\ &\quad\ \text{145 calories} \end{aligned}$$

You can calculate the percentage of total calories from carbohydrate, protein, and fat in the soup by dividing the number of calories supplied by each energy nutrient by the total number of calories in the soup, and then multiplying the results by 100:

- Carbohydrate: $\frac{\text{60 calories}}{\text{145 calories}} = 0.41 \times 100 = 41\%$
- Protein: $\frac{\text{40 calories}}{\text{145 calories}} = 0.28 \times 100 = 28\%$
- Fat: $\frac{\text{45 calories}}{\text{145 calories}} = 0.31 \times 100 = \underline{31\%}$

$$100\%$$

Given this information about the caloric value of carbohydrates, proteins, and fats, which of the items in Illustration 8.6 would you expect to be highest in calories?

College students have been found to miss this question 47% of the time.[8] It's the margarine! Margarine contains the most fat; it is made primarily from oil. High-fat foods provide more calories ounce for ounce than foods that contain primarily carbohydrate or protein. That means bread and potatoes (rich in carbohydrates), catsup (rich in water

Felicia Martinez Photography/PhotoEdit

Illustration 8.5 If you have observed the flames produced by fat dripping from a steak or hamburger on a grill, you have seen the powerhouse of energy stored in fat. The carbohydrate and protein contents of grilled foods don't burn with nearly the same intensity. They have less energy to give.

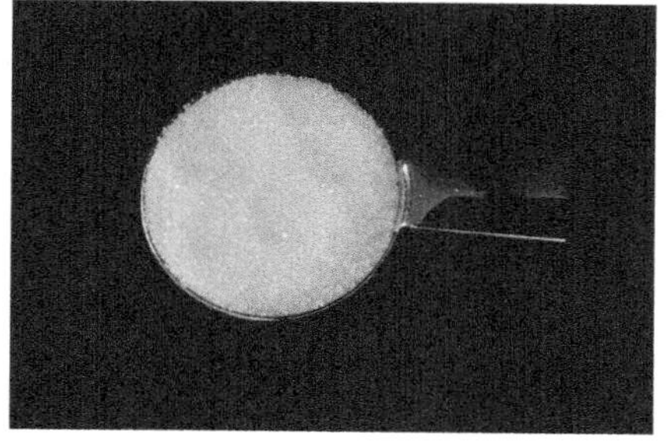

Richard Anderson

Illustration 8.6 Which contains the most calories—a tablespoon of margarine, sugar, or pork? (Calorie values are shown on page 8-8.)

and a low-calorie vegetable), lean meats, and other low-fat or no-fat foods provide fewer calories than equal amounts of foods that contain primarily fat.

Most Foods Are a Mixture

Some foods (such as oil and table sugar) consist almost exclusively of one energy nutrient, but most foods contain carbohydrates, proteins, and fats in varying amounts.

Bread is high in complex carbohydrates ("starch"), but it also contains protein and a small amount of fat. Likewise, steak is not all protein. Although protein constitutes about 32% of the total weight of lean sirloin steak, fats make up about 8%, and the largest single ingredient is water—60% of the weight. Yet we don't think of steak as a "high-water food"; instead, it's often thought of as pure protein.

So, even though some foods provide relatively more carbohydrate, protein, or fat than other foods, most foods contain a mixture of energy nutrients.

Resources for Estimating the Caloric Value of Food Some people say you can estimate the caloric value of food by how it tastes or by how appetizing it looks. Actually, it's not that simple. How many calories do you think are contained in a half cup of peanuts; a boiled, medium-sized potato; or a cup of rice (Illustration 8.7)?

Taste, appearance, and reputation do not make good criteria for determining the caloric value of foods. Here's an example:

I used to order the fish sandwich at fast food restaurants because I was trying to avoid all the calories in hamburgers. Then I found out the fish sandwich had about the same number of calories as the quarter-pound hamburger! Where did I get the idea that fried fish loaded with tartar sauce has fewer calories than a hamburger?

Richard Anderson

Illustration 8.7 What's the caloric value of these foods? One contains 420 calories, another 205, and the third 118 calories. Try matching the caloric values with the foods, and then check your answers on page 8-8.

Answer to Illustration 8.6
Margarine.
There are about 101 calories in one tablespoon of margarine, 46 in a tablespoon of sugar, and 40 in a tablespoon of relatively lean pork.

Answer to Illustration 8.7

	Calories
½ cup of peanuts	420
1 medium boiled potato	118
1 cup of white rice	205

This doesn't have to happen to you! By referring to Appendix A, you can determine that a typical fast food fish sandwich and a quarter-pound hamburger weigh in at about 400 calories each. If you're interested in the caloric value of foods you eat frequently, look them up in this appendix.

Energy Density

The obesity epidemic in the United States is related to increased calorie intake over recent decades.[9] Why are so many people consuming more calories than they need and gaining weight? Part of the answer appears to be related to the **energy density** (or calorie density) of foods that have become a regular part of the U.S. diet.[10,11]

Energy density represents the number of calories per gram of a food item. Twenty grams of potato chips (about 10 chips), for example, has 107 calories. Its energy density equals 107 calories divided by 20 grams, or 5.4. A 202-gram baked potato that provides 212 calories, on the other hand, has an energy density of 1.0 (212 calories divided by 202 grams).

Diets that regularly include energy-dense foods are associated with overeating, excess calorie intake, weight gain, obesity, and type 2 diabetes.[12,13] Diets high in energy-dense foods may interfere with normal food-intake regulation mechanisms by overriding the onset of satiety.[14] In addition, foods high in energy density tend to be nutrient poor, and those of low energy density are often nutrient rich.[15] Because they are not energy dense, nutrient-rich foods can be consumed in higher quantities while keeping calorie intake in check. Regular consumption of nutrient-rich foods that are low in energy density is associated with favorable nutrient intake and reduced weight gain.[13] Vegetables, fruits, whole grains, and lean meats have lower energy densities than foods such as processed meats, fried foods, and high-fat, sweet desserts.

Small Differences in Energy Density Make a Big Difference Rather small differences in the energy density of foods can make a sizable difference to calorie intake. On an equal weight basis, there are 38% more calories in macaroni and cheese (energy density = 1.2 calories/gram) than in ramen noodles (energy density = 0.7 calories/gram). More examples of the energy density of foods are given in the Health Action feature.

How Are Food and Energy Intake Regulated by the Body?

Because energy is critical to survival, the body has a number of mechanisms that encourage regular caloric intake. It has less effective means of discouraging excessive intake of calories.[16]

Mechanisms that encourage food intake don't depend on weight status. Rather, they are keyed to encouraging eating on a regular basis so that food, if available, will be consumed and carry the body through times when food isn't around. Humans developed on a schedule of "feast and famine." Those who could store enough fat to see them through the times when food was scarce had an advantage. So, no matter how thin or fat a person is, she or he experiences **hunger** if food is not consumed several times throughout the day. The "hungry" signal is thought to be sent by a series of complex mechanisms when cells run low on energy nutrients supplied by the last meal or snack.[16]

When we eat, we reach a point when we feel full and are no longer interested in eating. The signal is due to hormones and internal sensors in the brain, stomach, intestines, liver, and fat cells that indicate **satiety**—the feeling that we've had enough to eat.[18] Satiety stems partly from food's reduced reward over time—particular foods no longer taste as good as they did when eating began. People can override a feeling of satiety and continue to eat highly desired foods. There are hedonic (pleasure) mechanisms driven by the flavor and pleasure of food, and these influence **appetite**, food intake, and energy balance. Individuals can continue to eat highly preferred foods past the feeling of fullness.[1] (If you have ever made room for a delicious-looking piece of pie or cake after a big meal,

energy density The number of calories per gram of food. It is calculated by dividing the number of calories in a portion of food by the food's weight in grams. May also be calculated as calories per 100 grams of food.

hunger Unpleasant physical and psychological sensations (weakness, stomach pain, irritability) that lead people to acquire and ingest food.

satiety A feeling of fullness or of having had enough to eat.

appetite The desire to eat; a pleasant sensation that is aroused by thoughts, taste, and enjoyment of food.

health action Moving Toward Foods Lower in Energy Density

EAT MORE. CONSUME FEWER CALORIES!

No, this isn't an ad for a weight-loss product. It's a public service announcement about the benefits of lower energy-dense foods. You *can* improve your nutrient intake while eating more food and reducing calorie intake. The key is to more often select foods that are relatively low in energy density. The list below compares high energy-dense foods (left column) with their lower energy-dense alternatives (right column).

Higher energy-dense food	Calories/g	Lower energy-dense food	Calories/g
Taco shell	4.7	Corn tortilla	2.2
Bologna	3.1	Sliced turkey breast	0.9
Fried chicken	2.8	Grilled chicken	1.7
Fried pork chop	2.8	Broiled pork chop	2.0
Cheeseburger	2.7	Bean burrito	1.9
Hash brown potatoes	2.2	Boiled potato	0.9
Fried fish	2.2	Broiled fish	1.2
Fried rice	1.6	Rice	1.3
Potato salad	1.4	Tossed salad with salad dressing	1.1
Frozen, sweetened strawberries	1.1	Fresh strawberries	0.3

you have experienced this.) Hedonic mechanisms can also signal an early end to eating when the appetite for food diminishes soon after eating begins.[19]

The Question of Energy Balance

- **Describe how the first law of thermodynamics applies to energy balance in humans.**

The first law of thermodynamics applies to energy balance in humans. This law states that energy can be transformed from one form to another, but cannot be created or destroyed. In humans, the amount of energy we consume from food equals the amount of energy the body has available for growth, physical activity, other body functions, and energy stores.[21] Unless you are currently losing or gaining weight, the number of calories you need is the number you usually consume in your diet. Adults who maintain their weight are in a state of energy balance (Illustration 8.8). Because they are not losing weight (using fat and

Illustration 8.8 The body's energy status.

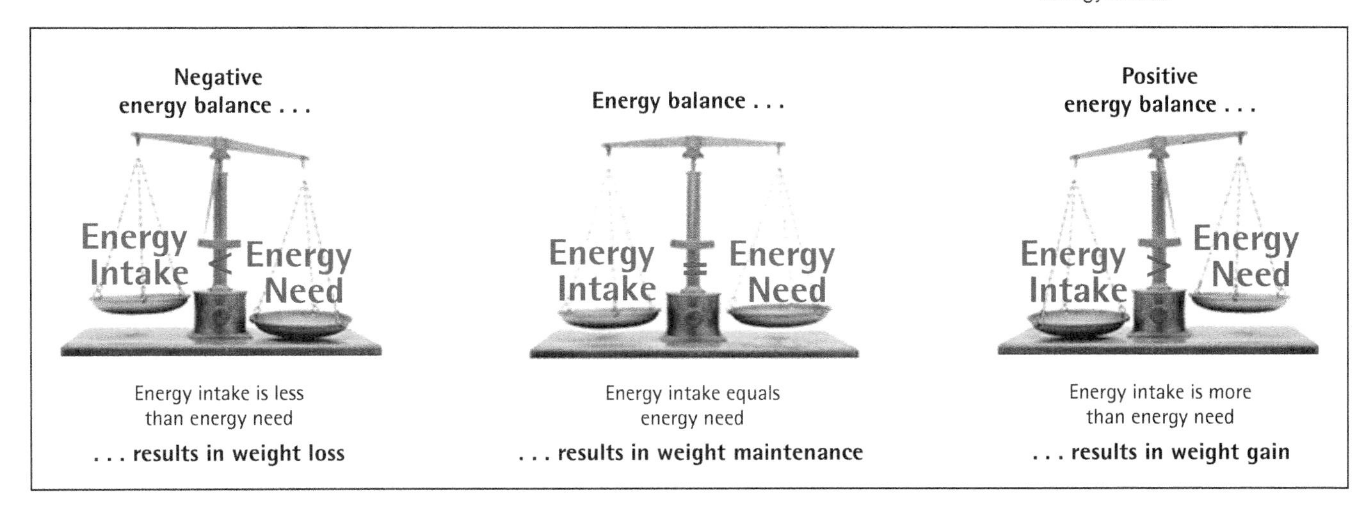

other energy stores) or gaining it (storing energy), their body's expenditure of energy and its intake of energy are balanced.

When energy intake is less than the amount of energy expended, people are in negative energy balance. In this case, energy stores are used and people lose weight. When a positive energy balance exists, weight and fat stores are gained because more energy is available from foods than is needed by the body. Small increases in calorie intake above calorie need may not amount to noticeable gains in body weight on a day-to-day basis, but they can add up to significant weight gain over the course of months or years.

Sometimes a positive energy balance is a healthy and normal circumstance. For example, a positive energy balance is normal when growth is occurring, as in childhood or pregnancy, or when a person is regaining weight lost during an illness.

Keep Calories in Perspective

Calorie is not a word that means fattening or bad for you. Calories represent our source of energy and are a life- and health-sustaining property of food.[20] Diets compatible with health contain a mixture of foods providing various amounts of calories. It's not the caloric content of individual foods that makes them good or bad. It's the sum of calories and nutrients in foods that make up our dietary pattern.

Howard Kingsnorth/The Image Bank/Getty Images

NUTRITION up close

Food as a Source of Calories

Focal Point: Examine the distribution of calories in the foods you eat.

What are the sources of calories in food? Determine the caloric contribution of the fat, carbohydrate, and protein content of the following snack foods using the composition data and calorie conversion factors listed. Then answer the question: Which snack is lowest in total calories?

PhotoDisc PhotoDisc

Calorie conversion factors

1g fat = 9 calories
1g carbohydrate = 4 calories
1g protein = 4 calories

Potato Chips

Serving size: 1 oz (about 20 chips)

Fat: 10g × ____ = ____ calories from fat
Carbohydrate: 15g × ____ = ____ calories from carbohydrate
Protein: 2g × ____ = ____ calories from protein
_____ Total calories

Percent of calories from fat: _____ ÷ __________ = _____ × 100 = _____ %

Mini Pretzels

Serving size: 1 oz (about 17 pieces)

Fat: 10g × ____ = ____ calories from fat
Protein: 3g × ____ = ____ calories from protein
Carbohydrate: 24g × ____ = ____ calories from carbohydrate
_____ Total calories

Percent of calories from fat: _____ ÷ __________ = _____ × 100 = _____ %

Feedback to the Nutrition Up Close is located in Appendix F.

REVIEW QUESTIONS

- **Estimate total calorie need using the formulas presented.**
- **Describe how the first law of thermodynamics applies to energy balance in humans.**

1. An adult's total calorie need can be estimated based on his or her energy expenditure from basal metabolism, physical activity, and dietary thermogenesis. **True/False**
2. Basal metabolic rate varies substantially among people due to differences in genetic traits and disease history. **True/False**
3. George, a city bus driver, weighs 220 pounds. The number of calories he expends for basal metabolism daily would equal about 2,420 calories. **True/False**
4. A common source of error in estimates of physical activity level is underestimation of the amount of time spent in physical activities. **True/False**
5. Dietary thermogenesis accounts for approximately 30% of a person's total calorie need. **True/False**
6. Appetite is related to the desire for food, whereas hunger refers to a physiological need for food. **True/False**
7. A person who is gaining weight is in positive energy balance. **True/False**

The next three questions refer to the following situation:

Sunita lost 10 pounds in the past few months and now weighs 162.

8. _____ What is the approximate difference in Sunita's calorie need for basal metabolism from before to after the 10-pound weight loss?
   a. 50 calories
   b. 100 calories
   c. 150 calories
   d. 200 calories
9. _____ Sunita continues to lose weight gradually until she reaches her target weight of 140 pounds. Her new calorie need for basal metabolism will be _____ calories lower than before her weight loss.
   a. 120 c. 320
   b. 220 d. 420
10. _____ Sunita starts strength training and increases her muscle mass by 10% while she maintains her weight. What happens to her calorie need for basal metabolism now?
    a. It will decrease, then gradually increase.
    b. It will stay the same.
    c. It will increase somewhat.
    d. It will decrease somewhat.
11. _____ You notice a customer buying a 20-ounce bottle of ginger ale in a store and wonder how many calories that represents. Use the information in Appendix A to determine the calorie value of 20 ounces of ginger ale. Your answer is:
    a. 248 calories c. 124 calories
    b. 186 calories d. 207 calories
12. _____ Use Appendix A to identify how many grams of white granulated sugar (table sugar) are in a teaspoon. Your answer is:
    a. 2 grams c. 6 grams
    b. 4 grams d. 8 grams
13. _____ If all of the calories provided by 20 ounces of ginger ale are in the form of table sugar, how many teaspoons of table sugar would be in the ginger ale?
    a. 32.1 c. 12.9
    b. 10.0 d. 51.5

- **Calculate the calorie value of a food from its content of carbohydrate, protein, fat, and alcohol (if present).**

14. A calorie is also referred to as a kilocalorie (kcal). **True/False**
15. Calories are a component of food that provides the body with energy. **True/False**
16. Say you ate a medium serving of french fries that contained 3 grams of protein, 38 grams of carbohydrate, and 14 grams of fat. The calorie value of the french fries would be 290. **True/False**
17. The serving of french fries referred to in question 16 would provide 60% of calories from fat. **True/False**
18. The composition of lean steak is nearly 100% protein. **True/False**

The next two questions refer to this scenario:

You are in your favorite coffee shop and notice the calorie value of the foods offered is listed right below the foods. You focus on the whole wheat English muffin, the oat bran muffin, and the mini cupcake because they all look delicious. The table below shows the calories listed, as well as the serving size and fat content of each food.

	Weight	Calories	Fat, grams
whole wheat English muffin	2 oz (57 grams)	127	1
Oat bran muffin	3 oz (84 grams)	231	6
Mini cupcake	2 oz (57 grams)	180	10

19. ______ Which of these menu items has the lowest energy density?

a. whole wheat English muffin
b. oat bran muffin
c. mini cupcake

20. ______ Which menu item has the highest percentage of total calories from fat?

a. whole wheat English muffin
b. oat bran muffin
c. mini cupcake

Answers to these questions can be found in Appendix F.

NUTRITION SCOREBOARD ANSWERS

1. Carbohydrates provide four calories per gram. Fat, on the other hand, provides nine calories per gram. **False**
2. Butter and margarine contain the same amount of fat, so they provide the same number of calories (35 per teaspoon). **False**
3. This is the conservation of energy principle, the first law of thermodynamics. **True**
4. People can feel they have had enough to eat of a particular food that has lost its flavor attraction, but appetite for another food that tastes differently is preserved.[1] **True**

Obesity to Underweight: The Highs and Lows of Weight Status

Ilene MacDonald/Alamy

NUTRITION SCOREBOARD

1 There is usually little difference between ideal body weights as defined by cultural norms and those defined by science. **True/False**

2 Fat stores located in the stomach area present a greater hazard to health than do fat stores located around the hips and thighs. **True/False**

3 Risk factors for obesity include interactions between genetic factors and an obesogenic environment, family history of obesity, and exaggerated, internal reward responses to appetizing foods. **True/False**

4 Healthy underweight people often find it nearly impossible to gain weight. **True/False**

Answers can be found at the end of the unit.

LEARNING OBJECTIVES

After completing Unit 9 you will be able to:

- Identify weight status based on weight, height, age (where appropriate), and sex using appropriate formulas and growth charts.
- Delineate the potential causes and health consequences of overweight and obesity.
- Delineate the potential causes and health consequences of being underweight.
- Comment on the awareness of the roles of weight bias on psychological and other aspects of the health status of children and adults.

Variations in Body Weight

- **Identify weight status based on weight, height, age (where appropriate), and sex using appropriate formulas and growth charts.**

For humans, one size does not fit all. Human bodies come in a range of sizes, varying from heavy and tall to thin and short. Over recent decades, however, the distribution of body sizes has become lopsided. The proportion of people in the United States and other countries classified as overweight or obese has increased, and rates of underweight are dropping. What lies behind these trends? Why are humans so susceptible to gaining weight?

Obesity appears to represent a weak link in the biological evolution of humans. Body processes that regulate food intake developed more than 40,000 years ago when "feast or famine" cycles were common and all foods were basic. Being underweight was a distinct disadvantage. The more body fat people could store after a feast, the better their chances of surviving the subsequent famines. Consequently, multiple mechanisms that favored food intake and body fat storage developed.[5] These body mechanisms, and other factors related to the modern environment, appear to encourage food intake even when food is constantly available and obesity poses a greater threat to survival than does famine.[6] In the words of Theodore Van Itallie, a noted obesity researcher, in environments with an abundant supply of food and no requirement for vigorous physical activity, "perhaps thin people are the ones who are abnormal."

Circumstances today are very different than they were 40,000 years ago. These differences may go a long way toward explaining why obesity rates are high and increasing in many countries.[7]

KEY NUTRITION CONCEPTS

Evidence addressed in this unit on weight status is related to the key nutrition concepts of:

1. Nutrition Concept #3: Health problems related to nutrition originate within cells.
2. Nutrition Concept #4: Poor nutrition can result from both inadequate and excessive levels of nutrient intake.
3. Nutrition Concept #8: Poor nutrition can influence the development of certain chronic diseases.

How Is Weight Status Defined?

Culture and science define the appropriateness of body size. Culturally defined body weight and shape preferences change with time and in ways that have little to do with health. In the fifteenth century, culturally defined body shape standards in Europe favored thin, muscular men and plump women.[8] In America in the early 1900s, moon-faced and pear-shaped women and muscular men were the rage. It was not until the latter part of the twentieth century that ideal body shape took a turn in Europe and America toward thinness, and at times, to unhealthy levels of extreme thinness. In contrast, the culturally defined ideal weight status in some poor nations is overweight or obese. People with ample fat stores are considered wealthier, higher class, and better able to produce children than thin people, who tend to be perceived as sickly and poor.[8] In the United States and some other developed countries, expectations of a socially accepted weight status tend to be different for women than for men. Women in the United States, for example, are more likely to underreport their body weight than men, who generally don't.[10]

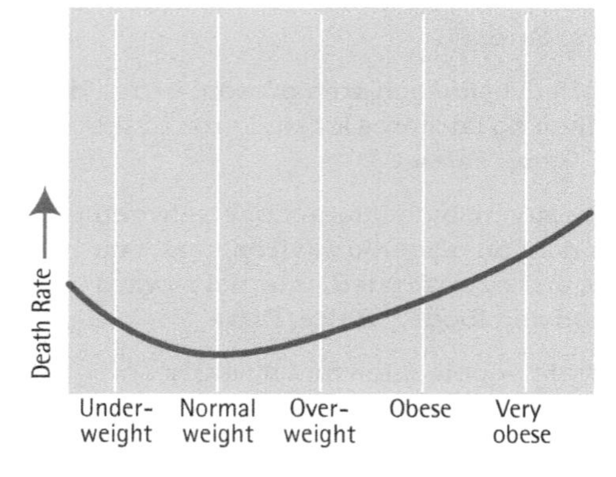

Illustration 9.1 The relationship between body weight status and deaths from all causes for adults. Note: For adults over the age of 65, the lowest mortality rates are associated with BMIs in the overweight range.[11]

Science defines standards for body weight for adults based primarily on the risk of death from all causes. Death rates are highest among adults who have very high body weights for their height; the next highest rates are for underweight adults; and the lowest rates are among adults who are normal weight for height (Illustration 9.1).[9] You can get an idea of what a normal weight for height is by following the steps detailed in the Health Action.

health action Estimating Normal Weight for Height

There is a quick way to estimate within ±10% what is a normal, or healthy, weight for height. It is called the Hamwi[24] method.

Women	Men
Begin with 5 feet equals 100 pounds, and then add 5 pounds for each additional inch of height. Here's an example of how you would estimate a healthy weight for a woman who is 5 feet, 7 inches tall:	For men, 5 feet equals 106 pounds, and each additional inch of height adds on 6 pounds. So, for example, to estimate a healthy weight for a man who is 5 feet 10 inches tall the following way:
5 feet = 100 pounds	5 feet = 106 pounds
7 inches × 5 pounds = 35 pounds	10 inches × 6 pounds = 60 pounds
100 pounds + 35 pounds = 135 pounds	100 pounds + 66 pounds = 166 pounds

In the past, standards used to identify healthy body weights for adults were developed by the insurance industry. Height and weight tables were created to estimate life and death expectancy and the risk of death for adults. These tables have been replaced by standards that employ the **body mass index (BMI)**.

body mass index (BMI) An indicator of the status of a person's weight for his or her height. It is calculated by dividing weight in kilograms by height in meters squared. It can also be calculated using inches and pounds, as shown in Table 9.2.

Body Mass Index

Most commonly referred to as BMI, body mass index is a measure of weight for height that provides a fairly good estimate of body fat content in most people.[12] Ranges of BMI are used to define weights for height that correspond to underweight, normal weight, overweight, and obesity in adults (Table 9.1). BMI has the advantage of being calculated the same way for adult females and males. Calculating BMI involves dividing weight in pounds by height in inches, then dividing this result by height in inches again, and then multiplying that result by 703. An example of BMI calculation is given in Table 9.2. You can use the chart shown in Illustration 9.2 to identify weight status based on BMI.

Table 9.1 Classifying adult weight status by body mass index[13,8]

	Body mass index
Underweight	Under 18.5 kg/m^2
Normal weight	18.5–24.9 kg/m^2
Overweight	25–29.9 kg/m^2
Obese	30 kg/m^2 or higher[a]

[a]Obesity is subdivided for some purposes into the BMI groups of moderate obesity (30–34.9 kg/m^2), severe obesity (35–39.9 kg/m^2), and very severe obesity (40+ kg/m^2).

Assessing Weight Status in Children and Adolescents Standards used to assess weight status in children and adolescents employ BMI percentile ranges for girls and boys (Illustration 9.3). Percentiles of BMI are based on the proportion of children and adolescents who have different levels of BMI at given ages. For example, if a child's BMI is at the 50th percentile for his or her age, then half of children will have BMIs below and half will have values above the 50th percentile. Table 9.3 shows BMI percentile ranges by age that correspond to underweight, healthy weight, overweight, and obesity in children and adolescents.

BMI percentile ranges for children and adolescents do not provide information on growth progress in terms of weight or height for age. Consequently, growth in height cannot be assessed using BMI. Other standards for assessment of weight and height for age in children and adolescents are available at cdc.gov/growthcharts.

Table 9.2 Calculating BMI: An example

Say Chris weighs 140 pounds and is 5 feet, 3 inches tall. To calculate BMI:
Figure out how many inches tall Chris is: 5 feet × 12 inches per foot = 60 inches; 60 inches + 3 inches = 63 inches
Divide Chris's weight by his height in inches: $\frac{140 \text{ pounds}}{63 \text{ inches}} = 2.22$
Divide the result (2.22) by Chris's height in inches again: $\frac{2.22}{63 \text{ inches}} = 0.035$
Multiply this result by 703: 0.035 × 703 = 24.8 This is Chris's BMI.

Overweight and Obesity

- **Delineate the potential causes and health consequences of overweight and obesity.**

Being overweight or obese is the norm for adults in the United States. The combined incidence of overweight and obesity among them is 70.2%.[9] For children and adolescents, the figure is 33.4%.[14,23] Rates of obesity have increased substantially since 1990, and vary substantially by state (Illustration 9.4 on page 9-6). In 2016, rates of obesity were highest in Mississippi (37.3%) and West Virginia (37.7%), and lowest in Colorado (22.3%) and Hawaii (21.8%). Rates of overweight and obesity in the United States have stabilized recently in many states, and rates for children aged 2 to 5 years have declined a bit. The high prevalence of overweight and obesity in the United States and other countries is being matched by increased rates of obesity-related health issues.[7]

Body Mass Index (BMI)

	18	19	20	21	22	23	24	25	26	27	28	29	30	31	32	33	34	35	36	37	38	39	40
Height	Body Weight (pounds)																						
4′10″	86	91	96	100	105	110	115	119	124	129	134	138	143	148	153	158	162	167	172	177	181	186	191
4′11″	89	94	99	104	109	114	119	124	128	133	138	143	148	153	158	163	168	173	178	183	188	193	198
5′0″	92	97	102	107	112	118	123	128	133	138	143	148	153	158	163	168	174	179	184	189	194	199	204
5′1″	95	100	106	111	116	122	127	132	137	143	148	153	158	164	169	174	180	185	190	195	201	206	211
5′2″	98	104	109	115	120	126	131	136	142	147	153	158	164	169	175	180	186	191	196	202	207	213	218
5′3″	102	107	113	118	124	130	135	141	146	152	158	163	169	175	180	186	191	197	203	208	214	220	225
5′4″	105	110	116	122	128	134	140	145	151	157	163	169	174	180	186	192	197	204	209	215	221	227	232
5′5″	108	114	120	126	132	138	144	150	156	162	168	174	180	186	192	198	204	210	216	222	228	234	240
5′6″	112	118	124	130	136	142	148	155	161	167	173	179	186	192	198	204	210	216	223	229	235	241	247
5′7″	115	121	127	134	140	146	153	159	166	172	178	185	191	198	204	211	217	223	230	236	242	249	255
5′8″	118	125	131	138	144	151	158	164	171	177	184	190	197	203	210	216	223	230	236	243	249	256	262
5′9″	122	128	135	142	149	155	162	169	176	182	189	196	203	209	216	223	230	236	243	250	257	263	270
5′10″	126	132	139	146	153	160	167	174	181	188	195	202	209	216	222	229	236	243	250	257	264	271	278
5′11″	129	136	143	150	157	165	172	179	186	193	200	208	215	222	229	236	243	250	257	265	272	279	286
6′0″	132	140	147	154	162	169	177	184	191	199	206	213	221	228	235	242	250	258	265	272	279	287	294
6′1″	136	144	151	159	166	174	182	189	197	204	212	219	227	235	242	250	257	265	272	280	288	295	302
6′2″	141	148	155	163	171	179	186	194	202	210	218	225	233	241	249	256	264	272	280	287	295	303	311
6′3″	144	152	160	168	176	184	192	200	208	216	224	232	240	248	256	264	272	279	287	295	303	311	319
6′4″	148	156	164	172	180	189	197	205	213	221	230	238	246	254	263	271	279	287	295	304	312	320	328
6′5″	151	160	168	176	185	193	202	210	218	227	235	244	252	261	269	277	286	294	303	311	319	328	336
6′6″	155	164	172	181	190	198	207	216	224	233	241	250	259	267	276	284	293	302	310	319	328	336	345
	Underweight (<18.5)	Healthy Weight (18.5-24.9)						Overweight (25-29.9)					Obese (30)										

Find your height along the left-hand column and look across the row until you find the number that is closest to your weight. The number at the top of that column identifies your BMI.

Illustration 9.2 BMI chart.

metabolism The chemical changes that take place in the body. The conversion of glucose to energy or to body fat is an example of a metabolic process.

C-reactive protein (CRP) A key inflammatory factor produced in the liver in response to infection or inflammation. Elevated concentrations of CRP are associated with heart disease, obesity, diabetes, inactivity, infection, smoking, and inadequate antioxidant intake.

subcutaneous fat (sub-q-tain-e-ous) Fat located under the skin.

visceral fat (vis-sir-el) Fat located under the skin and muscle of the abdomen.

chronic inflammation Low-grade inflammation that lasts weeks, months, or years. Inflammation is the first response of the body's immune system to infection or irritation. Inflammation triggers the release of biologically active substances that promote oxidation and other potentially harmful reactions in the body.

The Influence of Obesity on Health People who are obese are more likely to experience diabetes, heart disease, certain types of cancer, hypertension, and other health problems than are people of normal weight (Table 9.4).[7] Many factors are related to increased disease risk in people with obesity. A primary link is represented by the higher incidence of **metabolic** abnormalities such as:

- Hypertension
- Elevated triglycerides, glucose, and/or insulin
- Excess liver fat
- High **C-reactive protein** (a key marker of inflammation)[18–20]

About 53% of normal-weight adults, and 73% of obese adults, have one or more genetic or metabolic abnormality that increases disease risk.[21] Weight loss paired with exercise reduces metabolic abnormalities and disease risk, lessens sleep disturbances and depression, and improves physical mobility and quality of life.[15,22]

Obesity and Psychological Well-Being An important aspect of obesity is the prejudice to which obese people may be subjected. Children who are obese are more likely to suffer unfair or indifferent treatment from teachers and to experience more isolation, rejection, and feelings of inferiority than other children. Obese adults are likely to be discriminated against in hiring and promotion decisions and to be thought of as lazy or lacking in self-control—even by health professionals.[25–27] Society's prejudice against people who don't conform to the cultural ideal of body size may be one of the most injurious aspects of being obese. It is also a preventable part of the reduced health and well-being experienced by many obese people.[28]

Body Fat and Health: Location, Location, Location

It is clear that many of the health problems associated with obesity are directly related to where excess fat is stored. Humans store fat in two major locations: under the skin over the hips, upper arms, and thighs; and in the abdomen. Fat stored under the skin is called **subcutaneous fat**, and that stored in the abdomen under the skin and a layer of muscle is called **visceral fat** (see Illustration 9.5). People who store fat primarily in their hips, upper arms, and thighs are said to have a "pear shape," and those who store fat principally in the abdomen have an "apple shape." These body shapes are shown in Illustration 9.6. You may be better off health-wise if you're a "pear" rather than an "apple."

Visceral fat is much more metabolically active, and more strongly related to disease risk than subcutaneous fat.[29] Metabolic processes initiated by visceral fat can produce a state of **chronic inflammation** that disrupt normal body functions. These disruptions promote the development of **insulin resistance**, **metabolic syndrome**, elevated blood glucose and triglyceride levels, increased liver fat content, high blood pressure, and hardening of the arteries. These changes, in turn, can lead to the development of heart disease, some types of cancer, **type 2 diabetes**, hypertension, **fatty liver disease**, and other disorders.[30,31] Normal-weight and overweight individuals with excessive visceral fat deposits are also at increased risk of metabolic abnormalities and diseases associated with them.[32]

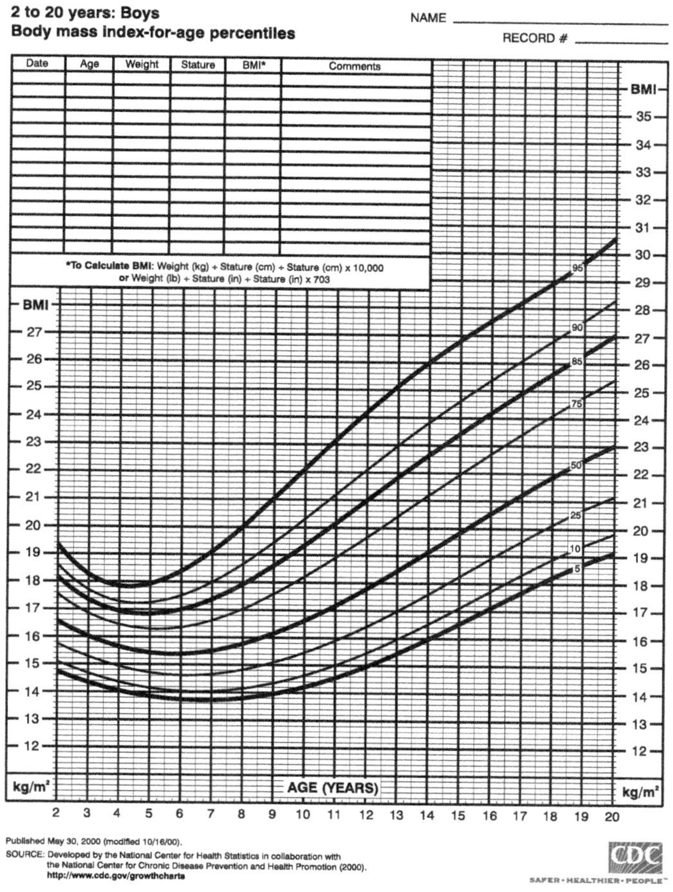

Illustration 9.3 An example of the CDC's BMI charts by age and sex, including examples of their interpretation for a 10-year-old boy. *Copies of all growth charts can be obtained from www.cdc.gov/growthcharts.*

Metabolic abnormalities and disease risk associated with visceral fat can be reduced by regular exercise and improved physical fitness level. Higher levels of health benefits are achieved if both aerobic (walking, jogging, swimming, gardening) and resistance (strength training) exercises are part of the program.[33] Weight loss combined with exercise is the bonus pack for large reductions in risks.[34]

Visceral Fat and Waist Circumference The size of visceral fat deposits can be closely estimated by measuring waist circumference (Illustration 9.7).[35] In men, waist circumferences over 40 inches (102 cm), and in women, over 35 inches (88 cm) are related to excess visceral fat.[36] Waist circumference may not accurately estimate visceral fat content in large, muscular individuals, however.[37]

Because waist circumference is such a strong indicator of disease risk, its measurement in clinical practice is being encouraged.[38] This is particularly true in Japan due to the exaggerated effects of waist circumference on health risk in Asians.[39] Japan is attempting

Table 9.3 Weight status standards for 2- to 19-year-olds based on BMI for age and sex

	Percentile range(s)
Underweight	<5th
Healthy weight	5th–85th
Overweight	85th–95th
Obese	95th or higher

insulin resistance A condition in which cells "resist" the action of insulin in facilitating the passage of glucose into cells.

metabolic syndrome A constellation of metabolic abnormalities generally characterized by insulin resistance, abdominal obesity, high blood pressure and triglyceride levels, low levels of HDL cholesterol, and impaired glucose tolerance. Metabolic syndrome predisposes people to the development of type 2 diabetes, heart disease, hypertension, and other disorders.

type 2 diabetes A disease characterized by high blood glucose levels due to the body's inability to use insulin normally or to produce enough insulin (previously called adult-onset diabetes).

fatty liver disease A reversible condition characterized by fat infiltration of the liver (10% or more by weight). If not corrected, fatty liver disease can produce liver damage and other disorders. The condition is associated primarily with obesity, diabetes, and excess alcohol consumption. The disease is called steatohepatitis when accompanied by inflammation.

Table 9.4 Health problems associated with obesity and those experienced by obese people because they are obese[14–17]

• Chronic inflammation
• Type 2 diabetes
• Hypertension
• Stroke
• Some types of cancer
• Heart disease
• Fatty liver disease
• Decreased life expectancy
• Skin disorders
• Sleep disorders
• Discrimination in social, health care, and work environments
• Depression
• Victimization, being bullied (by school children)
• Lower-quality health care
• Disordered eating behaviors

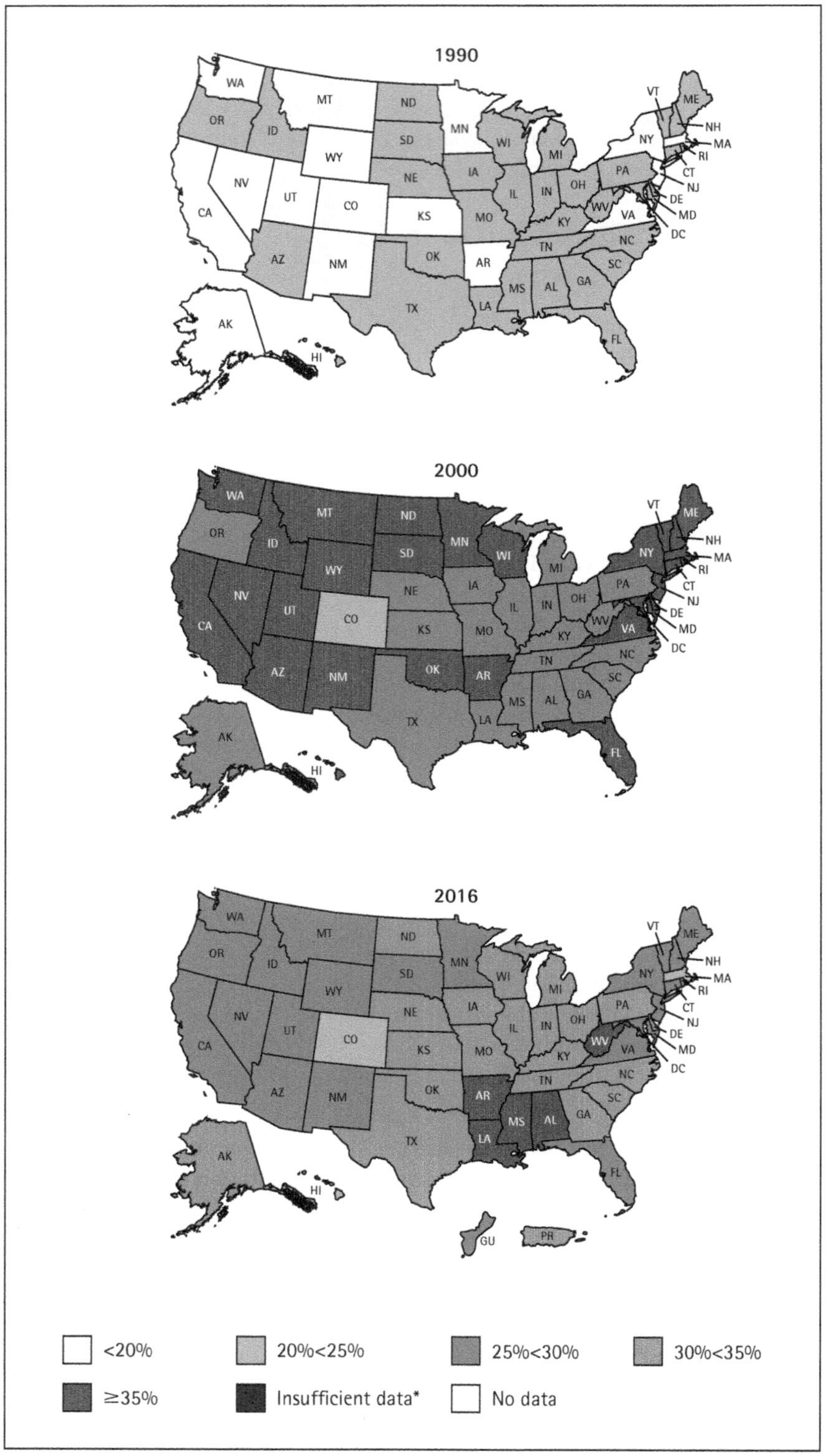

1990

2000

2016

<20% | 20%<25% | 25%<30% | 30%<35%

≥35% | Insufficient data* | No data

*Sample size is insufficient for reliable results.

Illustration 9.4 Percent of obese adults by state 2014-2015, and 2016.[9]

to prevent and control the country's quickly growing rates of diabetes and health care expenditures by population-wide screening of central body fat. The country has instituted a policy that requires all citizens aged 50–74 years (that's 56 million people) to have their waist circumference measured yearly (Illustration 9.8). Individuals with large circumferences and weight-related health problems are given dieting guidance.[40]

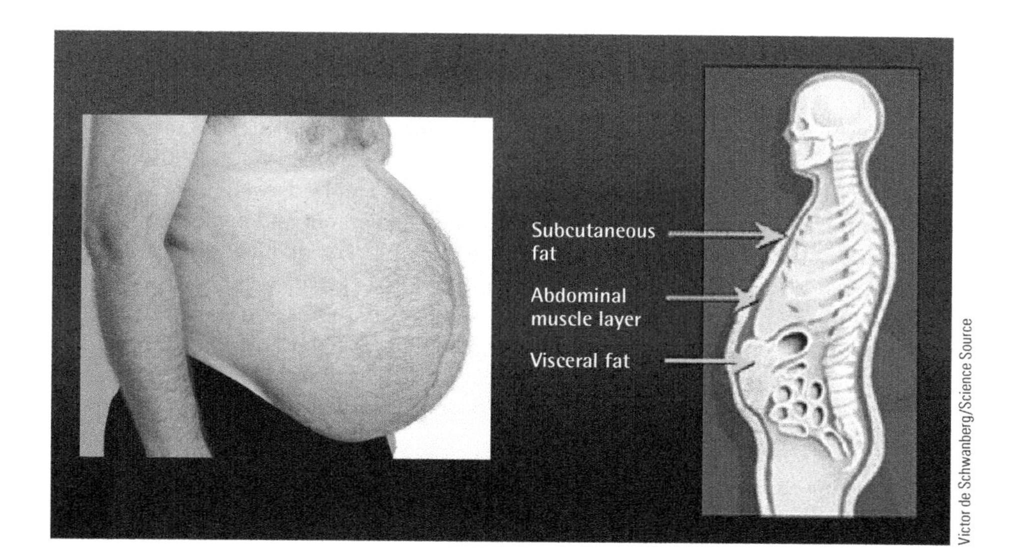

Illustration 9.5 A diagram of the location of subcutaneous and visceral fat.

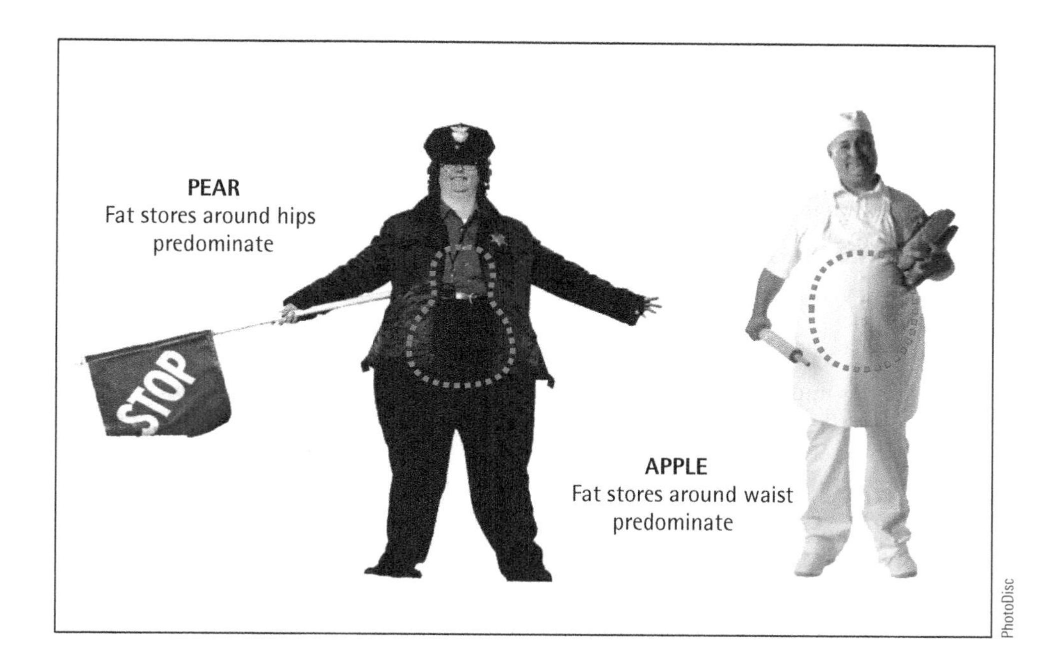

Illustration 9.6 Basic body shapes. *The pear normally has narrow shoulders, a small chest, and an average-size waist. Fat is concentrated in the hips, upper arms, and thighs. Apples are round in the middle. The apple shape is riskier than the pear shape due to the presence of visceral fat.*

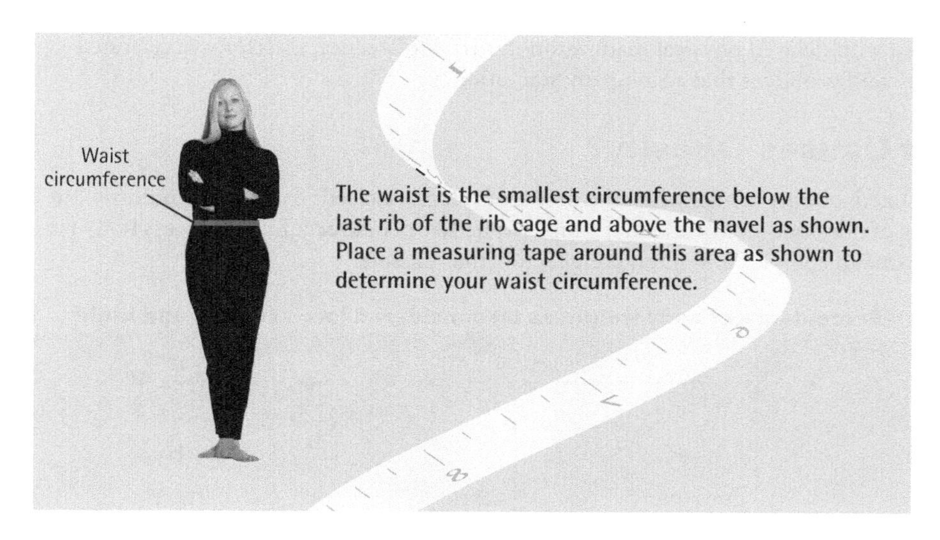

Illustration 9.7 Determining your waist circumference.

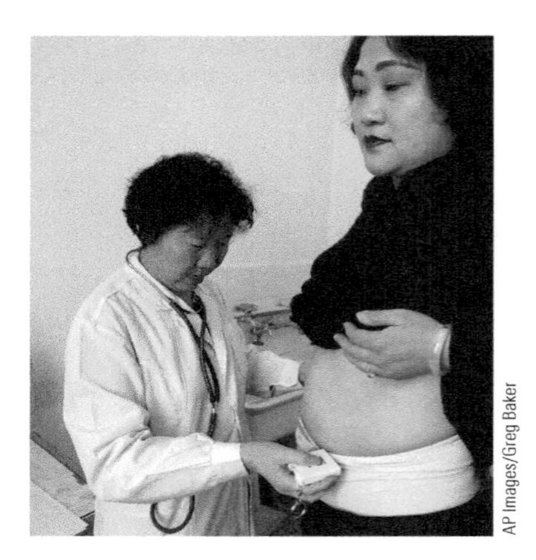

AP Images/Greg Baker

Illustration 9.8 Waist circumference measurements are required annually in Japan for adults ages 50–74 years.

Assessment of Body Fat Content

Body mass index is commonly used to approximate body fat content because the two measures correspond closely in groups of people. However, BMI measures alone cannot be used to quantify fat content or location in individuals.[32] Take a 165-pound, 30-year-old woman who is 5 feet, 6 inches tall and weight trains heavily. Her body weight for height would indicate obesity, but she may have a normal body fat content and be healthy. If BMI were used to assess body fat levels in professional football players, many of them would be wrongly categorized as obese.[37] Sometimes people who are classified as normal weight or underweight by BMI have too much body fat and are at risk of disease because of it.[75] Certain medications make people retain fluid. Their weight for height may qualify them as overweight, but their body fat content may actually be very low. Obviously, measures of body fat content and location are better estimators of health status than are measures related to weight for height.

Methods for Assessing Body Fat Content Accurate and inexpensive tools for assessing body fat content are available, and their use is spreading. Standards for classifying percent body fat, or the percent of weight that consists of fat, have been developed (Table 9.5) and will be refined as additional studies on the relationships between body fat and health risk are conducted.

Here are the most common methods for determining body fat content:

- Skinfold thickness measures
- Bioelectrical impedance analysis (BIA)
- Underwater weighing
- Magnetic resonance imaging (MRI)
- Dual-energy X-ray absorptiometry (DEXA)
- Whole-body air displacement

The theories underlying each of these methods and their advantages and limitations are presented in Table 9.6. The tests are most likely to provide accurate results when they are performed by skilled, experienced technicians using proper, well-maintained equipment, and when the measurements are converted into percent body fat by the appropriate formulas.

Everybody Needs Some Body Fat A certain amount of body fat—3 to 5% for men and 10 to 12% for women—is needed for survival. Body fat serves essential roles in the manufacture of hormones; it's a required component of every cell in the body, and it provides a cushion for internal organs. Fat that serves these purposes is not available for energy formation no matter how low energy reserves become. Low body fat levels are associated with delayed physical maturation during adolescence, infertility, accelerated bone loss, and problems that accompany starvation.[42]

What Causes Obesity?

Simply stated, obesity results when the intake of calories exceeds caloric expenditure. But the cause of obesity is not nearly that simple. Whether people accumulate excess body fat or not is due to complex and interacting factors that include:

Table 9.5 Percentages of body weight as fat considered low, average, and high[41]

	Percent body fat		
	Low	Average	High
Women	Less than 12	32	35 or more
Men	Less than 5	22	25 or more

Table 9.6 Commonly used methods for assessing body fat content

	Skinfold measurement	
Michael Weber/imagebroker/Alamy	Body fat content can be estimated by measuring the thickness of fat folds that lie underneath the skin. Calipers are used to measure the thickness of fat folds, preferably over several sites on the body. Body fat content is estimated by plugging the thicknesses into the appropriate formula.	*Advantages*: Calipers are relatively inexpensive (they cost about $250–$350). The procedure is painless if done correctly and can yield a fairly good estimate of percent body fat. *Limitations*: This measurement is often performed by untrained people. Skinfolds may be difficult to isolate in some individuals.
	Bioelectrical impedance analysis (BIA)	
May/Science Source	Because fat is a poor conductor of electricity and water and muscles are good conductors, body fat content can be estimated by determining how quickly electrical current passes from the ankle to the wrist.	*Advantages*: The equipment required is portable, and the test is easy to do and painless. The results are more accurate than skinfold measures for people who are not at the extremes of weight for height. *Limitations*: Equipment may be expensive; inferior equipment produces poor results. Hydration status and meal ingestion may affect electrical conductivity a bit. May cost $20 or more at fitness centers.
	Underwater weighing	
David Madison/Getty Images	The subject is first weighed on dry land; next he or she is submerged in water and exhales completely; then his or her weight is measured. The less the person weighs under water compared to the weight on dry land, the higher the percent body fat. (Fat, but not muscle or bone, floats in water.)	*Advantages*: If undertaken correctly and if appropriate formulas are used in calculations, this technique gives an accurate value for percent body fat. *Limitations*: The equipment required is expensive and not easily moved. The test doesn't work well for people who don't swim or who are ill or disabled in some way. A body fat assessment costs $50 or more.
	Magnetic resonance imaging (MRI)	
Living Art Enterprises/Science Source Custom Medical Stock Photo	MRI and CT or CAT scans provide similar results. Using this technology, a person's body fat and muscle mass can be photographed from cross-sectional images obtained when the body, or a part of it, is exposed to a magnetic field (or, in the case of CAT scans, to radiation). Based on the volume of fat and muscle observed, total fat and muscle content can be determined. Liver fat content is shown in the MRI scan in the bottom photo on the left.	*Advantages*: Provides highly accurate assessment of fat and muscle mass. *Limitations*: The test is expensive (over $1,000 per assessment) and largely used for clinical and research purposes.

(continued)

Table 9.6 (continued)

	Dual-energy X-ray absorptiometry (DEXA)	
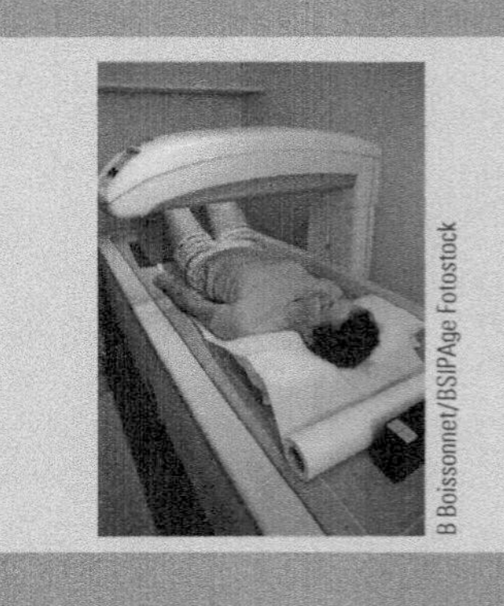 B Boissonnet/BSIP/Age Fotostock	DEXA (or DXA) is based on the principle that various body tissues can be differentiated by the level of X-ray absorption. The measurement is made by scanning the body with a small dose of X-rays (similar to the level of exposure from a transcontinental flight) and then calculating body fat content based on the level of X-ray absorption.	*Advantages*: Provides highly accurate results when measurements are undertaken correctly. DEXA is safe and user friendly for people being measured, and it can also be used to assess bone mineral content and lean tissue mass. *Limitations*: The DEXA machine is expensive; the cost of individual assessments is about $125 per person. The machine must be operated by a trained and certified radiation technologist in many states.
	Whole-body air displacement	
 Courtesy, Life Measurement, Inc. www.bodpod.com	This is an established method that has become practical for broader use due to development of the BOD POD. The method is similar to that of underwater weighing but uses air displacement for determining percent body fat. Individuals sit in an enclosed "cabin" for about five minutes while wearing a tight-fitting swim suit and cap. Computerized sensors determine body weight and the amount of air that is displaced by the body. The PEA POD is used for assessing body fat in infants.	*Advantages*: This method provides a quick, comfortable, automated, and reasonably accurate way to assess body fat content. It is suitable for disabled individuals, the elderly, and children. *Limitations*: Results can be modified by drinking, eating, or exercising before testing; a full bladder or failure to adhere to test procedures may lead to erroneous results. The test is costly but less expensive than DEXA. A BOD POD assessment costs around $50.

- An individual's genetic traits. Obesity tends to run in families due in part to shared types of genes that influence food intake by increasing the feelings of reward and pleasure associated with eating.[7]
- Gene–environment interactions. Obesity is influenced by interactions between gene types and an environment that offers an abundant, inexpensive, and readily available supply of highly palatable foods. People who tend to overeat and gain weight in this environment appear to have an exaggerated reward response to eating energy-dense, sweet, or other appetizing foods.[4,7,43]
- High added sugar intake. High intake of sugar-sweetened beverages has been found to promote the development of obesity in genetically susceptible people.[44]
- Poverty and low educational levels. These two factors are strongly associated with the development of obesity. The increased risk of obesity appears to stem from the limited variety of types of food available in nearby stores, food cost, time constraints, low levels of physical activity, and lack of transportation and related factors.[45,46]
- Lack of sleep. Getting 4–5 hours of sleep per night versus 8–10 hours a night is related to increased food intake and weight.[48]
- Exposure to PCBs (polychlorinated biphenyls). PCBs are used as coolants and lubricants. Exposure to them during pregnancy and early childhood is related to metabolic changes that encourage the development of obesity in children. PCBs are no longer used, but they remain in the environment and can contaminate food supplies.[47]
- Smoking cessation. Smoking cessation is related to appetite and weight gain in most people who quit.[49]
- Medications. A number of commonly prescribed medications—such as insulin, metformin, beta-blockers, sleeping pills, and hormone therapies—are related to weight gain.[50]

- Physical inactivity. Reduced levels of energy expenditure for physical activity at work and at home are part of our obesity-promoting environment. This situation encourages weight gain.

Many factors exist that increase an individual's biological susceptibility to obesity. It is not simply a matter of personal life choices. No study has found that obesity is due to a lack of will power or self-control.

Do Obese Children Become Obese Adults? The link between weight status very early in life and adult obesity appears to be weak or nonexistent. The relationship of early weight status to later obesity becomes meaningful among older children and adolescents, especially if one or both parents are obese.[51] Only 8% of obese children who are heavy at 1 to 2 years of age and who do not have an obese parent are obese as adults. However, nearly 80% of children who are obese between the ages of 10 and 14 and have at least one obese parent have been found to be obese as adults.[52]

Dietary Pattern and Obesity Development Overweight and obesity are prevalent in the United States and elsewhere largely because people are habitually consuming too many calories for their physical activity levels.[53] Large portions of inexpensive, palate-pleasing, high-calorie foods are readily available, whereas vegetables, whole grains, and fruits are more expensive and less readily available.[54,55] We are becoming accustomed to large portion sizes. Adults who consume large portions of food consume 30–50% more food than when presented with smaller food portions.[56]

The occasional fast food meal or bowl of ice cream does not lay the foundation for obesity development; unhealthy dietary patterns do. Dietary patterns least likely to promote obesity are based on plant foods and regularly include vegetables, fruits, legumes, whole grains, nuts and seeds, lean meats and fish, and low-fat dairy products.[74]

Obesity: The Future Lies in Its Prevention

Whether obesity is related to genetic predisposition, environmental factors, or a combination of the two, certain steps can be taken to help prevent it.

Preventing Obesity in Children For children, the prevention of obesity includes the early development of healthy eating and activity habits. Parents should offer a nutritious selection of food, but children themselves should be allowed to decide how much they eat. Physical activities that are fun for every child—not just those who show athletic promise—should be routine in schools and summer programs.[57,58]

Interactions between parents and children around eating and body weight can set the stage for the prevention or promotion of overweight in children. Parents who overreact to a child's weight by focusing on it, restricting food access, and making negative comments to the child may increase the likelihood that eating and weight problems will develop or endure. Lifestyle changes for the whole family—such as incorporating fun physical activities into daily schedules; making a wide assortment of nutritious foods available in the home; and decreasing a focus on eating, foods, and weight—are some of the positive changes families can make to promote healthy eating and exercise habits and normal weight in children.[57]

Preventing Obesity in Adults Action needs to be taken to prevent weight gain during the adult years. Many adults gain weight at a slow pace (about a pound per year) as they age, whereas others gain substantial amounts of weight over short periods of time.[59] Data from a national nutrition and health survey indicate that major gains in weight are most likely to occur in adults between the ages of 25 and 34.[60] Regular exercise may prevent or lessen the amount of weight gain that occurs with age,[61] as may decreased portion sizes at home and in restaurants.[62] Paying attention to the "I'm hungry" and "I'm full" signals can help moderate food intake.[76] For some people, regularly getting eight hours of sleep at night appears to reduce weight gain.[63]

© Jane Holmes/Shutterstock.com

Illustration 9.9 Some underweight people are genetically thin but still healthy.

Changing the Environment Environmental changes in food portion sizes; increased availability of energy-dense, inexpensive foods; and sedentary lifestyles are among the key factors that underlie the current obesity epidemic. From a public health point of view, it is reasoned that if environmental changes got us into the obesity epidemic, then environmental changes will help get us out of it.[65]

Many communities are taking action to promote healthy environments. These actions include limiting access to "junk" foods in schools, building exercise parks, and developing community gardens. Sidewalks, bicycle and walking paths, and nature trails are being planned or added in urban residential areas.[32] Fast food and other restaurants are gradually making changes toward offering smaller portions of energy-dense foods and offering more nutrient-dense items. Some restaurants are adding a wider selection of non-fried entrées and half portions to their menus. The general trend in weight- and health-friendly environmental change is toward providing individuals and families with widespread opportunities for making healthful choices.[66]

Underweight

- **Delineate the potential causes and health consequences of being underweight.**

In contrast to the desperate situations faced by many people in economically emerging countries, underweight in developed nations largely results from illnesses such as HIV/AIDS, pneumonia, cancer, or an eating disorder (anorexia nervosa). An important cause of underweight in some poor nations is poverty.[5,67]

Underweight Defined

People who are underweight have too little body fat—less than 12% body fat in adult females and 5% body fat in males.[68] They have BMIs below 18.5 kg/m^2, as indicated in Table 9.1. A portion of the 2% of U.S. adults classified as underweight by BMI will not actually be underweight, just as a subset of people categorized as obese by BMI are not really obese.

Some people assessed as underweight for height are healthy and have a normal body composition. Like the person in Illustration 9.9, they are probably genetically thin. People who are naturally thin often have as much difficulty gaining weight as obese people have losing it.[5] People who are thin and unhealthy are more likely than others to experience apathy, fatigue, and illnesses frequently, and they take longer to recover from illness. They tend to have reduced bone mineral density and more bone fractures, be intolerant of cold temperatures, and have impaired concentration.[69]

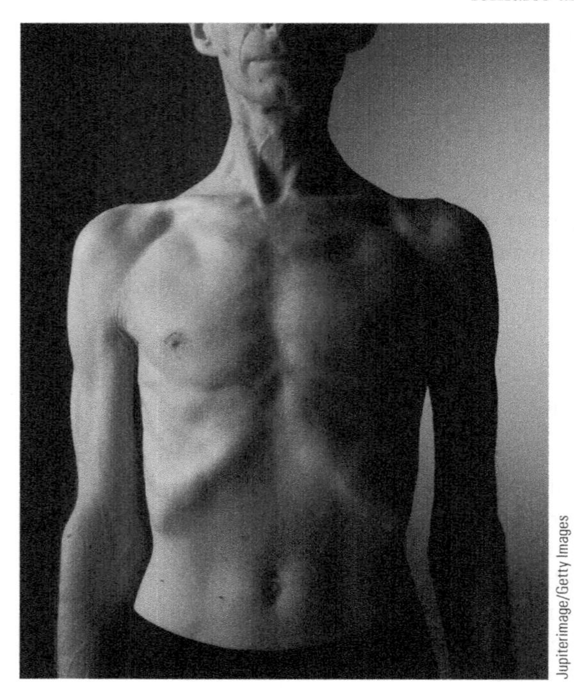

Jupiterimage/Getty Images

Illustration 9.10 The appearance of one man who follows a calorie-restricted diet.

Underweight and Longevity in Adults

Longevity can be extended in adult mice by feeding them a nutritious, calorie-restricted diet that produces underweight.[55] Underfeeding adult monkeys a basic diet does not appear to increase life expectancy; however, it appears to improve metabolic health.[70] Although it is not known whether caloric restriction and underweight would serve as a fountain of youth for humans,[71] there are groups of adults who believe that it will. Devotees of calorie restriction for longer life tightly control their food intake and activity level to maintain a lean body. Calorie restrictors tend to consume 1,600 to 1,700 calories per day, emphasize nutrient-rich foods, and exercise regularly.[72] In general, they appear thin but are healthy (see Illustration 9.10).

Adults following calorie-restricted diets tend not to have chronic inflammation nor the health problems associated with it. However, they are at risk of iron deficiency, osteoporosis, infertility, infections, and of becoming irritable. Life expectancy has been found to be longest, on average, among individuals with BMIs of 20 to 25 kg/m^2. Whether a lower BMI purposeful achieved by maintaining a low-calorie diet enhances life expectancy more than being on the lower end of the normal weight range or by other factors is unclear.[73]

Jim West/Alamy

Illustration 9.11 People come in many different sizes and shapes.

Toward a Realistic View of Body Weight

- **Comment on the awareness of the roles of weight bias on psychological and other aspects of the health status of children and adults.**

A widespread belief among Americans is that individuals can achieve any body weight or shape they desire if they just diet and exercise enough. It's a myth. People naturally come in different weights and shapes, and these can only be modified so much (Illustration 9.11). Half of all women in the United States wear sizes 14 to 26, yet many clothing models are very underweight. Many men, no matter how hard they work out, will never have a washboard stomach or fit into slim jeans.

Size Acceptance

The war is on obesity, not on the obese.[78] The U.S. obsession with body weight and shape is spreading to other industrialized countries as part of popular culture. Ironically, strong social bias against certain body sizes may contribute broadly to weight and health problems. Intolerance of overweight and obese children and adults tends to increase discrimination against them. This type of bias lowers the individual's feeling of self-worth and may promote eating disorders, including the consumption of too much food. Females are hardest hit by negative attitudes about body size. Although the incidence of overweight and obesity in females and males is similar,[14] obesity in females carries with it many more negative stereotypes.[8,23]

Acceptance of people of different sizes and a more realistic view of obtainable body weights and shapes may be two of the most important things society can do to ameliorate the harmful effects of obesity. Some changes are underway. Unproductive attitudes and harmful biases against heavier people by clinical practitioners are being addressed by various institutions. Health professionals are being urged to examine their own biases against individuals of various sizes and shapes by increasing their awareness of their personal biases.[26] They are being asked to think about whether they do any of the following:

- Judge a person's competence or worth by their size.
- Blame people for their size even though many factors are involved and there are no easy or simple solutions.
- Attempt to change their attitudes to become sensitive to the needs of obese people.[28]

The nearby Reality Check feature addresses the issue of weight bias.

The Health at Every Size Program Kinder and more effective approaches to health improvement in obese people have been developed. One, called the "Health at Every Size"

REALITY CHECK

A Weighty Topic

On their way through the park Carlos and Sandra see two kids playing. One is pulling a wagon while the other is riding in it. The child pulling the cart is thin and the child riding it is plump. Here's the thought that flashed through Carlos's and Sandra's heads when they saw the kids:

Answers are below.

Michael Newman/PhotoEdit

Carlos: The heavier kid should be pulling the wagon.

Michael Newman/PhotoEdit

Sandra: I wonder what they're doing. They're probably taking turns pulling the wagon. We used to do that.

program, is gaining acceptance among consumers and health professionals in the United States, Canada, and other countries.[8] Such programs emphasize:

- a focus on physical and mental health outcomes rather than weight
- engaging in enjoyable, life-enhancing physical activities that promote fitness
- use of interventions that include services based on personalized data on the metabolic consequences of genetic makeup, and dietary and physical activity preferences
- a focus on biological and mental health outcomes.[26,27]

Health at Every Size program participation is associated with reduced blood pressure and LDL-cholesterol, increased HDL-cholesterol levels, improved self-esteem, decreased disordered eating behaviors, and improved body image.[26] The Health at Every Size philosophy and program components could help future generations of children, men, and women achieve healthful eating patterns and high levels of well-being and quality of life, as well as improved long-term health.[17,26]

ANSWERS TO REALITY CHECK

A Weighty Topic

You can't know by looking why one child is pulling a wagon while another child rides in it. There are many possible reasons for this besides the children's weight statuses.

Michael Newman/PhotoEdit

Carlos:

Michael Newman/PhotoEdit

Sandra:

Ilene MacDonald/Alamy

NUTRITION up close

Are You an Apple?

Focal Point: Determining your waist circumference.

Follow the directions given in Illustration 9.7 for determining your waist circumference. If you don't have a measuring tape, use a string. Mark the string where the two ends intersect when placed around your waist. Then use a ruler to measure the distance between the end and the mark on the string. Are you an apple?

Feedback to the Nutrition Up Close is located in Appendix F.

REVIEW QUESTIONS

- **Identify weight status based on weight, height, age (where appropriate), and sex using appropriate formulas and growth charts.**

1. Death rates are lowest among adults who remain obese. **True/False**
2. According to the Hamwi method for estimating normal weight in adults, a 6-foot, 1-inch male should weigh approximately 204 pounds. **True/False**
3. A child with a body mass index higher than the 95th percentile on the CDC growth charts is considered obese. **True/False**
4. The prevalence of overweight and obesity in the United States has declined over the past decade. **True/False**
5. Culturally defined standards for ideal body shape and size have not changed through the course of history. **True/False**
6. Rates of obesity are substantially higher for females than for males in the United States. **True/False**
7. ____ A 57-year-old person weighs 176 pounds and is 5 feet, 10 inches tall. The person's BMI would qualify as:
   a. underweight
   b. normal weight
   c. overweight
   d. obese
8. ____ A 6-year-old child's weight was measured and her BMI percentile was plotted on the appropriate growth chart. Her BMI is at the 25th percentile. She would be considered:
   a. underweight
   b. healthy weight
   c. overweight

- **Delineate the potential causes and health consequences of overweight and obesity.**

9. Excess visceral fat poses higher risks to health than excess subcutaneous fat. **True/False**
10. Features of metabolic syndrome can include excess abdominal fat and insulin resistance. **True/False**
11. Women but not men require a minimal amount of body fat stores for health. **True/False**
12. If undertaken correctly, and if appropriate formulas are used, underwater weighing produces accurate values of a person's percent body fat. **True/False**
13. The use of medications that cause weight gain is the leading reason for the obesity epidemic in the United States. **True/False**
14. Vegetables, fruits, and fiber are often consumed in low amounts by people who regularly eat energy-dense, high-calorie foods. **True/False**
15. Adults tend to eat 30–50% more food when given large portions of food than when given smaller portions. **True/False**
16. ____ Which of the following statements about the potential causes of obesity is true?
   a. Obesity is caused primarily by a person's genes.
   b. Obesity is related primarily to lack of self-esteem and control.
   c. Low levels of physical activity are the main cause of obesity.
   d. Obesity is related primarily to genetic predispositions and an obesogenic environment.

17. ____ Which of the following statements does *not* represent an effective approach to the prevention of weight gain in adults?
    a. Consume smaller portion sizes.
    b. Pay attention to satiety signals.
    c. Sleep at least five hours daily.
    d. Get regular physical activity.

- **Delineate the potential causes and health consequences of being underweight.**

18. Calorie-restricted diets have been shown to prolong life in mice, monkeys, and humans. **True/False**
19. Unlike obesity, underweight has been related to specific genetic traits. **True/False**
20. Underweight is associated with an increased risk of illness and delayed recovery from illnesses. **True/False**

- **Comment on the awareness of the roles of weight bias on psychological and other aspects of the health status of children and adults.**

21. Health care professionals intentionally show bias against obese people because this behavior has been shown to encourage people to lose weight. **True/False**
22. People who lack an ideal body shape and size lack the will power to stay on a diet and exercise program. **True/False**
23. A healthy diet and regular exercise help improve the health status of people of all sizes. **True/False**

Answers to these questions can be found in Appendix F.

NUTRITION SCOREBOARD ANSWERS

1. Cultural norms of ideal body weights are often at odds with the "healthy" weights defined by science. **False**
2. Health is affected by the location of body fat stores as well as by obesity.[1] Adults who store body fat in the stomach, or central area of the body, are at higher risk for a number of health problems than are adults who store fat primarily in their hips and thighs. **True**
3. These are some of the factors associated with obesity.[2–4] **True**
4. Gaining weight is as difficult for many underweight people as losing weight is for many overweight people.[5] **True**

Weight Control: The Myths and Realities

Jose Luis Pelaez Inc/Getty Images

NUTRITION SCOREBOARD

1 Anybody who really wants to can lose weight and keep it off. **True/False**

2 Weight loss is the cure for obesity. **True/False**

3 Weight loss can be accomplished using many types of "quick and strict" fad diets. Such diets rarely help people maintain the weight loss in the long run. **True/False**

4 Weight-loss products and services must be shown to be safe and effective before they can be marketed. **True/False**

5 Small and acceptable improvements in eating and exercise behaviors are more likely to produce weight loss and weight-loss maintenance than are large and unpleasant changes in behaviors. **True/False**

Answers can be found at the end of the unit.

LEARNING OBJECTIVES

After completing Unit 10 you will be able to:

- Describe weight-control methods that have been demonstrated to be successful and those that have not.
- Discuss how an effective weight-control plan can be developed based on small and acceptable changes in dietary intake and physical activity level.

Weight Control Nation

- **Describe weight-control methods that have been demonstrated to be successful and those that have not.**

Americans in general are preoccupied with their weight. On any given day, 57% of women and 40% of men are trying to lose weight.[3] To help get the weight off, Americans spend about *$20 billion* annually on weight-loss products and services.[4] Yet many Americans are gaining weight faster than they are losing it. Consumers are paying handsomely for weight loss without experiencing the desired cosmetic changes or health benefits that come when weight loss is maintained.

Why do so many Americans fail at weight control? For some people, achieving permanent weight reduction on their own may be truly impossible. For others, the problem is the ineffective methods employed, not the people who use them.

KEY NUTRITION CONCEPTS

Information presented in this unit on weight control closely relates to these three key nutrition concepts:

1. Nutrition Concept #2: Foods provide energy (calories), nutrients, and other substances needed for growth and health.
2. Nutrition Concept #8: Poor nutrition can influence the development of certain chronic diseases.
3. Nutrition Concept #9: Adequacy, variety, and balance are key characteristics of healthful diets.

Weight Loss versus Weight Control

Dietary torture and quick weight loss are not the cure for overweight. Any popularized approach to quick weight loss (and some get pretty spectacular) that calls for a reduction in caloric intake can produce weight loss in the short run. These methods fail in the long run, however, because they become too unpleasant. Feelings of hunger and deprivation that tend to occur while on a rapid weight-loss diet can lead to a return to previous habits and weight regain. Humans are creatures of pleasure and not pain. Any painful approach to weight control is bound to fail. Individualized, enjoyable, and sustainable eating and exercise experiences are needed to keep excess weight off.[6] Quick weight-loss approaches don't promote sustainable lifestyle changes required for weight control for the long term.[5,6]

Steve Vidler/Alamy

Illustration 10.1 Weight-loss books and products.

The Business of Weight Loss Thousands of weight-loss products and services are available. Some of these are shown in Illustration 10.1. Most of them either don't work at all or don't prevent weight regain. The demand for such products and services is so great, however, that many are successful—financially.

One reason so many weight-loss products and services are available is that almost none of them work. If any widely advertised approach helped people lose weight and keep it off, manufacturers of bogus methods would go out of business. The weight-loss industry also thrives because of the social pressure to be thin. Many people try new weight-loss methods even though they sound strange or too good to be true. Often people believe that a product or service must be effective or it wouldn't be allowed on the market. Although reasonable, this belief is incorrect.

Table 10.1 A brief history of discontinued weight-loss methods[9–13]

Year	Method	Reasons for discontinuation
1940s–1960s	Amphetamines	Highly addictive, heart and blood pressure problems
	Vibrating machines	Did not work
1950s	Jejunoileal; bypass surgery	Often caused chronic diarrhea, vitamin and mineral deficiencies, kidney stones, liver failure, arthritis
	Tapeworms	Ingestion of tapeworms can eventually lead to malnutrition and poor health, cyst formation in various tissues, and death
1960s	Liquid protein diet	Poor-quality protein caused heart failure, deaths
1980s	Intestinal bypass surgery	Excessive risk of serious health problems
1990	Oprah Winfrey liquid diet success (she lost 67 pounds)	
1991	Oprah Winfrey gains 67 pounds and declares "No more diets!"	
1997	Phen-Fen (Redux)	Heart valve defects, hypertension in lung vessels
2004	Ephedra	Excessive risk of stroke, heart attack, and psychiatric illnesses
2010	Sibutramine (Merida)	Excessive risk of stroke and heart attack
2014	Fat shaming	Blaming people for obesity does not lead to weight loss and may encourage weight gain

The Lack of Consumer Protection The truth is that general societal standards for consumer protection do not apply to the weight-loss industry.[7] No laws require products to be effective; in most cases, companies do not have to show that weight-loss products or services are safe and actually work before they can be sold. That's why products such as herbal remedies, weight-loss skin patches, blue-tinted glasses, cellulite-dissolving creams, green coffee bean extract, and mud and plastic wraps are available for sale. These, like many other weight-loss products, do not work, and there is no reason why they should. Promotions for the products just have to make it *seem* as though they might work wonders on appetite or body fat.

Furthermore, weight-loss products and services are usually not tested for safety before they reach the market. The history of the weight-loss industry is littered with abject failures: fiber pills that can cause obstructions in the digestive tract, very low-calorie diets and intestinal bypass surgeries that may lead to nutrient-deficiency diseases and serious health problems, amphetamines that can produce physical addiction, diet pills that have caused heart valve problems, and liquid protein diets that have led to heart problems and death from heart failure.[8] A brief history of weight-loss method failures is chronicled in Table 10.1.

Pulling the Rug Out Fraudulent weight-loss products and services can be investigated and taken off the market. Currently, the Federal Trade Commission (FTC) monitors deceptive practices and weight-loss claims on a case-by-case basis. The FTC has filed suits against companies that make false or exaggerated claims and has identified the most common dubious claims made by the industry (Table 10.2).[13] The FTC performs most investigations in response to consumer complaints; the investigations are not proactive.[14] Illustration 10.2 shows three examples of products that were taken off the market and explains why.

Requirements for truth in labeling keep many bogus products from including false or misleading information on their labels. These laws, however, do not keep outrageous claims from being made in infomercials, on the Internet, or printed in newspapers, books, magazines, and advertisements.[16]

Table 10.2 The top eight features of weight-loss ads that make false or misleading claims[15]

1. Use testimonials, before-and-after photos.
2. Promise rapid weight loss.
3. Require no special diet or exercise.
4. Guarantee long-term weight loss.
5. Include a "clinically proven," "guaranteed," "scientific breakthrough," or "doctor-approved" statement.
6. Make a "safe," "natural," or "easy" claim.
7. Claim that use of a wearable product or lotion rubbed into the skin will produce rapid weight loss.
8. Causes a weight loss of 2 or more pounds a week without dieting.

Fat magnet pills were purported to break into thousands of magnetic particles once swallowed. When loaded with fat, the particles simply flushed themselves out of the body. The FTC found the product too hard to swallow. The company took the product off the market and made $750,000 available for customer refunds. © Scott Goodwin Photography

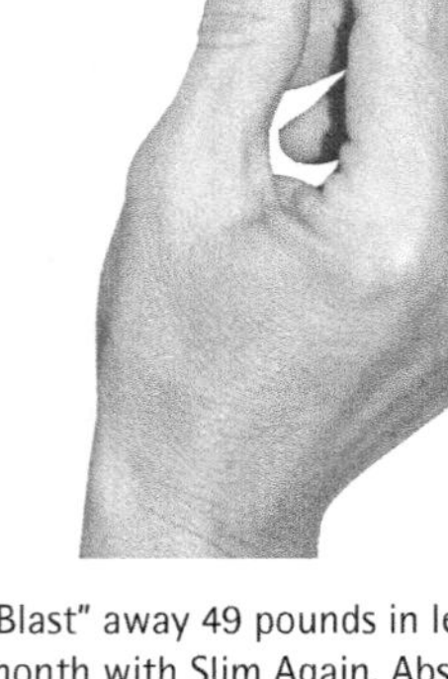

"Blast" away 49 pounds in less than a month with Slim Again, Absorbit-All, and Absorbit-AllPlus pills, claimed ads in magazines, newspapers, and on the Internet. The company got blasted by the FTC to the tune of $8 million for making fraudulent claims. Stan Maddock

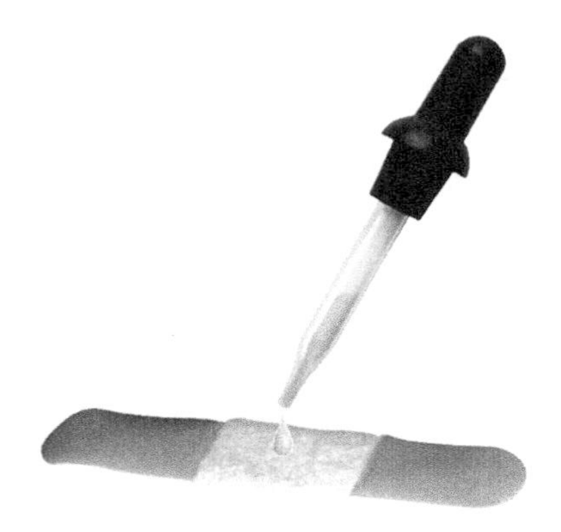

Take a few drops of herbal liquid, put them on a bandage patch, and voilà—a new weight-loss product! The herbal liquid was supposed to reach the appetite center of the brain and turn the appetite off. Federal marshals weren't impressed. They seized $22 million worth of patch kits and banned the sale of others. © Scott Goodwin Photography

Illustration 10.2 Examples of bogus weight-loss products.

What if the Truth Had to Be Told? Suppose the weight-loss industry had to inform consumers about the results of scientific tests of the effectiveness and safety of weight-loss products and services. What if this information had to be routinely included on weight-loss product labels (Illustration 10.3)? What do you think would happen to consumer choices and the weight-loss industry? Increased federal enforcement of truth-in-advertising laws may help put an element of honesty into promotions for many weight-loss products. In addition, the FTC has proposed that claims about long-term weight loss be based on the experience of patients followed over time after they complete the program.[17] If this proposal ever becomes law, it will change the weight-loss industry in the United States.

© Scott Goodwin Photography

Illustration 10.3 Just imagine what would happen if weight-loss approaches were required to divulge their effectiveness and risks.

Weight Control

- **Discuss how an effective weight-control plan can be developed based on small and acceptable changes in dietary intake and physical activity level.**

People usually begin weight-loss efforts with the goal of losing weight and maintaining the new, lower body weight forever. What programs help individuals do that?

The only way to know whether a weight-control method, drug, product, or surgery works is to test its safety and effectiveness in scientific studies. Weight-control methods found to be safe and that lead to an average of 10% loss of initial body weight over a one-year period are identified as being successful.[18] Translated into pounds of weight loss, this means that a group starting a weight-loss program with an average weight of 200 pounds would need to lose an average of 20 pounds or more in a year to obtain a 10% or greater average weight loss. Weight-control approaches identified as successful do not guarantee that everyone who uses the approach will lose 10% or more of their body weight and keep it off for two years, but about 20% will.[18] Effective weight control programs can also benefit people by decreasing or eliminating additional weight gain.[19]

Loss of 5% or more of body weight in obese individuals is considered clinically significant. Losses of this magnitude are associated with lowered blood concentrations of glucose, insulin, triglycerides, cholesterol, and markers of inflammation. They are also associated with reduced blood

Table 10.3 Examples of fad diets

1. *The Popcorn Diet.* Like popcorn? You'll get to eat all the unbuttered, unsalted popcorn on this diet you want. The diet also emphasizes fruits, vegetables, smaller portions, and exercise.
2. *The Grapefruit Diet.* This diet is based on the myth that, when combined with protein, grapefruit triggers fat burning and weight loss. The high grapefruit–high protein diet has been around since the 1930s.
3. *The Chocolate Diet.* A weight-loss diet for chocolate addicts, this nonsense diet categorizes chocolate lovers into types and provides a diet plan for each type. Includes a liquid chocolate diet shake.
4. *The Metabolism Diet.* As the name implies, this diet purports to produce weight loss by speeding up metabolism. The low-carbohydrate, low-calorie diet recommended can only be used for seven days.
5. *The Three-Day Diet.* If you follow the food prescriptions for what to eat, how much to eat, and when to eat, you'll supposedly be rewarded with ramped-up metabolism and a 10-pound weight loss in three days. But wait, there's more. The diet also provides internal cleansing, cholesterol reduction, and more energy.
6. *The Cabbage Soup Diet.* It's a seven-day weight-loss plan based on cabbage soup with other vegetables.
7. *Japanese Morning Banana Diet.* You get to eat all the bananas you want on this diet—and nothing else.
8. *The Cleanse Diet.* This diet targets people who believe their bodies contain "toxins" that slow down metabolism. Liquid cleanses made with hot water, lemon juice, cayenne pepper, maple syrup, or pureed vegetables can clean you out and increase your energy.

pressure, risk of type 2 diabetes, fatty liver disease, and heart disease. The magnitude of risk reduction increases as body weight becomes closer to normal weight.[20,21]

Weight-Control Methods: The Evidence

Identifying solutions to obesity is a major health care priority in the United States. A large part of the reason for a high level of interest in effective weight control programs stems from the fact that obesity is now classified as a disease. Disease management requires evidence that a procedure or practice be demonstrated to be effective by scientific studies in order to be reimbursed by health care insurance.[22,23] Given this reality, the number of research projects aimed at identifying weight-control approaches that work or don't work is expanding. Much more is known about successful and unsuccessful weight control methods now than in the past.

Fad diets, such as those listed in Table 10.3, often promise quick weight loss if a person follows strict rules about the types and amounts of food to eat. They all share the property of being limited in calories and often provide inadequate amounts of certain nutrients.[27]

REALITY CHECK

Do Some Foods Have Negative Calories?

You're looking for a way to shed the 10 pounds you gained during your freshman year and are seriously thinking about adding grapefruit and vinegar to your diet. You have heard these foods have "negative calories" because they make the body burn fat.

Answers on page 10-6.

PhotoDisc

Natalie: I know vinegar cleans the grease off windows. Maybe it will melt away my fat too.

PhotoDisc

Chuck: My grandfather lives in Florida and eats a lot of grapefruit. He's as thin as a rail.

ANSWERS TO REALITY CHECK

Do Some Foods Have Negative Calories?

The negative calorie myth surrounding vinegar or grapefruit lives on because it strikes many people as reasonable. Neither vinegar nor grapefruit, nor any other food, has negative calories or causes the body to gear up metabolism and burn fat.[25]

PhotoDisc

PhotoDisc

Natalie: 

Chuck: 

Quick and strict weight loss approaches used in fad diets have given "dieting" a bad name. Almost everyone gains weight after they go off a fad diet and many blame themselves for the failure even though it was due to the weight-loss approach and not them.[24]

Fad diets have a very slim chance of working in the long run, primarily because they don't change behavior in acceptable ways. Although any approach that calls for calorie restriction can lead to weight loss in the short run,[24] that is not the measure of a successful weight-control method. Successful measures help people keep the weight off.[18]

Popular Diets A number of popular diets have been tested by research studies. These include the Atkins, Weight Watchers, South Beach, Zone, Ornish, and Rosemary Conley diets. The diets tend to vary in protein, carbohydrate, and fat content and allow for individual food choices with certain groups. They also include calorie intake, behavioral, and exercise goals to various extents. The Atkins, South Beach, and Zone diets limit carbohydrate intake and are relatively high in protein, for example, whereas the Rosemary Conley and Ornish diets are low in fat. Individuals who adhere to these diets tend to achieve modest sustained weight loss and an improvement in metabolic risk factors.[31,32] Some individuals experience greater success with weight-control diets that are high in protein, whereas others find low-fat diets easier to follow in the long run.[33] Overall, higher-protein, lower-carbohydrate diets tend to achieve slightly higher weight loss amounts, likely due to the higher acceptability of, and longer adherence to, higher-protein diets.[34] People who adhere to these diets for a year lose an average of about 7–8 pounds (3.2–3.6 kg).[35] Most of the people in the studies were unable to stick to the particular diet for a year, primarily because it was too hard to follow.[26]

Federal Trade Commission/www.ftc.gov/bcp/edu/pubs/business/adv/bus60.pdf

Illustration 10.4 The Internet is home to hundreds of bogus weight-loss sites. File a complaint at www.ftc.gov/bcp/edu/microsites/redflag/beyond.html.

Internet Weight-Loss Products Weight-loss products sold over the Internet are particularly untrustworthy. Other than an Food and Drug Administration (FDA)-approved weight-loss drug (orlistat, also known as Alli or Xenical) that can be purchased without a prescription, there is no dietary supplement or medication sold over the Internet that is currently approved by the FDA as safe or effective for weight loss.[28,29] Some of the products sold as orlistat over the Internet have been found to contain another, potentially dangerous drug.[29] A number of the fraudulent products advertised on the Internet are also available in stores.[30]

Efforts are underway to limit the availability of bogus weight-loss products. The FTC issues red flag alerts for bogus weight-loss products advertised over the Internet and is requesting consumer help in identifying such products (Illustration 10.4).

Successful Weight-Control Programs

Two approaches to weight control are related to weight loss and weight-loss maintenance. One approach involves lifestyle-based programs; the other is weight-loss surgery. Prescription diet drugs may be used to facilitate weight loss. It is difficult to evaluate their effects on weight loss fully because it is required that they be used with diet, exercise, and behavioral counseling.[28]

Lifestyle Programs Weight-control programs based on individually tailored, sustainable lifestyle changes that focus on behavioral strategies for reducing calorie intake and increasing physical activity have been demonstrated to be successful.[21] Client-centered lifestyle programs generally offer positive and supportive counseling that encourages and motivates participants to identify and adopt small and acceptable behavioral changes related to healthy dietary and physically active patterns.[6,26] An example of such a program is one that offered monthly lifestyle coaching sessions with goal setting, behavior change strategy development, and follow-up sessions to evaluate and fine-tune personal approaches.[37]

Studies of client-centered lifestyle programs for weight control show that they lead to an average loss of 13.8 pounds (6.8 kg) over 18 months.[36] For many participants, the weight loss achieved is sufficient to improve blood cholesterol, glucose, and blood pressure levels, and to decrease inflammatory markers.[31]

Physical Activity and Weight Control The use of weight-control programs that include both dietary and physical activity components leads to greater weight loss than if only one is included.[36] Regular physical activity promotes health by decreasing abdominal fat stores, improving blood cholesterol and glucose levels, and decreasing blood pressure. A regular program of strength building increases lean muscle mass and reduces fat mass without weight loss.[40] Regular exercise fine-tunes appetite regulation mechanisms in some people.[27]

Recommendations for physical activity in overweight and obese adults for the prevention of additional weight gain, weight loss, and maintenance of weight loss have been developed by the American College for Sports Medicine.[38] These recommendations are summarized in Table 10.4. The minutes of physical activity listed in the table refer to moderate-intensity activities such as brisk walking, soccer, jogging, tennis, calisthenics, rope skipping, and aerobic dancing. Low-intensity physical activities are also related to improvements in health risks and reductions in weight gain, but to a lesser extent than are moderate-intensity activities.[39] Although uncommon, some people have health problems requiring medical supervision of any exercise program. Individuals with health concerns should get an "all clear" from their health care provider before undertaking a higher level of physical activity than usual.

Diet Pills Four types of diet pills are currently approved for use in the United States. Three of these have yet to be widely incorporated into clinical practice. The other is Orlistat, also know by the names Alli and Xenical.[28,41]

Orlistat works by partially blocking fat absorption in the intestines. The primary side effect associated with the use of orlistat is oily stools, which can be particularly bothersome if high-fat meals are consumed. The malabsorption of fat caused by orlistat also

Table 10.4 American College of Sports Medicine's recommendation for moderate physical activity and weight management in overweight and obese adults[40]

Weight outcome for most people	Average number of minutes per day of moderate physical activity
Prevention of weight gain	21 minutes or more
Weight loss	32 to 60 minutes
Weight-loss maintenance	29 to 43 minutes

reduces absorption of a number of fat-soluble nutrients such as vitamin D, vitamin E, and beta-carotene.[43]

Like all approved weight-loss drugs, orlistat is intended for use in conjunction with a reduced-calorie diet, exercise, and behavioral counseling. Use of orlistat as part of a weight-reduction program is related to a loss of five pounds, up to a total of about 10% of initial body weight.[42] Along with a reduced-calorie diet, the over-the-counter version of orlistat is associated with weight loss of up to 5% of initial body weight.[44] Weight regain after pill use stops is common.[43,45]

Many other diet drugs are under development or are being tested. Much activity currently centers around medications that decrease appetite and increase satiety. Even though people using prescribed diet drugs are carefully screened and monitored by medical professionals, serious side effects can develop. No diet drug is absolutely safe, and none is known to cure obesity forever.[41]

Obesity Surgery

bariatrics The field of medicine concerned with weight loss.

Weight-loss surgery may be needed by individuals whose disease risks do not improve enough with lifestyle interventions. This type of surgery is called **bariatric** surgery. Candidates for the surgery must be assessed as to their understanding of the surgery and the lifestyle changes required after surgery. Bariatric surgery usually leads to weight losses in the range of 75 to 90 pounds, and most of the weight loss is maintained over the years. It is the most effective method of weight control available.[46] A beneficial side effect of weight-loss surgery is a reduction in appetite due to surgery-induced changes in appetite mechanisms located in the stomach.[47]

A number of bariatric surgical techniques are used clinically, but the two most common techniques by far are gastric bypass surgery and adjustable gastric banding (lap-band).[47] Gastric bypass surgery is not reversible, but the lap-band procedure is.[50] The use of surgery for weight loss and weight-loss maintenance is expanding worldwide.[48]

Gastric Bypass Surgery Gastric bypass surgery is the most effective method for weight loss and weight maintenance available. On average, individuals undergoing this surgery lose 60% of their excess body weight (the amount of weight they would need to lose to achieve normal weight), and they often maintain much of the loss over the long term.[49] Gastric bypass surgery is approved for use by the FDA for adults with BMIs over 40 kg/m^2 and for others with BMIs over 35 who have diabetes or other serious health problems related to obesity.[50] Although a higher-risk procedure, gastric bypass surgery is associated with greater levels of weight loss than the less-risky lap-band procedure.[51] Health status generally improves dramatically as a result of the weight loss. Resolution of type 2 diabetes, hypertension, sleep disorders, and elevated LDL cholesterol blood levels often follow gastric bypass surgery.[49] The surgery comes at a high cost ($14,000 to $17,000) and may be accompanied by complications.[51] There is a relatively low (0.04 to 0.38%) chance of death from the operation.[52] People who undergo the surgery must be committed to long-term dietary and other lifestyle changes and follow-up care.[55]

Lap-Band Surgery Adults with BMIs over 35, or with BMIs of 30–35 and an obesity-related health problem, can qualify for lap-band surgery.[53] Lap-band surgery is performed laparoscopically by inserting a tube through small incisions made in the abdomen. The surgery produces a small stomach pouch by constricting the upper part of the stomach with a band (Illustration 10.5). The band can be adjusted, or inflated by the injection of saline water to control the amount of food allowed to enter the stomach. Individuals receiving this surgery tend to lose less weight (48% of excess body weight on average) than people having gastric bypass surgery.[54]

Concerns Related to Bariatric Surgery Some of the complications arising from bariatric surgery are related to nutrient deficiencies. The small stomach and changes in absorption that result from the surgery lead to malabsorption of a number of vitamins and minerals, most notably vitamin D, vitamin B_{12}, folate, calcium, and iron. Multivitamin

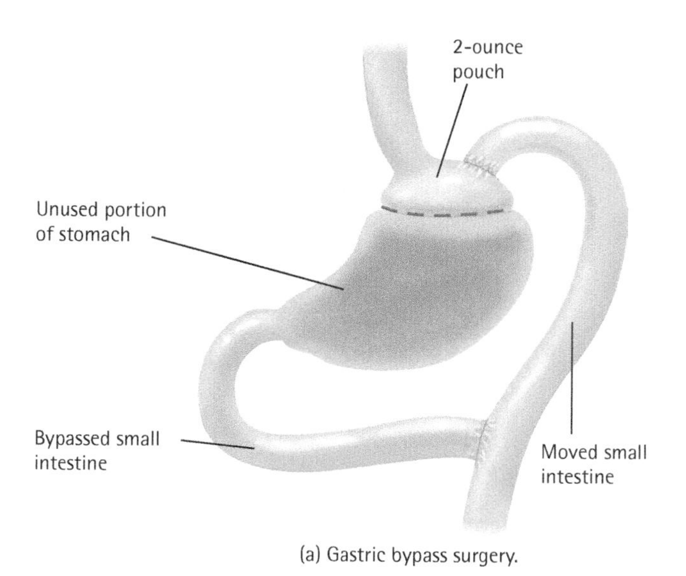

(a) Gastric bypass surgery.

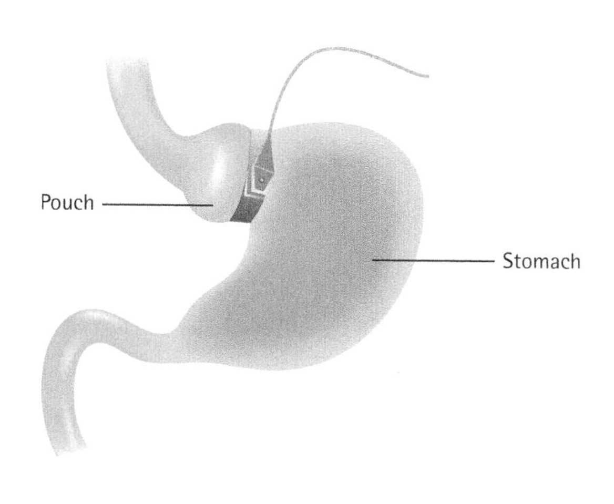

(b) Lap-band surgery.

Illustration 10.5 Gastric bypass surgery (*left*); lap-band surgery (*right*).

and mineral supplements are a routine component of care after bypass surgery.[55] Other complications related to adverse consequences of anesthesia—such as infection, nausea, vomiting, dehydration, and gallstones—can also develop.[49,51]

Some amount of lost weight is usually regained in the years following bariatric surgery.[49] Increasing food intake can stretch people's stomachs. The two-tablespoon pouch left after the surgery can be enlarged to hold half to two-thirds of a cup of food.[56] (A regular-size stomach has a capacity of about four cups.)

Bariatric surgery leaves some people with folds of excess skin where fat stores were lost. Skin tissue previously stretched by high levels of fat stores does not retract on its own after fat stores are lowered. The only known way of removing it is body-contouring surgery. The surgery may have to be repeated a number of times and can cost tens of thousands of dollars. It can be associated with healing complications due to poor nutritional status.[57]

Liposuction At a cost of $2,000 to $8,000 per surgery (prices vary by fat deposit site, practice location, and surgeon), fat deposits in the thighs, hips, arms, back, or chin can be partially removed by liposuction (Illustration 10.6). The procedure is the most common type of cosmetic surgery performed in the United States.[59] Only fat present underneath the skin, and not fat that lies underneath muscles or that surrounds organs, is removed by liposuction.

Liposuction is intended for body shaping like that shown in Illustration 10.7. It is not recommended for weight loss.[58] Surgical standards require that no more than eight pounds of fat be removed by liposuction.[59] Although considered a cosmetic procedure, liposuction that removes fat underneath abdominal skin may decrease plasma triglyceride levels in people who entered the surgery with high triglyceride levels.[60] Weight and fat losses from liposuction will not be permanent if weight is gained. With weight gain, fat deposits will increase at the surgical sites and other locations in the body.[61] In addition, surgery always carries a risk of infection and other complications, so it cannot be taken lightly.

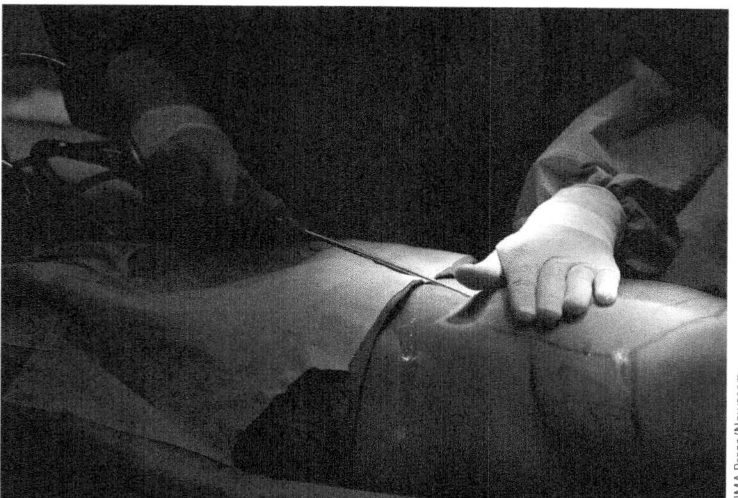
ZUMA Press/Newscom

Illustration 10.6 Liposuction surgery. In some cases, lasers are used during surgery to heat the fat, essentially turning the semi-soft fat in to a liquid to facilitate removal.

Illustration 10.7 The body-contouring effects of liposuction. The photo on the left shows the "spare tire" of a man before liposuction, and the photo on the right shows the same man's waist after liposuction.

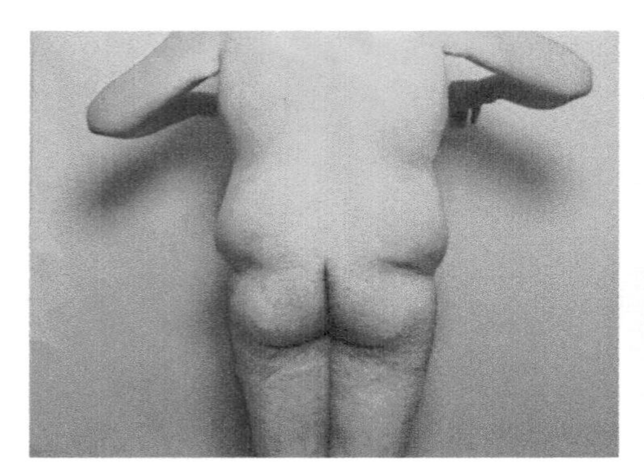

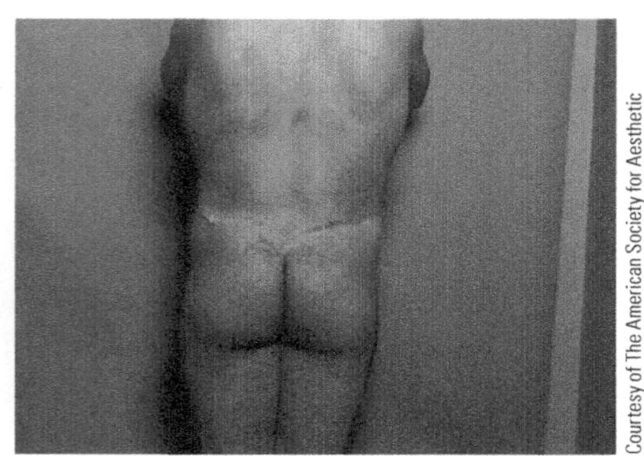

Courtesy of The American Society for Aesthetic Plastic Surgery

Weight Loss: Making It Last

People who lose weight and maintain their weight afterward tend to watch portion sizes carefully and to exercise regularly (Illustration 10.8). These are the major commonalities among people who maintain their weight after weight loss.[62] Other characteristics of people who lose weight and keep it off and characteristics of people who regain lost weight are listed in the Health Action feature. Changes in diet and physical activity likely to be maintained tend to be small, easy to implement, and acceptable—even preferable—to existing behaviors.[36]

Calorie Intake Reduction To lose weight a person has to decrease calorie intake. Decreases in calorie intake don't have to be large; they have to be enough to continue a gradual weight loss until a weight loss goal is met. In the recent past it was estimated that weight loss of a pound would require a reduction in calorie intake of 3,500 calories. Recent research indicates that the relationship between calorie reduction and weight is not that straightforward. A person's calorie needs decrease over time as weight is lost. Calorie intake levels that compensate for decreases in calorie needs related to weight loss must be incorporated into a weight-control plan if weight loss is to continue.[66,67] A number of Internet resources are available to help people track their calorie intake. The information can be used for decision making about additional reductions in food intake that will support the continuation of gradual weight loss.[20,67]

Identifying Small, Acceptable Changes To identify changes in eating and activity that have staying power, first list the weak points in your diet and activity level. Dietary weak points might include the consumption of high-fat foods due to eating out often or relying on high-fat convenience foods that can be heated up in seconds. Another weak point might be skipping breakfast and overeating later in the day because of extreme hunger. Weak points in physical activity might include driving instead of walking, not engaging in sports, or spending too little time playing outside.

Illustration 10.8 Small, acceptable changes are the key to a successful weight-loss/weight-maintenance program. Try to find activities that you enjoy; get out there and play!

Photodisc

health action Weight-Loss Maintainers versus Weight Regainers[63–65]

Weight-Loss Maintainers	Weight Regainers
• Exercise regularly.	• Exercise little.
• Make small and comfortable changes in diet and exercise.	• Use popular diets.
• Eat breakfast.	• Make drastic and unpleasant changes in their diets and physical activity levels.
• Choose low-fat foods.	• Take diet pills.
• Keep track of their weight, dietary intake, and physical activity level.	• Cope with problems and stress by eating.

For each weak point, identify options that seem acceptable and enjoyable. A person who enjoys broiled chicken with barbecue sauce might not mind eating that at a restaurant instead of fried chicken. That's a change people can make if they plan ahead. A person who gets too full from a large serving of fries might be happier not eating as many and ordering a small serving. A breakfast-skipper might find grabbing a piece of fruit and a slice of cheese for breakfast acceptable and doable. People who enjoy walking may not mind leaving the car or bus behind and letting their feet carry them to class, the grocery store, or a friend's house.

Many acceptable options for making small improvements in diet and activity may be available (see "Take Action—Small Steps Can Make a Big Difference"). The easiest changes to accomplish are those that should be incorporated into the overall lifestyle improvement plan. Some people, for example, lose weight and keep it off simply by consciously cutting down on portion sizes. Others avoid eating too much at any meal and walk more. Increasingly, people are losing weight and keeping it off by eating more nutrient-dense foods such as vegetables and fruits, and fewer high-fat, high-sugar, energy-dense foods.[26,68] This change doesn't require that people eat less food, but rather that they select and prepare foods that are nutrient dense rather than calorie dense. The more acceptable and easier the changes are to follow, the more likely they are to succeed.

Individualized plans that don't work out often include unacceptable or unenjoyable changes. The changes may be too large or too different from your usual activity. In these cases, go back to the drawing board and modify the plan to include small changes that are acceptable in the long run. Perhaps the original plan included jogging, but it turns out that you don't enjoy jogging. In that case, take jogging out of the plan! Replace it with another physical activity that you would enjoy. Midcourse corrections should be expected. Some experimentation may be required to identify small changes that will last.

What to Expect for Weight Loss If you're happy with the changes and have found the right levels of calorie intake and physical activity, weight loss will be gradual but lasting. The pattern of loss should be somewhat like that graphed in Illustration 10.9, where the person lost 18 pounds over 30 weeks. The pattern will include peaks, valleys, and plateaus—not a straight downward curve.[69] Sometimes a bit of weight will be gained, and other times more weight than expected will be lost. It's more important to enjoy and continue improved eating and activity patterns than to concentrate on the number of pounds lost.[24]

If diet and exercise behaviors are improved in acceptable ways, there may be little need to be preoccupied with the number of calories consumed, the number of calories burned off in a bout of exercise, or the number of pounds lost last week. The goal is reached when the small changes become an enjoyable part of life on a day-to-day basis. Improved and rewarding eating and activity patterns offer many benefits. Weight control is one of them.

take action Small Steps Can Make a Big Difference

Trying to lose weight or keep it off? Pick one eating and one physical activity option from the lists below that you find attractive. Give it a try for a day or two and see how it works out for you. Write down the changes you made and how you felt about making them. Could you see yourself making the changes for a week? The rewards offered by small steps may be greater than you expect.

Small Steps for Healthy Eating	Small Steps for Physical Activity
1. Eat cereal for breakfast.	1. Walk an additional 2,000 steps (20 minutes).
2. Eat a piece of fruit before dinner.	2. Park farther from the store than usual.
3. Eat larger portions of vegetables than of meat.	3. Take a 10-minute walk in a park and look for flowers.
4. Drink skim milk.	4. Lift a weight (or bottle of water, your textbook, a phone book) for 10 minutes while watching TV.
5. Switch from soft drinks to flavored water. You can make your own by diluting fruit juice with water.	5. Do 10 sit-ups in the morning.
6. Stop eating as soon as you start to feel full.	6. Use the stairs.
7. Eat only when you feel hungry.	7. Take your neighbor's dog for a walk.
8. Dish up a smaller than usual portion of dessert.	8. Jump rope for two minutes.
9. When your appetite tells you to eat another piece of cake (or other tempting food), put down the fork and start another activity.	9. Dance for five minutes while no one is watching.
10. Devote your full attention to eating. Chew your food thoroughly.	

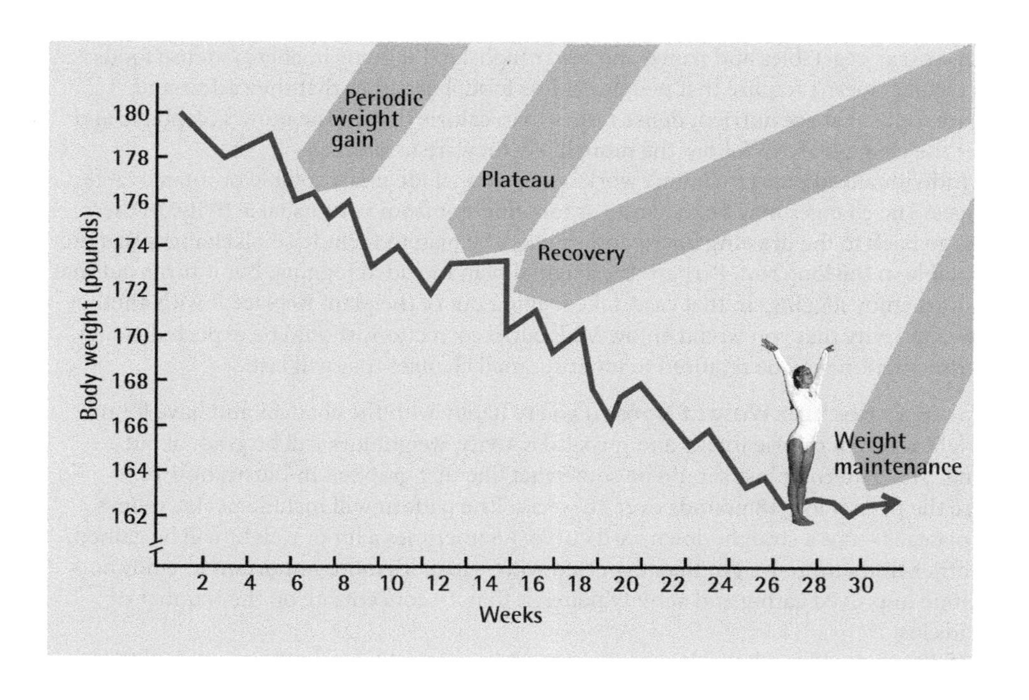

Illustration 10.9 The weight loss graphed here averages half a pound per week.

NUTRITION up close

Setting Small Behavior Change Goals

Focal Point: Small changes practice

Small changes in diet and exercise can return large benefits to health over time. What's the best way to get started on small and important behavioral changes? For many people, it begins with setting small but concrete goals.

This activity asks you to practice developing two small behavioral change goals, one related to diet and the other related to physical activity. Each small change goal should encompass the following:

a. State an activity.
b. State when the activity will be performed. A physical activity goal should also:
c. State how long the activity will be performed.

Here are two examples of small and concrete behavioral change goals:

"I will eat my vegetables first at dinner three times a week."

"I will work in the garden twice a week for 20 minutes."

Your Small Change Goals

Goal 1—Diet change:

Goal 2—Physical activity change:

Checklist: Do your goals meet the criteria listed above? Do they represent small rather than large changes? If not, rework the goals so they do.

Feedback to the Nutrition Up Close is located in Appendix F.

REVIEW QUESTIONS

- **Describe weight-control methods that have been demonstrated to be successful and those that have not.**

1. Claims made for the effectiveness of weight-loss products in advertisements must, by federal regulation, be true. **True/False**
2. Commercial weight-loss services and products must be tested for safety before they are made available to the public. **True/False**
3. Some foods, including celery and grapefruit, have negative calories and help overweight people burn fat. **True/False**
4. Many different types of quick weight-loss diets lead to weight loss if followed. **True/False**
5. Liposuction is a recommended method for weight loss and weight-loss maintenance. **True/False**
6. A person who weighs 250 pounds and has 95 pounds of excess body weight before gastric bypass surgery and loses the average amount of excess body weight would be expected to weigh 170 pounds two years after the procedure. **True/False**
7. ____Your friend Amoni developed her own weight-loss method. It's simple: Just eat fish, papaya, and broccoli. She lost five pounds in two weeks using her method and recommends it to everyone she hears talking about weight loss. This diet:
   a. Is nutrient-rich and likely meets people's need for nutrients.
   b. Would likely change people's eating behavior in the long run.
   c. Would be followed by weight regain in most people who try it.
   d. Would help people get started on losing weight forever.

8. ____While flipping through an in-flight magazine you notice an ad for a natural weight-loss supplement that guarantees users can safely lose at least 10 pounds in 10 days. Which term or word used in the above sentence should serve as the first clue to the fact that the weight-loss product is bogus?
   a. safely
   b. natural
   c. guarantees
   d. 10 pounds in 10 days

- **Discuss how an effective weight-control plan can be developed based on small and acceptable changes in dietary intake and physical activity level.**

9. It is recommended that people using a prescription diet pill for weight loss should also follow a reduced-calorie diet and exercise routine. **True/False**
10. The major problem shared by quick weight-loss approaches is that they do not help people maintain weight in the long run. **True/False**
11. A successful way to lose weight and keep it off is by making small and acceptable changes in diet and physical activity. **True/False**
12. Lifestyle approaches to weight control include individually based and sustainable behavioral changes. **True/False**
13. ____Your cousin weighs 183 pounds and wants to lose 5% of his current weight. How many pounds would your cousin have to lose to meet the goal of a 5% weight loss?
    a. 5
    b. 9
    c. 15
    d. 18

The next two questions refer to the following situation.

A research study finds that participants in a meal-replacement weight-control program lost an average of 4.4 pounds over two years. The control group (similar people who did not receive the program) lost 2 pounds over the same period. The average weight of people in the meal-replacement program prior to the onset of the study was 213 pounds, and the average weight of the control group was 210 pounds.

14. ____Which of the following statements about the results of this study is accurate?
    a. Neither program can be considered successful.
    b. Both programs can be considered successful.
    c. Only the meal-replacement program was successful.
    d. If participants in the control group had lost an average of 3 pounds more, the control group would have achieved a 5% or greater weight loss.
15. ____Assume the meal-replacement program was not successful. What is the most likely reason for its failure?
    a. Participants lacked the will power to stay on the diet.
    b. The meal replacements provided too many calories.
    c. The program did not focus on increasing physical activity.
    d. The program did not achieve sustainable and acceptable lifestyle changes.

Answers to these questions can be found in Appendix F.

NUTRITION SCOREBOARD ANSWERS

1. If this claim were true, hardly anybody would be obese. **False**
2. Maintenance of weight loss is the cure for obesity. **False**
3. Diets that have strict rules for weight loss can lead to lost pounds, but none help people keep the weight off.[1] **True**
4. Unfortunately, many weight-loss products and services on the market have not been shown to be safe or effective. Laws and regulations do not fully protect the consumer from the introduction of bogus products and services. **False**
5. Small and acceptable behavioral changes are easier to live with over time than drastic and disliked changes.[2] **True**

Disordered Eating: Anorexia and Bulimia Nervosa, Binge-Eating Disorder, and Pica

Jupiterimages/Stockbyte/Getty Images

NUTRITION SCOREBOARD

1. The United States has one of the world's highest rates of anorexia nervosa. **True/False**
2. Eating disorders result from psychological, and not biological, causes. **True/False**
3. "Food addiction" now qualifies as a substance abuse disorder in the *Diagnostic and Statistical Manual of Mental Disorders* (DSM-V). **True/False**
4. People in many different cultures may consume clay, dirt, and other nonfood substances. **True/False**

Answers can be found at the end of the unit.

LEARNING OBJECTIVES

After completing Unit 11 you will be able to:

- Discuss the potential causes, effects, and effective treatment approaches of eating disorders.

purging The use of self-induced vomiting, laxatives, or diuretics (water pills) to rid the body of food.

Eating Disorders

- **Discuss the potential causes, effects, and effective treatment approaches of eating disorders.**

Three square meals a day, an occasional snack or missed meal, and caloric intakes that average out to match the body's need for calories—this set of practices is considered "orderly" eating. Self-imposed semistarvation, feast and famine cycles, binge eating, **purging**, and the regular consumption of nonfood substances such as paint chips and clay—these behaviors are symptoms of disordered eating. Eating disorders are psychiatric illnesses for which diagnostic criteria have been established based on the presence of specific psychological, behavioral, and physiological factors. Like obesity, they are not due to personal choices or other factors that are easily within a person's control.[3]

Four specific types of disordered eating patterns are officially recognized as eating disorders and have been assigned diagnostic and treatment criteria. They are (1) anorexia nervosa, (2) bulimia nervosa, (3) binge eating disorder, and (4) pica.[3] Other forms of eating disorders such as compulsive overeating and nighttime eating syndrome have been proposed. They are classified as Eating Disorders Not Otherwise Specified (EDNOS). A variety of other forms of disordered eating have been designated to be eating disorders but as of yet lack treatment protocols.[3,4]

KEY NUTRITION CONCEPTS

Fundamental knowledge of nutrition science supports this content on eating disorders through these three key nutrition concepts:

1. Nutrition Concept #1: Food is a basic need of humans.
2. Nutrition Concept #6: Malnutrition can result from poor diets and from disease states, genetic factors, or combinations of these factors.
3. Nutrition Concept #9: Adequacy, variety, and balance are key characteristics of healthful diets.

Anorexia Nervosa

It's about 9:30 on a Tuesday night. You're at the grocery store picking up sandwich fixings and some milk. Although your grocery list contains only four items, you arrive at the checkout line with a half-filled cart. The woman in front of you has only five items: a bag with about 10 green beans, an apple, a bagel, a green pepper, and a 4-ounce carton of nonfat yogurt (Illustration 11.1). As she carefully places each item into her shopping bag, you notice that she is dreadfully thin.

Richard Anderson

Illustration 11.1 A day's diet for Alison. The foods shown provide approximately 562 calories.

The woman is Alison. She has just spent half an hour selecting the food she will eat tomorrow. Alison knows a lot about the caloric value of foods and makes only low-calorie choices. Otherwise, she will never get rid of her excess fat. To Alison, weight is everything—she cannot see the skeleton-like appearance others see when they look at her.

Alison has an intense fear of gaining weight and of being considered fat by others. She is annoyed when her parents and friends express their concern about her weight. You didn't know this about Alison when you saw her. There is much more to anorexia nervosa than meets the eye.

Individuals with **anorexia nervosa** such as described in the real-life case example of Alison, starve themselves. They can never be too thin—no matter how emaciated they may be. As shown in Illustration 11.2, people with anorexia nervosa look extraordinarily thin from the neck down. The face and the rest of the head usually look normal because the head is the last part of the body to be affected by starvation.

Women with anorexia nervosa have relatively little body fat (15% of body weight, compared to 27% in control women).[5] They often keep their body fat low by consuming fewer than 1,200 calories a day. They become cold easily and have unusually low heart rates and sometimes an irregular heartbeat, dry skin, and low blood pressure. Women with anorexia nervosa can experience absent or irregular menstrual cycles, infertility, and poor pregnancy outcomes (Table 11.1).[6] Men with anorexia nervosa tend to see themselves as being too fat and want less body fat and a more muscular body.[7] Anorexia nervosa is generally identified later in males than in females, largely because it is broadly thought to be a "woman's" disorder.[8] Low testosterone levels in males with this eating disorder produce diminished sexual drive and impaired fertility.[9]

Tony Freeman/PhotoEdit

Illustration 11.2 Eating disorders occur in males as well as in females, but females make up approximately 90% of all cases.

Approximately 9 in 10 women with anorexia nervosa have significant bone loss, and 38% have osteoporosis. The extent of bone loss correlates strongly with undernutrition: the lower the body weight, the lower the bone density. Males with anorexia nervosa lose bone mass, too. Improving calcium and vitamin D intake is recommended, along with nutritional rehabilitation.[6]

Table 11.1 Common features of anorexia nervosa[3,6]

- An intense fear of gaining weight
- Excessive dieting and exercise behaviors
- Severe weight loss
- Distorted body image
- Lack of normal menstrual cycles (females)
- Decreased testosterone levels, sex drive, fertility (males)
- Depression
- Low estrogen levels, low bone density
- Increased susceptibility to infections and injuries
- Periodic food binges or purging (in individuals with a certain form of anorexia nervosa)

The Female Athlete Triad Pediatricians, nutritionists, and coaches are beginning to be on the lookout for eating disorders, menstrual cycle dysfunction, and decreased bone mineral density in young female athletes. Calorie intake below calorie need and underweight related to eating disorders can lower estrogen levels and disrupt menstrual cycles. The lack of estrogen decreases calcium deposition in bones and reduces bone density at a time when peak bone mass is accumulating.[6] Adverse effects of low-calorie intakes occur in approximately 20% female as well as 8% of adult male athletes. Consequently, a more descriptive term for this eating disorder has been established as **"relative energy deficiency in sports" (RED-S)**. It is most common in athletes engaged in gymnastics, competitive cycling, distance running, figure skating, diving, and ballet.[11]

Irregular or absent menstrual cycles used to be thought of as no big deal. That attitude has changed, however, due to research results indicating that abnormal cycles in young females are related to delayed healing of bone and connective tissue injuries, and to bone fractures and osteoporosis later in life.[12]

Motivations Underlying Anorexia Nervosa The overwhelming desire to become and remain thin drives people with anorexia nervosa to not eat much and to exercise intensely. Half of the people with anorexia turn to **binge eating** and purging—features of bulimia nervosa—in their efforts to lose weight.[13] Preoccupied with food, people with anorexia may prepare wonderful meals for others but eat very little of the food themselves. Family members and friends, distressed by their failure to persuade the person with anorexia to eat, report high levels of anxiety. Although adults often describe people with anorexia as "model students" or "ideal children," their personal lives are usually marred by low self-esteem, social isolation, and unhappiness.[14]

anorexia nervosa An eating disorder characterized by extreme weight loss, poor body image, and irrational fears of weight gain and obesity.

binge eating The consumption of a large amount of food in a small amount of time.

Richard Anderson

Illustration 11.3 There is a need for the use of more realistic body shapes and sizes in the media.

What Causes Anorexia Nervosa? The cause of anorexia nervosa isn't yet clear. It is likely that many different conditions, both psychological and biological, predispose an individual to become totally dedicated to extreme thinness. The value that Western societies place on thinness, the need to conform to society's expectations of acceptable body weight and shape, low self-esteem, and a need to control some aspect of one's life completely are commonly offered as potential causes for this disorder (Illustration 11.3).[15,16]

Recent studies have identified potential biological factors associated with development of anorexia nervosa. Scans of brain activity among individuals with anorexia nervosa have identified a lack of the expected pleasure response to sugar or eating and a lower sensitivity to hunger. In some clinics, results of scans of people affected by anorexia nervosa are referred to as "anorexia nervosa brain."[17]

relative energy deficiency is sports (RED-S) A deficiency of energy intake compared to need that impairs metabolic rate, menstrual function, bone health, immunity, protein synthesis, and cardiovascular health. It may lead to a gradual reduction in physical performance, increased risk of injury, decreased strength, and other conditions.[10]

How Common Is Anorexia Nervosa? It is estimated that 1% of adolescent and young women in the Western world and around 0.3% of young males have anorexia nervosa.[6] The disorder has been reported in girls as young as 5 and in women through their 40s;[16] however, it usually begins during adolescence. It is estimated that 1 in 10 females between the ages of 16 and 25 has "subclinical" anorexia nervosa, or exhibits some of the symptoms of the disorder.[6]

Certain groups of people are at greater risk of developing anorexia nervosa than others (Table 11.2). People at risk come from all segments of society, but they tend to be overly concerned about their weight and food and have attempted weight loss from an early age.[18]

Table 11.2 Risk groups for anorexia nervosa[6]

• Teenagers
• Ballet dancers
• Competitive athletes (gymnasts, figure skaters, runners, wrestlers, dancers)
• Fitness instructors
• Dietetics majors
• People with type 1 (insulin-dependent) diabetes

Treatment There is no magic bullet treatment that cures anorexia nervosa quickly and completely. Patient-centered, individualized care delivered in a hospital by a team of specialists in mental health, nutrition, and medicine is recommended as the initial treatment for many people with anorexia nervosa.[3,6] Care generally focuses on the prompt restoration of nutritional health and body weight, psychological counseling to improve self-esteem and attitudes about body weight and shape, medically supervised use of antidepressants or other medications (if needed), family therapy, and normalizing eating and exercise behaviors.[3] Outpatient care follows hospitalization and weight gain in most cases.[3,19] Such programs may include the provision of a manual for families that discusses what they can do to help and how to do it, and regular counseling and follow-up sessions with care team members over the course of about nine months.[19] These programs are fully successful in 50% of people and partially successful in most other cases.[19] Eight years after diagnosis, 3% of people with anorexia nervosa die from the disorder.[16] Results of treatment are often excellent when the disorder is treated early.[20] Unfortunately, many people with the condition deny that problems exist and postpone treatment for years. Initiation of treatment is often prompted by a relative, coach, or friend.[21]

Bulimia Nervosa

Finally home alone, Lisa heads to the pantry and then to the freezer. She has carefully controlled her eating for the last day and a half and is ready to eat everything in sight.

It's a bittersweet time for her. Lisa knows the eating binge she is preparing will be pleasurable, but that she'll hate herself afterward. Her stomach will ache from the volume of food she'll consume, she'll feel enormous guilt over losing control, and she'll be horrified that she may gain weight and will have to starve herself all over again. Lisa is so preoccupied with her weight and body shape that she doesn't see the connection between her severe dieting and her bouts of uncontrolled eating. To get rid of all the food she is about to eat, she will do what she has done several times a week for the last year. Lisa avoids the horrible feelings that come after a binge by "tossing" everything she has eaten as soon as she can.

In just 10 minutes, Lisa devours 10 peanut butter cups (the regular size), a 12-ounce bag of chocolate chip cookies, and a quart of ice cream. Before five more minutes have passed, Lisa will have emptied her stomach, taken a few deep breaths, thrown on her shorts, and started the five-mile route she jogs most days. As she jogs, she obsesses about getting her 138-pound, 5-foot 5-inch frame down to 115 pounds. She will fast tomorrow and see what news the bathroom scale brings.

Lisa is not alone. **Bulimia nervosa** occurs in 1 to 3% of young women and in about 0.5% of young males in the United States.[16] The disorder is characterized by regular episodes of dieting, binge eating (see Illustration 11.4), and attempts to prevent weight gain by purging or using laxatives, diuretics, excessive exercise, or enemas. In most cases, bulimia nervosa starts with voluntary dieting to lose weight. At some point, voluntary control over dieting is lost, and people may engage in binge eating and vomiting.[16] The behaviors become cyclic: food binges are followed by guilt, purging, and dieting. Dieting leads to a

bulimia nervosa An eating disorder characterized by recurrent episodes of rapid, uncontrolled eating of large amounts of food in a short period of time. Episodes of binge eating are followed by compensatory behaviors such as self-induced vomiting, dieting, excessive exercise, or misuse of laxatives to prevent weight gain.

Illustration 11.4 Bulimia nervosa is characterized by the consumption of a large amount of food (such as shown here) followed by purging and dieting, or other behavior such as excessive exercise.

Table 11.3 Features of bulimia nervosa[6,13,23]

• Feeling of loss of control over how much food is eaten
• Overwhelming urges to overeat
• Binge eating, or eating larger amounts of food than most people would in a similar situation in a short period of time, at least once a week for 3 months
• Inappropriate use of vomiting, fasting, exercise, laxatives, or diuretics to compensate for food binges
• A feeling of being ashamed of overeating and very fearful of gaining weight
• Self-evaluation unduly influenced by body weight or shape

feeling of deprivation and intense hunger, which leads to binge eating, and so on. Once a food binge starts, it is hard to stop.

Table 11.3 lists the features of bulimia nervosa. Approximately 86% of people with this condition vomit to prevent weight gain and to avoid post-binge anguish. A smaller proportion of people use laxatives, ipecac (a vomiting-inducing medication), diuretics (water pills), or enemas, alone or in combination with vomiting.[4] These approaches do not prevent weight gain, however, and their regular use can be harmful. The habitual use of laxatives and enemas causes "laxative dependency"—these products become necessary for bowel movements. Long-term use of ipecac may damage heart muscle, and diuretics can cause illnesses by disturbing the body's fluid balance.[16]

The lives of people with bulimia nervosa are usually dominated by conflicts about eating and weight. Some affected individuals are so preoccupied with food that they spend days securing food, bingeing, and purging. Others experience only occasional episodes of binge eating, purging, and fasting.[13]

Unlike those with anorexia nervosa, people with bulimia usually are not underweight or emaciated. They tend to be normal weight or overweight. Like anorexia nervosa, bulimia nervosa is more common among athletes (including gymnasts, weight lifters, wrestlers, jockeys, figure skaters, physical trainers, and distance runners) and ballet dancers than other groups.[16]

Bulimia nervosa leads to major changes in metabolism. The body must constantly adjust to feast and famine cycles and mineral and fluid losses. Salivary glands become enlarged, and teeth may erode due to frequent vomiting of highly acidic foods from the stomach.[13] There is evidence that satiety signals are decreased among people with this eating disorder, and that may support overeating behaviors.[22]

Is the Cause of Bulimia Nervosa Known? The cause of bulimia nervosa is not known with certainty, but the scientific finger is pointing at depression, abnormal mechanisms for regulating food intake, and feast and famine cycles as possible causes. Fasts and **restrained eating** may prompt feelings of deprivation and hunger that may trigger binge eating.[16] The ideal thinness may become more and more difficult to achieve as the feast and famine cycles continue.

restrained eating The purposeful restriction of food intake below desired amounts in order to control body weight.

binge-eating disorder An eating disorder characterized by periodic binge eating, which normally is not followed by vomiting or the use of laxatives.

Treatment The goal of bulimia treatment is to break the feast and famine cycles via nutrition and psychological counseling. Replacing the disordered pattern of eating with regular meals and snacks often reduces the urge to binge and the need to purge. Psychological counseling aimed at improving self-esteem and attitudes toward body weight and shape goes hand in hand with nutrition counseling. In some cases, antidepressants are a useful component of treatment.[23] The rates of full recovery for women with bulimia nervosa is higher than that for women with anorexia nervosa. Nearly all women with bulimia achieve partial recovery, but one-third will relapse into bingeing and purging within seven years.[24] Bulimia nervosa usually improves substantially during pregnancy; about 70% of pregnant women with the condition will improve their eating habits for the sake of their unborn baby.[25]

Binge-Eating Disorder

Table 11.4 Common features of binge-eating disorder[3,6]

1. Rapid consumption of extremely large amounts of food (several thousand calories) in a short period of time
2. One or more episodes of binge eating per week over a period of three months
3. Binge eating by oneself
4. Lack of control over eating or an inability to stop eating during a binge
5. Post–binge-eating feelings of self-hatred, guilt, and depression or disgust
6. Purging, fasting, excessive exercise, or other compensation for high-calorie intake not present

Features of **binge-eating disorder** are shown in Table 11.4. People with this condition tend to be overweight or obese, and it affects an equal number of males and females.[3] Like individuals with bulimia nervosa, people with binge-eating disorder eat several thousand calories worth of food within a short period of time during a solitary binge, feel a lack of control over the binge, and experience distress or depression after the binge occurs. People must experience eating binges once a week or more on average over a period of three months to qualify for the diagnosis. Unlike individuals with bulimia nervosa, however, people with binge-eating disorder don't vomit, use laxatives, fast, or exercise excessively in an attempt to control weight gain.[6]

It is estimated that 9 to 30% of people in weight-control programs[22] and 2 to 3.5% of U.S. adults overall experience binge-eating disorder.[6] Stress, depression, anger, anxiety, and other negative emotions appear to prompt binge-eating episodes. Preliminary evidence indicates that binge-eating disorder runs in families and has both genetic and environmental orgins.[16]

Treatment The treatment of binge-eating disorder focuses on both the disordered eating and the underlying psychological issues. Persons with this condition will often be asked to record their food intake, indicate bingeing episodes, and note feelings, circumstances, and thoughts related to each eating event (Illustration 11.5). This information is used to identify circumstances that prompt binge eating and practical alternative behaviors that may prevent it. Individuals being treated for binge-eating disorder are usually given information about it, attend individual and group therapy sessions, and participate in nutrition counseling on mindful eating. Components of the mindful eating approach include paying attention to hunger and satiety cues, slowing down the pace of eating, and identifying triggers to eating. Antidepressants may be part of the treatment. Treatment is successful in 85% of women treated for binge-eating disorder.[16,26,27]

Daily Food Record

Date________

Time	Type and amount of food and beverage	Meal, Snack, Binge?	Eating triggers (feelings, situation)
7:30 am	coffee, 2 cups sugar 2 tsp cornflakes, 2 cups skim milk, 1 cup	M	Hunger!
11:30 am	tuna sandwich ice tea, 2 cups	M	Bored, hungry
7:30 pm	3 hamburgers 2 large fries 24 oreos 1/2 gallon ice cream	B	Stressed out, angry at my coach

Illustration 11.5 Example of a food diary of a person with binge-eating disorder.

Pica

When did I start eating clay? (Miss Williams repeated the question she had just been asked.) I know it might sound strange to you, but I started craving clay in the summer of '68. It was a beautiful spring morning—it had just rained. I smelled something really sweet in the breeze coming in my bedroom window. I went outside and knew instantly where the sweet smell was coming from. It was the wet clay that lies all around my house. I scooped some up and tasted it. That's when and how I started my craving for that sweet-smelling clay. I keep some in the fridge now because it tastes even better cold.

A most intriguing type of eating disorder, **pica** has been observed in chimpanzees and in humans in many different cultures since ancient times.[33] The history and persistence of pica might suggest that the practice has its rewards. Nevertheless, important health risks are associated with eating many types of nonfood substances.

The characteristics of pica are summarized in Table 11.5. Young children and pregnant women are most likely to engage in the practice; for unknown reasons, it rarely occurs in men.[34] It most commonly takes the form of **geophagia** (clay or dirt eating), **pagophagia** (ice eating), **amylophagia** (laundry starch and cornstarch eating), or **plumbism** (lead eating). A potpourri of nonfood substances, listed in Table 11.6, may be consumed. It is not clear why pica exists, although several theories have been proposed.

pica (pike-eh) The regular consumption of nonfood substances such as clay or laundry starch.

geophagia (ge-oh-phag-ah) Clay or dirt eating.

pagophagia (pa-go-phag-ah) Ice eating.

amylophagia (am-e-low-phag-ah) Laundry starch or cornstarch eating.

plumbism Lead eating (primarily from old paint flakes).

Geophagia Some people like to eat certain types of clay or dirt. Those who do often report that the clay or dirt tastes or smells good, quells a craving, or helps relieve nausea or an upset stomach. The belief that certain types of clay provide relief from stomach upset may have some validity: a component of some types of clay is used in nausea and diarrhea medicines. There is no evidence that geophagia is motivated by a need for minerals found in clay or dirt, however.[35]

Although the reasons given for clay and dirt ingestion make the practice understandable and help explain its acceptance in some cultures, the consequences for health outweigh the benefits. Clay and dirt consumption can block the intestinal tract and cause parasitic and bacterial infections.[13] The practice is also associated with iron deficiency and sickle-cell anemia in some individuals.[35]

Table 11.5 Characteristics of pica

A. **Essential features:** Regular ingestion of nonfood substances such as clay, paint chips, laundry starch, paste, plaster, dirt, or hair
B. **Other common features:** Occurs primarily in young children and pregnant women in the southern United States

Table 11.6 A partial list of nonfood substances reported to be consumed by individuals with pica

Animal droppings	Coffee grounds	Leaves	Plaster
Baking soda	Cornstarch	Mothballs	Sand
Burnt matches	Crayons	Nylon stockings	String
Cigarette butts	Dirt	Paint chips	Wool
Clay	Foam rubber	Paper	
Cloth	Hair	Paste	
Coal	Laundry starch	Pebbles	

PhotoDisc

PhotoDisc

Pagophagia Have you ever known somebody who constantly crunches on ice? That person may have a 9-in-10 chance of being iron deficient. Regular ice eating, to the extent of one or more trays of ice cubes a day, is closely associated with an iron-deficient state. Ice eating usually stops completely when the iron deficiency is treated.[36]

Ice eating may be common during pregnancy. In one study of women from low-income households in Texas, 54% of pregnant women reported eating large amounts of ice regularly. Ice eaters had poorer iron status than other pregnant women who did not eat ice.[37]

Amylophagia The sweet taste and crunchy texture of flaked laundry starch are attractive to a small number of women, especially during pregnancy. If the laundry starch preferred is not available, cornstarch is sometimes eaten in its place. Laundry starch is made from unrefined cornstarch. The taste for starch almost always disappears after pregnancy.[34]

Laundry starch and cornstarch have the same number of calories per gram as do other carbohydrates (4 calories per gram). Consequently, starch eating provides calories and may reduce the intake of nutrient-dense foods. In addition, starch may contain contaminants because it is not intended for consumption. Starch eaters' diets are generally inferior to the diets of pregnant women who don't consume starch, and their infants are more likely to be born in poor health.[34]

Plumbism The consumption of lead-containing paint chips poses a major threat to the health of children in the United States and many other countries (Illustration 11.6). Many older homes and buildings, especially those found in substandard housing areas, may be covered with lead-based paint and its dried-up flakes. Children may develop lead poisoning if they eat the sweet-tasting paint flakes or inhale lead from contaminated dust and soil near the buildings. Approximately 2.6% of children aged 1 to 5 years in the United States have elevated blood lead levels.[38]

High levels of exposure to lead can cause profound mental retardation and death in young children. Low levels of exposure can lead to hearing problems, growth retardation, reduced intelligence, and poor classroom performance. Children with lead poisoning are more likely to fail or drop out of school than children not exposed to lead in their environment.[38,39] Cases of lead poisoning due to plumbism in adults have been observed among people who consume ground-up clay pottery, clay, lead-containing nontraditional medicines, and lead-containing water and cooking utensils.[40]

Andre Kudyusov/Getty Images

Illustration 11.6 Regular exposure to dust and soil contaminated with lead can lead to lead poisoning.

Proposed Eating Disorders

Several other patterns of abnormal eating behaviors, referred to as EDNOS—eating disorders not otherwise specified—have been described and may become official diagnoses in the future.[3] These tentative eating disorders include:

- Nighttime eating syndrome—high food consumption during the night accompanied by sleep disturbances and psychological distress
- Compulsive overeating—the uncontrolled ingestion of large amounts of food, as found in binge-eating disorder
- Purging disorder—frequent purging without binge eating
- Restrained eating—the consistent limitation of food intake to avoid weight gain

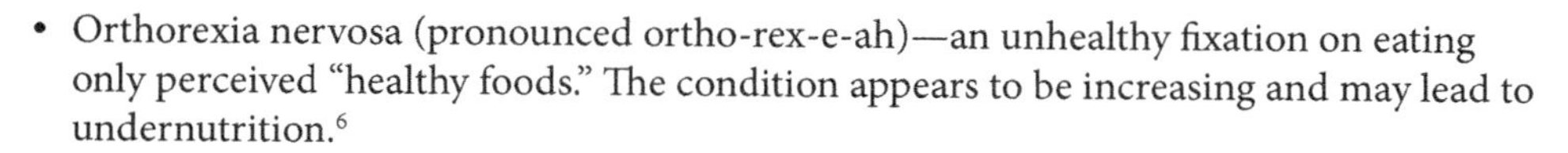

- Orthorexia nervosa (pronounced ortho-rex-e-ah)—an unhealthy fixation on eating only perceived "healthy foods." The condition appears to be increasing and may lead to undernutrition.[6]
- Selective eating disorder—children and adults who are picky eaters; they consume a very limited variety of food.

Individuals displaying these eating characteristics have been identified, but the core features are not sufficiently understood to develop reliable diagnostic criteria and treatment approaches.[16]

Although not yet proposed as a potential eating disorder, the topic of food addiction has captured the public's attention. It has been declared as fact in popular media articles and is thought to help explain why people overeat.[41]

Food Addiction The term *food addiction* is often applied to a strong desire for and binge eating of high-sugar or high-fat foods such as chocolate, fried foods, salty snack foods, sweets, ice cream, pizza, or desserts. The craving for specific foods can be intense and may encourage overeating, as has been identified in bulimia nervosa and binge-eating disorder.[42] However, it is unlikely that food addiction qualifies as a distinct eating disorder given what is currently known. Food addiction-like behaviors appear to exist in some people who have a strong urge to consume a lot of certain foods and experience pleasure from eating the foods.[2]

REALITY CHECK
Close to Home

Although she hides it, you are sure your sister has bulimia nervosa and that she is not getting help. You are deeply concerned for her health and well-being but don't know what to do about it.

Here's what Heather and Crystal say they would do if it were their sister. Who do you think has the better idea?

Answers on page 11-10.

PhotoDisc

Heather: I'd talk with her about getting help.

PhotoDisc

Crystal: I'd spend more time with her to let her know I love her.

ANSWERS TO REALITY CHECK

Close to Home

Both ideas are admirable and deserve a thumbs-up. Heather's idea is aimed directly at helping her sister consider treatment and may sometimes be the appropriate action to take. There are appropriate ways to talk to relatives or friends about your concerns for them. Learn more about it from the information presented in Table 11.7.

PhotoDisc

Heather:

PhotoDisc

Crystal:

Food items or ingredients have not been found to have addictive properties; the brain does not respond to foods like sugar and fat as it does to addictive drugs, and withdrawal symptoms associated with the consumption of particular foods or ingredients have not been identified. There are no diagnostic criteria or standardized treatments for food addiction. Much more research is needed before it can be known whether certain foods qualify as being truly addictive. Until then it is clear that we are all addicted to food and eating.

Resources for Eating Disorders

Information and services related to eating disorders are available from a variety of sources. Services are best delivered by health care teams that specialize in the treatment of eating disorders. Contact with a primary care physician, dietitian, or nurse practitioner is often a good start to the process of identifying qualified health care teams. Reliable sources of information about eating disorders can be located at the Academy of Eating Disorders website (www.aedweb.org).

One of the most important resources for people with an eating disorder may be a trusted friend or relative. This unit's Reality Check explores this resource in a very personal way—by putting you in the shoes of a person whose sister has bulimia.

Table 11.7 Helping a family member or friend with an eating disorder[4]

Whether at work, home, or play, many of us experience anxiety and a sense of helplessness when someone we love is living with an eating disorder. We may feel compelled to take action to help but aren't sure what to do or how to do it. Here are some tips on how to express your concerns to a friend or relative with an eating disorder:
1. Gather information about services for people with eating disorders to share with your friend or relative.
2. Talk with your friend or relative privately when there is enough time to discuss the issue fully. Tell them you are worried and that they may need to seek help.
3. Encourage your friend or relative to express his or her feelings and then listen intently. Be accepting of the feelings that are expressed. Be ready to talk to the friend or relative more about it in the future.
4. Do not argue with your friend or relative about whether she or he has an eating disorder. Let your friend or relative know you heard what was said but that you are concerned that he or she may not get better without treatment.
5. Seek emergency medical help in life-threatening situations.
6. Understand that affected individuals would rather not have an eating disorder and that parents are not to be blamed for them.
Only individuals with an eating disorder can make the decision to get help. Knowledge that people who love them will be around to support them and their decision to seek treatment may help encourage people with an eating disorder to take action.

Prevention of Eating Disorders

The pressure to conform to society's standard of beauty and acceptability is thought to be an important force underlying the development of eating disorders.[6] Children acquire prevailing cultural values of beauty before adolescence. As early as age 5, U.S. children learn to associate negative characteristics with people who are overweight and positive characteristics with those who are thin.[28] Standards of beauty as defined by models and movie and television stars often include thinness, but the body shape portrayed as most desirable is often unhealthfully thin and unattainable by many.[29] The disparity between this ideal and what people normally weigh can foster low self-esteem and body image dissatisfaction. Approximately 50% of normal-weight adult women are dissatisfied with their weight; many diet, binge, purge, or fast occasionally in an attempt to reach the standard of beauty set for them.[30] Men are more likely to express discontent with their body shape and fat content. They may strive obsessively for muscular bodies with little visible body fat.[31]

A movement toward acceptance of body size, fashionable attire for larger people, full-size models, and a more realistic view of individual differences in body shapes is emerging in the United States and Europe (Illustration 11.7). Adoption of the "health at every size" concepts and similar programs that stress body size acceptance, health and happiness rather than dieting and weight loss, and eating in response to physical rather than emotional cues could help reduce the drive for thinness at any cost.[6,32]

Richard Anderson

Richard Anderson

Illustration 11.7 The trend toward size acceptance.

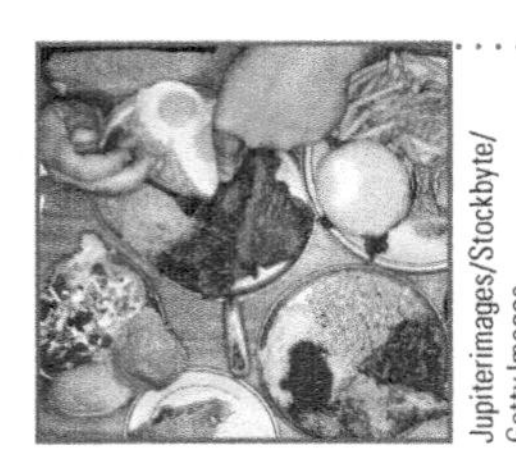
Jupiterimages/Stockbyte/Getty Images

NUTRITION up close

Eating Attitudes Test

Focal Point: Assess whether your eating attitudes and behaviors are likely to be within a normal range.

Date ________ Age ________ Gender ________

Height ________ Present weight ________ How long at present weight? ________

Highest past weight ________ How long ago? ________

Lowest past weight ________ How long ago? ________

Answer the following questions using these responses:

A = always S = sometimes U = usually R = rarely O = often N = never

____ 1. I am terrified of being overweight.
____ 2. I avoid eating when I am hungry.
____ 3. I find myself preoccupied with food.
____ 4. I have gone on eating binges where I feel that I may not be able to stop.
____ 5. I cut my food into very small pieces.
____ 6. I am aware of the calorie content of the foods I eat.
____ 7. I particularly avoid foods with a high carbohydrate content.
____ 8. I feel that others would prefer that I ate more.
____ 9. I vomit after I have eaten.
____ 10. I feel extremely guilty after eating.
____ 11. I am preoccupied with a desire to be thinner.
____ 12. I think about burning up calories when I exercise.
____ 13. Other people think I am too thin.
____ 14. I am preoccupied with the thought of having fat on my body.
____ 15. I take longer than other people to eat my meals.
____ 16. I avoid foods with sugar in them.
____ 17. I eat diet foods.
____ 18. I feel that food controls my life.
____ 19. I display self-control around food.
____ 20. I feel that others pressure me to eat.
____ 21. I give too much time and thought to food.
____ 22. I feel uncomfortable after eating sweets.
____ 23. I engage in dieting behavior.
____ 24. I like my stomach to be empty.
____ 25. I enjoy trying new rich foods.
____ 26. I have the impulse to vomit after meals.

Source: From Garner, D. M., Olmsted, M. P., Bohr, Y., and Garfinkel, P. E. (1982). "The Eating Attitudes Test: Psychometric features and clinical correlates," *Psychological Medicine*, 12, 871–878.

The questions are not intended to diagnose eating disorders but rather to screen for them. Feedback to the Nutrition Up Close is located in Appendix F.

REVIEW QUESTIONS

- **Discuss the potential causes, effects, and effective treatment approaches of eating disorders.**

1. Critical aspects of anorexia nervosa include low body fat and low bone density. **True/False**
2. Individuals with anorexia nervosa are at risk of anemia. **True/False**
3. The adverse effects of the female athlete triad on health are primarily related to low calorie intake and underweight. **True/False**
4. More individuals have anorexia nervosa than any other recognized eating disorder. **True/False**
5. Essential features of bulimia nervosa are binge eating and excessive sleeping. **True/False**
6. Obese people are more likely to experience binge-eating disorder than are people who are not obese. **True/False**
7. Treatment is effective for 85% of women treated for binge-eating disorder. **True/False**
8. *Pica* refers to the regular consumption of raw whole grains, grasses, and other foods generally used to feed livestock. **True/False**
9. People who consume lead, such as from lead-based paints, have a condition called *geophagia*. **True/False**
10. Nutritional rehabilitation and eating behavior counseling are key components of the treatment of eating disorders. **True/False**

Answers to these questions can be found in Appendix F.

NUTRITION SCOREBOARD ANSWERS

1. Anorexia nervosa is most common in the United States and other Westernized countries.[1] **True**
2. The causes of eating disorders are not known with certainty. Both psychological and biological factors play a role. **False**
3. The DSM-V does not currently include food addiction as a substance abuse, or any other type, of disorder.[2] **False**
4. Although not recommended for health reasons, people in many different cultures practice pica—the regular ingestion of nonfood items such as clay and dirt. **True**

Useful Facts about Sugars, Starches, and Fiber

John E. Kelly/Photolibrary/Getty Images

NUTRITION SCOREBOARD

1 Pasta, bread, and dried beans (legumes) are good sources of complex carbohydrates. **True/False**

2 Ounce for ounce, presweetened breakfast cereals and unsweetened cereals provide about the same number of calories. **True/False**

3 A 12-ounce can of soft drink contains about 3 tablespoons (9 teaspoons) of sugar. **True/False**

4 All foods high in fiber are fibrous. **True/False**

5 Cooking vegetables destroys their fiber content. **True/False**

6 Sugar consumption is strongly related to the development of tooth decay. **True/False**

Answers can be found at the end of the unit.

LEARNING OBJECTIVES

After completing Unit 12 you will be able to:

- Describe the food sources and functions of simple sugars, complex carbohydrates, fiber, and alcohol sugars.
- Explain relationships between carbohydrates and dental health.
- Discuss the foods and beverages that can be consumed by individuals with lactose maldigestion to assure adequate intake of calcium and vitamin D.

Carbohydrates

- **Describe the food sources and functions of simple sugars, complex carbohydrates, fiber, and alcohol sugars.**

Carbohydrates are the major source of energy for people throughout the world. They are the primary ingredient of staple foods such as pasta, rice, cassava, beans, and bread. On average, Americans consume fewer carbohydrates than people in much of the world, who consume approximately half of total calories from carbohydrates.[1]

The carbohydrate family consists of three types of chemical substances:

1. Simple sugars
2. Complex carbohydrates ("starch")
3. Dietary fiber

Some food sources of these different types of carbohydrates are shown in Illustration 12.1. Carbohydrates consist of carbon, hydrogen, and oxygen. They perform a number of functions in the body, but their primary function is to serve as an energy source. Simple sugars and complex carbohydrates supply the body with four calories per gram. Dietary fiber, on average, supplies two calories per gram. Although humans cannot digest fiber, bacteria in the colon can digest some types of fiber. These bacteria excrete fatty acids as a waste product from fiber digestion. The fatty acids are absorbed and used as a source of energy.[2] The total contribution of fiber to our energy intake is modest (around 50 calories), and supplying energy is not a major function of fiber. Certain carbohydrates perform roles in the functioning of the immune system, reproductive system, and blood clotting. One simple sugar (ribose) is a key a component of the genetic material DNA.

carbohydrates Chemical substances in foods that consist of a simple sugar molecule or multiples of them in various forms.

KEY NUTRITION CONCEPTS

The various forms of carbohydrates are important components of our diets and influence health in many ways. Knowledge about carbohydrates presented here relates to these three key nutrition concepts:

1. Nutrition Concept #2: Foods provide energy (calories), nutrients, and other substances needed for growth and health.
2. Nutrition Concept #4: Poor nutrition can result from both inadequate and excessive levels of nutrient intake.
3. Nutrition Concept #9: Adequacy, variety, and balance are key characteristics of healthful diets.

Illustration 12.1 The carbohydrate family. Some food sources of simple sugars (*left*), and some food sources of starch and dietary fiber (*right*) are shown.

Scott Goodwin Photography

Felicia Martinez Photography/PhotoEdit

Glucose Fructose Xylitol Ethanol

Illustration 12.2 The similar chemical structures of glucose, fructose, xylitol (an alcohol sugar), and ethanol (an alcohol).

Table 12.1 The monosaccharides and the disaccharides they form

Monosaccharides	Disaccharide formed
glucose + glucose	maltose
glucose + fructose	sucrose
glucose + galactose	lactose

Alcohol sugars and alcohol (ethanol) are similar in chemical structure to carbohydrates (Illustration 12.2). The alcohol sugars are presented in this unit and alcohol is discussed in Unit 14. Carbohydrates and carbohydrate-containing foods are also classified by their glycemic index, or the extent to which they increase blood glucose levels. This topic is introduced in this unit and explored further in Unit 13 on diabetes.

Simple Sugar Facts

Simple sugars are considered "simple" because they are small molecules that require little or no digestion before they can be used by the body. They come in two types: **monosaccharides** and **disaccharides**. Monosaccharides consist of one molecule; they include glucose (blood sugar or dextrose), fructose (fruit sugar), and galactose. Disaccharides consist of two monosaccharide molecules (see Table 12.1). The combination of a glucose molecule and a fructose molecule makes sucrose (or table sugar); maltose (malt sugar) is made from two glucose molecules; and lactose (milk sugar) consists of a glucose molecule plus a galactose molecule. Honey, by the way, is a disaccharide. It is composed of glucose and fructose just as sucrose is, but it's a liquid rather than a solid because of the way the two molecules of sugar are chemically linked together. Disaccharides are broken down into their monosaccharide components during digestion; only glucose, fructose, and galactose are absorbed into the bloodstream.

High-fructose corn syrup—a liquid sweetener used in many soft drinks, fruit drinks, breakfast cereals, and other products—is also considered a simple sugar. It generally consists of 55% fructose and 45% glucose, compared to sucrose, which contains 50% glucose and 50% fructose.[3] Most of the simple sugars have a distinctively sweet taste.

The simple sugars the body uses directly to form energy are glucose and fructose. Galactose is readily converted by the body to glucose. When the body has more glucose than it needs for energy formation, it converts the excess to fat and to **glycogen**, the body's storage form of glucose. Glycogen is a type of complex carbohydrate, and storage of it is limited. It consists of chains of glucose units linked together in long strands. Glycogen is produced only by animals and is stored in the liver (Illustration 12.3) and muscles. When the body needs additional glucose, glycogen is broken down, making glucose available for energy formation. Glucose can also be derived from certain amino acids and the glycerol component of fats. Illustration 12.4 shows the various ways glucose becomes available to the body. A constant supply of glucose is needed because the brain, red blood cells, white blood cells, and specific cells in the kidneys require glucose as an energy source.[4]

Simple Sugar Intake Most of the simple sugar in our diet comes from foods and beverages sweetened with sucrose and high-fructose corn syrup. As a matter of fact, simple sugars are the most commonly used food additive.[5] Average consumption of sugar increased dramatically from 1875 through 2000 (Illustration 12.5). Currently, average sugar intake appears to be stabilizing or decreasing globally,[9] but remains high. **Added sugars** make up an average 13.4% of total calorie intake among Americans, which is above the recommended level of less than 10% of total calories.[6]

simple sugars Carbohydrates that consist of a glucose, fructose, or galactose molecule, or a combination of glucose and either fructose or galactose. High-fructose corn syrup and alcohol sugars are also considered simple sugars. Simple sugars are often referred to as sugars.

monosaccharides (mono = one, saccharide = sugar): Simple sugars consisting of one sugar molecule. Glucose, fructose, and galactose are common examples of monosaccharides.

disaccharides (di = two, saccharide = sugar): Simple sugars consisting of two molecules of monosaccharides linked together. Sucrose, maltose, and lactose are disaccharides.

glycogen The body's storage form of glucose. Glycogen is stored in the liver and muscles.

added sugars Sugars that are either added during the processing of foods, or are packaged as such and include sugars (free, mono-, and disaccharides), syrups, naturally occurring sugars that are isolated from a whole food and concentrated, and other caloric sweeteners.

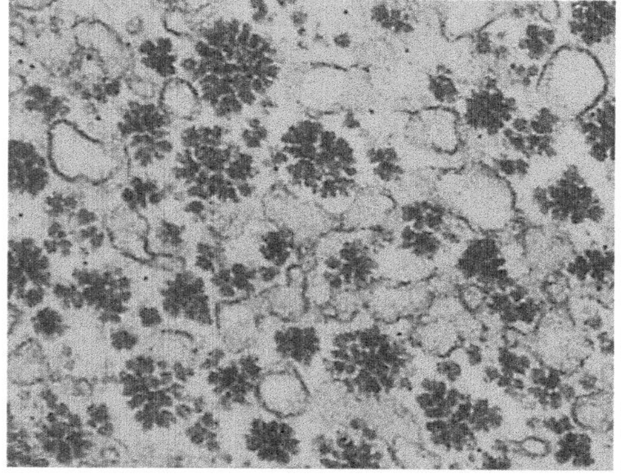

Illustration 12.3 Glycogen in a liver cell. The black "rosettes" are aggregates of glycogen molecules. This cell was photographed under an electron microscope at a magnification of 65,000×.

take action Lower Your Sugar Intake

Concerned about your sugar intake? Want to consider some ways to lower it and to protect your teeth?

Consider these actions and check the options you'd be willing to try.

_______ 1. When you want something sweet to eat, try:
- _______ sweet cherries
- _______ melon
- _______ a banana
- _______ a mango
- _______ unsweetened applesauce

_______ 2. Replace a serving of soft drink or fruit drink with water or a no-added-sugar, 100% juice serving such as:
- _______ tomato juice
- _______ vegetable juice
- _______ dark grape juice
- _______ apple cider/juice
- _______ pineapple juice
- _______ cranberry juice
- _______ grapefruit-tangerine juice
- _______ other juice

_______ 3. Taste-test beverages or gum sweetened with alcohol sugars or artificial sweeteners for acceptability. Try:
- _______ iced tea sweetened with aspartame or sucralose
- _______ soft drinks sweetened with Reb A or aspartame
- _______ gum sweetened with xylitol or other alcohol sugar

_______ 4. Keep sugar off your teeth. After you eat a food with added sugar:
- _______ rinse your mouth with water
- _______ brush your teeth
- _______ floss in between your teeth

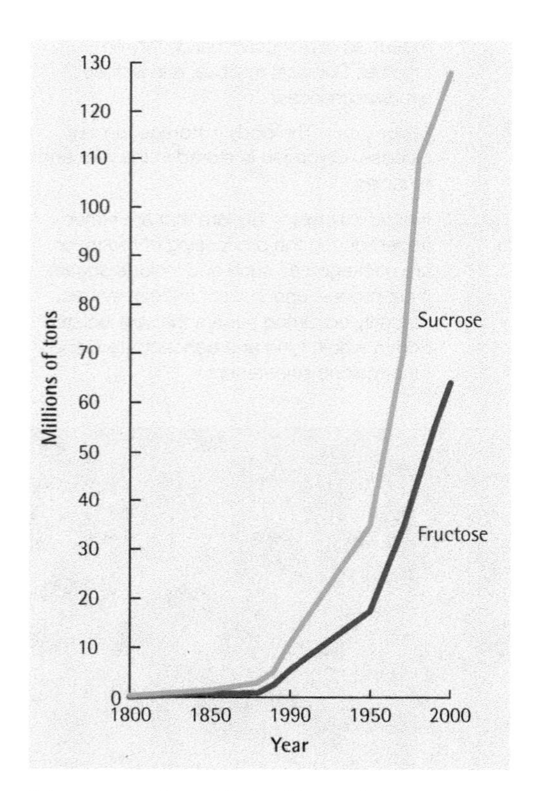

Illustration 12.5 Worldwide trends in sucrose and fructose consumption between 1800 and 2000.

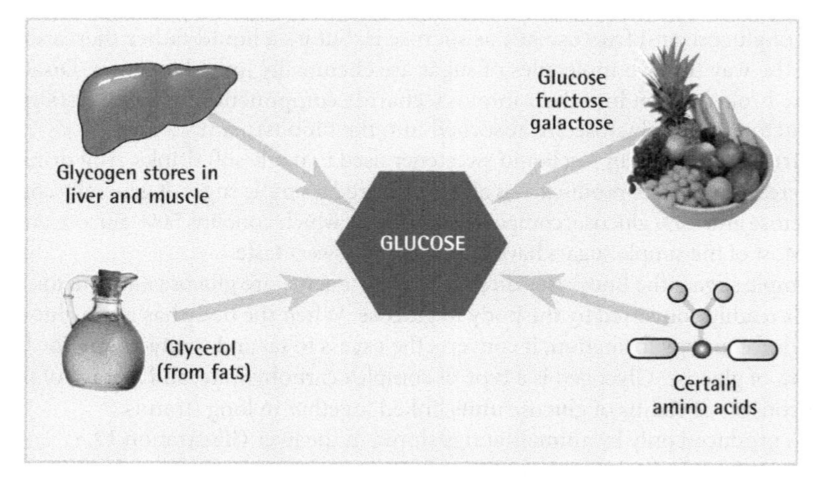

Illustration 12.4 The body's sources of glucose.

Almost all of the added sugar in diets comes from soft drinks, fruits drinks, energy and sports drinks, sweet snacks, candy, and desserts. One 12-ounce serving of a soft drink contains about 9 teaspoons of simple sugar. Simple sugars are also present in fruits, and small amounts are found in some vegetables (Table 12.2). Fruits and vegetables, however, provide an array of vitamins, minerals, fiber, and beneficial phytochemicals that are absent from many foods high in added sugar.[6] Milk is the only animal product that contains significant amounts of a simple sugar (lactose).

Nutrition Labeling of Sugars Nutrition labels must list the total amount of mono- and diglycerides per serving of food under the heading "sugars"

Table 12.2 The simple sugar content of some common foods

	Amount	Simple sugars (grams)[a]	Percent of total calories from simple sugars
Sweeteners:			
Corn syrup	1 tsp	5	100%
Honey	1 tsp	6	100
Maple syrup	1 tsp	4	100
Table sugar	1 tsp	4	100
Fruits:			
Apple	1 medium	16	91
Peach	1 medium	8	91
Watermelon	1 wedge (4″ × 8″)	25	87
Orange	1 medium	14	86
Banana	1 medium	21	85
Vegetables:			
Broccoli	½ cup	2	40
Corn	½ cup	3	30
Potato	1 medium	1	4
Beverages:			
Fruit drinks	1 cup	29	100
Soft drinks	12 oz	38	100
Skim milk	1 cup	12	53
Whole milk	1 cup	11	28
Candy:			
Gumdrops	1 oz	25	100
Hard candy	1 oz	28	100
Caramels	1 oz	21	73
Fudge	1 oz	21	73
Milk chocolate	1 oz	16	44
Breakfast cereals:			
Apple Jacks	1 oz	13	52
Raisin Bran	1 oz	19	40
Cheerios	1 oz	14	4

[a]4 grams sucrose = 1 teaspoon.

(Illustration 12.6). In addition, in the ingredient list, all simple sugars contained in the product must be listed in order of weight. Nutrition labels contain information on total sugars per serving and do not distinguish between sugars naturally present in foods and added sugars. Revised nutrition labels (due out around 2017) will include "added sugars" in the Nutrition Facts panel under "sugars." It is proposed that added sugars be defined as sugar or sugar-containing ingredients that are added during processing.[8]

Added Sugars and Health Foods to which simple sugars have been added are often not among the top sources of nutrients. By themselves, simple sugars are among the few foods that provide only calories. Many foods high in simple sugars, such as cakes, sweet rolls, cookies, pies, and ice cream, are also high in fat and calories. The likelihood that diets will provide insufficient amounts of vitamins and minerals increases along with

Illustration 12.6 Labeling the sugar content of breakfast cereal.

	% Daily Value**	
Total Fat 1 g*	2%	2%
Saturated Fat 0 g	0%	0%
Monounsaturated Fat 0 g		
Polyunsaturated Fat 0.5 g		
Trans Fat 0 g		
Cholesterol 0 mg	0%	0%
Sodium 5 mg	0%	3%
Potassium 200 mg	6%	12%
Total Carbohydrate 48 g	16%	18%
Dietary Fiber 6 g	24%	24%
Sugars 12 g		
Other Carbohydrate 30 g		
Protein 6 g		
Vitamin A	0%	4%
Vitamin C	0%	0%
Calcium	0%	15%
Iron	90%	90%
Thiamin	25%	30%
Riboflavin	25%	35%
Niacin	25%	25%
Vitamin B_6	25%	25%
Folic Acid	25%	25%
Vitamin B_{12}	25%	35%
Phosphorus	15%	25%
Magnesium	15%	20%
Zinc	10%	15%
Copper	10%	10%

		Calories	2,000	2,500
Total Fat	Less than	65 g	80 g
Saturated Fat	Less than	20 g	25 g
Cholesterol	Less than	300 mg	300 mg
Sodium	Less than	2,400 mg	2,400 mg
Potassium		3,500 mg	3,500 mg
Total Carbohydrate		300 g	375 g
Dietary Fiber		25 g	30 g

Calories Per gram: Fat 9 • Carbohydrate 4 • Protein 4

INGREDIENTS: WHOLE GRAIN WHEAT, SUGAR, HIGH FRUCTOSE CORN SYRUP, GELATIN,
VITAMINS AND MINERALS: REDUCED IRON, NIACINAMIDE, ZINC OXIDE, PYRIDOXINE HYDROCHLORIDE (VITAMIN B_6), RIBOFLAVIN (VITAMIN B_2), THIAMIN HYDROCHLORIDE (VITAMIN B_1), FOLIC ACID AND VITAMIN B_{12}, TO MAINTAIN QUALITY, BHT HAS BEEN ADDED TO THE PACKAGING.

sugar intake.[6] Neither sucrose nor high-fructose corn syrup has been found to cause obesity, type 2 diabetes, or attention-deficit hyperactivity disorder. However, high intake of these sweeteners is associated with increased blood triglyceride levels, liver fat accumulation, and greater body weight and an elevated risk of type 2 diabetes, hypertension, and heart disease. It is clear that the frequent consumption of sugary foods is directly related to the development of tooth decay.[6,10,11]

What's the bottom line on eating sugary foods? Enjoy them as treats as part of a healthy dietary pattern that includes less than 10% of total calories from added sugars per day.[6] If you are concerned about the amount of sugar in your diet, don't wait until it's time for a New Year's resolution. Get some suggestions for small changes that will have a positive impact in this unit's Take Action feature.

The Alcohol Sugars—What Are They?

alcohol sugars Simple sugars containing an alcohol group in their molecular structure. The most common are xylitol, mannitol, and sorbitol. They are a subgroup of chemical substances called polyols.

Nonalcoholic in the beverage sense, the **alcohol sugars** (or polyols) are like simple sugars except that they include a chemical component of alcohol. Like simple sugars, the alcohol sugars have a sweet taste. Xylitol is by far the sweetest alcohol sugar—it's much sweeter than the other two common alcohol sugars, mannitol and sorbitol.

Alcohol sugars are found naturally in very small amounts in some fruits. They are mostly used as sweetening agents in gums and candy (Illustration 12.7). Unlike the simple sugars, xylitol, mannitol, and sorbitol do not promote tooth decay because bacteria in the mouth that cause tooth decay cannot digest them.[5] Foods sweetened with alcohol sugars can use the health claim "Does not promote tooth decay" and "Sugar-free" on labels.[8]

Like dietary fiber, the alcohol sugars are slowly and incompletely broken down in the gastrointestinal tract and provide fewer calories per gram than other carbohydrates.[11] On average, alcohol sugars provide two calories per gram, so foods labeled sugar-free will not be calorie-free. High intake of alcohol sugars can cause diarrhea. This characteristic limits their use in foods. The "diarrhea dose" for alcohol sugars defined by the Food and Drug Administration (FDA) equals 50 grams of sorbitol and 20 grams of mannitol.[11] A food product's content of alcohol sugars per serving must be listed on the nutrition label as Sugar Alcohols or by the name of the sugar alcohol.[7]

Scott Goodwin Photography

Illustration 12.7 Examples of products sweetened with xylitol, mannitol, and sorbitol.

Artificial Sweetener Facts

Unwanted calories in simple sugars, the connection of sucrose with tooth decay, the need for a sugar substitute for people with diabetes, and sugar shortages such as occurred during the two world wars have all provided incentives for developing sugar substitutes.[12] Six artificial sweeteners are marketed in the United States (Table 12.3) and more are being developed. None of the artificial sweeteners that are currently approved for use exactly mimic the taste and properties of sugar.[13] Artificial sweeteners can have a bitter or metallic taste for people who are genetically sensitive to bitter tastes.[14]

Artificial sweeteners are also known as nonnutritive sweeteners because they are not a significant source of energy or nutrients.[13] They have chemical properties that invoke an intensely sweet taste on the tongue. Gram for gram, artificial sweeteners are 160 to 13,000 times sweeter than sucrose, and only small amounts are needed to sweeten food products. Illustration 12.8 shows the relative sweetening power of various artificial sweeteners and naturally occurring sugars. Of the artificial sweeteners, only aspartame provides calories (4 calories/gram).[5]

The artificial sweeteners currently on the market do not promote tooth decay because they are not utilized by bacteria in the mouth that cause decay.[5] They do not appear to promote weight loss without calorie restriction.[12] Artificial sweeteners are used to sweeten thousands of products, a few of which are shown in Illustration 12.9. Artificial sweeteners are not considered added sugar, and products containing them can be labeled sugar-free. Their presence in food must be noted in the ingredient label.[11]

Saccharin Saccharin was the first artificial sweetener to be developed. Did you know that it was discovered in a laboratory in the late 1800s? That's right—saccharin is over 100 years old. The availability of this artificial sweetener, which is 300 times as sweet as sucrose, helped relieve the sugar shortages that occurred during World Wars I and II.

Table 12.3 Artificial sweeteners currently approved for use in the United States

Trade name	Product name	Calories/gram
Saccharin	Sweet and Low	0
Aspartame	NutraSweet, Equal Sugar Twin	4
Sucralose	Splenda	0
Acesulfame potassium	Acesulfame K, Sunnette, Sweet One	0
Neotame	—	0
Stevia	Reb-A, Rebiana Truvia, PureVia	0

Scott Goodwin Photography

Illustration 12.8 Some of the thousands of foods that contain artificial sweeteners.

In 1977, saccharin was taken off the market after very high doses were found to cause cancer in laboratory animals. Saccharin was deemed safe in 2000 after scientists concluded there was no evidence that it causes cancer in humans.[15] It is used in some types of toothpastes, diet sodas, mouthwash, pill coatings, juices, and jellies.[16]

Aspartame Early in the 1980s, the artificial sweetener aspartame was approved for use in the United States and more than 90 other countries. Primarily known as NutraSweet, this artificial sweetener is about 200 times sweeter than sucrose.

Aspartame is made from two amino acids (phenylalanine and aspartame). Although both are found in nature, it took chemists to arrange their chemical partnership. Because aspartame is made from amino acids (the building blocks of protein), it supplies four calories per gram. Aspartame is so sweet, however, that very little is needed to sweeten products. Aspartame is used in more than 6,000 products worldwide, including soft drinks, whipped toppings, jellies, cereals, puddings, and some medicines. Products containing aspartame must carry a label warning people with **phenylketonuria (PKU)** (an inherited disease) and others with certain liver conditions about the presence of phenylalanine. People with these disorders are unable to utilize the amino acid phenylalanine, causing it to build up in the blood. Because high temperatures tend to break down aspartame, it is not used in baked or heated products.[11]

Is Aspartame Safe? A safe level of aspartame intake is defined as 50 milligrams per kilogram of body weight per day in the United States and as 40 milligrams per kilogram of body weight in Canada.[17] In food terms, the limit in the United States is equivalent to approximately 20 aspartame-sweetened soft drinks or 55 desserts per day (Illustration 12.10). The average intake of aspartame in the United States, Canada, Germany, and Finland, for example, ranges from 2 to 10 milligrams per day, well below the level of intake considered safe.[18,19]

Some individuals report that they develop headaches, dizziness, or anxiety when small amounts of aspartame are consumed. Studies have failed to confirm these effects. Aspartame has not been found to promote cancer, nerve disorders, or other health problems in humans.[19,20]

Sucralose This noncaloric, intense sweetener is made from sucrose, is safe, and is very sweet (600 times sweeter than sucrose). Known primarily as Splenda on product labels, it is used in both hot and cold food products, including soft drinks, baked goods, frosting, pudding, and chewing gum.

phenylketonuria (feen-ol-key-tone-u-re-ah) (PKU) A rare genetic disorder related to lack of the enzyme phenylalanine hydroxylase. Lack of this enzyme causes the essential amino acid phenylalanine to build up in blood.

- NEOTAME
- SUCRALOSE
- SACCHARIN
- REB-A
- ACESULFAME K
- ASPARTAME
- FRUCTOSE
- SUCROSE
- XYLITOL
- GLUCOSE
- SORBITOL
- MANNITOL
- GALACTOSE
- MALTOSE
- LACTOSE

Illustration 12.9 A ranking of various types of artificial sweeteners and naturally occurring sugars in order of sweetness.[11,12]

Illustration 12.10 You would have to consume more than 20 cans of soft drinks sweetened with aspartame (NutraSweet) a day to exceed the safe limit set for this artificial sweetener.

Acesulfame Potassium Also known as acesulfame K, Sunette, and Sweet One, acesulfame potassium was approved by the FDA in 1988. It is added to at least 4,000 foods and is used in food production in about 90 countries. It is 200 times as sweet as sucrose, provides zero calories, and does not break down when heated.

Neotame Neotame is derived from the same amino acids as aspartame and is extraordinarily sweet. Its sweetness potency is 7,000 to 13,000 times that of sucrose. Only minute amounts of neotame are absorbed when it is consumed, and it is not considered harmful to individuals with PKU. It is marketed as having a sweet taste with little bitter or metallic aftertaste.[5]

Stevia In December 2008, the FDA approved stevia for use as an artificial sweetener. This sweetener is derived from the herb stevia (Illustration 12.11), which grows in subtropical and tropical areas. It has been known in these areas as "sugar leaf" for centuries.

Illustration 12.11 The stevia plant is the source of the artificial sweetener rebiana.

Illustration 12.12 Examples of beverages sweetened with rebiana (Reb-A).

Scott Goodwin Photography

complex carbohydrates The form of carbohydrate found in starchy vegetables, grains, and dried beans and in many types of dietary fiber. The most common form of starch is made of long chains of interconnected glucose units.

polysaccharides (poly = many, saccharide = sugar): Carbohydrates containing many molecules of monosaccharides linked together. Starch, glycogen, and dietary fiber are the three major types of polysaccharides. Polysaccharides consisting of 3 to 10 monosaccharides may be referred to as oligosaccharides.

Stevia leaves contain rebina (Reb A), which imparts the sweet taste and is used in the purified form as an artificial sweetener. It is currently used to sweeten beverages (Illustration 12.12) and reportedly tastes best with citrus flavors.[13]

Complex Carbohydrate Facts

Starches, glycogen, and dietary fiber constitute the **complex carbohydrates**. They are known as **polysaccharides**. Only plant foods such as grains, potatoes, dried beans, and corn that contain starch and dietary fiber are considered dietary sources of complex carbohydrates (Table 12.4). Very little glycogen is available from animal products.

Table 12.4 The complex carbohydrate content of some common foods

	Amount	Complex carbohydrate (grams)	Percent of total calories from complex carbohydrate
Grain and grain products:			
Rice (white), cooked	½ cup	21	83%
Pasta, cooked	½ cup	15	81
Cornflakes	1 cup	11	76
Oatmeal, cooked	½ cup	12	74
Cheerios	1 cup	11	68
Whole wheat bread	1 slice	7	60
Dried beans (cooked):			
Lima beans	½ cup	11	64
White beans	½ cup	13	63
Kidney beans	½ cup	12	59
Vegetables:			
Potato	1 medium	30	85
Corn	½ cup	10	67
Broccoli	½ cup	2	40

Which Foods Have Carbohydrates? Food sources of complex carbohydrates include whole-grain breads, cereals, pastas, and crackers, as well as these same foods produced from refined grains. Whole-grain products provide more fiber and beneficial substances naturally present in grains than do refined grain products. Regular intake of whole-grain foods reduces the risk of heart disease, type 2 diabetes, and some types of cancer.[21]

Are They Fattening? Starchy foods are caloric bargains (Illustration 12.13). A medium baked potato weighs in at only 122 calories, a half-cup of corn at 85 calories, and a slice of whole wheat bread at 70 calories. You can expand the caloric value of complex carbohydrates quite easily by adding fat, sauces, and cheese. One cup of macaroni (about 200 calories) gains around 180 calories when it comes as macaroni and cheese. Adding a quarter cup of gravy to potatoes elevates calories by 150.

Dietary Fiber What is very low in calories, helps prevent constipation, encourages the growth of beneficial microorganisms in the gut, lowers the risk of a number of health disorders, and is generally under-consumed by people in the United States? The answer is **dietary fiber**.[22,23]

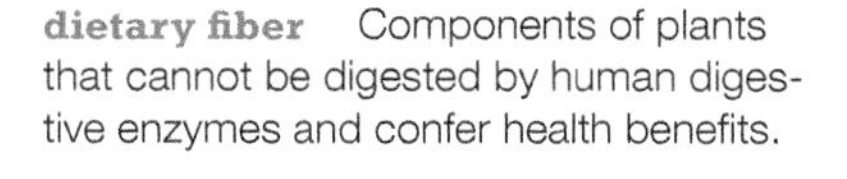

dietary fiber Components of plants that cannot be digested by human digestive enzymes and confer health benefits.

PhotoDisc

©Mark Herreid/Shutterstock.com

Richard Anderson

©Kawia Scharle/Shutterstock.com

Richard Anderson

Richard Anderson

Illustration 12.13 Which has more calories?

Check your answers below.

One medium baked potato (four ounces) *or* three ounces of lean hamburger?

One slice of bread *or* a half cup of low-fat cottage cheese?

One cup of spaghetti noodles *or* 17 french fries (three ounces)?

Answers
Potato = 122 calories;
Lean hamburger = 239 calories.
Bread = 70 calories;
Cottage cheese = 102 calories.
Spaghetti = 197 calories;
French fries = 265 calories.

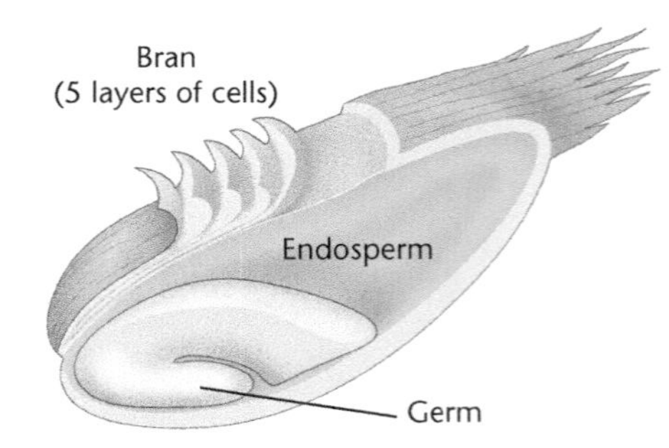

Illustration 12.14 Diagram of a grain of wheat showing the bran, which is a rich source of dietary fiber. The germ contains protein, unsaturated fats, thiamin, niacin, riboflavin, iron, and other nutrients. (The bran and germ are removed in the refining process.) The endosperm primarily contains starch, the storage form of glucose in plants.

The definition of dietary fiber has evolved over time as analytic methods and research on its health effects have advanced. Dietary fibers are now considered to be components of plants that cannot be digested by human digestive enzymes and that confer health benefits.[24,25] They have been categorized into two basic types: soluble and insoluble. Soluble fibers are not fibrous and form gels when combined with water.[26] They are found in foods such as oats, barley, fruit pulp, dried beans, and psyllium husks. Although there are crossovers in functions between soluble and insoluble fiber, soluble fibers have been shown to slow glucose absorption and reduce blood levels of cholesterol.[27] Insoluble fibers do not combine chemically with water; they decrease constipation by increasing stool bulk.[28] This type of fiber is found primarily in oat and wheat bran (Illustration 12.14), the coating on seeds, and the skins of fruits and vegetables. Both types of fiber are important to health. Total dietary fiber intake is related to the risk of heart disease, some cancers, and type 2 diabetes.[6,21,30]

Recommendations for Dietary Fiber Intake Total dietary fiber intake by U.S. adults (15 grams per day on average) is well below the amount recommended (28 grams for women and 35 grams for men).[29] People who consume the recommended amount of dietary fiber tend to select whole-grain breads, high-fiber cereal, and dried beans most days and eat at least five servings of vegetables and fruits daily.[30] Food sources of fiber are listed in Table 12.5. It doesn't matter whether the fiber foods are mashed, chopped, cooked, or raw. They retain their fiber value through it all.

Table 12.5 Examples of good sources of fiber

	Amount	Fiber (grams)
Grain and grain products:		
Bran Buds	½ cup	12.0
All Bran	½ cup	11.0
Raisin Bran	1 cup	7.0
Granola (homemade)	½ cup	6.0
Bran Flakes	¾ cup	5.0
Oatmeal	1 cup	4.0
Spaghetti noodles	1 cup	4.0
Shredded Wheat	1 biscuit	2.7
Whole wheat bread	1 slice	2.0
Bran (dry; wheat, oat)	2 Tbsp	2.0
Fruits:		
Raspberries	1 cup	8.0
Avocado	½ medium	7.0
Mango	1 medium	4.0
Pear (with skin)	1 medium	4.0
Apple (with skin)	1 medium	3.3
Banana	6" long	3.1
Orange (no peel)	1 medium	3.0
Peach (with skin)	1 medium	2.3
Strawberries	10 medium	2.1
Vegetables:		
Lima beans	½ cup	6.6
Green peas	½ cup	4.4
Potato (with skin)	1 medium	3.5
Brussels sprouts	½ cup	3.0
Broccoli	½ cup	2.8

Table 12.5 **Continued**

	Amount	Fiber (grams)
Carrots	½ cup	2.8
Green beans	½ cup	2.7
Collard greens	½ cup	2.7
Cauliflower	½ cup	2.5
Corn	½ cup	2.0
Nuts:		
Almonds	¼ cup	4.5
Peanuts	¼ cup	3.3
Peanut butter	2 Tbsp	2.3
Dried beans (cooked):		
Pinto beans	½ cup	10.0
Peas, split	½ cup	8.2
Black beans (turtle beans)	½ cup	8.0
Lentils	½ cup	7.8
Kidney or navy beans	½ cup	6.9
Black-eyed peas	½ cup	5.3
Fast foods:		
Big Mac	1	3
French fries	1 regular serving	3
Whopper	1	3
Cheeseburger	1	2
Taco	1	2
Chicken sandwich	2	1
Egg McMuffin	1	1
Fried chicken, drumstick	1	1

It was previously assumed that fiber had no calorie value because it is not broken down by human digestive enzymes. It is now known they do supply calories—an average of 2 calories per gram.[32] Bacteria in the colon are able to break down many types of fiber. The bacteria excrete fatty acids as a waste product from fiber digestion and the fatty acids are used as an energy source by the colon and the rest of the body. Dietary fiber, and the fatty acids produced by bacteria from fiber breakdown, are involved in signaling pathways that influence inflammation, **insulin resistance**, fat storage, and other processes related to health outcomes.[38]

insulin resistance A condition is which cell membranes have reduced sensitivity to insulin so that more insulin than normal is required to transport a given amount of glucose into cells.

REALITY CHECK
Sugar Addiction

Have you ever pined or whined for something sweet? Do you think you can get addicted to carbohydrates?

Who gets a thumbs-up?

Answers on page 12-14.

PhotoDisc

Terry: Maybe you could really love to eat them, but addicted? I don't think so.

PhotoDisc

Wolfgang: The more I don't eat candy, the more I want to eat some. I swear, I'm addicted.

ANSWERS TO REALITY CHECK

Sugar Addiction

The theory that consumption of sweet foods may become addictive in some people is being intensely studied, especially among individuals with restrictive followed by binge-eating patterns. Although it is the subject of ongoing research, sweet foods and sugar have *not* been found to be addictive in humans.[33,34]

PhotoDisc

Terry: 👍

PhotoDisc

Wolfgang:

type 2 diabetes A condition characterized by high blood glucose levels due to the body's inability to use insulin normally, or to produce enough insulin. This type of diabetes was called *adult-onset diabetes* in the past.

glycemic index A measure of the extent to which blood glucose is raised by a 50 gram portion of a carbohydrate-containing food compared to 50 grams of glucose or white bread.

Be Cautious When Adding More Fiber to Your Diet Newcomers to adequate fiber diets often experience diarrhea, bloating, and gas for the first week or so of increased fiber intake. These side effects can be avoided. They occur when too much fiber is added to the diet too quickly. Adding sources of dietary fiber to the diet gradually can prevent these side effects. In addition, dietary fiber can be constipating if consumed with too little fluid. Your fluid intake should increase along with your intake of dietary fiber.[35] You know you've got the right amount of fiber in your diet when stools float and are soft and well formed.

The Health Action feature suggests some food choices that will put fiber into your meals and snacks. Fiber should be added carefully to children's diets, because the large volume of some high-fiber diets can cause diarrhea and make them feel full too soon.[36]

Glycemic Index of Carbohydrates

For a long time it was assumed that "a carbohydrate is a carbohydrate is a carbohydrate." It was thought that all types of carbohydrates had the same effect on blood glucose levels and health, so it didn't matter what type was consumed. As is the case with many untested assumptions, this one ultimately fell by the wayside. Some types of simple and complex carbohydrates in foods elevate blood glucose levels more than do others. Such differences may be particularly important to people with disorders such as insulin resistance, hypertension, and **type 2 diabetes**.[37,38]

Carbohydrates and carbohydrate-containing foods are now being classified by the extent to which they increase blood glucose levels. This classification system is called the **glycemic index**. Carbohydrates that are digested and absorbed quickly have a high glycemic index; they raise blood glucose levels to a greater extent than do those with lower glycemic index values. Carbohydrates and carbohydrate-containing foods with high glycemic index values include glucose, white bread, baked potatoes, and jelly beans. Fructose, dried beans, bananas, and oatmeal are examples of carbohydrates and carbohydrate-containing foods with low glycemic indexes. Table 12.6 lists examples of the glycemic index of foods with high and low glycemic indices. You can find additional glycemic index values for foods on the Internet by searching the term *international table of glycemic index.*

Table 12.6 Examples of food sources of carbohydrates with high or low glycemic index[39]

	Glycemic Index
High glycemic index	
Glucose	103
Potatoes, instant mashed	87
Rice crackers, crisps	87
Cornflakes	81
Potato, boiled	78
White bread	75
White rice	73
Watermelon	76
Low glycemic index	
Spaghetti, white	49
Yogurt, fruit	41
Carrots, boiled	39
Milk, skim	37
Soy milk	34
Lentils	32
Chickpeas	28
Barley	28

Compared with high glycemic index carbohydrate foods, intake of low glycemic index carbohydrate foods is associated with:

- decreased blood glucose and insulin levels
- increased satiety (feeling of fullness)
- decreased food intake
- decreased activation of brain regions that motivate people to eat
- increased weight loss
- decreased blood levels of glucose and insulin[40,41]

health action Putting the Fiber into Meals and Snacks

HIGH-FIBER OPTIONS FOR BREAKFAST

Food	Amount	Fiber
Whole grain toast		2 g per slice
Bran cereal:		
Bran flakes	1 cup	7 g
All Bran	1/2 cup	11 g
Raisin bran	1 cup	7 g
Oat bran	1/3 cup	5 g
Bran muffin, with fruit:	1 small	3 g
Strawberries	10	2 g
Raspberries	1/2 cup	3 g
Bananas	1 medium	2 g

LUNCHES THAT INCLUDE FIBER

Food	Amount	Fiber
Whole grain bread		2 g per slice
Baked beans	1/2 cup	10 g
Carrot	1 medium	2 g
Raisins	1/4 cup	2 g
Peas	1/2 cup	4 g
Peanut butter	2 tablespoons	2 g

FIBER ON THE MENU FOR SUPPER

Food	Amount	Fiber
Brown rice	1/2 cup	2 g
Potato	1 medium	3 g
Dried cooked beans	1/2 cup	8 g
Broccoli	1/2 cup	3 g
Corn	1/2 cup	3 g
Tomato	1 medium	2 g
Green beans	1/2 cup	3 g

FIBER-FILLED SNACKS

Food	Amount	Fiber
Peanuts	1/4 cup	3 g
Apple	1 medium	2 g
Pear	1 medium	4 g
Orange	1 medium	3 g
Prunes	3	2 g[a]
Sunflower seeds	1/4 cup	2 g
Popcorn	2 cups	2 g

PhotoDisc

[a]Prunes contain fiber, but their laxative effect is primarily due to a naturally occurring chemical substance that causes an uptake of fluid into the intestines and the contraction of muscles that line the intestines.

Much more needs to be learned about the effects of various forms of carbohydrates on health. The importance of glycemic index to health as now measured is still somewhat controversial.[43] It is clear, however, that different types of carbohydrates influence bodily processes differently. Healthy dietary patterns now broadly recommended include low glycemic index sources of carbohydrates.[6]

Carbohydrates and Dental Health

- **Explain relationships between carbohydrates and dental health.**

The relationship between sugar and **tooth decay** is very close, and the history of tooth decay closely parallels the availability of sugar. The incidence of tooth decay is estimated to have been very low (less than 5%) among hunter-gatherers, who had minimal access to sugars.[44] Tooth decay did not become a widespread problem until the late

tooth decay The disintegration of teeth due to acids produced by bacteria in the mouth that feed on sugar. Also called dental caries or cavities.

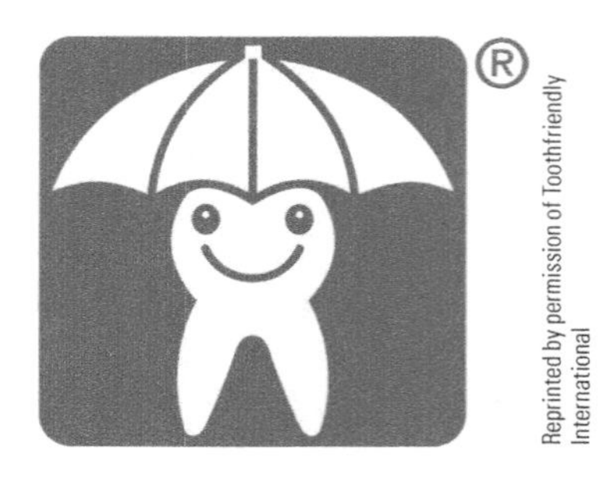

Illustration 12.15 Switzerland's "Tooth Friendly" symbol. It has become an internationally used symbol.

seventeenth century, when great quantities of sucrose were exported from the New World to Europe and other parts of the world. When sugar shortages occurred in the United States and Europe during World Wars I and II, rates of tooth decay declined; they rebounded when sugar became available again.[44] Rates of tooth decay in children vary substantially among countries, but the highest rates are in countries where sugar is widely available in processed foods and beverages. Tooth decay is spreading rapidly in developing countries where sugar, candy, soft drinks, and fruit drinks are becoming widely available.[45]

Sweets are not the only culprit. Simple sugars that promote tooth decay can also come from starchy foods, especially pretzels, crackers, and breads that stick to your gums and teeth. Some of the starch is broken down to simple sugars by enzymes in the mouth.

To reduce the incidence of tooth decay, a number of countries have adopted campaigns to help inform consumers about cavity-promoting foods. Switzerland and other countries label foods that are safe for the teeth with a "Tooth Friendly" symbol (Illustration 12.15) and encourage the use of alcohol sugars (which don't promote tooth decay) in gums and candies.[45] Other countries recommend that sweets be consumed with meals or that teeth be brushed after sweets are eaten.

There's More to Tooth Decay Than Sugar Per Se

How frequently sugary and starchy foods are consumed and how long they stick to gums and teeth make a difference in their tooth decay-promoting effects. Marshmallows, caramels, and taffy, for example, are much more likely to promote tooth decay than are apples and milk chocolate. Nevertheless, all of the foods listed in Table 12.7 can promote tooth decay if allowed to remain in contact with the gums and teeth. Drinking coffee or tea with sugar throughout the day or consuming three or more regular soft drinks between meals hastens tooth decay (more than if these beverages are consumed with meals).[46] Candy, cookies, and crackers eaten between meals are much more likely to promote tooth decay than are the same foods consumed as part of a meal. Chewing as few as two sticks of sugar-containing gum a day also significantly increases tooth decay.[47]

Why Does Sugar Promote Tooth Decay? Sugar promotes tooth decay because it is the sole food for certain bacteria that live in the mouth and excrete acid that dissolves the enamel covering on teeth. In the presence of sugar, bacteria in the mouth multiply rapidly and form a sticky white material called **plaque**. Tooth areas covered by plaque are prime locations for tooth decay because they are dense in acid-producing bacteria. Acid

plaque A soft, sticky, white material on teeth; formed by bacteria.

Table 12.7 The "stickiness" value of some foods[46,49]
The stickier the food, the worse it is for your teeth.

Very sticky	Sticky	Somewhat sticky	Barely sticky
Caramels	Doughnuts	Bagels	Apples
Chewy cookies	Figs	Cake	Bananas
Crackers	Frosting	Cereal	Fruit drinks
Cream-filled cookies	Fudge	Dry cookies	Fruit juices
Granola bars	Hard candy	Milk chocolate	Ice cream
Marshmallows	Honey	Rolls	Oranges
Pretzels	Jelly beans	White bread	Peaches
Taffy	Pastries		Pears
	Raisins		
	Syrup		

production by bacteria increases within five minutes of exposure to sugar. It continues for 20 to 30 minutes after the bacteria ingest the sugar.[48] If teeth are frequently exposed to sugar, the acid produced by bacteria may erode the enamel, producing a cavity. If the erosion continues, the cavity can extend into the tooth and allow bacteria to enter the inside of the tooth. That can cause an infection and loss of the tooth. It can be prevented if the plaque is removed before the acid erodes much of the enamel. Teeth are capable of replacing small amounts of minerals lost from enamel.[50] Table 12.8 lists foods that do not promote tooth decay.

Table 12.8 Foods that don't promote tooth decay[46,49]

Artificial sweeteners
Gum and candy sweetened with alcohol sugars
Peanut butter
Cheese
Tea
Coffee (no sugar)
Meats
Water
Eggs
Milk
Yogurt (plain)
Fats and oils
Nuts
Vegetables
Fresh fruit

Water Fluoridation In the early 1930s, lower rates of tooth decay were observed among children living in areas where water naturally contained fluoride. This provided the initial evidence that led to the fluoridation of many community water supplies. Fluoridated water reduces the incidences of tooth decay by 50% or more and is primarily responsible for declining rates of tooth decay and loss.[46]

Credit for declines in tooth decay and loss in the United States is shared by fluoride supplements, toothpastes, rinses and gels, protective sealants, and improved dental hygiene. Further improvements in rates of dental caries will occur with reduced intake of sugars and sticky carbohydrates, and behaviors such as rinsing the mouth with water after consuming sugar and chewing sugar-free gum after eating sweet foods. Fluoridation is a safe, effective, and inexpensive method of controlling dental disease.[46] Despite the advantages of fluoride, 31% of Americans consume water from a less than optimally fluoridated water supply.[51] Few bottled waters contain fluoride.

Baby Bottle Caries A startling example of the effect that frequent and prolonged exposure to sugary foods can have is *baby bottle caries* (Illustration 12.16). Infants and young children who routinely fall asleep while sucking a bottle of sugar water, fruit drink, milk, or formula—or while breastfeeding—may develop severe decay. After the child falls asleep, the fluid may continue to drip into the mouth. A pool of the fluid collects between the tongue and the front teeth, bathing the teeth in the sweet fluid for as long as the child sleeps. The upper front teeth become decayed first because the tongue protects the lower teeth. Baby bottle caries occur in 5 to 10% of infants and young children and can lead to the destruction of all the baby teeth.[52]

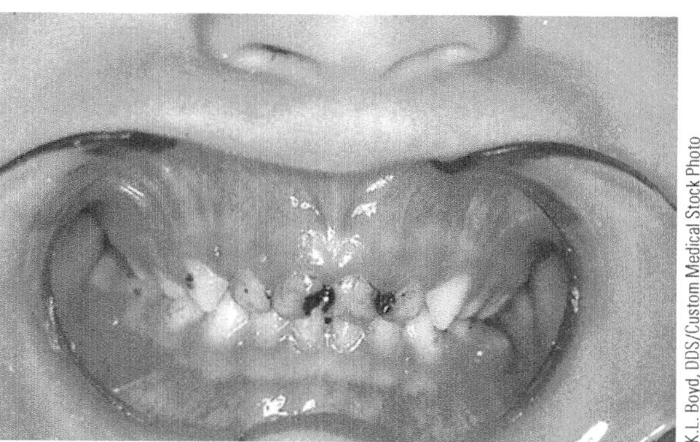

K.L. Boyd, DDS/Custom Medical Stock Photo

Illustration 12.16 "Baby bottle caries" (also called "nursing bottle syndrome") occurs in infants who habitually receive sweet fluids or milk in bottles when they go to sleep. Cavities occur first in the upper front teeth because that's where fluid pools when babies sleep.

Lactose Maldigestion and Intolerance

- **Discuss the foods and beverages that can be consumed by individuals with lactose maldigestion to assure adequate intake of calcium and vitamin D.**

A very common digestive disorder is **lactose maldigestion**. The lactose found in milk and milk products presents a problem for most of the world's adults, who cannot digest it, either partially or completely (Table 12.9). The condition occurs more commonly in population groups that have no historical links to dairy farming and milk drinking.[53] Early humans in central and northwestern Europe and the regions of Africa and China highlighted in Illustration 12.17 tended to raise dairy animals and drink milk.

Lactose maldigestion is caused by a genetically determined low production of lactase, the enzyme that digests lactose. People who lack this enzyme end up with free lactose in their large intestine after they consume milk or milk products. The presence of lactose in the large intestine produces the symptoms of **lactose intolerance**: a bloated feeling, diarrhea, gas, and abdominal cramping.[53]

Lactose maldigestion is rare in very young children and affects adults to various degrees. Some adults produce little or no lactase and develop symptoms of lactose intolerance when they consume even small amounts of milk or milk products. Others produce

lactose maldigestion A disorder characterized by reduced digestion of lactose due to low availability of the enzyme lactase.

lactose intolerance The term for gastrointestinal symptoms (flatulence, bloating, abdominal pain, diarrhea, and "rumbling in the bowel") resulting from consumption of more lactose than can be digested with available lactase.

Table 12.9 Estimated incidence of lactose maldigestion among older children and adults in different population groups[53,54]

	Incidence of lactose maldigestion
Asians	90%
Africans	70
Native Americans	60–80%
Mexicans	70%
U.S. adults (overall)	25
Northern Europeans	20

some lactase and can tolerate limited amounts of lactose-containing milk and milk products, such as a cup of milk at a time or two cups of milk consumed with meals during the day.[55] Regular consumption of milk may improve lactose digestion due to enhanced bacterial breakdown of lactose in the gut.[56] Lactase tablets can help improve lactose digestion, but the benefit is limited due to the inactivation of lactase by digestive processes in the stomach. Once lost, the body's ability to produce sufficient lactase cannot be restored.[57]

Many people who are lactose maldigesters have no trouble eating yogurt and other fermented milk products such as cultured buttermilk, kefir, and aged cheese. The bacteria used to culture yogurt can digest half or more of the lactose. This reduction in lactose content is sufficient to prevent adverse effects in many people with lactose maldigestion.[53]

Milk solids, milk, and other lactose-containing components of milk may be added to foods you wouldn't expect. Consequently, it's best to examine food ingredient labels when in doubt. Milk, for instance, is a primary ingredient in some types of sherbet, and milk solids are added to many types of candy.

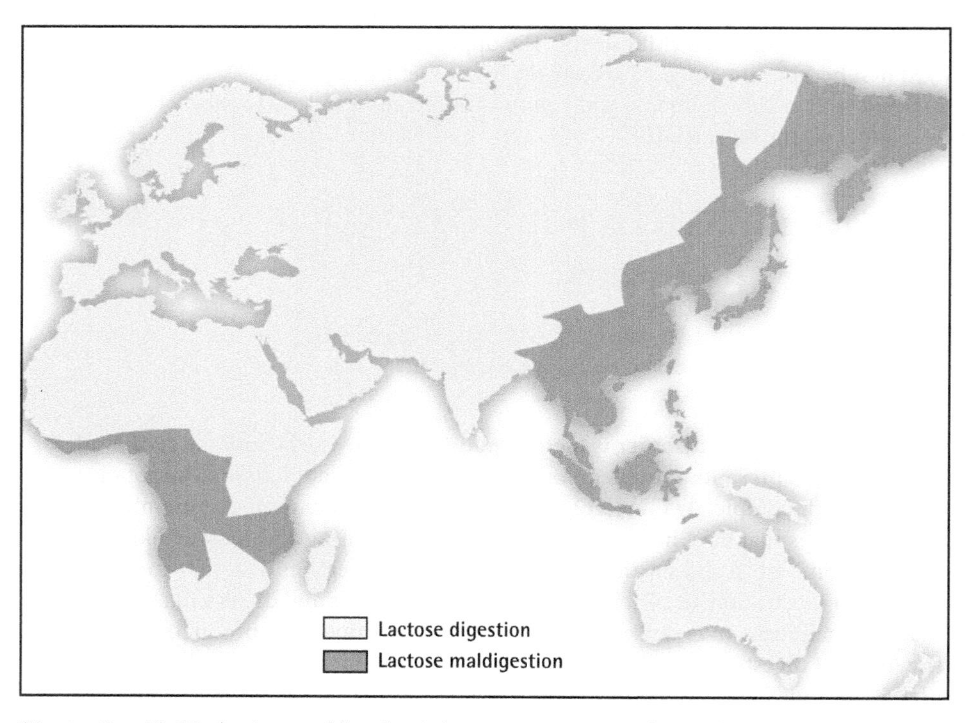

Illustration 12.17 Lactose maldigestion is less common among descendants of people who consumed milk from domesticated animals during prehistoric times (light areas) than among people whose early ancestors did not drink milk (dark areas).[58]

The single most reliable indicator of lactose maldigestion is the occurrence of lactose intolerance within hours after consuming lactose.[53] If you consistently experience symptoms of lactose maldigestion, visit your health care provider for a diagnosis. The symptoms could be due to lactose, other substances in milk, or another problem.

How Is Lactose Maldigestion Managed?

Lactose maldigestion should *not* be managed by omitting milk and milk products from the diet! Doing so would exclude a food group that contributes a variety of nutrients that cannot easily be replaced by other foods. The omission of milk and milk products from the diet of people with lactose intolerance promotes the development of osteoporosis.[53] Rather, fortified soy or rice milk, low-lactose cow's milk, milk pretreated with lactase drops, yogurt and other fermented milk products (if tolerated), as well as ready-to-eat cereals and fruit juices fortified with calcium and vitamin D should be consumed. Illustration 12.18 shows a variety of dairy products that are generally well tolerated by people with lactose maldigestion.

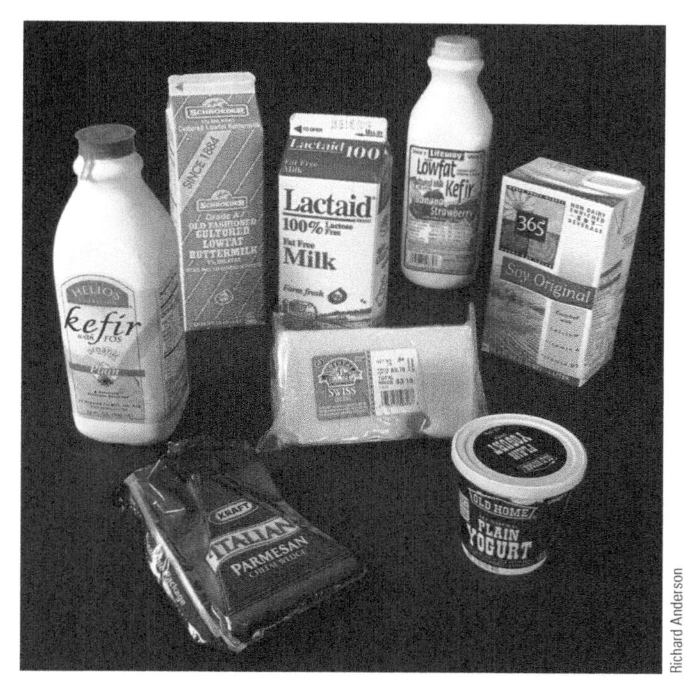

Illustration 12.18 Dairy and fortified products generally well tolerated by individuals with lactose maldigestion.

John E. Kelly/Photolibrary/
Getty Images

NUTRITION up close

Does Your Fiber Intake Measure Up?

Focal Point: Approximate the amount of fiber your diet contains.

Are you meeting your fiber quota, or do you consume the typical low-fiber American diet? To determine if your fiber intake is adequate, award yourself the allotted number of points for each serving of the following foods that you eat in a typical day. For example, if you normally eat one slice of whole-grain bread each day, give yourself 2 points. If you eat two slices daily, give yourself 4 points. After tallying your score, refer to the Feedback section for the results.

High-fiber food choices	Number of points
Fruits: 2 points for each serving	
1 whole fruit (e.g., apple, banana)	
½ cup cooked fruit	
¼ cup dried fruit	
Grains: 2 points for each serving	
½ cup cooked brown rice	
1 whole-grain slice of bread, roll, muffin, or tortilla	
½ cup hot whole-grain cereal (e.g., oatmeal)	
¾ cup cold whole-grain cereal (e.g., Cheerios)	
2 cups popcorn	
Nuts and seeds: 2 points for each serving	
¼ cup seeds (e.g., sunflower)	
2 tablespoons peanut butter	
Vegetables: 3 points for each serving	
1 whole vegetable (e.g., potato)	
½ cup cooked vegetable (e.g., green beans)	
Bran cereals: 7 points for each serving	
½ cup cooked oat bran cereal	
1 cup cold bran cereal	
Legumes: 8 points for each serving	
½ cup cooked beans (e.g., baked beans, pinto beans)	
Total score	

Feedback to the Nutrition Up Close can be found in Appendix F.

REVIEW QUESTIONS

- **Describe the food sources and functions of simple sugars, complex carbohydrates, fiber, and alcohol sugars.**

1. Excluding fiber, one gram of carbohydrate provides four calories. **True/False**
2. Fructose is a monosaccharide, maltose is a disaccharide, and glycogen is a trisaccharide. **True/False**
3. When the body has more glucose than it needs for energy formation, the excess glucose is converted to glycogen and fat. **True/False**
4. Sugars are the most commonly used food additive. **True/False**
5. Excess sugar intake is the primary cause of obesity in developed countries. **True/False**
6. Xylitol and sorbitol are examples of alcohol sugars. **True/False**
7. Excess intake of aspartame (also called NutraSweet) causes headaches in a majority of children and adults. **True/False**
8. Complex carbohydrates are also known as polysaccharides. **True/False**
9. All types of fiber share the characteristic of providing four calories per gram. **True/False**
10. Carbohydrates and carbohydrate-containing foods are classified by their glycemic index, or the extent to which they increase blood glucose levels. **True/False**
11. ______ Which of the following statements represents a documented relationship between simple sugar intake and health?
    a. High simple sugar intake is related to the risk of developing type 2 diabetes and obesity.
    b. Excessive simple sugar intake increases the onset of attention-deficit hyperactivity disorder in children.
    c. High sugar intake is related to increased blood lead levels.
    d. It is recommended that Americans consume ≤5% of total calories as sugar.

- **Explain relationships between carbohydrates and dental health.**

12. Declining rates of dental caries in the United States are primarily related to increased access to fluoridated water supplies. **True/False**
13. Children who consume water at home from a fluoridated water supply are protected from developing dental caries if they consume candy and other high-sugar foods regularly. **True/False**
14. ______ Which of the following measures is associated with the prevention of dental caries?
    a. Frequent intake of dried fruits.
    b. Regular consumption of sweetened beverages.
    c. Use of chewing gum sweetened with xylitol.
    d. Consumption of milk right before bedtime.

- **Discuss the foods and beverages that can be consumed by individuals with lactose maldigestion to assure adequate intake of calcium and vitamin D.**

15. Lactose intolerance refers to the disorder associated with low or absent lactase availability, whereas lactose maldigestion refers to the symptoms of gas and bloating experienced by people who cannot digest lactose. **True/False**
16. Individuals who lack the enzyme lactase are at increased risk of developing osteoporosis. **True/False**
17. ______ When he was young, Hank remembers hearing his father say, "I can't drink milk. It tears me up inside." This means:
    a. His father has lactose maldigestion.
    b. Hank will develop lactose maldigestion, too.
    c. His father is lactose intolerant.
    d. Unless diagnosed, you cannot know why his father has a problem he associates with milk drinking.
18. ______ Individuals with lactose maldigestion can replace the calcium and vitamin D they do not get from milk with food such as:
    a. Ice cream, cream, and coffee creamer.
    b. Fortified soy and rice milk, low-lactose milk.
    c. Fortified margarine, fruit drinks, and salad dressings.
    d. Raisins, cheese, and whole-grain bread.

Answers to these questions can be found in Appendix F.

NUTRITION SCOREBOARD ANSWERS

1. If you get this right, you may be in the minority. In one study of college students, only 38% could identify good sources of complex carbohydrates. Rice, crackers, grits, dried beans, corn, peas, tortillas, biscuits, and oatmeal are also good sources of complex carbohydrates. **True**
2. That's true. Both sweetened and unsweetened cereals consist primarily of carbohydrates. A gram of carbohydrate provides four calories, whether the source is sugar or flakes of corn. **True**
3. That's a lot of sugar! **True**
4. Not all foods high in fiber are fibrous, or crunchy when you eat them. Some form gels. **False**
5. Cooking doesn't destroy dietary fiber. **False**
6. Rates of tooth decay increase in populations as sugar intakes increases. **True**

Diabetes Now

Rosenfeld/AGE Fotostock

NUTRITION SCOREBOARD

1 Excess sugar consumption is the primary cause of diabetes. **True/False**

2 Diabetes generally develops over the course of many years. **True/False**

3 It is almost impossible to prevent or postpone the development of type 2 diabetes. **True/False**

4 Dietary recommendations for individuals with type 1 and type 2 diabetes are substantially different than the recommendations for a healthy dietary pattern. **True/False**

Answers can be found at the end of the unit.

LEARNING OBJECTIVES

Upon completing Unit 13 you will be able to:

- Describe risk factors for development of diabetes and approaches to its management and prevention.

diabetes A disease characterized by abnormal utilization of carbohydrates by the body and elevated blood glucose levels. There are three main types of diabetes: type 1, type 2, and gestational diabetes. The word *diabetes* in this unit refers to type 2, which is by far the most common form of diabetes.

type 1 diabetes A disease characterized by high blood glucose levels resulting from destruction of the insulin-producing cells of the pancreas. This type of diabetes was called juvenile-onset diabetes and insulin-dependent diabetes in the past, and its official medical name is type 1 diabetes mellitus.

type 2 diabetes A disease characterized by high blood glucose levels due to the body's inability to use insulin normally, or to produce enough insulin. This type of diabetes was called adult-onset diabetes and non-insulin-dependent diabetes in the past, and its official medical name is type 2 diabetes mellitus.

gestational diabetes Diabetes first discovered during pregnancy.

The Diabetes Epidemic

- **Describe risk factors for development of diabetes and approaches to its management and prevention.**

It's not the plague, yellow fever, or heart disease. The latest worldwide disease epidemic is **diabetes**, and the rising rates are directly related to the global increase in obesity (Illustration 13.1). Diabetes affects 8.5% of adults worldwide,[5] and 9.4% of the U.S. population has been diagnosed with dabetes. Less than 1% of U.S. adults were diagnosed with type 2 diabetes in 1960. One-fourth of U.S. adults with diabetes have yet to be diagnosed and are not receiving health care services for the disorder.[6]

KEY NUTRITION CONCEPTS

Two key nutrition concepts are directly related to this content on carbohydrates and other components of diet on the development of diabetes:

1. Nutrition Concept #3: Health problems related to nutrition originate within cells.
2. Nutrition Concept #8: Poor nutrition can influence the development of certain chronic diseases.

There are three major forms of diabetes: **type 1**, **type 2**, and **gestational diabetes**. Type 2 diabetes is the most common by far and is fueling the diabetes epidemic. Table 13.1 summarizes key features of type 1 and type 2 diabetes. Both types are diagnosed when fasting levels of blood glucose are 126 milligrams/deciliter (mg/dl) and higher; these types generally take years to develop.[2] In all cases of diabetes, the central problem is an elevated blood glucose level caused by an inadequate supply of insulin, ineffective utilization of insulin, or both.[7]

Insulin is a hormone produced by the pancreas that performs many functions, one of which is to reduce blood glucose levels after meals. By facilitating the passage of glucose into cells, insulin keeps a steady supply of glucose going into cells. Glucose is needed by cells as a source of energy for thousands of chemical reactions that participate in the maintenance of ongoing body functions and health. If insulin is produced in insufficient amounts, or if cell membranes are not sensitive to the action of insulin, cells become starved for glucose. Functional levels of multiple tissues and organs in the body degrade as a result. High levels of blood glucose are related to adverse side effects in the body, as well, such as elevated blood levels of triglycerides, **chronic inflammation**, increased blood pressure, and hardening of the arteries.[8,9]

Symptoms associated with high blood glucose levels vary somewhat among individuals. Common signs of elevated glucose level include:

- Increased thirst
- Increased hunger
- Headaches
- Difficulty concentrating
- Blurred vision
- Frequent urination
- Fatigue (weak, tired feeling)
- Weight loss[10]

HEALTH

The Global Epidemic of Diabesity: A Major Threat to Human Health

Associated Press

WORLD & NATION

Diabetes Threat on the Rise among U.S. Children, Specialists Say

Associated Press

WASHINGTON, D.C. —

CDC's Forecast for Diabetes is Grim

By Maria Elena Baca
Star Tribune Staff Writer

In a recent government study, the Centers for Disease Control and Prevention (CDC) estimated that obesity is fast approaching tobacco as the top underlyi- -reventable

percent of U.S. adults are considered overweight, and an additional 31 percent are obese.

Anyone with a body mass index (a ratio between your height and weight) of 25 or above -- that's someone, for example, who

considered overweight, according to the National Institutes of Health. Anyone with a body mass index of 30 or above -- such as someone who is 5-foot-6 and 186 pounds -- is considered obese. Check your bod- mass index here!

Illustration 13.1 Diabetes headlines say it all.

Table 13.1 Key characteristics of type 1 and type 2 diabetes[6]

Characteristic	Type 1 diabetes	Type 2 diabetes
Insulin deficiency?	Yes	Occurs in advanced stages of the disease
Proportion of cases	5–10%	90–95%
Risk factors	Genetic predisposition to effects of early life exposure to complex dietary proteins or other environmental factors that initiate an autoimmune reaction that destroys the part of the pancreas that produces insulin	Obesity (especially abdominal fatness), sedentary lifestyle, insulin resistance, low weight at birth, genetic traits, older age
Treatment	Insulin, individualized healthy dietary pattern, exercise	Weight loss (in most cases), exercise, individualized healthy dietary pattern, sometimes oral medications, and/or insulin

chronic inflammation Low-grade inflammation that lasts weeks, months, or years. Inflammation is the first response of the body's immune system to infection or irritation. Inflammation triggers the release of biologically active substances that promote oxidation and other potentially harmful reactions in the body.

Health Consequences of Diabetes

Health effects of diabetes vary depending on how well blood glucose levels are controlled and on the presence of other health problems such as hypertension or heart disease. In the short run, poorly controlled and untreated diabetes produces blurred vision, frequent urination, weight loss, increased susceptibility to infection, delayed wound healing, and extreme hunger and thirst. In the long run, diabetes contributes to heart disease, hypertension, nerve damage, blindness, kidney failure, stroke, and the loss of limbs due to poor circulation. The number one cause of death among people with diabetes is heart disease.[11] Many of the side effects of diabetes can be diminished or their onset delayed if blood glucose levels are well controlled.[12]

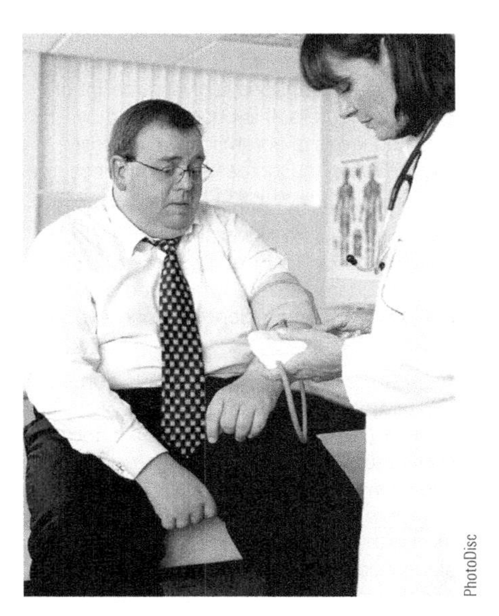

Illustration 13.2 Obesity characterized by central body fat stores and physical inactivity is a strong risk factor for type 2 diabetes.

Type 2 Diabetes

The development of this common type of diabetes is most likely to occur in overweight and obese, inactive people (Illustration 13.2). The term *diabesity* is being used to describe the close relationship between obesity and type 2 diabetes, and the current surge in worldwide rates of both is being called the diabesity epidemic.[13] Approximately 60% of people who develop type 2 diabetes are obese, and 30% are overweight.[12] Illustration 13.3 shows the close relationship between rising rates of obesity and type 2 diabetes in the United States.

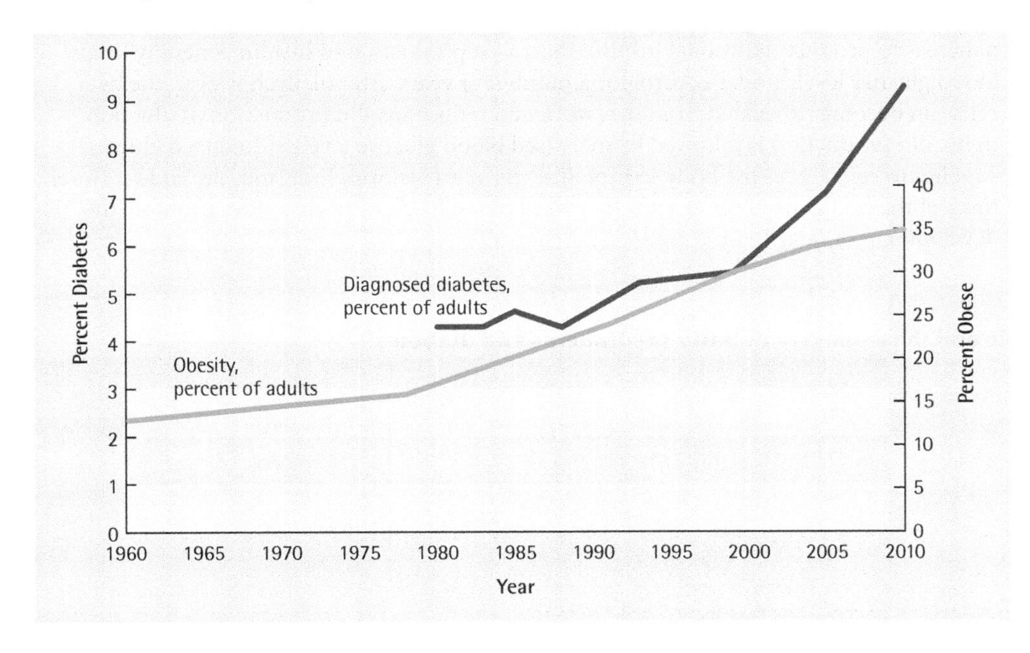

Illustration 13.3 The incidence of type 2 diabetes increases as rates of obesity increase.[6]

Although most often diagnosed in people over the age of 40, type 2 diabetes is becoming increasingly common in children and adolescents.[6] There are genetic components to this disease, as evidenced by the fact that it tracks in families and is more likely to occur in certain groups (Hispanic American, African American, Asian and Pacific Islanders, and Native Americans) than others.[6] Rather than inheriting genes that cause diabetes, certain individuals inherit or acquire very early in life multiple genetic traits that increase the likelihood that diabetes will develop. These genetic traits lead to metabolic changes that promote the development of diabetes when a person is exposed to certain environmental triggers, such as an obesity-promoting environment that offers a wide assortment of palate-pleasing and affordable foods (such as high-sugar and high-fat foods), and little requirement for physical activity in everyday life.[15,16]

The development of type 2 diabetes is characterized by a number of metabolic changes that can be identified before blood levels of glucose become high enough to qualify for the diagnosis. Type 2 diabetes usually begins with **insulin resistance**, which is followed by increased blood glucose levels. Small increases in blood glucose levels indicate the presence of **prediabetes**. If left unmanaged, prediabetes can proceed over time to become type 2 diabetes.[6]

insulin resistance A condition in which cell membranes have reduced sensitivity to insulin so that more insulin than normal is required to transport a given amount of glucose into cells. It is characterized by elevated levels of serum insulin, glucose, triglycerides, and increased blood pressure.

prediabetes A condition in which blood glucose levels are higher than normal but not high enough for the diagnosis of diabetes. It is characterized by impaired glucose tolerance, or fasting blood glucose levels of 100 to ≤125 mg/dL.

fatty liver disease A reversible condition characterized by fat infiltration of the liver (10% or more by weight). If not corrected, advanced forms of fatty liver disease can produce liver damage and other disorders. The condition is primarily associated with obesity, diabetes, and excess alcohol consumption. Also referred to as nonalcoholic fatty liver disease (NAFLD).

hemoglobin A1c A measure of the percentage of hemoglobin proteins in red blood cells that are attached to glucose. The higher the blood glucose levels, the more glucose will become attached to hemoglobin. Once glucose binds with hemoglobin, it will stay there for the lifespan of the hemoglobin (normally about 120 days). Consequently, hemoglobin A1c represents blood glucose levels over several months.

Prediabetes Individuals at risk of developing type 2 diabetes based on somewhat elevated blood glucose levels are considered to have prediabetes, which is defined by the blood glucose cut-points listed in Table 13.2. This table also shows blood glucose levels considered normal and those used to diagnose diabetes. Approximately 35% of the U.S. population is at risk of type 2 diabetes due to the existence of prediabetes.[6] Abdominal obesity, physical inactivity, unhealthy dietary habits, and genetic predisposition are common risk factors for the development of prediabetes, and insulin resistance is also thought to play a major role.[17,19] Medical nutrition therapy that successfully helps people with prediabetes follow a healthy dietary pattern, lose 7% of their initial body weight (if needed), and exercise 150 minutes per week is associated with a reversal of prediabetes or a substantial reduction in its progression to type 2 diabetes.[4,20,21]

Insulin Resistance Insulin resistance occurs due to abnormalities in the way the body uses insulin. Normally, insulin is able to lower blood glucose levels after meals by binding to cell membrane receptors in muscle, fat, and liver cells, enabling glucose to pass into the cells. With insulin resistance, cell membrane receptors "resist" the effects of insulin, and that lowers the amount of glucose that is transported into cells. Blood glucose that cannot make its way into cells is converted to glycogen and fat by the liver. If present in great enough amounts, fat produced by the excess supply of glucose will be stored in the liver and transferred for storage in muscle and fat tissue in the form of triglycerides.[7]

The body responds to elevated levels of blood glucose by signaling the beta cells of the pancreas to produce additional insulin. Increased production of insulin generally keeps blood glucose levels under control for a number of years. Eventually, however, the beta cells can become exhausted from overwork and reduce insulin production. Reduction in insulin production is followed by increased blood glucose levels.[6,7] Reduced glucose supplies to cells forces the body to mobilize triglycerides from liver, muscle, and fat stores to meet the body's need for energy. Increased blood levels of triglycerides promote the development of **fatty liver disease**.[22]

Table 13.2 Normal blood glucose levels and those used to identify prediabetes and diabetes[76,77]

Blood glucose test[a]	Normal values	Prediabetes	Diabetes
Fasting blood glucose	<100 mg/dL	100–≤125 mg/dL	≥126 mg/dL
Blood glucose measured 2 hr post glucose load	<140 mg/dL	140–≤199 mg/dL	≥200 mg/dL
Random blood glucose	—	—	≥200 mg/dL
Hemoglobin A1c	4–6%	5.7–≤6.4%	≥6.5%

[a]It is recommended that blood glucose measures be repeated unless very high glucose levels are found.

Fatty Liver Disease Fatty liver disease is characterized by an excess accumulation of fat in the liver. Officially called *nonalcoholic fatty liver disease* (NAFLD), it refers to a spectrum of liver damage ranging from levels of fat accumulation that pose few health risks to extensive fat accumulation and liver damage.[22] Susceptibility to the development of fatty liver disease varies among individuals based on genetic traits.[23] It is estimated that 21% of the U.S. population has fatty liver disease.[24] It occurs in people of all weight statuses, but it is most likely to occur in individuals with certain genetic traits, men, Mexican Americans, and people with insulin resistance, type 2 diabetes, and obesity.[24] Habitually high intake of sugar-sweetened beverages, whether sweetened with sucrose or fructose, is a risk factor for fatty liver disease in susceptible individuals.[25,26]

The first line of treatment for fatty liver disease is lifestyle modification. This disease can be diminished or reversed in most cases by a 5 to 10% reduction in body weight, consumption of a healthy dietary pattern, and increased physical activity. Reduced calorie diets based on whole grain foods, colorful vegetables and fruits, fish, seafood, poultry, and low in added sugars have been found to reverse fatty liver disease.[22,27]

Metabolic Syndrome Insulin resistance is also related to the development of a spectrum of metabolic abnormalities that have far-reaching effects. Physicians have known for decades that obese people with hypertension and type 2 diabetes are at high risk of heart disease. What they didn't know was why. Over time, research studies discovered that part of the answer was insulin resistance.[28] There is currently no simple and inexpensive test for insulin resistance. Clinically, it is assumed to be present in people with high waist circumferences.[29,30] Collectively, the adverse effects of insulin resistance are included in a disorder called **metabolic syndrome** that increases the risk of type 2 diabetes as well as heart disease. Symptoms related to the metabolic syndrome include:

metabolic syndrome A constellation of metabolic abnormalities that increase the risk of heart disease and type 2 diabetes. Metabolic syndrome is characterized by insulin resistance, abdominal obesity, high blood pressure and triglyceride levels, low levels of HDL cholesterol, and impaired glucose tolerance. It is also called *syndrome X* and *insulin resistance syndrome*.

- Waist circumference of 40″ or more in males, 35″ or more in females
- High blood pressure (130/85 mm Hg or higher)
- Elevated blood triglycerides levels (150 mg/dL or higher)
- Low levels of protective HDL cholesterol (less than 50 mg/dL in women and 40 mg/dL in men)
- Elevated fasting blood glucose levels (100 mg/dL or higher)

The diagnosis of metabolic syndrome is made when three or more abnormalities are identified.[30] Individuals with four or five metabolic abnormalities have a 3.7 times greater risk of heart disease, and a 25 times higher risk for diabetes than do people with no abnormalities.[8] It is estimated that 32% of men and women in the United States and 20% of Canadians have metabolic syndrome.[31] Most, but not all, people who develop metabolic syndrome are overweight or obese and are physically inactive.[31] Weight loss, a healthy dietary pattern, exercise, and other factors listed in Table 13.3 are key components of lifestyles related to prevention and management of metabolic syndrome.

Table 13.3 Key components of the prevention and management of metabolic syndrome[32–34]

1. Weight loss if overweight
2. Regular physical activity
3. Healthy dietary pattern that emphasizes:
• Whole grains, high-fiber foods • Low added sugar foods and fluids • Vegetables, fruits • Fish, seafood, poultry • No- or low-fat dairy products • Legumes, nuts, and seeds
4. Vitamin D adequacy

Managing Type 2 Diabetes

As with all forms of diabetes, diet is a cornerstone of the treatment of type 2 diabetes.[4] Modest weight loss alone (5–10% of body weight) has repeatedly been shown to significantly improve blood glucose control in overweight and obese people with type 2 diabetes.[35]

Dietary intake recommendations for individuals with type 2 diabetes are based on a healthy dietary pattern that is characterized by the regular intake of:

- vegetables and fruits
- whole grains and whole-grain products
- legumes
- fish and shellfish

- poultry and lean meats
- low-fat dairy products
- nuts

Moderate alcohol intake (one to two drinks per day) by adults who choose to drink is part of this dietary pattern. Small portions of sugar and sweet desserts can be planned into the diet but should be considered special treats.[36] Healthy dietary patterns are low in foods and beverages with added sugar and salt, emphasize unsaturated fats, and supply adequate fiber.[1,4,17]

The sources of carbohydrates included within the dietary recommendations are those with moderate and low **glycemic index.**[4]

glycemic index (GI) A measure of the extent to which blood glucose levels are raised by consumption of an amount of food that contains 50 grams of carbohydrate compared to 50 grams of glucose. A portion of white bread containing 50 grams of carbohydrate is sometimes used for comparison instead of 50 grams of glucose.

resistant starch Starches that do not release glucose within the small intestine but are consumed or fermented by bacteria in the colon released as fatty acids.

Not all types of complex carbohydrates are absorbed in the intestines as glucose. Some starches, called **resistant starch**, are not absorbed as glucose but rather as fatty acids that do not raise blood glucose levels. Food sources of resistant starch include bananas, oats and oatmeal, sweet potatoes, green peas, and dried beans. They also foster the growth of "friendly" bacteria in the lower gut that appear to play important roles in regulatory processes that affect health.[42,44]

In some cases, weight-loss surgery will be recommended for overweight or obese individuals with type 2 diabetes. This surgery successfully lowers or eliminates high blood glucose levels in many people, and generally reduces blood pressure and triglyceride levels.[4]

Regular physical activity is an important component of the management of type 2 diabetes. In addition to facilitating weight loss, physical activity reduces insulin resistance, decreases blood pressure and body fat content, and improves blood lipid and glucose levels.[4] Moderate-intensity aerobic and strength-building activities, such as weight lifting, jogging, fast walking, aerobic dancing, and swimming are recommended. Exercise is particularly beneficial after weight loss surgery in individuals with type 2 diabetes because it helps reduce insulin resistance, and improves physical fitness and weight-loss maintenance. The target generally set for the duration of physical activity is 150 minutes or more per week, or an average of 21 minutes per day or more.[4]

Carbohydrate Counting Dietary carbohydrates have the greatest effect on blood glucose level and insulin need of the energy nutrients. Consequently, managing carbohydrate intake is a key part managing diabetes. Several methods for managing carbohydrate intake are utilized clinically and include carbohydrate counting, portion control, and food lists of carbohydrate food choices (Table 13.4).[4] A number of tools (such as Nutrition Facts panels) and carbohydrate counting apps are available online.

Carbohydrate counting basically involves the following three steps:

1. Keeping a food log that includes amounts of food and beverages consumed.
2. Identifying the significant sources of carbohydrate in the food log.
3. Quantifying the carbohydrate content of the foods consumed.

With the assistance of a Registered Dietitian/Nutritionist or Certified Diabetes Educator, the results of assessments of carbohydrate intake and blood glucose levels are then used to adjust insulin to amounts that correspond to insulin need.[4,66]

Table 13.4 **Carbohydrate food lists**[4,66,67]

A. Foods that provide very little or no carbohydrate	
Meats	Fish, seafood
Seeds	Most cheeses
Nuts	Oils and other fats
Salad greens, broccoli, celery, peppers	
B. Foods that provide carbohydrates	
Grains and grain products	Fruits and fruit juices
Milk, yogurt	Potatoes
Corn	Sweet potatoes
Soft drinks	Fruit drinks
Energy drinks	Sports drinks
Dried beans	Sweets, candy, desserts
C. Carbohydrate content of common foods	
Food, amount	Carbohydrates, grams
Honey, 1 tsp	5
Table sugar, 1 tsp	4
Fruits	
Apple, 1 medium	19
Grapes, ½ cup	18
Peach, 1 medium	14
Watermelon, ½ cup	6
Canetloupe, ½ cup	15
Orange juice, ½ cup	13
Orange, 1 medium	14
Banana, 1 medium	27
Beverages	
Fruit drinks, 1 cup	29
Soft drinks, 12 oz	40
Skin milk, 1 cup	12
Whole milk, 1 cup	11
Candy	
Jelly Beans, 15	40
Hard candy, 1 oz	28
Snickers bar, 2 oz	35
M&M's, 2 oz	34
Breakfast cereals	
Apple Jacks, 1 oz	30
Cheerios, 1 oz	22
Corn Flakes, 1 cup	24
Corn Pops, 1 cup	28
Oatmeal, cooked, 1 cup	28
Grits, cooked ½ cup	16
Granola, ½ cup	33
Grain and grain products	
Rice, white, ½ cup cooked	22
Pasta, ½ cup cooked	24
Whole wheat bread, 1 slice	24
White bread, 1 slice	15
Dried, cooked beans, vegetables	
Dried beans, cooked, ½ cup	20
Vegetables:	
Potato, ½ cup	20
Corn, ½ cup	16

REALITY CHECK

Will the Real Whole Grain Please Stand Up?

Which bread is made from whole grains?

Who gets a thumbs-up?

Answers on page 13-8.

Richard Anderson

Wheat Bread

Richard Anderson

Whole Wheat Bread

ANSWERS TO REALITY CHECK

Will the Real Whole Grain Please Stand Up?

You can't tell by looking, but the slice of bread on the right contains more fiber (4 vs. 2 grams). Want to find whole-grain breads? Look for the term *whole grain* or *whole wheat* on the label. Bread products carrying that label contain 51% or more whole grains. The ingredient label on whole-grain products will list one or more whole grains before listing enriched flour.

Richard Anderson

Wheat Bread

Richard Anderson

Whole Wheat Bread

Prevention of Type 2 Diabetes

The effects of weight loss and exercise in preventing type 2 diabetes can be quite dramatic. In one large study that took place over a three-year period, people with prediabetes reduced their risk of developing type 2 diabetes by 58% by losing around 7% of body weight and exercising for 150 minutes a week in a structured lifestyle program.[47] Reduction in the risk of developing type 2 diabetes related to participation in the program was 71% for people over the age of 60 years. These reductions in risk were greater than that achieved by use of a common medication (metformin) for blood glucose control (31% reduction in risk). The type 2 diabetes preventive effects of participation in the program and weight-loss maintenance have persisted for at least 15 years.[48–50]

Healthy dietary patterns rich in whole-grain and high-fiber foods, vegetables, and fruits are protective against the development of type 2 diabetes and appear to aid weight loss (Illustration 13.4). Components of high-fiber, whole-grain foods raise blood glucose levels marginally and appear to provide nutrients and other biologically active substances that lessen the risk of this disease.[51] Consumption of regular or decaffeinated coffee (1 to 4 cups daily) and moderate alcohol intake (one to two drinks per day if appropriate) also appear to increase insulin sensitivity and decrease the risk of developing type 2 diabetes.[52,53]

Type 2 diabetes may also be prevented or postponed by not gaining weight during the young adult years. Weight gains of 10 to 15 pounds between the ages of 25 and 40 have been found to increase the risk of type 2 diabetes sevenfold compared to not gaining weight or gaining it after the age of 40.[55]

It is anticipated that 800,000 new cases of type 2 diabetes may develop each year in the United States due to rising rates of obesity.[56] The additional burden such an increase would place on individuals and health care costs clearly convey the message that prevention is urgently needed. Public health campaigns are now underway to encourage people to lose weight if overweight, to exercise regularly, and to follow a healthy dietary pattern.[57]

Are you at risk of developing type 2 diabetes? Find out if you are, and what people at risk can do to help prevent it in the accompanying Health Action feature.

PhotoDisc

Illustration 13.4 Whole-grain products, other high-fiber foods, and ample servings of vegetables and fruits can help prevent type 2 diabetes.

health action Preventing Type 2 Diabetes

Are You at Risk?

Check each category that applies to you:

_____ I have a brother, sister, mother, or father with type 2 diabetes.

_____ My BMI is 25 or higher.

_____ My waist circumference is over 35 inches (if female) or 40 inches (if male).

_____ I have been diagnosed with prediabetes or insulin resistance.

_____ I am habitually inactive.

_____ I have been diagnosed with high blood pressure.

_____ I have been diagnosed as having high blood triglyceride levels.

If you checked one or more of the categories above, you may be at increased risk of developing type 2 diabetes.

What to Do Now?

1. Recognize that there are no quick fixes for the prevention or management of type 2 diabetes.
2. If you are overweight or obese, gradually cut back on your calorie intake by making small and acceptable changes in your diet. Aim for a reduction in caloric intake of 100 to 200 calories a day.
3. Gradually increase your physical activity level until you are moderately to vigorously active for at least 150 minutes per week (21 minutes per day). Select activities you enjoy and will keep doing in the long run.
4. Choose a healthy dietary pattern such as offered by the Mediterranean Dietary Pattern, the DASH Healthy Eating Plan, or the Healthy Vegetarian Dietary pattern.[18]
5. Consume enough fiber (28 grams daily).
6. Eat your favorite vegetables and fruits often.
7. Choose low-fat dairy products.
8. Select and consume whole-grain products rather than refined grain products.
9. See your health care provider if you think you have a problem with your blood glucose levels.

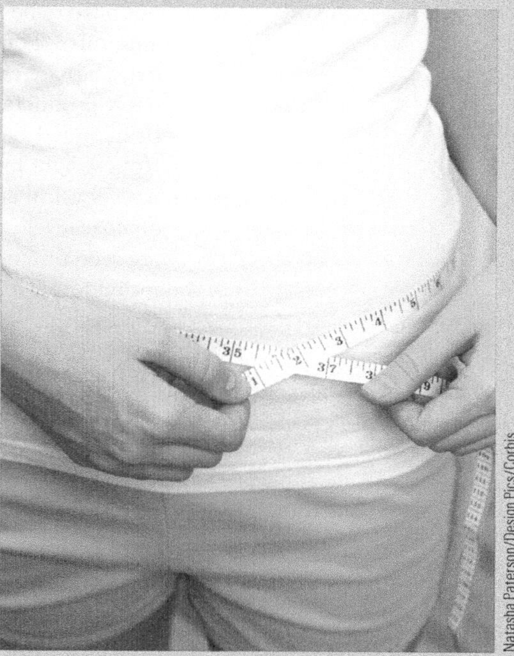

Natasha Paterson/Design Pics/Corbis

Type 1 Diabetes

Type 1 diabetes is an **autoimmune disease** that produces a deficiency of insulin. It accounts for 5–10% of cases of diabetes.[6] The onset of type 1 peaks around the ages of 11 to 12 years and usually occurs before age 40.[58] Although individuals may inherit or acquire genetic tendencies toward the development of type 1 diabetes, most people who have the disease have no family history of it. Like type 2 diabetes, worldwide rates of type 1 diabetes are increasing.[59]

The insulin deficiency that marks type 1 diabetes appears to develop when a person's own **immune system** destroys beta cells in the pancreas that produce insulin. The destruction may be triggered by exposure to complex dietary proteins during the first year of life.[60] Although the specific cause or causes of type 1 diabetes are not yet understood, breastfeeding infants for the first four months or more of life may confer some level of protection against the development of type 1 diabetes.[61]

autoimmune disease A disease initiated by the destruction of the body's own cells by components of the immune system that mistakenly recognize the cells as harmful.

immune system Body tissues that provide protection against bacteria, viruses, foreign proteins, and other substances identified by cells as harmful.

Managing Type 1 Diabetes The main goals of the management of type I diabetes are blood glucose control, prevention of hypoglycemia, and health maintenance.[63] Blood glucose levels are managed primarily by regular, healthful meals controlled in carbohydrate content that are consumed at planned times and in specific amounts. Diets are designed to match insulin dosage so that blood glucose levels remain within normal ranges. They are controlled in carbohydrate content because carbohydrates raise blood glucose levels and increase insulin need to a greater extent than protein or fats.[64]

It is generally recommended that individuals with type 1 diabetes limit intake of added sugars and foods high in refined carbohydrates. Sugar and sweetened foods do not have to be eliminated from the diet; rather, they should be consumed in limited amounts.[65] Foods low in glycemic index and high in fiber (especially soluble fiber such as oatmeal) are encouraged, as are brightly colored fruits and vegetables, low-fat meat and dairy products, fish, dried beans, and nuts and seeds.[66] Reduced-calorie diet plans should be included as part of the care for individuals with type 1 diabetes who would benefit from weight loss.

Physical activity is generally part of a diabetes care plan because it improves blood glucose levels, physical fitness, and insulin utilization. Both strength and aerobic exercises are recommended.[63] Individualized meal and physical activity plans and follow-up care for individuals with type I diabetes should be provided by an experienced health care team that includes a registered dietitian nutritionist.[67]

Insulin and New Technologies in the Management of Type 1 Diabetes People with type 1 diabetes require insulin to control blood glucose levels. The amount and type of insulin required depends on diet, physical activity level, and the presence of conditions such as pregnancy, stress, illness, and physical activity. To get the insulin dose right, individuals with type 1 diabetes measure their blood glucose level a number of times daily. Blood glucose levels have traditionally been tested using a pinprick blood sample and a device that measures blood glucose level in a droplet of blood. The glucose measurement would be followed by the appropriate amount of insulin delivered in an injection.[68] Although nearly painless if very small needles are used, many people hate the idea of getting pricked by a needle.

Technological advances are taking some of the perceived sting out of type 1 diabetes management. Insulin pumps that deliver programmed doses of insulin are now combined with continuous glucose monitors that assess the body's glucose level (Illustration 13.5). Continuous glucose monitors issue a warning signal if glucose levels are rising or dropping too much. The signal gives the individual enough advance warning to take appropriate action, such as eating or using insulin, before a problem develops. It must be used with self-monitoring of blood glucose levels. New therapies, including pancreatic beta cell transplants and stem cell therapies that increase the mass of cells in the pancreas that produce insulin, are under development. These techniques are being intensively studied now, and preliminary results are promising.[70]

AP Images/Tom Strattma

Illustration 13.5 An insulin pump with a built-in glucose sensor can warn individuals about potentially harmful changes in blood glucose levels.

Gestational Diabetes

It is estimated that up to 9.2% of women in the United States develop gestational diabetes during pregnancy,[71] though the incidence varies a good deal based on ethnicity. Native Americans, African Americans, and Asians are at higher risk for gestational diabetes than other groups. Women over the age of 35 years, obese women, and those with habitually low levels of physical activity are at greater risk than other women.[72] A number of dietary risk factors have been associated with the development of gestational diabetes, including diets high in red and processed meats, low fiber intake, and diets high in glycemic load.[73] Gestational diabetes is likely caused by the same basic environmental and genetic predisposition factors as type 2 diabetes.[64] The prevalence of gestational diabetes is increasing in many countries, paralleling increases in obesity and type 2 diabetes.[72] Infants born to women with poorly controlled diabetes may be excessively fat at birth, require cesarean delivery, and have blood glucose control problems after delivery. They are at greater risk for developing diabetes later in life.[74]

As with type 2 diabetes, women with gestational diabetes are insulin resistant and can often control their blood glucose levels with an individualized diet and exercise plan. Some women require daily oral medication or insulin injections for blood glucose control.[75] Gestational diabetes often disappears after delivery, but type 2 diabetes may appear after delivery or later in life. Exercise, maintenance of normal weight, and consumption of a healthy dietary pattern reduce the risk that diabetes will return.[63]

Hypoglycemia

Hypoglycemia, or low blood glucose level (<70 mg/dL), is a fairly common symptom associated with type 1 diabetes treatment. It is often mild and accompanied by a feeling of weakness, a negative mood, nervousness, hunger, shakiness, and confusion. Severe cases of hypoglycemia can lead to coma and death. Symptoms of hypoglycemia can usually be resolved by eating 5 or 6 pieces of hard candy or eating a tablespoon of sugar or honey.[76,77] Blood glucose level should be checked 15 minutes after sugar consumption.

hypoglycemia A disorder resulting from abnormally low blood glucose levels (<70 mg/dL). Symptoms of hypoglycemia include irritability, nervousness, weakness, sweating, and hunger.

Diabetes in the Future

The anticipated surge in the worldwide incidence of type 2 diabetes shown in Illustration 13.6 is not inevitable. It could be lowered substantially by individuals making environmental and lifestyle changes that reduce the risk for, and incidence of, overweight and obesity. Increased awareness of the connection between diabetes and body weight may help. The desired future of diabetes is one negating the dire forecasts of the experts.

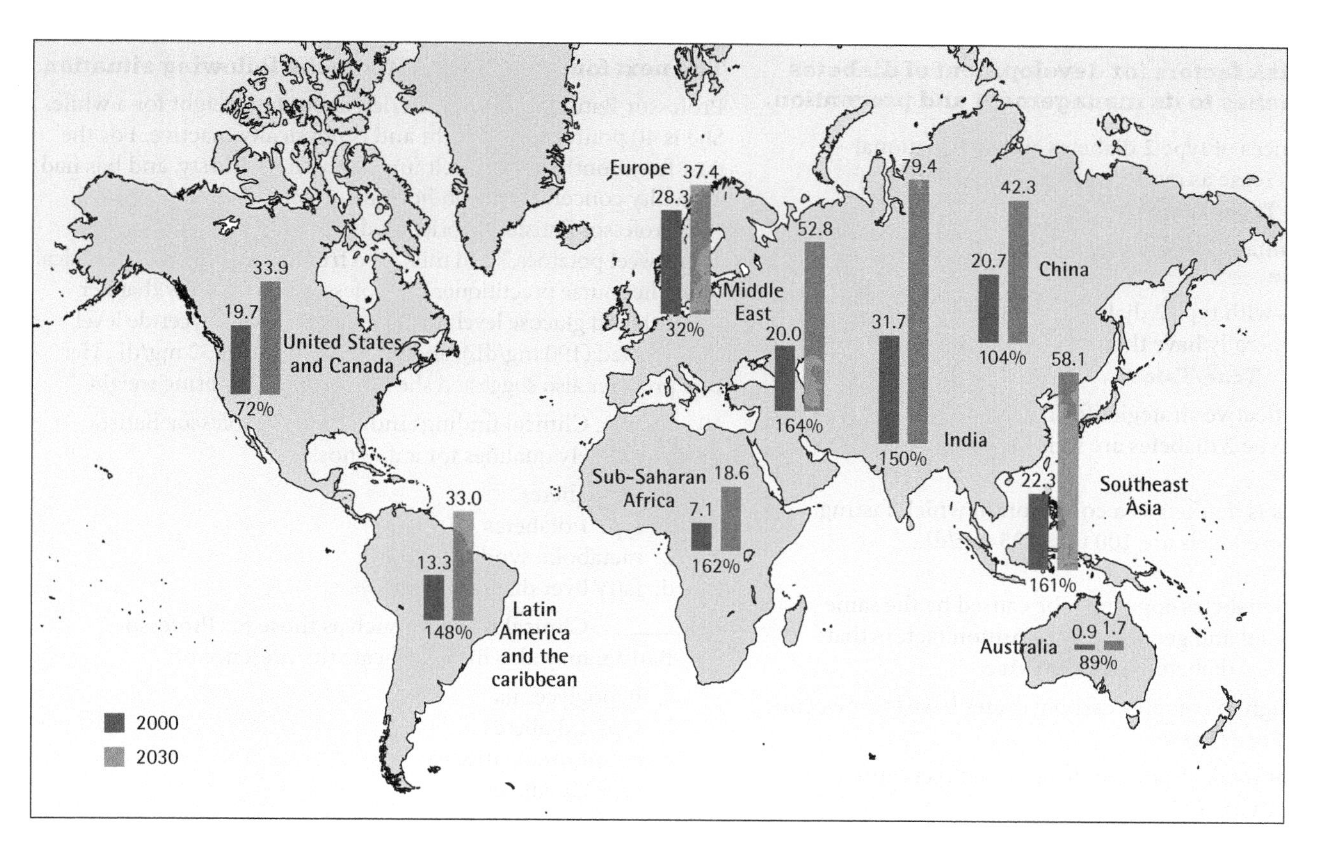

Illustration 13.6 The red bars indicate millions of cases of type 2 diabetes in 2000 and the green bars the projections for 2030. The projected percentage increases in cases of type 2 diabetes are shown below the bars.[69]

NUTRITION up close

Calculating Carbohydrates

Focal Point: To increase understanding of the carbohydrate content of foods in ways that help individuals with diabetes manage their diet and blood glucose levels.

Bart has been recently diagnosed with type 2 diabetes and is working with his health care team to learn how to estimate the carbohydrate content of his daily diet. His food log for one day indicates that Bart consumed the following carbohydrate-containing foods:

Breakfast:
Corn flakes, 1 cup
Skim milk, 1 cup
Orange juice, 1 cup

Mid-morning snack:
Banana, 1

Lunch:
Dried beans, cooked, 1 cup
White rice, 1 cup
Apple, 1 medium

Dinner:
Pork chop, 3 oz.
Corn, 1 cup
Fruit drink, 1 cup

Using the information in Table 13.4, calculate the total amount of carbohydrates in Bart's diet that day. ____ grams carbohydrate

REVIEW QUESTIONS

- **Describe risk factors for development of diabetes and approaches to its management and prevention.**

1. The incidences of type 2 diabetes and of gestational diabetes increase as rates of overweight and obesity increase. **True/False**
2. Insulin facilitates the passage of vitamins into cells. **True/False**
3. Individuals with type 2 diabetes and gestational diabetes generally have the condition known as insulin resistance. **True/False**
4. The most effective strategies for the prevention and control of type 2 diabetes are weight loss and exercise. **True/False**
5. Prediabetes is defined as a condition in which fasting blood glucose levels are 100 to ≤125 mg/dL. **True/False**
6. Gestational diabetes appears to be caused by the same environmental and genetic predisposition factors that apply to type 2 diabetes. **True/False**
7. All foods high in complex carbohydrates have low glycemic indices. **True/False**
8. Excess sugar intake from soft drinks is a direct cause of type 2 diabetes. **True/False**
9. Hypoglycemia in people with diabetes is most often caused by excess consumption of sugar. **True/False**

The next four questions refer to the following situation.

Professor Batista has been worried about her weight for a while. She is 40 pounds overweight and is physically inactive. For the past few months she has felt unusually tired, thirsty, and has had difficulty concentrating on her class preparation notes.

Professor Batista's favorite foods are white rice, pinto beans, corn, sweet potatoes, skim milk, and fruit drink. At her last physical exam, her nurse practitioner let Professor Batista know that her fasting blood glucose level was 136 mg/dL, her triglyceride level was elevated (190 mg/dL), and her HDL level was 32 mg/dL. Her care provider also suggested she should consider losing weight.

10. ______ Clinical findings indicate that Professor Batista most likely qualifies for a diagnosis of:
    a. prediabetes
    b. type 1 diabetes
    c. metabolic syndrome
    d. fatty liver disease
11. ______ Clinical findings, such as those for Professor Batista, may also likely indicate the presence of:
    a. hypoglycemia
    b. type 1 diabetes
    c. autoimmune disease
    d. type 2 diabetes

12. ______ Which of the following sets of foods preferred by Professor Batista would be expected to raise blood glucose levels the most:

a. white rice
b. fruit drink
c. pinto beans
d. skim milk

13. ______ Over the following year, Professor Batista lost 20 pounds (12% of her initial body weight), switched to whole-grain tortillas and brown rice, substituted unsweetened iced coffee for fruit punch, and walked vigorously for 30 minutes daily. Which of the following improvements in the clinical findings at her next checkup would *not* be expected?

a. increased insulin sensitivity
b. reduced blood glucose level
c. improved blood lipid levels
d. increased insulin resistance

14. A female with a waist circumference of 29 inches, high blood pressure, and low levels of LDL cholesterol would be diagnosed as having metabolic syndrome. **True/False**

15. Roger has a BMI of 35 mg/kg^2 and is considered clinically obese. His fasting blood glucose level is 97 mg/dL. Roger may have diabetes. **True/False**

NUTRITION SCOREBOARD ANSWERS

1. Simple sugar intake does not directly cause diabetes.[1] **False**
2. Disease processes underlying the development of diabetes exist for years before the onset of diabetes in most cases.[2] **True**
3. Yes you can.[1] **False**
4. Dietary recommendations for individuals with type 1 and type 2 diabetes are substantially different than the recommendations for a healthy dietary pattern. **False**

Alcohol: The Positives and Negatives

Image Source/Getty Images

NUTRITION SCOREBOARD

1. Alcohol is produced by the fermentation of carbohydrates. **True/False**
2. You can protect your body from the harmful effects of consuming excessive amounts of alcohol by eating a nutritious diet. **True/False**
3. Alcohol abuse plays a major role in injuries and deaths in the United States. **True/False**
4. Dark beer provides more calories than the same amount of regular beer. **True/False**

Answers can be found at the end of the unit.

LEARNING OBJECTIVES

After completing Unit 14 you will be able to:

- Summarize the effects of the quantity of alcohol intake on behavior, health, and disease.

Alcohol Facts

- **Summarize the effects of the quantity of alcohol intake on behavior, health, and disease.**

Alcohol is both a food and a drug. It's a food because alcohol is made from carbohydrates, and the body uses it as an energy source. Alcohol is a drug because it modifies various body functions.

The type of alcohol people consume in beverages is ethanol. (We refer to ethanol by the broader term *alcohol* in this unit.) Alcohol is produced from carbohydrates in grains, fruits, and other foods by the process of **fermentation**. Wines, brews made from grains, and other alcohol-containing beverages are a traditional part of the food supply of many cultural groups.[1] In high doses, however, alcohol is harmful to the body and can cause a wide variety of nutritional, social, and physical health problems.

KEY NUTRITION CONCEPTS

Content and activities included in this unit on alcohol are related to the following key nutrition concepts:

1. Nutrition Concept #3: Health problems related to nutrition originate within cells.
2. Nutrition Concept #6: Malnutrition can result from poor diets and from disease states, genetic factors, or combinations of these factors.

fermentation The process by which carbohydrates are converted to ethanol by the action of the enzymes in yeast.

chronic inflammation Inflammation that is low grade and lasts weeks, months, or years. Inflammation is the first response of the body's immune system to infection or irritation. It triggers the release of biologically active substances that promote oxidation and other potentially harmful reactions in the body.

The Positive

Whether alcohol has harmful effects on health depends on how much is consumed. The consumption of moderate amounts of alcohol by healthy adults who are not pregnant appears to cause no harm. In fact, moderate alcohol consumption versus abstinence is associated with a significant level of protection against heart disease, type 2 diabetes, hypertension, stroke, dementia (cognitive decline), and all causes of mortality.[2,3] In the United States, a moderate level of alcohol consumption is considered to be one standard-sized drink per day for women and two drinks for men (Illustration 14.1).[4] On a given day, 21% of women and 36% of men in the United States consume alcohol. Although most adults who drink consume moderate amounts of alcohol, 3% of women and 8% of men consume five or more drinks a day and are considered heavy alcohol drinkers.[4,17]

Illustration 14.1 Standard serving sizes of alcohol-containing beverages. Standard servings shown each contain 0.6 oz (14 g) of alcohol.

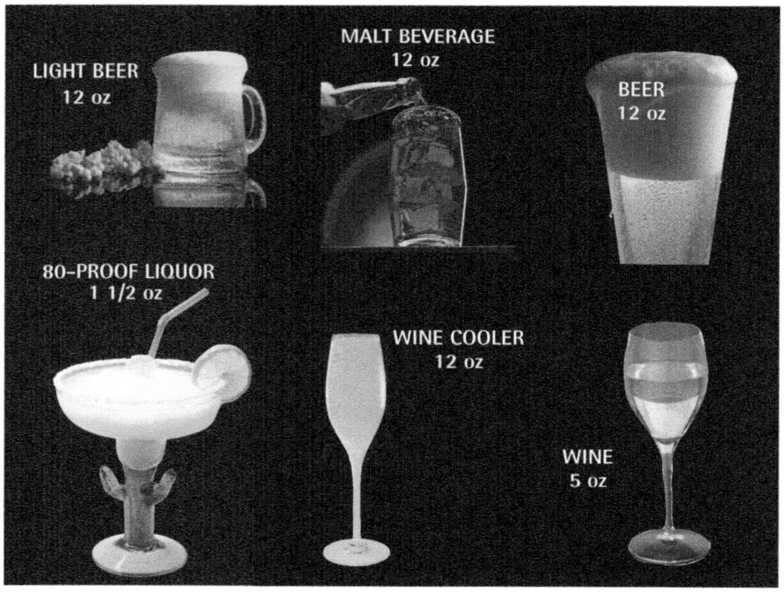

Alcohol at moderate doses increases HDL cholesterol levels (the "good" cholesterol) and decreases **chronic inflammation**. Decreased inflammation helps prevent the formation of plaque in arteries and improves circulatory function and maintenance of normal cell health. Moderate amounts of alcohol improve the body's utilization of insulin and glucose, lower post-meal blood glucose levels somewhat, and improve cognitive function.[5–7]

Some of the beneficial effects of alcohol on health are related to the phytochemical content of the fruit, vegetable, or grain fermented to produce it. Red wine, and to a lesser extent beer and white wine, contain pigments and other phytochemicals that act as powerful antioxidants and decrease inflammation and arterial plaque formation.[8] Purple grape juice and other purple- and blue-colored fruit juices also provide

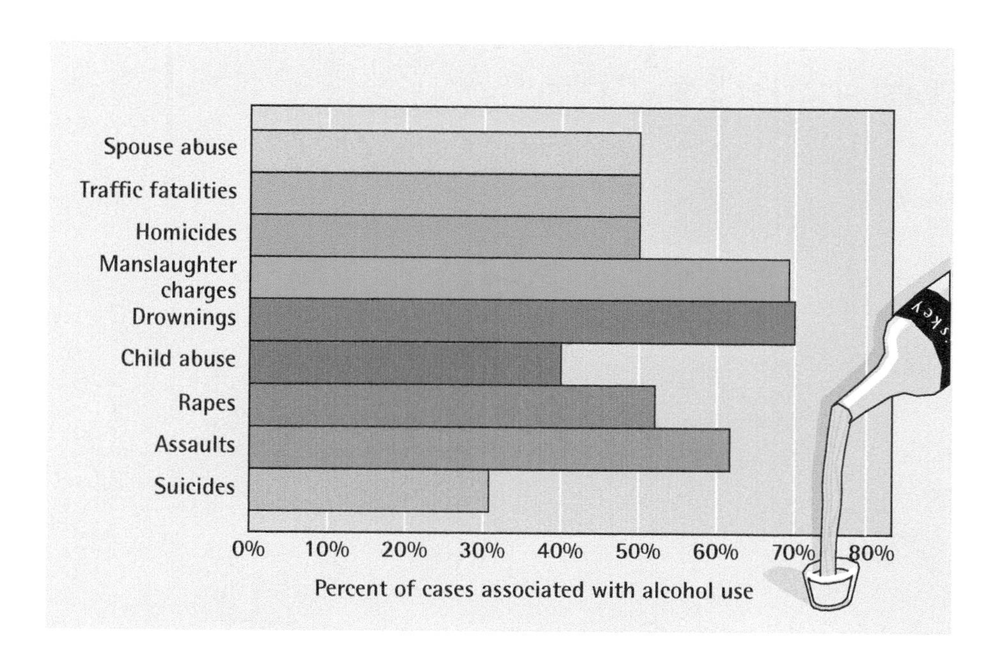

Illustration 14.2 Violence and injuries associated with alcohol.

Source: National Institute on Alcohol Abuse and Alcoholism, 2001, 2006.

antioxidants and have anti-inflammatory effects that benefit health, although to a lesser degree than red wine.[9]

People don't have to consume alcohol to reduce their risk of heart disease. Healthy dietary pattern, ample physical activity, and not smoking also reduce the risk of heart disease.[10]

The Negative

Heavy drinking poses a number of threats to the health of individual drinkers and often to other people as well. Although health can be damaged by the regular consumption of large amounts of alcohol, the ill effects of alcohol are most obvious in people with alcohol dependency, or **alcoholism**.

Habitually high alcohol intake and alcoholism increase the risk of developing high blood pressure, stroke, and dementia; throat, stomach, and bladder cancer; central nervous system disorders; and vitamin and mineral deficiency diseases. Alcohol abuse is associated with a high proportion of deaths from homicide, drowning, fires, traffic accidents, and suicide (Illustration 14.2). It is also involved in a large proportion of rapes and assaults, and can devastate families.[11] **Alcohol poisoning** from the consumption of a large amount of alcohol in a short period of time can cause death—and does to about 1,700 college students each year.[12] You can read more about the important topic of alcohol poisoning in the Health Action feature.

Long-term, excessive alcohol intake is related to the development of **fatty liver disease**, a condition marked by the build-up of fat in the liver, inflammation, and eventually **cirrhosis** (Illustration 14.3). As a result of cirrhosis, scar tissue forms in

alcoholism An illness characterized by dependence on alcohol and by a level of alcohol intake that interferes with health, family and social relations, and job performance. Also referred to as alcohol dependency.

alcohol poisoning A condition characterized by mental confusion, vomiting, seizures, slow or irregular breathing, and low body temperature due to the effects of excess alcohol consumption. It is life-threatening and requires emergency medical help.

fatty liver disease A reversible condition characterized by fat infiltration of the liver (10% or more by weight). If not corrected, fatty liver disease can produce liver damage and other disorders. The condition is primarily associated with obesity, type 2 diabetes, and alcohol consumption. The disease is called steatohepatitis when accompanied by inflammation.

cirrhosis (sear-row-sis): A disease of the liver characterized by widespread fibrous tissue buildup and disruption of normal liver structure and function. It can be caused by a number of chronic conditions that affect the liver.

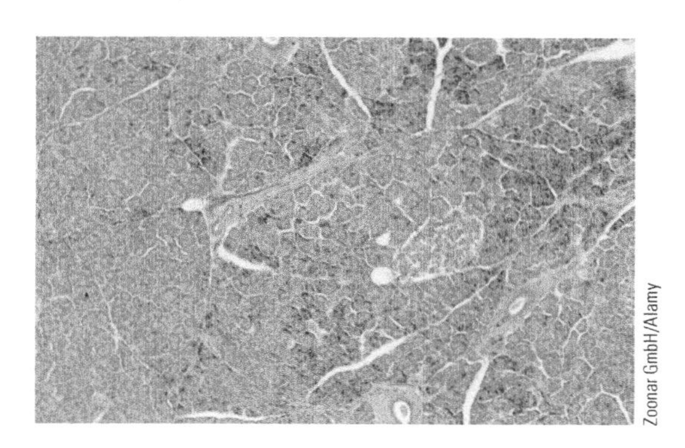

Zoonar GmbH/Alamy

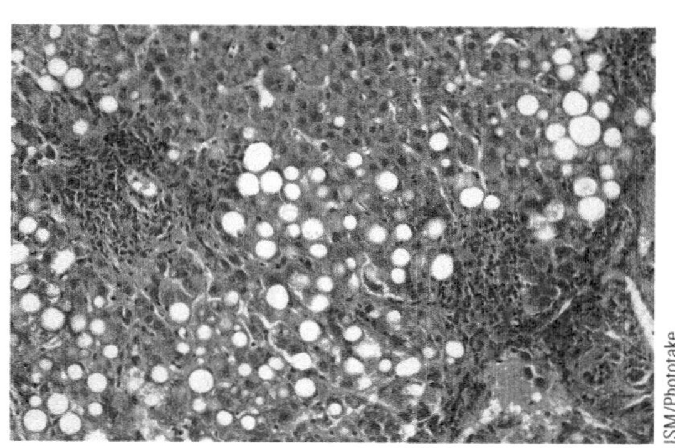

ISM/Phototake

Illustration 14.3 Normal liver tissue is shown in the left-hand photo. The photo on the right shows a liver with inflammation and fatty tissue, or steatohepatitis.

health action Alcohol Poisoning: Facts and Action[14]

Facts

- Alcohol poisoning is a serious and sometimes deadly result of drinking excessive amounts of alcohol, usually in an episode of binge drinking.
- A high level of alcohol in the blood depresses breathing, lowers body temperature, promotes choking and vomiting, and slows heart rate.
- Blood alcohol levels can continue to rise after a person passes out.

CRITICAL SIGNS of alcohol poisoning include the following:

1. Mental confusion and loss of consciousness; sometimes the person cannot be roused.
2. Vomiting
3. Seizures
4. Slow or irregular breathing (10 seconds or more between breaths)
5. Off-colored skin (paleness or bluish skin color)

ACTION

1. CALL 911 for help if there is any suspicion of an alcohol overdose.
2. DO NOT wait for all the symptoms to appear.
3. DO NOT assume a person can safely "sleep it off" or that a cold shower will help.

Bill Roth/Anchorage Daily News/MCT/Newscom

Illustration 14.4 Children with FASD experience physical and developmental impairments to varying degrees. This photograph shows facial characteristics associated with a severe form of FASD. Alcohol-containing beverages must show a warning statement on labels.

Table 14.1 Caloric value of common alcohol-containing beverages

	Serving	Calories
Beer, regular	12 oz	150
Beer, light	12 oz	110
Beer, dark	12 oz	168
Malt beverage	12 oz	225
80 proof liquor	1.5 oz	100
Wine, red	5 oz	105
Wine, white	5 oz	100
Wine cooler	12 oz	215

the liver and decreases the liver's ability to function normally in synthesizing protein, combating infection, and processing the nutrients it receives from the bloodstream. Cirrhosis can lead to easy bruising, bleeding, or nosebleeds; swelling of the abdomen or legs; kidney failure; and other health problems.[15]

Alcohol consumption by pregnant women can cause lifetime physical and developmental problems in their offspring. Alcohol is readily transferred from mother to fetus. Tissues and organs undergoing development are particularly sensitive to the effects of alcohol.

Fetal Alcohol Spectrum Disorder Once referred to as fetal alcohol syndrome, the term *fetal alcohol spectrum disorder* (FASD)[16] is now used to refer to the range of physical and developmental effects on offspring induced by alcohol consumption during pregnancy. Signs and symptoms present in individuals affected by various levels of severity of FASD range from mild to severe. Some individuals affected by FASD may exhibit physical characteristics such as those shown in Illustration 14.4. These characteristics include a smooth ridge between the nose and upper lip, small head size, short stature, and low weight for height. Other problems observed range from poor coordination, hyperactive behavior, and learning difficulties to hearing, vision, heart, and kidney problems. Some individuals affected to some degree by FASD are physically normal, experience some level of interrupted and impaired development, and become accomplished adults.[15]

The severity of the effects of alcohol exposure during pregnancy depends on how much alcohol is consumed, the time during pregnancy of alcohol consumption, and genetic traits of the mother that influence alcohol metabolism. No "safe" level of alcohol consumption during pregnancy has been identified, largely due to the fact that any alcohol intake by the mother exposes the rapidly developing fetus to alcohol. Dietary deficiency of nutrients and low body mass index may also contribute to the development of FASD. Liver damage induced by excess alcohol contributes to poor nutritional status by hampering the liver's ability to utilize nutrients.[16] Because there is no known safe level of alcohol intake, it is recommended that women who are or may become pregnant not drink.[4,15]

Alcohol Intake, Diet Quality, and Nutrient Status

Alcohol provides seven calories per gram, making alcohol-containing beverages rather high in caloric content (Table 14.1). Because many alcohol-containing

REALITY CHECK

Do Alcohol Calories Count?

Perhaps you've heard the popular opinion that alcohol intake does not increase the risk of obesity. Is that the same as the opinion of scientists?

Do your thoughts side with Pedro or Erik?

Answers on page 14-6.

Photodisc

Pedro: I started drinking a beer at night over the summer, and my weight never changed.

Photodisc

Erik: The six-pack around my abdomen is really a six-pack.

beverages provide calories and few or no nutrients, they are considered energy-dense, empty-calorie foods.[17] On average, alcohol accounts for 3–9% of the caloric intake of U.S. adults who drink. The average goes up to around 50% among heavy drinkers.[18] Although beer, wine, and mixed drinks are known to contain alcohol and to provide calories, some people are still confused about whether calories from alcohol contribute to weight gain. This issue is addressed in this unit's Reality Check.

Although diet quality tends to be better than average in moderate drinkers, as caloric consumption from alcohol-containing beverages increases, the quality of the diet generally decreases. Heavy drinkers are often malnourished, and their diets often provide too little thiamin, niacin, vitamins B_{12}, A, and C, and folate.[19,20] Deficiencies of nutrients, as well as direct toxic effects of high levels of alcohol ingestion on liver function, nutrient absorption, and cellular utilization of nutrients, produce most of the physical health problems associated with alcoholism.[16] The lack of thiamin, for example, impairs the brain's utilization of glucose. When people with alcoholism initially withdraw from alcohol, thiamin deficiency may be expressed and result in *delirium tremens*, a condition called the DTs by people who staff detoxification centers. People with delirium tremens experience convulsions and hallucinations, and are severely confused. Thiamin injections are a key component of treatment for delirium tremens.[21] Because alcohol in excess is directly toxic to body tissues, consuming an adequate diet protects heavy drinkers from only some of the harmful effects of alcoholism.[22] Nutritional rehabilitation is a cornerstone of the treatment for people recovering from alcoholism.[23]

How the Body Handles Alcohol

Alcohol (ethanol) is easily and rapidly absorbed in the stomach and small intestine. Within minutes after it is consumed, alcohol enters the circulatory system, where it increases blood alcohol concentration, and is on its way to the liver, brain, and other tissues throughout the body. The extent to which blood alcohol concentrations rise after alcohol consumption depends not only on the amount of alcohol consumed but also on whether the alcohol is consumed without food on an empty stomach or with food. Alcohol consumed without food is more rapidly absorbed into the blood stream and leads to higher blood alcohol levels than is the case when alcohol is consumed with food (especially fatty foods).[26] Illustration 14.5 shows the effect of alcohol (ethanol) consumption with and without food on blood alcohol concentration over time.

Alcohol cannot be stored in the body, so it remains in body tissues until it is used for energy formation or converted to fatty acids. The process of alcohol breakdown takes several hours or more, depending on the amount of alcohol consumed and a person's size and body composition. Because of the lag time between alcohol intake

ANSWERS TO REALITY CHECK

Do Alcohol Calories Count?

Alcohol provides 7 calories per gram, so it counts toward total calorie intake. However, there is no direct association between all levels of alcohol intake and body weight. People affected by alcoholism tend not to be overweight, whereas light and moderate drinkers appear to have a somewhat elevated risk of weight gain and overweight. Alcohol decreases the utilization of fat stores for energy and redirects fat stores to the abdominal area.[27–29]

Pedro:

Erik:

Illustration 14.5 Blood ethanol (alcohol) concentrations over time in response to ethanol administration with water or with a meal.

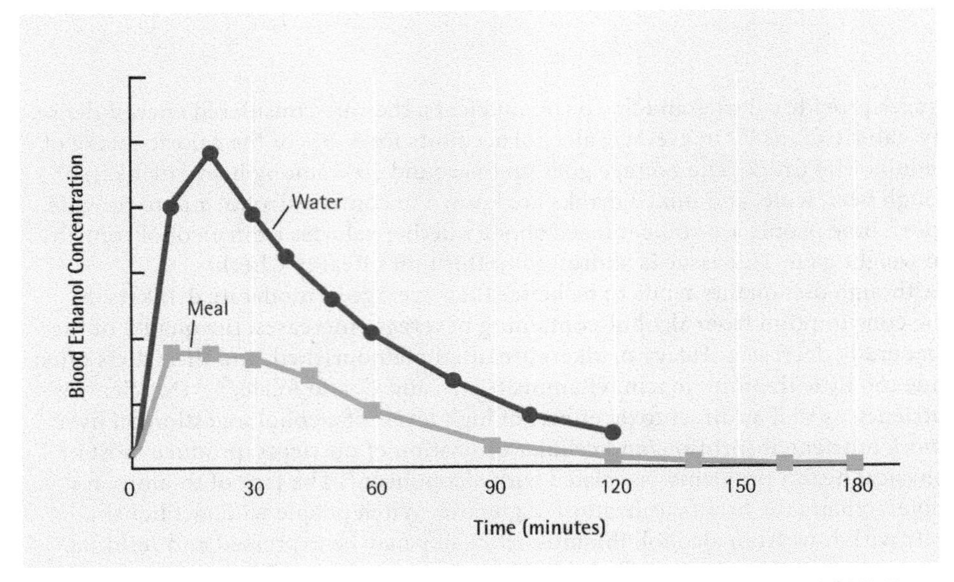

Source: Adapted from Figure 4, Levitt M D, et al. Am J Physiol Gastrointest Liver Physiol. 1997; 273:G951–7. Reference crediting: Levitt MD, et al. Use of measurements of ethanol absorption from stomach and intestine to assess human ethanol metabolism. Am J Physiol Gastrointestinal Liver Physiol. 1997;273(4):G951–57.

and the limited rate at which the body can metabolize it, blood levels of alcohol build up as drinking continues (Table 14.2).[30]

Individuals vary in the in the rate at which blood alcohol levels can be reduced due to differences in genetic traits.[16] Some individuals have gene types that direct the production of enzymes that quickly and thoroughly break down alcohol, whereas other people have gene types that lead to the production of enzymes that break it down slowly or incompletely. Individuals who completely break down alcohol quickly are more

Table 14.2 Alcohol doses and estimated percent blood alcohol levels

Number of drinks[a]	Percent blood alcohol by body weight					
	100 lb	120 lb	140 lb	160 lb	180 lb	200 lb
1	0.04	0.03	0.03	0.02	0.02	0.02
2	0.04	0.03	0.03	0.03	0.02	0.02
3	0.07	0.06	0.05	0.05	0.04	0.04
4	0.11	0.09	0.08	0.07	0.06	0.06
5	0.14	0.12	0.10	0.09	0.08	0.07
6	0.18	0.15	0.13	0.11	0.10	0.09

[a]Taken within an hour or so; each drink equal to ½ ounce of pure ethanol. Effects may vary based on food intake, sex, and other factors.
Source: University of Oklahoma Police Dept., Blood Alcohol Calculator. Available at: www.ou.edu/oupd/bac.htm. Accessed August 2006.

tolerant of alcohol and more prone to excess alcohol intake than are individuals who break it down slowly. People who break down alcohol slowly experience nausea, a rapid heart rate, and develop facial flushing when they drink alcohol.[30] The experience is unpleasant, so they tend to drink less. These and other differences in genetic traits account for approximately 50% of the heritable risk for alcohol dependency.[31]

Approximately 90% of the alcohol consumed is metabolized by the body. The remaining 10% is lost in sweat, urine, or breath. The excretion of alcohol through a person's breath is the reason you can smell alcohol on the breath of someone who has been drinking recently, and why blood alcohol concentration can be estimated by a breathalyzer test.[34]

Alcohol Intake and Blood Alcohol Concentration The intoxicating effects of alcohol correspond to blood alcohol levels (Illustration 14.6). It doesn't matter if the alcohol comes from beer, wine, or hard liquor; the intoxicating effects are the same. A drink or two in an hour raises blood levels of alcohol to approximately 0.03% in most people who weigh about 140 pounds. Blood alcohol levels of 0.03% correspond to mild intoxication. At this level, people lose some control over muscle movement and have slowed reaction times and impaired thought processes. A person's ability to drive or operate equipment in a safe manner is decreased at this level of blood alcohol content. Blood alcohol levels of around 0.06% are associated with increased involvement in traffic accidents. The legal limit for intoxication according to all states' highway safety ordinances is 0.08%—beyond the point where driving is impaired. When blood alcohol content increases to 0.13%, speech becomes slurred, double vision occurs, reflexes are dulled, and body movements become unsteady (Illustration 14.7). If blood alcohol level continues to increase, drowsiness occurs and people may lose consciousness. Levels of blood alcohol above 0.35% can cause death.[25]

A given amount of alcohol intake among women produces higher blood levels of alcohol than it does for men of the same body weight. Pound for pound, women's bodies contain less water than men's bodies, so blood alcohol levels in women increase faster than in men.[32]

Over 150 medications, including sleeping pills, antidepressants, and painkillers, interact harmfully with alcohol. Combining three or more drinks per day with aspirin or nonaspirin pain relievers (acetaminophen, ibuprofen) may cause stomach ulcers or liver damage.[33]

A Note about Alcohol Proof Alcohol proof is an old measure of how much alcohol is contained in alcohol-containing liquids. The term was derived hundreds of years ago by testing whether an alcohol containing liquid would ignite. If it did, that was "proof" of a high enough content of alcohol.[35] The proof of an alcohol-containing liquid is twice its alcohol content. So a beer labeled 3.5% alcohol by volume would be 7 degrees proof, or a liquor labeled 35% alcohol by volume would be 70 degrees proof. In the United States and in many other countries, it is required that alcohol-containing liquids be labeled with the percent alcohol by volume.[36]

How to Drink Safely if You Drink Many of the problems related to alcohol intake can be prevented by not drinking or by drinking responsibly. That means:

As BAC Increases, So Does Impairment

Blood Alcohol Content (BAC)

0.31–0.45%
0.16–0.30%
0.06–0.15%
0.0–0.05%

Life-threatening
- Loss of consciousness
- Danger of life-threatening alcohol poisoning
- Significant risk of death in most drinkers due to suppression of vital life functions

Severe Impairment
- Speech, memory, coordination, attention, reaction time, balance significantly impaired
- All driving-related skills dangerously impaired
- Judgment and decision making dangerously impaired
- Blackouts (amnesia)
- Vomiting and other signs of alcohol poisoning common
- Loss of consciousness

Increased Impairment
- Perceived beneficial effects of alcohol, such as relaxation, give way to increasing intoxication
- Increased risk of aggression in some people
- Speech, memory, attention, coordination, balance further impaired
- Significant impairments in all driving skills
- Increased risk of injury to self and others
- Moderate memory impairments

Mild Impairment
- Mild speech, memory, attention, coordination, balance impairments
- Perceived beneficial effects, such as relaxation
- Sleepiness can begin

Illustration 14.6 Dose-response effects of blood alcohol content (BAC) on body processes.[25]

PhotoDisc

Illustration 14.7 The legal limit for intoxication is 0.08%—beyond the point where driving is impaired.

- Not drinking if you are or could become pregnant.
- Not drinking on an empty stomach (which can make you intoxicated surprisingly fast).
- Slowly sipping rather than gulping drinks.
- Limiting alcohol to an amount that doesn't make you lose control over your mind and body.
- Never driving a car or boat, hunting, or operating heavy equipment while under the influence of alcohol.

Caffeine does not counteract the effects of alcohol on impaired judgment, reaction time, or motor skills.[37] Rather, the combination of caffeine with alcohol masks the intoxicating effects of alcohol while not affecting blood alcohol concentrations.[38]

What Causes Alcohol Dependency?

One in ten adults in the United States abuse alcohol or has alcohol dependency.[31] Alcoholism tends to run in families (about half of alcohol-dependent people have a family history of the disease), and there are documented genetic components to alcoholism.[39] Its development is also influenced by environmental factors. In general, the younger individuals are when they begin to drink, the greater the likelihood that they will develop a drinking problem at some point in life.[40] Individuals who begin drinking before the age of 15, for example, are four times more likely to become alcohol dependent than people who do not drink before age 21 (Illustration 14.8). Close association with friends or peers who drink, high levels of stress, and availability of alcohol may also increase the risk of alcoholism. Social media posts depicting youth-oriented parties, fun, and alcohol may encourage underage drinking.[41]

Alcohol Use Among Adolescents is increasing, and the age when teens begin drinking is going down. Underage drinking accounts for 11% of all the alcohol consumed in the United States. The average age when teens begin drinking is 14 years.[40] Reduction in alcohol intake by adolescents is a major public health initiative of the Health Objectives for the Nation.

PhotoDisc

Illustration 14.8 The younger a person is when drinking begins, the greater the probability that a drinking problem will develop.

Help for Alcohol Dependence Alcohol dependency is a chronic disease that can be successfully managed but not always cured. Many options for treatment are available. Treatments generally involve behavioral therapy, medications, or both. Behavioral therapy is successful in about one-third of people with alcoholism. Medications now available successfully treat alcohol dependency in certain individuals with a genetic predisposition toward developing the disease. Other medications that are under development would act by reducing the craving for and intake of alcohol among individuals without a genetic predisposition for the disease.[1]

Image Source/Getty Images

NUTRITION up close

Effects of Alcohol Intake

Focal Point: Estimating blood alcohol levels and side effects.

Scenario: Ligia and Mark attend a wedding reception. Prior to the meal, they both drink a glass of champagne to toast the bride. Fifteen minutes later, they drink another glass to toast the groom. Ligia weighs 140 pounds and Mark weighs 180.

Questions: Using the information in Table 14.2 and the information given on how the body handles alcohol, answer the following questions:

A. After two glasses of champagne, what would be Ligia's estimated blood alcohol level? What is her percent blood alcohol? What would be Mark's estimated blood alcohol level? What is his estimated percent blood alcohol?

B. List three side effects of 0.03% blood alcohol content:

1. 2. 3.

Feedback to the Nutrition Up Close can be found in Appendix F.

REVIEW QUESTIONS

- **Summarize the effects of the quantity of alcohol intake on behavior, health, and disease.**

1. Moderate consumption of alcohol-containing beverages decreases the risk of heart disease. **True/False**
2. In the United States, a "moderate" intake of alcohol-containing beverages is considered to be two standard servings daily for men and one for women. **True/False**
3. Alcohol is considered a food because it is a good source of a variety of vitamins and minerals. **True/False**
4. Alcohol is used by the body for energy or is converted to glycogen. **True/False**
5. Alcohol poisoning represents an emergency situation requiring medical care. **True/False**
6. The idea that alcohol is absorbed more slowly if you drink after you eat or while eating is a myth. **True/False**
7. Some people are genetically susceptible to developing alcoholism. **True/False**
8. Caffeinated energy drinks counteract the effects of alcohol on the body. **True/False**
9. Alcohol is converted to glucose before it is absorbed into the bloodstream. **True/False**
10. Alcohol is absorbed in the stomach. **True/False**
11. ____ The alcohol content of 12 ounces of regular beer is equivalent to the alcohol content of:
    a. 1 ounce of liquor
    b. 16 ounces of "light" beer
    c. 3½ ounces of wine
    d. 1½ ounces of liquor
12. ____ Moderate alcohol intake is related to:
    a. increased ability of cells to utilize insulin
    b. decreased HDL cholesterol levels
    c. increased chronic inflammation
    d. increased risk of type 2 diabetes
13. ____ Heavy drinking is defined as the daily consumption of ____ or more drinks per day.
    a. 3
    b. 5
    c. 7
    d. 9
14. ____ Which of the following is not usually related to heavy drinking?
    a. obesity
    b. traffic accidents
    c. assaults
    d. high blood pressure

Answers to these questions can be found in Appendix F.

NUTRITION SCOREBOARD ANSWERS

1. Alcohol (actually ethanol) is produced by the fermentation of carbohydrates in grains, fruits, and other plant foods. **True**
2. High intake of alcohol is harmful to the body, regardless of the quality of the diet. **False**
3. The statistics on alcohol abuse, injury, and death are startling. Alcohol abuse is a major personal, social, and public health problem in the United States. **True**
4. Dark beers provide more calories than regular beer. **True**

Proteins and Amino Acids

NUTRITION SCOREBOARD

1. The primary function of protein is to provide energy. **True/False**
2. *Nonessential* amino acids are not required for normal body processes. Only *essential* amino acids are. **True/False**
3. High-protein diets and amino acid supplements by themselves increase muscle mass and strength. **True/False**
4. High-protein diets increase the risk of heart disease. **True/False**

Answers can be found at the end of the unit.

Elena Veselova/Shutterstock.com

Protein

LEARNING OBJECTIVES

After completing Unit 15 you will be able to:

- Summarize the functions, structures, and food sources of proteins.

- **Summarize the functions, structures, and food sources of proteins.**

The term **protein** is derived from the Greek word *protos*, meaning "first." The derivation indicates the importance ascribed to this substance when it was first recognized. Protein is an essential structural component of all living matter and is involved in virtually every biological process that occurs in cells. Protein has a very positive image (Illustration 15.1). The perception is so positive that you don't have to talk about the importance of protein—people are already convinced of it.[3]

KEY NUTRITION CONCEPTS

Content covered in Unit 15 on protein and amino acids relates to the following key nutrition concepts:

1. Nutrition Concept #2: Foods provide energy (calories), nutrients, and other substances needed for growth and health.
2. Nutrition Concept #9: Adequacy, variety, and balance are key characteristics of healthful diets.

protein Chemical substance in foods made up of chains of amino acids.

hormone A substance, usually a protein or steroid (a cholesterol-derived chemical), produced by one tissue and conveyed by the bloodstream to another. Hormones affect the body's metabolic processes such as glucose utilization and fat deposition.

immunoproteins Blood proteins such as antibodies that play a role in the functioning of the immune system (the body's disease defense system). Antibodies attack foreign proteins.

Nearly all people in the United States get enough protein in their diets. The average intake of protein by adults in the United States is 98 grams per day, approximately twice the recommended daily allowance (RDA) for both men (56 grams) and women (46 grams). Approximately 15% of total calories in the average U.S. adult diet are supplied by protein.[4]

High-protein diets are often accompanied by high fat and low fiber intakes. That's because foods high in protein, such as hamburger, cheese, nuts, and eggs, tend to be high in fat and contain little or no fiber. Even lean meats provide a considerable proportion of their total calories as fat (Illustration 15.2).

Functions of Protein

Proteins perform structural and functional roles in the body (Table 15.1). They are an integral structural component of skeletal muscle, bone, connective tissues (skin, collagen, and cartilage), organs (such as the heart, liver, and kidneys), red blood cells and hemoglobin, hair, and fingernails. Proteins are the basic substances that make up thousands of enzymes in the human body; they are a major component of **hormones** such as insulin and growth hormone, and they serve as other substances that perform important biological functions. Tissue maintenance and the repair of organs and tissues damaged by illness or injury are functions of different types of protein. Albumin, a protein made by the liver, is the blood's "tramp steamer." It attaches to and transports fatty acids, calcium, and other substances through the circulatory system to cells throughout the body.[5] Protein serves as an energy source at the level of four calories per gram.

Illustration 15.1 The protein perception.

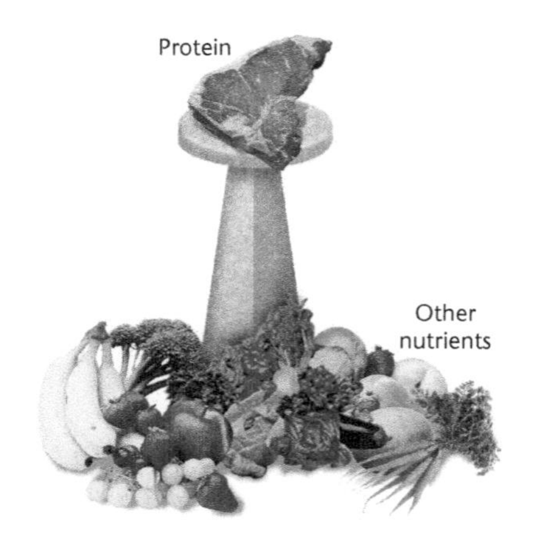

The body of a 154-pound man contains approximately 24 pounds of protein. Nearly half of the protein is found in muscle; the rest is present in the skin, collagen, blood, enzymes, and **immunoproteins**, and in organs such as the heart, liver, intestines, and other body parts. All protein in the body is continually being turned over, or broken down and rebuilt. This process helps maintain protein tissues in optimal condition so they continue to function normally. The process of protein turnover utilizes roughly nine ounces of protein each day. Yet we consume only two to three ounces of protein daily. Most of the protein used for maintenance is recycled from muscle and other protein tissues being turned over. Proteins play key roles in the repair of body tissues by serving as substances—such as fibrin—that help blood clot (Illustration 15.3) and by replacing tissue proteins damaged by illness or injury.[6]

Protein serves as a source of energy in healthy people, but not nearly to the extent that carbohydrates and fats usually do. Protein is unlike carbohydrate and fat in that it

©Erika Cross/Shutterstock.com

(a) Hamburger (90% lean): 45%

iStockphoto.com/svariophoto

(b) Tenderloin: 43%

Felicia Martinez Photography/PhotoEdit

(c) Sirloin: 33%

Felicia Martinez Photography/PhotoEdit

(d) Pork chop, lean: 48%

Felicia Martinez/PhotoEdit

(e) Pork loin roast: 36%

Felicia Martinez/PhotoEdit

(f) Pork tenderloin: 28%

Richard Anderson

(g) Chicken thigh, no skin: 47%

Richard Anderson

(h) Baked chicken breast, no skin: 19%

Illustration 15.2 The fat content of three-ounce portions of "lean" meats. The percentage of calories from fat is indicated for each portion. (A three-ounce portion of meat is about the size of a deck of cards.) Each portion of meat provides approximately 21 grams of protein.

contains nitrogen and does not have a storage form in the body. To use protein for energy, amino acids that make up proteins must first be stripped of their nitrogen by the liver. The free nitrogen can be used as a component of protein formation within the body, or, if present in excess, it is largely excreted in urine. Excretion of nitrogen requires water, so high intake of protein increases water need. Amino acids missing their nitrogen component are converted to glucose or fat that can then be used to form energy. A small amount of protein (1%) can be obtained from the liver and blood and used to cover occasional deficits in protein intake.[7]

Nitrogen Balance **Nitrogen balance** represents the balance between protein intake and protein utilization by the body. Protein intake is estimated as the nitrogen content of protein consumed. The nitrogen content of protein is estimated as 16% of the weight of protein consumed. Nitrogen excretion is assessed as the amount of nitrogen excreted in the form of **urea**. Nitrogen balance is measured as the difference between nitrogen intake and nitrogen excreted.[8]

- A person is considered to be in nitrogen balance when her or his intake of nitrogen equals nitrogen excreted. Example: A person consumes 10 grams of nitrogen and excretes 10 grams of nitrogen: 10 grams nitrogen − 10 grams nitrogen = 0 grams nitrogen. People who are in nitrogen balance are consuming as much protein as they are utilizing.

Table 15.1 Examples of functions of protein

Structural
• Serves as a structural material in muscles, connective tissue, organs, and hemoglobin
• Maintains and repairs protein-containing tissues
Functional
• Serves as the basic component of enzymes, hormones, and other biologically important chemicals
• Serves as an energy source
• Helps maintain body fluid balance
• Helps maintain acid–base balance in body fluids
• Contributes to a feeling of satiety

nitrogen balance The difference between nitrogen intake and excretion. It is assessed as the difference between nitrogen intake and nitrogen excreted.

urea Nitrogen released from the breakdown of proteins for energy is largely excreted in the urine in the form of urea. It can be measured in urine, or in blood as blood urea nitrogen (BUN).

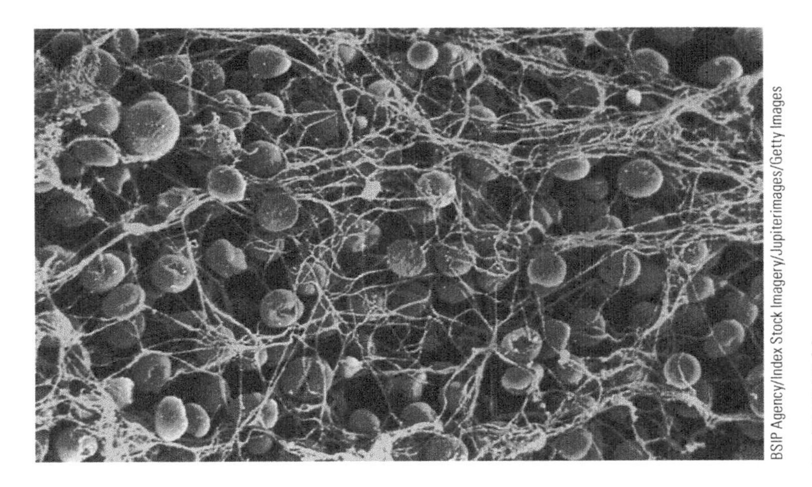
BSIP Agency/Index Stock Imagery/Jupiterimages/Getty Images

Illustration 15.3 Red blood cells enmeshed in fibrin in a color-enhanced microphotograph. Red blood cells and fibrin (which helps stop bleeding by causing blood to clot) are made primarily from protein.

- A person who consumes less nitrogen than excreted is in negative nitrogen balance. Example: A person consumes 10.2 grams nitrogen and excretes 14.0 grams: 10.2 grams nitrogen − 14.0 grams nitrogen = −3.8 grams nitrogen. A negative nitrogen balance means the person is consuming less protein than the body is utilizing. A negative nitrogen balance may occur with undernutrition, fasting, burns and other serious injuries, fever, and other illnesses.
- A person is considered to be in positive nitrogen balance when his or her intake of nitrogen is greater than nitrogen excretion. Example: A person consumes 18 grams of nitrogen and excretes 14.5 grams: 18 grams nitrogen − 14.5 grams nitrogen = −3.5 grams nitrogen. A positive nitrogen balance indicates that some of the protein consumed is being retained and used to build up body protein tissues. Positive nitrogen balance occurs during growth, pregnancy, breastfeeding, and recovery from illness or injury.

Results of nitrogen balance studies help determine protein need and are sometimes used clinically to adjust a person's protein intake to meet protein need.[8]

Amino Acids

The building blocks of protein are amino acids, which share the characteristic of containing nitrogen. Illustration 15.4 shows an example of the basic structure of an amino acid and its nitrogen component. Twenty common amino acids (Table 15.2) form proteins, and every protein in the body is composed of unique combinations of amino acids. Nine of the 20 common amino acids are considered **essential**, and 11 are **nonessential**. The essential amino acids are called *essential* because the body cannot produce them or produce enough of them, so they must be provided in the diet. Healthy individuals can produce the nonessential amino acids, so we don't require a dietary source of them. Despite the labels, all 20 amino acids are required to build and maintain protein tissues. Proteins in foods contain both essential and nonessential amino acids.

essential amino acids Amino acids that cannot be synthesized in adequate amounts by humans and therefore must be obtained from the diet. They are sometimes referred to as "indispensable amino acids."

nonessential amino acids Amino acids that can readily be produced by humans from components of the diet. Also referred to as "dispensable amino acids."

DNA (deoxyribonucleic acid) Genetic material contained in cells that directs the production of proteins in the body.

Amino Acids and Protein Structure The assembly of amino acids into proteins is directed by **DNA**, the genetic material within each cell. Some proteins are made of only a few amino acids, while other proteins contain over 25,000. The arrangement of amino acids determines whether a protein functions as an enzyme or a hormone, or it becomes a component of red blood cells, muscle, or some other substance. Human genes are able to produce around two million proteins from the raw material provided by amino acids, and each protein performs a specific function in the body.[9]

Proteins vary in size and complexity based on their role in cellular processes. They are classified by their properties as having primary, secondary, tertiary, or quaternary structures (Illustration 15.5). Proteins with primary structure consist of a linear arrangement of linked amino acids, whereas proteins with secondary structures have folded chains of amino acids. Some hormones and chemical messengers that initiate cellular processes have these structures. Tertiary structures represent the three-dimensional structure of more complex and larger proteins. These proteins have elaborate folding patterns such as

Table 15.2 Essential and nonessential amino acids

Essential		Nonessential	
Histidine	Threonine	Alanine	Glutamine
Isoleucine	Tryptophan	Arginine	Glycine
Leucine	Valine	Asparagine	Proline
Lysine		Aspartic acid	Serine
Methionine		Cysteine	Tyrosine
Phenylalanine		Glutamic acid	

Illustration 15.4 The basic chemical structure of alanine, a nonessential amino acid.

that found in insulin (Illustration 15.6). Proteins with quaternary structure are the largest and consist of multiple, linked chains of amino acids folded and formed in such a way that they can perform specific functions. Hemoglobin, a component of red blood cells that transports oxygen to cells and removes carbon dioxide from cells, is an example of this type of protein structure.[7]

Protein Quality

The ability of proteins to support protein tissue construction in the body varies depending on their content of essential amino acids. How well dietary proteins support protein tissue construction is captured by tests of the protein's quality.

Proteins of high quality contain all the essential amino acids in the amounts needed to support protein synthesis (formation) by the body. If any of the essential amino acids are missing from the diet, proteins are not formed or are formed incorrectly and do not function normally. Shutting off or modifying protein production for want of an amino

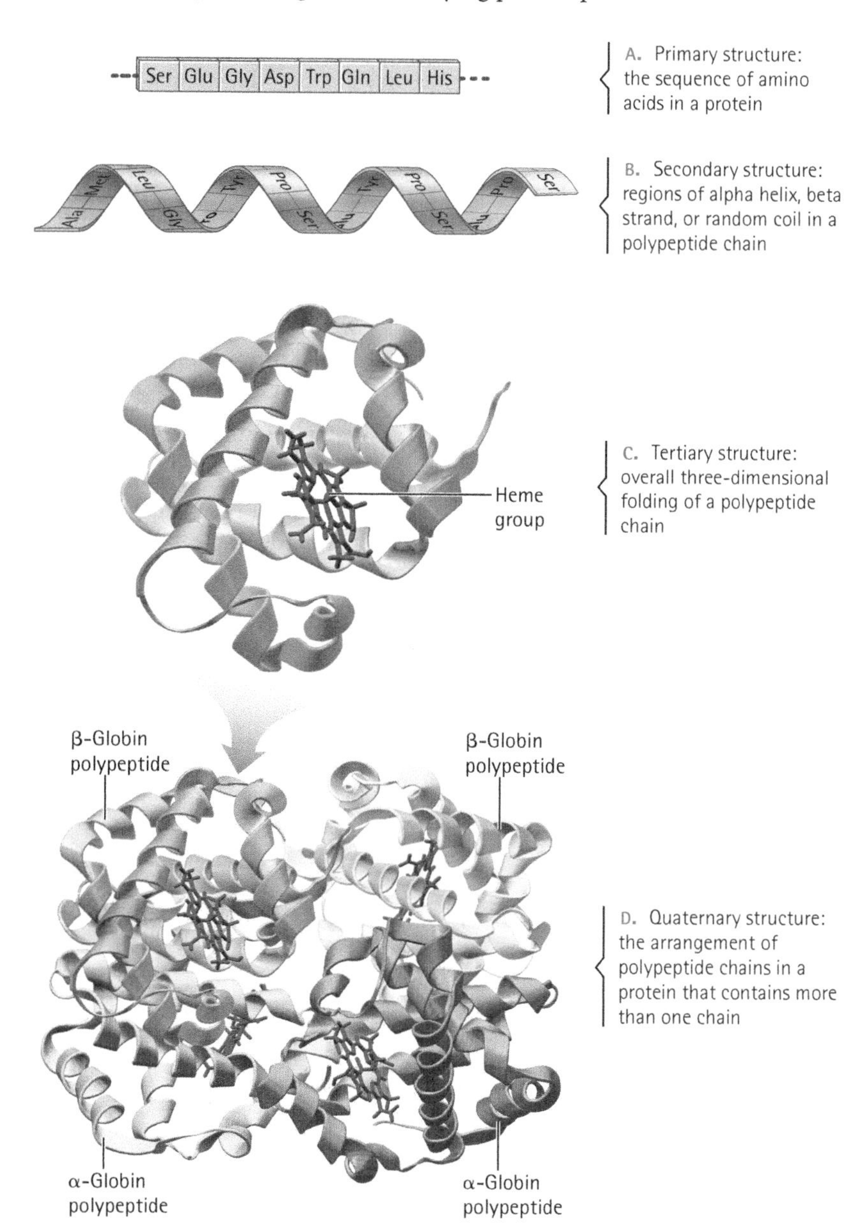

Illustration 15.5 The primary, secondary, tertiary, and quaternary structures of proteins. Source: Russell, Biology: The Dynamic Science, 2nd ed., chap. 3, fig. 3.18, p. 58

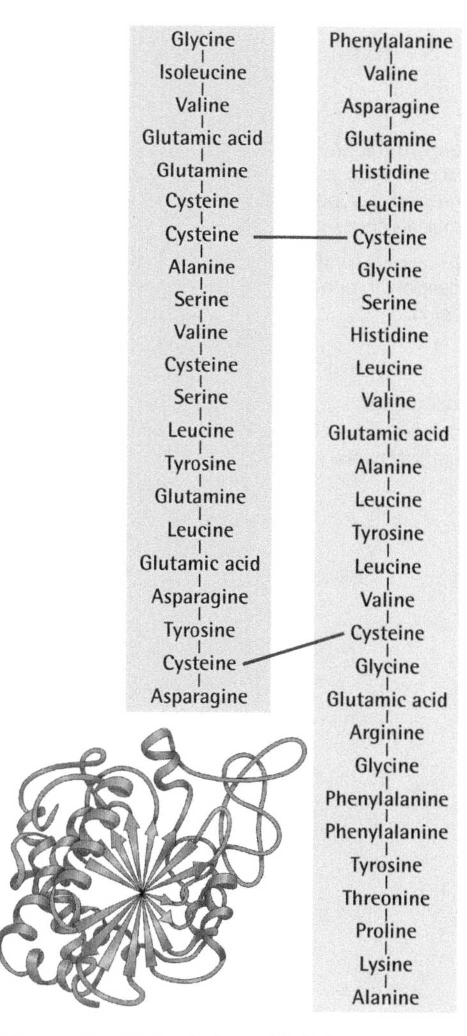

Illustration 15.6 Amino acid chains and folding of insulin, a tertiary protein.

Illustration 15.7 Animal sources of protein shown here supply complete proteins.

acid or two may seem inefficient. But if that did not happen, cells would end up with an assortment of proteins that could broadly disrupt critical body functions.[10] When the required level of an essential amino acid is lacking, the remaining amino acids are used primarily for energy. Amino acids cannot be stored very long in the body, so we need a fresh supply of essential amino acids daily.

Complete Proteins Food sources of high-quality protein (meaning they contain all the essential amino acids in the amounts needed to support protein formation) are called **complete proteins.** Proteins in this category include those found in animal products such as meat, milk, and eggs (Illustration 15.7). **Incomplete proteins** are deficient in one or more essential amino acid. Proteins in plants are incomplete, with the exception that soybeans are considered a complete source of protein for adults.[11] (Soybeans may not meet the essential amino acid requirements of young infants.) You can complement the essential amino acid composition of plant sources of protein by combining them to form a complete source of protein. Illustration 15.8 shows a few complementary plant food combinations that produce complete proteins.

Vegetarian Diets Dietary patterns consisting only of plant foods can provide an adequate amount of complete proteins. A key to success is eating a variety of complementary sources of protein each day. (The 2015 Dietary Guidelines scientific advisory committee identified a "healthy vegetarian-style" dietary pattern to be one that supports health and helps prevent disease.[12]) Unit 16 on vegetarian diets expands on this topic.

complete proteins Proteins that contain all of the essential amino acids in amounts needed to support growth and tissue maintenance.

incomplete proteins Proteins that are deficient in one or more essential amino acids.

Amino Acid Supplements Because amino acids occur naturally in foods, people may assume that amino acid supplements are harmless, no matter how much is taken (Illustration 15.9). Researchers have known for decades, however, that high intake of individual amino acids can harm health. Large amounts may disrupt normal protein production by overwhelming cells with a surplus of some amino acids and a relative deficit of others.[13] Excess consumption of the amino acid methionine, in particular, causes a host of problems. Intake levels two to five times above the amount normally consumed from foods worsen the symptoms of schizophrenia, promote hardening of the arteries,

Illustration 15.8 Each of these combinations of plant foods provides complementary protein sources.

impair fetal and infant development, and lead to nausea, vomiting, bad breath, and constipation.[13,14] Adverse health effects associated with supplemental cysteine and phenylalanine have also been reported.[13,15] Use of large amounts of amino acid supplements should be supervised by a physician.

Safe doses of certain amino acids or their derivatives are being used to manage certain conditions. For example, melatonin (a derivative of the essential amino acid tryptophan), is used to promote sleep.[16] Melatonin supplements appear to help reset the sleep clock in shift workers, pilots, jetlagged travelers, and people with sleep disorders. The biggest users of amino acid supplements are athletes who believe they will increase muscle mass and strength.[17] Additional information on the topic of physical performance aids is provided in Unit 28.

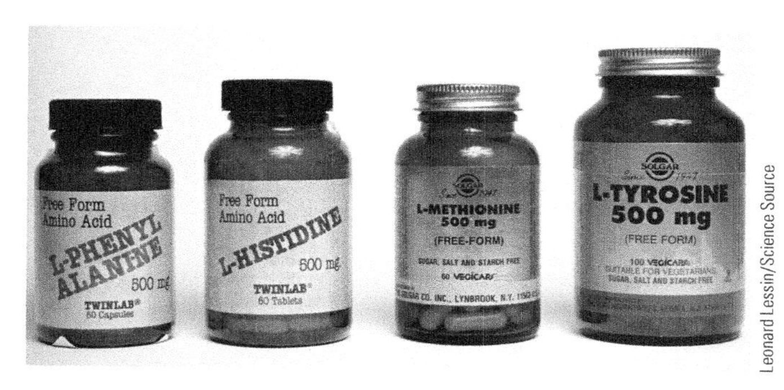

Leonard Lessin/Science Source

Illustration 15.9 A wide variety of amino acid supplements are available on the Web or in stores.

Amino Acid Supplements, Protein Powders, Muscle Mass, and Strength You can't just consume amino acids or protein powders and watch your muscles grow—no matter how convincing the ads that sell such products are. If that happened, anyone who wanted a ripped stomach and bulging triceps could have them. Neither essential amino acids nor protein supplements by themselves increase muscle size and strength.[1] Muscle size and strength are built over time and primarily from the raw ingredients of a healthy dietary pattern, including high-quality protein and resistance training (Illustration 15.10).[21] Current research indicates that young adults can enhance muscle protein synthesis, muscle mass, and muscle recovery by consuming 20 grams of high-quality protein shortly after resistance exercise workouts.[2] Older adults may need 40 grams of high-quality protein to obtain similar results.[22] Skim milk, lean meat, fish, egg white, and beans and rice are examples of high-quality protein foods; casein and whey are examples of protein concentrates that could be consumed. Protein intake over 20 grams following exercise does not appear to offer additional benefits in young adults. The excess protein intake is largely used for energy formation or storage.[22]

As with other dietary supplements, the purity, dosage, and safety of supplements available over the Internet and in stores is not guaranteed.[23] Even if the amino acid supplements are pure, safe, and the dose reported on the label is correct, amino acid supplements have not been found to work better for muscle and strength building than high-quality protein from foods.[1]

Paul Bradbury/The Image Bank/Getty Images

Illustration 15.10 Building muscles like these takes time, a healthy dietary pattern including high-quality protein sources, and resistance exercise.

Food as a Source of Protein

Approximately 70% of the protein consumed by Americans comes from meats, milk, and other animal products.[4] Dried beans and grains are not as well known for their protein, but are nevertheless good sources (Table 15.3). Plant sources of

REALITY CHECK

The Protein Halo Effect

Browsing the grocery store shelves for a box of breakfast cereal, you notice a new one with a front-of-package label announcing "a good source of protein." Is this cereal superior in nutrient content to cereals not so labeled?

Carole: I look for the protein label because I think more protein is really good for me.

Lauren: I'm looking to get more out of my morning bowl of cereal than protein. I already pour skim milk over it.

Who gets the thumbs-up?

Answers on page 15-8

Photodisc

Carole: Lean meats are the best source. They're pure protein!

Photodisc

Lauren: Pure protein? Even the driest, toughest meats contain more than protein.

ANSWERS TO REALITY CHECK

The Protein Halo Effect

Protein has a positive image and is important. But that doesn't mean foods labeled as having protein are more nutritious than other foods not so labeled.

Carole:

Lauren:

Table 15.3 Food sources of protein

The adult RDA is 46 grams for women aged 19 to 30 years and 56 grams for men aged 19 to 30 years.

Food	Amount	Grams	Protein content as percentage of total calories
Animal products			
Tuna (water packed)	3 oz	24	89%
Shrimp	3 oz	11	84
Cottage cheese (low-fat)	½ cup	14	69
Beef steak (lean)	3 oz	26	60
Chicken (no skin)	3 oz	24	60
Pork chop (lean)	3 oz	20	59
Beef roast (lean)	3 oz	23	45
Skim milk	1 cup	9	40
Fish (haddock)	3 oz	19	38
Leg of lamb	3 oz	22	37
Yogurt (low-fat)	1 cup	13	34
Hamburger (lean)	3 oz	24	34
Egg	1 medium	6	32
Swiss cheese	1 oz	8	30
Sausage (pork links)	3 oz	17	28
2% milk	1 cup	8	26
Cheddar cheese	1 oz	7	25
Whole milk	1 cup	8	23
Dried beans and nuts			
Tofu	½ cup	14	38
Soybeans (cooked)	½ cup	10	33
Split peas (cooked)	½ cup	5	31
Lima beans (cooked)	½ cup	6	27
Dried beans (cooked)	½ cup	8	26
Peanuts	½ cup	9	17
Peanut butter	1 Tbsp	4	17

(continued)

Table 15.3 **(continued)**

Food	Amount	Grams	Protein content as percentage of total calories
Grains			
Corn	½ cup	3	29
Egg noodles	½ cup	4	25
Oatmeal (cooked)	½ cup	3	15
Whole wheat bread	1 slice	2	15
Macaroni (cooked)	½ cup	3	13
White bread	1 slice	2	13
White rice (cooked)	½ cup	2	11
Brown rice (cooked)	½ cup	2	10

protein are generally low in calories, making them a wise choice for people who are trying to limit their calorie intake. Nearly all food sources of protein provide an assortment of vitamins and minerals as well. Beef and pork are particularly good sources of iron, a mineral often lacking in the diets of women (Table 15.4).

Protein Deficiency Protein deficiency has been found to occur in combination with a deficiency of calories and nutrients.[18] Because food sources of protein generally contain essential nutrients such as iron, zinc, vitamin B_{12}, and niacin, diets that produce protein deficiency usually cause a variety of other deficiencies, as well. Protein does not generally serve as an important source of energy, but body protein will be used as a major energy source during starvation. To meet the need for energy, the body will extract protein from the liver, intestines, heart, muscles, and other organs and tissues. Loss of more than about 30% of body protein results in reduced body strength for breathing, susceptibility to infection, abnormal organ functions, and death. Inadequate protein intake is related to decreased growth in children and loss of muscle mass and strength in adults.[6]

In the past it was thought that **kwashiorkor**, a devastating disease that can affect severely undernourished children, was primarily due to a protein deficiency. Although kwashiorkor is related to protein-calorie malnutrition, the disease does not appear to be due only to a lack of protein.[18] This conclusion is supported by studies that show that improving protein intake in children with kwashiorkor does not correct the disease. It appears that some children with protein-calorie malnutrition develop kwashiorkor due to an inability to utilize protein and fat normally during starvation.[19,20]

How Much Protein Is Too Much? Adults can consume a substantial amount of protein without ill effects.[2] This conclusion is largely based on studies of the diets of Eskimos, explorers, trappers, and hunters in northern America.[18] Consumption of 45% of total calories from protein is considered too high and is related to the development of symptoms such as nausea, weakness, and diarrhea. Diets very high in protein may result in death after several weeks. A complex disease resulting from excess protein intake was termed "rabbit fever" and "rabbit starvation syndrome" after it occurred in trappers attempting to exist exclusively on wild rabbit.[6,24]

High-protein diets have been implicated in the development of weak bones, kidney stones, cancer, heart disease, and obesity. The National Academy of Sciences has concluded that the risk of such disorders does not appear to be increased among individuals consuming 10–35% of total calories from protein, and adults consume 15% on average.[4,6]

Table 15.4 **Iron content in a 3-ounce serving of various meats**
The RDA for women aged 19 to 30 years is 15 milligrams. The RDA for men aged 19 to 30 years is 10 milligrams.

Meat	Iron content (mg)
Pork chop (lean)	3.4
Round steak (lean)	3.1
Hamburger (lean)	3.0
Shrimp	2.6
Tuna	1.6
Baked chicken (no skin)	1.4
Lamb (lean)	1.3

kwashiorkor A form of severe protein-energy malnutrition in young children. It is characterized by swelling, fatty liver, susceptibility to infection, profound apathy, and poor appetite.

Envision/Encyclopedia/Corbis

NUTRITION up close

My Protein Intake

Focal Point: Determine the amount of protein in your diet yesterday.

For *each serving* of a food item you ate yesterday, write the grams of protein the food contains in the corresponding blank. For example, a standard serving of meat is three ounces (about the size of the palm of your hand or a deck of cards). If you had *one* three-ounce pork chop yesterday, write *20 grams* in the corresponding blank. If you had *two* three-ounce pork chops, write *40 grams.* If a protein food you ate yesterday is not included, choose the item on the list closest to it. Then total the grams of protein you ate yesterday from both plant and animal sources. Finally, compare your protein intake with the RDA of 46 grams for women or 56 grams for men.

Food	One serving	Protein in one serving (grams)	Protein you ate (grams)
Animal products			
Milk (whole)	1 c (8 oz)	8	
Yogurt	1 c (8 oz)	13	
Cottage cheese	½ c (4 oz)	14	
Hard cheese	1 oz	7	
Hamburger (lean)	3 oz	24	
Beef steak (lean)	3 oz	26	
Chicken (no skin)	3 oz	24	
Pork chop (lean)	3 oz	20	
Fish	3 oz	19	
Hot dog	1	6	
Sausage	3 oz	17	
		Subtotal from animal foods:	
Plant products			
Bread	1 slice	2	
Rice	½ c (4 oz)	2	
Pasta	½ c (4 oz)	3	
Cereals	½ c (4 oz)	3	
Vegetables	½ c (4 oz)	2	
Peanut butter	1 Tbsp	4	
Nuts	¼ c (2 oz)	7	
Cooked beans (legumes)	½ c (4 oz)	8	
		Subtotal from plant foods:	
		Total grams of protein from plant and animal foods:	
		Amount above/below RDA:	

Special note: You can also calculate your protein intake using the Diet & Wellness Plus software. From the home page of Diet & Wellness Plus, select the Track Diet tab and input your food intake for one day. Then from the Reports tab, select the Source Analysis report. From the Source Analysis page, choose Protein from the drop-down menu. View your results.

Feedback to the Nutrition Up Close can be found in Appendix F.

REVIEW QUESTIONS

- **Summarize the functions, structures, and food sources of proteins.**

1. The only known function of protein is to serve as a structural component of muscle, bones, and other solid tissues. **True/False**
2. Low protein intake is a fairly common problem among adults and children in the United States. **True/False**
3. Immunoproteins are a part of the body's immune system. **True/False**
4. Fibrin is a protein that helps blood clot. **True/False**
5. The nitrogen component of amino acids must be removed before they can be used for energy. **True/False**
6. There are 22 amino acids, and half of them are *essential*. **True/False**
7. DNA provides the blueprint for the production of specific proteins by the body. **True/False**
8. People need to eat animal sources of protein in order to consume enough high-quality complete proteins. **True/False**
9. Individual amino acid supplements may be hazardous to health. **True/False**
10. Kwashiorkor is the name of a protein deficiency disease. **True/False**
11. Protein intake should constitute 10–35% of total calorie intake. **True/False**
12. Proteins with primary structure consist of a linear arrangement of linked amino acids, whereas proteins with secondary structures have folded chains of amino acid. **True/False**
13. Amino acid supplements plus resistance exercise are the keys to building muscle mass and strength. **True/False**

The next three questions refer to Bertie's resistance training and protein consumption.

14. _______ Bertie decides to consume 20 grams of high-quality protein after her resistance training workout. How many glasses of skim milk would Bertie need to consume to get about 20 grams of protein?
    a. 1
    b. 2
    c. 3
    d. 4
15. _______ Bertie consumes 20 grams of high-quality protein after each resistance training session. Over time, the protein would likely _______ Bertie's muscle mass.
    a. have no effect on
    b. decrease
    c. increase
    d. increases then decrease
16. _______ Bertie decides to consume 20 grams of high-quality protein daily as a supplement to her diet but to give up on resistance training. What do you expect would happen to her muscle mass?
    a. The extra protein would maintain her muscle mass.
    b. The extra protein would increase her muscle mass.
    c. Her muscle mass would decline.
    d. Her muscle mass would stay the same.

Answers to these questions can be found in Appendix F.

NUTRITION SCOREBOARD ANSWERS

1. Energy is a function of protein, but it's not the primary one. **False**
2. Nonessential amino acids are required by the body, but they are not required components of our diet. **False**
3. Muscles contain protein, but you can't increase muscle mass and strength by consuming a high-protein diet, amino acid supplements, or protein powders alone.[1] **False**
4. High-protein diets have not been found to increase the risk of heart disease.[2] **False**

Vegetarian Dietary Patterns

NUTRITION SCOREBOARD

1 There are a number of types of vegetarian dietary patterns. **True/False**

2 Dietary patterns based on plant foods are associated with a lower risk of obesity and heart disease. **True/False**

3 Macrobiotic diets cure some types of cancer. **True/False**

4 In order to consume enough high-quality protein, vegetarians need to consume combinations of plant foods that provide a complete source of protein at every meal. **True/False**

Answers can be found at the end of the unit.

MIB Pictures/UpperCut Images/Getty Images

LEARNING OBJECTIVES

After completing Unit 16 you will be able to:

- Discuss the health benefits and limitations of the types of vegetarian dietary patterns.

Perspectives on Vegetarian Dietary Patterns

- **Discuss the health benefits and limitations of the types of vegetarian dietary patterns.**

Vegetarianism in the United States, Canada, and other economically developed countries is moving from the realm of counterculture to the mainstream.[3] Yet even with the increased acceptance of vegetarian diets, people tend to be for vegetarianism or against it, often without knowing much about it. Few of those opposed to vegetarianism have tried to learn about the vegetarian way of life or understand that well-planned vegetarian dietary patterns are health promoting and environmentally sustainable.[1]

Vegetarian-style healthy dietary patterns are similar in many ways to the DASH, Mediterranean, and USDA's MyPlate healthy dietary patterns. Each of these patterns emphasizes whole grains, vegetables, fruits, legumes, fish, poultry, nuts, and vegetable oils and de-emphasize high-fat meats, processed meats, high-fat dairy products, and sugar-sweetened foods and beverages. Multiple studies have demonstrated that adherence to these types of dietary patterns reduces the risk of heart disease and obesity. Compared to a Western-style dietary pattern, other health benefits attributed to healthy vegetarian-style dietary patterns include a reduced risk of obesity, heart disease, hypertension, type 2 diabetes, and some types of cancer.[1,3] As with the other dietary patterns designated as "healthy," well-planned vegetarian dietary patterns provide adequate calories and essential nutrients and a good supply of fiber and beneficial phytochemicals.[1]

Vegetarians tend to be health conscious; they often avoid using alcohol, tobacco, and illicit drugs, and engage in regular physical activity. Diet is usually one of several characteristics shared by people practicing particular types of vegetarianism.[1]

KEY NUTRITION CONCEPTS

Key nutrition concepts that underlie this content on vegetarian diets include:

1. Nutrition Concept #4: Poor nutrition can result from both inadequate and excessive levels of nutrient intake.
2. Nutrition Concept #7: Some groups of people are at higher risk of becoming inadequately nourished than others.
3. Nutrition Concept #8: Poor nutrition can influence the development of certain chronic diseases.
4. Nutrition Concept #9: Adequacy, variety, and balance are key characteristics of healthful diets.

©Jamie Rogers/Shutterstock.com

Illustration 16.1 What's for lunch? Tofu and vegetables.

Reasons for Vegetarianism

Worldwide, vegetarians number in the hundreds of millions. In the United States, it is estimated that 3.3% of the population are vegetarians, and there is a pronounced trend toward incorporating more meatless meals and plant foods into dietary patterns among non-vegetarians (Illustration 16.1). A part of the world's population subsists on vegetarian diets because meat and other animal products are scarce or too expensive. In other societies, people have the luxury of choosing a healthy assortment of food from an abundant and affordable food supply that includes a wide variety of items acceptable to vegetarians. When food availability is not an issue, people tend to adopt vegetarian diets because of a desire to cause no harm to animals, a personal commitment to preserve the environment and the world's food supply, or a belief that animal products are unhealthful or unsafe. They may avoid animal products as part of a value or religious belief system. Others follow vegetarian diets to keep their weight down or to lower the risk of developing a particular disease or condition or to help manage it.[11] A summary of the primary reasons why people choose a vegetarian dietary pattern is provided in Table 16.1.

Vegetarian Diets Come in Many Types There is no one vegetarian diet. People who consider themselves to be vegetarians range from those who eat all foods except red meat (mainly beef) to those who exclude all foods from animal sources, including honey. Individuals who follow a vegetarian dietary pattern may include certain types of meats in small portions or consume them infrequently, or they may consume fish, eggs, cheese, or other dairy products. Only strict vegetarians (vegans) exclude all animal products from their diets. Various types of vegetarian dietary patterns are listed and defined in Table 16.2.[1]

Macrobiotic Diets Macrobiotic diets were a strict vegetarian type of dietary pattern in the past; they were totally based on whole grains and vegetables. Because the diet was inadequate in a number of nutrients, it was expanded to include legumes and bean products, sea vegetables, fish and seafood, certain fruits, nuts, and rice syrup. Organic and locally produced foods are strongly favored over other foods (Illustration 16.2). Some individuals who follow this dietary pattern believe it can prevent or cure cancer and lead to a long life. Research results currently do not support this belief.[4]

Persons adhering to the macrobiotic philosophy place value on balancing the intake of "yin" and "yang" foods. Foods are classed as yin or yang based on beliefs about the food's relationship to the emotions and the physical condition of the body. Yin foods such

Table 16.1 Reasons for vegetarianism[1]

- Lack of availability or affordability of animal products
- Desire not to cause harm to animals
- Religious beliefs or cultural values
- Desire to consume environmentally sustainable foods
- Taste preferences
- Personal health
- Desire to avoid hormones, antibiotics, and possible contaminants in meats and other foods

Table 16.2 Vegetarian dietary patterns and types of food generally included in each pattern[1,5]

	Foods included					
Type of diet	Beef, lamb, pork ("red meat")	Poultry	Fish	Eggs	Milk and milk products	Plant foods
Semi-vegetarian[a]		x	x	x	x	x
Pesco (fish)-vegetarian[a]			x			x
Lacto-ovo vegetarian[b]				x	x	x
Lacto-vegetarian					x	x
Macrobiotic				x		x
Strict vegetarian (vegan)						x
Raw food						x

[a]Semi-vegetarian dietary pattern tends to vary in the type and amount of animal products consumed.

REALITY CHECK

Reintroducing Meat

Larry has been a vegetarian for the last seven years and wants to eat a hamburger again to see if it tastes as good as he remembers. He's a bit nervous about doing it because he thinks meat might make him feel sick.

Who gets the thumbs-up?

Answers on page 16-4.

Photodisc

Susan: Larry should eat the hamburger if he wants to. It's a food, not an indigestion time bomb.

Photodisc

Doug: Larry's stomach isn't used to the heaviness of meat. It will make him nauseated.

ANSWERS TO REALITY CHECK

Reintroducing Meat

The thought that meat could make one sick may be enough to trigger indigestion. Many self-defined vegetarians consume meat on rare occasion without reported ill effects.[25]

Photodisc

Susan:

Photodisc

Doug:

Boston Globe/Contributor/Getty Images

Illustration 16.2 Organic and locally grown foods are part of a macrobiotic dietary pattern.

as corn, seeds, nuts, fruits, and leafy vegetables are considered negative, dark, cold, and feminine. Yang foods represent opposing positive forces of light, warmth, and masculinity. Poultry, fish, eggs, and cereal grains such as buckwheat are yang foods.[6]

Raw Food Diets Diets consisting primarily of raw foods have come in and out of fashion for hundreds of years. Although there is no formal definition, raw food diets are usually described as an uncooked vegan diet. The diet consists of vegetables, fruits, nuts, seeds, sprouted grains, and beans. Uncooked foods usually make up 50% or more of the foods consumed in the diet.[7]

The modern raw food diet emphasizes the consumption of organic plant foods that have not been processed, are minimally processed, or are raw based on a belief that nutrients and enzymes in food can be lost, or "killed," during processing. Individuals who adhere to a raw food diet are generally deeply concerned about animal rights, climate change, and sustainable agriculture. Consumption of a raw food diet is associated with low body mass index, a risk of vitamin B_{12} deficiency, low bone mineral density, and impaired growth in children.[9]

Other Vegetarian Dietary Patterns For a number of other vegetarian regimes, the spiritual or emotional importance assigned to certain foods supersedes consideration of their contribution to an adequate diet.[16] Vegetarians adhering to the "living foods diet" consume uncooked and fermented plant foods only. This diet is inadequate in a number of nutrients, including vitamin B_{12}. **Fruitarians** consume only fruit and olive oil. People adopting this type of diet rarely stick with it for long—it does not sustain health.[9]

fruitarian A form of vegetarian diet in which fruits are the major ingredient. Such diets provide inadequate amounts of a variety of nutrients.

Health Implications of Vegetarian Dietary Patterns

Health benefits of vegetarian-style dietary patterns vary somewhat based on the types and amounts of food consumed. Strict vegetarians, for example, are less likely to be overweight or obese than semi-vegetarians, although both groups tend to have lower rates of overweight and obesity than non-vegetarians. The more restrictive the vegetarian dietary pattern, the more likely it is that intake of certain nutrients will be lower than recommended, and that it may not meet the needs of pregnant and breastfeeding women, young children, and ill people.[5]

Vegetarians who consume fish, dairy products, or eggs can reduce their reliance on legumes, nuts, and grains as protein and nutrient sources. Animal products are a major source of **complete proteins**, vitamin B_{12}, vitamin D, and calcium, as well as EPA (eicosapentaenoic acid) and DHA (docosahexaenoic acid), two important fatty acids found primarily in fish and seafood. Vitamin B_{12}, vitamin D, calcium, EPA, and DHA are the nutrients most likely to be lacking in the diets of vegetarians overall.[5] Diets of

complete proteins Proteins that contain all of the nine essential amino acids in amounts sufficient to support protein tissue construction by the body.

non-vegetarians, on the other hand, are most likely to be lacking in vitamin D, calcium, potassium, and fiber.[10] Both vegetarian and non-vegetarian dietary patterns take knowledge and planning to implement optimally. The availability of a large assortment of vegan foods in the United States and other developed countries is making it easier to find healthful vegetarian food options.[1,6]

Dietary Recommendations for Vegetarians

Because vegetarian dietary patterns can exclude one or more type of food, it's important that the foods included provide sufficient calories and the assortment and quantity of nutrients needed for health. No matter what the motivation underlying the assortment of foods included, vegetarian diets that fail to provide all the nutrients humans need in required amounts will not sustain health.

There are a number of types of vegetarian dietary patterns, so there is no one dietary recommendation that is appropriate for all of them. Dietary guidelines for vegetarians generally recommend a variety of foods that includes grains, legumes, vegetables, fruits, fats and oils, nuts and seeds, and occasional sweets or other treats (Illustration 16.3).[1] Additional guidance for vegetarian diets is available from ChooseMyPlate.gov in the form of 10 healthy eating tips (Illustration 16.4).[11] USDA's current food guide highlights the role of plant foods in the diet and emphasizes that vegetables, fruits, and grains should make up about three-fourths of a plate of food. It is generally recommended that Americans consume more plant foods and fewer animal products.[10]

A specific example of the food and nutrient composition of one day of a vegetarian's diet are shown in Table 16.3. The diet provides small amounts of animal fat, is high in fiber, and provides adequate amounts of protein and most vitamins and minerals. These are fairly typical results for a vegetarian diet.[12] Of the four key nutrients most likely to be missing in the diet of vegetarians, this day's diet provides the recommended amount of

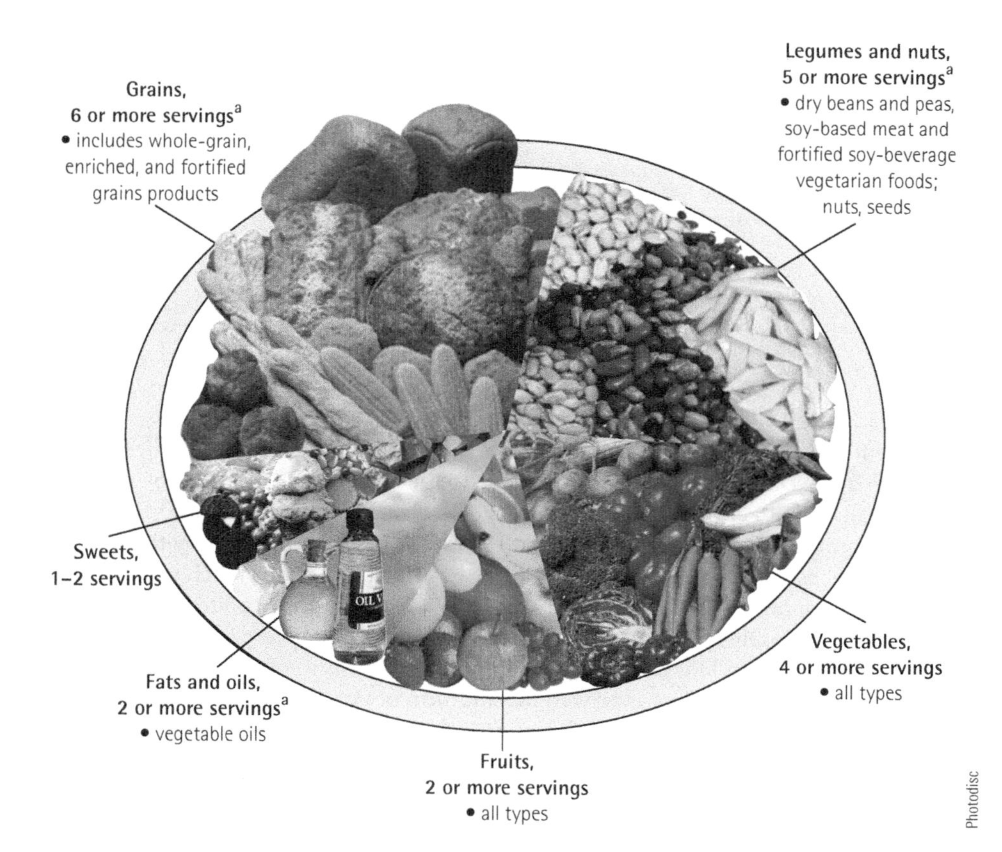

Illustration 16.3 Foundation foods for well-planned vegetarian diets.[11,12]

Vegan Diets

Grains: 6 or more servings[a]
- includes whole-grain, enriched, and fortified grain products

Legumes and nuts: 5 or more servings[a]
- dry beans and peas, soy-based meat and fortified soy-based beverages; vegetarian foods; nuts, seeds

Vegetables: 4 or more servings
- all types

Fruits: 2 or more servings
- all types

Fats and oils: 2 or more servings[a]
- vegetable oils

Sweets: 1–2 servings

[a]Number of servings depends on calorie need.

Illustration 16.4 ChooseMyPlate.gov's recommendations for healthy vegetarian diets.

10 tips
Nutrition Education Series

healthy eating for vegetarians

10 tips for vegetarians

A vegetarian eating pattern can be a healthy option. The key is to consume a variety of foods and the right amount of foods to meet your calorie and nutrient needs.

1 think about protein
Your protein needs can easily be met by eating a variety of plant foods. Sources of protein for vegetarians include beans and peas, nuts, and soy products (such as tofu, tempeh). Lacto-ovo vegetarians also get protein from eggs and dairy foods.

2 bone up on sources of calcium
Calcium is used for building bones and teeth. Some vegetarians consume dairy products, which are excellent sources of calcium. Other sources of calcium for vegetarians include calcium-fortified soymilk (soy beverage), tofu made with calcium sulfate, calcium-fortified breakfast cereals and orange juice, and some dark-green leafy vegetables (collard, turnip, and mustard greens; and bok choy).

3 make simple changes
Many popular main dishes are or can be vegetarian—such as pasta primavera, pasta with marinara or pesto sauce, veggie pizza, vegetable lasagna, tofu-vegetable stir-fry, and bean burritos.

4 enjoy a cookout
For barbecues, try veggie or soy burgers, soy hot dogs, marinated tofu or tempeh, and fruit kabobs. Grilled veggies are great, too!

5 include beans and peas
Because of their high nutrient content, consuming beans and peas is recommended for everyone, vegetarians and non-vegetarians alike. Enjoy some vegetarian chili, three bean salad, or split pea soup. Make a hummus-filled pita sandwich.

6 try different veggie versions
A variety of vegetarian products look—and may taste—like their non-vegetarian counterparts but are usually lower in saturated fat and contain no cholesterol. For breakfast, try soy-based sausage patties or links. For dinner, rather than hamburgers, try bean burgers or falafel (chickpea patties).

7 make some small changes at restaurants
Most restaurants can make vegetarian modifications to menu items by substituting meatless sauces or non-meat items, such as tofu and beans for meat, and adding vegetables or pasta in place of meat. Ask about available vegetarian options.

8 nuts make great snacks
Choose unsalted nuts as a snack and use them in salads or main dishes. Add almonds, walnuts, or pecans instead of cheese or meat to a green salad.

9 get your vitamin B_{12}
Vitamin B_{12} is naturally found only in animal products. Vegetarians should choose fortified foods such as cereals or soy products, or take a vitamin B_{12} supplement if they do not consume any animal products. Check the Nutrition Facts label for vitamin B_{12} in fortified products.

10 find a vegetarian pattern for you
Go to www.dietaryguidelines.gov and check appendices 8 and 9 of the *Dietary Guidelines for Americans, 2010* for vegetarian adaptations of the USDA food patterns at 12 calorie levels.

Go to www.ChooseMyPlate.gov for more information.

DG TipSheet No. 8
Nov. 2017
USDA is an equal opportunity provider and employer.

ChooseMyPlate.gov/USDA

vitamin D, but not of vitamin B_{12}, calcium, or EPA/DHA. Vitamin D intake reaches 6 mcg (240 IU) and B_{12} levels are 1.9 mcg in this diet due to the inclusion of fortified soy milk.

Vegetarians can meet their needs for vitamin B_{12}, vitamin D, calcium, and EPA/DHA by consuming fortified foods (such as those shown in Table 16.4) or by the use of B_{12}-fortified yeast, DHA algae supplements, or fortified plant foods (such as DHA-fortified soy or rice milk). EPA can be obtained, to some extent, from sources of DHA. EPA and DHA are closely related chemicals, and the body can convert one to the other to some degree.[15] Vitamin D is becoming easier for vegetarians to obtain. The Food and Drug Administration (FDA) has approved the addition of vitamin D to soy-based foods such as soy beverages, tofu, burgers, and desserts. The change is reflected on food labels of soy vegetarian products.[10] Direct exposure of the face, arms, and hands to direct sunlight for about 5 to 15 minutes a day is one of the best ways to obtain vitamin D.

Table 16.3 Example of food intake and nutrient content in one day for the vegetarian diet of a 31-year-old, 130-pound female

One day's diet:	
Oatmeal, 1 cup	Salad dressing, 3 Tbsp
Banana, 1 medium	Soy milk, 1 cup
Soy milk, 1 cup	Veggie burger, 1 patty
Brown sugar, 1 Tbsp	Bun, 1
Almonds, 1 oz (22 almonds)	French fries, 1 cup
Black beans, 2 cups	Herbal tea, 1 cup
Brown rice, 1 cup	Hummus, ½ cup
Lettuce salad with mixed vegetables, 1 cup	Baby carrots, 10

Selected nutrient analysis results: (calories: 2,170)		
Nutrients	Amount consumed	Recommended intake
Protein, g	87	46
Total fiber, g	53	25
Total fat, g	100	53–92
Saturated fat, g	20	<26
Cholesterol, mg	27	<300
EPA + DHA, mg[a]	50	300–600
Vitamin A, mcg	723	700
Vitamin C, mg	54	75
Vitamin E, mg	14	15
Vitamin B_6, mg	3.2	1.3
Vitamin B_{12}, mcg	1.9	2.4
Thiamin, mg	3.1	1.1
Riboflavin, mg	1.6	1.1
Niacin, mg	19	14
Folate, mcg	561	400
Vitamin D, mcg[a]	6	5
Calcium, mg	491	1,000
Magnesium, mg	614	320
Iron, mg	18.5	18
Zinc, mg	12.2	8
Selenium, mg	37	55

[a]EPA and DHA availability are estimated based on alpha-linolenic acid intake of 2.8 g; vitamin D intake is estimated based on fortified soymilk intake. Analysis is from mypyramidtracker.gov, September 2006.

Vitamin and mineral supplements can be, and often are, used by vegetarians to meet specific nutrient needs. If needed, supplementary intake levels of vitamins B_{12} and D, calcium, and EPA/DHA should approximate those shown in Table 16.5. The best way to know if supplemental nutrients are needed is to carefully perform a dietary assessment

Table 16.4 Key nutrient levels in fortified and other vegetarian foods[a]

	Key nutrients			
	Vitamin B_{12}	Vitamin D	Calcium	DHA
RDA for adults:	2.4 mcg	5 mcg	1,000 mg	250–500 mg[b]
Rice Dream, 1 cup	1.5 mcg	5 mcg	1,000 mg	—
Soy milk, 1 cup	2.6 mcg	3 mcg	300 mg	—
Meat analogs, 1 serv.	0–1.4 mcg	—	—	—
Tofu w/ calcium sulfate, ½ cup	—	—	130 mg	—
Fully fortified breakfast cereal, 1 cup	6 mcg	2.5 mcg	1,000 mg	—
Other breakfast cereals, ¾ cup	1.5 mcg	1 mcg	0–100 mg	—
Nutritional yeast, 2 Tbsp	7.8 mcg	—	—	—
Algae, 1 capsule	—	—	—	180 mg
DHA-fortified fruit juice, 1 cup	—	—	—	32 mg
Collard greens, 1 cup	—	—	357	—
Kale, 1 cup	—	—	180	—
Turnip greens, 1 cup	—	—	107	—

[a]Check out product labels for product-specific information on nutrient content.
[b]DRIs for DHA are yet to be established. The figure is from Harris et al., 2009.[15]

that covers several typical days of food intake. The Diet & Wellness Plus program available in MindTap is a good tool to use for dietary assessment.

Complementary Protein Foods Most well-planned vegetarian diets provide adequate amounts of protein,[1] but the quality of the protein varies depending on which plant foods are consumed.

Animal products such as meat, eggs, and milk provide all of the nine essential amino acids in sufficient quantity to qualify as complete source of protein. In addition, tests have shown soy proteins to be complete protein sources for adults.[23] The diet must include complete proteins because the body needs a sufficient supply of each essential amino acid if it is to build and replace protein substances such as red blood cells and enzymes. If any of the essential amino acids are missing from the diet, protein tissue construction stops or becomes disrupted, and the available amino acids will largely be used for energy instead. Essential amino acids consumed in foods are not stored for long, so the body needs a fresh supply each day or so.

Vegetarians who don't consume animal products can meet their need for essential amino acids by combining plant foods to yield complete proteins. This is done by consuming plant foods that *together* provide all the essential amino acids, although each individual food is missing some of these essential nutrients. The goal is to "complement" plant sources of essential amino acids, or to consume **complementary protein sources** from plant foods regularly.

Many different combinations of plant foods yield complete proteins. Basically, complete sources of protein can be obtained by combining grains such as rice, bulgur (whole wheat), millet, or barley with dried beans, tofu, or green peas; or corn with lima beans or dried beans; or seeds with dried beans. Some examples of complementary sources of plant proteins are shown in Illustration 16.5. Milk, meat, and eggs contain complete proteins and will complement the essential amino acids profile of any plant source of protein.

complementary protein source Plant sources of protein that together provide sufficient quantities of the nine essential amino acids.

Table 16.5 Amounts of key nutrients that may be needed by vegetarians daily[a]

Vitamin B_{12}	2–6 mcg
Vitamin D	200–400 IU (5–10 mcg)
Calcium	500 mg
EPA and DHA	300 mg

[a]Fortified foods may also supply these nutrients.

PhotoDisc

Rice and dried beans

Kelly Cline/Vetta/Getty Images

Hummus and bread

Michael Newman/PhotoEdit

Corn and black-eyed peas

PhotoDisc

Bulgur (whole wheat) and lentils

PhotoDisc

Tofu and rice

Jacobs, Martin/StockFood/Age Fotostock

Corn and lima beans (succotash)

Gkrphoto/Shutterstock.com

Tortilla with refried beans (e.g., a bean burrito)

PhotoDisc

Pea soup and bread

Illustration 16.5 Some combinations of plant foods that provide complete protein.

Meat and Dairy Analogs A wide array of alternatives to meat and dairy products are available in grocery stores (Illustration 16.6). Many of these products enjoy wide acceptance among vegetarians and non-vegetarians alike. Meat and dairy analogs are generally based on plant proteins such as gluten and soybean derivatives, and may contain egg whites or whey. These compounds add texture to products and pick up the flavorings that are added to them.[17] Some of the products are fortified with nutrients so that they more closely resemble the composition of meat and dairy products, or to supply nutrients that may be needed in the diet. Others may be high in sodium or calories. Careful examination of nutrition information labels on meat and dairy analog products is important for individuals who are looking for vegetarian sources of particular nutrients and those who are sensitive to gluten or MSG, or allergic to soy or egg protein.

Richard Anderson

Illustration 16.6 The growing selection of vegetarian foods in supermarkets.

Where to Go for More Information on Vegetarian Diets

You can find information about vegetarian diets on a variety of Internet sites that post science-based information (such as government educational sites, universities, journals) and in cookbooks that offer information about vegetarian nutrition and delicious recipes, like those pictured in Illustration 16.7. Unfortunately, some of the information available on vegetarianism is wrong or misconstrued. Some vegetarian organizations are more committed to selling particular beliefs, memberships, or books and magazines than to promoting healthy vegetarian diets. Beware of vegetarian diets that claim to cure cancer, AIDS, or other serious illnesses, or that promise you'll experience inner peace or spiritual renewal. Appropriately planned vegetarian diets are health promoting, but they are not magic bullets that will cure all the ills of body and soul.

Illustration 16.7 Many excellent vegetarian cookbooks are available; most provide information on diet as well as recipes. The books pictured here are *Moosewood Restaurant Low-Fat Favorites: Flavorful Recipes for Healthful Meals*, edited by Pam Krauss; *Vegetarian Times Complete Cookbook*, by the editors of *Vegetarian Times*; *Vegetarian Cooking for Everyone*, by Deborah Madison; *The Whole Soy Cookbook*, by Patricia Greenberg; and *Vegetable Heaven*, by Mollie Katzen.

MIB Pictures/UpperCut Images/Getty Images

NUTRITION up close

Vegetarian Main Dish Options

Focal Point: Serving up vegetarian alternatives to meat dishes.

Assume you are a member of the "vegetarian option" planning committee for your college. You are asked to identify three meatless main dishes that could be served in the dining halls for lunch and another three that could be served for dinner. The one stipulation is that they should be main dishes you would enjoy eating.

What meatless dishes would you identify as options for lunch and for dinner?

Lunch Dishes	Dinner Dishes
1.	1.
2.	2.
3.	3.

Feedback to the Nutrition Up Close is located in Appendix F.

REVIEW QUESTIONS

- **Discuss the health benefits and limitations of the types of vegetarian dietary patterns.**

1. The emotional and spiritual importance of certain foods in some vegetarian regimes supersedes any contribution the foods may make to an adequate diet. **True/False**
2. Raw food diets supply ample amounts of all vitamins and required minerals and improve indigestion in people with digestive problems. **True/False**
3. Compared to a Western-style omnivore dietary pattern, vegetarian dietary patterns are associated with lower rates of obesity and heart disease. **True/False**
4. Vitamin B_{12} and EPA/DHA are found largely in animal foods. **True/False**
5. The four nutrients most likely to be lacking in the diets of vegetarians are vitamin B_{12}, EPA/DHA, magnesium, and vitamin E. **True/False**
6. Well-planned vegetarian diets provide adequate amounts of iron and protein. **True/False**
7. When forming a new protein tissue, the body will respond to deficiency of an essential amino acid by replacing it with another essential amino acid that is available. **True/False**
8. Corn combined with cooked dried beans is an example of a complementary protein source. **True/False**

The next five questions refer to the following situation.

James and Manda have both gained weight since they met and decide it's time to lose the weight. Manda was a vegetarian for 14 years, liked that type of diet, and had switched to eating meat only a year ago because she missed the taste of steak. James has always been a meat eater, but he needs to lose weight, too, and wants to lose it. They both decide to go on a vegan diet and eat only plant foods.

9. ________ Which food could they include in their vegan diet that would provide vitamin D?
   a. dried beans
   b. rice
   c. spinach
   d. fully fortified breakfast cereal
10. ________ Which food could they consume to help meet their need for vitamin B_{12}?
    a. nutritional yeast
    b. sweet potatoes
    c. cabbage
    d. hummus
11. ________ While on the vegan diet, intake of which of the following nutrients will likely increase?
    a. fiber
    b. calcium
    c. protein
    d. iron

12. ________ Which of the following represents a complementary protein source Manda and James could consume to get a complete source of protein?
    a. navy and pinto beans
    b. carrots and corn
    c. hummus and black beans
    d. tofu and rice

13. ________ Which of the following sets of foods could be used to provide good sources of calcium in the vegan diet?
    a. soy milk and bread
    b. veggie burgers and squash
    c. collard greens and soy milk
    d. tomato juice and rice

Answers to these questions can be found in Appendix F.

NUTRITION SCOREBOARD ANSWERS

1. **True**[1]
2. **True**
3. Macrobiotic dietary patterns have not been found to prevent or cure cancer.[2] **False**
4. Vegetarians do need to plan their diets to ensure they obtain enough high-quality protein, but they need to consume complete sources of protein daily, not at every meal. Appropriately planned vegetarian diets provide adequate amounts of high-quality protein. **False**

Food Allergies and Intolerances

NUTRITION SCOREBOARD

1 About one in every three Americans is allergic to at least one food. **True/False**

2 Food ingredient labels must indicate the presence of major food allergens in food products. **True/False**

3 Skin prick tests are an accurate way to diagnose specific food allergies. **True/False**

4 Food intolerances cause less severe reactions than food allergies do. **True/False**

Answers can be found at the end of the unit.

Creatas/Fotosearch

LEARNING OBJECTIVES

After completing Unit 17 you will be able to:

- Explain food allergy as a primary reason for adverse reactions to food in humans.
- Explain food intolerance as a primary reason for adverse reactions to food in humans.

Food Allergy

- **Explain food allergy as a primary reason for adverse reactions to food in humans.**

In many circles, food allergy is a topic of heated debate and misconceptions. People use the term *food allergy* to refer to virtually any type of problem they have with food. At one extreme are people who believe allergies to milk, wheat, and sugar are to blame for hyperactivity and a host of other behavioral problems in children. At the other extreme are some health professionals and others who think people who complain of food allergies need to have their heads examined. In the middle is the fact that food allergies are increasingly common and are an important health concern in the United States and other Western countries.[2]

Food allergies are caused by genetic and environmental factors and can be very serious. At a minimum, true food allergies can cause a rash or an upset stomach. At the maximum, they can result in death. Unreal food allergies can cause problems, too. They can lead people to eliminate nutritious foods from their diet unnecessarily, resulting in inadequate diets and eventually in health problems. One of the most intriguing aspects of food allergies is the frequency and ease with which foods are falsely blamed for a variety of mental and physical health problems.

KEY NUTRITION CONCEPTS

This unit's content on food allergies and intolerance is directly relevant to two key nutrition concepts:

1. Nutrition Concept #3: Health problems related to nutrition originate within cells.
2. Nutrition Concept #6: Malnutrition can result from poor diets and from disease states, genetic factors, or combinations of these factors.

Prevalence of Food Allergy

The incidence of food allergies is estimated to be to 1% in adults and 4% in children.[2] However, around 30% of the general public believe they are allergic to one or more foods. The vast majority of complaints of food allergy fail to be confirmed by testing.[2] Here are two real-life examples of self-diagnosed food allergies that went awry:

1. Isaiah, a lover of blueberries and blueberry pie, hasn't touched a blueberry since 2007. That year, Isaiah had a piece of blueberry pie in a restaurant and later became violently sick to his stomach. Bingo! Isaiah decided he must be allergic to blueberries.

   Actually, Isaiah gave up blueberries for no good reason. He was coming down with the flu when he ate the pie. Nevertheless, to this day when he thinks of blueberries, he gets a bit queasy.

2. For 11 years, Emilia rigidly avoided even small amounts of cow's milk. She was convinced that just a few drops of cow's milk would cause pressure in her head, blurred vision, dizziness, cramps, and nausea.

   Finally, Emilia's presumed allergy to cow's milk was put to the test. She was given liquid through a dark tube inserted into her stomach from her mouth. When she was told the fluid was cow's milk—even though it was actually water—the familiar symptoms appeared within 10 minutes. When she was given milk but was told it was water, no symptoms appeared.[4]

   Emilia was shocked. For 11 years she had scrutinized nearly everything she ate and had wasted hundreds of dollars on special food products and supplements. Finishing a frosted brownie with a glass of milk, Emilia contemplated the practical realities of the power of suggestion.

food allergy Adverse reaction to a normally harmless substance in food that involves the body's immune system.

immune system Body tissues that provide protection against bacteria, viruses, and other substances identified by cells as harmful.

Adverse Reactions to Foods

People experience adverse reactions to food for three primary reasons. One is food poisoning. The other two involve the body's reaction to a specific protein in food that are normally harmless. With **food allergies**, the body's **immune system** reacts to a substance in food it identifies as harmful and attempts to fight off (Illustration 17.1). The immune system does not initiate **food intolerances**, which encompass other adverse reactions to normally harmless substances in food.

Food Allergies and the Immune System It may seem odd that a system that helps the body conquer bacteria and viruses is involved in protecting us from normal constituents of food. In people with allergies, however, the cells recognize some food components as harmful—just as they recognize bacteria and viruses. Components of food that trigger the immune system are called **food allergens**.[2]

Among susceptible people, exposure to trace amounts of an allergen triggers an allergic reaction. Highly sensitive people can develop an allergic reaction simply by inhaling vapors from allergenic food that is being prepared or cooked, or by touching the food.[4] Allergic reactions to peanuts have been triggered by kissing someone who has recently consumed peanuts or even by inhaling fumes in a confined space with people who are eating them.[5] Because food allergies are "hard wired," reactions are reproducible: the same reaction will occur following exposure to the same food or food ingredient.[2]

In response to the allergen, the immune system prompts the formation of **antibodies** (Illustration 17.2).[3] The antibodies attach to cells located in the nose, throat, lungs, skin, eyes, and other areas of the body. When the allergen appears again, the body is ready. The previously formed antibodies recognize the allergen, attach to it, and signal the body to secrete **histamine** and other substances that cause the physical signs of an allergic reaction. Most commonly, allergic reactions cause a rash (Illustration 17.3), diarrhea, congestion, or wheezing (Table 17.1), but the symptoms may be much more serious.

Exposure to trace amounts of an allergen in nuts, peanuts (peanuts are actually a legume, not a nut), fish, and shellfish can cause **anaphylactic shock**. This massive reaction of the immune system can result in death, caused by a cutoff of the blood supply to tissues throughout the body. People experiencing anaphylactic shock can be revived by

Felicia Martinez Photography/PhotoEdit

Illustration 17.1 What do these foods have in common? Each food shown, plus many more not shown, is related to food allergies in some individuals.

food intolerance Adverse reaction to a normally harmless substance in food that does not directly involve the body's immune system.

food allergen A substance in food (almost always a protein) that is identified as harmful by the body and elicits an allergic reaction from the immune system.

antibodies In the case of allergies, proteins the body makes to combat allergens.

histamine (hiss-tah-mean) A substance released in allergic reactions. It causes dilation of blood vessels, itching, hives, and a drop in blood pressure and stimulates the release of stomach acids and other fluids. Antihistamines neutralize the effects of histamine and are used in the treatment of some cases of allergies.

anaphylactic shock (an-ah-fa-lac-tic) A serious allergic reaction that is rapid in onset and can cause death. Symptoms of anaphylactic shock (or "anaphylaxis") may include abdominal cramps, chest tightness, difficulty breathing, cough, hives, flushing, swelling, and/or itching.

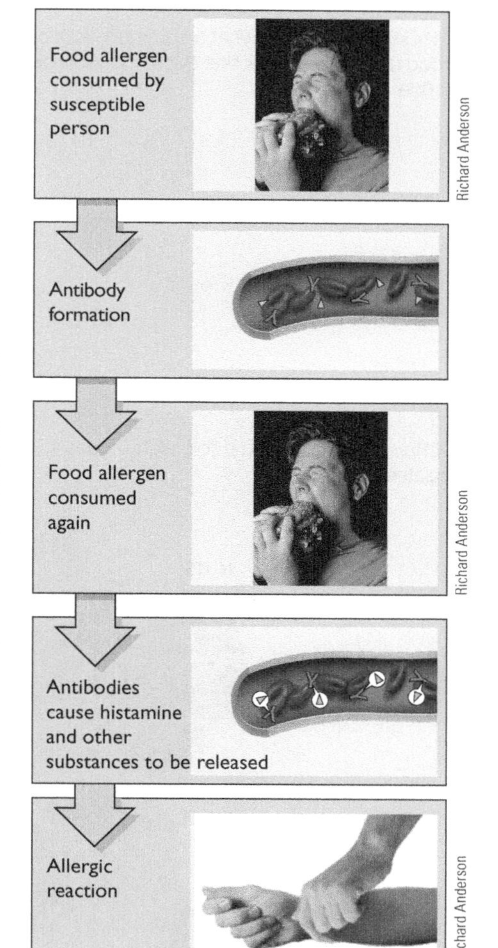

Illustration 17.2 The development of allergic reactions to foods.

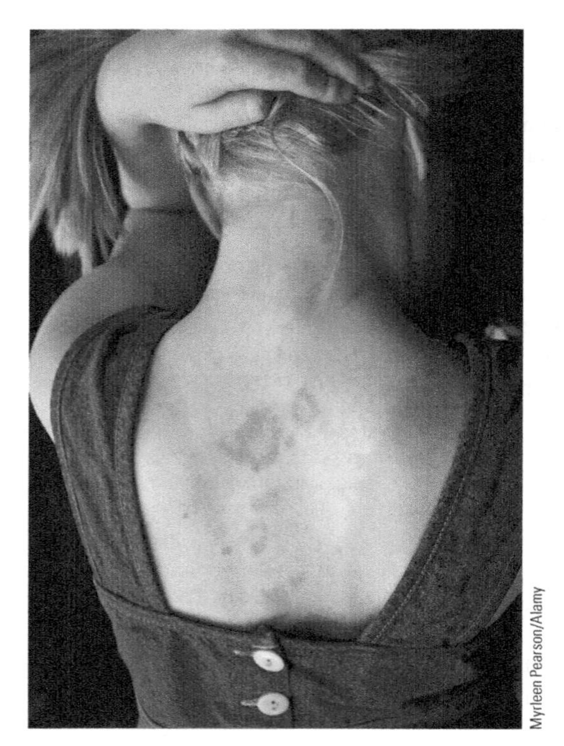
Myrleen Pearson/Alamy

Illustration 17.3 What a common reaction to a food allergen can look like. (Most food allergies cause milder symptoms.)

Table 17.1 The three most common types of symptoms caused by food allergies
Some people experience more than one type of reaction.[5,7]

Symptom	Percentage of people with food allergies who develop the symptom
• Skin eruptions: rash, hives	84%
• Gastrointestinal upsets: diarrhea, vomiting, cramps, nausea	52
• Respiratory and other problems: congestion, swelling of the tongue and throat, runny nose, cough, wheezing, asthma	32

an injection of epinephrine.[7] Follow-up care should be provided because symptoms of anaphylactic shock may reoccur a number of hours later.[2]

Foods That Are Most Likely to Cause Allergic Reactions Many foods can cause allergic reactions. Nevertheless, approximately 90% of all food allergies are related to the consumption of eight foods: nuts, eggs, wheat, milk, peanuts, soy, shellfish, and fish (Illustration 17.4).[8] Passage of the Food Allergen Labeling and Consumer Protection Act of 2004 is helping consumers with food allergies find offending foods and ingredients. Food ingredient labels are now required to state clearly whether the food contains one or more of the big eight allergenic foods.[8]

Wheat Allergy (Celiac Disease) Minnie is 46 years old and has three sisters and two brothers, all of whom are taller than she is. (She considers herself the "runt of the litter.") Since childhood, Minnie has had problems with diarrhea and cramps. Over the past two years, these problems have become worse, and she has been chronically tired. Assuming that she was lactose intolerant, Minnie stopped eating all dairy products. Her problems persisted, however. When she almost lost her job due to her frequent bathroom breaks and her obvious fatigue, she decided to see a doctor again.

Diagnosing anemia, the doctor sent Minnie to a registered dietitian/nutritionist. Hearing Minnie's story, the dietitian suspected celiac disease caused by an allergy to a component of wheat and other grains. After two weeks on an allergen-free diet, Minnie started to regain her strength and spent much less time in the bathroom. Diagnostic tests subsequently confirmed that Minnie had celiac disease, a condition that likely had existed since childhood.

Illustration 17.4 The "big eight" foods related to allergies.

PhotoDisc

An allergy to wheat, or more specifically to one or more components of gluten found in wheat and other grains such as barley, rye, and triticale, has been underdiagnosed in the past. Called **celiac disease**, it affects approximately 1% of the population.[9] It differs somewhat from other allergies in that an immune system reaction to gluten occurs that is localized in the lining of the small intestine.[10] It occurs in genetically and environmentally susceptible people who have been exposed to certain environmental triggers early in life that modify gene functions.[12,13] Environmental triggers for celiac disease have yet to be clearly identified.[13]

celiac disease An autoimmune disease characterized by inflammation of the small intestine lining related to genetically based intolerance to gluten. The inflammation produces diarrhea, fatty stools, weight loss, and vitamin and mineral deficiencies. (Also called *celiac sprue* and *gluten-sensitive enteropathy*.)

Celiac disease can be difficult to diagnose because the symptoms (diarrhea, weight loss, cramps, fatigue, anemia) are similar to those of other diseases, and because symptoms may be silent (meaning not obvious) in some individuals. In the United States, only about 17% of individuals with celiac disease are diagnosed with the condition.[14,15] With new awareness of the disorder, doctors are ordering diagnostic tests more frequently, and the rate of confirmed cases of celiac disease is increasing.[9] The gold standard diagnostic test for celiac disease is small bowel biopsy and examination of cells for signs of damage due to the disease. The test must be undertaken while individuals are consuming their normal diet, and not a gluten-free one. Removal of gluten from the diet corrects intestinal cell damage and may lead to an incorrect diagnosis.[8] Laboratory tests that identify the presence of gene types or specific antibodies related to the presence of celiac disease are also used to diagnose celiac disease.[16,17]

Celiac disease is treated with a lifelong, but rewarding, gluten-free diet. Because gluten is an inexpensive ingredient added to many products to improve food texture or cooking properties, it is found in a variety of food products that would not be expected to contain gluten.[18] It often takes the help of an experienced, registered dietitian/nutritionist to design and monitor gluten-free diets.[19] Table 17.2 provides examples of some foods that contain gluten and some that don't.

Table 17.2 Which foods contain gluten? Listed here are examples of foods that do or do not contain gluten[10,20]

Gluten-containing foods	Gluten-free foods[a]
Beer, ale	Fruit
Barley	Vegetables
Broth, bouillon powder/cubes	Dried beans
Brown rice syrup	Buckwheat
Bulgur	Cassava
Commercial soups	Grits
Bread, other wheat/rye flour products	Cornmeal
Imitation seafood	Nuts
Cakes, pies, cookies	Quinoa
Processed meats	Fresh meats, fish
Soy sauce	Soy flour, cereals
Wheat starch	Wild rice
Pizza	Eggs
Macaroni and cheese	Nuts, seeds
Seasonings	Cheese (not processed)
Marinades	Popcorn
Rye-containing products	Milk
Vegetarian meat substitutes	Chips (100% corn, potato)

[a]Assumes foods have not been contaminated with gluten during processing and are free of gluten-containing ingredients.

iStockphoto.com/Jamesbenet

Illustration 17.5 Food products containing less than 20 parts per million gluten can be labeled as gluten-free, without gluten, no gluten, or free of gluten.

The Food and Drug Administration (FDA) has defined the term "gluten-free" for food labeling purposes. Foods containing less than 20 parts per million gluten, the lowest level that can be detected in foods using scientifically valid analytical methods, can be labeled as "gluten-free," "no gluten," "free of gluten," or "without gluten" (Illustration 17.5).[21] This amount of gluten is not enough to cause a reaction in the vast majority of individuals with celiac disease and is generally considered safe for people with this disorder.[19] Unless a food label declares a product to be "gluten-free," it may contain very small amounts of gluten that can precipitate an adverse reaction. With or without the gluten-free phrase, foods with ingredient labels listing wheat or products made from wheat, rye, barley, or triticale should not be consumed, nor should foods labeled as containing, or potentially containing, these ingredients.

Non-Celiac Gluten Sensitivity A small but significant percentage of U.S. adults report adverse effects from gluten consumption and follow a gluten-free diet for symptom relief.[16] Symptoms related to the ingestion of gluten vary, but include gastrointestinal pain, headache, joint pain, foggy mind, and numbness in the arms, legs, or fingers.[22,23] Because individuals who report such symptoms following wheat ingestion do not have celiac disease, it has been suggested that the title of the disorder be changed to "wheat intolerance syndrome." Although most individuals with this condition report an improvement of symptoms with a gluten-free diet, they rarely respond positively to an oral gluten challenge test.[26] The cause of wheat intolerance is unknown, but it appears to be related to a component of wheat and to be more common than celiac disease.[25]

Diagnosis: Is It a Food Allergy?

A variety of tests are used to diagnose food allergies, but the gold standard test is the **double-blind, placebo-controlled food challenge.**[24] Suppose you suspect you are allergic to grapes because you got an annoying rash twice after eating them. Your doctor arranges an appointment for you at an allergy clinic. At the clinic, you will be given grapes, either in a concentrated pill form or blended into a liquid, and placebo pills (or a liquid) without being told which are the grapes and which is the placebo. After an appropriate amount of time, your reaction to the grapes and to the placebo will be noted. If the rash appears after the grapes but not after the placebo, you have an adverse reaction to grapes that may be an allergy.

double-blind, placebo-controlled food challenge A test used to determine the presence of a food allergy or other adverse reaction to a food. In this test, neither the patient nor the care provider knows whether a suspected offending food or a placebo is being tested.

Food challenges have to be undertaken under medical supervision. True food allergies can cause serious reactions, and immediate help may be needed. Foods suspected of causing anaphylactic shock in the past should not be given in food challenge tests. In fact, they should be totally omitted from the diet.[2,3]

Other tests for food allergy are used and are reliable to varying degrees. Immunoglobulin E (IgE) is a protein produced by the immune system in response to an allergen. Identification of specific types of IgE in a person's blood may identify the presence of a food allergy, but often the results do not correctly identify a specific food allergy. Skin prick tests are also used in the diagnosis of food allergies. For this test, a few drops of food extract are placed on the skin, and the skin is then pricked with a needle. At approximately the same time, another area of skin is pricked using only water. If the area around the food skin prick becomes redder and wider than the area pricked with water (Illustration 17.6), it is concluded that the person *might* be allergic to the food tested. A positive skin prick test result isn't proof positive that an allergy exists, however, because positive test results may not be reproducible and may not identify specific food allergens.[2,3]

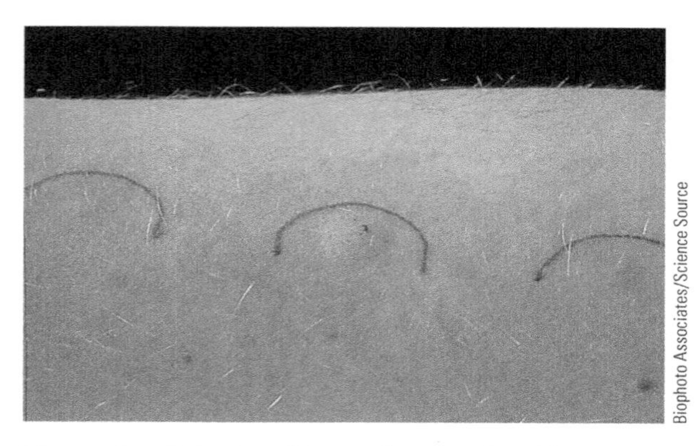

Biophoto Associates/Science Source

Illustration 17.6 A positive reaction to a skin prick test for a food allergy usually looks like this. The result indicates that an allergy to the food tested may be present but doesn't guarantee it.

Bogus Tests A number of companies offer food allergy tests through the mail, but they tend to be unreliable. In one study, researchers sent five of these companies the required samples, nine from adults who were allergic to fish

REALITY CHECK

I Think I'm Allergic to . . .

You and your coworkers decide to go to a Chinese restaurant for lunch. Trevor wants to go but hesitates because, as he says, "I must be allergic to Chinese food. Every time I eat it I feel dizzy, sweat, and get this ringing in my ears."

Who's got the better idea?

Answers on page 17-8.

Photodisc

Haley: Trevor, that's not an allergy. It's all in your head.

Photodisc

Darrel: Have you ever asked them to leave out the MSG?

and nine from adults who had no allergies. All five companies missed the fish allergy. When duplicate samples from the same adults were sent separately to each company, reports came back with different results for each of the pairs of samples. In addition, the companies reported their laboratories had identified various food allergies that none of the adults actually had.[6]

Treatment of Food Allergies

After a food allergy is confirmed, the food is eliminated from the diet. This is the *only* treatment currently available for food allergies. If the eliminated food is an important source of nutrients or is found in many food products, consultation with a registered dietitian/nutritionist is recommended.[26] Individuals with a history of anaphylaxis should be well informed about its signs and symptoms, and of the correct use of epinephrine self-injectors. Medical ID bracelets or necklaces are recommended.[23] Allergy shots used to overcome some nonfood allergies are not ready for use for food allergy treatment, but progress is being made. Immunotherapy, which involves slowly increasing the dose of allergenic foods given over time, shows promise for diminishing or eliminating a person's sensitivity to specific food allergens such as those in peanuts, milk, eggs, and peaches. It is being clinically used to some extent now and will be used increasingly in the future.[2,3]

Some food allergies don't last forever. Many infants and young children outgrow allergies to cow's milk (64% by age 12), wheat (66% by age 12), and eggs (37% by age 10).[4] Food allergies that develop in adults, however, tend to persist.[26] Children and adults with a severe allergy to peanuts, nuts, fish, or shellfish will likely have to eliminate the food from their diet for their lifetime.[4]

Prevention of Food Allergies

Several other factors appear to be related to the growing prevalence of food allergies. These include changes in food processing methods and diets, decreased exposure to microbes early in life that modify the gut's microbiome, and medical advice to avoid wheat, eggs, and peanuts during pregnancy, breastfeeding, and through the first year of life. Studies have demonstrated that this advice does not prevent food allergies and may encourage their development. Infants can be given these foods when solid foods are introduced into an infant's diet around 6 months of age.[2,3]

Food Intolerances

- **Explain food intolerance as a primary reason for adverse reactions to food in humans.**

Food intolerances produce some of the same reactions as food allergies, but the reactions develop by different mechanisms. Food intolerance reactions do not directly involve the immune system; they are due to genetic predispositions or environmental factors.[31]

ANSWERS TO REALITY CHECK

I Think I'm Allergic to . . .

Some people, like Trevor, may be sensitive to MSG. It doesn't elicit an immune system reaction, so it's not a food allergy.

Photodisc

Haley:

Photodisc

Darrel:

Table 17.3 Foods and substances in them linked to food intolerance reactions

• Aged cheese
• Anchovies
• Beer
• Catsup
• Chocolate
• Dried beans
• Food coloring
• Lactose
• Mushrooms
• Pineapple
• Red wine
• Sausage (hard, cured)
• Soy sauce
• Spinach
• Tomatoes
• Yeast

They are more common than food allergies, and the onset of symptoms is usually slower. Unlike food allergies, individuals with a food intolerance can generally tolerate a small to moderate amount of the specific food, and develop symptoms only when the food is eaten frequently or in large amounts. Symptoms caused by food intolerances vary; they include fatigue, joint pain, diarrhea, vomiting, bloating, and rashes.[27]

It is clear that some people are intolerant of lactose, sulfite, histamine (a component of red wine and aged cheese), and other foods (Table 17.3). It is also clear, however, that not all food intolerances are real. The best way to separate the real from the unreal is the double-blind, placebo-controlled food challenge. True food intolerances produce predictable reactions. Problems such as headache, diarrhea, swelling, or stomach pain will occur every time a person consumes a sufficient amount of the suspected food.

Lactose Maldigestion and Intolerance

Lactose maldigestion, or the inability to break down lactose in dairy products due to lack of the enzyme lactase, results in the condition known as **lactose intolerance**. Symptoms of lactose intolerance, such as flatulence, bloating, abdominal pain, diarrhea, and "rumbling in the bowel," occur in lactose maldigesters within several hours of consuming more lactose than can be broken down by the available lactase. These symptoms are due to the breakdown of undigested lactose by bacteria in the lower intestine (which produces gas as a by-product of lactose ingestion) and by fluid accumulation. Lactose maldigestion is a common disorder. Around 65% of the human population has a reduced ability to digest lactose after infancy, and the incidence varies based on ancestry. Only about 5% of people of Northern European descent are lactose intolerant.[27–29]

Reduced intake of lactose-containing dairy products (e.g., milk, ice cream, and cottage cheese) may prevent lactose intolerance symptoms. Foods such as hard cheese, low- or no-lactose milk, buttermilk, and yogurt without added milk solids contain low amounts of lactose and can generally be consumed. Small amounts of milk or other lactose-containing dairy products are generally well tolerated by people with lactose maldigestion.[28] Care should be taken to ensure adequate intake of calcium and vitamin D if milk and dairy product intake is restricted.

lactose maldigestion A disorder characterized by reduced digestion of lactose due to low availability of the enzyme lactase.

lactose intolerance The term for gastrointestinal symptoms (flatulence, bloating, abdominal pain, diarrhea, and "rumbling in the bowel") resulting from consumption of more lactose than can be digested with available lactase.

Sulfite Sensitivity

Sulfite is a food additive that has been used by food manufacturers to keep vegetables and fruits looking fresh and to prevent mold growth. It is added to some beers, wines, processed foods, and medications as a preservative (see Table 17.4). Very small amounts of sulfite can cause anaphylactic shock and bring on an asthma attack in sensitive people.[30] Because a variety of foods have added sulfites, people sensitive to sulfite should read food ingredient labels carefully. The FDA requires that processed foods containing sulfites list them on the ingredients label. It also prohibits the use of sulfite on fresh vegetables and fruits.

take action To Prevent Anaphylaxis

Are you at risk for anaphylactic shock due to a food allergy and unsure how to manage it? If so, medical experts have advice to share with you about how to be prepared.[4] The advice can be translated into action by completing the following activities.

1. Scrupulously learn about and avoid eating foods that may trigger anaphylaxis. Make an appointment with a registered dietitian/nutritionist if needed.
2. Talk with your doctor about the symptoms of anaphylaxis and what you should do if it happens.
3. Get a prescription for two self-injection epinephrine pens (EpiPens) and any other medications that may be needed from your doctor.
4. Ask your doctor to teach you how to use injectable epinephrine and other medications needed when anaphylaxis develops.
5. Obtain these medications and carry the EpiPen and other prescribed emergency medications with you at all times.
6. Get and wear a medical ID bracelet or necklace that contains information on the allergy.

True Images/Alamy

Red Wine, Aged Cheese, and Migraines

Some people develop migraine headaches when they drink red wine. The presence of histamine in wine has been blamed for the headaches. People with this intolerance are unable to break down histamine during digestion, so it accumulates in the blood and causes headaches.[31,32] Histamine is also found in beer, sardines, anchovies, hard cured sausage, pickled cabbage, spinach, and catsup. Tyramine, a compound closely related to histamine and found in aged cheese, soy sauce, and other fermented products, also causes migraine headaches in sensitive people. Other chemical compounds in red wine, hard cheese, and other foods may also contribute to the development of migraine headaches.[31,32]

MSG and the "Chinese Restaurant Syndrome"

Some people appear to be sensitive to MSG (monosodium glutamate), a flavor enhancer used on meats and in soups, stews, and many Chinese food items.[33] Because symptoms are reported to occur shortly after sensitive individuals eat Chinese food, sensitivity to MSG has been dubbed the "Chinese restaurant syndrome." Symptoms associated with MSG sensitivity include dizziness, sweating, flushing, a rapid heartbeat, and a ringing sound in the ears.[34]

Food Allergy and Intolerance Precautions

People with food allergies or intolerances must be very careful about what they eat and should have a plan of action ready in case they develop a serious reaction. Becoming highly knowledgeable about which foods and food products contain ingredients that cause an adverse reaction is essential. When eating out, people with allergies and intolerances should ask a lot of questions to make sure they know what is being served. They must become students of food ingredient labels.

Planning ahead is crucial for individuals with a history of anaphylactic shock. Although preparedness is the key factor related to quick treatment of anaphylaxis, individuals at risk are often not adequately prepared. Physicians may not have prescribed EpiPens (injectable epinephrine) to their patients with a history of anaphylactic shock or have instructed patients on their use. It is highly recommended that individuals with a history of anaphylactic shock have a plan for handling an emergency situation.[21] This topic is the subject of the unit's Take Action feature.

One final thought about food allergies and intolerances—they are never caused by studying about them.

Table 17.4 Some foods that may contain sulfite

Some foods that may contain sulfite
• Wine
• Beer
• Hard cider
• Tea
• Fruit juices
• Vegetable juices
• Guacamole
• Dried fruit
• Potato products
• Canned vegetables
• Baked goods
• Spices
• Gravy
• Soup mixes
• Jam
• Trail mix
• Fish and seafood

Creatas/Fotosearch

NUTRITION up close

Gluten-Free Cuisine

Focal Point: How to serve gluten-free meals and snacks.

You're having a ball game–viewing party and want to offer snacks and a meal that can also be enjoyed by your best friend, who was diagnosed recently with celiac disease. What should you serve?

Place a check mark in front of the foods that are naturally gluten free and could be served. Assume you prepare the foods without adding gluten-containing ingredients. (*Hint*: Refer to Table 17.2.)

____ fresh meats, fish
____ tomatoes
____ pizza
____ macaroni and cheese
____ eggs
____ corn grits
____ corn chips
____ sourdough rolls
____ baked apples with sugar
____ hamburgers
____ grapes
____ tossed salad with oil and vinegar dressing
____ apple pie
____ black beans and rice
____ canned soup
____ sausage
____ popcorn
____ cheddar cheese
____ veggie burgers
____ lemonade
____ broccoli

Feedback to the Nutrition Up Close is located in Appendix F.

REVIEW QUESTIONS

- **Explain food allergy as a primary reason for adverse reactions to food in humans.**

1. Peanuts, shellfish, cow's milk, and eggs are among the "big eight" foods related to food allergies. **True/False**
2. The prevalence of food allergies is increasing in the United States. **True/False**
3. The Food Allergen Labeling and Consumer Protection Act requires that manufacturers remove all potentially allergic ingredients from their foods. **True/False**
4. Skin prick tests definitively identify allergies to specific foods. **True/False**
5. Symptoms of food allergy can be triggered by exposure to a very small amount of a food allergen. **True/False**
6. Food allergies can be treated with allergy shots. **True/False**
7. Allergies to peanuts, nuts, fish, and shellfish are likely to last a lifetime. **True/False**
8. ____ Food allergy is initiated by the ingestion of ________.
   a. antibodies
   b. histamine
   c. an antigen
   d. sulfite
9. ____ Food allergies can be prevented by:
   a. consuming only small amounts of allergenic foods
   b. keeping potentially allergenic foods out of the diets of infants
   c. allergy shots
   d. avoiding allergen-containing foods

- **Explain food intolerance as a primary reason for adverse reactions to food in humans.**

The next three questions refer to the following scenario.

You're at dinner with your next-door neighbors when gluten becomes the topic of conversation. It started when Devin announced that his workmate, Bob, was diagnosed with celiac disease at the age of 48. "It hit him like wham! Right out of the blue," Devin said.

10. ____ Which of the following statements about Bob's celiac disease is most likely true?
    a. Bob suddenly developed celiac disease.
    b. Bob may have had few or no symptoms of celiac disease before he was diagnosed.
    c. The diagnosis is incorrect. Celiac disease develops only in people younger than age 30.
    d. He must have had a parent who had celiac disease as a young child.
11. ____ Bob's development of celiac disease is likely due to:
    a. excess consumption of refined wheat products.
    b. a viral infection.
    c. consumption of foods that use gluten as a food additive.
    d. predisposing genetic traits combined with exposure to specific environmental triggers.
12. Adverse reactions to sulfite in foods are prompted by immune system-mediated processes. **True/False**
13. Use of the term *gluten-free* on labels of qualifying food is regulated. **True/False**
14. Recent research proves that the "Chinese restaurant syndrome" does *not* actually exist. **True/False**
15. ____ Food intolerance is defined as:
    a. an adverse reaction to a normally harmless substance in food that involves the body's immune system.
    b. a condition that causes the regurgitation of food
    c. a limited ability to ingest large amounts of specific types of food.
    d. an adverse reaction to a normally harmless substance in food that does *not* involve the body's immune system.
16. ____ Which of the following phrases about food intolerance is true?
    a. Food intolerance is *not* initiated by the immune system.
    b. Symptoms of food intolerance are different from those of food allergy.
    c. Food intolerance reactions are "hit and miss"; they sometimes are experienced when a specific food is consumed and sometimes not.
    d. Food intolerance is never related to anaphylactic shock.

Answers to these questions can be found in Appendix F.

NUTRITION SCOREBOARD ANSWERS

1. About one in three Americans *believes* he or she is allergic to at least one food.[1] Approximately 1 in 100 adults and 4 in 100 children actually are.[2] **False**
2. Listing potential, relatively common food allergens on food ingredient labels is required for packaged foods. **True**
3. Skin prick test results do not accurately identify the food allergen that causes the allergic reaction.[2] **False**
4. Food allergies and food intolerances have different causes, but both can produce life-threatening reactions.[3] **False**

Fats and Health

NUTRITION SCOREBOARD

1 Dietary fats increase the flavor and palatability of foods. **True/False**

2 Fat cells store fat and fat-soluble vitamins, and produce chemical messengers that affect a number of important body processes. **True/False**

3 Dietary fats serve as the source of two essential fatty acids. **True/False**

4 Good food sources of polyunsaturated fats include nuts, seeds, vegetables oils, and fish. **True/False**

5 Low-fat diets and foods help people lose weight. **True/False**

Answers can be found at the end of the unit.

James And James/Photolibrary/Getty Images

LEARNING OBJECTIVES

After completing Unit 18 you will be able to:

- Summarize the major functions, food sources, recommended intake, and health effects of fats.

Fat Intake and Health

- **Summarize the major functions, types, food sources, recommended intake, and health effects of fats.**

As understanding of the effects of dietary fats on health evolves, so do recommendations about fat intake. In the past, it was recommended that Americans aim for diets providing less than 30% of total calories from fat. However, recent evidence indicates that the type of fat consumed may be more important to health than total fat intake.[23] American adults are being urged to select food sources of "healthy" fats while keeping fat intake within the range of 20–35% of total caloric intake. Concerns that high-fat diets encourage the development of obesity have been eased by studies demonstrating that excessive caloric intake and dietary patterns high in refined carbohydrates and added sugars—and not diets high in fat—are primarily related to weight gain.[2]

New recommendations regarding fat intake do not encourage increased fat consumption. Rather, they focus on the consumption of certain types of fat. Dietary patterns providing as low as 20% of calories from fat, or those providing up to 35%, can be healthy—depending on the types of fat consumed and the quality of the rest of the dietary pattern. This unit provides facts about fats, explains the reasons behind recent changes in recommendations for fat intake, and addresses the practical meaning of it all.

KEY NUTRITION CONCEPTS

Three key nutrition concepts are particularly relevant to the content covered in this unit on fats and cholesterol. They are the following:

1. Nutrition Concept #4: Poor nutrition can result from both inadequate and excessive levels of nutrient intake.
2. Nutrition Concept #8: Poor nutrition can influence the development of certain chronic diseases.
3. Nutrition Concept #9: Adequacy, variety, and balance are key characteristics of healthful diets.

Facts about Fats

Fats are a group of substances found in food. They have one major property in common: they are not soluble (or, in other words, will not dissolve) in water. If you have ever tried to mix vinegar and oil when making salad dressing, you have observed the principle of water and fat insolubility firsthand.

Fats are actually a subcategory of the fat-soluble substances known as **lipids**. Lipids include fats, oils, and cholesterol. Dietary fats such as butter, margarine, and shortening are often distinguished from oils by their property of being solid at room temperature. This physical difference between fats and oils is due to their chemical structures.

Functions of Fats Just like carbohydrates, fat molecules consist of carbon, hydrogen, and oxygen. The arrangement and number of carbon, hydrogen, and oxygen atoms determine the fat's structure and how it functions in the body.[1,3] Table 18.1 summarizes the major functions of dietary and body fats.

Fats in Foods Supply Energy and Fat-Soluble Nutrients Dietary fats are a concentrated source of energy. Each gram of fat consumed supplies the body with 9 calories worth of energy. That's enough energy for a 160-pound person to walk casually for a little over two minutes or to jog at a slow pace for about a minute. Fats in food supply the **essential**

lipids Compounds that are insoluble in water and soluble in fat. Triglycerides, saturated and unsaturated fats, and essential fatty acids are examples of lipids, or "fats."

essential fatty acids Components of fats (linoleic acid, pronounced *lynn-oh-lay-ick*, and alpha-linolenic acid, pronounced *lynn-oh-len-ick*) required in the diet.

adipose tissue A connective tissue that primarily contains adipocytes (fat cells) that store fat in the form of triglycerides.

adipocyte A fat cell. (Pronounced add-dip-o-site)

white fat The primary type of fat in the body composed of triglycerides and cholesterol particles stored within large lipid droplets. White adipocytes are largely used not only for energy storage and production, but also produce a number of chemical messengers that influence inflammation, appetite, and that affect regulation of blood glucose levels.

brown fat The type of fat primarily located above the shoulder blades, around the spinal cord, chest, and neck areas, and around certain organs. Proteins found in lipid droplets within these cells activate reactions that modify energy production from glucose and fat, and produce heat upon cold exposure.

fatty acids (linoleic acid and alpha-linolenic acid) and provide the fat-soluble vitamins D, E, K, and A (the "deka" vitamins). So part of the reason we need fats in our diets is to obtain a supply of the fat-soluble essential nutrients they "carry" in foods. Diets containing little fat (less than 20% of total calories) often fall short on delivering adequate amounts of essential fatty acids and fat-soluble vitamins.[12]

Fat Contributes to the Body's Energy Stores Fat consumed as part of a dietary intake that exceeds calorie need is converted to triglycerides and stored in **adipose tissue** within fat cells (**adipocytes**). A pound of body fat can provide approximately 3,500 calories of energy to the body when needed. Much of our body fat is located under two layers of skin tissue (the dermis and epidermis) and above muscle tissue (Illustration 18.1).

Body Fat Helps Cushion and Insulate Body Tissues Some body fat is located around organs such as the kidneys and heart. It's there to cushion and protect the organs, and for insulation. Cold-water swimmers can attest to the effectiveness of fat as an insulation material. They purposefully build up body fat stores because they need the extra layer of insulation (Illustration 18.2).

Body Fat Comes in Two Basic Types: White and Brown Adipose Tissue The two major forms of body fat, **white adipose tissue** (WAT) and **brown adipose tissue** (BAT), are metabolically active, producing and secreting a variety of hormones and other chemical messengers that effect energy and heat production, appetite, inflammation, and insulin sensitivity and blood glucose levels. White fat is, by far, the most abundant form of body fat and the major source of the body's energy reserves. Brown adipose fat is capable of forming energy from fat and glucose, and producing heat needed to maintain body temperature of around 98 degrees F (37 degrees C). It has a brownish tint due to the presence of a high concentration of mitochondria, which are tiny, energy-producing factories present in all cells (Illustration 18.1).

Brown fat stores are limited in adults and are primarily located between the shoulder blades and around the heart and kidneys. Higher amounts of brown fat have been identified in infants, physically active individuals, and females. Expansion of the body's brown

Table 18.1 Functions of dietary and body fats[9,11]

- Provide a concentrated source of energy
- Contribute to the body's energy reserves (fat stores)
- Carry the essential fatty acids, the fat-soluble vitamins, and certain phytochemicals
- Increase the flavor and palatability of foods
- Provide relief from hunger
- Cushion organs and other tissues
- Insulate organs and tissues
- Produce hormones, such as adiponectin and leptin, and other signaling substances that participate in the regulation of body processes
- Serve as a component of cell membranes that act as a barrier between the outside and inside of cells and in the maintenance of the structural integrity of cells.
- Serve as precursors to vitamin D, estrogen, testosterone

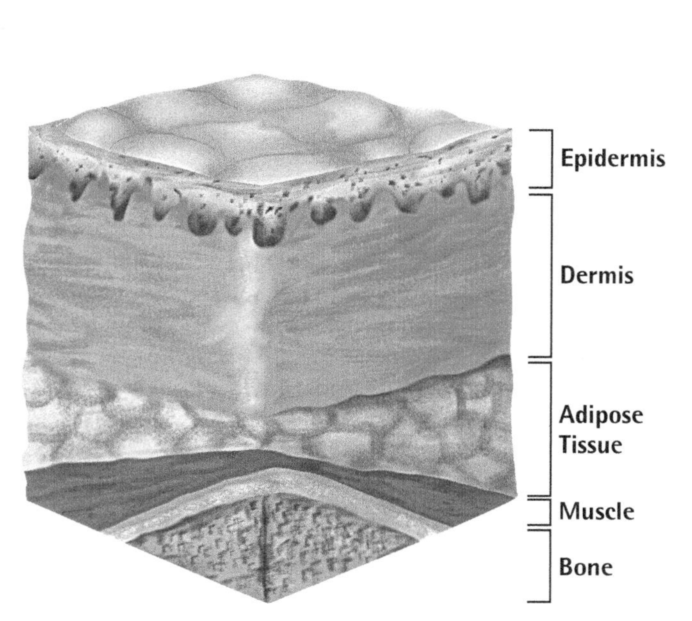

Illustration 18.1 Location of adipose tissue.

Illustration 18.2 Although their body fat stores don't fit the image of the superb athlete, cold-water swimmers need the fat to help stay warm. Pictured here is the English swimmer Mike Read, who swam the English Channel 20 times by age 39. The narrowest width of the English Channel is 22 miles, or 35 km.

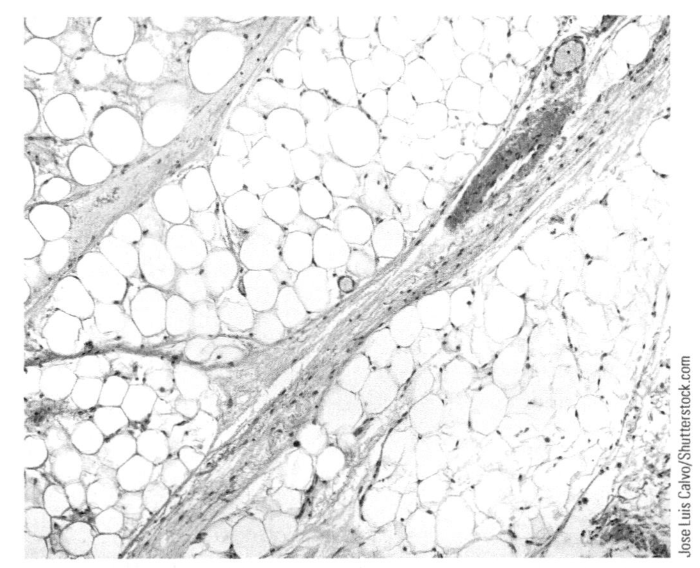

Jose Luis Calvo/Shutterstock.com

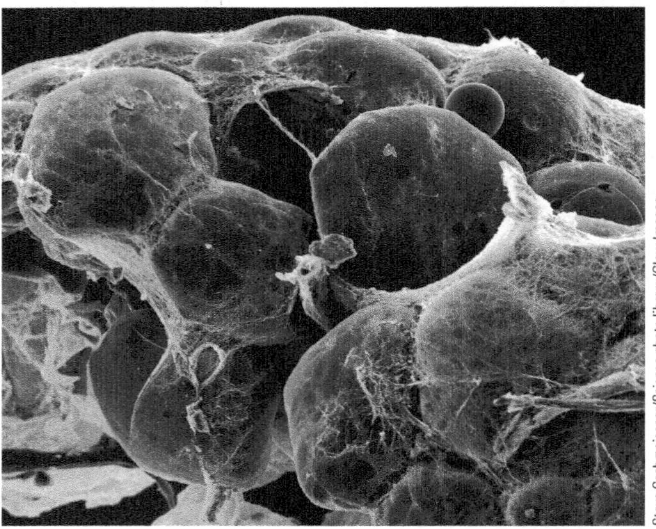

Steve Gschmeissner/Science photo library/Glow Images

Illustration 18.3 Scanning electron micrographs of white fat (on the top) and brown fat (on the bottom).

fat content is associated with improved insulin sensitivity and blood glucose levels[13,27]

Fats Increase the Flavor and Palatability of Foods Although "pure" fats by themselves tend to be tasteless, they absorb and retain the flavor of substances that surround them. Fats in meats, for example, pick up flavors from the meat and extend the meaty flavor to the fat. This characteristic of fat is why butter, if placed close enough to garlic in the refrigerator, tastes like garlic.

Fats Contribute to the Sensation of Feeling Full As they should, at nine calories per gram! Fats tend to stay in the stomach longer than carbohydrates or proteins and are absorbed over a longer period of time. Their presence in the stomach and small intestine triggers satiety signals and a feeling of fullness.[14]

Fats Are a Component of Cell Membranes, Vitamin D, and Sex Hormones Fats are found in every cell in the body, and are a key component of cell membranes. Fats help make cell membranes flexible and are involved in the selective transfer of nutrients and other substances into and out of cells. Other types of fat serve as precursors to vitamin D and sex hormones, including estrogen and testosterone.[12]

Fat Tissue Produces Hormones and Other Signaling Substances Two important hormones produced by fat cells are **adiponectin** and **leptin**. Both hormones play important roles in the development of type 2 diabetes and obesity. Adiponectin promotes the utilization of fat stores, increases insulin sensitivity (which lowers blood glucose levels and insulin need), and reduces inflammation (which helps protect tissues from injury). Leptin is involved in the regulation of energy balance and body weight. It initiates signals that suppress appetite and increase the utilization of fat stores. Levels of adiponectin are often low in people with high levels of body fat and type 2 diabetes. Obese individuals tend to be leptin resistant. Although high levels of leptin that would normally decrease food intake are available, the appetite-regulating mechanisms fail to recognize it.[16,17]

Table 18.2 Basic facts about the types of fat

Fats can be:
• Monoglycerides
• Diglycerides
• Triglycerides
• Saturated
• Monounsaturated
• Polyunsaturated
Unsaturated fats come in:
• *Cis* forms
• *Trans* forms

Types of Fat

There are many types of fat in food and in our bodies (Table 18.2). Of primary importance are *triglycerides* (or triacylglycerols), *saturated and unsaturated fats, cholesterol*, and *trans* fats (for definitions, see Table 18.3). The different types of fats have different effects on health.

Triglycerides, which consist of one *glycerol* unit (a glucose-like substance) and three fatty acids (Illustration 18.4), make up 98% of our dietary fat intake and the vast majority of our body's fat stores. Triglycerides are transported in the blood attached to protein carriers and are used by cells for energy formation and tissue maintenance. A minority of fats have the form of *diglycerides* (glycerol plus two fatty acids) and *monoglycerides* (glycerol and one fatty acid). Diglycerides are present in some oils and small amounts are used in food products as emulsifiers—or to increase the blending of fat- and water-soluble substances. Monoglycerides are present in small amounts in some oils; we don't consume very much of them in foods.

As far as health is concerned, the glycerol component of fat is relatively unimportant. It's the fatty acids that influence what the body does with the fat we eat; they are responsible, in part, for how fat affects health. Many different types of fatty acids are found in triglycerides. You've heard of the major ones: those that make fat *saturated* or *unsaturated.*

Table 18.3 A glossary of fats

Triglycerides: Fats in which the glycerol molecule has three fatty acids attached to it; also called triacylglycerides. Triglycerides are the most common type of fat in foods and in body fat stores.
Saturated fats: Molecules of fat in which adjacent carbons within fatty acids are linked only by single bonds. The carbons are "saturated" with hydrogens; that is, they are attached to the maximum possible number of hydrogens. Saturated fats tend to be solid at room temperature. Animal fat and palm and coconut oil are sources of saturated fats.
Unsaturated fats: Molecules of fat in which adjacent carbons are linked by one or more double bonds. The carbons are not saturated with hydrogens; that is, they are attached to fewer than the maximum possible number of hydrogens. Unsaturated fats tend to be liquid at room temperature and are found in plants, vegetable oils, meats, and dairy products.
Glycerol: A syrupy, colorless liquid component of fats that is soluble in water. It is similar to glucose in chemical structure.
Cholesterol: A fat-soluble, colorless liquid found primarily in animals. Cholesterol is produced by the liver and is used by the body to form hormones such as testosterone and estrogen. It is a component of cell membranes. Cholesterol is present in plant cell membranes, but the quantity is small and plants are not considered to be a significant dietary source of cholesterol.
Diglyceride: A fat in which the glycerol molecule has two fatty acids attached to it; also called diacylglyceride.
Monoglyceride: A fat in which the glycerol molecule has one fatty acid attached to it; also called monoacylglyceride.
Monounsaturated fats: Fats that contain a fatty acid in which one carbon–carbon bond is not saturated with hydrogen.
Polyunsaturated fats: Fats that contain a fatty acid in which two or more carbon–carbon bonds are not saturated with hydrogen.
***Trans* fats:** A type of unsaturated fatty acid produced by the addition of hydrogen to liquid vegetable oils to make them more solid. Small amounts of naturally occurring *trans* fat are found in milk and meat.

Saturated and Unsaturated Fats Fatty acids found in fats consist primarily of hydrogen atoms attached to carbon atoms (Illustration 18.5). When the carbons are attached to as many hydrogens as possible, the fatty acid is "saturated"—that is, saturated with hydrogen. Saturated fats tend to be solid at room temperature. Except for palm and coconut oil, only animal products are rich in saturated fats. Fatty acids that contain fewer hydrogens than the maximum are "unsaturated." They tend to be liquid at room temperature. By and large, plant foods are the best sources of unsaturated fats.

Glycerol + 3 fatty acids = Triglyceride

Illustration 18.4 A triglyceride.

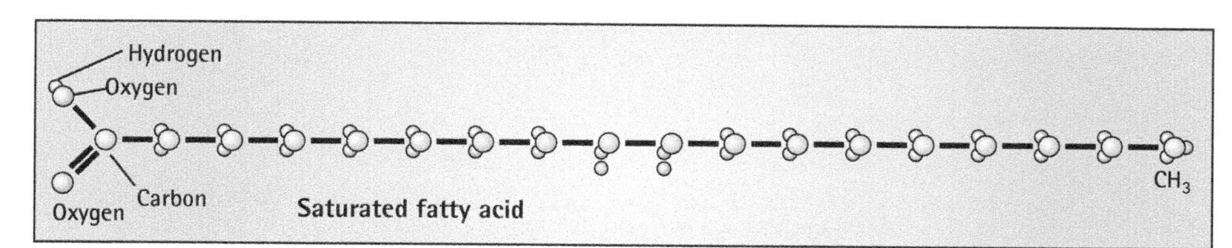

CH CH
Monounsaturated fatty acid

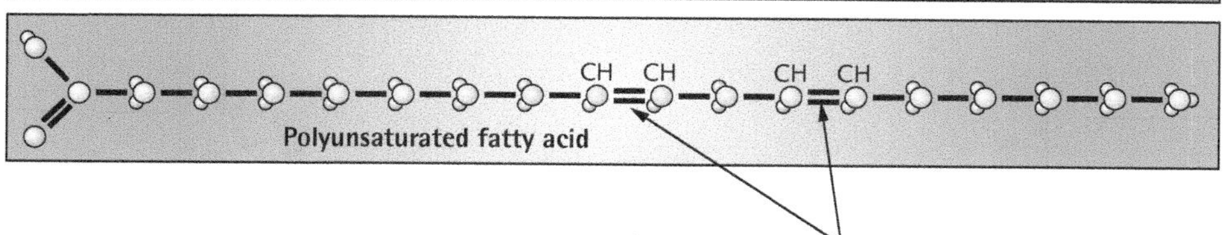

Illustration 18.5 A look at the difference between a saturated and an unsaturated fatty acid.

adiponectin (pronounced add-dip-o-neck-tin) A hormone produced by fat cells that increases insulin sensitivity and utilization of fat stores, and decreases inflammation.

leptin A hormone produced by fat cells involved in the regulation of energy balance and body weight. It initiates signals that suppress appetite and increase the utilization of fat stores.

Illustration 18.6 Fat profiles of selected foods.

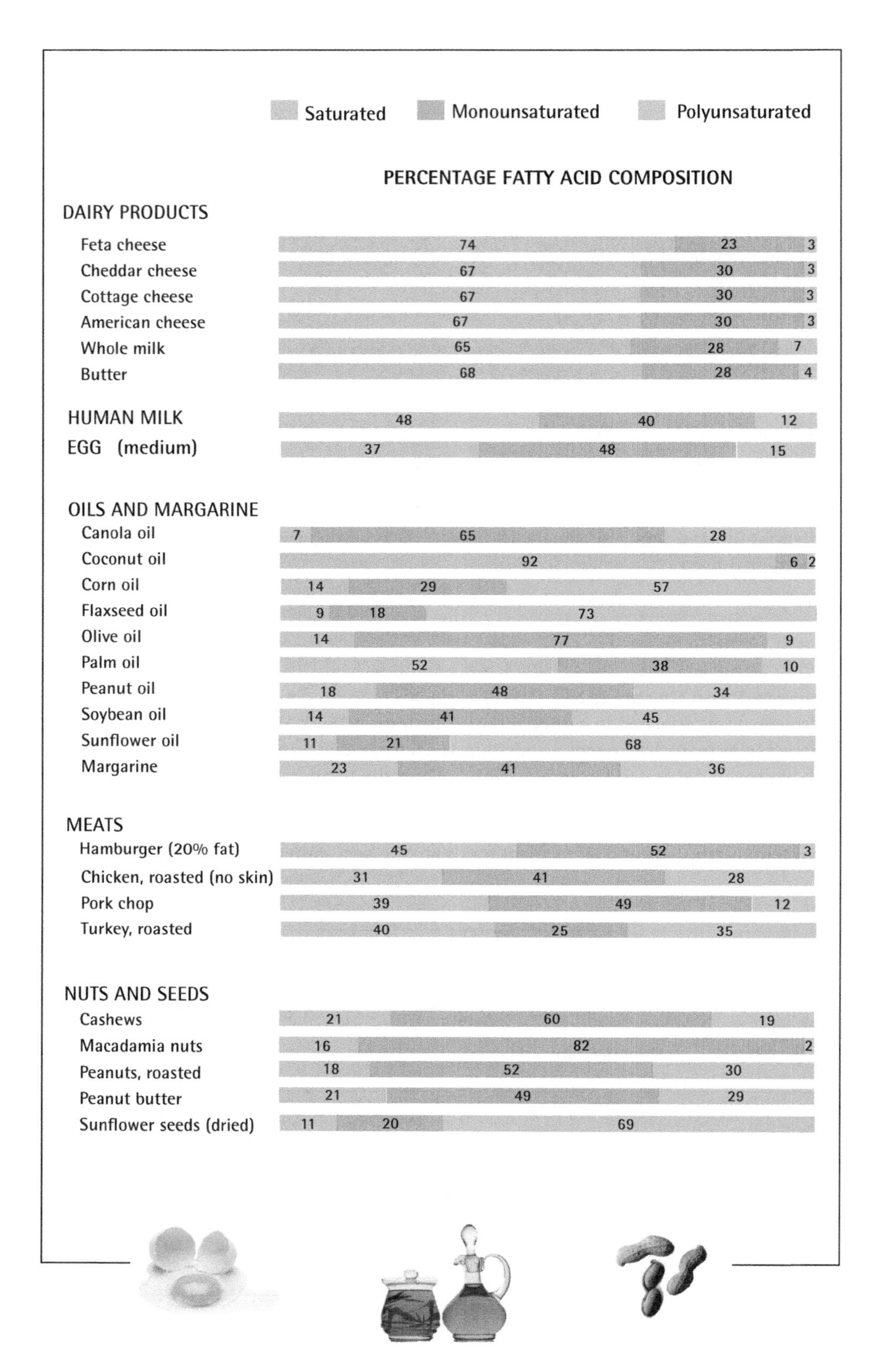

The saturated, monounsaturated, and polyunsaturated fat contents of a number of foods are shown in Illustration 18.6.

Unsaturated fats are classified by their degree of unsaturation. If only one carbon–carbon bond in the fatty acid is unsaturated, the fat is called *monounsaturated.* If two or more carbon–carbon bonds are unsaturated with hydrogen, the fat qualifies as *polyunsaturated.*

The Essential Fatty Acids

There are two essential fatty acids: linoleic acid and alpha-linolenic acid. Linoleic acid is a member of the omega-6 fatty acid family (also called n-6 fatty acids), and alpha-linolenic acid is an omega-3 fatty acid (also known as n-3 fatty acids). Both are polyunsaturated, can be used as a source of energy, and are stored in fat tissue. Because they are essential, both linoleic and alpha-linolenic acid are required in the diet.

Linoleic acid is required for growth, maintenance of healthy skin, and normal functioning of the reproductive system. It is a component of all cell membranes and is found in particularly large amounts in nerves and the brain. A number of biologically active compounds that are produced in the body and that participate in regulation of blood pressure, blood clotting, and pro-inflammatory reactions are derived from linoleic acid. Inflammation is a central component of many chronic diseases, including heart disease, type 2 diabetes, osteoporosis, cancer, Alzheimer's disease, and rheumatoid arthritis. It is part of the body's response to the presence of infectious agents or irritants. By-products of the body's inflammatory processes, however, are oxidation reactions that can harm cells and tissues. The major food sources of linoleic acid include vegetable oils, nuts, grains, and meats.

Alpha-linolenic acid is a structural component of cell membranes and is also found in high amounts in the brain and other nervous system tissues. It forms biologically active compounds used in the regulation of blood pressure, blood clotting, and anti-inflammatory reactions.[12] Alpha-linolenic acid is found in walnuts; dark, leafy green vegetables; and flaxseed, canola, and soybean oils.

Other biologically important omega-3 fatty acids exist; the two primary ones are EPA (eicosapentaenoic acid, pronounced *e-co-sah-pent-tah-no-ick*) and DHA (docosahexaenoic acid, pronounced *dough-cos-ah-hex-ah-no-ick*). These omega-3 fatty acids can be produced from alpha-linolenic acid, but the conversion process is slow and results in relatively small amounts of EPA and DHA being made available.[12]

EPA and DHA EPA and DHA are found primarily in fish oils and perform a number of important functions in the body. DHA is a structural component of the brain and is found in large amounts in the retina of the eye. During the last three months of pregnancy and during infancy, DHA accumulates in these developing tissues and promotes intellectual and visual development. These fatty acids are also associated with decreased inflammation and may reduce the risk of heart disease.[20] EPA serves as a precursor to a number of biologically active compounds involved in blood pressure regulation, blood clotting, and anti-inflammatory reactions. Derivatives of EPA limit the harmful effects of inflammatory and oxidation reactions.[21]

Adequate intake of EPA and DHA for adults is considered to be 250–500 mg per day.[22] These levels of intake may be achieved by consuming 8 ounces of the types of fish and shellfish included in the "best sources" list shown in Table 18.4, while excluding those listed as "choices to avoid" based on mercury content.[19] Mercury can cause nerve tissue damage and learning problems if consumed in excessive amounts. Pregnant and breastfeeding women can safely consume two to three servings of low-mercury fish and seafood weekly. Fish that are high in mercury, such as tilefish, swordfish, shark, and king mackerel, should not be consumed. The Food and Drug Administration (FDA) recommends that consumers ingest no more than a total of three grams of EPA and DHA daily and limit supplementary intakes to two grams a day. Fish liver oils should be used with caution because they contain relatively large amounts of vitamins A and D. Fish oils, made from the body of the fish, do not.[21]

EPA and DHA Fortified Foods Fatty fish are clearly the richest sources of EPA and DHA. But what if you (like many other people) don't like fish? In that case you can turn to seafoods like shrimp or EPA- and DHA-fortified foods.[24] (For more information on specific EPA- and DHA-fortified foods, see this unit's Take Action feature.) Purified fish oils with no fishy taste are being added to products from fruit juice to yogurt, as well as to animal feeds. Consequently, beef, pork, eggs, milk, and milk products consumed from animals fortified with EPA and DHA are fortified, as well. Some animal feeds contain DHA from algae and provide only DHA in their food products.

Table 18.4 "Best Choices" and "Choices to Avoid" for fish and shellfish based on methyl mercury content.[19]

Best choices	Choices to avoid
Anchovy	King mackerel
Atlantic mackerel	Marlin
Black sea bass	Orange roughy
Butterfish	Shark
Catfish	Swordfish
Clams	Tilefish
Cod	(Gulf of Mexico)
Crab	Tuna, bigeye
Crawfish	
Flounder Haddock	
Hake	
Herring	
Lobster,	
American and spiny	
Mullet Oyster	
Pacific chub mackerel	
Perch	
Pickerel	
Plaice	
Pollock	
Salmon	
Sardine	
Scallop	
Shad	
Shrimp	
Skate	
Smelt	
Sole	
Squid	
Tilapia	
Trout, freshwater	
Tuna, canned light	
Whitefish	
Whiting	

take action To Consume Enough EPA and DHA from Foods Other Than Fish

If you would like to increase your EPA and DHA intake, here are some of your non-fish food choices. Indicate with a check mark foods you would be likely to consume. At the bottom of the list, answer this question: Based on your choices, about how much EPA and DHA would you be adding to your daily diet by consuming the foods you checked?

EPA and DHA/DHA content, mg	
____ DHA fortified eggs, 1 150	
____ Healthy Heart Omega-3 orange juice, 8 oz.	50
____ Omega Farm nonfat yogurt, 8 oz.	75
____ Egg Creations Liquid, ¼ cup	260
____ Smart Balance Omega Buttery Spread 1 Tbsp	32
____ Italica Omega-3 Olive Oil, 1 Tbsp	120
____ Omega Farms Low-Fat Milk, 1 cup	75
____ Omega Farms Mild Cheddar, 1 oz	75
____ Smart Balance Lactose-Free Milk with Omega-3, 1 cup	32
____ Minute Maid 100% Fruit Juice Blend with Omega 3/DHA, 1 cup	50
____ Shrimp, 3 oz	268
____ Clams, 3 oz	241
____ Crab, 3 oz	375
____ Scallops, 3 oz	161

The approximate amount of EPA and DHA I would be adding to my diet daily based on my food choices is ____ mg. If your answer is not between 250 to 500 mg, at least you would be getting closer!

To make sure you're choosing foods with both EPA and DHA, or with DHA alone, confirm that the label specifies that these fatty acids are contained in the product. Just because a product announces "Omega-3" on the label doesn't mean it contains EPA and DHA. It may contain the omega-3 fatty acid alpha-linolenic acid from ingredients such as flaxseed, flaxseed oil, or walnut oil rather than EPA and DHA.[22]

Hydrogenated Fats

Unsaturated fats aren't as stable as saturated fats. They are more likely to turn rancid with time and exposure to air (oxygen) and heat than are saturated fats. Additionally, solid fats are preferable to oils for some cooking applications. These problems with unsaturated fats have a solution: it's called hydrogenation.

What's Hydrogenation? **Hydrogenation** is a process that adds hydrogen to liquid unsaturated fats, thereby making them more saturated and solid. The shelf life and cooking properties of vegetable oils are improved in the process.[11] Hydrogenation has two drawbacks, however. Hydrogenated vegetable oils contain more saturated fat than the original oil. Corn oil, for example, contains only 6% saturated fats; corn oil margarine has 17%. The other negative is that hydrogenation causes a change in the structure of the unsaturated fatty acids. Specifically, hydrogenation converts some unsaturated fats into ***trans* fats**.

hydrogenation The addition of hydrogen to unsaturated fatty acids.

***trans* fats** Unsaturated fatty acids in fats that contain atoms of hydrogen attached to opposite sides of carbons joined by a double bond:

```
  H
—C=C—              H H
     H            —C=C—
Trans fatty acid   Cis fatty acid
```

Fats containing fatty acids in the *trans* form are generally referred to as *trans* fats. *Cis* fatty acids are the most common, naturally occurring form of unsaturated fatty acids. They contain hydrogens located on the same side of doubly bonded carbons.

***Trans* Fatty Acids** The bulk of *trans* fats in our diets comes from industrially hydrogenated vegetable oils. Hydrogenation causes some of the unsaturated fatty acids to be converted from their naturally occurring *cis* to the *trans* form. Ruminant animals such as cows, goats, and sheep form a small amount of *trans* fats in their stomachs that are distributed within body tissues. Consequently, milk and milk products from these animals will contain some *trans* fats.[11]

The repositioned hydrogen molecules in *trans* fatty acids appear to be responsible for the specific adverse effects of these fatty acids on disease risk. *Trans* fatty acids have adverse effects on the functions and levels of blood lipids and increase the risk of heart disease, stroke, sudden death from heart disease, and type 2 diabetes. Sufficiently high levels of *trans* fatty acids promote the development of inflammation.[19] Regular intake of approximately 4 grams of *trans* fat daily places a person consuming 2,000 calories a day at increased risk of heart disease.[23]

It is recommended that Americans consume as little *trans* fats as possible,[1] and nutrition information labeling requirements are making that easier to accomplish. Nutrition Facts panels must include the *trans* fat content of food products (Illustration 18.7). The percent of daily value column (%DV) is not used for *trans* fats because there is no recommended level of intake. Products labeled "*trans* fat-free" (Illustration 18.8) must contain less than 0.5 gram of both *trans* and saturated fats. Products reporting "0" *trans* fats on Nutrition Facts panels must contain less than 0.5 g of *trans* fats per serving, but they may contain 0.1 to 0.5 g of *trans* fat per serving. If the ingredient label lists partially hydrogenated vegetable oil, shortening, palm oil, or other sources of *trans* fat, then you know the product contains some *trans* fat.[24]

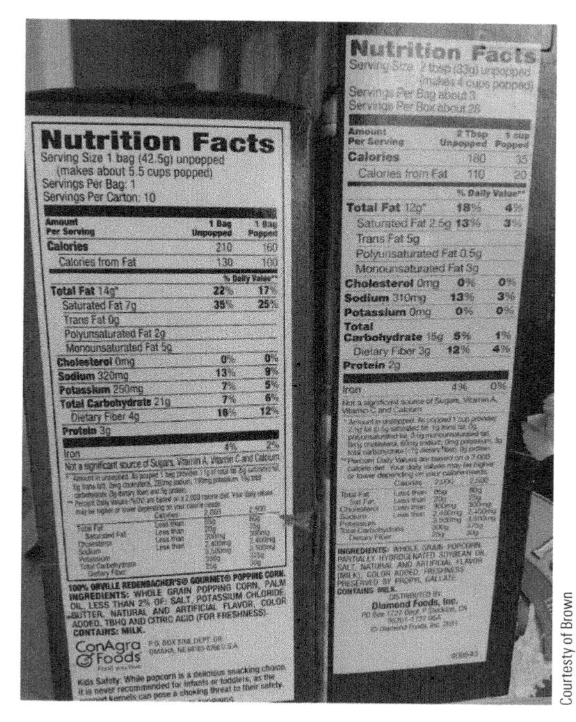

Courtesy of Brown

Illustration 18.7 Examples of Nutrition Facts panels on two brands of microwave popcorn. Is it likely that the popcorn labeled "0 *trans* fat" contains 0.5 grams or less *trans* fat? The ingredient label will let you know.

The requirement to label the *trans* fat content of food products, regulations by various states and cities that ban their use in restaurant foods, and increased consumer awareness of its adverse effects have encouraged food producers to take *trans* fats out of many prepared, processed, and fast foods.[11] Many major brands of potato and tortilla chips, bakery products, and frozen meals and desserts have taken the *trans* fats out of their products, and the presence of *trans* fats in prepared foods will continue to decrease.[25] The decline in the presence of *trans* fats in processed foods is continuing due to FDA's decision to end the use of partially hydrogenated oils in foods sold in the United States by 2018. In addition, *trans* fat is no longer considered a "generally recognized as safe" food ingredient, making it much more difficult to include in foods.[24,25]

Richard Anderson

Illustration 18.8 Products that feature "no *trans* fats" and "*trans* fat-free" labels.

Cholesterol

Cholesterol is a lipid found primarily in animal products. It is tasteless and odorless and is contained in both the lean and fat parts of animal products. It is also present in plant cell membranes, but the quantity is small and plants are not considered to be a significant dietary source of cholesterol.[24] About 75% of the cholesterol in the body is produced by the liver and the brain (which makes its own supply of cholesterol). The rest comes from the diet. Production of cholesterol deceases when cholesterol intake is high, and increases when it is low.[26] Because the liver produces cholesterol from other substances in our diet, it does not qualify as an essential nutrient.

Functions of Cholesterol Cholesterol is involved in a number of body functions (Table 18.5). It is used in the synthesis of vitamin D and of estrogen, testosterone, and other hormones. It is a major component of the lipid-rich sheath that coats nerve cells. The cholesterol-rich coating helps protect and insulate nerve cells, and enhances the transmission of nerve impulses. Cholesterol is a major component of cell membranes, where it functions to modulate the flexibility and permeability of cell membranes. The brain, our central nervous system's headquarters, is rich in cholesterol. It contains more cholesterol than any other organ—roughly a quarter of the total free cholesterol in the

Table 18.5 How the body uses cholesterol[28]

- Cholesterol is a major component of cell membranes and nerve cells.
- Cholesterol is used in the production of vitamin D, estrogen, testosterone, and other hormones.
- Cholesterol is a major constituent of bile.

human body. Cholesterol is a major constituent of bile. Bile is produced in the liver, stored in the gallbladder, and released into the small intestine during the digestion of fat. Bile is needed for the digestion and absorption of dietary fats.[28]

Recommendations for Cholesterol Intake The 2015 Dietary Guidelines Advisory Committee did not make any statement about recommended cholesterol intake. Instead, the committee focused its recommendations for heart disease prevention on consuming healthy dietary patterns. It appears that previous estimates of the effects of cholesterol intake on heart disease risk exaggerated its influence in most people. Cholesterol intake reduction may benefit "hyper-cholesterol responders," whose blood cholesterol level is highly responsive to cholesterol intake. An individual's particular gene types involved in cholesterol utilization by the body appear to influence heart disease risk in hyper-responders to cholesterol intake.[23,29]

Food Sources of Fat

Table 18.6 lists the fat content of common food sources of fat, including candy. Vegetables, fruits (except avocado and coconut), and grains are not listed because they don't contain much fat. The Food Composition Table provided in Appendix A, as well as product nutrition labels, can be used to identify the fat content of foods not listed in the table.

Table 18.6 The fat content of some foods

Food	Amount	Grams	Percentage of total calories from fat
Fats and oils			
Butter	1 tsp	4.0	100%
Margarine	1 tsp	4.0	100
Oil	1 tsp	4.7	100
Mayonnaise	1 Tbsp	11.0	99
Heavy cream	1 Tbsp	5.5	93
Salad dressing	1 Tbsp	6.0	83
Meats and fast foods			
Hot dog, 1	2 oz	17.0	83
Bologna	1 oz	8.0	80
Sausage	4 links	18.0	77
Bacon	3 pieces	9.0	74
Salami	2 oz	11.0	68
Hamburger, regular (20% fat)	3 oz	16.5	62
Chicken, fried with skin	3 oz	14.0	53
Big Mac	6.6 oz	31.4	52
Quarter Pounder with cheese	6.8 oz	28.6	50
Whopper	8.9 oz	32.0	48
Steak (rib eye)	3 oz	9.9	47
Veggie pita	1	17.0	38
Chicken, baked without skin	3 oz	4.0	25
Flounder, baked	3 oz	1.0	13

Table 18.6 **(continued)**

Food	Amount	Grams	Percentage of total calories from fat
Shrimp, boiled	3 oz	1.0	10
Milk and milk products			
Cheddar cheese	1 oz	9.5	74
American cheese	1 oz	6.0	66
Milk, whole	1 cup	8.5	49
Cottage cheese, regular	½ cup	5.1	39
Milk, 2%	1 cup	5.0	32
Milk, 1%	1 cup	2.7	24
Cottage cheese, 1% fat	½ cup	1.2	13
Milk, skim	1 cup	0.4	4
Yogurt, frozen	¾ cup	0.0–6.6	0–3
Other			
Olives	4 medium	1.5	90
Avocado	½	15.0	84
Almonds	1 oz	15.0	80
Sunflower seeds	¼ cup	17.0	77
Peanuts	¼ cup	17.5	75
Cashews	1 oz	13.2	73
Flax seed	¼ cup	17.7	71
Egg	1	6.0	61
Potato chips	1 oz (13 chips)	11.0	61
French fries	14 fries	9.5	46
Taco chips	1 oz (10 chips)	6.2	41
Candy			
Peanut butter cups, 2 regular	1.6 oz	15.0	54
Milk chocolate	1.6 oz	14.0	53
Almond Joy	1.8 oz	14.0	50
Kit Kat	1.5 oz	12.0	47
M&M's, peanut	1.7 oz	13.0	47

Knowledge of the caloric and fat content of a food can be used to calculate the percentage of calories that is provided by fat in foods or in a diet. For example, suppose that your diet provides 2,400 calories per day and 80 grams of fat. To calculate the percentage of fat calories, multiply 80 grams by 9 (the number of calories in each gram of fat), divide the result by 2,400 calories, and then multiply this result by 100:

$$80 \text{ grams fat} \times 9 \text{ calories/gram} = 720 \text{ calories}$$
$$720 \text{ calories} \div 2{,}400 = 0.30$$
$$0.30 \times 100 = 30\% \text{ of total calories from fat}$$

Fat Labeling Nutrition labeling regulations for fat require that food manufacturers adhere to standard definitions of *low-fat*, *fat-free*, and related terms on food labels.

Table 18.7 What claims about the cholesterol content of foods that normally contain cholesterol must mean

- No cholesterol or cholesterol-free: Contains less than three milligrams of cholesterol per serving
- Low cholesterol: Contains 20 milligrams or less of cholesterol per serving
- Reduced cholesterol: Contains at least 75% less cholesterol than normal
- Less cholesterol: Contains at least 25% less cholesterol than normal; the percentage less must be stated on the label

Similarly, claims made about the cholesterol content of food products must comply with standard definitions (Table 18.7). If a claim is made about the fat content of a food, the Nutrition Facts panel must specify the food's fat, saturated fat, *trans* fat, and cholesterol content. If a claim is made about cholesterol content (and claims can be made only for products that normally contain a meaningful amount of cholesterol), the nutrition panel must also reveal the product's fat and saturated fat content. To prevent the use of unrealistically small serving sizes as a way to appear to cut down on a product's fat content, realistic serving sizes must also be used on food labels.[30]

The Low-Fat Era Is Over

Low-fat diets and fat-free and low-fat food products have lost their luster because they are frequently associated with increased intake of refined carbohydrates and added sugars.[2] Low-fat diets are no more effective for weight control than other diets that limit calorie intake, and high-fat diets do not appear to promote weight gain. In fact, they are associated with the development of obesity, type 2 diabetes, and heart disease if dietary fat is replaced with refined carbohydrates and added sugar.

Currently, it is recommended that the types of fat-containing foods included in healthy dietary patterns—such as the Healthy Mediterranean-Style Dietary Pattern and the Healthy Vegetarian-Style Dietary Pattern—be consumed. Illustration 18.9 shows an example of the assortment of foods included in a Mediterranean-type dietary pattern. This and other healthy dietary patterns do not regularly include foods high in refined carbohydrates, sugar, or highly processed foods. They emphasize basic foods like vegetables, fruits, whole grains, fish, seafood, nuts, seeds, poultry, and vegetable oils. Healthy intake of fat is within the previously established range of 20–35% of total calories.[23] Average fat consumption in the United States is within this range (33% of total calories).[23]

PhotoDisc

Illustration 18.9 A look at the cuisine of the Mediterranean diet.

REALITY CHECK

Low Fat-Foods and Health

Are low-fat foods healthy?

Who gets the thumbs up?

Answers on page 18-13.

Photodisc

Kristen: How can I be wrong? Low-fat food products are best for healthy diets because they contain almost no fat!

Photodisc

Butch: Why would they be? Processed foods contain other ingredients besides "no fat."

The Adequate Intake (AI) for the essential fatty acid linoleic acid is set at 17 grams a day for men and 12 grams for women. The AI for the other essential fatty acid, alpha-linolenic acid, is 1.6 grams per day for men and 1.1 grams for women. It is recommended that intake of *trans* fats be kept as low as possible while consuming a nutritionally adequate diet. Americans are being encouraged to increase consumption of EPA and DHA by eating two servings of fish (the types low in mercury) weekly.

Recommendations for Fat Intake: What's next Scientific understanding of the roles played by fats in our diets has changed dramatically over time. During World War I when the troops of the United States and those of its allies were running short on food, the U.S. Food Administration Department developed a program that increased the amount of food available to the troops. The program developed encouraged Americans to eat more nutrient-dense foods such as fish and vegetables, and lower amounts of higher-calorie foods, including meats and fats, so that more of the high calorie foods could be sent to the troops (Illustration 18.9).[32] In those days, fat and sugar were valued for their calorie value.

In a nation and world where obesity has become wide-spread across most of the globe, the caloric value of fats has become more of a negative characteristic than a positive one. In addition, evidence related to the roles of the different types of fat (eg. saturated and polyunsaturated fats, cholesterol) in heart disease development has weakened. Research aimed at identifying the roles of multiple dietary, genetic, and environmental factors that contribute to heart disease development and progression is steaming ahead.[33]

Advances in knowledge will be studied and scrutinized by the 2020 Dietary Guidelines Committee and new recommendations for fat intake will emerge as part of the set of recommendations produced. Although scientific facts serve as the basis for new recommendations regarding diet and health, there are far-flung ramifications associated with changes in recommendations about the types of food people should consume for their health. To stay informed, stay tuned to the 2020 Dietary Guidelines. They will be released in early 2021.

Pictures Now/Alamy Stock Photo

ANSWERS TO REALITY CHECK

Low-Fat Foods and Health

Low- and no-fat foods contain less fat than the regular versions of the foods, but they may be high in refined carbohydrates and added sugar. The elimination of a particular, unpopular component of a processed food with another ingredient doesn't automatically mean the food is necessarily "better" for you. Healthy dietary patterns aren't based on individual foods; they are based on the overall pattern of dietary intake.

Photodisc

Kristen:

Photodisc

Butch:

James And James/Photolibrary/
Getty Images

NUTRITION up close

A Focus on DHA and EPA Safe Food Sources

Focal Point: Identify recommended dietary sources of DHA and EPA.

Assume your boss comes to you for advice because she knows you are taking a nutrition class. She wants to increase her consumption of omega-3 fatty acids EPA and DHA and asks you what the good sources would be.

Place a "✓" after the options listed that represent recommended choices.

1. Tilefish	______
2. Catfish	______
3. Clams	______
4. Fish liver oil	______
5. DHA-fortified eggs	______
6. Fish oil	______
7. Ketchup	______
8. Coconut oil	______
9. Perch	______

Feedback for the Nutrition Up Close is located in Appendix F.

REVIEW QUESTIONS

- **Summarize the major functions, food sources, recommended intake, and health effects of fats.**

1. Animal products are by far the leading source of saturated fats in the American diet. **True/False**
2. Flaxseed oil and dark, leafy green vegetables are good sources of the essential fatty acid alpha-linolenic acid. **True/False**
3. The best food sources of EPA and DHA are fish and shellfish. **True/False**
4. Cholesterol is not a required nutrient because it serves no essential, or life-sustaining, function in the body. **True/False**
5. Thanks in part to nutrition labeling requirements, there are lower amounts of *trans* fats in food products now than in the past. **True/False**
6. High *trans* fat intake is related to increased risk of heart disease. **True/False**
7. If a claim is made on a food package about the product's fat content, the Nutrition Facts panel must list the food's fat, saturated fat, *trans* fat, and cholesterol content. **True/False**
8. White adipose tissue is the major source of the body's energy reserves. **True/False**
9. Body fat functions are limited to serving as the body's major source of energy reserves. **True/False**
10. It is recommended that adults consume 20–25% of total calories from fat. **True/False**
11. ____ Assume Tao has a daily calorie need of 2,300 and decides to check out his diet to see if his fat intake is within the range of 20–35% of his total calorie intake. What would be the range of fat intake in grams that would meet his goal of consuming between 20 and 35% of total calories from fat?
    a. 46 to 80 grams
    b. 20 to 35 grams
    c. 51 to 89 grams
    d. 65 to 77 grams

Answers to these questions can be found in Appendix F.

NUTRITION SCOREBOARD ANSWERS

1. See Table 18.1. **True**
2. **True**
3. **True, alpha-linolenic and linoleic acids**
4. **True**
5. **False**[2]

Nutrition and Heart Disease

NUTRITION SCOREBOARD

1 Heart disease is the leading cause of death worldwide. **True/False**

2 Dietary patterns high in refined grain products and added sugar are related to an increased risk of heart disease. **True/False**

3 Plaque that causes hardening of the arteries develops under the endometrial lining of arteries. **True/False**

4 Chronic inflammation is an important contributor to the development and progression of heart disease. **True/False**

Answers can be found at the end of the unit.

LEARNING OBJECTIVES

After completing Unit 19 you will be able to:

- List dietary, lifestyle, and biological factors that contribute to heart disease development.
- Evaluate the components of your diet and lifestyle that help prevent the development of heart disease.

Nutrition and Heart Disease

- **List dietary, lifestyle, and biological factors that contribute to heart disease development.**

Suspicions that dietary fat may be related to **heart disease** were first raised over 200 years ago. During the late eighteenth century, physicians noted that people who died of heart attacks had fatty streaks and deposits in the arteries that led to the heart. Later it was discovered that people with heart disease tended to have high blood levels of cholesterol and high intakes of saturated fats and cholesterol. It was concluded that high intake of these fats increased the risk of heart disease by raising blood cholesterol levels, and that diets low in saturated fat, cholesterol, and total fat may lower blood cholesterol and help prevent and treat heart disease.[1,3]

iStockphoto.com/Creativeye99

Results of a large number of studies have demonstrated that this approach to heart disease prevention and treatment is not effective.[3,5] It is clear that there are multiple important dietary and other risk factors for heart disease, and that they all need to be addressed. Revised recommendations for heart disease prevention and management feature the consumption of overall healthy dietary patterns and lifestyles.[2] State-of-the-science dietary and other lifestyle recommendations for the prevention and treatment of heart disease are presented in this unit.

heart disease One of a number of disorders that result when circulation of blood to parts of the heart is inadequate. Also called *coronary heart disease*. (*Coronary* refers to the blood vessels at the top of the heart. They look somewhat like a crown.)

KEY NUTRITION CONCEPTS

Content in this unit on nutrition and heart disease directly relates to a number of key nutrition concepts, including the following:

1. Nutrition Concept #3: Health problems related to nutrition originate within cells.
2. Nutrition Concept #4: Poor nutrition can result from both inadequate and excessive levels of nutrient intake.
3. Nutrition Concept #8: Poor nutrition can influence the development of certain chronic diseases.
4. Nutrition Concept #9: Adequacy, variety, and balance are key characteristics of healthful diets.

A Primer on Heart Disease

There is no bigger health problem in the United States and other developed countries than heart disease. Heart disease is the leading cause of death worldwide, accounting for one out of every four deaths. It is an "equal opportunity" disease, striking as many women as men, although women on average die 10 years later from heart disease than do men.[6] Deaths from heart disease are distributed unequally across the United States (Illustration 19.1), disproportionately affecting African and Hispanic Americans.[7]

Overall, though, both the incidence of and death rate from heart disease are declining in the United States. The decrease in incidence is related to risk factor reduction; the decrease in death rate primarily to improved treatment.[7] Gains in life expectancy, quality of life, and reduced health care expenditures would be generated by further reduction of risk factors for heart disease.[8]

What Is Heart Disease? Heart disease, also called coronary heart disease, refers to several disorders that result from inadequate blood circulation to parts of the heart. It begins with damage to the lining of the coronary (heart) arteries and continues with the formation of cholesterol-rich **plaque** under the **endothelium** lining of the

plaque Deposits of cholesterol, other fats, white blood cells, calcium, and cell materials in the lining of the inner wall of arteries.

endothelium (en-dough-thiel-e-um) The layer of cells lining the inside of blood vessels.

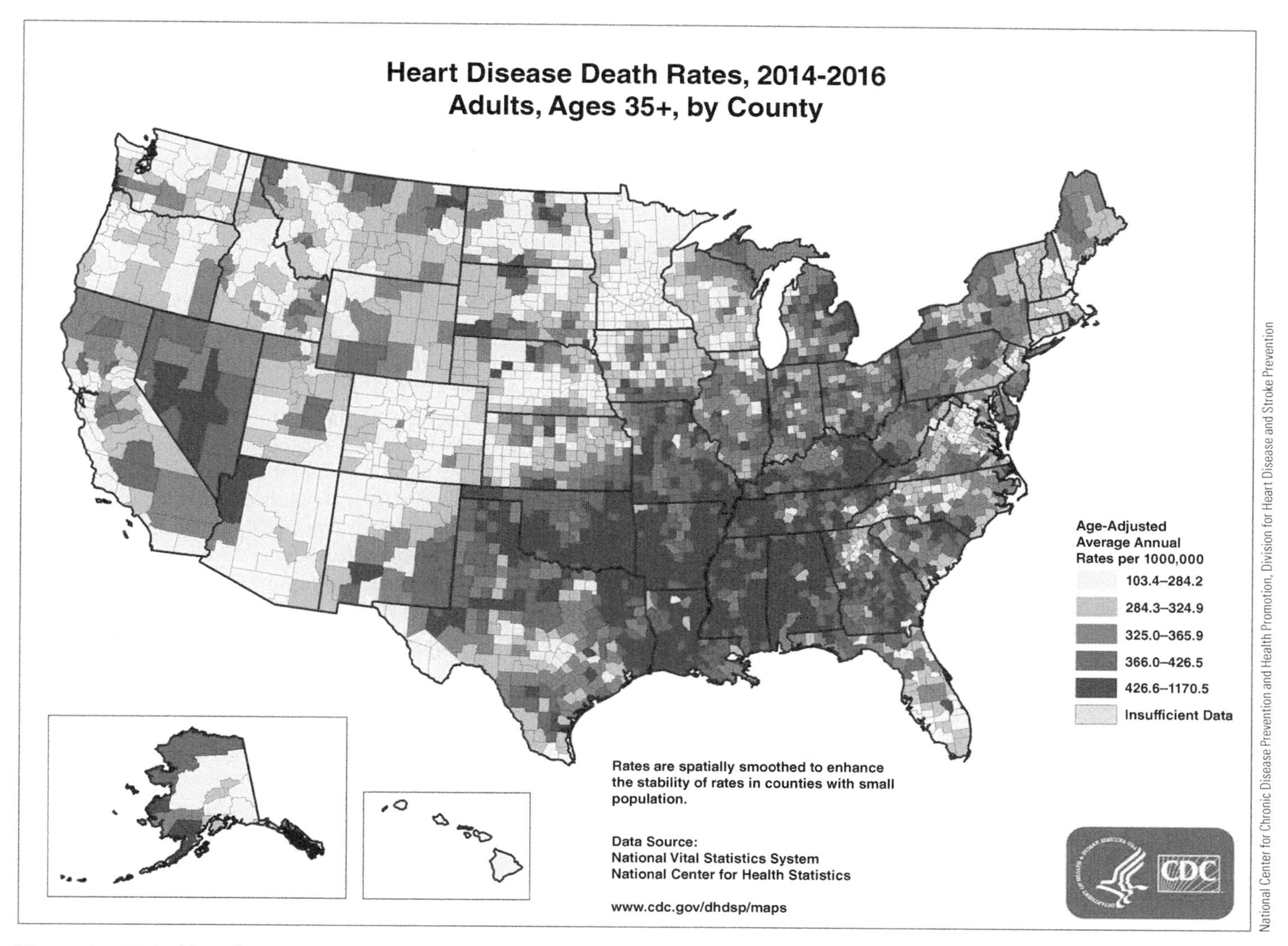

National Center for Chronic Disease Prevention and Health Promotion, Division for Heart Disease and Stroke Prevention

Illustration 19.1 Heart disease death rates, 2013–2015, adults age 35+ by county.

artery (Illustration 19.2). Plaque buildup eventually narrows blood vessels enough to reduce blood flow to the heart (Illustration 19.3). People with narrowed arteries have **atherosclerosis**, or "hardening of the arteries" as it is often called. Heart disease develops silently over time (usually decades).[4]

atherosclerosis "Hardening of the arteries" due to a buildup of plaque. Atherosclerosis is characterized by chronic inflammation of the arteries and develops over decades.

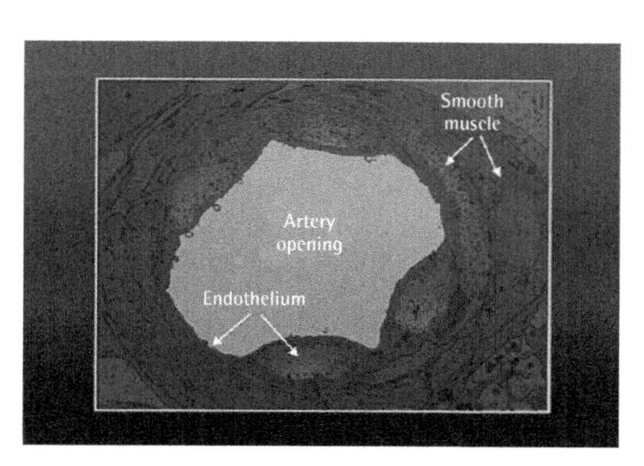

Illustration 19.2 The endothelium consists of cells that form the inside surface of arteries.

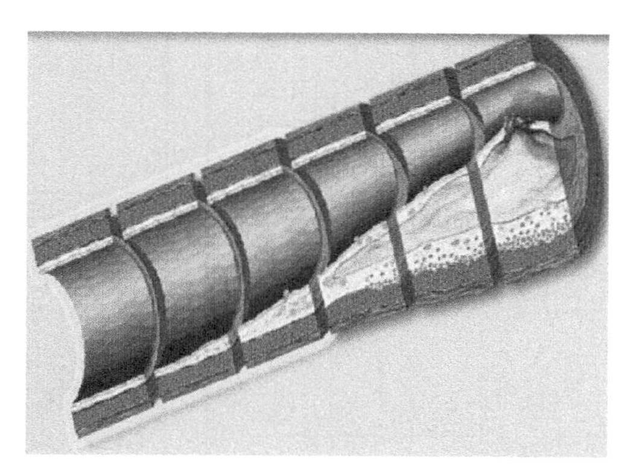

Illustration 19.3 Plaque (shown in yellow) forms under the endothelium and thickens, hardens, and narrows the artery.

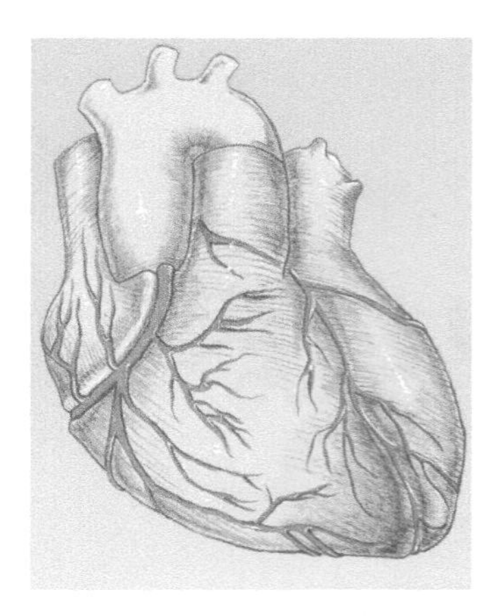

Illustration 19.4 The heart after a heart attack. The dark portion at the base of the heart is affected by the blockage in blood flow.

cardiovascular disease Disorders related to plaque buildup in blood vessels in the heart, brain, legs, and other tissues and organs. Other forms of cardiovascular disease exist, including disorders related to the malformation of the heart and blood vessels.

LDL cholesterol Low-density lipoprotein cholesterol. The function of LDL is to transport cholesterol in blood to cells, where it is used as a component of cell membranes, or for the synthesis of testosterone, estrogen, and other hormones. It is called "bad cholesterol" because high levels are associated with the development of heart disease in most people, and because it supplies the cholesterol found in plaque.

HDL cholesterol High-density lipoprotein cholesterol. The primary function of HLD cholesterol is to transport cholesterol from body tissues to the liver for incorporation into bile and excretion from the body. HDL cholesterol is sometimes referred to as "good cholesterol" or "healthy cholesterol" because high blood levels are related to a reduced risk of heart disease in most people.

chronic inflammation Low-grade inflammation that lasts weeks, months, or years. Inflammation is the first response of the body's immune system to infectious agents, toxins, or irritants. It triggers the release of biologically active substances that promote oxidation and other reactions to counteract the infection, toxin, or irritant. A side effect of chronic inflammation is that it also damages lipids, cells, and tissues.

gene variant An alteration in the normal nucleotide sequence of a gene that affects the gene's functions. Different forms of the same genes are considered "alleles."

When arteries are narrowed by 50% or more, the shortage of blood to the heart can produce chest pain (called *angina*). A heart attack occurs when an artery leading to the heart becomes clogged by a piece of plaque released by a ruptured portion of the artery wall or by a blood clot (Illustration 19.4).[38] Although heart disease primarily affects individuals over the age of 55, it's a progressive disease that may begin in childhood.[9]

Arteries leading to the heart aren't the only ones affected by atherosclerosis. Plaque can also build up in the arteries in the legs, neck, brain, and other body parts. If the blood supply to the legs is reduced, pain and muscle cramps may result after brief periods of exercise. Plaque buildup in the arteries of the brain contributes to stroke—an event that occurs when the blood supply to a part of the brain is inadequate.[14] Health problems due to atherosclerosis in arteries of the heart, brain, neck, and legs are collectively referred to as **cardiovascular disease**.

What Causes Atherosclerosis? A number of conditions are known to increase plaque formation in arteries leading to the heart. Four of the most important identified to date are elevated **LDL cholesterol** (LDLc) blood levels, low levels of **HDL cholesterol** (HDLc), **chronic inflammation**, and specific **gene variants**. (The composition and functions of LDLc and HDLc are presented as follows.) Part of the increased risk related to high concentrations of LDLc appears to be due to gene variants that affect the body's ability to excrete cholesterol. Low blood levels of HDLc reduce the delivery of cholesterol to the liver where it is incorporated into bile and excreted, thereby lowering total blood cholesterol levels. Chronic inflammation can damage the endothelial lining of arteries leading to the heart, which promotes plaque formation.[12] Increased emphasis is being placed on genetic screening for heart disease risk so that preventive efforts and treatments can be effectively targeted.

Understanding LDL and HDL Cholesterol Cholesterol is soluble in fat, but blood is mostly water. Therefore, cholesterol must be bound to a substance, such as protein, that mixes with water. (Otherwise, cholesterol would simply float in blood.) The resulting combination is called a *lipoprotein.* LDL cholesterol and HDL cholesterol are two types of lipoproteins. The composition and roles of both types of lipoproteins are summarized in Illustration 19.5.

LDL cholesterol generally carries more cholesterol than does HDL. It can be oxidized into reactive LDL particles that make it more likely to enter plaque and contribute to inflammation and plaque buildup. The higher the LDL cholesterol level, the greater the chances that atherosclerosis will develop and progress into heart disease.[13] HDL cholesterol gets its reputation as the good cholesterol because it helps remove cholesterol from the blood. An avid cholesterol acceptor, HDL escorts cholesterol to the liver for eventual excretion from the body. It also reduces the formation of plaque, speeds repair of the endothelial lining, and helps prevent oxidation and inflammation in plaque.[14]

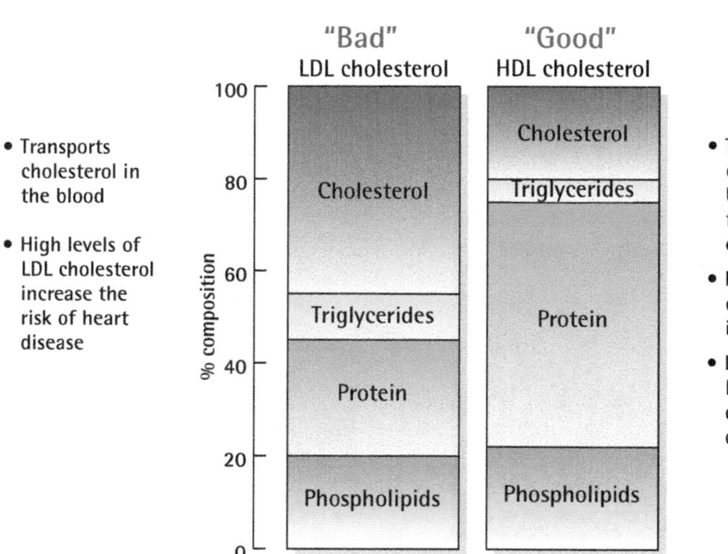

Illustration 19.5 Composition and functions of LDL cholesterol and HDL cholesterol.

Several types of lipoprotein particles are found in LDL cholesterol, and the different types vary in size and function. These differences influence the effects of LDL cholesterol on heart disease risk. Smaller LDL particles, for example, transport less cholesterol than larger ones. Consequently, the amount of cholesterol transported by LDL cholesterol and the resulting heart disease risk vary somewhat depending on the type and function of lipoprotein particles it contains.[15] A different indicator of heart disease risk that includes the amount of cholesterol in all of the heart disease-promoting lipoproteins and particles is becoming widely used. The indicator, which is derived by subtracting HDL cholesterol from total cholesterol, is called **non-HDL cholesterol**. This measure provides a more accurate assessment of heart disease risk than does LDL cholesterol.[34]

non-HDL cholesterol The difference between total cholesterol concentration and HDL cholesterol concentration in blood.

Although LDL and HDL cholesterol are related to the risk of heart disease and blood levels can be modified by dietary changes, there is no convincing evidence that lowering LDL cholesterol or raising HDL cholesterol through low-fat, low-saturated fat, or low-cholesterol diets deceases the risk of heart disease in most people.[3,13,15] Heart disease risk is influenced by many factors and by interactions among them.

Who's at Risk for Heart Disease?

Frankly, you may be. Take a look at Illustration 19.6 to get an idea if you are. Many factors are related to increased risk of heart disease, and a number of the risk factors are fairly common among the U.S. adult population. The major ones are:

- Family history of early onset of heart disease and genetic traits for heart disease
- High blood pressure
- Elevated LDL cholesterol and low levels of HDL cholesterol
- Smoking
- Physical inactivity
- Age (over 55 years)
- Diabetes
- Low intake of vegetables, fruits, whole grains, and other sources of fiber and antioxidants
- Obesity, specifically high amounts of visceral (abdominal) fat
- High intake of *trans* fats, refined carbohydrates, and added sugars
- **Insulin resistance** and chronic inflammation[11,12,17,18]

insulin resistance A condition in which cell membranes have reduced sensitivity to insulin so that more insulin is needed to transport glucose into cells. It is characterized by elevated levels of serum insulin, glucose, triglycerides, and increased blood pressure.

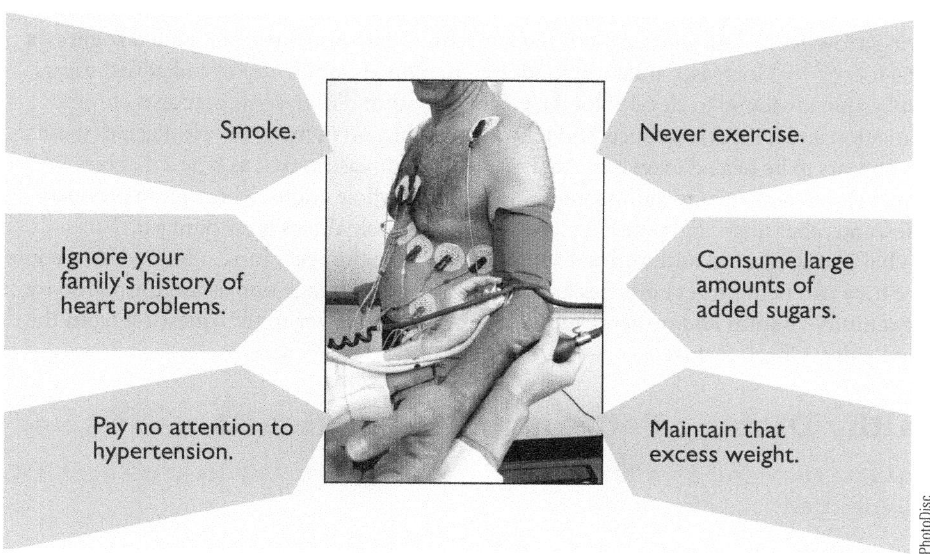

Illustration 19.6 How to have a heart attack.

Risk factors for heart disease apply to both women and men, but the symptoms associated with heart disease vary somewhat between the sexes.[19] Both sexes report feeling chest-related pain, but women are more likely than men to perceive pain in the neck, jaw, and throat.[20]

Genetic Factors Related Heart Disease Risk Several genetic disorders and gene types are known to directly affect blood cholesterol levels and the risk of heart disease. The effect of dietary cholesterol intake on blood cholesterol levels varies depending on whether a person is genetically a *hypo-* or *hyper-*responder to cholesterol in foods.[21,22] About one in three individuals are *cholesterol hyper-responders.*[23] Inherited defects in LDL cholesterol receptors are related to very high LDL cholesterol levels.[24]

Some of the genetic risks for heart disease are due to gene–environment interactions early in life. Certain environmental exposures can turn specific genes "on" or "off" for the rest of one's life.[38] Malnutrition during pregnancy, for example, can change the long-term functional status of certain genes in ways that increase the risk of heart disease later in life.[25]

Dietary Guidelines for Americans are issued every five years, but do not account for individual differences in genetic traits that affect diet and health relationships. Knowledge about the interactions of gene types, food intake, and health is rapidly evolving and will be used in the future to personalize dietary recommendations.[28] Researchers continue to look for additional risk factors because existing ones do not fully account for heart disease development. Some people identified as being at low risk for heart disease due to low LDL cholesterol or high HDL cholesterol levels, for example, develop heart disease.[27]

Dietary Risk Factors Dietary patterns that regularly include the consumption of red and processed meats, sugars, and other sources of refined carbohydrates, and limit intake of colorful vegetables and fruits, whole grains, fish and poultry, nuts, and legumes promote chronic inflammation and heart disease development. Such dietary patterns tend to provide low amounts of antioxidants, fiber, and essential nutrients needed to moderate inflammatory processes.[2]

Several types of dietary patterns have been found to reduce heart disease risk in multiple, large studies. These consist of the Mediterranean Dietary Pattern, the DASH Dietary Pattern, the Healthy U.S. Dietary Pattern, and the Healthy Vegetarian Dietary Pattern. Each of these patterns is based on plant foods such as vegetables and fruits, unsaturated oils, whole grains, whole grain products, and other sources of complex carbohydrates. Healthy dietary patterns share the characteristics of being low in processed meats, deep-fried foods, snack foods, sodium, saturated fat, sugars, and refined grain products. They provide healthy amounts of antioxidants, fiber, vitamins and minerals, and protein.[2]

Carbohydrate Intake and Heart Disease Risk It is beginning to look like the rising rates of obesity and type 2 diabetes that have occurred in the United States over the past few decades were, to some extent, related to the popularity of low-fat diets.[2,3] It turns out that many people following a low-fat diet substituted foods high in **refined carbohydrates** and **added sugars** for high-fat foods.[3,5,31,35] Increased intake of foods rich in refined carbohydrates and added sugars were subsequently found to elevate blood levels of glucose and triglycerides, trigger chronic inflammation and insulin resistance, and promote weight gain in many people. Each of these factors appears to be related to an increased risk of heart disease as well as type 2 diabetes.[2,3,5,36] Low-fat diets are no longer recommended, and intake of whole grains, whole-grain products, legumes, and other fiber- and nutrient-rich sources of carbohydrates is encouraged.[2]

refined carbohydrates Carbohydrate-containing foods that have been processed to remove most or all of the bran coating and germ from the grain. Refined grains are lower in fiber, beneficial phytonutrients, and vitamins and minerals than are whole grains.

added sugars Sugars that are either added during the processing of foods, or are packaged as such; they include sugars (free, mono-, and disaccharides), syrups, naturally occurring sugars that are isolated from a whole food and concentrated, and other caloric sweeteners.

What about vitamin and mineral supplements? Are they recommended? Many people believe they can help protect arteries and the heart from damage due to inflammation by taking a multi-vitamin and mineral supplement. Get an answer to this question from the nearby Reality Check feature.

Healthy Dietary Patterns for Heart Health

- **Evaluate the components of your diet and lifestyle that help prevent the development of heart disease.**

Recommendations made in the current edition of the Dietary Guidelines for Americans are largely based on the types of food included in dietary patterns associated with a low risk of developing heart disease and other disorders. Healthy dietary patterns encourage consumption of a variety of vegetables and fruits such as broccoli, turnip and collard greens, tomatoes, papaya, berries, and grapes (Illustration 19.7). To support this dietary pattern, the Take Action feature in this unit sends you on a treasure hunt for your favorite vegetables and fruits on your next trip to your local farmer's market.

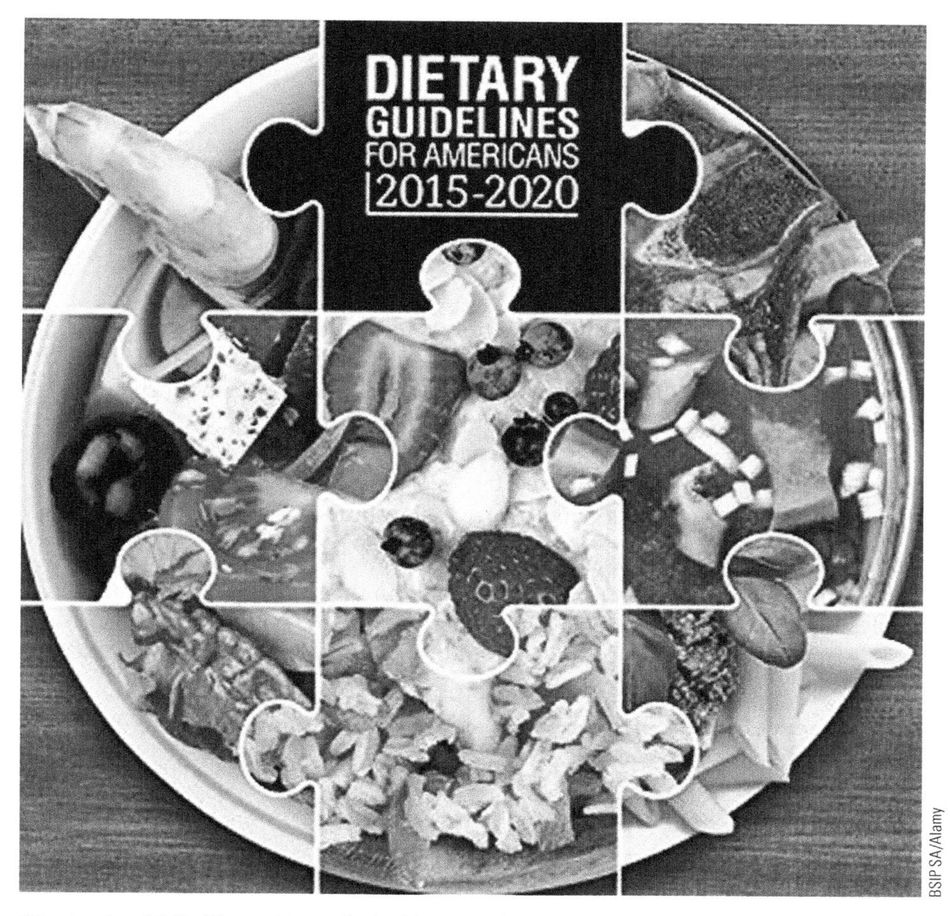

BSIP SA/Alamy

Illustration 19.7 The makings of a health-promoting dietary pattern.

REALITY CHECK

Are Supplements Heart-Healthy?

Can a daily multi-vitamin and mineral supplement reduce the risk of heart disease?

Who gets the thumbs up?

Answers on page 19-8.

Photodisc

Lupe: People used to think that vitamin and mineral supplements helped reduce the risk of heart disease, but that doesn't happen.

PhotoDisc

Sharon: A multivitamin and mineral supplement that contains vitamin C and E and selenium do help reduce the risk of heart disease. My Uncle Marty takes a vitamin pill every day because he already had a heart attack.

take action Farmer's Market Treasure Hunt

Many of the ingredients of a heart healthy dietary pattern are available at your local farmer's markets. Identify your favorite fresh vegetables and fruits from the list below, or add your own favorites to the list, and look for them first when you stroll through the farmers' offerings.

____ vine-ripe tomatoes	____ peaches	____ blueberries
____ squash	____ cucumbers	____ strawberries
____ apples	____ colorful peppers	____ raspberries
____ tangerines	____ melons	____ cherries
____ avocados	____ green beans	____ spinach

Dietary recommendations for the prevention of heart disease are summarized in Table 19.1. To help clarify two of the recommendations, Table 9.2 cites examples of foods considered to be high in refined carbohydrates and food products considered to be processed meats.

The important contribution of regular physical activity to heart disease risk reduction is also highlighted in the Dietary Guidelines. It is recommended that adults participate in 40 minutes of moderate to vigorous activity three to four times per week.[2]

Treatment of Heart Disease Statins, drugs that lower LDLc, are recommended for people with a risk of heart disease of 10% or more in the next 10 years. They are particularly effective in people with gene variants related to the body's utilization of cholesterol.[32] Statins have been used for years and, with the exception of an increased

Table 19.1 Dietary recommendations for the prevention of heart disease[2,26]

- Adhere to a healthy dietary pattern that emphasizes intake of vegetables, fruits, whole grains, fish, shellfish, legumes, olive and vegetable oils, nuts and seeds, poultry, and low-fat dairy products.
- Limit intake of foods and beverages high in refined carbohydrates and added sugars (including fructose).
- Keep sodium intake at or below 2,400 mg per day.
- Select food sources of polyunsaturated fats such as nuts, seeds, olive oil, and avocado.
- Limit intake of processed meats.
- Limit the intake of *trans* fats to as little as possible.
- Consume moderate amounts of alcohol-containing beverages, if any. (Moderate is defined as two drinks per day for men and one drink a day for women.)

ANSWERS TO REALITY CHECK

You got it, Lupe. Use of vitamin and mineral supplements do not decrease the risk of heart disease.[30]

PhotoDisc

Lupe:

PhotoDisc

Sharon:

Table 19.2 Food sources of refined carbohydrates, and foods considered "processed meats"

Food sources of:

Refined carbohydrates		
White bread	Flour tortillas	Maple syrup
White rice	Cakes, cookies	Jams, jellies
Pasta	Soda crackers	Agave syrup
Biscuits	Sucrose	Soft drinks
Pizza dough	Fructose	Fruit drinks
Breakfast cereals (cold)	Agave syrup	Some breakfast cereals

Food considered to be "processed meats"[a]			
Bologna	Hot dogs	Liver paté	Pastrami
Sausage	Pepperoni	Canned corned beef	Potted meat
Salami	Liverwurst	Fermented meats (e.g., Naem)	Bacon
Chorizo	Beef jerky	Ham loaf	Chicken nuggets

[a]Source: Categories of processed meats. www.fao.org/docrep/010/ai407e/AI407E09.htm, accessed 1/29/15.

risk of type 2 diabetes and side effects including liver injury and muscle weakness and stiffness, are considered to be safe and effective for most adults under the age of 65 years.[38,39] The effectiveness of statins for reducing the risk of heart attack and death among individuals aged 65 and older who are at risk of heart disease is not well established.[39]

Statins alone are not the answer to heart disease. Many factors are related to the development of heart disease, so approaches to reducing its impact on health must be multifaceted. It is expected that new information about optimal prevention and management strategies will result from the expansion of research related to the influence of individual genetic profiles on heart disease development. Once this is determined, personalized approaches for the prevention and management of heart disease can be development and implemented.

©JIL Photo/Shutterstock.com

NUTRITION up close

Evaluate Your Dietary and Lifestyle Strengths for Heart Disease Prevention

Focal Point: Assess the positive aspects of your diet and lifestyle on the risk of heart disease.

A number of dietary characteristics and lifestyle behaviors decrease heart disease risk. Give yourself credit by checking those characteristics you call your own.

____ I exercise at least 21 minutes daily.
____ I eat three or more servings of vegetables a day.
____ I eat a serving of fruit at least twice a day.
____ I don't smoke.
____ I am normal weight or on my way to becoming that way.
____ I eat more than two servings of whole grains or whole-grain foods daily.
____ I consume low- or no-fat milk when I drink milk.
____ I eat fish or seafood twice a week or more.
____ I eat nuts for snacks or otherwise consume them regularly.
____ I choose lean meats if I eat meat.
____ I generally drink water rather than sugar-sweetened beverages.
____ I check the *trans* fat content of packaged foods on Nutrition Facts panels.

Feedback to the Nutrition Up Close is located in Appendix F.

REVIEW QUESTIONS

- **List dietary, lifestyle, and biological factors that contribute to heart disease development.**

1. The term *cardiovascular disease* refers to disorders related to plaque buildup in arteries of the heart, brain, and other organs and tissues. **True/False**
2. High levels of LDL cholesterol increase the risk of heart disease just as low levels of HDL cholesterol do. **True/False**
3. High levels of total cholesterol in the blood always represent a major risk factor for heart disease. **True/False**
4. Risk factors for heart disease differ substantially between men and women. **True/False**
5. The major drawback to heart-healthy diets is that they help prevent heart disease only. **True/False**
6. Healthful diets include ample vegetables and fruits, whole-grain products, fish, and lean, unprocessed meats. **True/False**
7. A low-fat diet is the best diet to reduce the risk of heart disease. **True/False**
8. Physical inactivity is associated with increased risk of heart disease. **True/False**
9. Nutritional deprivation during pregnancy can modify the function of genes that influence blood lipid levels in offspring. **True/False**
10. ____ Food sources of processed meat include:
    a. chicken breast
    b. turkey thigh
    c. chicken nuggets
    d. lamb chops
11. ____ Foods sources of refined carbohydrates include:
    a. legumes and avocados
    b. white rice and pasta
    c. corn tortillas and oatmeal
    d. hot dogs and mayonnaise
12. ____ Healthy dietary patterns are characterized by the regular consumption of foods such as:
    a. olive oil and vinegar
    b. potatoes and flaxseeds
    c. whole-grain breakfast cereal and nuts
    d. deep-fried fish and seafood

- **Evaluate the components of your diet and lifestyle that help prevent the development of heart disease.**

13. _____ Which of the following dietary behaviors is related to a decreased risk of heart disease?
    a. eliminating saturated fats from the diet
    b. consuming beverages labeled as having "less sugar"
    c. selecting low-cholesterol foods
    d. consuming whole-grain breakfast cereals

14. Nuts are high in fats and should be consumed infrequently. **True False**

15. Following a healthy dietary pattern means you can never eat a hamburger or fried fish. **True False**

Answers to these questions can be found in Appendix F.

NUTRITION SCOREBOARD ANSWERS

1. About one in four deaths of men and women across the globe are related to heart disease.[1] **True**
2. **True**[2,3]
3. See Illustrations 19.2 and 19.3. **True**
4. **True**[4]

Vitamins and Your Health

NUTRITION SCOREBOARD

1. The only documented benefit of consuming sufficient amounts of vitamins is protection against deficiency diseases. **True/False**
2. Vitamins provide energy. **True/False**
3. Vitamin C is found only in citrus fruits. **True/False**
4. Nearly all cases of illness due to excessive intake of vitamins result from the overuse of vitamin supplements. **True/False**

Answers can be found at the end of the unit.

Fertnig/Getty Images

LEARNING OBJECTIVES

After completing Unit 20 you will be able to:

- Describe the functions and food sources of vitamins and their effects on health.

vitamins Components of food required by the body in small amounts for growth and health maintenance.

Vitamins: They're on Center Stage

- **Describe the functions and food sources of vitamins and their effects on health.**

These are exciting times for people interested in **vitamins**. The first vitamin (vitamin A) was identified at the University of Wisconsin in 1913, a year before World War I began and at the height of the women's suffrage movement. It came during a time when disease was thought to be caused by "germs" and the only essential components of food were proteins, carbohydrates, fats, and minerals. During the next 35 years, 13 additional vitamins would be discovered and utilized for the treatment and prevention of deficiency diseases. The discovery of vitamins established the fact that disease can be caused by vitamin deficiencies, which started a wave of scientific research on the health effects of vitamins that continues today.[2] This unit addresses the functions of vitamins in the body, their food sources, and health consequences related to the consumption of too little or too much of individual vitamins.

KEY NUTRITION CONCEPTS

A number of key nutrition concepts underlie the content presented on vitamins, including:

1. Nutrition Concept #2: Foods provide energy (calories), nutrients, and other substances needed for growth and health.
2. Nutrition Concept #4: Poor nutrition can result from both inadequate and excessive levels of nutrient intake.
3. Nutrition Concept #5: Humans have adaptive mechanisms for managing fluctuations in nutrient intake.

Vitamin Facts

Vitamins are chemical substances that perform specific functions in the body. They are essential nutrients because, in general, the body cannot produce them or produce sufficient amounts of them. If we fail to consume enough of any of the vitamins, specific deficiency diseases develop. Fourteen vitamins have been discovered so far, and they are listed in Table 20.1.

Table 20.1 Fourteen vitamins are known to be essential for health. Ten are water-soluble and four are fat-soluble.

The water-soluble vitamins	The fat-soluble vitamins
B-complex vitamins	Vitamin A (retinol) (provitamin is beta-carotene)
Thiamin (B_1)	Vitamin D (1,25-dihydroxy-cholecalciferol)
Riboflavin (B_2)	Vitamin E (tocopherol)
Niacin (B_3)	Vitamin K (phylloquinone, menaquinone)
Vitamin B_6 (pyridoxine)	
Folate (folacin, folic acid)	
Vitamin B_{12} (cyanocobalamin)	
Biotin	
Pantothenic acid (pantothenate)	
Choline[a]	
Vitamin C (ascorbic acid)	

[a]Choline is a vitamin-like essential nutrient.

Water- and Fat-Soluble Vitamins Vitamins come in two basic types—those soluble in water (the B-complex vitamins and vitamin C) and those that dissolve in fat (vitamins D, E, K, and A, or the *deka* vitamins). Their key features are summarized in Table 20.2. With the exception of vitamin B_{12}, the water-soluble vitamins can be stored in the body only in small amounts. Consequently, deficiency symptoms generally develop within a few weeks to several months after the diet becomes deficient in water-soluble vitamins. Vitamin B_{12} is unique in that the body can build up stores that last for a year or more after intake of the vitamin stops. Of the water-soluble vitamins, niacin, vitamin B_6, choline, and vitamin C are known to produce ill effects if consumed in excessive amounts.

The fat-soluble vitamins are primarily stored in body fat and the liver. Because the body is better able to store these vitamins, deficiencies of fat-soluble vitamins generally take longer to develop than deficiencies of water-soluble vitamins when intake from food is too low.

Bogus Vitamins Some substances are called "vitamins" even though they are not actually vitamins. A number of them are listed in Table 20.3. These bogus vitamins are called vitamins by some manufacturers of supplements, weight-loss products, and cosmetics. Although the label may help the product to sell, the ingredients aren't essential and therefore cannot be considered vitamins. People do not develop deficiency diseases when they consume too little of the bogus vitamins.

What Do Vitamins Do?

For starters, vitamins neither provide energy nor serve as components of body tissues such as muscle and bone. A number of vitamins do play critical roles as **coenzymes** in the conversion of proteins, carbohydrates, and fats into energy. Coenzymes are also involved in reactions that build and maintain body tissues such as bone, muscle, and red blood cells. Thiamin, for example, is needed for reactions that convert glucose into energy. People who are thiamin deficient tire easily and feel weak (among other things). Folate, another B-complex vitamin, is required for reactions that build body proteins. Without enough folate, proteins such as those found in red blood cells form abnormally and function poorly. Vitamin A is needed for reactions that generate new cells to replace worn-out cells lining the mouth, esophagus, intestines, and eyes. Without enough vitamin A, old cells aren't replaced, and the affected tissues are damaged. Vitamin C is required for reactions that build and maintain collagen, a protein found in skin, bones, blood vessels, gums, ligaments, and cartilage. Approximately 30% of the total amount of protein in the body is collagen. With vitamin C deficiency, collagen becomes weak, causing tissues that contain collagen to weaken and bleed easily.

These examples all relate to the physical effects of vitamins. Vitamins participate in reactions that affect behavior, too. Alterations in behaviors such as reduced attention span, poor appetite, irritability, depression, or paranoia often precede the physical signs of vitamin deficiency.[2] Vitamins are truly "vital" for health.

Gene Variants and Vitamin Functions and Status Specific functions of individual vitamins are highlighted in this unit. However, it should be noted that the requirement, functions, and effects of individual vitamins differ to some extent based on the presence of **gene variants**. Some people with low serum levels of vitamin D or folate, for example, do not respond to supplementation due to gene variants that modify the body's utilization and status of vitamin D or folate.[6,7] Other gene variants interfere with the body's ability to utilize vitamin B_{12} in ways that increase the risk of developing B_{12} deficiency.[8]

The Antioxidant Vitamins Beta-carotene (a **precursor** to vitamin A), vitamin E, and vitamin C function as **antioxidants**. This means they prevent or repair damage to components of cells caused by exposure to **free radicals**. A free radical results primarily when an atom of oxygen loses an electron. Without the electron, there is an imbalance between the atom's positive and negative charges. This makes the atom reactive—it needs to steal an electron from a nearby atom or molecule to reestablish a balance between its positive and negative charges. Free radicals play a number of roles in the body, so they are always present. They are produced during energy formation, by breathing, and by the immune

coenzymes Chemical substances, including many vitamins, that activate specific enzymes. Activated enzymes increase the rate at which reactions take place in the body, such as the breakdown of fats or carbohydrates in the small intestine and the conversion of glucose and fatty acids into energy within cells.

gene variant An alteration in the normal nucleotide sequence of a gene that affect the gene's functions. Different forms of the same genes are considered "alleles."

precursor In nutrition, a nutrient that can be converted into another nutrient (also called provitamin). Beta-carotene is a precursor of vitamin A.

antioxidants Chemical substances that prevent or repair damage to cells caused by exposure to free radicals. Beta-carotene, vitamin E, and vitamin C function as antioxidants.

free radicals Chemical substances (usually oxygen) that are missing an electron. The absence of the electron makes the chemical substances reactive and prone to oxidizing nearby atoms or molecules by stealing an electron from them.

Table 20.2 An intensive course on vitamins

	The water-soluble vitamins	
	Primary functions	**Consequences of deficiency**
Thiamin (vitamin B_1) AI[a] women: 1.1 mg men: 1.2 mg	• Helps body release energy from carbohydrates ingested • Facilitates growth and maintenance of nerve and muscle tissues • Promotes normal appetite	• Fatigue, weakness • Nerve disorders, mental confusion, apathy • Impaired growth • Swelling • Heart irregularity and failure
Riboflavin (vitamin B_2) AI women: 1.1 mg men: 1.3 mg	• Helps body capture and use energy released from carbohydrates, proteins, and fats • Aids in cell division • Promotes growth and tissue repair • Promotes normal vision	Biophoto Associates/Science Source • Reddened lips, cracks at both corners of the mouth • Fatigue
Niacin (vitamin B_3) RDA women: 14 mg men: 16 mg UL: 35 mg (from supplements and fortified foods)	• Helps body capture and use energy released from carbohydrates, proteins, and fats • Assists in the manufacture of body fats • Helps maintain normal nervous system functions	Dr. M.A. Ansary/Science Source Pellagra: the niacin-deficiency disease. • Skin disorders • Nervous and mental disorders • Diarrhea, indigestion • Fatigue
Vitamin B_6 (pyridoxine) AI women: 1.3 mg men: 1.3 mg UL: 100 mg	• Needed for reactions that build proteins and protein tissues • Assists in the conversion of tryptophan to niacin • Needed for normal red blood cell formation • Promotes normal functioning of the nervous system	• Irritability, depression • Convulsions, twitching • Muscular weakness • Dermatitis near the eyes • Anemia • Kidney stones

[a] (Adequate Intakes) and RDAs (Recommended Dietary Allowances) are for 19–30-year-olds; UL (Upper Limits) are for 19–70-year-olds.

The water-soluble vitamins		
Consequences of overdose	**Primary food sources**	**Highlights and comments**
• High intakes of thiamin are rapidly excreted by the kidneys. Oral doses of 500 mg/day or less are considered safe.	• Grains and grain products (cereals, rice, pasta, bread) • Pork • Nuts	• Need increases with carbohydrate intake • There is no "e" on the end of thiamin! • Deficiency rare in the United States; may occur in people with alcoholism • Enriched grains and cereals prevent thiamin deficiency
• None known. High doses are rapidly excreted by the kidneys.	• Milk, yogurt, cheese • Grains and grain products (cereals, rice, pasta, bread) • Liver, fish, beef • Eggs	• Destroyed by exposure to light (that's why milk comes in opaque containers)
• Flushing, headache, cramps, rapid heartbeat, nausea, diarrhea, decreased liver function with doses above 0.5 g per day	• Meats (all types), fish • Grains and grain products (cereals, rice, pasta, bread) • Nuts	• Niacin has a precursor: tryptophan. Tryptophan, an amino acid, is converted to niacin by the body. Much of our niacin intake comes from tryptophan. • High doses raise HDL cholesterol levels, lower LDL cholesterol and triglycerides.
• Bone pain, loss of feeling in fingers and toes, muscular weakness, numbness, loss of balance (mimicking multiple sclerosis)	• Meats (all types) • Breakfast cereals • Bananas, avocados • Potatoes, brussels sprouts, sweet peppers	• Vitamins go from B_3 to B_6 because B_4 and B_5 were found to be duplicates of vitamins already identified. • Requirement affected by gene variants.

(*continued*)

Table 20.2 **(continued)**

	The water-soluble vitamins	
	Primary functions	**Consequences of deficiency**
Folate (folacin, folic acid) RDA women: 400 mcg men: 400 mcg UL: 1,000 mcg (from supplements and fortified foods)	• Needed for reactions that utilize amino acids (the building blocks of protein) for protein tissue formation • Promotes the normal formation of red blood cells	• Megaloblastic anemia • Diarrhea • Red, sore tongue 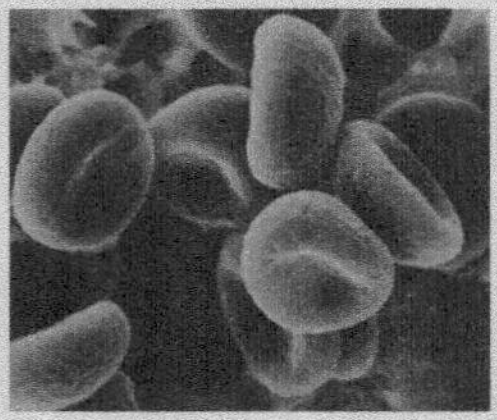 Normal red blood cells. © Dennis Kunkel Microscopy, Inc./Phototake • Increased rise of neural tube defects and other malformations, preterm delivery • Elevated blood levels of homocysteine 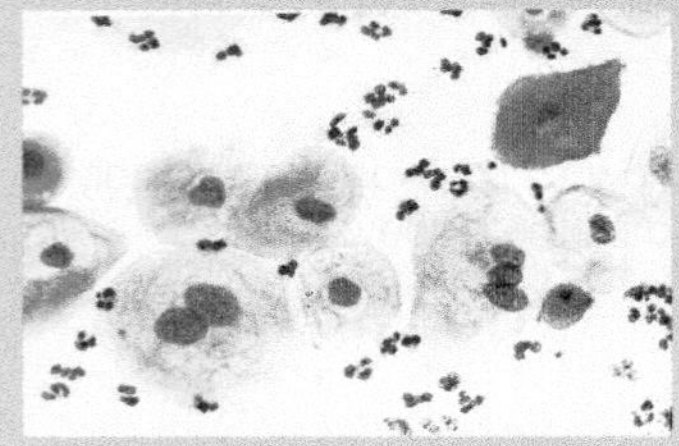 Red blood cells in megaloblastic anemia. Science Source
Vitamin B_{12} (cyanocobalamin) AI women: 2.4 mcg men: 2.4 mcg	• Helps maintain nerve tissues • Aids in reactions that build up protein and bone tissues • Needed for normal red blood cell development	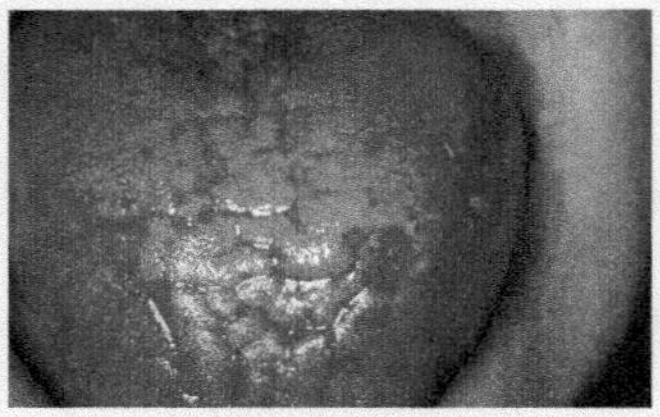 A look at beriberi, a thiamin-deficiency disease. Biophoto Associates/Science Source • Neurological disorders (nervousness, tingling sensations and numbness in fingers and toes, brain degeneration) • Pernicious anemia characterized by large, oval-shaped red blood cells • Fatigue • Sore, beefy red, smooth tongue • Requirement affected by gene variants.
Biotin AI women: 30 mcg men: 30 mcg	• Needed for the body's manufacture of fats, proteins, and glycogen	• Seizures • Vision problems • Muscular weakness • Hearing loss
Pantothenic acid (pantothenate) AI women: 5 mg men: 5 mg	• Needed for release of energy from fat and carbohydrates	• Fatigue, sleep disturbances, impaired coordination, vomiting, nausea
Vitamin C (ascorbic acid) RDA women: 75 mg men: 90 mg UL: 2,000 mg	• Needed for the manufacture of collagen • Helps the body fight infections, repair wounds • Acts as an antioxidant • Enhances iron absorption	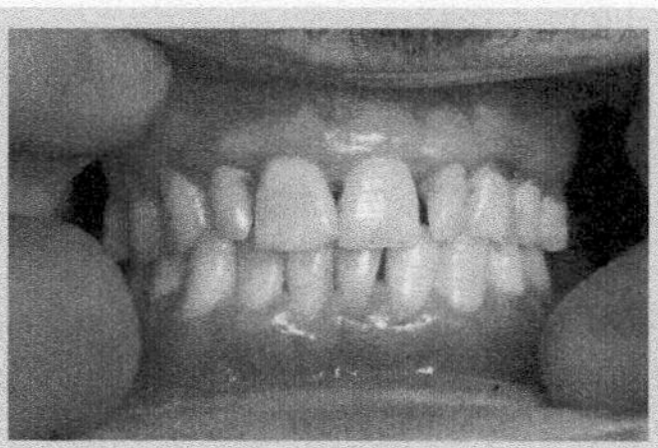 Gums that are swollen and bleed easily are signs of scurvy, the vitamin C–deficiency disease. Biophoto Associates/Science Source • Bleeding and bruising easily due to weakened blood vessels, cartilage, and other tissues containing collagen • Slow recovery from infections and poor wound healing • Fatigue, depression
Choline[a] AI women: 425 mg men: 550 mg UL: 3.5 g	• Serves as a structural and signaling component of cell membranes • Required for normal development of fetal spinal cord and brain, as well as attention and memory processes. • Serves as a source of methyl groups needed for many metabolic reactions.	• Fatty liver • Infertility • Hypertension

[a]Choline is a vitamin-like essential nutrient.

The water-soluble vitamins		
Consequences of overdose	**Primary food sources**	**Highlights and comments**
• May mask signs of vitamin B_{12} deficiency (pernicious anemia)	• Fortified, refined grain products (cereals, bread, pasta) • Dark green vegetables (spinach, collards, romaine) • Dried beans	• *Folate* means "foliage." It was first discovered in leafy green vegetables. • This vitamin is easily destroyed by heat. • Synthetic form (folic acid) added to fortified grain products is better absorbed than naturally occurring folates. • Some people have genetic traits that increase their need for folates. • May protect against colon cancer in early stages of development. • Fortification of whole grain flours with folic acid is being recommended. • Requirement affected by gene variants.
• None known. Excess vitamin B_{12} is rapidly excreted by the kidneys or is not absorbed into the bloodstream. • Vitamin B_{12} injections may cause a temporary feeling of heightened energy.	• Fish, seafood • Meat • Milk and cheese • Ready-to-eat cereals	• Older people, those who have had stomach surgery, and vegans are at risk for vitamin B_{12} deficiency. • Some people become vitamin B_{12} deficient because they are unable to absorb it. • Vitamin B_{12} is found in animal products and microorganisms only. • Requirement affected by gene variants.
• None known. Excesses are rapidly excreted.	• Grain and cereal products • Meats, dried beans, cooked eggs • Vegetables	• Deficiency is extremely rare. May be induced by the overconsumption of raw eggs. • Supplemental biotin is used in the treatment of multiple sclerosis.
• None known. Excesses are rapidly excreted.	• Many foods, including meats, grains, vegetables, fruits, and milk	• Deficiency is very rare.
• High intakes of 1 g or more per day can cause nausea, cramps, and diarrhea and may increase the risk of kidney stones.	• Fruits: guava, oranges, lemons, limes, strawberries, cantaloupe, grapefruit, kiwi fruit • Vegetables: broccoli, green and red peppers, collards, tomato, potatoes • Ready-to-eat cereals	• Need increases among smokers (to 110–125 mg per day). • Is fragile; easily destroyed by heat and exposure to air. • Supplements may decrease the severity of colds in some people. • Deficiency may develop within three weeks of very low intake.
• Low blood pressure • Sweating, diarrhea • Fishy body odor • Liver damage	• Meat (all types) • Eggs • Dried beans • Milk	• Most of the choline we consume from foods comes from its location in animal cell membranes. • Lecithin, an additive commonly found in processed foods, is a rich source of choline. • Choline is produced by the liver but in inadequate amounts to meet need • It is considered a B-complex vitamin. • Requirement for choline increases during pregnancy and lactation.

(*continued*)

Table 20.2 (continued)

	The fat-soluble vitamins		
	Primary functions	**Consequences of deficiency**	
Vitamin A **1. Retinol** RDA women: 700 mcg men: 900 mcg UL: 3,000 mcg	• Needed for the formation and maintenance of mucous membranes, skin, bone • Needed for vision in dim light	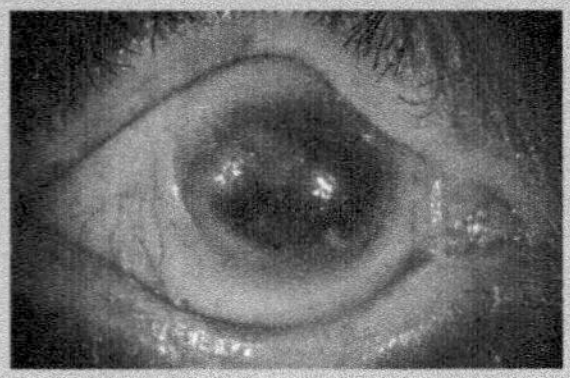 ISM/Phototake Xerophthalmia. Vitamin A deficiency is the leading cause of blindness in developing countries.	• Increased incidence and severity of infectious diseases • Impaired vision, blindness • Inability to see in dim light • Blindness
2. Beta-carotene (a vitamin A precursor or "provitamin") No RDA; suggested intake: 6 mg	• Acts as an antioxidant; prevents damage to cell membranes and the contents of cells by repairing damage caused by free radicals		• Deficiency disease related only to lack of vitamin A
Vitamin E (alpha-tocopherol) RDA women: 15 mg men: 15 mg UL: 1,000 mg	• Acts as an antioxidant, prevents damage to cell membranes in blood cells, lungs, and other tissues by repairing damage caused by free radicals • Participates in the regulation of gene expression		• Muscle loss, nerve damage • Anemia • Weakness
Vitamin D (vitamin D_2 = ergocalciferol, vitamin D_3 = cholecalciferol) RDA women: 15 mcg (600 IU) men: 15 mcg (600 IU) UL: 100 mcg (4,000 IU)	• Needed for the absorption of calcium and phosphorus, and for their utilization in bone formation, nerve and muscle activity. • Inhibits inflammation • Involved in insulin secretion and blood glucose level maintenance	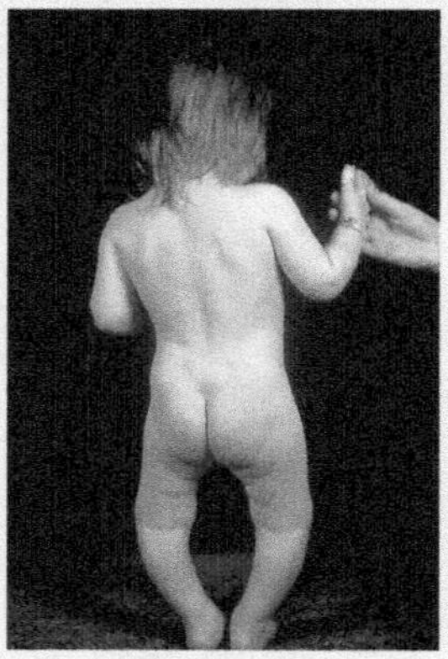 Biophoto Associates/Science Source The vitamin D–deficiency disease: rickets	• Weak, deformed bones (children) • Loss of calcium from bones (adults), osteoporosis • Increased risk of chronic inflammation • Increased risk of death from all causes

The fat-soluble vitamins		
Consequences of overdose	**Primary food sources**	**Highlights and comments**
• Vitamin A toxicity (hypervitaminosis A) with acute doses of 500,000 IU, or long-term intake of 50,000 IU per day. • Nausea, irritability, blurred vision, weakness, headache • Increased pressure in the skull, hip fracture • Liver damage • Hair loss, dry skin • Birth defects	• Vitamin A is found in animal products only. • Liver, clams, low-fat milk, eggs • Ready-to-eat cereals	• Symptoms of vitamin A toxicity may mimic those of brain tumors and liver disease. Vitamin A toxicity is sometimes misdiagnosed because of the similarities in symptoms. • 1 mcg vitamin A = 3.33 IU • Forms of retinol are used in the treatment of acne and skin wrinkles due to overexposure to the sun. American Journal of Clinical Nutrition, Vol. 71, No 4, 878-884, April 2000, Robert M. Russell Brittle hair and dry, rough, scaly, and cracked skin from vitamin A overdose.
• High intakes from supplements may increase lung damage in smokers. • With high intakes and supplemental doses (over 12 mg/day for months), skin may turn yellow-orange.	• Deep orange and dark green vegetables. • Carrots, sweet potatoes, pumpkin, spinach, collards, cantaloupe, apricots, vegetable juice	• The body converts beta-carotene to vitamin A. Other carotenes are also present in food, and some are converted to vitamin A. Beta-carotene and vitamin A perform different roles in the body. • High intakes decrease sunburn.
• Intakes of up to 800 IU per day are unrelated to toxic side effects; over 800 IU per day may increase bleeding (blood-clotting time). • Avoid supplement use if aspirin, anticoagulants, or fish oil supplements are taken regularly.	• Nuts and seeds • Vegetable oils • Salad dressings, mayonnaise • Whole grains, wheat germ • Leafy, green vegetables, asparagus	• Vitamin E is destroyed by exposure to oxygen and heat. • Oils naturally contain vitamin E. It's there to protect the fat from breakdown due to free radicals. • Eight forms of vitamin E exist, and each has different antioxidant strength. • 1 mg vitamin E = 1.49 IU • Intakes in the United States tend to be low.
• Mental retardation in young children • Abnormal bone growth and formation • Nausea, diarrhea, irritability, weight loss • Deposition of calcium in organs such as the kidneys, liver, and heart • Toxicity possible with long-term use of 10,000 IU daily	• Vitamin D–fortified milk, cereals, and other foods • Fish and shellfish	• Vitamin D_3, the most active form of the vitamin, is manufactured from a form of cholesterol in skin cells upon exposure to ultraviolet rays from the sun. • Most of our supply comes from sun exposure on the skin. • Inadequate vitamin D status is common. • Breast-fed infants with little sun exposure benefit from vitamin D supplements. • 1 mcg vitamin D = 40 IU. • Requirement affected by gene variants. Some people with low serum levels of vitamin D have adequate vitamin D status. • Most of our supply comes from sun exposure on the skin.

(continued)

Table 20.2 (continued)

	The fat-soluble vitamins	
	Primary functions	**Consequences of deficiency**
Vitamin K (phylloquinone, menaquinone) AI women: 90 mcg men: 120 mcg	• Is an essential component of mechanisms that cause blood to clot when bleeding occurs • Aids in the incorporation of calcium into bones	Science Source The long-term use of antibiotics can cause vitamin K deficiency. People with vitamin K deficiency bruise easily. • Bleeding, bruises • Decreased calcium in bones • Deficiency is rare. May be induced by the long-term use (months or more) of antibiotics.

Table 20.3 Nonvitamins
The "real" vitamins are listed in Table 20.1. These are some of the more popular nonvitamins.

Bioflavonoids (vitamin P)
Coenzyme Q_{10}
Gerovital H-3
Hesperidin
Inositol
Laetrile (vitamin B_{17})
Lecithin
Lipoic acid
Nucleic acids
Pangamic acid (vitamin B_{15})
Para-amino benzoic acid (PABA)
Provitamin B_5 complex
Rutin

Scott Goodwin Photography

Illustration 20.1 Apple slices exposed to air turn brown due to oxidation. Coating the slices with lemon juice, a rich source of vitamin C, reduces the oxidation.

system to help destroy bacteria and viruses that enter the body. They can also be formed when the body is exposed to alcohol, radiation emitted by the sun, smoke, ozone, smog, and other environmental pollutants.

Atoms and molecules that have lost electrons to free radicals are said to be *oxidized.* Oxidized substances can damage lipids, cell membranes, DNA, and other cell components. Antioxidants such as beta-carotene, vitamin E, and vitamin C donate electrons to stabilize oxidized molecules or repair them in other ways.[3] Illustration 20.1 shows visible effects of an oxidation and an antioxidation reaction.

Consumption of foods rich in beta-carotene, vitamin C, and vitamin E decrease the risk of heart disease, stroke, and cancer. Intake of these antioxidant vitamins in supplements, however, does not have the same protective effects.[1,3–5] Fruits, vegetables, whole grains, and other plant foods contain a variety of naturally occurring antioxidants that work together with the antioxidant vitamins and other nutrients in disease prevention.[4]

Controlling oxidation reactions is a very important process that does not rely solely on our intake of antioxidant vitamins. The body's sources of antioxidants also include colorful pigments in vegetables and fruits, and enzymes produced by the body that function as antioxidants.[3]

Vitamins and the Prevention and Treatment of Disorders

Current research on vitamins centers around their effects on disease prevention and treatment. It is a very active area of research and, no doubt, new and important advances in knowledge about the vitamins will be gained. Here are a few examples of developments in research on vitamins and disease prevention and treatment.

Folate and Neural Tube Defects Folate plays a key role during pregnancy in the synthesis of proteins needed for the normal development of fetal tissues, including the spinal cord and brain. When the folate status of women early in pregnancy is poor, the neural tube (which develops into the spinal cord and brain) may form abnormally and incompletely. Illustration 20.2 shows a photograph of spina bifida, an example of a potential outcome of inadequate folate status early in pregnancy. Daily consumption of 400 micrograms (mcg) of folic acid (the synthetic form of folate added to refined grain products) before and early in pregnancy significantly reduces the incidence of neural tube defects.[8]

In 1998, manufacturers started fortifying refined grain products such as bread, pasta, and rice with folic acid. The addition of folic acid to these foods has produced major gains

The fat-soluble vitamins		
Consequences of overdose	**Primary food sources**	**Highlights and comments**
• Toxicity is only a problem when synthetic forms of vitamin K are taken in excessive amounts. That may cause liver disease.	• Leafy green vegetables • Grain products • Dairy products (non low-fat)	• Vitamin K is produced by bacteria in the gut. About half of our vitamin K supply comes from these bacteria. • Newborns are given vitamin K because they have "sterile" guts and consequently no vitamin K–producing bacteria. • Deficiency is rare.

in people's folate status, and the prevalence of poor folate status and of neural tube defects has been reduced substantially.[9–11]

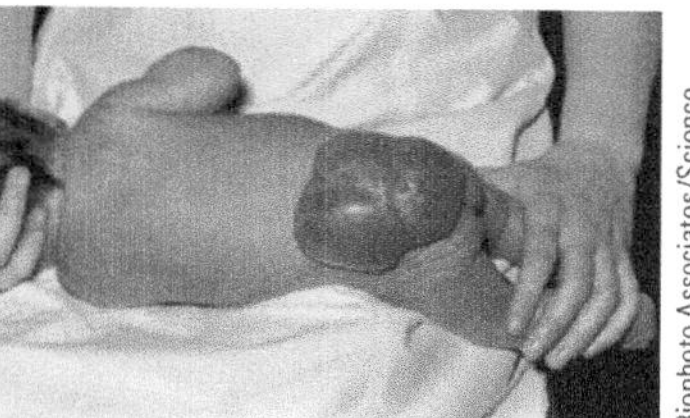

Biophoto Associates/Science Source

Illustration 20.2 A baby with spina bifida, a form of neural tube defect associated with poor folate status early in pregnancy.

Vitamin A: From Infectious Diseases to Acne and Wrinkles Studies undertaken both in developing countries and in the United States indicate that adequate vitamin A status helps prevent and decreases the severity of measles and other infectious diseases. It has been known for decades that adequate vitamin A intake also prevents blindness, an all-too-common consequence of vitamin A deficiency in developing nations.[12] Vitamin A is needed for the synthesis of substances that keep the outer layer of the eye moist and resistant to infection. Without sufficient vitamin A, eyes dry out, become susceptible to infection, and cloud over.

Forms of vitamin A are used in the treatment of acne and of skin wrinkles and blotches due to overexposure to the sun.[13,14] Very high intakes of vitamin A as retinol early in pregnancy (but not high intakes of the vitamin A precursor beta-carotene) are related to the development of specific birth defects. Women who are or may become pregnant should not use vitamin A–derived medications.[15]

chronic inflammation Low-grade inflammation that lasts weeks, months, or years. Inflammation is the first response of the body's immune system to infection or irritation. Inflammation triggers the release of biologically active substances that promote oxidation and other potentially harmful reactions in the body.

Vitamin D: From Osteoporosis to Chronic Inflammation Vitamin D is best known as the sunshine vitamin; it helps build strong bones and prevent osteoporosis by facilitating the absorption and utilization of calcium. Vitamin D does much more than that, however. It plays key roles as a hormone in combating **chronic inflammation**. Low-grade, chronic inflammation is at the core of the development of disorders such as type 2 diabetes, cardiovascular disease, multiple sclerosis, certain cancers, and rheumatoid arthritis.[16] Vitamin D reduces inflammation by entering cells and turning genes that produce inflammatory substances "off" and those that produce substances that reduce inflammation "on."[17] It also functions in the regulation of insulin secretion and blood glucose level.[18]

Recommended Intake of Vitamin D Based on research results related to optimal doses of vitamin D for bone formation and maintenance, the recommended intake of vitamin D was increased in 2011 from 5 mcg (200 IU) to 15 mcg (600 IU) daily for adult women and men.[19] The recommended levels of intake of vitamin D reflect amounts needed from the diet and do not include vitamin D manufactured in the skin from exposure to the ultraviolet rays from the sun (Illustration 20.3). Most people do not consume the recommended amount of vitamin D in their diet, even if fortified foods are consumed, but get the majority of their vitamin D

Dean Conger/Getty Images

Illustration 20.3 Russian children are exposed to a quartz UV lamp to prevent vitamin D deficiency during the long winter.

take action To Improve Your Vitamin D Status

How do you become a person who stands out from the average in terms of vitamin D intake? If you're not that person already, here are some tips that will get you on your way.

Choose and check at least two of the following options for getting more vitamin D that appeal to you.

I would:

_____ substitute a cup of skim milk for a sweetened beverage at one meal or snack a day.

_____ eat salmon once a week at dinner.

_____ select a vitamin D–fortified orange juice when I buy orange juice.

_____ buy or select and consume vitamin D–fortified breakfast cereals.

_____ consume swordfish, salmon, or trout several times a week.

_____ Enjoy exposing your arms and legs to direct sunshine for 10 to 15 minutes about three times a week in warm weather

Illustration 20.4 More foods are now fortified with Vitamin D than in the past. Check the labels on these foods for vitamin D fortification levels.

from exposure of their skin to direct sunlight.[1] Vitamin D is produced in the skin when the energy from ultraviolet rays from the sun is absorbed in skin cells. The energy absorbed initiates the conversion of a derivative of cholesterol in cells to an active form of vitamin D.[20]

Meeting the Need for Vitamin D through Foods and the Sun With the exception of fish, few foods are naturally rich in vitamin D. Consequently, most of the vitamin D in our diets comes from vitamin D–fortified foods, such as those shown in Illustration 20.4. The availability of food products fortified with vitamin D is steadily increasing, and the increased availability of these foods will help increase vitamin D intake.[21] Because most types of yogurt, cheese, cottage cheese, ice cream, and dairy products other than milk are not fortified with vitamin D, it's important to look at product labels. Vitamin D–fortified foods can be identified from the nutrition information label on food packages. A good source of vitamin D would provide 10% or more of the % Daily Value in a serving. There are several ways individuals can increase their intakes of vitamin D and production of vitamin D in the skin. The Take Action feature in this unit provides a list of options for doing that.

The amount of vitamin D produced in the skin from exposure to direct sunlight varies depending on the lightness or darkness of skin color and the intensity of the sun's ultraviolet rays that reach the skin.[22] Individuals with darker skin produce less vitamin D given the same circumstances than people with lighter skin because ultraviolet rays are less able to penetrate darker skin. Production of vitamin D in the skin is very low to zero in parts of the world during winter when sunlight is indirect, and higher in warmer parts of the world closer to the equator. Used properly, broad-spectrum sunscreen lotions with SPF factors of 15 and higher successfully block vitamin D production by absorbing the energy from the ultraviolet rays. Energy from ultraviolet rays does not pass through glass, windows, or plastic.[20,23]

For most people, exposing the arms and legs to direct sunlight for about 15 minutes between 10:00 a.m. and 5:00 p.m. daily is recommended as a sensible approach to getting enough vitamin D from the sun.[23] Maximum production of vitamin D in the skin is achieved before changes in skin color occur. You cannot get too much vitamin D from the sun. Production of the vitamin stops when adequate amounts have been produced for use and for storage in body fat.[24] Vitamin D–rich foods, and vitamin D supplements if needed, are recommended for individuals whose skin is sensitive to even short durations of sun exposure, and for individuals who, for other reasons, do not expose their skin to direct sunlight.[19]

Vitamin C and the Common Cold Do vitamin C supplements prevent colds or reduce the severity of cold symptoms? A review of 72 research studies on this topic concluded that vitamin C supplements of 200 mg per day or more reduce the incidence of colds by only 3%. Vitamin C supplements were found to decrease the severity of cold symptoms by 50% in athletes undergoing high levels of physical stress in very cold weather. For others, vitamin C supplement use was related to a decrease in the severity of cold symptoms in 8% of adults and 14% of children.[25]

Preserving the Vitamin Content of Foods

As demonstrated in Table 20.4, the vitamin content of foods can be affected by food preparation and cooking methods. (See this unit's Reality Check for an example of when this does not happen.) For example, food preparation methods that involve heat and

Table 20.4 Percent of original vitamin content lost in fruits by food storage method and in dried beans by cooking method

	Fruits			Dried beans boiled 2–2.5 hours	
	Canned (%)	Frozen (%)	Dried (%)	Water drained (%)	Water used (%)
Vitamin C	50%	30%	80%	35%	30%
Thiamin	15	10	10	60	55
Riboflavin	5	5	5	25	20
Niacin	10	5	5	45	40
Vitamin B_6	10	10	5	50	45
Folate	35	25	15	70	65
Choline	0	0	0	0	0
Vitamin B_{12}	0	0	0	0	0
Vitamin A	5	5	10	15	10

Source: Table prepared by author from data presented in USDA's nutrient retention in foods tables.

REALITY CHECK

To Peel or Not to Peel?

Jolene is peeling potatoes for dinner when she gets a tap on her shoulder from her mother. "Thank you for helping with dinner, sweetie, but quit peeling the potatoes! That's where the vitamins are!"

Who gets the thumbs-up?

Answer appears on page 20-14.

Gabriela Medina/Blend Images/Getty Images

Jolene: "Are you sure, mom? The peel is just fiber."

iStockphoto.com/MichaelDeLeon

Mom: "Of course I'm sure, honey."

ANSWERS TO REALITY CHECK

To Peel or Not to Peel?

Keeping the peel on potatoes does not appear to make a meaningful difference in the vitamin content of potatoes. On an equal-weight basis (5.4 ounces each), the vitamin content of peeled and boiled versus non-peeled and boiled potatoes is equivalent for thiamin, riboflavin, niacin, vitamin B_6, and folate. The content of vitamin C is approximately 60% higher in the non-peeled and boiled potatoes (18 mg) versus the peeled and boiled potatoes (11 mg). This difference is likely due to vitamin C loss in water in the peeled and boiled potatoes. There is little difference in the fiber content of the potatoes.

Gabriela Medina/Blend Images/Getty Images

Jolene:

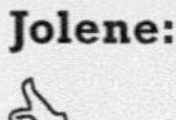

iStockphoto.com/MichaelDeLeon

Mom:

oxygen exposure lead to greater losses of vitamins such as vitamin C and folate and little or no loss of vitamins A and E, which are much less sensitive to heat and oxygen.[27] Vitamins in foods can be released into cooking water to some extent and lost down the drain if the water is thrown out. In general, boiling or steaming foods using a small amount of water, using the cooking water in soups, stews, or sauces, or stir-frying lead to superior vitamin retention. Cooking and softening vegetables like tomatoes and carrots increases the availability of beta-carotene and several beneficial phytochemicals.[27]

Vitamins: Getting Enough Without Getting Too Much

Table 20.5 lists good food sources of vitamins. Adequate amounts of vitamins can be obtained from diets that include a variety of basic foods including whole-grain products, low-fat dairy products, vegetables, fruits, fish, and other foods recommended in MyPlate.gov food guidance materials. Most fruits and vegetables are good sources of vitamins, and eating five or more servings a day is one way for individuals to get an assortment of vitamins. Fortified foods such as ready-to-eat cereals, fruit juices, and dairy products contribute to adequate intakes of vitamins.[1]

Recommended Intake Levels of Vitamins Updated recommendations for vitamin intakes associated with the prevention of deficiency and chronic diseases are represented by standards called Dietary Reference Intakes (DRIs). DRIs include the Recommended Dietary Allowances (RDAs) for vitamins, which convincing scientific data establishes for intake standards. Adequate Intakes (AIs) are assigned to vitamins for which scientific information about levels of intake associated with chronic disease prevention is less convincing. Tolerable Upper Levels of Intake (ULs) are also assigned to vitamins to indicate levels of vitamin intake from foods, fortified foods, and supplements that should *not* be exceeded. The RDAs or AIs and the ULs for the vitamins are given in Table 20.2.

Although people can get all the vitamins they need from supplements, it makes more sense to get them from foods. Foods offer fiber, minerals, beneficial phytochemicals, and other nutrients that don't come in supplements. Nutrients in foods interact to produce greater positive effects on health maintenance than do vitamin supplements in general.[28]

Table 20.5 **Food sources of vitamins**

Thiamin Food		Serving Size	Thiamin (mg)
Meats:	© Nordling/Shutterstock.com		
Ham		3 oz	0.6
Pork		3 oz	0.5
Beef		3 oz	0.4
Liver		3 oz	0.2
Nuts and seeds:	© Isak55/Shutterstock.com		
Pistachios		1/4 cup	0.3
Macadamia nuts		1/4 cup	0.2
Peanuts, dry roasted		1/4 cup	0.2
Grains:	© Svetlana Lukienko/Shutterstock.com		
Breakfast cereals		1 cup	0.3–1.4
Flour tortilla		1	0.2
Macaroni		1/2 cup	0.2
Rice		1/2 cup	0.2
Bread		1 slice	0.1
Vegetables:			
Peas		1/2 cup	0.2
Lima beans		1/2 cup	0.2
Corn		1/2 cup	0.2
Fruits:			
Orange juice		1 cup	0.2
Orange		1	0.1
Avocado		1/2	0.1
Riboflavin Food		**Serving size**	**Riboflavin (mg)**
Milk and milk products:	© Elena Elisseeva/Shutterstock.com		
Milk		1 cup	0.5
2% milk		1 cup	0.5
Yogurt, low-fat		1 cup	0.5
Skim milk		1 cup	0.4
Yogurt		1 cup	0.4
American cheese		1 oz	0.1
Cheddar cheese		1 oz	0.1
Meats:			
Liver		3 oz	3.6
Pork chop		3 oz	0.3
Beef		3 oz	0.2
Tuna		3 oz	0.1
Vegetables:			
Collard greens		1/2 cup	0.3
Spinach, cooked		1/2 cup	0.2
Broccoli		1/2 cup	0.1
Eggs:			
Egg		1	0.2

(continued)

Table 20.5 **(continued)**

Riboflavin Food		Serving size	Riboflavin (mg)
Grains:			
Breakfast cereals		1 cup	0.1–1.7
Macaroni		½ cup	0.1
Bread		1 slice	0.1

© Svetlana Lukienko/Shutterstock.com

Niacin Food		Serving size	Niacin (mg)
Meats:			
Liver		3 oz	14.0
Tuna		3 oz	7.0
Turkey		3 oz	4.0
Chicken		3 oz	11.0
Salmon		3 oz	6.9
Veal		3 oz	6.4
Beef (round steak)		3 oz	4.0
Pork		3 oz	4.0
Haddock		3 oz	3.9
Shrimp		3 oz	2.2
Nuts and seeds:			
Peanuts, dry roasted		½ cup	4.9
Almonds		½ cup	1.3
Vegetables:			
Asparagus		½ cup	1.2
Corn		½ cup	1.2
Green beans		½ cup	1.2
Grains:			
Breakfast cereals		1 cup	5.0–20.0
Brown rice		½ cup	1.5
Noodles, enriched		½ cup	1.0
Rice, white, enriched		½ cup	1.2
Bread, enriched		1 slice	1.1

© Gts/Shutterstock.com

Vitamin B_6 Food		Serving size	Vitamin B_6 (mg)
Meats:			
Liver		3 oz	0.8
Fish		3 oz	0.3–0.6
Chicken		3 oz	0.4
Ham		3 oz	0.4
Hamburger		3 oz	0.4
Veal		3 oz	0.4
Pork		3 oz	0.3
Beef		3 oz	0.2
Grains:			
Breakfast cereals		1 cup	0.5–7.0

Vitamin B_6 Food		Serving size	Vitamin B_6 (mg)
Fruits:			
Banana		1	0.4
Avocado		1/2	0.3
Watermelon		1 cup	0.3
	© Maks Narodenko/Shutterstock.com		
Vegetables:			
Brussels sprouts		1/2 cup	0.2
Potato		1/2 cup	0.4
Sweet potato		1/2 cup	0.3
Carrots		1/2 cup	0.2
Sweet peppers		1/2 cup	0.2

Folate Food		Serving size	Folate (mcg)
Vegetables:			
Garbanzo beans		1/2 cup	141
Spinach, cooked		1/2 cup	131
Navy beans		1/2 cup	128
Asparagus		1/2 cup	120
Lima beans		1/2 cup	76
Collard greens, cooked		1/2 cup	65
Romaine lettuce		1 cup	65
Peas		1/2 cup	47
	iStockphoto.com/loooby		
	© Nikola Bilic/Shutterstock.com		
Grains:[a]			
Ready-to-eat cereals		1 cup/1 oz	100–400
Rice		1/2 cup	77
Noodles		1/2 cup	45
Wheat germ		2 Tbsp	40

Vitamin B_{12} Food		Serving size	Vitamin B_{12} (mcg)
Fish and seafood:			
Oysters		3 oz	13.8
Scallops		3 oz	3.0
Salmon		3 oz	2.3
Clams		3 oz	2.0
Crab		3 oz	1.8
Tuna		3 oz	1.8

[a]Fortified, refined grain products such as bread, rice, pasta, and crackers provide approximately 60 micrograms of folic acid per standard serving.

(continued)

Table 20.5 **(continued)**

Vitamin B_{12}			
Food		**Serving size**	**Vitamin B_{12} (mcg)**
Meats:	© Robyn Mackenzie/ Shutterstock.com		
Liver		3 oz	6.8
Beef		3 oz	2.2
Veal		3 oz	1.7
Milk and milk products:			
Skim milk		1 cup	1.0
Milk		1 cup	0.9
Yogurt		1 cup	0.8
Cottage cheese		1/2 cup	0.7
American cheese		1 oz	0.2
Cheddar cheese		1 oz	0.2
Grains:			
Breakfast cereals		1 cup	0.6–12.0
Eggs:			
Egg		1	0.6
Vitamin C			
Food		**Serving size**	**Vitamin C (mg)**
Fruits:	iStockphoto.com/joe Biafore		
Guava		1/2 cup	180
Orange juice, vitamin C-fortified		1 cup	108
Kiwi fruit		1	108
Grapefruit juice, fresh		1 cup	94
Cranberry juice cocktail		1 cup	90
Orange		1	85
Strawberries, fresh		1 cup	84
Cantaloupe		1/4 whole	63
Grapefruit		1 medium	51
Raspberries, fresh		1 cup	31
Watermelon		1 cup	15
Vegetables:	iStockphoto.com/NoDerog		
Sweet red peppers		1/2 cup	142
Cauliflower, raw		1/2 cup	75
Broccoli		1/2 cup	70
Brussels sprouts		1/2 cup	65
Green peppers		1/2 cup	60
Collard greens		1/2 cup	48
Vegetable (V-8) juice		3/4 cup	45
Tomato juice		3/4 cup	33
Cauliflower, cooked		1/2 cup	30
Potato		1 medium	29
Tomato		1 medium	23

Choline Food	Serving size	Choline (mg)
Meats:		
Beef	3 oz	111
Pork chop	3 oz	94
Lamb	3 oz	89
Ham	3 oz	87
Beef	3 oz	85
Turkey	3 oz	70
Salmon	3 oz	56
Eggs:		
Egg	1 large	126
Vegetables:		
Baked beans	1/2 cup	50
Navy beans, boiled	1/2 cup	41
Collards, cooked	1/2 cup	39
Black-eyed-peas (cowpeas)	1/2 cup	39
Chickpeas (garbanzo beans)	1/2 cup	35
Brussels sprouts	1/2 cup	32
Broccoli	1/2 cup	32
Collard greens	1/2 cup	30
Refried beans	1/2 cup	29
Milk and milk products		
Milk, 2%	1 cup	40
Cottage cheese, low-fat	1/2 cup	37
Yogurt, low-fat	1 cup	35
Vitamin A Food	**Serving Size**	**Vitamin A (retinol) (mcg)**
Meats:		
Liver	3 oz	9,124
Clams	3 oz	145
Fortified breakfast cereals	3/4 cup	150
Milk and milk products:		
American cheese	1 oz	114
Fat-free/low-fat milk	1 cup	100
Whole milk	1 cup	58
Egg	1	84
Beta-Carotene Food	**Serving Size**	**Beta-carotene (mcg retinol equivalents, RE)**
Vegetables:		
Sweet potatoes	1/2 cup	961
Pumpkin, canned	1/2 cup	953
Carrots, raw	1/2 cup	665
Spinach, cooked	1/2 cup	524
Collard greens, cooked	1/2 cup	489
Kale, cooked	1/2 cup	478
Turnip greens, cooked	1/2 cup	441
Beet greens, cooked	1/2 cup	276
Swiss chard, cooked	1/2 cup	268
Winter squash, cooked	1/2 cup	268
Vegetable juice	1/2 cup	200
Romaine lettuce	1 cup	162

(continued)

Table 20.5 (continued)

Beta-Carotene Food	Serving Size	Beta-carotene (mcg retinol equivalents, RE)	
Fruit:			
Cantaloupe	½ cup	135	
Apricots, fresh	4	134	
Vitamin E **Food**	**Serving Size**	**Vitamin E (mg)**	
Nuts and seeds:			
Sunflower seeds	1 oz	7.4	
Almonds	1 oz	7.3	
Hazelnuts (filberts)	1 oz	4.3	
Mixed nuts	1 oz	3.1	
Pine nuts	1 oz	2.6	
Peanut butter	2 Tbsp	2.5	
Peanuts	1 oz	2.2	
Vegetable oil:			
Sunflower oil	1 Tbsp	5.6	
Safflower oil	1 Tbsp	5.6	
Canola oil	1 Tbsp	2.4	
Peanut oil	1 Tbsp	2.1	
Corn oil	1 Tbsp	1.9	
Olive oil	1 Tbsp	1.9	
Salad dressing	2 Tbsp	1.5	
Fish and seafood:			
Crab	3 oz	4.5	
Shrimp	3 oz	3.7	
Fish	3 oz	2.4	
Grains:			
Wheat germ	2 Tbsp	4.2	
Whole wheat bread	1 slice	2.5	
Vegetables:			
Spinach, cooked	½ cup	3.4	
Yellow bell pepper	1	2.8	
Turnip greens, cooked	½ cup	2.2	
Swiss chard, cooked	½ cup	1.7	
Asparagus	½ cup	1.5	
Sweet potato	½ cup	1.5	
Vitamin D **Food**	**Serving size**	**Vitamin D** **(mcg)**	**IU**
Fish and seafoods:			
Swordfish	3 oz	14	566
Trout	3 oz	13	502
Salmon	3 oz	11	447
Tuna, light, canned in oil	3 oz	5.7	228
Halibut	3 oz	4.9	196
Tuna, light, canned in water	3 oz	3.8	152
Tuna, white, canned in water	3 oz	1.7	68

iStockphoto.com/FotografiaBasica

Ray Kachatorian/Photodisc/Getty Images

Vitamin D			Vitamin D	
Food		Serving size	(mcg)	IU
Vitamin D-fortified breakfast cereals:				
Whole grain Total		1 cup	3.3	132
Total Raisin Bran		1 cup	2.6	104
Corn Pops, Kellogg's		1 cup	1.2	48
Crispix, Kellogg's		1 cup	1.2	48
Other vitamin D-fortified foods:				
Orange juice		1 cup	2.5	100
Rice milk		1 cup	2.5	100
Soy milk		1 cup	2.5	100
Yogurt		1 cup	2.0	80
Margarine		2 tsp	1.2	48
Milk:				
Milk, whole		1 cup	3.2	128
Milk, 2%		1 cup	2.9	116
Milk, 1%		1 cup	2.9	116
Milk, skim		1 cup	2.9	116

Vitamin K			
Food		Serving Size	Vitamin K (mcg)
Kale, cooked		½ cup	531
Spinach, cooked		½ cup	444
Turnip greens, cooked		½ cup	426
Broccoli, cooked		½ cup	110
Brussels sprouts, cooked		½ cup	109
Mustard greens, cooked		½ cup	105
Cabbage, cooked		½ cup	82
Spinach, raw		½ cup	73
Lettuce, leafy green		1 cup	71
Asparagus, cooked		4 spears	48
Kiwifruit		½ cup	37
Berries, blue or black		1 cup	29
Okra, cooked		½ cup	23
Peas, cooked		½ cup	21
Leeks		1	16

Fertnig/Getty Images

NUTRITION up close

Antioxidant Vitamins: How Adequate Is Your Diet?

Focal Point: Determine if you eat enough antioxidant-rich foods.

Vitamin C, beta-carotene, and vitamin E, the antioxidant vitamins, help maintain cellular integrity in the body. Good food sources of these antioxidants reduce the risk of heart disease, certain cancers, and other ailments. Check below to find out how frequently you consume foods containing these important, health-promoting nutrients from foods.

How often do you eat:	Seldom or never	1–2 times per week	3–5 times per week	Almost daily
Vitamin C food sources:				
1. Grapefruit, lemons, oranges, or pineapple?				
2. Strawberries, kiwi, or honeydew melon?				
3. Orange juice, cranberry juice cocktail, or tomato juice?				
4. Green, red, or chili peppers?				
5. Broccoli, Chinese cabbage, or cauliflower?				
6. Asparagus, tomatoes, or potatoes?				
Beta-carotene food sources:				
7. Carrots, sweet potatoes, or winter squash?				
8. Spinach, collard greens, or Swiss chard?				
9. Cantaloupe, papayas, or mangoes?				
10. Nectarines, peaches, or apricots?				
Vitamin E food sources:				
11. Whole-grain breads, whole-grain cereals, or wheat germ?				
12. Crab, shrimp, or fish?				
13. Peanuts, almonds, or sunflower seeds?				
14. Oils, margarine, butter, mayonnaise, or salad dressing?				

Feedback to the Nutrition Up Close is located in Appendix F.

REVIEW QUESTIONS

- **Describe the functions and food sources of vitamins and their effects on health.**

1. Vitamins are essential. Specific deficiency diseases develop if we fail to consume enough of them. **True/False**
2. Vitamins D, E, C, and B_6 are fat soluble. **True/False**
3. The niacin deficiency disease is called pellagra. **True/False**
4. Some people are deficient in vitamin B_{12} because they are genetically unable to absorb it. **True/False**
5. Vitamin A toxicity causes brain tumors. **True/False**
6. Three good sources of vitamin D are sunshine, milk, and seafood. **True/False**
7. Vitamins E and C, and beta-carotene supplements decrease the risk of heart disease. **True/False**
8. Vitamin D acts as a hormone. **True/False**
9. Adequate vitamin D status reduces chronic inflammation. **True/False**

10. Very high intakes of each of the vitamins have been found to cause toxicity disease. **True/False**
11. With the exception of vitamin B_{12}, the body is able to store high amounts of the water-soluble vitamins. **True/False**
12. A number of vitamins act as coenzymes by activating specific enzymes. **True/False**
13. Vitamin C functions in the replacement of cells that line the esophagus and eyes. **True/False**
14. The sun is the primary source of vitamin D for people in general. **True/False**
15. It is recommended that women who are or may become pregnant consume 400 mcg of folic acid daily to reduce the risk of fetal development of vision problems. **True/False**
16. ___ Which of the following is not considered a vitamin?
    a. pantothenic acid
    b. coenzyme Q_{10}
    c. biotin
    d. choline
17. ___ A person who never eats fish, seafood, nuts, seeds, or vegetable oil is at risk of developing a deficiency of ________.
    a. vitamin A
    b. vitamin K
    c. vitamin D
    d. vitamin E
18. ___ Which of the following is a consequence of vitamin A deficiency?
    a. rough, dry skin
    b. tingling sensation in the finger tips
    c. headache
    d. night blindness
19. ___ Assume you ate the following foods as snacks yesterday: an orange, ¼ cup sunflower seeds, 1 cup yogurt, and a carrot. Which of these foods would provide the most vitamin E?
    a. orange
    b. sunflower seeds
    c. yogurt
    d. carrot
20. ___ Which of the following foods is not a good source of vitamin A (retinol)?
    a. milk
    b. eggs
    c. sweet peppers
    d. fortified breakfast cereal
21. ___ Which of the following vitamins is needed for blood to clot when bleeding occurs?
    a. vitamin K
    b. thiamin
    c. vitamin B_6
    d. vitamin B_{12}

Answers to these questions can be found in Appendix F.

NUTRITION SCOREBOARD ANSWERS

1. For some vitamins, intake levels above those known to prevent deficiency diseases help protect humans from certain cancers, heart disease, osteoporosis, depression, and other disorders. **False**
2. Nope—only carbohydrates, proteins, and fats provide energy to the body. Vitamins are needed, however, to convert the energy in food into energy the body can use. **False**
3. Citrus fruits are good sources of vitamin C, but so are red sweet peppers, strawberries, and other noncitrus fruits and vegetables. **False**
4. True. Nearly all cases of illness due to vitamin overdoses result from excessive intake of vitamin supplements.[1] **True**

Phytochemicals

Scott Goodwin Photography

NUTRITION SCOREBOARD

1 Phytochemicals are found only in plants. **True/False**

2 Phytochemicals are also called *phytonutrients* because they are biologically active in the body and have beneficial effects on health. **True/False**

3 Brightly colored vegetables and fruits are the only sources of phytochemicals. **True/False**

4 Some chemical substances that occur naturally in food or result from food preparation may be harmful to health. **True/False**

Answers can be found at the end of the unit.

LEARNING OBJECTIVES

After completing Unit 21 you will be able to

- Describe the functions and food sources of key phytochemicals.

Phytochemicals: The "What Else" in Your Food

- **Describe the functions and food sources of key phytochemicals.**

As recently as 25 years ago, the science of nutrition focused on the study of the functions and health effects of protein, fats, carbohydrates, vitamins, minerals, and water. These classes of essential nutrients have been extensively studied and a good deal is known about their effects on growth, reproduction, and health. However, essential nutrients do not account for all the benefits associated with healthy diets. There are other components in food that influence health.

We have ample evidence that diets rich in vegetables, fruits, whole grains, and other plant foods support health and reduce the risk of developing a number of diseases. It was largely assumed that the health benefits came from the vitamin and mineral content of fruits and vegetables. That conclusion turned out to be incorrect because supplementation with specific vitamins and minerals failed to yield the same health benefits as did diets rich in fruits and vegetables. In addition, use of individual vitamin and mineral supplements was found to increase health risks in some studies.[1] Now nutrition and other scientists are investigating the health effects of thousands of other substances in food and their interactions with essential nutrients and genetic traits that affect their utilization.[2] Illustration 21.1 shows some of the hundreds of chemical substances found in two plant foods.[3]

The subjects of many current studies are plant chemicals, known as **phytochemicals** or *phytonutrients*. Phytochemicals are not considered essential nutrients because deficiency diseases do not develop when we fail to consume them. They are considered to be nutrients, however, because they are biologically active and perform health-promoting functions in the body. Meats, eggs, dairy products, and other foods of animal origin also contain biologically active substances that affect body processes. Much less is known about these **zoochemicals**, and their effects on health are not yet as clear as those of some of the phytochemicals. Most bioactive food constituents are derived from plants.[4]

This unit presents information on the functions and health benefits of the most extensively studied phytochemicals and identifies their major food sources. It also highlights substances in foods that are considered to be natural toxins because they can be harmful to health if consumed in excess.

phytochemicals (phyto = *plant*) Biologically active, or "bioactive," substances in plants that have positive effects on health. Also called *phytonutrients*.

zoochemicals Chemical substances in animal foods, some of which may be biologically active in the body.

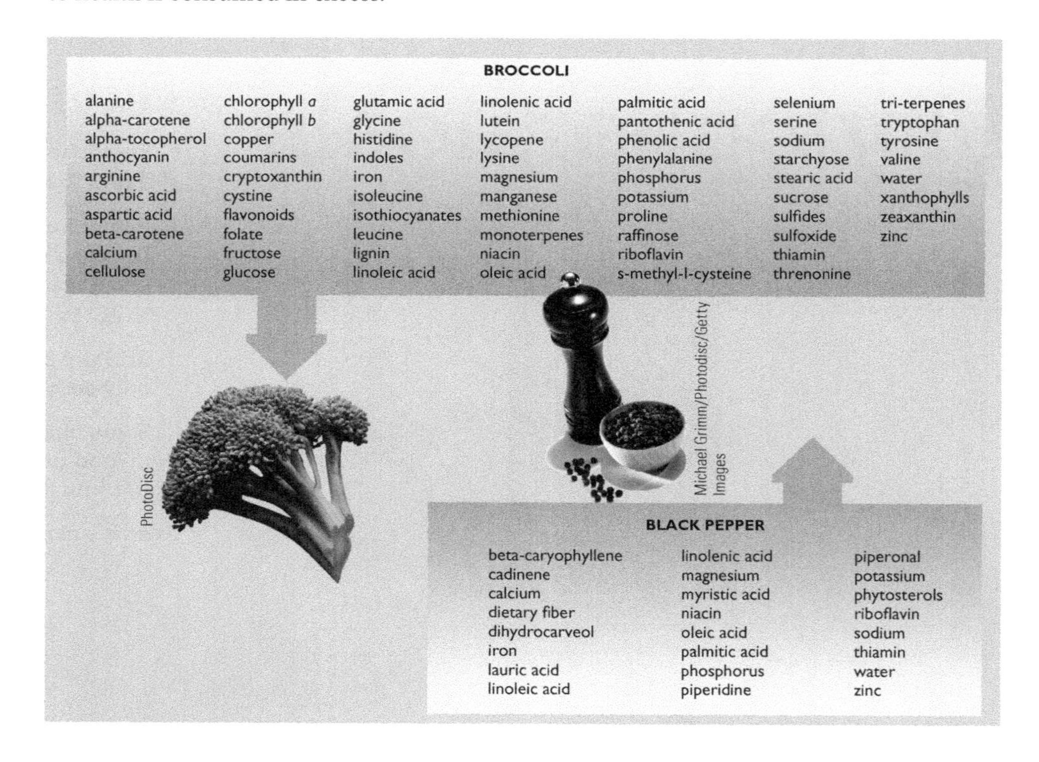

Illustration 21.1 A sampling of the chemical substances in two foods. There are hundreds more.[3]

KEY NUTRITION CONCEPTS

The key nutrition concepts underlying content presented on phytochemicals relate to the importance of substances in food that are beneficial to health but are not considered essential nutrients. These concepts are as follows:

1. Nutrition Concept #2: Foods provide energy (calories), nutrients, and other substances needed for growth and health.
2. Nutrition Concept #9: Adequacy, variety, and balance are key characteristics of healthy dietary patterns.

Characteristics of Phytochemicals

Phytochemicals play a variety of roles in plants. They provide protection against bacterial, viral, and fungal infection; ward off insects; and prevent tissue damage due to oxidation. Some act as plant hormones or participate in the regulation of gene function, while others provide plants with flavor and color.[5] Did you ever wonder where "brown" eggs get their color (see Illustration 21.2)? It comes from xanthophylls (pronounced zan-tho-fills), a yellow-orange pigment in plants. Brown eggs are laid by chickens known as Rhode Island Reds when they consume foods such as yellow corn or alfalfa that contain xanthophylls.[6]

More than 2,000 types of phytochemicals that act as pigments have been identified (Illustration 21.3). Specific types of phytochemicals, their color, and top food sources are listed in Table 21.1. Sometimes the color of phytochemicals contained in plants is obscured. Dark green vegetables, for example, are often good sources of orange and yellow carotenes, but the green chlorophyll obscures these colors. Many phytochemicals are colorless. You cannot identify rich sources of phytochemicals by their color alone. The Reality Check feature for this unit emphasizes this point using the example of white vegetables and fruits.

Although thousands of biologically active substances in plants have been identified, the effects on human health are

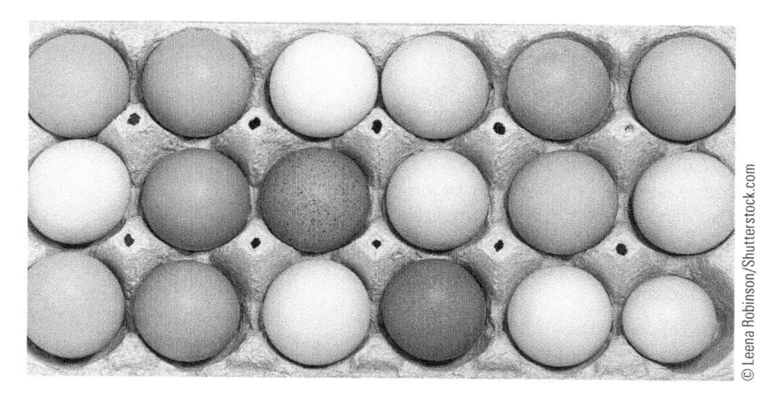

© Leena Robinson/Shutterstock.com

Illustration 21.2 Why are some eggshells brown?

REALITY CHECK

Color and the Phytonutrient Content of Vegetables and Fruits

Are white vegetables and fruits good sources of phytonutrients?

Who gets the thumbs-up?

Answers on page 21-4.

Photodisc

Hyde: I avoid eating white foods like onions, bananas, potatoes, and cauliflower. If you want phytonutrients, you have to eat the really colorful vegetables and fruits.

Photodisc

Ejay: I used to think that, too. But I just learned "white" doesn't mean that a white vegetable or fruit is a poor source of phytonutrients.

Keith Weller/ARS/USDA

Illustration 21.3 The colors in plant foods are created by phytochemicals.

currently known for only a small number of them. Rather than acting alone, phytochemicals appear to exert their beneficial effects primarily through complementary mechanisms of action and combined effects with other phytochemicals, nutrients in foods, and foods within dietary patterns. The specific functions of phytochemicals may be diminished or expanded based on these interactions.[7]

Functions of Phytochemicals in Humans

The beneficial effects of phytochemicals from plants on health promotion and disease prevention relates to their roles in combating oxidation reactions and chronic inflammation and to their effects on gene functions.[8,9] Various types of phytochemicals perform different functions in the body depending on their chemical characteristics and other factors. They have been found to play a role in preventing the development of cancer, heart disease, diabetes, hypertension, eye disease, and a variety of other diseases and disorders.[10–12]

Chronic Inflammation and the Antioxidant Roles of Phytochemicals Chronic inflammation, which plays an important role in the development of many common diseases and disorders, leads to the production of free radicals that trigger oxidation reactions within cells. Oxidation reactions can damage cell membranes and components of cells and impair cell functions in ways that encourage the development of disease. **Antioxidants**, which are both consumed in foods and produced by the body, help prevent and repair damage to cell functions by scavenging oxidized particles and neutralizing their destructive effects. A number of phytochemicals function as antioxidants and expand the body's arsenal of defense mechanisms against the harmful effects of chronic inflammation.[8,9] Table 21.2 provides a list of the top commonly consumed plant sources of antioxidants.

antioxidants Chemical substances that prevent or repair damage to cells caused by oxidizing agents such as pollutants, ozone, smoke, and reactive oxygen. Oxidation reactions are a normal part of cellular processes. Vitamins C and E and certain phytochemicals function as antioxidants.

age-related macular degeneration (AMD) Eye damage caused by oxidation of the macula, the central portion of the eye that allows you to see details clearly.

cruciferous family Sulfur-containing vegetables whose outer leaves form a cross (or crucifix). Vegetables in this family include broccoli, cabbage, cauliflower, brussels sprouts, mustard and collard greens, kale, bok choy, kohlrabi, rutabaga, turnips, broccoflower, and watercress.

Table 21.1 The color and top food sources of specific phytochemicals[13]

Phytochemical	Color	Top food sources
Beta-carotene	Orange	Carrots
Lycopene (like-oh-pene)	Red	Tomatoes
Anthocyanins (an-tho-sigh-ah-nins)	Blue to purple	Blueberries, grapes
Allicin (all-is-in)	White	Garlic
Lutein (loo-te-in) and zeaxanthin (ze-ah-zan-thin)	Yellow-green	Spinach
Xanthophylls (zan-tho-fills)	Yellow-orange	Corn, green vegetables

ANSWERS TO REALITY CHECK

Color and the Phytonutrient Content of Vegetables and Fruits

Ejay's got it right. White vegetables and fruits provide phytonutrients such as flavonoids and allicin. You can't identify rich sources of phytonutrients based solely on the color of vegetables and fruits. Not all phytonutrients that benefit health add bright colors to vegetables and fruits.

Photodisc

Hyde:

Photodisc

Ejay:

take action Hungry for a Snack? How About Fruit?

Fresh, frozen, dried, or canned—fruits are delicious. Read through this list of good sources of antioxidants and place a check in front of fruits you'd like to have as a snack.

_______ Pomegranate
_______ Blackberries
_______ Papaya
_______ Blueberries
_______ Strawberries
_______ Raspberries
_______ Cranberries
_______ Oranges
_______ Tangerines
_______ Pears
_______ Grapes
_______ Apples
_______ Kiwifruit
_______ Grapefruit
_______ Peach
_______ Cranberry juice
_______ Cherries, sour
_______ Pineapple juice
_______ Mango nectar

Antioxidant Roles of Lutein, Zeaxanthin, and Beta-Carotene Many pigments in plants act as antioxidants and participate in disease prevention. Lutein (pronounced loo-te-in) and zeaxanthin (pronounced ze-ah-zan-thin) are found together in plants and play key roles in the prevention and treatment of **age-related macular degeneration (AMD)**. AMD is the leading cause of blindness in people over the age of 50. It produces a loss of central vision due to degeneration of the macula (Illustration 21.4). The macula, located in the center of the inside wall of the eye, is yellow due to its content of lutein and zeaxanthin. When these two phytochemicals become depleted, vision declines. Increased intake of lutein and zeaxanthin helps prevent deterioration of the macula and improves the maintenance of vision.[14]

Beta-carotene, lutein, zeaxanthin, and other phytochemicals come in the form of dietary supplements. However, research has not shown that antioxidant supplements are superior to foods sources for disease prevention.[15] Plant sources of antioxidants are known to promote health and help prevent disease, and are preferred.[16]

Antioxidant Roles of Flavonoids The good news about flavonoids has made chocolate a health food. Cocoa, the main ingredient in chocolate, is a rich source of flavonoids. Flavonoids, which include anthocyanins and quercetin (pronounced queer-sah-tin), act as antioxidants and reduce inflammation. Regular intake of flavonoids (such as daily consumption of a cup of hot chocolate made with cocoa powder) is related to improved blood flow, reduced blood pressure, and decreased risk of heart disease, stroke, type 2 diabetes, and cancer.[17] The content of flavonoids in chocolate products increases with the amount of cocoa in the products. Chocolate products that contain 40% or more cocoa by weight have substantially higher amounts of flavonoids than other chocolate products.[18] Milk chocolate made in the United States tends to have less than 30% cocoa, whereas darker chocolate products usually have more than 40%. Flavonoids are also found in good amounts in foods such as blueberries, grapes, oranges, bananas, apples, wine, and tea.[18,19]

Because of their importance, this unit's Take Action feature attempts to remind you about specific fruits rich in flavonoids and other antioxidant phytochemicals that may have escaped your attention.

Phytochemicals and Gene Function Quercetin, a phytochemical present in apples, onions, and grapes, influences inflammation by modifying the expression (the turning on or off) of genes that promote the production of pro-inflammatory compounds. Its presence helps decrease chronic inflammation due to excess body fat, improves insulin utilization, and may decrease bone loss. Similarly, phytochemicals in the **cruciferous family** of

Table 21.2 **Top food sources of antioxidants**[20]

Pomegranate
Red cabbage
Blackberries
Pecans
Walnuts
Cloves, ground
Peanuts
Sunflower seeds
Blueberries
Strawberries
Chocolate, dark
Raspberries
Cranberry juice
Kale
Artichokes
Wine, red
Grape juice
Cranberries
Pineapple juice
Green tea
Guava nectar
Coffee
Mango nectar

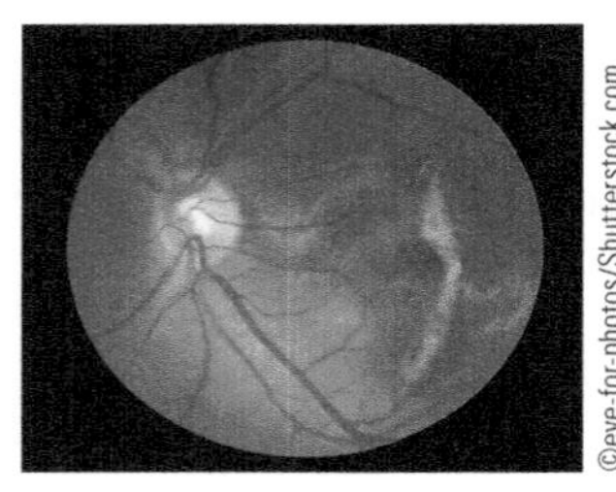

Illustration 21.4 Photograph of damage to the central part of the back of the eye due to macular degeneration.

vegetables, including broccoli and cauliflower (Illustration 21.5), appear to reduce the risk of certain types of cancer, particularly in people with specific gene types.[11,21] Consumption of one-half cup of a vegetable in the crucifer family daily is sufficient to decrease inflammation in many individuals.[22] Resveratrol (rez-ver-ah-trol) in the skin of red grapes also plays a role in the prevention of inflammation. This phytochemical is also found in peanuts, tea, blueberries, and cranberries.[19]

Anti-Infection Roles of Proanthocyanidins and Resveratrol Cranberries contain proanthocyanidins (pro-an-tho-sigh-an-ah-dins) that prevent Escherichia coli (*E. coli*) and other types of bacteria from sticking to the walls of the urinary tract and thus limit the spread of urinary tract infections. Cranberry juice by itself usually does not fully treat established urinary tract infections, but regular intake of cranberry juice (a cup daily) reduces the symptoms of urinary tract infections and helps prevent them from developing and reoccurring.[23,24] Resveratrol also helps prevent the spread of bacteria that cause infection.[24]

Illustration 21.5 Examples of cruciferous vegetables.

Caffeine: A Phytochemical with Multiple Functions Caffeine is an example of a phytochemical that is not easily classified by function because it has antioxidant, anti-inflammatory, and gene-regulating effects on body processes.[26,27] Most of the caffeine in diets comes from coffee (Illustration 21.6), one of the most commonly consumed beverages in the world. Caffeine sources also include tea, some soft drinks, and energy drinks (Table 21.3). Both positive and negative health effects are attributed to caffeine. On the positive side, caffeine or regular coffee intake has been found to:

- Decrease the risk of type 2 diabetes by lowering blood glucose levels, improving insulin sensitivity and insulin production, and decreasing glucose output by the liver.[28,29]
- Decrease the risk of death from heart disease and stroke.[30]
- Decrease the risk of estrogen-sensitive cancers, potentially by blocking the effects of estrogen and insulin on cancer development.[31]
- Decrease the risk of Parkinson's disease and Alzheimer's disease.[32]
- Increase mental alertness and energy by blocking the activity of a chemical messenger that decreases heart rate and increases drowsiness.[33]
- Improve mood; decrease depression.[33]

Intake of four cups of coffee daily does not appear to be related to an increased risk of disease.[30,34] Very high intakes of coffee, and particularly of caffeine, however, can have important negative effects on health in some people.

Illustration 21.6 Coffee is the leading source of caffeine in the diet, followed by tea.

Negative Effects of Excess Caffeine Some people may experience increased blood pressure if large amounts of coffee are consumed in combination with the use of some anti-hypertension medications.[35] Excess caffeine intake can lead to anxiety, nervousness, sleep problems, and a feeling of unusually strong heart beats (heart palpitations) in some people. Tolerance to caffeine occurs over time in most people, and abrupt withdrawal can lead to headache, fatigue, and irritability within 12 to 24 hours.[36]

Medical visits for children and adults for symptoms of excess caffeine intake have sharply increased in the United States in recent years. The increase is largely due to the availability and popularity of caffeinated energy drinks and alcohol plus caffeine beverages.[37,38] The mix of alcohol and caffeine gives drinkers the feeling of being wide awake and not very drunk. Caffeine, however, does not counteract the effects of alcohol on impaired judgment and reduced reaction time.[39] The Food and Drug Administration (FDA) has banned the sale of beverages containing both caffeine and alcohol, and both Canada and the United States have set limits on the amount of caffeine that can be added to energy drinks.[39,40]

Table 21.3 **Caffeine content of foods, beverages, and some drugs**

Source	Caffeine (mg)
Coffee (1 cup)	
Drip	115–175
Decaffeinated (ground or instant)	0.5–4.0
Instant	61–70
Percolated	97–140
Espresso (2 oz)	100
Tea (1 cup)	
Black, brewed 5 minutes, U.S. brands	32–144
Black, brewed 5 minutes, imported brands	40–176
Green, brewed 5 minutes	25
Instant	40–80
Soft drinks	
Coca-Cola (12 oz)	47
Cherry Coke (12 oz)	47
Diet Coke (12 oz)	47
Dr. Pepper (12 oz)	40
Ginger ale (12 oz)	0
Mountain Dew (12 oz)	54
Pepsi-Cola (12 oz)	38
Diet Pepsi (12 oz)	37
7-Up (12 oz)	0
Energy drinks	
Rockstar Energy Shot (8 oz)	229
5-Hour Energy (2 oz)	215
Red Bull (8 oz)	80
Full Throttle (8 oz)	210
Jolt (12 oz)	72
Chocolate	
Cocoa, chocolate milk (1 cup)	10–17
Milk chocolate candy (1 oz)	1–15
Chocolate syrup, one ounce (2 Tbsp)	4
Nonprescription drugs, two tablets	
Nodoz	200
Vivarin	200
Excedrin	130
Weight-control pills	150

Table 21.4 **The top five and other leading sources of phytochemicals**[41]

Top five sources
1. Tomatoes
2. Carrots
3. Oranges
4. Orange juice
5. Strawberries
Other leading sources
• Coffee
• Tea
• Spinach
• Corn
• Lettuce
• Collards
• Watermelon
• Grapes
• Blueberries
• Strawberries
• Bananas
• Onions
• Apples
• Raspberries

Food Sources of Phytochemicals

The amount and type of phytochemicals present in plants vary a good deal, depending on a number of factors. The content varies based on growing conditions, genetic strains used, storage, and processing and preparation methods. Some plant foods, such as spinach and carrots, which are considered rich sources of vitamins, contain a variety of phytochemicals. Celery, tea, and onions, foods determined by nutrient composition tables to be relative "vitamin weaklings," are actually good sources of a number of phytochemicals. Plant foods rich in phytochemicals may or may not be good sources of vitamins and minerals as well. However, intake of phytochemicals, and the beneficial effects of vegetable and fruit intake on health, increase as vegetable and fruit consumption increases.[16] Established health benefits related to plant food intake are reflected in the dietary recommendations developed for MyPlate.gov to "fill half your plate with vegetables and fruits." Similarly, the Dietary Guidelines recommend that people consume ample plant foods because of their benefits.[16] The top five and other leading sources of phytochemicals in the U.S. diet are listed in Table 21.4.

Naturally Occurring Toxins in Food

Some foods contain biologically active substances that can harm health if consumed in excess. These substances are considered naturally occurring toxins. Some naturally occurring toxins form in food during food processing and preparation.[42]

Spinach, collard greens, rhubarb, and other dark green, leafy vegetables contain oxalic acid. Eating too much of these foods can make your teeth feel as though they are covered with sand, and can give you a stomachache. Have you ever seen a potato that was partly colored green? (If not, take a look at Illustration 21.7.) The green area contains solanine, a bitter-tasting, insect-repelling phytochemical that is normally found only in the leaves and stalks of potato plants. Small amounts of solanine are harmless, but large quantities can interfere with the transmission of nerve impulses.

Phytate is present in whole grains, seeds, dried beans, and nuts. It tightly binds zinc, iron, calcium, magnesium, and copper, and reduces their absorption. Diets high in phytate have been found to produce mineral deficiency diseases.[42] Cassava, a root consumed daily in many parts of tropical Africa, can be very toxic if not prepared properly, because it contains cyanide. Soaking cassava roots in water for three nights will get rid of the cyanide, but soaking for shorter periods of time does not. When the soaking time is cut to one or two nights, as sometimes happens during periods of food shortage, enough of the toxin remains in the root to cause konzo, a disease caused by cyanide overdose.[43] Konzo is characterized by permanent, spastic paralysis.

Ackee fruit is another potential hazard to health (Illustration 21.8). If you're from Jamaica, chances are excellent that you love the taste of the core of ackee—and also know the fruit can be deadly. The national fruit of Jamaica, the yellow fleshy part around the seeds tastes like butter and looks like scrambled eggs. The rest of the fruit, however, is not edible. The fruit of unopened, unripe ackee contains high concentrations of phytochemicals that cause severe vomiting and a drastic drop in blood glucose levels. Ingestion of the fruit has caused hundreds of deaths in Jamaica. Its sale was banned in the United States until 2000, and imported ackee is routinely analyzed by the FDA for ripeness.[44]

Scott Goodwin Photography

Illustration 21.7 Potatoes grown partly above ground develop a green color in the part exposed to the sun. The green section contains solanine, a naturally occurring, potentially toxic phytochemical.

David Neil Madden/Getty Images

Illustration 21.8 Ackee fruit and seeds.

Scott Goodwin Photography

NUTRITION up close

Have You Had Your Phytochemicals Today?

Focal Point: Consuming good sources of the "other" beneficial components of food.

Good food sources of beneficial phytochemicals are listed below. Indicate foods you consumed at least twice last week and those you have never tried eating.

Foods eaten	Foods never tried	Twice last week
Broccoli		
Cabbage		
Brussels sprouts		
Cauliflower		
Carrots		
Celery		
Collard greens		
Turnip greens		
Kale		
Swiss chard		
Spinach		
Tomatoes		
Peanuts		
Walnuts		
Apple/apple juice		
Orange/orange juice		
Grapefruit/grapefruit juice		
Grapes/grape juice		
Strawberries		
Blueberries		
Papaya		
Banana		
Pear		
Peaches		

Feedback to the Nutrition Up Close is located in Appendix F.

REVIEW QUESTIONS

- **Describe the functions and food sources of key phytochemicals.**

1. Phytochemicals that benefit health are found in large amounts in fish and organ meats. **True/False**
2. Some plant foods that contain relatively small amounts of vitamins and minerals are rich sources of beneficial phytochemicals. **True/False**
3. The best way to achieve the health benefits of phytochemicals is by making vegetables and fruits a core component of your dietary pattern. **True/False**
4. If a chemical substance occurs naturally in a plant food, it can be considered harmless to health. **True/False**
5. Functions of phytochemicals that naturally occur in plants are often different than their functions in the body. **True/False**
6. MyPlate.gov food guidance recommends that half your plate consist of vegetables and fruits. **True/False**
7. Age-related macular degeneration is related to quercetin intake from plant foods. **True/False**
8. Caffeine's functions are limited to roles in combating inflammation. **True/False**
9. The combination of caffeine and alcohol prevents the intoxicating effects of alcohol from occurring. **True/False**
10. Most plant foods contain some level of phytochemicals, but onions and celery do not. **True/False**
11. Orange juice is not a leading source of phytochemicals. **True/False**
12. ____ Which of the following statements about antioxidants is true?
    a. Antioxidants decrease the body's utilization of oxygen.
    b. Lutein functions as an antioxidant in the body.
    c. Caffeine is a powerful antioxidant that increases the risk of heart disease.
    d. Vitamins E and C are the only components of plant foods that function as antioxidants.

The next three questions refer the following scenario.

Assume you eat a salad that consists of spinach, ripe red strawberries, cranberries, and walnuts with a salad dressing.

13. ____ Which of these plant foods would be the best source of lutein and zeaxanthin?
    a. salad dressing
    b. spinach
    c. strawberries
    d. walnuts
14. ____ The strawberries in the salad are considered to be a good source of:
    a. lutein
    b. zeaxanthin
    c. resveratrol
    d. lycopene
15. ____ The cranberries in the salad would contribute most to your intake of phytochemicals that play a role in:
    a. anti-inflammation processes
    b. anti-infection processes
    c. gene regulation processes
    d. antioxidation processes
16. Which of the following plant foods do *not* belong to the cruciferous vegetable family?
    a. cabbage
    b. broccoli
    c. potatoes
    d. cauliflower
17. ____ Which of the following biologically active substances is *not* considered a potential "natural toxin"?
    a. phytates
    b. resveratrol
    c. oxalic acid
    d. solanine

Answers to these questions can be found in Appendix F.

NUTRITION SCOREBOARD ANSWERS

1. The "phyto" in phytochemicals means plants. **True**
2. That's true. **True**
3. You cannot judge a plant's overall content of phytochemicals by looking at its color. **False**
4. Some foods contain naturally occurring toxins that can be harmful if consumed in excess. **True**

Diet and Cancer

NUTRITION SCOREBOARD

1 Some types of cancer are contagious. **True/False**

2 Cancer is primarily an inherited disease. **True/False**

3 People who regularly consume a variety of fruits and vegetables are less likely to develop cancer than people who don't. **True/False**

4 High levels of body fat contribute to the development of some types of cancer. **True/False**

5 Six out of 10 Americans never get cancer. **True/False**

Answers can be found at the end of the unit.

Digital Vision/Getty Images

LEARNING OBJECTIVES

After completing Unit 22 you will be able to:

- Determine the processes involved in the development and progression of cancer.
- Summarize the dietary patterns and lifestyle factors that affect cancer development and that decrease the risk of cancer.

What Is Cancer?

- **Determine the processes involved in the development and progression of cancer.**

Cancer, the second leading cause of death in the United States, is a group of conditions that result from the uncontrolled growth of abnormal cells. Although these cells can begin to grow in any tissue in the body, the lungs, colon, **prostate**, and breasts are the most common cancer sites (Illustration 22.1). Approximately 40% of the U.S. population will develop cancer at some point during life. Many forms of cancer can be successfully treated.[4]

cancer A group of diseases in which abnormal cells grow out of control and can spread throughout the body. Cancer is not contagious and has many causes.

prostate A gland located below the bladder in males. The prostate secretes a fluid that surrounds sperm.

acquired gene mutations Defects in genes that develop over time after conception due to exposure to smoke, toxins, radiation, various dietary components, excess body fat, or other factors; or that occur randomly during cell division. They are present only in certain cells, not in every cell in the body. Acquired mutations in somatic cells (cells other than sperm and egg cells) cannot be passed on to the next generation. Acquired mutations are involved in the development of cancer.

KEY NUTRITION CONCEPTS

The key nutrition concepts that underlie the complex relationships among diet and cancer are:

1. Nutrition Concept #3: Health problems related to nutrition originate within cells.
2. Nutrition Concept #8: Poor nutrition can influence the development of certain chronic diseases.
3. Nutrition Concept #9: Adequacy, variety, and balance are key characteristics of healthy dietary patterns.

How Does Cancer Develop?

Cancer is a complex group of diseases initiated primarily by **acquired gene mutations**. These mutations occur over time due to exposure to factors such as excess body fat, tobacco smoke, various components of the diet, physical inactivity, certain types of infections and hormones, and exposure to specific chemical pollutants and radiation. Acquired gene mutations also occur randomly during cell division. Some types of cancer, specifically prostate, breast, and colorectal, tend to cluster in families. Most cancers, however, are not clearly linked to the genes we inherit from our parents.[6–8]

Cancer prevention and treatment approaches in the past were primarily directed at organs most affected by the disease. It is now known that cancers that develop within many sites share specific gene mutations and that prevention and treatment should center on the types of gene mutations and not the location of the cancer.[9] Cancer development does not proceed in a straight line—it can progress two steps forward and then take a step or two back.

Illustration 22.2 summarizes the processes involved in the development of cancer, as currently understood. The risk of cancer development begins when DNA, the genetic material in cells that controls the body's production of proteins that regulate cell functions, becomes damaged. DNA can be damaged by reactive oxygen molecules (free radicals), radiation, toxins, and other reactive substances within cells. The structure of DNA is altered by the damage, and that changes the accuracy of DNA codes for protein synthesis. Cells have multiple and redundant systems that continuously work to repair DNA to maintain its proper structure. Most of the time, DNA is successfully repaired. Problems arise when the repair processes

Illustration 22.1 Percentage of new cancer cases by sites and sex. About 23% of deaths in the United States are due to cancer.[5]

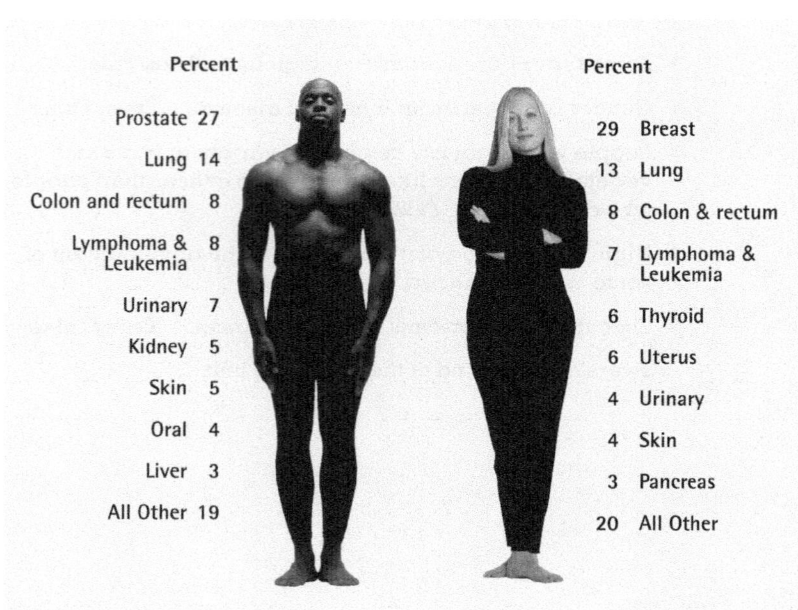

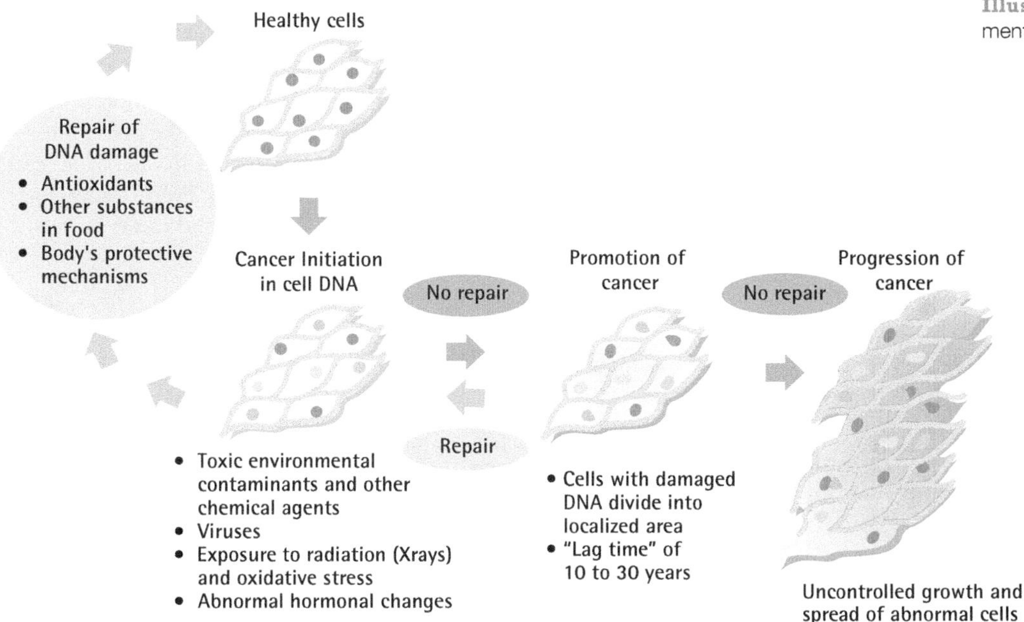

Illustration 22.2 Steps in the development of cancer.[11]

fail to keep up with the damage. When this happens, errors in DNA codes disrupt protein synthesis and normal cell functions in ways that can lead to the development of abnormal cell functions and structures. This situation can lead to the **initiation phase** of cancer development. Cancer development enters the **promotion phase** if damaged DNA is allowed to accumulate over the course of years (10 to 30 years in most cases).[10] Ultimately, the growth of abnormal cells becomes rampant and can spread throughout the body. Although there are many factors that can lead to DNA damage, all cancers share the hallmark of damaged DNA.[10]

initiation phase The start of the cancer process. It begins with damage to DNA.

promotion phase The period in cancer development when the number of cells with altered DNA increases.

What Causes DNA Damage?

About 80 to 90% of common forms of cancer are related to environmental factors that modify the structure and function of DNA.[11] Many environmental factors play a role in DNA damage development, and most are modifiable. The top nine modifiable risk factors for cancer development worldwide are listed in Table 22.1. Poor diet, excess alcohol intake, and obesity are three important environmental factors, accounting for 30 to 40% of cancer risk.[12,13]

Some of the most convincing evidence of the robust relationship between environmental factors and cancer comes from studies of cancer rates in people migrating to other countries.[2] Rates of breast cancer, for example, are low in rural Asia. When individuals from rural parts of Asia immigrate to the United States, their rates of breast cancer become the same as or higher than the U.S. rate by the third generation. Rates of prostate cancer similarly increase as men move from countries with low rates to countries with high rates. Rates of breast cancer in Japanese and Alaskan Native women have increased substantially as they have adopted Westernized diets and lifestyles.[14]

Some people have genetic predispositions toward cancer, which means they have a tendency to develop cancer if regularly exposed to certain substances in the diet or environment. A genetically based susceptibility to cancer can develop in a fetus during pregnancy and in infancy. Exposure to calorie deficits, certain viruses, and other specific substances during these periods of rapid growth and development can modify the function of genes that help protect people from developing cancer.[15]

Insulin resistance, oxidative stress, and chronic inflammation appear to be important mechanisms underlying acquired gene mutations and cancer development related to

Table 22.1 The nine leading modifiable risk factors related to cancer development worldwide[13]

Risk factor
1. Obesity
2. Low vegetable and fruit intake
3. Physical inactivity
4. Smoking
5. Excess alcohol intake
6. Unsafe sex
7. Air pollution
8. Indoor use of solid fuels
9. Hepatitis B or C viral infection

© Mark Lehigh/Shutterstock.com

Illustration 22.3 The charred and black coating on grilled or broiled meats is the part you shouldn't eat.

dietary and other exposures. These conditions are often associated with obesity, unhealthy dietary patterns, and physical inactivity.[16–18]

Given the high percentage of cancers related to diet and other environmental factors, cancer is considered a largely preventable disease. Increasing rates of new cases of cancer took a turn for the better after 1992 corresponding to declines in rates of tobacco use. Death rates from cancer in the United States are continuing to decline among both women and men, and among individuals from all major ethnic and racial groups.[19]

Fighting Cancer with a Fork

- **Summarize the dietary patterns and lifestyle factors that affect cancer development and that decrease the risk of cancer.**

Specific characteristics of dietary patterns linked to the development of cancer include low intake of vegetables, fruits, and whole grains and regular intake of charred meats (Illustration 22.3), red and processed meats, foods and beverages high in added sugars, and excess alcohol consumption.[16,18,20] Diets and lifestyles related to the prevention of cancer are represented by healthy dietary patterns and lifestyles, and not by hard rules about special diets or foods, dietary restrictions, or types of physical activities. Table 22.2 summarizes characteristics of dietary patterns and lifestyles that are related to a reduced risk of cancer.

Frequent consumption of certain types of food is sometimes more strongly related to particular cancers than to other types. For example, regular consumption of tomato products is related in particular to a decreased risk of prostate cancer, and regular intake of red and processed meats appears to increase the risk of colorectal cancer.[16,23]

How Do Healthy Dietary Patterns Help Prevent Cancer?

Foods contain a variety of vitamins and minerals, as well as fiber and phytochemicals that help prevent damage to DNA or assist in its repair. These substances in food, particularly plant foods, appear to work together in ways that provide this protection. Attempts to

Table 22.2 Dietary patterns and lifestyles related to reduced risk of cancer[16,18,20,21]

1. Rely on foods for nutrient needs.
2. Utilize dietary patterns that are plant-based: • 5+ servings of a variety of vegetables and fruits daily, including those that are dark green, orange, and red. • 3+ whole grains/products daily. • Regular consumption of dried beans, nuts, and seeds. • Include fish and seafood, chicken. • Limit intake of red and processed meats and highly processed foods. • Exclude charred meats. • Exclude excess alcohol intake. • Limit intake of foods and beverages with added sugars.
3. Exclude smoking.
4. Include 30–60 minutes daily of moderate-intensity physical activity.
5. Maintain normal weight.

REALITY CHECK

Kumquats and Cancer

"KUMQUATS PREVENT CANCER" reads the headline in the newspaper. Researchers speculate that *tartaphil*, the substance that makes kumquats tangy, may be the reason. Should you eat kumquats?

Who gets a thumbs-up?

Answers appear on page 22-6.

Photodisc

Sebastian: Kumquats are too sour. I'd take a tartaphil supplement though.

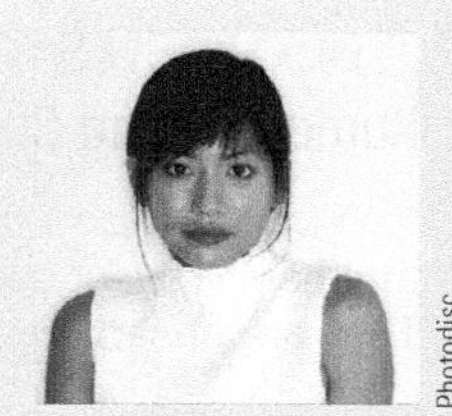
Photodisc

Flora: I already eat about three fruits every day. I'll include kumquats.

prevent cancer by giving large groups of people vitamin supplements or phytochemical extracts thought to account for the plant's beneficial effects on cancer development have not been successful.[23,24] In fact, a number of studies have noted that more harm than good results from the use of large amounts of individual supplements such as vitamin C, beta-carotene, selenium, and vitamin E.[25–27] Particular types of foods and dietary patterns clearly provide greater levels of protection against cancer than supplements. (The Reality Check feature for this unit addresses the issue of "magical" foods for cancer prevention. Take a look.) Some of the mechanisms underlying the relationships between intake of specific foods and cancer are fairly well described, whereas others are yet to be elucidated.

A major role plants foods play in reducing cancer risk appears to be related to the antioxidant and anti-inflammatory functions of certain vitamins and phytochemicals. These components of foods help neutralize reactive oxygen and other molecules, preventing them from damaging DNA. They also participate in the repair of DNA.[16,20] Many brightly colored vegetables and fruits contain phytochemicals that act as antioxidants, and their consumption is being encouraged. A list of some of the best sources of the antioxidant phytochemicals that make vegetables and fruits colorful is given in Table 22.3. In some individuals, vegetables from the cruciferous family (e.g., broccoli, brussels sprouts, cabbage, and cauliflower) appear to "turn off" genes that promote the oxidative stress and inflammation that can damage DNA.[27,28]

Table 22.3 Colorful vegetable and fruit antioxidant sources

 Photodisc	Red: tomatoes, red raspberries, watermelon, strawberries, red peppers, cherries
	Dark green: broccoli, brussels sprouts, kale, spinach, watercress, turnip greens, collard greens
	Orange: carrots, mangos, papayas, apricots, sweet potatoes, pumpkins, oranges, tangerines, peaches, cantaloupe

ANSWERS TO REALITY CHECK

Kumquats and Cancer

No individual food or specific phytochemical has been proved to prevent cancer. Diets that help prevent cancer are not based on one or even a few foods or supplements. They are based on day-to-day intake of a healthy array of foods and lifestyles.

P.S. The author made up the headline about kumquats.

Photodisc

Sebastian:

Photodisc

Flora:

Bogus Cancer Treatments

Cancer is a feared disease and, understandably, people want to identify changes they can make, or products they can use, that would ward off its development or help in its treatment. This situation leaves the area of cancer prevention and treatment open to fraudulent claims.[29] Unorthodox purported cancer cures such as macrobiotic diets; hydrogen peroxide ingestion; laetrile tablets; vitamin, mineral, and herbal supplements; and animal gland therapy have not been shown to be effective for the prevention or the treatment of cancer. Such remedies have been promoted since the early 1900s. They still exist because, although not proven to work, they offer some people hope. They should not be used as a substitute for evidence-based approaches to cancer prevention or treatment.[29,30]

Digital Vision/Getty Images

NUTRITION up close

A Cancer Risk Checkup

Focal Point: Reducing cancer risk.

A number of behaviors that help protect people from developing cancer are listed below. Check those that apply to you.

	Yes	No	Don't know
1. I eat a dark green, orange, or red vegetable or fruit at least daily.			
2. I consume whole-grain products daily.			
3. I eat broccoli, cauliflower, cabbage, or another vegetable of the cruciferous family twice a week.			
4. I avoid eating charred meats.			
5. I do not take large amounts of vitamin supplements.			
6. I do not smoke or chew tobacco.			
7. I exercise regularly.			
8. I do not have too much body fat.			

Feedback to the Nutrition Up Close is located in Appendix F.

REVIEW QUESTIONS

- **Determine the processes involved in the development and progression of cancer.**
- **Summarize the dietary patterns and lifestyle factors that affect cancer development and that decrease the risk of cancer.**

1. Cancer development is related to unrepaired damaged to DNA. **True/False**
2. Little can be done to prevent cancer. **True/False**
3. Diets providing five or more servings per day of vegetables and fruits reduce the risk of cancer. **True/False**
4. Cancer has many causes. **True/False**
5. Some of the strongest evidence of the lack of a relationship between environmental factors and cancer development come from studies demonstrating a lack of change in cancer rates among groups of people migrating from one country to another. **True/False**
6. Some people have genetic predispositions that increase their risk of developing cancer if regularly exposed to certain substances in the diet or the environment. **True/False**
7. Dietary patterns providing large amounts of fiber increase the risk of cancer development. **True/False**
8. A number of studies have shown that supplements of vitamin C, beta-carotene, and vitamin E reduce the risk of cancer. **True/False**
9. A major role of plant foods in reducing cancer risk appears to be related to the antioxidant function of certain phytochemicals. **True/False**
10. Although the reasons for the effect are not yet understood, it appears that hydrogen peroxide ingestion and laetrile tablets reverse DNA damage and prevent cancer from developing. **True/False**
11. _____ You get a call from your sister notifying you that your great uncle has been diagnosed with early stage cancer. Your sister tells you that your great uncle's daughter believes her father can get better if he drinks carrot juice often. What would be the evidence-based reaction to the daughter's belief about the effectiveness of carrot juice?
    a. Foods high in beta-carotene have been found to reverse the development of cancer.
    b. Carrots, but not carrot juice, reverse cancer development.
    c. Beta-carotene supplements, but not food high in beta-carotene, have been found to limit cancer progression.
    d. The daughter may believe that carrot juice is effective against cancer, but that has not been shown by scientific studies to be true.

12. _____ After hearing about your great uncle's cancer, you decide to look at your diet and make a change to lower your risk of cancer development. Which of the following would be reasonable changes to make?

    a. Reduce your intake of dairy products and spices.
    b. Eat the dark green, orange, and red vegetables and fruits you like often.
    c. Start taking a multivitamin and mineral supplement to enhance your intake of vitamins from food.
    d. Eat more food because having a few extra pounds of body weight helps prevent cancer.

Answers to these questions can be found in Appendix F.

NUTRITION SCOREBOARD ANSWERS

1. Cancer doesn't spread from person to person. **False**
2. Cancer development is primarily related to environmental factors including diet, smoking, and exposure to radioactive particles and toxins. Genetic traits play a role by placing some individuals at increased risk for developing cancer.[10,18] **False**
3. It's true.[3] **True**
4. High levels of body fat are related to the development of some types of cancer.[3] **True**
5. Most people don't develop cancer. **True**

Good Things to Know about Minerals

Mediablitzimages/Whitebox Media/Alamy

NUTRITION SCOREBOARD

1. The sole function of minerals is to serve as a component of body structures such as bones, teeth, and hair. **True/False**
2. Bone continue to mineralize for years after we reach adult height. **True/False**
3. Ounce for ounce, spinach provides more iron than beef. **True/False**
4. Worldwide, the most common nutritional deficiency is iron deficiency. **True/False**
5. More than one in four American adults has hypertension. **True/False**

Answers can be found at the end of the unit.

Mineral Facts

LEARNING OBJECTIVES

After completing Unit 23 you will be able to:

- Identify key functions and food sources of five essential minerals.

- **Identify key functions and food sources of five essential minerals.**

What substances are neither animal nor vegetable in origin, cannot be created or destroyed by living organisms (or by any other ordinary means), and provide the raw materials from which all things on earth are made? The answer is the **mineral** elements; they are displayed in full in the periodic table presented in Illustration 23.1. Minerals considered *essential*, or required in the diet, are highlighted.

The body contains 40 or more minerals. Only 15 are an essential part of our diets; we obtain the others through the air we breathe or from other essential nutrients in the diet such as protein and vitamins. Seven of the 15 essential minerals are required in trace amounts in the diet, and our need for them is generally measured in micrograms (mcg or μg). Recommended daily intake levels of the eight other required minerals reflect needs measured in gram (g) or milligram (mg) amounts. The recommended intake of calcium for adult women and men, for example, is 1,000 mg (1 g), whereas only 150 mcg of the trace mineral iodine is recommended for daily consumption.

Minerals are unlike the other essential nutrients in that they consist of single atoms. A single atom of a mineral typically does not have an equal number of protons (particles that carry a positive charge) and electrons (particles that carry a negative charge), and it therefore carries a charge. The charge makes minerals reactive. Many of the functions of minerals in the body are related to this property.

KEY NUTRITION CONCEPTS

Material covered in this unit on minerals relates to the following key nutrition concepts:

1. Nutrition Concept #2: Foods provide energy (calories), nutrients, and other substances needed for growth and health.
2. Nutrition Concept #5: Humans have adaptive mechanisms for managing fluctuations in nutrient intake.
3. Nutrition Concept #8: Poor nutrition can influence the development of certain chronic diseases.

Getting a Charge Out of Minerals

The charge carried by minerals allows them to combine with other minerals of the opposite charge and form fairly stable compounds that become part of bones, teeth, cartilage, and other tissues. In body fluids, charged minerals serve as a source of electrical energy that stimulates muscles to contract and nerves to react. The electrical current generated by charged minerals when performing these functions can be recorded by an electrocardiogram (abbreviated EKG or ECG) or an electroencephalogram (EEG). Abnormalities in the pattern of electrical activity in EKGs signal pending or past problems in the heart muscle. An EKG recording is shown in Illustration 23.2. Electroencephalograms similarly record electrical activity in the brain.

The charge minerals carry is related to many other functions. It helps maintain an adequate amount of water in the body and assists in neutralizing body fluids when they become too acidic or basic. Minerals that perform the roles of **cofactors** are components of proteins and enzymes, and they provide the "spark" that initiates enzyme activity.

Charge Problems Because minerals tend to be reactive, they may combine with other substances in food to form highly stable compounds that are not easily absorbed. Absorption of zinc from foods, for example, can vary from 0 to 100%, depending on what is attached to it. Zinc in whole-grain products is very poorly absorbed because it is bound tightly to a substance called phytate. In contrast, zinc in meats is readily available because

minerals In the context of nutrition, minerals are specific, single atoms that perform particular functions in the body. There are 15 essential minerals—or minerals required in the diet.

cofactors Individual minerals required for the activity of certain proteins. For example:

- Iron is needed for hemoglobin's function in oxygen and carbon dioxide transport.
- Zinc is needed to activate or is a structural component of more than 200 enzymes.
- Magnesium activates over 300 enzymes involved in the formation of energy and proteins.

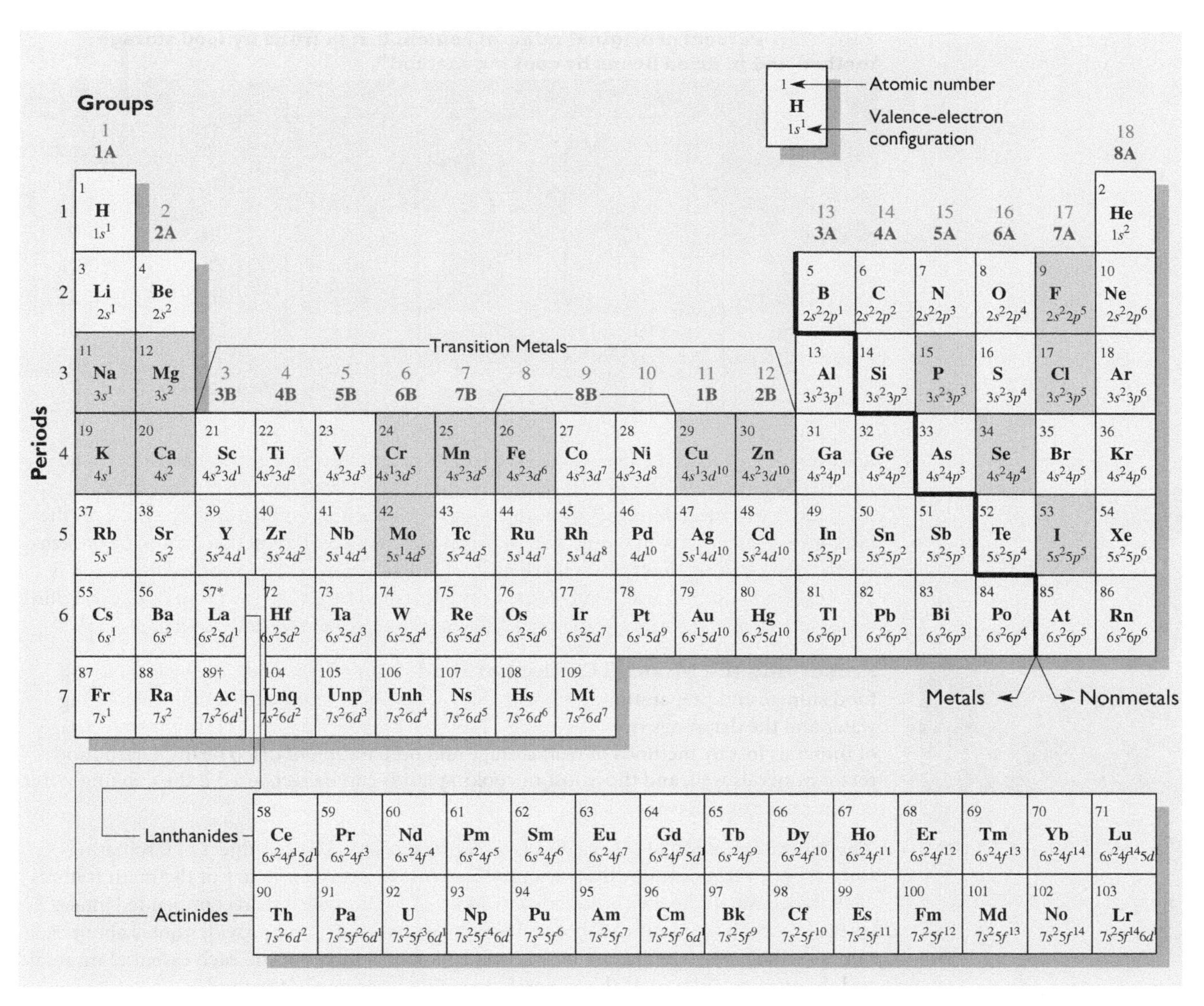

Illustration 23.1 The periodic table lists all known minerals. The highlighted minerals are required in the human diet.

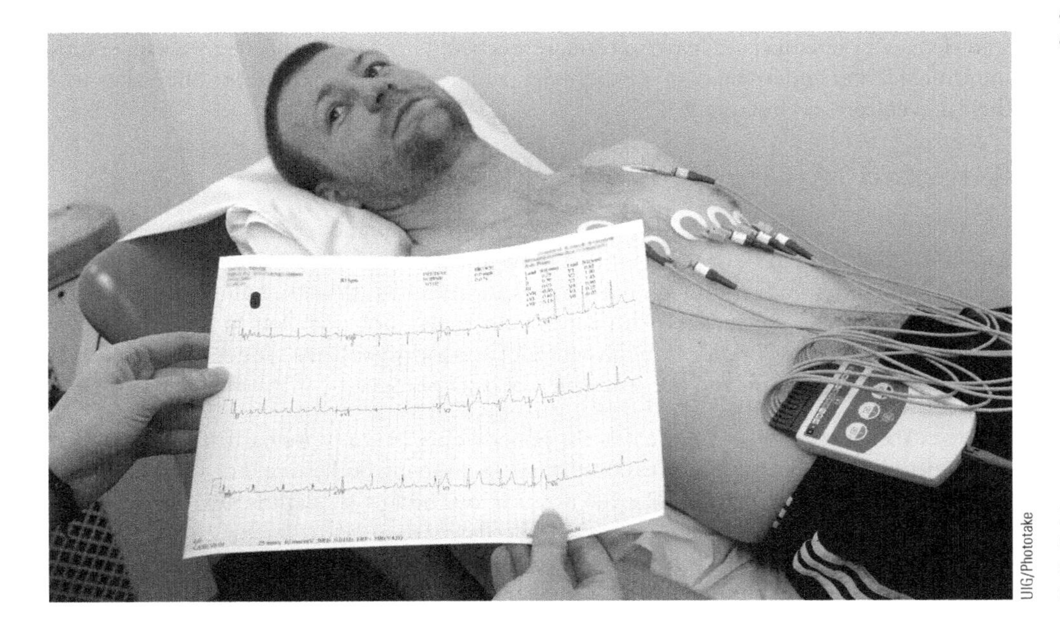

UIG/Phototake

Illustration 23.2 The electrical current measured by an EKG results from the movement of charged minerals across membranes of the muscle cells in the heart.

Table 23.1 Percent of original mineral content lost in fruits by food storage method and in dried beans by cooking method[51]

	Fruits			Dried beans boiled 2–2.5 hours	
	Canned (%)	Frozen (%)	Dried (%)	Water drained (%)	Water used (%)
Calcium	5	5	0	35	30
Iron	0	0	0	25	20
Magnesium	0	0	0	30	25
Potassium	10	10	0	35	30
Zinc	0	0	0	15	10
Copper	10	10	0	45	40

it is bound to protein. People whose sole source of zinc is whole grains have developed zinc deficiency, even though their intake of zinc is adequate.[3] The absorption of iron from foods in a meal decreases by as much as 50% if tea is consumed with the meal. In the intestines, iron binds with tannic acid in tea and forms a compound that cannot be broken down.[4] The calcium present in spinach and collard greens is poorly absorbed because it is firmly bound to oxalic acid. Many more examples could be given. The point is that you don't always get what you consume; the availability of minerals in food can vary a great deal.

Preserving the Mineral Content of Food Minerals in foods can be lost during food storage and preparation, primarily due to the leaching out of minerals in cooking water and the drippings from the meats. Table 23.1 gives two examples of the percentage of minerals lost by methods of fruit storage and preparation of dried beans. Dried foods retain minerals well, and those lost in cooking fluids can be recovered if the cooking water is minimized and consumed.

The Boundaries of This Unit All of the minerals could be the subject of fascinating stories, but in this unit we will concentrate on just three. A summary of the main features of all the essential minerals is provided in Table 23.2. This table lists recommended intake levels, functions, consequences of deficiency and overdose, and provides notes about each mineral. Table 23.10 at the end of this unit lists food sources of the each essential mineral and the average amount of the mineral present in a serving of the food.

The minerals highlighted in this unit are calcium, iron, and sodium. They have been selected primarily because they play important roles in the development of osteoporosis, iron-deficiency anemia, and hypertension, respectively. These disorders are widespread in the United States and in many other countries, and healthy dietary patterns offer a key to their prevention and treatment.

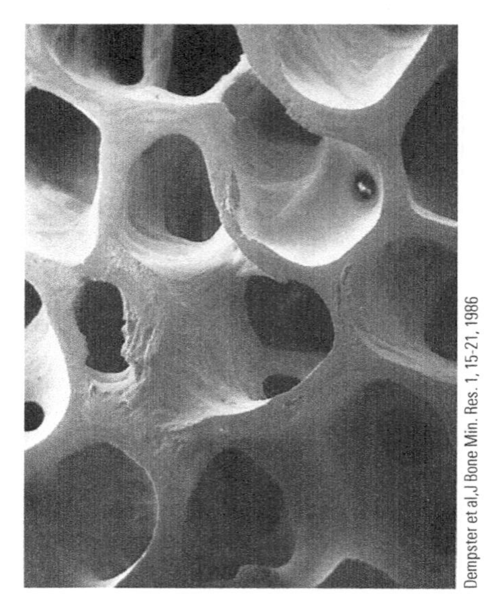

Dempster et al,J Bone Min. Res. 1, 15-21, 1986

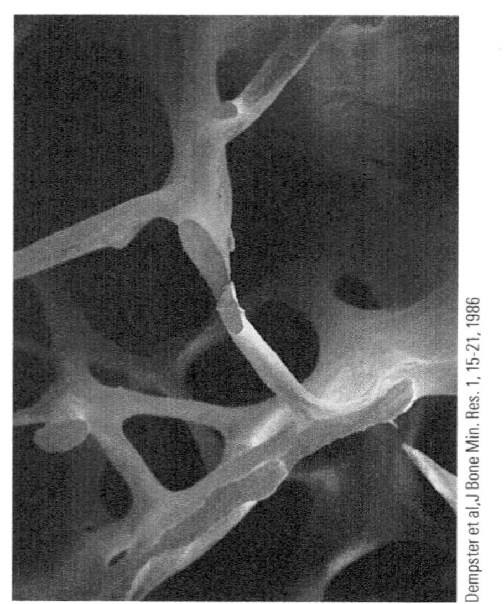

Dempster et al,J Bone Min. Res. 1, 15-21, 1986

Illustration 23.3 (top) Electron micrograph of healthy bone. (bottom) Electron micrograph of bone affected by osteoporosis.

Selected Minerals: Calcium

What you've heard about calcium is true: It's good for bones and teeth. About 99% of the 3 pounds of calcium in the body is located in bones and teeth. The remaining 1% is found in blood and other body fluids. We don't hear so much about this 1%, but it's very active. Every time a muscle contracts, a nerve sends out a signal, or blood clots to stop a bleeding wound, calcium in body fluids is involved. Calcium's most publicized function, however, is its role in bone formation and the prevention of osteoporosis.

A Short Primer on Bones Most of the bones we see or study are hard and dead. As a result, people often have the impression that bones in living bodies are that way. Nothing could be further from the truth. The 206 bones in our bodies are slightly flexible, living tissues infiltrated by blood vessels, nerves, and cells.

The solid parts of bones consist of networks of strong protein fibers (called the *protein matrix*) embedded with mineral crystals (Illustration 23.3). Calcium is the most abundant mineral found in bone, but many other minerals, such as phosphorus, magnesium, and carbon, are also incorporated into the protein matrix. The combination of water, the tough protein matrix, and mineral crystals makes bone very strong yet slightly flexible and capable of absorbing shocks.

Teeth have the same properties as other bone plus a hard outer covering called enamel, which is not infiltrated by blood vessels or nerves. Enamel serves to protect the teeth from destruction by bacteria and from mechanical wear and tear.

Illustration 23.4 This woman's stooped appearance is due to osteoporosis.

The Timing of Bone Formation Bones develop and mineralize throughout the first three decades of life. Even after the growth spurt occurs during adolescence and people think they are as tall as they will ever be, bones continue to increase in width and mineral content for 10 to 15 more years. Peak bone density, or the maximal level of mineral content in bones, is reached somewhere between the ages of 30 and 40. After that, bone mineral content no longer increases. The higher the peak bone mass, the less likely it is that osteoporosis will develop. People with higher peak bone mass simply have more calcium to lose before bones become weak and fracture easily. That also is the reason males experience osteoporosis less often than females do: they have more bone mass to lose.[5]

Bone size and density often remain fairly stable from age 30 to the mid-40s, but then bones tend to demineralize with increasing age. By the time women are 70, for example, their bones are 30–40% less dense on average than they once were.[6] A woman may lose an inch or more in height with age and develop the "dowager's hump" that is characteristic of osteoporosis in the spine (Illustration 23.4).

Bone Remodeling Bones slowly and continually go through a repair and replacement process known, appropriately enough, as **remodeling**. During remodeling, the old protein matrix is replaced and remineralized. If insufficient calcium is available to complete the remineralization, or if other conditions such as vitamin D inadequacy prevent calcium from being incorporated into the protein matrix, **osteoporosis** results.[7]

remodeling The breakdown and buildup of bone tissue.

osteoporosis (*osteo* = bones; *poro* = porous, *osis* = abnormal condition) A condition characterized by porous bones; it is due to the loss of minerals from the bones.

Osteoporosis

If you are female, you have a one in four chance of developing osteoporosis in your lifetime. If you are a Caucasian, Hispanic, or Asian female, you have a higher risk of developing osteoporosis than if you are an African American woman or a male. If you are male, your chance of developing osteoporosis is one in eight.[8] Approximately 49% of U.S. adults over the age of 50 are at are at risk of developing osteoporosis due to low bone mineral density, or *low bone mass* (Illustration 23.5).

Many factors including age, physical activity, dietary pattern, chronic inflammation, and body size influence the development of osteoporosis (Table 23.3). Key lifestyle characteristics associated with the premature development of osteoporosis include habitual consumption of poor quality diets (about half of U.S. adults), avoidance of weight-bearing exercise and sunshine, excessive alcohol intake, and smoking.[9-12,17]

Osteoporosis can be a disabling disease that reduces quality of life and dramatically increases the need for health care. Because the incidence of osteoporosis increases with age, its importance as a personal and public health problem is intensifying as the U.S. population ages. It currently appears, however, that a large percentage of the cases of osteoporosis can be prevented. The key to prevention is to build dense bones during childhood and the early adult years and then keep bones dense as you age.[16]

A number of factors affect the risk of osteoporosis (Table 23.3). The best approach to the prevention of osteoporosis is not to take vitamin, mineral or other types of dietary supplements but to live a healthy lifestyle.

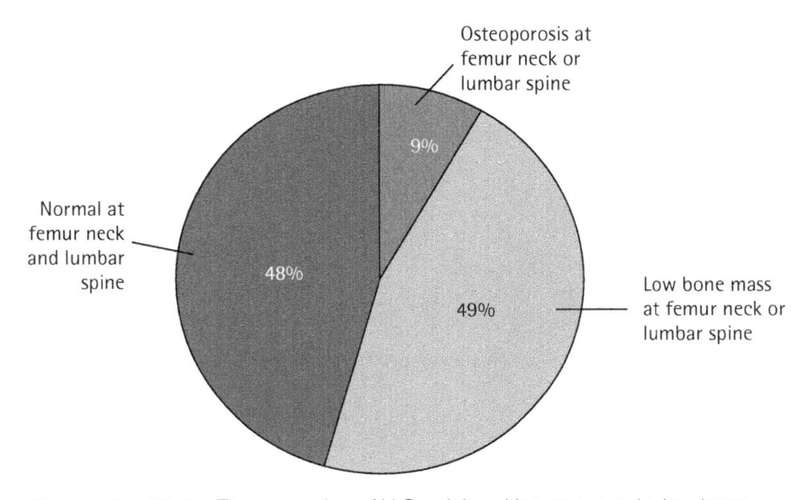

Illustration 23.5 The proportion of U.S. adults with osteoporosis, low bone mass, and normal bone mass.

Table 23.2 An intensive course on essential minerals

		Primary functions	Consequences of deficiency
Richard Anderson Calcium supplements are large because our daily need for calcium is high (1,000 milligrams/day). Four pills provide 800 milligrams of calcium, the amount in $2^2/_3$ cups of milk. An aspirin is shown for comparison.	**Calcium** AI[a] women: 1,000 mg men: 1,000 mg UL: 2,500 mg	• Component of bones and teeth • Needed for muscle and nerve activity, blood clotting	• Poorly mineralized, weak bones (osteoporosis) • Rickets in children • Osteomalacia (rickets in adults) • Stunted growth in children • Convulsions, muscle spasms
©Evan Lorne/Shutterstock.com Amphotos/Alamy	**Phosphorus** RDA women: 700 mg men: 700 mg UL: 4,000 mg	• Component of bones and teeth • Component of certain enzymes and other substances involved in energy formation • Needed to maintain the right acid–base balance of body fluids	• Loss of appetite • Nausea, vomiting • Weakness • Confusion • Loss of calcium from bones
©Adriana Nikolova/Shutterstock.com	**Magnesium** RDA women: 310 mg men: 400 mg UL: 350 mg (from supplements only)	• Component of bones and teeth • Needed for nerve activity • Activates hundreds of enzymes involved in energy and protein formation and other body processes	• Stunted growth in children • Weakness • Muscle spasms • Personality changes

[a]AIs (Adequate Intakes) and RDAs (Recommended Dietary Allowances) are for 19–30-year-olds; ULs (Upper Limits) are for 19–70-year-olds, 1997–2004.

Consequences of overdose	Primary food sources	Highlights and comments
• Drowsiness • Calcium deposits in kidneys, liver, and other tissues • Suppression of bone remodeling • Decreased zinc absorption	• Milk and milk products (cheese, yogurt) • Calcium-fortified foods (some juices, breakfast cereals, soy milk)  Bon Appetit/Deshong, Howard/Alamy	• The average intake of calcium among U.S. women is approximately 60% of the DRI. • One in four women and one in eight men in the United States develop osteoporosis. • Adequate calcium and vitamin D status must be maintained to prevent bone loss.
• Muscle spasms • Increased risk of cardiovascular disease and osteoporosis	• Milk and milk products (cheese, yogurt) • Meats • Seeds, nuts • Phosphates added to foods  © Christine Myaskovsky Phosphates are a common food additive.	• Deficiency is generally related to disease processes. • Phosphorus intake in the United States is increasing due to use of phosphates in processed foods.
• Diarrhea • Dehydration • Impaired nerve activity due to disrupted utilization of calcium	• Plant foods (dried beans, nuts, potatoes, green vegetables) • Ready-to-eat cereals	• Magnesium is found primarily in plant foods, where it is attached to chlorophyll. • Average intake among U.S. adults is below the RDA.  Photodisc

(continued)

Table 23.2 (continued)

		Primary functions	Consequences of deficiency
Food And Drink Photos/AGE Fotostock © Daniel Padavona/Shutterstock.com	**Iron** RDA women: 18 mg men: 8 mg UL: 45 mg Dr. Richard Kessel/Visuals Unlimited, Inc/Getty Images Dr. Gladden Willis/Visuals Unlimited, Inc/Getty Images Iron-deficiency anemia is characterized by microcytic anemia and small, pale red blood cells (bottom photo). Normal red blood cells are shown in the top photo.	• Transports oxygen as a component of hemoglobin in red blood cells • Component of myoglobin (a muscle protein) • Needed for certain reactions involving energy formation	• Iron deficiency • Iron-deficiency anemia • Weakness, fatigue • Pale appearance • Reduced attention span and resistance to infection • Hair loss • Mental retardation, developmental delay in children • Ice craving • Decreased resistance to infection
© Sbarabu/Shutterstock.com	**Zinc** RDA women: 8 mg men: 11 mg UL: 40 mg	• Required for the activation of many enzymes involved in the reproduction of proteins • Component of insulin, many enzymes	• Growth failure • Delayed sexual maturation • Slow wound healing • Loss of taste and appetite • In pregnancy, low-birth-weight infants and preterm delivery
© Marie C Fields/Shutterstock.com	**Fluoride** AI women: 3 mg men: 4 mg UL: 10 mg	• Component of bones and teeth (enamel) • Helps rebuild enamel that is beginning to decay	• Tooth decay and other dental diseases
CERTIFIED Red Label IODIZED SALT D.A. Weinstein/Custom Medical Stock Photo/Newscom	**Iodine** RDA women: 150 mcg men: 150 mcg UL: 1,100 mcg	• Required for the synthesis of thyroid hormones that help regulate energy production, growth, and development Mediscan/Alamy Iodine deficiency during pregnancy produces cretinism in the offspring.	• Goiter, thyroid disease • Cretinism (mental retardation, hearing loss, growth failure) Biophoto Associates/Getty Images

Consequences of overdose	Primary food sources	Highlights and comments
• Hemochromatosis ("iron poisoning") • Vomiting, abdominal pain, diarrhea • Blue coloration of skin • Iron deposition in liver and heart • Decreased zinc absorption • Oxidation-related damage to tissues and organs	• Liver, beef, pork • Dried beans • Iron-fortified cereals • Prunes, apricots, raisins • Spinach • Bread	• Cooking foods in iron and stainless steel pans increases the iron content of the foods. • Vitamin C, meat, and alcohol increase iron absorption. • Iron deficiency is the most common nutritional deficiency in the world. • Average iron intake of young children and women in the United States is low. • Inflammation and infection can affect the accuracy of results of some markers of iron status.
• Over 25 mg/day is associated with nausea, vomiting, weakness, fatigue, susceptibility to infection, copper deficiency, and metallic taste in mouth. • Increased blood lipids	• Meats (all kinds) • Dried beans • Grains • Nuts • Ready-to-eat cereals	• Like iron, zinc is better absorbed from meats than from plants. • Marginal zinc deficiency may be common, especially in children. • Zinc supplements taken within 24 hours of onset may decrease duration and severity of the common cold.
• Fluorosis • Brittle bones • Mottled teeth • Nerve abnormalities Dr. P. Marazzi/ Science Source "Mottled teeth" result from excessive fluoride.	• Fluoridated water and foods and beverages made with it • White grape juice • Raisins • Wine	• Toothpastes, mouth rinses, and other dental care products may provide fluoride. • Fluoride overdose has been caused by ingestion of fluoridated toothpaste. • Fluoridated water is not related to cancer. • Water fluoridation effectively decreases tooth decay in children and adults.
• Over 1 mg/day may produce pimples, goiter, decreased thyroid function, and thyroid disease.	• Iodized salt • Milk and milk products • Seaweed, seafoods • Bread from commercial bakeries	• Iodine deficiency was a major problem in the United States in the 1920s and 1930s. Deficiency remains a major health problem in some developing countries. • Amount of iodine in plants depends on iodine content of soil, season, and agriculture processing methods. • The need for iodine increases by 50% during pregnancy.

(*continued*)

Table 23.2 **(continued)**

		Primary functions	Consequences of deficiency
© Dani Vincek/ Shutterstock.com	**Selenium** RDA women: 55 mcg men: 55 mcg UL: 400 mcg	• Acts as an antioxidant in conjunction with vitamin E (protects cells from damage due to exposure to oxygen) • Needed for thyroid hormone production	• Anemia • Muscle pain and tenderness • Keshan disease (heart failure), Kashin-Beck disease (joint disease)
	Copper RDA women: 900 mcg men: 900 mcg UL: 10,000 mcg	• Component of enzymes involved in the body's utilization of iron and oxygen • Functions in growth, immunity, cholesterol and glucose utilization, brain development	• Anemia • Seizures • Nerve and bone abnormalities in children • Growth retardation
	Manganese AI women: 2.3 mg men: 1.8 mg	• Needed for the formation of body fat and bone	• Weight loss • Rash • Nausea and vomiting
	Chromium AI women: 35 mcg men: 25 mcg	• Required for the normal utilization of glucose and fat	• Elevated blood glucose and triglyceride levels • Weight loss
	Molybdenum RDA women: 45 mcg men: 45 mcg UL: 2,000 mcg	• Component of enzymes involved in the transfer of oxygen from one molecule to another	• Rapid heartbeat and breathing • Nausea, vomiting • Coma
© Jill Battaglia/ Shutterstock.com	**Sodium** AI women: 1,500 mg men: 1,500 mg UL: 2,300 mg	• Needed to maintain the right acid–base balance in body fluids • Helps maintain an appropriate amount of water in blood and body tissues • Needed for muscle and nerve activity	• Weakness • Apathy • Poor appetite • Muscle cramps • Headache • Swelling
© Dusan Zidar/ Shutterstock.com © Valentyn Volkov/ Shutterstock.com	**Potassium** AI women: 4,700 mg men: 4,700 mg UL: Not determined	• Same as for sodium	• Weakness • Irritability, mental confusion • Irregular heartbeat • Paralysis
	Chloride AI women: 2,300 mg men: 2,300 mg UL: 3,600 mg	• Component of hydrochloric acid secreted by the stomach (used in digestion) • Needed to maintain the right acid–base balance of body fluids • Helps maintain an appropriate water balance in the body	• Muscle cramps • Apathy • Poor appetite • Long-term mental retardation in infants

Consequences of overdose	Primary food sources	Highlights and comments
• "Selenosis," which includes symptoms of hair and fingernail loss, weakness, liver damage, irritability, and "garlic" or "metallic" breath.	• Fish • Eggs	• Content of foods depends on amount of selenium in soil, water, and animal feeds. • Selenium supplements have not been found to prevent cancer.
• Wilson's disease (excessive accumulation of copper in the liver and kidneys) • Vomiting, diarrhea • Tremors • Liver disease	• Potatoes • Grains • Dried beans • Nuts and seeds • Seafood • Ready-to-eat cereals	• Toxicity can result from copper pipes and cooking pans. • Average intake in the United States is below the RDA.
• Infertility in men • Disruptions in the nervous system, learning impairment • Muscle spasms	• Whole grains • Coffee, tea • Dried beans • Nuts	• Toxicity is related to overexposure to manganese dust in miners or contaminated groundwater.
• Kidney and skin damage	• Whole grains • Wheat germ • Liver, meat • Beer, wine • Oysters	• Toxicity usually results from exposure in chrome-making industries or overuse of supplements. • Supplements do not build muscle mass, increase endurance, or reduce blood glucose levels.
• Loss of copper from the body • Joint pain • Growth failure • Anemia • Gout	• Dried beans • Grains • Dark green vegetables • Liver • Milk and milk products	• Deficiency is extraordinarily rare.
• High blood pressure in susceptible people • Kidney disease • Heart problems	• Foods processed with salt • Cured foods (corned beef, ham, bacon, pickles, sauerkraut) • Table and sea salt • Bread • Milk, cheese • Salad dressing	• Very few foods naturally contain much sodium. • Processed foods are the leading source of dietary sodium. • High-sodium diets are associated with the development of hypertension in "salt-sensitive" people.
• Irregular heartbeat, heart attack	• Plant foods (potatoes, squash, lima beans, tomatoes, plantains, bananas, oranges, avocados), dried fruit • Fish, clams • Milk and milk products • Coffee • Dried beans, carrot juice	• Content of vegetables is often reduced in processed foods. • Diuretics (water pills) and other antihypertension drugs may deplete potassium. • Salt substitutes often contain potassium. • Potassium intake tracks with vegetable and fruit intake.
• Vomiting	• Same as for sodium. (Most of the chloride in our diets comes from salt.)	• Excessive vomiting and diarrhea may cause chloride deficiency. • Legislation regulating the composition of infant formulas was enacted in response to formula-related chloride deficiency and subsequent mental retardation in infants.

Table 23.3 Risk factors for osteoporosis[9–12]

- Female
- Menopause
- Poor overall dietary pattern
- Deficient calcium intake
- Caucasian or Asian heritage
- Thinness ("small bones")
- Cigarette smoking
- Poor vitamin D status
- Ovarectomy (ovaries removed) before age 45
- Chronic inflammation
- Deficient vitamin D status
- Genetic factors

A key lifestyle characteristic associated with the premature development of osteoporosis is adoption of a "Western"-style dietary pattern that predominately consists of fast and fried foods, sweetened beverages and food products, highly processed snack foods and prepared meals, meat, and refined carbohydrate foods. About half of U.S. adults consume this type of dietary pattern. Avoidance of weight-bearing exercise and sunshine, excessive alcohol intake, and smoking are additional risk factors for osteoporosis.[9-12,17]

The need for calcium can be met by consuming two to three servings of milk or other dairy products daily, and there are additional options for obtaining bioavailable calcium from other foods. Table 23.4 lists dairy and other sources of bioavailable calcium, and Illustrations 23.6 and 23.7 provide examples of calcium-fortified food products and labeling.

Healthy Dietary Patterns and Osteoporosis

Healthy dietary patterns are plant-based and include an array of colorful vegetables and fruits, nuts and seeds, whole grain products, fish and poultry, dairy products, and other basic foods. They provide limited amounts of red and processed meats, highly processed foods such as sweetened beverage and snack foods, refined grain products, and fried foods. Healthy dietary patterns contribute to osteoporosis prevention and management by decreasing chronic inflammation, and by supplying adequate amounts of calcium, vitamin D, antioxidants, fiber, and other nutrients needed for bone formation and maintenance.[9,12,14] Along with other aspects of healthy lifestyles, healthy dietary patterns promote bone health more safely and effectively than do supplements and unhealthy dietary patterns.[9–12,15,16,51]

Selected Minerals: Iron

hemoglobin The iron-containing protein in red blood cells.

myoglobin The iron-containing protein in muscle cells.

Most (80%) of the body's iron supply is found in **hemoglobin**. Small amounts are present in **myoglobin**, and free iron is involved in processes that capture energy released during the breakdown of proteins and fats.[23]

The Role of Iron in Hemoglobin and Myoglobin What happens to a car when its paint gets scratched? After a while, the exposed metal rusts. The iron in the metal combines with oxygen in the air, and the result is iron oxide, or *rust*. Iron readily combines with oxygen, and that property is put to good use in the body. From its location in hemoglobin in red blood cells, iron loosely attaches to oxygen when blood passes near the inner surface

Table 23.4 Caloric content, calcium level, and percentage of available calcium from different foods[20]

Food	Amount	Calories	Calcium (mg)	Calcium absorbed (%)	Available calcium (mg)
Yogurt, low fat	1 cup	143	13	32	132
Skim milk	1 cup	85	301	32	96
Soy milk (fortified)	1 cup	79	300	31	93
1% milk	1 cup	163	274	32	88
Tofu	1 cup	188	260	31	81
Cheese	1 oz	114	204	32	65
Kale, cooked	1 cup	42	94	49	46
Broccoli, cooked	1 cup	44	72	61	44
Bok choy, raw	1 cup	9	73	54	39
Dried beans	1 cup	209	120	24	29
Spinach, cooked	1 cup	42	244	5	12

Richard Anderson

Illustration 23.6 The number of calcium-fortified foods is increasing.

"Regular exercise and a healthy diet with enough calcium help maintain good bone health and may reduce the risk of osteoporosis later in life."

Nutrition Facts

Serving Size: 1 cup (240ml)
Servings per Container: 16

Amount per Serving	
Calories 110	Calories from Fat 20
	% Daily Value*
Total Fat 2.5g	4%
Saturated Fat 1.5g	8%
Trans Fat 0g	
Cholesterol 15mg	4%
Sodium 135mg	6%
Total Carbohydrate 13g	4%
Dietary Fiber 0g	0%
Sugars 12g	
Protein 8g	
Vitamin A 10% •	Vitamin C 4%
Calcium 30% Iron 0%	Vitamin D 25%
Phosphorus 10%	

*Percent Daily Values are based on a 2,000 calorie diet.

MILK
1% LOWFAT MILK

Illustration 23.7 Food products that are good sources of calcium can be labeled with a health claim.

of the lungs. The bright red, oxygenated blood is then delivered to cells throughout the body. When the oxygenated blood passes near cells that need oxygen for energy formation or for other reasons, oxygen is released from the iron and diffuses into cells (Illustration 23.8). The free iron in hemoglobin then picks up carbon dioxide, a waste product of energy formation. When carbon dioxide attaches to iron, blood turns from bright red to dark bluish red. Blood then circulates back to the lungs, where carbon dioxide is released from the iron and exhaled into the air. The free iron attaches again to oxygen that enters the lungs, and the cycle continues.

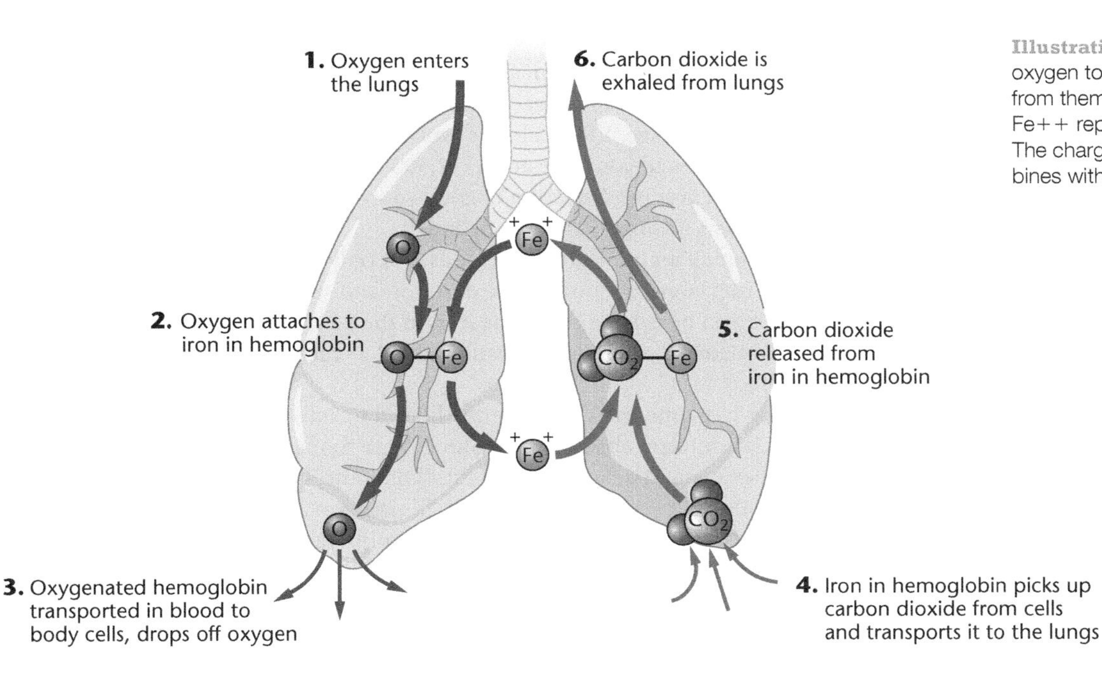

Illustration 23.8 Iron's role in carrying oxygen to cells and carbon dioxide away from them by way of the bloodstream. Fe++ represents charged particles of iron. The charge is neutralized when iron combines with oxygen or carbon dioxide.

Table 23.5 Incidence of iron deficiency[30]

	Population with iron deficiency (%)
Worldwide (children under 5 years):	
Developing countries	51
Developed countries	12
United States:	
Children, 1–3 years	8
Pregnant women	12
Females, 20–49 years	7
Males, 12–49 years	1 or less

Iron in myoglobin traps oxygen delivered by hemoglobin, stores it, and releases it as needed for energy formation for muscle activity. In effect, myoglobin boosts the supply of oxygen available to muscles.

The functions of iron just described operate smoothly when the body's supply of iron is sufficient. Unfortunately, that is often not the case.

iron deficiency A disorder that results from a depletion of iron stores in the body. It is characterized by weakness, fatigue, short attention span, poor appetite, increased susceptibility to infection, and irritability.

Iron Deficiency Is a Big Problem The most widespread nutritional deficiency in both developing and developed countries is **iron deficiency** (Table 23.5). It is estimated that a third of people in the world is iron deficient.[24] For the most part, iron deficiency affects very young children and women of childbearing age, who have a high need for iron and frequently consume too little of it.[25] Iron deficiency may develop in people who have lost blood due to injury, surgery, or ulcers. Donating blood more than three times a year can also precipitate iron deficiency.[26]

iron-deficiency anemia A condition that results when the content of hemoglobin in red blood cells is reduced due to a lack of iron. It is characterized by the signs of iron deficiency plus paleness, exhaustion, and a rapid heart rate.

Consequences of Iron Deficiency Many body processes sputter without sufficient oxygen. People with iron deficiency usually feel weak and tired. They have a shortened attention span and a poor appetite, are susceptible to infection, and become irritable easily. If the deficiency is serious enough, **iron-deficiency anemia** develops, and additional symptoms occur. People with iron-deficiency anemia look pale, are easily exhausted, and have rapid heart rates. Iron-deficiency anemia is a particular problem for infants and young children because it is related to lasting retardation in mental development.[27]

Food Sources of Iron "Enough" iron, according to the RDA, is 8 milligrams for men and 18 milligrams per day for women aged 19 to 50. Consuming that much iron can be difficult for women. On average, 1,000 calories' worth of food provides about 6 milligrams of iron. Women would have to consume around 2,500 calories per day to obtain even 15 milligrams of iron on average. Selection of good sources of iron has to be done on a better-than-average basis if women are to get enough.

Iron is found in small amounts in many foods, but only a few foods such as liver, beef, and prune juice are rich sources. Foods cooked in iron and stainless steel pans can be a significant source of iron because some of the iron in the pan leaches out during cooking. On average, approximately 1 milligram of iron is added to each 3-ounce serving of food cooked in these pans.[28]

Most of the iron in plants and eggs is tightly bound to substances such as phytates or oxalic acid, which limit iron absorption, making these foods relatively poor sources of iron even though they contain a fair amount of it. (Differences in the proportions of iron absorbed from various food sources are shown in Illustration 23.9.) A 3-ounce hamburger and a cup of asparagus both contain approximately 3 milligrams of iron, for example. But 20 times more iron can be absorbed from the hamburger than from the asparagus.[29] Absorption of iron from plants is increased substantially if foods containing vitamin C are included in the same meal. Iron absorption is also increased by low levels of iron stores in

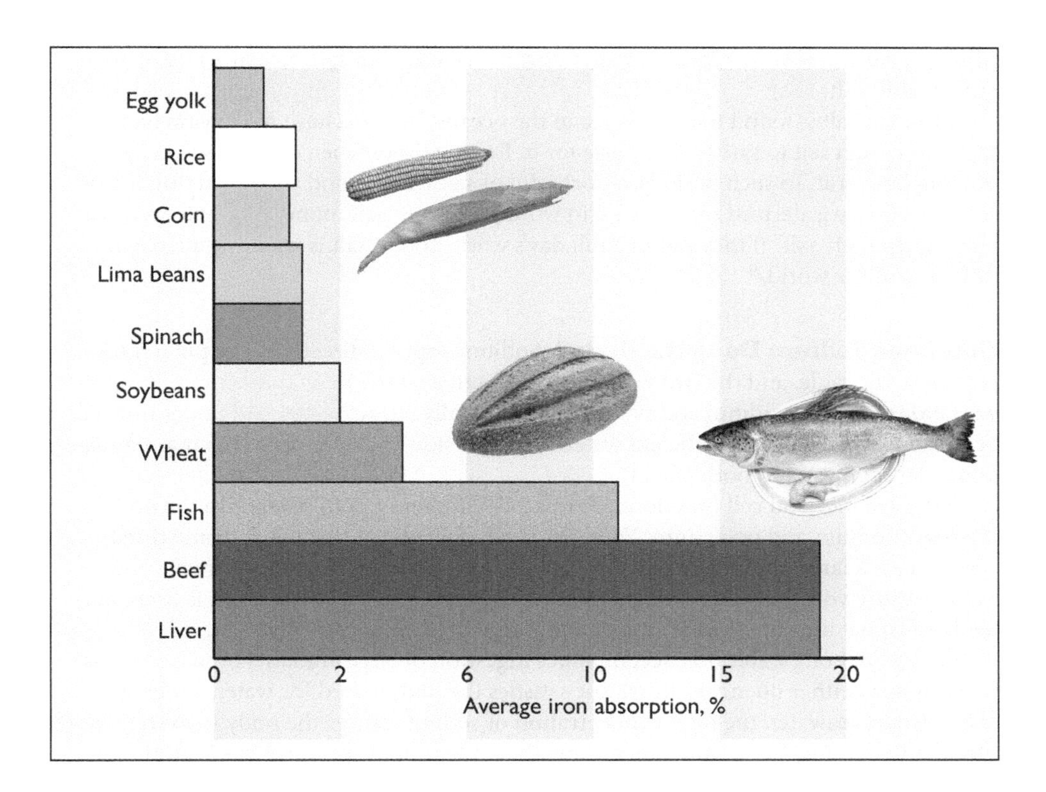

Illustration 23.9 Average percentage of iron absorbed from selected foods by healthy adults.[29]

the body (Illustration 23.10). In other words, if you're in need of iron, your body sets off mechanisms that allow more of it to be absorbed from foods or supplements. When iron stores are high, less iron is absorbed.[23]

The body's ability to regulate iron absorption provides considerable protection against iron deficiency and overdose. The protection is not complete, however, as evidenced by widespread iron deficiency and the occurrence of iron toxicity.

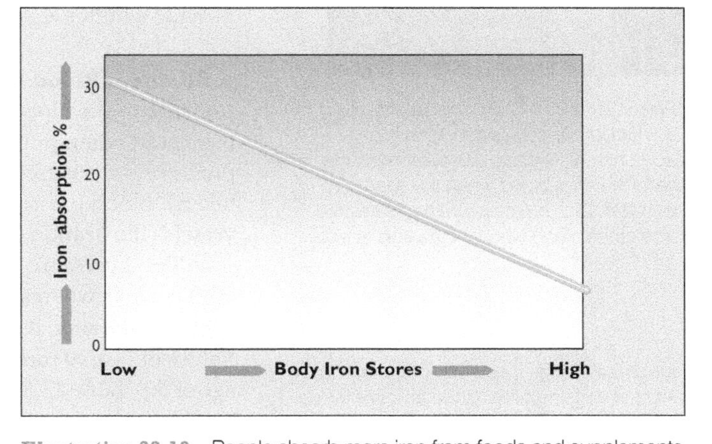

Illustration 23.10 People absorb more iron from foods and supplements when body stores of iron are low than when stores of iron are high.

Iron Toxicity Excess iron absorbed into the body cannot be easily excreted. Consequently, it is deposited in various tissues such as the liver, pancreas, and heart. There, the iron reacts with cells, causing damage that can result in liver disease, diabetes, and heart failure. One in every 200 people in the United States has an inherited tendency to absorb too much iron (a disorder called hemochromatosis (pronounced hem-oh-chrom-ah-toe-sis). Other people develop iron toxicity from consuming large amounts of iron with alcohol (alcohol increases iron absorption) or from very high iron intake, usually due to overdoses of iron supplements.[30]

Each year in the United States, more than 10,000 people accidentally overdose on iron supplements.[30] Victims are often young children who mistakenly think iron pills are candy (Illustration 23.11). The lethal dose of iron for a two-year-old child is about 3 grams,[31] the amount of iron present in 25 pills containing 120 milligrams of iron each. Iron supplements that are not being used should be thrown away or stored in a place where toddlers cannot get to them.

Illustration 23.11 If you were 3 years old, could you tell which "pills" are candy and which are the iron supplements? Overdoses of iron supplements are a leading cause of accidental poisoning in young children. (The iron supplements are the lightest green in color.)

Selected Minerals: Sodium

An extremely reactive mineral, sodium occurs in nature in combination with other elements. The most common chemical partner of sodium is chloride, and much of the sodium present on this planet is in the form of sodium chloride—table salt. Table salt is 40% sodium by weight; one teaspoon of salt contains about 2,300 milligrams of sodium. Visit the Reality Check for this unit to see how sea salt measures up against regular table salt.

Although salt is found in abundance in the oceans, humans have not always had access to enough salt to satisfy their taste for it. During times when salt was scarce, wars were fought over it. In such periods, a pocketful of salt was as good as a pocketful of cash. (The word *salary* is derived from the Latin word *salarium*, "salt money.") People were said to be "worth their salt" if they put in a full day's work. Today, salt is widely available in most parts of the world.[31]

What Does Sodium Do in the Body? Sodium appears directly above potassium in the periodic table, and the two work closely together in the body to maintain normal **water balance**. Both sodium and potassium chemically attract water, and under normal circumstances, each draws sufficient water to the outside or inside of cells to maintain an optimal level of water in both places.

Water balance and cell functions are upset when there's an imbalance in the body's supplies of sodium and potassium. You have probably noticed that you become thirsty when you eat a large amount of salted potato chips or popcorn. Salty foods make you thirsty because your body loses water when the high load of dietary sodium is excreted. The thirst signal indicates that you need water to replace what you have lost.

The loss of body water that accompanies ingestion of large amounts of salt explains why seawater neither quenches thirst nor satisfies the body's need for water. When a person drinks seawater, the high concentration of sodium causes the body to excrete more water than it retains. Rather than increasing the body's supply of water, the ingestion of seawater increases the need for water. In healthy people, the body's adaptive mechanisms provide a buffer against upsets in water balance due to low or high sodium intake.

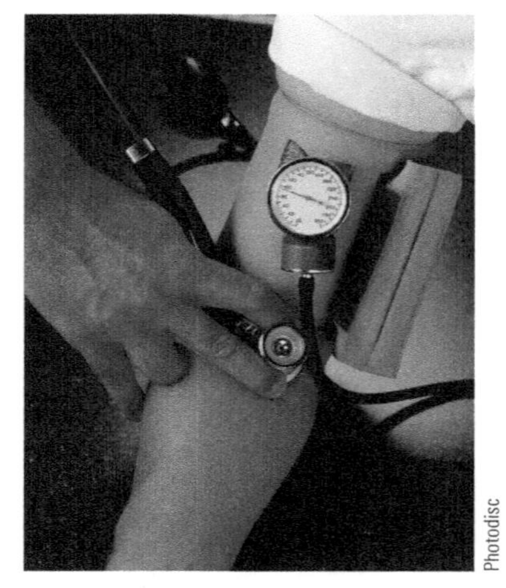

Photodisc

Illustration 23.12 A blood pressure test. Blood pressure is expressed as two numbers: Systolic pressure measures the force of the blood when the heart contracts. Diastolic pressure measures the force of the blood when the heart is at rest.

A Bit about Blood Pressure To circulate through the body, blood must exist under pressure in the blood vessels. The amount of pressure exerted on the walls of blood vessels is greatest when pulses of blood are passing through them (that's when *systolic* blood pressure is measured) and least between pulses (that's when *diastolic* blood pressure is taken). Blood pressure measurements note the highest and lowest pressure in blood vessels (Illustration 23.12).

Blood pressure levels less than 120/80 millimeters of mercury (mm Hg) are considered normal, whereas levels between 120/80 through 139/89 mm Hg are classified as **prehypertension**. It is estimated that 28% of U.S. adults qualify as having prehypertension. Values of 140/90 mm Hg and higher qualify as **hypertension** in adults younger then the age of 60 (Table 23.6), and approximately one-third of American adults have this disorder.[42] The blood pressure cut-off for hypertension in adults aged 60 and over is 150/90 mm Hg.[43] Several blood pressure measurements, taken while a person is relaxed, are needed to obtain an accurate measure of blood pressure. Even going to a clinic or doctor's office to have blood pressure measured can raise it for some people and lead to a false diagnosis of hypertension. This condition is referred to as "white coat hypertension."[44]

Although not considered a disease by itself, the presence of hypertension substantially increases the risk that a person will develop heart disease or kidney failure or will experience a heart attack or stroke.[45]

water balance The ratio of the amount of water outside cells to the amount inside cells; a proper balance is needed for normal cell functioning.

prehypertension In adults, typical blood pressure levels of 120/80 mm Hg through 139/89 mm Hg. In children and adolescents, prehypertension is defined as systolic or diastolic blood pressure equal to or greater than the 90th percentile but less than the 95th percentile of sex-, age- and height-specific blood pressure percentiles.

hypertension High blood pressure. It is defined as blood pressure exerted inside blood vessel walls that typically exceeds 140/90 millimeters of mercury in adults under 60 years of age and 150/90 in people 60 years of age and older. Hypertension in children and adolescents is defined as systolic or diastolic blood pressure equal to or greater than the 95th blood pressure percentile of sex-, age- and height-specific blood pressure percentiles.

Table 23.6 Hypertension categories[43]

Category	Systolic (mm Hg)[a]	Diastolic (mm Hg)
Prehypertension	120–139	80–89
Hypertension (adults through age 60)	≥140	≥90
Hypertension (adults 60 yrs. and over)	≥150	≥90

[a]mm Hg = millimeters of mercury.

What Causes Hypertension? Approximately 5–10% of all cases of hypertension can be directly linked to a cause. People who have hypertension with no identifiable cause (90–95% of all cases) are said to have **essential hypertension.**[45]

Risk Factors for Hypertension Table 23.7 summarizes common risk factors for hypertension. Obesity is a major risk factor for hypertension. For obese people with hypertension, the most effective treatment is weight loss. Excessive alcohol intake can prompt the development of hypertension, and moderation of intake (to two or fewer alcoholic drinks per day) can improve blood pressure. Physically inactive lifestyles also foster the development of hypertension. Diets low in potassium are a well-established risk factor for hypertension, and most Americans consume too little of it from foods. On average, U.S. women consume half of the RDA for potassium of 4,700 mg daily and men consume a third less than the RDA.[47] Regular intake of vegetables and fruits rich in potassium contributes to a reduction in risk for hypertension.[47] Diets containing adequate amounts of potassium from foods help lower blood pressure and appear to counteract the effects of high sodium intake on blood pressure.[35] The Take Action feature in this unit attempts to increase your awareness of good sources of potassium. If you are like most Americans, you are consuming too little of it.

Reduction of Sodium Intake Low sodium diets (<2,500 mg sodium per day) may improve blood pressure among a subset of individuals who are **salt-sensitive** and among people with habitually highsodium intake (over 5 grams per day).[32–35] Increasing evidence points to a relationship between the composition of diets of individuals who consume high amounts of salt rather than sodium intake by itself. High-salt foods tend to be processed, low in potassium, and lacking other essential nutrients compared to foods included in healthy dietary patterns.[35,36] The DASH healthy dietary pattern emphasizes lean sources of protein, including legumes, and low-fat dairy products, and whole grains, nuts, vegetables, and fruits, and has repeatedly been shown to improve blood pressure and lower the risk of hypertension and heart disease.[38–41]

Table 23.7 Risk factors for hypertension[46]

• Age
• Genetic predisposition
• High-sodium, low-potassium diet
• Obesity
• Physical inactivity
• Excessive alcohol consumption
• Smoking
• Frequent stress, anxiety

essential hypertension Hypertension of no known cause; also called primary or idiopathic hypertension, it accounts for 90–95% of all cases of hypertension.

salt sensitivity A genetically influenced condition in which a person's blood pressure rises when large amounts of salt or sodium are consumed. Such individuals are sometimes identified by blood pressure increases of 5 or 10% or more when switched from a low-salt to a high-salt diet. Approximately 51% of people with hypertension and 26% with normal blood pressure are salt sensitive.[37]

REALITY CHECK

Is Sea Salt a Better Choice Than Table Salt?

Matty and Miriam were shopping for snacks to serve when their friends came over for the game. They are now standing in front of the nut section at the store and can't decide if they should get nuts salted with sea salt or the regular salted nuts.

Who gets the thumbs up?

Answers on page 23-18.

iStockphoto.com/DRB Images, LLC

Matty: Sea salt is lower in sodium. Why don't we get the sea-salted nuts?

Wavebreakmedia Ltd/Getty Images Plus/Getty Images

Miriam: I agree. It's amazing, though. I've had them before and they still taste salty enough.

ANSWERS TO REALITY CHECK

Is Sea Salt a Better Choice Than Table Salt?

A survey by the American Heart Association of 1,000 adults discovered that 6 out of 10 thought sea salt was a low-sodium alternative to table salt. Equal amounts of sea salt and table salt contain the same amount of sodium (40%). Unlike table salt, sea salt (as well as Kosher salt) is generally not fortified with iodine. Iodized table salt is a leading source of iodine in U.S. diets.[48,49]

iStockphoto.com/DRB Images, LLC

Matty:

Wavebreakmedia Ltd/Getty Images Plus/Getty Images

Miriam:

take action To To Increase Foods Rich in Potassium in Your Diet

The Dietary Guidelines for Americans has designated potassium as a nutrient likely to be under-consumed by Americans in general. Foods listed below are good sources of potassium, as well as many other nutrients. Read the list and check the foods you consume at least twice a week. If you can't check four or more of the foods, consider taking action to include more good sources of potassium in your diet.

____ bran buds/flakes

____ avocado

____ dried beans

____ orange juice

____ lima beans

____ banana

____ baked potato

____ sweet potato

____ winter squash

____ spinach

____ tomato juice

____ orange juice

____ yogurt

____ fish

Table 28.8 Major sources of sodium in the American diet[49]

Sodium source	Contribution to sodium intake (%)
Processed foods	70
Fresh foods	14
Salt added at the table	10
Salt added during cooking and food preparation, other sources	5

The leading sources of salt (and therefore of sodium) in the U.S. diet are processed foods (Table 23.8). These foods account for over half of the total sodium intake by Americans, while approximately 10% of total sodium intake comes from salt added at the table.[50] Restricting the use of foods to which salt has been added during processing is the most effective way to lower salt intake.

High-salt processed foods include frozen meals, salad dressings, canned soups, ham, sausages, and biscuits. Only a small proportion of our total sodium intake enters our diet from fresh foods. Very few foods naturally contain much sodium—at least not until they are processed (Illustration 23.13).

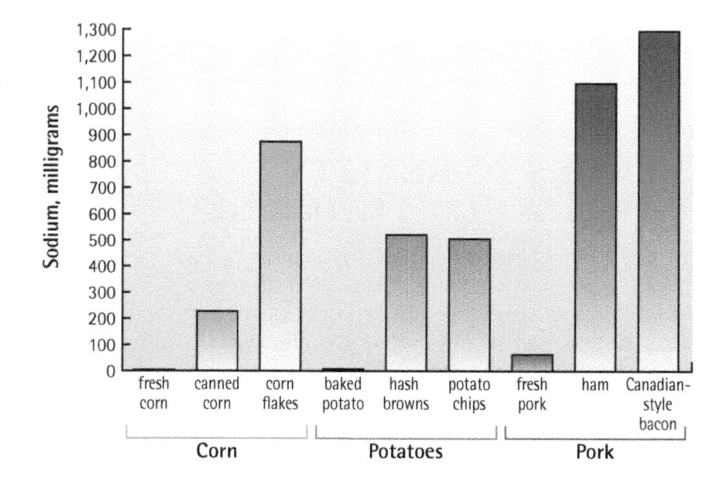

Illustration 23.13 Examples of how processing increases the sodium content of foods. Sodium values are for a 3-ounce serving of each food shown.

How Is Hypertension Treated? The recommended approach to treatment of all cases of hypertension consists of dietary and lifestyle changes and the use of medications if necessary.[43,44] Weight loss and smoking cessation (if needed), a reduced-sodium diet, if intake is high regular exercise, moderate alcohol consumption (if any), and the DASH eating plan are basic components of the approach to treatment (Table 23.9).[8,48] People who adhere to the DASH eating plan, shown in Illustration 23.14, often bring their blood pressure levels back into the normal range.[38–41]

The Dash Eating Plan

In the mid-1990s a revolutionary approach to the control of mild and moderate hypertension was tested, and the results have changed health professionals' thinking about high blood pressure prevention and management. Called the DASH (Dietary Approaches to Stop Hypertension) Eating Plan, it didn't focus on salt restriction; it was related to significant reductions in blood pressure within two weeks in most people tested. In some people, reductions in blood pressure were sufficient to erase the need for anti-hypertension medications, and for others the diet reduced the amount or variety of medications needed. A subsequent study showed that a low-sodium diet boosts the blood pressure–lowering effects of the DASH diet, especially in African Americans. This dietary pattern is also associated with reduced risk of heart disease and stroke,[50] and is recommended by the 2015 Dietary Guidelines Advisory Committee as a healthy dietary pattern for people in general.[38,40]

The DASH diet consists of eating patterns made up of the following food groups:

	Daily servings
Vegetables	4–5
Fruits	4–5
Grain products, mostly whole grain	7–8
Low-fat milk and dairy products	2–3
Lean meats, fish, poultry	6 oz
Nuts, seeds, dried beans	1

Although it does not work for all individuals with hypertension, the DASH diet is a mainstay in clinical efforts to reverse prehypertension and to manage hypertension.

A very useful Web site is available that gives practical tips on how to implement and follow the DASH eating pattern (www.nhlbi.nih.gov/health/public/heart/hbp/dash/new_dash.pdf). You can find the content quickly by searching the key term "DASH diet."

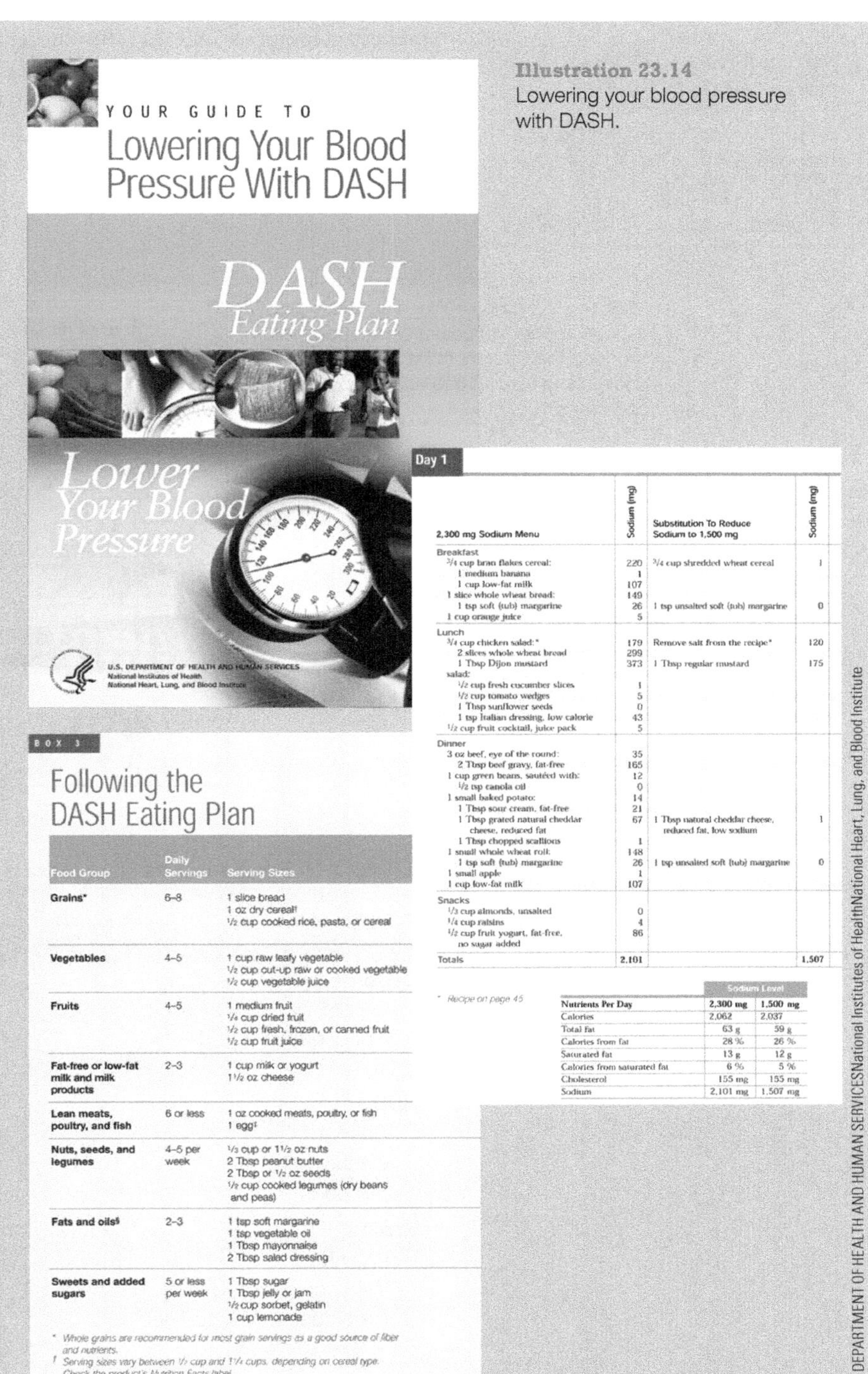

BOX 3

Following the DASH Eating Plan

Food Group	Daily Servings	Serving Sizes
Grains*	6–8	1 slice bread 1 oz dry cereal† ½ cup cooked rice, pasta, or cereal
Vegetables	4–5	1 cup raw leafy vegetable ½ cup cut-up raw or cooked vegetable ½ cup vegetable juice
Fruits	4–5	1 medium fruit ¼ cup dried fruit ½ cup fresh, frozen, or canned fruit ½ cup fruit juice
Fat-free or low-fat milk and milk products	2–3	1 cup milk or yogurt 1½ oz cheese
Lean meats, poultry, and fish	6 or less	1 oz cooked meats, poultry, or fish 1 egg‡
Nuts, seeds, and legumes	4–5 per week	⅓ cup or 1½ oz nuts 2 Tbsp peanut butter 2 Tbsp or ½ oz seeds ½ cup cooked legumes (dry beans and peas)
Fats and oils§	2–3	1 tsp soft margarine 1 tsp vegetable oil 1 Tbsp mayonnaise 2 Tbsp salad dressing
Sweets and added sugars	5 or less per week	1 Tbsp sugar 1 Tbsp jelly or jam ½ cup sorbet, gelatin 1 cup lemonade

* *Whole grains are recommended for most grain servings as a good source of fiber and nutrients.*
† *Serving sizes vary between ½ cup and 1¼ cups, depending on cereal type. Check the product's Nutrition Facts label.*

Day 1

2,300 mg Sodium Menu	Sodium (mg)	Substitution To Reduce Sodium to 1,500 mg	Sodium (mg)
Breakfast			
¾ cup bran flakes cereal:	220	¾ cup shredded wheat cereal	1
1 medium banana	1		
1 cup low-fat milk	107		
1 slice whole wheat bread:	149		
1 tsp soft (tub) margarine	26	1 tsp unsalted soft (tub) margarine	0
1 cup orange juice	5		
Lunch			
¾ cup chicken salad:*	179	Remove salt from the recipe*	120
2 slices whole wheat bread	299		
1 Tbsp Dijon mustard	373	1 Tbsp regular mustard	175
salad:			
½ cup fresh cucumber slices	1		
½ cup tomato wedges	5		
1 Tbsp sunflower seeds	0		
1 tsp Italian dressing, low calorie	43		
½ cup fruit cocktail, juice pack	5		
Dinner			
3 oz beef, eye of the round:	35		
2 Tbsp beef gravy, fat-free	165		
1 cup green beans, sautéed with:	12		
½ tsp canola oil	0		
1 small baked potato:	14		
1 Tbsp sour cream, fat-free	21		
1 Tbsp grated natural cheddar cheese, reduced fat	67	1 Tbsp natural cheddar cheese, reduced fat, low sodium	1
1 Tbsp chopped scallions	1		
1 small whole wheat roll:	148		
1 tsp soft (tub) margarine	26	1 tsp unsalted soft (tub) margarine	0
1 small apple	1		
1 cup low-fat milk	107		
Snacks			
⅓ cup almonds, unsalted	0		
¼ cup raisins	4		
½ cup fruit yogurt, fat-free, no sugar added	86		
Totals	**2,101**		**1,507**

* *Recipe on page 45*

Nutrients Per Day	Sodium Level 2,300 mg	Sodium Level 1,500 mg
Calories	2,062	2,037
Total fat	63 g	59 g
Calories from fat	28 %	26 %
Saturated fat	13 g	12 g
Calories from saturated fat	6 %	5 %
Cholesterol	155 mg	155 mg
Sodium	2,101 mg	1,507 mg

U.S. DEPARTMENT OF HEALTH AND HUMAN SERVICESNational Institutes of HealthNational Heart, Lung, and Blood Institute

Illustration 23.14
Lowering your blood pressure with DASH.

Table 23.9 Approaches to the treatment of hypertension[2,35,41,48]

- Weight loss (if needed)
- Sodium intake < 2,300 mg per day
- Moderate alcohol consumption (if any)
- Regular physical activity (≥30 minutes moderate intensity per day)
- The DASH or other health dietary pattern
- Antihypertension drugs (if needed)
- Meditation, yoga
- Smoking cessation (if applicable)

If blood pressure remains elevated after dietary and lifestyle changes have been implemented, or if blood pressure is quite high when diagnosed, anti-hypertension drugs are usually prescribed.[43]

Using spices and lemon juice to flavor foods, consuming fresh vegetables (rather than processed ones) and fresh fruits, checking out the sodium content of foods and selecting low-sodium ones, and repeated exposure to lower-salt foods can also help reduce sodium intake.[50]

Label Watch Not all processed foods with added sodium taste salty. To find out which processed foods are high in sodium, you have to examine the label. Increasingly, low-salt processed foods are entering the market and can be easily identified by the "low-salt" message on the label (Illustration 23.15). Terms used to identify low-salt (or low-sodium) foods are defined by the Food and Drug Administration (FDA). To be considered low sodium, foods must contain 140 milligrams or less of sodium per serving. Food manufacturers must adhere to the definitions when they make claims about the salt or sodium content of a food on the label.

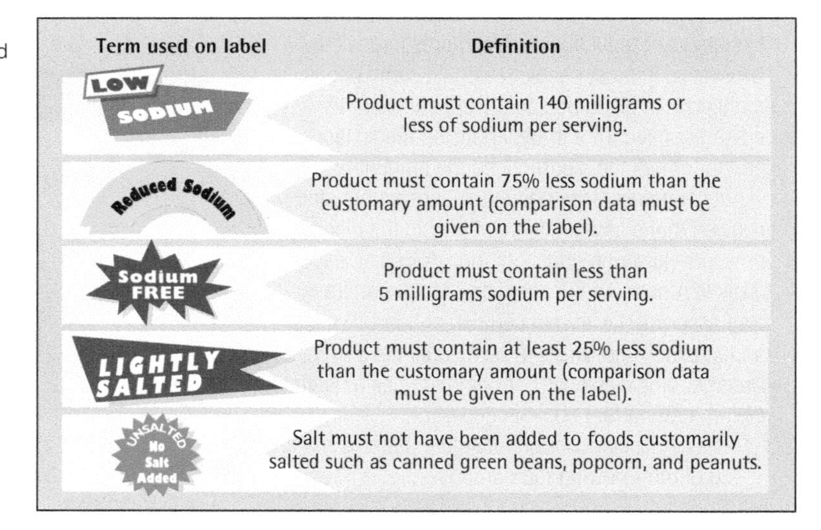

Term used on label	Definition
Low Sodium	Product must contain 140 milligrams or less of sodium per serving.
Reduced Sodium	Product must contain 75% less sodium than the customary amount (comparison data must be given on the label).
Sodium Free	Product must contain less than 5 milligrams sodium per serving.
Lightly Salted	Product must contain at least 25% less sodium than the customary amount (comparison data must be given on the label).
Unsalted No Salt Added	Salt must not have been added to foods customarily salted such as canned green beans, popcorn, and peanuts.

Illustration 23.15 When a label makes a claim about the sodium content of a food product, the label must list the amount of sodium in a serving and adhere to these definitions.

Table 23.10 Food sources of minerals

Magnesium		
Food	**Amount**	**Magnesium (mg)**
Legumes:		
Lentils, cooked	½ cup	134
Split peas, cooked	½ cup	134
Tofu	½ cup	130
Nuts:		
Peanuts	¼ cup	247
Cashews	¼ cup	93
Almonds	¼ cup	80
Grains:		
Bran buds	1 cup	240
Wild rice, cooked	½ cup	119
Breakfast cereal, fortified	1 cup	85
Wheat germ	2 Tbsp	45

Vegetables:		
Bean sprouts	½ cup	98
Black-eyed peas	½ cup	58
Spinach, cooked	½ cup	48
Lima beans	½ cup	32
Milk and milk products:		
Milk	1 cup	30
Cheddar cheese	1 oz	8
American cheese	1 oz	6
Meats:		
Chicken	3 oz	25
Beef	3 oz	20
Pork	3 oz	20

Calcium[a]

Food	Amount	Calcium (mg)
Milk and milk products:		
Yogurt, low-fat	1 cup	413
Milk shake (low-fat frozen yogurt)	1¼ cup	352
Yogurt with fruit, low-fat	1 cup	315
Skim milk	1 cup	301
1% milk	1 cup	300
2% milk	1 cup	298
3.25% milk (whole)	1 cup	288
Swiss cheese	1 oz	270
Milk shake (whole milk)	1¼ cup	250
Frozen yogurt, low-fat	1 cup	248
Frappuccino	1 cup	220
Cheddar cheese	1 oz	204
Frozen yogurt	1 cup	200
Cream soup	1 cup	186
Pudding	½ cup	185
Ice cream	1 cup	180
Ice milk	1 cup	180
American cheese	1 oz	175
Custard	½ cup	150
Cottage cheese	½ cup	70
Cottage cheese, low-fat	½ cup	69
Vegetables:		
Spinach, cooked	½ cup	122
Kale	½ cup	47
Broccoli	½ cup	36
Legumes:		
Tofu	½ cup	260
Dried beans, cooked	½ cup	60
Foods fortified with calcium:		
Orange juice	1 cup	350
Frozen waffles	2	300
Soy milk	1 cup	200–400
Breakfast cereals	1 cup	150–1000

[a]Actually, the richest source of calcium is alligator meat; 3½ ounces contain about 1,231 milligrams of calcium, but just try to find it on your grocer's shelf!

(continued)

Table 23.10 **(continued)**

Selenium		
Food	**Amount**	**Selenium (mcg)**
Seafood:		
Lobster	3 oz	66
Tuna	3 oz	60
Shrimp	3 oz	54
Oysters	3 oz	48
Fish	3 oz	40
Meats/Eggs:		
Liver	3 oz	56
Egg	1 medium	37
Ham	3 oz	29
Beef	3 oz	22
Bacon	3 oz	21
Chicken	3 oz	18
Lamb	3 oz	14
Veal	3 oz	10

Zinc		
Food	**Amount**	**Zinc (mg)**
Meats:		
Liver	3 oz	4.6
Beef	3 oz	4.0
Crab	½ cup	3.5
Lamb	3 oz	3.5
Turkey ham	3 oz	2.5
Pork	3 oz	2.4
Chicken	3 oz	2.0
Legumes:		
Dried beans, cooked	½ cup	1.0
Split peas, cooked	½ cup	0.9
Grains:		
Breakfast cereal, fortified	1 cup	1.5–4.0
Wheat germ	2 Tbsp	2.4
Oatmeal, cooked	1 cup	1.2
Bran flakes	1 cup	1.0
Brown rice, cooked	½ cup	0.6
White rice	½ cup	0.4
Nuts and seeds:		
Pecans	¼ cup	2.0
Cashews	¼ cup	1.8
Sunflower seeds	¼ cup	1.7
Peanut butter	2 Tbsp	0.9
Milk and milk products:		
Cheddar cheese	1 oz	1.1
Whole milk	1 cup	0.9
American cheese	1 oz	0.8

Sodium		
Food	**Amount**	**Sodium (mg)**
Miscellaneous:		
Salt	1 Tbsp	2,132
Dill pickle	1 (4½ oz)	1,930
Sea salt	1 Tbsp	1,716
Ravioli, canned	1 cup	1,065
Spaghetti with sauce, canned	1 cup	955
Baking soda	1 tsp	821
Beef broth	1 cup	810
Chicken broth	1 cup	770
Gravy	¼ cup	720
Italian dressing	2 Tbsp	720
Pretzels	5 (1 oz)	500
Green olives	5	465
Pizza with cheese	1 wedge	455
Soy sauce	1 tsp	444
Cheese twists	1 cup	329
Bacon	3 slices	303
French dressing	2 Tbsp	220
Potato chips	1 oz (10 pieces)	200
Catsup	1 Tbsp	155

Sodium		
Food	**Amount**	**Sodium (mg)**
Meats:		
Corned beef	3 oz	808
Ham	3 oz	800
Fish, canned	3 oz	735
Meat loaf	3 oz	555
Sausage	3 oz	483
Hot dog	1	477
Fish, smoked	3 oz	444
Bologna	1 oz	370
Milk and milk products:		
Cream soup	1 cup	1,070
Cottage cheese	½ cup	455
American cheese	1 oz	405
Cheese spread	1 oz	274
Parmesan cheese	1 oz	247
Gouda cheese	1 oz	232
Cheddar cheese	1 oz	175
Skim milk	1 cup	125
Whole milk	1 cup	120
Grains:		
Bran flakes	1 cup	363
Cornflakes	1 cup	325
Croissant	1 medium	270
Bagel	1	260
English muffin	1	203
White bread	1 slice	130
Whole wheat bread	1 slice	130
Saltine crackers	4 squares	125

(*continued*)

Table 23.10 **(continued)**

Iron		
Food	**Amount**	**Iron (mg)**
Meat and meat alternates:		
Liver	3 oz	7.5
Round steak	3 oz	3.0
Hamburger, lean	3 oz	3.0
Baked beans	½ cup	3.0
Pork	3 oz	2.7
White beans	½ cup	2.7
Soybeans	½ cup	2.5
Pork and beans	½ cup	2.3
Fish	3 oz	1.0
Chicken	3 oz	1.0
Grains:		
Breakfast cereal, iron-fortified	1 cup	8.0 (4–18)
Oatmeal, fortified, cooked	1 cup	8.0
Bagel	1	1.7
English muffin	1	1.6
Rye bread	1 slice	1.0
Whole-wheat bread	1 slice	0.8
White bread	1 slice	0.6
Fruits:		
Prune juice	1 cup	9.0
Apricots, dried	½ cup	2.5
Prunes	5 medium	2.0
Raisins	¼ cup	1.3
Plums	3 medium	1.1

Iron		
Food	**Amount**	**Iron (mg)**
Vegetables:		
Spinach, cooked	½ cup	2.3
Lima beans	½ cup	2.2
Black-eyed peas	½ cup	1.7
Peas	½ cup	1.6
Asparagus	½ cup	1.5

Phosphorus		
Food	**Amount**	**Phosphorus (mg)**
Milk and milk products:		
Yogurt	1 cup	327
Skim milk	1 cup	250
Whole milk	1 cup	250
Cottage cheese	½ cup	150
American cheese	1 oz	130
Meats:		
Pork	3 oz	275
Hamburger	3 oz	165
Tuna	3 oz	162
Lobster	3 oz	125
Chicken	3 oz	120
Nuts and seeds:		
Sunflower seeds	¼ cup	319
Peanuts	¼ cup	141
Pine nuts	¼ cup	106
Peanut butter	1 Tbsp	61

Grains:		
Bran flakes	1 cup	180
Shredded wheat	2 large biscuits	81
Whole-wheat bread	1 slice	52
Noodles, cooked	½ cup	47
Rice, cooked	½ cup	29
White bread	1 slice	24
Vegetables:		
Potato	1 medium	101
Corn	½ cup	73
Peas	½ cup	70
French fries	½ cup	61
Broccoli	½ cup	54
Other:		
Milk chocolate	1 oz	66
Cola	12 oz	51
Diet cola	12 oz	45

Potassium		
Food	**Amount**	**Potassium (mg)**
Vegetables:		
Potato	1 medium	780
Winter squash	½ cup	327
Tomato	1 medium	300
Celery	1 stalk	270
Carrots	1 medium	245
Broccoli	½ cup	205

Potassium		
Food	**Amount**	**Potassium (mg)**
Fruits:		
Avocado	½ medium	680
Orange juice	1 cup	469
Banana	1 medium	440
Raisins	¼ cup	370
Prunes	4 large	300
Watermelon	1 cup	158
Meats:		
Fish	3 oz	500
Hamburger	3 oz	480
Lamb	3 oz	382
Pork	3 oz	335
Chicken	3 oz	208
Grains:		
Bran buds	1 cup	1,080
Bran flakes	1 cup	248
Raisin bran	1 cup	242
Wheat flakes	1 cup	96
Milk and milk products:		
Yogurt	1 cup	531
Skim milk	1 cup	400
Whole milk	1 cup	370
Other:		
Salt substitutes	1 tsp	1,300–2,378

(continued)

Table 23.10 **(continued)**

Fluoride		
Food	**Amount**	**Fluoride (mcg)**
Grape juice, white	6 oz	350
Instant tea	1 cup	335
Raisins	3½ oz	234
Wine, white	3½ oz	202
Wine, red	3½ oz	105
French fries, McDonald's	1 medium	130
Dannon's Fluoride to Go	8 oz	178
Tap water, U.S.	8 oz	59
Municipal water, U.S.	8 oz	186
Bottled water, store brand	8 oz	37
Iodine		
Food	**Amount**	**Iodine (mcg)**
Seaweed	Sheet, 1 g	16–2,984
Iodized salt	1 tsp	400
Haddock	3 oz	125
Cod	3 oz	99
Yogurt, low-fat	1 cup	75
Milk, low-fat	1 cup	56
Cottage cheese	½ cup	50
Bread	2 oz (2 slices)	45
Shrimp	3 oz	30
Macaroni, cooked	1 cup	27
Egg	1	22
Cheddar cheese	1 oz	17

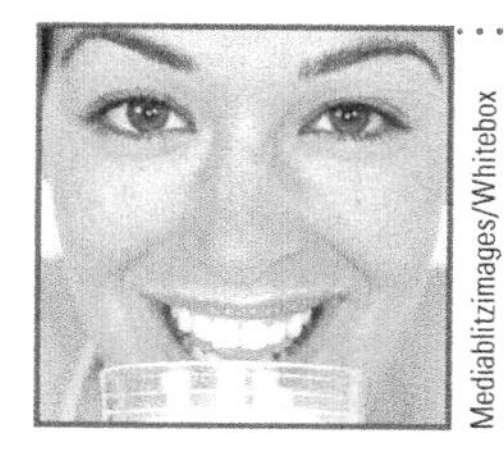
Mediabitzimages/Whitebox Media/Alamy

NUTRITION up close

Processed Foods and Your Diet

Focal Point: Are highly processed food keeping you from consuming a healthy dietary pattern?

The following is a list of some highly processed foods that are commonly consumed but don't easily fit into a healthy dietary pattern.

Place a "✓" next to the foods you commonly consume. In the blank lines at the end of the table, suggest basic foods you would enjoy eating instead.

_______ biscuits, refrigerated	_______ hot dogs
_______ bacon	_______ macaroni and cheese, boxed
_______ white bread	_______ beef jerky
_______ ham	_______ chicken nuggets
_______ white rice	_______ pork rind
_______ sausage	_______ pot pies, frozen
_______ cheese spread	_______ cheese crackers
_______ salami, pepperoni,	_______ taquitos
_______ frozen dinners	_______ potato chips

Feedback to the Nutrition Up Close is located in Appendix F.

REVIEW QUESTIONS

- **Identify key functions and food sources of five essential minerals.**

1. There are 40 essential minerals that perform specific functions in the body. Each must be obtained from the diet. **True/False**
2. Essential minerals are a structural component of teeth; they also serve as a source of electrical energy that stimulates muscles to contract and help maintain an appropriate balance of fluids in the body. **True/False**
3. The amount of certain minerals absorbed from foods varies a good deal based on the food source of the minerals. **True/False**
4. Calcium is the most abundant mineral found in bones, but other minerals such as magnesium, phosphorus, and carbon are also components of bone. **True/False**
5. Bones fully develop and mineralize during the first 18 years of life. **True/False**
6. Iodine deficiency and overdose are related to goiter and thyroid disease. **True/False**
7. Wilson's disease is related to magnesium deficiency. **True/False**
8. Vitamin D is required for the absorption and utilization of calcium by the body. **True/False**
9. People with iron deficiency generally feel weak and tired, have a poor appetite, and are susceptible to infection. **True/False**
10. Although iron deficiency is an important health problem in America, iron toxicity is not. **True/False**
11. High salt intake is the leading risk factor of hypertension. **True/False**
12. The major source of salt (or sodium) in our diet comes from salt added to food at the table. **True/False**

Pablo decides to lose 5 pounds in a week by going on a high-protein diet. The foods he'll eat consist of canned tuna, lean beef, skinless chicken, low-fat cottage cheese, and skim milk.

13. _____ Of the minerals listed below, which one is most likely to be lacking in his diet?
    a. iron
    b. calcium
    c. magnesium
    d. phosphorus

The next four questions refer to the following scenario.

Dorthea was just told by her adult nurse practitioner that she has prehypertension and recommends that she meet with a registered dietitian for consultation on the DASH eating plan. The appointment was made and Dorthea is speaking with the registered dietitian now.

14. _____ During the consultation, Dorthea will learn that the DASH eating pattern emphasizes:
    a. foods very low in sodium and high in fiber
    b. elimination of fast foods
    c. low-calorie, low-fat foods
    d. vegetables, fruits, and grain products
15. _____ Dorthea asks the dietitian how many vegetables a day are included in the DASH eating plan. The likely response from the dietitian is:
    a. 1–2 servings
    b. 2–3 servings
    c. 3–4 servings
    d. 4–5 servings
16. _____ How many servings of fruits would Dorthea be urged to include in her diet under the DASH eating plan?
    a. 1–2 servings
    b. 2–3 servings
    c. 3–4 servings
    d. 4–5 servings
17. _____ What other types of food would Dorthea be encouraged to include in her DASH eating plan?
    a. lean meats, fish, and poultry
    b. milk products of all types
    c. coffee and tea
    d. chocolate and molasses

Answers to these questions can be found in Appendix F.

NUTRITION SCOREBOARD ANSWERS

1. Minerals serve as structural components of the body, but they also play important roles in stimulating muscle and nerve activity and in other functions. **False**
2. Bones continue to grow and mineralize well after we reach adult height. Bone growth and development continues to about age 30. **True**
3. Spinach is a nutritious food, providing 0.6 mg iron per ounce. It contains less iron than beef (1 mg iron per ounce). The iron in spinach is poorly absorbed, but the iron in meat is well absorbed by the body. **False**
4. Iron deficiency is the most common nutritional deficiency in both developed and developing countries. Approximately one-fourth of the world's population is iron deficient.[1] **True**
5. Approximately one-third of American adults have hypertension (high blood pressure).[2] **True**

Dietary Supplements

NUTRITION SCOREBOARD

1. Products classified as *dietary supplements* consist of herbs and vitamin and mineral supplements only. **True/False**
2. Dietary supplements must be tested for safety and effectiveness before they can be sold. **True/False**
3. Herbal remedies have been used for over 1,000 years in Germany, so they must be safe and effective. **True/False**
4. *Probiotics* are "friendly bacteria" that benefit health. **True/False**

Answers can be found at the end of the unit.

Photodisc

LEARNING OBJECTIVES

After completing Unit 24 you will be able to:

- Make evidence-based decisions about the probable safety and effectiveness of specific dietary supplements.

dietary supplements Any products intended to supplement the diet, including vitamin and mineral supplements; proteins, enzymes, and amino acids; fish oils and fatty acids; hormones and hormone precursors; herbs and other plant extracts; and prebiotics and probiotics. Such products must be labeled "Dietary Supplement."

Dietary Supplements

- **Make evidence-based decisions about the probable safety and effectiveness of specific dietary supplements.**

What do vitamin E supplements, protein powders, energy drinks, homeopathic remedies, herbal remedies, and probiotics all have in common? They are members of the increasingly popular group of products called **dietary supplements**—and they are all discussed in this unit. Types of dietary supplements available to consumers are presented in Table 24.1. About half of U.S. adults use one or more of the over 54,000 dietary supplement products available on the market.[1] Dietary supplements are supposed to supplement the diet. They are not intended to prevent, treat, or cure disease. These disparate products are grouped together because they are regulated as food and not drugs.[2] You will learn from this unit that most dietary supplements have not been found to perform as claimed, and that they may have adverse, neutral, or beneficial effects on health. This "buyer beware" situation exists because of the loose rules that govern dietary supplements, and many consumers believe that they work or are worth a try.[3] You will also learn about exciting new developments related to intestinal fertilizers and friendly bacteria (no kidding).

KEY NUTRITION CONCEPTS

Foundational knowledge about nutrition that applies to the topic of dietary supplements is represented by the following three key nutrition concepts:

1. Nutrition Concept #1: Foods provide energy (calories), nutrients, and other substances needed for growth and health.
2. Nutrition Concept #4: Poor nutrition can result from both inadequate and excessive levels of nutrient intake.
3. Nutrition Concept #9: Adequacy, variety, and balance are key characteristics of healthful diets.

Regulation of Dietary Supplements

In 1994, Congress passed the Dietary Supplement Health and Education Act, which started the explosion in the availability of dietary supplements. Under the act, dietary supplements are minimally regulated by the Food and Drug Administration (FDA); they do not have to be tested prior to marketing or shown to be safe or effective.[2] Although often advertised to relieve certain ailments, they are not considered to be drugs. Dietary supplements are not subjected to rigorous testing to prove safety and effectiveness prior to sale, as drugs must be. Responsibility for evaluating the safety of dietary supplements lies with manufacturers and not the FDA. Supplements are deemed unsafe only when the FDA has proof they are harmful. Since few (about 0.3%) dietary supplements have been adequately tested, and because results of studies showing negative effects may never see the light of day, it is difficult to prove them to be unsafe.[4] The FDA largely relies on reports of ill effects from manufacturers, health professionals, and consumers to assess supplement safety. Since 1994, the FDA has received thousands of reports of adverse effects of supplements (primarily for herbal remedies), including several hundred deaths.[1,5,6] The FDA has taken action against hundreds of products because they contained hazardous drugs or ingredients harmful to health, or because they made false claims for products such as "helps replace medicine in the treatment of diabetes," and "cellulite fighter."[7]

The majority of adverse event complaints received by the FDA involve products with caffeine, yohimbe (marketed for male sexual enhancement), and stimulants; and those marketed for body building and weight loss.[1,5,22] Consumers are encouraged to report adverse effects of dietary supplements to the FDA and the National Institutes of Health at the website: safetyreporting.hhs.gov.

Table 24.1 Selected types of dietary supplements

Type	Example
1. Vitamins and minerals	Vitamins C and E, selenium
2. Herbs (botanicals)	Ginkgo, ginseng, St. John's wort
3. Proteins and amino acids	Shark cartilage, chondroitin sulfate, creatine
4. Hormones, hormone precursors	DHEA, vitamin D
5. Fats	Fish oils, DHA, lecithin
6. Other plant extracts	Garlic capsules, fiber, cranberry concentrate, echinacea
7. Prebiotics, probiotics	Psyllium, garlic, certain live bacteria

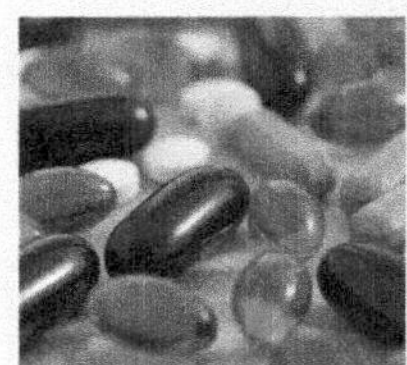
Photodisc

Vitamin, mineral, protein, and amino acid supplements.

Photodisc

Botanical supplements, such as dong quai, often come in gel caps.

Photodisc

Plant extracts can also be taken in liquid form, as tinctures.

Photodisc

According to FDA regulations (Table 24.2), dietary supplements must be labeled with a Supplemental Facts panel that lists serving size, ingredients, and percent Daily Value (% DV) of key nutrients, as well as a listing of ingredients.[8] Products can be labeled with a health claim, such as "high in calcium" or "low fat," if the product qualifies according to the nutrition labeling regulations. Supplements can also be labeled with structure/function claims, but these claims cannot refer to disease prevention or treatment effects. Claims such as "improves circulation," "helps prevents wrinkles," "supports the immune system," and "helps maintain mental health" can be used, whereas "prevents heart disease" or "cures depression" cannot be. If a function claim is made on the label or package inserts, the label or insert must include the FDA disclaimer, which states that the FDA does not support the claim. (This is done to reduce the FDA's liability for problems that may be caused by supplements.) Nonetheless, many people believe the health claims made for supplements.

Table 24.2 FDA regulations for dietary supplement labeling[7]

1. Product must be labeled "Dietary Supplement."
2. Product must have a Supplemental Facts label that includes serving size, amount of the product per serving, % Daily Value of key nutrients of public health significance, a list of other ingredients, and the manufacturer's name and address.
3. Nutrient claims (such as "low in sodium" and "high in fiber") can be made on labels of products that qualify based on nutrition labeling regulations.
4. A listing of ingredients in the supplement.
5. Structure/function claims about how the product affects normal body structures (such as "helps maintain strong bones") or functions ("enhances normal bowel function") can be made on product labels. If a structure/function claim is made, this FDA disclaimer must appear:

This statement has not been evaluated by the FDA. This product is not intended to diagnose, treat, cure, or prevent any disease.

Nutrition Facts
Serving size 1 Tablet

Amount Per Serving	% DV
Melatonin 3 mg	*

*Daily Value (DV) not established

Other Ingredients: Dicalcium Phosphate, Cellulose (Plant Origin), Vegetable Stearic Acid, Vegetable Magnesium Stearate, Silica, Croscarmellose.

GUARANTEED FREE OF: wheat, yeast, soy, corn, sugar, starch, milk, eggs. No artificial colors, flavors. No chemical additives. No preservatives. No animal derivatives.

Directions: As a dietary supplement for adults, take one (1) tablet, under the direction of a physician, only at bedtime as Melatonin may produce drowsiness. **DO NOT EXCEED 3 MG IN A 24 HOUR PERIOD.**

Warning: For Adults. Use only at bedtime. This product is not to be taken by pregnant or lactating women. If you are taking medication or have a medical condition such as an auto-immune condition or a depressive disorder, consult your physician before using this product. **NOT FOR USE BY CHILDREN 16 YEARS OF AGE OR YOUNGER.** Do not take this product when driving a motor vehicle, operating machinery or consuming alcoholic beverages.

In case of accidental overdose, seek professional assistance or contact a Poison Control Center immediately.

KEEP OUT OF REACH OF CHILDREN

Table 24.3 Consumer beliefs about dietary supplements versus reality[2,4,9–11]

Common consumer beliefs
Consumers tend to believe dietary supplements:
• Are not drugs.
• Have fewer side effects than prescription drugs.
• Have accurate health claims.
• Are approved for use by the FDA.
• Will improve health and help maintain health.
• Are safe, high quality, and effective.
• May replace conventional medicines and cost associated with health care.
Dietary supplement realities
• FDA does not approve, test, or regulate the manufacture or sale of dietary supplements.
• The FDA has limited power to keep potentially harmful dietary supplements off the market.
• Dietary supplements generally are not tested for safety or effectiveness before they are sold.
• Dietary supplements often do not list side effects, warnings, or drug or food interactions on product labels.
• Ingredients listed on dietary supplement labels often do not include all ingredients.
• Dietary supplements generally do not relieve problems or promote health and performance as advertised.
• Some dietary supplements contain potentially harmful or banned substances.

Table 24.3 provides a summary of common consumer beliefs about dietary supplements and then lists the realities of their safety and effectiveness.

The Federal Trade Commission (FTC) regulates claims for dietary supplements made in print and broadcast advertisements, including direct marketing, websites, infomercials, and mass e-mails. Claims made for dietary supplements in advertisements are supposed to be truthful, but often are not. Although some companies have been prosecuted for making false and misleading claims, neither the FDA nor the FTC has sufficient resources to fully monitor products and enforce laws related to dietary supplements.

The majority of adverse event complaints received by the FDA involve products with caffeine, yohimbe (marketed for male sexual enhancement), and stimulants, and those marketed for bodybuilding and weight loss.[1,5,22] Consumers are encouraged to report adverse effects of dietary supplements to the FDA and the National Institutes of Health at the website: safetyreporting.hhs.gov

Some dietary supplements are safe and effective, and may help people maintain their own health. Unfortunately, the for-profit-only manufacture and sale of dietary supplements taints products with suspicion.

Table 24.4 Who may benefit from vitamin and mineral supplements? Here are some examples:[12]

• People with diagnosed vitamin and/or mineral deficiency diseases
• Newborns (vitamin K)
• People living in areas without a fluoridated water supply (fluoride)
• Vegans (vitamins B_{12} and D)
• People experiencing blood loss (iron)
• Individuals with impaired absorption processes (B_{12})

Vitamin and Mineral Supplements: Enough Is as Good as a Feast

About one-third of U.S. consumers regularly use multiple vitamin/mineral supplements, and supplements of individual vitamins or minerals. They are the most popular of the dietary supplements.[13]

Supplements can have positive effects on health, as shown by the examples given in Table 24.4. This table lists examples of conditions for which specific vitamin and/or mineral supplements have been shown to be effective. Overall, a vitamin or mineral supplement is indicated when a deficiency exists, or when use is related to improved health among individuals at risk of developing a deficiency due to a particular disease or condition.[14]

People often take supplements as a sort of insurance policy against problems caused by poor diets or because of a "more is better" belief. Although multivitamin and mineral supplements can help fill in some of the nutrient gaps caused by poor food habits or low food intake, they can't turn a poor dietary pattern into a healthful one. Whether a dietary pattern is healthful or not is determined by more than its vitamin and mineral content. The healthfulness of a dietary pattern also depends on its content of essential fatty acids, protein, fiber, water, and other nutrients and phytochemicals from food.[15]

One of the most serious consequences of supplements results when they are used as a remedy for health problems that can be treated, but not by vitamins or minerals. For example, vitamin and mineral supplements have not been found to prevent or treat heart disease, cancer, diabetes, hypertension, premature death, behavioral problems, sexual dysfunction, hair loss, autism, chronic fatigue syndrome, obesity, cataracts, or stress. Some vitamin supplements, such as vitamin E, vitamin C, beta-carotene, and calcium, can be harmful to certain groups of people, especially in high doses.[1,3,13–19]

The Rational Use of Vitamin and Mineral Supplements Like all medications, vitamin and mineral supplements should be taken only if medically indicated. If they are taken, dosages should not be excessive.[19]

Herbal Remedies

Herbal remedies (also called *botanicals*) have been used in traditional medicine in China and India for over 5,000 years.[20] Discovery of the properties of these plant-based substances, and subsequent studies on specific chemical compounds contained in some of them, led to the development of about half of the drugs now used to treat diseases and disorders. Many people assume that herbal remedies are safe because they are natural components of plants and have been used in traditional medicine for a long time, but not all plants are safe: poison ivy, oleander, and mistletoe berries are toxic, for example. Modern medicine has developed safe and effective drugs that have decreased the historic reliance on herbal remedies.

Herbal remedies Extracts or other preparations made from ingredients of plants intended to prevent, alleviate, or treat disease, or to promote health.

Approximately 20% of U.S. adults use herbal dietary supplements each year.[21,24] The herb pharmacopoeia includes over 550 primary herbs known by at least 1,800 names (Illustration 24.1). Plant products known to effectively treat disease are considered drugs, and those that have not passed the scientific tests needed to demonstrate safety and effectiveness in disease treatment are often called herbs. Yet the truth is that many products sold as herbal remedies in the United States have drug-like effects on body functions

Illustration 24.1 Examples of plants used in the formulation of herbal supplements.

Table 24.5 Effectiveness and safety of a sampling of top-selling herbal remedies[a,24–26]

Herbal remedy	Effectiveness	Safety Concerns
Echinacea	May diminish upper respiratory infection	People allergic to ragweed, who have an autoimmune disorder, or are on drugs that affect liver function should not use this herb.
Garlic	May decrease blood pressure to a small extent	May decrease blood clotting, interact with blood thinners.
Ginkgo biloba	May improve cognitive and social function in people with Alzheimer's disease	Seeds are unsafe. Acts as an anticoagulant, should not be used by people with bleeding or seizure disorders.
St. John's wort	May relieve mild to moderate depression	Due to multiple interactions, should be used under medical supervision.
Ginger	May reduce motion sickness, nausea, and vomiting	May cause gastrointestinal upset, prolonged bleeding.

[a]Effects vary based on dose, purity, type, and duration of use of the herbal remedy. Safety and effectiveness have not consistently been tested in pregnant or breastfeeding women, or in infants and children. Herbal remedies are not considered "safe" until safety has been demonstrated.

but have not passed safety or effectiveness tests. Herbal products are regulated as dietary supplements and do not have to be scientifically demonstrated to be safe or effective prior to being sold.[24]

Effectiveness and Safety of Herbal Remedies Likely effects and identified adverse effects of five top-selling herbal products are listed in Table 24.5. Herbal remedies, like drugs, have biologically active ingredients that can have positive, negative, or neutral effects on body processes. Basically, an herbal remedy is considered valuable if it has beneficial effects on health and is safe. Knowledge of the risks and benefits of many herbal supplements remains incomplete. However, available evidence suggests that some herbal remedies are safe and effective, while others appear to be neither.

Which herbal remedies are likely ineffective or unsafe? Human studies with various herbal remedies have helped identify herbs and other dietary supplements that lack beneficial effects or have adverse side effects. Table 24.6 lists some of these products. The extent

Table 24.6 Examples of dietary supplements that may not be effective or safe[4,25,27]

Apricot pits (laetrile)	Ephedra	Pennyroyal
Androstenedione (Andro)	Eyebright	Pokeroot
Aristolochic acid	Ginkgo seed	Hoodia
Sassafras	Saw palmetto	Shark cartilage
Belladonna	Kava	Skullcap
Black cohosh	Blue cohosh	Licorice root
Star anise	Bitter orange	Liferoot
Vinca	Borage	Lily of the valley
Wild yam	Broom	Lobelia
Wormwood	Chaparral	Yohimbe
Chinese yew	Mandrake	Willow bark
Comfrey	Mistletoe	
Dong quai	Organ/glandular extracts	

health action Choosing and Using Dietary Supplements[1,19,32]

1. Undertake research on the supplement you are interested in taking at sites offered by the Office of Dietary Supplements and the National Institutes of Health.
2. Purchase supplements labeled *USP* or *CL*. They are tested for purity, ingredients, and dose.
3. Check the expiration date on supplements. Use unexpired supplements.
4. Choose vitamin and mineral supplements containing 100% of the Daily Value or less.
5. Take vitamin and mineral supplements with meals.
6. If you have a diagnosed need for a specific vitamin or mineral, take that individual vitamin or mineral and not a multiple supplement.
7. Store supplements where small children cannot get at them.
8. Don't use herbal remedies for serious, self-diagnosed conditions such as depression, persistent headaches, and memory loss. (You might benefit more from a different treatment.)
9. Don't use herbal remedies without medical advice if you are attempting to become pregnant or if you are pregnant or breastfeeding.
10. Don't mix herbal remedies.
11. If you are allergic to certain plants, make sure herbal remedies are not going to be a problem before you use them.
12. If you have a bad reaction to an herbal remedy, stop using it and report the reaction to the FDA using the site: www.fda.gov/Safety/MedWatch/HowToReport/ucm053074.htm.
13. Tell your health care provider about the dietary supplements you take or plan on taking.

to which the herbs included in the table pose a risk to health depends on the amount taken and the duration of use, the age and health status of the user, and other factors. Not everyone reacts the same way to different herbal supplements.[4,22]

Quality of Herbal Products Many herbal products available on the market are of poor quality. Some of the products have been found to contain ingredients other than those declared on the label, including banned drugs, and some contain contaminants such as bacteria, mold, mercury, and lead.[29,30] Analyses of the composition of 25 ginseng products, for example, found that concentrations of ginseng compounds in the supplements were up to 36 times different than labeled amounts.[31] High levels of contamination in herbal products were identified by another study. Of 260 Asian herbal products examined, 25% were contaminated with high levels of heavy metals, and 7% contained undeclared drugs purposefully and illegally added to produce a desired effect.[33] Studies of echinacea products found that 10% of samples contained no echinacea, and only half contained the labeled amount.[20] Male enhancement supplements are among the most common products recalled from the market due to contamination or because they contain unlisted prescription drugs.[1] No herbal or "all natural" substance has been shown to cure impotence.[35] Dietary supplements labeled "pure," "natural," or "quality assured" may or may not fit the description. You generally can't tell by the label.

There is no government body that monitors the contents of herbal supplements. Private groups, such as the U.S. Pharmacopeia (USP) and Consumer Laboratories (CL), offer testing services to ensure that herbal and other dietary supplements meet standards for disintegration, purity, potency, and labeling.[36,37] Products that pass these tests can display a USP logo or the CL symbol on product labels (Illustration 24.2). These symbols represent quality ingredients and labeling but do not address product safety or effectiveness.[38] Considerations for the use of dietary supplements are summarized in this unit's Health Action feature.

Due to the lack of studies and potential dangers, the FDA has advised dietary supplement manufacturers not to make claims related to pregnancy for herbs and other products, and to label products truthfully based on scientific evidence.[26,27]

Illustration 24.2 These symbols on the labels of dietary supplements certify quality ingredients and accurate labeling but do not address product safety or effectiveness.[39]

Prebiotics and Probiotics

The digestive tract, particularly the colon, is home to over 500 species of microorganisms representing 100 trillion bacteria (and billions of viruses and fungi, as well). Some species of bacteria such as *E. coli* 0157:H7 and *Salmonella* may cause disease, whereas others

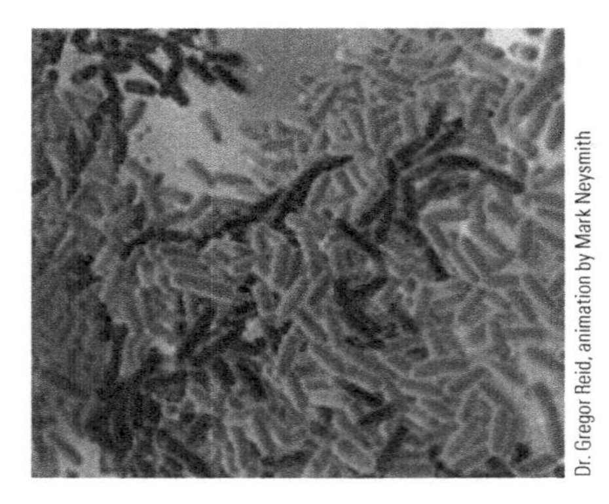
Dr. Gregor Reid, animation by Mark Neysmith

Illustration 24.3 A *Lactobacillus* species (blue) taking over harmful *E. coli* bacteria (red).

such as *Lactobacillus* and bifidobacteria (Illustration 24.3) help prevent disease through a variety of mechanisms.[40] The beneficial effects of the right species of microorganisms, and ways to increase their presence in the gut, are subjects of intense research and expanded knowledge. Knowledge gains center around the actions of **prebiotics** and **probiotics**. Pre- and probiotics are regulated by the same rules that govern other dietary supplements.[41]

Prebiotics are non-digestible dietary fibers that can be used as a food source by beneficial microorganisms (mainly bacteria) in the small intestine and colon. The breakdown products from microorganism digestion of prebiotics are released into the gut, where they foster the growth of beneficial bacteria (which is why prebiotics are referred to as "intestinal fertilizer") and diminish the population and effects of harmful microorganisms.[40,41] Probiotics is the term for live, beneficial—or "friendly"—bacteria that enter food through fermentation and aging processes and are resistant to digestion.[43] The term **synbiotics** is being used to classify combinations of prebiotics and probiotics. Table 24.7 lists food and other sources of pre- and probiotics. Many of the foods listed are central to a healthy dietary pattern.

Prebiotics, probiotics, and synbiotics have been found to benefit health by:

- Increasing the mass of beneficial, and decreasing the mass of harmful, microorganisms in the small intestines and colon, thereby limiting the effects of disease-causing microorganisms in the gut
- Decreasing insulin resistance and chronic inflammation by altering gene expression in microorganisms in ways that enhance glucose utilization
- Enhancing the immune functions of the small intestine and colon
- Decreasing the symptoms of irritable bowel syndrome and infantile colic

Table 24.7 Food and other sources of prebiotics and probiotics[43,44]

Prebiotics	Probiotics
Jerusalem artichokes	Yogurt with live culture
Wheat	Buttermilk
Barley	Kefir
Rye	Cottage cheese
Onions	Dairy products with added probiotics
Garlic	Soft cheeses
Leeks	Soy sauce
Prebiotics tablets and powders and nutritional beverages	Tempeh
Breastmilk	Sauerkraut
Psyllium husk	Miso soup
Onions	Probiotic powders, pills
Garlic	kimchee
Leeks	natto
Oats	pickled vegetables
Banana	
Tomatoes	
Dried beans	

prebiotics Non-digestible carbohydrates (various types of dietary fiber) that serve as food for and promote the growth of beneficial microorganisms in the small intestine and colon. Also called "intestinal fertilizer."

probiotics Live microorganisms which resist digestion, and when administered in adequate amounts of appropriate strains, confer health benefits to the host. Strains of *Lactobacillus* (lac-toe-bah-sil-us) and bifidobacteria (bif-id-dough bacteria) are the best-known probiotics. Also called "friendly bacteria."

synbiotics Combinations of prebotics and probiotics that interact in ways that generally benefit both and the health of the host.

- Enhancing the absorption of minerals such as calcium, magnesium, and iron
- Decreasing the symptoms and onset of vaginal and urinary tract infections
- Delaying the onset of allergy development in children
- Decreasing the duration of infection- and antibiotic use–related diarrhea
- Increasing stool bulk and reducing constipation[40–44]

Probiotics are considered safe and beneficial for healthy people but may not be appropriate for individuals with compromised immune status or who are critically ill. Table 24.8 reviews practical considerations for the use of probiotics.

Final Thoughts

From dietary supplements to friendly bacteria, the universe of substances considered dietary ingredients is expanding. Knowledge about potential benefits of pre- and probiotics is charging ahead, and advances are catching the attention of consumers and health care professionals. Perhaps you never thought that "intestinal fertilizer" or "friendly bacteria" would ever intentionally pass through your lips. But that may well be the nature of some dietary ingredients to come.

Table 24.8 Practical considerations related to the use of probiotics[40,42,44]

• Benefits to health are strain- and dose-specific.
• Products are not regulated by the FDA; quality, microorganism strains and doses, viability of the microorganisms, and purity vary considerably among products.
• Use may cause temporary gas and bloating.
• Effects appear to last as long as the probiotic is consumed.
• Appear to be safe for use by most people
• Individuals who are critically ill or have severe immune system disorders should not use them.

REALITY CHECK

Herbals on the Internet

Can you trust information on herbal products you see on the Web?

Who gets the thumbs-up?

Answers appears on page 24-10.

Photodisc

Sarah: When I'm sick, the first place I go is to the Internet to find an herb that will make me feel better.

Photodisc

Pablo: Herbal products I see advertised on the Internet look like they'll work for my problem. But I'm conflicted about buying them, because I'm not sure I can trust the information.

ANSWERS TO REALITY CHECK

Herbals on the Internet

Many people use the Internet as a source of information about illness remedies. Unfortunately, many of the sites selling herbal products illegally claim the products prevent or cure specific diseases as if they were real drugs and do not include the required FDA disclaimer and ingredient list.[25,45]

Take the worry out of decisions about herbal supplements. Check them out using scientifically reliable websites.

Photodisc

Sarah:

Photodisc

Pablo:

Photodisc

NUTRITION up close

Supplement Use and Misuse

Focal Point: Decide if a dietary supplement is warranted in these situations.

People take dietary supplements for many reasons, but is their use justified? Apply the information from this chapter to determine if you agree with the decisions made in each of the following scenarios.

1. Martha works part time and takes a full load of classes. Like many college students, she is always on the go, often grabbing something quick to eat at fast food restaurants or skipping meals altogether. Nevertheless, Martha feels confident her health will not suffer, because she takes a daily vitamin and mineral supplement.

   Is a supplement warranted in this case? Why or why not?

   ____

2. Sylvia is a 23-year-old student diagnosed with iron-deficiency anemia. She has learned in her nutrition class that it is preferable to get vitamins and minerals from food instead of supplements. Therefore, instead of taking the iron pills her doctor has prescribed, Sylvia has decided to counteract the anemia by increasing her consumption of iron-rich foods.

   Is a supplement warranted in this case? Why or why not?

   ____

3. John is a 21-year-old physical education major involved in collegiate sports. He is very aware that nutrition plays an important role in the way he feels, so he is careful to eat well-balanced meals. In addition, John takes megadoses of vitamins and minerals daily. He is convinced they enhance his physical performance.

   Is a supplement warranted in this case? Why or why not?

   ____

4. Roberto, a native Californian, is backpacking through Europe when he is slowed down by constipation. He visits a pharmacy where English is spoken and is given senna by the pharmacist. Roberto has never taken an herb before and is not sure how his body will react to it, or if it will work.

   Should Roberto try the senna, ask for a nonherbal drug, or take another action? (Assume they cost the same.) What's the rationale for this decision?

   ____

5. While shopping at the mall, Yuen notices a kiosk selling "Hypermetabolite," a weight-loss product that guarantees you'll lose 5 pounds a week without dieting. Having gained 10 pounds since she started working full time, Yuen decides to try it. Her examination of the product's label reveals that an ephedra derivative and Asian ginseng are major ingredients.

   Should Yuen take Hypermetabolite for weight loss? Why or why not?

   ____

Feedback to the Nutrition Up Close is located in Appendix F.

REVIEW QUESTIONS

- **Make evidence-based decisions about the probable safety and effectiveness of specific dietary supplements.**

1. Dietary supplements include vitamins, minerals, herbs, proteins, amino acids, and fish oil. **True/False**
2. The FDA must approve dietary supplements before they are allowed to enter the market. **True/False**
3. Dietary supplement labels must include a Supplemental Facts panel and may include qualifying nutrient claims and structure/function claims. **True/False**
4. Say you read a dietary supplement label that claims the product "helps enhance muscle tone or size." Because the statement appears on the label, it must be true. **True/False**
5. The Recommended Dietary Allowance (RDA) for vitamin A for breastfeeding women is 1,300 mcg per day. To make sure

you are getting enough vitamin A, you should select a supplement containing 3,300 mcg of vitamin A. **True/False**

6. Specific vitamin or mineral supplements appear to benefit individuals with diagnosed vitamin or mineral deficiency, and some adults at risk of osteoporosis. **True/False**
7. Vitamin and mineral supplements available over the counter can be safely consumed at any dose levels. **True/False**
8. A USP label displayed on dietary supplements indicated the product has been tested for safety and effectiveness. **True/False**
9. Prebiotics support the growth of beneficial bacteria in the colon. **True/False**
10. Prebiotics have been demonstrated to benefit health under certain circumstances, but probiotics have not. **True/False**
11. Prebiotics are non-digestible dietary fibers. **True/False**
12. Probiotics are microorganisms that, when consumed, resist digestive processes, occupy the small intestine and colon in a live state, and benefit the host. **True/False**
13. Herbal remedies, when demonstrated through scientific studies to be effective and safe, are generally considered to be drugs and regulated as such. **True/False**

The following four questions refer to the following scenario.

While reading a men's magazine you notice a full-page ad for a new dietary supplement that promises to "ignite sexual well-being." Containing a natural aphrodisiac, the supplement is specifically designed by scientists to encourage healthy, meaningful, and long-term relationships. The supplements have been tested on movie stars and college students, and you have heard from social media contacts that they work. Ingredients include vitamins, taurine, caffeine, dong quai, ginkgo, and ginseng.

14. ______ How many of the herbal ingredients listed have been shown to act as an aphrodisiac?
    a. 0
    b. 1
    c. 2
    d. 3
15. ______ Which herbal ingredient contained in the product does not appear to be effective for any health condition or be safe to consume?
    a. dong quai
    b. ginkgo
    c. ginseng
    d. taurine
16. ______ This product:
    a. has been tested for purity, safety, and effectiveness by the FDA before it became available.
    b. has probably not been tested for purity, safety, and effectiveness by the FDA before it became available.
    c. will be examined by the manufacturer for safety and effectiveness while on the market.
    d. will improve people's chances of finding a long-term relationship.
17. ______ Assume your friend bought and took the product and thinks it really works. What is the most likely reason for that?
    a. The vitamins in the product worked.
    b. Ginseng improves personality and the potential for long-term relationships.
    c. The caffeine provides extra energy for interpersonal relationships.
    d. Your friend believed the product would be effective (placebo effect).

Answers to these questions can be found in Appendix F.

NUTRITION SCOREBOARD ANSWERS

1. Dietary supplements include herbs, vitamin and mineral supplements, protein powders, amino acid and enzyme pills, fish oils and fatty acids, hormone extracts, and other products. **False**
2. Dietary supplements can be sold without proof of their safety or effectiveness. **False**
3. Results of clinical trials, and not historical use, are the gold standard for determining the safety and effectiveness of herbal remedies and other dietary supplements. **False**
4. Yes! There are healthful bacteria. **True**

Water Is an Essential Nutrient

Fotofeeling/Getty Images

NUTRITION SCOREBOARD

1 Bottled water is better for health than municipal water from your faucet. **True/False**

2 Drinking lots of water helps flush toxins out of your body. **True/False**

3 Few bottled water brands are fluoridated. **True/False**

4 You can't drink too much water. **True/False**

5 You should drink 8 or more glasses of water every day. **True/False**

Answers can be found at the end of the unit.

LEARNING OBJECTIVES

After completing Unit 25 you will be able to:

- List the main functions of water in the body and the consequences of water deficiency and toxicity.
- Describe the factors that affect the availability of safe water supplies and the potential consequences of inadequate and unsanitary water supplies on health.

Photodisc

Table 25.1 Key functions of water in the body[3]

Needed to maintain normal internal body temperature
Serves as the medium for chemical reactions that take place within the body
Serves as the medium for the transport of nutrients throughout the body in blood
Serves as the medium for the action of digestive enzymes
Required for the normal elimination of waste products in urine and stools
Participates in energy formation

Illustration 25.1 The body of a 160-pound person contains approximately 12 gallons of water. That's 96 pounds of water!

Water: Where Would We Be Without It?

- **List the main functions of water in the body and the consequences of water deficiency and toxicity.**

Ask any three people you know to name as many essential nutrients as they can. If they mention water, give them a prize. Our need for water is so obvious that it is often taken for granted. Water differs from other essential nutrients in that it is liquid, and our need for it is measured in cups rather than in grams or milligrams. It is required by every cell, tissue, and organ in the body. Without it, our days are limited to about six. The largest single component of our diet and body, water is a basic requirement of all living things.[3]

KEY NUTRITION CONCEPTS

Three key nutrition concepts are directly related to the essential nutrient water and its functions in the body:

1. Nutrition Concept #3: Health problems related to nutrition originate within cells.
2. Nutrition Concept #4: Poor nutrition can result from both inadequate and excessive levels of nutrient intake.
3. Nutrition Concept #5: Humans have adaptive mechanisms for managing fluctuations in nutrient intake.

Water's Role as an Essential Nutrient

Water qualifies in all respects as an essential nutrient. It is a required part of our diet; it performs specific functions in the body; and deficiency and toxicity signs develop when too little or too much is consumed. It is our body's main source of fluoride, an essential mineral needed for the formation and maintenance of enamel and resistance to tooth decay.[4] Fluoride is naturally present in some well waters and is added to 70% of all municipal water supplies in the United States.[5]

Water serves as the medium in which many chemical reactions take place within our body. Water plays key roles in digestion and energy formation. It provides the medium needed by digestive enzymes to access and break down carbohydrates, proteins, and fats during digestion. Water is produced as an end product of energy formation from carbohydrates, proteins, and fats. Water is needed to "carry" nutrients to cells and waste products away from them. Additionally, water acts as the body's cooling system. When our internal temperature gets too high, water transfers heat to the skin and releases it in perspiration. When we're too cool, less water—and heat—are released through the skin. Water's functions are summarized in Table 25.1.

Water has been given credit for other functions in the body, but undeservedly. Drinking more water than is normally needed does not prevent dry, wrinkled skin, lead to weight loss, or flush toxins out of the body. Nor will it cure chronic fatigue, arthritis, migraines, or hypertension.[1,5–7] This unit's Reality Check addresses another misconception about water.

Water, Water, Everywhere The body of a 160-pound person contains about 12 gallons of water (Illustration 25.1). Adults are approximately 60–65% water by weight.[3] Water is distributed through the body in blood, in the spaces between cells, and in all cells. The proportion of water in body tissues varies: blood is 83% water, muscle 75%, and bone 22%. Even fat cells are 10% water.[8]

REALITY CHECK

To Drink or Not to Drink Water with Meals

According to a health and wellness website, people should not drink water with a meal because it dilutes digestive juices and interferes with digestion. Really?

Who gets the thumbs-up?

Answer appears on page 25-4.

Paul Burns/Digital Vision/Getty Images

Barbara Ann: What website made that statement? Why would that happen? Do you know how watery food gets when it's in your stomach?

Jack Hollingsworth/Photodisc/Getty Images

Carole: I get a bloated feeling when I drink water with food. I think the water is slowing down the digestion of food in my stomach.

Most Foods Contain Lots of Water, Too Most beverages are more than 85% water, and fruits and vegetables are 75–90% (Illustration 25.2). Meats, depending on their type and how well done they are, contain 50–70% water. Consequently, the water content of foods makes an important contribution to our daily intake of water. Approximately half of our water intake comes from plain water and other beverages and half from foods.[9]

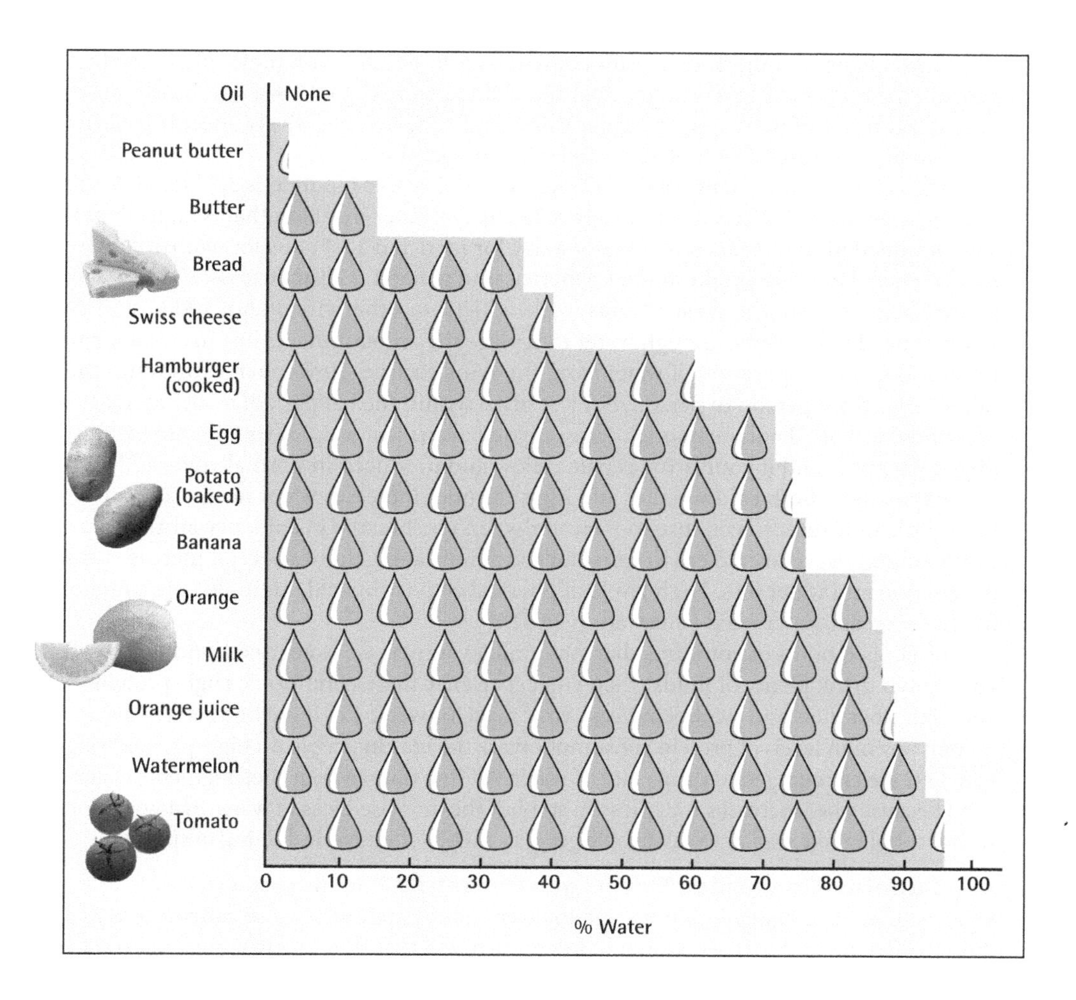

Illustration 25.2 The water content of some foods.

ANSWERS TO REALITY CHECK

To Drink or Not to Drink Water with Meals

Water intake during a meal does not interfere with digestion. On the contrary, it increases the exposure of digestive enzymes to food and facilitates the absorption of nutrients.[10]

Paul Burns/Digital Vision/Getty Images

Barbara Ann:

Jack Hollingsworth/Photodisc/Getty Images

Carole:

Meeting Our Need for Water

water balance The ratio between the water consumed and that lost from the body. Individuals are in water balance when water intake equals water output.

In general, individuals require enough water each day to maintain the normal functions of water in the body. When this is accomplished (which is normal), we are said to be in **water balance.** This means that we have consumed as much water from fluids and foods as we have released through urine (60%), respiration and perspiration (28%), sweat (6%), and stools (6%). Body processes that regulate thirst and re-absorption of body water tightly control water balance. We become thirsty when body fluids become too concentrated with various chemical substances to function normally. This state triggers mechanisms in the hypothalamus, which is located in the center of the brain, to signal the "I'm thirsty" feeling. Thirst disappears when sufficient water has been consumed to dilute body fluids back to the point where the concentration of chemical substances is normalized.[13]

Small changes in the body's water content have large effects on body functions. In general, the body's content of water does not change more than 0.2% a day. Losses of body fluid equal to 1% of body weight trigger thirst, and a 2% loss negatively affects physical performance. A water deficit of 20% can be life threatening.[11]

The recommended daily intake of water for 19–30-year-old men is 3.7 liters (L) a day and for women, it's 2.7 L. Since there are 4.23 cups of water in a liter, that is equivalent to a recommendation of 15.6 cups of water a day for men and 11.4 cups for women.[4] There is no evidence that water intake among Americans in general is either excessive or inadequate.[12] People who do strenuous work, athletic or otherwise, in hot and humid weather need to consume enough water to replace the amount that is lost in sweat, urine, respiration, and evaporation of water from the skin's surface. How much extra water this takes varies from person to person, but it is often within the ballpark of a 50% increase. You know you are drinking enough water if you haven't lost weight from before to after physical activity and if your urine is pale yellow and produced in normal volume.[13]

Exposure to both cold weather and high altitudes increases water need. Cold air holds little moisture, so you lose more water when you breathe in cold, dry air than warm, moist air. Exposure to cold environmental conditions at high altitudes can increase urine production and water loss. High levels of physical activity in cold, high-altitude climates further increase water need (Illustration 25.3).[14]

Prolonged bouts of vomiting, diarrhea, and fever increase water need, and that is why you should drink plenty of fluids when you experience these conditions. High-protein and high-fiber diets and alcohol increase your need for water. Losses in body water that accompany high levels of protein consumption are the reason people on high-protein weight-loss diets are encouraged to drink a lot of water. Adding fiber to your diet augments water need because fiber increases water loss in stools.[8] The increased loss of water that goes along with alcohol intake explains why people get very thirsty after overindulging in spirits.

Are Caffeine-Containing Beverages Hydrating? In the past it was widely believed that beverages containing caffeine were not hydrating because caffeine acts as a diuretic. Recent and better research has demonstrated that this conclusion is incorrect.

Illustration 25.3 Cold, high-altitude conditions increase water need.

Caffeine does not increase urine output in people accustomed to drinking coffee, tea, and other beverages that contain caffeine.[15] So count the coffee or tea you consume as contributing to your overall water intake.

Water Deficiency

A deficiency of water can lead to dehydration. Dehydrated people feel very sick. They are generally nauseated, have a fast heart rate and increased body temperature, feel dizzy, and may find it hard to move. The ingestion of fluids produces quick recovery in all but the most serious cases of dehydration. If it is not resolved, however, dehydration can lead to kidney failure and death.[16]

Water Toxicity

It is possible for people to overdose on water if they drink too much of it.[16] High intake of water can lead to a condition known as hyponatremia—or low blood sodium level and excessive water accumulation in the brain and lungs. The consequences can be devastating, including confusion, severe headache, nausea, vomiting, and even seizure, coma, and death.[2]

Rise in water intoxication found in babies on formula

Chicago, Ill.
Water intoxication, a potentially fatal condition, has increased dramatically among infants in part because poor parents give babies too much water when formula runs out, a study indicates.

Illustration 25.4 This headline is accurate—overdilution of infant formula can lead to water intoxication in infants.

Water intoxication is rare, but it has occurred in marathon runners who consumed too much water during an event, in infants given too much water or overdiluted formula (Illustration 25.4), and in people with schizophrenia, certain brain injuries, and among individuals who habitually drink lots of water (over 13 cups/day) due to a belief that is good for their health.[2] The drive for water created by antipsychotic medications can be so strong that access to water (even when showering) has to be limited.[17]

The Nature of Our Water Supply

- **Describe the factors that affect the availability of safe water supplies and the potential consequences of inadequate and unsanitary water supplies on health.**

Water covers about three-quarters of the earth's surface, yet very little of it is drinkable. Nearly 97% of the total supply is salt water, and only 3% is fresh.[18] Of the freshwater supply, only a fourth of the total amount is available for use. The rest is located in polar and glacier ice. Although fresh water is readily available in most locations in the United States, this is not the case in a number of other countries.

Water is so scarce in parts of Russia that drinking-water dispensers are coin operated. Fresh water is so highly prized in sections of the arid Middle East that having a decorative fountain in one's home is considered a sign of affluence. Drinking water is sampled, judged, and celebrated in Middle Eastern countries as ceremoniously as fine wines are in France. Although water is not a subject of envy or a cause for celebration in most parts of the United States, it is increasingly viewed as a precious resource that must be protected.[18,19]

Stringer/Getty Images

Illustration 25.5 Water distribution at a refugee camp.

Water scarcity has become a primary concern of many nations in the twenty-first century (Illustration 25.5). Demand for water worldwide is increasing due to population growth and expanding use of water-intensive energy production and farming techniques. Wasteful use of water, climate change, groundwater depletion, pollution, and leaky public water systems are contributing to water shortages and safety concerns (Illustration 25.6).[19,20] Without more effective protection of the world's water, the quantity of water available for agriculture and clean water for consumption may be insufficient, limiting food production and increasing the incidence of water-borne infectious diseases.[19–21]

The Environmental Protection Agency (EPA) in the United States is responsible for the safety of U.S. public water supplies and has set maximum allowable levels of contaminants. Water quality is monitored by local water utilities, and the results are reported to state and federal officials. Problems identified are remedied, and attempts are made to prevent future contamination in order to provide safe public water supplies.[21]

Photodisc

Illustration 25.6 Contaminated water supplies are a threat to the public's health.

Water Sources

About 40 years ago, the French made drinking "fine waters" very stylish. In Paris, boulevardiers crowded sidewalk cafés for hours sipping chilled Perrier served with thinly sliced lemon. The popularity of bottled mineral, spring, and sparkling waters skyrocketed and continues to grow in many countries. In the United States, mineral, spring, and seltzer waters became best sellers (Illustration 25.7). Bottled waters now outsell soft drinks in grocery stores, and new versions of waters appear often. These waters have a strong, positive image. But what *are* these waters? How do they differ from the water we get from the faucet?

True *mineral water* is taken from protected underground reservoirs that are lodged between layers of rock. The water dissolves some of the minerals found in the rocks, and as a result it contains higher amounts of minerals than most sources of surface water. Actually, most water contains some minerals and could be legitimately considered "mineral water." *Spring water* is taken from freshwater springs that form pools or streams on earth's surface. True seltzers (not the sweetened kind you often find for sale) are *sparkling waters* that are naturally carbonated. Most seltzers, however, become bubbly by the commercial addition of pressurized carbon dioxide.[22] The latest bottled waters to

Richard Anderson

Jerryb8/Dreamstime.com

Illustration 25.7 Is bottled water better for you than tap water? Consumers often perceive that bottled waters have fewer impurities and are better for health than tap water. Such beliefs are unfounded.

hit the shelves include artesian (taken from underground aquifers), alkaline (pH-elevated), caffeinated, flavored, and probiotic water (Illustration 25.7). Evidence that any of these bottled waters are better for health in some way than tap water does not exist.[22]

A new generation of bottled waters is coming to stores near you. The Dietary Guidelines for Americans recommends that people consume water rather than sugar-sweetened beverages, and water sales reflect the emphasis on increased intake. Many types of bottled water that may please consumer palates are being broadly promoted (Illustration 25.8). Most are calorie-free, but some are sweetened with no-cal sweeteners or sugar. Bottled waters sweetened with sugars tend to provide at least 75% less sugar than fruit or soft drinks.[24]

Most of the water consumed in the United States, including 45% of single-serve bottled waters, is water from a public municipal water supply, identified as "PWS" on labels.[22] Unlike most bottled water, tap water from most municipal water supplies is fluoridated (most bottled water is not), and the water is generally tested several times a day for bacterial and other contaminants. It is far cheaper than bottled water and does not contribute

Tony Freeman/Science Source

Illustration 25.8 Flavored waters are preferred by many consumers.

to landfill overflow as plastic bottles may.[25,26] In the recent past, the presence of bisphenol A (BPA) in plastic bottles was considered potentially unsafe, especially for pregnant women and young children. Use of BPA in plastic bottles has decreased substantially and, based on recent safety tests, the Food and Drug Administration (FDA) concluded that current levels of BPA in foods due to plastic containers is safe at current intake levels.[27]

Fluoridated Bottled Water Bottled waters have been criticized for their general lack of fluoride compared to fluoridated tap water. People of all ages need this mineral for bone health and the prevention of tooth decay. Children who regularly consume bottled water and infants who receive formula mixed with bottled water develop more tooth decay at an early age than do young children who receive fluoridated water.[29] The State of Ohio has formally recommended that all bottled waters sold in the state be fluoridated but, when tested, only 5% were.[26] Bottled waters with added fluoride must be labeled as *fluoridated.*[28] A quick look at the label can tell you if you're getting fluoride along with your bottled water or if you need to get fluoride from another source.

Fotofeeling/Getty Images

NUTRITION up close

Foods as a Source of Water

Focal Point: Water is a primary component of many foods.

About half of our requirement for water is met by fluids and solid foods. But how much water is in foods? You may be surprised.

Circle the food in each set that you think contains the highest percentage of water by weight.

1.	avocado	potato, boiled	corn, cooked
2.	egg	almonds	ripe olives (black)
3.	watermelon	celery	pineapple, fresh
4.	2% milk	Coke	cranberry juice, low calorie
5.	cheddar cheese	banana	refried beans
6.	hot dog	pork sausage, cooked	ham, extra lean
7.	apple	mushrooms, raw	orange
8.	onions, cooked	lettuce	okra, cooked
9.	peanut butter	butter	mayonnaise
10.	cake with frosting	bagel	Italian bread

Feedback to the Nutrition Up Close is located in Appendix F.

REVIEW QUESTIONS

- **List the main functions of water in the body and the consequences of water deficiency and toxicity.**

1. Water plays key roles in the prevention of acne and dry skin and in the dilution of toxins formed during digestion. **True/False**
2. Water is our major source of fluoride. **True/False**
3. The functions of water include serving as a medium for chemical reactions and for nutrient transport in blood. **True/False**
4. Adult females living in moderate climates require an average of 8 cups of water a day, and males require twice that amount. **True/False**
5. Most of a person's body weight consists of water. **True/False**
6. Beverages containing caffeine are not hydrating. **True/False**
7. ________ Water need is increased by three of the four following conditions. Which condition is *not* related to an increased need for water?
   a. dry skin
   b. hot, humid environment
   c. vomiting
   d. alcohol intake
8. ________ Dehydration is *not* characterized by:
   a. accumulation of water in the brain and lungs
   b. nausea and vomiting
   c. dizziness
   d. severe headache

- **Describe the factors that affect the availability of safe water supplies and the potential consequences of inadequate and unsanitary water supplies on health.**

9. The lack of adequate availability of safe water is a global health concern. **True/False**

10. The FDA requires that all bottled water be fluoridated. **True/False**
11. Increased incidence of infectious disease is one consequence of shortages of clean water. **True/False**
12. Approximately 20% of bottled waters contain tap water. **True/False**
13. Water balance exists when the amount of water consumed from foods and fluids equals the amount of water lost in urine in a day. **True/False**
14. ______ The Adequate Intake level for water for children aged 4 to 8 years is 1.7 liters. How many cups is that?
    a. 7.2
    b. 6.8
    c. 11.4
    d. 15.7

Answers to these questions can be found in Appendix F.

NUTRITION SCOREBOARD ANSWERS

1. Bottled waters have not been shown to be better for health or to be "purer" than tap water obtained from city water supplies. **False**
2. Drinking lots of water flushes toxic wastes from our bodies. **False**
3. Only a minority of bottled waters are fluoridated; these are labeled *fluoridated*. **True**
4. Yes, you can drink too much water, and it results in water intoxication.[1] **False**
5. This an urban myth. Drinking lots of water does not improve health.[2] **False**

Nutrient–Gene Interactions in Health and Disease

NUTRITION SCOREBOARD

1 Most chronic diseases are genetically caused. **True/False**

2 Differences in genetic traits are responsible for large differences in essential nutrient needs between individuals. **True/False**

3 Heart disease, cancer, obesity, and hypertension result primarily from interactions among environmental and genetic factors. **True/False**

4 Individual differences in inflammatory response to dietary fats can be explained largely by differences in the effects of dietary fats on gene expression. **True/False**

Answers can be found at the end of the unit.

LEARNING OBJECTIVES

After completing Unit 26 you will be able to:

- Explain how genetics and nutrition are interdependent.

Nutrition and Genomics

- **Explain how genetics and nutrition are interdependent.**

In the history of the relatively young science of nutrition, at least two breakthroughs have defined the future. The first resulted from experiments in the late 1800s demonstrating that some constituents of foods are essential for life. The second is represented by the mapping of the human *genome* and development of the field of **nutrigenomics**. (Find definitions and explanations for many of the genetic terms used in this unit in Table 26.1.) That breakthrough marked the beginning of a new era in the discovery of the underlying causes of a variety of common diseases and the specific roles of components of diets in disease prevention and treatment. It is being made possible, in part, by the identification of genetic codes for protein production and nutrients that affect, or are affected by, gene activity. Enzymes and other proteins produced as a result of genetic codes are central to life and health because they determine which chemical changes will take place within the body. These changes affect every body process, including growth, digestion, nutrient absorption, nutrient utilization, disease resistance, healing, and blood pressure control.[1]

KEY NUTRITION CONCEPTS

Key nutrition concepts are directly related to this unit's content on nutrient–gene interactions:

1. Nutrition Concept #3: Health problems related to nutrition originate within cells.
2. Nutrition Concept #6: Malnutrition can result from poor diets and from disease states, genetic factors, or combinations of these factors.

Genetic Secrets Unfolded

Our bodies may be new, but our genes have been around for over 40,000 years. Genes replicate themselves exactly over generations, and lasting modifications in them almost never occur. This means that changes in genetic traits do not account for increases or decreases in the incidence of disease. It is why genetic traits cannot be given as the cause of recent epidemics of obesity and diabetes, nor can they be given credit for major declines in heart disease and stroke rates.

©Muellek Josef/Shutterstock.com; ©Amst/Shutterstock.com; ©Volosina/Shutterstock.com; ©Valentyn Volkov/Shutterstock.com; Photodisc; ©Picsfive/Shutterstock.com; ©Stargazer/Shutterstock.com; ©Monticello/Shutterstock.com

Humans are united as a species because we all share 99.9% of the same DNA in our genes.[8] DNA is located in chromosomes and has four different chemical building blocks called "bases" abbreviated as A, T, C, and G. The sequence and order of the bases establish the blueprint for the specific amino acids needed to form a particular protein. When needed, each amino acid code contained in the blueprint is sent by messenger RNA to transfer RNA that then attaches to the appropriate amino acid. It brings the amino acid to ribosomal RNA. Ribosomal RNA is a component of ribosomes, which are considered to be the body's protein synthesis factories. Within the ribosomes, amino acids are assembled into chains that correspond to the coded protein. These amazing RNA processes are likely key to the origin of life.[6,7]

Chromosomes also contain noncoding regions that enact the important roles of maintaining chromosomal structural integrity and the regulation of where, when, and what quantity of proteins are made.[7]

Gene Variants Gene variants occur when base codes are altered to change the functional status of a gene. They may be inherited or formed during embryonic and fetal cell proliferation. They are set in place to help a developing fetus adapt to nutrient and other

Table 26.1 Definitions and explanations of genetic terms[8–10]

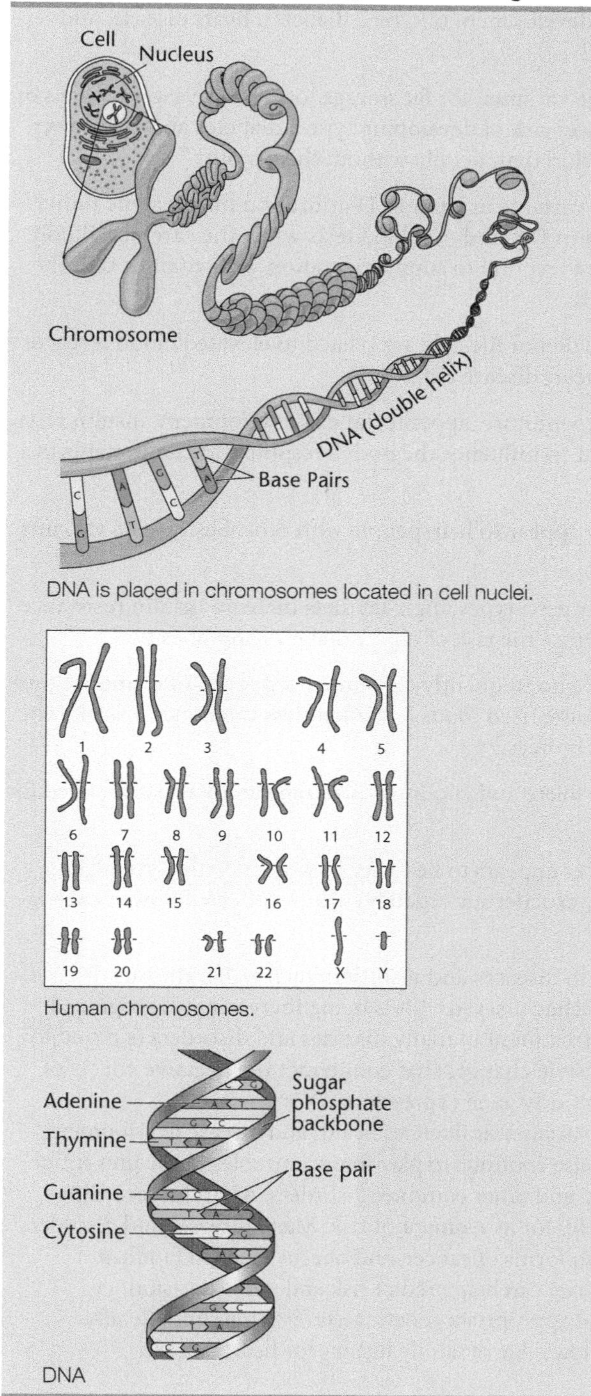

DNA is placed in chromosomes located in cell nuclei.

Human chromosomes.

DNA

- **DNA (deoxyribonucleic acid, pronounced de-oxy-rye-bow-new-clay-ic)** Segments of genes that provide instructions for the synthesis, or production, of specific enzymes and other proteins needed by cells to carry out life- and health-sustaining processes. Enlarged, DNA looks something like an immensely long ladder twisted into a coil. The sides of the ladder structure of DNA are formed by a backbone of sugar-phosphate molecules, and the "rungs" consist of base pairs (abbreviated A, T, C, and G) joined by weak chemical bonds. Bases are the "letters" that spell out the genetic code, and there are over 3 billion of them in human DNA. Some sections of DNA do not code for protein synthesis. They signal genes to turn on or off, or to increase or decrease activity. Characteristics of these non-protein coding segments of DNA vary among individuals, making it possible to identify individuals based on "DNA fingerprinting."
- **Genome** Combined term for "genes" and "chromosomes." It represents all the genes and DNA contained in an organism. It is estimated that the human genome consists of approximately 20,000 to 25,000 genes.
- **Chromosomes** Structures in the nuclei of cells that contain genes. Humans have 23 pairs of chromosomes (shown on the left); half of each pair comes from the mother and half from the father.
- **Genes** The basic units of heredity that occupy specific places (loci) on chromosomes. Genes consist of large DNA molecules, each of which contains the code for the manufacture of a specific enzyme or other protein.
- **Genotype** The specific genetic makeup of an individual as coded by DNA.
- **Genomics** The study of the functions and interactions of all genes in the genome. Genomics includes the scientific study of complex diseases such as heart disease, asthma, diabetes, and cancer. These diseases are typically caused by a combination of genetic and environmental (**epigenetic**) factors.
- **Epigenetic** Heritable changes in gene function that do not entail a change in DNA sequence. Epigenetic modifications play a crucial role in the silencing and expression of noncoding portions of genes.
- **Nucleotide** A nucleotide is one of the structural components, or building blocks, of DNA and RNA. A nucleotide consists of a base (one of four chemicals: adenine, thymine, guanine, and cytosine) plus a molecule of sugar and one of phosphoric acid.
- **Gene variant** A different form of a gene that has a different functional status than the original gene.
- **Nutrigenomics** The study of the interactions between genes, gene variants, and nutrients.

environmental exposures and thereby increase the chances of survival after birth.[15,21] Such adaptations change the functional status of certain genes by turning them "on" or "off." Not all gene variants affect health status, but some do. Here are some examples of gene variants that affect health:

- Exposure to famine during early pregnancy can promote the development of gene variants that modify glucose utilization in ways that increase the risk of type 2 diabetes.[15]

- Optimal nutrition before and throughout pregnancy may suppress the development of gene variants that promote the development of type 2 diabetes, heart disease, and obesity later in the offspring's life.[15]
- Individuals with 14 "favorable gene variants" for fat storage location have lower levels of central and liver fat, and a 40% lower risk of developing type 2 diabetes and 10% lower risk of heart disease and hypertension than people without the variants.[37]
- The presence of one or more gene variants in vitamin D utilization makes some individuals appear to be deficient in vitamin D based on blood tests when they are not. Blood concentrations of vitamin D do not respond to supplementation with vitamin D.[16] The same situation is true for vitamin B_{12}.[17]
- Specific gene variants unrelated to diet or lifestyle are related to elevated blood levels of triglycerides, inflammation, and heart disease risk.[18]
- Gene variants related to energy expenditure, appetite, fat cell development, insulin resistance, and lipid metabolism appear to influence the body's response to components of the diet and weight status.[4,19]
- Increased level of physical activity appear to help people with pro-obesity gene variants lose weight.[21]
- Among individuals with particular gene types, high-fat diets increase insulin resistance and fat stores in the liver, and increase the risk of type 2 diabetes and obesity.[23]
- Adults with specific gene variants who frequently consume fried foods (4 or more times per week) versus adults who consume fried foods less often (less than once a week) are associated with higher body mass indices.[16]
- Gene variants related to impaired folate and choline utilization can increase the need for folate and choline.[12,13]
- The risk of developing breast cancer appears to be lower in women with certain gene variants if they regularly consume cruciferous vegetables (broccoli, cauliflower, cabbage, for example).[25]

Other gene variants are related to diseases and disorders such as Parkinson's disease, late-onset Alzheimer's disease, and celiac disease.[1,11] It is being increasingly recognized that the path to the prevention and treatment of many diseases and disorders is paved by individualized dietary and other lifestyle changes that counteract the negative effects of gene variants, and techniques that modify gene expression.[1,4,5,15,20]

Although it is clear that genetic variation influences health and disease development, lifestyle factors such as diet and exercise continue to play important roles in the prevention and development of obesity, diabetes, and other common disorders. A careful recording of family disease history remains useful for assessment of risk. Many diseases and disorders such as diabetes, heart disease, certain forms of cancer, and obesity track in families. Knowledge gained from a family history can help predict risk and guide decision making about preventive services and appropriate genetic tests.[12,26] This Unit's Reality Check addresses the usefulness of knowledge of family history for heart disease.

Single-Gene Defects

Thousands of rare diseases related to a defect in a single gene have been identified, and many of these affect nutrient needs. Such defects can alter the absorption or utilization of nutrients such as amino acids, iron, zinc, and the vitamins B_{12}, B_6, or folate. Phenylketonuria (PKU), galacostemia, hemochromatosis, and sitosterolemia are four examples of well-characterized **single-gene defects** that substantially affect nutrient needs or utilization (Table 26.2). Due to its potentially severe affects on mental development if not treated with a low-phenylalanine diet starting after birth, all babies born in the United States and many other countries are tested for PKU after birth (Illustration 26.1).

single-gene defects Disorders resulting from one abnormal gene. Also called "inborn errors of metabolism." Over 6,000 single-gene disorders have been cataloged, and most are rare.

Table 26.2 Examples of single-gene disorders that affect nutrient need or utilization[27–30]

Disorder	Description
PKU (phenylketonuria, pronounced phen-el-key-tone-uria)	A rare disorder caused by lack of the enzyme phenylalanine hydroxylase. Lack of this enzyme causes phenylalanine, an essential amino acid, to build up in the blood. High blood levels of phenylalanine during growth lead to mental retardation, poor growth, and other problems. PKU is treated by low-phenylalanine diets.
Galactosemia (pronounced gal-lac-tos-see-me-ah)	Galactosemia is a single-gene defect disorder that interferes with the body's utilization of the sugar galactose found in lactose ("milk sugar"). The signs and symptoms of galactosemia result from an inability to use galactose to produce energy and the build-up of galactose in body tissues. If infants with classic galactosemia are not treated promptly with a low-galactose diet, life-threatening complications appear within a few days after birth. People with this condition must avoid all milk, milk-containing products (including dry milk), and other foods that contain galactose for life. It occurs in approximately 1 in 30,000 to 60,000 newborns.
Hemochromatosis (pronounced heme-oh-chrome-ah-toe-sis)	A fairly common single-gene defect disorder affecting 1 in 300 Caucasians. It occurs due to a defect in a gene that produces a protein that controls how much iron is absorbed from food. Individuals with hemochromatosis absorb more iron than normal and have excessive levels of body iron. High levels of body iron have toxic effects on tissues such as the liver and heart. Hemochromatosis is treated with medications and a diet low in iron and vitamin C. A high intake of vitamin C can make hemochromatosis worse because vitamin C increases the absorption of iron.
Sitosterolemia	A rare disorder (1 in 50,000 individuals) caused by a mutation in one of two genes that affect the ability of small intestine cells to dispose of plant sterols (a type of fat found in vegetables oils, nuts, and other plants) in the feces. The body cannot utilize plant sterols, and if they cannot be eliminated from the intestines they accumulate in the body and cause plaque buildup in arteries. That increases the risk of heart disease and stroke. It is treated by a diet low in plant sterols and other measures including drugs and surgery. Also called *phytosterolemia*.

Most diseases related to genetic traits are not as well defined as are single-gene defects. They are more likely to represent an interwoven mesh of gene variants and environmental factors.

Genetics of Taste

Food preferences are partly influenced by genetic traits related to taste sensitivity. Genetically influenced food preferences apply to dogs, cats, and other animals as well humans (Illustration 26.2). There are more than 80 genes that help people taste bitter foods, for example, and some people get the set of genes that make them highly sensitive to bitter-tasting foods. People born with a high sensitivity to bitter tastes tend to dislike cooked cabbage, collard greens, spinach, brussels sprouts, or other vegetables that taste bitter to them. People who tend to like these vegetables generally don't perceive them to be

BSIP/Getty Images

Illustration 26.1 A blood test for PKU is a routine part of newborn screening in most Western countries.[30]

REALITY CHECK

Changing Family History

Elena and Alfredo both come from families with a number of relatives who have died from heart disease.

Who gets the thumbs-up?

Answer appears on page 26-6.

PhotoDisc

Elena: Heart disease is in my genes. There's nothing I can do about it.

PhotoDisc

Alfredo: I already know my LDL cholesterol is high and my HDL cholesterol is low. I'm eating and exercising to beat the odds.

ANSWERS TO REALITY CHECK

Changing Family History

Alfredo recognizes that a family history of heart disease doesn't mean he is destined to die from it. A minority of diseases are due solely to genetic factors; many are primarily due to interactions between genetic traits and environmental factors. Disease-promoting effects of genetic traits can often be diminished by the right lifestyle and environmental changes.[12,31]

PhotoDisc

Elena:

PhotoDisc

Alfredo:

Dlillc/Cardinal/Corbis

Illustration 26.2 Dogs and many other animals have a taste for sweets, but cats don't. Somewhere during genetic evolution the gene in cats that enabled them to recognize sweetness was turned off.[36]

©Photographee.eu/Shutterstock.com

Illustration 26.3 In the future, advice given by registered dietitians/nutritionists and other health professionals will be tailored to the specific genetic traits of individuals.

bitter tasting. A genetic tendency to reject these vegetables is likely to limit intake and therefore may be linked to diseases associated with low vegetable intake.[32]

Genetic traits also affect a person's perception of sweet and savory (umami) tastes. People who prefer sweet and savory tastes tend to have higher intake of sugar and fats, and higher body mass index than people who don't.[33]

Nutrition Tomorrow

The promise of advances in knowledge of the genetic bases of disease is longer, healthier lives. No doubt drugs that counter the ill effects of genetic traits will continue to be developed, and attempts to fix abnormal genes by gene therapy will broaden.[34] Some of the most meaningful breakthroughs in the future will be in the area of disease prevention and treatment through nutritional changes.[13] Knowledge of genetic traits will increasingly be used to identify individual nutrient needs and responsiveness to dietary and other changes.[4]

Although used to some extent now, personalized modifications of dietary intake based on genotypes will become standard practice in clinical dietetics and medicine (Illustration 26.3).[12] It is already clear that lifestyle modifications, such as weight loss, increased physical activity, and increased intake of vegetables and fruits, for example, will be components of health improvement efforts stemming from knowledge of genetic risk factors.[31]

The sequencing of an individual human's genome, a DNA profile, cost $350,000 in 2008 but can now be done for $1,000.[35] Except for well-characterized single-gene defects, knowledge of an individual's genetic makeup does not provide enough information to reliably indicate disease risk. Even if specific genes related to disease risk are identified in a person's genome, that does not mean that the genes are functional. The function of genes is determined by epigenetic factors. It is estimated that 90% of DNA alterations associated with disease are due to gene variants rather than genes.[1,35]

Although tremendous advances in knowledge have been made, understanding the functions and interactions of the entire human genome is an extraordinarily complex undertaking and will evolve over time. When understood, it will represent the next revolution in breakthrough knowledge about nutrition and health.

©XiXinXing/Shutterstock.com

NUTRITION up close

Nature and Nurture

Focal Point: Gaining insight into your family's and your own dietary health history.

Families often share dietary factors that interact with genetic traits to influence chronic disease development. Check out your dietary health history and that of your family by completing the following family food tree activity. If you don't know all of your relatives included in the activity or their dietary behaviors, fill out the parts based on what you do know.

Check each dietary behavior that applies:

	Excess calorie consumption		Low vegetable and/or fruit intake		Low whole grains intake		Low fiber intake		Low fish intake		High alcohol intake	
Relative	Yes	No	Yes	No	Yes	No	Yes	No	Yes	No	Yes	No
A. On your mother's side:												
Grandmother												
Grandfather												
B. On your father's side:												
Grandmother												
Grandfather												
C. Your mother												
D. Your father												
E. Yourself												

Feedback to the Nutrition Up Close is located in Appendix F.

REVIEW QUESTIONS

- **Explain how genetics and nutrition are interdependent.**

1. Genomics is the study of the functions and interactions of genes. **True/False**
2. Nutrigenomics is the study of the interaction of body weight and genes. **True/False**
3. Genes provide the codes for the production of enzymes and other proteins in cells. **True/False**
4. Unfavorable changes in genetic traits are the primary cause of the current global epidemics of obesity and diabetes. **True/False**
5. Many chronic diseases result from single-gene defects. **True/False**
6. Many chronic diseases result from interactions among multiple genetic traits and environmental factors. **True/False**
7. Commercially available tests that identify an individual's genes are generally *not* helpful for identifying an individual's susceptibility to diseases. **True/False**
8. Some of the most meaningful breakthroughs in the prevention and management of chronic diseases in the future will be due to advances in nutrigenomics. **True/False**
9. Galactosemia is a single-gene defect that interferes with the digestion of lactose. **True/False**
10. Hemochromatosis is caused primarily by a defect in a gene that affects iron absorption. **True/False**
11. Gene variants represent different genes that code for the synthesis of the same specific protein. **True/False**
12. Availability of certain vitamins can affect the functional level of specific genes. **True/False**

The following three questions refer to the following scenario.

You're standing in a line to order Chinese takeout when you can't help but overhear the conversation about weight loss the gentleman behind you is having with his friend. The conversation goes like this. "You know, George, no matter what I do, I can't lose weight. I eat like a bird and I still gain weight. I swear I've inherited the obesity gene."

13. _____ Which of the following statements most likely represents the truth about the gentleman's problem with losing weight?
    a. He has probably inherited the gene known to cause obesity from his mother or father.
    b. He may have inherited genetic traits that increase the likelihood he will overeat in an environment where plenty of highly palatable foods are available.
    c. He has developed a single-gene defect that causes obesity.
    d. He has gene variants that decrease his calorie need substantially, so he will not be able to lose weight.

14. _____ If you could have turned around and talked with the gentleman about the likely reasons for obesity, what would you say that would represent state-of-the-science knowledge?
    a. Genetic traits do not influence a person's chances of becoming obese.
    b. Genetic defects in energy metabolism affect obesity development, not calorie consumption.
    c. It has become clear recently that multiple genetic traits make it impossible for some people to lose weight.
    d. Multiple genes may influence the development of obesity, but if you consume fewer calories than you use you will lose weight.

15. _____ The gentleman decides to have his genetic makeup tested to confirm that his genes produced his obesity. Which of the following statements about the results of the test is correct?
    a. The test will identify multiple genes that account for his obese state.
    b. The test will provide information needed to recommend a particular nutrient supplement that will enable him to lose weight.
    c. The test will be inconclusive because genes, environment, and behavior influence the development of obesity, and not genes alone.
    d. His DNA will likely indicate that he is obese because he is missing a gene that increases a person's level of physical activity.

Answers to these questions can be found in Appendix F.

NUTRITION SCOREBOARD ANSWERS

1. That's incorrect. Only a very small proportion of diseases are directly caused by a person's genes.[1] **False**
2. Individual genetic traits alter essential nutrient needs to a small extent in many people, and to a large extent in relatively few.[2] **False**
3. Correct. Common diseases result from interactions between multiple genetic traits and environmental factors such as nutrient intake.[3,4] **True**
4. The body's utilization of nutrients, including dietary fats, varies among individuals based on genetic makeup.[5] **True**

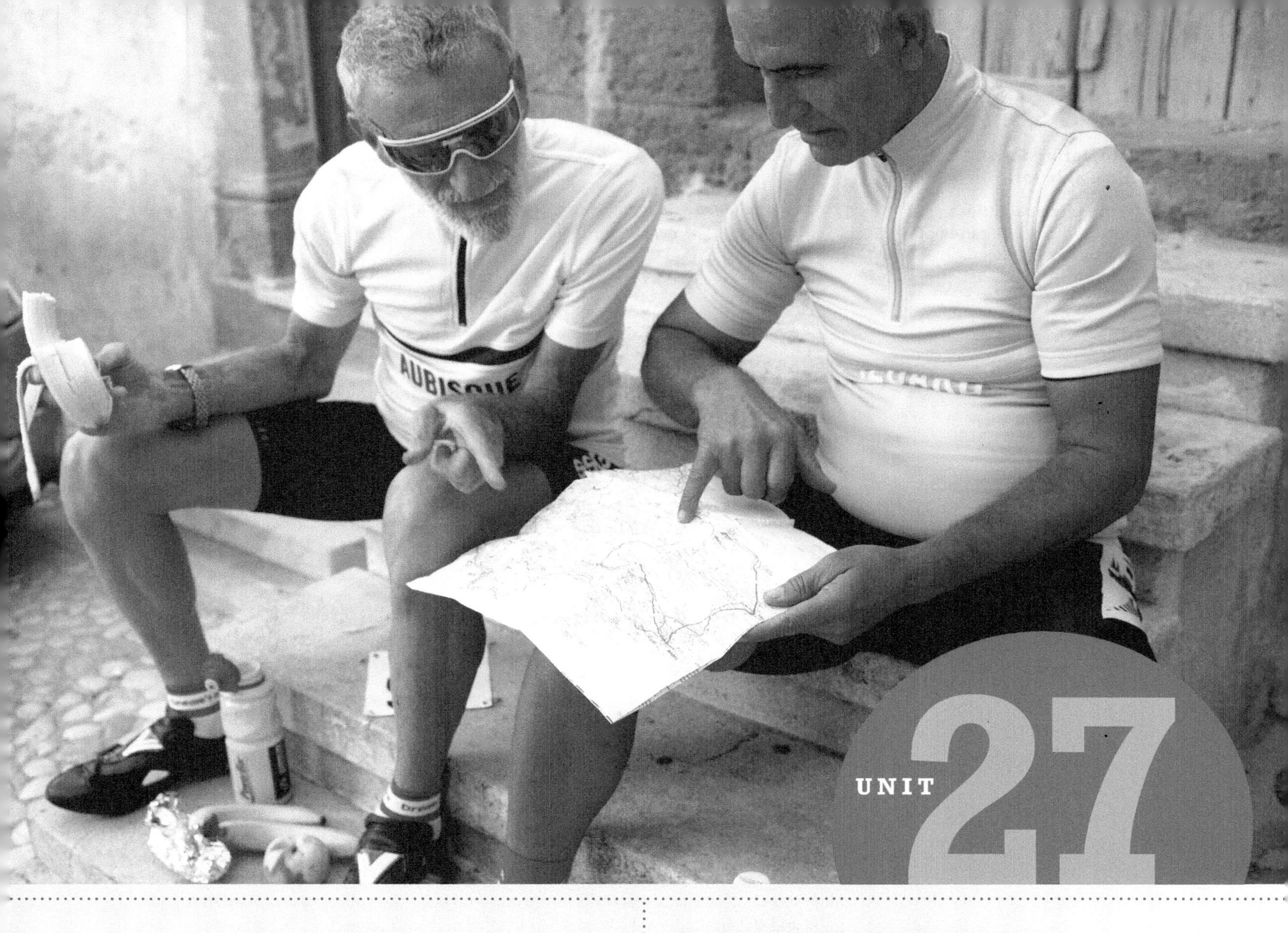

UNIT 27

Nutrition and Physical Fitness for Everyone

NUTRITION SCOREBOARD

1 Physical fitness means being very muscular. **True/False**

2 Overweight people can be physically fit. **True/False**

3 Exercise is more effective than diet in preventing heart disease. **True/False**

4 Physical fitness can only be achieved by exercising intensively for at least an hour every day. **True/False**

Answers can be found at the end of the unit.

Michael Blann/Iconica/Getty Images

Physical Activity: It Offers Something for Everyone

LEARNING OBJECTIVES

After completing Unit 27 you will be able to:

- Describe physical fitness and the health advantages of being physically fit.

I would not wish to imagine a world in which there were no games to play and no chance to satisfy the natural human impulse to run, to jump, to throw, to swim, to dance . . . . I see health as a happy consequence of sporting activity.

—Sir Roger Bannister, first person to run a mile in less than 4 minutes

- **Describe physical fitness and the health advantages of being physically fit.**

Physical activity has much to offer the athlete and non-athlete alike. It provides recreation, which is good for both the body and soul; it doesn't have to cost anything; and it benefits almost everyone. At its best, physical activity results from play and has positive consequences for health and well-being. As long as there are physical activities people enjoy doing, there's a physically fit "athlete" in everyone.

KEY NUTRITION CONCEPTS

Physical activity interacts with healthy diets in a number of ways to promote overall health and well-being. Key nutrition concepts that relate to the "good diet" portion of these relationships include:

1. Nutrition Concept #8: Poor nutrition can influence the development of certain chronic diseases.
2. Nutrition Concept #9: Adequacy, variety, and balance are key characteristics of healthy dietary patterns.

The "Happy Consequences" of Physical Activity

Ask people who are in good physical condition what they get from physical activity and you're likely to hear a variety of responses. Someone who is 20 years old may say he wants to look good and stay in shape. A 40-year-old individual may say that exercise helps keep her blood triglyceride levels and weight from increasing. People who are 80 might tell you they exercise so that they won't have to use a walker and will be able to maintain their independence. Ask children why they exercise, however, and they may not understand the question. Children engage in active play; they don't exercise. For them, fitness is truly an unintended consequence of play.

Regular physical activity benefits both physical and psychological health in people of all ages (Table 27.1). By improving a person's physical health, regular physical activity helps ward off heart disease, obesity, some types of cancer, hypertension and stroke, osteoporosis, and diabetes. It tends to increase a person's feeling of well-being and helps relieve depression, anxiety, and stress.[1]

The Bonus Pack: Physical Activity Plus a Healthy Dietary Pattern

Physical activity benefits health most when combined with a healthy dietary pattern and other healthy behaviors.[2] Regular physical activity helps build bone mass and reduce the risk of bone fractures.[9] But the risk is lowered to a greater extent if physical activity is combined with a diet that supplies adequate amounts of calcium and vitamin D.[3] Some other benefits of exercise, such as a reduced chance of developing colon and breast cancer, may be related to the effect of exercise on reducing the ill effects of high levels of body fat on health. Regular physical activity is one of the few factors we can identify that help people reduce weight gain with age.[4] From children to elders, to pregnant women and individuals with physical and cognitive disabilities, regular physical activity improves cardiorespiratory fitness, muscle function, mental health, and energy levels.[2]

physical activity Body movements produced by muscles that require energy expenditure. Exercise is a subcategory of physical activity. It is generally planned, structured, and repetitive. The terms *physical activity* and *exercise* are often used interchangeably.

physical fitness The health of the body as measured by muscular strength and endurance, cardiorespiratory fitness, and flexibility. Body composition, agility, and balance are sometimes included as components of physical fitness.

muscular strength The ability of a muscle or muscle group to exert force. It is assessed by the maximal amount of resistance or force that can be sustained in a single effort.

Table 27.1 **Benefits of regular physical activity**[1,14,36,37]

Reduced risk of certain diseases and disorders	Improved sense of well-being
• Heart disease	• Increases feeling of well-being
• Colon and breast cancer	• Decreases depression and anxiety
• Hypertension	• Helps relieve stress
• Stroke	• Decreases risk of dementia
• Osteoporosis	• Decreased feeling of fatigue
• Back and other injuries	
• Obesity, excess abdominal fat	
• Type 2 diabetes	
• Bone and joint diseases	
• Increased mobility	

Paul Bradbury/OJO Images/Getty Images

Exercise and Body Weight Combined with a moderate decrease in normal caloric intake (on the order of 200 calories per day), exercise helps people lose body fat, build muscle mass, and become physically fit.[5] Exercise alone may be an effective weight loss approach for the subset of individuals who carry the fat mass and obesity-associated (FTO) gene variant that reduces appetite in obese individuals in response to physical activity.[15] Exercise alone is generally ineffective, or less effective as a weight reduction measure, than is exercise plus a reduced calorie intake.[6,7] Because the body needs more calories to maintain muscle than fat, exercise that results in an increase in muscle mass leads to an increase in caloric requirements. For many people, this increase in calorie requirement makes it easier to maintain weight during adulthood and to keep lost weight off.[4,6]

Whether or not regular exercise helps overweight people lose weight and keep it off depends primarily on calorie intake. If it remains too high, weight will not be lost nor will weight loss be sustained. Calorie intake often fails to be adequately managed by individuals who perceive physical activity to be "work" or "exercise" that should be rewarded with something pleasurable like a savory snack or sweet dessert. People who enjoy physical activity because it is perceived as being fun or pleasurable are less likely to reward themselves with high-calorie foods than those who think it is hard.[7]

Table 27.2 lists the calorie cost of various types of exercise. The values are calculated based on body weight because an individual's weight affects the energy requirement for specific physical activities. If a 140-pound person cycles on a stationary bicycle at a moderate level of effort for an hour, and the calorie cost of the exercise is 3.4 calories per pound per hour, the person would use about 476 calories per hour with moderate cycling (3.4 calories per pound × 140 pounds = 476 calories). A person who weighs 212 pounds would use approximately 720 calories (3.4 calories per pound × 212 pounds). If the exercise is continued for 10 minutes, the calorie expenditure number would be divided by 6, or if done for 15 minutes, it would be divided by 4.

What Is Physical Fitness?

Many of the benefits of physical activity are related to the **physical fitness** it promotes. Physical fitness is not defined by bulging muscles, thin waistlines, or the amount of physical activity. Overweight as well as thin people can be physically fit or not. Physical fitness is a state of health measured by **muscular strength** and **endurance**, **cardiorespiratory fitness**, and **flexibility**.[2,9]

The strength component of physical fitness relates to the level of maximum force that muscles can produce. Muscular endurance refers to the length of time muscles can perform physical activities, and flexibility refers to a person's range of motion. Cardiorespiratory fitness relates to the functioning of the circulatory system, heart, and the lungs. Physical fitness exists when all four are present at health-promoting levels.

muscular endurance The ability of a muscle or group of muscles to sustain repeated contractions against a resistance for an extended period of time. Also called *muscular fitness*.

cardiorespiratory fitness The ability of the circulatory and respiratory systems to supply fuel during sustained physical activity and to eliminate fatigue products after supplying fuel. Cardiorespiratory fitness is also called *aerobic fitness* and *cardiorespiratory endurance*.

flexibility Range of motion of joints.

resistance exercise Physical activities that involve the use of muscles against a weight or force. Resistance exercise increases muscle strength, mass, power, and endurance. Also called resistance training, strength training, and weight training.

aerobic fitness A state of respiratory and circulatory health as measured by the ability to deliver oxygen to muscles and the capacity of muscles to use the oxygen for physical activity.

aerobic exercise Physical activity in which the body's large muscles move in a rhythmic manner for a sustained period of time. Aerobic exercise involves metabolic pathways that require oxygen for energy production and improve functioning of the cardiovascular and respiratory systems. Aerobic activities include walking, jogging, running, swimming, skiing, vacuuming, house cleaning, lawn mowing, raking, cycling, and aerobic dance.

Table 27.2 Average calorie output per pound of body weight for selected types of exercise

Exercise	Intensity	Calories (pound/hour)
Walking	3 mph (20 min/mi)	1.6
	3½ mph (17 min/mi)	1.8
	4 mph (15 min/mi)	2.7
	4½ mph (13 min/mi)	2.9
Jogging	5 mph (12 min/mi)	4.1
	5½ mph (11 min/mi)	4.5
	6 mph (10 min/mi)	4.9
	6½ mph (9 min/mi)	5.2
	7 mph (8½ min/mi)	5.6
	7½ mph (8 min/mi)	6.0
Running	8 mph (7½ min/mi)	6.3
	8½ mph (7 min/mi)	6.7
	9 mph (6⅔ min/mi)	7.1
	9½ mph (6⅓ min/mi)	7.4
	10 mph (6 min/mi)	7.6
	11 mph (5½ min/mi)	8.5
	12 mph (5 min/mi)	9.5
Cycling (stationary)	Mild effort	2.9
	Moderate effort	3.4
	Vigorous effort	4.3
Cycling (level)	6 mph (10 min/mi)	1.5
	8 mph (7½ min/mi)	1.8
	10 mph (6 min/mi)	2.0
	12 mph (5 min/mi)	2.8
	15 mph (4 min/mi)	3.9
	20 mph (3 min/mi)	5.7
Skating		2.9
Calisthenics	Moderate	2.4
	Vigorous	2.9
Rope skipping	Moderate	4.8
Bench stepping	12" high, 24 steps/min	3.2
Weight lifting		2.9
Wrestling		6.2

Exercise	Intensity	Calories (pound/hour)
Handball	Moderate	4.8
	Vigorous	6.2
Swimming	Resting strokes	1.4
	20 yd/min (mod)	2.9
	40 yd/min (vig)	4.8
Rowing (sculling or machine)		4.8
Downhill skiing		3.8
Cross-country skiing (level)	4 mph (15 min/mi)	4.3
	6 mph (10 min/mi)	5.7
	8 mph (7½ min/mi)	6.7
	10 mph (6 min/mi)	7.8
Aerobic dancing	Moderate	3.4
	Vigorous	4.3
Rebound trampoline	50–60 steps/min	4.1
Racquetball/squash	Moderate	4.3
	Vigorous	4.8
Tennis	Moderate	3.4
	Vigorous	4.3
Volleyball	Moderate	3.4
	Vigorous	3.8
Basketball	Moderate	3.8
	Vigorous	4.8
Football	Moderate	3.8
	Vigorous	4.3
Baseball/golf/woodcutting/horseback riding/badminton/canoeing		2.4
Soccer/hill climbing/fencing/judo/snowshoeing		5.3
Bowling/archery/pool		1.2

Source: Values calculated from *Guidelines for Graded Exercise Testing and Exercise Prescription*, 2nd ed. Philadelphia: American College of Sports Medicine, Lea and Febiger, 1984.

Victoria Snowber/Photodisc/Getty Images

Illustration 27.1 An example of a strength-building exercise.

Muscle Strength Muscle strength depends on the ability of a muscle or groups of muscles to lift, pull, push, or otherwise exert force against a weight or opposing force. Muscle-building activities are referred to as **resistance exercise**. Such activities require muscles to work harder than usual, which increases muscular strength over time. Strength-building exercise includes lifting weights, doing pull-ups and push-ups, and the use of stretch bands (Illustration 27.1).

Muscle Endurance and Cardiorespiratory Fitness Muscular endurance is a measure of the ability of a muscle or group of muscles to sustain repeated muscular contractions against a weight or force over time. The length of time muscles can work against a weight or force depends largely on cardiorespiratory fitness and is measured in terms of the amount of oxygen an individual is able to deliver to muscles. (Cardiorespiratory fitness is sometimes referred to as **aerobic fitness**.) Levels of cardiorespiratory fitness build up with time as individuals increase their level of **aerobic exercise**.

Aerobic exercises use oxygen for energy formation by muscles. Performing these types of activities enhances the functioning of the heart and lungs and increases their ability to deliver oxygen to muscles. Fat-burning, aerobic exercises increase cardiovascular fitness in three major ways:

1. They strengthen and expand the capacity of the lungs to deliver oxygen.
2. They increase the ability of the circulatory system to deliver blood and oxygen to muscles and other tissues throughout the body.
3. They strengthen the ability of the heart to move an increased volume of blood through the body.

Aerobic activities include jogging, basketball, swimming, soccer, bungee, aerobic dance, and other low- and moderate-intensity activities (Illustration 27.2). They give most of the body a workout.

Erik Isakson/Getty Images

Illustration 27.2 An example of aerobic exercise.

How Is Cardiorespiratory Fitness Determined? Cardiorespiratory fitness is assessed by measuring **maximal oxygen consumption** (abbreviated as VO_2 max) under supervision in a specially equipped laboratory (Illustration 27.3). In the lab, individuals are exercised at increasingly higher intensities, for example, by elevating the grade of a treadmill or increasing the resistance on a stationary cycle. The individual performing the exercise breathes through a tube that delivers air. Monitoring equipment measures the amount of oxygen from air that is used during the exercise. The maximal amount of oxygen a person delivers to working muscles is the amount used when the intensity of exercise can no longer be increased. The higher the level of oxygen used at the peak level of activity (or 100% VO_2 max), the higher the level of aerobic fitness and the longer physical activity can be performed.

maximal oxygen consumption The greatest amount of oxygen that can be delivered to, and used by, muscles for physical activity. Also called *VO_2 max* and *maximal volume of oxygen*.

People can perform physical activity at 100% of VO_2 max for only a few minutes. Consequently, aerobic fitness goals are set below that level. In general, it is recommended that beginners start a cardiorespiratory fitness program with a goal of exercising at 40 to 60% of VO_2 max and working up to a higher level.[9] Aerobically fit people generally train at 70 to 85% of VO_2 max. VO_2 max can be increased by exercising regularly at intensities that raise a person's heart rate.[11]

Physical Fitness and Heart Rate Heart rate, or the number of times your heart beats per minute (bpm), is used as an indicator of cardiorespiratory fitness and exercise intensity. It reflects the number of times your heart has to beat to send enough fuel, nutrients, and oxygen to your tissues to maintain their functions. The lower the heart rate, the more efficient your body is at delivering oxygen to cells and utilizing glucose and fat for energy formation.

Resting heart rate reflects the number of times per minute your heart must beat to maintain body functions while you are at complete rest. It can be used to help track progress in becoming physically fit. Although variable, the average heart rate of adults is between 60 and 80 beats per minute (bpm), whereas the rates for well-trained athletes are often between 40 to 60 bpm. A good time to assess resting heart rate is when you wake up after a good night's sleep. Before getting out of bed, gently press two fingertips on the artery located on the thumb side of your wrist (see Illustration 27.4). Count the number of pulses you feel while counting off or watching a clock for 10 seconds. Multiple that number by 6 to get beats per minute (bpm). This same technique for taking your pulse is used to measure heart rate during pauses in exercise.[13]

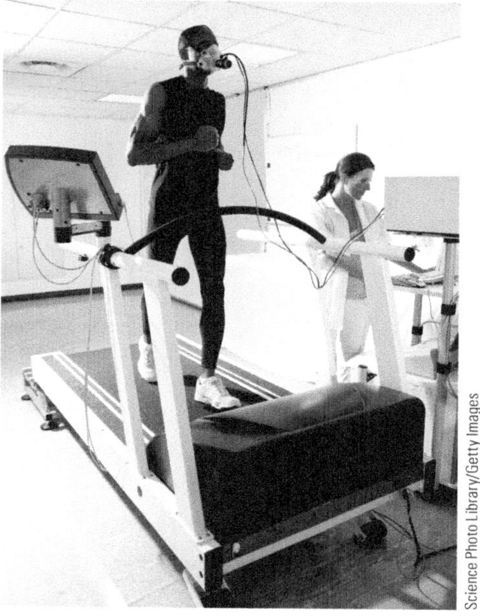

Science Photo Library/Getty Images

Illustration 27.3 Determining VO_2 max. VO_2 max is measured when a person's nose is plugged so she or he must breathe through a tube that delivers air. The tube is attached to an apparatus that measures the total amount of air breathed in and out, and calculates the difference between the amounts of oxygen inhaled and exhaled. Every few minutes the resistance of the cycle is increased, making the exercise more intense. The amount of oxygen a person uses at the point when exercise intensity can go no higher is considered "maximal oxygen consumption," or 100% VO_2 max.

Maximum Heart Rate The rate at which the highest level of oxygen delivery to working muscles can occur is considered the *maximum heart rate (MHR)* or 100% MHR. Moderate-intensity activities are considered those undertaken between 50 and 69% MHR, and intense activities fall between 70 and 90% MHR.[13] The most accurate way to determine MHR is to have it measured directly in a laboratory. Maximum heart rate varies based on sex, age, and other factors, and no one formula captures the true rate for most individuals.[14] However, identifying estimated MHR and using heart rate targets for physical activity can

Dorling Kindersley ltd/Alamy

Illustration 27.4 Finger location for measuring pulse on your wrist.

Table 27.3 Applying the 100% MHR formula to a 22-year-old who will exercise at a level of 60% MHR

100% MHR = 220 − age = 220 − 22 − 198 bpm
100% MHR = 198 bpm
60% MHR = 198 × 0.6 = 119 bpm

help individuals monitor their progress in meeting their aerobic fitness goals while staying within a safe range of MHR. The most commonly recommended and best-known formula for estimating MHR in adults is straightforward:

$$\text{MHR} = 220 - \text{age}$$ [13]

To obtain a target heart rate, or heart rate range for exercise, multiply your MHR by the desired percent of MHR (Table 27.3 shows an example). The result is your target heart rate for aerobic exercise. Table 27.4 lists maximal heart rate, 50% of the maximum, and 85% of the maximum rate for adults between the ages of 20 and 70 years. During aerobic exercise, take a break and measure your pulse; calculate your heart rate to see if you are within your target range.

Wearable Devices and Smart Phone Apps A number of wearable devices and smart phone apps are available that assess step counts, speed, and heart rate. Some of the wearable devices and their apps have been tested and found to calculate steps and speed with reasonable accurately 88 to 98% of the time, and smart phone sensors and apps are very close to being accurate at least 93% of the time.[16,17] Smart phones that detect heart beats or pulses appear to provide reasonably accurate results.[18] Activity monitors are also available that provide automated feedback and interactive behavioral change tools via smart phones or computers. Many of the available programs are based on established behavioral change methods for goal setting, self-monitoring, and feedback.[19]

Informing physical fitness progress by monitoring target heart rate or the use of smart technology helps motivate some people who want to improve their level of aerobic fitness.[17] Other people simply want to be more physically active, and although they may track time spent in physical activities, knowledge of heart rate and tracking behaviors does not provide inspiration or motivation. Is addition, reaching certain target heart rates is not for everyone. It may be dangerous for people with heart or other conditions that limit the amount of stress the heart can take.[20]

Have you been thinking about increasing your level of aerobic exercise? If you have, refer to the first of two Take Action features in this unit to help make a plan for doing just that.

Table 27.4 Target heart rates estimated at 100%, 50% (moderate-intensity exercise) and 85% (vigorous-intensity exercise) MHR for adults ages 20–70 years[17]

Age (years)	100% MHR	50% MHR	85% MHR
20	200	100	170
30	190	95	162
40	180	90	153
50	170	85	145
60	160	80	136
70	150	75	128

take action Increase Your Cardiorespiratory Fitness

Increase your level of cardiorespiratory fitness by choosing aerobic activities you enjoy. First, check the aerobic activities you like to do and can do. Then list how long and on how many days per week you will engage in the activity.

Aerobic activity	Duration of activity	Which days?
____ Swimming	____ minutes	________
____ Walking	____ minutes	________
____ Jogging	____ minutes	________
____ Running	____ minutes	________
____ House-cleaning	____ minutes	________
____ Table tennis	____ minutes	________
____ Treadmill	____ minutes	________
____ Basketball	____ minutes	________
____ Cycling	____ minutes	________
____ Aerobic dance	____ minutes	________
____ Zumba	____ minutes	________
____ Ballroom dancing	____ minutes	________
____ Pilates	____ minutes	________
____ Other:	____ minutes	________

Refer to these goals periodically to see how you're doing or modify your activities into a more workable plan. You can obtain forms for developing goals and a plan for aerobic and strengthening exercise, and for tracking your progress in meeting the goals from the Physical Activity Guidelines for Americans site: www.health.gov/paguidelines/pdf/adultguide.pdf. The Guidelines were endorsed by the 2015 Dietary Guidelines Advisory Committee (DGAC).

Flexibility

Flexibility refers to the range of motion of your muscles and connective tissues around your joints, or how "tight" or "loose" your body movements are. It affects your ability to stretch, react, bend, and maintain balance and agility. It is important for day-to-day activities such as walking, reaching, moving smoothly and quickly out of danger, and balance. Flexibility is maintained and increased by many types of physical activities. Stretching exercises, in particular, increase flexibility (Illustration 27.5).[21]

iStockphoto.com/Mark Bowden

Illustration 27.5 Stretching increases flexibility and reduces muscle stiffness when it is done before beginning to exercise.

Stretching helps improve flexibility by lengthening muscles and tendons. It involves extending large muscle groups, while breathing, to the point you feel muscle tightness rather than discomfort. This position is held for 10 to 30 seconds, released, and then repeated once or twice. You can stretch your shoulder muscles, for example, by extending as arm across your chest and placing a hand on your elbow and gently pulling the arm toward your chest. Or, you can stretch your upper leg muscles (quadriceps) by grabbing the bottom of one leg above the ankle and pulling the heel back toward the buttocks while pushing your hips out. Yoga, Pilates, and tai chi movements are good examples of exercises that improve flexibility.

Stretching before or after exercising does not appear to protect against muscle soreness or reduce muscle injury, and it is not yet clear whether it increases performance.[23] It does, however, improve flexibility and muscle stiffness when it is done before exercise begins.[22]

Delayed Onset Muscle Soreness Have you ever exercised really hard one day when you were out of shape and felt it the next day? You are certainly not alone. People who begin an exercise program and overdo it, or who perform physical activities they don't usually do, often hear back from their muscles within the next day or two. The muscle groups that were used most feel stiff and sore, and you feel every move that's made using them.

delayed onset muscle soreness (DOMS) Muscle pain, soreness, and stiffness that occurs a day or two after exercise of muscles that are not normally used to high levels of activity.

This condition is called **delayed onset muscle soreness (DOMS)**, and here's why it happens. Overused muscles develop microscopic tearing in muscle fibers that cause stiffness and soreness for a day or two. Within a few days your muscles recover and actually end up in a better state to be strengthened. Muscles rebuild and become ready for the next bout of use.[24]

DOMS is most likely to occur when you:

- Use muscles you don't generally use for vigorous physical activity.
- Engage in physical activity when you are out of shape.
- Dramatically increase exercise intensity or duration of physical activity.
- Run downstairs or downhill, or engage in other downward motions.[24]

Fueling Physical Activity

Muscles use fat, glucose, and amino acids for energy. The proportion and amount of each that is used depends on the intensity of activity (Illustration 27.6). When we're inactive, fat supplies between 85 and 90% of the total amount of energy needed by muscles. The rest is provided by glucose (about 10%) and amino acids (5% at most).[25] Fat is also the primary source of fuel for activities of low to moderate intensity such as jogging and swimming. Because oxygen is required to convert fat into energy, low- to moderate-intensity activities are aerobic—or "oxygen requiring." This unit's Reality Check addresses the topic of aerobic exercise and its relationship to body fat. If you are curious about that, take a look at the Reality Check.

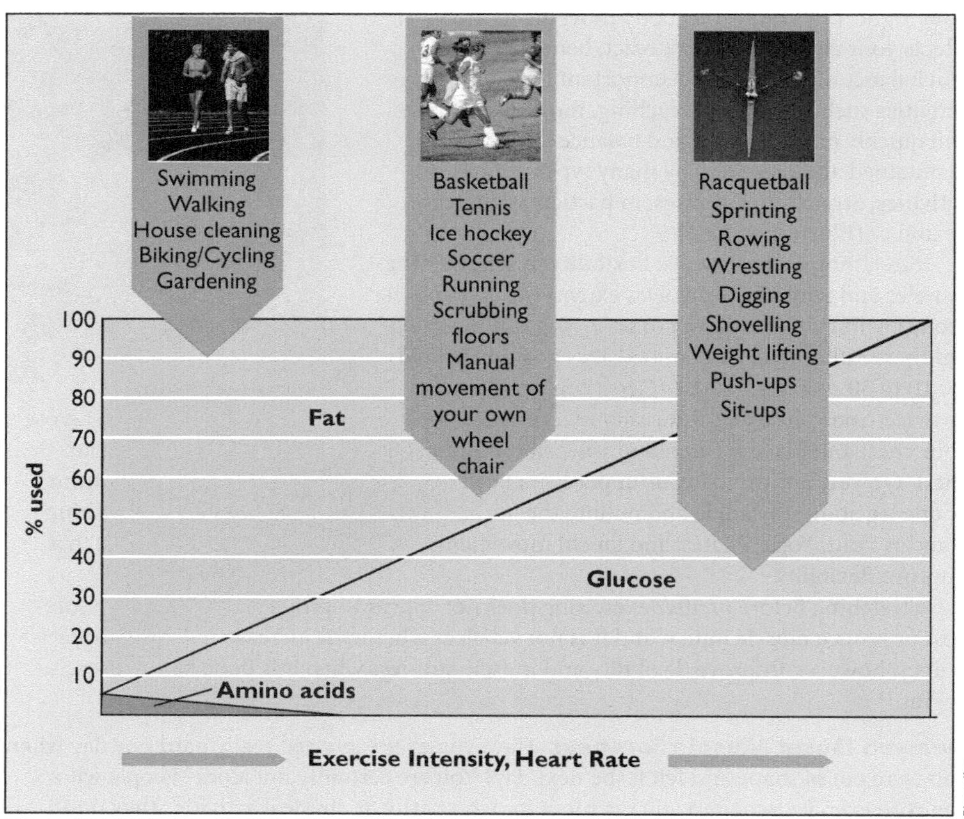

Illustration 27.6 Schematic representation showing the proportionate use of sources of energy for physical activities of various intensities. Most activities are fueled by both fat and glucose.

High-intensity, short-duration activities, like making a tackle, throwing a fastball, or sprinting down the block to catch a bus, are fueled primarily by glucose. Glucose is the primary fuel for these activities because metabolic processes exist within muscle cells that convert glucose to energy quickly. Our supply of glucose for intense activities comes principally from glycogen, the storage form of glucose. Glycogen is stored in muscles and the liver and can be rapidly converted to glucose when needed by working muscles. High-intensity, short-duration activities represent **anaerobic exercise** because the conversion of glucose to energy does not require oxygen (it's *an*aerobic). People can undertake very intense activity only as long as their stores of glycogen last.[26]

anaerobic exercise Short-duration, intense activities. Anaerobic exercise requires glucose for energy production. Energy formation from glucose does not require oxygen.

Activities such as basketball, hockey, tennis, football, and soccer, which involve walking, running, and high-intensity, quick moves, use both fat and glucose for energy.

A Reminder about Water Physical activity increases the body's need for water, and if the climate is hot and humid, this need increases even more. In general, people should drink in response to thirst and, overall, consume enough water to replace the amount lost in sweat, respiration, and urine during exercise.[27] (The amount of water lost during exercise is equivalent to the amount of weight that is lost during the exercise.) You are consuming the right amount of water if your urine is pale yellow and normal in volume. For exercise that lasts over an hour, consumption of a sports drink appears to improve hydration.[27] It's important to keep up fluid intake when exercising in cold weather. Cold air holds less water vapor than hot air, so you lose more water through breathing in cold weather.

Achieving Physical Fitness

Commenting on physical activity for the long run, Sir Roger Bannister said:

> *The best advice to those who are unfit is to take exercise unobtrusively. I am not an enthusiast for fitness and exercise schemes that are boring because they tend to fizzle out. . . . The vast majority of people, I think, can only be attracted for any length of time towards recreational activities if these are rewarding, enjoyable, and satisfying in themselves.*

Physical fitness is not something that, once achieved, lasts a lifetime. The beneficial effects of training on muscular endurance (aerobic fitness) diminish dramatically within 2 weeks after training stops. Muscle strength also decreases, but at a lower rate.[28] That makes it important to undertake physical activities that wear well—those are enjoyable and rewarding enough to fit into a busy schedule (Illustration 27.7). As the guidelines for becoming physically fit are presented in the next section, keep firmly in mind that a

REALITY CHECK

Does Aerobic Exercise Increase Body Fat Loss?

Who gets the thumbs-up?

Answers appears on page 27-10.

PhotoDisc

Tex: You bet. Aerobic exercise is fueled by body fat. If you do a lot of low-intensity aerobic exercise you'll burn off more of your fat than if you do high-intensity exercises.

PhotoDisc

Rose: I'm thinking you lose body fat when you expend more calories than you consume from food.

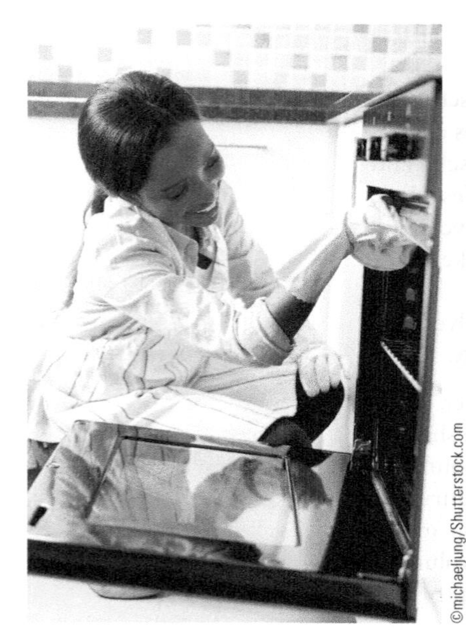
©michaeljung/Shutterstock.com

©Monkey Business Images/Shutterstock.com

© auremar/Shutterstock.com

Illustration 27.7 Physical activity can be fun. (Even joggers and runners smile sometimes.)

realistic and achievable plan is needed, one that feels right and that will last because it's good to you and for you.

Physical Activity Recommendations In response to the nationwide epidemic of inactivity, the Department of Health and Human Services developed recommendations aimed at improving the physical activity and fitness levels of Americans. Entitled *Physical Activity Guidelines for Americans*,[9] the recommendations urge Americans to undertake the following:

- At least 2 hours and 30 minutes (150 minutes) per week of moderate-intensity aerobic activity. The activity can be performed in bouts of 10 minutes each if desired. Moderate-intensity activities are those that increase heart and breathing rates. They include brisk walking and jogging, scrubbing floors, operating a wheelchair manually or playing a sport from a wheelchair, running around with children, fast dancing, tennis (doubles), and playing basketball.
- At least two sessions of strengthening activities per week. Strengthening activities should include activities like push-ups, sit-ups, weight lifting, or other resistance exercises. Exercises should work legs, hips, back, chest, stomach, shoulders, and arms. Each exercise should be repeated 8 to 12 times per session. Vigorous-intensity activities should be added to exercise programs a little at a time in order to reduce the risk of injuries.

ANSWERS TO REALITY CHECK

Does Aerobic Exercise Increase Body Fat Loss?

It's a myth that aerobic exercise is better for body fat loss than other types of exercise. Body fat levels decline when a person's calorie expenditure is greater than calorie intake.[35]

PhotoDisc

Tex:

PhotoDisc

Rose:

take action Ramping Up Exercise Intensity

Here are some options for increasing exercise intensity. Do you see an option or two that would work for you?

I can do that	No way	
______	______	1. Play "Beat your time." Trim seconds off the amount of time it takes you to walk a specific distance each time out for a month. Then maintain that level of walking intensity.
______	______	2. Gradually increase your cycling speed until you're safely traveling at least 1 mile in 6 minutes (10 mph).
______	______	3. Take the stairs two steps at a time. (Hold the handrail! Start with one floor of steps.)
______	______	4. Jog or run, rather than walk, whenever possible.
______	______	5. Carry light free weights with you when you walk or jog.

Source: Values calculated from *Guidelines for Graded Exercise Testing and Exercise Prescription*, 2nd ed. Philadelphia: American College of Sports Medicine, Lea and Febiger, 1984.

People who have a chronic disease should consult a physician before starting an exercise program. It may well turn out that exercise will be just what the doctor orders. Moderate exercise programs are beneficial and safe for almost everyone.[30]

Physical Activity Recommendations for Schoolchildren Regular physical activity is an important component of children's health and development. Yet only 30% or less of children and adolescents exercise at the recommended level of an hour or more each day at a moderate to vigorous level. Only one in three students are offered daily physical activity classes.[9,30]

Schools are being encouraged to include physical activity as a daily part of the curriculum for all children and adolescents. Physical activity should focus on meeting fitness goals rather than on competitive or sports performance goals.[31] Tests of physical fitness, such as the muscle strength and endurance assessment shown in Illustration 27.8, should be performed periodically on all children and the results used to modify instruction.

Some Exercise Is Better Than None Approximately 48% of U.S. adults met the Physical Activity Guidelines for Americans.[28] However, physical activity levels that are lower than the recommended benefit health and are superior to no exercise at all.[32,33] Sedentary people who take up walking, dancing, gardening, biking, golfing, or similar exercises tend to experience improvement in aerobic fitness, strength, and energy level. Short, interspersed bouts of moderate-intensity physical activity also count as health promoting. Bouts of running or jogging for 5 to 10 minutes during the day decrease heart disease risk, and mixing prolonged periods of sitting with short bouts of walking enhances blood glucose utilization.[5,33]

One way to improve levels of physical fitness without spending more time on exercise is to increase the intensity of physical activities. In general, 15 minutes of vigorous activity provides the same benefits as 30 minutes of moderate activity.[9] High-intensity physical activity, when combined with moderate-intensity activities, saves time while producing a range of cardiovascular and metabolic benefits that are comparable to those related to continuous aerobic activity.[35] The second Take Action feature for this unit provides some choices for increasing exercise intensity.

Stay tuned. The Physical Activity Guidelines for Americans were released in 2008 and are currently being updated.

Ilene MacDonald/Alamy

Illustration 27.8 An example of a flexibility exercise assessment in a child.

Michael Blann/Iconica/Getty Images

NUTRITION up close

Exercise: Your Options

Focal Point: Assess your level of physical activity.

About a third of U.S. adults are physically inactive. How much physical activity are you getting?

Physical Activity

1. Check the category that best describes your usual overall daily activity level. Then answer two questions about exercise.

Overall Daily Activity Level

- *Inactive*: Sitting most of the day, with less than 2 hours moving about slowly or standing.
- *Average*: Sitting most of the day, walking or standing 2 to 4 hours each day, but not engaging in strenuous activity.
- *Active*: Physically active 4 or more hours each day. Little sitting or standing, engaging in some physically strenuous activities.

Exercise

2. Do you exercise at moderate intensity a total of 150 minutes per week?
3. Do you perform strength-building exercises twice a week?

Feedback to the Nutrition Up Close is located in Appendix F.

REVIEW QUESTIONS

- **Describe physical fitness and the health advantages of being physically fit.**

1. Regular physical activity benefits psychological and physical health. **True/False**
2. The primary components of physical fitness are body fat composition, strength, and heart rate. **True/False**
3. Aerobic fitness is related to a person's ability to deliver oxygen to muscles and the capacity of muscles to use the oxygen for physical activity. **True/False**
4. Endurance is related to aerobic fitness. **True/False**
5. Maximal oxygen consumption is related to endurance. **True/False**
6. A 44-year-old individual who wants to exercise at 70% of maximal heart rate would aim for a target heart rate of around 123 beats per minute. **True/False**
7. Glycogen supplies the fuel to meet most of our energy needs while we are at rest. **True/False**
8. Fat is the major fuel for low- to moderate-intensity physical activities. **True/False**
9. Resistance training primarily benefits endurance. **True/False**
10. To build and maintain a state of physical fitness, it is recommended that adults engage in vigorous physical activity for 30 minutes daily. **True/False**
11. Increased physical activity plus reduced calorie intake is more effective for reducing body weight and for preventing obesity than is increased exercise alone. **True/False**

The next three questions refer to the following scenario.

Tacchi, now 45 years old, is well into an exercise program. He is exercising three times a week at 80% of maximum heart rate (MHR) for 30 minutes. During his exercise Tacchi takes a break to measure his heart rate. Yesterday he calculated his heart rate to be 155 beats per minute.

12. ______ Tacchi is:
    a. exercising at a level below 80% of his MHR.
    b. exercising at a level that is above 80% of his MHR.
    c. exercising at a level that is close to 80% of his MHR.
    d. exercising too short a time to accurately assess heart rate.
13. ______ Tacchi wants to expand his exercise program to include activities that would increase his flexibility. Which of the following types of physical activities should he choose?
    a. weight lifting
    b. aerobic dance
    c. roller-blading
    d. stretching

14. ______ Assume Tacchi adds strength training to his exercise program. On the first day of strength training, Tacchi decides to go for it in a big way and spends 45 minutes running up and down hills and doing sit-ups. He felt fine all day after that, but the next day his muscles felt really sore and stiff. What probably happened to Tacchi's muscles?

a. He probably seriously injured a muscle.
b. He's probably coming down with the flu.
c. He developed delayed onset muscle soreness. His muscles will feel better in a few days.
d. He caused serious damage to his muscles and will be weaker for months because of the damaged caused.

Answers to these questions can be found in Appendix F.

NUTRITION SCOREBOARD ANSWERS

1. Muscle strength is one of four components of physical fitness. **False**
2. You can be fit and overweight. You don't have to be thin to be physically fit.[1] **True**
3. Regular exercise does help reduce the risk of heart disease—but not as much as a combined program of exercise, weight loss, and consumption of a healthy diet.[2,3] **False**
4. You don't have to exercise to that extent to become physically fit.[1] **False**

Nutrition and Physical Performance

Bill Milne/StockFood Creative/Getty Images

NUTRITION SCOREBOARD

1 Nutritional status affects physical performance. **True/False**

2 Foods included as part of a healthy dietary pattern can supply all the energy and nutrients required by athletes for physical performance. **True/False**

3 Supplementation with protein and amino acids increases muscle strength to about the same extent as does resistance training. **True/False**

4 Losing your period is normal when you are a female athlete. **True/False**

Answers can be found at the end of the unit.

Sports Nutrition

LEARNING OBJECTIVES

After completing Unit 28 you will be able to:

- Describe how carbohydrates, proteins, and fats are utilized by muscles for energy formation.
- Explain hydration status and the nutritional concerns of athletes.
- Assess the safety and effectiveness of ergogenic aids offered to athletes.

- **Describe how carbohydrates, proteins, and fats are utilized by muscles for energy formation.**

BAM! Nobody heard it, but Lou felt it. She had "hit the wall." She was ahead of her planned pace, but now her legs felt like lead. Without a new supply of carbohydrates, she would have to finish the last two miles of the marathon at the slow pace her legs would allow.

From her carefully crafted and scrupulously followed training program to her refined shaping of mental attitude, Lou thought she had done everything right. She had left one thing out, however, and that may have cost her the race. Lou failed to pay attention to her diet while training and ran out of **glycogen** too soon (Illustration 28.1).[1]

Three major factors affect **physical performance**: genetics, training, and nutrition.[1] The first gives some people an innate edge in sprinting or endurance, and nothing can be done about it. The second is acknowledged as a basic truth. Most athletes know a good bit about proper training, and the trick is to follow the right plan. The third is often ignored or, when taken seriously, misunderstood.

glycogen The storage form of glucose. Glycogen is stored in muscles and the liver.

physical performance The ability to perform a physical task or sport at a desired or particular level.

KEY NUTRITION CONCEPTS

The roles played by carbohydrates, protein, fat, sodium, water, and other nutrients in physical performance relate to two key nutrition concepts.

1. Nutrition Concept #2: Foods provide energy (calories), nutrients, and other substances needed for growth and health.
2. Nutrition Concept #9: Adequacy, variety, and balance are key characteristics of a healthy dietary pattern.

Nutrition has important effects on physical performance, but the legitimate role of nutrition is often poorly understood by athletes and coaches alike.[5] Incorrect information about nutrition and physical performance can be heard in locker rooms, at neighborhood picnics, and in health food and sports stores. Misinformation about nutrition and physical performance can be found on dietary supplement labels, in online blogs, and in magazine articles. Buying into myths about nutrition and physical performance can cost you money or time, jeopardize health, or decrease performance. Common myths about nutrition and physical performance are listed in Table 28.1. Did you believe some of them? (If you did, you are not alone.[2]) Knowledge of the science supporting connections between nutrition and physical performance will help you make solid decisions about food, nutrients, dietary supplements, and performance.

ergogenic aids (*ergo* = work; *genic* = producing) In the context of sport, an ergogenic aid is broadly defined as a technique or substance used for the purpose of enhancing performance.

John Fryer/Alamy

Illustration 28.1 Hitting the wall. This runner ran out of muscle glycogen before the finish line.

Basic Components of Energy Formation during Exercise

Understanding the role of nutrition and performance and the potential effects of some **ergogenic aids** can be fostered by basic knowledge of how energy is formed within muscle cells. Illustration 28.2 summarizes the processes by which energy for muscle movement is formed.

There are two main substrates for energy formation in muscles: glucose from muscle and liver glycogen stores, and fatty acids released from fat stores. How much

Table 28.1 Eight common myths about nutrition and physical performance[1–5]

Myth	Reality
1. Very physically active people can eat all the fat and sugar they want.	Very active children and adults may stay thin no matter what they eat. However, that does not mean a poor diet won't affect their dental and heart health, or protect them from consuming low amounts of vitamins, minerals, and beneficial phytochemicals in plant foods. Calorie needs for athletes are best met through dietary patterns based on MyPlate.gov food groups that provide 45–65% of calories from carbohydrates, 10–35% from protein, and 20–35% from fat.
2. Protein and amino acid supplements improve strength and endurance.	Resistance exercise is the key ingredient in strength building. Athletes can get all the protein and other nutrients they need for building and strengthening muscles from exercise and high-quality proteins from food. Excess protein adds calories and increases the workload of the kidneys because they have to excrete the excess nitrogen that results from high levels of protein breakdown.
3. Saturation of glycogen stores works best if you start when glycogen stores are depleted.	Glycogen stores can be saturated by carbohydrate intake whether glycogen stores start out totally or partially depleted.
4. Athletes benefit from consuming a vitamin and mineral supplement.	Vitamins and minerals participate in energy formation but do not, by themselves, increase your ability to produce energy. Vitamin and mineral supplements benefit individuals with diagnosed deficiency diseases, but do not improve performance in well-nourished individuals, including athletes.
5. The more protein you consume after resistance training the more muscle protein synthesis will take place and the stronger you will become.	Consuming a 3-ounce portion of lean meat will support muscle protein synthesis to the same extent as consuming a 12-ounce portion. Muscles reach a "muscle full" level after resistance exercise that cannot be surpassed by additional protein intake.
6. Carbohydrate and protein intake should occur immediately after completion of a resistance exercise session in order to maximize muscle protein synthesis and glycogen formation.	Carbohydrate and protein intake does not have to occur immediately following resistance exercise to maximize muscle protein synthesis and glycogen formation.
7. Drinking water during exercise decreases performance.	Drinking water before, during, and after exercise keeps athletes hydrated and prevents dehydration.
8. Body weight is more important to athletes' performance than body composition.	Body composition can be more important than weight for some types of sports.

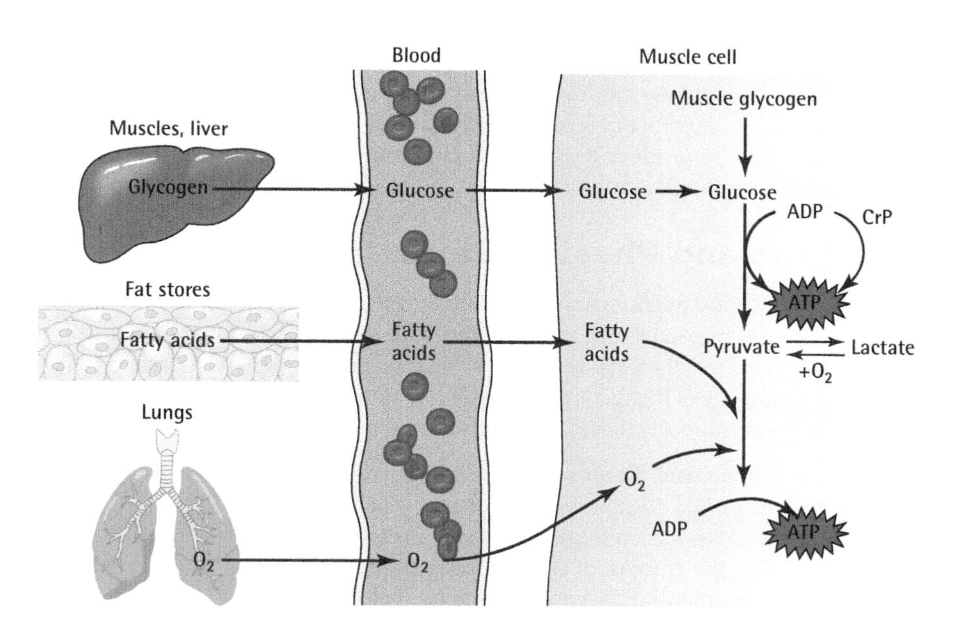

Illustration 28.2 Schematic representation of how ATP is formed for muscular movement.

PhotoDisc

ThinkStock/Jupiter Images

PhotoDisc

©Rena Schild/Shutterstock.com

PhotoDisc

Comstock Images/Getty Images

Illustration 28.3 Fat is the main source of energy for low- and moderate-intensity activities, whereas glycogen is the primary fuel for high-intensity activities.

ATP, ADP Adenosine triphosphate (ah-den-o-scene tri-phos-fate) and adenosine diphosphate. Molecules containing a form of phosphorus that can trap energy obtained from the macronutrients. ADP becomes ATP when it traps energy and returns to being ADP when it releases energy for muscular and other work.

of each is used depends on the intensity and duration of the exercise, as well as the body's ability to deliver each along with oxygen to muscle cells (Illustration 28.3). Each substrate is used to form **ATP** from **ADP**. ATP serves as the source of energy for muscle contraction.

Anaerobic Energy Formation Glucose is obtained from the liver, and muscle glycogen stores form ATP without oxygen. This route of energy formation is *anaerobic*, or "without oxygen," and it generates most of the energy used for intense muscular work (70% VO_2 max or higher).[1] Creatine phosphate (abbreviated CrP in Illustration 28.2), an amino acid containing a high-energy phosphate molecule in muscles, converts ADP to ATP to some extent. Creatine phosphate stores are limited and decrease rapidly during intensive exercise.

Glucose is converted to pyruvate during energy formation. In the absence of oxygen, pyruvate is converted to lactate. Lactate can build up in muscles and blood if not reconverted to pyruvate by the addition of oxygen. Pyruvate yields additional energy when it enters aerobic energy formation pathways along with fatty acids from fat stores.

Aerobic Energy Formation The conversion of pyruvate and fatty acids to ATP requires oxygen. Much more ATP is delivered by the breakdown of fatty acids than glucose (fats provide 9 calories per gram, glucose only 4). The rate of energy formation from fatty acids is four times slower than that from glucose, however. It's the reason fatty acids are used to fuel low- and moderate-intensity exercise, or those below 60% VO_2 max. Unlike glucose, energy formation from fatty acids is not limited by availability. Muscle cells can continue to produce energy from fatty acids as long as delivery of oxygen from the lungs and the circulation is sufficient.[6]

Nutrition and Physical Performance

The habitual diet and nutritional status of athletes should be considered first when improved performance is the goal. Glycogen stores, and foods and fluids consumed before, during, and after exercise are all related to energy formation and physical performance.[1] Table 28.2 provides a summary of the recommendations for nutrition and athletic performance, and the evidence that supports these recommendations is presented next.

Glycogen Stores and Performance Glycogen stores in muscles and the liver can deliver about 2,000 calories worth of energy, whereas adults have access to over 100,000 calories from fat. Consequently, a person's ability to perform endurance exercise can be limited by the amount of stored glycogen.[6] People who run out of liver and muscle glycogen during an endurance event such as cycling or long-distance running

Table 28.2 Summary of nutrition recommendations for athletic performance[1]

A. Healthy Dietary Pattern
- Healthy dietary patterns (including vegetarian) support physical health and performance. Athletes may require somewhat more protein (0.5–0.8 g/lb body weight, or 1.2–1.7 g/kg), than non-athletes (0.4 g/lb or 0.8g/kg per day). A nutritious assortment of foods and beverages can meet all of the energy and nutrient needs of athletes.

B. Endurance Events
- Top off glycogen stores with food or fluids providing about 60 g (240 calories) of carbohydrate 2–3 hours prior to endurance events.
- Consume a source of carbohydrate during the event (30–60 g of carbohydrate per hour) to help maintain blood glucose levels and conserve glycogen. Use of carbohydrate- and sodium-containing sports drinks is appropriate.
- Fluid intake should match fluid loss. About 2 cups of fluid are needed for every pound of body weight lost during exercise.

C. Resistance Exercise
- Consume 20 g of high-quality protein and carbohydrate in a meal or snack within 2 hours after exercise to support muscle recovery and growth, and to build up muscle glycogen stores.
  - Individuals aged 60 years and older may need up to 40 g protein after resistance training to maximize muscle protein synthesis.
- Maintain appropriate hydration with water during exercise sessions lasting less than an hour.

"hit the wall"—they have to slow down their pace substantially because they can no longer use glycogen as a fuel. The pace they are able to maintain will be dictated by the body's ability to use fat as fuel for muscular work. If athletes keep pushing themselves after glycogen runs out, they may end up "bonking." That's worse than hitting the wall, and the experience makes a profound impression on athletes. It is something they do not want to have happen again. Bonking is due to hypoglycemia (low blood sugar) and causes severe weakness, fatigue, confusion, and disorientation. In severe cases, it can cause an athlete to pass out.[7] Obviously, endurance athletes do not want to exhaust their glycogen stores too soon.

Most athletes consuming a typical U.S. diet normally have enough glycogen stores to fuel continuous, intense exercise for about two hours. Glycogen stores can be increased by consuming a meal or snack containing about 60 g of carbohydrate several hours before endurance exercise begins (Illustration 28.4), supplemented by consuming 30–60 g (120–240 calories) of carbohydrate an hour during endurance exercise. A variety of types of simple and complex carbohydrates that involve different metabolic processes are preferred over consumption of a single type.[1] Serving sizes of carbohydrate-rich foods and beverages that supply about 60 grams of carbohydrates are listed in Table 28.3.

Protein and Performance Many athletes require no more than their Recommended Dietary Allowance (RDA) of protein. Individuals undertaking strength or endurance training, however, may need 20–40 grams of additional protein daily to support muscle protein synthesis and repair.[9] This higher level of protein may already be included in their diets. Athletes tend to have higher protein intakes than non-athletes due to their higher calorie intake. On average, adult females in the United States consume 67 grams of protein daily (the RDA is 48 grams), and males 98 grams (versus the RDA of 56 grams).[8] Diets providing up to 35% of total calories from protein are compatible with health, but may provide too little carbohydrate and lead to early fatigue in athletes.[1]

Kim has been hitting the gym daily trying to get his body ripped when he started to experience muscle aches and a burning feel during urination. Thinking he had a venereal disease, Kim went to the clinic to get it checked out. During the visit, his doctor asked him if he had been taking a protein supplement. He had been. The amount of protein he was consuming from a muscle-building supplement (200 g per day) and his diet (80 grams per day) far exceeded his need for protein and his body's ability to utilize it. The problems resolved after Kim cut his total protein intake down by over half.

Table 28.3 Foods and beverages that provide about 60 grams carbohydrate

Quinoa, cooked	1 cup
Graham crackers	12 (3 oz)
Brown or white rice, cooked	1½ cups
Breakfast cereal	1–3 cups (check nutrition information label)
Chocolate milk	1¾ cups
Pasta	2 cups
Apple juice, cranberry juice	
cocktail	2⅓ cups
Banana	2 large
Raisins	½ cup
Carbohydrate gel pack	2–3 (varies)

©BLACKDAY/Shutterstock.com

Judith Brown

Illustration 28.4 A snack of 1¾ cups of chocolate milk or ½ cup of raisins provides approximately 60 grams of carbohydrate. The chocolate milk also provides about 17 grams of high-quality protein.

high-quality protein Proteins that contain all of the essential amino acids in amounts needed to support growth and tissue maintenance. Examples of high-quality proteins include eggs, soy milk, milk, meat, and beans and rice. Also referred to as "complete proteins."

Table 28.4 Foods and beverages that provide about 20 grams high-quality protein

Food	Amount
Tuna fish	3 oz
Chicken (no skin)	3 oz
Beef, lean	2½ oz
Pork, lean	3 oz
Egg whites	3
Yogurt, low-fat	1½ cups
Skim milk	2 cups
Skim milk powder	½ cup
Dried beans (cooked)	1½ cups
Protein bar	1–2 (varies)
Soybeans, cooked	¾ cup

The Protein–Muscle Connection Muscle fibers develop microscopic tears during training, and protein is needed to help limit muscle tissue breakdown and to repair and rebuild the muscle. Strength training, by itself, increases muscle mass and strength, and prepares muscles to continue increasing in mass and strength as training continues. Muscular strength has been shown to increase 25–35% and muscle mass by 9% after a 12-week, three session per week resistance training program, for example.[12] The question has been whether additional protein or amino acids would increase gains in muscle mass and strength achieved by resistance exercise.

Muscle is high in protein, and it has been assumed that high protein intake helps build muscle mass and strength. Protein increases muscle protein synthesis and repair and muscle strength only when combined with sufficient and regular intense resistance exercise.[4] Consumption of about 20 grams of **high-quality protein** within 2 hours after exercise sessions facilitates muscle protein synthesis and repair and enhances strength.[1] Illustration 28.5 shows an example of how this amount of protein can be supplied in a sandwich. The 20-gram protein level appears to represent the "muscle full" amount, or the limit on the amount of amino acids that can be incorporated into muscle cells post-exercise. Protein intake above this amount will be used for the synthesis of nonessential amino acids or converted to an energy source. Table 28.4 lists examples of foods and their amounts that provide 20 grams of high-quality protein.

Although not yet known with certainty, it appears that the addition of approximately 60 grams of carbohydrate to post-training protein intake facilitates repletion of muscle protein synthesis and glycogen stores. Carbohydrate intake stimulates insulin release, and insulin increases the uptake of glucose and amino acids by muscle cells.[10] Table 28.3 lists examples of types and amounts of food that provide approximately 60 grams of carbohydrate.

Nutrients such as amino acids and glucose obtained from food enter the blood stream and become available for uptake by cells about 1 to 2 hours after food is consumed. Consequently, intake of a high-quality protein and carbohydrate snack before resistance exercise that lasts about an hour means that amino acids will be available for use by muscle cells postexercise. Athletes who are not able to eat a meal or snack within 2 hours after training may benefit from consuming 20 grams of high-quality protein prior to the training.[10]

Protein powders providing high-quality protein, such as whey, and essential amino acid supplements can be used as a protein source before or after exercise. They have not been found to more effectively build muscle size or strength than food sources of high-quality protein, but they may be more convenient to use.[8] Protein supplements should be purchased from reliable companies and bear the NF or USP symbols for purity and quantity of ingredients.[11]

Although it is recommended that athletes consume specific types and amounts of food after training sessions, it is not always easy to do. Athletes don't always feel like eating, or have an appetite after exercising. The Reality Check feature in this unit addresses the issue of appetite postexercise. The content may ring a bell whether you feel really hungry or not hungry at all after exercise.

Michael Neelon(misc)/Alamy

Illustration 28.5 The sandwich supplies around 25 grams of high-quality protein from 3 ounces of turkey. It also provides about 25 grams of carbohydrate from the whole-wheat bread.

Hydration

- **Explain hydration status and the nutritional concerns of athletes.**

Hydration status is a major factor affecting physical performance and health. Adequate hydration during training and competition enhances performance, prevents excessive body temperature, delays fatigue, and helps prevent injuries.[1]

hydration status The state of the adequacy of fluid in the body tissues.

Hydration status during exercise is affected primarily by how much a person sweats—by how much water he or she loses through the skin while exercising. Muscular activity produces heat that must be eliminated to prevent the body from becoming overheated. To keep the body cool internally, the heat produced by muscles is collected in the blood and then released both through blood that circulates near the surface of the skin and in sweat. Sweating cools the body because heat is released when water evaporates on the skin. People sweat more during physical activity in hot, humid weather because it is harder to add moisture and heat to warm, moist air than to dry, cool air. To stay cool during exercise in hot, humid conditions, the body must release more heat and water than when exercise is undertaken in drier, cooler conditions.[12]

Estimating Fluid Needs: Sweat Rate The amount of fluid athletes need during an event can be estimated by calculating an hourly **sweat rate** for a specific activity (Table 28.5). A person preparing for a marathon who loses a pound in an hour of training and who drinks nothing during the hour would have a sweat rate of 1 pound (16 ounces). If that athlete drank 8 ounces of fluid in the hour and lost a pound (16 ounces), his or her sweat rate would be 24 ounces (16 ounces + 8 ounces). The sweat rate amount of fluid should be consumed per hour of the marathon event. Athletes who gain weight during an event have consumed too much water.[1]

sweat rate Fluid loss per hour of exercise. It equals the sum of body weight loss plus fluid intake.

REALITY CHECK

Does Strenuous Physical Activity Increase Appetite and Food Intake?

Who gets the thumbs-up?

Answer appears on page 28-8.

PhotoDisc

Junior: It must. You read about all these long-distance runners who chow down on a 3,000-calorie meal after the race is over.

PhotoDisc

Lakisha: I can't even think about eating after I've been working out on a hot day. I can't believe people would want to eat after that, either.

ANSWERS TO REALITY CHECK

Does Strenuous Physical Activity Increase Appetite and Food Intake?

Two thumbs-up for these responses because exercise increases appetite and food intake in some athletes and it doesn't in others. Part of the difference seems to be due to the temperature. People who exercise in the heat, in particular, tend to feel less like eating after exercise.[10,29]

PhotoDisc

Junior:

PhotoDisc

Lakisha:

Table 28.5 Calculating sweat rate: An example

1. Determine your body weight 1 hour before and 1 hour after exercise.
2. Subtract your postexercise weight from your pre-exercise weight.
3. Convert the number of pounds lost to ounces. (One pound equals 16 ounces.)
4. Add the number of ounces lost or gained to the number of ounces of fluid you consumed during the hour of exercise. The result is your sweat rate, and that's an approximation of the amount of fluid you need to consume during 1 hour of that exercise.
Example
1. Terrell weighed 172 pounds an hour before an hour-long bout of exercise and 171 pounds an hour after the exercise: 172 pounds − 171 pounds = 1 pound lost 1 pound × 16 ounce per pound = 16 ounces
2. He drank 16 ounces of fluid during the hour of exercise. 16 ounces + 16 ounces = 32 ounces, or "sweat rate"
Terrell would need to consume about 32 ounces of fluid per hour to remain hydrated.

Dehydration Loss of more than 2% of body weight (2 to 4 pounds generally) during an event indicates that the body is becoming dehydrated. It occurs when a person's intake of water and other fluids fails to replace water losses, primarily from sweat. Effects of **dehydration** range from mild to severe, depending on how much body water is lost. People who experience dehydration tend to feel thirsty, sweat less, and have reduced urine output. If dehydration progresses, they begin to feel confused and light-headed.[13] At the extreme, dehydration can lead to heat exhaustion or heat stroke (Table 28.6). People who over-exercise in hot weather when they are out of condition are most likely to suffer heat exhaustion or heat stroke, but these conditions occasionally occur among seasoned athletes, as well. Dehydration and heat exhaustion can be remedied by fluids and appropriate amounts of **electrolytes**, but heat stroke requires emergency medical care.[14] The Health Action feature for this unit highlights dehydration. The information may help you recognize and avoid it.

Water Intoxication **Water intoxication** is a problem that can occur during endurance events when insufficient sodium is consumed compared to water (Illustration 28.6). The body needs sodium to maintain a normal fluid balance within and around cells; its lack of availability alters fluid balance and cell functions. Also called **hyponatremia**, water intoxication can cause nausea, vomiting, confusion, seizures, and coma. It should be treated promptly.[15] It is most likely to occur in endurance athletes who consume a lot of water and too little sodium and thus gain weight, rather than replace weight, during an event.[8] In one Boston marathon, 13% of runners developed blood sodium levels that qualified as hyponatremia, and 0.6% became very ill due to extremely low levels of sodium.[18] The condition needs to be treated promptly. Water intoxication can generally be prevented by periodic consumption of a food or beverage containing about 100 mg sodium (<1/8 teaspoon salt).[1]

dehydration A condition that occurs when the body loses more water than it takes in. Prolonged exercise accompanied by profuse sweating in high temperatures, vomiting, diarrhea, certain drugs, and decreased water intake can lead to dehydration.

electrolytes Minerals such as sodium, magnesium, and potassium that carry a charge when in solution. Many electrolytes help the body maintain an appropriate amount of fluid.

water intoxication A condition that results when the body accumulates more water than it can excrete due to low availability of sodium. The condition can result in headache, fatigue, nausea and vomiting, confusion, irritability, seizures, and coma. It can occur in people who consume or receive too much water without sodium and in those with impaired kidney function. Also referred to as overhydration and hyponatremia.

hyponatremia A deficiency of sodium in the blood (135 mmol/L sodium or less).

health action Be Aware of the Signs and Symptoms of Dehydration!

The signs and symptoms of dehydration can include:

- Thirst
- Lightheadedness or fatigue
- Headache
- Inability to concentrate
- Feeing cold or having chills when it is hot out and you are sweating
- Collapse due to heat exhaustion or stroke
- Low amount of urine and dark yellow color

Table 28.6 A primer on heat exhaustion and heat stroke[14,15]

Heat exhaustion: A condition caused by low body water and sodium content due to excessive loss of water through sweat in hot weather. Symptoms include intense thirst, weakness, paleness, dizziness, nausea, fainting, and confusion. Fluids with electrolytes and a cool place are the remedy. Also called "heat prostration" and "heat collapse."

Heat stroke: A condition requiring emergency medical care. It is characterized by hot, dry skin, labored and rapid breathing, a rapid pulse, nausea, blurred vision, irrational behavior, and, often, coma. Internal body temperature exceeds 105°F due to a breakdown of the mechanisms for regulating body temperature. Heat stroke is caused by prolonged exposure to environmental heat or strenuous physical activity. The person affected by heat stroke should be kept cool by any means possible, such as removing clothing and soaking the person in ice-cold water. If conscious, the person should be given fluids. Also called "sunstroke."

Failure to replace lost body water

↓ Blood volume declines

↓ Volume of water in and around cells declines

↓ Sweat, flushing, dry mouth
↑ Body temperature rises
↓ Physical work capacity drops
↑ Electrolyte concentration in muscles increases (causes muscle cramps)

Heat exhaustion

↑↑ Body temperature rises
↑ Heart rate increases
↑ Hot, dry skin

Heat stroke

Illustration 28.6 Water intoxication can result from over-consumption of water being handed out during marathon events.

Maintaining Hydration Status during Exercise It is recommended that athletes engaged in events that last an hour or less drink water to maintain hydration status. Athletes undertaking longer events (over 1 hour in duration) should consume primarily water, beverages, and foods that provide sodium (about 100 mg sodium), and carbohydrate (4–8% by weight) during the exercise. The sodium is needed to replace what is lost in sweat and the carbohydrate helps maintain blood glucose levels.[1] Sodium is the only electrolyte that should be added to sports drinks or other fluids consumed by athletes. Potassium needs should be met by the foods such as vegetables and fruits in the athlete's diet.[16] Hydration status is maintained when athletes do not lose or gain weight during an event and when their urine remains pale yellow and is normal in volume.[1]

Fluids That Don't Hydrate Not all fluids help maintain hydration status. Fluids containing over 8% sugar don't quench a thirst and should not be consumed for fluid replacement. Their high sugar content may draw fluid from the blood into the intestines, thereby increasing the risk of dehydration, nausea, and bloating.[17] Alcohol-containing

Table 28.7 Incidence of irregular or absent menstrual cycles in female athletes and sedentary women[41,42]

Joggers (5 to 30 miles per week)	23%
Runners (over 30 miles per week)	34
Long-distance runners (over 70 miles per week)	43
Competitive bodybuilders	86
Noncompetitive bodybuilders	30
Volleyball players	48
Ballet dancers	44
Sedentary women	13

beverages such as beer, wine, and gin and tonics are not hydrating, nor is water when consumed by a sodium-depleted person.[1]

Nutrition-Related Concerns of Athletes

It is not uncommon for female athletes to experience irregular or absent menstrual periods or late onset of periods during adolescence (Table 28.7).[2] Female athletes engaged in "leanness" sports such as gymnastics, diving, and figure skating are particularly at risk for inadequate calorie intake.[21] These aberrations in menstrual periods appear to be related to specific effects on hormone production of deficits in caloric intake.[2] Female athletes with missing and irregular periods are at risk for developing low bone density, osteoporosis, and bone fractures. Women at particular risk of bone fractures are those with the female-athlete triad of disordered eating, amenorrhea (pronounced a-men-or-re-ah, meaning "no menstrual periods"), and osteoporosis. Abnormal menstrual cycles should not be dismissed as a normal part of training. Normal menstrual periods should be reinstated through increased caloric intake.[8] Suppressed testosterone levels or other metabolic changes due to caloric deficits in male athletes may be the parallel to menstrual irregularities in female athletes.[2]

PHOTOINKE/Alamy

Illustration 28.7 Dropping weight quickly in the days before a match can wreck a wrestler's chances and harm his health.

Wrestling: The Sport of Weight Cycling "Weight cutting," although not recommended, remains popular among amateur wrestlers.[22] Most wrestlers will lose 1 to 20 pounds 10 times or more per season (Illustration 28.7). Most wrestlers will "cut" 1 to 20 pounds over a period of days between 50 and 100 times during a high school or college career.[23] That makes wrestling more than a test of strength and agility. It makes it a contest of rapid weight loss.

For competitive reasons, wrestlers often want to stay in the lowest weight class possible and may go to great lengths to achieve it. They may fast, "sweat the weight off" in saunas or rubber suits, or vomit after eating to lose weight before the weigh-in. These practices can be dangerous if taken too far. After the match is over, the wrestlers may binge and regain the weight they lost.[22]

Wrestlers, like other athletes involved in intense exercise, perform better if they have a good supply of glycogen and a normal amount of body water. Fasting before a weigh-in dramatically reduces glycogen stores, and withholding fluids or losing water by sweating puts the wrestler at risk of becoming dehydrated. Trying to stay within a particular weight class too long may also stunt or delay a young wrestler's growth.[23]

The American Medical Association and the Association for Sports Medicine recommend that wrestling weight be determined after six weeks of training and normal eating. In addition, a minimum of 7% body fat should be used as a qualifier for assigning wrestlers to a particular weight class. In addition, weight classes are now based on normal weight for height and age, and weigh-ins are scheduled close to event times.[25]

Iron Status of Athletes Iron status is an important topic in sports nutrition because iron deficiency (or low iron stores) and iron-deficiency anemia (or low blood hemoglobin level) decrease endurance. Iron is a component of hemoglobin, a protein in blood that carries oxygen to cells throughout the body, and it works with enzymes involved in energy production. When iron stores or hemoglobin levels are low, less oxygen is delivered to cells, and less energy is produced than normal.[24]

Female athletes are at higher risk of iron-deficiency anemia than other females. One study of aerobically fit athletes found that 36% of women and 6% of males were iron deficient.[25] Consequently, it is recommended that female athletes especially pay attention to their iron status.[8] Iron deficiency should be diagnosed before it is treated with iron supplements. The International Olympic Committee recommends that female athletes be screened for iron deficiency so that females needing additional iron can be identified.[24]

Ergogenic Aids: The Athlete's Dilemma

- **Assess the safety and effectiveness of ergogenic aids offered to athletes.**

The quest for a competitive edge has drawn athletes to ergogenic aids throughout much of history. (See Table 28.8 for a historical review of ergogenic aids use.) Relatively few of the hundreds of currently available products have been tested for safety and effectiveness, and most are sold as dietary supplements so they do not have to be.[1,11] Those known to increase muscle mass, strength, or endurance are usually banned from use by competitive athletes.[27] Some ergogenic aids pose no particular risk to health (just wallets). Some clearly pose risks to health due to contamination, others have been found to contain banned substances, and some contain ingredients intended to appeal to athletes with certain opinions.[11,28] The National Athletic Trainers' Association supports a food-first philosophy to support health and performance among athletes.[11]

Table 28.9 lists dietary supplements and other ergogenic aids for which studies have not demonstrated effectiveness and those banned for use by athletes. Products that do not increase strength or endurance to a greater extent than high-quality protein from foods or other normal components of diets are considered ineffective.

Anabolic Steroids Substances derived from testosterone, a primary male sex hormone, are considered to be *anabolic steroids.* At least 51 testosterone derivatives are banned by the U.S. Olympic Committee because they increase strength and body weight, and have adverse effects on health. Depending on the dose and duration of use, anabolic steroids can lead to the development of acne, increased sex drive, increased body hair, impaired fertility, and mood changes ranging from depression to hostility. Male characteristics, such as facial hair and voice deepening, can occur in females who use anabolic steroids.[29]

Caffeine Caffeine is a mild stimulant consumed by many athletes to decrease fatigue and increase alertness during exercise. Coffee is a popular source of caffeine among athletes, and caffeine is also available to them in energy drinks and gels. Consumed in moderate doses of about 200 mg per day (about 2 cups of brewed coffee), caffeine appears to increase alertness, mood, and cognitive processes during exercise. Effects of caffeine on performance vary from one athlete to the next, so individuals should make their own decisions about its helpfulness. Caffeine is associated with few adverse side effects if consumed by adults in moderate doses. The World Anti-Doping Agency (WADA) has removed caffeine from their list of prohibited substances.[1,30]

The Path to Improved Performance Genetics, training, and nutrition—these are the real keys to physical performance. Although other aids will be sought, those that exceed the boundaries of what is considered fair and safe will not be approved for use by athletes. After all, athletic competition is not a test of drugs or performance aids. It's a test of an individual's ability to excel. Anything less wouldn't be sporting.

Table 28.8 Faster, higher, stronger, longer: A brief history of performance-enhancing substances used by athletes

B.C.	Large quantities of beef consumed by athletes in Greece to obtain "the strength of 10 men." Deer liver and lion heart consumed for stamina.
1880s	Morphine used to increase performance in (painful) endurance events.
1910s	Strychnine consumed for the same reason as morphine.
1930s	Amphetamines used to increase energy levels and endurance. Testosterone taken to increase muscle mass.
1980s	Blood doping, EPO used to increase endurance; ephedra to increase energy.
2009	HGH; gene doping to increase strength and endurance.

Table 28.9 Noneffective and banned dietary supplements and ergogenic aids[28,43–45]

Not shown to be effective	Banned substances
Glutamine	Testosterone
Isoflavones	Testosterone derivatives:
Sulfo-polysaccharides (myostatin inhibitors)	Erythropoietin (EPO)
Boron	Growth hormone (HGH)
Calcium pyruvate	Mechano growth factors (MGFs)
Chromium	Gonadotrophins (e.g., LH, hCG)
Creatine	Insulin
	Beta-2 agonists
	Anti-estrogenic substances
	Diuretics
	Stimulants
	Narcotics
	Cannabinoids (hash, marijuana)
	Glucocorticosteroids
	Alcohol
	Beta-blockers
	Blood doping (enhancing oxygen delivery)
	Gene doping
Gamma oryzanol (ferulic acid)	
Herbal diuretics	
Tribulus terrestris	
Vanadyl sulfate (vanadium)	
Chitosan	
Garcinia cambogia (HCA)	
L-carnitine	
Phosphates	
Glycerol	
Ribose	
Inosine	
Sodium bicarbonate	
β-HMB	
Conjugated linoleic acid	
Branched chain amino acids (BCAA)	
Medium-chain triglycerides (MCT)	
Omega-3 fatty acids	

Bill Milne/StockFood Creative/ Getty Images

NUTRITION up close

Testing Performance Aids

Focal Point: The critical examination of studies on performance aids.

Read the following summary of a fictitious study of the effects of a phosphorus supplement on strength and then answer the critical thinking questions.

- *Purpose:* To assess the effect of a phosphorus supplement on strength.
- *Methods:* Twenty volunteers from the crew team were given the phosphorus supplement for a week. Strength, assessed as the maximum number of push-ups a study participant could do in one session, was assessed before and after supplementation. Participants recorded any supplement side effects.
- *Results:* The number of push-ups increased by an average of 5% after supplementation. Diarrhea was the only side effect consistently noted by participants.

Critical Thinking Questions

1. What is a major limitation in the basic design of the study? ______________________
2. Does the study demonstrate that the supplement is safe?

Yes _____ No _____ Give reasons for your answer. _________________________________

3. Does the study demonstrate that the supplement increases strength?

Yes _____ No _____ Give reasons for your answer. _________________________________

Feedback to the Nutrition Up Close is located in Appendix F.

REVIEW QUESTIONS

- **Describe how carbohydrates, proteins, and fats are utilized by muscles.**
- **Explain hydration status and the nutritional concerns of athletes.**
- **Assess the safety and effectiveness of erogenic aids offered to athletes.**

1. The two main substrates for energy formation in muscles are amino acids and fatty acids. **True/False**
2. Glucose is converted to energy by muscles anaerobically. **True/False**
3. Much more ATP is produced from the breakdown of fatty acids than from amino acids or glucose. **True/False**
4. Hypoglycemia results when athletes use up glycogen stored in muscles and liver. **True/False**
5. Glucose is the major source of energy for low- and moderate-intensity activities. **True/False**
6. Fat intake improves performance in short-duration, high-intensity events if consumed within 1 hour before the events. **True/False**
7. Protein serves as the major source of fuel for resistance exercise. **True/False**
8. Water intoxication is caused primarily by drinking excessive amounts of water during prolonged exercise. **True/False**
9. An athlete who runs for an hour, loses 2 pounds in that hour, and drinks 12 ounces of fluid during the hour has a sweat rate of 60 ounces. **True/False**
10. Signs of dehydration can include thirst, lightheadedness, and confusion. **True/False**
11. After a resistance exercise session, the "muscle full" phenomenon is reached after 20 grams of high-quality protein has been consumed. **True/False**
12. For best results, athletes attempting to build muscle mass and strength should take both protein and amino acid supplements. **True/False**

The next three questions refer to the following scenario.

Assume you attend a sports competition and are offered a free sample of MuscleMaxX, a new protein powder that promises to help you build muscle after resistance exercise. You take a look at the ingredient label and this is what it contains:

triple-filtered water
starch
whey
glucose
lean lipids
growth peptides
micellar proteins
lactalbumins

13. ____Which ingredient in this product would most likely help rebuild muscles after resistance exercise?
    a. triple-filtered water
    b. growth peptides
    c. whey
    d. lean lipids

14. ____Why do you think the real reason is for giving ingredients titles such as "lean lipids" and "micellar proteins"?
    a. To ensure users know they are getting unique ingredients for muscle recovery.
    b. To increase sales.
    c. To assure users they are getting ingredients for which scientific studies have shown effectiveness.
    d. To assure users that the ingredients are highly technical and the accurate name for each is being used.

15. ____Is it likely this product would work better for muscle rebuilding after resistance exercise than consuming 2 cups of skim milk?
    a. Yes, because it also contains growth peptides.
    b. Yes, because it provides starch.
    c. No, because high-quality protein from food also works.
    d. No, because skim milk also contain lactalbumins.

The next two questions refer to the following scenario.

Tahmina comes home from her 1-hour weight training session and looks in the fridge to find foods that will provide about 60 grams of carbohydrate and 20 grams of high-quality protein. Voilà! She notices leftover black beans and rice and eats a cup of each.

16. ____Based on the carbohydrate and protein content of black beans and rice given in Appendix A, Tahmina consumed about:
    a. 20 grams of carbohydrate and 10 grams of protein
    b. 85 grams of carbohydrate and 19 grams of protein
    c. 44 grams of carbohydrate and 16 grams of protein
    d. 72 grams of carbohydrate and 32 grams of protein

17. Black beans and rice are a source of high quality protein. **True/False**

Answers to these questions can be found in Appendix F.

NUTRITION SCOREBOARD ANSWERS

1. **True**[1]
2. **True**[1]
3. Consumption of protein and/or amino acid supplements alone does not increase muscle strength.[1,32] **False**
4. It is not normal and reflects a low energy intake that is likely interfering with normal body functions.[20] **False**

Good Nutrition for Life: Pregnancy, Breastfeeding, and Infancy

Kablonk! RF/Golden Pixels LLC/Alamy

NUTRITION SCOREBOARD

1. The fetus is able to obtain needed nutrients from the mother regardless of the mother's diet or nutrient stores. **True/False**
2. The United States has the lowest rate of infant mortality in the world. **True/False**
3. Normal health, growth, and development of infants are defined by the experience of breastfed infants. **True/False**
4. Drinking alcoholic beverages is safe during breastfeeding because the alcohol doesn't pass into the milk. **True/False**

Answers can be found at the end of the unit.

LEARNING OBJECTIVES

After completing Unit 29 you will be able to:

- Discuss the importance of nutrition in improving the health of infants in the United States.
- Describe critical periods of fetal growth and development and the potential consequences of inadequate and excessive nutrient availability during these periods on future health status.
- Discuss the nutrition advantages of breastfeeding for infants and the dietary recommendations for breastfeeding mothers.
- Summarize the role of nutrition in the development of infants.

Nutrition and a Healthy Start in Life

The day has finally arrived, the one Crystal and Raymond have anticipated for a year. Crystal's pregnancy test is positive!

This is a planned pregnancy, and both Crystal and Raymond have done all they could to prepare. Crystal's diet has been the picture of perfection, she has kept up her regular exercise schedule, and she has quit drinking alcohol entirely. Now they are ready to have the baby of their dreams. They are convinced that not only will this baby be healthy and strong, but it will be above average in every respect.

All parents want to give their children every advantage in life that they can. Although no one can guarantee that a baby will be born healthy and strong—no matter what the parents do—there are steps parents can take to make the healthiest baby possible.

Unfortunately, many infants born in the United States do not have the advantage of good health at birth. The U.S. **infant mortality rate** continues to lag far behind that of 55 other developed countries, and maternal mortality is 76% higher (Illustration 29.1).[1,2] The relatively high rate of infant deaths in the United States is due primarily to the proportion of infants born **low birthweight** or **preterm** (Illustration 29.2). About 1 in 12 U.S. infants is born too small, and 1 in 10 is born too soon. Rates of low birth weight and preterm differ substantially among population groups, with the highest rates occurring in African Americans. These infants are at particular risk for requiring intensive and continuing care and of dying within the first year of life. In contrast, infants weighing between 7 pounds, 11 ounces and 8 pounds, 13 ounces (or 3,500 to 4,000 grams) are least likely to die within the first year of life.[3]

infant mortality rate Deaths that occur within the first year of life per 1,000 live births.

low-birthweight infants Infants weighing less than 2,500 grams (5.5 pounds) at birth.

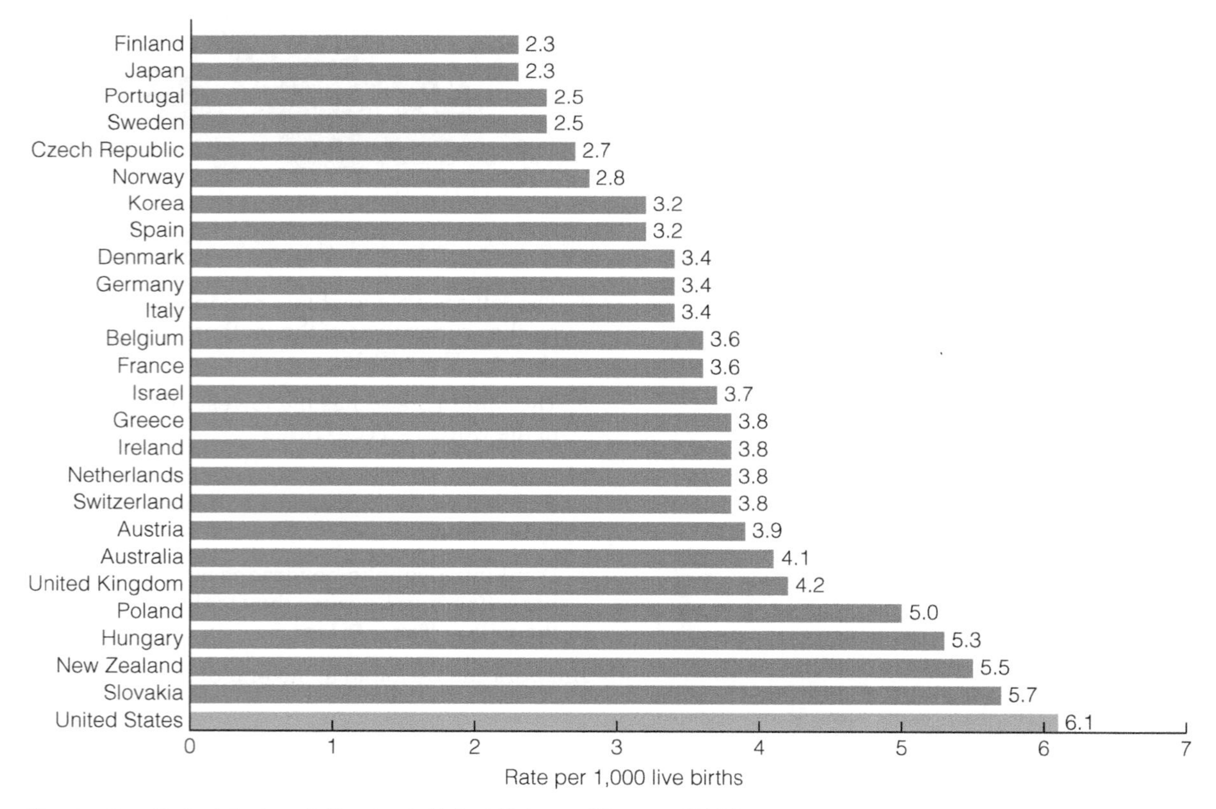

Illustration 29.1 Infant mortality rates in United States and Europe, 2010.[3]
Source: From the National Vital Statistics Reports, Volume 63, Number 5, September 24, 2014. MacDorman,MF et al. International Comparisons of Infant Mortality and Related Factors: United States and Europe, 2010. Available at: www.cdc.gov/nchs/data/nvsr/nvsr63/nvsr63_05.pdf.

KEY NUTRITION CONCEPTS

Six of the 10 key nutrition concepts presented in Unit 1 are directly relevant to advances in knowledge about nutrition during growth and development and health. Four are highlighted here:

1. Nutrition Concept #2: Foods provide energy (calories), nutrients, and other substances needed for growth and health.
2. Nutrition Concept #4: Poor nutrition can result from both inadequate and excessive levels of nutrient intake.
3. Nutrition Concept #7: Some groups of people are at higher risk of becoming inadequately nourished than others.
4. Nutrition Concept #9: Adequacy, variety, and balance are key characteristics of healthy dietary patterns.

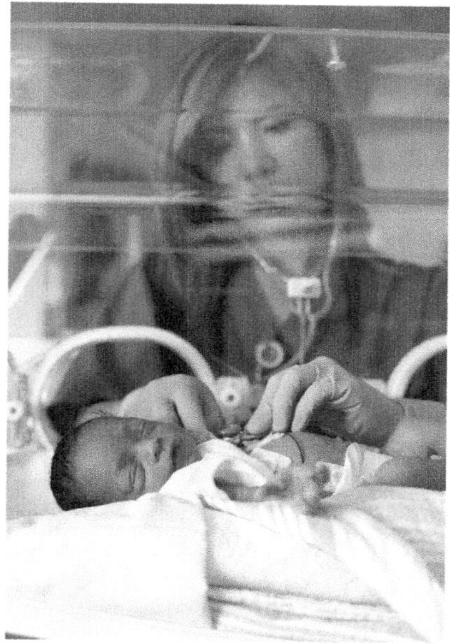

Blend Images · ERproductions Ltd/Brand X Pictures/Getty Images

Illustration 29.2 In 2013, 8% of infants born in the United States had a low birthweight, and 10% were born preterm.[1]

Improving the Health of U.S. Infants

A high proportion of poor infant outcomes in the United States is attributed to a combination of factors including poverty, poor nutrition, limited access to health care, and a maternal lifestyle that includes the use of illicit drugs, cigarettes, and excessive amounts of alcohol. Many infant deaths, preterm infants, and low-birthweight infants can be prevented, however, by optimizing environmental conditions and maternal behaviors that influence health prior to pregnancy. Of the various behaviors that affect maternal health, none offers potentially greater advantages to both pregnant women and their infants than good nutrition.[4–6]

PhotoDisc

Nutrition and Pregnancy

- **Discuss the importance of nutrition in improving the health of infants in the United States.**
- **Describe critical periods of fetal growth and development and the potential consequences of inadequate nutrient availability during these periods on future health status.**

Nutrition is important during pregnancy because the **fetus** depends on it. In almost all instances, the fetus is not protected from inadequate or excessive nutrient intake of mothers. Whether or not fetal needs for nutrients are met depends primarily on the availability of nutrients from the mother's diet and maternal nutrient stores.[4] This situation makes sense in terms of survival of the species. By meeting the mother's needs first, nature protects the reproducer.[7,8]

Rather striking evidence that maternal health is not placed in jeopardy by fetal nutrient utilization comes from studies showing that deficiencies of vitamin B_{12}, thiamin, iodine, folate, zinc, and other nutrients occur in newborns but not in their mothers. Similarly, infants born to women who consume excessive levels of vitamin or mineral supplements during pregnancy are more likely to display signs of nutrient overdose than are the mothers.[8,9]

Fetal growth and development can be compromised by insufficient and excessive availability of nutrients. The nature and extent of impairment depends primarily on when they occur during pregnancy and the extent of the deficiency or excess. The timing during pregnancy is important because organs and tissues growing and developing most rapidly when the nutritional deficiencies or overdoses occur are affected most.[7,10]

Critical Periods of Growth and Development

Fetal **growth** and **development** proceed in a series of critical periods. A **critical period** of growth and development is an interval of time during which cells of a tissue or organ are

preterm infants Infants born at or before 37 weeks of gestation (pregnancy).

fetus A baby in the womb from the eighth week of pregnancy until birth. (Before then, it is referred to as an embryo.)

growth A process characterized by increases in cell number and size.

development Processes involved in enhancing functional capabilities. For example, the brain grows, but the ability to reason develops.

critical period A specific interval of time during which cells of a tissue or organ are genetically programmed to multiply.

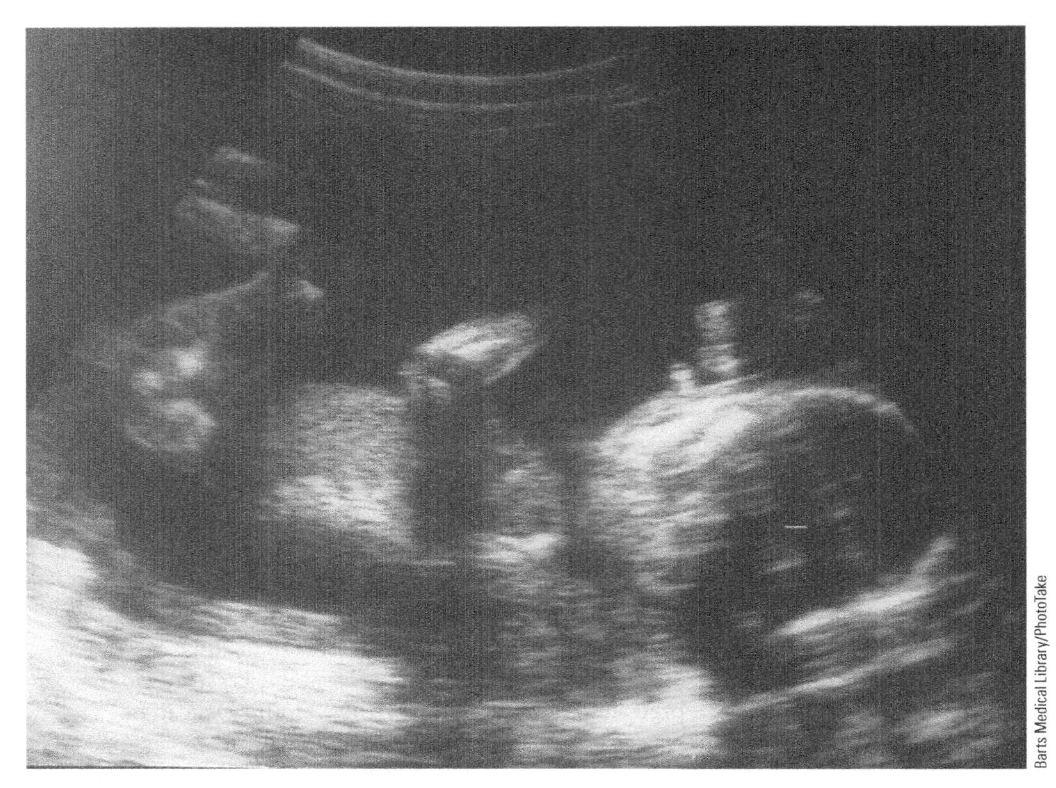
Barts Medical Library/PhotoTake

Illustration 29.3 Baby's first picture. An ultrasound image of a rapidly growing and developing 16-week-old fetus.

genetically programmed to multiply. The period is considered critical because if the cells do not multiply as programmed during this set time interval, they cannot make up the deficiency later. The level of nutrients required for cell multiplication to occur normally must be available during this specific time interval. If the nutrients are not available, the developing tissue or organ will contain fewer cells than normal, will form abnormally, or will function less than optimally.[4,7]

The roof of the mouth (the hard palate), for example, is formed early in the third month of pregnancy when two developing plates fuse together. This process can only occur early in the third month. If excessive amounts of vitamin A are present in fetal tissues during this period, the two plates may fail to combine, resulting in a cleft palate.[96] (The divided hard palate can be surgically corrected after birth.)

Critical periods of cell multiplication are most intensive in the first few months of pregnancy, when fetal tissues and organs are forming rapidly (Illustration 29.3)—hence the importance of the nutritional status of women at the time of conception and very early in pregnancy. For some organs and tissues, cell multiplication continues through the first two years after birth. Most of the growth that occurs in the fetus late in pregnancy and throughout the rest of the growing years, however, is due to increases in the size of cells within tissues and organs.[4,7]

Developmental Origins of Later Disease Risk

One of the most striking advances in research on pregnancy concerns the potential effects of maternal nutrition on the baby's risk of developing certain chronic diseases later in life.[13] It appears that an important element of increased susceptibility to heart disease, stroke, diabetes, obesity, hypertension, and other disorders may be "programmed," or gene functions modified, by inadequate or excessive supplies of energy or nutrients during specific time periods during pregnancy.[4,16]

Given optimal conditions, fetal growth and development proceed according to the genetic blueprint established at conception. Organs and tissues are well developed and ready to function optimally after birth. With less-than-optimal growing conditions such as those introduced by maternal weight loss, poor nutrient intake, or diseases in the mother that alter her ability to supply the fetus with energy or nutrients, fetal growth and development are modified. Fetal tissues undergoing critical phases of development at that time have to make adaptations to cope with the under- or oversupply of nutrients. These adaptations may produce long-term changes in the functional status of certain genes (called "**gene variants**"). Fetal exposure to low or high concentrations of nutrients such as vitamin B_{12}, iodine, and vitamin A, and high maternal blood glucose concentrations, for example, can modify the offspring's future growth, development, and disease risk due to the presence of specific gene variants.

Prepregnancy Weight Status, Pregnancy Weight Gain, and Pregnancy Outcomes

Women who enter pregnancy underweight or who fail to gain a certain minimum amount of weight during pregnancy are much more likely to deliver low-birthweight and preterm infants than women who enter pregnancy at or above normal weight and gain an appropriate amount of weight.[12] The risk of low birthweight can be reduced by healthy diets that lead to a desired rate and amount of weight gain during pregnancy. Along with the duration of pregnancy and smoking, prepregnancy weight status and weight gain in pregnancy are the major factors known to influence an infant's birthweight (Table 29.1).

Table 29.1 Major factors that directly influence birthweight[11,12]

- Duration of pregnancy
- Prenatal weight gain
- Prepregnancy weight status
- Smoking

What's the Right Amount of Weight to Gain During Pregnancy? Weight-gain recommendations for pregnancy vary depending on whether a woman enters pregnancy underweight, normal weight, overweight, or obese. A woman who is obese when she becomes pregnant will need to gain less weight than a woman who enters pregnancy underweight. It is recommended that underweight women gain 28 to 40 pounds, normal-weight women gain 25 to 35 pounds, overweight women gain 15 to 25 pounds, and obese women gain 11 to 20 pounds during pregnancy. Normal-weight women carrying twins should gain 37 to 54 pounds during pregnancy.[11] Table 29.2 can be used to identify prepregnancy weight status by a woman's body mass index (BMI) category and the recommended weight gain during pregnancy. After the goal is identified, a woman may want to plot her prenatal weight gain on a chart like the one in Illustration 29.4. The weight gained should be the result of a high-quality diet that leads to gradual and consistent gains in weight throughout pregnancy.

Where Does the Weight Gain Go during Pregnancy? If healthy infants tend to weigh about 8 pounds at birth, why do women need to gain more than that? Where does the rest of the weight go? Many pregnant women ask themselves these questions, especially when they don't relish the thought of gaining weight. Illustration 29.5 shows where the weight gain goes.

Table 29.2 Prepregnancy weight status and recommended weight gain during pregnancy[11]

Prepregnancy weight status, body mass index	Recommended weight gain
Underweight, <18.5 kg/m^2	28–40 pounds
Normal weight, 18.5–24.9 kg/m^2	25–35 pounds
Overweight, 25–29.9 kg/m^2	15–25 pounds
Obese, 30 kg/m^2 or higher	11–20 pounds
Twin pregnancy, normal weight status	37–54 pounds

Gene variants A different form of the same genes that direct specific protein production. Proteins produced affect every metabolic process in the body, from the digestion to glucose utilization.

trimester One-third of the normal duration of pregnancy. The first trimester is 0 to 13 weeks, the second is 13 to 26 weeks, and the third is 26 to 40 weeks.

neural tube defects Malformations of the spinal cord and brain. They are among the most common and severe fetal malformations, occurring in approximately 1 in every 1,000 pregnancies. Neural tube defects include spina bifida (spinal cord fluid protrudes through a gap in the spinal cord; shown in Illustration 29.7), anencephaly (absence of the brain or spinal cord), and encephalocele (protrusion of the brain through the skull).

Fetal growth is accompanied by marked increases in maternal blood volume, fat stores, and breast and uterus size, all of which contribute to weight gain. In addition, water accumulates in the amniotic fluid, which cushions and protects the fetus, and the volume of fluid that exists outside cells increases. The placenta (the tissue that transfers nutrients in the mother's blood supply to the fetus) also accounts for some of the weight that is gained during pregnancy.

Where Does the Weight Gain Go after Pregnancy? Delivery is one of the greatest weight-loss plans known to humankind. On average, women lose 15 pounds within the first week after delivery, and weight loss generally continues for up to a year.[13] Women who gain more than the recommended amount of weight, however, and those who gain weight after delivery may have extra pounds to lose to get back to prepregnancy weight.[14] The Reality Check for this unit addresses a common issue related to weight gain during pregnancy among obese women.

The Need for Calories and Key Nutrients during Pregnancy

Pregnant women need more calories, protein, and essential nutrients than nonpregnant women. All of these are important, but levels of folate, vitamin A, vitamin D, iron, and iodine intake are of particular concern because intake may be low. Dietary fiber performs important roles during pregnancy, and few women consume the recommended 28 grams daily. Although many pregnant women are concerned about protein, low protein intake is rare. Pregnant women consume, on average, around 70 grams of protein daily. The recommended amount of protein for pregnancy is 71 grams a day.[6,17]

According to dietary intake recommendations, pregnant women need approximately 15% more calories and up to 50% more of various nutrients than do

State of California Health and Human Services Agency California Department of Public Health

Name:

Weight Categories for Women According to Height and Pre-pregnancy Weight (lbs)[1]:

Height	Under Weight (BMI <18.5)	Normal Weight (BMI 18.5-24.9)	Over Weight (BMI 25-29.9)	Obese (BMI ≥ 30)
4'7"	< 80	80-107	108-128	> 128
4'8"	< 83	83-111	112-133	> 133
4'9"	< 86	86-115	116-138	> 138
4'10"	< 89	89-119	120-143	> 143
4'11"	< 92	92-123	124-148	> 148
5'	< 95	95-127	128-153	> 153
5'1"	< 98	98-132	133-158	> 158
5'2"	< 101	101-136	137-163	> 163
5'3"	< 105	105-140	141-168	> 168
5'4"	< 108	108-145	146-174	> 174
5'5"	< 111	111-149	150-179	> 179
5'6"	< 115	115-154	155-185	> 185
5'7"	< 118	118-159	160-191	> 191
5'8"	< 122	122-164	165-196	> 196
5'9"	< 125	125-168	169-202	> 202
5'10"	< 129	129-173	174-208	> 208
5'11"	< 133	133-178	179-214	> 214
6'	< 137	137-183	184-220	> 220
6'1"	< 140	140-189	190-227	> 227
6'2"	< 143	143-194	195-233	> 233
6'3"	< 148	148-199	200-239	> 239

BMI = Weight (lbs.)/Height (in.)^2 X 703

Recommended Weight Gain[1]:

Mark One:	Single	Twins
☐ Underweight	28-40 lbs.	N/A
☐ Normal	25-35 lbs.	37-54 lbs.
☐ Overweight	15-25 lbs.	31-50 lbs.
☐ Obese	11-20 lbs.	25-42 lbs.

Pre-pregnancy Weight:

Height:

[1] IOM, 2009. *Weight Gain During Pregnancy: Reexamining the Guidelines.* Washington, DC: National Academies Press.
[2] Per Personal Communication with the Committee to Reexamine IOM Pregnancy Weight Guidelines

Date	Wt	Wk Gest	Initials

Pre-pregnancy Normal Weight Range
Prenatal Weight Gain Grid[2]

Weight Gain (Pounds)

Weeks in Gestation

CDPH 4472B2 (03/10)

Illustration 29.4 An example of a prenatal weight-gain chart for women entering pregnancy at normal weight, from the California Department of Maternal and Child Health.

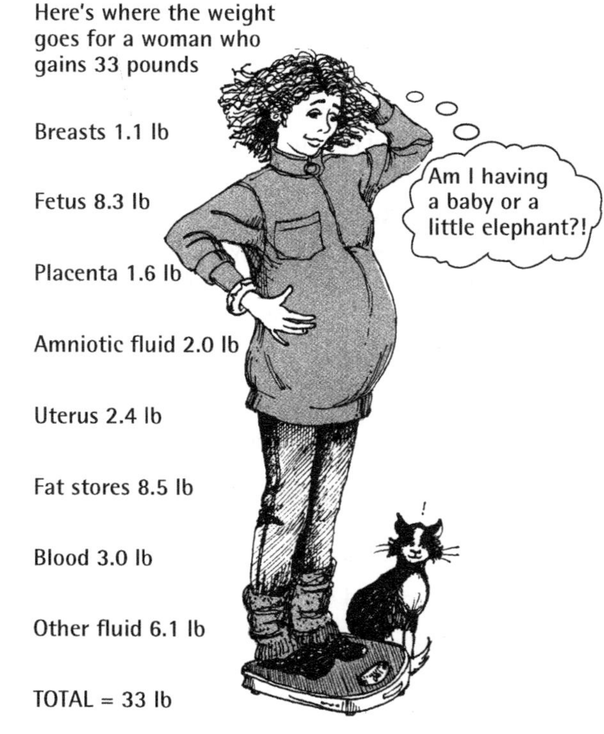

Illustration 29.5 Where does all the weight go?

nonpregnant women (Illustration 29.6). Relatively high nutrient requirements mean that pregnant women should increase their intake of nutrient-dense foods more than their consumption of calorie-rich foods.[18,19]

Calories On average, women need an additional 340 calories a day in the second **trimester**, and 450 calories in the third trimester of pregnancy.[18] Women entering pregnancy underweight will need more calories than this, and those entering overweight will need fewer. In addition, physically active pregnant women require higher caloric intake than average. Rather than counting calories, however, it is generally more practical to monitor the adequacy of caloric intake by tracking weight gain.

Folate Folate is required for protein tissue construction and therefore is in high demand during pregnancy. Inadequate folate status early in pregnancy has been found to be one of the most important causes of **neural tube defects** such as spina bifida (Illustration 29.7). Sufficient folate is particularly important in the first month of pregnancy when the neural tube, the future spinal cord, forms in a layer and then closes to encapsulate nerve tissue. Neural tube defects (NTDs) occur when the neural tube fails to close and damage to nerve tissue results.[20] Adequate folate intake throughout pregnancy is also related to a decreased incidence of infants born preterm and small for their gestational age.[21,22]

Prior to 1998, inadequate folate status was quite common in women before and during pregnancy, and combined with its known relationship to neural tube defects, led to the implementation of folic acid fortification of refined grain products (Illustration 29.8). The current fortification program calls for the addition of 140 μg of folic acid per 100 g (3½ oz) of refined grain products. (Folic acid is a highly available form of folate.) It is estimated that folic acid fortification provides 100–200 mcg folic acid per day to the diet of women of childbearing age in the United States, and the prevalence of poor folate status has plummeted. More than 53 countries mandate

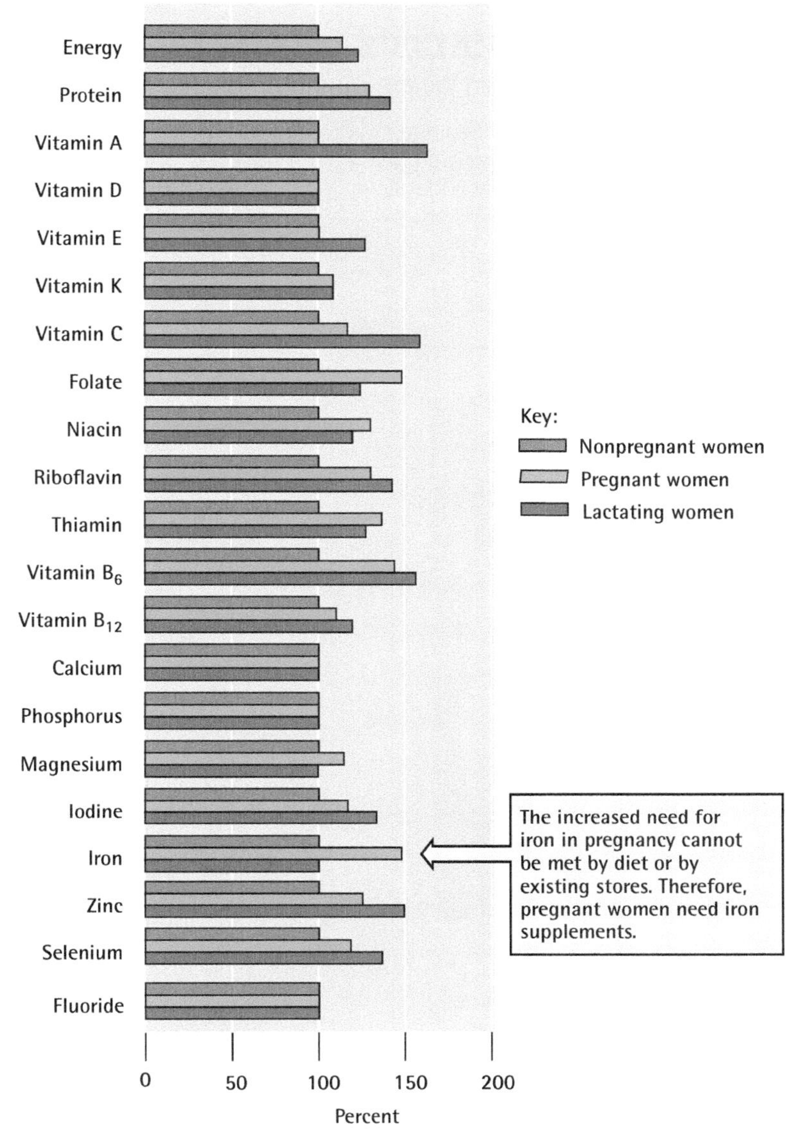

Illustration 29.6 Percentage increases in the RDAs or AIs for pregnant and breastfeeding women compared to other women.

REALITY CHECK

Should Obese Women Gain Weight During Pregnancy?

Elesha, who has been obese for 5 years, is 28 weeks pregnant. At her recent prenatal appointment she asked her health care provider about how much weight she should gain. Because she started pregnancy with a BMI indicating obesity, the health care provider tells her that she does not need to gain any weight during pregnancy. She returns home and asks her cousin Joanne if this is the right advice.

Who gets the thumbs-up?

Answers appear on page 29-8.

©Vasiliy Koval/Shutterstock.com

Joanne: Makes sense to me. You have a good bit of stored energy for the baby to grow on!

iStockphoto.com/Wmiami

Elesha: I thought all women were supposed to gain weight during pregnancy for the baby's sake.

ANSWERS TO REALITY CHECK

Should Obese Women Gain Weight During Pregnancy?

Yes, they should. Obese women need to gain less (11 to 20 pounds) than women entering pregnancy normal weight (25 to 35 pounds) or overweight (15 to 25 pounds). Infant outcomes are generally better if women gain weight during pregnancy than if they lose it.[11]

©Vasiliy Koval/Shutterstock.com

Joanne:

iStockphoto.com/Wmiami

Elesha:

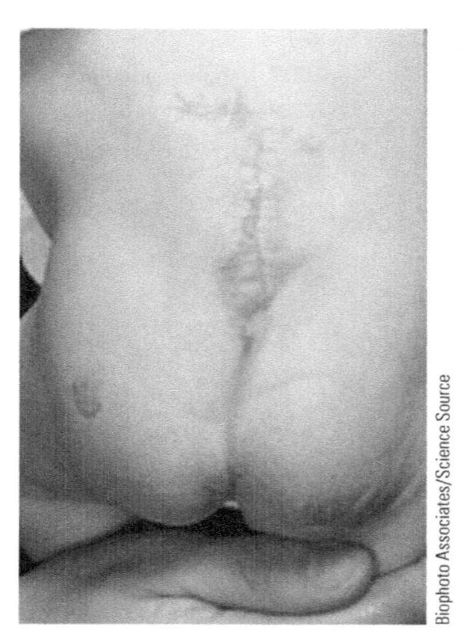

Biophoto Associates/Science Source

Illustration 29.7 Spina bifida: a neural tube defect. The interruption in the spinal cord results in paralysis below the injury. This photo shows the back after surgery to close the interruption.

Scott Goodwin

Scott Goodwin

Illustration 29.8 Foods fortified with folic acid.

folic acid fortification of refined grain products. The incidence of neural tube defects in countries that fortify decline by about 50%.[20,22]

Whole-grain products do not have to be fortified because they are not refined and the folate content contained in the germ is present. Whole-grain and enriched grain products provide similar amounts of folate. A slice of white bread, for example, contains about 28 mcg of folate, whereas a slice of rye or whole-wheat bread provides around 30 mcg folate.

The recommended daily intake of folic acid before pregnancy is 400 mcg (0.4 mg) from fortified foods or supplements. During pregnancy, it is recommended that women consume 600 mcg (0.6 mg) folate, including 400 mcg (0.4 mg) folic acid.[24] Women can generally obtain 400 mcg folic acid by consuming two servings of fortified ready-to-eat breakfast cereal, six servings of grain products, and three servings of vegetables daily.[25] Intake of high amounts of folic acid from supplements is no longer deemed advantageous. Excess folic acid may lead to genetic changes in the fetus that may be harmful.[29,30]

Vitamin A Both low and high intakes of vitamin A may cause problems during pregnancy. Too little vitamin A is associated with poor fetal growth. Too much vitamin A in the form of retinol from supplements can cause fetal malformations.[53]

The effects of vitamin A overdoses during pregnancy came to public attention in the early 1990s, when women taking Accutane or retinoic acid for acne were found to deliver far more than the expected number of infants with malformations. Use of these drugs before and very early in pregnancy increases the risk that babies will be born with malformations of facial features and the heart; intake of more than 10,000 to 15,000 IU of retinol daily during the same period may have the same effect. As a precaution, the American College of Obstetrics and Gynecology recommends that women who are or could become pregnant should limit their intake of retinol to less than 5,000 IU per day and should not use medicines containing vitamin A.[54] Beta-carotene, a precursor to vitamin A, is not harmful, however.[55]

Calcium Uptake of calcium by the fetus is especially high during the third trimester, when the fetus's bones are mineralizing. Pregnant women who regularly consume low-calcium diets lose calcium from their bones during pregnancy but can regain it from calcium intake after delivery. Because calcium losses are not from the teeth, the old saying, "for every baby a tooth" has no basis in fact.[57]

Vitamin D Vitamin D during pregnancy supports fetal growth and the addition of calcium to bone, and participates in the programming of genes in ways that may influence the development of chronic diseases such as rheumatoid arthritis and cancer later in life.[59] Insufficient vitamin D compromises fetal growth and bone development, and this appears to be happening during some pregnancies. About 42% of African American and 4% of Caucasian women have low blood levels of vitamin D.[58] Vegan women may also be at risk for poor vitamin D status because vitamin D is naturally present only in animal products.[59] This requirement can be met through food sources of vitamin D and by exposing the skin to sunshine for about 10 minutes on warm, sunny days. Current research questions the benefit of taking high-dose vitamin D supplements due to apparent lack of a beneficial response on vitamin D status.[26] An intake of 15 mcg (600 IU) of vitamin D daily is officially recommended during pregnancy. Intake of vitamin D from foods and supplements should not exceed 100 mcg (4,000 IU) daily.[60]

Iron Iron deficiency is the most common nutrient deficiency in pregnant women in the United States.[61] It develops when a woman enters pregnancy with low iron stores and fails to consume enough iron during her pregnancy. Iron requirements increase during pregnancy due to increases in hemoglobin production and storage of iron by the fetus. Inadequate iron availability during pregnancy is associated with an increased rate of iron-deficiency anemia in pregnant women and preterm delivery, low birthweight, and low iron stores in infants.[62] Iron deficiency is least likely to occur when iron stores are adequate prior to and throughout pregnancy.[4,62]

Recommendations for iron supplementation during pregnancy vary across the globe. Prior to supplementation, screening for iron deficiency is generally recommended to determine the need, a position held by the American College of Obstetrics and Gynecology.[29] High-dose iron supplements (45 mg/day plus) are no longer recommended for use in pregnancy because they surpass the tolerable upper limit for iron intake, are not associated with health benefits, and because there is evidence of harm from inflammation and gastrointestinal side effects due to the presence of high blood levels of reactive iron.[29] High doses of iron have limited effect on iron status because iron absorption decreases steadily as iron intake increases (Illustration 4.21).

Women and their pregnancies benefit more from healthy dietary patterns that include sufficient iron than from attempts to "catch up" on needed iron during pregnancy.

Iodine Iodine is needed for normal thyroid function and plays important roles in protein tissue construction and maintenance. A lack of iodine during pregnancy can interfere with fetal development and, in extreme cases, can cause mental and growth retardation, as well as malformations in children.[63]

It is recommended that pregnant women consume 220 mcg of iodine daily throughout pregnancy. The most reliable source of iodine is iodized salt. One teaspoon contains 400 mcg iodine. Salt with added iodine is clearly labeled as "iodized." Fish, shellfish, seaweed, and some types of tea also provide iodine. Women who consume iodized salt are not likely to need supplemental iodine.[45] Iodine supplements should be used only by women who are at risk of iodine deficiency due to the lack of iodine in the food supply, or because of diagnosed iodine deficiency. Excess blood concentration of iodine interferes with normal thyroid function.[27,28] Usual Normal iodine intake should not exceed 1,100 mcg daily during pregnancy.

Dietary Fiber Adequate intake of fiber, especially soluble fiber from dried beans, fruits, and vegetables, is particularly encouraged due to beneficial effects on gut microbiota, inflammation, glucose levels, and insulin resistance.[6,31] The adequate intake set for fiber is 28 grams per day for pregnant women (few women consume that much).

Other Dietary Components

Other components of diets of particular interest during pregnancy include alcohol, salt, artificial sweeteners, and omega-3 fatty acids.

Alcohol Knowledge is spreading about the potential, adverse effects of alcohol intake very early in and throughout pregnancy. Consumption of alcohol-containing beverages among most women attempting pregnancy or during pregnancy has dropped precipitously. Although 86% of females 18 years of age and older drink, only 8.5% drink some amount of alcohol during pregnancy. Around 3% of women binge drink during pregnancy.[31,32]

It is estimated that 1 to 5% of newborns experience adverse effects related to in utero exposure to alcohol.[35] The risk for development of adverse effects of alcohol exposure increases along with alcohol intake. Binge drinkers and women who drink consistently throughout pregnancy are at highest risk for delivering newborns with fetal alcohol syndrome (FAS).[36]

FAS has two major diagnostic categories:

1. FAS characterized by the presence of three specific facial features that can be observed in Illustration 29.9:
   - A smooth ridge between the nose and upper lip
   - Narrow openings between the upper and lower eyelids
   - A thin border on the upper edge of the lip. (See illustration 29.9)
2. Partial fetal alcohol syndrome (PFAS):
   - Presence of two of the three characteristic facial features

microbiome The totality of microorganisms and their collective genetic material present in or on the human body.

Important effects of fetal alcohol exposure are not visible. They are behavioral and intellectual impairments in offspring including central nervous system abnormalities, poor social and learning skills, and short attention span.[36,37] It is recommended that women do not drink alcohol-containing beverages during any part of pregnancy.[36]

©Rick's Photography/Shutterstock

Illustration 29.9 A child with fetal alcohol syndrome (FAS).

Artificial Sweeteners Regular consumption of artificial sweeteners appears to modify the gut **microbiome** in ways that influence metabolic mechanisms involved in glucose and insulin utilization, and satiety. Regular use of artificial sweeteners has been associated with decreased satiety, altered glucose homeostasis, and increased calorie consumption and weight gain in adults.[38] These changes, in turn, are associated with increased calorie consumption and weight gain in adults.[38,39] It now appears that regular intake of artificial sweeteners during pregnancy may have lasting effects on weight status of offspring. Preliminary evidence indicates that infants and children born to women with gestational diabetes who regularly consumed artificially sweetened beverages were more likely to be large for gestational age at birth and overweight or obese at age 7 than infants and children born to women with gestational diabetes who did not consume such drinks.[39]

Coffee Although it is often recommended that pregnant women avoid coffee during pregnancy, evidence that backs up this recommendation is weak. Eliminating coffee consumption during pregnancy does not appear to benefit pregnant women or their newborns.[40,41] Moderate amounts of coffee, such as 1 to 3 cups daily, do not appear to pose a risk to mom or baby.[42,43] Most U.S. pregnant women consume less than 100 mg caffeine daily.[44]

Salt Salt restriction during pregnancy was recommended in the past because it was thought that salt intake was related to the development of hypertension during pregnancy. However, there is no solid evidence that dietary salt reduction helps prevent or treat hypertension during pregnancy. Salt restriction during pregnancy for the prevention of hypertension during pregnancy is no longer recommended.[45]

Omega-3-Fally Acids Omega-3-fatty acids (EPA, DHA) are needed for fetal neural development, and the best way to get them is through food sources and not supplements. Fish and seafood are generally good sources of the omega-3 fatty acids. Table 29.3 shows a list of safe choices for fish and seafood. It is recommended that pregnant women consume two to three servings of cooked fish or seafood weekly.[46] Women who do not like fish can select DHA-fortified eggs, juices, and other foods. It does not appear that supplements of DHA or EPA during pregnancy significantly improve cognitive or neural development in children.[98,99]

Table 29.3 "Best choices" for fish and seafood for pregnant women.[46]

Anchovy	Perch
Atlantic mackerel	Pickerel
Black sea bass	Plaice
Butterfish	Pollock
Catfish	Salmon
Clams	Sardine
Cod	Scallop
Crab	Shad
Crawfish	Shrimp
Flounder	Smelt
Haddock	Sole
Hake	Squid
Herring	Tilapia
Lobster, American and spiny	Trout, freshwater
Mullet	Tuna, canned light
Oyster	Whitefish
Pacific chub mackerel	Whiting

What's a Good Diet for Pregnancy?

Recent research results clearly support the benefits of healthy dietary patterns (the Healthy Eating Index, Mediterranean Dietary pattern, the Healthy Vegetarian pattern, or the DASH Eating Plan) on the course and outcome pregnancy.[6,47] These dietary patterns capture the metabolic and health effects of multiple and complex interactions among nutrients and other components of food on health.[47] They are low in sugar and refined carbohydrates, highly processed foods, and red meats, and highlight vegetables, fruits, and low-fat dairy products; poultry, fish, and seafood; nuts and seeds; vegetables oils; and whole grains and whole grain products.

Table 29.4 lists recommendations for diet during pregnancy. Illustration 29.10 shows an example of food group recommendations for Dianne, a 30-year-old women who is 5 foot, 8 inches tall, weighed 140 pounds before pregnancy, is in her second trimester of pregnancy, and is physically active for 30–60 minutes daily. (This example of dietary intake recommendations for pregnancy was generated from the previous edition of MyPlate.gov and is no longer available at the site.)

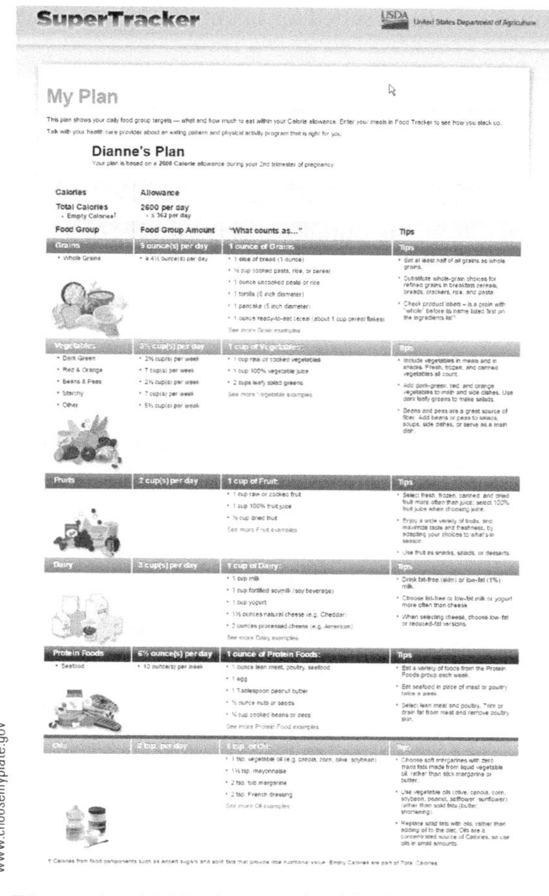

Illustration 29.10 An example of food group recommendations for a pregnant or breastfeeding woman based on MyPlate Plan for pregnancy.

Table 29.4 Dietary recommendations for pregnancy[6,16,97]

Consume sufficient calories for adequate weight gain.
Follow a healthy dietary pattern. Eat a variety of foods from each MyPlate food group.
Eat regular meals and snacks.
Consume sufficient dietary fiber (about 28 grams per day).
Consume 11 to 12 cups of water each day from fluids and foods.
Use salt to taste (within reason).
Do not drink alcoholic beverages.
Be careful not to contract a food-borne illness.
Eat foods you enjoy at pleasant mealtimes.

PhotoDisc

Illustration 29.11 Babies are born to be breastfed.

Food Safety during Pregnancy Hormonal changes during pregnancy lead to decreased immune function during pregnancy, which places pregnant women and their fetuses at increased risk of food-borne illness. Pregnant women should eat only appropriately cooked eggs, seafood, meat, and poultry, and should not consume raw sprouts or raw or unpasteurized milk or milk products. As a precaution, deli meats and hot dogs should be heated to the point of steaming (165 degrees F) to kill *Listeria* bacteria that may be present. *Listeria* is a particularly hazardous source of food-borne illness in pregnant women.[65]

Physical Activity Recommendations for Pregnancy Physical activity is encouraged during pregnancy because it enhances maternal health and fetal development. It decreases the risk of developing high blood glucose during pregnancy, reduces excessive weight gain during pregnancy, and helps women return to their prepregnancy weight. Unless not recommended for health reasons, pregnant women are encouraged to engage in moderate-intensity physical activity for 30 minutes most days of the week.[65] Moderate-intensity physical activities include brisk walking, aerobic dance and swimming, raking leaves, and tennis.

Breastfeeding

- **Discuss the nutrition advantages of breastfeeding for infants and the dietary recommendations for breastfeeding mothers.**

A woman's capacity to nourish a growing infant does not end at birth; it continues in the form of breastfeeding (Illustration 29.11). Breastmilk from healthy, well-nourished women is ideally suited for infant nutrition and health.[67]

What's So Special about Breastmilk?

Breastmilk is like a bonus pack. In addition to serving as a complete source of nutrition for infants for the first 4–6 months of life, breastmilk contains substances that convey a significant degree of protection against a variety of infectious and other diseases (Table 29.5). Breastmilk confers a degree of protection against the development of diarrhea and other gastrointestinal disorders, allergies, and infectious diseases during childhood. It also affects infants' taste preferences early in life due to exposure to food flavors transferred from mom's diet to her breastmilk.[48] The disease-preventing and development-promoting components of human milk are lifesaving assets in many developing countries, where a safe water supply and medical care may be unavailable.[68] Although breastfeeding doesn't protect infants from all infectious diseases and food allergies, it's the ounce of prevention that's worth a pound of cure.

Breastfeeding also offers other benefits—ones that may mean a lot to parents. Table 29.6 presents some of them.

Table 29.5 Health benefits of breastfeeding[67,71]

Breastfeeding decreases the risk that infants will develop:
• Respiratory Infections
• Ear infections
• Gastrointestinal tract infections
• Sudden infant death syndrome
• Allergic disease
• Celiac disease (if infants were exposed to breastfeeding at the time of gluten exposure)
• Inflammatory bowel disease
• Obesity
• Type 1 diabetes

Is Breastfeeding Best for All New Mothers and Infants?

Over 96% of women are biologically capable of breastfeeding, and the vast majority of infants thrive on breastmilk.[69] In the United States, about 80% of new mothers breastfeed their infants to some extent (Illustration 29.12).[70] Successful breastfeeding involves more than biology; it is heavily influenced by environmental and psychological conditions.

The increase in women returning to work soon after delivery, a lack of health care providers and emotional support for breastfeeding, embarrassment, early hospital discharge, and inadequate knowledge about how to breastfeed all appear to be deterring U.S. women from breastfeeding. If more women are to have the opportunity to

Table 29.6 Fifteen reasons to breastfeed

1. The milk container is easy to clean.
2. Breastmilk is a renewable resource.
3. There's no packaging to discard.
4. Breastmilk comes in an attractive container.
5. The temperature of breastmilk is always perfect right out of the container.
6. Breastmilk tastes really good.
7. There are no leftovers.
8. You don't have to go to the kitchen in the middle of the night to get breastmilk ready.
9. It takes just seconds to get a meal ready.
10. There's no bottle to repeatedly pick up off the floor.
11. The price is right.
12. The meal comes in a perfect serving size.
13. Meals and snacks are easy to bring along on a trip or outing.
14. Feeding units come in an assortment of beautiful colors and sizes.
15. One food makes a complete meal.

breastfeed, these and other barriers must be broken down.[100] A number of European countries that actively promote breastfeeding allow women to stay in the hospital after delivery until breastfeeding is going well. The usual practice in some African countries is to relieve a breastfeeding mother's workload so that she can devote nearly full time to feeding and caring for her young infant. A relative may move in with the family and take over household chores, or the mother and baby may live with her parents for a time.

Public health initiatives are in place to facilitate breastfeeding.[100] To meet national health objectives for improving rates of breastfeeding initiation and continuation in the United States (see Table 29.7), it is recommended that:

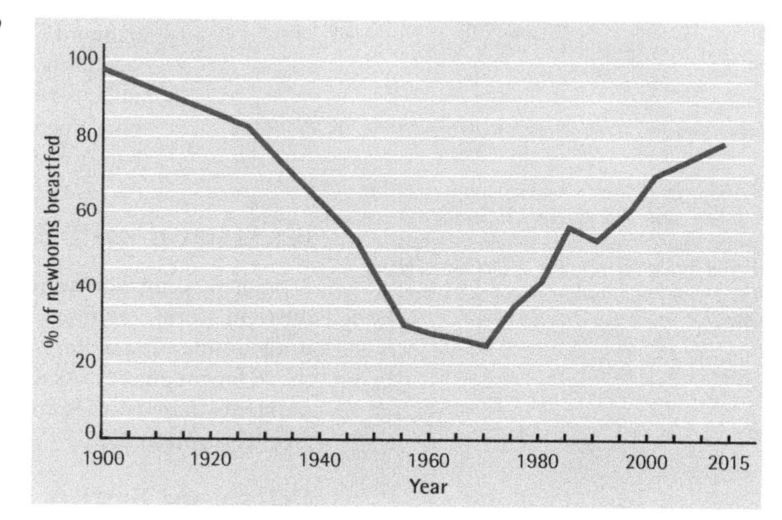

Illustration 29.12 Percentage of breastfed newborns in the United States.[72,73]

- Health care workers encourage and facilitate breastfeeding
- Hospitals do not routinely distribute infant formula to new moms
- "How-to" advice and problem-solving guidance be available for breastfeeding women from qualified health care staff
- Breastfeeding women returning to work should have easy access to breastmilk pumping private spaces and refrigeration for breastmilk

Breastfeeding is at its best when both the mother and infant benefit from the experience. If mutual benefit is not possible, then formula feeding may be necessary. Infant growth and development are well supported by commercially available infant formulas.[74]

Table 29.7 Actual versus 2020 national health goals for breastfeeding in the United States[72]

	Newborns	6 Months	12 Months
National goal	82%	60%	34%
Actual (all women)	79%	49%	27%

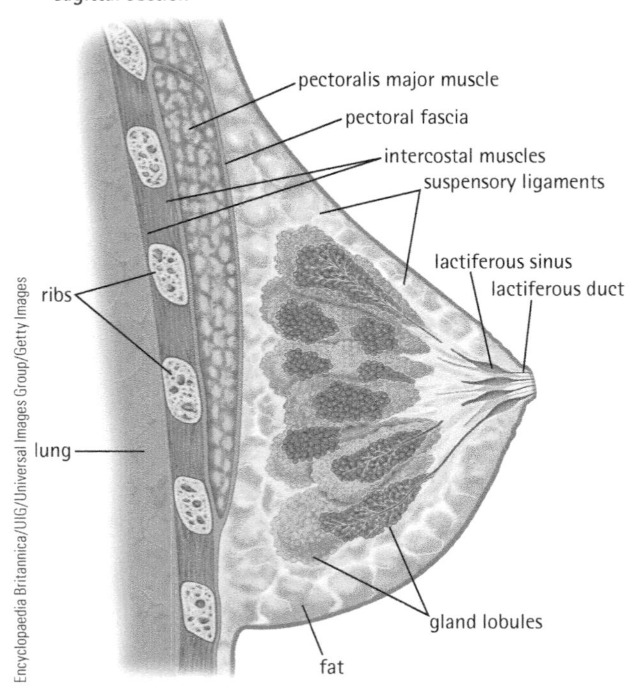

Illustration 29.13 A view of the interior of the breast.

colostrum The milk produced during the first few days after delivery. It contains more antibodies, protein, and certain minerals than the mature milk that is produced later. It is thicker than mature milk and has a yellowish color.

How Breastfeeding Works

The mother's body prepares for breastfeeding during pregnancy. Fat is deposited in breast tissue, and networks of blood vessels and nerves infiltrate the breasts. Ducts that will channel milk from the milk-producing cells forward to the nipple—the milk collection ducts—also mature (Illustration 29.13). Hormonal changes that occur at delivery signal milk production to begin.

Breastmilk produced during the first three days or so after delivery is different from the milk produced later. Called **colostrum**, this early milk contains higher levels of protein, minerals, and antibodies than "mature" milk.

While an infant is consuming one meal, she or he is "ordering" the next. The pressure produced inside the breast by the infant's sucking and the emptying of the breasts during a feeding cause a hormone to be released from specific cells in the brain. The hormone stimulates the production of milk so that more milk is produced for the next feeding. It generally takes about 2 hours for the milk-producing cells to manufacture enough milk for the next feeding. An important exception to this occurs when an infant enters a growth spurt and consumes more milk than usual. Then milk production takes longer, perhaps a day, to catch up with demand.

Only very rarely is a breastfeeding woman unable to produce enough milk. As long as an infant is allowed to satisfy her or his appetite by breastfeeding as often as desired, milk production will catch up with the baby's need.[75]

Nutrition and Breastfeeding

A breastfeeding woman needs an adequate and balanced diet to replenish her body's nutrient stores, maintain her health, and produce sufficient milk for her baby. Increases in recommended dietary intakes for breastfeeding women are generally higher than those for pregnancy. As during pregnancy, proportionately higher amounts of nutrients than calories are required, indicating the need for a nutrient-dense diet (see Illustration 29.6).

Calorie and Nutrient Needs The RDA for calories is about 15% higher for breastfeeding women than for other women. Actually, a breastfeeding woman needs about 30% more calories than the RDA for women who are not breastfeeding, but she does not have to consume that level of calories from food. Energy supplied from fat stores that normally accumulate during pregnancy contributes to meeting these needs during breastfeeding, so not all of the calories must come from the mother's diet.[65]

Nutrient levels in breastmilk are, in part, determined by the mother's diet. Low maternal blood levels of vitamin D and vitamin, B_{12}, for example, reflect maternal intake. The mother's diet also influences the development of a healthy microbiome in the baby's intestine.[49,50]

Table 29.8 Dietary recommendations for breastfeeding women[72,73]

• Diets should supply all of the nutrients breastfeeding women need. (The routine use of vitamin and mineral supplements is not recommended.)
• Fluids should be consumed to thirst.
• Weight loss should not exceed 6 to 8 pounds per month after the first month after delivery; caloric intakes should not fall below 1,500.
• Alcohol intake should be avoided.

What's a Good Diet for Breastfeeding Women? The calories and nutrients needed by breastfeeding women can be obtained from a healthy dietary pattern that is similar to that described for pregnancy in Illustration 29.10. Overall dietary recommendations for breastfeeding women are summarized in Table 29.8.

Increases in hunger and food intake that accompany breastfeeding generally take care of meeting caloric needs. Failure to consume enough calories from food can decrease milk production, however. Low-calorie diets (those providing fewer than 1,500 calories per day) and weight loss that exceeds 1.5 to 2 pounds per week—even in women with a good supply of fat stores—can reduce the amount of milk women produce. Weight loss of about a pound a week starting a month after delivery appears to be safe and helpful in women's attempts to return to prepregnancy weight.[76]

Illustration 29.14 Infants develop at a remarkable rate.

Are Supplements Recommended for Breastfeeding Women? Supplements are not recommended for breastfeeding women. Instead, breastfeeding women should try to get all of the nutrients they need from food.[67] Supplements, if needed, should be prescribed on an individual basis.[77]

Dietary Cautions for Breastfeeding Women Infants are much smaller than women, and it takes a smaller dose of alcohol, drugs, or environmental contaminants to have an effect on them than on an adult. Almost anything a woman consumes may end up in her breastmilk. Alcohol is transferred from a woman's body to breastmilk. The development of the brain and nervous system of infants breastfed by chronic, heavy drinkers appears to be impaired.[78] Breastfeeding women are advised to avoid alcohol or limit it to one drink when breastfeeding will not be undertaken for 3 to 4 hours. It takes that long to metabolize the alcohol in one standard drink and clear it from the blood and breastmilk.[79]

Infant Nutrition

- **Summarize the role of nutrition in the development of infants.**

At no other time during life outside the womb do growth and development proceed at a faster pace than during infancy. Infants grow out of clothes long before they wear them out. Each day infants learn new behaviors, and their minds absorb large chunks of information that will serve them well in the future (Illustration 29.14). The rate at which growth and development proceed in the first year of life is truly amazing. Just as infants need security, love and attention to flourish, so too do they need calories and essential nutrients.

Infant Growth

During the first week or two of life, it is not uncommon for infants to lose up to 10% of their birthweight while adjusting to the new surroundings.[80] After that, infants grow rapidly. Most infants double their birth weight by 4 months and triple it by 1 year. Length usually increases by 50% during the first year.[81] If this rate of growth were to continue, 10-year-old children would be about 10 stories high and weigh over 220 tons! After infancy, the growth rate declines and remains at a fairly low level until the adolescent growth spurt begins. Development proceeds in parallel with growth during the first year of life (Illustration 29.15).

Growth Charts for Infants The Centers for Disease Control and Prevention (CDC) recommends that the World Health Organization (WHO) growth standards be used to monitor the growth of infants and children ages 0 to 2 years of age in the United States. Examples of WHO growth charts for weight for age for newborns to 2 year olds are shown in Illustration 29.16. The charts should be carefully plotted based on accurate, periodic measures of an infant's size. They are best used for screening growth problems; confirmation of underweight or obesity requires assessment of body fat by skinfold thickness and other measures.[82]

Body Composition Changes with Growth Humans grow up and out, and their body composition and proportions change as growth progresses (Illustration 29.17). Although generally measured by gains in pounds and inches, growth also reflects changes in bone mass, organ size, body proportions, and composition (the proportionate amounts of water, muscle, and fat). Infants are not shaped like little adults. Their heads are very large in relation to the rest of their bodies and their arms are short, for example. Brain growth takes precedence over trunk and limb growth early in life, and the body eventually

Illustration 29.15 Growth and developmental characteristics from birth through 1 year.

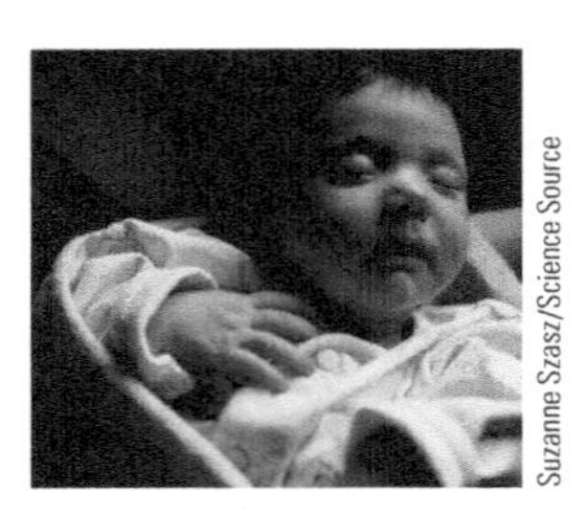

First Month
Generally weighs from 8 to 11 pounds; length is 20 to 23 inches. Head is relatively large and has soft spot on top. Startles and sneezes easily. Jaw may tremble. May hiccup and spit up. Eats every few hours.

One to Three Months
Lifts head briefly when placed on stomach; smiles, coos, and gurgles. Whole body moves when infant is touched or lifted. Eats every three to four hours.

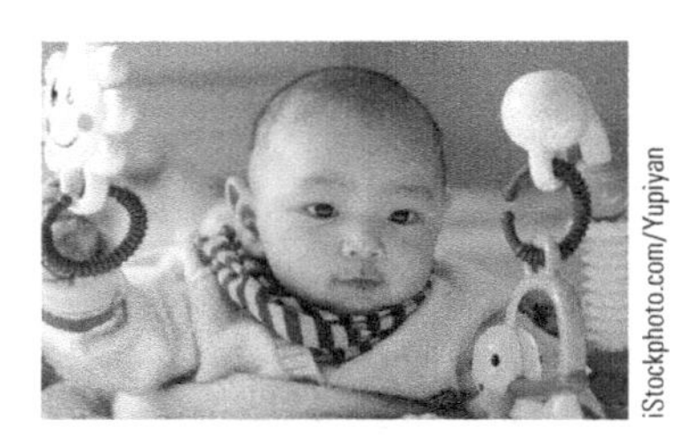

Four to Six Months
Weight nearly doubled. Has grown three to four inches. Follows objects with eyes. Reaches toward objects with both hands; puts fingers and objects into mouth. Turns over, sits unassisted. Awake longer at feeding time. Eats six to seven times per day. Sleeps six to seven hours at night.

Seven to Eleven Months
Gains in weight and height are less rapid, appetite has decreased. Stands up with help, hitches self along the floor. Reaches for, grasps, and examines objects with hands, eyes, and mouth. Has one or two teeth. Takes two naps a day.

Twelve Months
Usually has tripled birth weight and increased length by 50%. Grasps and releases objects with fingers. Holds spoon, but uses it poorly. Begins to walk unassisted.

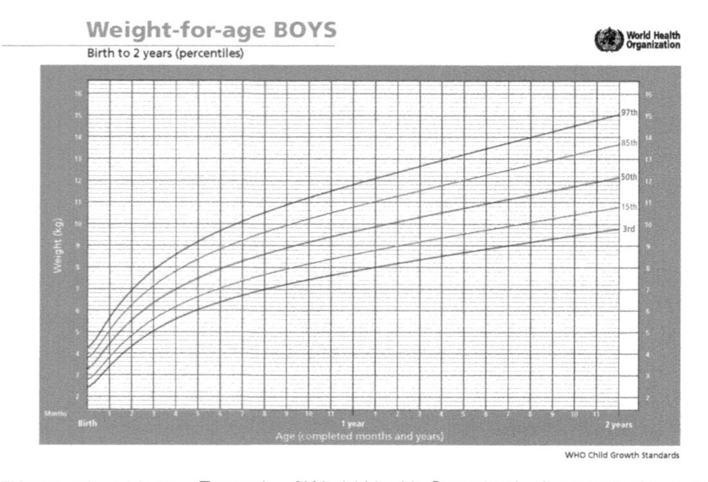

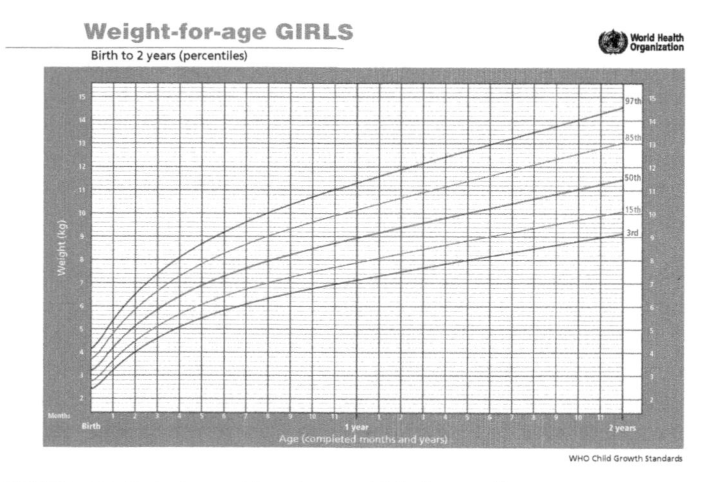

Illustration 29.16 Example of World Health Organization's growth charts for weight for age for males and females aged 0 to 2 years.[81]

Illustration 29.17 Infants' heads are proportionately large for the rest of the body. But, as you can see, there's much more growth to come.

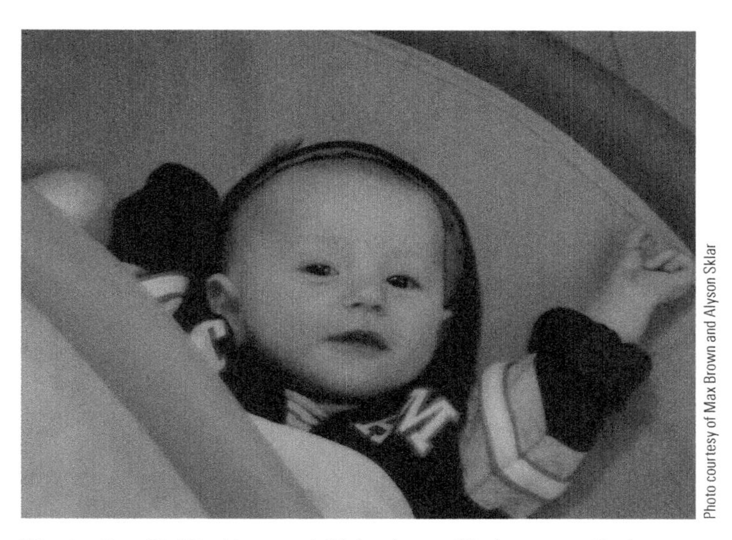

Illustration 29.18 How much higher is your lifted arm over the top of your head?

grows to "fit" the head (Illustration 29.18). During the first 10 years of life, infants who tend to be shaped like a loaf of bread normally take on the shape of a string bean.

Nutrition and Mental Development Malnutrition has the greatest impact on mental development when it is severe and occurs during the critical period for brain cell multiplication. For humans, this vulnerable period begins during pregnancy and ends after the first year of life. Impairment in mental development is less severe when malnutrition occurs only during pregnancy or only during infancy than if it occurs throughout both pregnancy and infancy.[83]

Children's mental development is also greatly influenced by the social and psychological environment in which they are raised. Because malnutrition is generally accompanied by both social and psychological deprivation, these factors often contribute jointly to poor mental development. For children in the United States, poverty, neglect, illness, and psychological problems appear to be the main cause of undernutrition and poor mental development.[84]

Infant Feeding Recommendations

Feeding recommendations for healthy infants are changing due to evidence that supports the introduction of small amounts of peanuts, eggs, milk, and wheat during the first year to help prevent the development of allergies to these foods. Newly published recommendations conclude that exclusive breastfeeding should continue throughout the first 4 to 6 months of life and continue until the baby is 12 months of age.[51,52] Infants should be fed "on demand," that is, when they indicate they are hungry, rather than on a rigid schedule set by the clock. In the first few months of life, most infants will get hungry every 3 or 4 hours. These and other recommendations for infant feeding are shown in Table 29.9.

Soft, solid foods should be added to an infant's diet of breastmilk or infant formula at around 6 months.[68] At one time, it was thought that infants should be offered solid foods within the first few months of life. Although such young infants are unable to swallow much of the food (most of it ends up on their face and bib) or to digest completely what they do swallow, solid foods were thought to help the baby grow and sleep through the night. This belief does not appear to be correct. Infants who receive cereal at bedtime before the age of 6 months are no more likely to grow normally or sleep through the night than are infants who start to receive solid foods around 6 months of age.[85] The age at which an infant begins to sleep for 6 or more hours during the night depends on other factors, including the infant's developmental level and how much he or she has slept during the day.[86] Neither the infant nor the infant's parents are likely to get a full night's sleep for at least 4 months.

Table 29.9 Overview of feeding recommendations for healthy infants.[51,52]

• Exclusive breastfeeding for the first 4–6 months of life and continuation of breastfeeding through the first year of life are preferred. Iron-fortified infant formulas may be used as a secondary option to breastfeeding.
• Introduce solid foods around 4–6 months of age.
• An infant's first food should be a good source of iron.
• Introduce a variety of non-sweetened solids foods around 4–6 months of age. Pay attention to the textures of baby's food, going from smooth to mashed to chopped to tiny pieces.
• Keep trying new foods—babies may need to try a new food 10 times or more before they like it!
• Use a small baby toothbrush with a tiny dot of fluoride toothpaste when the teeth start to come in.
• Infants should be fed "on demand" and not by a set schedule. Feeding should stop when the baby loses interest in eating.
• Infants not receiving fluoridated water should be given fluoride drops (dose depends on age).

Vance Duxbury/Shutterstock.com

Illustration 29.19 Older infants develop self-feeding skills. It requires practice and parental patience.

Table 29.10 Examples of foods that may cause choking in infants

Grapes
Hot dog pieces
Hard candy
Meat chunks
Nuts and seeds
Popcorn
Raw vegetables

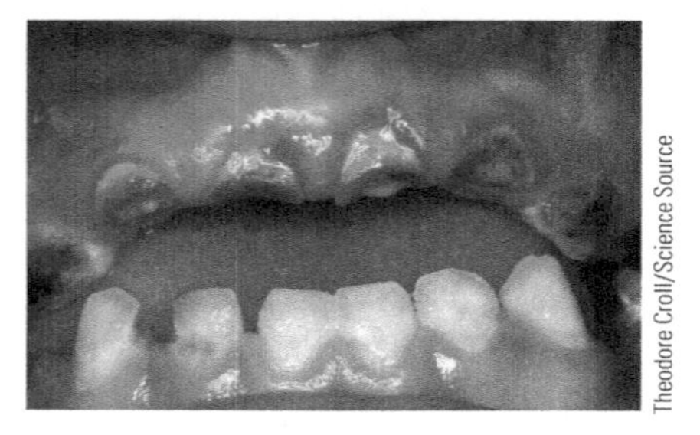
Theodore Croll/Science Source

Illustration 29.20 Baby-bottle tooth decay.

Introducing Solid Foods It is recommended that iron-rich foods be given to breastfed infants and to bottle-fed infants who are not receiving iron-fortified formula. The iron provided by iron-fortified rice cereal or formula helps restock the infant's iron stores, which have been drawn on since birth. Generally, solid foods prepared for infants should consist of single, basic foods such as strained vegetables, fruits, or meats.[51,52]

Solid foods for infants can be purchased as commercial baby food or prepared at home. When baby foods are prepared at home, care should be taken to avoid contamination and to achieve the right consistency. Variety is the key to achieving a healthy diet for infants in their second 6 months of life, and an assortment of basic, textured foods should be given.[88] By 9 months of age, infants are ready for mashed foods and foods such as yogurt, applesauce, ripe banana pieces, and grits. Most infants have several teeth by this time, and they are able to bite into and chew soft foods.

Infants graduate to adult-type foods after the age of 12 months. Although most foods still need to be mashed or cut up into small pieces for them, 1-year-olds are able to eat the same types of food as the rest of the family. They can drink from a cup and nearly feed themselves with a spoon (Illustration 29.19). Infants have come a long way in 12 months.

Foods to Avoid Foods like hot dog pieces, peanut butter, and pieces of raw vegetables should not be given to infants because they are difficult for infants to chew into small pieces or swallow (Table 29.10). Current research indicates that delaying the introduction of specific foods during infancy is likely not protective against the development of food allergies.[51,52]

Reduced-fat products are not recommended for infants. Infants need a relatively high-fat diet for brain and nervous system development. Unpasteurized honey is not recommended either, because it may cause botulism in infants due to their still maturing gastrointestinal defenses against bacteria. Beverages containing added sugar, such as fruit drinks, sweetened cereals and yogurt, cookies, and sweet snack foods, should not be given to infants because they promote tooth decay and provide empty calories. To help prevent the baby bottle tooth decay, shown in Illustration 29.20, infants should not be put to sleep with a bottle containing sweet fluids or formula.[87]

Do Infants Need Supplements? Two situations call for the use of supplements during infancy. Breastfed infants and infants living in households that do not receive fluoridated water should be given a fluoride supplement (0.25 mg daily) after 6 months of age.[78] Most infants do not consume the recommended 400 IU of vitamin D daily, and a vitamin D supplement is recommended to fill the gap between intake and need.[89] Increasing maternal vitamin D intake during breastfeeding increases the concentration of vitamin D in breastmilk by up to 20%.[49]

The Development of Healthy Eating Habits Begins in Infancy

Older infants and young children should be offered a wide variety of nutritious foods in a positive eating environment. They alone should make the decision about how much to eat of any food offered. Food preferences are primarily learned and are unique to every individual. The learning process begins early in life and is affected by a variety of influences. "Instincts" that draw infants to select certain nutritious foods are not one of the influences, however. Infants do not instinctively know what to eat. Rather, they are born with mechanisms that help regulate how much they eat. These basic characteristics of infant food intake were beautifully demonstrated by Clara Davis in experiments with 7- to 9-month-old infants in the 1920s and 1930s.[90] One of her most famous experiments is described in Illustration 29.21.

Subsequent research reinforced earlier conclusions that infants and young children should be offered a wide variety of nutritious foods and that decisions about how much to eat should be left up to the infant or child.[91] Beginning at birth, infants who are hungry eat enthusiastically. They stop eating when they are full. Coaxing, cajoling, and pleading by parents or caregivers can override infant's decisions about how much food to eat. That, unfortunately, may upset infants' built-in food intake regulating mechanisms and lead to over- or under-eating, and to "fussy" eaters.

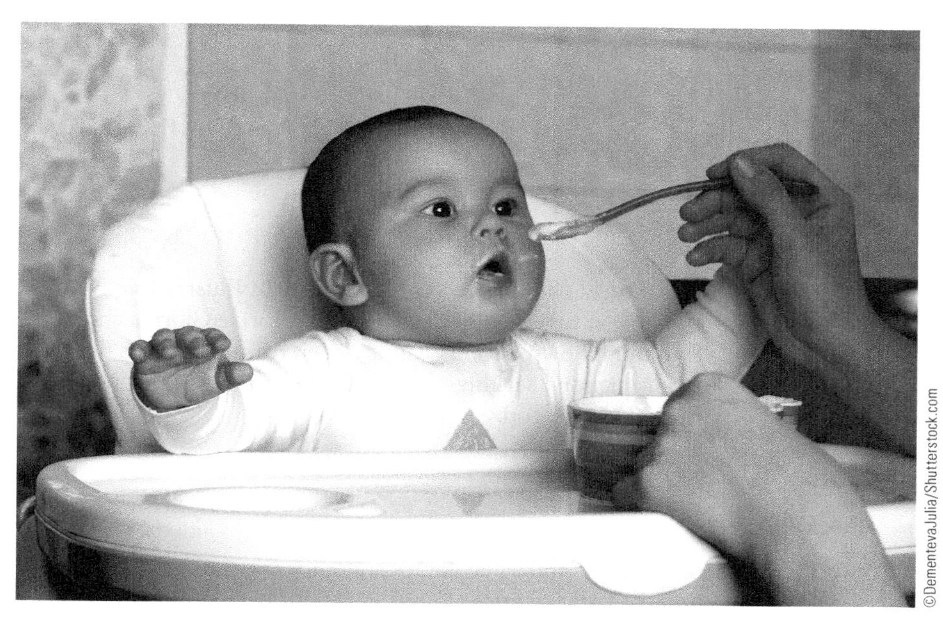

©Dementeva.Julia/Shutterstock.com

Illustration 29.21 Infant food preferences.

Clara Davis conducted her most quoted feeding experiment with older infants (7 to 9 months old) living in an orphanage connected to a large hospital in Chicago. Three times a day a nurse would offer the infants a tray of various foods. The infants would pick up or grab for the foods they wanted, and the nurse would bring the food selected to the infant's hands or mouth.

Altogether 32 different foods were offered, consisting of fresh, unprocessed, unseasoned, and simply prepared basic foods. Foods included milk, beef, kidney, bone marrow, liver, brain, thymus, fish, whole-wheat cereals, raw eggs, sea salt, 15 fresh fruits, and 10 fresh vegetables. Desserts, sugars, syrups, and other sweetened foods were not offered.

Infants quickly formed preferences and narrowed their choices to 14 of the 32 foods. Vegetables were the least liked, whereas milk, bone marrow, eggs, bananas, apple bits, oranges, and oatmeal were the best liked. They ate enough to grow normally and to remain in good health.

Dr. Davis concluded that appetite is the best guide to how much food an infant or young child needs, and that self-selection will have doubtful value only if the diet is chosen from inferior foods.[90,91]

Teaching Infants the Right Lessons about Food Recommendations for feeding infants are based primarily on energy and nutrient needs and the developmental readiness of infants for solid foods. But infant feeding recommendations also include a large educational component. Many of the lessons infants learn about food and eating make an impression that lasts a lifetime. Later food habits and preferences, appetite, and food intake regulation are all influenced by early learning experiences.[92] Table 29.11 provides a lesson plan for teaching your infant the right lessons about food and eating.

Making Feeding Time Pleasurable T. Berry Brazelton, an expert on child development, says that feeding is an arena in which parents and baby work out the continuing struggle between dependence (being fed) and independence (feeding oneself). Independence must win. Pushing a child to eat is the surest way to create problems. Feeding has to be pleasurable. Dr. Brazelton points out that there is much more to the feeding of infants than meets the eye. Infants learn about communication, relationships, and independence during pleasurable eating experiences. If undertaken in positive and supportive circumstances, mealtimes for infants can provide lessons that will support their physical and psychological health well into the future.[90,91]

Table 29.11 Tips for teaching infants the right lessons about food and eating[90,91]

1. Infants learn to eat a variety of healthy foods by being offered an assortment of nutritious food choices. There are no inborn mechanisms that direct babies to select a nutritious diet.
2. Infants must be allowed to eat when they are hungry and to stop eating when they are full. Infants, not parents, know when they are hungry and when they have had enough to eat.
3. Food should be offered in a pleasant environment with positive adult attention.
4. Food should not be used as a reward, punishment, or pacifier.
5. Infants or children should never be coerced into eating anything.
6. Food preferences change throughout infancy. Because an infant rejects a food one time doesn't mean she or he will not accept the food if offered later. Offering a food on a number of occasions often improves acceptance of the food. Babies still may not like strong-flavored vegetables until they are older, however.

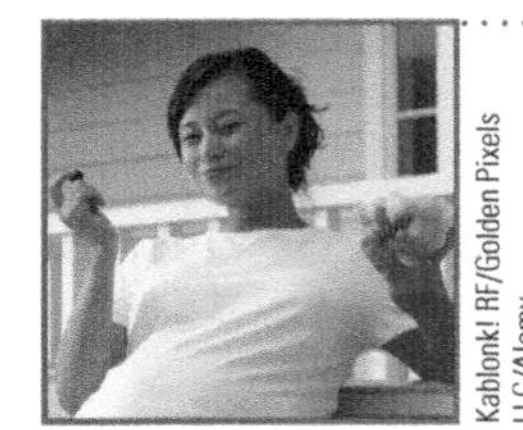

Kablonk! RF/Golden Pixels LLC/Alamy

NUTRITION up close

You Be the Judge!

Focal Point: Apply your knowledge of nutrition to the diet of a pregnant woman.

Gloria's baby is due in 5 months. Before becoming pregnant, she never gave much thought to what she ate, but now she is trying to eat healthy foods for her baby and herself. Compare her choices with the recommendations in Illustration 29.10 to find out how well she is doing. Categorize Gloria's choices into the following food groups to determine if she ate the recommended servings from MyPlate. The grains group has been done for you.

Gloria's one-day diet:

Breakfast: Bran muffin (2 ounces), cold cereal (1 cup) with strawberries (½ cup), and a glass of milk (1 cup)
Lunch: Vegetable soup (1 cup vegetables), carrot sticks (½ cup), and a glass of orange juice (1 cup)
Afternoon snack: Yogurt (1 cup)
Dinner: Chicken (3 ounces), turnip greens (½ cup), a tossed salad (2 cups), and iced tea (2 cups)

Feedback to Nutrition Up Close is located in Appendix F

Grains
bran muffin, 2 ounces
cold cereal, 1 cup
______
Gloria's intake: 3 ounces
Recommended intake: **8 ounces**

©David Kay/Shutterstock.com

Vegetables
______
______
______
Gloria's intake: ______
Recommended intake: **3 cups**

©Feng Yu/Shutterstock.com

Fruits
______
______
______
Gloria's intake: ______
Recommended intake: **2 cups**

©Tatiana Popova/Shutterstock.com

Milk and Milk Products
______
______
______
Gloria's intake: ______
Recommended intake: **3 cups**

©DenisNata/Shutterstock.com

Meat and Beans
______
______
______
Gloria's intake: ______
Recommended intake: **6 1/2 ounces**

©PHB.cz (Richard Semik)/Shutterstock.com

Oils
______
______
______
Gloria's intake: ______
Recommended intake: **7 teaspoons**

©Bonchan/Shutterstock.com

REVIEW QUESTIONS

- **Discuss the importance of nutrition in improving the health of infants in the United States.**

1. Relatively high rates of infant mortality, low-birthweight infants, and preterm delivery represent major public health problems in the United States. **True/False**
2. Critical periods of growth and development are marked by large increases in the size of cells within developing organs. **True/False**
3. Reduction in the size of organs related to the lack of nutrients during critical periods of growth and development can be corrected during the first year of life. **True/False**
4. Inadequate or excessive levels of nutrient availability during pregnancy can modify the function of fetal genes in ways that increase the risk of chronic disease later in life. **True/False**

- **Describe critical periods of fetal growth and development and the potential consequences of inadequate nutrient availability during these periods on future health status.**

5. A woman entering pregnancy with a BMI (body mass index) of 27 should gain 25 to 35 pounds during pregnancy. **True/False**
6. The critical period of growth and development for formation of the neural tube in a fetus occurs during the second trimester of pregnancy. **True/False**
7. Iodine and iron, EPA and DHA, and vitamin D are considered "risk nutrients" during pregnancy because a portion of women fail to consume enough of them. **True/False**
8. A daily multivitamin and mineral supplement is recommended for all breastfeeding women. **True/False**
9. Women are advised not to eat any types of fish or seafood during pregnancy because these foods contain excessively high levels of mercury. **True/False**
10. Moderate intake of alcohol in the second and third trimester of pregnancy is safe for the mother and fetus, but intake of alcohol in the first trimester is not. **True/False**

The next three questions refer to the following situation.

Simonetta is 6 months pregnant. One night while visiting her aunt and uncle for dinner her aunt gives her advice on what and how much she should eat for a healthy pregnancy.

11. ____ Which of the following pieces of Simonetta's aunt's advice about a healthy diet for pregnant women is correct?
    a. "Simonetta, you're eating for two and you need to eat more than that."
    b. "Don't eat so many vegetables. They will take up space in your stomach that needs to be filled with the meat. It's full of protein, you know."
    c. "Glad you enjoyed the salmon. I heard this kind of fish is good for pregnant women so I made it especially for you."
    d. "Just eat enough food, honey. Your baby will get enough of the vitamins and minerals it needs from your body."
12. ____ Simonetta decides to have a cup of coffee after dinner. Both her aunt and uncle look at her while her uncle asks, "Are you sure you should be drinking coffee? Isn't it bad for the baby?" What would be a reasonable response that Simonetta could make to that question?
    a. "Decaffeinated coffee is okay for the baby if you have only a cup a day."
    b. " Maybe coffee used to be forbidden during pregnancy, but times have changed. A few cups of coffee a day won't harm the baby."
    c. "You're right! What was I thinking?"
    d. "Do you have any tea?"
13. ____ Assume Simonetta delivers a boy who weighs 5 kg (11 pounds) when he reaches 2 months of age. Referring to WHO growth charts in Illustration 29.15, the range of weight for age percentiles that would correspond to the baby's weight at 2 months of age would be in the:
    a. 3rd to the 15th percentiles.
    b. 15th to 50th percentiles.
    c. 50th to 85th percentiles.
    d. 85th to 97th percentiles.

- **Discuss the nutrition advantages of breastfeeding for infants and the dietary recommendations for breastfeeding mothers.**

14. Breastfeeding promotes infants' health in part because breastmilk contains substances that prevent infections and reduce the risk of development of certain diseases later in life. **True/False**
15. Approximately one in four women is biologically unable to breastfeed successfully. **True/False**
16. A woman's milk supply automatically increases as an infant consumes more breastmilk. **True/False**

- **Summarize the role of nutrition in the development of infants.**

17. Infants should be given their first solid food around 4–6 months of age. **True/False**
18. Infants are able to self-regulate the amount of food they consume. They will eat when they are hungry and stop eating when they have had enough food. **True/False**
19. It has been demonstrated in multiple studies that infants will select and consume a well-balanced and adequate diet if given access to a wide variety of both "junk" and nutritious foods. **True/False**
20. Body proportions and composition of infants change a good deal with age. **True/False**

The next three questions refer to the following scenario.

Your cousin and her husband have a 10-month-old girl who refuses to try any vegetable whatsoever. The parents find it totally frustrating and want their daughter to at least taste them.

21. _____ Which type of parental behavior would be *least* likely to convince their daughter to taste vegetables?
    a. Continuing to offer her different types of vegetables at meals during pleasant mealtimes.
    b. Letting the baby decide if she doesn't want to eat the vegetable offered but keep offering small portions of vegetables with a positive attitude at future meals.
    c. Showing your enjoyment of vegetables when you eat together.
    d. Stopping the meal as soon as the baby refuses to try the vegetable.

22. _____ Finally the baby tries sweet potatoes and likes them. What should be a parent's appropriate reaction?
    a. Give the baby a reward for eating her vegetables.
    b. Smile and continue eating your meal.
    c. Clap your hands and shout hurray!
    d. Offer her sweet potatoes at every meal.

23. _____ A month later the daughter tries chopped spinach, immediately makes a face, and spits the spinach out of her mouth. What's the most likely reason this happened?
    a. She didn't like the taste of spinach.
    b. She is exerting her independence and decides not to eat what her parents want her to eat.
    c. She was too full to eat any more.
    d. She loves the look on her dad's face when she does something like that.

Answers to these questions can be found in Appendix F.

NUTRITION SCOREBOARD ANSWERS

1. In most all instances, the fetus does not harm the mother for its own gain. If there is a shortage of essential nutrients, the mother generally gets access to them before the fetus does.[7] **False**
2. More than 50 countries in the world have lower infant mortality rates than the United States.[1] **False**
3. **True**[82]
4. Alcohol does pass into breastmilk.[67] **False**

Nutrition for the Growing Years: Childhood through Adolescence

Jose Luis Pelaez Inc/Blend Images/Getty Images

NUTRITION SCOREBOARD

1 Gains in height and weight during childhood occur in spurts rather than gradually. **True/False**

2 Young children are able to associate the feeling of hunger with a need for food around the age of 5. **True/False**

3 The incidence of overweight and obesity in U.S. children tripled between 1990 and 2012. **True/False**

4 Childhood and adolescence represent "grace periods" during which the diet consumed does not influence future health. **True/False**

Answers can be found at the end of the unit.

LEARNING OBJECTIVES

After completing Unit 30 you will be able to:

- Describe connections between growth and development and eating behaviors in children and adolescents.
- Describe five key elements of healthy dietary patterns for children and adolescents.

The Span of Growth and Development

- **Describe connections between growth and development and eating behaviors in children and adolescents.**

Physical and mental development proceed at a high rate from infancy through adolescence (Illustration 30.1). Young children generally enter this phase of life able to take a step or two and to guide a spoon into their mouths sideways. They will likely leave adolescence with the ability to drive, work for pay, and solve complex problems. These are the formative years that lay the foundation for the rest of life.

KEY NUTRITION CONCEPTS

Content presented in this unit on nutrition during childhood and adolescence relates most directly to these key nutrition concepts:

1. Nutrition Concept #2: Foods provide energy (calories), nutrients, and other substances needed for growth and health.
2. Nutrition Concept #3: Poor nutrition can result from both inadequate and excessive levels of nutrient intake.
3. Nutrition Concept #7: Some groups of people are at higher risk of becoming inadequately nourished than others.
4. Nutrition Concept #8: Poor nutrition can influence the development of certain chronic diseases.
5. Nutrition Concept #9: Adequacy, variety, and balance are key characteristics of healthy dietary patterns.

The Nutritional Foundation

Good nutrition takes on particular importance during the growing years for many reasons. Growth is an energy- and nutrient-requiring process. It will not proceed normally unless the diet supplies enough of both.

Children learn about food and its importance to health and well-being during these early years of life, and they also establish food preferences and physical activity patterns that may endure into the adult years. Food intake regulatory mechanisms are affected by the lessons children learn early in life. Given control over decisions about how much to eat, children generally become responsive to internal cues that signal when they should eat and when they should stop eating.[1] If the lessons go well, they learn to eat for the right reason.

Eating well and learning the right lessons about food and health have implications that transcend the growing years. The types and amounts of foods and beverages children and adolescents normally consume can influence growth and development and the risk of specific diseases and disorders throughout life.[2] Table 30.1 summarizes the potential benefits of healthy dietary patterns during the growing years.

Table 30.1 Benefits of healthy dietary patterns in children[2,41]

- Promotion of optimal growth and physical and mental development
- Decreased risk of developing:
  - Obesity
  - Dental caries
  - Iron deficiency
  - Heart disease
  - Type 2 diabetes
  - Cancer
  - High blood pressure
  - Osteoporosis

Characteristics of Growth in Children

Between the ages of 2 and 10 years children normally gain somewhere around 5 pounds and 2–3 inches in height per year (Illustration 30.2). Gains in weight and height occur in "spurts" rather than continuously and gradually. Prior to a growth spurt, appetite and food intake increase (given an adequate food supply), and the child puts on a few pounds of fat stores (Illustration 30.3). During the growth spurt, these fat stores are normally used to supply energy needed for growth in height. When growth is not occurring, children are often disinterested in food and eat very little at times. They may go on "food jags,"

Illustration 30.1 Growth and development characteristics from 1 through 16 years.

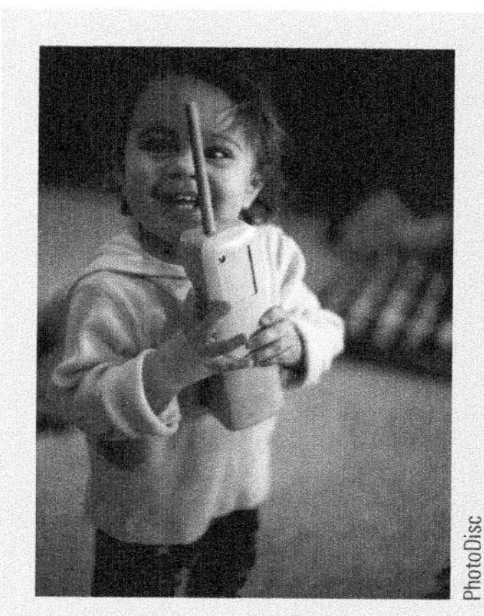
PhotoDisc

One to Two Years
Gains in height and weight continue at a lower rate; appetite is less. Uses finger and thumb to pick up things. Soft spot grows smaller and then disappears. Baby teeth continue to appear. Usually takes one long nap a day. Drinks from a cup, attempts to feed self with a spoon. Likes to eat with hands. Pulls self up to standing position. Walks alone.

Felicia Martinez/PhotoEdit

Two to Three Years
Slower and more irregular gains in height and weight. Has all 20 teeth. Runs and climbs, pushes, pulls, lugs, walks upstairs one step at a time. Feeds self using fingers, spoon, and cup; spills a lot. At times has one favorite food. By age 3, associates the sensation of hunger with a need for food.

Bridget Taylor/Getty Images

Three to Four Years
Gains 4–6 pounds and grows about 2–3 inches. Feeds self and drinks from a cup quite neatly, carries things without spilling. May give up sleep at naptime, substituting quiet play.

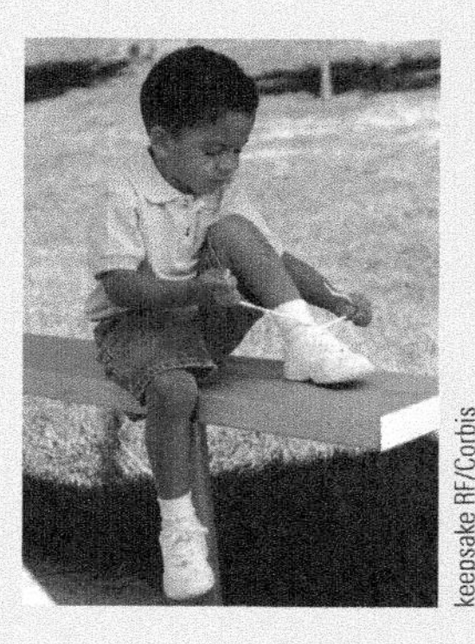
keepsake RF/Corbis

Four to Five Years
Gains in height and weight about same as previous year. Hops and skips, throws ball. Has increasingly good coordination, masters buttons and shoelaces. Can use knife and fork and is a good self-feeder.

Mary Kate Denny/PhotoEdit

Five to Six Years
Growth continues at about the same rate. Legs lengthen. Six-year permanent molars usually appear (new teeth, not replacing baby teeth). Begins to lose front baby teeth. Prefers plain, bland, and unmixed foods.

(*continued*)

Carl Purcell/Science Source

Six to Nine Years
Slow gains in height and weight (2–3 inches and 4–6 pounds a year). Some additional permanent teeth appear. Likely to have the childhood communicable diseases. Sleeps 11 to 13 hours.

©Blend Images/Shutterstock.com

Nine to Twelve Years
May be long legged and rangy, but health is generally sturdy. Permanent teeth continue to appear. Appetite good. Needs about 10 hours sleep. Growth spurt in girls usually begins. May get very irritable when hungry.

Myrleen Pearson/Alamy

Twelve to Fourteen Years
Wide differences in height and weight in children of either sex of same age. Menstruation usually begins (sometimes earlier). Girls develop breasts, are usually taller and heavier than boys of the same age. Growth spurt of boys begins. Muscle growth rapid. Appetite increases. Likes to spend time with friends.

Ghislain & Marie David de Lossy/The Image Bank/Getty Images

Fourteen to Sixteen Years
Boys are in period of growth spurt. Voice deepens. Girls have usually achieved maximum growth. Menstruation is established though may still be irregular. Enormous appetite. Pimples a common problem. Sleep reaches its adult pattern. All permanent teeth except wisdom teeth.

insisting on eating limited amounts of only a few favorite foods like peanut butter and jelly sandwiches or breakfast cereal.

The ups and downs of children's food intake can be unnerving for parents (see the Reality Check). But as long as growth continues normally and children remain in good health, there's little reason to worry about food jags and fluctuations in appetite and food intake.

CDC's Growth Charts for Children and Adolescents Growth progress during childhood and adolescence is generally monitored with the use of the Centers for Disease Control and Prevention (CDC) growth charts. (The World Health Organization [WHO] charts are recommended for children under the age of 2 years.[4]) Growth charts are available for females and males from 2 to 20 years. These growth charts consist of graphs of:

Weight for age (see Illustration 30.4)

Height (stature) for age

Weight for height

Body mass index (BMI) for age

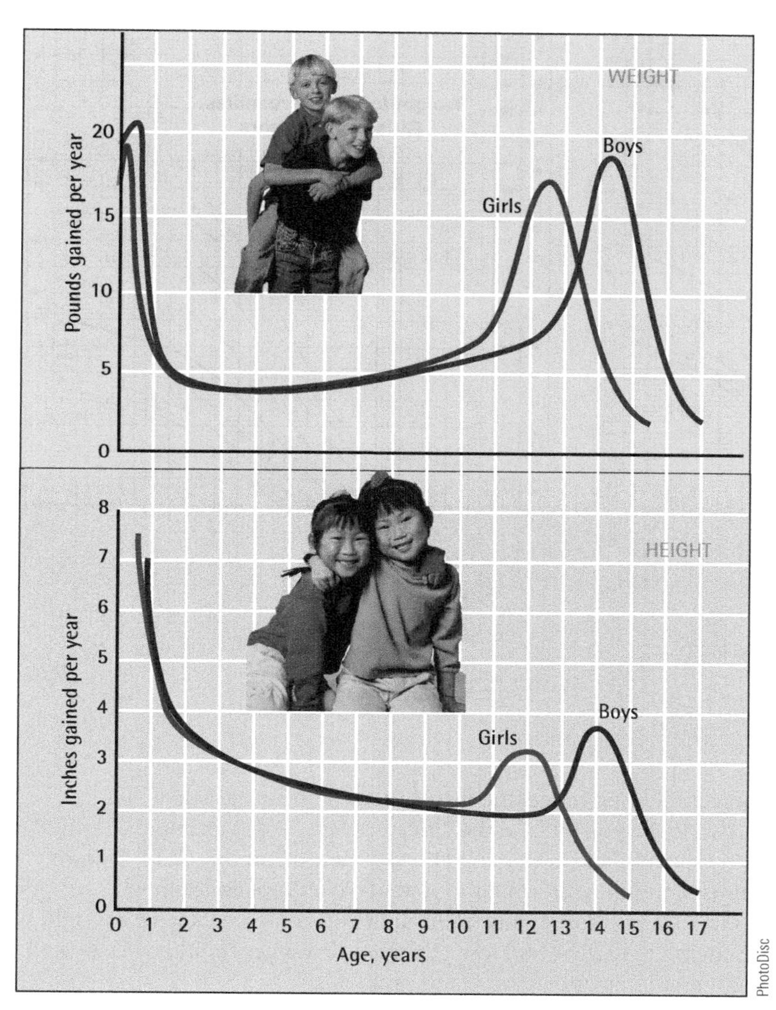

Illustration 30.2 Average yearly growth in weight and height during childhood and adolescence.

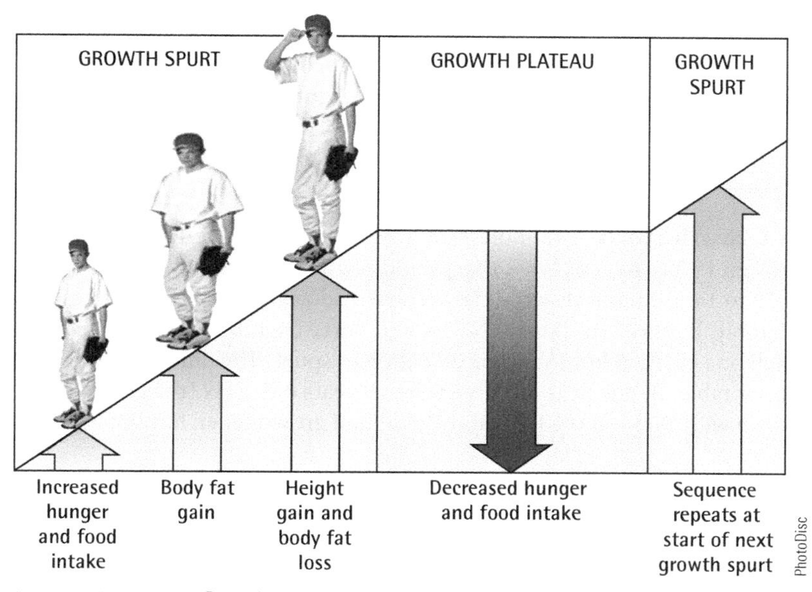

Illustration 30.3 Growth occurs in a series of "spurts." Each spurt follows this sequence.

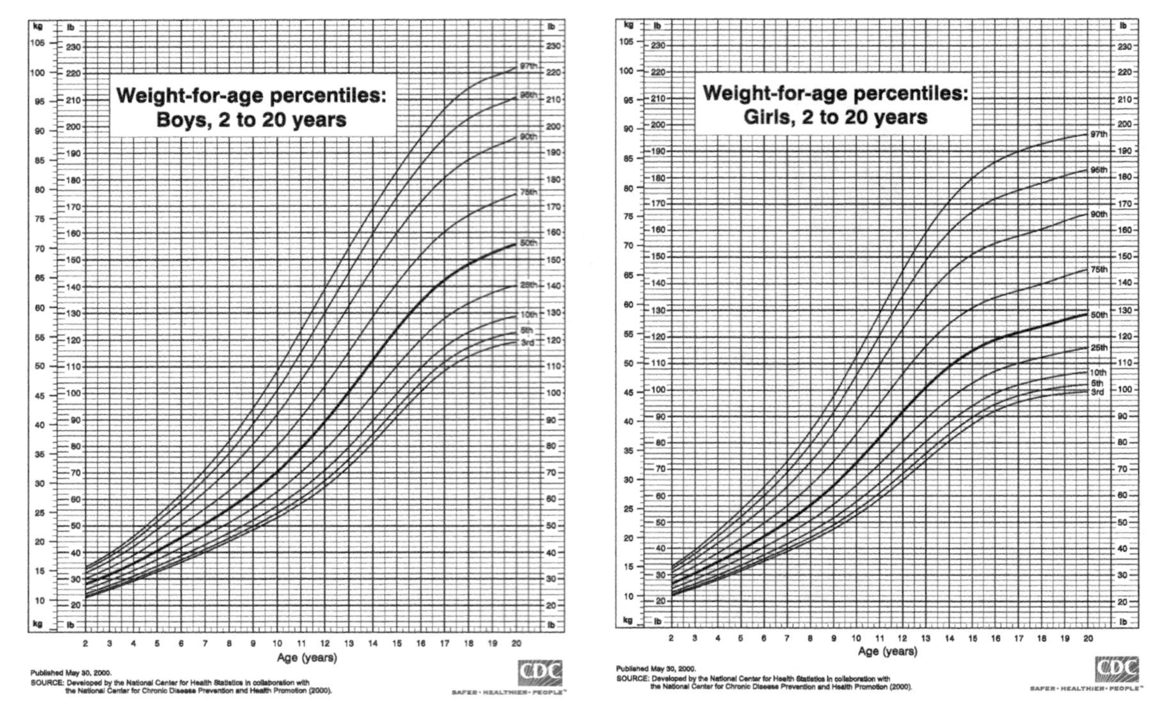

Illustration 30.4 An example of the CDC's growth charts: weight for age for boys and girls. The charts can be printed off the Internet using the address www.cdc.gov/growthcharts.

Each graph provides percentiles that reflect the distribution of these measures in a representative sample of 2- to 20-year-olds in the United States. So, for example, if a child's weight for age is between the 50th and 75th percentiles, it means that this child's weight is greater than that of most children and similar to that of the 25% of children who are represented in the 50th to 75th percentile range. Children and adolescents whose weight measurements place them in the highest and lowest percentile ranges should be evaluated further for potential underlying nutrition and health problems.[5]

Body Mass Index (BMI) for Age BMI charts by age and sex are a feature of the growth charts. Since BMI increases with age in children and adolescents, ranges of BMI used to classify weight status in adults cannot be used. BMIs for age and sex among 2- to 20-year-olds who are at or above the 85th percentile and lower than the 95th percentile are considered **overweight**. Those at or above the 95th percentile up to and including the 99th percentile are considered **obese**, and **severe obesity** is classified as having a BMI over the 99th percentile.[6] Now recommended for determining weight status in children and adolescents, use of the CDC's new BMI for age charts requires that BMI be calculated. Illustration 30.5 shows you how to calculate BMI and gives an example of a BMI chart for girls that shows where the BMI measure would be plotted.

overweight In children and adolescents, overweight is defined as a BMI at or above the 85th percentile and lower than the 95th percentile for children of the same age and sex.

obese In children and adolescents, a BMI at or above the 95th percentile for children of the same age and sex.

severe obesity In children and adolescents, severe obesity is defined as BMIs for age over the 99th percentile.

The Adolescent Growth Spurt The adolescent growth spurt usually occurs in girls between the ages of 9 and 12 years. For boys, this period of growth generally begins around the ages of 12 to 14. Actually, though, the age when adolescents start their growth spurt varies considerably. Pictured in Illustration 30.6 are three friends, all age 12. It's not hard to tell which one of them has experienced a growth spurt! The difference was also clear at the dinner table. By the time all three were 19 years old, Max (the one in the middle) had caught up with Ben (on the left), and David had grown taller, but not as tall as Max or Ben.

During these years of growth, teenagers gain approximately 50% of their adult weight, 20–25% of adult height, and 45% of their total bone mass. In the year of peak growth, girls gain 18 pounds on average, and boys gain 20 pounds.[7]

Can You Predict or Influence Adult Height? Ultimate height is difficult to predict. On average, children tend to achieve adult heights that are between the heights of their

How to calculate BMI

Lucy just turned 7 years old and has had her weight and height carefully measured. She weighs 57 pounds (lb), and her height (stature) is 4 feet (ft), 2 inches (in.). What's her BMI?

PhotoDisc

1. Convert her height of 4 ft 2 in. to inches:

$$4 \text{ ft} \times 12 \text{ in. per foot} = 48 \text{ in.}$$
$$+2 \text{ in.}$$
$$50 \text{ in.}$$

2. Square the inches:

$$50 \text{ in.} \times 50 \text{ in.} = 2500 \text{ in.}^2$$

3. Apply formula:

$$\text{BMI} = \frac{\text{kg}}{\text{m}^2} \quad \text{or} \quad \frac{\text{weight in lbs} \times 700}{\text{height in in.}^2}$$

$$\text{BMI} = \frac{57 \text{ lbs} \times 700}{2500 \text{ in.}^2} = \frac{39{,}900}{2500 \text{ in.}^2} = 15.96 \text{ kg/m}^2$$

In this example, Lucy's BMI for age of 15.96 kg/m² falls between the 50th and 75th percentiles, or well within the normal range.

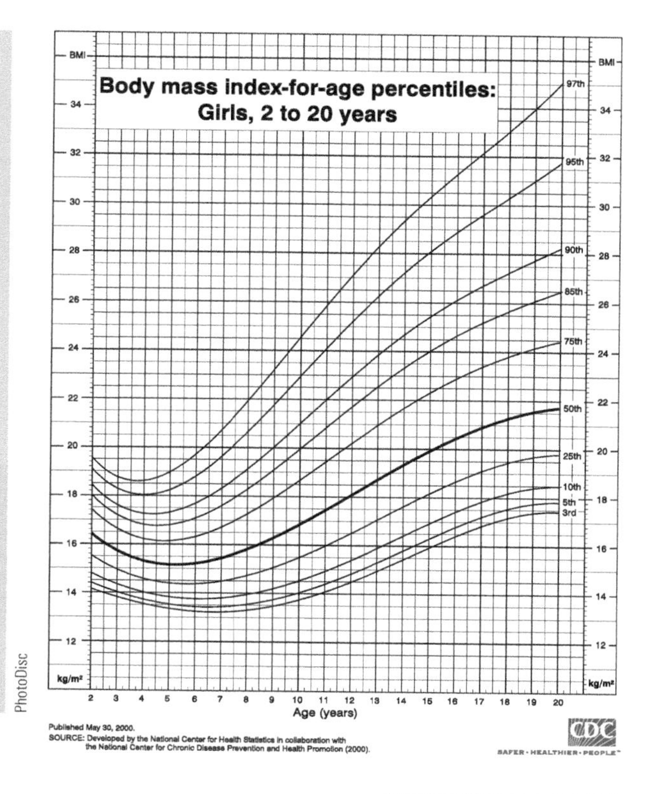

http://www.cdc.gov/nchs/data/series/sr_11/sr11_246.pdf

Illustration 30.5 How to calculate BMI, and an example of plotting BMI on a CDC BMI-by-sex-and-age growth chart.

biological parents.[8] There are many exceptions to this general finding, however, and that means that heredity is not the only influence on height. The dramatic increases in height of Japanese youth since World War II provide clear evidence that non-genetic factors have the strongest influence on height. Since the late 1940s, Japanese youth have grown an average of 2 inches taller each generation. (The increase in height has meant that everything from shoes to beds must be produced in larger sizes.) The increase in size of Japanese people is largely attributed to the availability of sufficient, nutritious foods.[9] Indeed, people in most economically developed countries continue to grow taller; a maximal genetically determined height has not yet been reached. If children are less well nourished than their parents, though, they tend to be shorter than their parents as adults.[10]

A healthy birthweight and diet during the growing years, and freedom from frequent bouts of illness, support growth in height.[11] There are no supplements, powders, or special diets that can be used to increase growth rate. Because treatment with growth hormone is

REALITY CHECK

The "Clean Your Plate" Club

Aiden and Zoe are eating lunch with a friend and his 4-year-old son, Jayden. During lunch, they notice Jayden has quit eating, become distracted, and wants to leave the table to play. Then their friend says, "Jayden, you can go out and play as soon as you finish all the food on your plate." Is this good advice?

Who gets the thumbs-up?

Answers appear on page 30-8.

iStockphoto.com/DRB Images, LLC

Aiden: I don't think it's good advice. Jayden isn't hungry any more.

iStockphoto.com/DRB Images, LLC

Zoe: It sounds like good advice to me. It's a shame to waste food.

ANSWERS TO REALITY CHECK

The "Clean Your Plate" Club

Children should not be encouraged to eat more food than they want. They should be allowed to decide for themselves when they are full. Serving small portions, and then adding additional food if the child does not feel full, can help reduce food waste.[3]

Aiden:

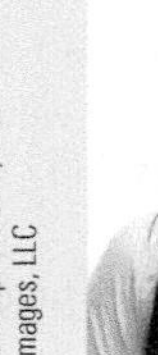

Zoe:

Illustration 30.6 This photo was taken when these boys, born 2 months apart, were 12 years old. The photo on the right shows the same boys at age 19.

Table 30.2 Key elements of healthy dietary patterns for children over the age of 2 and adolescents[14,15]

- Grain products emphasize whole-grain foods.
- Protein sources consist primarily of lean meats such as poultry, fish, and seafood, as well as legumes and tofu.
- Limit intake of highly processed meats and food products, and foods and beverages with added sugar.
- Emphasize water intake over sugar-sweetened beverages.
- Include sufficient dietary fiber from plant foods.
- Vegetables and fruits are consumed daily.

expensive and associated with a number of side effects, its use is limited to children with growth hormone deficiency and related disorders.[12]

Nutrition for the Growing Years

- **Describe five key elements of healthy dietary patterns for children and adolescents.**

Calorie and nutrient needs of children can be met by any of the four healthy dietary patterns with intake amounts adjusted to meet the needs of growing children (Table 30.2). Dietary recommendations for children differ from those for adolescents in respect to portion sizes. During the growing years, calorie need generally increases from 800 calories per day in 1-year-olds to around 3,100 calories for boys and 2,400 calories for girls by age 18. Illustration 30.7 shows portion sizes of foods and beverages for two meal and snack patterns generated by the previous edition of MyPlate.gov for a 4-year-old who needs 1,200 calories a day. Illustration 30.8 provides an example of a one-day diet plan appropriate for a 13-year-old girl who is moderately active 30 to 60 minutes a day and needs 2,000 calories a day. There is room in the dietary pattern for the occasional "empty-calorie" treat such as ice cream, cookies, or french fries.

Meal and Snack Patterns

for a 1200 calorie Daily Food Plan ...

These patterns are examples of how the Daily Food Plan can be divided into meals and snacks for a preschooler. There are many ways to divide the amounts recommended from each food group into daily meals and snacks.

Click on either pattern to see examples of food choices for meals and snacks.

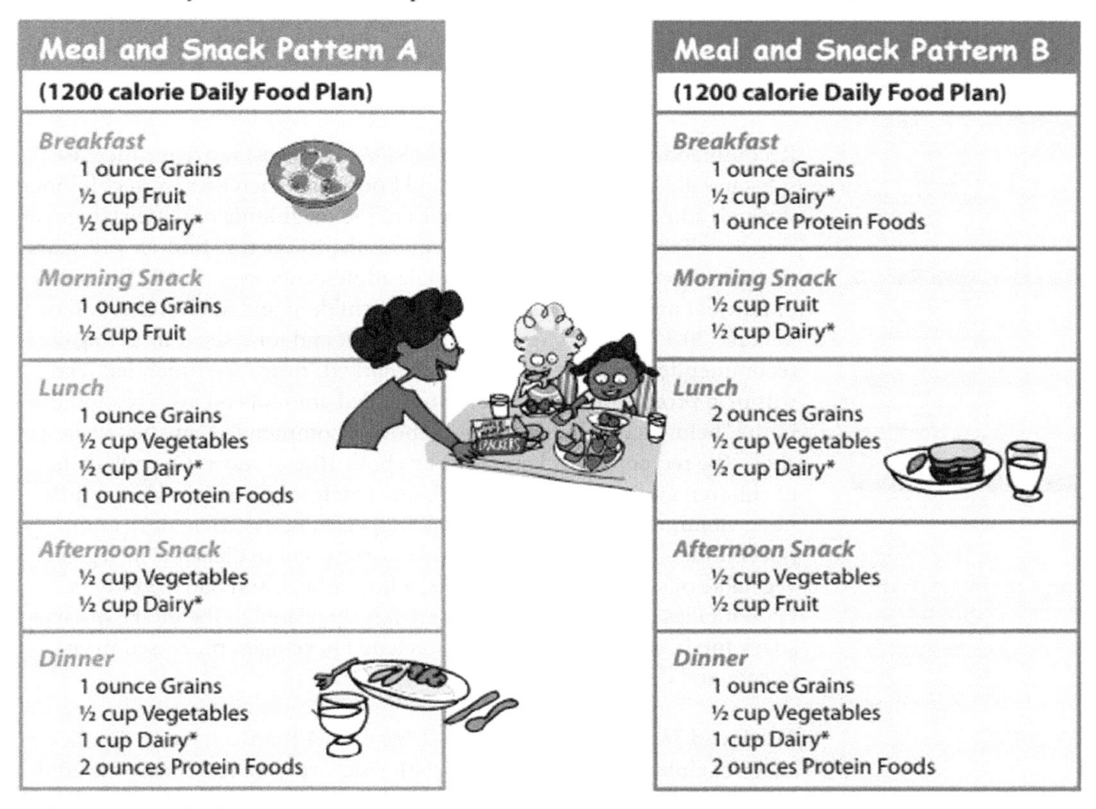

Meal and Snack Pattern A
(1200 calorie Daily Food Plan)
Breakfast 1 ounce Grains ½ cup Fruit ½ cup Dairy*
Morning Snack 1 ounce Grains ½ cup Fruit
Lunch 1 ounce Grains ½ cup Vegetables ½ cup Dairy* 1 ounce Protein Foods
Afternoon Snack ½ cup Vegetables ½ cup Dairy*
Dinner 1 ounce Grains ½ cup Vegetables 1 cup Dairy* 2 ounces Protein Foods

Meal and Snack Pattern B
(1200 calorie Daily Food Plan)
Breakfast 1 ounce Grains ½ cup Dairy* 1 ounce Protein Foods
Morning Snack ½ cup Fruit ½ cup Dairy*
Lunch 2 ounces Grains ½ cup Vegetables ½ cup Dairy*
Afternoon Snack ½ cup Vegetables ½ cup Fruit
Dinner 1 ounce Grains ½ cup Vegetables 1 cup Dairy* 2 ounces Protein Foods

*Offer your child fat-free or low-fat milk, yogurt, and cheese.

Daily Food Plan (1200 calories)	*Total amount for the day*
Grain Group	4 ounces
Vegetable Group	1½ cups
Fruit Group	1 cup
Dairy* Group	2½ cups
Protein Foods Group	3 ounces

Choosemyplate

Illustration 30.7 Portion sizes of foods and beverages for two meal and snack patterns generated by the previous version of MyPlate.gov for a 4-year-old who needs 1,200 calories a day.

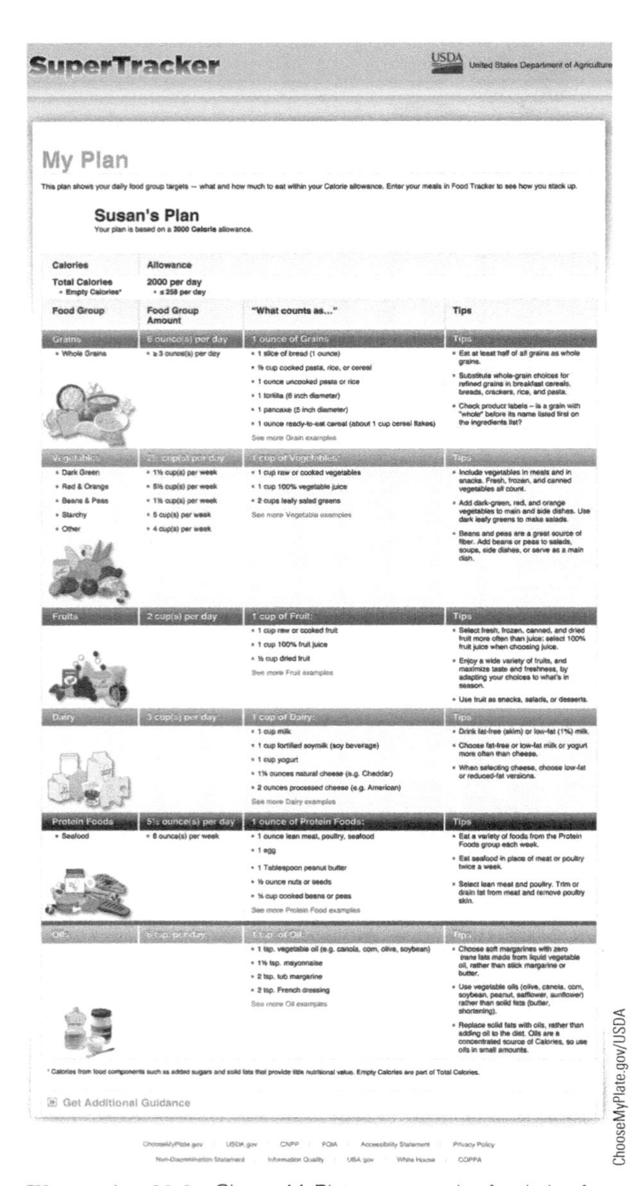

Illustration 30.8 ChooseMyPlate.gov one-day food plan for 13-year-old Susan, who is physically active for 30–60 minutes a day, weighs 115 pounds, and is 5 foot, 3 inches tall.

Diets of children and adolescents rarely match the ChooseMyPlate recommendations. "Sometimes" foods such as chips, fries, candy, desserts, and soft drinks can be consumed as "regular" foods, and diets of children are often high in energy-dense foods rather than foods like vegetables, fruits, lean meats, and whole-grain products that are nutrient dense.[14] Regular intake of energy-dense and high-sugar foods and beverages is associated with the development of overweight and obesity in children and teens, and to the beginning of metabolic abnormalities that contribute to the risk of diseases such as type 2 diabetes, hypertension, and heart disease.[16]

Nutrient Needs of Children and Adolescents

Recommended Dietary Allowances (RDAs) for most nutrients increase substantially as growth continues and body mass increases from childhood through adolescence. Some examples of the magnitude of these changes are shown in Table 30.3. This table compares the RDAs for children 4–8 years of age to those set for male and female adolescents ages 14–18. The status of nutrient and total sugar intake in U.S. children and adolescents is shown in Table 30.4. These tables compare nutrient and total sugar intake against recommended levels of intake. Those marked "near recommended" are within approximately 10% of recommended amounts on average, and those in the "below recommended" and "above recommended" are below or above the recommended amounts by about 10% or more. Overall, diets of children and adolescents in the United States tend to provide too little fiber, vitamin D, vitamin E, calcium, magnesium, iron (females), and potassium.[17] These low levels of intake correspond to diets low in vegetables, vegetable oils, nuts and seeds, fruits, whole grains, and dairy products. High intakes of sugar and sodium are largely related to the intake of savory snack foods such as chips, sugar-sweetened beverages, processed meats, sweets, and desserts.[3]

Diet and Health Status of Children and Adolescents The health status of children and teens, and health risks experienced during the adult years, are affected by dietary intake. Inadequate fluoride intake from water and excess consumption of sugary foods places children and teens at high risk of developing dental decay and related dental health problems.[18,19] Importantly, over-consumption of energy-dense and sugary foods and low levels of physical activity place children and adolescents at increased risk of becoming obese and developing disorders related to obesity later in life.[19]

Snacks are an important source of calories and nutrients in the diets of children and adolescents and should contribute to a healthy dietary pattern. Good choices for snack foods include the following:

Yogurt	Bananas	Carrots
Cheese	Oranges	Cucumbers
Milk	Apples	Popcorn
Nuts, seeds	Dried fruit	Peanuts
Pears	Mangos	Cherry tomatoes
Melons	Grapes	Peanut butter

Obesity and Type 2 Diabetes in Children and Adolescents Rates of obesity in U.S. children and adolescents have risen remarkably since 1980 (Illustration 30.9), and overweight- and obesity-related disorders have also increased. Conditions such as

Table 30.3 RDAs for selected nutrients for children (4–8 years) and adolescents (14–18 years)

Nutrient	RDA		
		Adolescents	
	Children	Males	Females
Protein, g	19	52	46
Vitamin A, mcg	400	900	700
Vitamin B_6, mg	0.6	1.3	1.2
Folate, mcg	200	400	400
Vitamin C, mg	25	75	65
Vitamin D, mcg	15	15	15
Vitamin E, mg	7	15	15
Calcium, mg	1,000	1,300	1,300
Magnesium, mg	130	410	360
Iron, mg	10	11	15
Zinc, mg	5	11	9
Potassium, mg[a]	3,800	4,700	4,700
Sodium, mg[a]	1,200	1,500	1,500

[a]Adequate Intake (AI)

Table 30.4 Average nutrient and total sugar intake compared to recommended levels of intake among children and adolescents in the United States ages 2 to 12[17]

Food constituent	Near recommended	Below recommended	Above recommended
Protein			√
Fat	√		
Total sugar			√
Fiber		√	
Vitamin A	√		
Vitamin B_6			√
Folate			√
Vitamin C			√
Vitamin D		√	
Vitamin E		√	
Calcium		√	
Magnesium		√	
Iron		√ (females)	√ (males)
Zinc	√		
Potassium		√	
Sodium			√

type 2 diabetes, bone and joint disorders, abnormal blood lipid levels, and elevated blood pressure that were very rarely observed in children and adolescents in the past are being diagnosed with increasing frequency.[19] The emergence of type 2 diabetes as a problem of the early years is of particular concern because diabetes generally worsens with time and causes long-term health impairments.

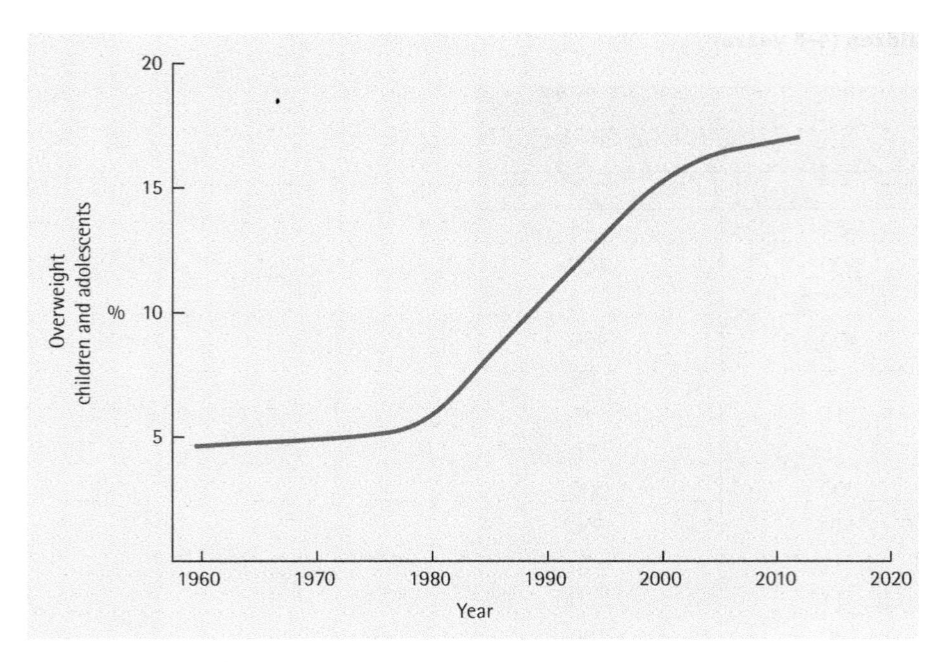

Illustration 30.9 Changes in the percent of obese children and adolescents in the United States.

Source: Developed by Judith Brown using data from Chartbook/Health, United States, 2008, and the CDC.[20]

Recent evidence indicated that rates of childhood and adolescent overweight and obesity are increasing in the United States. In 2015–16, the prevalence of obesity in youth aged 2 to 19 years was 18.5 percent.[6,21]

Causes of Overweight in Young People The development of obesity is related to both environmental factors and genetic susceptibility to gaining weight. Multiple inherited and acquired gene variants influence obesity risk by promoting inflammation, insulin resistance, and appetite. Individuals with these genetic susceptibilities tend to gain body fat when exposed to a diet loaded with energy-dense, highly processed, and sweetened foods and beverages. It is estimated that 40 to 70% of the risk for obesity stems from the presence of specific gene variants.[22–24] Compared to 20 or so years ago, children and adolescents now have fewer opportunities for physical activity and are exposed to a generally plentiful supply of energy-dense and sugary foods.[45] Most schools offer too few physical activity classes and opportunities for exercise.[46] Vending machines in many school districts are loaded with energy-dense, empty-calorie snack foods, and portion sizes of foods served to children in fast food and other restaurants are often excessive.[29,30] Many communities lack bike paths or lanes, sidewalks, and safe playgrounds that invite physical activity.[47]

Prevention of Overweight Increased opportunities for physical activity and the availability and consumption of nutrient-dense foods rather than highly processed snack foods and sweetened beverages are key to the prevention of obesity.[19,31,32] School-based gardening projects have been found to encourage healthier eating in children. Improving teachers' and parents' understanding of what constitutes a healthy dietary pattern and the "how to" of implementing one for children is also helpful.[31,32] Reducing the incidence of gestational diabetes, a disorder that can lead to the development of gene variants that favor obesity development in children, would help prevent childhood obesity.[33]

School-Based Programs for Obesity Prevention High rates of obesity, and increases in obesity-related diseases and disorders in school-aged children, have prompted federal and state governments to become involved in school-based obesity prevention programs. Nearly all schools in the United States are now involved in the delivery of nutrition education to parents, children, and teachers and in the inclusion in school meals of foods and beverages, snacks, and events that correspond to healthy dietary patterns.

New standards for foods served and sold in schools are aimed at reducing the availability of calorie-dense and high-sugar, salt, and saturated fat foods in school environments, and center on providing:

- Breads and other grain products that contain over 50% whole grains
- Meats such as poultry and fish, and meat substitutes such as tofu and veggie burgers
- A wide variety of vegetables and fruits
- Fat-free chocolate or 1% milk to children over the age of 2 years
- A source of drinking water at all times
- Foods and beverages low in sugar (less than 22 grams per 8 oz), sodium, and saturated fat[34]

Table 30.5 **Physical activity guidelines for children and adolescents**[36]

Activity	Duration	Examples
Moderate to vigorous aerobic exercise	60 minutes daily	Running, hopping, skipping, rope jumping, line dancing, cheerleading
Strength-building exercise	Included in the 60 minutes daily • 3 days a week	Tree climbing, playing on playground equipment, hill climbing, weight lifting
Bone strengthening	Included in the 60 minutes daily • 3 days a week	Tennis, basketball, rope jumping, skiing, push-ups, soccer

Syracuse Newspapers/Li-Hau Lan/The Image Works

AP Images/Douglas Healey

Instead of pizza or ice cream parties, schools are encouraged to offer board or card game time, or other fun activities. Rewards offered for good behavior or work should consist of prizes such as crayons, jump ropes, lunch with a teacher, stickers, or the chance to be the leader in the school lunch line. These new standards conform to the Dietary Guidelines for Americans.[35]

Physical Activity Guidelines for Children and Adolescents Regular exercise during childhood and adolescence helps prevent overweight and improves a child's chances of having a healthy adulthood. How much is enough? Recommendations for healthy levels of physical activity for children and adolescents (Table 30.5) call for 60 minutes or more of moderate- to vigorous-intensity aerobic activity daily. The 60 minutes daily should include activities that build muscle, bone strength, and bone density.[36,37] Although exercises that increase strength are recommended, power lifting and body building are not. These intense, high-impact exercises should not be undertaken before physical and skeletal maturity is reached.[36]

Helping Children Learn the Right Lessons about Food

When I was a teenager and babysat for the neighbor's kids, the parents would leave a frozen pizza for dinner. I'd always have to find a fruit or vegetable to serve with it. Just pizza for dinner didn't seem right. I never had a dinner at home with just pizza.

This real response from a teenager relays an important message. Food choices and preferences are a highly individual matter and in large part are determined by family, educational, and environmental experiences. Parents and caretakers may take an overly relaxed attitude toward poor food habits and inadvertently contribute to the development of health problems later in life. Informed parents should decide what types of food their child should be offered, but the decision about how much to eat should be left to the child. Food likes and dislikes and dietary habits appear to be shaped largely by the environment in which children learn about food. Which foods are offered, the way they are offered, and how frequently particular foods are offered all influence children's diets and long-term health outcomes. Children may need to get used to a new food before they trust it. Parents may have to offer a new food on five or more occasions before the child makes a decision.[3,38–41]

About one-third of children refuse to taste a new food, even given repeated attempts to get them to try it. For these children, combining a novel food with a liked food, such as cheese sauce on broccoli or cauliflower offered with a dip, may increase the chances they will taste and eventually like the food. Calling a kale smoothie a "green monster," fresh carrots "X-ray vision carrots," or cauliflower "power flowers" can encourage some children to try, and eventually like, novel vegetables.[42] Engaging children in gardening and the preparation of meals and snacks from vegetables and fruits they help grow can encourage healthier eating behaviors.[43]

Forcing a young child to eat foods she or he does not like, to "clean their plates," or totally restricting access to favorite foods can have lifelong negative effects on food preferences and health. When attempts to get a child to eat a particular food turn the dinner table into a battleground for control, nobody wins. Foods should be offered in an objective, nonthreatening way so that the child has a fair chance to try the food and make a decision about it. Restricting access to or prohibiting intake of children's favorite junk foods tends to strengthen their interest in the foods and consumption of those foods when they get a chance. Such prohibitions have the opposite effect of that intended because they make children want the foods even more.[3,41]

Healthy eating and physical activity levels during childhood and adolescence are clearly important to long-term health and quality of life. Improvements in both areas are needed. Families, schools, and communities—as well as children and adolescents themselves—can do many things to help improve eating and physical activity patterns. The rewards of successful efforts would be for life.

Jose Luis Pelaez Inc/Blend Images/Getty Images

NUTRITION up close

Obesity Prevention Close to Home

Focal Point: Tips for parents on preventing obesity in their children.

Debra and Dale have a 1-year-old daughter, and their second adopted infant will arrive in a few weeks. Both are acutely aware of the growing problem of obesity and type 2 diabetes in children and don't want their children to become obese.

Referring back to information presented in this unit, list four actions Debra and Dale can take to help their children develop healthful food and activity habits.

1. ______
2. ______
3. ______
4. ______

Feedback to Nutrition Up Close is located in Appendix F.

REVIEW QUESTIONS

- **Describe connections between growth and development, and eating behavior in children and adolescents.**

1. Children generally become good "self-feeders" between the ages of 4 to 5 years. **True/False**
2. As shown in Illustration 30.2, the rate of gain in height increases sharply between the ages of 10 to 12 years in girls and between 13 to 14 years in boys. **True/False**
3. Food likes and dislikes are strongly influenced by the environment in which children learn about food. **True/False**
4. Young children instinctively know what to eat. **True/False**
5. According to ChooseMyPlate food intake guidance, a 13-year-old girl should consume 3 cups of vegetables and 2.5 cups of fruits daily. **True/False**

The next two questions refer to the following scenario.

Deena and Britt are big football fans and want Greg, their 10-year-old, to grow up big and strong and play professional football. They figure that if they get him to eat enough he will grow tall and the extra calories will help put weight on his frame. At dinner, they encourage him to eat all the meat he can fit in his stomach.

6. _____ Will Greg grow taller if he eats a lot of meat?
   a. Maybe fatter, but overeating meat won't help him gain height.
   b. Yes, that's how adolescents grow taller and add muscle mass.
   c. If Greg eats more protein than he needs he will grow as tall as genetically possible.
   d. How tall Greg becomes depends on how many calories he consumes during the adolescent years only.
7. _____ What type of diet should Greg be offered to support his growth and physical development?
   a. A diet that will help him gain a pound week.
   b. A diet that includes the recommended assortment and amounts of foods recommended by ChooseMyPlate.gov food guidance materials for 10-year-olds of his activity level.
   c. A diet that is adequate in calories.
   d. A diet that eliminates oils and other sources of fat.

The next three questions refer to the following scenario.

Ender just had his 10th birthday. He weighs 105 pounds (47.7 kg).

8. _____ According to the CDC weight for age chart (Illustration 30.4), Ender's weight for age would be located between the _____ and _____ percentiles.
   a. 95th, 97th
   b. 50th, 95th
   c. 50th, 75th
   d. 25th, 50th

9. _____ Ender's weight for age means that he may be:

   a. underweight for age
   b. normal weight for age
   c. overweight for age
   d. obese for age

10. _____ Ender is 4 feet tall. Using the formula given in Illustration 30.5, Ender's BMI is:

   a. 26.3 kg/m^2
   b. 21.6 kg/m^2
   c. 31.9 kg/m^2
   d. 35.3 kg/m^2

- **Describe five key elements of a healthy dietary pattern for children and adolescents.**

11. Sugar-sweetened beverages increase the risk of obesity in some children. **True/False**
12. Children and adolescents in the United States tend to consume too much sodium and potassium. **True/False**
13. Diets of U.S. children and adolescents tend to provide adequate amounts of fiber and vitamin D. **True/False**
14. Average protein intakes of U.S. children and adolescents are below recommended levels. **True/False**

The next three questions refer to the following scenario and the follow-up scenario listed after question 15.

Your supervisor from work is having a cookout at his home in the country. You're in the kitchen with your supervisor and his two young children, who are having an early dinner. The two children are very active, running and playing most of the day, and appear to be thin. Your supervisor gives them their favorite dessert, which the children eat with relish, run out of the house, and go back to playing outside.

The parents make this dessert for their children often. Here is the recipe:

2 cups warm Rice Krispies marshmallow mix
4 fruity roll-ups
1 package gummy bears

The warm Rice Krispies mix is flattened out, lined with fruity rollups, stuffed with gummy bears, and rolled up.

15. _____ Is this dessert the type of food young children should eat often?

   a. Yes, if the children don't have a weight problem.
   b. Yes, because the Rice Krispies are probably fortified with vitamins.
   c. No, in part because sugary foods are related to dental decay in young children.
   d. No, because there aren't enough calories in the dessert to help the children gain weight.

Four months later you are chatting with your supervisor and he remarks that he's looking for a second job to pay for his children's dental care. He asks you if you know any way to help prevent cavities in children. The water supply where he lives is a well and the water contains no fluoride.

16. _____ What would *not* be a reasonable suggestion to make about preventing dental decay in his children?

   a. Cut down on the sugary foods in the children's diet.
   b. Investigate getting an appropriate fluoride supplement for the children.
   c. Stop offering the children dessert.
   d. Use bottled water fortified with fluoride for the children to drink.

17. _____Which of the following options represent nutrient-dense and tasty dessert options for children?

   a. Caramel-covered apples
   b. Cupcakes from a natural foods market
   c. Vanilla cake with whipped cream
   d. Fruit and yogurt

18. A number of chronic diseases such as heart disease, hypertension, and type 2 diabetes may begin to develop during childhood and adolescence. **True/False**
19. Inadequate intake of calcium and vitamin D during the growing years increases the risk of osteoporosis later in life. **True/False**
20. Overconsumption of energy-dense and sugary foods increases the risk of overweight and obesity during the adult years. **True/False**

Answers to these questions can be found in Appendix F.

NUTRITION SCOREBOARD ANSWERS

1. There really are "growth spurts." **True**
2. Most young children begin to equate the feeling of hunger with the need for food around age 3. **False**
3. The incidence of obesity in U.S. children has decreased over the past 5 years. **False**
4. Childhood and adolescent diets can affect future health.[44] **False**

Nutrition and Health Maintenance for Adults of All Ages

Photodisc

NUTRITION SCOREBOARD

1. Healthy dietary patterns contribute to longevity in part by delaying the age at which chronic disease develops. **True/False**
2. Aging is a kind of disease, and people can do little to preserve their health and quality of life as they age. **True/False**
3. Calorie restriction and a low body mass index increase longevity in humans. **True/False**
4. The need for protein and most vitamins and minerals decreases with age during the adult years. **True/False**

Answers can be found at the end of the unit.

LEARNING OBJECTIVES

After completing Unit 31 you will be able to:

- Identify three examples of interactions among diet, chronic disease development, and longevity.
- Explain why the need for calories and some nutrients changes with aging.

adult Generally defined as people 18 to 65 years old.

older adult Generally defined as individuals 65 or more years old.

life expectancy The average length of life of people of a given age or from birth.

Nutrition for Adults of All Ages

- **Identify three examples of interactions among diet, chronic disease development, and longevity.**

Eating right during the **adult** and **older adult** years is a wise practice. Healthy diets, exercise, and normal weight status help adults feel healthy and vigorous as they age and improve their overall sense of well-being. Adults who adhere to a healthy dietary pattern tend to develop heart disease, cancer, hypertension, and diabetes at older ages and have more life in their years and years of life than adults who do not.[7,8]

Aging is a normal process, not a disease. Although the incidence of many diseases increases with age, the causes of the diseases can be unrelated to normal aging processes. They may fully or partially result from the cumulative effects of diets low in vegetables, fruits, dairy products, and whole grains and of overeating and obesity, smoking, physical inactivity, excessive stress, or other habits that insidiously influence health on a day-to-day basis.[8,9] Aging cannot be prevented, but how healthy we are during aging can be influenced by what we do to our bodies.

Maintaining health as we age is becoming an increasingly important concern. There are more older adults in the United States than ever before, and the numbers are growing.[10]

KEY NUTRITION CONCEPTS

Information about nutrition during the adult years presented in this unit most directly relates to the key nutrition concepts of:

1. Nutrition Concept #2: Foods provide energy (calories), nutrients, and other substances needed for growth and health.
2. Nutrition Concept #3: Health problems related to nutrition originate within cells.
3. Nutrition Concept #4: Poor nutrition can influence the development of certain chronic diseases.

The Age Wave

The proportion of people in the United States aged 65 years and older is steadily increasing (Illustration 31.1). By the year 2050, approximately 20% of the U.S. population will be 65 years of age or older. Since 1900, **life expectancy** at birth has increased by over 40 years, from 47.3 to 78.6 years. Beginning in 2015, however, average life expectancy began to reverse the previous trend and declined.[11,12]

Tomas Rodriguez/Fancy/Corbis

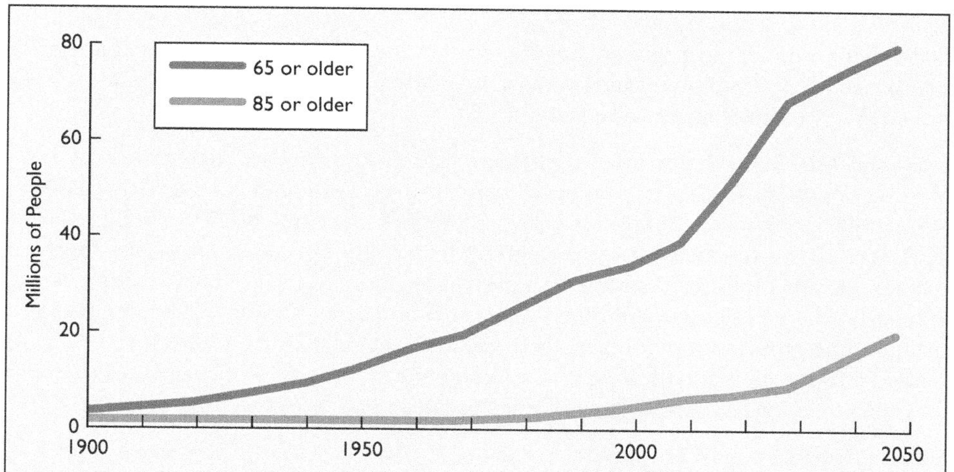

Note: Data for the years 2000 to 2050 are middle-series projections of the population.
Reference population: These data refer to the resident population.
Source: U.S. Census Bureau, *Decennial Census Data and Population Projections*, 2000.

Illustration 31.1 Expected growth in the U.S. population of people aged 65 and 85 years and older.[15,16]

Not all population groups in the United States have the same average life expectancy from birth. Life expectancy varies a good deal by sex and race. Table 31.1 shows overall life expectancy for Caucasian, African American, and Hispanic American men and women over time. African American men have the shortest life expectancy, and Hispanic women have the longest. As in most other countries, life expectancy of females exceeds that of males.[11,12] Reasons for the greater average life expectancy in women compared to men are not yet fully understood. It appears that behaviors in men that accelerate disease progression (especially heart disease) are partly related to the four- to six-year shorter average life span in men. Female hormones do not appear to account for differences in life expectancy.[12,13]

Factors That Influence Life Expectancy Before World War II, most of the gains in life expectancy in the United States were due to public health advances such as improved diet and nutrition, safe water and milk supplies, sanitation measures, better control of infectious diseases, and improved housing. After World War II, new drugs and the development of effective medical devices and procedures accounted for most of the gains in life expectancy. Advances in medicine currently account for 80% of the increase in life expectancy of people aged 65 years and older.[6] Reductions in infant mortality also increase a population's life expectancy.[17] In 1900, 1 out of 10 newborns died in the first year of life, but now less than 1 newborn in 100 does.

Table 31.1 Average length of life for males and females by population group in the United States in 1900, 1960, and 2012[18,19]

	Life expectancy (years)		
	1900	1960	2012
Caucasian Americans			
Males	46.6	67.4	76.4
Females	48.7	74.1	81.1
African Americans			
Males	32.5	61.1	71.4
Females	33.5	63.3	77.7
Hispanic Americans			
Males	—	—	78.5
Females	—	—	83.8

Due to the high and rising costs of medical care, innovations in health and medical care that promote health and quality of life at lower cost than is the current case are urgently needed.[20] The United States ranks 42nd in the world in life expectancy, yet spends more on medical care than any other country.[21]

Genes and Life Expectancy Many people gauge how long they will likely live by looking at how long their parents and grandparents lived. Although not a totally reliable method, it is true that genetic characteristics influence longevity. Children of long-lived parents tend to live somewhat longer as a group than children whose parents had shorter life spans. This relationship strengthens as parents live beyond the age of 80.[22] It appears that long-lived people tend to inherit types of genes and gene variants that confer disease resistance. Life expectancy of individuals from families that are not long-lived can be extended most by diets and lifestyles that promote the maintenance of health. Longevity is also related to living life with happiness, enjoyment, and a sense of well-being.[2,22,23]

Calorie Restriction and Longevity For decades it has been known that some species of animals fed low-calorie diets have increased life expectancies. Why this happens isn't clear, but the proposed reasons relate to reductions in calorie need, modifications in nutrient utilization, reduced metabolic rate, and decelerated aging processes. Human studies have not shown that calorie restriction increases longevity.[3,4]

Individuals who follow low-calorie diets during the adult years report being hungry most of the time, experience decreased libido, and feel cold. Adults aged 20–65 years who tend to live the longest have normal weights for height, whereas people over the age of 65 years are more likely to live the longest if they have weights for height within the overweight body mass index (BMI) range of 26–29 kg/m^2.[4,24] This unit's Reality Check expands on the topic of weight status and life expectancy.

Diet and Health Promotion for Adults of All Ages

- **Explain why the need for calories and some nutrients changes with aging.**

Researchers and others interested in nutrition and health during middle age and beyond have tended to focus their attention on the cumulative effects of diet on chronic disease. Nutrition exerts its effects on chronic disease development over time, and therefore diseases related to poor dietary patterns are most likely to express themselves during the adult years. The occurrence of diseases related to behavioral traits such as smoking and physical inactivity also increases among adults as they age.[9,25] In addition to the health effects of behaviors, people age biologically.[26]

The combined effects of poor diets, other risky lifestyle behaviors, and biological aging increase the rates of serious illness during adulthood. How soon a disease develops depends largely on the intensity and duration of exposure to behavioral risks that contribute to disease development.

Nutrient Needs

Calorie and nutrient needs of adults and elderly are affected by hormonal changes that occur with **menopause** in women and **andropause** in men. Symptoms related to these normal processes begin to appear after the age of 40 and include decreased energy levels, loss of lean muscle mass, and redistribution of body fat (Table 31.2).[48] Effects of these changes on health vary among individuals, and can be influenced by healthy lifestyles. However, people who remain physically active into their older years maintain more muscle mass, experience less muscle and bone pain, and gain less body fat than people who are inactive.[27] Reduction in calorie need with age increases the importance of nutrient-dense diets.

menopause Cessation of the menstrual cycle and reproductive capacity in females.

andropause The time during aging when testosterone levels as well as muscle and bone mass decline.

Table 31.2 **Examples of biological changes during aging and nutritional consequences**[29–32]

Biological change	Nutritional consequences
• Lowered stomach acidity	• Decreased absorption of vitamin B_{12}
• Decreased lean muscle mass	• Increased protein need
• Redistribution of fat stores from the limbs to the trunk	• Decreased calorie need
• Decreased sensation of thirst	• Dehydration risk

Nutrient needs of older adults can be affected by physiological changes that may develop with increasing age (Table 31.3). Decreased stomach acidity, for example, reduces absorption of vitamin B_{12}.

Nutrients of Public Health Concern among Adults and the Elderly Among adults and the elderly in the United States, average dietary intake of vitamins A, D, and E, calcium, potassium, and dietary fiber are below the recommended intake levels, and intakes of sodium are substantially higher than recommended. In addition, iron intake tends to be low in females aged 19–50 years.[33] Inadequate intake of these nutrients can largely be traced back to dietary patterns that contain low amounts of vegetables, fruits, whole grains, legumes, lean meats, and milk products and to high intake of salty snacks and other processed foods. Low levels of intake of fortified foods such as some types of breakfast cereals and fortified soy or almond milk, also contribute to low nutrient intake. Absent a diagnosed deficiency, foods rather than vitamin and mineral supplements are recommended for individuals as they age.[34]

Dietary Recommendations for Adults of All Ages

For the most part, the development of chronic disease in middle-aged and older adults can be viewed as a chain that represents the accumulation over time of problems that impair the functions of cells. Each link that is added to the chain, or each additional insult to cellular function, increases the risk that a chronic disease will develop. The presence of a disease indicates that the chain has gotten too long—that the accumulation of problems is sufficient to interfere noticeably with the normal functions of cells and tissues.[14,35]

Normal cell functions and health promotion are facilitated by healthful dietary and other behaviors. For example:

- Correcting obesity and stabilizing weight during the adult years tends to lengthen life expectancy.
- Entering aging with good levels of muscle mass and maintaining muscular mass and strength throughout the adult years enhance mobility, flexibility, and independence.

REALITY CHECK

Do People Live Longer if They Are Underweight?

While golfing, old friends Evelyn and Dorothy discuss the news story they had read in the morning paper about underweight and increased longevity in laboratory animals. As a result of the news story, Evelyn has become convinced that if she goes on a low-calorie diet, loses 30 pounds, and becomes and stays underweight she will live longer. Dorothy wants to live long, too, but research on laboratory animals isn't enough to convince her to go on a lifelong diet.

Who gets the thumbs-up?

Answers appear on page 31-6.

Purestock/Getty Images

Evelyn: It's worth a try. It's probably true if it happens in animals.

iStockphoto.com/Yuri

Dorothy: Hmmm . . . I'm thinking results of some animal studies are not enough to convince me it's true for humans.

Table 31.3 RDAs for selected nutrients for adults 19–50 and over 70 years old[41]

Nutrient	RDA			
	Adults 19–50		Adults >70	
	Male	Female	Male	Female
Protein, g	56	46	56	46
Vitamin A, mcg	900	700	900	700
Vitamin B_6, mg	1.3	1.3	1.7	1.5
Vitamin B_{12}, mcg	2.4	2.4	2.4	2.4
Folate, mcg	400	400	400	400
Vitamin C, mg	90	75	90	75
Vitamin D, mcg	15	15	20	20
Vitamin E, mg	15	15	15	15
Calcium, mg	1,000	1,000	1,200	1,200
Iron, mg	8	18	8	8
Zinc, mg	11	8	11	8
Potassium, mg[a]	4,700	4,700	4,700	4,700
Sodium, mg[a]	1,500	1,500	1,200	1,200

[a]Adequate Intake (AI)

- Dietary intake patterns that correspond to the Dietary Guidelines for Americans are related to longer life expectancy.
- Maintaining adequate calcium, vitamin D, and protein intake and engaging in regular physical activity during the adult years may prevent or postpone the development of osteoporosis and help maintain muscle mass and strength.
- Above average intake of vegetables, fruits, and whole-grain foods may delay the development or help prevent a number of types of cancer, heart disease, hypertension, and cataracts.[36–38]

The health status of adults is not necessarily "fixed" by age; it can change for the better or the worse, or not much at all.[39]

ChooseMyPlate.gov Food Intake Guidance Diets that consist of foods recommended by ChooseMyPlate are nutrient dense and low in saturated fat. They provide healthy amounts of calories, protein, vitamins, minerals, fiber, and beneficial phytonutrients. Illustration 31.2 shows an example of a one-day diet plan generated from the previous version of ChooseMyPlate.gov materials for a 56-year-old male. Diet plans generated are based on age, sex, weight, height, and usual physical activity level.

Foods identified by the ChooseMyPlate plan can be used to guide food selection. Many different types of foods in the grain, dairy, vegetable, fruit, and protein groups can be selected to meet individual food preferences.

ANSWERS TO REALITY CHECK

Do People Live Longer if They Are Underweight?

Dorothy's thinking is on the right track. Research results in laboratory animals on low-calorie diets and longevity are inconclusive and may not apply to humans.[15] Recent evidence indicates that adults 65 years of age and older who have a BMI in the overweight range may live the longest.[4]

Purestock/Getty Images

Evelyn:

iStockphoto.com/Yuri

Dorothy:

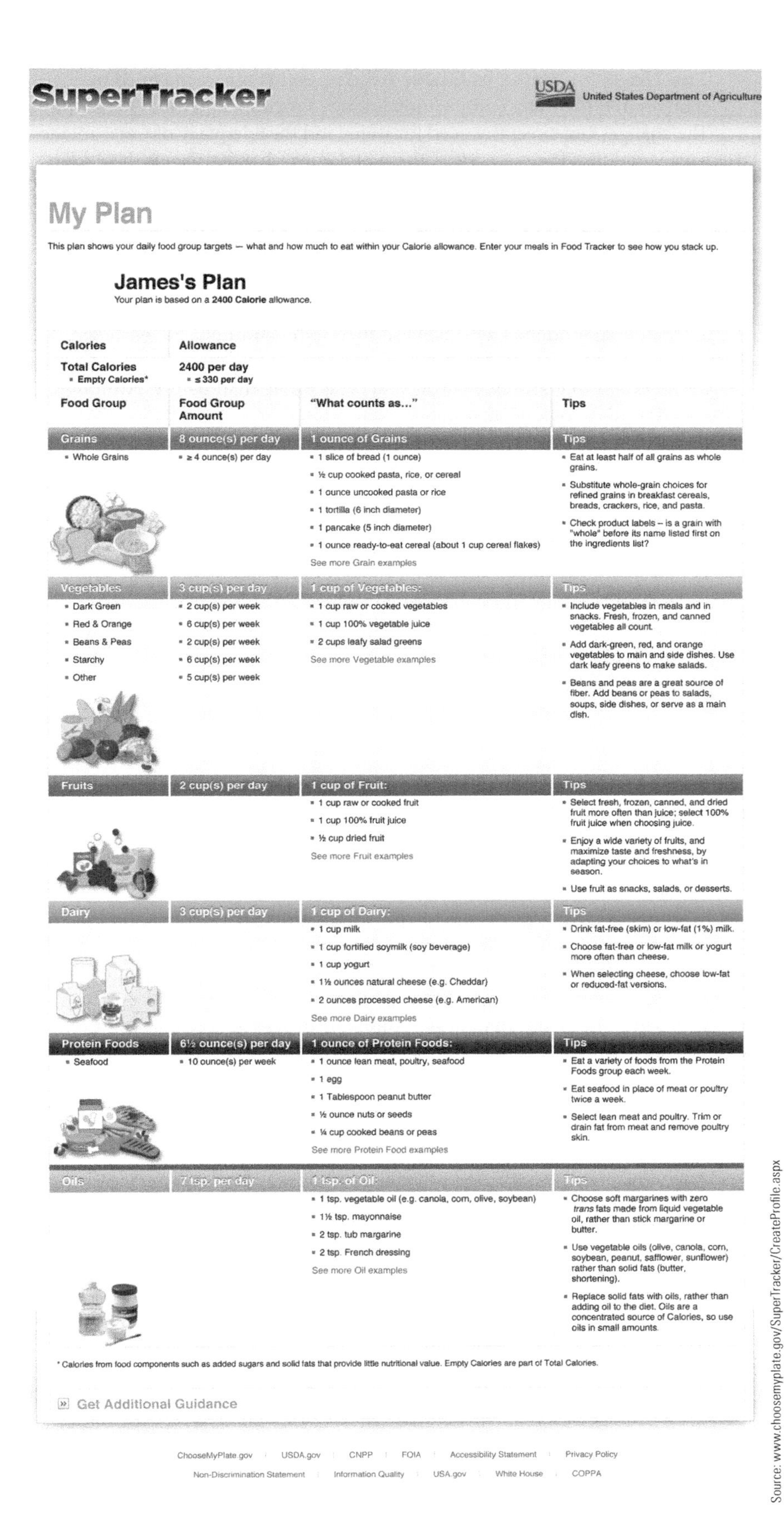

SuperTracker

USDA United States Department of Agriculture

My Plan

This plan shows your daily food group targets — what and how much to eat within your Calorie allowance. Enter your meals in Food Tracker to see how you stack up.

James's Plan

Your plan is based on a **2400 Calorie** allowance.

Calories	Allowance
Total Calories	**2400 per day**
• Empty Calories*	• ≤ 330 per day

Food Group	Food Group Amount	"What counts as..."	Tips
Grains	**8 ounce(s) per day**	**1 ounce of Grains**	**Tips**
• Whole Grains	• ≥ 4 ounce(s) per day	• 1 slice of bread (1 ounce) • ½ cup cooked pasta, rice, or cereal • 1 ounce uncooked pasta or rice • 1 tortilla (6 inch diameter) • 1 pancake (5 inch diameter) • 1 ounce ready-to-eat cereal (about 1 cup cereal flakes) See more Grain examples	• Eat at least half of all grains as whole grains. • Substitute whole-grain choices for refined grains in breakfast cereals, breads, crackers, rice, and pasta. • Check product labels – is a grain with "whole" before its name listed first on the ingredients list?
Vegetables	**3 cup(s) per day**	**1 cup of Vegetables:**	**Tips**
• Dark Green • Red & Orange • Beans & Peas • Starchy • Other	• 2 cup(s) per week • 6 cup(s) per week • 2 cup(s) per week • 6 cup(s) per week • 5 cup(s) per week	• 1 cup raw or cooked vegetables • 1 cup 100% vegetable juice • 2 cups leafy salad greens See more Vegetable examples	• Include vegetables in meals and in snacks. Fresh, frozen, and canned vegetables all count. • Add dark-green, red, and orange vegetables to main and side dishes. Use dark leafy greens to make salads. • Beans and peas are a great source of fiber. Add beans or peas to salads, soups, side dishes, or serve as a main dish.
Fruits	**2 cup(s) per day**	**1 cup of Fruit:**	**Tips**
		• 1 cup raw or cooked fruit • 1 cup 100% fruit juice • ½ cup dried fruit See more Fruit examples	• Select fresh, frozen, canned, and dried fruit more often than juice; select 100% fruit juice when choosing juice. • Enjoy a wide variety of fruits, and maximize taste and freshness, by adapting your choices to what's in season. • Use fruit as snacks, salads, or desserts.
Dairy	**3 cup(s) per day**	**1 cup of Dairy:**	**Tips**
		• 1 cup milk • 1 cup fortified soymilk (soy beverage) • 1 cup yogurt • 1½ ounces natural cheese (e.g. Cheddar) • 2 ounces processed cheese (e.g. American) See more Dairy examples	• Drink fat-free (skim) or low-fat (1%) milk. • Choose fat-free or low-fat milk or yogurt more often than cheese. • When selecting cheese, choose low-fat or reduced-fat versions.
Protein Foods	**6½ ounce(s) per day**	**1 ounce of Protein Foods:**	**Tips**
• Seafood	• 10 ounce(s) per week	• 1 ounce lean meat, poultry, seafood • 1 egg • 1 Tablespoon peanut butter • ½ ounce nuts or seeds • ¼ cup cooked beans or peas See more Protein Food examples	• Eat a variety of foods from the Protein Foods group each week. • Eat seafood in place of meat or poultry twice a week. • Select lean meat and poultry. Trim or drain fat from meat and remove poultry skin.
Oils	**7 tsp. per day**	**1 tsp. of Oil:**	**Tips**
		• 1 tsp. vegetable oil (e.g. canola, corn, olive, soybean) • 1½ tsp. mayonnaise • 2 tsp. tub margarine • 2 tsp. French dressing See more Oil examples	• Choose soft margarines with zero *trans* fats made from liquid vegetable oil, rather than stick margarine or butter. • Use vegetable oils (olive, canola, corn, soybean, peanut, safflower, sunflower) rather than solid fats (butter, shortening). • Replace solid fats with oils, rather than adding oil to the diet. Oils are a concentrated source of Calories, so use oils in small amounts.

* Calories from food components such as added sugars and solid fats that provide little nutritional value. Empty Calories are part of Total Calories.

» Get Additional Guidance

ChooseMyPlate.gov | USDA.gov | CNPP | FOIA | Accessibility Statement | Privacy Policy
Non-Discrimination Statement | Information Quality | USA.gov | White House | COPPA

Source: www.choosemyplate.gov/SuperTracker/CreateProfile.aspx

Illustration 31.2 A one-day diet plan for James, who is 56 years old, weighs 160 pounds (normal weight), is 5 foot, 10 inches tall, and is physically active 30–60 minutes a day.

Jeff Greenberg/PhotoEdit

Illustration 31.3 Taste is less affected by age than are some other senses.

The Need for Water in Older Adults Aging is accompanied by changes in body water composition, kidney function, and thirst perception. The changes can affect the need for water and the adequacy of water intake in older adults. The presence of these biological changes means that older people may become dehydrated and need medical assistance. The need for water can be met by consumption of tap water, juices, teas, coffee, and other beverages. Women need about 11 cups of water a day from fluids and foods, and men need around 15 cups.[40,41]

Does Taste Change with Age? Poor diets observed in some middle-aged and older adults have been ascribed to "declining taste" with age. Besides being a dreadful thought, it's not true that taste declines with age to the extent that it makes eating less pleasurable than before. Although taste sensitivity does diminish somewhat with age, chances are your favorite foods will taste as good to you in your older years as they did in your youth (Illustration 31.3). Sight, smell, and hearing usually decline with age more than taste does.[42]

The senses of taste and smell are affected by medications such as antibiotics, antihistamines, some lipid-lowering drugs, and cancer treatments; diseases including Alzheimer's disease, cancer, and allergies; and surgeries that affect parts of the brain and nasal passages. These treatments and conditions are more prevalent in older adults than in middle-aged populations and are primarily responsible for pronounced declines in the senses of taste and smell. Changes in these senses increase the likelihood that people will experience food poisoning or consume an inadequate diet due to low food intake.[43]

Physical Activity Recommendations for Adults

Regular physical activity is a core part of a healthy lifestyle that maintains health and delays aging processes. It increases oxygen delivery to organs and tissues, decreases biologic aging, improves cognitive functions, decreases chronic inflammation, and helps lower central body fat.[26,44] Substantial health benefits result from performing moderate-intensity exercise for 150 minutes per week, or from 75 minutes of vigorous-intensity exercise weekly. Resistance exercise performed twice a week or more further enhances the benefits of regular exercise while increasing strength, muscle mass, and balance.[27] Regular physical activity recommendations apply to individuals with physical and cognitive disabilities who are able to exercise.

Some exercise is better for health than none. Adults of all ages are encouraged to do "the best they can" when it comes to regular physical activity.[44] With proper training and spirit, physical fitness can be maintained during aging, or increased to the level of the "over 80" event winners pictured in Illustration 31.4.

Psychological and Social Aspects of Nutrition for Older Adults Preparing meals and eating right may not be as simple as it sounds for many older adults. Consuming an adequate and balanced diet may not be easy when you depend on someone else to take you shopping, when mealtimes involve little social life, or when you "don't feel up" to making a meal. Isolation, loneliness, depression, and poor health can be major contributors to poor diets in older adults.[45] A sense of humor, an active social life, optimism, and an ability to take life as it comes also appear to enhance well-being and longevity.[2]

Peter Muller/Getty Images

Cultura/Frank and Helena/Getty Images

Illustration 31.4 Athletes who are over 80.

Photodisc

NUTRITION up close

Does He Who Laughs, Last?

Focal Point: Critically thinking about factors that influence longevity.

The following is an adaptation of a letter printed in the "Dear Abby" newspaper column:

> Dear Abby,
>
> Since I've reached my 80s, my mail is full of ads for health products to help me live longer.
>
> I once had many friends, all of whom were health vigilantes. They shook their heads knowingly as I avoided all health food fads and exercise. They made liquid out of good vegetables and spent fortunes buying all the latest supplements. They argued that "organic" was better and "natural" was best. I would tell them that snake venom, poison ivy and manure were "natural." But they wouldn't listen and they didn't laugh.
>
> Now my friends are all dead and I have no one left to argue with.

Think critically about the contents of this letter and identify three alternate explanations for why this individual outlived his friends.

Feedback to the Nutrition Up Close is located in Appendix F.

REVIEW QUESTIONS

- **Identify three examples of interactions among diet, chronic disease development, and longevity.**

1. One of the main reasons the United States ranks first among all countries in life expectancy is the high-quality diet consumed by adults in general. **True/False**
2. Three characteristics of dietary patterns related to increased longevity are ample consumption of fruits and vegetables, above-average consumption of whole-grain products, and calorie restriction. **True/False**
3. Obesity is associated with increased life expectancy in elderly people. **True/False**
4. Vitamin D supplements during the adult years may prevent or postpone the development of osteoporosis and help build muscle mass and strength. **True/False**

- **Explain why the need for calories and some nutrients changes with aging.**

5. Elderly persons tend to have a lowered sensation of thirst. **True/False**
6. The RDAs for older adults are the same as those for younger adults. **True/False**
7. Reduction in calorie need with age increases the importance of nutrient-dense diets. **True/False**
8. Decreased stomach acidity in older adults reduces absorption of vitamin B_{12}. **True/False**

The next three questions refer to the following scenario.

Assume your favorite uncle is visiting and explains to you that he is trying to eat a healthy diet now that he has turned 50. This is his new plan for his daily diet:

Breakfast: grits, eggs, English muffin with margarine, coffee
Lunch: Deli pastrami or corned beef sandwich on rye bread, coleslaw, soft drink
Dinner: Frozen dinner that includes meat, a vegetable, and potatoes; coffee
Bedtime snack: cookies or peanuts

9. _____ How many ChooseMyPlate food groups are represented in your uncle's diet plan?
   a. 1
   b. 2
   c. 3
   d. 4

10. _____ Which ChooseMyPlate food groups are missing from his plan?
    a. grains and fruit
    b. protein and dairy
    c. vegetables and fish
    d. fruit and dairy

11. _____ Your uncle asks you if you have any suggestions for changes to his diet plan that would make it healthier. What would be the most appropriate changes to suggest?
    a. Substitute a blueberry muffin for the English muffin and leave off the margarine.
    b. Have a whole-grain cereal with low-fat milk and orange juice for breakfast.
    c. Switch the cookies at bedtime to ice cream and top the ice cream with sliced fruit.
    d. Eat egg whites instead of whole eggs and replace the peanuts with a lower-fat snack like popcorn.

Answers to these questions can be found in Appendix F.

NUTRITION SCOREBOARD ANSWERS

1. Healthy eating pays off.[1] **True**
2. Aging is not a disease! It's normal, and you can normally promote healthy aging through healthy diets and lifestyles.[2] **False**
3. Neither calorie restriction nor low BMI has not been shown to increase longevity in humans.[3,4] **False**
4. **False**[5,6]

The Multiple Dimensions of Food Safety

Garry Adams/Getty Images

NUTRITION SCOREBOARD

1 Most cases of food-borne illness outbreaks can be traced back to eggs. **True/False**

2 Freezing kills bacteria in food. **True/False**

3 Approximately one in six Americans becomes sick due to a food-borne illness each year. **True/False**

4 You can tell if food has gone bad by smelling it. **True/False**

Answers can be found at the end of the unit.

LEARNING OBJECTIVES

After completing Unit 32 you will be able to:

- Discuss the top causes of food-borne illness and the transmission and symptoms of each.
- Describe five consumer practices that help prevent food-borne illnesses.

Threats to the Safety of the Food Supply

- **Discuss the top causes of food-borne illness and the transmission and symptoms of each.**

Chia seed sprouts and chicken contaminated with *Salmonella*, raw milk tainted with *Campylobacter*, leafy green vegetables laced with *Norovirus*—it's enough to give you food fright. Actually, Americans have good reason to be concerned about the safety of some foods (Illustration 32.1). The Centers for Disease Control and Prevention (CDC) estimates that each year in the United States **food-borne illnesses** cause one in six Americans (or 48 million people) to get sick, 128,000 hospitalizations, and 3,000 deaths.[2]

Food-borne illnesses can be caused by bacteria, viruses, algae, fungi, and toxins, as well as by chemical contaminants in foods or water. They are spread by a wide assortment of foods, from contaminated eggs to hummus. Any food can become contaminated and transmit food-borne diseases. Table 32.1 lists the top six foods that were the source of the largest food-borne illness outbreaks during 2015.

Table 32.1 Top six foods associated with food-borne illness in the United States, 2015[1]

Food
Fish
Chicken
Pork
Cucumbers
Tomatoes
Leafy green vegetables

KEY NUTRITION CONCEPTS

Foodborne illness affects health and nutritional status. The following two nutrition concepts relate to the effects on nutritional status.

1. Nutrition Concept #3: Health problems related to nutrition originate within cells.
2. Nutrition Concept #6: Malnutrition can result from poor diets and from disease states, genetic factors, or combinations of these factors.

I visited my friend Joyce at her house. She asked if I would like some chips and salsa. Liking salsa, I replied "yes." On my fourth mouthful of salsa, Joyce asked me: "It lasts a long time in the fridge, right?"

How Good Foods Go Bad

Bacteria and viruses are the most common causes of food-borne illnesses; they enter the food supply mainly during food production, processing, storage, or preparation. They are transferred to humans from contaminated foods and water through many

food-borne illness An illness related to consumption of foods or beverages containing disease-causing bacteria, viruses, parasites, toxic chemicals, or other harmful substances.

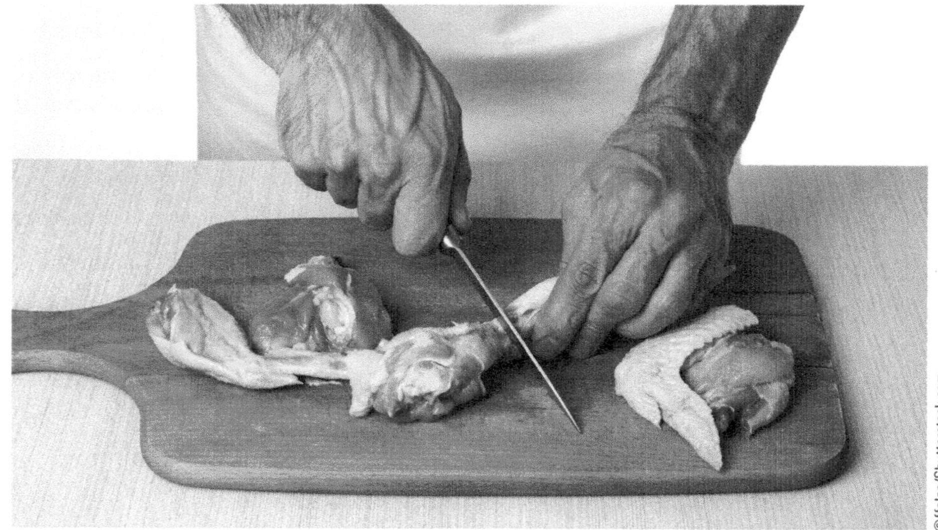

©ffolas/Shutterstock.com

Illustration 32.1 Most raw chicken sold in the United States is contaminated with potentially harmful bacteria.[3]

Dennis MacDonald/PhotoEdit

Illustration 32.2 Inappropriate restaurant food preparation methods are linked to food-borne illnesses.

different routes, a major one being feces.[4] The lower intestines of many healthy farm animals are colonized by bacteria that may be harmful to humans, and soil and water used to grow crops can become contaminated. Bacteria and other substances can contaminate food when unsanitary practices are used to grow or prepare vegetables and fruits, and during the processing and storage of meats and other foods. Once affected, humans can transfer bacterial, viral, and other types of food-borne illnesses to others when unsanitary practices are used to prepare foods (Illustration 32.2). Humans can also transfer certain harmful microorganisms to food when fluids from infected injuries or body secretions contact food.[5]

Bacteria and other contaminants may be present on the inside of plant and animal foods, as well as on food surfaces. Bacteria and other toxins can enter vegetables and fruits, for example, if the protective skin coatings are broken, allowing an entrance for bacteria, or if contaminated soil or water are used to grow crops. *Salmonella* bacteria can infect the ovaries of hens and cause them to lay normal-looking but infected eggs. Shellfish can concentrate microorganisms present in the surrounding water, and although the microbes may be harmless to the shellfish, they can provide large doses of harmful bacteria and viruses to humans.[6]

Cross-Contamination of Foods The source of many cases of food-borne illness is food that has come into contact with other contaminated food. This situation, referred to as **cross-contamination**, increases the reach of food-borne illnesses.

cross-contamination The spread of bacteria, viruses, or other harmful substances from one surface to another.

Bacteria, viruses, toxins, and other harmful substances can contaminate safe foods during production, processing, shipping, preparation, or storage. Opportunities for cross-contamination of foods during processing, for example, are plentiful. The hamburger you eat may contain meat from hundreds of different cows, an omelet in a restaurant may contain the eggs of hundreds of different chickens, and the chicken you baked may have been bathed with hundreds of others when they were all washed in the same vat of water at the meat-processing plant. Cross-contamination occurs on cutting boards in restaurants and in homes across America. The failure to routinely wash cutting boards between the preparation of different raw foods is a major route for the spread of food-borne illness.[6,7]

Table 32.2 High-risk groups for severe effects of food-borne illness[8]

- People with weakened immune systems due to HIV/AIDS, and others with weakened defenses against infections
- People with certain chronic illnesses such as diabetes and cancer
- Pregnant women
- Infants and young children
- Older adults
- Transplant recipients

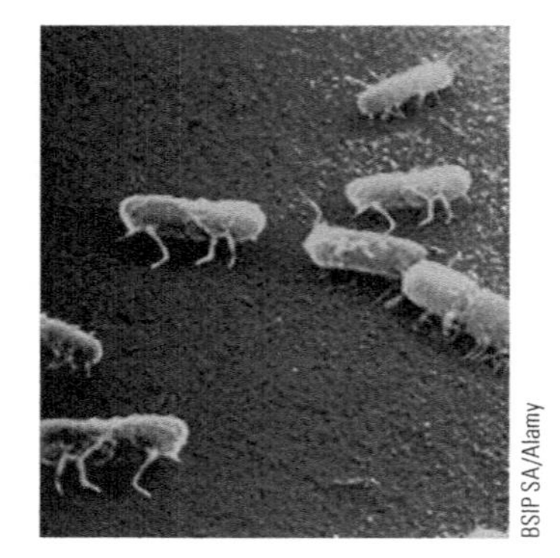

Illustration 32.3 A magnified view of *Salmonella* bacteria.

Causes and Consequences of Food-Borne Illness

More than 250 types of food-borne illnesses caused by infectious agents (bacteria, viruses, and parasites) and noninfectious agents (toxins and chemical contaminants) have been identified. Their impact on health ranges from a day or two of nausea and diarrhea to death within minutes. The most common symptoms of food-borne illness are nausea, vomiting, diarrhea, and fever.[9] Effects of food-borne illnesses are generally most severe in people with weakened immune systems or certain chronic illnesses, pregnant women, young children, and older persons (Table 32.2). Many cases of food-borne illness go undiagnosed and unreported, making it difficult to identify the true incidence and causes of food-borne illness. Estimates of the causes are usually based on confirmed cases reported by health care professionals to local health departments.[10] The following discussion and Table 32.3 summarize facts about five types of food-borne illnesses that result from bacteria or viruses.

Salmonella It is estimated that over 1,027,000 cases of *Salmonella* infection occur in the United States each year.[11] A photograph of *Salmonella* bacteria is presented in Illustration 32.3. An outbreak of *Salmonella* that affected approximately 224,000 people was linked to the transport of an ice cream mixture in a tanker that had previously transported *Salmonella*-infected eggs. The infected ice cream was distributed in 41 states.

Campylobacter This bacteria is a common contaminant on chicken and causes an estimated 845,025 cases of food-borne illness each year.[12] The bacteria have been found in two-thirds of raw chickens, and some of the bacteria represent a strain of *Campylobacter* resistant to antibiotics.[13,15] In addition to poultry, *Campylobacter* infection has been traced to other foods, such as unpasteurized milk and contaminated water.[14]

E. coli Many strains of *E. coli* are harmless, but four have been found to cause illness in humans. *E. coli* 0157:H7 is a relatively new strain of *E. coli* that has evolved into a potential killer. As few as 10 of these bacteria can lead to death by kidney failure in vulnerable people. About 2,000 cases of *E. coli* 0157:H7 occur each year in the United States, but the incidence has begun to decrease.[4] *E. coli* infections are particularly associated with consumption of undercooked ground beef; however, the bacteria can contaminate many types

Table 32.3 Description of selected bacteria-caused food-borne illnesses[4,16]

Bacteria	Onset	Illness duration	Symptoms	Foods most commonly affected	Usual source of contamination
Salmonella	12–74 hours	4–7 days	Diarrhea, abdominal pain, chills, fever, vomiting, dehydration	Uncooked or undercooked eggs, unpasteurized milk, raw meat and poultry, vegetables and fruits	Infected animals, human feces on food, contaminated water
Campylobacter	2–5 days	2–10 days	Diarrhea (may be bloody), abdominal cramps, fever, vomiting	Undercooked poultry, unpasteurized milk	Raw, undercooked poultry and other animals, raw milk and cheese made from it, contaminated water
E. coli 0157:H7	1–8 days	5–10 days	Watery, bloody diarrhea, abdominal cramps, little or no fever	Raw or undercooked beef, unpasteurized milk, raw vegetables and fruits, contaminated water	Infected cattle
Noroviruses (Norwalk-like viruses)	1–2 days	1–3 days	Nausea, explosive vomiting, diarrhea	Undercooked seafood, raw produce	Human feces contamination of oysters and other shellfish beds
Listeria	9–48 hours or longer in invasive cases	Variable	Fever, muscle aches, nausea, vomiting; in pregnant women, may lead to miscarriage or stillbirth delivery	Raw milk and cheeses made from it, cold cuts, hot dogs	Soil and water contaminated with *Listeria*

of food. A major outbreak of *E. coli* 0157:H7 infection occurred in the past when contaminated ground beef left over from one day's production was added to the next day's batch of beef. Use of the leftover, contaminated meat kept recontaminating subsequent batches. Some 25 million pounds of ground beef had to be recalled as a result.[17]

Noroviruses Noroviruses are the most common cause of food-borne illness in the United States, usually resulting in a bout of serious vomiting that resolves within two days.[18] Norovirus is spread primarily by infected kitchen workers and by fishermen who have dumped sewage waste into waters above oyster beds. Norovirus infection can also be spread from person to person. Over 5 million cases of illnesses related to noroviruses are estimated to occur in the United States each year.[4,6]

Listeria This type of bacteria represents a major hazard to pregnant women, older adults, and individuals with compromised immune systems. *Listeria* infection (called listeriosis) during pregnancy can cause miscarriage and stillbirth. Most outbreaks of listeriosis are linked to contaminated soft cheese, but cold cuts, hot dogs, sushi rolls, ice cream, cantaloupe, and other foods have been found to harbor this bacteria. An outbreak of listeriosis related to cantaloupe in Colorado in 2011 sickened 147 people and killed 33, and is considered to be in one of the nation's deadliest food-borne illness outbreaks.[7] Listeria is unlike other germs because it grows during refrigeration. It is killed by cooking and pasteurization.[19]

Other Causes of Food-Borne Illnesses

Of the hundreds of other causes of food-borne illness, seven have been selected for brief review here. They represent examples of food-borne illnesses ranging from mercury toxicity to mad cow disease.

Mercury Contamination Onset of symptoms of mercury poisoning usually occur 72 hours or more after ingestion; they include numbness, weakness of the legs, spastic paralysis, impairment of vision, blindness, and coma. Seafoods have come under fire as a potential source of food-borne illnesses due to mercury contamination of waters and fish by fungicides, fossil fuel exhaust, smelting plants, pulp and paper mills, leather-tanning facilities, and chemical manufacturing plants. High levels of mercury are most likely to be present in large, long-lived fish such as shark and swordfish. Because mercury can interfere with fetal brain development, pregnant women should limit their consumption of fish with high mercury content. Women who are pregnant or may become pregnant are advised to not eat shark, swordfish, king mackerel, or tilefish. Consumption of "safe" fish by pregnant women and adults in general benefits health.[20]

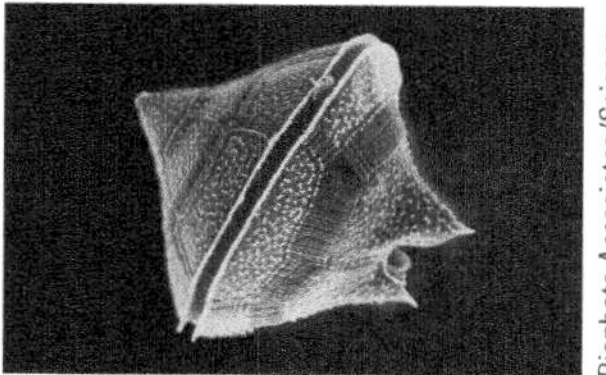

Biophoto Associates/Science Source

Ciguatera Ciguatera poisoning from fish is caused by a neurotoxin (ciguatoxin) present in microorganisms called dinoflagellates that live on reefs (Illustration 32.4). The toxin is transferred through herbivorous reef fish to carnivorous fish and then to humans who eat these fish. Over 200 types of fish may cause ciguatera poisoning, the most common being large groupers and snappers, amberjack, and barracuda. Primary areas affected by ciguatera poisoning include the southwestern and southern waters off Florida, and the Caribbean and South Pacific islands. Symptoms develop within 1 to 6 hours after ingestion of poisoned fish and can cause tingling and numbness, gastroenteritis, dizziness, dry mouth, muscle aches, dilated pupils, blurred vision, and paralysis. The toxin causing the poisoning is not destroyed by cooking, freezing, or digestive enzymes, and there is no effective treatment.[4]

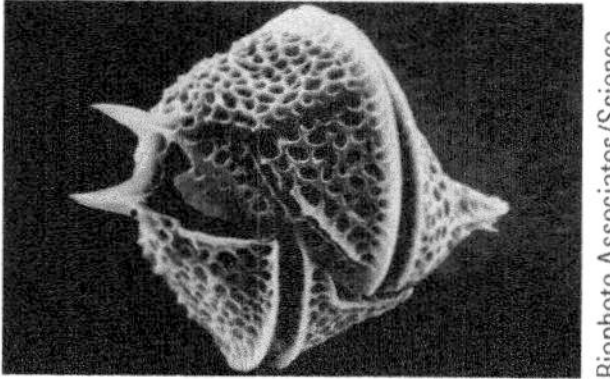

Biophoto Associates/Science Source

Red Tide "Red tide" may occur between June and October on the Pacific and Atlantic coasts. It is due to the accumulation of a microorganism in algae from pollution that produces a nerve toxin. Oysters and other shellfish that consume the microorganism become contaminated, as do humans who ingest the shellfish. Resistant to cooking, the toxin produced causes a burning or prickling sensation in the mouth from 5 to 30 minutes after contaminated shellfish are consumed. This symptom is followed by nausea, vomiting, muscle weakness, and a loss of feeling in the hands and feet. Recovery is usually complete, but it's easy to understand that the warning not to eat mussels, clams, oysters, scallops, or other shellfish from red-tide waters is for real.[21]

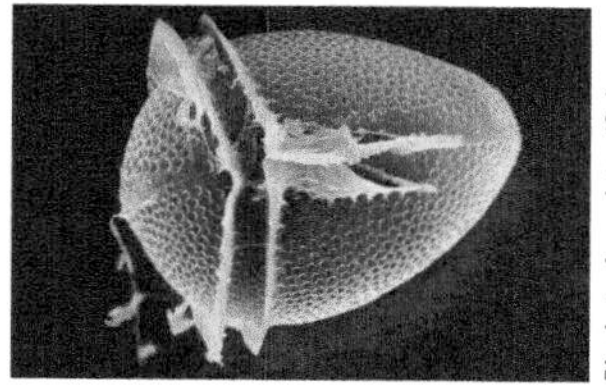

Biophoto Associates/Science Source

Illustration 32.4 Magnified view of dinoflagellates that cause ciguatera poisoning.

Illustration 32.5 Pressure caused by bacterial gases in this can of glucose drink made the can explode.

Botulism *C. botulinum* bacteria produce a toxin that is one of the deadliest known. Botulism infection causes dizziness, double or blurred vision; loss of reflex to light; difficulty in swallowing, speaking, and breathing; dry mouth; weakness; and respiratory paralysis. An antidote is available but must be given soon after the infection occurs. These bacteria are commonly present in soil and in ocean and lake sediment, and they may contaminate crops, honey, animals, and seafood. In humans, food-borne botulism usually results from eating under-heated, contaminated foods stored in airtight containers.[4] The bacteria thrive without oxygen and produce gases as they grow. The gases expand the food container. For a real example of a can that exploded from the inside due to gas production by bacteria (see Illustration 32.5). Foods and fluids in cans, plastic bags and wraps, and other airtight containers that have bulged-out areas should *not* be consumed.

For many years Alaska had the highest rate of botulism, primarily due to traditional, Native Alaskan methods for fermenting raw fish, blubber, and other seafood in airless plastic bags. Statewide public health and clinical educational programs have successfully reduced the rate of botulism among Native Alaskans.[22] In the past 50 years, the proportion of patients with botulism who die has fallen from about 50% to 3–5%.[23]

Parasites Various parasitic worms, such as tapeworms, flatworms, and roundworms, can enter food and water through fecal material and soil. Most parasites are too small for the naked eye to see. Many die a short time after ingestion and are excreted in stools, but some survive and grow in the gastrointestinal tract.[9] Once consumed, roundworms may attach to the lining of the intestine and feed on the person's blood. This can lead to anemia. One type of roundworm can bore a hole through the stomach within an hour after the worm's source—raw fish—is eaten. The severe pain that results sends people immediately to their doctors. Hundreds of Japanese people experience this food-borne illness every year.[24] Many people in the United States harbor parasites but do not develop symptoms because the immune symptoms keeps the ill effects in check. Immuno-compromised individuals are at highest risk of illness due to parasitic infection. Much less is known about the consequences of parasitic infections than most other food-borne diseases, and they are on the CDC's "watch list."[25]

Parasites are generally killed by freezing and always killed by high temperature. It's a benefit that most parasites were destroyed by freezing. One study in Seattle assessed the parasite content of raw fish used in sushi. About 40% of the raw fish samples contained roundworms, but since the fish had been deep frozen, all the worms were dead.[24]

Bisphenol A It is hard to drink from a bottle of water without wondering if the plastic contains bisphenol A (BPA). News about the potential effects of BPA on fetal and infant growth and development are popular topics for the media and a cause of concern among parents and others.[26] BPA is a chemical used primarily in the production of plastic and epoxy resins used in food packaging materials. It is still present in some can liners and food storage containers.[27]

The presence of BPA in human blood samples has been identified, but there is no clear scientific evidence that the presence of BPA in blood is related to health problems. As a precaution and due to consumer concerns, the Food and Drug Administration (FDA) has initiated a policy that prohibits the use of bisphenol A in baby bottles, sippy cups, or in coatings in packaging for infant formula. Based on the FDA's ongoing safety review of scientific evidence, the available information continues to support the safety of BPA for currently approved uses in food containers and packaging.[28] Health Canada recently concluded that current dietary exposure to BPA through food packaging is not expected to pose a health risk to the general population, including newborns and infants. The Canadian policy on BPA states that exposure to BPA should be limited in newborn and infant products as a precautionary measure.[27,29]

Due to lingering concerns about the safety of BPA, especially for pregnant women, infants, and young children, many people prefer not to use containers that contain it.[24] Individuals concerned about BPA can reduce their exposure by following the tips given in Table 32.4. Many plastic containers no longer contain BPA, and you can often identify which are BPA-free by looking at the recycling codes on plastic containers (Illustration 32.6).

Table 32.4 Avoiding BPA in plastic bottles and containers[30]

• Don't microwave food in plastic containers.
• Don't add hot liquids or foods to plastic containers.
• Check recycling codes on plastic bottles and containers (see Illustration 32.6). Plastic containers with the recycling code of 3 or 7 may contain BPA.
• Use glass bottles or containers.
• Consume clean, fresh, or frozen vegetables and fruits.

Mad Cow Disease Technically called **bovine spongiform encephalopathy** (BSE), this rare disease in cattle is suspected of causing at least 150 human deaths in Europe. Four cases of the disease in humans—caused by consumption of **prion**-infected beef—have been diagnosed in the United States, and each originated in another country.[31] The disease appears to have started when cows, which are herbivores by nature, were given sheep intestines and parts of the spinal cord in their feed. Some of the sheep harbored a protein called a prion that caused a deadly disease when consumed by cows. A prion is not a bacterium or other microorganism; it's a small, misfolded protein that can replicate itself and transmit disease when consumed by another species.[32] The only other known food-borne illness that is spread by the consumption of otherwise healthy body parts is kuru, and it is transmitted by cannibalism.[33]

Researchers concluded that mad cow disease was transferred to humans who ate the meat of prion-infected cows. In 1994, mad cow disease was discovered to be a new form of Creutzfeldt-Jakob disease (variant Creutzfeldt-Jakob disease, or VCJD), previously identified in humans. VCJD may take 5 to 50 years to develop. After many years, it inevitably leads to death in humans due to brain damage.[32] Needless to say, it is no longer legal to feed cows animal parts that may transmit the disease. Although the risk of consuming beef from cattle with mad cow disease is extremely small, the possibility that animals may develop this disease exists. Stringent monitoring efforts are in place in the United States, Canada, and Europe, and rates of VCJD are declining.[4]

Antibiotics and Pesticides in Foods Other potential food-borne illnesses of concern relate to antibiotic use in feeds and pesticides applied to crops. Overuse of antibiotics in farm animals (as well as humans) contributes to the emergence, persistence, and spread of antibiotic-resistant bacteria. Given repeated doses of antibiotics, bacteria adapt by modifying specific genes in a way that enables them to "resist" the effects of the antibiotics that are given. These antibiotic-resistant bacteria present in animals can end up in the meat, eggs, and milk they supply, and introduce antibiotic-resistant bacteria in to a person's body. Since some of the antibiotics given to livestock are the same as those used in humans, humans can become infected with antibiotic-resistant bacteria that are very difficult or impossible to treat.[9,34]

Current FDA regulations stipulate that antibiotics be used in livestock only for treating a veterinarian-diagnosed infection and disallows use to prevent infection or facilitate growth.[9] Nonetheless, the use of antibiotics in livestock feed and water for these prohibited uses continues in the United States (Illustration 32.7).[35,36] This situation makes it particularly

©koko-tewan/Shutterstock.com

Illustration 32.7 Overuse of antibiotics in livestock feeds increases the development of antibiotic-resistant bacteria in humans.

©Eltoro69/Shutterstock

Illustration 32.6 Recycling codes are generally imprinted near or on the bottom of plastic containers. (Not all plastic containers have recycling codes.) A code of 3 or 7 indicates the plastic may contain BPA.[30]

bovine spongiform encephalopathy (pronounced bow-vine sponge-ah-form en-cef-a-lop-pathy) A fatal neurodegenerative disease caused by a prion that mainly affects cattle. It can affect humans if animal tissue infected with the prior is consumed; that condition is called variant Creutzfeldt-Jakob disease.

prion An infectious, misfolded protein that has the capability of causing normal proteins to become misfolded, thereby producing disease. The resulting diseases are called spongiform encephalopathies.

health action Sleuthing a Food-Borne Illness

This is a true story about an outbreak of a food-borne illness. One day after having enjoyed several meals together at the house, 4 of 7 visiting relatives woke up with painful diarrhea and cramps. Everybody's first thought was it must be a food-borne illness, although the food tasted fine. Expiration dates of all the food in the fridge were okay, and the temperature reading on the thermostat was good: 38°F. What happened?

The mystery was finally solved when a regular thermometer was placed in the fridge and it gave a temperature of 44°F. Replacement of the thermostat fixed the problem. Rather small, unnoticeable increases in the temperature inside a refrigerator can cause food to spoil.

Photodisc

Lee Snider/The Image Works

Illustration 32.8 Protective gear—including a chemical-resistant suit, respirator, and goggles—is required for use with some of the most toxic pesticides but is not always worn.

important to follow food safety rules in restaurants and homes, and highlights the need for increased monitoring of livestock farms for antibiotic use.

Pesticides containing organophosphates, mercury-containing fungicides, and DDT remain causes of food-borne illnesses.[37] Use of DDT on plants for insect control was phased out in the United States over 20 years ago due to links with cancer, but these long-lasting chemicals still contaminate some land, lakes, and streams. Pesticide levels in crops are periodically measured by the FDA.[38] Fewer than 1% of the foods on the market contain an excessive level of pesticide. Agricultural workers who fail to take safety precautions when applying pesticides are most likely to experience the illnesses related to pesticide exposure (Illustration 32.8). Health problems due to pesticides, although not extensively examined in long-term human studies, appear to be rare in other groups of people.[39]

Exposure to antibiotics and synthetic pesticides and fertilizers can be reduced by consuming certified organic foods. Efforts are underway to increase the production and availability, and to decrease the cost, of these foods.[40]

Preventing Food-Borne Illnesses

- **Describe five consumer practices that help prevent food-borne illnesses.**

There are two major approaches to the prevention of food-borne illnesses. The first relies on food safety regulations that control food processing and handling practices, and the second involves consumer behaviors that lessen the risk of consuming contaminated foods. (Read what can happen when one of these principles is violated in this unit's Health Action.)

Food Safety Regulations

According to the federal Food, Drug, and Cosmetic Act, it is illegal to produce or dispense foods that are contaminated with substances that cause illness in humans. Foods are considered "safe" if there is a reasonable certainty that no harm will result from repeated exposure to any substance added to foods. The act governs all substances that are added intentionally or accidentally to foods—except pesticides. According to legislation passed in 1996, pesticides are permitted if their consumption is associated with a "negligible risk" of cancer or other health problems.[41]

Irradiation of Foods Increasing rates of food-borne illness have led some health officials to recommend the expanded use of food irradiation. Food irradiation is a safe process when conducted under specified conditions. It destroys bacteria, parasites, insects, and viruses present in or on foods. NASA astronauts have been consuming irradiated foods for years to prevent food-borne illness while in space. Irradiation used as approved on foods:

- reduces or eliminates disease causing germs;
- does not introduce radioactivity into foods;
- does not cause dangerous substances to appear in foods; and
- does not significantly impact the nutrient content or nutritional value of the food.

Availability of irradiated foods is increasing. Irradiation of beef, pork, poultry, shellfish, fruits, vegetables, leafy greens, sprouts, spices, and eggs has been approved in the United States by the FDA.[42]

Food irradiation, as well as other food sterilization techniques, are not silver bullets that will prevent all cases of food-borne illness. Prions, toxins, pesticides, and mercury, for example, are resistant to radiation. Once irradiated or sterilized, foods can later become contaminated in such places as a packing plant, grocery store, restaurant, or home. The absence of all microorganisms in irradiated food may enable individual types of bacteria that contact the food after it is sterilized to grow at unusually high rates.[42]

Food Safety Basics

Due in large measure to the demands of individuals and consumer groups, many mechanisms are in place to help ensure the safety of the food supply. However, existing safeguards in food production and processing, and government regulations and enforcement efforts, are insufficient to guarantee that all foods purchased by consumers will be free from contamination. This situation, along with the possibility that food can become contaminated in unanticipated ways, means that responsibility for food safety is shared by consumers.

Food-borne illness is twice as likely to originate from restaurant foods as home prepared foods, so the food safety rules outlined in Table 32.5 apply to both settings.[43] The first rule of food safety is to wash your hands thoroughly with soap and water before and after handling food (Illustration 32.9). According to the CDC, this is the single most important means of preventing the spread of food-borne illness caused by bacteria.

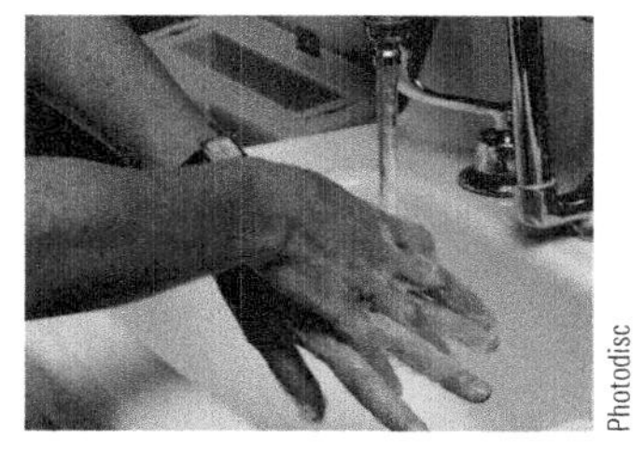
Photodisc

Illustration 32.9 There's a right way to wash your hands.

Table 32.5 Food safety rules[4,44]

Food safety rules
1. Clean hands, surfaces, utensils. • Wash your hands. It takes 20 seconds worth of washing with soap and water to really clean your hands of germs. • Repeat hand washing between handling of different foods. • Wash your hands after using the bathroom. • Wash counters, equipment, and utensils in between handling different foods.
2. Thoroughly wash raw fruits and vegetables under running water.
3. Separate raw and cooked foods. • Keep raw and cooked foods in separate containers. • Clean equipment, utensils, and counter tops in between uses for different foods.
4. Cook foods to proper temperatures (see Table 32.6). • Internal temperatures of meat are maintained or increase after being removed from heat. Let beef, lamb, and pork "rest" for 3 minutes before serving to allow them to keep cooking.
5. Store foods at the proper temperature. • Keep cold foods cold. Store foods at 40°F or less as soon as possible after purchase or use.
6. Don't consume unpasteurized milk or other dairy products, or raw or undercooked eggs.
7. When in doubt, throw it out! • If you are not sure that a food is safe to eat, don't eat it.

There's a Right Way to Wash Your Hands Ever watch a TV dramatic series about doctors and hospitals and wonder why they show surgeons scrubbing their hands and upper arms for so long before surgery? TV has it right. It takes about 20 seconds (or the time it takes to repeat the word "Mississippi" 20 times) to sanitize your hands. First you need to lather up with soap and very warm water. The soap makes germs lose their grip on your skin. Next you have to make sure to scrub all the crevices between your fingers and under your fingernails. Rinse your hands thoroughly with really warm water and dry them with a paper towel. (If you use a dishcloth, you may re-infect your hands.)[48]

Keep Hot Foods Hot and Cold Foods Cold Good foods can go bad if stored improperly. One of the most effective ways to prevent foods from spoiling is to store them at, and heat them to, the right temperatures (see Table 32.6 and Illustrations 32.10 and 32.11). When holding prepared foods before serving, hot foods should be kept hot, and cold foods cold. Freezing foods halts the growth of all of the main types of bacteria and other microorganisms that contaminate foods. Once the foods are thawed, however, microorganism growth may resume.

Bacteria Bacteria grow best at temperatures between 40 and 135°F (see Illustration 32.10). The room temperature of homes is within this range, and that's why foods that spoil should not be left outside the refrigerator for more than an hour. If a food has been improperly stored or kept past the "use by" date on the label, it should not be eaten even if it tastes, smells, and looks all right. Bacteria that most commonly cause food-borne illnesses usually do not change the taste, smell, or appearance of foods.[49] This general rule should apply: when in doubt, throw it out!

Table 32.6 A guide for cooking foods to safe internal temperatures[46,47]

Food	Internal temperature
Ground meats	
Beef, veal, lamb, pork	160°F
Chicken, turkey	165°F
Roasts, steaks, chops	
Beef, veal, lamb[a]	145°F
Chicken, turkey, duck	165°F
Stuffing	165°F
Pork[a]	145°F
Ham, fresh[a]	145°F
Ham, cooked, reheated	140°F
Eggs and other foods	
Fried, poached	Yolk and white are firm
Casseroles, leftovers	165°F
Sauces, custards	160°F
Fish should be cooked until the flesh is milky white and flakes easily with a fork.	
Shrimp, lobster, and crab should be cooked until shells are red and the flesh is milky white, firm, and not raw looking.	
Clams, oysters, and mussels should be cooked until the shells open, and not eaten if the shell is not open.	

[a]Let the meat rest for 3 minutes after cooking.

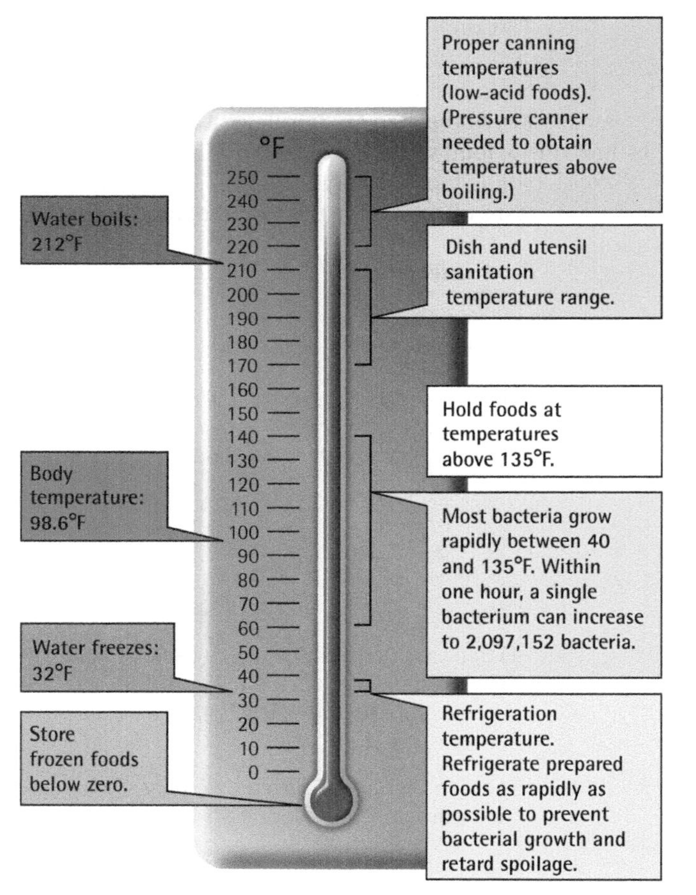

Illustration 32.10 Temperature guide for safe handling of food in the home.

Source: Adapted from Minnesota Extension Service, University of Minnesota, updated in 2004, based on data from USDA and reference 47.

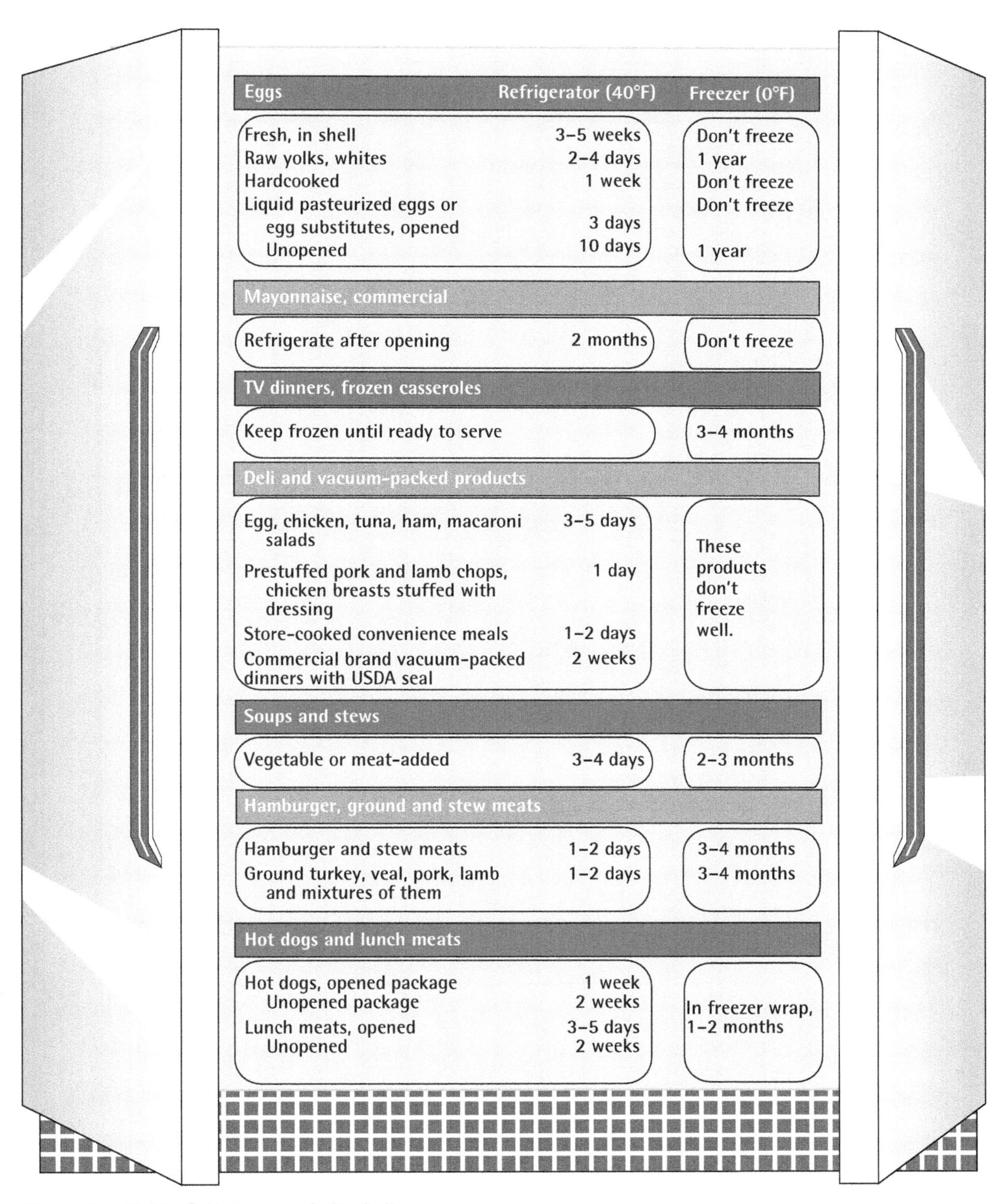

Eggs	Refrigerator (40°F)	Freezer (0°F)
Fresh, in shell	3–5 weeks	Don't freeze
Raw yolks, whites	2–4 days	1 year
Hardcooked	1 week	Don't freeze
Liquid pasteurized eggs or egg substitutes, opened	3 days	Don't freeze
Unopened	10 days	1 year
Mayonnaise, commercial		
Refrigerate after opening	2 months	Don't freeze
TV dinners, frozen casseroles		
Keep frozen until ready to serve		3–4 months
Deli and vacuum-packed products		
Egg, chicken, tuna, ham, macaroni salads	3–5 days	These products don't freeze well.
Prestuffed pork and lamb chops, chicken breasts stuffed with dressing	1 day	
Store-cooked convenience meals	1–2 days	
Commercial brand vacuum-packed dinners with USDA seal	2 weeks	
Soups and stews		
Vegetable or meat-added	3–4 days	2–3 months
Hamburger, ground and stew meats		
Hamburger and stew meats	1–2 days	3–4 months
Ground turkey, veal, pork, lamb and mixtures of them	1–2 days	3–4 months
Hot dogs and lunch meats		
Hot dogs, opened package	1 week	In freezer wrap, 1–2 months
Unopened package	2 weeks	
Lunch meats, opened	3–5 days	
Unopened	2 weeks	

Illustration 32.11 Cold storage: safe time limits.
These safe time limits will help keep refrigerated food from spoiling or becoming dangerous to eat. These time limits will keep frozen food at top quality.[49]

Bacon and Sausage	Refrigerator (40°F)	Freezer (0°F)
Bacon	7 days	1 month
Sausage, raw from pork, beef, turkey	2 days	1–2 months
Smoked breakfast links, patties	7 days	1–2 months
Hard sausage—pepperoni, jerky sticks	2–3 weeks	1–2 months
Ham, corned beef		
Corned beef, in pouch with pickling juices	5–7 days	Drained, wrapped 1 month
Ham, canned— label says keep refrigerated	6–9 months	Don't freeze
opened	3–5 days	1–2 months
Ham, fully cooked—whole	7 days	1–2 months
Ham, fully cooked—half	3–5 days	1–2 months
Ham, fully cooked—slices	3–4 days	1–2 months
Fresh meats		
Steaks, beef	3–5 days	6–12 months
Chops, pork	3–5 days	4–6 months
Chops, lamb	3–5 days	6–9 months
Roasts	3–5 days	4–12 months
Variety meats—tongue, brain, kidneys, liver, heart, chitterlings	2 days	3–4 months
Meat leftovers		
Cooked meat and meat dishes	3–4 days	2–3 months
Gravy and meat broth	1–2 days	2–3 months
Fresh poultry		
Chicken or turkey, whole	1–2 days	1 year
Chicken or turkey pieces	1–2 days	9 months
Giblets	1–2 days	3–4 months
Cooked poultry, leftover		
Fried chicken	3–4 days	4 months
Cooked poultry dishes	3–4 days	4–6 months
Pieces, plain	3–4 days	4 months
Pieces covered with broth, gravy	1–2 days	6 months
Chicken nuggets, patties	1–2 days	1–3 months

Illustration 32.11 (*Continued*) Cold storage: safe time limits.

Don't Eat Raw Milk, Eggs, or Meats Unpasteurized milk and cheese made from it are 150 times more likely to cause a food-borne illness than pasteurized milk and its products.[45] Raw and partially cooked eggs and raw and undercooked meats and fish may be contaminated with microorganisms that cause food-borne illness. Certain raw oysters may be contaminated, as well, and should be avoided.[50]

take action Limit Your Exposure to Potential Sources of Food-Borne Illness

Check the options from this list below that you do, or believe you could do, to reduce your risk of developing food-borne illness.

_______ If you are not sure about the safety of a food in your refrigerator, you don't take the risk. You toss it.
_______ Use a thermometer to measure the temperature of cooked foods such as meat, fish, and egg dishes.
_______ Check the "Sell By" and "Use By" dates printed on food packaging.
_______ Thoroughly wash raw fruits and vegetables, even if they look clean.
_______ Don't drink or eat from a container of refrigerated food and then put it back in the fridge.
_______ Don't drink raw milk or eat cheeses made from it.
_______ If you want a less than well done hamburger, you buy irradiated beef.
_______ Wash your hands, counter tops, utensils, and equipment frequently while preparing foods.

Food Handling and Storage Particular care needs to be taken when handling and storing foods. Dairy products such as milk and cheese may spoil in about a week or sooner if contaminated with bacteria from the air, surface areas, or hands. If you hold cheese by the wrapper when you take it out of the refrigerator to cut a slice off, it may last longer than if you touch it with dirty hands. (The Reality Check for this unit addresses the issue of the safety of cheese that has grown mold.) Handling foods with clean hands and using clean utensils on washed surface areas helps reduce the number of bacteria that come in contact with the food. Raw meats should be separated from other foods, and utensils and surface areas used to prepare meats should be thoroughly cleaned after each use.[44]

The Safety of Canned Foods Commercially canned foods are heated to the point where all bacteria are killed, so there should be no bacterial growth. Consequently, the contents of canned foods are sterile and will not spoil if left on the shelf for years. Nevertheless, canned foods may spoil if the can is dented, develop a pinhole leak, or if errors made during the canning process allow bacteria to enter the food. A cardinal sign that canned foods have spoiled is a buildup of pressure inside the can. The pressure will cause the top of the can to bulge out instead of curving in. Such a can of food should be thrown away or returned.

There Is a Limit to What the Consumer Can Do Keeping food safe from the farm, lake, or ocean to the table represents a major challenge to industry and government. Until contamination of food is prevented, consumers will continue to play an important role in ensuring food's safety. This unit's Take Action feature presents some options for reducing food-borne illness risk in today's food supply environment.

REALITY CHECK

Is Moldy Cheese Safe to Eat?

Tino decides to make tacos for dinner and finds that the shredded cheese he wants to use has some mold on it. Is it safe to eat the cheese?

Who gets the thumbs-up?

Answers appear on page 32-14

©AISPIX by Image Source/Shutterstock.com

Tino: I'm not sure if it's safe, but I think I'd better toss the cheese.

iStockphoto.com/Ranplett

Nancy: Just remove the moldy pieces. The rest is okay to eat.

ANSWERS TO REALITY CHECK

Is Moldy Cheese Safe to Eat?

Soft cheeses—such as cottage cheese, cream cheese, and brie—and sliced or shredded cheeses with mold should be thrown out. Mold can easily penetrate throughout these cheeses. Mold generally cannot penetrate very far in blocks of hard cheeses, such as cheddar, Swiss, and Colby. The cheese that lies an inch below the mold is considered safe to eat.[45]

Tino:

Nancy:

Garry Adams/Getty Images

NUTRITION up close

Food Safety Detective

Focal Point: Investigating paths to food-borne illness.

Assume an outbreak of food-borne illness occurred among people who ate foods from the onion and lettuce platter shown being prepared in Illustration 32.2.

Identify three potential causes of the outbreak related to the preparation of the foods as shown in the illustration.

1. ______________________________

2. ______________________________

3. ______________________________

Feedback to the Nutrition Up Close is located in Appendix F.

REVIEW QUESTIONS

- **Discuss the top causes of food-borne illness and the transmission and symptoms of each.**

1. All food-borne illnesses are caused by bacteria. **True/False**
2. Food-borne illnesses often begin with the transfer of harmful bacteria or viruses from feces of an infected animal or person to food. **True/False**
3. Cross-contamination can result from a failure to wash cutting boards after using them to cut raw meat but not raw vegetables. **True/False**
4. Antibiotics given to livestock can result in the development of strains of bacteria that resist treatment with antibiotics. **True/False**
5. Symptoms of food-borne illness most commonly consist of headache, blurred vision, and loss of balance. **True/False**
6. Infection with norovirus is an extremely common cause of food-borne illness. The illness is marked by episodes of vomiting that resolve within two days. **True/False**

- **Describe five consumer practices that help prevent food-borne illnesses.**

7. The first rule of food safety is to thoroughly wash your hands before and during food preparation. **True/False**
8. Bacteria grow best at temperatures between 100 and 200°F. **True/False**
9. Commercially canned vegetables, given the can is free from damage, can be safely consumed two years after purchase. **True/False**
10. Chicken, beef, and turkey roasts should be cooked until they reach a temperature of 160°F, but ground beef and chicken are safe to eat after they reach a temperature of 145°F. **True/False**

The following three questions refer to the following scenario.

Moshe comes home after an afternoon in the library and is really hungry. He hasn't shopped for a while, so he has to eat what he can find in the fridge. This is what he finds:

- **Half a package of deli turkey he bought 10 days ago**
  - Milk that expired five days ago
  - Beef stew left over from his dinner a week ago
  - Eggs he bought two weeks ago

11. ____ The milk smelled fine to Moshe. Should he drink it?
   a. If it smells fine, it is safe to drink.
   b. Moshe should look at it and taste it. If it looks and tastes okay, it would be safe to drink.
   c. It may be spoiled (contaminated). Moshe should throw it out.
   d. He could safely drink a small amount of it even if it's past the "use by" date.

12. ____ Which of the foods in Moshe's fridge is most likely to be safe to eat?
   a. deli turkey
   b. beef stew
   c. milk
   d. eggs

13. ____ Assume Moshe decides to eat the deli turkey on bread slices he had stored in the freezer. Two days later he develops diarrhea, abdominal cramps, fever, and vomiting. He has a food-borne illness. Which of the following causes of food-borne illness accounts best for his symptoms?
   a. *E. coli* 0157:H7
   b. *Campylobacter*
   c. Botulinum
   d. Ciguatera

Answers to these questions can be found in Appendix F.

NUTRITION SCOREBOARD ANSWERS

1. Leading foods associated with food-borne illnesses change every year. Fish, chicken, pork, and vegetables were the most common foods associated with recent cases of food-borne illnesses in the United States.[2] **False**
2. Freezing causes most bacteria to cease multiplying, but does not kill them. If a food is contaminated with bacteria before you freeze it, bacteria will be present when you thaw it. **False**
3. **True**[2]
4. Smell is not a foolproof indicator of contamination. Microorganisms that most commonly cause food-borne illnesses may not change the smell, taste, or appearance of foods. **False**

UNIT 33

Aspects of Global Nutrition

NUTRITION SCOREBOARD

1 Approximately 30% of the world's population does not have access to a safe supply of water. **True/False**

2 Many developing countries currently have rates of undernutrition and infectious disease that are similar to rates experienced in the United States and in many European countries 100 years ago. **True/False**

3 Many developing countries now face the dual public health problems of increasing rates of obesity as well as high a prevalence of underweight. **True/False**

4 Approximately one in nine people in the world is chronically undernourished. **True/False**

Answers can be found at the end of the unit.

LEARNING OBJECTIVES

After completing Unit 33 you will be able to:

- Identify the steps in the nutrition transition that occur when countries progress in terms of human development.

State of the World's Health

- **Identify the steps in the nutrition transition that occur when countries progress in terms of human development.**

If you represented the world's population with a group of 100 people, 61 would be Asian, 15 African, 10 European, 9 Latin American and Caribbean, and 5 North American. The group of 100 would include 65 people who live in countries where overweight and obesity represents a greater threat to health than undernutrition. Thirty-three members of the group would be Internet users, seven would have a college education, and about 33% would believe in ghosts or witchcraft. Large differences exist in standards of living and health status among people in the world's countries.[2,5]

Countries used to be classified as industrialized, developing, and least developed based on economic factors only. This classification system has changed. Countries are now classified by the United Nations based on a human development index that considers life expectancy, education, and per capita income.[2] Table 33.1 lists the top 20 countries in the highest and lowest categories of human development, and Table 32.2 reports life expectancy from birth, educational status, and per person income by level of human development. On average, individuals in countries with high human development scores live for over 70 years, while individuals living in countries with the lowest scores have life expectancies that are about 20 years less.[2]

KEY NUTRITION CONCEPTS

Two key nutrition concepts directly underlie this content on global nutrition:

1. Nutrition Concept #1: Food is a basic need of humans.
2. Nutrition Concept #4: Poor nutrition can result from both inadequate and excessive levels of nutrient intake.

The general state of the health of populations is monitored through evaluation of key environmental, health, and behavioral characteristics. Health outcomes, such as the number of low-birthweight newborns, prevalence of child underweight, rates of breastfeeding, and access to safe drinking water, are key indicators evaluated.[6] Worldwide, 17% of infants are born with low birthweights and only 36% of people have access to safe water supplies.[1,7] As can be seen in Illustration 33.1, rates of underweight, a measure of malnutrition, vary substantially worldwide. When percentages of low birthweight in newborns and underweight in young children are high, and rates of breastfeeding and access to a safe water supply are low, one can rightfully assume that undernutrition, short stature (low height for age), infection, and other health problems are common. When these key indicators show improvement, the health status and longevity of the population improve.[7]

Recent reports have identified leading problem areas related to undernutrition that should be addressed as part of global strategies aimed at improving health and well-being (Table 33.3). Important progress is being made, and rates of undernutrition, short stature, and infectious disease are decreasing worldwide. New threats posed by global warming-related diseases decreased availability of croplands and water and are currently reversing some of these gains. Rates of obesity and diabetes are increasing.[9]

The "Nutrition Transition"

nutrition transition A change in a population's diet from traditional foods such as grains, roots, beans, and rice to calorie-dense foods high in sugar, fat, sodium, and other components that characterize Western diets.

Throughout the history of modern humans, the typical diet and health status of population groups have changed along with advances in the human development level of countries. These changes tend to follow a pattern referred to as the **nutrition transition.**[10] Table 33.4 outlines changes in nutrition-related indicators and health status that accompany the various phases of the nutrition transition. The changes occur largely due to economic

Table 33.1 Countries with the highest and lowest human development scores based on average life expectancy, years of education, and income of people[2]

Top 20 countries based on level of human development	Bottom 20 countries based on level of human development
1. Norway	168. Haiti
2. Australia	169. Afghanistan
3. Switzerland	170. Djibouti
4. The Netherlands	171. Côte d'Ivoire
5. The United States	172. Gambia
6. Germany	173. Ethiopia
7. New Zealand	174. Malawi
8. Canada	175. Liberia
9. Singapore	176. Mali
10. Denmark	177. Guinea-Bissau
11. Ireland	178. Mozambique
12. Sweden	179. Guinea
13. Iceland	180. Burundi
14. The United Kingdom	181. Burkina Faso
15. Hong Kong, China	182. Eritrea
15. Korea (Republic of)	183. Sierra Leone
17. Japan	184. Chad
18. Liechtenstein	185. Central African Republic
19. Israel	186. Congo (Democratic Republic of the)
20. France	187. Niger

Table 33.2 Average scores for components of the human development ranking by level of human development[2]

	Number of countries	Life expectancy (years)	Education (years)	Income per person per year (US$)
Very high	49	80.2	11.7	$40,046
High	52	74.5	8.1	$13,231
Medium	44	67.9	5.5	$5,960
Low	42	59.4	4.2	$2,904

development and modifications in the food supply. In populations undergoing the nutrition transition, such as India and China, rates of underweight decline but rates of obesity, diabetes, hypertension, heart disease, dental disease, and cancer increase.[3,10,11] Illustration 33.2 shows WHO's projected increases in the incidence of chronic, noncommunicable diseases within countries by income level.[12]

Countries undergoing the nutrition transition often experience rising rates of obesity on top of existing high rates of underweight and undernutrition.[13] In 1980, 8% of adults worldwide were obese, and the figure that increased to 13% by 2014.[14] Children born to poorly nourished women in countries with low levels of human development may be biologically programmed in the womb to conserve energy. Adaptations are made by the fetus to adjust growth to correspond to the available energy and nutrient supplies, and these adaptations persist throughout life. When exposed to an improved food supply after birth, as happens when conditions improve in a country, children may gain fat more readily than

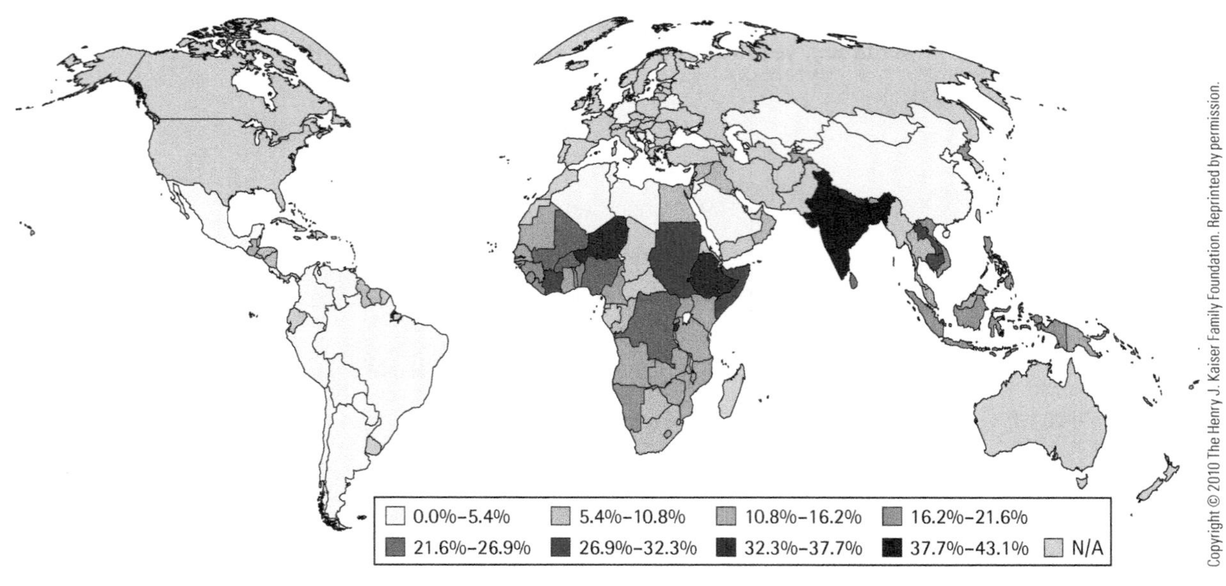

Illustration 33.1 Worldwide prevalence of child malnutrition assessed as the percentage of underweight among children under 5 years of age, 2000–2009.[16]

height.[3,15] Excess accumulation of fat places children at increased risk for hypertension, diabetes, heart disease, and a number of other "diseases of civilization" later in life. Underweight, short stature, undernutrition, and infections such as malaria and tuberculosis will continue to be top health problems in developing countries for many years to come, while diseases of Western civilization will continue to gain ground. This change is driven by the worldwide increased availability of inexpensive sugary, fast, and highly processed foods that replace foods traditionally consumed.[9,18]

Food and Nutrition: The Global Challenge

We have all heard about starvation in Somalia, Ethiopia, Sudan, and Bangladesh. The pictures of starving children with desperation in their big eyes and heads too large for their

Table 33.3 Priority problem areas related to malnutrition in developing countries[15,19,20]

1. Childhood protein-calorie malnutrition
2. Vitamin deficiencies: vitamin A, folate
3. Mineral deficiencies: iodine, iron, zinc
4. Lack of breastfeeding
5. Alcohol abuse
6. Overweight and obesity
7. Insufficient physical activity
8. Low vegetable and fruit intake
9. Malnutrition and increased complications from HIV/AIDS
10. Poor nutritional status of women of childbearing age

Table 33.4 Changes in health status associated with the nutrition transition in countries as human development advances[10,21]

Nutrition transition indicators	Health status indicators
• High prevalence of nutrient deficiencies, infectious disease, high levels of physical activity; diets tend to be high in wild game and plants	• High birth rates and low life expectancy
↓	↓
• Underweight, infectious diseases decline, diets characterized by high grain intake	• Decreased birth rate, increased life expectancy
↓	↓
• Nutrition-related chronic diseases increase, physical activity levels decline; intake of prepared and processed food intake increase.	• Age of population increases, birth rate continues to decrease
↓	↓
• Chronic diseases (obesity, heart disease diabetes, hypertension) predominate; diets characterized by energy-dense, high sugar and salt, low-fiber foods	• Focus changes to behavior changes, health care

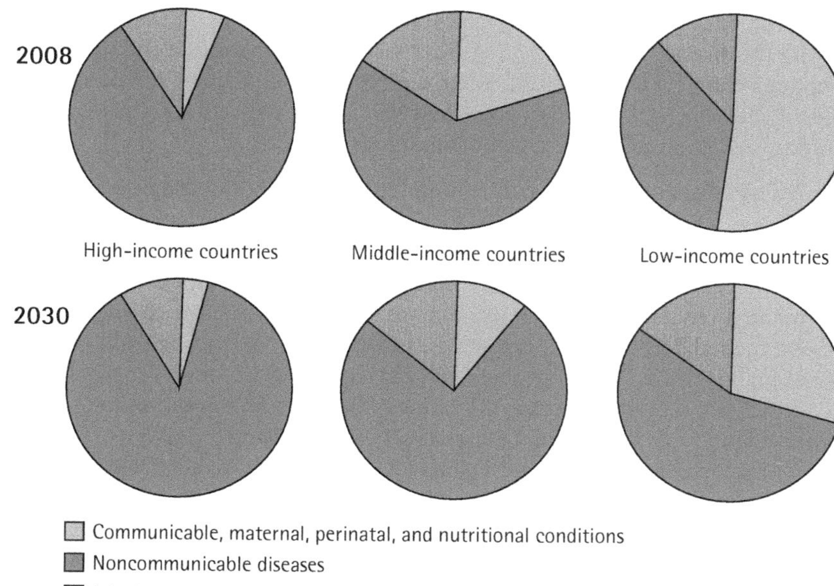

Illustration 33.2 Worldwide rates of chronic, noncommunicable disease, 2008, and projections for 2030.[12]

Source: www.nia.nig.gov/research/publication/global-health-and-aging/new-disease-patterns.

marasmus A severe form of malnutrition due primarily to a chronic lack of calories and protein. Also called protein-energy malnutrition.

kwashiorkor A severe form of protein-energy malnutrition in young children. It is characterized by swelling, fatty liver, susceptibility to infection, profound apathy, and poor appetite. The cause of kwashiorkor is unclear.

frail bodies burn an image in our minds and make many people wonder why it happens. (The Reality Check for this unit provides a follow-up discussion on this topic.)

What we don't hear about in the news, however, is the extent of starvation and malnutrition in the world. These are not just problems of a few isolated areas experiencing civil war or crop failures. In regions of the world with low levels of human development, undernutrition is an ongoing problem for up to 45% of children under the age of 5, and short stature affects an average of 24% of the population.[16,22] Approximately half of young children and pregnant women in the world are iron deficient, and 33% of young children are deficient in vitamin A. Vitamin A deficiency is related to frequent and severe infectious disease, decreased growth, and blindness. Deficiencies of iodine, zinc, and vitamin B_{12} are common in some developing countries.[3,23]

The children we usually see in pictures from famine areas are victims of **marasmus**, a disease caused primarily by a lack of calories and protein. These children look starved (Illustration 33.3) and consume far fewer calories and far less protein and other essential nutrients than they need. In victims of marasmus, the body uses its own muscle and other tissues as an energy source.

Protein-energy malnutrition produces a disease called **kwashiorkor** in some young children. It is similar to marasmus in that it is primarily due to a chronic lack of protein and calories. Kwashiorkor is unlike marasmus, however, in some ways. Young children with kwashiorkor appear to be genetically susceptible to developing faulty protein utilization mechanisms when chronically exposed to protein and calorie shortages.[24] These children

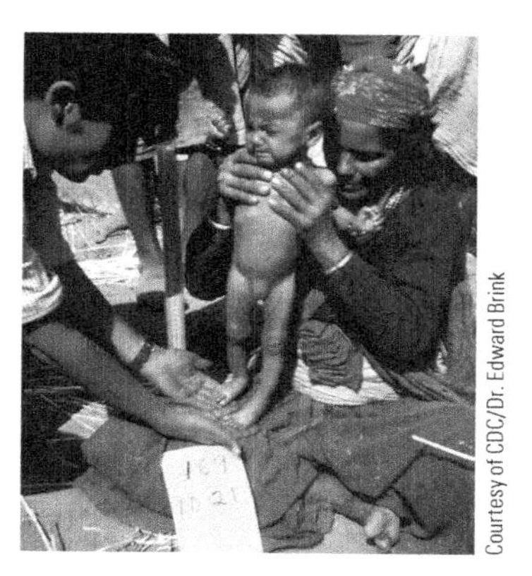

Courtesy of CDC/Dr. Edward Brink

Illustration 33.3 The look of marasmus. Children with marasmus look like "skin and bones."

REALITY CHECK

Can the World Produce Enough Food?

The image of starving children upset both Kai and Kari after they saw the image posted on YouTube today. Kai says to Kari that he thinks the major reason why children starve is that the world cannot produce enough food for everyone. Kari believes that's not the reason. She thinks starvation is mostly related to causes such as civil unrest, discrimination, and abject poverty.

Who gets the thumbs-up?

Answers appear on page 33-6.

©Szefei/Shutterstock.com

Kai: The planet is overpopulated, and we can't produce enough food for everyone.

©Ashwin/Shutterstock.com

Kari: There doesn't have to be starvation. Enough food can be produced.

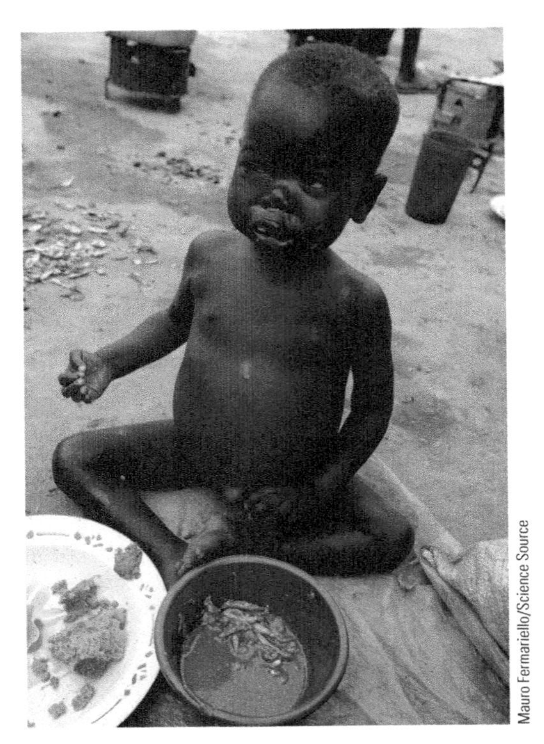

Illustration 33.4 The look of kwashiorkor. This child has the characteristic "moon face" (edema), swollen belly, and patchy dermatitis often seen with kwashiorkor.

may appear fat due to massive swelling related to abnormal protein utilization that occurs with the disease (Illustration 33.4). Most victims of kwashiorkor show some of the symptoms of marasmus as well. Children with either type of severe malnutrition are generally deficient in multiple vitamins and minerals and are at high risk of dying from infectious diseases due to weakened immune systems.[25,26]

Survivors of Malnutrition Malnutrition in the form of undernutrition that occurs before a child's brain has completely grown has lasting effects on development.[3] Brain growth occurs primarily during pregnancy through the age of 2 years. If children suffer severe malnutrition during these years, they will experience permanent reductions in growth and mental development. The severity of the growth and mental development impairments depends on the timing and duration of the malnutrition. The longer it persists, the harder it becomes for young children to achieve a brighter future.[3] The state of hunger and starvation in older children and adults can lead them to become self-centered and to lose any sense of well-being. The need for food for survival may prompt unethical behaviors such as stealing and injuring others to obtain food. The devastating psychological effects of starvation may follow children into adulthood and have a lasting effect on behavior throughout life. Persistent malnutrition saps the physical and mental energy of people and, ultimately, compromises the economic and social progress of nations.[4,20]

Malnutrition and Infection There is a very close relationship between malnutrition and infection: malnutrition weakens the immune system and increases the likelihood and severity of infection. Simultaneously, repeated infection and the bouts of diarrhea that often accompany it can produce undernutrition. Poor sanitary conditions, contaminated water supplies, and lack of refrigeration contribute to the spread of infectious diseases that promote the development of malnutrition. Most deaths of children under the age of 5 in countries with low levels of human development are related to both malnutrition and infection. Low rates of breastfeeding promote the development of infectious disease, especially in countries where contaminated water is used to prepare formula. In addition to providing optimal nutrition, breastmilk contains a number of substances that protect babies from infection.[20,27–29]

Vitamin A deficiency in children plays important roles in the spread and severity of infections. Children who are deficient in vitamin A are much more likely to die from a measles infection, for example, than are children whose vitamin A status is sufficient.[30] HIV infection progresses more rapidly into AIDS in malnourished individuals and shortens the time parents can work and support their families. Life expectancy in Sudan, for example, decreased from 53.0 years in 1996 to 51.4 years in 2009. Over 38 million people in the world are infected with HIV, including 3.8 million children.[21] Fortunately, transmission of HIV in most developed and developing countries is declining, and treatment is becoming increasingly available.[32]

Why Do Starvation and Malnutrition Happen? Malnutrition and starvation don't have to happen anywhere. The world's agricultural systems have the capacity to produce enough food to feed everyone on earth.[33] People become malnourished or starve primarily due to poverty, human-made disasters, discrimination against women and girls, the HIV/AIDS

ANSWERS TO REALITY CHECK

Can the World Produce Enough Food?

The world can and does produce enough food for everyone. Children in need of food may not have access to it for the reasons Kari gave and for other reasons listed in Table 33.5.

Kai:

Kari:

epidemic, racism, the use of agricultural land for biofuel crops, corrupt governance, and other factors (Table 33.5). In some instances, starvation and malnutrition are due to natural disasters that lead to crop failures or the inability to distribute food to those in need.[4]

The famine in Somalia that led to the death of over 1.5 million people was due to the collapse of the government, fighting among rival clans, and general lawlessness. Farmers and their families were driven off their land by bandits who stole their crops and possessions. Many had nowhere to go for help.[34] Mass starvation in Ethiopia and Sudan was similarly initiated by violent conflict within the countries.[25] Seizing the opposing side's food and preventing them from obtaining food aid are among the primary weapons used to fight these civil wars. In Bangladesh, where poor people have no choice but to live on floodplains, cyclones wipe out crops and regularly result in the death of thousands due to floods, starvation, and infectious disease.[15]

Women and female children are at particular risk for malnutrition in some societies because cultural practices call for food to be allocated to men and boys first (Illustration 33.5). If too little food is available, what remains after meals for women and daughters may be too little to support health and growth.[36] In many developing countries, discrimination against women in education and employment, sanctions against the use of birth control, and violence toward women place women at high risk of developing malnutrition and having a low quality of life.[19]

Table 33.5 Root causes of malnutrition in developing countries[8,19]

• Poverty
• Low levels of education
• Discrimination against females
• Inequitable distribution of the food supply
• Impact of climate change on cropland and water availability
• Lack of economic opportunities
• Racism, ethnocentrism
• Low agricultural productivity
• Poor and corrupt governance
• Unsafe water due primarily to the ravages of nature

Ending Malnutrition The long-term solution to malnutrition will depend on the ability of humans to work together to achieve educational and economic development, peace, control of population growth, improved sanitation and clean water supplies, social equity for women and children, and environmentally sound and productive agricultural policies and practices (Illustration 33.6). The people most affected by malnutrition must be active participants in the planning and implementation of improvement programs. Efforts to improve the nutritional status of populations can and have paid off.

Success Stories Examples of strikingly successful efforts to reduce malnutrition in various countries around the world are available for the telling. A repeated theme is the reduction in malnutrition as economic, educational, nutritional, and sanitary conditions improve. Improvements in these same areas were responsible for the dramatic declines in malnutrition and infectious diseases and the increases in life expectancy experienced in the United States, Canada, and many European countries around 100 years ago.[2] In other instances, specific nutrient deficiencies have been greatly reduced or eliminated by food fortification and nutrition education programs. Here are several specific examples:

- Vitamin A supplements and education on vitamin A–rich foods are related to a dramatic decrease in cases of severe and moderate vitamin A deficiency and infection in children in many countries.[30]
- Food supplements given to groups of infants in Russia, Brazil, South Africa, and China were associated with higher IQ scores at age 8.
- Iodization of salt has eliminated iodine deficiency in Bolivia and Ecuador.
- Worldwide, iodization of salt is associated with a drop in the number of iodine-deficient children.
- Fortification of flour with iron has led to a decrease in iron-deficiency anemia in the Philippines.[37,38]

In Thailand, a country with an iodine-deficiency problem, a creative approach to increasing iodine intake was implemented by the king in the 1990s. To celebrate his birthday, salt producers gave the king 10,000 tons of iodized salt. The king had the salt packaged in small plastic bags and, with the help of the Red Cross and the army, delivered a bag to every household in Thailand. A message from the king about the importance of using iodized salt was attached to each bag.[39]

Deaths of young children from malnutrition and its related diseases have dropped substantially in countries experiencing a resurgence of breastfeeding (Illustration 33.7).[40] Breastmilk protects infants and young children from a variety of infectious diseases,

Illustration 33.6 Safe water supplies represent a major advance in quality of life for people in developing countries.

Porterfield/Chickering/Science Source

supports the growth and health of infants, and protects them from the hazards of formulas reconstituted with contaminated water. Worldwide health initiatives have led to the development of the International Code of Marketing Breast Milk Substitutes and to Baby Friendly Hospitals. The international marketing code calls for the prohibition of free formula samples as well as the promotion of infant formulas by health care professionals, and staff training on breastfeeding support. More than 15,000 hospitals and health facilities have adopted Baby Friendly policies that promote and facilitate breastfeeding.[40]

Illustration 33.7 After a rooming-in policy that enabled mothers to breastfeed their infants on demand was introduced in one hospital in the Philippines, the incidence of early infant deaths due to infection dropped by 95%.[40]

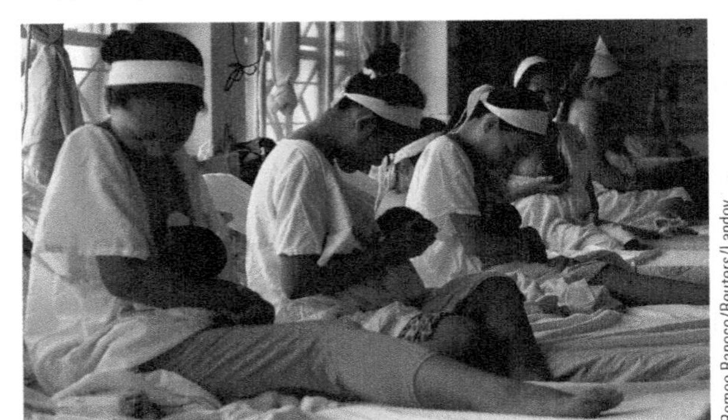
Romeo Ranoco/Reuters/Landov

The Future

Benefits derived from improving the nutritional status of the world's population are immense. Adequate nutrition and basic public assurances such as adequate supplies of affordable nutritious foods, clean water, vaccinations, and education are the bedrock on which present generations secure a future for themselves and the next ones. People who are well nourished, grow normally, and are free from infectious disease are more likely to be productive members of society, earn enough money to buy material objects, and make sure their children are educated.[20] In this type of environment, parents tend to have fewer children and are able to take better care of the children they have.[21,33] Members of families are more productive, happier, require less medical care, commit fewer crimes, and are more likely to be self-sufficient than undernourished people. When the undernutrition problem is solved, the world will likely wonder why it didn't happen sooner.

©Stephane Bidouze/ Shutterstock.com

NUTRITION up close

Ethnic Foods Treasure Hunt

Focal Point: Values placed on foods are culturally specific.

The following questions deal with ethnic food preferences. Answer any three of the questions. Resources that will help you find the answers include knowledgeable individuals, the Internet, travel and anthropology books, and encyclopedias.

1. Name three foods commonly served at celebrations by the Igbo in Nigeria (Igbos were formerly referred to as Biafrans).
   a. ______________________
   b. ______________________
   c. ______________________
2. List two special foods given to Hmong women during pregnancy.
   a. ______________________
   b. ______________________
3. List the five foods that make up the traditional Costa Rican breakfast.
   a. ______________________
   b. ______________________
   c. ______________________
   d. ______________________
   e. ______________________
4. List three foods commonly sold at soccer matches in Italy.
   a. ______________________
   b. ______________________
   c. ______________________
5. Name two traditional foods of the Zulu in South Africa.
   a. ______________________
   b. ______________________
6. Name one of the staple foods of people in Nentsi (it's near Siberia).
   a. ______________________
7. List two foods considered to be "yin" and two considered to be "yang" in Chinese culture.

Yin	Yang
a. ______________________	a. ______________________
b. ______________________	b. ______________________

Feedback to the Nutrition Up Close is located in Appendix F.

REVIEW QUESTIONS

- **Identify the steps in the nutrition transition that occur when countries progress in terms of human development.**

1. Due to the expanding adoption of Western lifestyles and dietary patterns, some developing countries are experiencing increasing rates of overweight and chronic diseases. **True/False**
2. The term *nutrition transition* is defined as improvements in a population's nutritional status and health that result from a country's economic development. **True/False**
3. Children who were undernourished during their mother's pregnancy and early in their own lives may develop a tendency to gain fat more readily than height given access to an abundant food supply. **True/False**

4. As countries develop, the leading causes of death generally change from infectious diseases to chronic diseases such as diabetes and heart disease. **True/False**
5. Malnutrition and infectious diseases are common health problems in countries with low life expectancy. **True/False**
6. Decreased availability of croplands and water are reversing some of these gains previously accomplished in reducing malnutrition. **True/False**
7. Deficiencies of iodine, iron, and vitamin A remain important public health problems in many developing countries. **True/False**
8. The need for food for survival may prompt behaviors such as stealing and injuring others to obtain food. **True/False**
9. In many developing countries, discrimination places women at high risk of malnutrition. **True/False**
10. Death rates for young children decrease in developing countries when rates of breastfeeding increase. **True/False**

The following three questions refer to this scenario:

In 2012, North Korea lowered the minimum height requirement for service in the military from 4 feet, 9 inches (145 cm) to 4 feet 8 inches (142 cm). The country is experiencing a shortage of taller men due to a famine in 1990 and continued undernutrition after that. The average height of the population is decreasing.[41]

11. _____ What is the most likely cause of the decreased stature of males?
    a. Their parents are short.
    b. Mothers lacked adequate health care during pregnancy.
    c. They experienced undernutrition during the growing years.
    d. They grew up lacking a safe and adequate supply of water.
12. _____ What other conditions are likely present in the North Korean population that are related to diminished growth in height?
    a. Increased agility due to low stature.
    b. Increased calorie need per pound of body weight.
    c. Decreased need for health care.
    d. Reduced life expectancy.
13. _____ Assume 30 years from now North Korea cuts back on military expenditures and the country grows economically. Peace prevails and people have jobs and money to spend. How would that likely change the health status or behaviors of the North Korean population?
    a. Individuals in future generations would gain height.
    b. Individuals would remain short because height is genetically determined.
    c. Heart disease would replace diabetes as the major cause of death.
    d. Couples would decide to have additional children because they can afford them.

Answers to these questions can be found in Appendix F.

NUTRITION SCOREBOARD ANSWERS

1. It is estimated that 36% of the world's population lack access to a safe water supply.[1] **True**
2. Many developing countries are evolving just as the United States and Europe began to do a century ago.[2] **True**
3. Rates of overweight are increasing in many developing countries.[3] **True**
4. **True**[4]

Appendix Table of Contents

Appendix A

Table of Food Composition

This table of food composition is updated periodically to reflect current nutrient data for foods, to remove outdated foods, and to add foods that are new to the marketplace.* The nutrient database for this appendix is compiled from a variety of sources, including the United States Department of Agriculture (USDA) Nutrient Database and manufacturers' data. The USDA database provides data for a wider variety of foods and nutrients than other sources. Because laboratory analysis for each nutrient can be quite costly, manufacturers tend to provide data only for those nutrients mandated on food labels. Consequently, data for their foods are often incomplete; any missing information on this table is designated as a dash. Keep in mind that a dash means only that the information is unknown and should not be interpreted as a zero. A zero means that the nutrient is not present in the food.

When using nutrient data, remember that many factors influence the nutrient contents of foods. These factors include the mineral content of the soil, the diet fed to the animal or the fertilizer used on the plant, the season of harvest, the method of processing, the length and method of storage, the method of cooking, the method of analysis, and the moisture content of the sample analyzed. With so many influencing factors, users should view nutrient data as a close approximation of the actual amount.

For updates, corrections, and a list of more than 35,000 foods and codes found in the diet analysis software that accompanies this text, visit www.cengagebrain.com and click on Diet Analysis Plus.

- *Fats* Total fats, as well as the breakdown of total fats to saturated, monounsaturated, and polyunsaturated, are listed in the table. The fatty acids seldom add up to the total in part due to rounding but also because values may include some nonfatty acids, such as glycerol, phosphate, or sterols. *Trans*-fatty acids are not listed separately in this edition because newer hydrogenated fats generally add less than 0.5 g *trans* fat to a serving of food, an amount often reported as 0.
- *Vitamin A, Vitamin E, and Folate* In keeping with the Dietary Reference Intake (DRI) values for vitamin A, this appendix presents data for vitamin A in micrograms (µg) RAE. Similarly, because the DRI intake values for vitamin E are based only on the alpha-tocopherol form of vitamin E, this appendix reports vitamin E data in milligrams (mg) alpha-tocopherol, listed on the table as Vit E (mg α). Folate values are listed in µg DFE, a unit that equalizes the bioavailability of naturally occurring folate and added folic acid in enriched foods.
- *Bioavailability* Keep in mind that the availability of nutrients from foods depends not only on the quantity provided by a food as reflected in this table, but also on the amount absorbed and used by the body.
- *Using the Table* The foods and beverages in this table are organized into several categories, which are listed at the head of each right-hand page. Page numbers are provided, and each group is color-coded to make it easier to find individual foods.

*This food composition table has been prepared by Cengage Learning. The nutritional data are supplied by Axxya Systems.

Table A–1 Table of Food Composition

(Computer code is for Cengage Diet Analysis Plus program)

DA+ Code	Food Description	QTY	Measure	Wt (g)	H₂O (g)	Ener (cal)	Prot (g)	Carb (g)	Fiber (g)	Fat (g)	Fat Breakdown (g)		
											Sat	Mono	Poly
BREADS, BAKED GOODS, CAKES, COOKIES, CRACKERS, CHIPS, PIES													
	Bagels												
8534	Cinnamon and raisin	1	item(s)	71	22.7	194	7.0	39.2	1.6	1.2	0.2	0.1	0.5
14395	Multi-grain	1	item(s)	61	—	170	6.0	35.0	1.0	1.5	0.5	0.1	0.4
8538	Oat bran	1	item(s)	71	23.4	181	7.6	37.8	2.6	0.9	0.1	0.2	0.3
4910	Plain, enriched	1	item(s)	71	25.8	182	7.1	35.9	1.6	1.2	0.3	0.4	0.5
4911	Plain, enriched, toasted	1	item(s)	66	18.7	190	7.4	37.7	1.7	1.1	0.2	0.3	0.6
	Biscuits												
25008	Biscuits	1	item(s)	41	15.8	121	2.6	16.4	0.5	4.9	1.4	1.4	1.8
16729	Scone	1	item(s)	42	11.5	148	3.8	19.1	0.6	6.2	2.0	2.5	1.3
25166	Wheat biscuits	1	item(s)	55	21.0	162	3.6	21.9	1.4	6.7	1.9	1.9	2.5
	Bread												
325	Boston brown, canned	1	slice(s)	45	21.2	88	2.3	19.5	2.1	0.7	0.1	0.1	0.3
8716	Bread sticks, plain	4	item(s)	24	1.5	99	2.9	16.4	0.7	2.3	0.3	0.9	0.9
25176	Cornbread	1	piece(s)	55	25.9	141	4.7	18.3	0.9	5.4	2.1	1.4	1.5
327	Cracked wheat	1	slice(s)	25	9.0	65	2.2	12.4	1.4	1.0	0.2	0.5	0.2
9079	Croutons, plain	¼	cup(s)	8	0.4	31	0.9	5.5	0.4	0.5	0.1	0.2	0.1
8582	Egg	1	slice(s)	40	13.9	113	3.8	19.1	0.9	2.4	0.6	0.9	0.4
8585	Egg, toasted	1	slice(s)	37	10.5	117	3.9	19.5	0.9	2.4	0.6	1.1	0.4
329	French	1	slice(s)	32	8.9	92	3.8	18.1	0.8	0.6	0.2	0.1	0.3
8591	French, toasted	1	slice(s)	23	4.7	73	3.0	14.2	0.7	0.5	0.1	0.1	0.2
42096	Indian fry, made with lard (Navajo)	3	ounce(s)	85	26.9	281	5.7	41.0	—	10.4	3.9	3.8	0.9
332	Italian	1	slice(s)	30	10.7	81	2.6	15.0	0.8	1.1	0.3	0.2	0.4
1393	Mixed grain	1	slice(s)	26	9.6	69	3.5	11.3	1.9	1.1	0.2	0.2	0.5
8604	Mixed grain, toasted	1	slice(s)	24	7.6	69	3.5	11.3	1.9	1.1	0.2	0.2	0.5
8605	Oat bran	1	slice(s)	30	13.2	71	3.1	11.9	1.4	1.3	0.2	0.5	0.5
8608	Oat bran, toasted	1	slice(s)	27	10.4	70	3.1	11.8	1.3	1.3	0.2	0.5	0.5
8609	Oatmeal	1	slice(s)	27	9.9	73	2.3	13.1	1.1	1.2	0.2	0.4	0.5
8613	Oatmeal, toasted	1	slice(s)	25	7.8	73	2.3	13.2	1.1	1.2	0.2	0.4	0.5
1409	Pita	1	item(s)	60	19.3	165	5.5	33.4	1.3	0.7	0.1	0.1	0.3
7905	Pita, whole wheat	1	item(s)	64	19.6	170	6.3	35.2	4.7	1.7	0.3	0.2	0.7
338	Pumpernickel	1	slice(s)	32	12.1	80	2.8	15.2	2.1	1.0	0.1	0.3	0.4
334	Raisin, enriched	1	slice(s)	26	8.7	71	2.1	13.6	1.1	1.1	0.3	0.6	0.2
8625	Raisin, toasted	1	slice(s)	24	6.7	71	2.1	13.7	1.1	1.2	0.3	0.6	0.2
10168	Rice, white, gluten free, wheat free	1	slice(s)	38	—	130	1.0	18.0	0.5	6.0	0	—	—
8653	Rye	1	slice(s)	32	11.9	83	2.7	15.5	1.9	1.1	0.2	0.4	0.3
8654	Rye, toasted	1	slice(s)	29	9.0	82	2.7	15.4	1.9	1.0	0.2	0.4	0.3
336	Rye, light	1	slice(s)	25	9.3	65	2.0	12.0	1.6	1.0	0.2	0.3	0.3
8588	Sourdough	1	slice(s)	25	7.0	72	2.9	14.1	0.6	0.5	0.1	0.1	0.2
8592	Sourdough, toasted	1	slice(s)	23	4.7	73	3.0	14.2	0.7	0.5	0.1	0.1	0.2
491	Submarine or hoagie roll	1	item(s)	135	40.6	400	11.0	72.0	3.8	8.0	1.8	3.0	2.2
8596	Vienna, toasted	1	slice(s)	23	4.7	73	3.0	14.2	0.7	0.5	0.1	0.1	0.2
8670	Wheat	1	slice(s)	25	8.9	67	2.7	11.9	0.9	0.9	0.2	0.2	0.4
8671	Wheat, toasted	1	slice(s)	23	5.6	72	3.0	12.8	1.1	1.0	0.2	0.2	0.4
340	White	1	slice(s)	25	9.1	67	1.9	12.7	0.6	0.8	0.2	0.2	0.3
1395	Whole wheat	1	slice(s)	46	15.0	128	3.9	23.6	2.8	2.5	0.4	0.5	1.4
	Cakes												
386	Angel food, prepared from mix	1	piece(s)	50	16.5	129	3.1	29.4	0.1	0.2	0	0	0.1
8772	Butter pound, ready to eat, commercially prepared	1	slice(s)	75	18.5	291	4.1	36.6	0.4	14.9	8.7	4.4	0.8
28517	Carrot	1	slice(s)	131	56.6	339	4.8	56.5	1.9	11.1	1.0	5.7	3.8
4931	Chocolate with chocolate icing, commercially prepared	1	slice(s)	64	14.7	235	2.6	34.9	1.8	10.5	3.1	5.6	1.2
8756	Chocolate, prepared from mix	1	slice(s)	95	23.2	352	5.0	50.7	1.5	14.3	5.2	5.7	2.6
393	Devil's food cupcake with chocolate frosting	1	item(s)	35	8.4	120	2.0	20.0	0.7	4.0	1.8	1.6	0.6
8757	Fruitcake, ready to eat, commercially prepared	1	piece(s)	43	10.9	139	1.2	26.5	1.6	3.9	0.5	1.8	1.4
1397	Pineapple upside down, prepared from mix	1	slice(s)	115	37.1	367	4.0	58.1	0.9	13.9	3.4	6.0	3.8
411	Sponge, prepared from mix	1	slice(s)	63	18.5	187	4.6	36.4	0.3	2.7	0.8	1.0	0.4

Chol (mg)	Calc (mg)	Iron (mg)	Magn (mg)	Pota (mg)	Sodi (mg)	Zinc (mg)	Vit A (µg)	Thia (mg)	Vit E (mg α)	Ribo (mg)	Niac (mg)	Vit B_6 (mg)	Fola (µg DFE)	Vit C (mg)	Vit B_{12} (µg)	Sele (µg)
0	13	2.69	19.9	105.1	228.6	0.80	14.9	0.27	0.22	0.19	2.18	0.04	123.5	0.5	0	22.0
0	60	1.08	—	—	310.0	—	0	—	—	—	—	—	—	0	—	—
0	9	2.18	22.0	81.7	360.0	0.63	0.7	0.23	0.23	0.24	2.10	0.03	95.1	0.1	0	24.3
0	63	4.29	15.6	53.3	318.1	1.34	0	0.42	0.07	0.18	2.82	0.04	160.5	0.7	0	16.2
0	65	2.97	15.8	56.1	316.8	0.85	0	0.39	0.07	0.17	2.88	0.04	134.0	0	0	16.6
0	38	0.94	6.0	47.4	206.0	0.20	—	0.16	0.01	0.12	1.20	0.01	46.8	0.1	0.1	7.1
49	79	1.35	7.1	48.7	277.2	0.29	64.7	0.14	0.42	0.15	1.19	0.02	49.1	0	0.1	10.9
0	57	1.21	16.1	81.0	321.1	0.42	—	0.19	0.01	0.14	1.65	0.03	63.2	0.1	0.1	0
0	32	0.94	28.4	143.1	284.0	0.22	11.3	0.01	0.14	0.05	0.50	0.03	6.3	0	0	9.9
0	5	1.02	7.7	29.8	157.7	0.21	0	0.14	0.24	0.13	1.26	0.01	61.0	0	0	9.0
21	94	0.91	10.5	71.5	209.8	0.48	—	0.14	0.32	0.15	1.03	0.04	78.7	1.7	0.2	6.2
0	11	0.70	13.0	44.3	134.5	0.31	0	0.09	—	0.06	0.92	0.08	19.0	0	0	6.3
0	6	0.30	2.3	9.3	52.4	0.06	0	0.04	—	0.02	0.40	0.00	15.7	0	0	2.8
20	37	1.21	7.6	46.0	196.8	0.31	25.2	0.17	0.10	0.17	1.93	0.02	52.0	0	0	12.0
21	38	1.23	7.8	46.6	199.8	0.31	25.5	0.14	0.10	0.16	1.77	0.02	47.7	0	0	12.2
0	14	1.16	9.0	41.0	208.0	0.29	0	0.13	0.05	0.09	1.52	0.03	73.6	0.1	0	8.7
0	11	0.89	7.1	32.2	165.6	0.24	0	0.10	0.04	0.09	1.24	0.02	49.9	0	0	6.8
6	48	3.43	15.3	65.5	279.8	0.29	0	0.36	0.00	0.18	3.91	0.03	166.7	—	0	15.8
0	23	0.88	8.1	33.0	175.2	0.25	0	0.14	0.08	0.08	1.31	0.01	91.2	0	0	8.2
0	27	0.65	20.3	59.8	109.2	0.44	0	0.07	0.09	0.03	1.05	0.06	19.5	0	0	8.6
0	27	0.65	20.4	60.0	109.7	0.44	0	0.06	0.10	0.03	1.05	0.07	16.8	0	0	8.6
0	20	0.93	10.5	44.1	122.1	0.26	0.6	0.15	0.13	0.10	1.44	0.02	36.0	0	0	9.0
0	19	0.92	9.2	33.2	121.0	0.28	0.5	0.12	0.13	0.09	1.29	0.01	28.1	0	0	8.9
0	18	0.72	10.0	38.3	161.7	0.27	1.4	0.10	0.13	0.06	0.84	0.01	23.5	0	0	6.6
0	18	0.74	10.3	38.5	162.8	0.28	1.3	0.09	0.13	0.06	0.77	0.02	18.7	0.1	0	6.7
0	52	1.57	15.6	72.0	321.6	0.50	0	0.35	0.18	0.19	2.77	0.02	99.0	0	0	16.3
0	10	1.95	44.2	108.8	340.5	0.97	0	0.21	0.39	0.05	1.81	0.17	22.4	0	0	28.2
0	22	0.91	17.3	66.6	214.7	0.47	0	0.10	0.13	0.09	0.98	0.04	40.0	0	0	7.8
0	17	0.75	6.8	59.0	101.4	0.18	0	0.08	0.07	0.10	0.90	0.01	40.6	0	0	5.2
0	17	0.76	6.7	59.0	101.8	0.19	0	0.07	0.07	0.09	0.81	0.02	35.5	0.1	0	5.2
0	100	1.08	—	—	140	—	—	0.15	—	0.10	1.20	—	47.5	0	—	—
0	23	0.90	12.8	53.1	211.2	0.36	0	0.13	0.10	0.10	1.21	0.02	48.3	0.1	0	9.9
0	23	0.89	12.5	53.1	210.3	0.36	0	0.11	0.10	0.09	1.09	0.02	42.9	0.1	0	9.9
0	20	0.70	3.9	51.0	175.0	0.18	0	0.10	—	0.08	0.80	0.01	5.3	0	0	8.0
0	11	0.91	7.0	32.0	162.5	0.23	0	0.11	0.05	0.07	1.19	0.03	57.5	0.1	0	6.8
0	11	0.89	7.1	32.2	165.6	0.24	0	0.10	0.04	0.09	1.24	0.02	49.9	0	0	6.8
0	100	3.80	—	128.0	683.0	—	0	0.54	—	0.33	4.50	0.04	—	0	—	42.0
0	11	0.89	7.1	32.2	165.6	0.24	0	0.10	0.04	0.09	1.24	0.02	49.9	0	0	6.8
0	36	0.87	12.0	46.0	130.3	0.30	0	0.09	0.05	0.08	1.30	0.03	24.8	0.1	0	7.2
0	38	0.94	13.6	51.3	140.5	0.34	0	0.10	0.06	0.09	1.44	0.04	23.0	0	0	7.7
0	38	0.94	5.8	25.0	170.3	0.19	0	0.11	0.06	0.08	1.10	0.02	42.8	0	0	4.3
0	15	1.42	37.3	144.4	159.2	0.69	0	0.13	0.35	0.10	1.83	0.09	35.9	0	0	17.8
0	42	0.11	4.0	67.5	254.5	0.06	0	0.04	0.01	0.10	0.08	0.00	14.5	0	0	7.7
166	26	1.03	8.3	89.3	298.5	0.34	111.8	0.10	—	0.17	0.98	0.03	46.5	0	0.2	6.6
0	65	2.18	23.0	279.6	367.7	0.44	—	0.25	0.01	0.19	1.73	0.10	98.2	4.6	0	14.7
27	28	1.40	21.8	128.0	213.8	0.44	16.6	0.01	0.62	0.08	0.36	0.02	14.7	0.1	0.1	2.1
55	57	1.53	30.4	133.0	299.3	0.65	38.0	0.13	—	0.20	1.08	0.03	37.1	0.2	0.2	11.3
19	21	0.70	—	46.0	92.0	—	—	0.04	—	0.05	0.30	—	8.1	0	—	2.0
2	14	0.89	6.9	65.8	116.1	0.11	3.0	0.02	0.38	0.04	0.34	0.02	13.8	0.2	0	0.9
25	138	1.70	15.0	128.8	366.9	0.35	71.3	0.17	—	0.17	1.36	0.03	44.9	1.4	0.1	10.8
107	26	0.99	5.7	88.8	143.6	0.37	48.5	0.10	—	0.19	0.75	0.03	33.4	0	0.2	11.7

											Fat Breakdown (g)		
DA+ Code	Food Description	QTY	Measure	Wt (g)	H_2O (g)	Ener (cal)	Prot (g)	Carb (g)	Fiber (g)	Fat (g)	Sat	Mono	Poly
BREADS, BAKED GOODS, CAKES, COOKIES, CRACKERS, CHIPS, PIES—continued													
8817	White with coconut frosting, prepared from mix	1	slice(s)	112	23.2	399	4.9	70.8	1.1	11.5	4.4	4.1	2.4
8819	Yellow with chocolate frosting, ready to eat, commercially prepared	1	slice(s)	64	14.0	243	2.4	35.5	1.2	11.1	3.0	6.1	1.4
8822	Yellow with vanilla frosting, ready to eat, commercially prepared	1	slice(s)	64	14.1	239	2.2	37.6	0.2	9.3	1.5	3.9	3.3
	Snack cakes												
8791	Chocolate snack cake, creme filled, with frosting	1	item(s)	50	9.3	200	1.8	30.2	1.6	8.0	2.4	4.3	0.9
25010	Cinnamon coffee cake	1	piece(s)	72	22.6	231	3.6	35.8	0.7	8.3	2.2	2.6	3.0
16777	Funnel cake	1	item(s)	90	37.6	276	7.3	29.1	0.9	14.4	2.7	4.7	6.1
8794	Sponge snack cake, creme filled	1	item(s)	43	8.6	155	1.3	27.2	0.2	4.8	1.1	1.7	1.4
	Snacks, chips, pretzels												
29428	Bagel chips, plain	3	item(s)	29	—	130	3.0	19.0	1.0	4.5	0.5	—	—
29429	Bagel chips, toasted onion	3	item(s)	29	—	130	4.0	20.0	1.0	4.5	0.5	—	—
38192	Chex traditional snack mix	1	cup(s)	45	—	197	3.0	33.3	1.5	6.1	0.8	—	—
654	Potato chips, salted	1	ounce(s)	28	0.6	155	1.9	14.1	1.2	10.6	3.1	2.8	3.5
8816	Potato chips, unsalted	1	ounce(s)	28	0.5	152	2.0	15.0	1.4	9.8	3.1	2.8	3.5
5096	Pretzels, plain, hard, twists	5	item(s)	30	1.0	114	2.7	23.8	1.0	1.1	0.2	0.4	0.4
4632	Pretzels, whole wheat	1	ounce(s)	28	1.1	103	3.1	23.0	2.2	0.7	0.2	0.3	0.2
4641	Tortilla chips, plain	6	item(s)	11	0.2	53	0.8	7.1	0.6	2.5	0.3	0.8	0.5
	Cookies												
8859	Animal crackers	12	item(s)	30	1.2	134	2.1	22.2	0.3	4.1	1.0	2.3	0.6
8876	Brownie, prepared from mix	1	item(s)	24	3.0	112	1.5	12.0	0.5	7.0	1.8	2.6	2.3
25207	Chocolate chip cookies	1	item(s)	30	3.7	140	2.0	16.2	0.6	7.9	2.1	3.3	2.1
8915	Chocolate sandwich cookie with extra creme filling	1	item(s)	13	0.2	65	0.6	8.9	0.4	3.2	0.7	2.1	0.3
14145	Fig Newtons cookies	1	item(s)	16	—	55	0.5	11.0	0.5	1.3	0	—	—
8920	Fortune cookie	1	item(s)	8	0.6	30	0.3	6.7	0.1	0.2	0.1	0.1	0
25208	Oatmeal cookies	1	item(s)	69	12.3	234	5.7	45.1	3.1	4.2	0.7	1.3	1.8
25213	Peanut butter cookies	1	item(s)	35	4.1	163	4.2	16.9	0.9	9.2	1.7	4.7	2.3
33095	Sugar cookies	1	item(s)	16	4.1	61	1.1	7.4	0.1	3.0	0.6	1.3	0.9
9002	Vanilla sandwich cookie with creme filling	1	item(s)	10	0.2	48	0.5	7.2	0.2	2.0	0.3	0.8	0.8
	Crackers												
9012	Cheese cracker sandwich with peanut butter	4	item(s)	28	0.9	139	3.5	15.9	1.0	7.0	1.2	3.6	1.4
9008	Cheese crackers (mini)	30	item(s)	30	0.9	151	3.0	17.5	0.7	7.6	2.8	3.6	0.7
33362	Cheese crackers, low sodium	1	serving(s)	30	0.9	151	3.0	17.5	0.7	7.6	2.9	3.6	0.7
8928	Honey graham crackers	4	item(s)	28	1.2	118	1.9	21.5	0.8	2.8	0.4	1.1	1.1
9016	Matzo crackers, plain	1	item(s)	28	1.2	112	2.8	23.8	0.9	0.4	0.1	0	0.2
9024	Melba toast	3	item(s)	15	0.8	59	1.8	11.5	0.9	0.5	0.1	0.1	0.2
9028	Melba toast, rye	3	item(s)	15	0.7	58	1.7	11.6	1.2	0.5	0.1	0.1	0.2
14189	Ritz crackers	5	item(s)	16	0.5	80	1.0	10.0	0	4.0	1.0	—	—
9014	Rye crispbread crackers	1	item(s)	10	0.6	37	0.8	8.2	1.7	0.1	0	0	0.1
9040	Rye wafer	1	item(s)	11	0.6	37	1.1	8.8	2.5	0.1	0	0	0
432	Saltine crackers	5	item(s)	15	0.8	64	1.4	10.6	0.5	1.7	0.2	1.1	0.2
9046	Saltine crackers, low salt	5	item(s)	15	0.6	65	1.4	10.7	0.5	1.8	0.4	1.0	0.3
9052	Snack cracker sandwich with cheese filling	4	item(s)	28	1.1	134	2.6	17.3	0.5	5.9	1.7	3.2	0.7
9054	Snack cracker sandwich with peanut butter filling	4	item(s)	28	0.8	138	3.2	16.3	0.6	6.9	1.4	3.9	1.3
9048	Snack crackers, round	10	item(s)	30	1.1	151	2.2	18.3	0.5	7.6	1.1	3.2	2.9
9050	Snack crackers, round, low salt	10	item(s)	30	1.1	151	2.2	18.3	0.5	7.6	1.1	3.2	2.9
9044	Soda crackers	5	tem(s)	15	0.8	64	1.4	10.6	0.5	1.7	0.2	1.1	0.2
9059	Wheat cracker sandwich with cheese filling	4	item(s)	28	0.9	139	2.7	16.3	0.9	7.0	1.2	2.9	2.6
9061	Wheat cracker sandwich with peanut butter filling	4	item(s)	28	1.0	139	3.8	15.1	1.2	7.5	1.3	3.3	2.5
9055	Wheat crackers	10	item(s)	30	0.9	142	2.6	19.5	1.4	6.2	1.6	3.4	0.8

Chol (mg)	Calc (mg)	Iron (mg)	Magn (mg)	Pota (mg)	Sodi (mg)	Zinc (mg)	Vit A (µg)	Thia (mg)	Vit E (mg α)	Ribo (mg)	Niac (mg)	Vit B_6 (mg)	Fola (µg DFE)	Vit C (mg)	Vit B_{12} (µg)	Sele (µg)
1	101	1.29	13.4	110.9	318.1	0.37	13.4	0.14	0.13	0.21	1.19	0.03	57.12	0.1	0.1	12.0
35	24	1.33	19.2	113.9	215.7	0.39	21.1	0.07	—	0.10	0.79	0.02	20.5	0	0.1	2.2
35	40	0.68	3.8	33.9	220.2	0.16	12.2	0.06	—	0.04	0.32	0.01	25.6	0	0.1	3.5
0	58	1.80	18.0	88.0	194.5	0.52	0.5	0.01	0.54	0.03	0.46	0.07	17.5	1.0	0	1.7
26	55	1.36	9.9	91.9	277.6	0.30	—	0.17	0.23	0.16	1.29	0.02	66.1	0.3	0.1	9.6
62	126	1.90	16.2	152.1	269.1	0.65	49.5	0.23	1.54	0.32	1.86	0.04	75.6	0	0.3	17.7
7	19	0.54	3.4	37.0	155.1	0.12	2.1	0.06	0.50	0.05	0.52	0.01	23.0	0	0	1.3
0	0	0.72	—	45.0	70.0	—	0	—	—	—	—	—	—	0	0	—
0	0	0.72	—	50.0	300.0	—	0	—	—	—	—	—	—	0	0	—
0	0	0.55	—	75.8	621.2	—	0	0.09	—	0.05	1.21	—	38.5	0	—	—
0	7	0.45	19.8	465.5	148.8	0.67	0	0.01	1.91	0.06	1.18	0.20	21.3	5.3	0	2.3
0	7	0.46	19.0	361.5	2.3	0.30	0	0.04	2.58	0.05	1.08	0.18	12.8	8.8	0	2.3
0	11	1.29	10.5	43.8	514.5	0.25	0	0.13	0.10	0.18	1.57	0.03	86.0	0	0	1.7
0	8	0.76	8.5	121.9	57.6	0.17	0	0.12	—	0.08	1.85	0.07	15.3	0.3	0	—
0	19	0.25	15.8	23.2	45.5	0.26	0	0.00	0.46	0.01	0.13	0.02	2.2	0	0	0.7
0	13	0.82	5.4	30.0	117.9	0.19	0	0.10	0.03	0.09	1.04	0.01	49.5	0	0	2.1
18	14	0.44	12.7	42.2	82.3	0.23	42.2	0.03	—	0.05	0.24	0.02	9.4	0.1	0	2.8
13	11	0.69	12.4	62.1	108.8	0.24	—	0.08	0.54	0.06	0.87	0.01	14.1	0	0	4.1
0	2	1.01	4.7	17.8	45.6	0.10	0	0.02	0.25	0.02	0.25	0.00	9.0	0	0	1.1
0	10	0.36	—	—	57.5	—	0	—	—	—	—	—	—	0	—	—
0	1	0.12	0.6	3.3	21.9	0.01	0.1	0.01	0.00	0.01	0.15	0.00	8.4	0	0	0.2
0	26	1.93	48.8	176.7	311.1	1.42	—	0.26	0.23	0.13	1.35	0.09	65.6	0.3	0	17.4
13	27	0.65	21.1	112.8	154.1	0.46	—	0.08	0.73	0.09	1.85	0.05	35.0	0.1	0.1	4.8
18	5	0.30	1.7	12.2	49.4	0.08	—	0.04	0.28	0.05	0.31	0.01	13.1	0	0	3.1
0	3	0.22	1.4	9.1	34.9	0.04	0	0.02	0.16	0.02	0.27	0.00	8.2	0	0	0.3
0	14	0.76	15.7	61.0	198.8	0.29	0.3	0.15	0.66	0.08	1.63	0.04	39.8	0	0.1	2.3
4	45	1.43	10.8	43.5	298.5	0.33	8.7	0.17	0.01	0.12	1.40	0.16	72.3	0	0.1	2.6
4	45	1.43	10.8	31.8	137.4	0.33	5.1	0.17	0.09	0.12	1.40	0.16	40.2	0	0.1	2.6
0	7	1.04	8.4	37.8	169.4	0.22	0	0.06	0.09	0.08	1.15	0.01	18.5	0	0	2.9
0	4	0.89	7.1	31.8	0.6	0.19	0	0.11	0.01	0.08	1.10	0.03	4.8	0	0	10.5
0	14	0.55	8.9	30.3	124.4	0.30	0	0.06	0.06	0.04	0.61	0.01	29.0	0	0	5.2
0	12	0.55	5.9	29.0	134.9	0.20	0	0.07	—	0.04	0.70	0.01	19.4	0	0	5.8
0	20	0.72	—	10.0	135.0	—	—	—	—	—	—	—	—	0	—	—
0	3	0.24	7.8	31.9	26.4	0.23	0	0.02	0.08	0.01	0.10	0.02	6.5	0	0	3.7
0	4	0.65	13.3	54.5	87.3	0.30	0	0.04	0.08	0.03	0.17	0.03	5.0	0	0	2.6
0	10	0.84	3.3	23.1	160.8	0.12	0	0.01	0.14	0.06	0.78	0.01	33.2	0	0	1.5
0	18	0.81	4.1	108.6	95.4	0.11	0	0.08	0.01	0.06	0.78	0.01	33.2	0	0	2.9
1	72	0.66	10.1	120.1	392.3	0.17	4.8	0.12	0.06	0.19	1.05	0.01	44.8	0	0	6.0
0	23	0.77	15.4	60.2	201.0	0.31	0.3	0.13	0.57	0.07	1.71	0.04	34.2	0	0	3.0
0	36	1.08	8.1	39.9	254.1	0.20	0	0.12	0.60	0.10	1.21	0.01	55.8	0	0	2.0
0	36	1.08	8.1	106.5	111.9	0.20	0	0.12	0.60	0.10	1.21	0.01	55.8	0	0	2.0
0	10	0.84	3.3	23.1	160.8	0.12	0	0.01	0.14	0.06	0.78	0.01	33.2	0	0	1.5
2	57	0.73	15.1	85.7	255.6	0.24	4.8	0.10	—	0.12	0.89	0.07	26.9	0.4	0	6.8
0	48	0.74	10.6	83.2	226.0	0.23	0	0.10	—	0.08	1.64	0.03	26.0	0	0	6.1
0	15	1.32	18.6	54.9	238.5	0.48	0	0.15	0.15	0.09	1.48	0.04	56.1	0	0	1.9

Table A–1 Table of Food Composition (*continued*) (Computer code is for Cengage Diet Analysis Plus program)

DA+ Code	Food Description	QTY	Measure	Wt (g)	H₂O (g)	Ener (cal)	Prot (g)	Carb (g)	Fiber (g)	Fat (g)	Fat Breakdown (g) Sat	Mono	Poly
BREADS, BAKED GOODS, CAKES, COOKIES, CRACKERS, CHIPS, PIES—continued													
9057	Wheat crackers, low salt	10	item(s)	30	0.9	142	2.6	19.5	1.4	6.2	1.6	3.4	0.8
9022	Whole wheat crackers	7	item(s)	28	0.8	124	2.5	19.2	2.9	4.8	1.0	1.6	1.8
	Pastry												
16754	Apple fritter	1	item(s)	17	6.4	61	1.0	5.5	0.2	3.9	0.9	1.7	1.1
41565	Cinnamon rolls with icing, refrigerated dough	1	serving(s)	44	12.3	145	2.0	23.0	0.5	5.0	1.5	—	—
4945	Croissant, butter	1	item(s)	57	13.2	231	4.7	26.1	1.5	12.0	6.6	3.1	0.6
9096	Danish, nut	1	item(s)	65	13.3	280	4.6	29.7	1.3	16.4	3.8	8.9	2.8
9115	Doughnut with creme filling	1	item(s)	85	32.5	307	5.4	25.5	0.7	20.8	4.6	10.3	2.6
9117	Doughnut with jelly filling	1	item(s)	85	30.3	289	5.0	33.2	0.8	15.9	4.1	8.7	2.0
4947	Doughnut, cake	1	item(s)	47	9.8	198	2.4	23.4	0.7	10.8	1.7	4.4	3.7
9105	Doughnut, cake, chocolate glazed	1	item(s)	42	6.8	175	1.9	24.1	0.9	8.4	2.2	4.7	1.0
437	Doughnut, glazed	1	item(s)	60	15.2	242	3.8	26.6	0.7	13.7	3.5	7.7	1.7
10617	Toaster pastry, brown sugar cinnamon	1	item(s)	50	5.3	210	3.0	35.0	1.0	6.0	1.0	4.0	1.0
30928	Toaster pastry, cream cheese	1	item(s)	54	—	200	3.0	23.0	0	11.0	4.5	—	—
	Muffins												
25015	Blueberry	1	item(s)	63	29.7	160	3.4	23.0	0.8	6.0	0.9	1.5	3.3
9189	Corn, ready to eat	1	item(s)	57	18.6	174	3.4	29.0	1.9	4.8	0.8	1.2	1.8
9121	English muffin, plain, enriched	1	item(s)	57	24.0	134	4.4	26.2	1.5	1.0	0.1	0.2	0.5
29582	English muffin, toasted	1	item(s)	50	18.6	128	4.2	25.0	1.5	1.0	0.1	0.2	0.5
9145	English muffin, wheat	1	item(s)	57	24.1	127	5.0	25.5	2.6	1.1	0.2	0.2	0.5
8894	Oat bran	1	item(s)	57	20.0	154	4.0	27.5	2.6	4.2	0.6	1.0	2.4
	Granola bars												
38161	Kudos milk chocolate granola bars w/fruit and nuts	1	item(s)	28	—	90	2.0	15.0	1.0	3.0	1.0	—	—
38196	Nature Valley banana nut crunchy granola bars	2	item(s)	42	—	190	4.0	28.0	2.0	7.0	1.0	—	—
38187	Nature Valley fruit 'n' nut trail mix bar	1	item(s)	35	—	140	3.0	25.0	2.0	4.0	0.5	—	—
1383	Plain, hard	1	item(s)	25	1.0	115	2.5	15.8	1.3	4.9	0.6	1.1	3.0
4606	Plain, soft	1	item(s)	28	1.8	126	2.1	19.1	1.3	4.9	2.1	1.1	1.5
	Pies												
454	Apple pie, prepared from home recipe	1	slice(s)	155	73.3	411	3.7	57.5	2.3	19.4	4.7	8.4	5.2
470	Pecan pie, prepared from home recipe	1	slice(s)	122	23.8	503	6.0	63.7	—	27.1	4.9	13.6	7.0
33356	Pie crust mix, prepared, baked	1	slice(s)	20	2.1	100	1.3	10.1	0.4	6.1	1.5	3.5	0.8
9007	Pie crust, ready to bake, frozen, enriched, baked	1	slice(s)	16	1.8	82	0.7	7.9	0.2	5.2	1.7	2.5	0.6
472	Pumpkin pie, prepared from home recipe	1	slice(s)	155	90.7	316	7.0	40.9	—	14.4	4.9	5.7	2.8
	Rolls												
8555	Crescent dinner roll	1	item(s)	28	9.7	78	2.7	13.8	0.6	1.2	0.3	0.3	0.6
489	Hamburger roll or bun, plain	1	item(s)	43	14.9	120	4.1	21.3	0.9	1.9	0.5	0.5	0.8
490	Hard roll	1	item(s)	57	17.7	167	5.6	30.0	1.3	2.5	0.3	0.6	1.0
5127	Kaiser roll	1	item(s)	57	17.7	167	5.6	30.0	1.3	2.5	0.3	0.6	1.0
5130	Whole wheat roll or bun	1	item(s)	28	9.4	75	2.5	14.5	2.1	1.3	0.2	0.3	0.6
	Sport bars												
37026	Balance original chocolate bar	1	item(s)	50	—	200	14.0	22.0	0.5	6.0	3.5	—	—
37024	Balance original peanut butter bar	1	item(s)	50	—	200	14.0	22.0	1.0	6.0	2.5	—	—
36580	Clif Bar chocolate brownie energy bar	1	item(s)	68	—	240	10.0	45.0	5.0	4.5	1.5	—	—
36583	Clif Bar crunchy peanut butter energy bar	1	item(s)	68	—	250	12.0	40.0	5.0	6.0	1.5	—	—
36589	Clif Luna Nutz over Chocolate energy bar	1	item(s)	48	—	180	10.0	25.0	3.0	4.5	2.5	—	—
12005	PowerBar apple cinnamon	1	item(s)	65	—	230	9.0	45.0	3.0	2.5	0.5	1.5	0.5
16078	PowerBar banana	1	item(s)	65	—	230	9.0	45.0	3.0	2.5	0.5	1.0	0.5
16080	PowerBar chocolate	1	item(s)	65	6.4	230	10.0	45.0	3.0	2.0	0.5	0.5	1.0
29092	PowerBar peanut butter	1	item(s)	65	—	240	10.0	45.0	3.0	3.5	0.5	—	—
	Tortillas												
1391	Corn tortillas, soft	1	item(s)	26	11.9	57	1.5	11.6	1.6	0.7	0.1	0.2	0.4
1669	Flour tortilla	1	item(s)	32	9.7	100	2.7	16.4	1.0	2.5	0.6	1.2	0.5

Chol (mg)	Calc (mg)	Iron (mg)	Magn (mg)	Pota (mg)	Sodi (mg)	Zinc (mg)	Vit A (µg)	Thia (mg)	Vit E (mg α)	Ribo (mg)	Niac (mg)	Vit B_6 (mg)	Fola (µg DFE)	Vit C (mg)	Vit B_{12} (µg)	Sele (µg)
0	15	1.32	18.6	60.9	84.9	0.48	0	0.15	0.15	0.09	1.48	0.04	21.6	0	0	10.1
0	14	0.86	27.7	83.2	184.5	0.60	0	0.05	0.24	0.02	1.26	0.05	7.8	0	0	4.1
14	9	0.26	2.2	22.4	6.8	0.09	7.1	0.03	0.07	0.04	0.23	0.01	9.2	0.2	0.1	2.6
0	—	0.72	—	—	340.1	—	0	—	—	—	—	—	—	—	—	—
38	21	1.15	9.1	67.3	424.1	0.42	117.4	0.22	0.47	0.13	1.24	0.03	74.1	0.1	0.1	12.9
30	61	1.17	20.8	61.8	236.0	0.56	5.9	0.14	0.53	0.15	1.49	0.06	79.3	1.1	0.1	9.2
20	21	1.55	17.0	68.0	262.7	0.68	9.4	0.28	0.24	0.12	1.90	0.05	92.7	0	0.1	9.2
22	21	1.49	17.0	67.2	249.1	0.63	14.5	0.26	0.36	0.12	1.81	0.08	88.4	0	0.2	10.6
17	21	0.91	9.4	59.7	256.6	0.25	17.9	0.10	0.90	0.11	0.87	0.02	32.9	0.1	0.1	4.4
24	89	0.95	14.3	44.5	142.8	0.23	5.0	0.01	0.08	0.02	0.19	0.01	27.3	0	0	1.7
4	26	0.36	13.2	64.8	205.2	0.46	2.4	0.53	—	0.04	0.39	0.03	13.2	0.1	0.1	5.0
0	0	1.80	—	70.0	190.0	—	—	0.15	—	0.17	2.00	0.20	21.0	0	0	—
10	100	1.80	—	—	220.0	—	—	0.15	—	0.17	2.00	—	21.0	0	0.6	—
20	56	1.02	7.8	70.2	289.4	0.28	—	0.17	0.75	0.15	1.25	0.02	62.5	0.4	0.1	8.8
15	42	1.60	18.2	39.3	297.0	0.30	29.6	0.15	0.45	0.18	1.16	0.04	63.8	0	0.1	8.7
0	30	1.42	12.0	74.7	264.5	0.39	0	0.25	—	0.16	2.21	0.02	57.0	0	0	—
0	95	1.36	11.0	71.5	252.0	0.38	0	0.19	0.16	0.14	1.90	0.02	62.5	0.1	0	13.5
0	101	1.63	21.1	106.0	217.7	0.61	0	0.24	0.25	0.16	1.91	0.05	46.7	0	0	16.6
0	36	2.39	89.5	289.0	224.0	1.04	0	0.14	0.37	0.05	0.23	0.09	79.2	0	0	6.3
0	200	0.36	—	—	60.0	—	0	—	—	—	—	—	—	0	0	—
0	20	1.08	—	120.0	160.0	—	0	—	—	—	—	—	—	0	—	—
0	0	0.00	—	—	95.0	—	0	—	—	—	—	—	—	0	—	—
0	15	0.72	23.8	82.3	72.0	0.50	0	0.06	—	0.03	0.39	0.02	5.8	0.2	0	4.0
0	30	0.72	21.0	92.3	79.0	0.42	0	0.08	—	0.04	0.14	0.02	6.8	0	0.1	4.6
0	11	1.73	10.9	122.5	327.1	0.29	17.1	0.22	—	0.16	1.90	0.05	58.9	2.6	0	12.1
106	39	1.80	31.7	162.3	319.6	1.24	100.0	0.22	—	0.22	1.03	0.07	41.5	0.2	0.2	14.6
0	12	0.43	3.0	12.4	145.8	0.07	0	0.06	—	0.03	0.47	0.01	22.2	0	0	4.4
0	3	0.36	2.9	17.6	103.5	0.05	0	0.04	0.42	0.06	0.39	0.01	16.3	0	0	0.5
65	146	1.96	29.5	288.3	348.8	0.71	660.3	0.14	—	0.31	1.21	0.07	43.4	2.6	0.1	11.0
0	39	0.93	5.9	26.3	134.1	0.18	0	0.11	0.02	0.09	1.16	0.02	47.6	0	0.1	5.5
0	59	1.42	9.0	40.4	206.0	0.28	0	0.17	0.03	0.13	1.78	0.03	73.5	0	0.1	8.4
0	54	1.87	15.4	61.6	310.1	0.53	0	0.27	0.23	0.19	2.41	0.02	86.1	0	0	22.3
0	54	1.86	15.4	61.6	310.1	0.53	0	0.27	0.23	0.19	2.41	0.01	86.1	0	0	22.3
0	30	0.69	24.1	77.1	135.5	0.57	0	0.07	0.26	0.04	1.04	0.06	8.5	0	0	14.0
3	100	4.50	40.0	160.0	180.0	3.75	—	0.37	—	0.42	5.00	0.50	102.0	60.0	1.5	17.5
3	100	4.50	40.0	130.0	230.0	3.75	—	0.37	—	0.42	5.00	0.50	102.0	60.0	1.5	17.5
0	250	4.50	100.0	370.0	150.0	3.00	—	0.37	—	0.25	3.00	0.40	80.0	60.0	0.9	14.0
0	250	4.50	100.0	230.0	250.0	3.00	—	0.37	—	0.25	3.00	0.40	80.0	60.0	0.9	14.0
0	350	5.40	80.0	190.0	190.0	5.25	—	1.20	—	1.36	16.00	2.00	400.0	60.0	6.0	24.5
0	300	6.30	140.0	125.0	100.0	5.25	—	1.50	—	1.70	20.00	2.00	400.0	60.0	6.0	—
0	300	6.30	140.0	190.0	100.0	5.25	0	1.50	—	1.70	20.00	2.00	400.0	60.0	6.0	—
0	300	6.30	140.0	200.0	95.0	5.25	0	1.50	—	1.70	20.00	2.00	400.0	60.0	6.0	5.1
0	300	6.30	140.0	130.0	120.0	5.25	0	1.50	—	1.70	20.00	2.00	400.0	60.0	6.0	—
0	21	0.32	18.7	48.4	11.7	0.34	0	0.02	0.07	0.02	0.39	0.06	1.3	0	0	1.6
0	41	1.06	7.0	49.6	203.5	0.17	0	0.17	0.06	0.08	1.14	0.01	64.3	0	0	7.1

DA+ Code	Food Description	QTY	Measure	Wt (g)	H₂O (g)	Ener (cal)	Prot (g)	Carb (g)	Fiber (g)	Fat (g)	Fat Breakdown (g)		
											Sat	Mono	Poly
BREADS, BAKED GOODS, CAKES, COOKIES, CRACKERS, CHIPS, PIES—continued													
	Pancakes, waffles												
8926	Pancakes, blueberry, prepared from recipe	3	item(s)	114	60.6	253	7.0	33.1	0.8	10.5	2.3	2.6	4.7
5037	Pancakes, prepared from mix with egg and milk	3	item(s)	114	60.3	249	8.9	32.9	2.1	8.8	2.3	2.4	3.3
1390	Taco shells, hard	1	item(s)	13	1.0	62	0.9	8.3	0.6	2.8	0.6	1.6	0.5
30311	Waffle, 100% whole grain	1	item(s)	75	32.3	200	6.9	25.0	1.9	8.4	2.3	3.3	2.1
9219	Waffle, plain, frozen, toasted	2	item(s)	66	20.2	206	4.7	32.5	1.6	6.3	1.1	3.2	1.5
500	Waffle, plain, prepared from recipe	1	item(s)	75	31.5	218	5.9	24.7	1.7	10.6	2.1	2.6	5.1
CEREAL, FLOUR, GRAIN, PASTA, NOODLES, POPCORN													
	Grain												
2861	Amaranth, dry	½	cup(s)	98	9.6	365	14.1	64.5	9.1	6.3	1.6	1.4	2.8
1953	Barley, pearled, cooked	½	cup(s)	79	54.0	97	1.8	22.2	3.0	0.3	0.1	0	0.2
1956	Buckwheat groats, cooked, roasted	½	cup(s)	84	63.5	77	2.8	16.8	2.3	0.5	0.1	0.2	0.2
1957	Bulgur, cooked	½	cup(s)	91	70.8	76	2.8	16.9	4.1	0.2	0	0	0.1
1963	Couscous, cooked	½	cup(s)	79	57.0	88	3.0	18.2	1.1	0.1	0	0	0.1
1967	Millet, cooked	½	cup(s)	120	85.7	143	4.2	28.4	1.6	1.2	0.2	0.2	0.6
1969	Oat bran, dry	½	cup(s)	47	3.1	116	8.1	31.1	7.2	3.3	0.6	1.1	1.3
1972	Quinoa, dry	½	cup(s)	85	11.3	313	12.0	54.5	5.9	5.2	0.6	1.4	2.8
	Rice												
129	Brown, long grain, cooked	½	cup(s)	98	71.3	108	2.5	22.4	1.8	0.9	0.2	0.3	0.3
2863	Brown, medium grain, cooked	½	cup(s)	98	71.1	109	2.3	22.9	1.8	0.8	0.2	0.3	0.3
37488	Jasmine, saffroned, cooked	½	cup(s)	280	—	340	8.0	78.0	0	0	0	0	0
30280	Pilaf, cooked	½	cup(s)	103	74.0	129	2.1	22.2	0.6	3.3	0.6	1.5	1.0
28066	Spanish, cooked	½	cup(s)	244	184.2	241	5.7	50.2	3.3	1.9	0.4	0.6	0.7
2867	White glutinous, cooked	½	cup(s)	87	66.7	84	1.8	18.3	0.9	0.2	0	0.1	0.1
484	White, long grain, boiled	½	cup(s)	79	54.1	103	2.1	22.3	0.3	0.2	0.1	0.1	0.1
482	White, long grain, enriched, instant, boiled	½	cup(s)	83	59.4	97	1.8	20.7	0.5	0.4	0	0.1	0
486	White, long grain, enriched, parboiled, cooked	½	cup(s)	79	55.6	97	2.3	20.6	0.7	0.3	0.1	0.1	0.1
1194	Wild brown, cooked	½	cup(s)	82	60.6	83	3.3	17.5	1.5	0.3	0	0	0.2
	Flour and grain fractions												
505	All purpose flour, self-rising, enriched	½	cup(s)	63	6.6	221	6.2	46.4	1.7	0.6	0.1	0	0.2
503	All purpose flour, white, bleached, enriched	½	cup(s)	63	7.4	228	6.4	47.7	1.7	0.6	0.1	0	0.2
1643	Barley flour	½	cup(s)	56	5.5	198	4.2	44.7	2.1	0.8	0.2	0.1	0.4
383	Buckwheat flour, whole groat	½	cup(s)	60	6.7	201	7.6	42.3	6.0	1.9	0.4	0.6	0.6
504	Cake wheat flour, enriched	½	cup(s)	69	8.6	248	5.6	53.5	1.2	0.6	0.1	0.1	0.3
426	Cornmeal, degermed, enriched	½	cup(s)	69	7.8	255	5.0	54.6	2.8	1.2	0.1	0.2	0.5
424	Cornmeal, yellow whole grain	½	cup(s)	61	6.2	221	4.9	46.9	4.4	2.2	0.3	0.6	1.0
1978	Dark rye flour	½	cup(s)	64	7.1	207	9.0	44.0	14.5	1.7	0.2	0.2	0.8
1644	Masa corn flour, enriched	½	cup(s)	57	5.1	208	5.3	43.5	5.5	2.1	0.3	0.6	1.0
1976	Rice flour, brown	½	cup(s)	79	9.4	287	5.7	60.4	3.6	2.2	0.4	0.8	0.8
1645	Rice flour, white	½	cup(s)	79	9.4	289	4.7	63.3	1.9	1.1	0.3	0.3	0.3
1980	Semolina, enriched	½	cup(s)	84	10.6	301	10.6	60.8	3.2	0.9	0.1	0.1	0.4
2827	Soy flour, raw	½	cup(s)	42	2.2	185	14.7	14.9	4.1	8.8	1.3	1.9	4.9
1990	Wheat germ, crude	2	tablespoon(s)	14	1.6	52	3.3	7.4	1.9	1.4	0.2	0.2	0.9
506	Whole wheat flour	½	cup(s)	60	6.2	203	8.2	43.5	7.3	1.1	0.2	0.1	0.5
	Breakfast bars												
39230	Atkins Morning Start apple crisp breakfast bar	1	item(s)	37	—	170	11.0	12.0	6.0	9.0	4.0	—	—
10571	Nutri-Grain apple cinnamon cereal bar	1	item(s)	37	—	140	2.0	27.0	1.0	3.0	0.5	2.0	0.5
10647	Nutri-Grain blueberry cereal bar	1	item(s)	37	5.4	140	2.0	27.0	1.0	3.0	0.5	2.0	0.5
10648	Nutri-Grain raspberry cereal bar	1	item(s)	37	5.4	140	2.0	27.0	1.0	3.0	0.5	2.0	0.5
10649	Nutri-Grain strawberry cereal bar	1	item(s)	37	5.4	140	2.0	27.0	1.0	3.0	0.5	2.0	0.5
	Breakfast cereals, hot												
1260	Cream of Wheat, instant, prepared	½	cup(s)	121	—	388	12.9	73.3	4.3	0	0	0	0
365	Farina, enriched, cooked w/water and salt	½	cup(s)	117	102.4	56	1.7	12.2	0.3	0.1	0	0	0

PAGE KEY: A-2 = Breads/Baked Goods A-8 = Cereal/Rice/Pasta A-12 = Fruit A-18 = Vegetables/Legumes A-28 = Nuts/Seeds A-30 = Vegetarian A-32 = Dairy A-40 = Eggs A-40 = Seafood A-44 = Meats A-48 = Poultry A-48 = Processed Meats A-50 = Beverages A-54 = Fats/Oils A-56 = Sweets A-58 = Spices/Condiments/Sauces A-62 = Mixed Foods/Soups/Sandwiches A-68 = Fast Food A-88 = Convenience A-90 = Baby Foods

Chol (mg)	Calc (mg)	Iron (mg)	Magn (mg)	Pota (mg)	Sodi (mg)	Zinc (mg)	Vit A (µg)	Thia (mg)	Vit E (mg α)	Ribo (mg)	Niac (mg)	Vit B_6 (mg)	Fola (µg DFE)	Vit C (mg)	Vit B_{12} (µg)	Sele (µg)
64	235	1.96	18.2	157.3	469.7	0.61	57.0	0.22	—	0.31	1.73	0.05	60.4	2.5	0.2	16.0
81	245	1.48	25.1	226.9	575.7	0.85	82.1	0.22	—	0.35	1.40	0.12	61.6	0.7	0.4	—
0	13	0.25	11.3	29.7	51.7	0.21	0.1	0.03	0.09	0.01	0.25	0.03	11.1	0	0	0.6
71	194	1.60	28.5	171.0	371.3	0.87	48.8	0.15	0.32	0.25	1.47	0.08	38.3	0	0.4	20.0
10	203	4.56	15.8	95.0	481.8	0.35	262.7	0.34	0.64	0.46	5.86	0.68	78.5	0	1.9	8.3
52	191	1.73	14.3	119.3	383.3	0.51	48.8	0.19	—	0.26	1.55	0.04	51.0	0.3	0.2	34.7
0	149	7.40	259.3	356.8	20.5	3.10	0	0.06	—	0.20	1.24	0.20	47.8	4.1	0	—
0	9	1.04	17.3	73.0	2.4	0.64	0	0.06	0.01	0.04	1.61	0.09	12.6	0	0	6.8
0	6	0.67	42.8	73.9	3.4	0.51	0	0.03	0.07	0.03	0.79	0.06	11.8	0	0	1.8
0	9	0.87	29.1	61.9	4.6	0.51	0	0.05	0.01	0.02	0.91	0.07	16.4	0	0	0.5
0	6	0.30	6.3	45.5	3.9	0.20	0	0.05	0.10	0.02	0.77	0.04	11.8	0	0	21.6
0	4	0.75	52.8	74.4	2.4	1.09	0	0.12	0.02	0.09	1.59	0.13	22.8	0	0	1.1
0	27	2.54	110.5	266.0	1.9	1.46	0	0.55	0.47	0.10	0.43	0.07	24.4	0	0	21.2
0	40	3.88	167.4	478.5	4.2	2.62	0.8	0.30	2.06	0.26	1.28	0.40	156.4	0	0	7.2
0	10	0.41	41.9	41.9	4.9	0.61	0	0.09	0.02	0.02	1.49	0.14	3.9	0	0	9.6
0	10	0.51	42.9	77.0	1.0	0.60	0	0.09	—	0.01	1.29	0.14	3.9	0	0	38.0
0	—	2.16	—	—	780.0	—	—	—	—	—	—	—	—	—	—	—
0	11	1.16	9.3	54.6	390.4	0.37	33.0	0.13	0.28	0.02	1.23	0.06	73.1	0.4	0	4.3
0	37	1.52	95.4	330.5	97.1	1.40	—	0.27	0.12	0.05	3.24	0.38	89.9	22.6	0	14.3
0	2	0.12	4.4	8.7	4.4	0.35	0	0.01	0.03	0.01	0.25	0.02	0.9	0	0	4.9
0	8	0.94	9.5	27.7	0.8	0.38	0	0.12	0.03	0.01	1.16	0.07	76.6	0	0	5.9
0	7	1.46	4.1	7.4	3.3	0.40	0	0.06	0.01	0.01	1.43	0.04	97.4	0	0	4.0
0	15	1.43	7.1	44.2	1.6	0.29	0	0.16	0.01	0.01	1.82	0.12	107.4	0	0	7.3
0	2	0.49	26.2	82.8	2.5	1.09	0	0.04	0.19	0.07	1.05	0.11	21.3	0	0	0.7
0	211	2.90	11.9	77.5	793.7	0.38	0	0.42	0.02	0.24	3.64	0.02	193.4	0	0	21.5
0	9	2.90	13.7	66.9	1.2	0.42	0	0.48	0.02	0.30	3.68	0.02	183.3	0	0	21.2
0	16	0.70	45.4	185.9	4.5	1.04	0	0.06	—	0.02	2.56	0.14	4.5	0	0	2.0
0	25	2.42	150.6	346.2	6.6	1.86	0	0.24	0.18	0.10	3.68	0.34	32.4	0	0	3.4
0	10	5.01	11.0	71.9	1.4	0.42	0	0.61	0.01	0.29	4.65	0.02	194.6	0	0	3.4
0	2	2.98	24.1	104.9	4.8	0.48	7.6	0.42	0.10	0.28	3.66	0.12	231.1	0	0	8.0
0	4	2.10	77.5	175.1	21.3	1.10	6.7	0.22	0.24	0.12	2.20	0.18	15.2	0	0	9.4
0	36	4.12	158.7	467.2	0.6	3.58	0.6	0.20	0.90	0.16	2.72	0.28	21.1	0	0	22.8
0	80	4.10	62.7	169.9	2.8	1.00	0	0.80	0.08	0.42	5.60	0.20	190.9	0	0	8.5
0	9	1.56	88.5	228.3	6.3	1.92	0	0.34	0.94	0.06	5.00	0.58	12.6	0	0	—
0	8	0.26	27.6	60.0	0	0.62	0	0.10	0.08	0.02	2.04	0.34	3.2	0	0	11.9
0	14	3.64	39.2	155.3	0.8	0.86	0	0.66	0.20	0.46	5.00	0.08	219.2	0	0	74.6
0	87	2.70	182.0	1067.0	5.5	1.65	2.5	0.24	0.82	0.48	1.83	0.18	146.4	0	0	3.2
0	6	0.90	34.4	128.2	1.7	1.76	0	0.27	—	0.07	0.97	0.18	40.4	0	0	11.4
0	20	2.32	82.8	243.0	3.0	1.74	0	0.26	0.48	0.12	3.82	0.20	26.4	0	0	42.4
0	200	—	—	90.0	70.0	—	—	0.22	—	0.25	3.00	—	—	9.0	—	—
0	200	1.80	8.0	75.0	110.0	1.50	—	0.37	—	0.42	5.00	0.50	40.0	0	—	—
0	200	1.80	8.0	75.0	110.0	1.50	—	0.37	—	0.42	5.00	0.50	40.0	0	0	—
0	200	1.80	8.0	70.0	110.0	1.50	—	0.37	—	0.42	5.00	0.50	40.0	0	0	—
0	200	1.80	8.0	55.0	110.0	1.50	—	0.37	—	0.42	5.00	0.50	40.0	0	0	—
0	862	34.91	21.4	150.8	732.6	0.86	—	1.59	—	1.47	21.55	2.15	122.2	0	0	—
0	5	0.58	2.3	15.1	383.3	0.09	0	0.07	0.01	0.05	0.57	0.01	139.2	0	0	10.6

											Fat Breakdown (g)		
DA+ Code	**Food Description**	**QTY**	**Measure**	**Wt (g)**	**H_2O (g)**	**Ener (cal)**	**Prot (g)**	**Carb (g)**	**Fiber (g)**	**Fat (g)**	**Sat**	**Mono**	**Poly**
CEREAL, FLOUR, GRAIN, PASTA, NOODLES, POPCORN—continued													
363	Grits, white corn, regular and quick, enriched, cooked w/water and salt	½	cup(s)	121	103.3	71	1.7	15.6	0.4	0.2	0	0.1	0.1
8636	Grits, yellow corn, regular and quick, enriched, cooked w/salt	½	cup(s)	121	103.3	71	1.7	15.6	0.4	0.2	0	0.1	0.1
8657	Oatmeal, cooked w/water	½	cup(s)	117	97.8	83	3.0	14.0	2.0	1.8	0.4	0.5	0.7
5500	Oatmeal, maple and brown sugar, instant, prepared	1	item(s)	198	150.2	200	4.8	40.4	2.4	2.2	0.4	0.7	0.8
5510	Oatmeal, ready to serve, packet, prepared	1	item(s)	186	158.7	112	4.1	19.8	2.7	2.0	0.4	0.7	0.8
	Breakfast cereals, ready to eat												
1197	All-Bran	1	cup(s)	62	1.3	160	8.1	46.0	18.2	2.0	0.4	0.4	1.3
1200	All-Bran Buds	1	cup(s)	91	2.7	212	6.4	72.7	39.1	1.9	0.4	0.5	1.2
1199	Apple Jacks	1	cup(s)	33	0.9	130	1.0	30.0	0.5	0.5	0	—	—
1204	Cap'n Crunch	1	cup(s)	36	0.9	147	1.3	30.7	1.3	2.0	0.5	0.4	0.3
1205	Cap'n Crunch Crunchberries	1	cup(s)	35	0.9	133	1.3	29.3	1.3	2.0	0.5	0.4	0.3
1206	Cheerios	1	cup(s)	30	1.0	110	3.0	22.0	3.0	2.0	0	0.5	0.5
3415	Cocoa Puffs	1	cup(s)	30	0.6	120	1.0	26.0	0.2	1.0	—	—	—
1207	Cocoa Rice Krispies	1	cup(s)	41	1.0	160	1.3	36.0	1.3	1.3	0.7	0	0
5522	Complete wheat bran flakes	1	cup(s)	39	1.4	120	4.0	30.7	6.7	0.7	—	—	—
1211	Corn Flakes	1	cup(s)	28	0.9	100	2.0	24.0	1.0	0	0	0	0
1247	Corn Pops	1	cup(s)	31	0.9	120	1.0	28.0	0.3	0	0	0	0
1937	Cracklin' Oat Bran	1	cup(s)	65	2.3	267	5.3	46.7	8.0	9.3	4.0	4.7	1.3
1220	Froot Loops	1	cup(s)	32	0.8	120	1.0	28.0	1.0	1.0	0.5	0	0
38214	Frosted Cheerios	1	cup(s)	37	—	149	2.5	31.1	1.2	1.2	—	—	—
372	Frosted Flakes	1	cup(s)	41	1.1	160	1.3	37.3	1.3	0	0	0	0
38215	Frosted Mini Chex	1	cup(s)	40	—	147	1.3	36.0	0	0	0	0	0
10268	Frosted Mini-Wheats	1	cup(s)	59	3.1	208	5.8	47.4	5.8	1.2	0	0	0.6
38216	Frosted Wheaties	1	cup(s)	40	—	147	1.3	36.0	0.3	0	0	0	0
1223	Granola, prepared	½	cup(s)	61	3.3	298	9.1	32.5	5.5	14.7	2.5	5.8	5.6
2415	Honey Bunches of Oats honey roasted	1	cup(s)	40	0.9	160	2.7	33.3	1.3	2.0	0.7	1.2	0.1
1227	Honey Nut Cheerios	1	cup(s)	37	0.9	149	3.7	29.9	2.5	1.9	0	0.6	0.6
2424	Honeycomb	1	cup(s)	22	0.3	83	1.5	19.5	0.8	0.4	0	—	—
10286	Kashi whole grain puffs	1	cup(s)	19	—	70	2.0	15.0	1.0	0.5	0	—	—
41142	Kellogg's Mueslix	1	cup(s)	83	7.2	298	7.6	60.8	6.1	4.6	0.7	2.4	1.5
1231	Kix	1	cup(s)	24	0.5	96	1.6	20.8	0.8	0.4	—	—	—
30569	Life	1	cup(s)	43	1.7	160	4.0	33.3	2.7	2.0	0.3	0.6	0.6
1233	Lucky Charms	1	cup(s)	24	0.6	96	1.6	20.0	0.8	0.8	—	—	—
38220	Multi Grain Cheerios	1	cup(s)	30	—	110	3.0	24.0	3.0	1.0	—	—	—
1201	Multi-Bran Chex	1	cup(s)	63	1.3	216	4.3	52.9	8.6	1.6	0	0	0.5
13633	Post Bran Flakes	1	cup(s)	40	1.5	133	4.0	32.0	6.7	0.7	0	—	—
1241	Product 19	1	cup(s)	30	1.0	100	2.0	25.0	1.0	0	0	0	0
32432	Puffed rice, fortified	1	cup(s)	14	0.4	56	0.9	12.6	0.2	0.1	0	—	—
32433	Puffed wheat, fortified	1	cup(s)	12	0.4	44	1.8	9.6	0.5	0.1	0	—	—
13334	Quaker 100% natural granola oats and honey	½	cup(s)	48	—	220	5.0	31.0	3.0	9.0	3.8	4.1	1.2
13335	Quaker 100% natural granola oats, honey, and raisins	½	cup(s)	51	—	230	5.0	34.0	3.0	9.0	3.6	3.8	1.1
2420	Raisin Bran	1	cup(s)	59	5.0	190	4.0	46.0	8.0	1.0	0	0.1	0.4
1244	Rice Chex	1	cup(s)	31	0.8	120	2.0	27.0	0.3	0	0	0	0
1245	Rice Krispies	1	cup(s)	26	0.8	96	1.6	23.2	0	0	0	0	0
5593	Shredded Wheat	1	cup(s)	49	0.4	177	5.8	40.9	6.9	1.1	0.1	0	0.2
1248	Smacks	1	cup(s)	36	1.1	133	2.7	32.0	1.3	0.7	—	—	—
1246	Special K	1	cup(s)	31	0.9	110	7.0	22.0	0.5	0	0	0	0
3428	Total corn flakes	1	cup(s)	23	0.6	83	1.5	18.0	0.6	0	0	0	0
1253	Total whole grain	1	cup(s)	40	1.1	147	2.7	30.7	4.0	1.3	—	—	—
1254	Trix	1	cup(s)	30	0.6	120	1.0	27.0	1.0	1.0	—	—	—
382	Wheat germ, toasted	2	tablespoon(s)	14	0.8	54	4.1	7.0	2.1	1.5	0.3	0.2	0.9
1257	Wheaties	1	cup(s)	36	1.2	132	3.6	28.8	3.6	1.2	—	—	—
	Pasta, noodles												
449	Chinese chow mein noodles, cooked	½	cup(s)	23	0.2	119	1.9	12.9	0.9	6.9	1.0	1.7	3.9
1995	Corn pasta, cooked	½	cup(s)	70	47.8	88	1.8	19.5	3.4	0.5	0.1	0.1	0.2

Chol (mg)	Calc (mg)	Iron (mg)	Magn (mg)	Pota (mg)	Sodi (mg)	Zinc (mg)	Vit A (µg)	Thia (mg)	Vit E (mg α)	Ribo (mg)	Niac (mg)	Vit B_6 (mg)	Fola (µg DFE)	Vit C (mg)	Vit B_{12} (µg)	Sele (µg)
0	4	0.73	6.1	25.4	269.8	0.08	0	0.10	0.02	0.07	0.87	0.03	46.0	0	0	3.8
0	4	0.73	6.1	25.4	269.8	0.08	2.4	0.10	0.02	0.07	0.87	0.03	44.8	0	0	3.3
0	11	1.05	31.6	81.9	4.7	1.17	0	0.09	0.09	0.02	0.26	0.01	7.0	0	0	6.3
0	26	6.83	49.9	126.4	403.5	1.03	0	1.02	—	0.05	1.56	0.30	42.2	0	0	11.1
0	21	3.96	44.7	112.4	240.9	0.92	0	0.60	—	0.04	0.77	0.18	18.7	0	0	3.8
0	241	10.90	224.4	632.4	150.0	3.00	300.1	1.40	—	1.68	9.16	7.44	1362.8	12.4	12.0	5.8
0	57	13.64	186.4	909.1	614.5	4.55	464.5	1.09	1.42	1.27	15.45	6.09	2054.8	18.2	18.2	26.3
0	0	4.50	8.0	30.0	130.0	1.50	150.2	0.37	—	0.42	5.00	0.50	196.0	15.0	1.5	2.4
0	5	6.80	20.0	73.3	266.7	5.00	2.5	0.51	—	0.57	6.68	0.67	946.8	0	0	6.7
0	7	6.53	18.7	73.3	240.0	5.13	2.4	0.51	—	0.57	6.68	0.67	910.7	0	0	6.7
0	100	8.10	40.0	95.0	280.0	3.75	150.3	0.37	—	0.42	5.00	0.50	493.2	6.0	1.5	11.3
0	100	4.50	8.0	50.0	170.0	3.75	0	0.37	—	0.42	5.00	0.50	165.9	6.0	1.5	2.0
0	53	6.00	10.7	66.7	253.3	2.00	200.1	0.49	—	0.56	6.67	0.67	442.8	20.0	2.0	5.8
0	0	24.00	53.3	226.7	280.0	20.00	300.1	2.00	—	2.27	26.67	2.67	909.1	80.0	8.0	4.1
0	0	8.10	3.4	25.0	200.0	0.16	149.8	0.37	—	0.42	5.00	0.50	221.8	6.0	1.5	1.4
0	0	1.80	2.5	25.0	120.0	1.50	150.0	0.37	—	0.42	5.00	0.50	174.7	6.0	1.5	2.0
0	27	2.40	80.0	293.3	200.0	2.00	299.9	0.49	—	0.56	6.67	0.67	217.1	20.0	2.0	14.4
0	0	4.50	8.0	35.0	150.0	1.50	150.1	0.37	—	0.42	5.00	0.50	166.1	15.0	1.5	2.3
0	124	5.60	19.9	68.4	261.3	4.67	—	0.46	—	0.52	6.22	0.62	444.4	7.5	1.9	—
0	0	6.00	3.7	26.7	200.0	0.20	200.1	0.49	—	0.56	6.67	0.67	260.4	8.0	2.0	1.8
0	133	12.00	—	33.3	266.7	4.00	—	0.49	—	0.56	6.67	0.67	266.7	8.0	2.0	—
0	0	16.66	69.4	196.7	5.8	1.74	0	0.43	—	0.49	5.78	0.58	193.5	0	1.7	2.4
0	133	10.80	0	46.7	266.7	10.00	—	1.00	—	1.13	13.33	1.33	901.2	8.0	4.0	—
0	48	2.58	106.8	329.4	15.3	2.45	0.6	0.44	6.77	0.17	1.30	0.17	50.0	0.7	0	17.0
0	0	10.80	21.3	0	253.3	0.40	—	0.49	—	0.56	6.67	0.67	549.6	0	2.0	—
0	124	5.60	39.8	112.0	336.0	4.67	—	0.46	—	0.52	6.22	0.62	444.4	7.5	1.9	8.8
0	0	2.03	6.0	26.3	165.4	1.13	—	0.28	—	0.32	3.74	0.37	126.1	0	1.1	—
0	0	0.36	—	60.0	0	—	0	0.03	—	0.03	0.80	0.00	—	0	—	—
0	48	6.83	74.2	363.3	257.5	5.67	136.7	0.67	6.00	0.67	8.33	3.08	1030.0	0.3	9.2	14.4
0	120	6.48	6.4	28.0	216.0	3.00	120.2	0.30	—	0.34	4.00	0.40	317.8	4.8	1.2	4.8
0	149	11.87	41.3	120.0	213.3	5.33	0.9	0.53	—	0.60	7.12	0.71	607.6	0	0	10.7
0	80	3.60	12.8	48.0	168.0	3.00	—	0.30	—	0.34	4.00	0.40	300.7	4.8	1.2	4.8
0	100	18.00	24.0	85.0	200.0	15.00	—	1.50	—	1.70	20.00	2.00	699.3	15.0	6.0	—
0	108	17.50	64.8	237.7	410.6	4.05	171.1	0.40	—	0.45	5.40	0.54	893.3	6.5	1.6	4.9
0	0	10.80	80.0	266.7	280.0	2.00	—	0.49	—	0.56	6.67	0.67	453.3	0	2.0	—
0	0	18.00	16.0	50.0	210.0	15.00	225.3	1.50	—	1.70	20.00	2.00	675.9	60.0	6.0	3.6
0	1	4.43	3.5	15.8	0.4	0.14	0	0.36	—	0.25	4.94	0.01	2.7	0	0	1.5
0	3	3.80	17.4	41.8	0.5	0.28	0	0.31	—	0.21	4.23	0.02	3.8	0	0	14.8
0	61	1.20	51.0	220.0	20.0	1.05	0.5	0.13	—	0.12	0.82	0.07	16.8	0.2	0.1	8.3
0	59	1.20	49.0	250.0	20.0	0.99	0.5	0.13	—	0.12	0.80	0.08	15.8	0.4	0.1	8.8
0	20	10.80	80.0	360.0	360.0	2.25	—	0.37	—	0.42	5.00	0.50	248.4	0	2.1	—
0	100	9.00	9.3	35.0	290.0	3.75	—	0.37	—	0.42	5.00	0.50	389.4	6.0	1.5	1.2
0	0	1.44	12.8	32.0	256.0	0.48	120.1	0.30	—	0.34	4.80	0.40	237.4	4.8	1.2	4.1
0	18	2.90	60.3	179.3	1.1	1.37	0	0.14	—	0.12	3.47	0.18	21.1	0	0	2.0
0	0	0.48	10.7	53.3	66.7	0.40	200.2	0.49	—	0.56	6.67	0.67	224.6	8.0	2.0	17.5
0	0	8.10	16.0	60.0	220.0	0.90	225.1	0.52	—	0.59	7.00	2.00	675.8	21.0	6.0	7.0
0	752	13.53	0	22.6	157.9	11.28	112.8	1.13	22.56	1.28	15.04	1.50	518.2	45.1	4.5	1.2
0	1333	24.00	32.0	120.0	253.3	20.00	200.4	2.00	31.32	2.27	26.67	2.67	901.2	80.0	8.0	1.9
0	100	4.50	0	15.0	190.0	3.75	150.3	0.37	—	0.42	5.00	0.50	155.4	6.0	1.5	6.0
0	6	1.28	45.2	133.8	0.6	2.35	0.7	0.23	2.25	0.11	0.79	0.13	49.7	0.8	0	9.2
0	24	9.72	38.4	126.0	264.0	9.00	180.4	0.90	—	1.02	12.00	1.20	403.6	7.2	3.6	1.7
0	5	1.06	11.7	27.0	98.8	0.31	0	0.13	0.78	0.09	1.33	0.02	31.1	0	0	9.7
0	1	0.18	25.2	21.7	0	0.44	2.1	0.04	—	0.02	0.39	0.04	4.2	0	0	2.0

DA+ Code	Food Description	QTY	Measure	Wt (g)	H₂O (g)	Ener (cal)	Prot (g)	Carb (g)	Fiber (g)	Fat (g)	Fat Breakdown (g) Sat	Mono	Poly
	CEREAL, FLOUR, GRAIN, PASTA, NOODLES, POPCORN—continued												
448	Egg noodles, enriched, cooked	½	cup(s)	80	54.2	110	3.6	20.1	1.0	1.7	0.3	0.5	0.4
1563	Egg noodles, spinach, enriched, cooked	½	cup(s)	80	54.8	106	4.0	19.4	1.8	1.3	0.3	0.4	0.3
440	Macaroni, enriched, cooked	½	cup(s)	70	43.5	111	4.1	21.6	1.3	0.7	0.1	0.1	0.2
2000	Macaroni, tricolor vegetable, enriched, cooked	½	cup(s)	67	45.8	86	3.0	17.8	2.9	0.1	0	0	0
1996	Plain pasta, fresh-refrigerated, cooked	½	cup(s)	64	43.9	84	3.3	16.0	—	0.7	0.1	0.1	0.3
1725	Ramen noodles, cooked	½	cup(s)	114	94.5	104	3.0	15.4	1.0	4.3	0.2	0.2	0.2
2878	Soba noodles, cooked	½	cup(s)	95	69.4	94	4.8	20.4	—	0.1	0	0	0
2879	Somen noodles, cooked	½	cup(s)	88	59.8	115	3.5	24.2	—	0.2	0	0	0.1
493	Spaghetti, al dente, cooked	½	cup(s)	65	41.6	95	3.5	19.5	1.0	0.5	0.1	0.1	0.2
2884	Spaghetti, whole wheat, cooked	½	cup(s)	70	47.0	87	3.7	18.6	3.2	0.4	0.1	0.1	0.1
	Popcorn												
476	Air popped	1	cup(s)	8	0.3	31	1.0	6.2	1.2	0.4	0	0.1	0.2
4619	Caramel	1	cup(s)	35	1.0	152	1.3	27.8	1.8	4.5	1.3	1.0	1.6
4620	Cheese flavored	1	cup(s)	36	0.9	188	3.3	18.4	3.5	11.8	2.3	3.5	5.5
477	Popped in oil	1	cup(s)	11	0.1	64	0.8	5.0	0.9	4.8	0.8	1.1	2.6
	FRUIT AND FRUIT JUICES												
	Apples												
952	Juice, prepared from frozen concentrate	½	cup(s)	120	105.0	56	0.2	13.8	0.1	0.1	0	0	0
225	Juice, unsweetened, canned	½	cup(s)	124	109.0	58	0.1	14.5	0.1	0.1	0	0	0
224	Slices	½	cup(s)	55	47.1	29	0.1	7.6	1.3	0.1	0	0	0
946	Slices without skin, boiled	½	cup(s)	86	73.1	45	0.2	11.7	2.1	0.3	0	0	0.1
223	Raw medium, with peel	1	item(s)	138	118.1	72	0.4	19.1	3.3	0.2	0	0	0.1
948	Dried, sulfured	¼	cup(s)	22	6.8	52	0.2	14.2	1.9	0.1	0	0	0
226	Applesauce, sweetened, canned	½	cup(s)	128	101.5	97	0.2	25.4	1.5	0.2	0	0	0.1
227	Applesauce, unsweetened, canned	½	cup(s)	122	107.8	52	0.2	13.8	1.5	0.1	0	0	0
38492	Crabapples	1	item(s)	35	27.6	27	0.1	7.0	0.9	0.1	0	0	0
	Apricot												
228	Fresh without pits	4	item(s)	140	120.9	67	2.0	15.6	2.8	0.5	0	0.2	0.1
229	Halves with skin, canned in heavy syrup	½	cup(s)	129	100.1	107	0.7	27.7	2.1	0.1	0	0	0
230	Halves, dried, sulfured	¼	cup(s)	33	10.1	79	1.1	20.6	2.4	0.2	0	0	0
	Avocado												
233	California, whole, without skin or pit	½	cup(s)	115	83.2	192	2.2	9.9	7.8	17.7	2.4	11.3	2.1
234	Florida, whole, without skin or pit	½	cup(s)	115	90.6	138	2.5	9.0	6.4	11.5	2.2	6.3	1.9
2998	Pureed	⅛	cup(s)	28	20.2	44	0.5	2.4	1.8	4.0	0.6	2.7	0.5
	Banana												
4580	Dried chips	¼	cup(s)	55	2.4	285	1.3	32.1	4.2	18.5	15.9	1.1	0.3
235	Fresh whole, without peel	1	item(s)	118	88.4	105	1.3	27.0	3.1	0.4	0.1	0	0.1
	Blackberries												
237	Raw	½	cup(s)	72	63.5	31	1.0	6.9	3.8	0.4	0	0	0.2
958	Unsweetened, frozen	½	cup(s)	76	62.1	48	0.9	11.8	3.8	0.3	0	0	0.2
	Blueberries												
959	Canned in heavy syrup	½	cup(s)	128	98.3	113	0.8	28.2	2.0	0.4	0	0.1	0.2
238	Raw	½	cup(s)	73	61.1	41	0.5	10.5	1.7	0.2	0	0	0.1
960	Unsweetened, frozen	½	cup(s)	78	67.1	40	0.3	9.4	2.1	0.5	0	0.1	0.2
	Boysenberries												
961	Canned in heavy syrup	½	cup(s)	128	97.6	113	1.3	28.6	3.3	0.2	0	0	0.1
962	Unsweetened, frozen	½	cup(s)	66	56.7	33	0.7	8.0	3.5	0.2	0	0	0.1
35576	**Breadfruit**	1	item(s)	384	271.3	396	4.1	104.1	18.8	0.9	0.2	0.1	0.3
	Cherries												
967	Sour red, canned in water	½	cup(s)	122	109.7	44	0.9	10.9	1.3	0.1	0	0	0
3000	Sour red, raw	½	cup(s)	78	66.8	39	0.8	9.4	1.2	0.2	0.1	0.1	0.1
3004	Sweet, canned in heavy syrup	½	cup(s)	127	98.2	105	0.8	26.9	1.9	0.2	0	0.1	0.1
969	Sweet, canned in water	½	cup(s)	124	107.9	57	1.0	14.6	1.9	0.2	0	0	0
240	Sweet, raw	½	cup(s)	73	59.6	46	0.8	11.6	1.5	0.1	0	0	0

Chol (mg)	Calc (mg)	Iron (mg)	Magn (mg)	Pota (mg)	Sodi (mg)	Zinc (mg)	Vit A (μg)	Thia (mg)	Vit E (mg α)	Ribo (mg)	Niac (mg)	Vit B_6 (mg)	Fola (μg DFE)	Vit C (mg)	Vit B_{12} (μg)	Sele (μg)
23	10	1.17	16.8	30.4	4.0	0.52	4.8	0.23	0.13	0.11	1.66	0.03	110.4	0	0.1	19.1
26	15	0.87	19.2	29.6	9.6	0.50	8.0	0.19	0.46	0.09	1.17	0.09	75.2	0	0.1	17.4
0	5	0.90	12.6	30.8	0.7	0.36	0	0.19	0.04	0.10	1.18	0.03	83.3	0	0	18.5
0	7	0.33	12.7	20.8	4.0	0.30	3.4	0.08	0.14	0.04	0.72	0.02	71.0	0	0	13.3
21	4	0.73	11.5	15.4	3.8	0.36	3.8	0.13	—	0.10	0.64	0.02	66.6	0	0.1	—
18	9	0.89	8.5	34.5	414.5	0.30	—	0.08	—	0.04	0.71	0.03	4.0	0.1	0	—
0	4	0.45	8.5	33.2	57.0	0.11	0	0.09	—	0.02	0.48	0.03	6.6	0	0	—
0	7	0.45	1.8	25.5	141.7	0.19	0	0.01	—	0.03	0.08	0.01	1.8	0	0	—
0	7	1.00	12.4	51.5	0.5	0.35	0	0.12	0.04	0.07	0.90	0.04	77.4	0	0	40.0
0	11	0.74	21.0	30.8	2.1	0.57	0	0.08	0.21	0.03	0.50	0.06	3.5	0	0	18.1
0	1	0.25	11.5	26.3	0.6	0.25	0.8	0.01	0.02	0.01	0.18	0.01	2.5	0	0	0
2	15	0.61	12.3	38.4	72.5	0.20	0.7	0.02	0.42	0.02	0.77	0.01	1.8	0	0	1.3
4	40	0.79	32.5	93.2	317.4	0.71	13.6	0.04	—	0.08	0.52	0.08	3.9	0.2	0.2	4.3
0	0	0.22	8.7	20.0	116.4	0.34	0.9	0.01	0.27	0.00	0.13	0.01	2.8	0	0	0.2
0	7	0.31	6.0	150.6	8.4	0.05	0	0.00	0.01	0.02	0.05	0.04	0	0.7	0	0.1
0	9	0.46	3.7	147.6	3.7	0.04	0	0.03	0.01	0.02	0.12	0.04	0	1.1	0	0.1
0	3	0.06	2.7	58.8	0.5	0.02	1.6	0.01	0.10	0.01	0.05	0.02	1.6	2.5	0	0
0	4	0.16	2.6	75.2	0.9	0.03	1.7	0.01	0.04	0.01	0.08	0.04	0.9	0.2	0	0.3
0	8	0.16	6.9	147.7	1.4	0.05	4.1	0.02	0.24	0.03	0.12	0.05	4.1	6.3	0	0
0	3	0.30	3.4	96.8	18.7	0.04	0	0.00	0.11	0.03	0.20	0.03	0	0.8	0	0.3
0	5	0.44	3.8	77.8	3.8	0.05	1.3	0.01	0.26	0.03	0.24	0.03	1.3	2.2	0	0.4
0	4	0.14	3.7	91.5	2.4	0.03	1.2	0.01	0.25	0.03	0.22	0.03	1.2	1.5	0	0.4
0	6	0.12	2.5	67.9	0.4	—	0.7	0.01	0.20	0.01	0.03	—	2.0	2.8	0	—
0	18	0.54	14.0	362.6	1.4	0.28	134.4	0.04	1.24	0.05	0.84	0.07	12.6	14.0	0	0.1
0	12	0.38	9.0	180.6	5.2	0.14	80.0	0.02	0.77	0.02	0.48	0.07	2.6	4.0	0	0.1
0	18	0.87	10.5	381.5	3.3	0.12	59.1	0.00	1.42	0.02	0.85	0.05	3.3	0.3	0	0.7
0	15	0.66	33.3	583.0	9.2	0.78	8.0	0.08	2.23	0.16	2.19	0.31	102.3	10.1	0	0.4
0	12	0.19	27.6	403.6	2.3	0.45	8.0	0.02	3.03	0.04	0.76	0.08	40.3	20.0	0	—
0	3	0.14	8.0	134.1	1.9	0.17	1.9	0.01	0.57	0.03	0.48	0.07	22.4	2.8	0	0.1
0	10	0.69	41.8	294.8	3.3	0.40	2.2	0.04	0.13	0.01	0.39	0.14	7.7	3.5	0	0.8
0	6	0.30	31.9	422.4	1.2	0.17	3.5	0.03	0.11	0.08	0.78	0.43	23.6	10.3	0	1.2
0	21	0.45	14.4	116.6	0.7	0.38	7.9	0.01	0.84	0.02	0.47	0.02	18.0	15.1	0	0.3
0	22	0.60	16.6	105.7	0.8	0.19	4.5	0.02	0.88	0.03	0.91	0.05	25.7	2.3	0	0.3
0	6	0.42	5.1	51.2	3.8	0.09	2.6	0.04	0.49	0.07	0.14	0.05	2.6	1.4	0	0.1
0	4	0.20	4.4	55.8	0.7	0.12	2.2	0.03	0.41	0.03	0.30	0.04	4.4	7.0	0	0.1
0	6	0.14	3.9	41.9	0.8	0.05	1.6	0.03	0.37	0.03	0.40	0.05	5.4	1.9	0	0.1
0	23	0.55	14.1	115.2	3.8	0.24	2.6	0.03	—	0.04	0.29	0.05	43.5	7.9	0	0.5
0	18	0.56	10.6	91.7	0.7	0.15	2.0	0.04	0.57	0.02	0.51	0.04	41.6	2.0	0	0.1
0	65	2.07	96.0	1881.6	7.7	0.46	0	0.42	0.38	0.11	3.45	0.38	53.8	111.4	0	2.3
0	13	1.67	7.3	119.6	8.5	0.09	46.4	0.02	0.28	0.05	0.22	0.05	9.8	2.6	0	0
0	12	0.25	7.0	134.1	2.3	0.08	49.6	0.02	0.05	0.03	0.31	0.03	6.2	7.8	0	0
0	11	0.44	11.4	183.4	3.8	0.12	10.1	0.02	0.29	0.05	0.50	0.03	5.1	4.6	0	0
0	14	0.45	11.2	162.4	1.2	0.10	9.9	0.03	0.29	0.05	0.51	0.04	5.0	2.7	0	0
0	9	0.26	8.0	161.0	0	0.05	2.2	0.02	0.05	0.02	0.11	0.04	2.9	5.1	0	0

DA+ Code	Food Description	QTY	Measure	Wt (g)	H_2O (g)	Ener (cal)	Prot (g)	Carb (g)	Fiber (g)	Fat (g)	Fat Breakdown (g) Sat	Mono	Poly
FRUIT AND FRUIT JUICES—continued													
	Cranberries												
3007	Chopped, raw	½	cup(s)	55	47.9	25	0.2	6.7	2.5	0.1	0	0	0
1717	Cranberry apple juice drink	½	cup(s)	123	102.6	77	0	19.4	0	0.1	0	0	0.1
1638	Cranberry juice cocktail	½	cup(s)	127	109.0	68	0	17.1	0	0.1	0	0	0.1
241	Cranberry juice cocktail, low calorie, with saccharin	½	cup(s)	119	112.8	23	0	5.5	0	0	0	0	0
242	Cranberry sauce, sweetened, canned	¼	cup(s)	69	42.0	105	0.1	26.9	0.7	0.1	0	0	0
	Dates												
244	Domestic, chopped	¼	cup(s)	45	9.1	125	1.1	33.4	3.6	0.2	0	0	0
243	Domestic, whole	¼	cup(s)	45	9.1	125	1.1	33.4	3.6	0.2	0	0	0
	Figs												
975	Canned in heavy syrup	½	cup(s)	130	98.8	114	0.5	29.7	2.8	0.1	0	0	0.1
974	Canned in water	½	cup(s)	124	105.7	66	0.5	17.3	2.7	0.1	0	0	0.1
973	Raw, medium	2	item(s)	100	79.1	74	0.7	19.2	2.9	0.3	0.1	0.1	0.1
	Fruit cocktail and salad												
245	Fruit cocktail, canned in heavy syrup	½	cup(s)	124	99.7	91	0.5	23.4	1.2	0.1	0	0	0
978	Fruit cocktail, canned in juice	½	cup(s)	119	103.6	55	0.5	14.1	1.2	0	0	0	0
977	Fruit cocktail, canned in water	½	cup(s)	119	107.6	38	0.5	10.1	1.2	0.1	0	0	0
979	Fruit salad, canned in water	½	cup(s)	123	112.1	37	0.4	9.6	1.2	0.1	0	0	0
	Gooseberries												
982	Canned in light syrup	½	cup(s)	126	100.9	92	0.8	23.6	3.0	0.3	0	0	0.1
981	Raw	½	cup(s)	75	65.9	33	0.7	7.6	3.2	0.4	0	0	0.2
	Grapefruit												
251	Juice, pink, sweetened, canned	½	cup(s)	125	109.1	57	0.7	13.9	0.1	0.1	0	0	0
249	Juice, white	½	cup(s)	124	111.2	48	0.6	11.4	0.1	0.1	0	0	0
3022	Pink or red, raw	½	cup(s)	114	100.8	48	0.9	12.2	1.8	0.2	0	0	0
248	Sections, canned in light syrup	½	cup(s)	127	106.2	76	0.7	19.6	0.5	0.1	0	0	0
983	Sections, canned in water	½	cup(s)	122	109.6	44	0.7	11.2	0.5	0.1	0	0	0
247	White, raw	½	cup(s)	115	104.0	38	0.8	9.7	1.3	0.1	0	0	0
	Grapes												
255	American, slip skin	½	cup(s)	46	37.4	31	0.3	7.9	0.4	0.2	0.1	0	0
256	European, red or green, adherent skin	½	cup(s)	76	60.8	52	0.5	13.7	0.7	0.1	0	0	0
3159	Grape juice drink, canned	½	cup(s)	125	106.6	71	0	18.2	0.1	0	0	0	0
259	Grape juice, sweetened, with added vitamin C, prepared from frozen concentrate	½	cup(s)	125	108.6	64	0.2	15.9	0.1	0.1	0	0	0
3060	Raisins, seeded, packed	¼	cup(s)	41	6.8	122	1.0	32.4	2.8	0.2	0.1	0	0.1
987	**Guava, raw**	1	item(s)	55	44.4	37	1.4	7.9	3.0	0.5	0.2	0	0.2
35593	**Guavas, strawberry**	1	item(s)	6	4.8	4	0	1.0	0.3	0	0	0	0
3027	**Jackfruit**	½	cup(s)	83	60.4	78	1.2	19.8	1.3	0.2	0	0	0.1
990	**Kiwi fruit or Chinese gooseberries**	1	item(s)	76	63.1	46	0.9	11.1	2.3	0.4	0	0	0.2
	Lemon												
262	Juice	1	tablespoon(s)	15	13.8	4	0.1	1.3	0.1	0	0	0	0
993	Peel	1	teaspoon(s)	2	1.6	1	0	0.3	0.2	0	0	0	0
992	Raw	1	item(s)	108	94.4	22	1.3	11.6	5.1	0.3	0	0	0.1
	Lime												
269	Juice	1	tablespoon(s)	15	14.0	4	0.1	1.3	0.1	0	0	0	0
994	Raw	1	item(s)	67	59.1	20	0.5	7.1	1.9	0.1	0	0	0
995	**Loganberries, frozen**	½	cup(s)	74	62.2	40	1.1	9.6	3.9	0.2	0	0	0.1
	Mandarin orange												
1038	Canned in juice	½	cup(s)	125	111.4	46	0.8	11.9	0.9	0	0	0	0
1039	Canned in light syrup	½	cup(s)	126	104.7	77	0.6	20.4	0.9	0.1	0	0	0
999	**Mango**	½	cup(s)	83	67.4	54	0.4	14.0	1.5	0.2	0.1	0.1	0
1005	**Nectarine, raw, sliced**	½	cup(s)	69	60.4	30	0.7	7.3	1.2	0.2	0	0.1	0.1

Chol (mg)	Calc (mg)	Iron (mg)	Magn (mg)	Pota (mg)	Sodi (mg)	Zinc (mg)	Vit A (μg)	Thia (mg)	Vit E (mg α)	Ribo (mg)	Niac (mg)	Vit B6 (mg)	Fola (μg DFE)	Vit C (mg)	Vit B12 (μg)	Sele (μg)
0	4	0.13	3.3	46.8	1.1	0.05	1.7	0.01	0.66	0.01	0.05	0.03	0.6	7.3	0	0.1
0	4	0.09	1.2	20.8	2.5	0.02	0	0.00	0.15	0.00	0.00	0.00	0	48.4	0	0
0	4	0.13	1.3	17.7	2.5	0.04	0	0.00	0.28	0.00	0.05	0.00	0	53.5	0	0.3
0	11	0.05	2.4	29.6	3.6	0.02	0	0.00	0.06	0.00	0.00	0.00	0	38.2	0	0
0	3	0.15	2.1	18.0	20.1	0.03	1.4	0.01	0.57	0.01	0.06	0.01	0.7	1.4	0	0.2
0	17	0.45	19.1	291.9	0.9	0.12	0	0.02	0.02	0.02	0.56	0.07	8.5	0.2	0	1.3
0	17	0.45	19.1	291.9	0.9	0.12	0	0.02	0.02	0.02	0.56	0.07	8.5	0.2	0	1.3
0	35	0.36	13.0	128.2	1.3	0.14	2.6	0.03	0.16	0.05	0.55	0.09	2.6	1.3	0	0.3
0	35	0.36	12.4	127.7	1.2	0.15	2.5	0.03	0.10	0.05	0.55	0.09	2.5	1.2	0	0.1
0	35	0.36	17.0	232.0	1.0	0.14	7.0	0.06	0.10	0.04	0.40	0.10	6.0	2.0	0	0.2
0	7	0.36	6.2	109.1	7.4	0.09	12.4	0.02	0.49	0.02	0.46	0.06	3.7	2.4	0	0.6
0	9	0.25	8.3	112.6	4.7	0.11	17.8	0.01	0.47	0.02	0.48	0.06	3.6	3.2	0	0.6
0	6	0.30	8.3	111.4	4.7	0.11	15.4	0.02	0.47	0.01	0.43	0.06	3.6	2.5	0	0.6
0	9	0.37	6.1	95.6	3.7	0.10	27.0	0.02	—	0.03	0.46	0.04	3.7	2.3	0	1.0
0	20	0.42	7.6	97.0	2.5	0.14	8.8	0.03	—	0.07	0.19	0.02	3.8	12.6	0	0.5
0	19	0.23	7.5	148.5	0.8	0.09	11.3	0.03	0.28	0.02	0.23	0.06	4.5	20.8	0	0.5
0	10	0.45	12.5	202.2	2.5	0.08	0	0.05	0.05	0.03	0.40	0.03	12.5	33.6	0	0.1
0	11	0.25	14.8	200.1	1.2	0.06	1.2	0.05	0.27	0.02	0.25	0.05	12.4	46.9	0	0.1
0	25	0.09	10.3	154.5	0	0.07	66.4	0.04	0.14	0.03	0.23	0.06	14.9	35.7	0	0.1
0	18	0.50	12.7	163.8	2.5	0.10	0	0.04	0.11	0.02	0.30	0.02	11.4	27.1	0	1.1
0	18	0.50	12.2	161.0	2.4	0.11	0	0.05	0.11	0.03	0.30	0.02	11.0	26.6	0	1.1
0	14	0.07	10.4	170.2	0	0.08	2.3	0.04	0.15	0.02	0.30	0.05	11.5	38.3	0	1.6
0	6	0.13	2.3	87.9	0.9	0.02	2.3	0.04	0.09	0.02	0.14	0.05	1.8	1.8	0	0
0	8	0.27	5.3	144.2	1.5	0.05	2.3	0.05	0.14	0.05	0.14	0.07	1.5	8.2	0	0.1
0	9	0.16	7.5	41.3	11.3	0.04	0	0.28	0.00	0.44	0.18	0.04	1.3	33.1	0	0.1
0	5	0.13	5.0	26.3	2.5	0.05	0	0.02	0.00	0.03	0.16	0.05	1.3	29.9	0	0.1
0	12	1.06	12.4	340.3	11.6	0.07	0	0.04	—	0.07	0.46	0.07	1.2	2.2	0	0.2
0	10	0.14	12.1	229.4	1.1	0.12	17.1	0.03	0.40	0.02	0.59	0.06	27.0	125.6	0	0.3
0	1	0.01	1.0	17.5	2.2	—	0.3	0.00	—	0.00	0.03	0.00	—	2.2	0	
0	28	0.49	30.5	250.0	2.5	0.35	12.4	0.02	—	0.09	0.33	0.09	11.5	5.5	0	0.5
0	26	0.23	12.9	237.1	2.3	0.10	3.0	0.02	1.11	0.01	0.25	0.04	19.0	70.5	0	0.2
0	1	0.00	0.9	18.9	0.2	0.01	0.2	0.00	0.02	0.00	0.02	0.01	2.0	7.0	0	0
0	3	0.01	0.3	3.2	0.1	0.01	0.1	0.00	0.01	0.00	0.01	0.00	0.3	2.6	0	0
0	66	0.75	13.0	156.6	3.2	0.10	2.2	0.05	—	0.04	0.21	0.11	—	83.2	0	1.0
0	2	0.02	1.2	18.0	0.3	0.01	0.3	0.00	0.03	0.00	0.02	0.01	1.5	4.6	0	0
0	22	0.40	4.0	68.3	1.3	0.07	1.3	0.02	0.14	0.01	0.13	0.02	5.4	19.5	0	0.3
0	19	0.47	15.4	106.6	0.7	0.25	1.5	0.04	0.64	0.03	0.62	0.05	19.1	11.2	0	0.1
0	14	0.34	13.7	165.6	6.2	0.64	53.5	0.10	0.12	0.04	0.55	0.05	6.2	42.6	0	0.5
0	9	0.47	10.1	98.3	7.6	0.30	52.9	0.07	0.13	0.06	0.56	0.05	6.3	24.9	0	0.5
0	8	0.10	7.4	128.7	1.7	0.03	31.4	0.05	0.92	0.04	0.48	0.11	11.6	22.8	0	0.5
0	4	0.19	6.2	138.7	0	0.12	11.7	0.02	0.53	0.02	0.78	0.02	3.5	3.7	0	0

Table A–1 Table of Food Composition (*continued*)

(Computer code is for Cengage Diet Analysis Plus program)

DA+ Code	Food Description	QTY	Measure	Wt (g)	H_2O (g)	Ener (cal)	Prot (g)	Carb (g)	Fiber (g)	Fat (g)	Fat Breakdown (g) Sat	Mono	Poly
FRUIT AND FRUIT JUICES—continued													
	Melons												
271	Cantaloupe	½	cup(s)	80	72.1	27	0.7	6.5	0.7	0.1	0	0	0.1
1000	Casaba melon	½	cup(s)	85	78.1	24	0.9	5.6	0.8	0.1	0	0	0
272	Honeydew	½	cup(s)	89	79.5	32	0.5	8.0	0.7	0.1	0	0	0
318	Watermelon	½	cup(s)	76	69.5	23	0.5	5.7	0.3	0.1	0	0	0
	Orange												
14412	Juice with calcium and vitamin D	½	cup(s)	120	—	55	1.0	13.0	0	0	0	0	0
29630	Juice, fresh squeezed	½	cup(s)	124	109.5	56	0.9	12.9	0.2	0.2	0	0	0
14411	Juice, not from concentrate	½	cup(s)	120	—	55	1.0	13.0	0	0	0	0	0
278	Juice, unsweetened, prepared from frozen concentrate	½	cup(s)	125	109.7	56	0.8	13.4	0.2	0.1	0	0	0
3040	Peel	1	teaspoon(s)	2	1.5	2	0	0.5	0.2	0	0	0	0
273	Raw	1	item(s)	131	113.6	62	1.2	15.4	3.1	0.2	0	0	0
274	Sections	½	cup(s)	90	78.1	42	0.8	10.6	2.2	0.1	0	0	0
	Papaya, raw												
16830	Dried, strips	2	item(s)	46	12.0	119	1.9	29.9	5.5	0.4	0.1	0.1	0.1
282	Papaya	½	cup(s)	70	62.2	27	0.4	6.9	1.3	0.1	0	0	0
35640	**Passion fruit, purple**	1	item(s)	18	13.1	17	0.4	4.2	1.9	0.1	0	0	0.1
	Peach												
285	Halves, canned in heavy syrup	½	cup(s)	131	103.9	97	0.6	26.1	1.7	0.1	0	0	0.1
286	Halves, canned in water	½	cup(s)	122	113.6	29	0.5	7.5	1.6	0.1	0	0	0
290	Slices, sweetened, frozen	½	cup(s)	125	93.4	118	0.8	30.0	2.3	0.2	0	0.1	0.1
283	Raw, medium	1	item(s)	150	133.3	59	1.4	14.3	2.3	0.4	0	0.1	0.1
	Pear												
8672	Asian	1	item(s)	122	107.7	51	0.6	13.0	4.4	0.3	0	0.1	0.1
293	D'Anjou	1	item(s)	200	168.0	120	1.0	30.0	5.2	1.0	0	0.2	0.2
294	Halves, canned in heavy syrup	½	cup(s)	133	106.9	98	0.3	25.5	2.1	0.2	0	0	0
1012	Halves, canned in juice	½	cup(s)	124	107.2	62	0.4	16.0	2.0	0.1	0	0	0
291	Raw	1	item(s)	166	139.0	96	0.6	25.7	5.1	0.2	0	0	0
1017	**Persimmon**	1	item(s)	25	16.1	32	0.2	8.4	—	0.1	0	0	0
	Pineapple												
3053	Canned in extra heavy syrup	½	cup(s)	130	101.0	108	0.4	28.0	1.0	0.1	0	0	0
1019	Canned in juice	½	cup(s)	125	104.0	75	0.5	19.5	1.0	0.1	0	0	0
296	Canned in light syrup	½	cup(s)	126	108.0	66	0.5	16.9	1.0	0.2	0	0	0.1
1018	Canned in water	½	cup(s)	123	111.7	39	0.5	10.2	1.0	0.1	0	0	0
299	Juice, unsweetened, canned	½	cup(s)	125	108.0	66	0.5	16.1	0.3	0.2	0	0	0.1
295	Raw, diced	½	cup(s)	78	66.7	39	0.4	10.2	1.1	0.1	0	0	0
1024	**Plantain, cooked**	½	cup(s)	77	51.8	89	0.6	24.0	1.8	0.1	0.1	0	0
300	**Plum, raw, large**	1	item(s)	66	57.6	30	0.5	7.5	0.9	0.2	0	0.1	0
1027	**Pomegranate**	1	item(s)	154	124.7	105	1.5	26.4	0.9	0.5	0.1	0.1	0.1
	Prunes												
5644	Dried	2	item(s)	17	5.2	40	0.4	10.7	1.2	0.1	0	0	0
305	Dried, stewed	½	cup(s)	124	86.5	133	1.2	34.8	3.8	0.2	0	0.1	0
306	Juice, canned	1	cup(s)	256	208.0	182	1.6	44.7	2.6	0.1	0	0.1	0
	Raspberries												
309	Raw	½	cup(s)	62	52.7	32	0.7	7.3	4.0	0.4	0	0	0.2
310	Red, sweetened, frozen	½	cup(s)	125	90.9	129	0.9	32.7	5.5	0.2	0	0	0.1
311	**Rhubarb, cooked with sugar**	½	cup(s)	120	81.5	140	0.5	37.5	2.7	0.1	0	0	0.1
	Strawberries												
313	Raw	½	cup(s)	72	65.5	23	0.5	5.5	1.4	0.2	0	0	0.1
315	Sweetened, frozen, thawed	½	cup(s)	128	99.5	99	0.7	26.8	2.4	0.2	0	0	0.1
16828	**Tangelo**	1	item(s)	95	82.4	45	0.9	11.2	2.3	0.1	0	0	0
	Tangerine												
1040	Juice	½	cup(s)	124	109.8	53	0.6	12.5	0.2	0.2	0	0	0
316	Raw	1	item(s)	88	74.9	47	0.7	11.7	1.6	0.3	0	0.1	0.1

Chol (mg)	Calc (mg)	Iron (mg)	Magn (mg)	Pota (mg)	Sodi (mg)	Zinc (mg)	Vit A (µg)	Thia (mg)	Vit E (mg α)	Ribo (mg)	Niac (mg)	Vit B_6 (mg)	Fola (µg DFE)	Vit C (mg)	Vit B_{12} (µg)	Sele (µg)
0	7	0.17	9.6	213.6	12.8	0.14	135.2	0.03	0.04	0.01	0.59	0.05	16.8	29.4	0	0.3
0	9	0.29	9.4	154.7	7.7	0.06	0	0.01	0.04	0.03	0.20	0.14	6.8	18.5	0	0.3
0	5	0.15	8.8	201.8	15.9	0.07	2.7	0.03	0.01	0.01	0.37	0.07	16.8	15.9	0	0.6
0	5	0.18	7.6	85.1	0.8	0.07	21.3	0.02	0.04	0.01	0.13	0.03	2.3	6.2	0	0.3
0	175	0.00	12.0	225.0	0	—	0	0.08	—	0.03	0.40	0.06	30.0	36.0	0	—
0	14	0.25	13.6	248.0	1.2	0.06	12.4	0.11	0.05	0.04	0.50	0.05	37.2	62.0	0	0.1
0	10	0.00	12.5	225.0	0	0.06	0	0.08	—	0.03	0.40	0.06	30.0	36.0	0	0.1
0	11	0.12	12.5	236.6	1.2	0.06	6.2	0.10	0.25	0.02	0.25	0.06	54.8	48.4	0	0.1
0	3	0.01	0.4	4.2	0.1	0.01	0.4	0.00	0.01	0.00	0.01	0.00	0.6	2.7	0	0
0	52	0.13	13.1	237.1	0	0.09	14.4	0.11	0.23	0.05	0.36	0.07	39.3	69.7	0	0.7
0	36	0.09	9.0	162.9	0	0.06	9.9	0.07	0.16	0.03	0.25	0.05	27.0	47.9	0	0.4
0	73	0.30	30.4	782.9	9.2	0.21	83.7	0.06	2.22	0.08	0.93	0.05	58.0	37.7	0	1.8
0	17	0.07	7.0	179.9	2.1	0.05	38.5	0.02	0.51	0.02	0.24	0.01	26.6	43.3	0	0.4
0	2	0.28	5.2	62.6	5.0	0.01	11.5	0.00	0.00	0.02	0.27	0.01	2.5	5.4	0	0.1
0	4	0.35	6.6	120.5	7.9	0.11	22.3	0.01	0.64	0.03	0.80	0.02	3.9	3.7	0	0.4
0	2	0.39	6.1	120.8	3.7	0.11	32.9	0.01	0.59	0.02	0.63	0.02	3.7	3.5	0	0.4
0	4	0.46	6.3	162.5	7.5	0.06	17.5	0.01	0.77	0.04	0.81	0.02	3.8	117.8	0	0.5
0	9	0.37	13.5	285.0	0	0.25	24.0	0.03	1.09	0.04	1.20	0.03	6.0	9.9	0	0.2
0	5	0.00	9.8	147.6	0	0.02	0	0.01	0.14	0.01	0.26	0.02	9.8	4.6	0	0.1
0	22	0.50	12.0	250.0	0	0.24	—	0.04	1.00	0.08	0.20	0.03	14.6	8.0	0	1.0
0	7	0.29	5.3	86.5	6.7	0.10	0	0.01	0.10	0.02	0.32	0.01	1.3	1.5	0	0
0	11	0.36	8.7	119.0	5.0	0.11	0	0.01	0.10	0.01	0.25	0.02	1.2	2.0	0	0
0	15	0.28	11.6	197.5	1.7	0.16	1.7	0.02	0.19	0.04	0.26	0.04	11.6	7.0	0	0.2
0	7	0.62	—	77.5	0.3	—	0	—	—	—	—	—	—	16.5	0	—
0	18	0.49	19.5	132.6	1.3	0.14	1.3	0.11	—	0.03	0.36	0.09	6.5	9.5	0	—
0	17	0.35	17.4	151.9	1.2	0.12	2.5	0.12	0.01	0.02	0.35	0.09	6.2	11.8	0	0.5
0	18	0.49	20.2	132.3	1.3	0.15	2.5	0.11	0.01	0.03	0.36	0.09	6.3	9.5	0	0.5
0	18	0.49	22.1	156.2	1.2	0.15	2.5	0.11	0.01	0.03	0.37	0.09	6.2	9.5	0	0.5
0	16	0.39	15.0	162.5	2.5	0.14	0	0.07	0.03	0.03	0.25	0.13	22.5	12.5	0	0.1
0	10	0.22	9.3	84.5	0.8	0.09	2.3	0.06	0.02	0.03	0.39	0.09	14.0	37.0	0	0.1
0	2	0.45	24.6	358.1	3.9	0.10	34.7	0.04	0.10	0.04	0.58	0.19	20.0	8.4	0	1.1
0	4	0.11	4.6	103.6	0	0.06	11.2	0.02	0.17	0.02	0.27	0.02	3.3	6.3	0	0
0	5	0.46	4.6	398.9	4.6	0.18	7.7	0.04	0.92	0.04	0.46	0.16	9.2	9.4	0	0.9
0	7	0.16	6.9	123.0	0.3	0.07	6.6	0.01	0.07	0.03	0.32	0.03	0.7	0.1	0	0
0	24	0.51	22.3	398.0	1.2	0.24	21.1	0.03	0.24	0.12	0.90	0.27	0	3.6	0	0.1
0	31	3.02	35.8	706.6	10.2	0.53	0	0.04	0.30	0.17	2.01	0.55	0	10.5	0	1.5
0	15	0.42	13.5	92.9	0.6	0.26	1.2	0.02	0.54	0.02	0.37	0.03	12.9	16.1	0	0.1
0	19	0.81	16.3	142.5	1.3	0.22	3.8	0.02	0.90	0.05	0.28	0.04	32.5	20.6	0	0.4
0	174	0.25	16.2	115.0	1.0	—	—	0.02	—	0.03	0.25	—	—	4.0	0	—
0	12	0.30	9.4	110.2	0.7	0.10	0.7	0.02	0.21	0.02	0.28	0.03	17.3	42.3	0	0.3
0	14	0.59	7.7	125.0	1.3	0.06	1.3	0.01	0.30	0.09	0.37	0.03	5.1	50.4	0	0.9
0	38	0.09	9.5	172.0	0	0.06	10.5	0.08	0.17	0.03	0.26	0.05	28.5	50.5	0	0.5
0	22	0.25	9.9	219.8	1.2	0.04	16.1	0.07	0.16	0.02	0.12	0.05	6.2	38.3	0	0.1
0	33	0.13	10.6	146.1	1.8	0.06	29.9	0.05	0.18	0.03	0.33	0.07	14.1	23.5	0	0.1

DA+ Code	Food Description	QTY	Measure	Wt (g)	H₂O (g)	Ener (cal)	Prot (g)	Carb (g)	Fiber (g)	Fat (g)	Fat Breakdown (g) Sat	Mono	Poly
	VEGETABLES, LEGUMES												
	Amaranth												
1043	Leaves, boiled, drained	½	cup(s)	66	60.4	14	1.4	2.7	—	0.1	0	0	0.1
1042	Leaves, raw	1	cup(s)	28	25.7	6	0.7	1.1	—	0.1	0	0	0
8683	**Arugula leaves, raw**	1	cup(s)	20	18.3	5	0.5	0.7	0.3	0.1	0	0	0.1
	Artichoke												
1044	Boiled, drained	1	item(s)	120	100.9	64	3.5	14.3	10.3	0.4	0.1	0	0.2
2885	Hearts, boiled, drained	½	cup(s)	84	70.6	45	2.4	10.0	7.2	0.3	0.1	0	0.1
	Asparagus												
566	Boiled, drained	½	cup(s)	90	83.4	20	2.2	3.7	1.8	0.2	0	0	0.1
568	Canned, drained	½	cup(s)	121	113.7	23	2.6	3.0	1.9	0.8	0.2	0	0.3
565	Tips, frozen, boiled, drained	½	cup(s)	90	84.7	16	2.7	1.7	1.4	0.4	0.1	0	0.2
	Bamboo shoots												
1048	Boiled, drained	½	cup(s)	60	57.6	7	0.9	1.2	0.6	0.1	0	0	0.1
1049	Canned, drained	½	cup(s)	66	61.8	12	1.1	2.1	0.9	0.3	0.1	0	0.1
	Beans												
1801	Adzuki beans, boiled	½	cup(s)	115	76.2	147	8.6	28.5	8.4	0.1	0	—	—
511	Baked beans with franks, canned	½	cup(s)	130	89.8	184	8.7	19.9	8.9	8.5	3.0	3.7	1.1
513	Baked beans with pork in sweet sauce, canned	½	cup(s)	127	89.3	142	6.7	26.7	5.3	1.8	0.6	0.6	0.5
512	Baked beans with pork in tomato sauce, canned	½	cup(s)	127	93.0	119	6.5	23.6	5.1	1.2	0.5	0.7	0.3
1805	Black beans, boiled	½	cup(s)	86	56.5	114	7.6	20.4	7.5	0.5	0.1	0	0.2
14597	Chickpeas, garbanzo beans or bengal gram, boiled	½	cup(s)	82	49.4	134	7.3	22.5	6.2	2.1	0.2	0.5	0.9
569	Fordhook lima beans, frozen, boiled, drained	½	cup(s)	85	62.0	88	5.2	16.4	4.9	0.3	0.1	0	0.1
1806	French beans, boiled	½	cup(s)	89	58.9	114	6.2	21.3	8.3	0.7	0.1	0	0.4
2773	Great northern beans, boiled	½	cup(s)	89	61.1	104	7.4	18.7	6.2	0.4	0.1	0	0.2
2736	Hyacinth beans, boiled, drained	½	cup(s)	44	37.8	22	1.3	4.0	—	0.1	0.1	0.1	0
570	Lima beans, baby, frozen, boiled, drained	½	cup(s)	90	65.1	95	6.0	17.5	5.4	0.3	0.1	0	0.1
515	Lima beans, boiled, drained	½	cup(s)	85	57.1	105	5.8	20.1	4.5	0.3	0.1	0	0.1
579	Mung beans, sprouted, boiled, drained	½	cup(s)	62	57.9	13	1.3	2.6	0.5	0.1	0	0	0
510	Navy beans, boiled	½	cup(s)	91	58.1	127	7.5	23.7	9.6	0.6	0.1	0.1	0.4
32816	Pinto beans, boiled, drained, no salt added	½	cup(s)	63	58.8	14	1.2	2.6	—	0.2	0	0	0.1
1052	Pinto beans, frozen, boiled, drained	½	cup(s)	47	27.3	76	4.4	14.5	4.0	0.2	0	0	0.1
514	Red kidney beans, canned	½	cup(s)	128	99.0	108	6.7	19.9	6.9	0.5	0.1	0.2	0.2
1810	Refried beans, canned	½	cup(s)	127	96.1	119	6.9	19.6	6.7	1.6	0.6	0.7	0.2
1053	Shell beans, canned	½	cup(s)	123	111.1	37	2.2	7.6	4.2	0.2	0	0	0.1
1670	Soybeans, boiled	½	cup(s)	86	53.8	149	14.3	8.5	5.2	7.7	1.1	1.7	4.4
1108	Soybeans, green, boiled, drained	½	cup(s)	90	61.7	127	11.1	9.9	3.8	5.8	0.7	1.1	2.7
1807	White beans, small, boiled	½	cup(s)	90	56.6	127	8.0	23.1	9.3	0.6	0.1	0.1	0.2
575	Yellow snap, string or wax beans, boiled, drained	½	cup(s)	63	55.8	22	1.2	4.9	2.1	0.2	0	0	0.1
576	Yellow snap, string or wax beans, frozen, boiled, drained	½	cup(s)	68	61.7	19	1.0	4.4	2.0	0.1	0	0	0.1
	Beets												
584	Beet greens, boiled, drained	½	cup(s)	72	64.2	19	1.9	3.9	2.1	0.1	0	0	0.1
2730	Pickled, canned with liquid	½	cup(s)	114	92.9	74	0.9	18.5	3.0	0.1	0	0	0
581	Sliced, boiled, drained	½	cup(s)	85	74.0	37	1.4	8.5	1.7	0.2	0	0	0.1
583	Sliced, canned, drained	½	cup(s)	85	77.3	26	0.8	6.1	1.5	0.1	0	0	0
580	Whole, boiled, drained	2	item(s)	100	87.1	44	1.7	10.0	2.0	0.2	0	0	0.1
585	**Cowpeas or black-eyed peas, boiled, drained**	½	cup(s)	83	62.3	80	2.6	16.8	4.1	0.3	0.1	0	0.1
	Broccoli												
588	Chopped, boiled, drained	½	cup(s)	78	69.6	27	1.9	5.6	2.6	0.3	0.1	0	0.1
590	Frozen, chopped, boiled, drained	½	cup(s)	92	83.5	26	2.9	4.9	2.8	0.1	0	0	0.1
587	Raw, chopped	½	cup(s)	46	40.6	15	1.3	3.0	1.2	0.2	0	0	0

PAGE KEY: A-2 = Breads/Baked Goods A-8 = Cereal/Rice/Pasta A-12 = Fruit A-18 = Vegetables/Legumes A-28 = Nuts/Seeds A-30 = Vegetarian A-32 = Dairy A-40 = Eggs A-40 = Seafood A-44 = Meats A-48 = Poultry A-48 = Processed Meats A-50 = Beverages A-54 = Fats/Oils A-56 = Sweets A-58 = Spices/Condiments/Sauces A-62 = Mixed Foods/Soups/Sandwiches A-68 = Fast Food A-88 = Convenience A-90 = Baby Foods

Chol (mg)	Calc (mg)	Iron (mg)	Magn (mg)	Pota (mg)	Sodi (mg)	Zinc (mg)	Vit A (μg)	Thia (mg)	Vit E (mg α)	Ribo (mg)	Niac (mg)	Vit B_6 (mg)	Fola (μg DFE)	Vit C (mg)	Vit B_{12} (μg)	Sele (μg)
0	138	1.49	36.3	423.1	13.9	0.58	91.7	0.01	—	0.09	0.37	0.12	37.6	27.1	0	0.6
0	60	0.65	15.4	171.1	5.6	0.25	40.9	0.01	—	0.04	0.18	0.05	23.8	12.1	0	0.3
0	32	0.29	9.4	73.8	5.4	0.09	23.8	0.01	0.09	0.02	0.06	0.01	19.4	3.0	0	0.1
0	25	0.73	50.4	343.2	72.0	0.48	1.2	0.06	0.22	0.10	1.33	0.09	106.8	8.9	0	0.2
0	18	0.51	35.3	240.2	50.4	0.33	0.8	0.04	0.16	0.07	0.93	0.06	74.8	6.2	0	0.2
0	21	0.81	12.6	201.6	12.6	0.54	45.0	0.14	1.35	0.12	0.97	0.07	134.1	6.9	0	5.5
0	19	2.21	12.1	208.1	347.3	0.48	49.6	0.07	1.47	0.12	1.15	0.13	116.2	22.3	0	2.1
0	16	0.50	9.0	154.8	2.7	0.36	36.0	0.05	1.08	0.09	0.93	0.01	121.5	22.0	0	3.5
0	7	0.14	1.8	319.8	2.4	0.28	0	0.01	—	0.03	0.18	0.06	1.2	0	0	0.2
0	5	0.21	2.6	52.4	4.6	0.43	0.7	0.02	0.41	0.02	0.09	0.09	2.0	0.7	0	0.3
0	32	2.30	59.8	611.8	9.2	2.03	0	0.13	—	0.07	0.82	0.11	139.2	0	0	1.4
8	62	2.24	36.3	304.3	556.9	2.42	5.2	0.08	0.21	0.07	1.17	0.06	38.9	3.0	0.4	8.4
9	75	2.08	41.7	326.4	422.5	1.73	0	0.05	0.03	0.07	0.44	0.07	10.1	3.5	0	6.3
9	71	4.09	43.0	373.2	552.8	6.93	5.1	0.06	0.12	0.05	0.62	0.08	19.0	3.8	0	5.9
0	23	1.80	60.2	305.3	0.9	0.96	0	0.21	—	0.05	0.43	0.05	128.1	0	0	1.0
0	40	2.36	39.4	238.6	5.7	1.25	0.8	0.09	0.28	0.05	0.43	0.11	141.0	1.1	0	3.0
0	26	1.54	35.7	258.4	58.7	0.62	8.5	0.06	0.24	0.05	0.90	0.10	17.9	10.9	0	0.5
0	56	0.95	49.6	327.5	5.3	0.56	0	0.11	—	0.05	0.48	0.09	66.4	1.1	0	1.1
0	60	1.88	44.3	346.0	1.8	0.77	0	0.14	—	0.05	0.60	0.10	90.3	1.2	0	3.6
0	18	0.33	18.3	114.0	0.9	0.16	3.0	0.02	—	0.03	0.20	0.01	20.4	2.2	0	0.7
0	25	1.76	50.4	369.9	26.1	0.49	7.2	0.06	0.57	0.04	0.69	0.10	14.4	5.2	0	1.5
0	27	2.08	62.9	484.5	14.5	0.67	12.8	0.11	0.11	0.08	0.88	0.16	22.1	8.6	0	1.7
0	7	0.40	8.7	62.6	6.2	0.29	0.6	0.03	0.04	0.06	0.51	0.03	18.0	7.1	0	0.4
—	63	2.14	48.2	354.0	0	0.93	0	0.21	0.01	0.06	0.59	0.12	127.4	0.8	0	2.6
0	9	0.41	11.3	61.7	32.1	0.10	0	0.04	—	0.03	0.45	0.03	18.3	3.8	0	0.4
0	24	1.27	25.4	303.6	39.0	0.32	0	0.12	—	0.05	0.29	0.09	16.0	0.3	0	0.7
0	32	1.62	35.8	327.7	330.2	2.09	0	0.13	0.02	0.11	0.57	0.10	25.6	1.4	0	0.6
10	44	2.10	41.7	337.8	378.2	1.48	0	0.03	0.00	0.02	0.39	0.18	13.9	7.6	0	1.6
0	36	1.21	18.4	133.5	409.2	0.33	13.5	0.04	0.04	0.07	0.25	0.06	22.1	3.8	0	2.6
0	88	4.42	74.0	442.9	0.9	0.98	0	0.13	0.30	0.24	0.34	0.20	46.4	1.5	0	6.3
0	131	2.25	54.0	485.1	12.6	0.82	7.2	0.23	—	0.14	1.13	0.05	99.9	15.3	0	1.3
0	65	2.54	60.9	414.4	1.8	0.97	0	0.21	—	0.05	0.24	0.11	122.6	0	0	1.2
0	29	0.80	15.6	186.9	1.9	0.23	2.5	0.05	0.28	0.06	0.38	0.04	20.6	6.1	0	0.3
0	33	0.59	16.2	85.1	6.1	0.32	4.1	0.02	0.03	0.06	0.26	0.04	15.5	2.8	0	0.3
0	82	1.36	49.0	654.5	173.5	0.36	275.8	0.08	1.30	0.20	0.35	0.09	10.1	17.9	0	0.6
0	12	0.46	17.0	168.0	299.6	0.29	1.1	0.01	—	0.05	0.28	0.05	30.6	2.6	0	1.1
0	14	0.67	19.6	259.3	65.5	0.30	1.7	0.02	0.03	0.03	0.28	0.05	68.0	3.1	0	0.6
0	13	1.54	14.5	125.8	164.9	0.17	0.9	0.01	0.02	0.03	0.13	0.04	25.5	3.5	0	0.4
0	16	0.79	23.0	305.0	77.0	0.35	2.0	0.02	0.04	0.04	0.33	0.06	80.0	3.6	0	0.7
0	106	0.92	42.9	344.9	3.3	0.85	33.0	0.08	0.18	0.12	1.15	0.05	104.8	1.8	0	2.1
0	31	0.52	16.4	228.5	32.0	0.35	60.1	0.04	1.13	0.09	0.43	0.15	84.2	50.6	0	1.2
0	30	0.56	12.0	130.6	10.1	0.25	46.9	0.05	1.21	0.07	0.42	0.12	51.5	36.9	0	0.6
0	21	0.33	9.6	143.8	15.0	0.19	14.1	0.03	0.36	0.05	0.29	0.08	28.7	40.6	0	1.1

Table A–1 Table of Food Composition (*continued*)

(Computer code is for Cengage Diet Analysis Plus program)

DA+ Code	Food Description	QTY	Measure	Wt (g)	H_2O (g)	Ener (cal)	Prot (g)	Carb (g)	Fiber (g)	Fat (g)	Fat Breakdown (g) Sat	Mono	Poly
	VEGETABLES, LEGUMES—continued												
16848	**Broccoflower, raw, chopped**	½	cup(s)	32	28.7	10	0.9	1.9	1.0	0.1	0	0	0
	Brussels sprouts												
591	Boiled, drained	½	cup(s)	78	69.3	28	2.0	5.5	2.0	0.4	0.1	0	0.2
592	Frozen, boiled, drained	½	cup(s)	78	67.2	33	2.8	6.4	3.2	0.3	0.1	0	0.2
	Cabbage												
595	Boiled, drained, no salt added	1	cup(s)	150	138.8	35	1.9	8.3	2.8	0.1	0	0	0
35611	Chinese (pak choi or bok choy), boiled with salt, drained	1	cup(s)	170	162.4	20	2.6	3.0	1.7	0.3	0	0	0.1
16869	Kim chee	1	cup(s)	150	137.5	32	2.5	6.1	1.8	0.3	0	0	0.2
594	Raw, shredded	1	cup(s)	70	64.5	17	0.9	4.1	1.7	0.1	0	0	0
596	Red, shredded, raw	1	cup(s)	70	63.3	22	1.0	5.2	1.5	0.1	0	0	0.1
597	Savoy, shredded, raw	1	cup(s)	70	63.7	19	1.4	4.3	2.2	0.1	0	0	0
35417	**Capers**	1	teaspoon(s)	4	—	2	0	0	0	0	0	0	0
	Carrots												
8691	Baby, raw	8	item(s)	80	72.3	28	0.5	6.6	2.3	0.1	0	0	0.1
601	Grated	½	cup(s)	55	48.6	23	0.5	5.3	1.5	0.1	0	0	0.1
1055	Juice, canned	½	cup(s)	118	104.9	47	1.1	11.0	0.9	0.2	0	0	0.1
600	Raw	½	cup(s)	61	53.9	25	0.6	5.8	1.7	0.1	0	0	0.1
602	Sliced, boiled, drained	½	cup(s)	78	70.3	27	0.6	6.4	2.3	0.1	0	0	0.1
32725	**Cassava or manioc**	½	cup(s)	103	61.5	165	1.4	39.2	1.9	0.3	0.1	0.1	0
	Cauliflower												
606	Boiled, drained	½	cup(s)	62	57.7	14	1.1	2.5	1.4	0.3	0	0	0.1
607	Frozen, boiled, drained	½	cup(s)	90	84.6	17	1.4	3.4	2.4	0.2	0	0	0.1
605	Raw, chopped	½	cup(s)	50	46.0	13	1.0	2.6	1.2	0	0	0	0
	Celery												
609	Diced	½	cup(s)	51	48.2	8	0.3	1.5	0.8	0.1	0	0	0
608	Stalk	2	item(s)	80	76.3	13	0.6	2.4	1.3	0.1	0	0	0.1
	Chard												
1057	Swiss chard, boiled, drained	½	cup(s)	88	81.1	18	1.6	3.6	1.8	0.1	0	0	0
1056	Swiss chard, raw	1	cup(s)	36	33.4	7	0.6	1.3	0.6	0.1	0	0	0
	Collard greens												
610	Boiled, drained	½	cup(s)	95	87.3	25	2.0	4.7	2.7	0.3	0	0	0.2
611	Frozen, chopped, boiled, drained	½	cup(s)	85	75.2	31	2.5	6.0	2.4	0.3	0.1	0	0.2
	Corn												
29614	Yellow corn, fresh, cooked	1	item(s)	100	69.2	107	3.3	25.0	2.8	1.3	0.2	0.4	0.6
615	Yellow creamed sweet corn, canned	½	cup(s)	128	100.8	92	2.2	23.2	1.5	0.5	0.1	0.2	0.3
612	Yellow sweet corn, boiled, drained	½	cup(s)	82	57.0	89	2.7	20.6	2.3	1.1	0.2	0.3	0.5
614	Yellow sweet corn, frozen, boiled, drained	½	cup(s)	82	63.2	66	2.1	15.8	2.0	0.5	0.1	0.2	0.3
618	**Cucumber**	¼	item(s)	75	71.7	11	0.5	2.7	0.4	0.1	0	0	0
16870	**Cucumber, kim chee**	½	cup(s)	75	68.1	16	0.8	3.6	1.1	0.1	0	0	0
	Dandelion greens												
620	Chopped, boiled, drained	½	cup(s)	53	47.1	17	1.1	3.4	1.5	0.3	0.1	0	0.1
2734	Raw	1	cup(s)	55	47.1	25	1.5	5.1	1.9	0.4	0.1	0	0.2
1066	**Eggplant, boiled, drained**	½	cup(s)	50	44.4	17	0.4	4.3	1.2	0.1	0	0	0
621	**Endive or escarole, chopped, raw**	1	cup(s)	50	46.9	8	0.6	1.7	1.5	0.1	0	0	0
8784	**Jicama or yambean**	½	cup(s)	65	116.5	49	0.9	11.4	6.3	0.1	0	0	0.1
	Kale												
623	Frozen, chopped, boiled, drained	½	cup(s)	65	58.8	20	1.8	3.4	1.3	0.3	0	0	0.2
29313	Raw	1	cup(s)	67	56.6	33	2.2	6.7	1.3	0.5	0.1	0	0.2
	Kohlrabi												
1072	Boiled, drained	½	cup(s)	83	74.5	24	1.5	5.5	0.9	0.1	0	0	0
1071	Raw	1	cup(s)	135	122.9	36	2.3	8.4	4.9	0.1	0	0	0.1

Chol (mg)	Calc (mg)	Iron (mg)	Magn (mg)	Pota (mg)	Sodi (mg)	Zinc (mg)	Vit A (µg)	Thia (mg)	Vit E (mg α)	Ribo (mg)	Niac (mg)	Vit B_6 (mg)	Fola (µg DFE)	Vit C (mg)	Vit B_{12} (µg)	Sele (µg)
0	11	0.23	6.4	96.0	7.4	0.20	2.6	0.02	0.01	0.03	0.23	0.07	18.2	28.2	0	0.2
0	28	0.93	15.6	247.3	16.4	0.25	30.4	0.08	0.33	0.06	0.47	0.13	46.8	48.4	0	1.2
0	20	0.37	14.0	224.8	11.6	0.18	35.7	0.08	0.39	0.08	0.41	0.22	78.3	35.4	0	0.5
0	72	0.24	22.5	294.0	12.0	0.30	6.0	0.08	0.20	0.04	0.36	0.16	45.0	56.2	0	0.9
0	158	1.76	18.7	630.7	459.0	0.28	360.4	0.04	0.14	0.10	0.72	0.28	69.7	44.2	0	0.7
0	144	1.26	27.0	379.5	996.0	0.36	288.0	0.06	0.36	0.10	0.80	0.32	88.5	79.6	0	1.5
0	28	0.33	8.4	119.0	12.6	0.12	3.5	0.04	0.10	0.02	0.16	0.08	30.1	25.6	0	0.2
0	31	0.56	11.2	170.1	18.9	0.15	39.2	0.04	0.07	0.05	0.29	0.14	12.6	39.9	0	0.4
0	24	0.28	19.6	161.0	19.6	0.18	35.0	0.05	0.11	0.02	0.21	0.13	56.0	21.7	0	0.6
0	0	0.00	—	—	140	—	0	—	—	—	—	—	—	0	—	—
0	26	0.71	8.0	189.6	62.4	0.13	552.0	0.02	—	0.02	0.44	0.08	21.6	2.1	0	0.7
0	18	0.16	6.6	176.0	37.9	0.13	459.2	0.03	0.36	0.03	0.54	0.07	10.4	3.2	0	0.1
0	28	0.54	16.5	344.6	34.2	0.21	1128.1	0.11	1.37	0.07	0.46	0.26	4.7	10.0	0	0.7
0	20	0.18	7.3	195.2	42.1	0.15	509.4	0.04	0.40	0.04	0.60	0.08	11.6	3.6	0	0.1
0	23	0.26	7.8	183.3	45.2	0.15	664.6	0.05	0.80	0.03	0.50	0.11	10.9	2.8	0	0.5
0	16	0.27	21.6	279.1	14.4	0.35	1.0	0.08	0.19	0.04	0.87	0.09	27.8	21.2	0	0.7
0	10	0.19	5.6	88.0	9.3	0.10	0.6	0.02	0.04	0.03	0.25	0.10	27.3	27.5	0	0.4
0	15	0.36	8.1	125.1	16.2	0.11	0	0.03	0.05	0.04	0.27	0.07	36.9	28.2	0	0.5
0	11	0.22	7.5	151.5	15.0	0.14	0.5	0.03	0.04	0.03	0.26	0.11	28.5	23.2	0	0.3
0	20	0.10	5.6	131.3	40.4	0.07	11.1	0.01	0.14	0.03	0.16	0.04	18.2	1.6	0	0.2
0	32	0.16	8.8	208.0	64.0	0.10	17.6	0.01	0.21	0.04	0.25	0.05	28.8	2.5	0	0.3
0	51	1.98	75.3	480.4	156.6	0.29	267.8	0.03	1.65	0.08	0.32	0.07	7.9	15.8	0	0.8
0	18	0.64	29.2	136.4	76.7	0.13	110.2	0.01	0.68	0.03	0.14	0.03	5.0	10.8	0	0.3
0	133	1.10	19.0	110.2	15.2	0.21	385.7	0.03	0.83	0.10	0.54	0.12	88.4	17.3	0	0.5
0	179	0.95	25.5	213.4	42.5	0.22	488.8	0.04	1.06	0.09	0.54	0.09	64.6	22.4	0	1.3
0	2	0.61	32.0	248.0	242.0	0.48	13.0	0.20	0.09	0.07	1.60	0.06	46.0	6.2	0	0.2
0	4	0.48	21.8	171.5	364.8	0.67	5.1	0.03	0.09	0.06	1.22	0.08	57.6	5.9	0	0.5
0	2	0.36	21.3	173.8	0	0.50	10.7	0.17	0.07	0.05	1.32	0.04	37.7	5.1	0	0.2
0	2	0.38	23.0	191.1	0.8	0.51	8.2	0.02	0.05	0.05	1.07	0.08	28.7	2.9	0	0.6
0	12	0.20	9.8	110.6	1.5	0.14	3.8	0.01	0.01	0.01	0.07	0.03	5.3	2.1	0	0.2
0	7	3.61	6.0	87.8	765.8	0.38	—	0.02	—	0.02	0.34	0.08	17.3	2.6	0	—
0	74	0.95	12.6	121.8	23.1	0.15	179.6	0.07	1.28	0.09	0.27	0.08	6.8	9.5	0	0.2
0	103	1.70	19.8	218.3	41.8	0.22	279.4	0.10	1.89	0.14	0.44	0.13	14.8	19.2	0	0.3
0	3	0.12	5.4	60.9	0.5	0.06	1.0	0.04	0.20	0.01	0.30	0.04	6.9	0.6	0	0
0	26	0.41	7.5	157.0	11.0	0.39	54.0	0.04	0.22	0.03	0.20	0.01	71.0	3.2	0	0.1
0	16	0.78	15.5	194.0	5.2	0.20	1.3	0.02	0.59	0.04	0.25	0.05	15.5	26.1	0	0.9
0	90	0.61	11.7	208.7	9.8	0.11	477.8	0.02	0.59	0.07	0.43	0.05	9.1	16.4	0	0.6
0	90	1.14	22.8	299.5	28.8	0.29	515.2	0.07	—	0.08	0.66	0.18	19.4	80.4	0	0.6
0	21	0.33	15.7	280.5	17.3	0.26	1.7	0.03	0.43	0.02	0.32	0.13	9.9	44.6	0	0.7
0	32	0.54	25.7	472.5	27.0	0.04	2.7	0.06	0.64	0.02	0.54	0.20	21.6	83.7	0	0.9

DA+ Code	Food Description	QTY	Measure	Wt (g)	H2O (g)	Ener (cal)	Prot (g)	Carb (g)	Fiber (g)	Fat (g)	Fat Breakdown (g) Sat	Mono	Poly
VEGETABLES, LEGUMES—continued													
	Leeks												
1074	Boiled, drained	½	cup(s)	52	47.2	16	0.4	4.0	0.5	0.1	0	0	0
1073	Raw	1	cup(s)	89	73.9	54	1.3	12.6	1.6	0.3	0	0	0.1
	Lentils												
522	Boiled	¼	cup(s)	50	34.5	57	4.5	10.0	3.9	0.2	0	0	0.1
1075	Sprouted	1	cup(s)	77	51.9	82	6.9	17.0	—	0.4	0	0.1	0.2
	Lettuce												
625	Butterhead leaves	11	piece(s)	83	78.9	11	1.1	1.8	0.9	0.2	0	0	0.1
624	Butterhead, Boston or Bibb	1	cup(s)	55	52.6	7	0.7	1.2	0.6	0.1	0	0	0.1
626	Iceberg	1	cup(s)	55	52.6	8	0.5	1.6	0.7	0.1	0	0	0
628	Iceberg, chopped	1	cup(s)	55	52.6	8	0.5	1.6	0.7	0.1	0	0	0
629	Looseleaf	1	cup(s)	36	34.2	5	0.5	1.0	0.5	0.1	0	0	0
1665	Romaine, shredded	1	cup(s)	56	53.0	10	0.7	1.8	1.2	0.2	0	0	0.1
	Mushrooms												
15585	Crimini (about 6)	3	ounce(s)	85	—	28	3.7	2.8	1.9	0	0	0	0
8700	Enoki	30	item(s)	90	79.7	40	2.3	6.9	2.4	0.3	0	0	0.1
1079	Mushrooms, boiled, drained	½	cup(s)	78	71.0	22	1.7	4.1	1.7	0.4	0	0	0.1
1080	Mushrooms, canned, drained	½	cup(s)	78	71.0	20	1.5	4.0	1.9	0.2	0	0	0.1
630	Mushrooms, raw	½	cup(s)	48	44.4	11	1.5	1.6	0.5	0.2	0	0	0.1
15587	Portabella, raw	1	item(s)	84	—	30	3.0	3.9	3.0	0	0	0	0
2743	Shiitake, cooked	½	cup(s)	73	60.5	41	1.1	10.4	1.5	0.2	0	0.1	0
	Mustard greens												
2744	Frozen, boiled, drained	½	cup(s)	75	70.4	14	1.7	2.3	2.1	0.2	0	0.1	0
29319	Raw	1	cup(s)	56	50.8	15	1.5	2.7	1.8	0.1	0	0	0
	Okra												
16866	Batter coated, fried	11	piece(s)	83	55.6	156	2.1	12.7	2.0	11.2	1.5	3.7	5.5
32742	Frozen, boiled, drained, no salt added	½	cup(s)	92	83.8	26	1.9	5.3	2.6	0.3	0.1	0	0.1
632	Sliced, boiled, drained	½	cup(s)	80	74.1	18	1.5	3.6	2.0	0.2	0	0	0
	Onions												
635	Chopped, boiled, drained	½	cup(s)	105	92.2	46	1.4	10.7	1.5	0.2	0	0	0.1
2748	Frozen, boiled, drained	½	cup(s)	106	97.8	30	0.8	7.0	1.9	0.1	0	0	0
1081	Onion rings, breaded and pan fried, frozen, heated	10	piece(s)	71	20.2	289	3.8	27.1	0.9	19.0	6.1	7.7	3.6
633	Raw, chopped	½	cup(s)	80	71.3	32	0.9	7.5	1.4	0.1	0	0	0
16850	Red onions, sliced, raw	½	cup(s)	57	50.7	24	0.5	5.8	0.8	0	0	0	0
636	Scallions, green or spring onions	2	item(s)	30	26.9	10	0.5	2.2	0.8	0.1	0	0	0
16860	**Palm hearts, cooked**	½	cup(s)	73	50.7	84	2.0	18.7	1.1	0.1	0	0	0.1
637	**Parsley, chopped**	1	tablespoon(s)	4	3.3	1	0.1	0.2	0.1	0	0	0	0
638	**Parsnips, sliced, boiled, drained**	½	cup(s)	78	62.6	55	1.0	13.3	2.8	0.2	0	0.1	0
	Peas												
639	Green peas, canned, drained	½	cup(s)	85	69.4	59	3.8	10.7	3.5	0.3	0.1	0	0.1
641	Green peas, frozen, boiled, drained	½	cup(s)	80	63.6	62	4.1	11.4	4.4	0.2	0	0	0.1
35694	Pea pods, boiled with salt, drained	½	cup(s)	80	71.1	32	2.6	5.2	2.2	0.2	0	0	0.1
1082	Peas and carrots, canned with liquid	½	cup(s)	128	112.4	48	2.8	10.8	2.6	0.3	0.1	0	0.2
1083	Peas and carrots, frozen, boiled, drained	½	cup(s)	80	68.6	38	2.5	8.1	2.5	0.3	0.1	0	0.2
2750	Snow or sugar peas, frozen, boiled, drained	½	cup(s)	80	69.3	42	2.8	7.2	2.5	0.3	0.1	0	0.1
640	Snow or sugar peas, raw	½	cup(s)	32	28.0	13	0.9	2.4	0.8	0.1	0	0	0
29324	Split peas, sprouted	½	cup(s)	60	37.4	77	5.3	16.9	—	0.4	0.1	0	0.2
	Peppers												
644	Green bell or sweet, boiled, drained	½	cup(s)	68	62.5	19	0.6	4.6	0.8	0.1	0	0	0.1
643	Green bell or sweet, raw	½	cup(s)	75	69.9	15	0.6	3.5	1.3	0.1	0	0	0
1664	Green hot chili	1	item(s)	45	39.5	18	0.9	4.3	0.7	0.1	0	0	0
1663	Green hot chili, canned with liquid	½	cup(s)	68	62.9	14	0.6	3.5	0.9	0.1	0	0	0
1086	Jalapeno, canned with liquid	½	cup(s)	68	60.4	18	0.6	3.2	1.8	0.6	0.1	0	0.3
8703	Yellow bell or sweet	1	item(s)	186	171.2	50	1.9	11.8	1.7	0.4	0.1	0	0.2

Chol (mg)	Calc (mg)	Iron (mg)	Magn (mg)	Pota (mg)	Sodi (mg)	Zinc (mg)	Vit A (µg)	Thia (mg)	Vit E (mg α)	Ribo (mg)	Niac (mg)	Vit B_6 (mg)	Fola (µg DFE)	Vit C (mg)	Vit B_{12} (µg)	Sele (µg)
0	16	0.56	7.3	45.2	5.2	0.02	1.0	0.01	—	0.01	0.10	0.04	12.5	2.2	0	0.3
0	53	1.86	24.9	160.2	17.8	0.10	73.9	0.05	0.81	0.02	0.35	0.20	57.0	10.7	0	0.9
0	9	1.65	17.8	182.7	1.0	0.63	0	0.08	0.05	0.04	0.52	0.09	89.6	0.7	0	1.4
0	19	2.47	28.5	247.9	8.5	1.16	1.5	0.17	—	0.09	0.86	0.14	77.0	12.7	0	0.5
0	29	1.02	10.7	196.4	4.1	0.16	137.0	0.04	0.14	0.05	0.29	0.06	60.2	3.1	0	0.5
0	19	0.68	7.1	130.9	2.7	0.11	91.3	0.03	0.09	0.03	0.19	0.04	40.1	2.0	0	0.3
0	10	0.22	3.8	77.5	5.5	0.08	13.7	0.02	0.09	0.01	0.07	0.02	15.9	1.5	0	0.1
0	10	0.22	3.8	77.5	5.5	0.08	13.7	0.02	0.09	0.01	0.07	0.02	15.9	1.5	0	0.1
0	13	0.31	4.7	69.8	10.1	0.06	133.2	0.02	0.10	0.02	0.13	0.03	13.7	6.5	0	0.2
0	18	0.54	7.8	138.3	4.5	0.13	162.4	0.04	0.07	0.03	0.17	0.04	76.2	13.4	0	0.2
0	0	0.67	—	—	32.6	—	0	—	—	—	—	—	—	0	0	—
0	1	0.98	14.4	331.2	2.7	0.54	0	0.16	0.01	0.14	5.31	0.07	46.8	0	0	2.0
0	5	1.35	9.4	277.7	1.6	0.67	0	0.05	0.01	0.23	3.47	0.07	14.0	3.1	0	9.3
0	9	0.61	11.7	100.6	331.5	0.56	0	0.06	0.01	0.01	1.24	0.04	9.4	0	0	3.2
0	1	0.24	4.3	152.6	2.4	0.25	0	0.04	0.01	0.19	1.73	0.05	7.7	1.0	0	4.5
0	39	0.35	—	—	9.9	—	0	—	—	—	—	—	—	0	0	—
0	2	0.31	10.2	84.8	2.9	0.96	0	0.02	0.02	0.12	1.08	0.11	15.2	0.2	0	18.0
0	76	0.84	9.8	104.3	18.8	0.15	265.5	0.03	1.01	0.04	0.19	0.08	52.5	10.4	0	0.5
0	58	0.81	17.9	198.2	14.0	0.11	294.0	0.04	1.12	0.06	0.45	0.10	104.7	39.2	0	0.5
2	54	1.13	32.2	170.8	109.7	0.44	14.0	0.16	1.50	0.12	1.29	0.11	43.7	9.2	0	3.6
0	88	0.61	46.9	215.3	2.8	0.57	15.6	0.09	0.29	0.11	0.72	0.04	134.3	11.2	0	0.6
0	62	0.22	28.8	108.0	4.8	0.34	11.2	0.10	0.21	0.04	0.69	0.15	36.8	13.0	0	0.3
0	23	0.24	11.5	174.3	3.1	0.21	0	0.03	0.02	0.02	0.17	0.12	15.7	5.5	0	0.6
0	17	0.32	6.4	114.5	12.7	0.06	0	0.02	0.01	0.02	0.14	0.06	13.8	2.8	0	0.4
0	22	1.20	13.5	91.6	266.3	0.29	7.8	0.19	—	0.09	2.56	0.05	73.1	1.0	0	2.5
0	18	0.16	8.0	116.8	3.2	0.13	0	0.03	0.01	0.02	0.09	0.09	15.2	5.9	0	0.4
0	13	0.10	5.7	82.4	1.7	0.09	0	0.02	0.01	0.01	0.04	0.08	10.9	3.7	0	0.3
0	22	0.44	6.0	82.8	4.8	0.11	15.0	0.01	0.16	0.02	0.15	0.01	19.2	5.6	0	0.2
0	13	1.23	7.3	1318.4	10.2	2.72	2.2	0.03	0.36	0.12	0.62	0.53	14.6	5.0	0	0.5
0	5	0.23	1.9	21.1	2.1	0.04	16.0	0.00	0.02	0.00	0.05	0.00	5.8	5.1	0	0
0	29	0.45	22.6	286.3	7.8	0.20	0	0.06	0.78	0.04	0.56	0.07	45.2	10.1	0	1.3
0	17	0.80	14.5	147.1	214.2	0.60	23.0	0.10	0.02	0.06	0.62	0.05	37.4	8.2	0	1.4
0	19	1.21	17.6	88.0	57.6	0.53	84.0	0.22	0.02	0.08	1.18	0.09	47.2	7.9	0	0.8
0	34	1.57	20.8	192.0	192.0	0.29	41.6	0.10	0.31	0.06	0.43	0.11	23.2	38.3	0	0.6
0	29	0.96	17.9	127.5	331.5	0.74	368.5	0.09	—	0.07	0.74	0.11	23.0	8.4	0	1.1
0	18	0.75	12.8	126.4	54.4	0.36	380.8	0.18	0.41	0.05	0.92	0.07	20.8	6.5	0	0.9
0	47	1.92	22.4	173.6	4.0	0.39	52.8	0.05	0.37	0.09	0.45	0.13	28.0	17.6	0	0.6
0	14	0.65	7.6	63.0	1.3	0.08	17.0	0.04	0.12	0.02	0.19	0.05	13.2	18.9	0	0.2
0	22	1.34	33.6	228.6	12.0	0.62	4.8	0.12	—	0.08	1.84	0.14	86.4	6.2	0	0.4
0	6	0.31	6.8	112.9	1.4	0.08	15.6	0.04	0.34	0.02	0.32	0.15	10.9	50.6	0	0.2
0	7	0.25	7.5	130.4	2.2	0.09	13.4	0.04	0.27	0.02	0.35	0.16	7.5	59.9	0	0
0	8	0.54	11.3	153.0	3.2	0.13	26.6	0.04	0.31	0.04	0.42	0.12	10.4	109.1	0	0.2
0	5	0.34	9.5	127.2	797.6	0.10	24.5	0.01	0.46	0.02	0.54	0.10	6.8	46.2	0	0.2
0	16	1.28	10.2	131.2	1136.3	0.23	57.8	0.03	0.47	0.03	0.27	0.13	9.5	6.8	0	0.3
0	20	0.85	22.3	394.3	3.7	0.31	18.6	0.05	—	0.04	1.65	0.31	48.4	341.3	0	0.6

DA+ Code	Food Description	QTY	Measure	Wt (g)	H₂O (g)	Ener (cal)	Prot (g)	Carb (g)	Fiber (g)	Fat (g)	Fat Breakdown (g)		
											Sat	Mono	Poly
VEGETABLES, LEGUMES—continued													
1087	**Poi**	½	cup(s)	120	86.0	134	0.5	32.7	0.5	0.2	0	0	0.1
	Potatoes												
1090	Au gratin mix, prepared with water, whole milk and butter	½	cup(s)	124	97.7	115	2.8	15.9	1.1	5.1	3.2	1.5	0.2
1089	Au gratin, prepared with butter	½	cup(s)	123	90.7	162	6.2	13.8	2.2	9.3	5.8	2.6	0.3
5791	Baked, flesh and skin	1	item(s)	202	151.3	188	5.1	42.7	4.4	0.3	0.1	0	0.1
645	Baked, flesh only	½	cup(s)	61	46.0	57	1.2	13.1	0.9	0.1	0	0	0
1088	Baked, skin only	1	item(s)	58	27.4	115	2.5	26.7	4.6	0.1	0	0	0
5795	Boiled in skin, flesh only, drained	1	item(s)	136	104.7	118	2.5	27.4	2.1	0.1	0	0	0.1
5794	Boiled, drained, skin and flesh	1	item(s)	150	115.9	129	2.9	29.8	2.5	0.2	0	0	0.1
647	Boiled, flesh only	½	cup(s)	78	60.4	67	1.3	15.6	1.4	0.1	0	0	0
648	French fried, deep fried, prepared from raw	14	item(s)	70	32.8	187	2.7	23.5	2.9	9.5	1.9	4.2	3.0
649	French fried, frozen, heated	14	item(s)	70	43.7	94	1.9	19.4	2.0	3.7	0.7	2.3	0.2
1091	Hashed brown	½	cup(s)	78	36.9	207	2.3	27.4	2.5	9.8	1.5	4.1	3.7
652	Mashed with margarine and whole milk	½	cup(s)	105	79.0	119	2.1	17.7	1.6	4.4	1.0	2.0	1.2
653	Mashed, prepared from dehydrated granules with milk, water, and margarine	½	cup(s)	105	79.8	122	2.3	16.9	1.4	5.0	1.3	2.1	1.4
2759	Microwaved	1	item(s)	202	145.5	212	4.9	49.0	4.6	0.2	0.1	0	0.1
2760	Microwaved in skin, flesh only	½	cup(s)	78	57.1	78	1.6	18.1	1.2	0.1	0	0	0
5804	Microwaved, skin only	1	item(s)	58	36.8	77	2.5	17.2	4.2	0.1	0	0	0
1097	Potato puffs, frozen, heated	½	cup(s)	64	38.2	122	1.3	17.8	1.6	5.5	1.2	3.9	0.3
1094	Scalloped mix, prepared with water, whole milk and butter	½	cup(s)	124	98.4	116	2.6	15.9	1.4	5.3	3.3	1.5	0.2
1093	Scalloped, prepared with butter	½	cup(s)	123	99.2	108	3.5	13.2	2.3	4.5	2.8	1.3	0.2
	Pumpkin												
1773	Boiled, drained	½	cup(s)	123	114.8	25	0.9	6.0	1.3	0.1	0	0	0
656	Canned	½	cup(s)	123	110.2	42	1.3	9.9	3.6	0.3	0.2	0	0
	Radicchio												
8731	Leaves, raw	1	cup(s)	40	37.3	9	0.6	1.8	0.4	0.1	0	0	0
2498	Raw	1	cup(s)	40	37.3	9	0.6	1.8	0.4	0.1	0	0	0
657	**Radishes**	6	item(s)	27	25.7	4	0.2	0.9	0.4	0	0	0	0
1099	**Rutabaga, boiled, drained**	½	cup(s)	85	75.5	33	1.1	7.4	1.5	0.2	0	0	0.1
658	**Sauerkraut, canned**	½	cup(s)	118	109.2	22	1.1	5.1	3.4	0.2	0	0	0.1
	Seaweed												
1102	Kelp	½	cup(s)	40	32.6	17	0.6	3.8	0.5	0.2	0.1	0	0
1104	Spirulina, dried	½	cup(s)	8	0.4	22	4.3	1.8	0.3	0.6	0.2	0.1	0.2
1106	**Shallots**	3	tablespoon(s)	30	23.9	22	0.8	5.0	—	0	0	0	0
	Soybeans												
1670	Boiled	½	cup(s)	86	53.8	149	14.3	8.5	5.2	7.7	1.1	1.7	4.4
2825	Dry roasted	½	cup(s)	86	0.7	388	34.0	28.1	7.0	18.6	2.7	4.1	10.5
2824	Roasted, salted	½	cup(s)	86	1.7	405	30.3	28.9	15.2	21.8	3.2	4.8	12.3
8739	Sprouted, stir fried	½	cup(s)	63	42.3	79	8.2	5.9	0.5	4.5	0.6	1.0	2.5
	Soy products												
1813	Soy milk	1	cup(s)	240	211.3	130	7.8	15.1	1.4	4.2	0.5	1.0	2.3
2838	Tofu, dried, frozen (koyadofu)	3	ounce(s)	85	4.9	408	40.8	12.4	6.1	25.8	3.7	5.7	14.6
13844	Tofu, extra firm	3	ounce(s)	85	—	86	8.6	2.2	1.1	4.3	0.5	0.9	2.8
13843	Tofu, firm	3	ounce(s)	85	—	75	7.5	2.2	0.5	3.2	0	0.9	2.3
1816	Tofu, firm, with calcium sulfate and magnesium chloride (nigari)	3	ounce(s)	85	72.2	60	7.0	1.4	0.8	3.5	0.7	1.0	1.5
1817	Tofu, fried	3	ounce(s)	85	43.0	230	14.6	8.9	3.3	17.2	2.5	3.8	9.7
13841	Tofu, silken	3	ounce(s)	85	—	42	3.7	1.9	0	2.3	0.5	—	—
13842	Tofu, soft	3	ounce(s)	85	—	65	6.5	1.1	0.5	3.2	0.5	1.1	2.2
1671	Tofu, soft, with calcium sulfate and magnesium chloride (nigari)	3	ounce(s)	85	74.2	52	5.6	1.5	0.2	3.1	0.5	0.7	1.8

Chol (mg)	Calc (mg)	Iron (mg)	Magn (mg)	Pota (mg)	Sodi (mg)	Zinc (mg)	Vit A (μg)	Thia (mg)	Vit E (mg α)	Ribo (mg)	Niac (mg)	Vit B_6 (mg)	Fola (μg DFE)	Vit C (mg)	Vit B_{12} (μg)	Sele (μg)
0	19	1.06	28.8	219.6	14.4	0.26	3.6	0.16	2.76	0.05	1.32	0.33	25.2	4.8	0	0.8
19	103	0.39	18.6	271.0	543.3	0.29	64.4	0.02	—	0.10	1.16	0.05	8.7	3.8	0	3.3
28	146	0.78	24.5	485.1	530.4	0.85	78.4	0.08	—	0.14	1.22	0.21	16.0	12.1	0	3.3
0	30	2.18	56.6	1080.7	20.2	0.72	2.0	0.12	0.08	0.09	2.84	0.62	56.6	19.4	0	0.8
0	3	0.21	15.3	238.5	3.1	0.18	0	0.06	0.02	0.01	0.85	0.18	5.5	7.8	0	0.2
0	20	4.08	24.9	332.3	12.2	0.28	0.6	0.07	0.02	0.06	1.77	0.35	12.8	7.8	0	0.4
0	7	0.42	29.9	515.4	5.4	0.40	0	0.14	0.01	0.02	1.95	0.40	13.6	17.7	0	0.4
0	13	1.27	34.1	572.0	7.4	0.46	0	0.14	0.01	0.03	2.13	0.44	15.0	18.4	0	—
0	6	0.24	15.6	255.8	3.9	0.21	0	0.07	0.01	0.01	1.02	0.21	7.0	5.8	0	0.2
0	16	1.05	30.8	567.0	8.4	0.39	0	0.08	0.09	0.03	1.34	0.37	16.1	21.2	0	0.4
0	8	0.51	18.2	315.7	271.6	0.26	0	0.09	0.07	0.02	1.55	0.12	19.6	9.3	0	0.1
0	11	0.43	27.3	449.3	266.8	0.37	0	0.13	0.01	0.03	1.80	0.37	12.5	10.1	0	0.4
1	23	0.27	19.9	344.4	349.6	0.31	43.0	0.09	0.44	0.04	1.23	0.25	9.4	11.0	0.1	0.8
2	36	0.21	21.0	164.8	179.5	0.26	49.3	0.09	0.53	0.09	0.90	0.16	8.4	6.8	0.1	5.9
0	22	2.50	54.5	902.9	16.2	0.72	0	0.24	—	0.06	3.46	0.69	24.2	30.5	0	0.8
0	4	0.31	19.4	319.0	5.4	0.25	0	0.10	—	0.01	1.26	0.25	9.3	11.7	0	0.3
0	27	3.44	21.5	377.0	9.3	0.29	0	0.04	0.01	0.04	1.28	0.28	9.9	8.9	0	0.3
0	9	0.41	10.9	199.7	307.2	0.21	0	0.08	0.15	0.02	0.97	0.08	9.0	4.0	0	0.4
14	45	0.47	17.4	252.2	423.7	0.31	43.5	0.02	—	0.06	1.28	0.05	12.4	4.1	0	2.0
15	70	0.70	23.3	463.1	410.4	0.49	0	0.08	—	0.11	1.29	0.22	16.0	13.0	0	2.0
0	18	0.69	11.0	281.8	1.2	0.28	306.3	0.03	0.98	0.09	0.50	0.05	11.0	5.8	0	0.2
0	32	1.70	28.2	252.4	6.1	0.20	953.1	0.02	1.29	0.06	0.45	0.06	14.7	5.1	0	0.5
0	8	0.23	5.2	120.8	8.8	0.25	0.4	0.01	0.90	0.01	0.10	0.02	24.0	3.2	0	0.4
0	8	0.23	5.2	120.8	8.8	0.25	0.4	0.01	0.90	0.01	0.10	0.02	24.0	3.2	0	0.4
0	7	0.09	2.7	62.9	10.5	0.07	0	0.00	0.00	0.01	0.06	0.01	6.8	4.0	0	0.2
0	41	0.45	19.6	277.1	17.0	0.30	0	0.07	0.27	0.04	0.61	0.09	12.8	16.0	0	0.6
0	35	1.73	15.3	200.6	780.0	0.22	1.2	0.03	0.17	0.03	0.17	0.15	28.3	17.3	0	0.7
0	67	1.12	48.4	35.6	93.2	0.48	2.4	0.02	0.32	0.04	0.16	0.00	72.0	1.2	0	0.3
0	9	2.14	14.6	102.2	78.6	0.15	2.2	0.18	0.38	0.28	0.96	0.03	7.1	0.8	0	0.5
0	11	0.36	6.3	100.2	3.6	0.12	18.0	0.02	—	0.01	0.06	0.09	10.2	2.4	—	0.4
0	88	4.42	74.0	442.9	0.9	0.98	0	0.13	0.30	0.24	0.34	0.20	46.4	1.5	0	6.3
0	120	3.39	196.1	1173.0	1.7	4.10	0	0.36	—	0.64	0.90	0.19	176.3	4.0	0	16.6
0	119	3.35	124.7	1264.2	140.2	2.70	8.6	0.08	0.78	0.12	1.21	0.17	181.5	1.9	0	16.4
0	52	0.25	60.4	356.6	8.8	1.32	0.6	0.26	—	0.12	0.69	0.10	79.9	7.5	0	0.4
0	60	1.53	60.0	283.2	122.4	0.28	0	0.14	0.26	0.16	1.23	0.18	43.2	0	0	11.5
0	310	8.27	50.2	17.0	5.1	4.16	22.1	0.42	—	0.27	1.01	0.24	78.2	0.6	0	46.2
0	65	1.16	84.1	—	0	—	0	—	—	—	—	—	—	0	0	—
0	108	1.16	56.1	—	0	—	0	—	—	—	—	—	—	0	0	—
0	171	1.36	31.5	125.9	10.2	0.70	0	0.05	0.01	0.05	0.08	0.06	16.2	0.2	0	8.4
0	316	4.14	51.0	124.2	13.6	1.69	0.9	0.14	0.03	0.04	0.08	0.08	23.0	0	0	24.2
0	56	0.34	33.1	—	4.7	—	0	—	—	—	—	—	—	0	1.7	—
0	108	1.16	35.5	—	0	—	0	—	—	—	—	—	—	0	1.9	—
0	94	0.94	23.0	102.1	6.8	0.54	0	0.04	0.01	0.03	0.45	0.04	37.4	0.2	0	7.6

DA+ Code	Food Description	QTY	Measure	Wt (g)	H2O (g)	Ener (cal)	Prot (g)	Carb (g)	Fiber (g)	Fat (g)	Fat Breakdown (g) Sat	Mono	Poly
VEGETABLES, LEGUMES—continued													
	Spinach												
663	Canned, drained	½	cup(s)	107	98.2	25	3.0	3.6	2.6	0.5	0.1	0	0.2
660	Chopped, boiled, drained	½	cup(s)	90	82.1	21	2.7	3.4	2.2	0.2	0	0	0.1
661	Chopped, frozen, boiled, drained	½	cup(s)	95	84.5	32	3.8	4.6	3.5	0.8	0.1	0	0.4
662	Leaf, frozen, boiled, drained	½	cup(s)	95	84.5	32	3.8	4.6	3.5	0.8	0.1	0	0.4
659	Raw, chopped	1	cup(s)	30	27.4	7	0.9	1.1	0.7	0.1	0	0	0
8470	Trimmed leaves	1	cup(s)	32	27.5	3	0.9	0	2.8	0.1	—	—	—
	Squash												
1662	Acorn winter, baked	½	cup(s)	103	85.0	57	1.1	14.9	4.5	0.1	0	0	0.1
29702	Acorn winter, boiled, mashed	½	cup(s)	123	109.9	42	0.8	10.8	3.2	0.1	0	0	0
29451	Butternut, frozen, boiled	½	cup(s)	122	106.9	47	1.5	12.2	1.8	0.1	0	0	0
1661	Butternut winter, baked	½	cup(s)	102	89.5	41	0.9	10.7	3.4	0.1	0	0	0
32773	Butternut winter, frozen, boiled, mashed, no salt added	½	cup(s)	121	106.4	47	1.5	12.2	—	0.1	0	0	0
29700	Crookneck and straightneck summer, boiled, drained	½	cup(s)	65	60.9	12	0.6	2.6	1.2	0.1	0	0	0.1
29703	Hubbard winter, baked	½	cup(s)	102	86.8	51	2.5	11.0	—	0.6	0.1	0	0.3
1660	Hubbard winter, boiled, mashed	½	cup(s)	118	107.5	35	1.7	7.6	3.4	0.4	0.1	0	0.2
29704	Spaghetti winter, boiled, drained, or baked	½	cup(s)	78	71.5	21	0.5	5.0	1.1	0.2	0	0	0.1
664	Summer, all varieties, sliced, boiled, drained	½	cup(s)	90	84.3	18	0.8	3.9	1.3	0.3	0.1	0	0.1
665	Winter, all varieties, baked, mashed	½	cup(s)	103	91.4	38	0.9	9.1	2.9	0.4	0.1	0	0.2
1112	Zucchini summer, boiled, drained	½	cup(s)	90	85.3	14	0.6	3.5	1.3	0	0	0	0
1113	Zucchini summer, frozen, boiled, drained	½	cup(s)	112	105.6	19	1.3	4.0	1.4	0.1	0	0	0.1
	Sweet potatoes												
666	Baked, peeled	½	cup(s)	100	75.8	90	2.0	20.7	3.3	0.2	0	0	0.1
667	Boiled, mashed	½	cup(s)	164	131.4	125	2.2	29.1	4.1	0.2	0.1	0	0.1
668	Candied, home recipe	½	cup(s)	91	61.1	132	0.8	25.4	2.2	3.0	1.2	0.6	0.1
670	Canned, vacuum pack	½	cup(s)	100	76.0	91	1.7	21.1	1.8	0.2	0	0	0.1
2765	Frozen, baked	½	cup(s)	88	64.5	88	1.5	20.5	1.6	0.1	0	0	0
1136	Yams, baked or boiled, drained	½	cup(s)	68	47.7	79	1.0	18.7	2.7	0.1	0	0	0
32785	**Taro shoots, cooked, no salt added**	½	cup(s)	70	66.7	10	0.5	2.2	—	0.1	0	0	0
	Tomatillo												
8774	Raw	2	item(s)	68	62.3	22	0.7	4.0	1.3	0.7	0.1	0.1	0.3
8777	Raw, chopped	½	cup(s)	66	60.5	21	0.6	3.9	1.3	0.7	0.1	0.1	0.3
	Tomato												
16846	Cherry, fresh	5	item(s)	85	80.3	15	0.7	3.3	1.0	0.2	0	0	0.1
671	Fresh, ripe, red	1	item(s)	123	116.2	22	1.1	4.8	1.5	0.2	0	0	0.1
675	Juice, canned	½	cup(s)	122	114.1	21	0.9	5.2	0.5	0.1	0	0	0
75	Juice, no salt added	½	cup(s)	122	114.1	21	0.9	5.2	0.5	0.1	0	0	0
1699	Paste, canned	2	tablespoon(s)	33	24.1	27	1.4	6.2	1.3	0.2	0	0	0.1
1700	Puree, canned	¼	cup(s)	63	54.9	24	1.0	5.6	1.2	0.1	0	0	0.1
1118	Red, boiled	½	cup(s)	120	113.2	22	1.1	4.8	0.8	0.1	0	0	0.1
3952	Red, diced	½	cup(s)	90	85.1	16	0.8	3.5	1.1	0.2	0	0	0.1
1120	Red, stewed, canned	½	cup(s)	128	116.7	33	1.2	7.9	1.3	0.2	0	0	0.1
1125	Sauce, canned	¼	cup(s)	61	55.6	15	0.8	3.3	0.9	0.1	0	0	0
8778	Sun dried	½	cup(s)	27	3.9	70	3.8	15.1	3.3	0.8	0.1	0.1	0.3
8783	Sun dried in oil, drained	¼	cup(s)	28	14.8	59	1.4	6.4	1.6	3.9	0.5	2.4	0.6
	Turnips												
678	Turnip greens, chopped, boiled, drained	½	cup(s)	72	67.1	14	0.8	3.1	2.5	0.2	0	0	0.1
679	Turnip greens, frozen, chopped, boiled, drained	½	cup(s)	82	74.1	24	2.7	4.1	2.8	0.3	0.1	0	0.1
677	Turnips, cubed, boiled, drained	½	cup(s)	78	73.0	17	0.6	3.9	1.6	0.1	0	0	0
	Vegetables, mixed												
1132	Canned, drained	½	cup(s)	82	70.9	40	2.1	7.5	2.4	0.2	0	0	0.1
680	Frozen, boiled, drained	½	cup(s)	91	75.7	59	2.6	11.9	4.0	0.1	0	0	0.1

Chol (mg)	Calc (mg)	Iron (mg)	Magn (mg)	Pota (mg)	Sodi (mg)	Zinc (mg)	Vit A (µg)	Thia (mg)	Vit E (mg α)	Ribo (mg)	Niac (mg)	Vit B_6 (mg)	Fola (µg DFE)	Vit C (mg)	Vit B_{12} (µg)	Sele (µg)
0	136	2.45	81.3	370.2	28.9	0.48	524.3	0.02	2.08	0.14	0.41	0.11	104.8	15.3	0	1.5
0	122	3.21	78.3	419.4	63.0	0.68	471.6	0.08	1.87	0.21	0.44	0.21	131.4	8.8	0	1.4
0	145	1.86	77.9	286.9	92.2	0.46	572.9	0.07	3.36	0.16	0.41	0.12	115.0	2.1	0	5.2
0	145	1.86	77.9	286.9	92.2	0.46	572.9	0.07	3.36	0.16	0.41	0.12	115.0	2.1	0	5.2
0	30	0.81	23.7	167.4	23.7	0.16	140.7	0.02	0.61	0.06	0.22	0.06	58.2	8.4	0	0.3
0	25	2.13	25.5	134.1	38.0	0.18	—	0.03	—	0.05	0.18	0.07	0	7.5	0	—
0	45	0.95	44.1	447.9	4.1	0.17	21.5	0.17	—	0.01	0.90	0.19	19.5	11.1	0	0.7
0	32	0.68	31.9	322.2	3.7	0.13	50.2	0.12	—	0.01	0.65	0.14	13.5	8.0	0	0.5
0	23	0.70	10.9	161.9	2.4	0.14	203.3	0.06	0.14	0.05	0.56	0.08	19.5	4.3	0	0.6
0	42	0.61	29.6	289.6	4.1	0.13	569.1	0.07	1.31	0.01	0.99	0.12	19.4	15.4	0	0.5
0	23	0.70	10.9	161.2	2.4	0.14	202.4	0.06	—	0.05	0.56	0.08	19.4	4.2	0	0.6
0	14	0.31	13.6	137.1	1.3	0.19	5.2	0.03	—	0.02	0.29	0.07	14.9	5.4	0	0.1
0	17	0.48	22.4	365.1	8.2	0.15	308.0	0.07	—	0.04	0.57	0.17	16.3	9.7	0	0.6
0	12	0.33	15.3	252.5	5.9	0.11	236.0	0.05	0.14	0.03	0.39	0.12	11.8	7.7	0	0.4
0	16	0.26	8.5	90.7	14.0	0.15	4.7	0.02	0.09	0.01	0.62	0.07	6.2	2.7	0	0.2
0	24	0.32	21.6	172.8	0.9	0.35	9.9	0.04	0.12	0.03	0.46	0.05	18.0	5.0	0	0.2
0	23	0.45	13.3	247.0	1.0	0.23	267.5	0.02	0.12	0.07	0.51	0.17	20.5	9.8	0	0.4
0	12	0.32	19.8	227.7	2.7	0.16	50.4	0.04	0.11	0.04	0.39	0.07	15.3	4.1	0	0.2
0	19	0.54	14.5	216.3	2.2	0.22	10	0.05	0.13	0.04	0.43	0.05	8.9	4.1	0	0.2
0	38	0.69	27.0	475.0	36.0	0.32	961.0	0.10	0.71	0.10	1.48	0.28	6.0	19.6	0	0.2
0	44	1.18	29.5	377.2	44.3	0.33	1290.7	0.09	1.54	0.08	0.88	0.27	9.8	21.0	0	0.3
7	24	1.03	10.0	172.6	63.9	0.13	0	0.01	—	0.03	0.36	0.03	10.0	6.1	0	0.7
0	22	0.89	22.0	312.0	53.0	0.18	399.0	0.04	1.00	0.06	0.74	0.19	17.0	26.4	0	0.7
0	31	0.47	18.4	330.1	7.0	0.26	913.3	0.05	0.67	0.04	0.49	0.16	19.3	8.0	0	0.5
0	10	0.35	12.2	455.6	5.4	0.13	4.1	0.06	0.23	0.01	0.37	0.15	10.9	8.2	0	0.5
0	10	0.28	5.6	240.8	1.4	0.37	2.1	0.02	—	0.03	0.56	0.07	2.1	13.2	0	0.7
0	5	0.42	13.6	182.2	0.7	0.15	4.1	0.03	0.25	0.02	1.25	0.03	4.8	8.0	0	0.3
0	5	0.41	13.2	176.9	0.7	0.15	4.0	0.03	0.25	0.02	1.22	0.04	4.6	7.7	0	0.3
0	9	0.22	9.4	201.5	4.3	0.14	35.7	0.03	0.45	0.01	0.50	0.06	12.8	10.8	0	0
0	12	0.33	13.5	291.5	6.2	0.20	51.7	0.04	0.66	0.02	0.73	0.09	18.5	15.6	0	0
0	12	0.52	13.4	278.2	326.8	0.18	27.9	0.06	0.39	0.04	0.82	0.14	24.3	22.2	0	0.4
0	12	0.52	13.4	278.2	12.2	0.18	27.9	0.06	0.39	0.04	0.82	0.14	24.3	22.2	0	0.4
0	12	0.97	13.8	332.6	259.1	0.20	24.9	0.02	1.41	0.05	1.00	0.07	3.9	7.2	0	1.7
0	11	1.11	14.4	274.4	249.4	0.22	16.3	0.01	1.23	0.05	0.91	0.07	6.9	6.6	0	0.4
0	13	0.82	10.8	261.6	13.2	0.17	28.8	0.04	0.67	0.03	0.64	0.10	15.6	27.4	0	0.6
0	9	0.24	9.9	213.3	4.5	0.15	37.8	0.03	0.48	0.01	0.53	0.07	13.5	11.4	0	0
0	43	1.70	15.3	263.9	281.8	0.22	11.5	0.06	1.06	0.04	0.91	0.02	6.4	10.1	0	0.8
0	8	0.62	9.8	201.9	319.6	0.12	10.4	0.01	0.87	0.04	0.59	0.06	6.7	4.3	0	0.1
0	30	2.45	52.4	925.3	565.7	0.53	11.9	0.14	0.00	0.13	2.44	0.09	18.4	10.6	0	1.5
0	13	0.73	22.3	430.4	73.2	0.21	17.6	0.05	—	0.10	0.99	0.08	6.3	28.0	0	0.8
0	99	0.58	15.8	146.2	20.9	0.10	274.3	0.03	1.35	0.05	0.30	0.13	85.0	19.7	0	0.6
0	125	1.59	21.3	183.7	12.3	0.34	441.2	0.04	2.18	0.06	0.38	0.06	32.0	17.9	0	1.0
0	26	0.14	7.0	138.1	12.5	0.09	0	0.02	0.02	0.02	0.23	0.05	7.0	9.0	0	0.2
0	22	0.86	13.0	237.2	121.4	0.33	475.1	0.04	0.24	0.04	0.47	0.06	19.6	4.1	0	0.2
0	23	0.74	20.0	153.8	31.9	0.44	194.7	0.06	0.34	0.10	0.77	0.06	17.3	2.9	0	0.3

Table A–1 Table of Food Composition (*continued*)

(Computer code is for Cengage Diet Analysis Plus program)

DA+ Code	Food Description	QTY	Measure	Wt (g)	H₂O (g)	Ener (cal)	Prot (g)	Carb (g)	Fiber (g)	Fat (g)	Fat Breakdown (g) Sat	Mono	Poly
VEGETABLES, LEGUMES—continued													
7489	V8 100% vegetable juice	½	cup(s)	120	—	25	1.0	5.0	1.0	0	0	0	0
7490	V8 low sodium vegetable juice	½	cup(s)	120	—	25	0	6.5	1.0	0	0	0	0
7491	V8 spicy hot vegetable juice	½	cup(s)	120	—	25	1.0	5.0	0.5	0	0	0	0
	Water chestnuts												
31073	Sliced, drained	½	cup(s)	75	70.0	20	0	5.0	1.0	0	0	0	0
31087	Whole	½	cup(s)	75	70.0	20	0	5.0	1.0	0	0	0	0
1135	**Watercress**	1	cup(s)	34	32.3	4	0.8	0.4	0.2	0	0	0	0
NUTS, SEEDS, AND PRODUCTS													
	Almonds												
32940	Almond butter with salt added	1	tablespoon(s)	16	0.2	101	2.4	3.4	0.6	9.5	0.9	6.1	2.0
1137	Almond butter, no salt added	1	tablespoon(s)	16	0.2	101	2.4	3.4	0.6	9.5	0.9	6.1	2.0
32886	Blanched	¼	cup(s)	36	1.6	211	8.0	7.2	3.8	18.3	1.4	11.7	4.4
32887	Dry roasted, no salt added	¼	cup(s)	35	0.9	206	7.6	6.7	4.1	18.2	1.4	11.6	4.4
29724	Dry roasted, salted	¼	cup(s)	35	0.9	206	7.6	6.7	4.1	18.2	1.4	11.6	4.4
29725	Oil roasted, salted	¼	cup(s)	39	1.1	238	8.3	6.9	4.1	21.7	1.7	13.7	5.3
508	Slivered	¼	cup(s)	27	1.3	155	5.7	5.9	3.3	13.3	1.0	8.3	3.3
1138	**Beechnuts, dried**	¼	cup(s)	57	3.8	328	3.5	19.1	5.3	28.5	3.3	12.5	11.4
517	**Brazil nuts, dried, unblanched**	¼	cup(s)	35	1.2	230	5.0	4.3	2.6	23.3	5.3	8.6	7.2
1166	**Breadfruit seeds, roasted**	¼	cup(s)	57	28.3	118	3.5	22.8	3.4	1.5	0.4	0.2	0.8
1139	**Butternuts, dried**	¼	cup(s)	30	1.0	184	7.5	3.6	1.4	17.1	0.4	3.1	12.8
	Cashews												
32931	Cashew butter with salt added	1	tablespoon(s)	16	0.5	94	2.8	4.4	0.3	7.9	1.6	4.7	1.3
32889	Cashew butter, no salt added	1	tablespoon(s)	16	0.5	94	2.8	4.4	0.3	7.9	1.6	4.7	1.3
1140	Dry roasted	¼	cup(s)	34	0.6	197	5.2	11.2	1.0	15.9	3.1	9.4	2.7
518	Oil roasted	¼	cup(s)	32	1.1	187	5.4	9.6	1.1	15.4	2.7	8.4	2.8
	Coconut, shredded												
32896	Dried, not sweetened	¼	cup(s)	23	0.7	152	1.6	5.4	3.8	14.9	13.2	0.6	0.2
1153	Dried, shredded, sweetened	¼	cup(s)	23	2.9	116	0.7	11.1	1.0	8.3	7.3	0.4	0.1
520	Shredded	¼	cup(s)	20	9.4	71	0.7	3.0	1.8	6.7	5.9	0.3	0.1
	Chestnuts												
1152	Chinese, roasted	¼	cup(s)	36	14.6	87	1.6	19.0	—	0.4	0.1	0.2	0.1
32895	European, boiled and steamed	¼	cup(s)	46	31.3	60	0.9	12.8	—	0.6	0.1	0.2	0.2
32911	European, roasted	¼	cup(s)	36	14.5	88	1.1	18.9	1.8	0.8	0.1	0.3	0.3
32922	Japanese, boiled and steamed	¼	cup(s)	36	31.0	20	0.3	4.5	—	0.1	0	0	0
32923	Japanese, roasted	¼	cup(s)	36	18.1	73	1.1	16.4	—	0.3	0	0.1	0.1
4958	**Flax seeds or linseeds**	¼	cup(s)	43	3.3	225	8.4	12.3	11.9	17.7	1.7	3.2	12.6
32904	**Ginkgo nuts, dried**	¼	cup(s)	39	4.8	136	4.0	28.3	—	0.8	0.1	0.3	0.3
	Hazelnuts or filberts												
32901	Blanched	¼	cup(s)	30	1.7	189	4.1	5.1	3.3	18.3	1.4	14.5	1.7
32902	Dry roasted, no salt added	¼	cup(s)	30	0.8	194	4.5	5.3	2.8	18.7	1.3	14.0	2.5
1156	**Hickory nuts, dried**	¼	cup(s)	30	0.8	197	3.8	5.5	1.9	19.3	2.1	9.8	6.6
	Macadamias												
32905	Dry roasted, no salt added	¼	cup(s)	34	0.5	241	2.6	4.5	2.7	25.5	4.0	19.9	0.5
32932	Dry roasted, with salt added	¼	cup(s)	34	0.5	240	2.6	4.3	2.7	25.5	4.0	19.9	0.5
1157	Raw	¼	cup(s)	34	0.5	241	2.6	4.6	2.9	25.4	4.0	19.7	0.5
	Mixed nuts												
1159	With peanuts, dry roasted	¼	cup(s)	34	0.6	203	5.9	8.7	3.1	17.6	2.4	10.8	3.7
32933	With peanuts, dry roasted, with salt added	¼	cup(s)	34	0.6	203	5.9	8.7	3.1	17.6	2.4	10.8	3.7
32906	Without peanuts, oil roasted, no salt added	¼	cup(s)	36	1.1	221	5.6	8.0	2.0	20.2	3.3	11.9	4.1
	Peanuts												
2807	Dry roasted	¼	cup(s)	37	0.6	214	8.6	7.9	2.9	18.1	2.5	9.0	5.7
2806	Dry roasted, salted	¼	cup(s)	37	0.6	214	8.6	7.9	2.9	18.1	2.5	9.0	5.7

Chol (mg)	Calc (mg)	Iron (mg)	Magn (mg)	Pota (mg)	Sodi (mg)	Zinc (mg)	Vit A (µg)	Thia (mg)	Vit E (mg α)	Ribo (mg)	Niac (mg)	Vit B_6 (mg)	Fola (µg DFE)	Vit C (mg)	Vit B_{12} (µg)	Sele (µg)
0	20	0.36	12.9	260.0	310.0	0.24	100.0	0.05	—	0.03	0.87	0.17	—	30.0	0	—
0	20	0.36	—	450.0	70.0	—	100.0	0.02	—	0.02	0.75	—	—	30.0	0	—
0	20	0.36	12.9	240.0	360.0	0.24	50.0	0.05	—	0.03	0.88	0.17	—	15.0	0	—
0	7	0.00	—	—	5.0	—	0	—	—	—	—	—	—	2.0	—	—
0	7	0.00	—	—	5.0	—	0	—	—	—	—	—	—	2.0	—	—
0	41	0.06	7.1	112.2	13.9	0.03	54.4	0.03	0.34	0.04	0.06	0.04	3.1	14.6	0	0.3
0	43	0.59	48.5	121.3	72.0	0.49	0	0.02	4.16	0.10	0.46	0.01	10.4	0.1	0	0.8
0	43	0.59	48.5	121.3	1.8	0.48	0	0.02	—	0.09	0.46	0.01	10.4	0.1	0	—
0	78	1.34	99.7	249.0	10.2	1.13	0	0.07	8.95	0.20	1.32	0.04	10.9	0	0	1.0
0	92	1.55	98.7	257.4	0.3	1.22	0	0.02	8.97	0.29	1.32	0.04	11.4	0	0	1.0
0	92	1.55	98.7	257.4	117.0	1.22	0	0.02	8.97	0.29	1.32	0.04	11.4	0	0	1.0
0	114	1.44	107.5	274.4	133.1	1.20	0	0.03	10.19	0.30	1.43	0.04	10.6	0	0	1.1
0	71	1.00	72.4	190.4	0.3	0.83	0	0.05	7.07	0.27	0.91	0.03	13.5	0	0	0.7
0	1	1.39	0	579.7	21.7	0.20	0	0.16	—	0.20	0.48	0.38	64.4	8.8	0	4.0
0	56	0.85	131.6	230.7	1.1	1.42	0	0.21	2.00	0.01	0.10	0.03	7.7	0.2	0	671.0
0	49	0.50	35.3	616.7	15.9	0.58	8.5	0.22	—	0.12	4.20	0.22	33.6	4.3	0	8.0
0	16	1.21	71.1	126.3	0.3	0.94	1.8	0.12	—	0.04	0.31	0.17	19.8	1.0	0	5.2
0	7	0.81	41.3	87.4	98.2	0.83	0	0.05	0.15	0.03	0.26	0.04	10.9	0	0	1.8
0	7	0.81	41.3	87.4	2.4	0.83	0	0.05	—	0.03	0.26	0.04	10.9	0	0	1.8
0	15	2.06	89.1	193.5	5.5	1.92	0	0.07	0.32	0.07	0.48	0.09	23.6	0	0	4.0
0	14	1.95	88.0	203.8	4.2	1.73	0	0.12	0.30	0.07	0.56	0.10	8.1	0.1	0	6.5
0	6	0.76	20.7	125.2	8.5	0.46	0	0.01	0.10	0.02	0.13	0.07	2.1	0.3	0	4.3
0	3	0.45	11.6	78.4	60.9	0.42	0	0.01	0.09	0.00	0.11	0.06	1.9	0.2	0	3.9
0	3	0.48	6.4	71.2	4.0	0.21	0	0.01	0.04	0.00	0.11	0.01	5.2	0.7	0	2.0
0	7	0.54	32.6	173.0	1.4	0.33	0	0.05	—	0.03	0.54	0.15	26.1	13.9	0	2.6
0	21	0.80	24.8	328.9	12.4	0.11	0.5	0.06	—	0.03	0.32	0.10	17.5	12.3	0	—
0	10	0.32	11.8	211.6	0.7	0.20	0.4	0.08	0.18	0.05	0.48	0.18	25.0	9.3	0	0.4
0	4	0.19	6.5	42.8	1.8	0.14	0.4	0.04	—	0.01	0.19	0.03	6.1	3.4	0	—
0	13	0.75	23.2	154.8	6.9	0.51	1.4	0.15	—	—	0.24	0.14	21.4	10.1	0	—
0	142	2.13	156.1	354.0	11.9	1.83	0	0.06	0.14	0.06	0.59	0.39	118.4	0.5	0	2.3
0	8	0.62	20.7	390.2	5.1	0.26	21.5	0.17	—	0.07	4.58	0.25	41.4	11.4	0	—
0	45	0.98	48.0	197.4	0	0.66	0.6	0.14	5.25	0.03	0.46	0.17	23.4	0.6	0	1.2
0	37	1.31	51.9	226.5	0	0.74	0.9	0.10	4.58	0.03	0.61	0.18	26.4	1.1	0	1.2
0	18	0.64	51.9	130.8	0.3	1.29	2.1	0.26	—	0.04	0.27	0.06	12.0	0.6	0	2.4
0	23	0.88	39.5	121.6	1.3	0.43	0	0.23	0.19	0.02	0.76	0.12	3.4	0.2	0	3.9
0	23	0.88	39.5	121.6	88.8	0.43	0	0.23	0.19	0.02	0.76	0.12	3.4	0.2	0	3.9
0	28	1.24	43.6	123.3	1.7	0.44	0	0.40	0.18	0.05	0.83	0.09	3.7	0.4	0	1.2
0	24	1.27	77.1	204.5	4.1	1.30	0.3	0.07	—	0.07	1.61	0.10	17.1	0.1	0	1.0
0	24	1.26	77.1	204.5	229.1	1.30	0	0.06	3.74	0.06	1.61	0.10	17.1	0.1	0	2.6
0	38	0.92	90.4	195.8	4.0	1.67	0.4	0.18	—	0.17	0.70	0.06	20.2	0.2	0	—
0	20	0.82	64.2	240.2	2.2	1.20	0	0.16	2.52	0.03	4.93	0.09	52.9	0	0	2.7
0	20	0.82	64.2	240.2	296.7	1.20	0	0.16	2.84	0.03	4.93	0.09	52.9	0	0	2.7

Table A–1 Table of Food Composition (*continued*) (Computer code is for Cengage Diet Analysis Plus program)

											Fat Breakdown (g)		
DA+ Code	Food Description	QTY	Measure	Wt (g)	H_2O (g)	Ener (cal)	Prot (g)	Carb (g)	Fiber (g)	Fat (g)	Sat	Mono	Poly
NUTS, SEEDS, AND PRODUCTS—continued													
1763	Oil roasted, salted	¼	cup(s)	36	0.5	216	10.1	5.5	3.4	18.9	3.1	9.4	5.5
1884	Peanut butter, chunky	1	tablespoon(s)	16	0.2	94	3.8	3.5	1.3	8.0	1.3	3.9	2.4
30303	Peanut butter, low sodium	1	tablespoon(s)	16	0.2	95	4.0	3.1	0.9	8.2	1.8	3.9	2.2
30305	Peanut butter, reduced fat	1	tablespoon(s)	18	0.2	94	4.7	6.4	0.9	6.1	1.3	2.9	1.8
524	Peanut butter, smooth	1	tablespoon(s)	16	0.3	94	4.0	3.1	1.0	8.1	1.7	3.9	2.3
2804	Raw	¼	cup(s)	37	2.4	207	9.4	5.9	3.1	18.0	2.5	8.9	5.7
	Pecans												
32907	Dry roasted, no salt added	¼	cup(s)	28	0.3	198	2.6	3.8	2.6	20.7	1.8	12.3	5.7
32936	Dry roasted, with salt added	¼	cup(s)	27	0.3	192	2.6	3.7	2.5	20.0	1.7	11.9	5.6
1162	Oil roasted	¼	cup(s)	28	0.3	197	2.5	3.6	2.6	20.7	2.0	11.3	6.5
526	Raw	¼	cup(s)	27	1.0	188	2.5	3.8	2.6	19.6	1.7	11.1	5.9
12973	**Pine nuts or pignolia, dried**	1	tablespoon(s)	9	0.2	58	1.2	1.1	0.3	5.9	0.4	1.6	2.9
	Pistachios												
1164	Dry roasted	¼	cup(s)	31	0.6	176	6.6	8.5	3.2	14.1	1.7	7.4	4.3
32938	Dry roasted, with salt added	¼	cup(s)	32	0.6	182	6.8	8.6	3.3	14.7	1.8	7.7	4.4
1167	**Pumpkin or squash seeds, roasted**	¼	cup(s)	57	4.0	296	18.7	7.6	2.2	23.9	4.5	7.4	10.9
	Sesame												
32912	Sesame butter paste	1	tablespoon(s)	16	0.3	94	2.9	3.8	0.9	8.1	1.1	3.1	3.6
32941	Tahini or sesame butter	1	tablespoon(s)	15	0.5	89	2.6	3.2	0.7	8.0	1.1	3.0	3.5
1169	Whole, roasted, toasted	3	tablespoon(s)	10	0.3	54	1.6	2.4	1.3	4.6	0.6	1.7	2.0
	Soy nuts												
34173	Deep sea salted	¼	cup(s)	28	—	119	11.9	8.9	4.9	4.0	1.0	—	—
34174	Unsalted	¼	cup(s)	28	—	119	11.9	8.9	4.9	4.0	0	—	—
	Sunflower seeds												
528	Kernels, dried	1	tablespoon(s)	9	0.4	53	1.9	1.8	0.8	4.6	0.4	1.7	2.1
29721	Kernels, dry roasted, salted	1	tablespoon(s)	8	0.1	47	1.5	1.9	0.7	4.0	0.4	0.8	2.6
29723	Kernels, toasted, salted	1	tablespoon(s)	8	0.1	52	1.4	1.7	1.0	4.8	0.5	0.9	3.1
32928	Sunflower seed butter with salt added	1	tablespoon(s)	16	0.2	93	3.1	4.4	—	7.6	0.8	1.5	5.0
	Trail mix												
4646	Trail mix	¼	cup(s)	38	3.5	173	5.2	16.8	2.0	11.0	2.1	4.7	3.6
4647	Trail mix with chocolate chips	¼	cup(s)	38	2.5	182	5.3	16.8	—	12.0	2.3	5.1	4.2
4648	Tropical trail mix	¼	cup(s)	35	3.2	142	2.2	23.0	—	6.0	3.0	0.9	1.8
	Walnuts												
529	Dried black, chopped	¼	cup(s)	31	1.4	193	7.5	3.1	2.1	18.4	1.1	4.7	11.0
531	English or Persian	¼	cup(s)	29	1.2	191	4.5	4.0	2.0	19.1	1.8	2.6	13.8
VEGETARIAN FOODS													
	Prepared												
34222	Brown rice and tofu stir-fry (vegan)	8	ounce(s)	227	244.4	302	16.5	18.0	3.2	21.0	1.7	4.7	13.4
34368	Cheese enchilada casserole (lacto)	8	ounce(s)	227	80.3	385	16.6	38.4	4.1	17.8	9.5	6.1	1.1
34247	Five bean casserole (vegan)	8	ounce(s)	227	175.8	178	5.9	26.6	6.0	5.8	1.1	2.5	1.9
34261	Lentil stew (vegan)	8	ounce(s)	227	227.9	188	11.5	35.9	11.0	0.7	0.1	0.1	0.3
34397	Macaroni and cheese (lacto)	8	ounce(s)	227	352.1	391	18.1	37.1	1.0	18.7	9.8	6.0	1.8
34238	Steamed rice and vegetables (vegan)	8	ounce(s)	227	222.9	587	11.2	87.9	5.8	23.1	4.1	8.7	9.1
34308	Tofu rice burgers (ovo-lacto)	1	piece(s)	218	77.6	435	22.4	68.6	5.6	8.4	1.7	2.4	3.5
34276	Vegan spinach enchiladas (vegan)	1	piece(s)	82	59.2	93	4.9	14.5	1.8	2.4	0.3	0.6	1.3
34243	Vegetable chow mein (vegan)	8	ounce(s)	227	163.3	166	6.5	22.1	2.0	6.4	0.7	2.7	2.5
34454	Vegetable lasagna (lacto)	8	ounce(s)	227	178.9	208	13.7	29.9	2.6	4.1	2.3	1.1	0.3
34339	Vegetable marinara (vegan)	8	ounce(s)	252	200.7	104	3.0	16.7	1.4	3.1	0.4	1.4	1.0
34356	Vegetable rice casserole (lacto)	8	ounce(s)	227	178.9	238	9.7	24.4	4.0	12.5	4.9	3.5	3.1
34311	Vegetable strudel (ovo-lacto)	8	ounce(s)	227	63.1	478	12.0	32.4	2.5	33.8	11.5	16.7	3.9
34371	Vegetable taco (lacto)	1	item(s)	85	46.5	117	4.2	13.6	2.9	5.6	2.1	1.9	1.3
34282	Vegetarian chili (vegan)	8	ounce(s)	227	191.4	115	5.6	21.4	7.1	1.5	0.2	0.3	0.7
34367	Vegetarian vegetable soup (vegan)	8	ounce(s)	227	257.9	111	3.2	16.0	3.2	5.0	1.0	2.1	1.6
	Boca burger												
32067	All American flamed grilled patty	1	item(s)	71	—	90	14.0	4.0	3.0	3.0	1.0	—	—
32074	Boca chik'n nuggets	4	item(s)	87	—	180	14.0	17.0	3.0	7.0	1.0	—	—

Chol (mg)	Calc (mg)	Iron (mg)	Magn (mg)	Pota (mg)	Sodi (mg)	Zinc (mg)	Vit A (µg)	Thia (mg)	Vit E (mg α)	Ribo (mg)	Niac (mg)	Vit B_6 (mg)	Fola (µg DFE)	Vit C (mg)	Vit B_{12} (µg)	Sele (µg)
0	22	0.54	63.4	261.4	115.2	1.18	0	0.03	2.49	0.03	4.97	0.16	43.2	0.3	0	1.2
0	7	0.30	25.6	119.2	77.8	0.45	0	0.02	1.01	0.02	2.19	0.07	14.7	0	0	1.3
0	6	0.29	25.4	107.0	2.7	0.47	0	0.01	1.23	0.02	2.14	0.07	11.8	0	0	1.2
0	6	0.34	30.6	120.4	97.2	0.50	0	0.05	1.20	0.01	2.63	0.06	10.8	0	0	1.4
0	7	0.30	24.6	103.8	73.4	0.47	0	0.01	1.44	0.02	2.14	0.09	11.8	0	0	0.9
0	34	1.67	61.3	257.3	6.6	1.19	0	0.23	3.04	0.04	4.40	0.12	87.6	0	0	2.6
0	20	0.78	36.8	118.3	0.3	1.41	1.9	0.12	0.35	0.03	0.32	0.05	4.5	0.2	0	1.1
0	19	0.75	35.6	114.5	103.4	1.36	1.9	0.11	0.34	0.03	0.31	0.05	4.3	0.2	0	1.1
0	18	0.68	33.3	107.8	0.3	1.23	1.4	0.13	0.70	0.03	0.33	0.05	4.1	0.2	0	1.7
0	19	0.69	33.0	111.7	0	1.23	0.8	0.18	0.38	0.04	0.32	0.06	6.0	0.3	0	1.0
0	1	0.47	21.6	51.3	0.2	0.55	0.1	0.03	0.80	0.02	0.37	0.01	2.9	0.1	0	0.1
0	34	1.29	36.9	320.4	3.1	0.71	4.0	0.26	0.59	0.05	0.44	0.39	15.4	0.7	0	2.9
0	35	1.34	38.4	333.4	129.6	0.73	4.2	0.26	0.61	0.05	0.45	0.40	16.0	0.7	0	3.0
0	24	8.48	303.0	457.4	10.2	4.22	10.8	0.12	0.00	0.18	0.99	0.05	32.3	1.0	0	3.2
0	154	3.07	57.9	93.1	1.9	1.17	0.5	0.04	—	0.03	1.07	0.13	16.0	0	0	0.9
0	21	0.66	14.3	68.9	5.3	0.69	0.5	0.24	—	0.02	0.85	0.02	14.7	0.6	0	0.3
0	94	1.40	33.8	45.1	1.0	0.68	0	0.07	—	0.02	0.43	0.07	9.3	0	0	0.5
0	59	1.07	—	—	148.1	—	0	—	—	—	—	—	—	0	—	—
0	59	1.07	—	—	9.9	—	0	—	—	—	—	—	—	0	—	—
0	7	0.47	29.3	58.1	0.8	0.45	0.3	0.13	2.99	0.03	0.75	0.12	20.4	0.1	0	4.8
0	6	0.30	10.3	68.0	32.8	0.42	0	0.01	2.09	0.02	0.56	0.06	19.0	0.1	0	6.3
0	5	0.57	10.8	41.1	51.3	0.44	0	0.03	—	0.02	0.35	0.07	19.9	0.1	0	5.2
0	20	0.76	59.0	11.5	83.2	0.85	0.5	0.05	—	0.05	0.85	0.13	37.9	0.4	0	—
0	29	1.14	59.3	256.9	85.9	1.20	0.4	0.17	—	0.07	1.76	0.11	26.6	0.5	0	—
2	41	1.27	60.4	243.0	45.4	1.17	0.8	0.15	—	0.08	1.65	0.09	24.4	0.5	0	—
0	20	0.92	33.6	248.2	3.5	0.41	0.7	0.15	—	0.04	0.51	0.11	14.7	2.7	0	—
0	19	0.97	62.8	163.4	0.6	1.05	0.6	0.01	0.56	0.04	0.14	0.18	9.7	0.5	0	5.3
0	29	0.85	46.2	129.0	0.6	0.90	0.3	0.10	0.20	0.04	0.32	0.15	28.7	0.4	0	1.4
0	353	6.34	118.3	501.4	142.2	2.03	—	0.23	0.07	0.14	1.49	0.36	51.8	24.8	0	14.8
39	441	2.44	34.6	191.2	1139.7	1.84	—	0.31	0.05	0.35	2.23	0.11	118.3	20.4	0.4	20.0
0	48	1.78	40.8	364.1	613.6	0.61	—	0.10	0.52	0.07	0.93	0.11	64.5	8.3	0	3.3
0	34	3.23	50.0	548.8	436.5	1.42	—	0.24	0.14	0.16	2.31	0.29	202.7	26.4	0	12.1
43	415	1.71	45.4	267.8	1641.0	2.32	—	0.32	0.27	0.48	2.18	0.13	162.7	0.9	0.8	33.3
0	91	3.31	153.1	810.1	3117.8	2.04	—	0.37	3.03	0.21	6.16	0.64	70.1	35.2	0	18.8
52	467	9.01	89.7	455.6	2449.5	2.06	—	0.27	0.12	0.26	3.43	0.29	167.7	2.0	0.1	43.0
0	117	1.13	40.4	170.5	134.2	0.68	—	0.07	—	0.07	0.53	0.10	20.3	1.8	0	5.1
0	189	3.70	28.0	310.3	372.7	0.76	—	0.13	0.05	0.11	1.43	0.14	76.8	8.0	0	6.5
10	176	1.86	41.9	470.0	759.4	1.14	—	0.26	0.05	0.25	2.49	0.22	124.5	19.0	0.4	21.8
0	17	0.94	19.1	189.9	439.6	0.42	—	0.15	0.55	0.08	1.36	0.12	88.4	23.5	0	10.8
17	190	1.28	29.3	414.2	626.0	1.24	—	0.16	0.35	0.29	2.00	0.19	154.8	56.0	0.2	5.8
29	200	2.15	24.5	181.0	512.1	1.24	—	0.28	0.20	0.31	2.88	0.11	111.4	17.4	0.2	19.7
7	77	0.88	26.3	174.1	280.7	0.59	—	0.08	0.04	0.06	0.49	0.08	38.7	4.6	0	3.0
0	65	1.98	41.0	543.1	390.7	0.74	—	0.14	0.15	0.10	1.31	0.18	47.7	20.3	0	4.4
0	46	1.87	34.9	550.3	729.5	0.56	—	0.13	0.55	0.09	1.99	0.27	49.9	29.9	0	1.4
5	150	1.80	—	—	280.0	—	0	—	—	—	—	—	—	0	—	—
0	40	1.44	—	—	500.0	—	—	—	—	—	—	—	—	0	—	—

DA+ Code	Food Description	QTY	Measure	Wt (g)	H₂O (g)	Ener (cal)	Prot (g)	Carb (g)	Fiber (g)	Fat (g)	Fat Breakdown (g) Sat	Mono	Poly
	VEGETARIAN FOODS—continued												
32075	Boca meatless ground burger	½	cup(s)	57	—	60	13.0	6.0	3.0	0.5	0	—	—
32072	Breakfast links	2	item(s)	45	—	70	8.0	5.0	2.0	3.0	0.5	—	—
32071	Breakfast patties	1	item(s)	38	—	60	7.0	5.0	2.0	2.5	0	—	—
35780	Cheeseburger meatless burger patty	1	item(s)	71	—	100	12.0	5.0	3.0	5.0	1.5	—	—
33958	Original meatless chik'n patties	1	item(s)	71	—	160	11.0	15.0	2.0	6.0	1.0	—	—
32066	Original patty	1	item(s)	71	—	70	13.0	6.0	4.0	0.5	0	—	—
32068	Roasted garlic patty	1	item(s)	71	—	70	12.0	6.0	4.0	1.5	0	—	—
37814	Roasted onion meatless burger patty	1	item(s)	71	—	70	11.0	7.0	4.0	1.0	0	—	—
	Gardenburger												
37810	BBQ chik'n with sauce	1	item(s)	142	—	250	14.0	30.0	5.0	8.0	1.0	—	—
39661	Black bean burger	1	item(s)	71	—	80	8.0	11.0	4.0	2.0	0	—	—
39666	Buffalo chik'n wing	3	item(s)	95	—	180	9.0	8.0	5.0	12.0	1.5	—	—
39665	Country fried chicken with creamy pepper gravy	1	item(s)	142	—	190	9.0	16.0	2.0	9.0	1.0	—	—
37808	Flamed grilled chik'n	1	item(s)	71	—	100	13.0	5.0	3.0	2.5	0	—	—
37803	Garden vegan	1	item(s)	71	—	100	10.0	12.0	2.0	1.0	—	—	—
39663	Homestyle classic burger	1	item(s)	71	—	110	12.0	6.0	4.0	5.0	0.5	—	—
37807	Meatless breakfast sausage	1	item(s)	43	—	50	5.0	2.0	2.0	3.5	0	—	—
37809	Meatless meatballs	6	item(s)	85	—	110	12.0	8.0	4.0	4.5	1.0	—	—
37806	Meatless riblets with sauce	1	item(s)	142	—	160	17.0	11.0	4.0	5.0	0	—	—
29913	Original	1	item(s)	71	—	90	10.0	8.0	3.0	2.0	0.5	—	—
39662	Sun-dried tomato basil burger	1	item(s)	71	—	80	10.0	11.0	3.0	1.5	0.5	—	—
29915	Veggie medley	1	item(s)	71	—	90	9.0	11.0	4.0	2.0	0	—	—
	Loma Linda												
9311	Big franks, canned	1	item(s)	51	—	110	11.0	3.0	2.0	6.0	1.0	1.5	3.5
9323	Fried chik'n with gravy	2	piece(s)	80	45.9	150	12.0	5.0	2.0	10	1.5	2.5	5.0
9326	Linketts, canned	1	item(s)	35	21.0	70	7.0	1.0	1.0	4.0	0.5	1.0	2.5
9336	Redi-Burger patties, canned	1	slice(s)	85	50.5	120	18.0	7.0	4.0	2.5	0.5	0.5	1.5
9350	Swiss Stake pattie with gravy, frozen	1	piece(s)	92	65.7	130	9.0	9.0	3.0	6.0	1.0	1.5	3.5
9354	Tender Rounds meatball substitute, canned in gravy	6	piece(s)	80	53.9	120	13.0	6.0	1.0	4.5	0.5	1.5	2.5
	Morningstar Farms												
33707	America's Original Veggie Dog links	1	item(s)	57	—	80	11.0	6.0	1.0	0.5	0	—	—
9362	Better'n Eggs egg substitute	¼	cup(s)	57	50.3	20	5.0	0	0	0	0	0	0
9371	Breakfast bacon strips	2	item(s)	16	6.8	60	2.0	2.0	0.5	4.5	0.5	1.0	3.0
9368	Breakfast sausage links	2	item(s)	45	26.8	80	9.0	3.0	2.0	3.0	0.5	1.5	1.0
33705	Chik'n nuggets	4	piece(s)	86	—	190	12.0	18.0	2.0	7.0	1.0	2.0	4.0
11587	Chik patties	1	item(s)	71	36.3	150	9.0	16.0	2.0	6.0	1.0	1.5	2.5
2531	Garden veggie patties	1	item(s)	67	40.1	100	10.0	9.0	4.0	2.5	0.5	0.5	1.5
33702	Spicy black bean veggie burger	1	item(s)	78	—	140	12.0	15.0	3.0	4.0	0.5	1.0	2.5
9412	Vegetarian chili, canned	1	cup(s)	230	172.6	180	16.0	25.0	10.0	1.5	0.5	0.5	0.5
	Worthington												
9424	Chili, canned	1	cup(s)	230	167.0	280	24.0	25.0	8.0	10.0	1.5	1.5	7.0
9436	Diced chik, canned	¼	cup(s)	55	42.7	50	9.0	2.0	1.0	0	0	0	0
9440	Dinner roast, frozen	1	slice(s)	85	53.2	180	14.0	6.0	3.0	11.0	1.5	4.5	5.0
9420	Meatless chicken slices, frozen	3	slice(s)	57	38.9	90	9.0	2.0	0.5	4.5	1.0	1.0	2.5
36702	Meatless chicken style roll, frozen	1	slice(s)	55	—	90	9.0	2.0	1.0	4.5	1.0	1.0	2.5
9428	Meatless corned beef, sliced, frozen	3	slice(s)	57	31.2	140	10.0	5.0	0	9.0	1.0	2.0	5.0
9470	Meatless salami, sliced, frozen	3	slice(s)	57	32.4	120	12.0	3.0	2.0	7.0	1.0	1.0	5.0
9480	Meatless smoked turkey, sliced	3	slice(s)	57	—	140	10.0	4.0	0	9.0	1.5	2.0	5.0
9462	Prosage links	2	item(s)	45	26.8	80	9.0	3.0	2.0	3.0	0.5	0.5	2.0
9484	Stakelets patty beef steak substitute, frozen	1	piece(s)	71	41.5	150	14.0	7.0	2.0	7.0	1.0	2.5	3.5
9486	Stripples bacon substitute	2	item(s)	16	6.8	60	2.0	2.0	0.5	4.5	0.5	1.0	3.0
9496	Vegetable Skallops meat substitute, canned	½	cup(s)	85	—	90	17.0	4.0	3.0	1.0	0	0	0.5
	DAIRY												
	Cheese												
1433	Blue, crumbled	1	ounce(s)	28	12.0	100	6.1	0.7	0	8.1	5.3	2.2	0.2
884	Brick	1	ounce(s)	28	11.7	105	6.6	0.8	0	8.4	5.3	2.4	0.2

Chol (mg)	Calc (mg)	Iron (mg)	Magn (mg)	Pota (mg)	Sodi (mg)	Zinc (mg)	Vit A (µg)	Thia (mg)	Vit E (mg α)	Ribo (mg)	Niac (mg)	Vit B_6 (mg)	Fola (µg DFE)	Vit C (mg)	Vit B_{12} (µg)	Sele (µg)
0	60	1.80	—	—	270.0	—	0	—	—	—	—	—	—	0	—	—
0	20	1.44	—	—	330.0	—	0	—	—	—	—	—	—	0	—	—
0	20	1.08	—	—	280.0	—	0	—	—	—	—	—	—	0	—	—
5	80	1.80	—	—	360.0	—	—	—	—	—	—	—	—	0	—	—
0	40	1.80	—	—	430.0	—	—	—	—	—	—	—	—	0	—	—
0	60	1.80	—	—	280.0	—	0	—	—	—	—	—	—	0	—	—
0	60	1.80	—	—	370.0	—	0	—	—	—	—	—	—	0	—	—
0	100	2.70	—	—	300.0	—	—	—	—	—	—	—	—	0	—	—
0	150	1.08	—	—	890.0	—	—	—	—	—	—	—	—	0	—	—
0	40	1.44	—	—	330.0	—	—	—	—	—	—	—	—	0	—	—
0	40	0.72	—	—	1000.0	—	—	—	—	—	—	—	—	0	—	—
5	40	1.44	—	—	550.0	—	—	—	—	—	—	—	—	0	—	—
0	60	3.60	—	—	360.0	—	—	—	—	—	—	—	—	0	—	—
0	40	4.50	—	—	230.0	—	—	—	—	—	—	—	—	0	—	—
0	80	1.44	—	—	380.0	—	—	—	—	—	—	—	—	0	—	—
0	20	0.72	—	—	120.0	—	—	—	—	—	—	—	—	0	—	—
0	60	1.80	—	—	400.0	—	—	—	—	—	—	—	—	0	—	—
0	60	1.80	—	—	720.0	—	—	—	—	—	—	—	—	3.6	—	—
0	80	1.08	30.4	193.4	490.0	0.89	—	0.10	—	0.15	1.08	0.08	10.1	1.2	0.1	7.0
5	60	1.44	—	—	260.0	—	—	—	—	—	—	—	—	3.6	—	—
0	40	1.44	27.0	182.0	290.0	0.46	—	0.07	—	0.08	0.90	0.09	10.6	9.0	0	4.0
0	0	0.77	—	50.0	220.0	—	0	0.22	—	0.10	2.00	0.70	—	0	2.4	—
0	20	1.80	—	70.0	430.0	0.33	0	1.05	—	0.34	4.00	0.30	—	0	2.4	—
0	0	0.36	—	20.0	160.0	0.46	0	0.12	—	0.20	0.80	0.16	—	0	0.9	—
0	0	1.06	—	140.0	450.0	—	0	0.15	—	0.25	4.00	0.40	—	0	1.2	—
0	0	0.72	—	200.0	430.0	—	0	0.45	—	0.25	10.00	1.00	—	0	5.4	—
0	20	1.08	—	80.0	340.0	0.66	0	0.75	—	0.17	2.00	0.16	—	0	1.2	—
0	0	0.72	—	60.0	580.0	—	0	—	—	—	—	—	—	0	—	—
0	20	0.72	—	75.0	90.0	0.60	37.5	0.03	—	0.34	0.00	0.08	24.0	—	0.6	—
0	0	0.36	—	15.0	220.0	0.05	0	0.75	—	0.04	0.40	0.07	—	0	0.2	—
0	0	1.80	—	50.0	300.0	—	0	0.37	—	0.17	7.00	0.50	—	0	3.0	—
0	20	2.70	—	320.0	490.0	—	0	0.52	—	0.25	5.00	0.30	—	0	1.5	—
0	0	1.80	—	210.0	540.0	—	0	1.80	—	0.17	2.00	0.20	—	0	1.2	—
0	40	0.72	—	180.0	350.0	—	—	—	—	—	—	—	—	0	—	—
0	40	1.80	—	320.0	470.0	—	0	—	—	—	0.00	—	—	0	—	—
0	40	3.60	—	660.0	900.0	—	—	—	—	—	—	—	—	0	—	—
0	40	3.60	—	330.0	1130.0	—	0	0.30	—	0.13	2.00	0.70	—	0	1.5	—
0	0	1.08	—	100.0	220.0	0.24	0	0.06	—	0.10	4.00	0.08	—	0	0.2	—
0	20	1.80	—	120.0	580.0	0.64	0	1.80	—	0.25	6.00	0.60	—	0	1.5	—
0	250	1.80	—	250.0	250.0	0.26	0	0.37	—	0.13	4.00	0.30	—	0	1.8	—
0	100	1.08	—	240.0	240.0	—	0	0.37	—	0.13	4.00	0.30	—	0	1.8	—
0	0	1.80	—	130.0	460.0	0.26	0	0.45	—	0.17	5.00	0.30	—	0	1.8	—
0	0	1.08	—	95.0	800.0	0.30	0	0.75	—	0.17	4.00	0.20	—	0	0.6	—
0	60	2.70	—	60.0	450.0	0.23	0	1.80	—	0.17	6.00	0.40	—	0	3.0	—
0	0	1.44	—	50.0	320.0	0.36	0	1.80	—	0.17	2.00	0.30	—	0	3.0	—
0	40	1.08	—	130.0	480.0	0.50	0	1.20	—	0.13	3.00	0.30	—	0	1.5	—
0	0	0.36	—	15.0	220.0	0.05	0	0.75	—	0.03	0.40	0.08	—	0	0.2	—
0	0	0.36	—	10.0	390.0	0.67	0	0.03	—	0.03	0.00	0.01	—	0	0	—
21	150	0.08	6.5	72.6	395.5	0.75	56.1	0.01	0.07	0.10	0.28	0.04	10.2	0	0.3	4.1
27	191	0.12	6.8	38.6	158.8	0.73	82.8	0.00	0.07	0.10	0.03	0.01	5.7	0	0.4	4.1

Table A–1 Table of Food Composition (*continued*) (Computer code is for Cengage Diet Analysis Plus program)

DA+ Code	Food Description	QTY	Measure	Wt (g)	H_2O (g)	Ener (cal)	Prot (g)	Carb (g)	Fiber (g)	Fat (g)	Fat Breakdown (g) Sat	Mono	Poly
DAIRY—continued													
885	Brie	1	ounce(s)	28	13.7	95	5.9	0.1	0	7.8	4.9	2.3	0.2
34821	Camembert	1	ounce(s)	28	14.7	85	5.6	0.1	0	6.9	4.3	2.0	0.2
5	Cheddar, shredded	¼	cup(s)	28	10.4	114	7.0	0.4	0	9.4	6.0	2.7	0.3
888	Cheddar or colby	1	ounce(s)	28	10.8	112	6.7	0.7	0	9.1	5.7	2.6	0.3
32096	Cheddar or colby, low fat	1	ounce(s)	28	17.9	49	6.9	0.5	0	2.0	1.2	0.6	0.1
889	Edam	1	ounce(s)	28	11.8	101	7.1	0.4	0	7.9	5.0	2.3	0.2
890	Feta	1	ounce(s)	28	15.7	75	4.0	1.2	0	6.0	4.2	1.3	0.2
891	Fontina	1	ounce(s)	28	10.8	110	7.3	0.4	0	8.8	5.4	2.5	0.5
8527	Goat cheese, soft	1	ounce(s)	28	17.2	76	5.3	0.3	0	6.0	4.1	1.4	0.1
893	Gouda	1	ounce(s)	28	11.8	101	7.1	0.6	0	7.8	5.0	2.2	0.2
894	Gruyere	1	ounce(s)	28	9.4	117	8.5	0.1	0	9.2	5.4	2.8	0.5
895	Limburger	1	ounce(s)	28	13.7	93	5.7	0.1	0	7.7	4.7	2.4	0.1
896	Monterey jack	1	ounce(s)	28	11.6	106	6.9	0.2	0	8.6	5.4	2.5	0.3
13	Mozzarella, part skim milk	1	ounce(s)	28	15.2	72	6.9	0.8	0	4.5	2.9	1.3	0.1
12	Mozzarella, whole milk	1	ounce(s)	28	14.2	85	6.3	0.6	0	6.3	3.7	1.9	0.2
897	Muenster	1	ounce(s)	28	11.8	104	6.6	0.3	0	8.5	5.4	2.5	0.2
898	Neufchatel	1	ounce(s)	28	17.6	74	2.8	0.8	0	6.6	4.2	1.9	0.2
14	Parmesan, grated	1	tablespoon(s)	5	1.0	22	1.9	0.2	0	1.4	0.9	0.4	0.1
17	Provolone	1	ounce(s)	28	11.6	100	7.3	0.6	0	7.5	4.8	2.1	0.2
19	Ricotta, part skim milk	¼	cup(s)	62	45.8	85	7.0	3.2	0	4.9	3.0	1.4	0.2
18	Ricotta, whole milk	¼	cup(s)	62	44.1	107	6.9	1.9	0	8.0	5.1	2.2	0.2
20	Romano	1	tablespoon(s)	5	1.5	19	1.6	0.2	0	1.3	0.9	0.4	0
900	Roquefort	1	ounce(s)	28	11.2	105	6.1	0.6	0	8.7	5.5	2.4	0.4
21	Swiss	1	ounce(s)	28	10.5	108	7.6	1.5	0	7.9	5.0	2.1	0.3
	Imitation cheese												
42245	Imitation American cheddar cheese	1	ounce(s)	28	15.1	68	4.7	3.3	0	4.0	2.5	1.2	0.1
53914	Imitation cheddar	1	ounce(s)	28	15.1	68	4.7	3.3	0	4.0	2.5	1.2	0.1
	Cottage cheese												
9	Low fat, 1% fat	½	cup(s)	113	93.2	81	14.0	3.1	0	1.2	0.7	0.3	0
8	Low fat, 2% fat	½	cup(s)	113	89.6	102	15.5	4.1	0	2.2	1.4	0.6	0.1
	Cream cheese												
11	Cream cheese	2	tablespoon(s)	29	15.6	101	2.2	0.8	0	10.1	6.4	2.9	0.4
17366	Fat-free cream cheese	2	tablespoon(s)	30	22.7	29	4.3	1.7	0	0.4	0.3	0.1	0
10438	Tofutti Better than Cream Cheese	2	tablespoon(s)	30	—	80	1.0	1.0	0	8.0	2.0	—	6.0
	Processed cheese												
24	American cheese food, processed	1	ounce(s)	28	12.3	94	5.2	2.2	0	7.1	4.2	2.0	0.3
25	American cheese spread, processed	1	ounce(s)	28	13.5	82	4.7	2.5	0	6.0	3.8	1.8	0.2
22	American cheese, processed	1	ounce(s)	28	11.1	106	6.3	0.5	0	8.9	5.6	2.5	0.3
9110	Kraft deluxe singles pasteurized process American cheese	1	ounce(s)	28	—	108	5.4	0	0	9.5	5.4	—	—
23	Swiss cheese, processed	1	ounce(s)	28	12.0	95	7.0	0.6	0	7.1	4.5	2.0	0.2
	Soy cheese												
10437	Galaxy Foods vegan grated parmesan cheese alternative	1	tablespoon(s)	8	—	23	3.0	1.5	0	0	0	0	0
10430	Nu Tofu cheddar flavored cheese alternative	1	ounce(s)	28	—	70	6.0	1.0	0	4.0	0.5	2.5	1.0
	Cream												
26	Half and half cream	1	tablespoon(s)	15	12.1	20	0.4	0.6	0	1.7	1.1	0.5	0.1
32	Heavy whipping cream, liquid	1	tablespoon(s)	15	8.7	52	0.3	0.4	0	5.6	3.5	1.6	0.2
28	Light coffee or table cream, liquid	1	tablespoon(s)	15	11.1	29	0.4	0.5	0	2.9	1.8	0.8	0.1
30	Light whipping cream, liquid	1	tablespoon(s)	15	9.5	44	0.3	0.4	0	4.6	2.9	1.4	0.1
34	Whipped cream topping, pressurized	1	tablespoon(s)	3	1.8	8	0.1	0.4	0	0.7	0.4	0.2	0
	Sour cream												
30556	Fat-free sour cream	2	tablespoon(s)	32	25.8	24	1.0	5.0	0	0	0	0	0
36	Sour cream	2	tablespoon(s)	24	17.0	51	0.8	1.0	0	5.0	3.1	1.5	0.2
	Imitation cream												
3659	Coffeemate nondairy creamer, liquid	1	tablespoon(s)	15	—	20	0	2.0	0	1.0	0	0.5	0
40	Cream substitute, powder	1	teaspoon(s)	2	0	11	0.1	1.1	0	0.7	0.7	0	0
904	Imitation sour cream	2	tablespoon(s)	29	20.5	60	0.7	1.9	0	5.6	5.1	0.2	0

Chol (mg)	Calc (mg)	Iron (mg)	Magn (mg)	Pota (mg)	Sodi (mg)	Zinc (mg)	Vit A (μg)	Thia (mg)	Vit E (mg α)	Ribo (mg)	Niac (mg)	Vit B_6 (mg)	Fola (μg DFE)	Vit C (mg)	Vit B_{12} (μg)	Sele (μg)
28	52	0.14	5.7	43.1	178.3	0.67	49.3	0.02	0.06	0.14	0.10	0.06	18.4	0	0.5	4.1
20	110	0.09	5.7	53.0	238.7	0.67	68.3	0.01	0.06	0.14	0.18	0.06	17.6	0	0.4	4.1
30	204	0.19	7.9	27.7	175.4	0.87	74.9	0.01	0.08	0.10	0.02	0.02	5.1	0	0.2	3.9
27	194	0.21	7.4	36.0	171.2	0.87	74.8	0.00	0.07	0.10	0.02	0.02	5.1	0	0.2	4.1
6	118	0.11	4.5	18.7	173.5	0.51	17.0	0.00	0.01	0.06	0.01	0.01	3.1	0	0.1	4.1
25	207	0.12	8.5	53.3	273.6	1.06	68.9	0.01	0.06	0.11	0.02	0.02	4.5	0	0.4	4.1
25	140	0.18	5.4	17.6	316.4	0.81	35.4	0.04	0.05	0.23	0.28	0.12	9.1	0	0.5	4.3
33	156	0.06	4.0	18.1	226.8	0.99	74.0	0.01	0.07	0.05	0.04	0.02	1.7	0	0.5	4.1
13	40	0.53	4.5	7.4	104.3	0.26	81.6	0.02	0.05	0.10	0.12	0.07	3.4	0	0.1	0.8
32	198	0.06	8.2	34.3	232.2	1.10	46.8	0.01	0.06	0.09	0.01	0.02	6.0	0	0.4	4.1
31	287	0.04	10.2	23.0	95.3	1.10	76.8	0.01	0.07	0.07	0.03	0.02	2.8	0	0.5	4.1
26	141	0.03	6.0	36.3	226.8	0.59	96.4	0.02	0.06	0.14	0.04	0.02	16.4	0	0.3	4.1
25	211	0.20	7.7	23.0	152.0	0.85	56.1	0.00	0.07	0.11	0.02	0.02	5.1	0	0.2	4.1
18	222	0.06	6.5	23.8	175.5	0.78	36.0	0.01	0.04	0.08	0.03	0.02	2.6	0	0.2	4.1
22	143	0.12	5.7	21.5	177.8	0.82	50.7	0.01	0.05	0.08	0.02	0.01	2.0	0	0.6	4.8
27	203	0.11	7.7	38.0	178.0	0.79	84.5	0.00	0.07	0.09	0.02	0.01	3.4	0	0.4	4.1
22	21	0.07	2.3	32.3	113.1	0.14	84.5	0.00	—	0.05	0.03	0.01	3.1	0	0.1	0.9
4	55	0.04	1.9	6.3	76.5	0.19	6.0	0.00	0.01	0.02	0.01	0.00	0.5	0	0.1	0.9
20	214	0.14	7.9	39.1	248.3	0.91	66.9	0.01	0.06	0.09	0.04	0.02	2.8	0	0.4	4.1
19	167	0.27	9.2	76.9	76.9	0.82	65.8	0.01	0.04	0.11	0.04	0.01	8.0	0	0.2	10.3
31	127	0.23	6.8	64.6	51.7	0.71	73.8	0.01	0.06	0.12	0.06	0.02	7.4	0	0.2	8.9
5	53	0.03	2.1	4.3	60	0.12	4.8	0.00	0.01	0.01	0.00	0.00	0.4	0	0.1	0.7
26	188	0.15	8.5	25.8	512.9	0.59	83.3	0.01	—	0.16	0.20	0.03	13.9	0	0.2	4.1
26	224	0.05	10.8	21.8	54.4	1.23	62.4	0.01	0.10	0.08	0.02	0.02	1.7	0	0.9	5.2
10	159	0.08	8.2	68.6	381.3	0.73	32.3	0.01	0.07	0.12	0.03	0.03	2.0	0	0.1	4.3
10	159	0.09	8.2	68.6	381.3	0.73	32.3	0.01	0.07	0.12	0.04	0.03	2.0	0	0.1	4.3
5	69	0.15	5.7	97.2	458.8	0.42	12.4	0.02	0.01	0.18	0.14	0.07	13.6	0	0.7	10.2
9	78	0.18	6.8	108.5	458.8	0.47	23.7	0.02	0.02	0.20	0.16	0.08	14.7	0	0.8	11.5
32	23	0.34	1.7	34.5	85.8	0.15	106.1	0.01	0.08	0.05	0.02	0.01	3.8	0	0.1	0.7
2	56	0.05	4.2	48.9	163.5	0.26	83.7	0.01	0.00	0.05	0.04	0.01	11.1	0	0.2	1.5
0	0	0.00	—	—	135.0	—	0	—	—	—	—	—	—	0	—	—
23	162	0.16	8.8	82.5	358.6	0.90	57.0	0.01	0.06	0.14	0.04	0.02	2.0	0	0.4	4.6
16	159	0.09	8.2	68.6	381.3	0.73	49.0	0.01	0.05	0.12	0.03	0.03	2.0	0	0.1	3.2
27	156	0.05	7.7	47.9	422.1	0.80	72.0	0.01	0.07	0.10	0.02	0.02	2.3	0	0.2	4.1
27	338	0.00	0	33.8	459.0	1.22	114.0	—	—	0.14	—	—	—	0	0.2	—
24	219	0.17	8.2	61.2	388.4	1.02	56.1	0.00	0.09	0.07	0.01	0.01	1.7	0	0.3	4.5
0	60	0.00	—	75.0	97.5	—	—	—	—	—	—	—	—	—	—	—
0	200	0.36	—	—	190.0	—	—	—	—	—	—	—	—	0	—	—
6	16	0.01	1.5	19.5	6.2	0.08	14.6	0.01	0.05	0.02	0.01	0.01	0.5	0.1	0	0.3
21	10	0.00	1.1	11.3	5.7	0.03	61.7	0.00	0.15	0.01	0.01	0.00	0.6	0.1	0	0.1
10	14	0.01	1.4	18.3	6.0	0.04	27.2	0.01	0.08	0.02	0.01	0.01	0.3	0.1	0	0.1
17	10	0.00	1.1	14.6	5.1	0.03	41.9	0.00	0.13	0.01	0.01	0.00	0.6	0.1	0	0.1
2	3	0.00	0.3	4.4	3.9	0.01	5.6	0.00	0.01	0.00	0.00	0.00	0.1	0	0	0
3	40	0.00	3.2	41.3	45.1	0.16	23.4	0.01	0.00	0.04	0.02	0.01	3.5	0	0.1	1.7
11	28	0.01	2.6	34.6	12.7	0.06	42.5	0.01	0.14	0.03	0.01	0.00	2.6	0.2	0.1	0.5
0	0	0.00	—	30.0	0	—	0	0.01	—	0.01	0.20	—	—	0	—	—
0	0	0.02	0.1	16.2	3.6	0.01	0	0.00	0.01	0.00	0.00	0.00	0	0	0	0
0	1	0.11	1.7	46.3	29.3	0.34	0	0.00	0.21	0.00	0.00	0.00	0	0	0	0.7

Table A–1 Table of Food Composition (*continued*)

(Computer code is for Cengage Diet Analysis Plus program)

DA+ Code	Food Description	QTY	Measure	Wt (g)	H₂O (g)	Ener (cal)	Prot (g)	Carb (g)	Fiber (g)	Fat (g)	Fat Breakdown (g)		
											Sat	Mono	Poly
	DAIRY—continued												
35972	Nondairy coffee whitener, liquid, frozen	1	tablespoon(s)	15	11.7	21	0.2	1.7	0	1.5	0.3	1.1	0
35976	Nondairy dessert topping, frozen	1	tablespoon(s)	5	2.4	15	0.1	1.1	0	1.2	1.0	0.1	0
35975	Nondairy dessert topping, pressurized	1	tablespoon(s)	4	2.7	12	0	0.7	0	1.0	0.8	0.1	0
	Fluid milk												
60	Buttermilk, low fat	1	cup(s)	245	220.8	98	8.1	11.7	0	2.2	1.3	0.6	0.1
54	Low fat, 1%	1	cup(s)	244	219.4	102	8.2	12.2	0	2.4	1.5	0.7	0.1
55	Low fat, 1%, with nonfat milk solids	1	cup(s)	245	220.0	105	8.5	12.2	0	2.4	1.5	0.7	0.1
57	Nonfat, skim or fat free	1	cup(s)	245	222.6	83	8.3	12.2	0	0.2	0.1	0.1	0
58	Nonfat, skim or fat free with nonfat milk solids	1	cup(s)	245	221.4	91	8.7	12.3	0	0.6	0.4	0.2	0
51	Reduced fat, 2%	1	cup(s)	244	218.0	122	8.1	11.4	0	4.8	3.1	1.4	0.2
52	Reduced fat, 2%, with nonfat milk solids	1	cup(s)	245	217.7	125	8.5	12.2	0	4.7	2.9	1.4	0.2
50	Whole, 3.3%	1	cup(s)	244	215.5	146	7.9	11.0	0	7.9	4.6	2.0	0.5
	Canned milk												
62	Nonfat or skim evaporated	2	tablespoon(s)	32	25.3	25	2.4	3.6	0	0.1	0	0	0
63	Sweetened condensed	2	tablespoon(s)	38	10.4	123	3.0	20.8	0	3.3	2.1	0.9	0.1
61	Whole evaporated	2	tablespoon(s)	32	23.3	42	2.1	3.2	0	2.4	1.4	0.7	0.1
	Dried milk												
64	Buttermilk	¼	cup(s)	30	0.9	117	10.4	14.9	0	1.8	1.1	0.5	0.1
65	Instant nonfat with added vitamin A	¼	cup(s)	17	0.7	61	6.0	8.9	0	0.1	0.1	0	0
5234	Skim milk powder	¼	cup(s)	17	0.7	62	6.1	9.1	0	0.1	0.1	0	0
907	Whole dry milk	¼	cup(s)	32	0.8	159	8.4	12.3	0	8.5	5.4	2.5	0.2
909	**Goat milk**	1	cup(s)	244	212.4	168	8.7	10.9	0	10.1	6.5	2.7	0.4
	Chocolate milk												
33155	Chocolate syrup, prepared with milk	1	cup(s)	282	227.0	254	8.7	36.0	0.8	8.3	4.7	2.1	0.5
33184	Cocoa mix with aspartame, added sodium and vitamin A, no added calcium or phosphorus, prepared with water	1	cup(s)	192	177.4	56	2.3	10.8	1.2	0.4	0.3	0.1	0
908	Hot cocoa, prepared with milk	1	cup(s)	250	206.4	193	8.8	26.6	2.5	5.8	3.6	1.7	0.1
69	Low fat	1	cup(s)	250	211.3	158	8.1	26.1	1.3	2.5	1.5	0.8	0.1
68	Reduced fat	1	cup(s)	250	205.4	190	7.5	30.3	1.8	4.8	2.9	1.1	0.2
67	Whole	1	cup(s)	250	205.8	208	7.9	25.9	2.0	8.5	5.3	2.5	0.3
70	**Eggnog**	1	cup(s)	254	188.9	343	9.7	34.4	0	19.0	11.3	5.7	0.9
	Breakfast drinks												
10093	Carnation Instant Breakfast classic chocolate malt, prepared with skim milk, no sugar added	1	cup(s)	243	—	142	11.1	21.3	0.7	1.3	0.7	—	—
10092	Carnation Instant Breakfast classic French vanilla, prepared with skim milk, no sugar added	1	cup(s)	273	—	150	12.9	24.0	0	0.4	0.4	—	—
10094	Carnation Instant Breakfast stawberry sensation, prepared with skim milk, no sugar added	1	cup(s)	243	—	142	11.1	21.3	0	0.4	0.4	—	—
10091	Carnation Instant Breakfast strawberry sensation, prepared with skim milk	1	cup(s)	273	—	220	12.5	38.8	0	0.4	0.4	—	—
1417	Ovaltine rich chocolate flavor, prepared with skim milk	1	cup(s)	258	—	170	8.5	31.0	0	0	0	0	0
8539	**Malted milk, chocolate mix, fortified, prepared with milk**	1	cup(s)	265	215.8	223	8.9	28.9	1.1	8.6	5.0	2.2	0.5
	Milkshakes												
73	Chocolate	1	cup(s)	227	164.0	270	6.9	48.1	0.7	6.1	3.8	1.8	0.2
3163	Strawberry	1	cup(s)	226	167.8	256	7.7	42.8	0.9	6.3	3.9	—	—
74	Vanilla	1	cup(s)	227	169.2	254	8.8	40.3	0	6.9	4.3	2.0	0.3

Chol (mg)	Calc (mg)	Iron (mg)	Magn (mg)	Pota (mg)	Sodi (mg)	Zinc (mg)	Vit A (µg)	Thia (mg)	Vit E (mg α)	Ribo (mg)	Niac (mg)	Vit B_6 (mg)	Fola (µg DFE)	Vit C (mg)	Vit B_{12} (µg)	Sele (µg)
0	1	0.00	0	28.9	12.0	0.00	0.2	0.00	0.12	0.00	0.00	0.00	0	0	0	0.2
0	0	0.00	0.1	0.9	1.2	0.00	0.3	0.00	0.05	0.00	0.00	0.00	0	0	0	0.1
0	0	0.00	0	0.8	2.8	0.00	0.2	0.00	0.04	0.00	0.00	0.00	0	0	0	0.1
10	284	0.12	27.0	370.0	257.3	1.02	17.2	0.08	0.12	0.37	0.14	0.08	12.3	2.5	0.5	4.9
12	290	0.07	26.8	366.0	107.4	1.02	141.5	0.04	0.02	0.45	0.22	0.09	12.2	0	1.1	8.1
10	314	0.12	34.3	396.9	127.4	0.98	144.6	0.09	—	0.42	0.22	0.11	12.3	2.5	0.9	5.6
5	306	0.07	27.0	382.2	102.9	1.02	149.5	0.11	0.02	0.44	0.23	0.09	12.3	0	1.3	7.6
5	316	0.12	36.8	419.0	129.9	1.00	149.5	0.10	0.00	0.42	0.22	0.11	12.3	2.5	1.0	5.4
20	285	0.07	26.8	366.0	100.0	1.04	134.2	0.09	0.07	0.45	0.22	0.09	12.2	0.5	1.1	6.1
20	314	0.12	34.3	396.9	127.4	0.98	137.2	0.09	—	0.42	0.22	0.11	12.3	2.5	0.9	5.6
24	276	0.07	24.4	348.9	97.6	0.97	68.3	0.10	0.14	0.44	0.26	0.08	12.2	0	1.1	9.0
1	93	0.09	8.6	105.9	36.7	0.28	37.6	0.01	0.00	0.09	0.05	0.01	2.9	0.4	0.1	0.8
13	109	0.07	9.9	141.9	48.6	0.36	28.3	0.03	0.06	0.16	0.08	0.02	4.2	1.0	0.2	5.7
9	82	0.06	7.6	95.4	33.4	0.24	20.5	0.01	0.04	0.10	0.06	0.01	2.5	0.6	0.1	0.7
21	359	0.09	33.3	482.5	156.7	1.21	14.9	0.11	0.03	0.48	0.27	0.10	14.2	1.7	1.2	6.2
3	209	0.05	19.9	289.9	93.3	0.75	120.5	0.07	0.00	0.30	0.15	0.06	8.5	1.0	0.7	4.6
3	214	0.05	20.3	296.0	95.3	0.76	123.1	0.07	0.00	0.30	0.15	0.06	8.7	1.0	0.7	4.7
31	292	0.15	27.2	425.6	118.7	1.06	82.2	0.09	0.15	0.38	0.20	0.09	11.8	2.8	1.0	5.2
27	327	0.12	34.2	497.8	122.0	0.73	139.1	0.11	0.17	0.33	0.67	0.11	2.4	3.2	0.2	3.4
25	251	0.90	50.8	408.9	132.5	1.21	70.5	0.11	0.14	0.46	0.38	0.09	14.1	0	1.1	9.6
0	92	0.74	32.6	405.1	138.2	0.51	0	0.04	0.00	0.20	0.16	0.04	1.9	0	0.2	2.5
20	263	1.20	57.5	492.5	110.0	1.57	127.5	0.09	0.07	0.45	0.33	0.10	12.5	0.5	1.1	6.8
8	288	0.60	32.5	425.0	152.5	1.02	145.0	0.09	0.05	0.41	0.31	0.10	12.5	2.3	0.9	4.8
20	273	0.60	35.0	422.5	165.0	0.97	160.0	0.11	0.10	0.45	0.41	0.06	5.0	0	0.8	8.5
30	280	0.60	32.5	417.5	150.0	1.02	65.0	0.09	0.15	0.40	0.31	0.10	12.5	2.3	0.8	4.8
150	330	0.50	48.3	419.1	137.2	1.16	116.8	0.08	0.50	0.48	0.26	0.12	2.5	3.8	1.1	10.7
9	444	4.00	88.9	631.1	195.6	3.38	—	0.33	—	0.45	4.44	0.44	4.0	26.7	1.3	8.0
9	500	4.50	100.0	665.0	192.0	3.75	—	0.37	—	0.51	5.00	0.49	100.0	30.0	1.5	9.0
9	444	4.00	88.9	568.9	186.7	3.38	—	0.33	—	0.45	4.44	0.44	88.9	26.7	1.3	8.0
9	500	4.47	100.0	665.0	288.0	3.75	—	0.37	—	0.51	5.07	0.50	100.0	30.0	1.5	8.8
5	350	3.60	100.0	—	270.0	3.75	—	0.37	—	—	4.00	0.40	—	12.0	1.2	—
27	339	3.76	45.1	577.7	230.6	1.16	903.7	0.75	0.15	1.31	11.08	1.01	13.3	31.8	1.1	12.5
25	300	0.70	36.4	508.9	252.2	1.09	40.9	0.10	0.11	0.50	0.28	0.05	11.4	0	0.7	4.3
25	256	0.24	29.4	412.0	187.9	0.81	58.9	0.10	—	0.44	0.39	0.10	6.8	1.8	0.7	4.8
27	332	0.22	27.3	415.8	215.8	0.88	56.8	0.06	0.11	0.44	0.33	0.09	15.9	0	1.2	5.2

											Fat Breakdown (g)		
DA+ Code	Food Description	QTY	Measure	Wt (g)	H_2O (g)	Ener (cal)	Prot (g)	Carb (g)	Fiber (g)	Fat (g)	Sat	Mono	Poly
DAIRY—continued													
	Ice cream												
4776	Chocolate	½	cup(s)	66	36.8	143	2.5	18.6	0.8	7.3	4.5	2.1	0.3
12137	Chocolate fudge, no sugar added	½	cup(s)	71	—	100	3.0	16.0	2.0	3.0	1.5	—	—
16514	Chocolate, soft serve	½	cup(s)	87	49.9	177	3.2	24.1	0.7	8.4	5.2	2.4	0.3
16523	Sherbet, all flavors	½	cup(s)	97	63.8	139	1.1	29.3	3.2	1.9	1.1	0.5	0.1
4778	Strawberry	½	cup(s)	66	39.6	127	2.1	18.2	0.6	5.5	3.4	—	—
76	Vanilla	½	cup(s)	72	43.9	145	2.5	17.0	0.5	7.9	4.9	2.1	0.3
12146	Vanilla chocolate swirl, fat-free, no sugar added	½	cup(s)	71	—	100	3.0	14.0	2.0	3.0	2.0	—	—
82	Vanilla, light	½	cup(s)	76	48.3	125	3.6	19.6	0.2	3.7	2.2	1.0	0.2
78	Vanilla, light, soft serve	½	cup(s)	88	61.2	111	4.3	19.2	0	2.3	1.4	0.7	0.1
	Soy desserts												
10694	Tofutti low fat vanilla fudge nondairy frozen dessert	½	cup(s)	70	—	140	2.0	24.0	0	4.0	1.0	—	—
15721	Tofutti premium chocolate supreme nondairy frozen dessert	½	cup(s)	70	—	180	3.0	18.0	0	11.0	2.0	—	—
15720	Tofutti premium vanilla nondairy frozen dessert	½	cup(s)	70	—	190	2.0	20.0	0	11.0	2.0	—	—
	Ice milk												
16517	Chocolate	½	cup(s)	66	42.9	94	2.8	16.9	0.3	2.1	1.3	0.6	0.1
16516	Flavored, not chocolate	½	cup(s)	66	41.4	108	3.5	17.5	0.2	2.6	1.7	0.6	0.1
	Pudding												
25032	Chocolate	½	cup(s)	144	109.7	155	5.1	22.7	0.7	5.4	3.1	1.7	0.2
1923	Chocolate, sugar free, prepared with 2% milk	½	cup(s)	133	—	100	5.0	14.0	0.3	3.0	1.5	—	—
1722	Rice	½	cup(s)	113	75.6	151	4.1	29.9	0.5	1.9	1.1	0.5	0.1
4747	Tapioca, ready to eat	1	item(s)	142	102.0	185	2.8	30.8	0	5.5	1.4	3.6	0.1
25031	Vanilla	½	cup(s)	136	109.7	116	4.7	17.6	0	2.8	1.6	0.9	0.2
1924	Vanilla, sugar free, prepared with 2% milk	½	cup(s)	133	—	90	4.0	12.0	0.2	2.0	1.5	—	—
	Frozen yogurt												
4785	Chocolate, soft serve	½	cup(s)	72	45.9	115	2.9	17.9	1.6	4.3	2.6	1.3	0.2
1747	Fruit varieties	½	cup(s)	113	80.5	144	3.4	24.4	0	4.1	2.6	1.1	0.1
4786	Vanilla, soft serve	½	cup(s)	72	47.0	117	2.9	17.4	0	4.0	2.5	1.1	0.2
	Milk substitutes												
	Lactose free												
16081	Fat-free, calcium fortified [milk]	1	cup(s)	240	—	80	8.0	13.0	0	0	0	0	0
36486	Low fat milk	1	cup(s)	240	—	110	8.0	13.0	0	2.5	1.5	—	—
36487	Reduced fat milk	1	cup(s)	240	—	130	8.0	12.0	0	5.0	3.0	—	—
36488	Whole milk	1	cup(s)	240	—	150	8.0	12.0	0	8.0	5.0	—	—
	Rice												
10083	Rice Dream carob rice beverage	1	cup(s)	240	—	150	1.0	32.0	0	2.5	0	—	—
17089	Rice Dream original rice beverage, enriched	1	cup(s)	240	—	120	1.0	25.0	0	2.0	0	—	—
10087	Rice Dream vanilla enriched rice beverage	1	cup(s)	240	—	130	1.0	28.0	0	2.0	0	—	—
	Soy												
34750	Soy Dream chocolate enriched soy beverage	1	cup(s)	240	—	210	7.0	37.0	1.0	3.5	0.5	—	—
34749	Soy Dream vanilla enriched soy beverage	1	cup(s)	240	—	150	7.0	22.0	0	4.0	0.5	—	—
13840	Vitasoy light chocolate soymilk	1	cup(s)	240	—	100	4.0	17.0	0	2.0	0.5	0.5	1.0
13839	Vitasoy light vanilla soymilk	1	cup(s)	240	—	70	4.0	10.0	0	2.0	0.5	0.5	1.0
13836	Vitasoy rich chocolate soymilk	1	cup(s)	240	—	160	7.0	24.0	1.0	4.0	0.5	1.0	2.5
13835	Vitasoy vanilla delite soymilk	1	cup(s)	240	—	120	7.0	13.0	1.0	4.0	0.5	1.0	2.5
	Yogurt												
3615	Custard style, fruit flavors	6	ounce(s)	170	127.1	190	7.0	32.0	0	3.5	2.0	—	—
3617	Custard style, vanilla	6	ounce(s)	170	134.1	190	7.0	32.0	0	3.5	2.0	0.9	0.1
32101	Fruit, low fat	1	cup(s)	245	184.5	243	9.8	45.7	0	2.8	1.8	0.8	0.1

Chol (mg)	Calc (mg)	Iron (mg)	Magn (mg)	Pota (mg)	Sodi (mg)	Zinc (mg)	Vit A (µg)	Thia (mg)	Vit E (mg α)	Ribo (mg)	Niac (mg)	Vit B_6 (mg)	Fola (µg DFE)	Vit C (mg)	Vit B_{12} (µg)	Sele (µg)
22	72	0.61	19.1	164.3	50.2	0.38	77.9	0.02	0.19	0.12	0.14	0.03	10.6	0.5	0.2	1.7
10	100	0.36	—	—	65.0	—	—	—	—	—	—	—	—	0	—	—
22	103	0.32	19.0	192.0	43.3	0.45	66.6	0.03	0.22	0.13	0.11	0.03	4.3	0.5	0.3	2.5
0	52	0.13	7.7	92.6	44.4	0.46	9.7	0.02	0.02	0.08	0.07	0.02	6.8	5.6	0.1	1.3
19	79	0.13	9.2	124.1	39.6	0.22	63.4	0.03	—	0.16	0.11	0.03	7.9	5.1	0.2	1.3
32	92	0.06	10.1	143.3	57.6	0.49	85.0	0.03	0.21	0.17	0.08	0.03	3.6	0.4	0.3	1.3
10	100	0.00	—	—	65.0	—	—	—	—	—	—	—	—	0	—	—
21	122	0.14	10.6	158.1	56.2	0.55	97.3	0.04	0.09	0.19	0.10	0.03	4.6	0.9	0.4	1.5
11	138	0.05	12.3	194.5	61.6	0.46	25.5	0.04	0.05	0.17	0.10	0.04	4.4	0.8	0.4	3.2
0	0	0.00	—	8.0	90.0	—	0	—	—	—	—	—	—	0	—	—
0	0	0.00	—	7.0	180.0	—	0	—	—	—	—	—	—	0	—	—
0	0	0.00	—	2.0	210.0	—	0	—	—	—	—	—	—	0	—	—
6	94	0.15	13.1	155.2	40.6	0.36	15.7	0.03	0.05	0.11	0.08	0.02	3.9	0.5	0.3	2.2
16	76	0.05	9.2	136.2	48.5	0.47	90.4	0.02	0.05	0.11	0.06	0.01	3.3	0.1	0.2	1.3
35	149	0.46	31.3	226.7	137.0	0.71	—	0.05	0.00	0.22	0.15	0.06	8.3	1.2	0.5	4.9
10	150	0.72	—	330.0	310.0	—	—	0.06	—	0.26	—	—	—	0	—	—
7	113	0.28	15.8	201.4	66.4	0.52	41.6	0.03	0.05	0.17	0.34	0.06	4.5	0.2	0.2	4.8
1	101	0.15	8.5	130.6	205.9	0.31	0	0.03	0.21	0.13	0.09	0.03	4.3	0.4	0.3	0
35	146	0.17	17.2	188.9	136.4	0.52	—	0.04	0.00	0.22	0.10	0.05	8.0	1.2	0.5	4.6
10	150	0.00	—	190.0	380.0	—	—	0.03	—	0.17	—	—	—	0	—	—
4	106	0.90	19.4	187.9	70.6	0.35	31.7	0.02	—	0.15	0.22	0.05	7.9	0.2	0.2	1.7
15	113	0.52	11.3	176.3	71.2	0.31	55.4	0.04	0.10	0.20	0.07	0.04	4.5	0.8	0.1	2.1
1	103	0.21	10.1	151.9	62.6	0.30	42.5	0.02	0.07	0.16	0.20	0.05	4.3	0.6	0.2	2.4
3	500	0.00	—	—	125.0	—	100.0	—	—	—	—	—	—	0	0	—
10	300	0.00	—	—	125.0	—	100.0	—	—	—	—	—	—	0	—	—
20	300	0.00	—	—	125.0	—	98.2	—	—	—	—	—	—	0	—	—
35	300	0.00	—	—	125.0	—	58.1	—	—	—	—	—	—	0	—	—
0	20	0.72	—	82.5	100.0	—	—	—	—	—	—	—	—	1.2	—	—
0	300	0.00	13.3	60.0	90.0	0.24	—	0.06	—	0.00	0.84	0.07	—	0	1.5	—
0	300	0.00	—	53.0	90.0	—	—	—	—	—	—	—	—	0	1.5	—
0	300	1.80	60.0	350.0	160.0	0.60	33.3	0.15	—	0.06	0.80	0.12	60.0	0	3.0	—
0	300	1.80	40.0	260.0	140.0	0.60	33.3	0.15	—	0.06	0.80	0.12	60.0	0	3.0	—
0	300	0.72	24.0	200.0	140.0	0.90	—	0.09	—	0.34	—	—	24.0	0	0.9	—
0	300	0.72	24.0	200.0	120.0	0.90	—	0.09	—	0.34	—	—	24.0	0	0.9	—
0	300	1.08	40.0	320.0	150.0	0.90	—	0.15	—	0.34	—	—	60.0	0	0.9	—
0	40	0.72	—	320.0	115.0	—	0	—	—	—	—	—	—	0	—	—
15	300	0.00	16.0	310.0	100.0	—	—	—	—	0.25	—	—	—	0	—	—
15	300	0.00	16.0	310.0	100.0	—	—	—	—	0.25	—	—	—	0	—	—
12	338	0.14	31.9	433.7	129.9	1.64	27.0	0.08	0.04	0.39	0.21	0.09	22.1	1.5	1.1	6.9

DA+ Code	Food Description	QTY	Measure	Wt (g)	H₂O (g)	Ener (cal)	Prot (g)	Carb (g)	Fiber (g)	Fat (g)	Fat Breakdown (g) Sat	Mono	Poly
	DAIRY—continued												
29638	Fruit, nonfat, sweetened with low-calorie sweetener	1	cup(s)	241	208.3	123	10.6	19.4	1.2	0.4	0.2	0.1	0
93	Plain, low fat	1	cup(s)	245	208.4	154	12.9	17.2	0	3.8	2.5	1.0	0.1
94	Plain, nonfat	1	cup(s)	245	208.8	137	14.0	18.8	0	0.4	0.3	0.1	0
32100	Vanilla, low fat	1	cup(s)	245	193.6	208	12.1	33.8	0	3.1	2.0	0.8	0.1
5242	Yogurt beverage	1	cup(s)	245	199.8	172	6.2	32.8	0	2.2	1.4	0.6	0.1
38202	Yogurt smoothie, nonfat, all flavors	1	item(s)	325	—	290	10.0	60.0	6.0	0	0	0	0
	Soy yogurt												
34617	Stonyfield Farm O'Soy strawberry-peach pack organic cultured soy yogurt	1	item(s)	113	—	100	5.0	16.0	3.0	2.0	0	—	—
34616	Stonyfield Farm O'Soy vanilla organic cultured soy yogurt	1	item(s)	170	—	150	7.0	26.0	4.0	2.0	0	—	—
10453	White Wave plain silk cultured soy yogurt	8	ounce(s)	227	—	140	5.0	22.0	1.0	3.0	0.5	—	—
	EGGS												
	Eggs												
99	Fried	1	item(s)	46	31.8	90	6.3	0.4	0	7.0	2.0	2.9	1.2
100	Hard boiled	1	item(s)	50	37.3	78	6.3	0.6	0	5.3	1.6	2.0	0.7
101	Poached	1	item(s)	50	37.8	71	6.3	0.4	0	5.0	1.5	1.9	0.7
97	Raw, white	1	item(s)	33	28.9	16	3.6	0.2	0	0.1	0	0	0
96	Raw, whole	1	item(s)	50	37.9	72	6.3	0.4	0	5.0	1.5	1.9	0.7
98	Raw, yolk	1	item(s)	17	8.9	54	2.7	0.6	0	4.5	1.6	2.0	0.7
102	Scrambled, prepared with milk and butter	2	item(s)	122	89.2	204	13.5	2.7	0	14.9	4.5	5.8	2.6
	Egg substitute												
4028	Egg Beaters	¼	cup(s)	61	—	30	6.0	1.0	0	0	0	0	0
920	Frozen	¼	cup(s)	60	43.9	96	6.8	1.9	0	6.7	1.2	1.5	3.7
918	Liquid	¼	cup(s)	63	51.9	53	7.5	0.4	0	2.1	0.4	0.6	1.0
	SEAFOOD												
	Cod												
6040	Atlantic cod or scrod, baked or broiled	3	ounce(s)	85	64.6	89	19.4	0	0	0.7	0.1	0.1	0.2
1573	Atlantic cod, cooked, dry heat	3	ounce(s)	85	64.6	89	19.4	0	0	0.7	0.1	0.1	0.2
2905	**Eel, raw**	3	ounce(s)	85	58.0	156	15.7	0	0	9.9	2.0	6.1	0.8
	Fish fillets												
25079	Baked	3	ounce(s)	84	79.9	99	21.7	0	0	0.7	0.1	0.1	0.3
8615	Batter coated or breaded, fried	3	ounce(s)	85	45.6	197	12.5	14.4	0.4	10.5	2.4	2.2	5.3
25082	Broiled fish steaks	3	ounce(s)	85	68.1	128	24.2	0	0	2.6	0.4	0.9	0.8
25083	Poached fish steaks	3	ounce(s)	85	67.1	111	21.1	0	0	2.3	0.3	0.8	0.7
25084	Steamed	3	ounce(s)	85	72.2	79	17.2	0	0	0.6	0.1	0.1	0.2
25089	**Flounder, baked**	3	ounce(s)	85	64.4	113	14.8	0.4	0.1	5.5	1.1	2.2	1.4
1825	**Grouper, cooked, dry heat**	3	ounce(s)	85	62.4	100	21.1	0	0	1.1	0.3	0.2	0.3
	Haddock												
6049	Baked or broiled	3	ounce(s)	85	63.2	95	20.6	0	0	0.8	0.1	0.1	0.3
1578	Cooked, dry heat	3	ounce(s)	85	63.1	95	20.6	0	0	0.8	0.1	0.1	0.3
1886	**Halibut, Atlantic and Pacific, cooked, dry heat**	3	ounce(s)	85	61.0	119	22.7	0	0	2.5	0.4	0.8	0.8
1582	**Herring, Atlantic, pickled**	4	piece(s)	60	33.1	157	8.5	5.8	0	10.8	1.4	7.2	1.0
1587	**Jack mackerel, solids, canned, drained**	2	ounce(s)	57	39.2	88	13.1	0	0	3.6	1.1	1.3	0.9

Chol (mg)	Calc (mg)	Iron (mg)	Magn (mg)	Pota (mg)	Sodi (mg)	Zinc (mg)	Vit A (μg)	Thia (mg)	Vit E (mg α)	Ribo (mg)	Niac (mg)	Vit B_6 (mg)	Fola (μg DFE)	Vit C (mg)	Vit B_{12} (μg)	Sele (μg)
5	369	0.62	41.0	549.5	139.8	1.83	4.8	0.10	0.16	0.44	0.49	0.10	31.3	26.5	1.1	7.0
15	448	0.19	41.7	573.3	171.5	2.18	34.3	0.10	0.07	0.52	0.27	0.12	27.0	2.0	1.4	8.1
5	488	0.22	46.6	624.8	188.7	2.37	4.9	0.11	0.00	0.57	0.30	0.13	29.4	2.2	1.5	8.8
12	419	0.17	39.2	536.6	161.7	2.03	29.4	0.10	0.04	0.49	0.26	0.11	27.0	2.0	1.3	12.0
13	260	0.22	39.2	399.4	98.0	1.10	14.7	0.11	0.00	0.51	0.30	0.14	29.4	2.1	1.5	—
5	300	2.70	100.0	580.0	290.0	2.25	—	0.37	—	0.42	5.00	0.50	100.0	15.0	1.5	—
0	100	1.08	24.0	5.0	20.0	—	0	0.22	—	0.10	—	0.04	—	0	0	—
0	150	1.44	40.0	15.0	40.0	—	—	0.30	—	0.13	—	0.08	—	0	0	—
0	400	1.44	—	0	30.0	—	0	—	—	—	—	—	—	0	—	—
210	27	0.91	6.0	67.6	93.8	0.55	91.1	0.03	0.56	0.23	0.03	0.07	23.5	0	0.6	15.7
212	25	0.59	5.0	63.0	62.0	0.52	84.5	0.03	0.51	0.25	0.03	0.06	22.0	0	0.6	15.4
211	27	0.91	6.0	66.5	147.0	0.55	69.5	0.02	0.48	0.20	0.03	0.06	17.5	0	0.6	15.8
0	2	0.02	3.6	53.8	54.8	0.01	0	0.00	0.00	0.14	0.03	0.00	1.3	0	0	6.6
212	27	0.91	6.0	67.0	70.0	0.55	70.0	0.03	0.48	0.23	0.03	0.07	23.5	0	0.6	15.9
210	22	0.46	0.9	18.5	8.2	0.39	64.8	0.03	0.43	0.09	0.00	0.06	24.8	0	0.3	9.5
429	87	1.46	14.6	168.4	341.6	1.22	174.5	0.06	1.33	0.53	0.09	0.14	36.6	0.2	0.9	27.5
0	20	1.08	4.0	85.0	115.0	0.60	112.5	0.15	—	0.85	0.20	0.08	60.0	0	1.2	—
1	44	1.18	9.0	127.8	119.4	0.58	6.6	0.07	0.95	0.23	0.08	0.08	9.6	0.3	0.2	24.8
1	33	1.32	5.6	207.1	111.1	0.82	11.3	0.07	0.17	0.19	0.07	0.00	9.4	0	0.2	15.6
47	12	0.41	35.7	207.5	66.3	0.49	11.9	0.07	0.68	0.06	2.13	0.24	6.8	0.8	0.9	32.0
47	12	0.41	35.7	207.5	66.3	0.49	11.9	0.07	0.68	0.06	2.13	0.24	6.8	0.9	0.9	32.0
107	17	0.42	17.0	231.3	43.4	1.37	887.0	0.13	3.40	0.03	2.97	0.05	12.8	1.5	2.6	5.5
44	8	0.31	29.1	489.0	86.1	0.48	—	0.02	—	0.05	2.47	0.46	8.1	3.0	1.0	44.3
29	15	1.79	20.4	272.2	452.5	0.37	9.4	0.09	—	0.09	1.78	0.08	17.0	0	0.9	7.7
37	55	0.97	96.7	524.3	62.9	0.49	—	0.05	—	0.08	6.47	0.36	12.6	0	1.2	42.5
32	48	0.85	84.0	455.6	54.7	0.42	—	0.05	—	0.07	5.92	0.33	11.5	0	1.1	37.0
41	12	0.29	24.7	319.3	41.7	0.34	—	0.06	—	0.06	1.89	0.21	6.1	0.8	0.8	32.0
44	19	0.34	47.3	224.7	280.2	0.20	—	0.06	0.40	0.07	2.02	0.18	7.4	2.8	1.6	33.5
40	18	0.96	31.5	404.0	45.1	0.43	42.5	0.06	—	0.01	0.32	0.29	8.5	0	0.6	39.8
63	36	1.15	42.5	339.4	74.0	0.40	16.2	0.03	0.42	0.03	3.94	0.29	6.8	0	1.2	34.4
63	36	1.14	42.5	339.3	74.0	0.40	16.2	0.03	—	0.03	3.93	0.29	11.1	0	1.2	34.4
35	51	0.91	91.0	489.9	58.7	0.45	45.9	0.05	—	0.07	6.05	0.33	11.9	0	1.2	39.8
8	46	0.73	4.8	41.4	522.0	0.31	154.8	0.02	1.02	0.08	1.98	0.10	1.2	0	2.6	35.1
45	137	1.15	21.0	110.0	214.9	0.57	73.7	0.02	0.58	0.12	3.50	0.11	2.8	0.5	3.9	21.4

Table A–1 Table of Food Composition (*continued*) (Computer code is for Cengage Diet Analysis Plus program)

DA+ Code	Food Description	QTY	Measure	Wt (g)	H_2O (g)	Ener (cal)	Prot (g)	Carb (g)	Fiber (g)	Fat (g)	Fat Breakdown (g)		
											Sat	Mono	Poly
SEAFOOD—continued													
8580	**Octopus, common, cooked, moist heat**	3	ounce(s)	85	51.5	139	25.4	3.7	0	1.8	0.4	0.3	0.4
1831	**Perch, mixed species, cooked, dry heat**	3	ounce(s)	85	62.3	100	21.1	0	0	1.0	0.2	0.2	0.4
1592	**Pacific rockfish, cooked, dry heat**	3	ounce(s)	85	62.4	103	20.4	0	0	1.7	0.4	0.4	0.5
	Salmon												
2938	Coho, farmed, raw	3	ounce(s)	85	59.9	136	18.1	0	0	6.5	1.5	2.8	1.6
1594	Broiled or baked with butter	3	ounce(s)	85	53.9	155	23.0	0	0	6.3	1.2	2.3	2.3
29727	Smoked chinook (lox)	2	ounce(s)	57	40.8	66	10.4	0	0	2.4	0.5	1.1	0.6
154	**Sardine, Atlantic with bones, canned in oil**	3	ounce(s)	85	50.7	177	20.9	0	0	9.7	1.3	3.3	4.4
	Scallops												
155	Mixed species, breaded, fried	3	item(s)	47	27.2	100	8.4	4.7	—	5.1	1.2	2.1	1.3
1599	Steamed	3	ounce(s)	85	64.8	90	13.8	2.0	0	2.6	0.4	1.0	0.8
1839	**Snapper, mixed species, cooked, dry heat**	3	ounce(s)	85	59.8	109	22.4	0	0	1.5	0.3	0.3	0.5
	Squid												
1868	Mixed species, fried	3	ounce(s)	85	54.9	149	15.3	6.6	0	6.4	1.6	2.3	1.8
16617	Steamed or boiled	3	ounce(s)	85	63.3	89	15.2	3.0	0	1.3	0.4	0.1	0.5
1570	**Striped bass, cooked, dry heat**	3	ounce(s)	85	62.4	105	19.3	0	0	2.5	0.6	0.7	0.9
1601	**Sturgeon, steamed**	3	ounce(s)	85	59.4	111	17.0	0	0	4.3	1.0	2.0	0.7
1840	**Surimi, formed**	3	ounce(s)	85	64.9	84	12.9	5.8	0	0.8	0.2	0.1	0.4
1842	**Swordfish, cooked, dry heat**	3	ounce(s)	85	58.5	132	21.6	0	0	4.4	1.2	1.7	1.0
1846	**Tuna, yellowfin or ahi, raw**	3	ounce(s)	85	60.4	92	19.9	0	0	0.8	0.2	0.1	0.2
	Tuna, canned												
159	Light, canned in oil, drained	2	ounce(s)	57	33.9	112	16.5	0	0	4.6	0.9	1.7	1.6
355	Light, canned in water, drained	2	ounce(s)	57	42.2	66	14.5	0	0	0.5	0.1	0.1	0.2
33211	Light, no salt, canned in oil, drained	2	ounce(s)	57	33.9	112	16.5	0	0	4.7	0.9	1.7	1.6
33212	Light, no salt, canned in water, drained	2	ounce(s)	57	42.6	66	14.5	0	0	0.5	0.1	0.1	0.2
2961	White, canned in oil, drained	2	ounce(s)	57	36.3	105	15.0	0	0	4.6	0.7	1.8	1.7
351	White, canned in water, drained	2	ounce(s)	57	41.5	73	13.4	0	0	1.7	0.4	0.4	0.6
33213	White, no salt, canned in oil, drained	2	ounce(s)	57	36.3	105	15.0	0	0	4.6	0.9	1.4	1.9
33214	White, no salt, canned in water, drained	2	ounce(s)	57	42.0	73	13.4	0	0	1.7	0.4	0.4	0.6
	Yellowtail												
8548	Mixed species, cooked, dry heat	3	ounce(s)	85	57.3	159	25.2	0	0	5.7	1.4	2.2	1.5
2970	Mixed species, raw	2	ounce(s)	57	42.2	83	13.1	0	0	3.0	0.7	1.1	0.8
	Shellfish, meat only												
1857	Abalone, mixed species, fried	3	ounce(s)	85	51.1	161	16.7	9.4	0	5.8	1.4	2.3	1.4
16618	Abalone, steamed or poached	3	ounce(s)	85	40.7	177	28.8	10.1	0	1.3	0.3	0.2	0.2
	Crab												
1851	Blue crab, canned	2	ounce(s)	57	43.2	56	11.6	0	0	0.7	0.1	0.1	0.2
1852	Blue crab, cooked, moist heat	3	ounce(s)	85	65.9	87	17.2	0	0	1.5	0.2	0.2	0.6
8562	Dungeness crab, cooked, moist heat	3	ounce(s)	85	62.3	94	19.0	0.8	0	1.1	0.1	0.2	0.3
1860	**Clams, cooked, moist heat**	3	ounce(s)	85	54.1	126	21.7	4.4	0	1.7	0.2	0.1	0.5
1853	**Crayfish, farmed, cooked, moist heat**	3	ounce(s)	85	68.7	74	14.9	0	0	1.1	0.2	0.2	0.4

Chol (mg)	Calc (mg)	Iron (mg)	Magn (mg)	Pota (mg)	Sodi (mg)	Zinc (mg)	Vit A (µg)	Thia (mg)	Vit E (mg α)	Ribo (mg)	Niac (mg)	Vit B_6 (mg)	Fola (µg DFE)	Vit C (mg)	Vit B_{12} (µg)	Sele (µg)
82	90	8.11	51.0	535.8	391.2	2.85	76.5	0.04	1.02	0.06	3.21	0.55	20.4	6.8	30.6	76.2
98	87	0.98	32.3	292.6	67.2	1.21	8.5	0.06	—	0.10	1.61	0.11	5.1	1.4	1.9	13.7
37	10	0.45	28.9	442.3	65.5	0.45	60.4	0.03	1.32	0.07	3.33	0.22	8.5	0	1.0	39.8
43	10	0.29	26.4	382.7	40.0	0.36	47.6	0.08	—	0.09	5.79	0.56	11.1	0.9	2.3	10.7
40	15	1.02	26.9	376.6	98.6	0.56	—	0.13	1.14	0.05	8.33	0.18	4.2	1.8	2.3	41.0
13	6	0.48	10.2	99.2	1134.0	0.17	14.7	0.01	—	0.05	2.67	0.15	1.1	0	1.8	21.6
121	325	2.48	33.2	337.6	429.5	1.10	27.2	0.04	1.70	0.18	4.43	0.14	10.2	0	7.6	44.8
28	20	0.38	27.4	154.8	215.8	0.49	10.7	0.02	—	0.05	0.70	0.06	23.3	1.1	0.6	12.5
27	20	0.22	45.9	238.0	358.7	0.78	32.3	0.01	0.16	0.05	0.84	0.11	10.2	2.0	1.1	18.2
40	34	0.20	31.5	444.0	48.5	0.37	29.8	0.04	—	0.00	0.29	0.39	5.1	1.4	3.0	41.7
221	33	0.85	32.3	237.3	260.3	1.48	9.4	0.04	—	0.39	2.21	0.04	11.9	3.6	1.0	44.1
227	31	0.62	28.9	192.1	356.2	1.49	8.5	0.01	1.17	0.32	1.69	0.04	3.4	3.2	1.0	43.7
88	16	0.91	43.4	279.0	74.8	0.43	26.4	0.09	—	0.03	2.17	0.29	8.5	0	3.8	39.8
63	11	0.59	29.8	239.7	388.5	0.35	198.9	0.06	0.52	0.07	8.30	0.19	14.5	0	2.2	13.3
26	8	0.22	36.6	95.3	121.6	0.28	17.0	0.01	0.53	0.01	0.18	0.02	1.7	0	1.4	23.9
43	5	0.88	28.9	313.8	97.8	1.25	34.9	0.03	—	0.09	10.02	0.32	1.7	0.9	1.7	52.5
38	14	0.62	42.5	377.6	31.5	0.44	15.3	0.37	0.42	0.04	8.33	0.77	1.7	0.8	0.4	31.0
10	7	0.79	17.6	117.3	200.6	0.51	13.0	0.02	0.49	0.07	7.03	0.06	2.8	0	1.2	43.1
17	6	0.87	15.3	134.3	191.5	0.43	9.6	0.01	0.19	0.04	7.52	0.19	2.3	0	1.7	45.6
10	7	0.78	17.6	117.4	28.3	0.51	0	0.02	—	0.06	7.03	0.06	2.8	0	1.2	43.1
17	6	0.86	15.3	134.4	28.3	0.43	0	0.01	—	0.04	7.52	0.19	2.3	0	1.7	45.6
18	2	0.36	19.3	188.8	224.5	0.26	2.8	0.01	1.30	0.04	6.63	0.24	2.8	0	1.2	34.1
24	8	0.55	18.7	134.3	213.6	0.27	3.4	0.00	0.48	0.02	3.28	0.12	1.1	0	0.7	37.2
18	2	0.36	19.3	188.8	28.3	0.26	0	0.01	—	0.04	6.63	0.24	2.8	0	1.2	34.1
24	8	0.54	18.7	134.4	28.3	0.27	3.4	0.00	—	0.02	3.28	0.12	1.1	0	0.7	37.3
60	25	0.53	32.3	457.6	42.5	0.56	26.4	0.14	—	0.04	7.41	0.15	3.4	2.5	1.1	39.8
31	13	0.28	17.0	238.1	22.1	0.29	16.4	0.08	—	0.02	3.86	0.09	2.3	1.6	0.7	20.7
80	31	3.23	47.6	241.5	502.6	0.80	1.7	0.18	—	0.11	1.61	0.12	11.9	1.5	0.6	44.1
144	50	4.84	68.9	295.0	980.1	1.38	3.4	0.28	6.74	0.12	1.89	0.21	6.0	2.6	0.7	75.6
50	57	0.47	22.1	212.1	188.8	2.27	1.1	0.04	1.04	0.04	0.77	0.08	24.4	1.5	0.3	18.0
85	88	0.77	28.1	275.6	237.3	3.58	1.7	0.08	1.56	0.04	2.80	0.15	43.4	2.8	6.2	34.2
65	50	0.36	49.3	347.0	321.5	4.65	26.4	0.04	—	0.17	3.08	0.14	35.7	3.1	8.8	40.5
57	78	23.78	15.3	534.1	95.3	2.32	145.4	0.12	—	0.36	2.85	0.09	24.7	18.8	84.1	54.4
117	43	0.94	28.1	202.4	82.5	1.25	12.8	0.03	—	0.06	1.41	0.11	9.4	0.4	2.6	29.1

Table A–1 Table of Food Composition (*continued*)

(Computer code is for Cengage Diet Analysis Plus program)

DA+ Code	Food Description	QTY	Measure	Wt (g)	H_2O (g)	Ener (cal)	Prot (g)	Carb (g)	Fiber (g)	Fat (g)	Fat Breakdown (g)		
											Sat	Mono	Poly
	SEAFOOD—continued												
	Oysters												
8720	Baked or broiled	3	ounce(s)	85	68.6	89	5.6	3.2	0	5.8	1.3	2.1	1.9
152	Eastern, farmed, raw	3	ounce(s)	85	73.3	50	4.4	4.7	0	1.3	0.4	0.1	0.5
8715	Eastern, wild, cooked, moist heat	3	ounce(s)	85	59.8	117	12.0	6.7	0	4.2	1.3	0.5	1.6
8584	Pacific, cooked, moist heat	3	ounce(s)	85	54.5	139	16.1	8.4	0	3.9	0.9	0.7	1.5
1865	Pacific, raw	3	ounce(s)	85	69.8	69	8.0	4.2	0	2.0	0.4	0.3	0.8
1854	**Lobster, northern, cooked, moist heat**	3	ounce(s)	85	64.7	83	17.4	1.1	0	0.5	0.1	0.1	0.1
1862	**Mussel, blue, cooked, moist heat**	3	ounce(s)	85	52.0	146	20.2	6.3	0	3.8	0.7	0.9	1.0
	Shrimp												
158	Mixed species, breaded, fried	3	ounce(s)	85	44.9	206	18.2	9.8	0.3	10.4	1.8	3.2	4.3
1855	Mixed species, cooked, moist heat	3	ounce(s)	85	65.7	84	17.8	0	0	0.9	0.2	0.2	0.4
	BEEF, LAMB, PORK												
	Beef												
4450	Breakfast strips, cooked	2	slice(s)	23	5.9	101	7.1	0.3	0	7.8	3.2	3.8	0.4
174	Corned beef, canned	3	ounce(s)	85	49.1	213	23.0	0	0	12.7	5.3	5.1	0.5
33147	Cured, thin siced	2	ounce(s)	57	32.9	100	15.9	3.2	0	2.2	0.9	1.0	0.1
4581	Jerky	1	ounce(s)	28	6.6	116	9.4	3.1	0.5	7.3	3.1	3.2	0.3
	Ground beef												
5898	Lean, broiled, medium	3	ounce(s)	85	50.4	202	21.6	0	0	12.2	4.8	5.3	0.4
5899	Lean, broiled, well done	3	ounce(s)	85	48.4	214	23.8	0	0	12.5	5.0	5.7	0.3
5914	Regular, broiled, medium	3	ounce(s)	85	46.1	246	20.5	0	0	17.6	6.9	7.7	0.6
5915	Regular, broiled, well done	3	ounce(s)	85	43.8	259	21.6	0	0	18.4	7.5	8.5	0.5
	Beef rib												
4241	Rib, small end, separable lean, 0″ fat, broiled	3	ounce(s)	85	53.2	164	25.0	0	0	6.4	2.4	2.6	0.2
4183	Rib, whole, lean and fat, ¼″ fat, roasted	3	ounce(s)	85	39.0	320	18.9	0	0	26.6	10.7	11.4	0.9
	Beef roast												
16981	Bottom round, choice, separable lean and fat, ⅛″ fat, braised	3	ounce(s)	85	46.2	216	27.9	0	0	10.7	4.1	4.6	0.4
16979	Bottom round, separable lean and fat, ⅛″ fat, roasted	3	ounce(s)	85	52.4	185	22.5	0	0	9.9	3.8	4.2	0.4
16924	Chuck, arm pot roast, separable lean and fat, ⅛″ fat, braised	3	ounce(s)	85	42.9	257	25.6	0	0	16.3	6.5	7.0	0.6
16930	Chuck, blade roast, separable lean and fat, ⅛″ fat, braised	3	ounce(s)	85	40.5	290	22.8	0	0	21.4	8.5	9.2	0.8
5853	Chuck, blade roast, separable lean, 0″ trim, pot roasted	3	ounce(s)	85	47.4	202	26.4	0	0	9.9	3.9	4.3	0.3
4296	Eye of round, choice, separable lean, 0″ fat, roasted	3	ounce(s)	85	56.5	138	24.4	0	0	3.7	1.3	1.5	0.1
16989	Eye of round, separable lean and fat, ⅛″ fat, roasted	3	ounce(s)	85	52.2	180	24.2	0	0	8.5	3.2	3.6	0.3
	Beef steak												
4348	Short loin, t-bone steak, lean and fat, ¼″ fat, broiled	3	ounce(s)	85	43.2	274	19.4	0	0	21.2	8.3	9.6	0.8
4349	Short loin, t-bone steak, lean, ¼″ fat, broiled	3	ounce(s)	85	52.3	174	22.8	0	0	8.5	3.1	4.2	0.3
4360	Top loin, prime, lean and fat, ¼″ fat, broiled	3	ounce(s)	85	42.7	275	21.6	0	0	20.3	8.2	8.6	0.7
	Beef variety												
188	Liver, pan fried	3	ounce(s)	85	52.7	149	22.6	4.4	0	4.0	1.3	0.5	0.5
4447	Tongue, simmered	3	ounce(s)	85	49.2	242	16.4	0	0	19.0	6.9	8.6	0.6
	Lamb chop												
3275	Loin, domestic, lean and fat, ¼″ fat, broiled	3	ounce(s)	85	43.9	269	21.4	0	0	19.6	8.4	8.3	1.4

PAGE KEY: A-2 = Breads/Baked Goods A-8 = Cereal/Rice/Pasta A-12 = Fruit A-18 = Vegetables/Legumes A-28 = Nuts/Seeds A-30 = Vegetarian A-32 = Dairy A-40 = Eggs A-40 = Seafood A-44 = Meats A-48 = Poultry A-48 = Processed Meats A-50 = Beverages A-54 = Fats/Oils A-56 = Sweets A-58 = Spices/Condiments/Sauces A-62 = Mixed Foods/Soups/Sandwiches A-68 = Fast Food A-88 = Convenience A-90 = Baby Foods

Chol (mg)	Calc (mg)	Iron (mg)	Magn (mg)	Pota (mg)	Sodi (mg)	Zinc (mg)	Vit A (µg)	Thia (mg)	Vit E (mg α)	Ribo (mg)	Niac (mg)	Vit B_6 (mg)	Fola (µg DFE)	Vit C (mg)	Vit B_{12} (µg)	Sele (µg)
43	36	5.30	37.4	125.0	403.8	72.22	60.4	0.07	0.98	0.06	1.04	0.04	7.7	2.8	14.7	50.7
21	37	4.91	28.1	105.4	151.3	32.23	6.8	0.08	—	0.05	1.07	0.05	15.3	4.0	13.8	54.1
89	77	10.19	80.8	239.0	358.9	154.45	45.9	0.16	—	0.15	2.11	0.10	11.9	5.1	29.8	60.9
85	14	7.82	37.4	256.8	180.3	28.27	124.2	0.10	0.72	0.37	3.07	0.07	12.8	10.9	24.5	131.0
43	7	4.34	18.7	142.9	90.1	14.13	68.9	0.05	—	0.20	1.70	0.04	8.5	6.8	13.6	65.5
61	52	0.33	29.8	299.4	323.2	2.48	22.1	0.01	0.85	0.05	0.91	0.06	9.4	0	2.6	36.3
48	28	5.71	31.5	227.9	313.8	2.27	77.4	0.25	—	0.35	2.55	0.08	64.6	11.6	20.4	76.2
150	57	1.07	34.0	191.3	292.4	1.17	0	0.11	—	0.11	2.60	0.08	33.2	1.3	1.6	35.4
166	33	2.62	28.9	154.8	190.5	1.32	57.8	0.02	1.17	0.02	2.20	0.10	3.4	1.9	1.3	33.7
27	2	0.71	6.1	93.1	509.2	1.44	0	0.02	0.06	0.05	1.46	0.07	1.8	0	0.8	6.1
73	10	1.76	11.9	115.7	855.6	3.03	0	0.01	0.12	0.12	2.06	0.11	7.7	0	1.4	36.5
23	6	1.53	10.8	243.2	815.9	2.25	0	0.04	0.00	0.10	2.98	0.19	6.2	0	1.5	16.0
14	6	1.53	14.5	169.2	627.4	2.29	0	0.04	0.13	0.04	0.49	0.05	38.0	0	0.3	3.0
58	6	2.00	17.9	266.2	59.5	4.63	0	0.05	—	0.23	4.21	0.23	7.6	0	1.8	16.0
69	12	2.21	18.4	250.0	62.4	5.86	0	0.08	—	0.23	5.10	0.16	9.4	0	1.7	19.0
62	9	2.07	17.0	248.3	70.6	4.40	0	0.02	—	0.16	4.90	0.23	7.6	0	2.5	16.2
71	12	2.30	18.5	242.4	72.4	5.18	0	0.08	—	0.23	4.93	0.17	8.5	0	1.6	18.0
65	16	1.59	21.3	319.8	51.9	4.64	0	0.06	0.34	0.12	7.15	0.53	8.5	0	1.4	29.2
72	9	1.96	16.2	251.7	53.6	4.45	0	0.06	—	0.14	2.85	0.19	6.0	0	2.1	18.7
68	6	2.29	17.9	223.7	35.7	4.59	0	0.05	0.41	0.15	5.05	0.36	8.5	0	1.7	29.3
64	5	1.83	14.5	182.0	29.8	3.76	0	0.05	0.34	0.12	3.92	0.29	6.8	0	1.3	23.0
67	14	2.15	17.0	205.8	42.5	5.93	0	0.05	0.45	0.15	3.63	0.25	7.7	0	1.9	24.1
88	11	2.66	16.2	198.2	55.3	7.15	0	0.06	0.17	0.20	2.06	0.22	4.3	0	1.9	20.9
73	11	3.12	19.6	223.7	60.4	8.73	0	0.06	—	0.23	2.27	0.24	5.1	0	2.1	22.7
49	5	2.16	16.2	200.7	32.3	4.28	0	0.05	0.30	0.15	4.69	0.34	8.5	0	1.4	28.0
54	5	1.98	15.3	193.1	31.5	3.95	0	0.05	0.34	0.13	4.37	0.31	7.7	0	1.5	25.2
58	7	2.56	17.9	233.9	57.8	3.56	0	0.07	0.18	0.17	3.29	0.27	6.0	0	1.8	10.0
50	5	3.11	22.1	278.1	65.5	4.34	0	0.09	0.11	0.21	3.93	0.33	6.8	0	1.9	8.5
67	8	1.88	19.6	294.3	53.6	3.85	0	0.06	—	0.15	3.96	0.31	6.0	0	1.6	19.5
324	5	5.24	18.7	298.5	65.5	4.44	6586.3	0.15	0.39	2.91	14.86	0.87	221.1	0.6	70.7	27.9
112	4	2.22	12.8	156.5	55.3	3.47	0	0.01	0.25	0.25	2.96	0.13	6.0	1.1	2.7	11.2
85	17	1.53	20.4	278.1	65.5	2.96	0	0.08	0.11	0.21	6.03	0.11	15.3	0	2.1	23.3

DA+ Code	Food Description	QTY	Measure	Wt (g)	H₂O (g)	Ener (cal)	Prot (g)	Carb (g)	Fiber (g)	Fat (g)	Fat Breakdown (g) Sat	Mono	Poly
BEEF, LAMB, PORK—continued													
	Lamb leg												
3264	Domestic, lean and fat, ¼″ fat, cooked	3	ounce(s)	85	45.7	250	20.9	0	0	17.8	7.5	7.5	1.3
	Lamb rib												
182	Domestic, lean and fat, ¼″ fat, broiled	3	ounce(s)	85	40.0	307	18.8	0	0	25.2	10.8	10.3	2.0
183	Domestic, lean, ¼″ fat, broiled	3	ounce(s)	85	50.0	200	23.6	0	0	11.0	4.0	4.4	1.0
	Lamb shoulder												
186	Shoulder, arm and blade, domestic, choice, lean and fat, ¼″ fat, roasted	3	ounce(s)	85	47.8	235	19.1	0	0	17.0	7.2	6.9	1.4
187	Shoulder, arm and blade, domestic, choice, lean, ¼″ fat, roasted	3	ounce(s)	85	53.8	173	21.2	0	0	9.2	3.5	3.7	0.8
3287	Shoulder, arm, domestic, lean and fat, ¼″ fat, braised	3	ounce(s)	85	37.6	294	25.8	0	0	20.4	8.4	8.7	1.5
3290	Shoulder, arm, domestic, lean, ¼″ fat, braised	3	ounce(s)	85	41.9	237	30.2	0	0	12.0	4.3	5.2	0.8
	Lamb variety												
3375	Brain, pan fried	3	ounce(s)	85	51.6	232	14.4	0	0	18.9	4.8	3.4	1.9
3406	Tongue, braised	3	ounce(s)	85	49.2	234	18.3	0	0	17.2	6.7	8.5	1.1
	Pork, cured												
29229	Bacon, Canadian style, cured	2	ounce(s)	57	37.9	89	11.7	1.0	0	4.0	1.3	1.8	0.4
161	Bacon, cured, broiled, pan fried or roasted	2	slice(s)	16	2.0	87	5.9	0.2	0	6.7	2.2	3.0	0.7
35422	Breakfast strips, cured, cooked	3	slice(s)	34	9.2	156	9.8	0.4	0	12.5	4.3	5.6	1.9
189	Ham, cured, boneless, 11% fat, roasted	3	ounce(s)	85	54.9	151	19.2	0	0	7.7	2.7	3.8	1.2
29215	Ham, cured, extra lean, 4% fat, canned	2	2 ounce(s)	57	41.7	68	10.5	0	0	2.6	0.9	1.3	0.2
1316	Ham, cured, extra lean, 5% fat, roasted	3	ounce(s)	85	57.6	123	17.8	1.3	0	4.7	1.5	2.2	0.5
16561	Ham, smoked or cured, lean, cooked	1	slice(s)	42	27.6	66	10.5	0	0	2.3	0.8	1.1	0.3
	Pork chop												
32671	Loin, blade, chops, lean and fat, pan fried	3	ounce(s)	85	42.5	291	18.3	0	0	23.6	8.6	10	2.6
32672	Loin, center cut, chops, lean and fat, pan fried	3	ounce(s)	85	45.1	236	25.4	0	0	14.1	5.1	6.0	1.6
32682	Loin, center rib, chops, boneless, lean and fat, braised	3	ounce(s)	85	49.5	217	22.4	0	0	13.4	5.2	6.1	1.1
32603	Loin, center rib, chops, lean, broiled	3	ounce(s)	85	55.4	158	21.9	0	0	7.1	2.4	3.0	0.8
32478	Loin, whole, lean and fat, braised	3	ounce(s)	85	49.6	203	23.2	0	0	11.6	4.3	5.2	1.0
32481	Loin, whole, lean, braised	3	ounce(s)	85	52.2	174	24.3	0	0	7.8	2.9	3.5	0.6
	Pork leg or ham												
32471	Pork leg or ham, rump portion, lean and fat, roasted	3	ounce(s)	85	48.3	214	24.6	0	0	12.1	4.5	5.4	1.2
32468	Pork leg or ham, whole, lean and fat, roasted	3	ounce(s)	85	46.8	232	22.8	0	0	15.0	5.5	6.7	1.4
	Pork ribs												
32693	Loin, country style, lean and fat, roasted	3	ounce(s)	85	43.3	279	19.9	0	0	21.6	7.8	9.4	1.7
32696	Loin, country style, lean, roasted	3	ounce(s)	85	49.5	210	22.6	0	0	12.6	4.5	5.5	0.9
	Pork shoulder												
32626	Shoulder, arm picnic, lean and fat, roasted	3	ounce(s)	85	44.3	270	20.0	0	0	20.4	7.5	9.1	2.0
32629	Shoulder, arm picnic, lean, roasted	3	ounce(s)	85	51.3	194	22.7	0	0	10.7	3.7	5.1	1.0
	Rabbit												
3366	Domesticated, roasted	3	ounce(s)	85	51.5	168	24.7	0	0	6.8	2.0	1.8	1.3
3367	Domesticated, stewed	3	ounce(s)	85	50.0	175	25.8	0	0	7.2	2.1	1.9	1.4

Chol (mg)	Calc (mg)	Iron (mg)	Magn (mg)	Pota (mg)	Sodi (mg)	Zinc (mg)	Vit A (μg)	Thia (mg)	Vit E (mg α)	Ribo (mg)	Niac (mg)	Vit B_6 (mg)	Fola (μg DFE)	Vit C (mg)	Vit B_{12} (μg)	Sele (μg)
82	14	1.59	19.6	263.7	61.2	3.79	0	0.08	0.11	0.21	5.66	0.11	15.3	0	2.2	22.5
84	16	1.59	19.6	229.5	64.6	3.40	0	0.07	0.10	0.18	5.95	0.09	11.9	0	2.2	20.3
77	14	1.87	24.7	266.1	72.3	4.47	0	0.08	0.15	0.21	5.56	0.12	17.9	0	2.2	26.4
78	17	1.67	19.6	213.4	56.1	4.44	0	0.07	0.11	0.20	5.22	0.11	17.9	0	2.2	22.3
74	16	1.81	21.3	225.3	57.8	5.13	0	0.07	0.15	0.22	4.89	0.12	21.3	0	2.3	24.2
102	21	2.03	22.1	260.3	61.2	5.17	0	0.06	0.12	0.21	5.66	0.09	15.3	0	2.2	31.6
103	22	2.29	24.7	287.5	64.6	6.20	0	0.06	0.15	0.23	5.38	0.11	18.7	0	2.3	32.1
2130	18	1.73	18.7	304.5	133.5	1.70	0	0.14	—	0.31	3.87	0.19	6.0	19.6	20.5	10.2
161	9	2.23	13.6	134.4	57.0	2.54	0	0.06	—	0.35	3.13	0.14	2.6	6.0	5.4	23.8
28	5	0.38	9.6	195.0	798.9	0.78	0	0.42	0.11	0.09	3.53	0.22	2.3	0	0.4	14.2
18	2	0.22	5.3	90.4	369.6	0.56	1.8	0.06	0.04	0.04	1.76	0.04	0.3	0	0.2	9.9
36	5	0.67	8.8	158.4	713.7	1.25	0	0.25	0.08	0.12	2.58	0.11	1.4	0	0.6	8.4
50	7	1.13	18.7	347.7	1275.0	2.09	0	0.62	0.26	0.28	5.22	0.26	2.6	0	0.6	16.8
22	3	0.53	9.6	206.4	711.6	1.09	0	0.47	0.09	0.13	3.00	0.25	3.4	0	0.5	8.2
45	7	1.25	11.9	244.1	1023.1	2.44	0	0.64	0.21	0.17	3.42	0.34	2.6	0	0.6	16.6
23	3	0.39	9.2	132.7	557.3	1.07	0	0.28	0.10	0.10	2.10	0.19	1.7	0	0.3	10.7
72	26	0.74	17.9	282.4	57.0	2.71	1.7	0.52	0.17	0.25	3.35	0.28	3.4	0.5	0.7	29.7
78	23	0.77	24.7	361.5	68.0	1.96	1.7	0.96	0.21	0.25	4.76	0.39	5.1	0.9	0.6	33.2
62	4	0.78	14.5	329.1	34.0	1.76	1.7	0.44	—	0.20	3.66	0.26	3.4	0.3	0.4	28.4
56	22	0.57	21.3	291.7	48.5	1.91	0	0.48	0.08	0.18	6.68	0.57	0	0	0.4	38.6
68	18	0.91	16.2	318.1	40.8	2.02	1.7	0.53	0.20	0.21	3.75	0.31	2.6	0.5	0.5	38.5
67	15	0.96	17.0	329.1	42.5	2.10	1.7	0.56	0.17	0.22	3.90	0.32	3.4	0.5	0.5	41.0
82	10	0.89	23.0	318.1	52.7	2.39	2.6	0.63	0.18	0.28	3.95	0.26	2.6	0.2	0.6	39.8
80	12	0.85	18.7	299.4	51.0	2.51	2.6	0.54	0.18	0.26	3.89	0.34	8.5	0.3	0.6	38.5
78	21	0.90	19.6	292.6	44.2	2.00	2.6	0.75	—	0.29	3.67	0.37	4.3	0.3	0.7	31.6
79	25	1.09	20.4	296.8	24.7	3.24	1.7	0.48	—	0.29	3.96	0.37	4.3	0.3	0.7	36.0
80	16	1.00	14.5	276.4	59.5	2.93	1.7	0.44	—	0.25	3.33	0.29	3.4	0.2	0.6	28.6
81	8	1.20	17.0	298.5	68.0	3.46	1.7	0.49	—	0.30	3.66	0.34	4.3	0.3	0.7	32.7
70	16	1.93	17.9	325.7	40.0	1.93	0	0.07	—	0.17	7.17	0.40	9.4	0	7.1	32.7
73	17	2.01	17.0	255.1	31.5	2.01	0	0.05	0.37	0.14	6.09	0.28	7.7	0	5.5	32.7

Table A–1 Table of Food Composition (*continued*) (Computer code is for Cengage Diet Analysis Plus program)

DA+ Code	Food Description	QTY	Measure	Wt (g)	H₂O (g)	Ener (cal)	Prot (g)	Carb (g)	Fiber (g)	Fat (g)	Fat Breakdown (g) Sat	Mono	Poly
	BEEF, LAMB, PORK—continued												
	Veal												
3391	Liver, braised	3	ounce(s)	85	50.9	163	24.2	3.2	0	5.3	1.7	1.0	0.9
3319	Rib, lean only, roasted	3	ounce(s)	85	55.0	151	21.9	0	0	6.3	1.8	2.3	0.6
1732	Deer or venison, roasted	3	ounce(s)	85	55.5	134	25.7	0	0	2.7	1.1	0.7	0.5
	POULTRY												
	Chicken												
29562	Flaked, canned	2	ounce(s)	57	39.3	97	10.3	0.1	0	5.8	1.6	2.3	1.3
	Chicken, fried												
29632	Breast, meat only, breaded, baked or fried	3	ounce(s)	85	44.3	193	25.3	6.9	0.2	6.6	1.6	2.7	1.7
35327	Broiler breast, meat only, fried	3	ounce(s)	85	51.2	159	28.4	0.4	0	4.0	1.1	1.5	0.9
36413	Broiler breast, meat and skin, flour coated, fried	3	ounce(s)	85	48.1	189	27.1	1.4	0.1	7.5	2.1	3.0	1.7
36414	Broiler drumstick, meat and skin, flour coated, fried	3	ounce(s)	85	48.2	208	22.9	1.4	0.1	11.7	3.1	4.6	2.7
35389	Broiler drumstick, meat only, fried	3	ounce(s)	85	52.9	166	24.3	0	0	6.9	1.8	2.5	1.7
35406	Broiler leg, meat only, fried	3	ounce(s)	85	51.5	177	24.1	0.6	0	7.9	2.1	2.9	1.9
35484	Broiler wing, meat only, fried	3	ounce(s)	85	50.9	179	25.6	0	0	7.8	2.1	2.6	1.8
29580	Patty, fillet or tenders, breaded, cooked	3	ounce(s)	85	40.2	256	14.5	12.2	0	16.5	3.7	8.4	3.7
	Chicken, roasted, meat only												
35409	Broiler leg, meat only, roasted	3	ounce(s)	85	55.0	162	23.0	0	0	7.2	1.9	2.6	1.7
35486	Broiler wing, meat only, roasted	3	ounce(s)	85	53.4	173	25.9	0	0	6.9	1.9	2.2	1.5
35138	Roasting chicken, dark meat, meat only, roasted	3	ounce(s)	85	57.0	151	19.8	0	0	7.4	2.1	2.8	1.7
35136	Roasting chicken, light meat, meat only, roasted	3	ounce(s)	85	57.7	130	23.1	0	0	3.5	0.9	1.3	0.8
35132	Roasting chicken, meat only, roasted	3	ounce(s)	85	57.3	142	21.3	0	0	5.6	1.5	2.1	1.3
	Chicken, stewed												
1268	Gizzard, simmered	3	ounce(s)	85	57.8	124	25.8	0	0	2.3	0.6	0.4	0.3
1270	Liver, simmered	3	ounce(s)	85	56.8	142	20.8	0.7	0	5.5	1.8	1.2	1.7
3174	Meat only, stewed	3	ounce(s)	85	56.8	151	23.2	0	0	5.7	1.6	2.0	1.3
	Duck												
1286	Domesticated, meat and skin, roasted	3	ounce(s)	85	44.1	287	16.2	0	0	24.1	8.2	11.0	3.1
1287	Domesticated, meat only, roasted	3	ounce(s)	85	54.6	171	20.0	0	0	9.5	3.5	3.1	1.2
	Goose												
35507	Domesticated, meat and skin, roasted	3	ounce(s)	85	44.2	259	21.4	0	0	18.6	5.8	8.7	2.1
35524	Domesticated, meat only, roasted	3	ounce(s)	85	48.7	202	24.6	0	0	10.8	3.9	3.7	1.3
1297	Liver pate, smoked, canned	4	tablespoon(s)	52	19.3	240	5.9	2.4	0	22.8	7.5	13.3	0.4
	Turkey												
3256	Ground turkey, cooked	3	ounce(s)	85	50.5	200	23.3	0	0	11.2	2.9	4.2	2.7
3263	Patty, batter coated, breaded, fried	1	item(s)	94	46.7	266	13.2	14.8	0.5	16.9	4.4	7.0	4.4
219	Roasted, dark meat, meat only	3	ounce(s)	85	53.7	159	24.3	0	0	6.1	2.1	1.4	1.8
222	Roasted, fryer roaster breast, meat only	3	ounce(s)	85	58.2	115	25.6	0	0	0.6	0.2	0.1	0.2
220	Roasted, light meat, meat only	3	ounce(s)	85	56.4	134	25.4	0	0	2.7	0.9	0.5	0.7
1303	Turkey roll, light and dark meat	2	slice(s)	57	39.8	84	10.3	1.2	0	4.0	1.2	1.3	1.0
1302	Turkey roll, light meat	2	slice(s)	57	42.5	56	8.4	2.9	0	0.9	0.2	0.2	0.1
	PROCESSED MEATS												
	Beef												
1331	Corned beef loaf, jellied, sliced	2	slice(s)	57	39.2	87	13.0	0	0	3.5	1.5	1.5	0.2
	Bologna												
13459	Beef	1	slice(s)	28	15.1	90	3.0	1.0	0	8.0	3.5	4.3	0.3
13461	Light, made with pork and chicken	1	slice(s)	28	18.2	60	3.0	2.0	0	4.0	1.0	2.0	0.4
13458	Made with chicken and pork	1	slice(s)	28	15.0	90	3.0	1.0	0	8.0	3.0	4.1	1.1
13565	Turkey bologna	1	slice(s)	28	19.0	50	3.0	1.0	0	4.0	1.0	1.1	1.0
	Chicken												
7125	Breast, smoked	1	slice(s)	10	—	10	1.8	0.3	0	0.2	0	—	—

Chol (mg)	Calc (mg)	Iron (mg)	Magn (mg)	Pota (mg)	Sodi (mg)	Zinc (mg)	Vit A (µg)	Thia (mg)	Vit E (mg α)	Ribo (mg)	Niac (mg)	Vit B_6 (mg)	Fola (µg DFE)	Vit C (mg)	Vit B_{12} (µg)	Sele (µg)
435	5	4.34	17.0	279.8	66.3	9.55	8026	0.15	0.57	2.43	11.18	0.78	281.5	0.9	72.0	16.4
98	10	0.81	20.4	264.5	82.5	3.81	0	0.05	0.30	0.24	6.37	0.23	11.9	0	1.3	9.4
95	6	3.80	20.4	284.9	45.9	2.33	0	0.15	—	0.51	5.70	—	—	0	—	11.0
35	8	0.89	6.8	147.4	408.2	0.79	19.3	0.01	—	0.07	3.58	0.19	2.3	0	0.2	—
67	19	1.05	24.7	222.6	450.2	0.84	—	0.08	—	0.09	10.97	0.46	4.3	0	0.3	—
77	14	0.96	26.4	234.7	67.2	0.91	6.0	0.06	0.35	0.10	12.57	0.54	3.4	0	0.3	22.3
76	14	1.01	25.5	220.3	64.6	0.93	12.8	0.06	0.39	0.11	11.68	0.49	6.0	0	0.3	20.3
77	10	1.13	19.6	194.8	75.7	2.45	21.3	0.06	0.65	0.19	5.13	0.29	9.4	0	0.3	15.6
80	10	1.12	20.4	211.8	81.6	2.73	15.3	0.06	—	0.20	5.22	0.33	7.7	0	0.3	16.7
84	11	1.19	21.3	216.0	81.6	2.53	17.0	0.07	0.38	0.21	5.68	0.33	7.7	0	0.3	16.0
71	13	0.96	17.9	176.9	77.4	1.80	15.3	0.03	0.40	0.10	6.15	0.50	3.4	0	0.3	21.6
49	11	0.75	19.6	244.8	411.4	0.79	4.3	0.09	1.04	0.12	5.99	0.24	35.7	0	0.2	13.9
80	10	1.11	20.4	205.8	77.4	2.43	16.2	0.06	0.22	0.19	5.37	0.31	6.8	0	0.3	18.8
72	14	0.98	17.9	178.6	78.2	1.82	15.3	0.03	0.22	0.10	6.21	0.50	3.4	0	0.3	21.0
64	9	1.13	17.0	190.5	80.8	1.81	13.6	0.05	—	0.16	4.87	0.26	6.0	0	0.2	16.7
64	11	0.91	19.6	200.7	43.4	0.66	6.8	0.05	0.22	0.07	8.90	0.45	2.6	0	0.3	21.9
64	10	1.02	17.9	194.8	63.8	1.29	10.2	0.05	—	0.12	6.70	0.34	4.3	0	0.2	20.9
315	14	2.71	2.6	152.2	47.6	3.75	0	0.02	0.17	0.17	2.65	0.06	4.3	0	0.9	35.0
479	9	9.89	21.3	223.7	64.6	3.38	3385.8	0.24	0.69	1.69	9.39	0.64	491.6	23.7	14.3	70.1
71	12	0.99	17.9	153.1	59.5	1.69	12.8	0.04	0.22	0.13	5.20	0.22	5.1	0	0.2	17.8
71	9	2.29	13.6	173.5	50.2	1.58	53.6	0.14	0.59	0.22	4.10	0.15	5.1	0	0.3	17.0
76	10	2.29	17.0	214.3	55.3	2.21	19.6	0.22	0.59	0.39	4.33	0.21	8.5	0	0.3	19.1
77	11	2.40	18.7	279.8	59.5	2.22	17.9	0.06	1.47	0.27	3.54	0.31	1.7	0	0.3	18.5
82	12	2.44	21.3	330.0	64.6	2.69	10.2	0.07	—	0.33	3.47	0.39	10.2	0	0.4	21.7
78	36	2.86	6.8	71.8	362.4	0.47	520.5	0.04	—	0.15	1.30	0.03	31.2	0	4.9	22.9
87	21	1.64	20.4	229.6	91.0	2.43	0	0.04	0.28	0.14	4.09	0.33	6.0	0	0.3	31.6
71	13	2.06	14.1	258.5	752.0	1.35	9.4	0.09	0.87	0.17	2.16	0.18	57.3	0	0.2	20.8
72	27	1.98	20.4	246.6	67.2	3.79	0	0.05	0.54	0.21	3.10	0.30	7.7	0	0.3	34.8
71	10	1.30	24.7	248.3	44.2	1.48	0	0.03	0.07	0.11	6.37	0.47	5.1	0	0.3	27.3
59	16	1.14	23.8	259.4	54.4	1.73	0	0.05	0.07	0.11	5.81	0.45	5.1	0	0.3	27.3
31	18	0.76	10.2	153.1	332.3	1.13	0	0.05	0.19	0.16	2.72	0.15	2.8	0	0.1	16.6
19	4	0.21	10.8	242.1	590.8	0.50	0	0.01	0.07	0.08	4.05	0.23	2.3	0	0.2	7.4
27	6	1.15	6.2	57.3	540.4	2.31	0	0.00	—	0.06	0.99	0.06	4.5	0	0.7	9.8
20	0	0.36	3.9	47.0	310.0	0.56	0	0.01	—	0.03	0.67	0.04	3.6	0	0.4	—
20	40	0.36	5.6	45.6	300.0	0.45	0	—	—	—	—	—	—	0	—	—
30	20	0.36	5.9	43.1	300.0	0.39	0	—	—	—	—	—	—	0	—	—
20	40	0.36	6.2	42.6	270.0	0.51	0	—	—	—	—	—	—	0	—	—
4	0	0.00	—	—	100.0	—	0	—	—	—	—	—	—	0	—	—

											Fat Breakdown (g)		
DA+ Code	Food Description	QTY	Measure	Wt (g)	H_2O (g)	Ener (cal)	Prot (g)	Carb (g)	Fiber (g)	Fat (g)	Sat	Mono	Poly
PROCESSED MEATS—continued													
	Ham												
7127	Deli-sliced, honey	1	slice(s)	10	—	10	1.7	0.3	0	0.3	0.1	—	—
7126	Deli-sliced, smoked	1	slice(s)	10	—	10	1.7	0.2	0	0.3	0.1	—	—
8614	**Beef and pork mortadella, sliced**	2	slice(s)	46	24.1	143	7.5	1.4	0	11.7	4.4	5.2	1.4
1323	**Pork olive loaf**	2	slice(s)	57	33.1	133	6.7	5.2	0	9.4	3.3	4.5	1.1
1324	**Pork pickle and pimento loaf**	2	slice(s)	57	34.2	128	6.4	4.8	0.9	9.1	3.0	4.0	1.6
	Sausages and frankfurters												
37296	Beerwurst beef, beer salami (bierwurst)	1	slice(s)	29	16.6	74	4.1	1.2	0	5.7	2.5	2.7	0.2
37257	Beerwurst pork, beer salami	1	slice(s)	21	12.9	50	3.0	0.4	0	4.0	1.3	1.9	0.5
35338	Berliner, pork and beef	1	ounce(s)	28	17.3	65	4.3	0.7	0	4.9	1.7	2.3	0.4
37298	Bratwurst pork, cooked	1	piece(s)	74	42.3	181	10.4	1.9	0	14.3	5.1	6.7	1.5
37299	Braunschweiger pork liver sausage	1	slice(s)	15	8.2	51	2.0	0.3	0	4.5	1.5	2.1	0.5
1329	Cheesefurter or cheese smokie, beef and pork	1	item(s)	43	22.6	141	6.1	0.6	0	12.5	4.5	5.9	1.3
1330	Chorizo, beef and pork	2	ounce(s)	57	18.1	258	13.7	1.1	0	21.7	8.2	10.4	2.0
8600	Frankfurter, beef	1	item(s)	45	23.4	149	5.1	1.8	0	13.3	5.3	6.4	0.5
202	Frankfurter, beef and pork	1	item(s)	45	25.2	137	5.2	0.8	0	12.4	4.8	6.2	1.2
1293	Frankfurter, chicken	1	item(s)	45	28.1	100	7.0	1.2	0.2	7.3	1.7	2.7	1.7
3261	Frankfurter, turkey	1	item(s)	45	28.3	100	5.5	1.7	0	7.8	1.8	2.6	1.8
37275	Italian sausage, pork, cooked	1	item(s)	68	32.0	234	13.0	2.9	0.1	18.6	6.5	8.1	2.2
37307	Kielbasa or kolbassa, pork and beef	1	slice(s)	30	18.5	67	5.0	1.0	0	4.7	1.7	2.2	0.5
1333	Knockwurst or knackwurst, beef and pork	2	ounce(s)	57	31.4	174	6.3	1.8	0	15.7	5.8	7.3	1.7
37285	Pepperoni, beef and pork	1	slice(s)	11	3.4	51	2.2	0.4	0.2	4.4	1.8	2.1	0.3
37313	Polish sausage, pork	1	slice(s)	21	11.4	60	2.8	0.7	0	5.0	1.8	2.3	0.5
206	Salami, beef, cooked, sliced	2	slice(s)	52	31.2	136	6.5	1.0	0	11.5	5.1	5.5	0.5
37272	Salami, pork, dry or hard	1	slice(s)	13	4.6	52	2.9	0.2	0	4.3	1.5	2.0	0.5
40987	Sausage, turkey, cooked	2	ounce(s)	57	36.9	111	13.5	0	0	5.9	1.3	1.7	1.5
8620	Smoked sausage, beef and pork	2	ounce(s)	57	30.6	181	6.8	1.4	0	16.3	5.5	6.9	2.2
8619	Smoked sausage, pork	2	ounce(s)	57	32.0	178	6.8	1.2	0	16.0	5.3	6.4	2.1
37273	Smoked sausage, pork link	1	piece(s)	76	29.8	295	16.8	1.6	0	24.0	8.6	11.1	2.8
1336	Summer sausage, thuringer, or cervelat, beef and pork	2	ounce(s)	57	25.6	205	9.9	1.9	0	17.3	6.5	7.4	0.7
37294	Vienna sausage, cocktail, beef and pork, canned	1	piece(s)	16	10.4	37	1.7	0.4	0	3.1	1.1	1.5	0.2
	Spreads												
1318	Ham salad spread	¼	cup(s)	60	37.6	130	5.2	6.4	0	9.3	3.0	4.3	1.6
32419	Pork and beef sandwich spread	4	tablespoon(s)	60	36.2	141	4.6	7.2	0.1	10.4	3.6	4.6	1.5
	Turkey												
13604	Breast, fat free, oven roasted	1	slice(s)	28	—	25	4.0	1.0	0	0	0	0	0
13606	Breast, hickory smoked fat free	1	slice(s)	28	—	25	4.0	1.0	0	0	0	0	0
16049	Breast, hickory smoked slices	1	slice(s)	56	—	50	11.0	1.0	0	0	0	0	0
16047	Breast, honey roasted slices	1	slice(s)	56	—	60	11.0	3.0	0	0	0	0	0
16048	Breast, oven roasted slices	1	slice(s)	56	—	50	11.0	1.0	0	0	0	0	0
7124	Breast, oven roasted	1	slice(s)	10	—	10	1.8	0.3	0	0.1	0	0	0
13567	Turkey ham, 10% water added	2	slice(s)	56	40.9	70	10.0	2.0	0	3.0	0	0.4	0.6
37270	Turkey pastrami	1	slice(s)	28	20.3	35	4.6	1.0	0	1.2	0.3	0.4	0.3
3262	Turkey salami	2	slice(s)	57	39.1	98	10.9	0.9	0.1	5.2	1.6	1.8	1.4
37318	Turkey salami, cooked	1	slice(s)	28	20.4	43	4.3	0.1	0	2.7	0.8	0.9	0.7
BEVERAGES													
	Beer												
866	Ale, mild	12	fluid ounce(s)	360	332.3	148	1.1	13.3	0.4	0	0	0	0
686	Beer	12	fluid ounce(s)	356	327.7	153	1.6	12.7	0	0	0	0	0
16886	Beer, nonalcoholic	12	fluid ounce(s)	360	328.1	133	0.8	29.0	0	0.4	0.1	0	0.2
31609	Bud Light beer	12	fluid ounce(s)	355	335.5	110	0.9	6.6	0	0	0	0	0
31608	Budweiser beer	12	fluid ounce(s)	355	327.7	145	1.3	10.6	0	0	0	0	0

Chol (mg)	Calc (mg)	Iron (mg)	Magn (mg)	Pota (mg)	Sodi (mg)	Zinc (mg)	Vit A (µg)	Thia (mg)	Vit E (mg α)	Ribo (mg)	Niac (mg)	Vit B_6 (mg)	Fola (µg DFE)	Vit C (mg)	Vit B_{12} (µg)	Sele (µg)
4	0	0.12	—	—	100.0	—	0	—	—	—	—	—	—	0.6	—	—
4	0	0.12	—	—	103.3	—	0	—	—	—	—	—	—	0.6	—	—
26	8	0.64	5.1	75.0	573.2	0.96	0	0.05	0.10	0.07	1.23	0.06	1.4	0	0.7	10.4
22	62	0.30	10.8	168.7	842.9	0.78	34.1	0.16	0.14	0.14	1.04	0.13	1.1	0	0.7	9.3
33	62	0.75	19.3	210.7	740.7	0.95	44.3	0.22	0.22	0.06	1.41	0.23	21.0	4.4	0.3	4.5
18	3	0.44	3.5	66.5	264.9	0.71	0	0.02	0.05	0.03	0.98	0.04	0.9	0	0.6	4.7
12	2	0.15	2.7	53.3	261.0	0.36	0	0.11	0.03	0.04	0.68	0.07	0.6	0	0.2	4.4
13	3	0.32	4.3	80.2	367.7	0.70	0	0.10	—	0.06	0.88	0.05	1.4	0	0.8	4.0
44	33	0.95	11.1	156.9	412.2	1.70	0	0.37	0.01	0.13	2.36	0.15	1.5	0.7	0.7	15.7
24	1	1.42	1.7	27.5	131.5	0.42	641.0	0.03	0.05	0.23	1.27	0.05	6.7	0	3.1	8.8
29	25	0.46	5.6	88.6	465.3	0.96	20.2	0.10	0.10	0.06	1.24	0.05	1.3	0	0.7	6.8
50	5	0.90	10.2	225.7	700.2	1.93	0	0.35	0.12	0.17	2.90	0.30	1.1	0	1.1	12.0
24	6	0.67	6.3	70.2	513.0	1.10	0	0.01	0.09	0.06	1.06	0.04	2.3	0	0.8	3.7
23	5	0.51	4.5	75.2	504.0	0.82	8.1	0.09	0.11	0.05	1.18	0.05	1.8	0	0.6	6.2
43	33	0.52	9.0	90.9	379.8	0.50	0	0.02	0.09	0.11	2.10	0.14	3.2	0	0.2	10.4
35	67	0.66	6.3	176.4	485.1	0.82	0	0.01	0.27	0.08	1.65	0.06	4.1	0	0.4	6.8
39	14	0.97	12.2	206.7	820.8	1.62	6.8	0.42	0.17	0.15	2.83	0.22	3.4	0.1	0.9	15.0
20	13	0.44	4.9	84.4	283.0	0.61	0	0.06	0.06	0.06	0.87	0.05	1.5	0	0.5	5.4
34	6	0.37	6.2	112.8	527.3	0.94	0	0.19	0.32	0.07	1.55	0.09	1.1	0	0.7	7.7
13	2	0.15	2.0	34.7	196.7	0.30	0	0.05	0.00	0.02	0.59	0.04	0.7	0.1	0.2	2.4
15	2	0.29	2.9	37.3	199.3	0.40	0	0.10	0.04	0.03	0.71	0.03	0.4	0.2	0.2	3.7
37	3	1.14	6.8	97.8	592.8	0.92	0	0.04	0.08	0.08	1.68	0.08	1.0	0	1.6	7.6
10	2	0.16	2.8	48.4	289.3	0.53	0	0.11	0.02	0.04	0.71	0.07	0.3	0	0.4	3.3
52	12	0.84	11.9	169.0	377.1	2.19	7.4	0.04	0.10	0.14	3.24	0.18	3.4	0.4	0.7	0
33	7	0.42	7.4	101.5	516.5	0.71	7.4	0.10	0.07	0.06	1.66	0.09	1.1	0	0.3	0
35	6	0.33	6.2	273.9	468.9	0.74	0	0.12	0.14	0.10	1.59	0.10	0.6	0	0.4	10.4
52	23	0.87	14.4	254.6	1136.6	2.13	0	0.53	0.18	0.19	3.43	0.26	3.8	1.5	1.2	16.4
42	5	1.15	7.9	147.4	737.1	1.45	0	0.08	0.12	0.18	2.44	0.14	1.1	9.4	3.1	11.5
14	2	0.14	1.1	16.2	155.0	0.25	0	0.01	0.03	0.01	0.25	0.01	0.6	0	0.2	2.7
22	5	0.35	6.0	90.0	547.2	0.66	0	0.26	1.04	0.07	1.25	0.09	0.6	0	0.5	10.7
23	7	0.47	4.8	66.0	607.8	0.61	15.6	0.10	1.04	0.08	1.03	0.07	1.2	0	0.7	5.8
10	0	0.00	—	—	340.0	—	0	—	—	—	—	—	—	0	—	—
10	0	0.00	—	—	300.0	—	0	—	—	—	—	—	—	0	—	—
25	0	0.72	—	—	720.0	—	0	—	—	—	—	—	—	0	—	—
20	0	0.72	—	—	660.0	—	0	—	—	—	—	—	—	0	—	—
20	0	0.72	—	—	660.0	—	0	—	—	—	—	—	—	0	—	—
4	0	0.06	—	—	103.3	—	0	—	—	—	—	—	—	0	—	—
40	0	0.72	12.3	162.4	700.0	1.44	0	—	—	—	—	—	—	0	—	—
19	3	1.19	4.0	97.8	278.1	0.61	1.1	0.01	0.06	0.07	1.00	0.07	1.4	4.6	0.1	4.6
43	23	0.70	12.5	122.5	569.3	1.31	1.1	0.24	0.13	0.17	2.25	0.24	5.7	0	0.6	15.0
22	11	0.35	6.2	61.2	284.6	0.65	0.6	0.12	0.06	0.08	1.12	0.12	2.8	0	0.3	7.5
0	18	0.07	21.6	90.0	14.4	0.03	0	0.03	0.00	0.10	1.62	0.18	21.6	0	0.1	2.5
0	14	0.07	21.4	96.2	14.3	0.03	0	0.01	0.00	0.08	1.82	0.16	21.4	0	0.1	2.1
0	25	0.21	25.2	28.8	46.8	0.07	—	0.07	0.00	0.18	3.99	0.10	50.4	1.8	0.1	4.3
0	18	0.14	17.8	63.9	9.0	0.10	0	0.03	—	0.10	1.39	0.12	14.6	0	0	4.0
0	18	0.10	21.3	88.8	9.0	0.07	0	0.02	—	0.09	1.60	0.17	21.3	0	0.1	4.0

Table A–1 Table of Food Composition (*continued*) (Computer code is for Cengage Diet Analysis Plus program)

DA+ Code	Food Description	QTY	Measure	Wt (g)	H_2O (g)	Ener (cal)	Prot (g)	Carb (g)	Fiber (g)	Fat (g)	Fat Breakdown (g) Sat	Mono	Poly
BEVERAGES—continued													
869	Light beer	12	fluid ounce(s)	354	335.9	103	0.9	5.8	0	0	0	0	0
31613	Michelob beer	12	fluid ounce(s)	355	323.4	155	1.3	13.3	0	0	0	0	0
31614	Michelob Light beer	12	fluid ounce(s)	355	329.8	134	1.1	11.7	0	0	0	0	0
	Gin, rum, vodka, whiskey												
857	Distilled alcohol, 100 proof	1	fluid ounce(s)	28	16.0	82	0	0	0	0	0	0	0
687	Distilled alcohol, 80 proof	1	fluid ounce(s)	28	18.5	64	0	0	0	0	0	0	0
688	Distilled alcohol, 86 proof	1	fluid ounce(s)	28	17.8	70	0	0	0	0	0	0	0
689	Distilled alcohol, 90 proof	1	fluid ounce(s)	28	17.3	73	0	0	0	0	0	0	0
856	Distilled alcohol, 94 proof	1	fluid ounce(s)	28	16.8	76	0	0	0	0	0	0	0
	Liqueurs												
33187	Coffee liqueur, 53 proof	1	fluid ounce(s)	35	10.8	113	0	16.3	0	0.1	0	0	0
3142	Coffee liqueur, 63 proof	1	fluid ounce(s)	35	14.4	107	0	11.2	0	0.1	0	0	0
736	Cordials, 54 proof	1	fluid ounce(s)	30	8.9	106	0	13.3	0	0.1	0	0	0
	Wine												
861	California red wine	5	fluid ounce(s)	150	133.4	125	0.3	3.7	0	0	0	0	0
858	Domestic champagne	5	fluid ounce(s)	150	—	105	0.3	3.8	0	0	0	0	0
690	Sweet dessert wine	5	fluid ounce(s)	147	103.7	235	0.3	20.1	0	0	0	0	0
1481	White wine	5	fluid ounce(s)	148	128.1	121	0.1	3.8	0	0	0	0	0
1811	Wine cooler	10	fluid ounce(s)	300	267.4	159	0.3	20.2	0	0.1	0	0	0
	Carbonated												
31898	7 Up	12	fluid ounce(s)	360	321.0	140	0	39.0	0	0	0	0	0
692	Club soda	12	fluid ounce(s)	355	354.8	0	0	0	0	0	0	0	0
12010	Coca-Cola Classic cola soda	12	fluid ounce(s)	360	319.4	146	0	40.5	0	0	0	0	0
693	Cola	12	fluid ounce(s)	368	332.7	136	0.3	35.2	0	0.1	0	0	0
2391	Cola or pepper-type soda, low calorie with saccharin	12	fluid ounce(s)	355	354.5	0	0	0.3	0	0	0	0	0
9522	Cola soda, decaffeinated	12	fluid ounce(s)	372	333.4	153	0	39.3	0	0	0	0	0
9524	Cola, decaffeinated, low calorie with aspartame	12	fluid ounce(s)	355	354.3	4	0.4	0.5	0	0	0	0	0
1415	Cola, low calorie with aspartame	12	fluid ounce(s)	355	353.6	7	0.4	1.0	0	0.1	0	0	0
1412	Cream soda	12	fluid ounce(s)	371	321.5	189	0	49.3	0	0	0	0	0
31899	Diet 7 Up	12	fluid ounce(s)	360	—	0	0	0	0	0	0	0	0
12031	Diet Coke cola soda	12	fluid ounce(s)	360	—	2	0	0.2	0	0	0	0	0
29392	Diet Mountain Dew soda	12	fluid ounce(s)	360	—	0	0	0	0	0	0	0	0
29389	Diet Pepsi cola soda	12	fluid ounce(s)	360	—	0	0	0	0	0	0	0	0
12034	Diet Sprite soda	12	fluid ounce(s)	360	—	4	0	0	0	0	0	0	0
695	Ginger ale	12	fluid ounce(s)	366	333.9	124	0	32.1	0	0	0	0	0
694	Grape soda	12	fluid ounce(s)	372	330.3	160	0	41.7	0	0	0	0	0
1876	Lemon lime soda	12	fluid ounce(s)	368	330.8	147	0.2	37.4	0	0.1	0	0	0
29391	Mountain Dew soda	12	fluid ounce(s)	360	314.0	170	0	46.0	0	0	0	0	0
3145	Orange soda	12	fluid ounce(s)	372	325.9	179	0	45.8	0	0	0	0	0
1414	Pepper-type soda	12	fluid ounce(s)	368	329.3	151	0	38.3	0	0.4	0.3	0	0
29388	Pepsi regular cola soda	12	fluid ounce(s)	360	318.9	150	0	41.0	0	0	0	0	0
696	Root beer	12	fluid ounce(s)	370	330.0	152	0	39.2	0	0	0	0	0
12044	Sprite soda	12	fluid ounce(s)	360	321.0	144	0	39.0	0	0	0	0	0
	Coffee												
731	Brewed	8	fluid ounce(s)	237	235.6	2	0.3	0	0	0	0	0	0
9520	Brewed, decaffeinated	8	fluid ounce(s)	237	234.3	5	0.3	1.0	0	0	0	0	0
16882	Cappuccino	8	fluid ounce(s)	240	224.8	79	4.1	5.8	0.2	4.9	2.3	1.0	0.2
16883	Cappuccino, decaffeinated	8	fluid ounce(s)	240	224.8	79	4.1	5.8	0.2	4.9	2.3	1.0	0.2
16880	Espresso	8	fluid ounce(s)	237	231.8	21	0	3.6	0	0.4	0.2	0	0.2
16881	Espresso, decaffeinated	8	fluid ounce(s)	237	231.8	21	0	3.6	0	0.4	0.2	0	0.2
732	Instant, prepared	8	fluid ounce(s)	239	236.5	5	0.2	0.8	0	0	0	0	0
	Fruit drinks												
29357	Crystal Light sugar-free lemonade drink	8	fluid ounce(s)	240	—	5	0	0	0	0	0	0	0
6012	Fruit punch drink with added vitamin C, canned	8	fluid ounce(s)	248	218.2	117	0	29.7	0.5	0	0	0	0
31143	Gatorade Thirst Quencher, all flavors	8	fluid ounce(s)	240	—	50	0	14.0	0	0	0	0	0
260	Grape drink, canned	8	fluid ounce(s)	250	210.5	153	0	39.4	0	0	0	0	0

Chol (mg)	Calc (mg)	Iron (mg)	Magn (mg)	Pota (mg)	Sodi (mg)	Zinc (mg)	Vit A (μg)	Thia (mg)	Vit E (mg α)	Ribo (mg)	Niac (mg)	Vit B_6 (mg)	Fola (μg DFE)	Vit C (mg)	Vit B_{12} (μg)	Sele (μg)
0	14	0.10	17.7	74.3	14.2	0.03	0	0.01	0.00	0.05	1.38	0.12	21.2	0	0.1	1.4
0	18	0.10	21.3	88.8	9.0	0.07	0	0.02	—	0.09	1.60	0.17	21.3	0	0.1	4.0
0	18	0.14	17.8	63.9	9.0	0.10	0	0.03	—	0.10	1.39	0.12	14.6	0	0	4.0
0	0	0.01	0	0.6	0.3	0.01	0	0.00	—	0.00	0.00	0.00	0	0	0	0
0	0	0.01	0	0.6	0.3	0.01	0	0.00	0.00	0.00	0.00	0.00	0	0	0	0
0	0	0.01	0	0.6	0.3	0.01	0	0.00	0.00	0.00	0.00	0.00	0	0	0	0
0	0	0.01	0	0.6	0.3	0.01	0	0.00	0.00	0.00	0.00	0.00	0	0	0	0
0	0	0.01	0	0.6	0.3	0.01	0	0.00	—	0.00	0.00	0.00	0	0	0	0
0	0	0.02	1.0	10.4	2.8	0.01	0	0.00	0.00	0.00	0.05	0.00	0	0	0	0.1
0	0	0.02	1.0	10.4	2.8	0.01	0	0.00	—	0.00	0.05	0.00	0	0	0	0.1
0	0	0.02	0.6	4.5	2.1	0.01	0	0.00	0.00	0.00	0.02	0.00	0	0	0	0.1
0	12	1.43	16.2	170.6	15.0	0.14	0	0.01	0.00	0.04	0.11	0.05	1.5	0	0	—
0	—	—	—	—	—	—	—	—	—	—	—	—	—	—	0	—
0	12	0.34	13.2	135.4	13.2	0.10	0	0.01	0.00	0.01	0.30	0.00	0	0	0	0.7
0	13	0.39	14.8	104.7	7.4	0.18	0	0.01	0.00	0.01	0.15	0.06	1.5	0	0	0.1
0	18	0.75	15.0	129.0	24.0	0.18	—	0.01	0.03	0.03	0.13	0.03	3.0	5.4	0	0.6
0	—	—	—	0.6	75.0	—	—	—	—	—	—	—	—	—	—	—
0	18	0.03	3.5	7.1	74.6	0.35	0	0.00	0.00	0.00	0.00	0.00	0	0	0	0
0	—	—	—	0	49.5	—	0	—	—	—	—	—	—	0	—	—
0	7	0.41	0	7.4	14.7	0.06	0	0.00	0.00	0.00	0.00	0.00	0	0	0	0.4
0	14	0.06	3.5	14.2	56.8	0.11	0	0.00	0.00	0.00	0.00	0.00	0	0	0	0.3
0	7	0.08	0	11.2	14.9	0.03	0	0.00	0.00	0.00	0.00	0.00	0	0	0	0.4
0	11	0.06	0	24.9	14.2	0.03	0	0.02	0.00	0.08	0.00	0.00	0	0	0	0.3
0	11	0.39	3.5	28.4	28.4	0.03	0	0.02	0.00	0.08	0.00	0.00	0	0	0	0
0	19	0.18	3.7	3.7	44.5	0.26	0	0.00	0.00	0.00	0.00	0.00	0	0	0	0
0	—	—	—	77.0	45.0	—	—	—	—	—	—	—	—	—	—	—
0	—	—	—	18.0	42.0	—	0	—	—	—	—	—	—	0	—	—
0	—	—	—	70.0	35.0	—	—	—	—	—	—	—	—	—	—	—
0	—	—	—	30.0	35.0	—	—	—	—	—	—	—	—	—	—	—
0	—	—	—	109.5	36.0	—	0	—	—	—	—	—	—	0	—	—
0	11	0.65	3.7	3.7	25.6	0.18	0	0.00	0.00	0.00	0.00	0.00	0	0	0	0.4
0	11	0.29	3.7	3.7	55.8	0.26	0	0.00	—	0.00	0.00	0.00	0	0	0	0
0	7	0.41	3.7	3.7	33.2	0.14	0	0.00	0.00	0.00	0.05	0.00	0	0	0	0
0	—	—	—	0	70.0	—	—	—	—	—	—	—	—	—	—	—
0	19	0.21	3.7	7.4	44.6	0.36	0	0.00	—	0.00	0.00	0.00	0	0	0	0
0	11	0.14	0	3.7	36.8	0.14	0	0.00	—	0.00	0.00	0.00	0	0	0	0.4
0	—	—	—	0	35.0	—	—	—	—	—	—	—	—	—	—	—
0	18	0.18	3.7	3.7	48.0	0.26	0	0.00	0.00	0.00	0.00	0.00	0	0	0	0.4
0	—	—	—	0	70.5	—	0	—	—	—	—	—	—	0	—	—
0	5	0.02	7.1	116.1	4.7	0.04	0	0.03	0.02	0.18	0.45	0.00	4.7	0	0	0
0	7	0.14	11.8	108.9	4.7	0.00	0	0.00	0.00	0.03	0.66	0.00	0	0	0	0.5
12	144	0.19	14.4	232.8	50.4	0.50	33.6	0.04	0.09	0.27	0.13	0.04	7.2	0	0.4	4.6
12	144	0.19	14.4	232.8	50.4	0.50	33.6	0.04	0.09	0.27	0.13	0.04	7.2	0	0.4	4.6
0	5	0.30	189.6	272.6	33.2	0.11	0	0.00	0.04	0.42	12.34	0.00	2.4	0.5	0	0
0	5	0.30	189.6	272.6	33.2	0.11	0	0.00	0.04	0.42	12.34	0.00	2.4	0.5	0	0
0	10	0.09	9.5	71.6	9.5	0.01	0	0.00	0.00	0.00	0.56	0.00	0	0	0	0.2
0	0	0.00	—	160.0	40.0	—	0	—	—	—	—	—	—	0	—	—
0	20	0.22	7.4	62.0	94.2	0.02	5.0	0.05	0.04	0.05	0.05	0.02	9.9	89.3	0	0.5
0	0	0.00	—	30.0	110.0	—	0	—	—	—	—	—	—	0	—	—
0	130	0.17	2.5	30.0	40.0	0.30	0	0.00	0.00	0.01	0.02	0.01	0	78.5	0	0.3

DA+ Code	Food Description	QTY	Measure	Wt (g)	H₂O (g)	Ener (cal)	Prot (g)	Carb (g)	Fiber (g)	Fat (g)	Fat Breakdown (g) Sat	Mono	Poly
BEVERAGES—continued													
17372	Kool-Aid (lemonade/punch/fruit drink)	8	fluid ounce(s)	248	220.0	108	0.1	27.8	0.2	0	0	0	0
17225	Kool-Aid sugar free, low calorie tropical punch drink mix, prepared	8	fluid ounce(s)	240	—	5	0	0	0	0	0	0	0
266	Lemonade, prepared from frozen concentrate	8	fluid ounce(s)	248	221.6	99	0.2	25.8	0	0.1	0	0	0
268	Limeade, prepared from frozen concentrate	8	fluid ounce(s)	247	212.6	128	0	34.1	0	0	0	0	0
14266	Odwalla strawberry C monster smoothie blend	8	fluid ounce(s)	240	—	160	2.0	38.0	0	0	0	0	0
10080	Odwalla strawberry lemonade quencher	8	fluid ounce(s)	240	—	110	0	28.0	0	0	0	0	0
10099	Snapple fruit punch fruit drink	8	fluid ounce(s)	240	—	110	0	29.0	0	0	0	0	0
10096	Snapple kiwi strawberry fruit drink	8	fluid ounce(s)	240	211.2	110	0	28.0	0	0	0	0	0
	Slim Fast ready-to-drink shake												
16054	French vanilla ready to drink shake	11	fluid ounce(s)	325	—	220	10.0	40.0	5.0	2.5	0.5	1.5	0.5
40447	Optima rich chocolate royal ready-to-drink shake	11	fluid ounce(s)	330	—	180	10.0	24.0	5.0	5.0	1.0	3.5	0.5
16055	Strawberries n cream ready to drink shake	11	fluid ounce(s)	325	—	220	10.0	40.0	5.0	2.5	0.5	1.5	0.5
	Tea												
33179	Decaffeinated, prepared	8	fluid ounce(s)	237	236.3	2	0	0.7	0	0	0	0	0
1877	Herbal, prepared	8	fluid ounce(s)	237	236.1	2	0	0.5	0	0	0	0	0
735	Instant tea mix, lemon flavored with sugar, prepared	8	fluid ounce(s)	259	236.2	91	0	22.3	0.3	0.2	0	0	0
734	Instant tea mix, unsweetened, prepared	8	fluid ounce(s)	237	236.1	2	0.1	0.4	0	0	0	0	0
733	Tea, prepared	8	fluid ounce(s)	237	236.3	2	0	0.7	0	0	0	0	0
	Water												
1413	Mineral water, carbonated	8	fluid ounce(s)	237	236.8	0	0	0	0	0	0	0	0
33183	Poland spring water, bottled	8	fluid ounce(s)	237	237.0	0	0	0	0	0	0	0	0
1821	Tap water	8	fluid ounce(s)	237	236.8	0	0	0	0	0	0	0	0
1879	Tonic water	8	fluid ounce(s)	244	222.3	83	0	21.5	0	0	0	0	0
FATS AND OILS													
	Butter												
104	Butter	1	tablespoon(s)	14	2.3	102	0.1	0	0	11.5	7.3	3.0	0.4
2522	Butter Buds, dry butter substitute	1	teaspoon(s)	2	—	5	0	2.0	0	0	0	0	0
921	Unsalted	1	tablespoon(s)	14	2.5	102	0.1	0	0	11.5	7.3	3.0	0.4
107	Whipped	1	tablespoon(s)	9	1.5	67	0.1	0	0	7.6	4.7	2.2	0.3
944	Whipped, unsalted	1	tablespoon(s)	11	2.0	82	0.1	0	0	9.2	5.9	2.4	0.3
	Fats, cooking												
2671	Beef tallow, semisolid	1	tablespoon(s)	13	0	115	0	0	0	12.8	6.4	5.4	0.5
922	Chicken fat	1	tablespoon(s)	13	0	115	0	0	0	12.8	3.8	5.7	2.7
5454	Household shortening with vegetable oil	1	tablespoon(s)	13	0	115	0	0	0	13.0	3.4	5.5	2.7
111	Lard	1	tablespoon(s)	13	0	115	0	0	0	12.8	5.0	5.8	1.4
	Margarine												
114	Margarine	1	tablespoon(s)	14	2.3	101	0	0.1	0	11.4	2.1	5.5	3.4
5439	Soft	1	tablespoon(s)	14	2.3	103	0.1	0.1	0	11.6	1.7	4.4	2.1
32329	Soft, unsalted, with hydrogenated soybean and cottonseed oils	1	tablespoon(s)	14	2.5	101	0.1	0.1	0	11.3	2.0	5.4	3.5
928	Unsalted	1	tablespoon(s)	14	2.6	101	0.1	0.1	0	11.3	2.1	5.2	3.5
119	Whipped	1	tablespoon(s)	9	1.5	64	0.1	0.1	0	7.2	1.2	3.2	2.5
	Spreads												
54657	I Can't Believe It's Not Butter!, tub, soya oil (non-hydrogenated)	1	tablespoon(s)	14	2.3	103	0.1	0.1	0	11.6	2.8	2.0	5.1
2708	Mayonnaise with soybean and safflower oils	1	tablespoon(s)	14	2.1	99	0.2	0.4	0	11.0	1.2	1.8	7.6
16157	Promise vegetable oil spread, stick	1	tablespoon(s)	14	4.2	90	0	0	0	10.0	2.5	2.0	4.0

Chol (mg)	Calc (mg)	Iron (mg)	Magn (mg)	Pota (mg)	Sodi (mg)	Zinc (mg)	Vit A (μg)	Thia (mg)	Vit E (mg α)	Ribo (mg)	Niac (mg)	Vit B_6 (mg)	Fola (μg DFE)	Vit C (mg)	Vit B_{12} (μg)	Sele (μg)
0	14	0.45	5.0	49.6	31.0	0.19	—	0.03	—	0.05	0.04	0.01	4.3	41.6	0	1.0
0	0	0.00	—	10.1	10.1	—	0	—	—	—	—	—	—	6.0	—	—
0	10	0.39	5.0	37.2	9.9	0.05	0	0.01	0.02	0.05	0.04	0.01	2.5	9.7	0	0.2
0	5	0.00	4.9	24.7	7.4	0.02	0	0.01	0.00	0.01	0.02	0.01	2.5	7.7	0	0.2
0	20	0.72	—	0	20.0	—	0	—	—	—	—	—	—	600.0	0	—
0	0	0.00	—	70.0	10.0	—	0	—	—	—	—	—	—	54.0	0	—
0	0	0.00	—	20.0	10.0	—	0	—	—	—	—	—	—	0	0	—
0	0	0.00	—	40.0	10.0	—	0	—	—	—	—	—	—	0	0	—
5	400	2.70	140.0	600.0	220.0	2.25	—	0.52	—	0.59	7.00	0.70	120.0	60.0	2.1	17.5
5	1000	2.70	140.0	600.0	220.0	2.25	—	0.52	—	0.59	7.00	0.70	120.0	30.0	2.1	17.5
5	400	2.70	140.0	600.0	220.0	2.25	—	0.52	—	0.59	7.00	0.70	120.0	60.0	2.1	17.5
0	0	0.04	7.1	87.7	7.1	0.04	0	0.00	0.00	0.03	0.00	0.00	11.9	0	0	0
0	5	0.18	2.4	21.3	2.4	0.09	0	0.02	0.00	0.01	0.00	0.00	2.4	0	0	0
0	5	0.05	2.6	38.9	5.2	0.02	0	0.00	0.00	0.00	0.02	0.00	0	0	0	0.3
0	7	0.02	4.7	42.7	9.5	0.02	0	0.00	0.00	0.01	0.07	0.00	0	0	0	0
0	0	0.04	7.1	87.7	7.1	0.04	0	0.00	0.00	0.03	0.00	0.00	11.9	0	0	0
0	33	0.00	0	0	2.4	0.00	0	0.00	—	0.00	0.00	0.00	0	0	0	0
0	2	0.02	2.4	0	2.4	0.00	0	0.00	—	0.00	0.00	0.00	0	0	0	0
0	7	0.00	2.4	2.4	7.1	0.00	0	0.00	0.00	0.00	0.00	0.00	0	0	0	0
0	2	0.02	0	0	29.3	0.24	0	0.00	0.00	0.00	0.00	0.00	0	0	0	0
31	3	0.00	0.3	3.4	81.8	0.01	97.1	0.00	0.32	0.01	0.01	0.00	0.4	0	0	0.1
0	0	0.00	0	1.6	120.0	0.00	0	0.00	0.00	0.00	0.00	0.00	0	0	0	—
31	3	0.00	0.3	3.4	1.6	0.01	97.1	0.00	0.32	0.01	0.01	0.00	0.4	0	0	0.1
21	2	0.01	0.2	2.4	77.7	0.01	64.3	0.00	0.21	0.00	0.00	0.00	0.3	0	0	0.1
25	3	0.00	0.2	2.7	1.3	0.01	78.0	0.00	0.26	0.00	0.00	0.00	0.3	0	0	0.1
14	0	0.00	0	0	0	0.00	0	0.00	0.34	0.00	0.00	0.00	0	0	0	0
11	0	0.00	0	0	0	0.00	0	0.00	0.34	0.00	0.00	0.00	0	0	0	0
0	0	0.00	0	0	0	0.00	0	0.00	—	0.00	0.00	0.00	0	0	0	—
12	0	0.00	0	0	0	0.01	0	0.00	0.07	0.00	0.00	0.00	0	0	0	0
0	4	0.01	0.4	5.9	133.0	0.00	115.5	0.00	1.26	0.01	0.00	0.00	0.1	0	0	0
0	4	0.00	0.3	5.5	155.4	0.00	142.7	0.00	1.00	0.00	0.00	0.00	0.1	0	0	0
0	4	0.00	0.3	5.4	3.9	0.00	103.1	0.00	0.98	0.00	0.00	0.00	0.1	0	0	0
0	2	0.00	0.3	3.5	0.3	0.00	115.5	0.00	1.80	0.00	0.00	0.00	0.1	0	0	0
0	2	0.00	0.2	3.4	97.1	0.00	73.7	0.00	0.45	0.00	0.00	0.00	0.1	0	0	0
0	4	0.00	0.3	5.5	155.3	0.00	142.6	0.00	0.72	0.00	0.00	0.00	0.1	0	0	0
8	2	0.06	0.1	4.7	78.4	0.01	11.6	0.00	3.03	0.00	0.00	0.08	1.1	0	0	0.2
0	10	0.18	—	8.7	90.0	—	—	0.00	—	0.00	0.00	—	—	0.6	—	

DA+ Code	Food Description	QTY	Measure	Wt (g)	H₂O (g)	Ener (cal)	Prot (g)	Carb (g)	Fiber (g)	Fat (g)	Fat Breakdown (g) Sat	Mono	Poly
FATS AND OILS—continued													
	Oils												
2681	Canola	1	tablespoon(s)	14	0	120	0	0	0	13.6	1.0	8.6	3.8
120	Corn	1	tablespoon(s)	14	0	120	0	0	0	13.6	1.8	3.8	7.4
122	Olive	1	tablespoon(s)	14	0	119	0	0	0	13.5	1.9	9.9	1.4
124	Peanut	1	tablespoon(s)	14	0	119	0	0	0	13.5	2.3	6.2	4.3
2693	Safflower	1	tablespoon(s)	14	0	120	0	0	0	13.6	0.8	10.2	2.0
923	Sesame	1	tablespoon(s)	14	0	120	0	0	0	13.6	1.9	5.4	5.7
128	Soybean, hydrogenated	1	tablespoon(s)	14	0	120	0	0	0	13.6	2.0	5.8	5.1
130	Soybean, with soybean and cottonseed oil	1	tablespoon(s)	14	0	120	0	0	0	13.6	2.4	4.0	6.5
2700	Sunflower	1	tablespoon(s)	14	0	120	0	0	0	13.6	1.8	6.3	5.0
357	**Pam original no stick cooking spray**	1	serving(s)	0	0.2	0	0	0	0	0	0	0	0
	Salad dressing												
132	Blue cheese	2	tablespoon(s)	30	9.7	151	1.4	2.2	0	15.7	3.0	3.7	8.3
133	Blue cheese, low calorie	2	tablespoon(s)	32	25.4	32	1.6	0.9	0	2.3	0.8	0.6	0.8
1764	Caesar	2	tablespoon(s)	30	10.3	158	0.4	0.9	0	17.3	2.6	4.1	9.9
29654	Creamy, reduced calorie, fat-free, cholesterol-free, sour cream and/or buttermilk and oil	2	tablespoon(s)	32	23.9	34	0.4	6.4	0	0.9	0.2	0.2	0.5
29617	Creamy, reduced calorie, sour cream and/or buttermilk and oil	2	tablespoon(s)	30	22.2	48	0.5	2.1	0	4.2	0.6	1.0	2.4
134	French	2	tablespoon(s)	32	11.7	146	0.2	5.0	0	14.3	1.8	2.7	6.7
135	French, low fat	2	tablespoon(s)	32	17.4	74	0.2	9.4	0.4	4.3	0.4	1.9	1.6
136	Italian	2	tablespoon(s)	29	16.6	86	0.1	3.1	0	8.3	1.3	1.9	3.8
137	Italian, diet	2	tablespoon(s)	30	25.4	23	0.1	1.4	0	1.9	0.1	0.7	0.5
139	Mayonnaise-type	2	tablespoon(s)	29	11.7	115	0.3	7.0	0	9.8	1.4	2.6	5.3
942	Oil and vinegar	2	tablespoon(s)	32	15.2	144	0	0.8	0	16.0	2.9	4.7	7.7
1765	Ranch	2	tablespoon(s)	30	11.6	146	0.1	1.6	0	15.8	2.3	5.2	7.6
3666	Ranch, reduced calorie	2	tablespoon(s)	30	20.5	62	0.1	2.2	0	6.1	1.1	1.8	2.9
940	Russian	2	tablespoon(s)	30	11.6	107	0.5	9.3	0.7	7.8	1.2	1.8	4.4
939	Russian, low calorie	2	tablespoon(s)	32	20.8	45	0.2	8.8	0.1	1.3	0.2	0.3	0.7
941	Sesame seed	2	tablespoon(s)	30	11.8	133	0.9	2.6	0.3	13.6	1.9	3.6	7.5
142	Thousand Island	2	tablespoon(s)	32	14.9	118	0.3	4.7	0.3	11.2	1.6	2.5	5.8
143	Thousand Island, low calorie	2	tablespoon(s)	30	18.2	61	0.3	6.7	0.4	3.9	0.2	1.9	0.8
	Sandwich spreads												
138	Mayonnaise with soybean oil	1	tablespoon(s)	14	2.1	99	0.1	0.4	0	11.0	1.6	2.7	5.8
140	Mayonnaise, low calorie	1	tablespoon(s)	16	10.0	37	0	2.6	0	3.1	0.5	0.7	1.7
141	Tartar sauce	2	tablespoon(s)	28	8.7	144	0.3	4.1	0.1	14.4	2.2	3.8	7.7
SWEETS													
4799	**Butterscotch or caramel topping**	2	tablespoon(s)	41	13.1	103	0.6	27.0	0.4	0	0	0	0
	Candy												
1786	Almond Joy candy bar	1	item(s)	45	4.3	220	2.0	27.0	2.0	12.0	8.0	3.3	0.7
1785	Bit-O-Honey candy	6	item(s)	40	—	190	1.0	39.0	0	3.5	2.5	—	—
33375	Butterscotch candy	2	piece(s)	12	0.6	47	0	10.8	0	0.4	0.2	0.1	0
1701	Chewing gum, stick	1	item(s)	3	0.1	7	0	2.0	0.1	0	0	0	0
33378	Chocolate fudge with nuts, prepared	2	piece(s)	38	2.9	175	1.7	25.8	1.0	7.2	2.5	1.5	2.9
1787	Jelly beans	15	item(s)	43	2.7	159	0	39.8	0.1	0	0	0	0
1784	Kit Kat wafer bar	1	item(s)	42	0.8	210	3.0	27.0	0.5	11.0	7.0	3.5	0.3
4674	Krackel candy bar	1	item(s)	41	0.6	210	2.0	28.0	0.5	10.0	6.0	3.9	0.4
4934	Licorice	4	piece(s)	44	7.3	154	1.1	35.1	0	1.0	0	0.1	0
1780	Life Savers candy	1	item(s)	2	—	8	0	2.0	0	0	0	0	0
1790	Lollipop	1	item(s)	28	—	108	0	28.0	0	0	0	0	0
4679	M & Ms peanut chocolate candy, small bag	1	item(s)	49	0.9	250	5.0	30.0	2.0	13.0	5.0	5.4	2.1
1781	M & Ms plain chocolate candy, small bag	1	item(s)	48	0.8	240	2.0	34.0	1.0	10.0	6.0	3.3	0.3
4673	Milk chocolate bar, Symphony	1	item(s)	91	0.9	483	7.7	52.8	1.5	27.8	16.7	7.2	0.6

PAGE KEY: A-2 = Breads/Baked Goods A-8 = Cereal/Rice/Pasta A-12 = Fruit A-18 = Vegetables/Legumes A-28 = Nuts/Seeds A-30 = Vegetarian A-32 = Dairy A-40 = Eggs A-40 = Seafood A-44 = Meats A-48 = Poultry A-48 = Processed Meats A-50 = Beverages A-54 = Fats/Oils A-56 = Sweets A-58 = Spices/Condiments/Sauces A-62 = Mixed Foods/Soups/Sandwiches A-68 = Fast Food A-88 = Convenience A-90 = Baby Foods

Chol (mg)	Calc (mg)	Iron (mg)	Magn (mg)	Pota (mg)	Sodi (mg)	Zinc (mg)	Vit A (µg)	Thia (mg)	Vit E (mg α)	Ribo (mg)	Niac (mg)	Vit B_6 (mg)	Fola (µg DFE)	Vit C (mg)	Vit B_{12} (µg)	Sele (µg)
0	0	0.00	0	0	0	0.00	0	0.00	2.37	0.00	0.00	0.00	0	0	0	0
0	0	0.00	0	0	0	0.00	0	0.00	1.94	0.00	0.00	0.00	0	0	0	0
0	0	0.07	0	0.1	0.3	0.00	0	0.00	1.93	0.00	0.00	0.00	0	0	0	0
0	0	0.00	0	0	0	0.00	0	0.00	2.11	0.00	0.00	0.00	0	0	0	0
0	0	0.00	0	0	0	0.00	0	0.00	4.63	0.00	0.00	0.00	0	0	0	0
0	0	0.00	0	0	0	0.00	0	0.00	0.19	0.00	0.00	0.00	0	0	0	0
0	0	0.00	0	0	0	0.00	0	0.00	1.10	0.00	0.00	0.00	0	0	0	0
0	0	0.00	0	0	0	0.00	0	0.00	1.64	0.00	0.00	0.00	0	0	0	0
0	0	0.00	0	0	0	0.00	0	0.00	5.58	0.00	0.00	0.00	0	0	0	0
0	0	0.00	0	0.3	1.5	0.01	0.1	0.00	0.00	0.00	0.00	0.00	0	0	0	0
5	24	0.06	0	11.1	328.2	0.08	20.1	0.00	1.80	0.03	0.03	0.01	7.8	0.6	0.1	0.3
0	28	0.16	2.2	1.6	384.0	0.08	—	0.01	0.08	0.03	0.01	0.01	1.0	0.1	0.1	0.5
1	7	0.05	0.6	8.7	323.4	0.03	0.6	0.00	1.56	0.00	0.01	0.00	0.9	0	0	0.5
0	12	0.08	1.6	42.6	320.0	0.05	0.3	0.00	0.21	0.01	0.01	0.01	1.9	0	0	0.5
0	2	0.03	0.6	10.8	306.9	0.01	—	0.00	0.71	0.00	0.01	0.01	0	0.1	0	0.5
0	8	0.25	1.6	21.4	267.5	0.09	7.4	0.01	1.60	0.01	0.06	0.00	0	0	0	0
0	4	0.27	2.6	34.2	257.3	0.06	8.6	0.01	0.09	0.01	0.14	0.01	0.6	0	0	0.5
0	2	0.18	0.9	14.1	486.3	0.03	0.6	0.00	1.47	0.01	0.00	0.01	0	0	0	0.6
2	3	0.19	1.2	25.5	409.8	0.05	0.3	0.00	0.06	0.00	0.00	0.02	0	0	0	2.4
8	4	0.05	0.6	2.6	209.0	0.05	6.2	0.00	0.60	0.01	0.00	0.01	1.8	0	0.1	0.5
0	0	0.00	0	2.6	0.3	0.00	0	0.00	1.46	0.00	0.00	0.00	0	0	0	0.5
1	4	0.03	1.2	8.4	354.0	0.01	5.4	0.00	1.84	0.01	0.00	0.00	0.3	0.1	0	0.1
0	5	0.01	1.5	8.4	413.7	0.01	0.9	0.00	0.72	0.01	0.00	0.00	0.3	0.1	0	0.1
0	6	0.20	3.0	51.9	282.3	0.06	13.2	0.01	0.98	0.01	0.16	0.02	1.5	1.4	0	0.5
2	6	0.18	0	50.2	277.8	0.02	0.6	0.00	0.12	0.00	0.00	0.00	1.0	1.9	0	0.5
0	6	0.18	0	47.1	300.0	0.02	0.6	0.00	1.50	0.00	0.00	0.00	0	0	0	0.5
8	5	0.37	2.6	34.2	276.2	0.08	4.5	0.46	1.28	0.01	0.13	0.00	0	0	0	0.5
0	5	0.27	2.1	60.6	249.3	0.05	4.8	0.01	0.30	0.01	0.13	0.00	0	0	0	0
5	1	0.03	0.1	1.7	78.4	0.02	11.2	0.01	0.72	0.01	0.00	0.08	0.7	0	0	0.2
4	0	0.00	0	1.6	79.5	0.01	0	0.00	0.32	0.00	0.00	0.00	0	0	0	0.3
8	6	0.20	0.8	10.1	191.5	0.05	20.2	0.00	0.97	0.00	0.01	0.07	2.0	0.1	0.1	0.5
0	22	0.08	2.9	34.4	143.1	0.07	11.1	0.01	—	0.03	0.01	0.01	0.8	0.1	0	0
0	18	0.33	30.3	126.5	65.0	0.36	0	0.01	—	0.06	0.21	—	—	0	—	—
0	20	0.00	—	—	150.0	—	0	—	—	—	—	—	—	0	—	—
1	0	0.00	0	0.4	46.9	0.01	3.4	0.00	0.01	0.00	0.00	0.00	0	0	0	0.1
0	0	0.00	0	0.1	0	0.00	0	0.00	0.00	0.00	0.00	0.00	0	0	0	0
5	22	0.74	20.9	69.5	14.8	0.54	14.4	0.02	0.09	0.03	0.12	0.03	6.1	0.1	0	1.1
0	1	0.05	0.9	15.7	21.3	0.02	0	0.00	0.00	0.01	0.00	0.00	0	0	0	0.5
3	60	0.36	16.4	126.0	30.0	0.51	0	0.07	—	0.22	1.07	0.05	59.6	0	0.1	2.0
3	40	0.36	—	168.8	50.0	—	0	—	—	—	—	—	—	0	—	—
0	0	0.22	2.6	28.2	126.3	0.07	0	0.01	0.07	0.01	0.04	0.00	0	0	0	—
0	0	0.00	—	0	0	—	0	0.00	—	0.00	0.00	—	—	0	—	0
0	0	0.00	—	—	10.8	—	0	0.00	—	0.00	0.00	—	—	0	—	1.0
5	40	0.36	36.5	170.6	25.0	1.13	14.8	0.03	—	0.06	1.60	0.04	17.3	0.6	0.1	1.9
5	40	0.36	19.6	127.4	30.0	0.46	14.8	0.02	—	0.06	0.10	0.01	2.9	0.6	0.1	1.4
22	228	0.82	61.0	398.6	91.9	1.00	0	0.06	—	0.25	0.14	0.10	10.9	2.0	0.4	—

Table A–1 Table of Food Composition (*continued*) (Computer code is for Cengage Diet Analysis Plus program)

DA+ Code	Food Description	QTY	Measure	Wt (g)	H₂O (g)	Ener (cal)	Prot (g)	Carb (g)	Fiber (g)	Fat (g)	Fat Breakdown (g) Sat	Mono	Poly
SWEETS—continued													
1783	Milky Way bar	1	item(s)	58	3.7	270	2.0	41.0	1.0	10.0	5.0	3.5	0.3
1788	Peanut brittle	1½	ounce(s)	43	0.3	207	3.2	30.3	1.1	8.1	1.8	3.4	1.9
1789	Reese's peanut butter cups	2	piece(s)	51	0.8	280	6.0	19.0	2.0	15.5	6.0	7.2	2.7
4689	Reese's pieces candy, small bag	1	item(s)	43	1.1	220	5.0	26.0	1.0	11.0	7.0	0.9	0.4
33399	Semisweet chocolate candy, made with butter	½	ounce(s)	14	0.1	68	0.6	9.0	0.8	4.2	2.5	1.4	0.1
1782	Snickers bar	1	item(s)	59	3.2	280	4.0	35.0	1.0	14.0	5.0	6.1	2.9
4694	Special dark chocolate bar	1	item(s)	41	0.4	220	2.0	25.0	3.0	12.0	8.0	4.6	0.4
4695	Starburst fruit chews, original fruits	1	package(s)	59	3.9	240	0	48.0	0	5.0	1.0	2.1	1.8
4698	Taffy	3	piece(s)	45	2.2	179	0	41.2	0	1.5	0.9	0.4	0.1
4699	Three Musketeers bar	1	item(s)	60	3.5	260	2.0	46.0	1.0	8.0	4.5	2.6	0.3
4702	Twix caramel cookie bars	2	item(s)	58	2.4	280	3.0	37.0	1.0	14.0	5.0	7.7	0.5
4705	York peppermint pattie	1	item(s)	39	3.9	160	0.5	32.0	0.5	3.0	1.5	1.2	0.1
	Frosting, icing												
4760	Chocolate frosting, ready to eat	2	tablespoon(s)	31	5.2	122	0.3	19.4	0.3	5.4	1.7	2.8	0.6
4771	Creamy vanilla frosting, ready to eat	2	tablespoon(s)	28	4.2	117	0	19.0	0	4.5	0.8	1.4	2.2
17291	Dec-A-Cake variety pack candy decoration	1	teaspoon(s)	4	—	15	0	3.0	0	0.5	0	—	—
536	White icing	2	tablespoon(s)	40	3.6	162	0.1	31.8	0	4.2	0.8	2.0	1.2
	Gelatin												
13697	Gelatin snack, all flavors	1	item(s)	99	96.8	70	1.0	17.0	0	0	0	0	0
2616	Sugar free, low calorie mixed fruit gelatin mix, prepared	½	cup(s)	121	—	10	1.0	0	0	0	0	0	0
548	**Honey**	1	tablespoon(s)	21	3.6	64	0.1	17.3	0	0	0	0	0
	Jams, jellies												
550	Jam or preserves	1	tablespoon(s)	20	6.1	56	0.1	13.8	0.2	0	0	0	0
42199	Jams, preserves, dietetic, all flavors, w/sodium saccharin	1	tablespoon(s)	14	6.4	18	0	7.5	0.4	0	0	0	0
552	Jelly	1	tablespoon(s)	21	6.3	56	0	14.7	0.2	0	0	0	0
545	**Marshmallows**	4	item(s)	29	4.7	92	0.5	23.4	0	0.1	0	0	0
4800	**Marshmallow cream topping**	2	tablespoon(s)	40	7.9	129	0.3	31.6	0	0.1	0	0	0
555	**Molasses**	1	tablespoon(s)	20	4.4	58	0	14.9	0	0	0	0	0
4780	**Popsicle or ice pop**	1	item(s)	59	47.5	47	0	11.3	0	0.1	0	0	0
	Sugar												
559	Brown sugar, packed	1	teaspoon(s)	5	0.1	17	0	4.5	0	0	0	0	0
563	Powdered sugar, sifted	⅓	cup(s)	33	0.1	130	0	33.2	0	0	0	0	0
561	White granulated sugar	1	teaspoon(s)	4	0	16	0	4.2	0	0	0	0	0
	Sugar substitute												
1760	Equal sweetener, packet size	1	item(s)	1	—	0	0	0.9	0	0	0	0	0
13029	Splenda granular no calorie sweetener	1	teaspoon(s)	1	—	0	0	0.5	0	0	0	0	0
1759	Sweet N Low sugar substitute, packet	1	item(s)	1	0.1	4	0	0.5	0	0	0	0	0
	Syrup												
3148	Chocolate syrup	2	tablespoon(s)	38	11.6	105	0.8	24.4	1.0	0.4	0.2	0.1	0
29676	Maple syrup	¼	cup(s)	80	25.7	209	0	53.7	0	0.2	0	0.1	0.1
4795	Pancake syrup	¼	cup(s)	80	30.4	187	0	49.2	0	0	0	0	0
SPICES, CONDIMENTS, SAUCES													
	Spices												
807	Allspice, ground	1	teaspoon(s)	2	0.2	5	0.1	1.4	0.4	0.2	0	0	0
1171	Anise seeds	1	teaspoon(s)	2	0.2	7	0.4	1.1	0.3	0.3	0	0.2	0.1
729	Bakers' yeast, active	1	teaspoon(s)	4	0.3	12	1.5	1.5	0.8	0.2	0	0.1	0
683	Baking powder, double acting with phosphate	1	teaspoon(s)	5	0.2	2	0	1.1	0	0	0	0	0
1611	Baking soda	1	teaspoon(s)	5	0	0	0	0	0	0	0	0	0
8552	Basil	1	teaspoon(s)	1	0.8	0	0	0	0	0	0	0	0
34959	Basil, fresh	1	piece(s)	1	0.5	0	0	0	0	0	0	0	0

PAGE KEY: A-2 = Breads/Baked Goods A-8 = Cereal/Rice/Pasta A-12 = Fruit A-18 = Vegetables/Legumes A-28 = Nuts/Seeds A-30 = Vegetarian A-32 = Dairy A-40 = Eggs A-40 = Seafood A-44 = Meats A-48 = Poultry A-48 = Processed Meats A-50 = Beverages A-54 = Fats/Oils A-56 = Sweets A-58 = Spices/Condiments/Sauces A-62 = Mixed Foods/Soups/Sandwiches A-68 = Fast Food A-88 = Convenience A-90 = Baby Foods

Chol (mg)	Calc (mg)	Iron (mg)	Magn (mg)	Pota (mg)	Sodi (mg)	Zinc (mg)	Vit A (µg)	Thia (mg)	Vit E (mg α)	Ribo (mg)	Niac (mg)	Vit B_6 (mg)	Fola (µg DFE)	Vit C (mg)	Vit B_{12} (µg)	Sele (µg)
5	60	0.18	19.8	140.1	95.0	0.41	15.1	0.02	—	0.06	0.20	0.02	5.8	0.6	0.2	3.3
5	11	0.51	17.9	71.4	189.2	0.37	16.6	0.05	1.08	0.01	1.12	0.03	19.6	0	0	1.1
3	40	0.72	45.4	217.4	180.0	0.93	0	0.12	—	0.08	2.35	0.07	28.1	0	0.1	2.3
0	20	0.00	18.9	169.9	80.0	0.32	0	0.04	—	0.06	1.22	0.03	12.0	0	0.1	0.8
3	5	0.44	16.3	51.7	1.6	0.23	0.4	0.01	—	0.01	0.06	0.01	0.4	0	0	0.5
5	40	0.36	42.3	—	140.0	1.37	15.3	0.03	—	0.06	1.60	0.05	23.5	0.6	0.1	2.7
0	0	1.80	45.5	136.0	50.0	0.59	0	0.01	—	0.02	0.16	0.01	0.8	0	0	1.2
0	10	0.18	0.6	1.2	0	0.00	—	0.00	—	0.00	0.00	0.00	0	30	0	0.5
4	4	0.00	0	1.4	23.4	0.09	12.2	0.01	0.04	0.01	0.00	0.00	0	0	0	0.3
5	20	0.36	17.5	80.3	110.0	0.33	14.5	0.01	—	0.03	0.20	0.01	0	0.6	0.1	1.5
5	40	0.36	18.5	116.8	115.0	0.45	15.0	0.09	—	0.13	0.69	0.01	13.9	0.6	0.1	1.2
0	0	0.33	23.4	66.1	10.0	0.28	0	0.01	—	0.03	0.31	0.01	1.5	0	0	—
0	2	0.44	6.4	60.0	56.1	0.09	0	0.00	0.48	0.00	0.03	0.00	0.3	0	0	0.2
0	1	0.04	0.3	9.5	51.5	0.01	0	0.00	0.43	0.08	0.06	0.00	2.2	0	0	0
0	0	0.00	—	—	15.0	—	0	—	—	—	—	—	—	0	—	—
0	4	0.01	0.4	5.6	76.4	0.01	44.4	0.00	0.32	0.01	0.00	0.00	0	0	0	0.3
0	0	0.00	—	0	40.0	—	0	—	—	—	—	—	—	0	—	—
0	0	0.00	0	0	50.0	0.00	0	0.00	0.00	0.00	0.00	0.00	0	0	0	—
0	1	0.08	0.4	10.9	0.8	0.04	0	0.00	0.00	0.01	0.02	0.01	0.4	0.1	0	0.2
0	4	0.10	0.8	15.4	6.4	0.01	0	0.00	0.02	0.02	0.01	0.00	2.2	1.8	0	0.4
0	1	0.56	0.7	9.7	0	0.01	0	0.00	0.01	0.00	0.00	0.00	1.3	0	0	0.2
0	1	0.04	1.3	11.3	6.3	0.01	0	0.00	0.00	0.01	0.01	0.00	0.4	0.2	0	0.1
0	1	0.06	0.6	1.4	23.0	0.01	0	0.00	0.00	0.00	0.02	0.00	0.3	0	0	0.5
0	1	0.08	0.8	2.0	32.0	0.01	0	0.00	0.00	0.00	0.03	0.00	0.4	0	0	0.7
0	41	0.94	48.4	292.8	7.4	0.05	0	0.01	0.00	0.00	0.18	0.13	0	0	0	3.6
0	0	0.31	0.6	8.9	4.1	0.08	0	0.00	0.00	0.00	0.00	0.00	0	0.4	0	0.1
0	4	0.03	0.4	6.1	1.3	0.00	0	0.00	0.00	0.00	0.01	0.00	0	0	0	0.1
0	0	0.01	0	0.7	0.3	0.00	0	0.00	0.00	0.00	0.00	0.00	0	0	0	0.2
0	0	0.00	0	0.1	0	0.00	0	0.00	0.00	0.00	0.00	0.00	0	0	0	0
0	0	0.00	0	0	0	0.00	0	0.00	0.00	0.00	0.00	0.00	0	0	0	0
0	0	0.00	—	—	0	—	—	0.00	—	0.00	0.00	—	—	0	0	—
0	0	0.00	—	—	0	—	0	—	0.00	—	—	—	—	0	—	—
0	5	0.79	24.4	84.0	27.0	0.27	0	0.00	0.01	0.01	0.12	0.00	0.8	0.1	0	0.5
0	54	0.96	11.2	163.2	7.2	3.32	0	0.01	0.00	0.01	0.02	0.00	0	0	0	0.5
0	2	0.02	1.6	12.0	65.6	0.06	0	0.01	0.00	0.01	0.00	0.00	0	0	0	0
0	13	0.13	2.6	19.8	1.5	0.01	0.5	0.00	—	0.00	0.05	0.00	0.7	0.7	0	0.1
0	14	0.77	3.6	30.3	0.3	0.11	0.3	0.01	—	0.01	0.06	0.01	0.2	0.4	0	0.1
0	3	0.66	3.9	80.0	2.0	0.25	0	0.09	0.00	0.21	1.59	0.06	93.6	0	0	1.0
0	339	0.51	1.8	0.2	363.1	0.00	0	0.00	0.00	0.00	0.00	0.00	0	0	0	0
0	0	0.00	0	0	1258.6	0.00	0	0.00	0.00	0.00	0.00	0.00	0	0	0	0
0	2	0.02	0.6	2.6	0	0.01	2.3	0.00	0.01	0.00	0.01	0.00	0.6	0.2	0	0
0	1	0.01	0.4	2.3	0	0.00	1.3	0.00	—	0.00	0.00	0.00	0.3	0.1	0	0

Table A–1 Table of Food Composition (*continued*)

(Computer code is for Cengage Diet Analysis Plus program)

DA+ Code	Food Description	QTY	Measure	Wt (g)	H_2O (g)	Ener (cal)	Prot (g)	Carb (g)	Fiber (g)	Fat (g)	Fat Breakdown (g)		
											Sat	Mono	Poly
SPICES, CONDIMENTS, SAUCES—continued													
808	Basil, ground	1	teaspoon(s)	1	0.1	4	0.2	0.9	0.6	0.1	0	0	0
809	Bay leaf	1	teaspoon(s)	1	0	2	0	0.5	0.2	0.1	0	0	0
11720	Betel leaves	1	ounce(s)	28	—	17	1.8	2.4	0	0	—	—	—
730	Brewers' yeast	1	teaspoon(s)	3	0.1	8	1.0	1.0	0.8	0	0	0	0
11710	Capers	1	teaspoon(s)	5	—	0	0	0	0	0	0	0	0
1172	Caraway seeds	1	teaspoon(s)	2	0.2	7	0.4	1.0	0.8	0.3	0	0.2	0.1
1173	Celery seeds	1	teaspoon(s)	2	0.1	8	0.4	0.8	0.2	0.5	0	0.3	0.1
1174	Chervil, dried	1	teaspoon(s)	1	0	1	0.1	0.3	0.1	0	0	0	0
810	Chili powder	1	teaspoon(s)	3	0.2	8	0.3	1.4	0.9	0.4	0.1	0.1	0.2
8553	Chives, chopped	1	teaspoon(s)	1	0.9	0	0	0	0	0	0	0	0
51420	Cilantro (coriander)	1	teaspoon(s)	0	0.3	0	0	0	0	0	0	0	0
811	Cinnamon, ground	1	teaspoon(s)	2	0.2	6	0.1	1.9	1.2	0	0	0	0
812	Cloves, ground	1	teaspoon(s)	2	0.1	7	0.1	1.3	0.7	0.4	0.1	0	0.1
1175	Coriander leaf, dried	1	teaspoon(s)	1	0	2	0.1	0.3	0.1	0	0	0	0
1176	Coriander seeds	1	teaspoon(s)	2	0.2	5	0.2	1.0	0.8	0.3	0	0.2	0
1706	Cornstarch	1	tablespoon(s)	8	0.7	30	0	7.3	0.1	0	0	0	0
1177	Cumin seeds	1	teaspoon(s)	2	0.2	8	0.4	0.9	0.2	0.5	0	0.3	0.1
11729	Cumin, ground	1	teaspoon(s)	5	—	11	0.4	0.8	0.8	0.4	—	—	—
1178	Curry powder	1	teaspoon(s)	2	0.2	7	0.3	1.2	0.7	0.3	0	0.1	0.1
1179	Dill seeds	1	teaspoon(s)	2	0.2	6	0.3	1.2	0.4	0.3	0	0.2	0
1180	Dill weed, dried	1	teaspoon(s)	1	0.1	3	0.2	0.6	0.1	0	0	0	0
34949	Dill weed, fresh	5	piece(s)	1	0.9	0	0	0.1	0	0	0	0	0
4949	Fennel leaves, fresh	1	teaspoon(s)	1	0.9	0	0	0.1	0	0	—	—	—
1181	Fennel seeds	1	teaspoon(s)	2	0.2	7	0.3	1.0	0.8	0.3	0	0.2	0
1182	Fenugreek seeds	1	teaspoon(s)	4	0.3	12	0.9	2.2	0.9	0.2	0.1	—	—
11733	Garam masala, powder	1	ounce(s)	28	—	107	4.4	12.8	0	4.3	—	—	—
1067	Garlic clove	1	item(s)	3	1.8	4	0.2	1.0	0.1	0	0	0	0
813	Garlic powder	1	teaspoon(s)	3	0.2	9	0.5	2.0	0.3	0	0	0	0
1068	Ginger root	2	teaspoon(s)	4	3.1	3	0.1	0.7	0.1	0	0	0	0
1183	Ginger, ground	1	teaspoon(s)	2	0.2	6	0.2	1.3	0.2	0.1	0	0	0
35497	Leeks, bulb and lower-leaf, freeze-dried	¼	cup(s)	1	0	3	0.1	0.6	0.1	0	0	0	0
1184	Mace, ground	1	teaspoon(s)	2	0.1	8	0.1	0.9	0.3	0.6	0.2	0.2	0.1
1185	Marjoram, dried	1	teaspoon(s)	1	0	2	0.1	0.4	0.2	0	0	0	0
1186	Mustard seeds, yellow	1	teaspoon(s)	3	0.2	15	0.8	1.2	0.5	0.9	0	0.7	0.2
814	Nutmeg, ground	1	teaspoon(s)	2	0.1	12	0.1	1.1	0.5	0.8	0.6	0.1	0
2747	Onion flakes, dehydrated	1	teaspoon(s)	2	0.1	6	0.1	1.4	0.2	0	0	0	0
1187	Onion powder	1	teaspoon(s)	2	0.1	7	0.2	1.7	0.1	0	0	0	0
815	Oregano, ground	1	teaspoon(s)	2	0.1	5	0.2	1.0	0.6	0.2	0	0	0.1
816	Paprika	1	teaspoon(s)	2	0.2	6	0.3	1.2	0.8	0.3	0	0	0.2
817	Parsley, dried	1	teaspoon(s)	0	0	1	0.1	0.2	0.1	0	0	0	0
818	Pepper, black	1	teaspoon(s)	2	0.2	5	0.2	1.4	0.6	0.1	0	0	0
819	Pepper, cayenne	1	teaspoon(s)	2	0.1	6	0.2	1.0	0.5	0.3	0.1	0	0.2
1188	Pepper, white	1	teaspoon(s)	2	0.3	7	0.3	1.6	0.6	0.1	0	0	0
1189	Poppy seeds	1	teaspoon(s)	3	0.2	15	0.5	0.7	0.3	1.3	0.1	0.2	0.9
1190	Poultry seasoning	1	teaspoon(s)	2	0.1	5	0.1	1.0	0.2	0.1	0	0	0
1191	Pumpkin pie spice, powder	1	teaspoon(s)	2	0.1	6	0.1	1.2	0.3	0.2	0.1	0	0
1192	Rosemary, dried	1	teaspoon(s)	1	0.1	4	0.1	0.8	0.5	0.2	0.1	0	0
11723	Rosemary, fresh	1	teaspoon(s)	1	0.5	1	0	0.1	0.1	0	0	0	0
2722	Saffron powder	1	teaspoon(s)	1	0.1	2	0.1	0.5	0	0	0	0	0
11724	Sage	1	teaspoon(s)	1	—	1	0	0.1	0	0	—	—	—
1193	Sage, ground	1	teaspoon(s)	1	0.1	2	0.1	0.4	0.3	0.1	0	0	0
30189	Salt substitute	¼	teaspoon(s)	1	—	0	0	0	0	0	0	0	0
30190	Salt substitute, seasoned	¼	teaspoon(s)	1	—	1	0	0.1	0	0	0	—	—
822	Salt, table	¼	teaspoon(s)	2	0	0	0	0	0	0	0	0	0
1194	Savory, ground	1	teaspoon(s)	1	0.1	4	0.1	1.0	0.6	0.1	0	—	—
820	Sesame seed kernels, toasted	1	teaspoon(s)	3	0.1	15	0.5	0.7	0.5	1.3	0.2	0.5	0.6
11725	Sorrel	1	teaspoon(s)	3	—	1	0.1	0.1	0	0	0	—	—
11721	Spearmint	1	teaspoon(s)	2	1.6	1	0.1	0.2	0.1	0	0	0	0
35498	Sweet green peppers, freeze-dried	¼	cup(s)	2	0	5	0.3	1.1	0.3	0	0	0	0
11726	Tamarind leaves	1	ounce(s)	28	—	33	1.6	5.2	0	0.6	—	—	—
11727	Tarragon	1	ounce(s)	28	—	14	1.0	1.8	0	0.3	—	—	—

Chol (mg)	Calc (mg)	Iron (mg)	Magn (mg)	Pota (mg)	Sodi (mg)	Zinc (mg)	Vit A (µg)	Thia (mg)	Vit E (mg α)	Ribo (mg)	Niac (mg)	Vit B6 (mg)	Fola (µg DFE)	Vit C (mg)	Vit B12 (µg)	Sele (µg)
0	30	0.58	5.9	48.1	0.5	0.08	6.6	0.00	0.10	0.00	0.09	0.03	3.8	0.9	0	0
0	5	0.25	0.7	3.2	0.1	0.02	1.9	0.00	—	0.00	0.01	0.01	1.1	0.3	0	0
0	110	2.29	—	155.9	2.0	—	—	0.04	—	0.07	0.19	—	—	0.9	0	—
0	6	0.46	6.1	50.7	3.3	0.21	0	0.41	—	0.11	1.00	0.06	104.3	0	0	0
0	—	—	—	—	105.0	—	—	—	—	—	—	—	—	—	0	—
0	14	0.34	5.4	28.4	0.4	0.11	0.4	0.01	0.05	0.01	0.07	0.01	0.2	0.4	0	0.3
0	35	0.89	8.8	28.0	3.2	0.13	0.1	0.01	0.02	0.01	0.06	0.01	0.2	0.3	0	0.2
0	8	0.19	0.8	28.4	0.5	0.05	1.8	0.00	—	0.00	0.03	0.01	1.6	0.3	0	0.2
0	7	0.37	4.4	49.8	26.3	0.07	38.6	0.01	0.75	0.02	0.20	0.09	2.6	1.7	0	0.2
0	1	0.01	0.4	3.0	0	0.01	2.2	0.00	0.00	0.00	0.01	0.00	1.1	0.6	0	0
0	0	0.01	0.1	1.7	0.2	0.00	1.1	0.00	0.01	0.00	0.00	0.00	0.2	0.1	0	0
0	23	0.19	1.4	9.9	0.2	0.04	0.3	0.00	0.05	0.00	0.03	0.00	0.1	0.1	0	0.1
0	14	0.18	5.5	23.1	5.1	0.02	0.6	0.00	0.17	0.01	0.03	0.01	2.0	1.7	0	0.1
0	7	0.25	4.2	26.8	1.3	0.02	1.8	0.01	0.01	0.01	0.06	0.00	1.6	3.4	0	0.2
0	13	0.29	5.9	22.8	0.6	0.08	0	0.00	—	0.01	0.03	—	0	0.4	0	0.5
0	0	0.03	0.2	0.2	0.7	0.01	0	0.00	0.00	0.00	0.00	0.00	0	0	0	0.2
0	20	1.39	7.7	37.5	3.5	0.10	1.3	0.01	0.07	0.01	0.09	0.01	0.2	0.2	0	0.1
0	20	—	—	43.6	4.8	—	—	—	—	—	—	—	—	—	—	—
0	10	0.59	5.1	30.9	1.0	0.08	1.0	0.01	0.44	0.01	0.06	0.02	3.1	0.2	0	0.3
0	32	0.34	5.4	24.9	0.4	0.10	0.1	0.01	—	0.01	0.05	0.01	0.2	0.4	0	0.3
0	18	0.48	4.5	33.1	2.1	0.03	2.9	0.00	—	0.00	0.02	0.01	1.5	0.5	0	—
0	2	0.06	0.6	7.4	0.6	0.01	3.9	0.00	0.01	0.00	0.01	0.00	1.5	0.9	0	—
0	1	0.02	—	4.0	0.1	—	—	0.00	—	0.00	0.01	0.00	—	0.3	0	—
0	24	0.37	7.7	33.9	1.8	0.07	0.1	0.01	—	0.01	0.12	0.01	—	0.4	0	—
0	7	1.24	7.1	28.5	2.5	0.09	0.1	0.01	—	0.01	0.06	0.02	2.1	0.1	0	0.2
0	215	9.24	93.6	411.1	27.5	1.07	—	0.09	—	0.09	0.70	—	0	0	0	—
0	5	0.05	0.8	12.0	0.5	0.03	0	0.01	0.00	0.00	0.02	0.03	0.1	0.9	0	0.4
0	2	0.07	1.6	30.8	0.7	0.07	0	0.01	0.01	0.00	0.01	0.08	0.1	0.5	0	1.1
0	1	0.02	1.7	16.6	0.5	0.01	0	0.00	0.01	0.00	0.02	0.01	0.4	0.2	0	0
0	2	0.20	3.3	24.2	0.6	0.08	0.1	0.00	0.32	0.00	0.09	0.01	0.7	0.1	0	0.7
0	3	0.06	1.3	19.2	0.3	0.01	0.1	0.01	—	0.00	0.02	0.01	2.9	0.9	0	0
0	4	0.23	2.8	7.9	1.4	0.03	0.7	0.01	—	0.01	0.02	0.00	1.3	0.4	0	0
0	12	0.49	2.1	9.1	0.5	0.02	2.4	0.00	0.01	0.00	0.02	0.01	1.6	0.3	0	0
0	17	0.32	9.8	22.5	0.2	0.18	0.1	0.01	0.09	0.01	0.26	0.01	2.5	0.1	0	4.4
0	4	0.06	4.0	7.7	0.4	0.04	0.1	0.01	0.00	0.00	0.02	0.00	1.7	0.1	0	0
0	4	0.02	1.5	27.1	0.4	0.03	0	0.01	0.00	0.00	0.01	0.02	2.8	1.3	0	0.1
0	8	0.05	2.6	19.8	1.1	0.04	0	0.01	0.01	0.00	0.01	0.02	3.5	0.3	0	0
0	24	0.66	4.1	25.0	0.2	0.06	5.2	0.01	0.28	0.01	0.09	0.01	4.1	0.8	0	0.1
0	4	0.49	3.9	49.2	0.7	0.08	55.4	0.01	0.62	0.03	0.32	0.08	2.2	1.5	0	0.1
0	4	0.29	0.7	11.4	1.4	0.01	1.5	0.00	0.02	0.00	0.02	0.00	0.5	0.4	0	0.1
0	9	0.60	4.1	26.4	0.9	0.03	0.3	0.00	0.01	0.01	0.02	0.01	0.2	0.4	0	0.1
0	3	0.14	2.7	36.3	0.5	0.04	37.5	0.01	0.53	0.01	0.15	0.04	1.9	1.4	0	0.2
0	6	0.34	2.2	1.8	0.1	0.02	0	0.00	—	0.00	0.01	0.00	0.2	0.5	0	0.1
0	41	0.26	9.3	19.6	0.6	0.28	0	0.02	0.03	0.01	0.02	0.01	1.6	0.1	0	0
0	15	0.53	3.4	10.3	0.4	0.04	2.0	0.00	0.02	0.00	0.04	0.02	2.1	0.2	0	0.1
0	12	0.33	2.3	11.3	0.9	0.04	0.2	0.00	0.01	0.00	0.03	0.01	0.9	0.4	0	0.2
0	15	0.35	2.6	11.5	0.6	0.03	1.9	0.01	—	0.01	0.01	0.02	3.7	0.7	0	0.1
0	2	0.04	0.6	4.7	0.2	0.01	1.0	0.00	—	0.00	0.01	0.00	0.8	0.2	0	—
0	1	0.07	1.8	12.1	1.0	0.01	0.2	0.00	—	0.00	0.01	0.01	0.7	0.6	0	0
0	4	—	1.1	2.7	0	0.01	—	0.00	—	—	—	—	—	—	0	—
0	12	0.19	3.0	7.5	0.1	0.03	2.1	0.01	0.05	0.00	0.04	0.01	1.9	0.2	0	0
0	7	0.00	0	603.6	0.1	—	0	—	—	—	—	—	—	0	—	—
0	0	0.00	—	476.3	0.1	—	0	—	—	—	—	—	—	0	—	—
0	0	0.01	0	0.1	581.4	0.00	0	0.00	0.00	0.00	0.00	0.00	0	0	0	0
0	30	0.53	5.3	14.7	0.3	0.06	3.6	0.01	—	—	0.05	0.02	—	0.7	0	0.1
0	3	0.21	9.2	10.8	1.0	0.27	0.1	0.03	0.01	0.01	0.15	0.00	2.6	0	0	0
0	—	—	—	—	0.1	—	—	—	—	—	—	—	—	—	—	—
0	4	0.22	1.2	8.7	0.6	0.02	3.9	0.00	—	0.00	0.01	0.00	2.0	0.3	0	—
0	2	0.16	3.0	50.7	3.1	0.03	4.5	0.01	0.06	0.01	0.11	0.03	3.7	30.4	0	0.1
0	85	1.48	20.2	—	—	—	—	0.06	—	0.02	1.16	—	—	0.9	0	—
0	48	—	14.5	128.1	2.6	0.17	—	0.04	—	—	—	—	—	0.6	0	—

											Fat Breakdown (g)		
DA+ Code	Food Description	QTY	Measure	Wt (g)	H_2O (g)	Ener (cal)	Prot (g)	Carb (g)	Fiber (g)	Fat (g)	Sat	Mono	Poly
SPICES, CONDIMENTS, SAUCES—continued													
1195	Tarragon, ground	1	teaspoon(s)	2	0.1	5	0.4	0.8	0.1	0.1	0	0	0.1
11728	Thyme, fresh	1	teaspoon(s)	1	0.5	1	0	0.2	0.1	0	0	0	0
821	Thyme, ground	1	teaspoon(s)	1	0.1	4	0.1	0.9	0.5	0.1	0	0	0
1196	Turmeric, ground	1	teaspoon(s)	2	0.3	8	0.2	1.4	0.5	0.2	0.1	0	0
11995	Wasabi	1	tablespoon(s)	14	10.7	10	0.7	2.3	0.2	0	—	—	—
	Condiments												
674	Catsup or ketchup	1	tablespoon(s)	15	10.4	15	0.3	3.8	0	0	0	0	0
703	Dill pickle	1	ounce(s)	28	26.7	3	0.2	0.7	0.3	0	0	0	0
138	Mayonnaise with soybean oil	1	tablespoon(s)	14	2.1	99	0.1	0.4	0	11.0	1.6	2.7	5.8
140	Mayonnaise, low calorie	1	tablespoon(s)	16	10.0	37	0	2.6	0	3.1	0.5	0.7	1.7
1682	Mustard, brown	1	teaspoon(s)	5	4.1	5	0.3	0.3	0	0.3	—	—	—
700	Mustard, yellow	1	teaspoon(s)	5	4.1	3	0.2	0.3	0.2	0.2	0	0.1	0
706	Sweet pickle relish	1	tablespoon(s)	15	9.3	20	0.1	5.3	0.2	0.1	0	0	0
141	Tartar sauce	2	tablespoon(s)	28	8.7	144	0.3	4.1	0.1	14.4	2.2	3.8	7.7
	Sauces												
685	Barbecue sauce	2	tablespoon(s)	31	18.9	47	0	11.3	0.2	0.1	0	0	0.1
834	Cheese sauce	¼	cup(s)	63	44.4	110	4.2	4.3	0.3	8.4	3.8	2.4	1.6
32123	Chili enchilada sauce, green	2	tablespoon(s)	57	53.0	15	0.6	3.1	0.7	0.3	0	0	0.1
32122	Chili enchilada sauce, red	2	tablespoon(s)	32	24.5	27	1.1	5.0	2.1	0.8	0.1	0	0.4
29688	Hoisin sauce	1	tablespoon(s)	16	7.1	35	0.5	7.1	0.4	0.5	0.1	0.2	0.3
1641	Horseradish sauce, prepared	1	teaspoon(s)	5	3.3	10	0.1	0.2	0	1.0	0.6	0.3	0
16670	Mole poblano sauce	½	cup(s)	133	102.7	156	5.3	11.4	2.7	11.3	2.6	5.1	3.0
29689	Oyster sauce	1	tablespoon(s)	16	12.8	8	0.2	1.7	0	0	0	0	0
1655	Pepper sauce or Tabasco	1	teaspoon(s)	5	4.8	1	0.1	0	0	0	0	0	0
347	Salsa	2	tablespoon(s)	32	28.8	9	0.5	2.0	0.5	0.1	0	0	0
52206	Soy sauce, tamari	1	tablespoon(s)	18	12.0	11	1.9	1.0	0.1	0	0	0	0
839	Sweet and sour sauce	2	tablespoon(s)	39	29.8	37	0.1	9.1	0.1	0	0	0	0
1613	Teriyaki sauce	1	tablespoon(s)	18	12.2	16	1.1	2.8	0	0	0	0	0
25294	Tomato sauce	½	cup(s)	150	132.8	63	2.2	11.9	2.6	1.8	0.2	0.4	0.9
728	White sauce, medium	¼	cup(s)	63	46.8	92	2.4	5.7	0.1	6.7	1.8	2.8	1.8
1654	Worcestershire sauce	1	teaspoon(s)	6	4.5	4	0	1.1	0	0	0	0	0
	Vinegar												
30853	Balsamic	1	tablespoon(s)	15	—	10	0	2.0	0	0	0	0	0
727	Cider	1	tablespoon(s)	15	14.0	3	0	0.1	0	0	0	0	0
1673	Distilled	1	tablespoon(s)	15	14.3	2	0	0.8	0	0	0	0	0
12948	Tarragon	1	tablespoon(s)	15	13.8	2	0	0.1	0	0	0	0	0
MIXED FOODS, SOUPS, SANDWICHES													
	Mixed dishes												
16652	Almond chicken	1	cup(s)	242	186.8	281	21.8	15.8	3.4	14.7	1.8	6.3	5.6
25224	Barbecued chicken	1	serving(s)	177	99.3	327	27.1	15.7	0.5	17.1	4.8	6.8	3.8
25227	Bean burrito	1	item(s)	149	81.8	326	16.1	33.0	5.6	14.8	8.3	4.7	0.9
9516	Beef and vegetable fajita	1	item(s)	223	143.9	397	22.4	35.3	3.1	18.0	5.9	8.0	2.5
16796	Beef or pork egg roll	2	item(s)	128	85.2	225	9.9	18.4	1.4	12.4	2.9	6.0	2.6
177	Beef stew with vegetables, prepared	1	cup(s)	245	201.0	220	16.0	15.0	3.2	11.0	4.4	4.5	0.5
30233	Beef stroganoff with noodles	1	cup(s)	256	190.1	343	19.7	22.8	1.5	19.1	7.4	5.7	4.4
16651	Cashew chicken	1	cup(s)	242	186.8	281	21.8	15.8	3.4	14.7	1.8	6.3	5.6
30274	Cheese pizza with vegetables, thin crust	2	slice(s)	140	76.6	298	12.7	35.4	2.5	12.0	4.9	4.7	1.6
30330	Cheese quesadilla	1	item(s)	54	18.3	190	7.7	15.3	1.0	10.8	5.2	3.6	1.3
215	Chicken and noodles, prepared	1	cup(s)	240	170.0	365	22.0	26.0	1.3	18.0	5.1	7.1	3.9
30239	Chicken and vegetables with broccoli, onion, bamboo shoots in soy based sauce	1	cup(s)	162	125.5	180	15.8	9.3	1.8	8.6	1.7	3.0	3.1
25093	Chicken cacciatore	1	cup(s)	244	175.7	284	29.9	5.7	1.3	15.3	4.3	6.2	3.3
28020	Chicken fried turkey steak	3	ounce(s)	492	276.2	706	77.1	68.7	3.6	12.0	3.4	2.9	3.9
218	Chicken pot pie	1	cup(s)	252	154.6	542	22.6	41.4	3.5	31.3	9.8	12.5	7.1
30240	Chicken teriyaki	1	cup(s)	244	158.3	364	51.0	15.2	0.7	7.0	1.8	2.0	1.7
25119	Chicken waldorf salad	½	cup(s)	100	67.2	179	14.0	6.8	1.0	10.8	1.8	3.1	5.2
25099	Chili con carne	¾	cup(s)	215	174.4	198	13.7	21.4	7.5	6.9	2.5	2.8	0.5
1062	Coleslaw	¾	cup(s)	90	73.4	70	1.2	11.2	1.4	2.3	0.3	0.6	1.2

Chol (mg)	Calc (mg)	Iron (mg)	Magn (mg)	Pota (mg)	Sodi (mg)	Zinc (mg)	Vit A (µg)	Thia (mg)	Vit E (mg α)	Ribo (mg)	Niac (mg)	Vit B_6 (mg)	Fola (µg DFE)	Vit C (mg)	Vit B_{12} (µg)	Sele (µg)
0	18	0.51	5.6	48.3	1.0	0.06	3.4	0.00	—	0.02	0.14	0.03	4.4	0.8	0	0.1
0	3	0.14	1.3	4.9	0.1	0.01	1.9	0.00	—	0.00	0.01	0.00	0.4	1.3	0	—
0	26	1.73	3.1	11.4	0.8	0.08	2.7	0.01	0.10	0.01	0.06	0.01	3.8	0.7	0	0.1
0	4	0.91	4.2	55.6	0.8	0.09	0	0.00	0.06	0.01	0.11	0.04	0.9	0.6	0	0.1
0	13	0.11	—	—	—	—	—	0.02	—	0.01	0.07	—	—	11.2	0	—
0	3	0.07	2.9	57.3	167.1	0.03	7.1	0.00	0.21	0.02	0.21	0.02	1.5	2.3	0	0
0	12	0.10	2.0	26.1	248.1	0.03	2.6	0.01	0.02	0.01	0.03	0.01	0.3	0.2	0	0
5	1	0.03	0.1	1.7	78.4	0.02	11.2	0.01	0.72	0.01	0.00	0.08	0.7	0	0	0.2
4	0	0.00	0	1.6	79.5	0.01	0	0.00	0.32	0.00	0.00	0.00	0	0	0	0.3
0	6	0.09	1.0	6.8	68.1	0.01	0	0.00	0.09	0.00	0.01	0.00	0.2	0.1	0	—
0	3	0.07	2.5	6.9	56.8	0.03	0.2	0.01	0.01	0.00	0.02	0.00	0.4	0.1	0	1.6
0	0	0.13	0.8	3.8	121.7	0.02	9.2	0.00	0.08	0.01	0.03	0.00	0.2	0.2	0	0
8	6	0.20	0.8	10.1	191.5	0.05	20.2	0.00	0.97	0.00	0.01	0.07	2.0	0.1	0.1	0.5
0	4	0.06	3.8	65.0	349.7	0.04	3.8	0.00	0.20	0.01	0.15	0.01	0.6	0.2	0	0.4
18	116	0.13	5.7	18.9	521.6	0.61	50.4	0.00	—	0.07	0.01	0.01	2.5	0.3	0.1	2.0
0	5	0.36	9.5	125.7	61.9	0.11	—	0.02	0.00	0.02	0.63	0.06	5.7	43.9	0	0
0	7	1.05	11.1	231.3	113.8	0.14	—	0.01	0.00	0.21	0.61	0.34	6.6	0.3	0	0.3
0	5	0.16	3.8	19.0	258.4	0.05	0	0.00	0.04	0.03	0.18	0.01	3.7	0.1	0	0.3
2	5	0.00	0.5	6.7	14.6	0.01	8.0	0.00	0.02	0.01	0.00	0.00	0.5	0.1	0	0.1
1	38	1.81	58.3	280.9	304.8	1.15	13.3	0.06	1.72	0.08	1.84	0.09	15.9	3.4	0.1	1.1
0	5	0.02	0.6	8.6	437.3	0.01	0	0.00	0.00	0.02	0.23	0.00	2.4	0	0.1	0.7
0	1	0.05	0.6	6.4	31.7	0.01	4.1	0.00	0.00	0.00	0.01	0.01	0.1	0.2	0	0
0	9	0.14	4.8	95.0	192.0	0.11	4.8	0.01	0.37	0.01	0.02	0.05	1.3	0.6	0	0.3
0	4	0.43	7.3	38.7	1018.9	0.08	0	0.01	0.00	0.03	0.72	0.04	3.3	0	0	0.1
0	5	0.20	1.2	8.2	97.5	0.01	0	0.00	—	0.01	0.11	0.03	0.2	0	0	—
0	5	0.30	11.0	40.5	689.9	0.01	0	0.01	0.00	0.01	0.22	0.01	1.4	0	0	0.2
0	23	1.24	28.9	536.8	268.6	0.36	—	0.08	0.52	0.08	1.64	0.20	23.2	32.0	0	1.0
4	74	0.20	8.8	97.5	221.3	0.25	—	0.04	—	0.11	0.25	0.02	3.1	0.5	0.2	—
0	6	0.30	0.7	45.4	55.6	0.01	0.3	0.00	0.00	0.01	0.03	0.00	0.5	0.7	0	0
0	0	0.00	—	—	0	—	0	—	—	—	—	—	—	0	—	—
0	1	0.03	0.7	10.9	0.7	0.01	0	0.00	0.00	0.00	0.00	0.00	0	0	0	0
0	1	0.09	0	2.3	0.1	0.00	0	0.00	0.00	0.00	0.00	0.00	0	0	0	5.0
0	0	0.07	—	2.3	0.7	—	—	0.07	—	0.07	0.07	—	—	0.3	0	—
41	68	1.86	58.1	539.7	510.6	1.50	31.5	0.07	4.11	0.22	9.57	0.43	26.6	5.1	0.3	13.6
120	26	1.70	32.4	419.7	500.9	2.67	—	0.09	0.01	0.24	6.87	0.40	15.0	7.9	0.3	19.5
38	333	3.01	52.5	447.6	510.6	1.98	—	0.28	0.01	0.30	1.92	0.19	134.4	8.2	0.3	15.9
45	85	3.65	37.9	475.0	756.0	3.52	17.8	0.38	0.80	0.29	5.33	0.39	102.6	23.4	2.1	28.3
74	31	1.68	20.5	248.3	547.8	0.89	25.6	0.32	1.28	0.24	2.55	0.18	38.4	4.0	0.3	17.5
71	29	2.90	—	613.0	292.0	—	—	0.15	0.51	0.17	4.70	—	—	17.0	0	15.0
74	69	3.25	35.8	391.7	816.6	3.63	69.1	0.21	1.25	0.30	3.80	0.21	69.1	1.3	1.8	27.9
41	68	1.86	58.1	539.7	510.6	1.50	31.5	0.07	4.11	0.22	9.57	0.43	26.6	5.1	0.3	13.6
17	249	2.78	28.0	294.0	739.2	1.42	47.6	0.29	1.05	0.33	2.84	0.14	89.6	15.3	0.4	18.6
23	190	1.04	13.5	75.6	469.3	0.86	58.3	0.11	0.43	0.15	0.89	0.02	32.4	2.4	0.1	9.2
103	26	2.20	—	149.0	600.0	—	—	0.05	—	0.17	4.30	—	75.5	0	—	29.0
42	28	1.19	22.7	299.7	620.5	1.32	81.0	0.07	1.11	0.14	5.28	0.36	16.2	22.5	0.2	12.0
109	47	1.97	40.0	489.3	492.1	2.13	—	0.11	0.00	0.20	9.81	0.57	25.3	14.0	0.3	22.6
156	423	8.79	110.3	1182.9	880.4	6.18	—	0.72	0.00	1.05	20.16	1.20	180.1	2.7	1.3	97.6
68	66	3.32	37.8	390.6	652.7	1.94	259.6	0.39	1.05	0.39	7.25	0.23	113.4	10.3	0.2	27.0
156	51	3.26	68.3	588.0	3208.6	3.75	31.7	0.15	0.58	0.36	16.68	0.88	24.4	2.0	0.5	36.1
42	20	0.82	23.9	202.5	246.5	1.13	—	0.05	0.62	0.09	4.06	0.25	15.8	2.5	0.2	10.7
27	42	2.83	50.6	636.8	864.8	2.36	—	0.15	0.01	0.22	3.18	0.19	58.1	10.3	0.6	7.3
7	41	0.53	9.0	162.9	20.7	0.18	47.7	0.06	—	0.05	0.24	0.11	24.3	29.4	0	0.6

Table A–1 Table of Food Composition (*continued*)

(Computer code is for Cengage Diet Analysis Plus program)

DA+ Code	Food Description	QTY	Measure	Wt (g)	H_2O (g)	Ener (cal)	Prot (g)	Carb (g)	Fiber (g)	Fat (g)	Fat Breakdown (g)		
											Sat	Mono	Poly
	MIXED FOODS, SOUPS, SANDWICHES—continued												
1574	Crab cakes, from blue crab	1	item(s)	60	42.6	93	12.1	0.3	0	4.5	0.9	1.7	1.4
32144	Enchiladas with green chili sauce (enchiladas verdes)	1	item(s)	144	103.8	207	9.3	17.6	2.6	11.7	6.4	3.6	1.0
2793	Falafel patty	3	item(s)	51	17.7	170	6.8	16.2	—	9.1	1.2	5.2	2.1
28546	Fettuccine alfredo	1	cup(s)	244	88.7	279	13.1	46.1	1.4	4.2	2.2	1.0	0.4
32146	Flautas	3	item(s)	162	78.0	438	24.9	36.3	4.1	21.6	8.2	8.8	2.3
29629	Fried rice with meat or poultry	1	cup(s)	198	128.5	333	12.3	41.8	1.4	12.3	2.2	3.5	5.7
16649	General Tso chicken	1	cup(s)	146	91.0	296	18.7	16.4	0.9	17.0	4.0	6.3	5.3
1826	Green salad	¾	cup(s)	104	98.9	17	1.3	3.3	2.2	0.1	0	0	0
1814	Hummus	½	cup(s)	123	79.8	218	6.0	24.7	4.9	10.6	1.4	6.0	2.6
16650	Kung pao chicken	1	cup(s)	162	87.2	434	28.8	11.7	2.3	30.6	5.2	13.9	9.7
16622	Lamb curry	1	cup(s)	236	187.9	257	28.2	3.7	0.9	13.8	3.9	4.9	3.3
25253	Lasagna with ground beef	1	cup(s)	237	158.4	284	16.9	22.3	2.4	14.5	7.5	4.9	0.8
442	Macaroni and cheese, prepared	1	cup(s)	200	122.3	390	14.9	40.6	1.6	18.6	7.9	6.4	2.9
29637	Meat filled ravioli with tomato or meat sauce, canned	1	cup(s)	251	198.7	208	7.8	36.5	1.3	3.7	1.5	1.4	0.3
25105	Meat loaf	1	slice(s)	115	84.5	245	17.0	6.6	0.4	16.0	6.1	6.9	0.9
16646	Moo shi pork	1	cup(s)	151	76.8	512	18.9	5.3	0.6	46.4	6.9	15.8	21.2
16788	Nachos with beef, beans, cheese, tomatoes and onions	1	serving(s)	551	253.5	1576	59.1	137.5	20.4	90.8	32.6	41.9	9.4
6116	Pepperoni pizza	2	slice(s)	142	66.1	362	20.2	39.7	2.9	13.9	4.5	6.3	2.3
29601	Pizza with meat and vegetables, thin crust	2	slice(s)	158	81.4	386	16.5	36.8	2.7	19.1	7.7	8.1	2.2
655	Potato salad	½	cup(s)	125	95.0	179	3.4	14.0	1.6	10.3	1.8	3.1	4.7
25109	Salisbury steaks with mushroom sauce	1	serving(s)	135	101.8	251	17.1	9.3	0.5	15.5	6.0	6.7	0.8
16637	Shrimp creole with rice	1	cup(s)	243	176.6	309	27.0	27.7	1.2	9.2	1.7	3.6	2.9
497	Spaghetti and meatballs with tomato sauce, prepared	1	cup(s)	248	174.0	330	19.0	39.0	2.7	12.0	3.9	4.4	2.2
28585	Spicy thai noodles (pad thai)	8	ounce(s)	227	73.3	221	8.9	35.7	3.0	6.4	0.8	3.3	1.8
33073	Stir fried pork and vegetables with rice	1	cup(s)	235	173.6	348	15.4	33.5	1.9	16.3	5.6	6.9	2.6
28588	Stuffed shells	2½	item(s)	249	157.5	243	15.0	28.0	2.5	8.1	3.1	3.0	1.3
16821	Sushi with egg in seaweed	6	piece(s)	156	116.5	190	8.9	20.5	0.3	7.9	2.2	3.2	1.5
16819	Sushi with vegetables and fish	6	piece(s)	156	101.6	218	8.4	43.7	1.7	0.6	0.2	0.1	0.2
16820	Sushi with vegetables in seaweed	6	piece(s)	156	110.3	183	3.4	40.6	0.8	0.4	0.1	0.1	0.1
25266	Sweet and sour pork	¾	cup(s)	249	205.9	265	29.2	17.1	1.0	8.1	2.6	3.5	1.5
16824	Tabouli, tabbouleh or tabuli	1	cup(s)	160	123.7	198	2.6	15.9	3.7	14.9	2.0	10.9	1.6
25276	Three bean salad	½	cup(s)	99	82.2	95	1.9	9.7	2.6	5.9	0.8	1.4	3.5
160	Tuna salad	½	cup(s)	103	64.7	192	16.4	9.6	0	9.5	1.6	3.0	4.2
25241	Turkey and noodles	1	cup(s)	319	228.5	270	24.0	21.2	1.0	9.2	2.4	3.5	2.3
16794	Vegetable egg roll	2	item(s)	128	89.8	201	5.1	19.5	1.7	11.6	2.5	5.7	2.6
16818	Vegetable sushi, no fish	6	piece(s)	156	99.0	226	4.8	49.9	2.0	0.4	0.1	0.1	0.1
	Sandwiches												
1744	Bacon, lettuce and tomato with mayonnaise	1	item(s)	164	97.2	341	11.6	34.2	2.3	17.6	3.8	5.5	6.7
30287	Bologna and cheese with margarine	1	item(s)	111	45.6	345	13.4	29.3	1.2	19.3	8.1	7.0	2.4
30286	Bologna with margarine	1	item(s)	83	33.6	251	8.1	27.3	1.2	12.1	3.7	5.0	2.1
16546	Cheese	1	item(s)	83	31.0	261	9.1	27.6	1.2	12.7	5.4	4.2	2.1
8789	Cheeseburger, large, plain	1	item(s)	185	78.9	564	32.0	38.5	2.6	31.5	12.5	10.2	1.0
8624	Cheeseburger, large, with bacon, vegetables, and condiments	1	item(s)	195	91.4	550	30.8	36.8	2.5	30.9	11.9	10.6	1.3
1745	Club with bacon, chicken, tomato, lettuce, and mayonnaise	1	item(s)	246	137.5	546	31.0	48.9	3.0	24.5	5.3	7.5	9.4
1908	Cold cut submarine with cheese and vegetables	1	item(s)	228	131.8	456	21.8	51.0	2.0	18.6	6.8	8.2	2.3
30247	Corned beef	1	item(s)	130	74.9	265	18.2	25.3	1.6	9.6	3.6	3.5	1.0
25283	Egg salad	1	item(s)	126	72.1	278	10.7	28.0	1.4	13.5	2.9	4.2	5.0
16686	Fried egg	1	item(s)	96	49.7	226	10.0	26.2	1.2	8.6	2.3	3.2	1.9
16547	Grilled cheese	1	item(s)	83	27.5	291	9.2	27.9	1.2	15.8	6.0	5.7	3.0
16659	Gyro with onion and tomato	1	item(s)	105	68.3	163	12.0	20.0	1.1	3.5	1.3	1.3	0.5

Chol (mg)	Calc (mg)	Iron (mg)	Magn (mg)	Pota (mg)	Sodi (mg)	Zinc (mg)	Vit A (µg)	Thia (mg)	Vit E (mg α)	Ribo (mg)	Niac (mg)	Vit B_6 (mg)	Fola (µg DFE)	Vit C (mg)	Vit B_{12} (µg)	Sele (µg)
90	63	0.64	19.8	194.4	198.0	2.45	34.2	0.05	—	0.04	1.74	0.10	36.6	1.7	3.6	24.4
27	266	1.07	38.5	251.4	276.3	1.26	—	0.07	0.02	0.16	1.27	0.17	44.6	59.3	0.2	6.0
0	28	1.74	41.8	298.4	149.9	0.76	0.5	0.07	—	0.08	0.53	0.06	47.4	0.8	0	0.5
9	218	1.83	38.4	163.9	472.9	1.24	—	0.41	0.00	0.35	2.85	0.09	225.0	1.6	0.4	38.4
73	146	2.66	61.3	222.9	885.7	3.43	0	0.10	0.10	0.16	3.00	0.26	95.7	0	1.2	36.7
103	38	2.77	33.7	196.0	833.6	1.34	41.6	0.33	1.60	0.18	4.17	0.27	146.5	3.4	0.3	22.0
66	26	1.46	23.4	248.2	849.7	1.40	29.2	0.10	1.62	0.18	6.28	0.28	23.4	12.0	0.2	19.9
0	13	0.65	11.4	178.0	26.9	0.21	59.0	0.03	—	0.05	0.56	0.08	38.3	24.0	0	0.4
0	60	1.91	35.7	212.8	297.7	1.34	0	0.10	0.92	0.06	0.49	0.49	72.6	9.7	0	3.0
65	50	1.96	63.2	427.7	907.2	1.50	38.9	0.15	4.32	0.14	13.22	0.58	42.1	7.5	0.3	23.0
90	38	2.95	40.1	493.2	495.6	6.60	—	0.08	1.29	0.28	8.03	0.21	28.3	1.4	2.9	30.4
68	233	2.22	40.1	420.1	433.6	2.70	—	0.21	0.21	0.29	3.06	0.22	91.1	15.0	0.8	21.4
34	310	2.06	40.0	258.0	784.0	2.06	180.0	0.27	0.72	0.43	2.18	0.08	100.0	0	0.5	30.6
15	35	2.10	20.1	283.6	1352.9	1.28	27.6	0.19	0.70	0.16	2.77	0.14	60.2	21.6	0.4	13.3
85	59	1.87	21.8	300.8	411.7	3.40	—	0.08	0.00	0.27	3.72	0.13	18.7	0.9	1.6	17.9
172	32	1.57	25.7	333.7	1052.5	1.82	49.8	0.49	5.39	0.36	2.88	0.31	21.1	8.0	0.8	30.0
154	948	7.32	242.4	1201.2	1862.4	10.68	259.0	0.29	7.71	0.81	6.39	1.09	148.8	16.0	2.6	44.1
28	129	1.87	17.0	305.3	533.9	1.03	105.1	0.26	—	0.46	6.09	0.11	76.7	3.3	0.4	26.1
36	258	3.14	31.6	352.3	971.7	2.02	49.0	0.37	1.13	0.37	3.77	0.19	91.6	15.6	0.6	22.8
85	24	0.81	18.8	317.5	661.3	0.38	40.0	0.09	—	0.07	1.11	0.17	8.8	12.5	0	5.1
60	74	1.94	23.8	314.5	360.5	3.45	—	0.10	0.00	0.27	3.95	0.13	20.8	0.7	1.6	17.4
180	102	4.68	63.2	413.1	330.5	1.72	94.8	0.29	2.06	0.11	4.75	0.21	121.5	12.9	1.2	49.3
89	124	3.70	—	665.0	1009.0	—	81.5	0.25	—	0.30	4.00	—	—	22.0	—	22.0
37	31	1.56	49.4	181.3	591.6	1.05	—	0.18	0.35	0.13	1.82	0.17	55.8	22.6	0.1	3.2
46	38	2.71	33.0	396.9	569.5	2.08	—	0.51	0.38	0.20	5.07	0.30	162.4	18.8	0.4	22.8
30	188	2.26	49.4	403.0	471.5	1.41	—	0.26	0.00	0.26	3.83	0.24	161.2	17.9	0.2	28.9
214	45	1.84	18.7	135.7	463.3	0.98	106.1	0.13	0.67	0.28	1.35	0.13	82.7	1.9	0.7	20.3
11	23	2.15	25.0	202.8	340.1	0.78	45.2	0.26	0.24	0.07	2.76	0.14	121.7	3.6	0.3	13.9
0	20	1.54	18.7	96.7	152.9	0.68	25.0	0.19	0.12	0.03	1.86	0.13	118.6	2.3	0	9.8
74	40	1.76	35.6	619.9	621.8	2.53	—	0.81	0.20	0.37	6.69	0.66	14.7	11.9	0.7	49.6
0	30	1.21	35.2	249.6	796.8	0.48	54.4	0.07	2.43	0.04	1.11	0.11	30.4	26.1	0	0.5
0	26	0.96	15.5	144.8	224.2	0.30	—	0.02	0.88	0.04	0.26	0.04	32.2	10.0	0	2.7
13	17	1.02	19.5	182.5	412.1	0.57	24.6	0.03	—	0.07	6.86	0.08	8.2	2.3	1.2	42.2
77	69	2.56	33.2	400.8	577.1	2.51	—	0.23	0.28	0.30	6.41	0.29	109.5	1.4	1.1	33.4
60	29	1.65	17.9	193.3	549.1	0.48	25.6	0.15	1.28	0.20	1.59	0.09	46.1	5.5	0.2	11.3
0	23	2.38	21.8	157.6	369.7	0.82	48.4	0.28	0.15	0.05	2.44	0.12	135.7	3.7	0	8.1
21	79	2.36	27.9	351.0	944.6	1.08	44.3	0.32	1.16	0.24	4.36	0.21	105.0	9.7	0.2	27.1
40	258	2.38	24.4	215.3	941.3	1.88	102.1	0.31	0.55	0.35	2.97	0.14	91.0	0.2	0.8	20.4
17	100	2.24	16.6	138.6	579.3	1.02	44.8	0.29	0.49	0.21	2.92	0.12	88.8	0.2	0.5	16.0
22	233	2.04	19.9	127.0	733.7	1.24	97.1	0.25	0.47	0.30	2.25	0.05	88.8	0	0.3	13.1
104	309	4.47	44.4	401.5	986.1	5.75	0	0.35	—	0.77	8.26	0.49	129.48	0	2.8	38.9
98	267	4.03	44.9	464.1	1314.3	5.20	0	0.33	—	0.67	8.25	0.47	122.9	1.4	2.4	6.6
71	157	4.57	46.7	464.9	1087.3	1.82	41.8	0.54	1.52	0.40	12.82	0.61	172.2	6.4	0.4	42.3
36	189	2.50	68.4	394.4	1650.7	2.57	70.7	1.00	—	0.79	5.49	0.13	109.4	12.3	1.1	30.8
46	81	3.04	19.5	127.4	1206.4	2.26	2.6	0.23	0.20	0.24	3.42	0.11	88.4	0.3	0.9	31.2
219	85	2.25	18.8	159.0	423.1	0.87	—	0.27	0.12	0.43	2.06	0.15	112.1	0.9	0.6	29.9
206	104	2.79	17.3	117.1	438.7	0.92	89.3	0.26	0.66	0.40	2.26	0.10	110.4	0	0.6	24.2
22	235	2.05	19.9	128.7	763.6	1.26	129.5	0.19	0.72	0.28	2.05	0.05	58.1	0	0.2	13.2
28	47	1.77	22.1	218.4	235.2	2.33	9.5	0.23	0.26	0.20	3.12	0.13	63.0	3.2	0.9	18.3

Table A–1 Table of Food Composition (*continued*) (Computer code is for Cengage Diet Analysis Plus program)

DA+ Code	Food Description	QTY	Measure	Wt (g)	H₂O (g)	Ener (cal)	Prot (g)	Carb (g)	Fiber (g)	Fat (g)	Fat Breakdown (g) Sat	Mono	Poly
	MIXED FOODS, SOUPS, SANDWICHES—continued												
1906	Ham and cheese	1	item(s)	146	74.2	352	20.7	33.3	2.0	15.5	6.4	6.7	1.4
31890	Ham with mayonnaise	1	item(s)	112	56.3	271	13.0	27.9	1.9	11.6	2.8	4.0	4.0
756	Hamburger, double patty, large, with condiments and vegetables	1	item(s)	226	121.5	540	34.3	40.3	—	26.6	10.5	10.3	2.8
8793	Hamburger, large, plain	1	item(s)	137	57.7	426	22.6	31.7	1.5	22.9	8.4	9.9	2.1
8795	Hamburger, large, with vegetables and condiments	1	item(s)	218	121.4	512	25.8	40.0	3.1	27.4	10.4	11.4	2.2
25134	Hot chicken salad	1	item(s)	98	48.4	242	15.2	23.8	1.3	9.2	2.9	2.5	3.0
25133	Hot turkey salad	1	item(s)	98	50.1	224	15.6	23.8	1.3	6.9	2.3	1.6	2.5
1411	Hotdog with bun, plain	1	item(s)	98	52.9	242	10.4	18.0	1.6	14.5	5.1	6.9	1.7
30249	Pastrami	1	item(s)	134	71.2	328	13.4	27.8	1.6	17.7	6.1	8.3	1.2
16701	Peanut butter	1	item(s)	93	23.6	345	12.2	37.6	3.3	17.4	3.4	7.7	5.2
30306	Peanut butter and jelly	1	item(s)	93	24.2	330	10.3	41.9	2.9	14.7	2.9	6.5	4.4
1909	Roast beef submarine with mayonnaise and vegetables	1	item(s)	216	127.4	410	28.6	44.3	—	13.0	7.1	1.8	2.6
1910	Roast beef, plain	1	item(s)	139	67.6	346	21.5	33.4	1.2	13.8	3.6	6.8	1.7
1907	Steak with mayonnaise and vegetables	1	item(s)	204	104.2	459	30.3	52.0	2.3	14.1	3.8	5.3	3.3
25288	Tuna salad	1	item(s)	179	102.2	415	24.5	28.4	1.6	22.4	3.5	6.2	11.4
30283	Turkey submarine with cheese, lettuce, tomato, and mayonnaise	1	item(s)	277	168.0	529	30.4	49.4	3.0	22.8	6.8	6.0	8.6
31891	Turkey with mayonnaise	1	item(s)	143	74.5	329	28.7	26.4	1.3	11.2	2.6	2.6	4.8
	Soups												
25296	Bean	1	cup(s)	301	253.1	191	13.8	29.0	6.5	2.3	0.7	0.8	0.5
711	Bean with pork, condensed, prepared with water	1	cup(s)	253	215.9	159	7.3	21.0	7.3	5.5	1.4	2.0	1.7
713	Beef noodle, condensed, prepared with water	1	cup(s)	244	224.9	83	4.7	8.7	0.7	3.0	1.1	1.2	0.5
825	Cheese, condensed, prepared with milk	1	cup(s)	251	206.9	231	9.5	16.2	1.0	14.6	9.1	4.1	0.5
826	Chicken broth, condensed, prepared with water	1	cup(s)	244	234.1	39	4.9	0.9	0	1.4	0.4	0.6	0.3
25297	Chicken noodle soup	1	cup(s)	286	258.4	117	10.8	10.9	0.9	2.9	0.8	1.1	0.7
827	Chicken noodle, condensed, prepared with water	1	cup(s)	241	226.1	60	3.1	7.1	0.5	2.3	0.6	1.0	0.6
724	Chicken noodle, dehydrated, prepared with water	1	cup(s)	252	237.3	58	2.1	9.2	0.3	1.4	0.3	0.5	0.4
823	Cream of asparagus, condensed, prepared with milk	1	cup(s)	248	213.3	161	6.3	16.4	0.7	8.2	3.3	2.1	2.2
824	Cream of celery, condensed, prepared with milk	1	cup(s)	248	214.4	164	5.7	14.5	0.7	9.7	3.9	2.5	2.7
708	Cream of chicken, condensed, prepared with milk	1	cup(s)	248	210.4	191	7.5	15.0	0.2	11.5	4.6	4.5	1.6
715	Cream of chicken, condensed, prepared with water	1	cup(s)	244	221.1	117	3.4	9.3	0.2	7.4	2.1	3.3	1.5
709	Cream of mushroom, condensed, prepared with milk	1	cup(s)	248	215.0	166	6.2	14.0	0	9.6	3.3	2.0	1.8
716	Cream of mushroom, condensed, prepared with water	1	cup(s)	244	224.6	102	1.9	8.0	0	7.0	1.6	1.3	1.7
25298	Cream of vegetable	1	cup(s)	285	250.7	165	7.2	15.2	1.9	8.6	1.6	4.6	1.9
16689	Egg drop	1	cup(s)	244	228.9	73	7.5	1.1	0	3.8	1.1	1.5	0.6
25138	Golden squash	1	cup(s)	258	223.9	145	7.6	20.4	0.4	4.1	0.8	2.2	0.9
16663	Hot and sour	1	cup(s)	244	209.7	161	15.0	5.4	0.5	7.9	2.7	3.4	1.1
28054	Lentil chowder	1	cup(s)	244	202.8	153	11.4	27.7	12.6	0.5	0.1	0.1	0.2
28560	Macaroni and bean	1	cup(s)	246	138.8	146	5.8	22.9	5.1	3.7	0.5	2.2	0.6
714	Manhattan clam chowder, condensed, prepared with water	1	cup(s)	244	225.1	73	2.1	11.6	1.5	2.1	0.4	0.4	1.2
28561	Minestrone	1	cup(s)	241	185.4	103	4.5	16.8	4.8	2.3	0.3	1.4	0.4
717	Minestrone, condensed, prepared with water	1	cup(s)	241	220.1	82	4.3	11.2	1.0	2.5	0.6	0.7	1.1

Chol (mg)	Calc (mg)	Iron (mg)	Magn (mg)	Pota (mg)	Sodi (mg)	Zinc (mg)	Vit A (µg)	Thia (mg)	Vit E (mg α)	Ribo (mg)	Niac (mg)	Vit B_6 (mg)	Fola (µg DFE)	Vit C (mg)	Vit B_{12} (µg)	Sele (µg)
58	130	3.24	16.1	290.5	770.9	1.37	96.4	0.30	0.29	0.48	2.68	0.20	78.8	2.8	0.5	23.1
34	91	2.47	23.5	210.6	1097.6	1.13	5.6	0.57	0.50	0.26	3.80	0.25	90.7	2.2	0.2	20.2
122	102	5.85	49.7	569.5	791.0	5.67	0	0.36	—	0.38	7.57	0.54	110.7	1.1	4.1	25.5
71	74	3.57	27.4	267.2	474.0	4.11	0	0.28	—	0.28	6.24	0.23	80.8	0	2.1	27.1
87	96	4.92	43.6	479.6	824.0	4.88	0	0.41	—	0.37	7.28	0.32	115.5	2.6	2.4	33.6
39	115	1.88	20.0	172.4	505.1	1.19	—	0.23	0.28	0.22	4.84	0.19	79.5	0.5	0.2	19.9
37	114	1.99	21.3	189.0	494.5	1.07	—	0.22	0.28	0.20	4.26	0.22	79.6	0.5	0.2	23.4
44	24	2.31	12.7	143.1	670.3	1.98	0	0.23	—	0.27	3.64	0.04	60.7	0.1	0.5	26.0
51	80	3.02	22.8	182.2	1364.1	2.70	2.7	0.28	0.26	0.26	4.97	0.14	89.8	0.3	1.0	14.3
0	110	2.92	66.0	226.0	580.3	1.32	0	0.31	2.39	0.23	6.72	0.18	130.2	0	0	13.2
0	94	2.50	56.7	198.1	492.9	1.12	—	0.26	2.01	0.20	5.66	0.15	110.7	0.1	0	11.2
73	41	2.80	67.0	330.5	844.6	4.38	30.2	0.41	—	0.41	5.96	0.32	88.6	5.6	1.8	25.7
51	54	4.22	30.6	315.5	792.3	3.39	11.1	0.37	—	0.30	5.86	0.26	68.1	2.1	1.2	29.2
73	92	5.16	49.0	524.3	797.6	4.52	0	0.40	—	0.36	7.30	0.36	128.5	5.5	1.6	42.0
59	78	2.97	35.9	316.3	724.7	1.02	—	0.27	0.34	0.26	12.07	0.46	99.6	1.8	2.4	76.9
64	307	4.59	49.9	534.6	1759.0	2.74	74.8	0.49	1.19	0.65	3.91	0.31	171.7	10.5	0.6	42.1
67	100	3.46	34.3	304.6	564.9	3.00	5.7	0.28	0.74	0.32	6.84	0.46	94.4	0	0.3	40
5	79	3.05	61.8	588.8	689.0	1.41	—	0.27	0.02	0.15	3.63	0.23	140.1	3.6	0.2	7.9
3	78	1.89	43.0	371.9	883.0	0.96	43.0	0.08	1.08	0.03	0.52	0.03	30.4	1.5	0	7.8
5	20	1.07	7.3	97.6	929.6	1.51	12.2	0.06	1.22	0.05	1.03	0.03	29.3	0.5	0.2	7.3
48	289	0.80	20.1	341.4	1019.1	0.67	358.9	0.06	—	0.33	0.50	0.07	10.0	1.3	0.4	7.0
0	10	0.51	2.4	209.8	775.9	0.24	0	0.01	0.04	0.07	3.34	0.02	4.9	0	0.2	0
24	25	1.38	16.4	340.1	774.5	0.77	—	0.15	0.02	0.16	5.57	0.13	37.2	1.8	0.3	10.2
12	14	1.59	9.6	53.0	638.7	0.38	26.5	0.13	0.07	0.10	1.30	0.04	28.9	0	0	11.6
10	5	0.50	7.6	32.8	577.1	0.20	2.5	0.20	0.12	0.07	1.08	0.02	27.7	0	0.1	9.6
22	174	0.86	19.8	359.6	1041.6	0.91	62.0	0.10	—	0.27	0.88	0.06	29.8	4.0	0.5	8.0
32	186	0.69	22.3	310.0	1009.4	0.19	114.1	0.07	—	0.24	0.43	0.06	7.4	1.5	0.5	4.7
27	181	0.67	17.4	272.8	1046.6	0.67	178.6	0.07	—	0.25	0.92	0.06	7.4	1.2	0.5	8.0
10	34	0.61	2.4	87.8	985.8	0.63	163.5	0.02	—	0.06	0.82	0.01	2.4	0.2	0.1	7.0
10	164	1.36	19.8	267.8	823.4	0.79	81.8	0.10	1.01	0.29	0.62	0.05	7.4	0.2	0.6	6.0
0	17	1.31	4.9	73.2	775.9	0.24	9.8	0.05	0.97	0.05	0.50	0.00	2.4	0	0	2.9
1	80	1.20	17.5	340.7	787.9	0.56	—	0.12	1.05	0.18	3.32	0.12	39.5	10.7	0.3	4.3
102	22	0.75	4.9	219.6	729.6	0.48	41.5	0.02	0.29	0.19	3.02	0.05	14.6	0	0.5	7.6
4	262	0.78	42.4	542.2	515.6	0.88	—	0.16	0.52	0.30	1.14	0.16	32.7	12.5	0.7	6.0
34	29	1.24	19.5	373.3	1561.6	1.43	—	0.26	0.12	0.24	4.96	0.20	14.6	0.5	0.4	19.3
0	48	4.38	59.3	626.2	26.7	1.57	—	0.24	0.06	0.12	1.87	0.32	192.7	16.1	0	3.5
0	59	1.90	32.4	275.9	531.0	0.51	—	0.16	0.37	0.12	1.44	0.10	92.3	9.1	0	8.8
2	27	1.56	9.8	180.6	551.4	0.87	48.8	0.02	1.22	0.03	0.77	0.09	9.8	3.9	3.9	9.0
0	62	1.70	29.9	287.5	442.7	0.42	—	0.09	0.23	0.09	0.70	0.07	62.0	13.3	0	3.5
2	34	0.91	7.2	313.3	911.0	0.74	118.1	0.05	—	0.04	0.94	0.09	50.6	1.2	0	8.0

											Fat Breakdown (g)		
DA+ Code	Food Description	QTY	Measure	Wt (g)	H_2O (g)	Ener (cal)	Prot (g)	Carb (g)	Fiber (g)	Fat (g)	Sat	Mono	Poly
MIXED FOODS, SOUPS, SANDWICHES—continued													
28038	Mushroom and wild rice	1	cup(s)	244	199.7	86	4.7	13.2	1.7	0.3	0	0	0.2
828	New England clam chowder, condensed, prepared with milk	1	cup(s)	248	212.2	151	8.0	18.4	0.7	5.0	2.1	0.7	0.6
28036	New England style clam chowder	1	cup(s)	244	227.5	61	3.8	8.8	1.8	0.2	0.1	0	0
28566	Old country pasta	1	cup(s)	252	183.3	146	6.5	18.3	3.6	4.5	2.0	2.4	0.9
725	Onion, dehydrated, prepared with water	1	cup(s)	246	235.7	30	0.8	6.8	0.7	0	0	0	0
16667	Shrimp gumbo	1	cup(s)	244	207.2	166	9.5	18.2	2.4	6.7	1.3	2.9	2.0
28037	Southwestern corn chowder	1	cup(s)	244	217.8	98	4.9	17.0	2.4	0.5	0.1	0.1	0.2
30282	Soybean (miso)	1	cup(s)	240	218.6	84	6.0	8.0	1.9	3.4	0.6	1.1	1.4
25140	Split pea	1	cup(s)	165	119.7	72	4.5	16.0	1.6	0.3	0.1	0	0.2
718	Split pea with ham, condensed, prepared with water	1	cup(s)	253	206.9	190	10.3	28.0	2.3	4.4	1.8	1.8	0.6
726	Tomato vegetable, dehydrated, prepared with water	1	cup(s)	253	238.4	56	2.0	10.2	0.8	0.9	0.4	0.3	0.1
710	Tomato, condensed, prepared with milk	1	cup(s)	248	213.4	136	6.2	22.0	1.5	3.2	1.8	0.9	0.3
719	Tomato, condensed, prepared with water	1	cup(s)	244	223.0	73	1.9	16.0	1.5	0.7	0.2	0.2	0.2
28595	Turkey noodle	1	cup(s)	244	216.9	114	8.1	15.1	1.9	2.4	0.3	1.1	0.7
28051	Turkey vegetable	1	cup(s)	244	220.8	96	12.2	6.6	2.0	1.1	0.3	0.2	0.3
25141	Vegetable	1	cup(s)	252	228.1	82	5.2	16.5	4.5	0.3	0	0	0.1
720	Vegetable beef, condensed, prepared with water	1	cup(s)	244	224.0	76	5.4	9.9	2.0	1.9	0.8	0.8	0.1
28598	Vegetable gumbo	1	cup(s)	252	184.5	170	4.4	28.9	3.6	4.7	0.7	3.2	0.5
721	Vegetarian vegetable, condensed, prepared with water	1	cup(s)	241	222.7	67	2.1	11.8	0.7	1.9	0.3	0.8	0.7
FAST FOOD													
	Arby's												
36094	Au jus sauce	1	serving(s)	85	—	43	1.0	7.0	0	1.3	0.4	—	—
751	Beef 'n cheddar sandwich	1	item(s)	195	—	445	22.0	44.0	2.0	21.0	6.0	—	—
9279	Cheddar curly fries	1	serving(s)	198	—	631	8.0	73.0	7.0	37.4	6.8	—	—
34770	Chicken breast fillet sandwich, grilled	1	item(s)	233	—	414	32.0	36.0	3.0	17.0	3.0	—	—
36131	Chocolate shake, regular	1	serving(s)	397	—	507	13.0	83.0	0	13.0	8.0	—	—
36045	Curly fries, large size	1	serving(s)	198	—	631	8.0	73.0	7.0	37.0	7.0	—	—
36044	Curly fries, medium size	1	serving(s)	128	—	406	5.0	47.0	5.0	24.0	4.0	—	—
752	Ham 'n cheese sandwich	1	item(s)	167	—	304	23.0	35.0	1.0	7.0	2.0	—	—
36048	Homestyle fries, large size	1	serving(s)	213	—	566	6.0	82.0	6.0	37.0	7.0	—	—
36047	Homestyle fries, medium size	1	serving(s)	142	—	377	4.0	55.0	4.0	25.0	4.0	—	—
33465	Homestyle fries, small size	1	serving(s)	113	—	302	3.0	44.0	3.0	20.0	4.0	—	—
9249	Junior roast beef sandwich	1	item(s)	125	—	272	16.0	34.0	2.0	10.0	4.0	—	—
9251	Large roast beef sandwich	1	item(s)	281	—	547	42.0	41.0	3.0	28.0	12.0	—	—
39640	Market Fresh chicken salad with pecans sandwich	1	item(s)	322	—	769	30.0	79.0	9.0	39.0	10.0	—	—
39641	Market Fresh Martha's Vineyard salad, without dressing	1	serving(s)	330	—	277	26.0	24.0	5.0	8.0	4.0	—	—
34769	Market Fresh roast turkey and Swiss sandwich	1	serving(s)	359	—	725	45.0	75.0	5.0	30.0	8.0	—	—
9267	Market Fresh roast turkey ranch and bacon sandwich	1	serving(s)	382	—	834	49.0	75.0	5.0	38.0	11.0	—	—
39642	Market Fresh Santa Fe salad, without dressing	1	serving(s)	372	—	499	30.0	42.0	7.0	23.0	8.0	—	—
39650	Market Fresh Southwest chicken wrap	1	serving(s)	251	—	567	36.0	42.0	4.0	29.0	9.0	—	—
37021	Market Fresh Ultimate BLT sandwich	1	item(s)	294	—	779	23.0	75.0	6.0	45.0	11.0	—	—
750	Roast beef sandwich, regular	1	item(s)	154	—	320	21.0	34.0	2.0	14.0	5.0	—	—
36132	Strawberry shake, regular	1	serving(s)	397	—	498	13.0	81.0	0	13.0	8.0	—	—
2009	Super roast beef sandwich	1	item(s)	198	—	398	21.0	40.0	2.0	19.0	6.0	—	—
36130	Vanilla shake, regular	1	serving(s)	369	—	437	13.0	66.0	0	13.0	8.0	—	—
	Auntie Anne's												
35371	Cheese dipping sauce	1	serving(s)	35	—	100	3.0	4.0	0	8.0	4.0	—	—
35353	Cinnamon sugar soft pretzel	1	item(s)	120	—	350	9.0	74.0	2.0	2.0	0	—	—

Chol (mg)	Calc (mg)	Iron (mg)	Magn (mg)	Pota (mg)	Sodi (mg)	Zinc (mg)	Vit A (µg)	Thia (mg)	Vit E (mg α)	Ribo (mg)	Niac (mg)	Vit B_6 (mg)	Fola (µg DFE)	Vit C (mg)	Vit B_{12} (µg)	Sele (µg)
0	30	1.38	27.3	376.9	283.8	1.00	—	0.06	0.07	0.23	3.21	0.14	27.3	3.7	0.1	4.7
17	169	3.00	29.8	456.3	887.8	0.99	91.8	0.20	0.54	0.43	1.96	0.17	22.3	5.2	11.9	10.9
3	89	1.30	29.2	503.7	256.8	0.52	—	0.06	0.02	0.10	1.30	0.15	24.2	11.8	3.0	3.8
5	57	2.43	51.9	500.4	355.7	0.78	—	0.21	0.01	0.14	2.64	0.20	105.2	22.0	0.1	10.3
0	22	0.12	9.8	76.3	851.2	0.12	0	0.03	0.02	0.03	0.15	0.06	0	0.2	0	0.5
51	105	2.85	48.8	461.2	441.6	0.90	80.5	0.18	1.90	0.12	2.52	0.19	102.5	17.6	0.3	14.9
1	83	1.03	26.3	434.4	217.4	0.57	—	0.08	0.09	0.13	1.81	0.21	32.1	39.6	0.2	1.7
0	65	1.87	36.0	362.4	988.8	0.86	232.8	0.06	0.96	0.16	2.61	0.15	57.6	4.6	0.2	1.0
0	28	1.26	29.8	328.1	602.4	0.52	—	0.10	0.00	0.07	1.50	0.16	49.5	8.1	0	0.7
8	23	2.27	48.1	399.7	1006.9	1.31	22.8	0.14	—	0.07	1.47	0.06	2.5	1.5	0.3	8.0
0	20	0.60	10.1	169.5	334.0	0.20	10.1	0.06	0.43	0.09	1.26	0.06	12.7	3.0	0.1	2.0
10	166	1.36	29.8	466.2	711.8	0.84	94.2	0.09	0.44	0.31	1.35	0.15	5.0	15.6	0.6	9.2
0	20	1.31	17.1	273.3	663.7	0.29	24.4	0.04	0.41	0.07	1.23	0.10	0	15.4	0	6.1
26	28	1.40	24.1	223.8	395.9	0.75	—	0.21	0.01	0.12	2.85	0.15	65.1	7.1	0.1	13.7
21	38	1.48	25.0	423.4	348.7	0.99	—	0.09	0.01	0.09	3.64	0.27	23.9	10.6	0.2	10.3
0	40	2.08	39.1	681.5	670.3	0.67	—	0.16	0.00	0.09	2.70	0.26	36.3	22.4	0	2.2
5	20	1.09	7.3	168.4	773.5	1.51	190.3	0.03	0.58	0.04	1.00	0.07	9.8	2.4	0.3	2.7
0	56	1.85	38.9	360.2	518.4	0.64	—	0.18	0.64	0.08	1.77	0.17	107.5	21.9	0	4.1
0	24	1.06	7.2	207.3	814.6	0.45	171.1	0.05	1.39	0.04	0.90	0.05	9.6	1.4	0	4.3
0	0	—	—	—	1510.0	—	—	—	—	—	—	—	—	—	—	—
51	80	3.96	—	—	1274.0	—	—	—	—	—	—	—	—	1.8	—	—
0	80	3.24	—	—	1476.0	—	—	—	—	—	—	—	—	9.6	—	—
9	90	3.06	—	—	913.0	—	—	—	—	—	—	—	—	10.8	—	—
34	510	0.54	—	—	357.0	—	—	—	—	—	—	—	—	5.4	—	—
0	80	3.24	—	—	1476.0	—	—	—	—	—	—	—	—	9.6	—	—
0	50	1.98	—	—	949.0	—	—	—	—	—	—	—	—	6.0	—	—
35	160	2.70	—	—	1420.0	—	—	—	—	—	—	—	—	1.2	—	—
0	50	1.62	—	—	1029.0	—	—	—	—	—	—	—	—	12.6	—	—
0	30	1.08	—	—	686.0	—	—	—	—	—	—	—	—	8.4	—	—
0	30	0.90	—	—	549.0	—	—	—	—	—	—	—	—	6.6	—	—
29	60	3.06	—	—	740.0	—	0	—	—	—	—	—	—	0	—	—
102	70	6.30	—	—	1869.0	—	0	—	—	—	—	—	—	0.6	—	—
74	180	4.32	—	—	1240.0	—	—	—	—	—	—	—	—	30.0	—	—
72	200	1.62	—	—	454.0	—	—	—	—	—	—	—	—	33.6	—	—
91	360	5.22	—	—	1788.0	—	—	—	—	—	—	—	—	10.2	—	—
109	330	5.40	—	—	2258.0	—	—	—	—	—	—	—	—	11.4	—	—
59	420	3.60	—	—	1231.0	—	—	—	—	—	—	—	—	36.6	—	—
88	240	4.50	—	—	1451.0	—	—	—	—	—	—	—	—	7.8	—	—
51	170	4.68	—	—	1571.0	—	—	—	—	—	—	—	—	16.8	—	—
44	60	3.60	—	—	953.0	—	0	—	—	—	—	—	—	0	—	—
34	510	0.72	—	—	363.0	—	—	—	—	—	—	—	—	6.6	—	—
44	70	3.78	—	—	1060.0	—	—	—	—	—	—	—	—	6.0	—	—
34	510	0.36	—	—	350.0	—	—	—	—	—	—	—	—	5.4	—	—
10	100	0.00	—	—	510.0	—	—	—	—	—	—	—	—	0	—	—
0	20	1.98	—	—	410.0	—	0	—	—	—	—	—	—	0	—	—

DA+ Code	Food Description	QTY	Measure	Wt (g)	H_2O (g)	Ener (cal)	Prot (g)	Carb (g)	Fiber (g)	Fat (g)	Fat Breakdown (g)		
											Sat	Mono	Poly
	FAST FOOD—continued												
35354	Cinnamon sugar soft pretzel with butter	1	item(s)	120	—	450	8.0	83.0	3.0	9.0	5.0	—	—
35372	Marinara dipping sauce	1	serving(s)	35	—	10	0	4.0	0	0	0	0	0
35357	Original soft pretzel	1	serving(s)	120	—	340	10.0	72.0	3.0	1.0	0	—	—
35358	Original soft pretzel with butter	1	item(s)	120	—	370	10.0	72.0	3.0	4.0	2.0	—	—
35359	Parmesan herb soft pretzel	1	item(s)	120	—	390	11.0	74.0	4.0	5.0	2.5	—	—
35360	Parmesan herb soft pretzel with butter	1	item(s)	120	—	440	10.0	72.0	9.0	13.0	7.0	—	—
35361	Sesame soft pretzel	1	item(s)	120	—	350	11.0	63.0	3.0	6.0	1.0	—	—
35362	Sesame soft pretzel with butter	1	item(s)	120	—	410	12.0	64.0	7.0	12.0	4.0	—	—
35364	Sour cream and onion soft pretzel	1	item(s)	120	—	310	9.0	66.0	2.0	1.0	0	—	—
35366	Sour cream and onion soft pretzel with butter	1	item(s)	120	—	340	9.0	66.0	2.0	5.0	3.0	—	—
35373	Sweet mustard dipping sauce	1	serving(s)	35	—	60	0.5	8.0	0	1.5	1.0	—	—
35367	Whole wheat soft pretzel	1	item(s)	120	—	350	11.0	72.0	7.0	1.5	0	—	—
35368	Whole wheat soft pretzel with butter	1	item(s)	120	—	370	11.0	72.0	7.0	4.5	1.5	—	—
	Boston Market												
34978	Butternut squash	¾	cup(s)	143	—	140	2.0	25.0	2.0	4.5	3.0	—	—
35006	Caesar side salad	1	serving(s)	71	—	40	3.0	3.0	1.0	20.0	2.0	—	—
35013	Chicken Carver sandwich with cheese and sauce	1	item(s)	321	—	700	44.0	68.0	3.0	29.0	7.0	—	—
34979	Chicken gravy	4	ounce(s)	113	—	15	1.0	4.0	0	0.5	0	—	—
35053	Chicken noodle soup	¾	cup(s)	283	—	180	13.0	16.0	1.0	7.0	2.0	—	—
34973	Chicken pot pie	1	item(s)	425	—	800	29.0	59.0	4.0	49.0	18.0	—	—
35054	Chicken tortilla soup with toppings	¾	cup(s)	227	—	340	12.0	24.0	1.0	22.0	7.0	—	—
35007	Cole slaw	¾	cup(s)	125	—	170	2.0	21.0	2.0	9.0	2.0	—	—
35057	Cornbread	1	item(s)	45	—	130	1.0	21.0	0	3.5	1.0	—	—
34980	Creamed spinach	¾	cup(s)	191	—	280	9.0	12.0	4.0	23.0	15.0	—	—
34998	Fresh vegetable stuffing	1	cup(s)	136	—	190	3.0	25.0	2.0	8.0	1.0	—	—
34991	Garlic dill new potatoes	¾	cup(s)	156	—	140	3.0	24.0	3.0	3.0	1.0	—	—
34983	Green bean casserole	¾	cup(s)	170	—	60	2.0	9.0	2.0	2.0	1.0	—	—
34982	Green beans	¾	cup(s)	91	—	60	2.0	7.0	3.0	3.5	1.5	—	—
34984	Homestyle mashed potatoes	¾	cup(s)	221	—	210	4.0	29.0	3.0	9.0	6.0	—	—
34985	Homestyle mashed potatoes and gravy	1	cup(s)	334	—	225	5.0	33.0	3.0	9.5	6.0	—	—
34988	Hot cinnamon apples	¾	cup(s)	145	—	210	0	47.0	3.0	3.0	0	—	—
34989	Macaroni and cheese	¾	cup(s)	221	—	330	14.0	39.0	1.0	12.0	7.0	—	—
51193	Market chopped salad with dressing	1	item(s)	563	—	580	11.0	31.0	9.0	48.0	9.0	—	—
34970	Meatloaf	1	serving(s)	218	—	480	29.0	23.0	2.0	33.0	13.0	—	—
39383	Nestle Toll House chocolate chip cookie	1	item(s)	78	—	370	4.0	49.0	2.0	19.0	9.0	—	—
34965	Quarter chicken, dark meat, no skin	1	item(s)	134	—	260	30.0	2.0	0	13.0	4.0	—	—
34966	Quarter chicken, dark meat, with skin	1	item(s)	149	—	280	31.0	3.0	0	15.0	4.5	—	—
34963	Quarter chicken, white meat, no skin or wing	1	item(s)	173	—	250	41.0	4.0	0	8.0	2.5	—	—
34964	Quarter chicken, white meat, with skin and wing	1	item(s)	110	—	330	50.0	3.0	0	12.0	4.0	—	—
34968	Roasted turkey breast	5	ounce(s)	142	—	180	38.0	0	0	3.0	1.0	—	—
35011	Seasonal fresh fruit salad	1	serving(s)	142	—	60	1.0	15.0	1.0	0	0	0	0
51192	Spinach with garlic butter sauce	1	serving(s)	170	—	130	5.0	9.0	5.0	9.0	6.0	—	—
34969	Spiral sliced holiday ham	8	ounce(s)	227	—	450	40.0	13.0	0	26.0	10.0	—	—
35003	Steamed vegetables	1	cup(s)	136	—	50	2.0	8.0	3.0	2.0	0	—	—
35005	Sweet corn	¾	cup(s)	176	—	170	6.0	37.0	2.0	4.0	1.0	—	—
35004	Sweet potato casserole	¾	cup(s)	198	—	460	4.0	77.0	3.0	17.0	6.0	—	—
	Burger King												
29731	Biscuit with sausage, egg, and cheese	1	item(s)	191	—	610	20.0	33.0	1.0	45.0	15.0	—	—
14249	Cheeseburger	1	item(s)	133	—	330	17.0	31.0	1.0	16.0	7.0	—	—
14251	Chicken sandwich	1	item(s)	219	—	660	24.0	52.0	4.0	40.0	8.0	—	—
3808	Chicken Tenders, 8 pieces	1	serving(s)	123	—	340	19.0	21.0	0.5	20.0	5.0	—	—
14259	Chocolate shake, small	1	item(s)	315	—	470	8.0	75.0	1.0	14.0	9.0	—	—
29732	Croissanwich with sausage and cheese	1	item(s)	106	37.2	370	14.0	23.0	0.5	25.0	9.0	12.7	3.3

Chol (mg)	Calc (mg)	Iron (mg)	Magn (mg)	Pota (mg)	Sodi (mg)	Zinc (mg)	Vit A (μg)	Thia (mg)	Vit E (mg α)	Ribo (mg)	Niac (mg)	Vit B_6 (mg)	Fola (μg DFE)	Vit C (mg)	Vit B_{12} (μg)	Sele (μg)
25	30	2.34	—	—	430.0	—	—	—	—	—	—	—	—	0	—	—
0	0	0.00	—	—	180.0	—	0	—	—	—	—	—	—	0	—	—
0	30	2.34	—	—	900.0	—	0	—	—	—	—	—	—	0	—	—
10	30	2.16	—	—	930.0	—	—	—	—	—	—	—	—	0	—	—
10	80	1.80	—	—	780.0	—	—	—	—	—	—	—	—	1.2	—	—
30	60	1.80	—	—	660.0	—	—	—	—	—	—	—	—	1.2	—	—
0	20	2.88	—	—	840.0	—	0	—	—	—	—	—	—	0	—	—
15	20	2.70	—	—	860.0	—	—	—	—	—	—	—	—	0	—	—
0	30	1.98	—	—	920.0	—	—	—	—	—	—	—	—	0	—	—
10	40	2.16	—	—	930.0	—	—	—	—	—	—	—	—	0	—	—
40	0	0.00	—	—	120.0	—	0	—	—	—	—	—	—	0	—	—
0	30	1.98	—	—	1100.0	—	0	—	—	—	—	—	—	0	—	—
10	30	2.34	—	—	1120.0	—	—	—	—	—	—	—	—	0	—	—
10	59	0.80	—	—	35.0	—	—	—	—	—	—	—	—	22.2	—	—
0	60	0.43	—	—	75.0	—	—	—	—	—	—	—	—	5.4	—	—
90	211	2.85	—	—	1560.0	—	—	—	—	—	—	—	—	15.8	—	—
0	0	0.00	—	—	570.0	—	0	—	—	—	—	—	—	0	—	—
55	0	1.07	—	—	220.0	—	—	—	—	—	—	—	—	1.8	—	—
115	40	4.50	—	—	800.0	—	—	—	—	—	—	—	—	1.2	—	—
45	123	1.32	—	—	1310.0	—	—	—	—	—	—	—	—	18.4	—	—
10	41	0.48	—	—	270.0	—	—	—	—	—	—	—	—	24.5	—	—
5	0	0.71	—	—	220.0	—	0	—	—	—	—	—	—	0	—	—
70	264	2.84	—	—	580.0	—	—	—	—	—	—	—	—	9.5	—	—
0	41	1.48	—	—	580.0	—	—	—	—	—	—	—	—	2.5	—	—
0	0	0.85	—	—	120.0	—	0	—	—	—	—	—	—	14.3	—	—
5	20	0.72	—	—	620.0	—	—	—	—	—	—	—	—	2.4	—	—
0	43	0.38	—	—	180.0	—	—	—	—	—	—	—	—	5.1	—	—
25	51	0.46	—	—	660.0	—	—	—	—	—	—	—	—	19.2	—	—
25	100	0.59	—	—	1230.0	—	—	—	—	—	—	—	—	24.9	—	—
0	16	0.28	—	—	15.0	—	—	—	—	—	—	—	—	0	—	—
30	345	1.65	—	—	1290.0	—	—	—	—	—	—	—	—	0	—	—
10	—	—	—	—	2010.0	—	—	—	—	—	—	—	—	—	—	—
125	140	3.77	—	—	970.0	—	—	—	—	—	—	—	—	1.8	—	—
20	0	1.32	—	—	340.0	—	—	—	—	—	—	—	—	0	—	—
155	0	1.52	—	—	260.0	—	0	—	—	—	—	—	—	0	—	—
155	0	2.14	—	—	660.0	—	0	—	—	—	—	—	—	0	—	—
125	0	0.89	—	—	480.0	—	0	—	—	—	—	—	—	0	—	—
165	0	0.78	—	—	960.0	—	0	—	—	—	—	—	—	0	—	—
70	20	1.80	—	—	620.0	—	0	—	—	—	—	—	—	0	—	—
0	16	0.29	—	—	20.0	—	—	—	—	—	—	—	—	29.5	—	—
20	—	—	—	—	200.0	—	—	—	—	—	—	—	—	—	—	—
140	0	1.73	—	—	2230.0	—	0	—	—	—	—	—	—	0	—	—
0	53	0.46	—	—	45.0	—	—	—	—	—	—	—	—	24.0	—	—
0	0	0.43	—	—	95.0	—	—	—	—	—	—	—	—	5.8	—	—
20	44	1.18	—	—	210.0	—	—	—	—	—	—	—	—	9.8	—	—
210	250	2.70	—	—	1620.0	—	89.9	—	—	—	—	—	—	0	—	—
55	150	2.70	—	—	780.0	—	—	0.24	—	0.31	4.17	—	—	1.2	—	—
70	64	2.89	—	—	1440.0	—	—	0.50	—	0.32	10.29	—	—	0	—	—
55	20	0.72	—	—	960.0	—	—	0.14	—	0.11	10.93	—	—	0	—	—
55	333	0.79	—	—	350.0	—	—	0.11	—	0.61	0.26	—	—	2.7	0	—
50	99	1.78	20.1	217.3	810.0	1.51	—	0.34	1.03	0.33	4.33	—	—	0	0.6	22.2

DA+ Code	Food Description	QTY	Measure	Wt (g)	H_2O (g)	Ener (cal)	Prot (g)	Carb (g)	Fiber (g)	Fat (g)	Fat Breakdown (g)		
											Sat	Mono	Poly
FAST FOOD—continued													
14261	Croissanwich with sausage, egg, and cheese	1	item(s)	159	71.4	470	19.0	26.0	0.5	32.0	11.0	15.8	6.1
3809	Double cheeseburger	1	item(s)	189	—	500	30.0	31.0	1.0	29.0	14.0	—	—
14244	Double Whopper sandwich	1	item(s)	373	—	900	47.0	51.0	3.0	57.0	19.0	—	—
14245	Double Whopper with cheese sandwich	1	item(s)	398	—	990	52.0	52.0	3.0	64.0	24.0	—	—
14250	Fish Filet sandwich	1	item(s)	250	—	630	24.0	67.0	4.0	30.0	6.0	—	—
14255	French fries, medium, salted	1	serving(s)	116	—	360	4.0	41.0	4.0	20.0	4.5	—	—
14262	French toast sticks, 5 pieces	1	serving(s)	112	37.6	390	6.0	46.0	2.0	20.0	4.5	10.6	2.9
14248	Hamburger	1	item(s)	121	—	290	15.0	30.0	1.0	12.0	4.5	—	—
14263	Hash brown rounds, small	1	serving(s)	75	27.1	230	2.0	23.0	2.0	15.0	4.0	—	—
14256	Onion rings, medium	1	serving(s)	91	—	320	4.0	40.0	3.0	16.0	4.0	—	—
39000	Tendercrisp chicken sandwich	1	item(s)	286	—	780	25.0	73.0	4.0	43.0	8.0	—	—
37514	TenderGrill chicken sandwich	1	item(s)	258	—	450	37.0	53.0	4.0	10.0	2.0	—	—
14258	Vanilla shake, small	1	item(s)	296	—	400	8.0	57.0	0	15.0	9.0	—	—
1736	Whopper sandwich	1	item(s)	290	—	670	28.0	51.0	3.0	39.0	11.0	—	—
14243	Whopper with cheese sandwich	1	item(s)	315	—	760	33.0	52.0	3.0	47.0	16.0	—	—
	Carl's Jr												
33962	Carl's bacon Swiss crispy chicken sandwich	1	item(s)	268	—	750	31.0	91.0	—	28.0	28.0	—	—
10801	Carl's Catch fish sandwich	1	item(s)	215	—	560	19.0	58.0	2.0	27.0	7.0	—	1.9
10862	Carl's Famous Star hamburger	1	item(s)	254	—	590	24.0	50.0	3.0	32.0	9.0	—	—
10785	Charbroiled chicken club sandwich	1	item(s)	270	—	550	42.0	43.0	4.0	23.0	7.0	—	2.9
10866	Charbroiled chicken salad	1	item(s)	437	—	330	34.0	17.0	5.0	7.0	4.0	—	1.0
10855	Charbroiled Santa Fe chicken sandwich	1	item(s)	266	—	610	38.0	43.0	4.0	32.0	8.0	—	—
10790	Chicken stars, 6 pieces	1	serving(s)	85	—	260	13.0	14.0	1.0	16.0	4.0	—	1.6
34864	Chocolate shake, small	1	serving(s)	595	—	540	15.0	98.0	0	11.0	7.0	—	—
10797	Crisscut fries	1	serving(s)	139	—	410	5.0	43.0	4.0	24.0	5.0	—	—
10799	Double Western Bacon cheeseburger	1	item(s)	308	—	920	51.0	65.0	2.0	50	21.0	—	6.6
14238	French fries, small	1	serving(s)	92	—	290	5.0	37.0	3.0	14.0	3.0	—	—
10798	French toast dips without syrup, 5 pieces	1	serving(s)	155	—	370	8.0	49.0	0	17.0	5.0	—	1.4
10802	Onion rings	1	serving(s)	128	—	440	7.0	53.0	3.0	22.0	5.0	—	0.8
34858	Spicy chicken sandwich	1	item(s)	198	—	480	14.0	48.0	2.0	26.0	5.0	—	—
34867	Strawberry shake, small	1	serving(s)	595	—	520	14.0	93.0	0	11.0	7.0	—	—
10865	Super Star hamburger	1	item(s)	348	—	790	41.0	52.0	3.0	47.0	14.0	—	—
38925	The Six Dollar burger	1	item(s)	429	—	1010	40.0	60.0	3.0	66.0	26.0	—	—
10818	Vanilla shake, small	1	item(s)	398	—	314	10.0	51.5	0	7.4	4.7	—	—
10770	Western Bacon cheeseburger	1	item(s)	225	—	660	32.0	64.0	2.0	30.0	12.0	—	4.8
	Chick Fil-A												
38746	Biscuit with bacon, egg, and cheese	1	item(s)	163	—	470	18.0	39.0	1.0	26.0	9.0	—	—
38747	Biscuit with egg	1	item(s)	135	—	350	11.0	38.0	1.0	16.0	4.5	—	—
38748	Biscuit with egg and cheese	1	item(s)	149	—	400	14.0	38.0	1.0	21.0	7.0	—	—
38753	Biscuit with gravy	1	item(s)	192	—	330	5.0	43.0	1.0	15.0	4.0	—	—
38752	Biscuit with sausage, egg, and cheese	1	item(s)	212	—	620	22.0	39.0	2.0	42.0	14.0	—	—
38771	Carrot and raisin salad	1	item(s)	113	—	170	1.0	28.0	2.0	6.0	1.0	—	—
38761	Chargrilled chicken Cool Wrap	1	item(s)	245	—	390	29.0	54.0	3.0	7.0	3.0	—	—
38766	Chargrilled chicken garden salad	1	item(s)	275	—	180	22.0	9.0	3.0	6.0	3.0	—	—
38758	Chargrilled chicken sandwich	1	item(s)	193	—	270	28.0	33.0	3.0	3.5	1.0	—	—
38742	Chicken biscuit	1	item(s)	145	—	420	18.0	44.0	2.0	19.0	4.5	—	—
38743	Chicken biscuit with cheese	1	item(s)	159	—	470	21.0	45.0	2.0	23.0	8.0	—	—
38762	Chicken Caesar Cool Wrap	1	item(s)	227	—	460	36.0	52.0	3.0	10.0	6.0	—	—
38757	Chicken deluxe sandwich	1	item(s)	208	—	420	28.0	39.0	2.0	16.0	3.5	—	—
38764	Chicken salad sandwich on wheat bun	1	item(s)	153	—	350	20.0	32.0	5.0	15.0	3.0	—	—
38756	Chicken sandwich	1	item(s)	170	—	410	28.0	38.0	1.0	16.0	3.5	—	—
38768	Chick-n-Strip salad	1	item(s)	327	—	400	34.0	21.0	4.0	20.0	6.0	—	—
38763	Chick-n-Strips	4	item(s)	127	—	300	28.0	14.0	1.0	15.0	2.5	—	—
38770	Cole slaw	1	item(s)	128	—	260	2.0	17.0	2.0	21.0	3.5	—	—
38776	Diet lemonade, small	1	cup(s)	255	—	25	0	5.0	0	0	0	0	0

Chol (mg)	Calc (mg)	Iron (mg)	Magn (mg)	Pota (mg)	Sodi (mg)	Zinc (mg)	Vit A (µg)	Thia (mg)	Vit E (mg α)	Ribo (mg)	Niac (mg)	Vit B_6 (mg)	Fola (µg DFE)	Vit C (mg)	Vit B_{12} (µg)	Sele (µg)
180	146	2.63	28.6	313.2	1060.0	2.08	—	0.38	1.66	0.51	4.72	0.28	—	0	1.1	38.0
105	250	4.50	—	—	1030.0	—	—	0.26	—	0.44	6.37	—	—	1.2	—	—
175	150	8.07	—	—	1090.0	—	—	0.39	—	0.59	11.05	—	—	9.0	—	—
195	299	8.08	—	—	1520.0	—	—	0.39	—	0.66	11.03	—	—	9.0	—	—
60	101	3.62	—	—	1380.0	—	—	—	—	—	—	—	—	3.6	—	—
0	20	0.71	—	—	590.0	—	0	0.15	—	0.48	2.30	—	—	8.9	—	—
0	60	1.80	21.3	124.3	440.0	0.57	—	0.31	0.98	0.19	2.88	0.05	—	0	0	13.7
40	80	2.70	—	—	560.0	—	—	0.25	—	0.28	4.25	—	—	1.2	—	—
0	0	0.36	—	—	450.0	—	0	0.11	0.83	0.06	1.35	0.17	—	1.2	—	—
0	100	0.00	—	—	460.0	—	0	0.14	—	0.09	2.32	—	—	0	—	—
75	79	4.43	—	—	1730.0	—	—	—	—	—	—	—	—	8.9	—	—
75	57	6.82	—	—	1210.0	—	—	—	—	—	—	—	—	5.7	—	—
60	348	0.00	—	—	240.0	—	—	0.11	—	0.63	0.21	—	—	2.4	0	—
51	100	5.38	—	—	1020.0	—	—	0.38	—	0.43	7.30	—	—	9.0	—	—
115	249	5.38	—	—	1450.0	—	—	0.38	—	0.51	7.28	—	—	9.0	—	—
80	200	5.40	—	—	1900.0	—	—	—	—	—	—	—	—	2.4	—	—
80	150	2.70	—	—	990.0	—	60.0	—	—	—	—	—	—	2.4	—	—
70	100	4.50	—	—	910.0	—	—	—	—	—	—	—	—	6.0	—	—
95	200	3.60	—	—	1330.0	—	—	—	—	—	—	—	—	9.0	—	—
75	200	1.80	—	—	880.0	—	—	—	—	—	—	—	—	30.0	—	—
100	200	3.60	—	—	1440.0	—	—	—	—	—	—	—	—	9.0	—	—
35	19	1.02	—	—	470.0	—	0	—	—	—	—	—	—	0	—	—
45	600	1.08	—	—	360.0	—	0	—	—	—	—	—	—	0	—	—
0	20	1.80	—	—	950.0	—	0	—	—	—	—	—	—	12.0	—	—
155	300	7.20	—	—	1730.0	—	—	—	—	—	—	—	—	1.2	—	—
0	0	1.08	—	—	170.0	—	0	—	—	—	—	—	—	21.0	—	—
3	0	0.00	—	—	470.0	—	0	0.25	—	0.23	2.00	—	—	0	—	—
0	20	0.72	—	—	700.0	—	0	—	—	—	—	—	—	3.6	—	—
40	100	3.60	—	—	1220.0	—	—	—	—	—	—	—	—	6.0	—	—
45	600	0.00	—	—	340.0	—	0	—	—	—	—	—	—	0	—	—
130	100	7.20	—	—	980.0	—	—	—	—	—	—	—	—	9.0	—	—
145	279	4.29	—	—	1960.0	—	—	—	—	—	—	—	—	16.7	—	—
30	401	0.00	—	—	234.0	—	0	—	—	—	—	—	—	0	—	—
85	200	5.40	—	—	1410.0	—	60.0	—	—	—	—	—	—	1.2	—	—
270	150	2.70	—	—	1190.0	—	—	—	—	—	—	—	—	0	—	—
240	80	2.70	—	—	740.0	—	—	—	—	—	—	—	—	0	—	—
255	150	2.70	—	—	970.0	—	—	—	—	—	—	—	—	0	—	—
5	60	1.80	—	—	930.0	—	0	—	—	—	—	—	—	0	—	—
300	200	3.60	—	—	1360.0	—	—	—	—	—	—	—	—	0	—	—
10	40	0.36	—	—	110.0	—	—	—	—	—	—	—	—	4.8	—	—
65	200	3.60	—	—	1020.0	—	—	—	—	—	—	—	—	6.0	—	—
65	150	0.72	—	—	620.0	—	—	—	—	—	—	—	—	30	—	—
65	80	2.70	—	—	940.0	—	—	—	—	—	—	—	—	6.0	—	—
35	60	2.70	—	—	1270.0	—	0	—	—	—	—	—	—	0	—	—
50	150	2.70	—	—	1500.0	—	—	—	—	—	—	—	—	0	—	—
80	500	3.60	—	—	1350.0	—	—	—	—	—	—	—	—	1.2	—	—
60	100	2.70	—	—	1300.0	—	—	—	—	—	—	—	—	2.4	—	—
65	150	1.80	—	—	880.0	—	—	—	—	—	—	—	—	0	—	—
60	100	2.70	—	—	1300.0	—	—	—	—	—	—	—	—	0	—	—
80	150	1.44	—	—	1070.0	—	—	—	—	—	—	—	—	6.0	—	—
65	40	1.44	—	—	940.0	—	—	—	—	—	—	—	—	0	—	—
25	60	0.36	—	—	220.0	—	—	—	—	—	—	—	—	36.0	—	—
0	0	0.36	—	—	5.0	—	0	—	—	—	—	—	—	15.0	—	—

DA+ Code	Food Description	QTY	Measure	Wt (g)	H₂O (g)	Ener (cal)	Prot (g)	Carb (g)	Fiber (g)	Fat (g)	Fat Breakdown (g)		
											Sat	Mono	Poly
	FAST FOOD—continued												
38755	Hashbrowns	1	serving(s)	84	—	260	2.0	25.0	3.0	17.0	3.5	—	—
38765	Hearty breast of chicken soup	1	cup(s)	241	—	140	8.0	18.0	1.0	3.5	1.0	—	—
38741	Hot buttered biscuit	1	item(s)	79	—	270	4.0	38.0	1.0	12.0	3.0	—	—
38778	IceDream, small cone	1	item(s)	135	—	160	4.0	28.0	0	4.0	2.0	—	—
38774	IceDream, small cup	1	serving(s)	227	—	240	6.0	41.0	0	6.0	3.5	—	—
38775	Lemonade, small	1	cup(s)	255	—	170	0	41.0	0	0.5	0	—	—
38777	Nuggets	8	item(s)	113	—	260	26.0	12.0	0.5	12.0	2.5	—	—
38769	Side salad	1	item(s)	108	—	60	3.0	4.0	2.0	3.0	1.5	—	—
38767	Southwest chargrilled salad	1	item(s)	303	—	240	25.0	17.0	5.0	8.0	3.5	—	—
40481	Spicy chicken cool wrap	1	serving(s)	230	—	380	30.0	52.0	3.0	6.0	3.0	—	—
38772	Waffle potato fries, small, salted	1	serving(s)	85	—	270	3.0	34.0	4.0	13.0	3.0	—	—
	Cinnabon												
39572	Caramellata Chill w/whipped cream	16	fluid ounce(s)	480	—	406	10.0	61.0	0	14.0	8.0	—	—
39571	Cinnabon Bites	1	serving(s)	149	—	510	8.0	77.0	2.0	19.0	5.0	—	—
39570	Cinnabon Stix	5	item(s)	85	—	379	6.0	41.0	1.0	21.0	6.0	—	—
39567	Classic roll	1	item(s)	221	—	813	15.0	117.0	4.0	32.0	8.0	—	—
39568	Minibon	1	item(s)	92	—	339	6.0	49.0	2.0	13.0	3.0	—	—
39573	Mochalatta Chill w/whipped cream	16	fluid ounce(s)	480	—	362	9.0	55.0	0	13.0	8.0	—	—
39569	Pecanbon	1	item(s)	272	—	1100	16.0	141.0	8.0	56.0	10.0	—	—
	Dairy Queen												
1466	Banana split	1	item(s)	369	—	510	8.0	96.0	3.0	12.0	8.0	—	—
38552	Brownie Earthquake®	1	serving(s)	304	—	740	10.0	112.0	0	27.0	16.0	—	—
38561	Chocolate chip cookie dough blizzard,® small	1	item(s)	319	—	720	12.0	105.0	0	28.0	14.0	—	—
1464	Chocolate malt, small	1	item(s)	418	—	640	15.0	111.0	1.0	16.0	11.0	—	—
38541	Chocolate shake, small	1	item(s)	397	—	560	13.0	93.0	1.0	15.0	10.0	—	—
17257	Chocolate soft serve	½	cup(s)	94	—	150	4.0	22.0	0	5.0	3.5	—	—
1463	Chocolate sundae, small	1	item(s)	163	—	280	5.0	49.0	0	7.0	4.5	—	—
1462	Dipped cone, small	1	item(s)	156	—	340	6.0	42.0	1.0	17.0	9.0	4.0	3.0
38555	Oreo cookies blizzard, small	1	item(s)	283	—	570	11.0	83.0	0.5	21.0	10.0	—	—
38547	Royal Treats Peanut Buster® Parfait	1	item(s)	305	—	730	16.0	99.0	2.0	31.0	17.0	—	—
17256	Vanilla soft serve	½	cup(s)	94	—	140	3.0	22.0	0	4.5	3.0	—	—
	Domino's												
31606	Barbeque buffalo wings	1	item(s)	25	—	50	6.0	2.0	0	2.5	0.5	—	—
31604	Breadsticks	1	item(s)	30	—	115	2.0	12.0	0	6.3	1.1	—	—
37551	Buffalo Chicken Kickers	1	item(s)	24	—	47	4.0	3.0	0	2.0	0.5	—	—
37548	CinnaStix	1	item(s)	30	—	123	2.0	15.0	1.0	6.1	1.1	—	—
37549	Dot, cinnamon	1	item(s)	28	7.6	99	1.9	14.9	0.7	3.7	0.7	—	—
31605	Double cheesy bread	1	item(s)	35	—	123	4.0	13.0	0	6.5	1.9	—	—
31607	Hot buffalo wings	1	item(s)	25	—	45	5.0	1.0	0	2.5	0.5	—	—
	Domino's Classic hand tossed pizza												
31573	America's favorite feast, 12″	1	slice(s)	102	—	257	10.0	29.0	2.0	11.5	4.5	—	—
31574	America's favorite feast, 14″	1	slice(s)	141	—	353	14.0	39.0	2.0	16.0	6.0	—	—
37543	Bacon cheeseburger feast, 12″	1	slice(s)	99	—	273	12.0	28.0	2.0	13.0	5.5	—	—
37545	Bacon cheeseburger feast, 14″	1	slice(s)	137	—	379	17.0	38.0	2.0	18.0	8.0	—	—
37546	Barbeque feast, 12″	1	slice(s)	96	—	252	11.0	31.0	1.0	10.0	4.5	—	—
37547	Barbeque feast, 14″	1	slice(s)	131	—	344	14.0	43.0	2.0	13.5	6.0	—	—
31569	Cheese, 12″	1	slice(s)	55	—	160	6.0	28.0	1.0	3.0	1.0	—	—
31570	Cheese, 14″	1	slice(s)	75	—	220	8.0	38.0	2.0	4.0	1.0	—	—
37538	Deluxe feast, 12″	1	slice(s)	201	101.8	465	19.5	57.4	3.5	18.2	7.7	—	—
37540	Deluxe feast, 14″	1	slice(s)	273	138.4	627	26.4	78.3	4.7	24.1	10.2	—	—
31685	Deluxe, 12″	1	slice(s)	100	—	234	9.0	29.0	2.0	9.5	3.5	—	—
31694	Deluxe, 14″	1	slice(s)	136	—	316	13.0	39.0	2.0	12.5	5.0	—	—
31686	Extravaganzza, 12″	1	slice(s)	122	—	289	13.0	30.0	2.0	14.0	5.5	—	—
31695	Extravaganzza, 14″	1	slice(s)	165	—	388	17.0	40.0	3.0	18.5	7.5	—	—
31575	Hawaiian feast, 12″	1	slice(s)	102	—	223	10.0	30.0	2.0	8.0	3.5	—	—
31576	Hawaiian feast, 14″	1	slice(s)	141	—	309	14.0	41.0	2.0	11.0	4.5	—	—
31687	Meatzza, 12″	1	slice(s)	108	—	281	13.0	29.0	2.0	13.5	5.5	—	—

Chol (mg)	Calc (mg)	Iron (mg)	Magn (mg)	Pota (mg)	Sodi (mg)	Zinc (mg)	Vit A (μg)	Thia (mg)	Vit E (mg α)	Ribo (mg)	Niac (mg)	Vit B_6 (mg)	Fola (μg DFE)	Vit C (mg)	Vit B_{12} (μg)	Sele (μg)
5	20	0.72	—	—	380.0	—	—	—	—	—	—	—	—	0	—	—
25	40	1.08	—	—	900.0	—	—	—	—	—	—	—	—	0	—	—
0	60	1.80	—	—	660.0	—	0	—	—	—	—	—	—	0	—	—
15	100	0.36	—	—	80.0	—	—	—	—	—	—	—	—	0	—	—
25	200	0.36	—	—	105.0	—	—	—	—	—	—	—	—	0	—	—
0	0	0.36	—	—	10.0	—	0	—	—	—	—	—	—	15.0	—	—
70	40	1.08	—	—	1090.0	—	0	—	—	—	—	—	—	0	—	—
10	100	0.00	—	—	75.0	—	—	—	—	—	—	—	—	15.0	—	—
60	200	1.08	—	—	770.0	—	—	—	—	—	—	—	—	24.0	—	—
60	200	3.60	—	—	1090.0	—	—	—	—	—	—	—	—	3.6	—	—
0	20	1.08	—	—	115.0	—	0	—	—	—	—	—	—	1.2	—	—
46	—	—	—	—	187.0	—	—	—	—	—	—	—	—	—	—	—
35	—	—	—	—	530.0	—	—	—	—	—	—	—	—	—	—	—
16	—	—	—	—	413.0	—	—	—	—	—	—	—	—	—	—	—
67	—	—	—	—	801.0	—	—	—	—	—	—	—	—	—	—	—
27	—	—	—	—	337.0	—	—	—	—	—	—	—	—	—	—	—
46	—	—	—	—	252.0	—	—	—	—	—	—	—	—	—	—	—
63	—	—	—	—	600.0	—	—	—	—	—	—	—	—	—	—	—
30	250	1.80	—	—	180.0	—	—	—	—	—	—	—	—	15.0	—	—
50	250	1.80	—	—	350.0	—	—	—	—	—	—	—	—	0	—	—
50	350	2.70	—	—	370.0	—	—	—	—	—	—	—	—	1.2	—	—
55	450	1.80	—	—	340.0	—	—	—	—	—	—	—	—	2.4	—	—
50	450	1.44	—	—	280.0	—	—	—	—	—	—	—	—	2.4	—	—
15	100	0.72	—	—	75.0	—	—	—	—	—	—	—	—	0	—	—
20	200	1.08	—	—	140.0	—	—	—	—	—	—	—	—	0	—	—
20	200	1.08	—	—	130.0	—	—	—	—	—	—	—	—	0	—	—
40	350	2.70	—	—	430.0	—	—	—	—	—	—	—	—	1.2	—	—
35	300	1.80	—	—	400.0	—	—	—	—	—	—	—	—	1.2	—	—
15	150	0.72	—	—	70.0	—	—	—	—	—	—	—	—	1.2	—	—
														0		
26	10	0.36	—	—	175.5	—	—	—	—	—	—	—	—	0	—	—
0	0	0.72	—	—	122.1	—	—	—	—	—	—	—	—	0	—	—
9	0	0.00	—	—	162.5	—	—	—	—	—	—	—	—	0	—	—
0	0	0.72	—	—	111.4	—	—	—	—	—	—	—	—	0	—	—
0	6	0.59	—	—	85.7	—	—	—	—	—	—	—	—	0	—	—
6	40	0.72	—	—	162.3	—	—	—	—	—	—	—	—	0	—	—
26	10	0.36	—	—	254.5	—	—	—	—	—	—	—	—	1.2	—	—
																—
22	100	1.80	—	—	625.5	—	—	—	—	—	—	—	—	0.6	—	—
31	140	2.52	—	—	865.5	—	—	—	—	—	—	—	—	0.6	—	—
27	140	1.80	—	—	634.0	—	—	—	—	—	—	—	—	0	—	—
38	190	2.52	—	—	900.0	—	—	—	—	—	—	—	—	0	—	—
20	140	1.62	—	—	600.0	—	—	—	—	—	—	—	—	0.6	—	—
27	190	2.16	—	—	831.5	—	—	—	—	—	—	—	—	0.6	—	—
0	0	1.80	—	—	110.0	—	0	—	—	—	—	—	—	0	—	—
0	0	2.70	—	—	150.0	—	0	—	—	—	—	—	—	0	—	—
40	199	3.56	—	—	1063.1	—	—	—	—	—	—	—	—	1.4	—	—
53	276	4.84	—	—	1432.2	—	—	—	—	—	—	—	—	1.8	—	—
17	100	1.80	—	—	541.5	—	—	—	—	—	—	—	—	0.6	—	—
23	130	2.34	—	—	728.5	—	—	—	—	—	—	—	—	1.2	—	—
28	140	1.98	—	—	764.0	—	—	—	—	—	—	—	—	0.6	—	—
37	190	2.70	—	—	1014.0	—	—	—	—	—	—	—	—	1.2	—	—
16	130	1.62	—	—	546.5	—	—	—	—	—	—	—	—	1.2	—	—
23	180	2.34	—	—	765.0	—	—	—	—	—	—	—	—	1.2	—	—
28	130	1.80	—	—	739.5	—	—	—	—	—	—	—	—	0	—	—

DA+ Code	Food Description	QTY	Measure	Wt (g)	H₂O (g)	Ener (cal)	Prot (g)	Carb (g)	Fiber (g)	Fat (g)	Fat Breakdown (g) Sat	Mono	Poly
	FAST FOOD—continued												
31696	Meatzza, 14″	1	slice(s)	146	—	378	17.0	39.0	2.0	18.0	7.5	—	—
31571	Pepperoni feast, extra pepperoni and cheese, 12″	1	slice(s)	98	—	265	11.0	28.0	2.0	12.5	5.0	—	—
31572	Pepperoni feast, extra pepperoni and cheese, 14″	1	slice(s)	135	—	363	16.0	39.0	2.0	17.0	7.0	—	—
31577	Vegi feast, 12″	1	slice(s)	102	—	218	9.0	29.0	2.0	8.0	3.5	—	—
31578	Vegi feast, 14″	1	slice(s)	139	—	300	13.0	40.0	3.0	11.0	4.5	—	—
	Domino's thin crust pizza												
31583	America's favorite, 12″	1	slice(s)	72	—	208	8.0	15.0	1.0	13.5	5.0	—	—
31584	America's favorite, 14″	1	slice(s)	100	—	285	11.0	20.0	2.0	18.5	7.0	—	—
31579	Cheese, 12″	1	slice(s)	49	—	137	5.0	14.0	1.0	7.0	2.5	—	—
31580	Cheese, 14″	1	slice(s)	68	27.0	214	8.8	19.0	1.4	11.4	4.6	2.9	2.5
31688	Deluxe, 12″	1	slice(s)	70	—	185	7.0	15.0	1.0	11.5	4.0	—	—
31697	Deluxe, 14″	1	slice(s)	94	—	248	10.0	20.0	2.0	15.0	5.5	—	—
31689	Extravaganzza, 12″	1	slice(s)	92	—	240	11.0	16.0	1.0	15.5	6.0	—	—
31698	Extravaganzza, 14″	1	slice(s)	123	—	320	14.0	21.0	2.0	20.5	8.0	—	—
31585	Hawaiian, 12″	1	slice(s)	71	—	174	8.0	16.0	1.0	9.5	3.5	—	—
31586	Hawaiian, 14″	1	slice(s)	100	—	240	11.0	21.0	2.0	13.0	5.0	—	—
31690	Meatzza, 12″	1	slice(s)	78	—	232	11.0	15.0	1.0	15.0	6.0	—	—
31699	Meatzza, 14″	1	slice(s)	104	—	310	14.0	20.0	2.0	20	8.0	—	—
31581	Pepperoni, extra pepperoni and cheese, 12″	1	slice(s)	68	—	216	9.0	14.0	1.0	14.0	5.5	—	—
31582	Pepperoni, extra pepperoni and cheese, 14″	1	slice(s)	93	—	295	13.0	20.0	1.0	19.0	7.5	—	—
31587	Vegi, 12″	1	slice(s)	71	—	168	7.0	15.0	1.0	9.5	3.5	—	—
31588	Vegi, 14″	1	slice(s)	97	—	231	10.0	21.0	2.0	13.5	5.0	—	—
	Domino's Ultimate deep dish pizza												
31596	America's favorite, 12″	1	slice(s)	115	—	309	12.0	29.0	2.0	17.0	6.0	—	—
31702	America's favorite, 14″	1	slice(s)	162	—	433	17.0	42.0	3.0	23.5	8.0	—	—
31590	Cheese, 12″	1	slice(s)	90	—	238	9.0	28.0	2.0	11.0	3.5	—	—
31591	Cheese, 14″	1	slice(s)	128	53.9	351	14.5	41.0	2.9	13.2	5.2	3.8	2.5
31589	Cheese, 6″	1	item(s)	215	—	598	22.9	68.4	3.9	27.6	9.9	—	—
31691	Deluxe, 12″	1	slice(s)	122	—	287	11.0	29.0	2.0	15.0	5.0	—	—
31700	Deluxe, 14″	1	slice(s)	156	—	396	15.0	42.0	3.0	20.0	7.0	—	—
31692	Extravaganzza, 12″	1	slice(s)	136	—	341	14.0	30.0	2.0	19.0	7.0	—	—
31701	Extravaganzza, 14″	1	slice(s)	186	—	468	20.0	43.0	3.0	25.5	9.5	—	—
31599	Hawaiian, 12″	1	slice(s)	114	—	275	12.0	30.0	2.0	13.0	5.0	—	—
31600	Hawaiian, 14″	1	slice(s)	162	—	389	17.0	43.0	3.0	18.0	6.5	—	—
31693	Meatzza, 12″	1	slice(s)	121	—	333	14.0	29.0	2.0	19.0	7.0	—	—
31703	Meatzza, 14″	1	slice(s)	167	—	458	19.0	42.0	3.0	25.0	9.5	—	—
31593	Pepperoni, extra pepperoni and cheese, 12″	1	slice(s)	110	—	317	13.0	29.0	2.0	17.5	6.5	—	—
31594	Pepperoni, extra pepperoni and cheese, 14″	1	slice(s)	155	—	443	18.0	42.0	3.0	24.0	9.0	—	—
31602	Vegi, 12″	1	slice(s)	114	—	270	11.0	30.0	2.0	13.5	5.0	—	—
31603	Vegi, 14″	1	slice(s)	159	—	380	15.0	43.0	3.0	18.0	6.5	—	—
31598	With ham and pineapple tidbits, 6″	1	item(s)	430	—	619	25.2	69.9	4.0	28.3	10.2	—	—
31595	With Italian sausage, 6″	1	item(s)	430	—	642	24.8	69.6	4.2	31.1	11.3	—	—
31592	With pepperoni, 6″	1	item(s)	430	—	647	25.1	68.5	3.9	32.0	11.7	—	—
31601	With vegetables, 6″	1	item(s)	430	—	619	23.4	70.8	4.6	28.7	10.1	—	—
	In-n-Out Burger												
34391	Cheeseburger with mustard and ketchup	1	serving(s)	268	—	400	22.0	41.0	3.0	18.0	9.0	—	—
34374	Cheeseburger	1	serving(s)	268	—	480	22.0	39.0	3.0	27.0	10.0	—	—

Chol (mg)	Calc (mg)	Iron (mg)	Magn (mg)	Pota (mg)	Sodi (mg)	Zinc (mg)	Vit A (μg)	Thia (mg)	Vit E (mg α)	Ribo (mg)	Niac (mg)	Vit B_6 (mg)	Fola (μg DFE)	Vit C (mg)	Vit B_{12} (μg)	Sele (μg)
37	190	2.52	—	—	983.5	—	—	—	—	—	—	—	—	0	—	—
24	130	1.62	—	—	670.0	—	70.9	—	—	—	—	—	—	0	—	—
33	180	2.34	—	—	920.0	—	104.7	—	—	—	—	—	—	0	—	—
13	130	1.62	—	—	489.0	—	—	—	—	—	—	—	—	0.6	—	—
18	180	2.34	—	—	678.0	—	—	—	—	—	—	—	—	0.6	—	—
23	100	0.90	—	—	533.0	—	—	—	—	—	—	—	—	2.4	—	—
32	140	1.26	—	—	736.5	—	—	—	—	—	—	—	—	3.0	—	—
10	90	0.54	—	—	292.5	—	60.0	—	—	—	—	—	—	1.8	—	—
14	151	0.48	17.7	125.1	338.0	0.02	64.6	0.05	1.01	0.07	0.69	—	—	2.4	0.5	24.1
19	100	0.90	—	—	449.0	—	—	—	—	—	—	—	—	2.4	—	—
24	130	1.08	—	—	601.0	—	—	—	—	—	—	—	—	3.6	—	—
29	140	1.08	—	—	671.5	—	—	—	—	—	—	—	—	2.4	—	—
38	190	1.44	—	—	886.5	—	—	—	—	—	—	—	—	3.6	—	—
17	130	0.72	—	—	454.0	—	—	—	—	—	—	—	—	3.0	—	—
24	180	0.90	—	—	637.5	—	—	—	—	—	—	—	—	3.6	—	—
29	140	0.90	—	—	647.0	—	—	—	—	—	—	—	—	1.8	—	—
38	190	1.26	—	—	865.5	—	—	—	—	—	—	—	—	2.4	—	—
26	130	0.72	—	—	577.0	—	80.0	—	—	—	—	—	—	1.8	—	—
35	80	1.08	—	—	792.5	—	105.8	—	—	—	—	—	—	2.4	—	—
14	130	0.72	—	—	396.5	—	—	—	—	—	—	—	—	2.4	—	—
19	180	1.08	—	—	550.5	—	—	—	—	—	—	—	—	3.0	—	—
25	120	2.34	—	—	796.5	—	—	—	—	—	—	—	—	0.6	—	—
34	170	3.24	—	—	1110.0	—	—	—	—	—	—	—	—	0.6	—	—
11	110	1.98	—	—	555.5	—	70.0	—	—	—	—	—	—	0	—	—
18	189	3.78	32.0	209.9	718.1	1.75	99.8	0.29	1.13	0.31	5.44	—	—	0	0.6	45.6
36	295	4.67	—	—	1341.4	—	174.0	—	—	—	—	—	—	0.5	—	—
20	120	2.16	—	—	712.0	—	—	—	—	—	—	—	—	1.2	—	—
26	170	3.06	—	—	974.5	—	—	—	—	—	—	—	—	1.2	—	—
31	160	2.52	—	—	934.5	—	—	—	—	—	—	—	—	1.2	—	—
40	220	3.42	—	—	1260.0	—	—	—	—	—	—	—	—	1.2	—	—
19	150	1.98	—	—	717.0	—	—	—	—	—	—	—	—	1.2	—	—
26	210	2.88	—	—	1011.0	—	—	—	—	—	—	—	—	1.8	—	—
31	160	2.34	—	—	910.5	—	—	—	—	—	—	—	—	0	—	—
40	220	3.24	—	—	1230.0	—	—	—	—	—	—	—	—	0.6	—	—
27	150	2.16	—	—	840.5	—	86.5	—	—	—	—	—	—	0	—	—
37	220	3.06	—	—	1166.0	—	115.4	—	—	—	—	—	—	0.6	—	—
15	150	2.16	—	—	659.5	—	—	—	—	—	—	—	—	0.6	—	—
21	220	3.06	—	—	924.0	—	—	—	—	—	—	—	—	1.2	—	—
43	298	4.84	—	—	1497.8	—	—	—	—	—	—	—	—	1.5	—	—
45	302	4.89	—	—	1478.1	—	—	—	—	—	—	—	—	0.6	—	—
47	299	4.81	—	—	1523.7	—	167.9	—	—	—	—	—	—	0.6	—	—
36	307	5.10	—	—	1472.5	—	—	—	—	—	—	—	—	4.7	—	—
60	200	3.60	—	—	1080.0	—	—	—	—	—	—	—	—	12.0	—	—
60	200	3.60	—	—	1000.0	—	—	—	—	—	—	—	—	9.0	—	—

DA+ Code	Food Description	QTY	Measure	Wt (g)	H₂O (g)	Ener (cal)	Prot (g)	Carb (g)	Fiber (g)	Fat (g)	Fat Breakdown (g) Sat	Mono	Poly
FAST FOOD—continued													
34390	Cheeseburger, lettuce leaves instead of buns	1	serving(s)	300	—	330	18.0	11.0	3.0	25.0	9.0	—	—
34377	Chocolate shake	1	serving(s)	425	—	690	9.0	83.0	0	36.0	24.0	—	—
34375	Double-Double cheeseburger	1	serving(s)	330	—	670	37.0	39.0	3.0	41.0	18.0	—	—
34393	Double-Double cheeseburger with mustard and ketchup	1	serving(s)	330	—	590	37.0	41.0	3.0	32.0	17.0	—	—
34392	Double-Double cheeseburger, lettuce leaves instead of buns	1	serving(s)	362	—	520	33.0	11.0	3.0	39.0	17.0	—	—
34376	French fries	1	serving(s)	125	—	400	7.0	54.0	2.0	18.0	5.0	—	—
34373	Hamburger	1	item(s)	243	—	390	16.0	39.0	3.0	19.0	5.0	—	—
34389	Hamburger with mustard and ketchup	1	serving(s)	243	—	310	16.0	41.0	3.0	10.0	4.0	—	—
34388	Hamburger, lettuce leaves instead of buns	1	serving(s)	275	—	240	13.0	11.0	3.0	17.0	4.0	—	—
34379	Strawberry shake	1	serving(s)	425	—	690	9.0	91.0	0	33.0	22.0	—	—
34378	Vanilla shake	1	serving(s)	425	—	680	9.0	78.0	0	37.0	25.0	—	—
	Jack in the Box												
30392	Bacon ultimate cheeseburger	1	item(s)	338	—	1090	46.0	53.0	2.0	77.0	30.0	—	—
1740	Breakfast Jack	1	item(s)	125	—	290	17.0	29.0	1.0	12.0	4.5	—	—
14074	Cheeseburger	1	item(s)	131	—	350	18.0	31.0	1.0	17.0	8.0	—	—
14106	Chicken breast strips, 4 pieces	1	serving(s)	201	—	500	35.0	36.0	3.0	25.0	6.0	—	—
37241	Chicken club salad, plain, without salad dressing	1	serving(s)	431	—	300	27.0	13.0	4.0	15.0	6.0	—	—
14064	Chicken sandwich	1	item(s)	145	—	400	15.0	38.0	2.0	21.0	4.5	—	—
14111	Chocolate ice cream shake, small	1	serving(s)	414	—	880	14.0	107.0	1.0	45.0	31.0	—	—
14073	Hamburger	1	item(s)	118	—	310	16.0	30.0	1.0	14.0	6.0	—	—
14090	Hash browns	1	serving(s)	57	—	150	1.0	13.0	2.0	10.0	2.5	—	—
14072	Jack's Spicy Chicken sandwich	1	item(s)	270	—	620	25.0	61.0	4.0	31.0	6.0	—	—
1468	Jumbo Jack hamburger	1	item(s)	261	—	600	21.0	51.0	3.0	35.0	12.0	—	—
1469	Jumbo Jack hamburger with cheese	1	item(s)	286	—	690	25.0	54.0	3.0	42.0	16.0	—	—
14099	Natural cut french fries, large	1	serving(s)	196	—	530	8.0	69.0	5.0	25.0	6.0	—	—
14098	Natural cut french fries, medium	1	serving(s)	133	—	360	5.0	47.0	4.0	17.0	4.0	—	—
1470	Onion rings	1	serving(s)	119	—	500	6.0	51.0	3.0	30.0	6.0	—	—
33141	Sausage, egg, and cheese biscuit	1	item(s)	234	—	740	27.0	35.0	2.0	55.0	17.0	—	—
14095	Seasoned curly fries, medium	1	serving(s)	125	—	400	6.0	45.0	5.0	23.0	5.0	—	—
14077	Sourdough Jack	1	item(s)	245	—	710	27.0	36.0	3.0	51.0	18.0	—	—
37249	Southwest chicken salad, plain, without salad dressing	1	serving(s)	488	—	300	24.0	29.0	7.0	11.0	5.0	—	—
14112	Strawberry ice cream shake, small	1	serving(s)	417	—	880	13.0	105.0	0	44.0	31.0	—	—
14078	Ultimate cheeseburger	1	item(s)	323	—	1010	40.0	53.0	2.0	71.0	28.0	—	—
14110	Vanilla ice cream shake, small	1	serving(s)	379	—	790	13.0	83.0	0	44.0	31.0	—	—
	Jamba Juice												
31645	Aloha Pineapple smoothie	24	fluid ounce(s)	730	—	500	8.0	117.0	4.0	1.5	1.0	—	—
31646	Banana Berry smoothie	24	fluid ounce(s)	719	—	480	5.0	112.0	4.0	1.0	0	—	—
31656	Berry Lime Sublime smoothie	24	fluid ounce(s)	728	—	460	3.0	106.0	5.0	2.0	1.0	—	—
31647	Carribean Passion smoothie	24	fluid ounce(s)	730	—	440	4.0	102.0	4.0	2.0	1.0	—	—
38422	Carrot juice	16	fluid ounce(s)	472	—	100	3.0	23.0	0	0.5	0	—	—
31648	Chocolate Moo'd smoothie	24	fluid ounce(s)	634	—	720	17.0	148.0	3.0	8.0	5.0	—	—
31649	Citrus Squeeze smoothie	24	fluid ounce(s)	727	—	470	5.0	110.0	4.0	2.0	1.0	—	—
31651	Coldbuster smoothie	24	fluid ounce(s)	724	—	430	5.0	100.0	5.0	2.5	1.0	—	—
31652	Cranberry Craze smoothie	24	fluid ounce(s)	793	—	460	6.0	104.0	4.0	0.5	0	—	—
31654	Jamba Powerboost smoothie	24	fluid ounce(s)	738	—	440	6.0	105.0	6.0	1.0	0	—	—
38423	Lemonade	16	fluid ounce(s)	483	—	300	1.0	75.0	0	0	0	0	0
31657	Mango-a-go-go smoothie	24	fluid ounce(s)	690	—	440	3.0	104.0	4.0	1.5	0.5	—	—
38424	Orange juice, freshly squeezed	16	fluid ounce(s)	496	—	220	3.0	52.0	0.5	1.0	0	—	—
38426	Orange/carrot juice	16	fluid ounce(s)	484	—	160	3.0	37.0	0	1.0	0	—	—
31660	Orange-a-peel smoothie	24	fluid ounce(s)	726	—	440	8.0	102.0	5.0	1.5	0	—	—
31662	Peach Pleasure smoothie	24	fluid ounce(s)	720	—	460	4.0	108.0	4.0	2.0	1.0	—	—
31665	Protein Berry Pizzaz smoothie	24	fluid ounce(s)	710	—	440	20.0	92.0	5.0	1.5	0	—	—
31668	Razzmatazz smoothie	24	fluid ounce(s)	730	—	480	3.0	112.0	4.0	2.0	1.0	—	—
31669	Strawberries Wild smoothie	24	fluid ounce(s)	725	—	450	6.0	105.0	4.0	0.5	0	—	—
38421	Strawberry Tsunami smoothie	24	fluid ounce(s)	740	—	530	4.0	128.0	4.0	2.0	1.0	—	—

Chol (mg)	Calc (mg)	Iron (mg)	Magn (mg)	Pota (mg)	Sodi (mg)	Zinc (mg)	Vit A (µg)	Thia (mg)	Vit E (mg α)	Ribo (mg)	Niac (mg)	Vit B_6 (mg)	Fola (µg DFE)	Vit C (mg)	Vit B_{12} (µg)	Sele (µg)
60	200	2.70	—	—	720.0	—	—	—	—	—	—	—	—	12.0	—	—
95	300	0.72	—	—	350.0	—	—	—	—	—	—	—	—	0	—	—
120	350	5.40	—	—	1440.0	—	—	—	—	—	—	—	—	9.0	—	—
115	350	5.40	—	—	1520.0	—	—	—	—	—	—	—	—	12.0	—	—
120	350	4.50	—	—	1160.0	—	—	—	—	—	—	—	—	12.0	—	—
0	20	1.80	—	—	245.0	—	0	—	—	—	—	—	—	0	—	—
40	40	3.60	—	—	650.0	—	—	—	—	—	—	—	—	9.0	—	—
35	40	3.60	—	—	730.0	—	—	—	—	—	—	—	—	12.0	—	—
40	40	2.70	—	—	370.0	—	—	—	—	—	—	—	—	12.0	—	—
85	300	0.00	—	—	280.0	—	—	—	—	—	—	—	—	0	—	—
90	300	0.00	—	—	390.0	—	—	—	—	—	—	—	—	0	—	—
140	308	7.38	—	540.0	2040.0	—	—	—	—	—	—	—	—	0.6	—	—
220	145	3.48	—	210.0	760.0	—	—	—	—	—	—	—	—	3.5	—	—
50	151	3.61	—	270.0	790.0	—	40.2	—	—	—	—	—	—	0	—	—
80	18	1.60	—	530.0	1260.0	—	—	—	—	—	—	—	—	1.1	—	—
65	280	3.35	—	560.0	880.0	—	—	—	—	—	—	—	—	50.4	—	—
35	100	2.70	—	240.0	730.0	—	—	—	—	—	—	—	—	4.8	—	—
135	460	0.47	—	840.0	330.0	—	—	—	—	—	—	—	—	0	—	—
40	100	3.60	—	250.0	600.0	—	0	—	—	—	—	—	—	0	—	—
0	10	0.18	—	190.0	230.0	—	0	—	—	—	—	—	—	0	—	—
50	150	1.80	—	450.0	1100.0	—	—	—	—	—	—	—	—	9.0	—	—
45	164	4.92	—	380.0	940.0	—	—	—	—	—	—	—	—	9.8	—	—
70	234	4.20	—	410.0	1310.0	—	—	—	—	—	—	—	—	8.4	—	—
0	20	1.42	—	1240.0	870.0	—	0	—	—	—	—	—	—	8.9	—	—
0	19	1.01	—	840.0	590.0	—	0	—	—	—	—	—	—	5.6	—	—
0	40	2.70	—	140.0	420.0	—	40.0	—	—	—	—	—	—	18.0	—	—
280	88	2.36	—	310.0	1430.0	—	—	—	—	—	—	—	—	0	—	—
0	40	1.80	—	580.0	890.0	—	—	—	—	—	—	—	—	0	—	—
75	200	4.50	—	430.0	1230.0	—	—	—	—	—	—	—	—	9.0	—	—
55	274	4.10	—	670.0	860.0	—	—	—	—	—	—	—	—	43.8	—	—
135	466	0.00	—	750.0	290.0	—	—	—	—	—	—	—	—	0	—	—
125	308	7.39	—	480.0	1580.0	—	—	—	—	—	—	—	—	0.6	—	—
135	532	0.00	—	750.0	280.0	—	—	—	—	—	—	—	—	0	—	—
5	200	1.80	60.0	1000.0	30.0	0.30	—	0.37	—	0.34	2.00	0.60	60.0	102.0	0	1.4
0	200	1.44	40.0	1010.0	115.0	0.60	—	0.09	—	0.25	0.80	0.70	24.0	15.0	0.2	1.4
5	200	1.80	16.0	510.0	35.0	0.30	—	0.06	—	0.25	6.00	0.70	140.0	54.0	0	1.4
5	100	1.80	24.0	810.0	60.0	0.30	—	0.09	—	0.25	5.00	0.50	100.0	78.0	0	1.4
0	150	2.70	80.0	1030.0	250.0	0.90	—	0.52	—	0.25	5.00	0.70	80.0	18.0	0	5.6
30	500	1.08	60.0	810.0	380.0	1.50	—	0.22	—	0.76	0.40	0.16	16.0	6.0	1.5	4.2
5	100	1.80	80.0	1170.0	35.0	0.30	—	0.37	—	0.34	1.90	0.60	100.0	180.0	0	1.4
5	100	1.08	60.0	1260.0	35.0	16.50	—	0.37	—	0.34	3.00	0.40	121.5	1302.0	0	1.4
0	250	1.44	16.0	500.0	50.0	0.30	—	0.03	—	0.25	5.00	0.60	120.0	54.0	0	1.4
0	1200	1.80	480.0	1070.0	45.0	16.50	—	5.55	—	6.12	68.00	7.40	640.0	288.0	10.8	77.0
0	20	0.00	8.0	200.0	10.0	0.00	—	0.03	—	0.17	14.00	1.80	320.0	36.0	0	0
5	100	1.08	24.0	780.0	50.0	0.30	—	0.15	—	0.25	5.00	0.70	120.0	72.0	0	1.4
0	60	1.08	60.0	990.0	0	0.30	—	0.45	—	0.13	2.00	0.20	160.0	246.0	0	0
0	100	1.80	60.0	1010.0	125.0	0.60	—	0.45	—	0.25	3.00	0.50	120.0	132.0	0	2.8
0	250	1.80	80.0	1380.0	160.0	0.90	—	0.45	—	0.42	2.00	0.50	140.0	240.0	0.6	1.4
5	100	0.72	32.0	740.0	60.0	0.30	—	0.06	—	0.25	4.00	0.60	80.0	18.0	0	1.4
0	1100	2.62	60.0	650.0	240.0	0.58	—	0.08	—	0.17	1.20	0.70	58.3	60.0	0	5.6
5	150	1.80	32.0	810.0	70.0	0.30	—	0.09	—	0.34	6.00	1.00	160.0	60.0	0	1.4
5	250	1.80	40.0	1050.0	180.0	0.90	—	0.12	—	0.34	0.80	0.40	40.0	60.0	0.6	1.4
5	100	1.08	24.0	480.0	10.0	0.30	—	0.06	—	0.34	14.00	1.80	320.0	90.0	0	1.4

DA+ Code	Food Description	QTY	Measure	Wt (g)	H₂O (g)	Ener (cal)	Prot (g)	Carb (g)	Fiber (g)	Fat (g)	Fat Breakdown (g)		
											Sat	Mono	Poly
	FAST FOOD—continued												
38427	Vibrant C juice	16	fluid ounce(s)	448	—	210	2.0	50.0	1.0	0	0	0	0
38428	Wheatgrass juice, freshly squeezed	1	ounce(s)	28	—	5	0.5	1.0	0	0	0	0	0
	Kentucky Fried Chicken (KFC)												
31850	BBQ baked beans	1	serving(s)	136	—	220	8.0	45.0	7.0	1.0	0	—	—
31853	Biscuit	1	item(s)	57	—	220	4.0	24.0	1.0	11.0	2.5	—	—
51223	Boneless Fiery Buffalo Wings	6	item(s)	211	—	530	30.0	44.0	3.0	26.0	5.0	—	—
39386	Boneless Honey BBQ Wings	6	item(s)	213	—	570	30.0	54.0	5.0	26.0	5.0	—	—
51224	Boneless Sweet & Spicy Wings	6	item(s)	203	—	550	30.0	50.0	3.0	26.0	5.0	—	—
31851	Cole slaw	1	serving(s)	130	—	180	1.0	22.0	3.0	10.0	1.5	—	—
31842	Colonel's Crispy Strips	3	item(s)	151	—	370	28.0	17.0	1.0	20.0	4.0	—	—
31849	Corn on the cob	1	item(s)	162	—	150	5.0	26.0	7.0	3.0	1.0	—	—
51221	Double Crunch sandwich	1	item(s)	213	—	520	27.0	39.0	3.0	29.0	5.0	—	—
3761	Extra Crispy chicken, breast	1	item(s)	162	—	370	33.0	10.0	2.0	22.0	5.0	—	—
3762	Extra Crispy chicken, drumstick	1	item(s)	60	—	150	12.0	4.0	0	10.0	2.5	—	—
3763	Extra Crispy chicken, thigh	1	item(s)	114	—	290	17.0	16.0	1.0	18.0	4.0	—	—
3764	Extra Crispy chicken, whole wing	1	item(s)	52	—	150	11.0	11.0	1.0	7.0	1.5	—	—
51218	Famous Bowls mashed potatoes with gravy	1	serving(s)	531	—	720	26.0	79.0	6.0	34.0	9.0	—	—
51219	Famous Bowls rice with gravy	1	serving(s)	384	—	610	25.0	67.0	5.0	27.0	8.0	—	—
31841	Honey BBQ chicken sandwich	1	item(s)	147	—	290	23.0	40.0	2.0	4.0	1.0	—	—
31833	Honey BBQ wing pieces	6	item(s)	157	—	460	27.0	26.0	3.0	27.0	6.0	—	—
10859	Hot wings pieces	6	piece(s)	134	—	450	26.0	19.0	2.0	30.0	7.0	—	—
42382	KFC Snacker sandwich	1	serving(s)	119	—	320	14.0	29.0	2.0	17.0	3.0	—	—
31848	Macaroni and cheese	1	serving(s)	136	—	180	8.0	18.0	0	8.0	3.5	—	—
31847	Mashed potatoes with gravy	1	serving(s)	151	—	140	2.0	20.0	1.0	5.0	1.0	—	—
10825	Original Recipe chicken, breast	1	item(s)	161	—	340	38.0	9.0	2.0	17.0	4.0	—	—
10826	Original Recipe chicken, drumstick	1	item(s)	59	—	140	13.0	3.0	0	8.0	2.0	—	—
10827	Original Recipe chicken, thigh	1	item(s)	126	—	350	19.0	7.0	1.0	27.0	7.0	—	—
10828	Original Recipe chicken, whole wing	1	item(s)	47	—	140	10.0	4.0	0	9.0	2.0	—	—
51222	Oven roasted Twister chicken wrap	1	item(s)	269	—	520	30.0	46.0	4.0	23.0	3.5	—	—
31844	Popcorn chicken, small or individual	1	item(s)	114	—	370	19.0	21.0	2.0	24.0	4.5	—	—
31852	Potato salad	1	serving(s)	128	—	180	2.0	22.0	2.0	9.0	1.5	—	—
10845	Potato wedges, small	1	serving(s)	102	—	250	4.0	32.0	3.0	12.0	2.0	—	—
31839	Tender Roast chicken sandwich with sauce	1	item(s)	236	—	430	37.0	29.0	2.0	18.0	3.5	—	—
	Long John Silver												
39392	Baked cod	1	serving(s)	101	—	120	22.0	1.0	0	4.5	1.0	—	—
3777	Batter dipped fish sandwich	1	item(s)	177	—	470	18.0	48.0	3.0	23.0	5.0	—	—
37568	Battered fish	1	item(s)	92	—	260	12.0	17.0	0.5	16.0	4.0	—	—
37569	Breaded clams	1	serving(s)	85	—	240	8.0	22.0	1.0	13.0	2.0	—	—
37566	Chicken plank	1	item(s)	52	—	140	8.0	9.0	0.5	8.0	2.0	—	—
39404	Clam chowder	1	item(s)	227	—	220	9.0	23.0	0	10.0	4.0	—	—
39398	Cocktail sauce	1	ounce(s)	28	—	25	0	6.0	0	0	0	0	0
3770	Coleslaw	1	serving(s)	113	—	200	1.0	15.0	3.0	15.0	2.5	1.8	4.1
39400	French fries, large	1	item(s)	142	—	390	4.0	56.0	5.0	17.0	4.0	—	—
3774	Fries, regular	1	serving(s)	85	—	230	3.0	34.0	3.0	10.0	2.5	—	—
3779	Hushpuppy	1	piece(s)	23	—	60	1.0	9.0	1.0	2.5	0.5	—	—
3781	Shrimp, batter-dipped, 1 piece	1	piece(s)	14	—	45	2.0	3.0	0	3.0	1.0	—	—
39399	Tartar sauce	1	ounce(s)	28	—	100	0	4.0	0	9.0	1.5	—	—
39395	Ultimate Fish sandwich	1	item(s)	199	—	530	21.0	49.0	3.0	28.0	8.0	—	—
	McDonald's												
50828	Asian salad with grilled chicken	1	item(s)	362	—	290	31.0	23.0	6.0	10.0	1.0	—	—
2247	Barbecue sauce	1	item(s)	28	—	45	0	11.0	0	0	0	0	0
737	Big Mac hamburger	1	item(s)	219	—	560	25.0	47.0	3.0	30.0	10.0	—	—
29777	Caesar salad dressing	1	package(s)	44	—	150	1.0	5.0	0	13.0	2.5	—	—
38391	Caesar salad with grilled chicken, no dressing	1	serving(s)	278	230.6	181	26.4	10.5	3.1	6.0	2.9	1.7	0.8
38393	Caesar salad without chicken, no dressing	1	serving(s)	190	170.4	84	6.0	8.1	3.0	3.9	2.2	0.9	0.3
738	Cheeseburger	1	item(s)	119	—	310	15.0	35.0	1.0	12.0	6.0	—	—
29775	Chicken McGrill sandwich	1	item(s)	213	—	400	27.0	38.0	3.0	16.0	3.0	—	—

Chol (mg)	Calc (mg)	Iron (mg)	Magn (mg)	Pota (mg)	Sodi (mg)	Zinc (mg)	Vit A (μg)	Thia (mg)	Vit E (mg α)	Ribo (mg)	Niac (mg)	Vit B_6 (mg)	Fola (μg DFE)	Vit C (mg)	Vit B_{12} (μg)	Sele (μg)
0	20	1.08	40.0	720.0	0	0.30	—	0.30	—	0.10	1.60	0.40	80.0	678.0	0	0
0	0	1.80	8.0	80.0	0	0.00	0	0.03	—	0.03	0.40	0.04	16.0	3.6	0	2.8
0	100	2.70	—	—	730.0	—	—	—	—	—	—	—	—	1.2	—	—
0	40	1.80	—	—	640.0	—	—	—	—	—	—	—	—	0	—	—
65	40	1.80	—	—	2670.0	—	—	—	—	—	—	—	—	1.2	—	—
65	40	1.80	—	—	2210.0	—	—	—	—	—	—	—	—	1.2	—	—
65	60	1.80	—	—	2000.0	—	—	—	—	—	—	—	—	1.2	—	—
5	40	0.72	—	—	270.0	—	—	—	—	—	—	—	—	12.0	—	—
65	40	1.44	—	—	1220.0	—	0	—	—	—	—	—	—	1.2	—	—
0	60	1.08	—	—	10.0	—	—	—	—	—	—	—	—	6.0	—	—
55	100	2.70	—	—	1220.0	—	—	—	—	—	—	—	—	6.0	—	—
85	20	2.70	—	—	1020.0	—	—	—	—	—	—	—	—	1.2	—	—
55	0	1.44	—	—	300.0	—	0	—	—	—	—	—	—	0	—	—
95	20	2.70	—	—	700.0	—	—	—	—	—	—	—	—	—	—	—
45	20	1.08	—	—	340.0	—	—	—	—	—	—	—	—	0	—	—
35	200	5.40	—	—	2330.0	—	—	—	—	—	—	—	—	6.0	—	—
35	200	4.50	—	—	2130.0	—	—	—	—	—	—	—	—	6.0	—	—
60	80	2.70	—	—	710.0	—	—	—	—	—	—	—	—	2.4	—	—
140	40	1.80	—	—	970.0	—	—	—	—	—	—	—	—	21.0	—	—
115	40	1.44	—	—	990.0	—	—	—	—	—	—	—	—	1.2	—	—
25	60	2.70	—	—	690.0	—	—	—	—	—	—	—	—	2.4	—	—
15	150	0.72	—	—	800.0	—	—	—	—	—	—	—	—	1.2	—	—
0	40	1.44	—	—	560.0	—	—	—	—	—	—	—	—	1.2	—	—
135	20	2.70	—	—	960.0	—	—	—	—	—	—	—	—	6.0	—	—
70	20	1.08	—	—	340.0	—	—	—	—	—	—	—	—	0	—	—
110	20	2.70	—	—	870.0	—	—	—	—	—	—	—	—	1.2	—	—
50	20	1.44	—	—	350.0	—	0	—	—	—	—	—	—	1.2	—	—
60	40	6.30	—	—	1380.0	—	—	—	—	—	—	—	—	15.0	—	—
25	40	1.80	—	—	1110.0	—	0	—	—	—	—	—	—	0	—	—
5	0	0.36	—	—	470.0	—	—	—	—	—	—	—	—	6.0	—	—
0	20	1.08	—	—	700.0	—	0	—	—	—	—	—	—	0	—	—
80	80	2.70	—	—	1180.0	—	—	—	—	—	—	—	—	9.0	—	—
90	20	0.72	—	—	240.0	—	—	—	—	—	—	—	—	0	—	—
45	60	2.70	—	—	1210.0	—	—	—	—	—	—	—	—	2.4	—	—
35	20	0.72	—	—	790.0	—	—	—	—	—	—	—	—	4.8	—	—
10	20	1.08	—	—	1110.0	—	0	—	—	—	—	—	—	0	—	—
20	0	0.72	—	—	480.0	—	0	—	—	—	—	—	—	2.4	—	—
25	150	0.72	—	—	810.0	—	—	—	—	—	—	—	—	0	—	—
0	0	0.00	—	—	250.0	—	—	—	—	—	—	—	—	0	—	—
20	40	0.36	—	222.7	340.0	0.70	—	0.07	—	0.08	2.34	—	—	18.0	—	—
0	0	0.00	—	—	580.0	—	0	—	—	—	—	—	—	24.0	—	—
0	0	0.00	—	370.0	350.0	0.30	0	0.09	—	0.01	1.60	—	—	15.0	—	—
0	20	0.36	—	—	200.0	—	0	—	—	—	—	—	—	0	—	—
15	0	0.00	—	—	160.0	—	0	—	—	—	—	—	—	1.2	—	—
15	0	0.00	—	—	250.0	—	0	—	—	—	—	—	—	0	—	—
60	150	2.70	—	—	1400.0	—	—	—	—	—	—	—	—	4.8	—	—
65	150	3.60	—	—	890.0	—	—	—	—	—	—	—	—	54.0	—	—
0	0	0.00	—	55.0	260.0	—	—	—	—	—	—	—	—	0	—	—
80	250	4.50	—	400.0	1010.0	—	—	—	—	—	—	—	—	1.2	—	—
10	40	0.18	—	30.0	400.0	—	—	—	—	—	—	—	—	0.6	—	—
67	178	1.77	—	708.9	767.3	—	—	0.15	—	0.19	10.62	—	127.9	29.2	0.2	—
10	163	1.15	17.1	410.4	157.7	—	—	0.08	—	0.07	0.40	—	102.6	26.8	0	0.4
40	200	2.70	—	240.0	740.0	—	60.0	—	—	—	—	—	—	1.2	—	—
70	150	2.70	—	510.0	1010.0	—	—	—	—	—	—	—	—	6.0	—	—

Table A–1 Table of Food Composition (*continued*) (Computer code is for Cengage Diet Analysis Plus program)

DA+ Code	Food Description	QTY	Measure	Wt (g)	H₂O (g)	Ener (cal)	Prot (g)	Carb (g)	Fiber (g)	Fat (g)	Fat Breakdown (g) Sat	Mono	Poly
FAST FOOD—continued													
1873	Chicken McNuggets, 6 piece	1	serving(s)	96	—	250	15.0	15.0	0	15.0	3.0	—	—
3792	Chicken McNuggets, 4 piece	1	serving(s)	64	—	170	10.0	10.0	0	10.0	2.0	—	—
29774	Crispy chicken sandwich	1	item(s)	232	121.8	500	27.0	63.0	3.0	16.0	3.0	5.7	7.4
743	Egg McMuffin	1	item(s)	139	76.8	300	17.0	30.0	2.0	12.0	4.5	3.8	2.5
742	Filet-O-Fish sandwich	1	item(s)	141	—	400	14.0	42.0	1.0	18.0	4.0	—	—
2257	French fries, large	1	serving(s)	170	—	570	6.0	70.0	7.0	30.0	6.0	—	—
1872	French fries, small	1	serving(s)	74	—	250	2.0	30.0	3.0	13.0	2.5	—	—
33822	Fruit 'n Yogurt Parfait	1	item(s)	149	111.2	160	4.0	31.0	1.0	2.0	1.0	0.2	0.1
739	Hamburger	1	item(s)	105	—	260	13.0	33.0	1.0	9.0	3.5	—	—
2003	Hash browns	1	item(s)	53	—	140	1.0	15.0	2.0	8.0	1.5	—	—
2249	Honey sauce	1	item(s)	14	—	50	0	12.0	0	0	0	0	0
38397	Newman's Own creamy caesar salad dressing	1	item(s)	59	32.5	190	2.0	4.0	0	18.0	3.5	4.6	9.6
38398	Newman's Own low fat balsamic vinaigrette salad dressing	1	item(s)	44	29.1	40	0	4.0	0	3.0	0	1.0	1.2
38399	Newman's Own ranch salad dressing	1	item(s)	59	30.1	170	1.0	9.0	0	15.0	2.5	9.0	3.7
1874	Plain Hotcakes with syrup and margarine	3	item(s)	221	—	600	9.0	102.0	2.0	17.0	4.0	—	—
740	Quarter Pounder hamburger	1	item(s)	171	—	420	24.0	40.0	3.0	18.0	7.0	—	—
741	Quarter Pounder hamburger with cheese	1	item(s)	199	—	510	29.0	43.0	3.0	25.0	12.0	—	—
2005	Sausage McMuffin with egg	1	item(s)	165	82.4	450	20.0	31.0	2.0	27.0	10.0	10.9	4.6
50831	Side salad	1	item(s)	87	—	20	1.0	4.0	1.0	0	0	0	0
	Pizza Hut												
39009	Hot chicken wings	2	item(s)	57	—	110	11.0	1.0	0	6.0	2.0	—	—
14025	Meat Lovers hand tossed pizza	1	slice(s)	118	—	300	15.0	29.0	2.0	13.0	6.0	—	—
14026	Meat Lovers pan pizza	1	slice(s)	123	—	340	15.0	29.0	2.0	19.0	7.0	—	—
31009	Meat Lovers stuffed crust pizza	1	slice(s)	169	—	450	21.0	43.0	3.0	21.0	10.0	—	—
14024	Meat Lovers thin 'n crispy pizza	1	slice(s)	98	—	270	13.0	21.0	2.0	14.0	6.0	—	—
14031	Pepperoni Lovers hand tossed pizza	1	slice(s)	113	—	300	15.0	30.0	2.0	13.0	7.0	—	—
14032	Pepperoni Lovers pan pizza	1	slice(s)	118	—	340	15.0	29.0	2.0	19.0	7.0	—	—
31011	Pepperoni Lovers stuffed crust pizza	1	slice(s)	163	—	420	21.0	43.0	3.0	19.0	10.0	—	—
14030	Pepperoni Lovers thin 'n crispy pizza	1	slice(s)	92	—	260	13.0	21.0	2.0	14.0	7.0	—	—
10834	Personal Pan pepperoni pizza	1	slice(s)	61	—	170	7.0	18.0	0.5	8.0	3.0	—	—
10842	Personal Pan supreme pizza	1	slice(s)	77	—	190	8.0	19.0	1.0	9.0	3.5	—	—
39013	Personal Pan Veggie Lovers pizza	1	slice(s)	69	—	150	6.0	19.0	1.0	6.0	2.0	—	—
14028	Veggie Lovers hand tossed pizza	1	slice(s)	118	—	220	10.0	31.0	2.0	6.0	3.0	—	—
14029	Veggie Lovers pan pizza	1	slice(s)	119	—	260	10.0	30.0	2.0	12.0	4.0	—	—
31010	Veggie Lovers stuffed crust pizza	1	slice(s)	172	—	360	16.0	45.0	3.0	14.0	7.0	—	—
14027	Veggie Lovers thin 'n crispy pizza	1	slice(s)	101	—	180	8.0	23.0	2.0	7.0	3.0	—	—
39012	Wing blue cheese dipping sauce	1	item(s)	43	—	230	2.0	2.0	0	24.0	5.0	—	—
39011	Wing ranch dipping sauce	1	item(s)	43	—	210	0.5	4.0	0	22.0	3.5	—	—
	Starbucks												
38052	Cappuccino, tall	12	fluid ounce(s)	360	—	120	7.0	10.0	0	6.0	4.0	—	—
38053	Cappuccino, tall nonfat	12	fluid ounce(s)	360	—	80	7.0	11.0	0	0	0	0	0
38054	Cappuccino, tall soymilk	12	fluid ounce(s)	360	—	100	5.0	13.0	0.5	2.5	0	—	—
38059	Cinnamon spice mocha, tall nonfat w/o whipped cream	12	fluid ounce(s)	360	—	170	11.0	32.0	0	0.5	—	—	—
38057	Cinnamon spice mocha, tall w/whipped cream	12	fluid ounce(s)	360	—	320	10.0	31.0	0	17.0	11.0	—	—
38051	Espresso, single shot	1	fluid ounce(s)	30	—	5	0	1.0	0	0	0	0	0
38088	Flavored syrup, 1 pump	1	serving(s)	10	—	20	0	5.0	0	0	0	0	0
32562	Frappuccino bottled coffee drink, mocha	9½	fluid ounce(s)	298	—	190	6.0	39.0	3.0	3.0	2.0	—	—
32561	Frappuccino coffee drink, all bottled flavors	9½	fluid ounce(s)	281	—	190	7.0	35.0	0	3.5	2.5	—	—
38073	Frappuccino, mocha	12	fluid ounce(s)	360	—	220	5.0	44.0	0	3.0	1.5	—	—
38067	Frappuccino, tall caramel w/o whipped cream	12	fluid ounce(s)	360	—	210	4.0	43.0	0	2.5	1.5	—	—
38070	Frappuccino, tall coffee	12	fluid ounce(s)	360	—	190	4.0	38.0	0	2.5	1.5	—	—
39894	Frappuccino, tall coffee, light blend	12	fluid ounce(s)	360	—	110	5.0	22.0	2.0	1.0	0	—	—

PAGE KEY: A-2 = Breads/Baked Goods A-8 = Cereal/Rice/Pasta A-12 = Fruit A-18 = Vegetables/Legumes A-28 = Nuts/Seeds A-30 = Vegetarian A-32 = Dairy A-40 = Eggs A-40 = Seafood A-44 = Meats A-48 = Poultry A-48 = Processed Meats A-50 = Beverages A-54 = Fats/Oils A-56 = Sweets A-58 = Spices/Condiments/Sauces A-62 = Mixed Foods/Soups/Sandwiches A-68 = Fast Food A-88 = Convenience A-90 = Baby Foods

Chol (mg)	Calc (mg)	Iron (mg)	Magn (mg)	Pota (mg)	Sodi (mg)	Zinc (mg)	Vit A (µg)	Thia (mg)	Vit E (mg α)	Ribo (mg)	Niac (mg)	Vit B_6 (mg)	Fola (µg DFE)	Vit C (mg)	Vit B_{12} (µg)	Sele (µg)
35	20	0.72	—	240.0	670.0	—	—	—	—	—	—	—	—	1.2	—	—
25	0	0.36	—	160.0	450.0	—	—	—	—	—	—	—	—	1.2	—	—
60	80	3.60	62.6	526.6	1380.0	1.53	41.8	0.46	2.27	0.39	12.85	—	94.2	6.0	0.4	—
230	300	2.70	26.4	218.2	860.0	1.59	—	0.36	0.82	0.51	4.31	0.20	109.8	1.2	0.9	—
40	150	1.80	—	250.0	640.0	—	36.2	—	—	—	—	—	—	0	—	—
0	20	1.80	—	—	330.0	—	0	—	—	—	—	—	—	9.0	—	—
0	20	0.72	—	—	140.0	—	0	—	—	—	—	—	—	3.6	—	—
5	150	0.67	20.9	248.8	85.0	0.53	0	0.06	—	0.17	0.35	—	19.4	9.0	0.3	—
30	150	2.70	—	210.0	530.0	—	5.0	—	—	—	—	—	—	1.2	—	—
0	0	0.36	—	210.0	290.0	—	0	—	—	—	—	—	—	1.2	—	—
0	0	0.00	—	0	0	—	0	—	—	—	—	—	—	0	—	—
20	61	0.00	3.0	16.0	500.0	0.20	—	0.01	15.43	0.02	0.01	0.64	2.4	0	0.1	0.1
0	4	0.00	1.3	8.8	730.0	0.01	—	0.00	0.00	0.00	0.00	0.00	0	2.4	0	0
0	40	0.00	1.8	70.4	530.0	0.03	0	0.01	—	0.08	0.01	0.02	0.6	0	0	0.2
20	150	2.70	—	280.0	620.0	—	—	—	—	—	—	—	—	0	—	—
70	150	4.50	—	390.0	730.0	—	10.0	—	—	—	—	—	—	1.2	—	—
95	300	4.50	—	440.0	1150.0	—	100.0	—	—	—	—	—	—	1.2	—	—
255	300	3.60	29.7	282.2	950.0	2.01	—	0.43	0.82	0.56	4.83	0.24	—	0	1.2	—
0	20	0.72	—	—	10.0	—	—	—	—	—	—	—	—	15.0	—	—
70	0	0.36	—	—	450.0	—	—	—	—	—	—	—	—	0	—	—
35	150	1.80	—	—	760.0	—	—	—	—	—	—	—	—	6.0	—	—
35	150	2.70	—	—	750.0	—	—	—	—	—	—	—	—	6.0	—	—
55	250	2.70	—	—	1250.0	—	—	—	—	—	—	—	—	9.0	—	—
35	150	1.44	—	—	740.0	—	—	—	—	—	—	—	—	6.0	—	—
40	200	1.80	—	—	710.0	—	57.7	—	—	—	—	—	—	2.4	—	—
40	200	2.70	—	—	700.0	—	57.7	—	—	—	—	—	—	2.4	—	—
55	300	2.70	—	—	1120.0	—	—	—	—	—	—	—	—	3.6	—	—
40	200	1.44	—	—	690.0	—	58.0	—	—	—	—	—	—	2.4	—	—
15	80	1.44	—	—	340.0	—	38.5	—	—	—	—	—	—	1.4	—	—
20	80	1.86	—	—	420.0	—	—	—	—	—	—	—	—	3.6	—	—
10	80	1.80	—	—	280.0	—	—	—	—	—	—	—	—	3.6	—	—
15	150	1.80	—	—	490.0	—	—	—	—	—	—	—	—	9.0	—	—
15	150	2.70	—	—	470.0	—	—	—	—	—	—	—	—	9.0	—	—
35	250	2.70	—	—	980.0	—	—	—	—	—	—	—	—	9.0	—	—
15	150	1.44	—	—	480.0	—	—	—	—	—	—	—	—	9.0	—	—
25	20	0.00	—	—	550.0	—	0	—	—	—	—	—	—	0	—	—
10	0	0.00	—	—	340.0	—	0	—	—	—	—	—	—	0	—	—
25	250	0.00	—	—	95.0	—	—	—	—	—	—	—	—	1.2	0	—
3	200	0.00	—	—	100.0	—	—	—	—	—	—	—	—	0	0	—
0	250	0.72	—	—	75.0	—	—	—	—	—	—	—	—	0	0	—
5	300	0.72	—	—	150.0	—	—	—	—	—	—	—	—	0	0	—
70	350	1.08	—	—	140.0	—	—	—	—	—	—	—	—	2.4	0	—
0	0	0.00	—	—	0	—	0	—	—	—	—	—	—	0	0	—
0	0	0.00	—	—	0	—	0	—	—	—	—	—	—	0	0	—
12	219	1.08	—	530.0	110.0	—	—	—	—	—	—	—	—	0	—	—
15	250	0.36	—	510.0	105.0	—	—	—	—	—	—	—	—	0	—	—
10	150	0.72	—	—	180.0	—	—	—	—	—	—	—	—	0	0	—
10	150	0.00	—	—	180.0	—	—	—	—	—	—	—	—	0	0	—
10	150	0.00	—	—	180.0	—	—	—	—	—	—	—	—	0	0	—
0	150	0.00	—	—	220.0	—	—	—	—	—	—	—	—	0	—	—

DA+ Code	Food Description	QTY	Measure	Wt (g)	H₂O (g)	Ener (cal)	Prot (g)	Carb (g)	Fiber (g)	Fat (g)	Fat Breakdown (g) Sat	Mono	Poly
	FAST FOOD—continued												
38071	Frappuccino, tall espresso	12	fluid ounce(s)	360	—	160	4.0	33.0	0	2.0	1.5	—	—
39897	Frappuccino, tall mocha, light blend	12	fluid ounce(s)	360	—	140	5.0	28.0	3.0	1.5	0	—	—
39887	Frappuccino, tall Strawberries and Creme, w/o whipped cream	12	fluid ounce(s)	360	—	330	10.0	65.0	0	3.5	1.0	—	—
38063	Frappuccino, tall Tazo chai creme w/o whipped cream	12	fluid ounce(s)	360	—	280	10.0	52.0	0	3.5	1.0	—	—
38066	Frappuccino, tall Tazoberry	12	fluid ounce(s)	360	—	140	0.5	36.0	0.5	0	0	0	0
38065	Frappuccino, tall Tazoberry Crème	12	fluid ounce(s)	360	—	240	4.0	54.0	0.5	1.0	0	—	—
38080	Frappuccino, tall vanilla w/o whipped cream	12	fluid ounce(s)	360	—	270	10.0	51.0	0	3.5	1.0	—	—
39898	Frappuccino, tall white chocolate mocha, light blend	12	fluid ounce(s)	360	—	160	6.0	32.0	2.0	2.0	1.0	—	—
38074	Frappuccino, tall white chocolate w/o whipped cream	12	fluid ounce(s)	360	—	240	5.0	48.0	0	3.5	2.5	—	—
39883	Java Chip Frappuccino, tall w/o whipped cream	12	fluid ounce(s)	360	—	270	5.0	51.0	1.0	7.0	4.5	—	—
33111	Latte, tall w/nonfat milk	12	fluid ounce(s)	360	335.3	120	12.0	18.0	0	0	0	0	0
33112	Latte, tall w/whole milk	12	fluid ounce(s)	360	—	200	11.0	16.0	0	11.0	7.0	—	—
33109	Macchiato, tall caramel w/nonfat milk	12	fluid ounce(s)	360	—	170	11.0	30.0	0	1.0	0	—	—
33110	Macchiato, tall caramel w/whole milk	12	fluid ounce(s)	360	—	240	10.0	28.0	0	10.0	6.0	—	—
33107	Mocha coffee drink, tall nonfat, w/o whipped cream	12	fluid ounce(s)	360	—	170	11.0	33.0	1.0	1.5	0	—	—
38089	Mocha syrup	1	serving(s)	17	—	25	1.0	6.0	0	0.5	0	—	—
33108	Mocha, tall mocha w/whole milk	12	fluid ounce(s)	360	—	310	10.0	32.0	1.0	17.0	10.0	—	—
38042	Steamed apple cider, tall	12	fluid ounce(s)	360	—	180	0	45.0	0	0	0	0	0
38087	Tazo chai black tea, soymilk, tall	12	fluid ounce(s)	360	—	190	4.0	39.0	0.5	2.0	0	—	—
38084	Tazo chai black tea, tall	12	fluid ounce(s)	360	—	210	6.0	36.0	0	5.0	3.5	—	—
38083	Tazo chai black tea, tall nonfat	12	fluid ounce(s)	360	—	170	6.0	37.0	0	0	0	0	0
38076	Tazo iced tea, tall	12	fluid ounce(s)	360	—	60	0	16.0	0	0	0	0	0
38077	Tazo tea, grande lemonade	16	fluid ounce(s)	480	—	120	0	31.0	0	0	0	0	0
38045	Vanilla crème steamed nonfat milk, tall w/whipped cream	12	fluid ounce(s)	360	—	260	11.0	33.0	0	8.0	5.0	—	—
38046	Vanilla crème steamed soymilk, tall w/whipped cream	12	fluid ounce(s)	360	—	300	8.0	37.0	1.0	12.0	6.0	—	—
38044	Vanilla crème steamed whole milk, tall w/whipped cream	12	fluid ounce(s)	360	—	330	10.0	31.0	0	18.0	11.0	—	—
38090	Whipped cream	1	serving(s)	27	—	100	0	2.0	0	9.0	6.0	—	—
38062	White chocolate mocha, tall nonfat w/o whipped cream	12	fluid ounce(s)	360	—	260	12.0	45.0	0	4.0	3.0	—	—
38061	White chocolate mocha, tall w/ whipped cream	12	fluid ounce(s)	360	—	410	11.0	44.0	0	20.0	13.0	—	—
38048	White hot chocolate, tall nonfat w/o whipped cream	12	fluid ounce(s)	360	—	300	15.0	51.0	0	4.5	3.5	—	—
38050	White hot chocolate, tall soymilk w/ whipped cream	12	fluid ounce(s)	360	—	420	11.0	56.0	1.0	16.0	9.0	—	—
38047	White hot chocolate, tall w/whipped cream	12	fluid ounce(s)	360	—	460	13.0	50.0	0	22.0	15.0	—	—
	Subway												
15842	Cheese steak sandwich, 6″, wheat bread	1	item(s)	250	—	360	24.0	47.0	5.0	10.0	4.5	—	—
40478	Chicken and bacon ranch sandwich, 6″, white or wheat bread	1	serving(s)	297	—	540	36.0	47.0	5.0	25.0	10.0	—	—
38622	Chicken and bacon ranch wrap with cheese	1	item(s)	257	—	440	41.0	18.0	9.0	27.0	10.0	—	—
32045	Chocolate chip cookie	1	item(s)	45	—	210	2.0	30.0	1.0	10.0	6.0	—	—
32048	Chocolate chip M&M cookie	1	item(s)	45	—	210	2.0	32.0	0.5	10.0	5.0	—	—
32049	Chocolate chunk cookie	1	item(s)	45	—	220	2.0	30.0	0.5	10.0	5.0	—	—
4024	Classic Italian B.M.T. sandwich, 6″, white bread	1	item(s)	236	—	440	22.0	45.0	2.0	21.0	8.5	—	—

Chol (mg)	Calc (mg)	Iron (mg)	Magn (mg)	Pota (mg)	Sodi (mg)	Zinc (mg)	Vit A (μg)	Thia (mg)	Vit E (mg α)	Ribo (mg)	Niac (mg)	Vit B_6 (mg)	Fola (μg DFE)	Vit C (mg)	Vit B_{12} (μg)	Sele (μg)
10	100	0.00	—	—	160.0	—	—	—	—	—	—	—	—	0	0	—
0	150	0.72	—	—	220.0	—	—	—	—	—	—	—	—	0	—	—
3	350	0.00	—	—	270.0	—	—	—	—	—	—	—	—	21.0	—	—
3	350	0.00	—	—	270.0	—	—	—	—	—	—	—	—	3.6	0	—
0	0	0.00	—	—	30.0	—	0	—	—	—	—	—	—	0	0	—
0	150	0.00	—	—	125.0	—	0	—	—	—	—	—	—	1.2	0	—
3	350	0.00	—	—	370.0	—	—	—	—	—	—	—	—	3.6	0	—
3	150	0.00	—	—	250.0	—	—	—	—	—	—	—	—	0	—	—
10	150	0.00	—	—	210.0	—	—	—	—	—	—	—	—	0	0	—
10	150	1.44	—	—	220.0	—	—	—	—	—	—	—	—	0	—	—
5	350	0.00	39.8	—	170.0	1.35	—	0.12	—	0.47	0.36	0.13	17.5	0	1.3	—
45	400	0.00	46.6	—	160.0	1.28	—	0.12	—	0.54	0.34	0.14	16.8	2.4	1.2	—
5	300	0.00	—	—	160.0	—	—	—	—	—	—	—	—	1.2	—	—
30	300	0.00	—	—	135.0	—	—	—	—	—	—	—	—	2.4	—	—
5	300	2.70	—	—	135.0	—	—	—	—	—	—	—	—	0	—	—
0	0	0.72	—	—	0	—	0	—	—	—	—	—	—	0	0	—
55	300	2.70	—	—	115.0	—	—	—	—	—	—	—	—	0	—	—
0	0	1.08	—	—	15.0	—	0	—	—	—	—	—	—	0	0	—
0	200	0.72	—	—	70.0	—	—	—	—	—	—	—	—	0	0	—
20	200	0.36	—	—	85.0	—	—	—	—	—	—	—	—	1.2	0	—
5	200	0.36	—	—	95.0	—	—	—	—	—	—	—	—	0	0	—
0	0	0.00	—	—	0	—	0	—	—	—	—	—	—	0	0	—
0	0	0.00	—	—	15.0	—	0	—	—	—	—	—	—	4.8	0	—
35	350	0.00	—	—	170.0	—	—	—	—	—	—	—	—	0	0	—
30	400	1.44	—	—	130.0	—	—	—	—	—	—	—	—	0	0	—
65	350	0.00	—	—	140.0	—	—	—	—	—	—	—	—	0	0	—
40	0	0.00	—	—	10.0	—	—	—	—	—	—	—	—	0	0	—
5	400	0.00	—	—	210.0	—	—	—	—	—	—	—	—	0	0	—
70	400	0.00	—	—	210.0	—	—	—	—	—	—	—	—	2.4	0	—
10	450	0.00	—	—	250.0	—	—	—	—	—	—	—	—	0	0	—
35	500	1.44	—	—	210.0	—	—	—	—	—	—	—	—	0	0	—
75	500	0.00	—	—	250.0	—	—	—	—	—	—	—	—	3.6	0	—
35	150	8.10	—	—	1090.0	—	—	—	—	—	—	—	—	18.0	—	—
90	250	4.50	—	—	1400.0	—	—	—	—	—	—	—	—	21.0	—	—
90	300	2.70	—	—	1680.0	—	—	—	—	—	—	—	—	9.0	—	—
15	0	1.08	—	—	150.0	—	—	—	—	—	—	—	—	0	—	—
10	20	1.00	—	—	100.0	—	—	—	—	—	—	—	—	0	—	—
10	0	1.00	—	—	100.0	—	—	—	—	—	—	—	—	0	—	—
55	150	2.70	—	—	1770.0	—	—	—	—	—	—	—	—	16.8	—	—

DA+ Code	Food Description	QTY	Measure	Wt (g)	H₂O (g)	Ener (cal)	Prot (g)	Carb (g)	Fiber (g)	Fat (g)	Fat Breakdown (g) Sat	Mono	Poly
	FAST FOOD—continued												
15838	Classic tuna sandwich, 6″, wheat bread	1	item(s)	250	—	530	22.0	45.0	4.0	31.0	7.0	—	—
15837	Classic tuna sandwich, 6″, white bread	1	item(s)	243	—	520	21.0	43.0	2.0	31.0	7.5	—	—
16397	Club salad, no dressing and croutons	1	item(s)	412	—	160	18.0	15.0	4.0	4.0	1.5	—	—
3422	Club sandwich, 6″, white bread	1	item(s)	250	—	310	23.0	45.0	2.0	6.0	2.5	—	—
4030	Cold cut combo sandwich, 6″, white bread	1	item(s)	242	—	400	20.0	45.0	2.0	17.0	7.5	—	—
34030	Ham and egg breakfast sandwich	1	item(s)	142	—	310	16.0	35.0	3.0	13.0	3.5	—	—
3885	Ham sandwich, 6″, white bread	1	item(s)	238	—	310	17.0	52.0	2.0	5.0	2.0	—	—
3888	Meatball marinara sandwich, 6″, wheat bread	1	item(s)	377	—	560	24.0	63.0	7.0	24.0	11.0	—	—
4651	Meatball sandwich, 6″, white bread	1	item(s)	370	—	550	23.0	61.0	5.0	24.0	11.5	—	—
15839	Melt sandwich, 6″, white bread	1	item(s)	260	—	410	25.0	47.0	4.0	15.0	5.0	—	—
32046	Oatmeal raisin cookie	1	item(s)	45	—	200	3.0	30.0	1.0	8.0	4.0	—	—
16379	Oven-roasted chicken breast sandwich, 6″, wheat bread	1	item(s)	238	—	330	24.0	48.0	5.0	5.0	1.5	—	—
32047	Peanut butter cookie	1	item(s)	45	—	220	4.0	26.0	1.0	12.0	5.0	—	—
4655	Roast beef sandwich, 6″, wheat bread	1	item(s)	224	—	290	19.0	45.0	4.0	5.0	2.0	—	—
3957	Roast beef sandwich, 6″, white bread	1	item(s)	217	—	280	18.0	43.0	2.0	5.0	2.5	—	—
16378	Roasted chicken breast, 6″, white bread	1	item(s)	231	—	320	23.0	46.0	3.0	5.0	2.0	—	—
34028	Southwest steak and cheese sandwich, 6″, Italian bread	1	item(s)	271	—	450	24.0	48.0	6.0	20.0	6.0	—	—
4032	Spicy Italian sandwich, 6″, white bread	1	item(s)	220	—	470	20.0	43.0	2.0	25.0	9.5	—	—
4031	Steak and cheese sandwich, 6″, white bread	1	item(s)	243	—	350	23.0	45.0	3.0	10.0	5.0	—	—
32050	Sugar cookie	1	item(s)	45	—	220	2.0	28.0	0.5	12.0	6.0	—	—
40477	Sweet onion chicken teriyaki sandwich, 6″, white or wheat bread	1	serving(s)	281	—	370	26.0	59.0	4.0	5.0	1.5	—	—
38623	Turkey breast and bacon melt wrap with chipotle sauce	1	item(s)	228	—	380	31.0	20.0	9.0	24.0	7.0	—	—
15834	Turkey breast and ham sandwich, 6″, white bread	1	item(s)	227	—	280	19.0	45.0	2.0	5.0	2.0	—	—
16376	Turkey breast sandwich, 6″, white bread	1	item(s)	217	—	270	17.0	44.0	2.0	4.5	2.0	—	—
15841	Veggie Delite sandwich, 6″, wheat bread	1	item(s)	167	—	230	9.0	44.0	4.0	3.0	1.0	—	—
16375	Veggie Delite, 6″, white bread	1	item(s)	160	—	220	8.0	42.0	2.0	3.0	1.5	—	—
32051	White chip macadamia nut cookie	1	item(s)	45	—	220	2.0	29.0	0.5	11.0	5.0	—	—
	Taco Bell												
29906	7-Layer burrito	1	item(s)	283	—	490	17.0	65.0	9.0	18.0	7.0	—	—
744	Bean burrito	1	item(s)	198	—	340	13.0	54.0	8.0	9.0	3.5	—	—
749	Beef burrito supreme	1	item(s)	248	—	410	17.0	51.0	7.0	17.0	8.0	—	—
33417	Beef Chalupa Supreme	1	item(s)	153	—	380	14.0	30.0	3.0	23.0	7.0	—	—
34474	Beef Gordita Baja	1	item(s)	153	—	340	13.0	29.0	4.0	19.0	5.0	—	—
29910	Beef Gordita Supreme	1	item(s)	153	—	310	14.0	29.0	3.0	16.0	6.0	—	—
2014	Beef soft taco	1	item(s)	99	—	200	10.0	21.0	3.0	9.0	4.0	—	—
10860	Beef soft taco supreme	1	item(s)	135	—	250	11.0	23.0	3.0	13.0	6.0	—	—
34472	Chicken burrito supreme	1	item(s)	248	—	390	20.0	49.0	6.0	13.0	6.0	—	—
33418	Chicken Chalupa Supreme	1	item(s)	153	—	360	17.0	29.0	2.0	20.0	5.0	—	—
34475	Chicken Gordita Baja	1	item(s)	153	—	320	17.0	28.0	3.0	16.0	3.5	—	—
29909	Chicken quesadilla	1	item(s)	184	—	520	28.0	40.0	3.0	28.0	12.0	—	—
29907	Chili cheese burrito	1	item(s)	156	—	390	16.0	40.0	3.0	18.0	9.0	—	—
10794	Cinnamon twists	1	serving(s)	35	—	170	1.0	26.0	1.0	7.0	0	—	—
29911	Grilled chicken Gordita Supreme	1	item(s)	153	—	290	17.0	28.0	2.0	12.0	5.0	—	—
14463	Grilled chicken soft taco	1	item(s)	99	—	190	14.0	19.0	1.0	6.0	2.5	—	—

PAGE KEY: A-2 = Breads/Baked Goods A-8 = Cereal/Rice/Pasta A-12 = Fruit A-18 = Vegetables/Legumes A-28 = Nuts/Seeds A-30 = Vegetarian A-32 = Dairy A-40 = Eggs A-40 = Seafood A-44 = Meats A-48 = Poultry A-48 = Processed Meats A-50 = Beverages A-54 = Fats/Oils A-56 = Sweets A-58 = Spices/Condiments/Sauces A-62 = Mixed Foods/Soups/Sandwiches A-68 = Fast Food A-88 = Convenience A-90 = Baby Foods

Chol (mg)	Calc (mg)	Iron (mg)	Magn (mg)	Pota (mg)	Sodi (mg)	Zinc (mg)	Vit A (µg)	Thia (mg)	Vit E (mg α)	Ribo (mg)	Niac (mg)	Vit B_6 (mg)	Fola (µg DFE)	Vit C (mg)	Vit B_{12} (µg)	Sele (µg)
45	100	5.40	—	—	1030.0	—	—	—	—	—	—	—	—	21.0	—	—
45	100	3.60	—	—	1010.0	—	—	—	—	—	—	—	—	16.8	—	—
35	60	3.60	—	—	880.0	—	—	—	—	—	—	—	—	30.0	—	—
35	60	3.60	—	—	1290.0	—	—	—	—	—	—	—	—	13.8	—	—
60	150	3.60	—	—	1530.0	—	—	—	—	—	—	—	—	16.8	—	—
190	80	4.50	—	—	720.0	—	66.7	—	—	—	—	—	—	3.6	—	—
25	60	2.70	—	—	1375.0	—	—	—	—	—	—	—	—	13.8	—	—
45	200	7.20	—	—	1610.0	—	—	—	—	—	—	—	—	36.0	—	—
45	200	5.40	—	—	1590.0	—	—	—	—	—	—	—	—	31.8	—	—
45	150	5.40	—	—	1720.0	—	—	—	—	—	—	—	—	24.0	—	—
15	20	1.08	—	—	170.0	—	—	—	—	—	—	—	—	0	—	—
45	60	4.50	—	—	1020.0	—	—	—	—	—	—	—	—	18.0	—	—
15	20	0.72	—	—	200.0	—	—	—	—	—	—	—	—	0	—	—
20	60	6.30	—	—	920.0	—	—	—	—	—	—	—	—	18.0	—	—
20	60	4.50	—	—	900.0	—	—	—	—	—	—	—	—	13.8	—	—
45	60	2.70	—	—	1000.0	—	—	—	—	—	—	—	—	13.8	—	—
45	150	8.10	—	—	1310.0	—	—	—	—	—	—	—	—	21.0	—	—
55	60	2.70	—	—	1650.0	—	—	—	—	—	—	—	—	16.8	—	—
35	150	6.30	—	—	1070.0	—	—	—	—	—	—	—	—	13.8	—	—
15	0	0.72	—	—	140.0	—	—	—	—	—	—	—	—	0	—	—
50	80	4.50	—	—	1220.0	—	—	—	—	—	—	—	—	24.0	—	—
50	200	2.70	—	—	1780.0	—	—	—	—	—	—	—	—	6.0	—	—
25	60	2.70	—	—	1210.0	—	—	—	—	—	—	—	—	13.8	—	—
20	60	2.70	—	—	1000.0	—	—	—	—	—	—	—	—	13.8	—	—
0	60	4.50	—	—	520.0	—	—	—	—	—	—	—	—	18.0	—	—
0	60	2.70	—	—	500.0	—	—	—	—	—	—	—	—	13.8	—	—
15	20	0.72	—	—	160.0	—	—	—	—	—	—	—	—	0	—	—
25	250	5.40	—	—	1350.0	—	—	—	—	—	—	—	—	15.0	—	—
5	200	4.50	—	—	1190.0	—	5.9	—	—	—	—	—	—	4.8	—	—
40	200	4.50	—	—	1340.0	—	9.9	—	—	—	—	—	—	6.0	—	—
40	150	2.70	—	—	620.0	—	—	—	—	—	—	—	—	3.6	—	—
35	100	2.70	—	—	780.0	—	—	—	—	—	—	—	—	2.4	—	—
40	150	2.70	—	—	620.0	—	—	—	—	—	—	—	—	3.6	—	—
25	100	1.80	—	—	630.0	—	—	—	—	—	—	—	—	1.2	—	—
40	150	2.70	—	—	650.0	—	—	—	—	—	—	—	—	3.6	—	—
45	200	4.50	—	—	1360.0	—	—	—	—	—	—	—	—	9.0	—	—
45	100	2.70	—	—	650.0	—	—	—	—	—	—	—	—	4.8	—	—
40	100	1.80	—	—	800.0	—	—	—	—	—	—	—	—	3.6	—	—
75	450	3.60	—	—	1420.0	—	—	—	—	—	—	—	—	1.2	—	—
40	300	1.80	—	—	1080.0	—	—	—	—	—	—	—	—	0	—	—
0	0	0.37	—	—	200.0	—	0	—	—	—	—	—	—	0	—	—
45	150	1.80	—	—	650.0	—	—	—	—	—	—	—	—	4.8	—	—
30	100	1.08	—	—	550.0	—	14.6	—	—	—	—	—	—	1.2	—	—

DA+ Code	Food Description	QTY	Measure	Wt (g)	H₂O (g)	Ener (cal)	Prot (g)	Carb (g)	Fiber (g)	Fat (g)	Fat Breakdown (g) Sat	Mono	Poly
FAST FOOD—continued													
29912	Grilled Steak Gordita Supreme	1	item(s)	153	—	290	15.0	28.0	2.0	13.0	5.0	—	—
29904	Grilled steak soft taco	1	item(s)	128	—	270	12.0	20.0	2.0	16.0	4.5	—	—
29905	Grilled steak soft taco supreme	1	item(s)	135	—	235	13.0	21.0	1.0	11.0	6.0	—	—
2021	Mexican pizza	1	serving(s)	216	—	530	20.0	42.0	7.0	30.0	8.0	—	—
29894	Mexican rice	1	serving(s)	131	—	170	6.0	23.0	1.0	11.0	3.0	—	—
10772	Meximelt	1	serving(s)	128	—	280	15.0	22.0	3.0	14.0	7.0	—	—
2011	Nachos	1	serving(s)	99	—	330	4.0	32.0	2.0	21.0	3.5	—	—
2012	Nachos Bellgrande	1	serving(s)	308	—	770	19.0	77.0	12.0	44.0	9.0	—	—
2023	Pintos 'n cheese	1	serving(s)	128	—	150	9.0	19.0	7.0	6.0	3.0	—	—
34473	Steak burrito supreme	1	item(s)	248	—	380	18.0	49.0	6.0	14.0	7.0	—	—
33419	Steak Chalupa Supreme	1	item(s)	153	—	360	15.0	28.0	2.0	21.0	6.0	—	—
747	Taco	1	item(s)	78	—	170	8.0	13.0	3.0	10.0	3.5	—	—
2015	Taco salad with salsa, with shell	1	serving(s)	548	—	840	30.0	80.0	15.0	45.0	11.0	—	—
14459	Taco supreme	1	item(s)	113	—	210	9.0	15.0	3.0	13.0	6.0	—	—
748	Tostada	1	item(s)	170	—	230	11.0	27.0	7.0	10.0	3.5	—	—
CONVENIENCE MEALS													
	Banquet												
29961	Barbeque chicken meal	1	item(s)	281	—	330	16.0	37.0	2.0	13.0	3.0	—	—
14788	Boneless white fried chicken meal	1	item(s)	286	—	310	10.0	21.0	4.0	20.0	5.0	—	—
29960	Fish sticks meal	1	item(s)	207	—	470	13.0	58.0	1.0	20.0	3.5	—	—
29957	Lasagna with meat sauce meal	1	item(s)	312	—	320	15.0	46.0	7.0	9.0	4.0	—	—
14777	Macaroni and cheese meal	1	item(s)	340	—	420	15.0	57.0	5.0	14.0	8.0	—	—
1741	Meatloaf meal	1	item(s)	269	—	240	14.0	20.0	4.0	11.0	4.0	—	—
39418	Pepperoni pizza meal	1	item(s)	191	—	480	11.0	56.0	5.0	23.0	8.0	—	—
33759	Roasted white turkey meal	1	item(s)	255	—	230	14.0	30.0	5.0	6.0	2.0	—	—
1743	Salisbury steak meal	1	item(s)	269	196.9	380	12.0	28.0	3.0	24.0	12.0	—	—
	Budget Gourmet												
1914	Cheese manicotti with meat sauce entrée	1	item(s)	284	194.0	420	18.0	38.0	4.0	22.0	11.0	6.0	1.3
1915	Chicken with fettucini entrée	1	item(s)	284	—	380	20.0	33.0	3.0	19.0	10.0	—	—
3986	Light beef stroganoff entrée	1	item(s)	248	177.0	290	20.0	32.0	3.0	7.0	4.0	—	—
3996	Light sirloin of beef in herb sauce entrée	1	item(s)	269	214.0	260	19.0	30.0	5.0	7.0	4.0	2.3	0.3
3987	Light vegetable lasagna entrée	1	item(s)	298	227.0	290	15.0	36.0	4.8	9.0	1.8	0.9	0.6
	Healthy Choice												
9425	Cheese French bread pizza	1	item(s)	170	—	340	22.0	51.0	5.0	5.0	1.5	—	—
9306	Chicken enchilada suprema meal	1	item(s)	320	251.5	360	13.0	59.0	8.0	7.0	3.0	2.0	2.0
3821	Familiar Favorites lasagna bake with meat sauce entrée	1	item(s)	255	—	270	13.0	38.0	4.0	7.0	2.5	—	—
13744	Familiar Favorites sesame chicken with vegetables and rice entrée	1	item(s)	255	—	260	17.0	34.0	4.0	6.0	2.0	2.0	2.0
9316	Lemon pepper fish meal	1	item(s)	303	—	280	11.0	49.0	5.0	5.0	2.0	1.0	2.0
9322	Traditional salisbury steak meal	1	item(s)	354	250.3	360	23.0	45.0	5.0	9.0	3.5	4.0	1.0
9359	Traditional turkey breasts meal	1	item(s)	298	—	330	21.0	50.0	4.0	5.0	2.0	1.5	1.5
	Stouffers												
2313	Cheese French bread pizza	1	serving(s)	294	—	380	15.0	43.0	3.0	16.0	6.0	—	—
11138	Cheese manicotti with tomato sauce entrée	1	item(s)	255	—	360	18.0	41.0	2.0	14.0	6.0	—	—
2366	Chicken pot pie entrée	1	item(s)	284	—	740	23.0	56.0	4.0	47.0	18.0	12.4	10.5
11116	Homestyle baked chicken breast with mashed potatoes and gravy entrée	1	item(s)	252	—	270	21.0	21.0	2.0	11.0	3.5	—	—
11146	Homestyle beef pot roast and potatoes entrée	1	item(s)	252	—	260	16.0	24.0	3.0	11.0	4.0	—	—
11152	Homestyle roast turkey breast with stuffing and mashed potatoes entrée	1	item(s)	273	—	290	16.0	30.0	2.0	12.0	3.5	—	—
11043	Lean Cuisine Comfort Classics baked chicken and whipped potatoes and stuffing entrée	1	item(s)	245	—	240	15.0	34.0	3.0	4.5	1.0	2.0	1.0
11046	Lean Cuisine Comfort Classics honey mustard chicken with rice pilaf entrée	1	item(s)	227	—	250	17.0	37.0	1.0	4.0	1.0	1.0	1.0

Chol (mg)	Calc (mg)	Iron (mg)	Magn (mg)	Pota (mg)	Sodi (mg)	Zinc (mg)	Vit A (µg)	Thia (mg)	Vit E (mg α)	Ribo (mg)	Niac (mg)	Vit B_6 (mg)	Fola (µg DFE)	Vit C (mg)	Vit B_{12} (µg)	Sele (µg)
40	100	2.70	—	—	530.0	—	—	—	—	—	—	—	—	3.6	—	—
35	100	2.70	—	—	660.0	—	—	—	—	—	—	—	—	3.6	—	—
35	120	1.44	—	—	565.0	—	29.2	—	—	—	—	—	—	3.6	—	—
40	350	3.60	—	—	1000.0	—	—	—	—	—	—	—	—	4.8	—	—
15	100	1.44	—	—	790.0	—	—	—	—	—	—	—	—	3.6	—	—
40	250	2.70	—	—	880.0	—	—	—	—	—	—	—	—	2.4	—	—
3	80	0.71	—	—	530.0	—	0	—	—	—	—	—	—	0	—	—
35	200	3.60	—	—	1280.0	—	—	—	—	—	—	—	—	4.8	—	—
15	150	1.44	—	—	670.0	—	—	—	—	—	—	—	—	3.6	—	—
35	200	4.50	—	—	1250.0	—	9.9	—	—	—	—	—	—	9.0	—	—
40	100	2.70	—	—	530.0	—	—	—	—	—	—	—	—	3.6	—	—
25	80	1.08	—	—	350.0	—	—	—	—	—	—	—	—	1.2	—	—
65	450	7.20	—	—	1780.0	—	—	—	—	—	—	—	—	12.0	—	—
40	100	1.08	—	—	370.0	—	—	—	—	—	—	—	—	3.6	—	—
15	200	1.80	—	—	730.0	—	—	—	—	—	—	—	—	4.8	—	—
50	40	1.08	—	—	1210.0	—	0	—	—	—	—	—	—	4.8	—	—
45	80	1.44	—	—	1200.0	—	—	—	—	—	—	—	—	18.0	—	—
55	20	1.44	—	—	710.0	—	—	—	—	—	—	—	—	0	—	—
20	100	2.70	—	—	1170.0	—	—	—	—	—	—	—	—	0	—	—
20	150	1.44	—	—	1330.0	—	0	—	—	—	—	—	—	0	—	—
30	0	1.80	—	—	1040.0	—	0	—	—	—	—	—	—	0	—	—
35	150	1.80	—	—	870.0	—	0	—	—	—	—	—	—	0	—	—
25	60	1.80	—	—	1070.0	—	—	—	—	—	—	—	—	3.6	—	—
60	40	1.44	—	—	1140.0	—	0	—	—	—	—	—	—	0	—	—
85	300	2.70	45.4	484.0	810.0	2.29	—	0.45	—	0.51	4.00	0.22	30.7	0	0.7	—
85	100	2.70	—	—	810.0	—	—	0.15	—	0.42	6.00	—	—	0	—	—
35	40	1.80	38.9	280.0	580.0	4.71	—	0.17	—	0.36	4.28	0.27	18.9	2.4	2.5	—
30	40	1.80	57.7	540.0	850.0	4.81	—	0.15	—	0.29	5.53	0.37	38.4	6.0	1.6	—
15	283	3.03	78.5	420.0	780.0	1.39	—	0.22	—	0.45	3.13	0.32	74.8	59.1	0.2	—
10	350	3.60	—	—	600.0	—	—	—	—	—	—	—	—	0	—	—
30	40	1.44	—	—	580.0	—	—	—	—	—	—	—	—	3.6	—	—
20	100	1.80	—	—	600.0	—	—	—	—	—	—	—	—	0	—	—
35	18	0.72	—	—	580.0	—	—	—	—	—	—	—	—	12.0	—	—
35	20	0.36	—	—	580.0	—	—	—	—	—	—	—	—	30.0	—	—
45	80	2.70	—	—	580.0	—	—	—	—	—	—	—	—	21.0	—	—
35	40	1.80	—	—	600.0	—	—	—	—	—	—	—	—	0	—	—
30	200	1.80	—	230.0	660.0	—	—	—	—	—	—	—	—	2.4	—	—
70	250	1.44	—	550.0	920.0	—	—	—	—	—	—	—	—	6.0	—	—
65	150	2.70	—	—	1170.0	—	—	—	—	—	—	—	—	2.4	—	—
55	20	0.72	—	490.0	770.0	—	0	—	—	—	—	—	—	0	—	—
35	20	1.80	—	800.0	960.0	—	—	—	—	—	—	—	—	6.0	—	—
45	40	1.08	—	490.0	970.0	—	—	—	—	—	—	—	—	3.6	—	—
25	40	1.16	—	500.0	650.0	—	—	—	—	—	—	—	—	3.6	—	—
30	64	0.38	—	370.0	650.0	—	—	—	—	—	—	—	—	0	—	—

DA+ Code	Food Description	QTY	Measure	Wt (g)	H_2O (g)	Ener (cal)	Prot (g)	Carb (g)	Fiber (g)	Fat (g)	Fat Breakdown (g)		
											Sat	Mono	Poly
CONVENIENCE MEALS—continued													
9479	Lean Cuisine Deluxe French bread pizza	1	item(s)	174	—	310	16.0	44.0	3.0	9.0	3.5	0.5	0.5
360	Lean Cuisine One Dish Favorites chicken chow mein with rice	1	item(s)	255	—	190	13.0	29.0	2.0	2.5	0.5	1.0	0.5
11054	Lean Cuisine One Dish Favorites chicken enchilada Suiza with Mexican-style rice	1	serving(s)	255	—	270	10.0	47.0	3.0	4.5	2.0	1.5	1.0
9467	Lean Cuisine One Dish Favorites fettucini alfredo entrée	1	item(s)	262	—	270	13.0	39.0	2.0	7.0	3.5	2.0	1.0
11055	Lean Cuisine One Dish Favorites lasagna with meat sauce entrée	1	item(s)	298	—	320	19.0	44.0	4.0	7.0	3.0	2.0	0.5
	Weight Watchers												
11164	Smart Ones chicken enchiladas suiza entrée	1	item(s)	255	—	340	12.0	38.0	3.0	10.0	4.5	—	—
39763	Smart Ones chicken oriental entrée	1	item(s)	255	—	230	15.0	34.0	3.0	4.5	1.0	—	—
11187	Smart Ones pepperoni pizza	1	item(s)	198	—	400	22.0	58.0	4.0	9.0	3.0	—	—
39765	Smart Ones spaghetti bolognese entrée	1	item(s)	326	—	280	17.0	43.0	5.0	5.0	2.0	—	—
31512	Smart Ones spicy szechuan style vegetables and chicken	1	item(s)	255	—	220	11.0	34.0	4.0	5.0	1.0	—	—
BABY FOODS													
787	Apple juice	4	fluid ounce(s)	127	111.6	60	0	14.8	0.1	0.1	0	0	0
778	Applesauce, strained	4	tablespoon(s)	64	56.7	26	0.1	6.9	1.1	0.1	0	0	0
779	Bananas with tapioca, strained	4	tablespoon(s)	60	50.4	34	0.2	9.2	1.0	0	0	0	0
604	Carrots, strained	4	tablespoon(s)	56	51.7	15	0.4	3.4	1.0	0.1	0	0	0
770	Chicken noodle dinner, strained	4	tablespoon(s)	64	54.8	42	1.7	5.8	1.3	1.3	0.4	0.5	0.3
801	Green beans, strained	4	tablespoon(s)	60	55.1	16	0.7	3.8	1.3	0.1	0	0	0
910	Human milk, mature	2	fluid ounce(s)	62	53.9	43	0.6	4.2	0	2.7	1.2	1.0	0.3
760	Mixed cereal, prepared with whole milk	4	ounce(s)	113	84.6	128	5.4	18.0	1.5	4.0	2.2	1.2	0.4
772	Mixed vegetable dinner, strained	2	ounce(s)	57	50.3	23	0.7	5.4	0.8	0	—	—	0
762	Rice cereal, prepared with whole milk	4	ounce(s)	113	84.6	130	4.4	18.9	0.1	4.1	2.6	1.0	0.2
758	Teething biscuits	1	item(s)	11	0.7	44	1.0	8.6	0.2	0.6	0.2	0.2	0.1

Chol (mg)	Calc (mg)	Iron (mg)	Magn (mg)	Pota (mg)	Sodi (mg)	Zinc (mg)	Vit A (μg)	Thia (mg)	Vit E (mg α)	Ribo (mg)	Niac (mg)	Vit B_6 (mg)	Fola (μg DFE)	Vit C (mg)	Vit B_{12} (μg)	Sele (μg)
20	150	2.70	—	300.0	700.0	—	—	—	—	—	—	—	—	15.0	—	—
25	40	0.72	—	380.0	650.0	—	—	—	—	—	—	—	—	2.4	—	—
20	150	0.72	—	350.0	510.0	—	—	—	—	—	—	—	—	2.4	—	—
15	200	0.72	—	290.0	690.0	—	0	—	—	—	—	—	—	0	—	—
30	250	1.47	—	610.0	690.0	—	—	—	—	—	—	—	—	2.4	—	—
40	200	0.72	—	—	800.0	—	—	—	—	—	—	—	—	2.4	—	—
35	40	0.72	—	—	790.0	—	—	—	—	—	—	—	—	6.0	—	—
15	200	1.08	—	401.0	700.0	—	69.1	—	—	—	—	—	—	4.8	—	—
15	150	3.60	—	—	670.0	—	—	—	—	—	—	—	—	9.0	—	—
10	40	1.44	—	—	890.0	—	—	—	—	—	—	—	—	0	—	—
0	5	0.72	3.8	115.4	3.8	0.03	1.3	0.01	0.76	0.02	0.10	0.03	0	73.4	0	0.1
0	3	0.12	1.9	45.4	1.3	0.01	0.6	0.01	0.36	0.02	0.04	0.02	1.3	24.5	0	0.2
0	3	0.12	6.0	52.8	5.4	0.04	1.2	0.01	0.36	0.02	0.08	0.04	3.6	10.0	0	0.4
0	12	0.20	5.0	109.8	20.7	0.08	320.9	0.01	0.29	0.02	0.25	0.04	8.4	3.2	0	0.1
10	17	0.40	9.0	89.0	14.7	0.32	70.4	0.03	0.12	0.04	0.44	0.04	8.3	0	0	2.4
0	23	0.40	12.0	87.6	3.0	0.12	10.8	0.02	0.04	0.04	0.20	0.02	14.4	0.2	0	0
9	20	0.02	1.8	31.4	10.5	0.10	37.6	0.01	0.04	0.02	0.10	0.01	3.1	3.1	0	1.1
12	249	11.82	30.6	225.7	53.3	0.80	28.4	0.49	—	0.65	6.54	0.07	10.2	1.4	0.3	
—	12	0.18	6.2	68.6	4.5	0.08	77.1	0.01	—	0.02	0.28	0.04	4.5	1.6	0	0.4
12	271	13.82	51.0	215.5	52.2	0.72	24.9	0.52	—	0.56	5.90	0.12	6.8	1.4	0.3	4.0
0	11	0.39	3.9	35.5	28.4	0.10	3.1	0.02	0.02	0.05	0.47	0.01	7.6	1.0	0	2.6

Appendix B

Reliable Sources of Nutrition Information

Many sources of nutrition information are available to consumers, but the quality of the information they provide varies widely. All of the sources listed here provide scientifically based information.

Expert Advice

Registered dietitians (hospitals and the yellow pages)

Public health nutritionists (public health departments)

College nutrition instructors/professors (colleges and universities)

Extension Service home economists (state and county U.S. Department of Agriculture Extension Service offices)

Consumer affairs staff of the Food and Drug Administration (national, regional, and state FDA offices)

You can find hundreds of toll-free telephone numbers for health information through the following website: www.healthfinder.gov. After you are connected, search the term toll-free numbers.

U.S. Government

- **Federal Trade Commission (FTC)**
  Public Reference Branch
  (202) 326-2222
  www.ftc.gov
- **Food and Drug Administration (FDA)**
  Office of Consumer Affairs, HFE 1
  Division of Communication and Consumer Affairs Office of Communication, Outreach and Development Food and Drug Administration
  10903 New Hampshire Avenue
  Building 71, Room 3103
  Silver Spring, MD 20993-0002
  (240) 402-7800
  www.fda.gov
- **FDA Consumer Information Line**
  1-888-INFO-FDA
  (1-888-463-6332)
- **FDA Office of Food Labeling,**
  Office of Nutrition, Labeling, and Dietary Supplements
  HFS-800
  Center for Food Safety and Applied Nutrition
  Food and Drug Administration
  5100 Paint Branch Parkway
  College Park, MD 20740
  (240) 402-2373
  www.fda.gov/FoodLabelingGuide
  www.cfsan.fda.gov
- **FDA Office of Plant and Dairy Foods and Beverages**
  HFS-316
  Office of Food Safety
  Center for Food Safety and Applied Nutrition
  5100 Paint Branch Parkway
  College Park, MD 20740
  CAP-OFS-CFSAN@fda.hhs.gov
- **FDA Office of Special Nutritionals,**
  HFS 450
  200 C Street SW
  Washington, DC 20204
  (202) 205-4168; fax (202) 205-5295
- **Food and Nutrition Information Center**
  National Agricultural Library
  10301 Baltimore Avenue, Room 108
  Beltsville, MD 20705
  (301) 504-5719; fax (301) 504-6409
  www.nal.usda.gov/fnic
- **Food Research Action Center (FRAC)**
  1200 18th Street NW, Suite 400
  Washington, DC 20036
  (202) 986-2200; fax (202) 986-2525
- **Superintendent of Documents**
  U.S. Government Printing Office
  732 North Capitol Street NW
  Washington, DC 20401-0001
  (202) 512-1800
  www.access.gpo.gov/su_docs
- **U.S. Department of Agriculture (USDA)**
  U.S. Department of Agriculture
  1400 Independence Ave. SW
  Washington, DC 20250
  (202) 720-2791
  www.usda.gov/fcs

- **USDA Center for Nutrition Policy and Promotion**
  3101 Park Center Drive, 10th Floor
  Alexandria, VA 22302-1594
  (202) 208-2417
  www.usda.gov/fcs/cnpp.htm
- **USDA Food Safety and Inspection Service**
  Food Safety Education Office,
  Room 1180-S
  Washington, DC 20250
  (202) 690-0351
  www.usda.gov/fsis
- **U.S. Department of Education (DOE)**
  Accreditation Agency Evaluation Branch
  7th and D Street SW
  ROB 3, Room 3915
  Washington, DC 20202-5244
  (202) 708-7417
  www2.ed.gov/admins/finaid/accred
  /accreditation_pg6.html
- **U.S. Department of Health and Human Services**
  200 Independence Avenue SW
  Washington, DC 20201
  1-877-696-6775
  www.os.dhhs.gov
- **U.S. Environmental Protection Agency (EPA)**
  1200 Pennsylvania Avenue NW
  Washington, DC 20460
  http://www2.epa.gov/aboutepa/mailing
  -addresses-and-phone-numbers#HQ
  (202) 260-2090
  www.epa.gov
- **U.S. Public Health Service**
  Assistant Secretary of Health
  Humphrey Building, Room 725-H
  200 Independence Avenue SW
  Washington, DC 20201
  (202) 690-7694

Health Canada

Headquarters

- **Health Canada**
  A.L. 0900C2
  Ottawa, Canada
  K1A 0K9
  Telephone: (613) 957-2991
  Toll free: 1-866-225-0709
  http://www.hc-sc.gc.ca/

Regional Headquarters

- **British Columbia/Yukon**
  Federal Building Sinclair Centre
  420-757 Hastings St. W
  Vancouver, BC
  All info found at http://www.hc-sc.gc.ca
  /contact/ahc-asc/branch-eng.php
  V6C 1A1
  Tel: (604) 666-2083
  Fax: (604) 666-2258
- **Alberta/NWT**
  Suite 710, Canada Place
  9700 Jasper Avenue
  Edmonton, AB
  T5J 4C3
  Tel: (780) 495-6815
  Fax: (780) 495-5551
- **Manitoba**
  391 York Avenue, Suite 300
  Winnipeg, MB
  R3C 4W1
  Tel: (204) 983-4199
  Fax: (204) 983-5325
- **Saskatchewan**
  1st floor, 2045 Broad Street
  Regina, SK
  S4P 3T7
  Tel 306-780-5449 or 306-780-5038
  Fax: 204-983-5325
- **Ontario/Nunavut**
  180 Queen Street West
  Toronto, ON
  M5V 3L7
  Tel: 1-800-O Canada
  Fax: (416) 973-1423
- **Quebec**
  Room 218, Complexe Guy-Favreau
  East Tower
  200 René Lévesque Blvd. West
  Montreal, QC
  H2Z 1X4
  Tel: (450) 646-1353
  Toll free: 1-800-561-3350
  Fax: (514) 283-6739
- **Atlantic**
  Suite 1525, 15th Floor,
  Maritime Centre
  1505 Barrington Street
  Halifax, NS B3J 3Y6
  Tel: 1-866-225-0709
  Fax: (902) 426-6659

International Agencies

- **Food and Agriculture Organization of the United Nations (FAO)**
  Liaison Office for North America
  2121 K Street NW, Suite 800B
  Washington, DC 20037 USA
  Tel: +1 (202) 653-2400
  Fax: +1 (202) 653-5760
  www.fao.org
- **International Food Information Council Foundation**
  1100 Connecticut Avenue NW,
  Suite 430
  Washington, DC 20036
  (202) 296-6540
  ificinfo.health.org
- **UNICEF**
  3 United Nations Plaza
  New York, NY 10017
  (212) 326-7000
  www.unicef.com
- **World Health Organization (WHO)**
  Regional Office
  525 23rd Street NW
  Washington, DC 20037
  (202) 974-3000
  www.who.org

Professional Nutrition Organizations

- **Academy of Nutrition and Dietetics (formerly American Dietetics Association)**
  120 South Riverside Plaza, Suite 2000
  Chicago, IL 60606-6995
  (800) 877-1600; (312) 899-0040
  www.eatright.org
- **AND, The Nutrition Hotline**
  (800) 366-1655
- **American Society for Clinical Nutrition**
  9650 Rockville Pike
  Bethesda, MD 20814-3998
  Tel: (301) 634-7050
  Fax: (301) 634-7894
  www.faseb.org/ascn
- **Dietitians of Canada**
  480 University Avenue, Suite 604
  Toronto, Ontario M5G 1V2, Canada
  (416) 596-0857; fax (416) 596-0603
  www.dietitians.ca
- **Human Nutrition Institute (INACG)**
  1126 Sixteenth Street NW
  Washington, DC 20036
  (202) 659-0789
  www.ilsi.org
- **National Academy of Sciences/ National Research Council (NAS/NRC)**
  2101 Constitution Avenue, NW
  Washington, DC 20418
  (202) 334-2000
  www.nas.edu
- **National Institute of Nutrition**
  265 Carling Avenue, Suite 302
  Ottawa, Ontario K1S 2E1
  (613) 235-3355; fax (613) 235-7032
  www.nin.ca
- **Society for Nutrition Education and Behavior (SNEB)**
  9100 Purdue Road, Suite 200
  Indianapolis, IN 46268
  Phone: (317) 328-4627
  or 1-800-235-6690
  (301) 656-4938

Aging

- **Administration on Aging**
  One Massachusetts Avenue NW
  Washington, DC 20001
  Office of the Administrator,
  Administration for Community Living:
  (202) 401-4634
  (202) 619-0724
  www.aoa.dhhs.gov

- **American Association of Retired Persons (AARP)**
  601 E Street NW
  Washington, DC 20049
  Toll-Free Nationwide:1-888-OUR-AARP (1-888-687-2277)
  Toll-Free TTY: 1-877-434-7598
  Toll-Free Spanish: 1-877-342-2277
  www.aarp.org
- **National Aging Information Center**
  330 Independence Avenue SW
  Washington, DC 20201
  (202) 619-7501
  www.aoa.dhhs.gov/naic
- **National Institute on Aging**
  Public Information Office
  31 Center Drive, MSC 2292
  Bethesda, MD 20892
  (301) 496-1752
  www.nih.gov/nia

Alcohol and Drug Abuse

- **Al-Anon Family Group Headquarters, Inc.**
  1600 Corporate Landing Parkway
  Virginia Beach, VA 23454-5617
  Telephone: (757) 563-1600
  Fax: (757) 563-1656
- **Alateen**
  1600 Corporate Landing Parkway
  Virginia Beach, VA 23454-5617
  (800) 356-9996
  www.al-anon.alateen.org
- **Alcoholics Anonymous (AA)**
  General Service Office
  475 Riverside Drive
  New York, NY 10115
  (212) 870-3400
  www.aa.org
- **Narcotics Anonymous (NA)**
  P.O. Box 9999
  Van Nuys, CA 91409
  (818) 773-9999; fax (818) 700-0700
  www.wsoinc.com
- **National Clearinghouse for Alcohol and Drug Information (NCADI)**
  P.O. Box 2345
  Rockville, MD 20847-2345
  (800) 729-6686
  www.health.org
- **National Council on Alcoholism and Drug Dependence (NCADD)**
  217 Broadway, Suite 712
  New York, NY 10007
  (212) 269-7797
  Fax: (212) 269-7510
  HOPE LINE: 800-NCA-CALL
  (800-622-2255) 24 Hour Affiliate Referral
  www.ncadd.org
- **U.S. Center for Substance Abuse Prevention**
  1010 Wayne Avenue, Suite 850
  Silver Spring, MD 20910
  (301) 459-1591 ext. 244;
  fax (301) 495-2919
  www.covesoft.com/csap.html

Consumer Organizations

- **Center for Science in the Public Interest (CSPI)**
  1220 L St. NW, Suite 300
  Washington, DC 20005
  (202) 332-9110; fax (202) 265-4954
  www.cspinet.org
- **Choice in Dying, Inc.**
  1035 30th Street NW
  Washington, DC 20007
  (202) 338-9790; fax (202) 338-0242
  www.choices.org
- **Consumer Information Center**
  Pueblo, CO 81009
  (888) 8 PUEBLO or (888) 878-3256
  www.pueblo.gsa.gov
- **Consumers Union of US Inc.**
  101 Truman Avenue
  Yonkers, NY 10703-1057
  (914) 378-2000
  www.consunion.org
- **National Council Against Health Fraud, Inc. (NCAHF)**
  P.O. Box 1276
  Loma Linda, CA 92354
  (909) 824-4690
  www.ncahf.org

Fitness

- **American College of Sports Medicine**
  American College of Sports Medicine
  401 W. Michigan Street
  Indianapolis, IN 46202-3233
  (317) 637-9200
  www.acsm.org/sportsmed
- **American Council on Exercise (ACE)**
  4851 Paramount Drive
  San Diego, California 92123
  Toll Free: (888) 825-3636
  Phone: (858) 576-6500
  Fax: (858) 576-6564
  www.acefitness.org
- **President's Council on Physical Fitness and Sports**
  President's Council on Fitness, Sports & Nutrition
  1101 Wootton Parkway, Suite 560
  Rockville, MD 20852
  Phone: (240) 276-9567
  Fax: (240) 276-9860
  http://www.fitness.gov
- **Shape Up America!**
  6707 Democracy Boulevard, Suite 306
  Bethesda, MD 20817
  (301) 493-5368
  www.shapeup.org
- **Sport Medicine and Science Council of Canada**
  1600 James Naismith Drive, Suite 314
  Gloucester, Ontario K1B 5N4, Canada
  (613) 748-5671; fax (613) 748-5729
  www.smscc.ca

Food Safety

- **Alliance for Food & Fiber Food Safety Hotline**
  (800) 266-0200
- **FDA Center for Food Safety and Applied Nutrition**
  Center for Food Safety and Applied Nutrition
  Food and Drug Administration
  5100 Paint Branch Parkway
  College Park, MD 20740
  1-888-SAFEFOOD
  1-888-723-3366
  http://www.fda.gov/AboutFDA/CentersOffices/OfficeofFoods/CFSAN/
- **National Lead Information Center**
  (800) LEAD-FYI or (800) 532-3394
  (800) 424-LEAD or (800) 424-5323
- **National Pesticide Telecommunications Network (NPTN)**
  National Pesticide Information Center
  Oregon State University
  310 Weniger Hall
  Corvallis, OR 97331-6502
  1-800-858-7378
  http://npic.orst.edu
  _www.ace.orst.edu/info/nptn
- **USDA Meat and Poultry Hotline**
  1-888-674-6854
- **U.S. EPA Safe Drinking Water Hotline**
  (800) 426-4791

Health and Disease

- **Alzheimer's Disease Education and Referral Center**
  P. O. Box 8250
  Silver Spring, MD 20907-8250
  (800) 438-4380
  www.alzheimers.org
- **Alzheimer's Disease Information and Referral Service**
  919 North Michigan Avenue, Suite 1000
  Chicago, IL 60611
  (800) 272-3900
  www.alz.org

- **American Academy of Allergy, Asthma, and Immunology**
  555 East Wells Street
  Suite 1100
  Milwaukee, WI 53202-3823
  Phone: (414) 272-6071
  www.aaaai.org
- **American Cancer Society National Home Office**
  250 Williams Street NW
  Atlanta, GA 30303
  (800) ACS-2345 or (800) 227-2345
  www.cancer.org
- **American Council on Science and Health**
  1995 Broadway, 2nd Floor
  New York, NY 10023-5882
  (212) 362-7044; fax (212) 362-4919
  www.acsh.org
- **American Dental Association**
  211 East Chicago Avenue
  Chicago, IL 60611
  (312) 440-2500
  www.ada.org
- **American Diabetes Association**
  1701 North Beauregard Street
  Alexandria, VA 22311
  1-800-DIABETES (800-342-2383)
  www.diabetes.org
- **American Heart Association**
  Box BHG, National Center
  7320 Greenville Avenue
  Dallas, TX 75231
  1-800-AHA-USA-1
  1-800-242-8721
  http://www.heart.org
- **American Institute for Cancer Research**
  1759 R Street NW
  Washington, DC 20009
  (800) 843-8114 or (202) 328-7226
  fax (202) 328-7226
  www.aicr.org
- **American Medical Association**
  American Medical Association
  AMA Plaza
  330 N. Wabash Ave.
  Chicago, IL 60611-5885
  1-800-621-8335
  https://apps.ama-assn.org
- **American Public Health Association (APHA)**
  800 I Street, NW
  Washington, DC 20001
  (202) 777-2742
  www.apha.org
- **American Red Cross**
  National Headquarters
  2025 E Street NW
  Washington, DC 20006
  1-800-RED CROSS
  (1-800-733-2767)
  www.redcross.org
- **Canadian Diabetes Association**
  15 Toronto Street, Suite 800
  Toronto, ON M5C 2E3
  (800) BANTING or (800) 226-8464
  (416) 363-3373
  www.diabetes.ca
- **Canadian Public Health Association**
  Canadian Public Health Association
  404-1525 Carling Ave
  Ottawa ON K1Z 8R9
  Canada
  (613) 725-3769; fax (613) 725-9826
  www.cpha.ca
- **Centers for Disease Control and Prevention (CDC)**
  1600 Clifton Road NE
  Atlanta, GA 30329-4027
  800-CDC-INFO
  (800-232-4636)
  www.cdc.gov
- **The Food Allergy Network**
  7925 Jones Branch Drive
  Suite 1100
  McLean, VA 22102
  (800) 929-4040 or (703) 691-3179
  www.foodallergy.org
- **Internet Health Resources**
  www.ihr.com
- **National AIDS Hotline (CDC)**
  (800) 342-AIDS (English)
  (800) 344-SIDA (Spanish)
  (800) 2437-TTY (Deaf)
  (900) 820-2437
- **National Cancer Institute**
  Office of Cancer Communications
  Building 31, Room 10824
  Bethesda, MD 20892
  (800) 4-CANCER or (800) 422-6237
  www.nci.nih.gov
- **National Diabetes Information Clearinghouse**
  1 Information Way
  Bethesda, MD 20892-3560
  (301) 654-3327
  www.niddk.nih.gov
- **National Digestive Disease Information Clearinghouse (NDDIC)**
  2 Information Way
  Bethesda, MD 20892-3570
  (301) 654-3810
  www.niddk.nih.gov
- **National Health Information Center (NHIC)**
  Office of Disease Prevention and Health Promotion
  (800) CDC-INFO (232-4636)
  (English/Spanish)
  http://www.health.gov/nhic/
- **National Heart, Lung, and Blood Institute**
  Information Center
  P.O. Box 30105
  Bethesda, MD 20824-0105
  (301) 251-1222
  www.nhlbi.nih.gov/nhlbi/nhlbi.htm
- **National Institute of Allergy and Infectious Diseases**
  NIAID Office of Communications and Government Relations
  5601 Fishers Lane, MSC 9806
  Bethesda, MD 20892-9806
  (301) 496-5717
  www.niaid.nih.gov
- **National Institute of Dental Research (NIDR)**
  National Institutes of Health
  Bethesda, MD 20892-2190
  1-866-232-4528
  www.nidr.nih.gov
- **National Institutes of Health (NIH)**
  9000 Rockville Pike
  Bethesda, MD 20892
  (301) 496-2433
  www.nih.gov
- **National Osteoporosis Foundation**
  1150 17th Street NW Suite 850
  Washington, DC 20036
  (202) 223-2226
  www.nof.org
- **Office of Disease Prevention and Health Promotion**
  http://health.gov/
- **Office on Smoking and Health (OSH)**
  http://www.cdc.gov/tobacco/about/osh/index.htm

Infancy and Childhood

- **American Academy of Pediatrics**
  141 Northwest Point Boulevard
  Elk Grove Village, IL 60007-1098
  847-434-4000 (tel)
  800-433-9016 (toll-free tel)
  www.aap.org
- **Association of Birth Defect Children, Inc.**
  976 Lake Baldwin Lane, Suite 104
  Orlando FL 32814
  (407) 895-0802
  www.birthdefects.org
- **Canadian Paediatric Society**
  2305 St. Laurent Blvd.
  Ottawa, ON Canada K1G 4J8
  (613) 526-9397; fax (613) 526-3332
  www.cps.ca
- **National Center for Education in Maternal & Child Health**
  2115 Wisconsin Avenue NW, Suite 601
  Washington, DC 20007-2292
  (703) 524-7802
  www.ncemch.org

Pregnancy and Lactation

- **American College of Obstetricians and Gynecologists Resource Center**
  409 12th Street SW
  Washington, DC 20024-2188
  (202) 638-5577
  www.acog.org
- **La Leche International, Inc.**
  La Leche League International
  35 E. Wacker Drive, Suite 850
  Chicago, IL 60601
  Voice: (312) 646-6260
  1-800-LALECHE (525-3243)
  http://www.llli.org
- **March of Dimes Birth Defects Foundation**
  1275 Mamaroneck Avenue
  White Plains, NY 10605
  (914) 997-4488
  http://www.marchofdimes.org

World Hunger

- **Bread for the World**
  425 3rd Street SW, Ste. 1200
  Washington, DC 20024
  phone: (202) 639-9400
  toll-free: (800) 822-7323
  www.bread.org
- **Center on Hunger, Poverty and Nutrition Policy**
  Tufts University School of Nutrition
  11 Curtis Avenue
  Medford, MA 02155
  (617) 627-3956
- **Freedom from Hunger**
  1460 Drew Avenue, Suite 300
  Davis, CA 95618
  (530) 758-6200
  www.freefromhunger.org
- **Oxfam America**
  226 Causeway Street, 5th Floor
  Boston, MA 02114
  (800) 776-9326
  www.oxfamamerica.org
- **SEEDS Magazine**
  P.O. Box 6170
  Waco, TX 76706
  (254) 755-7745
  www.helwys.com/seedhome.htm
- **Worldwatch Institute**
  1400 16th St. NW, Ste. 430
  Washington, DC 20036
  (202) 745-8092
  www.worldwatch.org

Scientific Literature

Nutrition Journals

American Journal of Clinical Nutrition
British Journal of Nutrition
Human Nutrition, Applied Nutrition
Journal of the American Academy of Nutrition and Dietetics
Journal of the Canadian Dietetic Association
Journal of Food Composition and Analysis
Journal of Nutrition
Journal of Nutrition Education
Nutrition Abstracts and Reviews
Nutrition and Metabolism
Nutrition Reports International
Nutrition Research
Nutrition Reviews

Other Journals

American Journal of Epidemiology
American Journal of Nursing
American Journal of Public Health
Annals of Internal Medicine
Annals of Surgery
Canadian Journal of Public Health
Caries Research
Food Technology
Gastroenterology
International Journal of Obesity
Journal of the American Dental Association
Journal of the American Medical Association
Journal of Clinical Investigation
Journal of Food Science
Journal of Home Economics
Journal of Pediatrics
Lancet
New England Journal of Medicine
Pediatrics
Science
The Scientist

Appendix C

Aids to Calculations

Mathematical problems have been worked out for you as examples at appropriate places in the text. This appendix aims to help with the use of the metric system and with those problems not fully explained elsewhere.

Conversion Factors

Conversion factors are useful mathematical tools in every day calculations, like the ones encountered in the study of nutrition. A conversion factor is a fraction in which the numerator (top) and the denominator (bottom) express the same quantity in different units. For example, 2.2 pounds (lb) and 1 kilogram (kg) are equivalent; they express the same weight. The conversion factor used to change pounds to kilograms or vice versa is:

$$\frac{2.2 \text{ lb}}{1 \text{ kg}} \quad \text{or} \quad \frac{1 \text{ kg}}{2.2 \text{ lb}}$$

Because both factors equal 1, measurements can be multiplied by the factor without changing the value of the measurement. Thus, the units can be changed.

The correct factor to use in a problem is the one with the unit you are seeking in the numerator (top) of the fraction. Following are some examples of problems commonly encountered in nutrition study; they illustrate the usefulness of conversion factors.

Example 1
Convert the weight of 130 pounds to kilograms.

1. Choose the conversion factor in which the unit you are seeking is on top:

$$\frac{1 \text{ kg}}{2.2 \text{ lb}}$$

2. Multiply 130 pounds by the factor:

$$130 \text{ lb} \times \frac{1 \text{ kg}}{2.2 \text{ lb}} = \frac{130 \text{ kg}}{2.2}$$

$$= 59 \text{ kg (rounded off to the nearest whole number)}$$

Example 2
How many grams (g) of saturated fat are contained in a 3-ounce (oz) hamburger?

1. Appendix A shows that a 4-ounce hamburger contains 7 grams of saturated fat. You are seeking grams of saturated fat; therefore, the conversion factor is:

$$\frac{7 \text{ g saturated fat}}{4 \text{ oz hamburger}}$$

2. Multiply 3 ounces of hamburger by the conversion factor:

$$3 \text{ oz hamburger} \times \frac{7 \text{ g saturated fat}}{4 \text{ oz hamburger}} = \frac{3 \times 7}{4} = \frac{21}{4}$$

$$= 5 \text{ g saturated fat (rounded off to the nearest whole number)}$$

Energy Units
1 calorie* (cal) = 4.2 kilojoules
1 millijoule (MJ) = 240 cal
1 kilojoule (kJ) = 0.24 cal
1 gram (g) carbohydrate = 4 cal = 17 kJ
1 g fat = 9 cal = 37 kJ
1 g protein = 4 cal = 17 kJ
1 g alcohol = 7 cal = 29 kJ

Nutrient Unit Conversions

Sodium
To convert milligrams of sodium to grams of salt:

mg sodium ÷ 400 = g of salt

The reverse is also true:

g salt × 400 = mg sodium

Folate
To convert micrograms (μg) of synthetic folate in supplements and enriched foods to Dietary Folate Equivalents (μg DFE):

μg synthetic folate × 1.7 = μg DFE

For naturally occurring folate, assign each microgram folate a value of 1 μg DFE:

μg folate = μg DFE

Example 3
Consider a pregnant woman who takes a supplement and eats a bowl of fortified cornflakes, 2 slices of fortified bread, and a cup of fortified pasta.

1. From the supplement and fortified foods, she obtains synthetic folate:

Supplement	100 μg folate
Fortified cornflakes	100 μg folate
Fortified bread	40 μg folate
Fortified pasta	60 μg folate
	300 μg folate

2. To calculate the DFE, multiply the amount of synthetic folate by 1.7:

   300 μg × 1.7 = 510 μg DFE

3. Now add the naturally occurring folate from the other foods in her diet—in this example, another 90 μg of folate.

   510 μg DFE + 90 μg = 600 μg DFE

Notice that if we had not converted synthetic folate from supplements and fortified foods to DFE, then this woman's intake would appear to fall short of the 600 μg recommendation for pregnancy (300 μg + 90 μg = 390 μg). But as this example shows, her intake does meet the recommendation. At this time, supplement and fortified food labels list folate in μg only, not μg DFE, making such calculations necessary.

*Throughout this book and in the appendixes, the term calorie is used to mean kilocalorie. Thus, when converting calories to kilojoules, do not enlarge the calorie values—they are kilocalorie values.

Vitamin A
Equivalencies for vitamin A:

1 μg RAE = 1 μg retinol
= 12 μg beta-carotene
= 24 μg other vitamin A carotenoids

1 international unit (IU) = 0.3 μg retinol
= 3.6 μg beta-carotene
= 7.2 μg other vitamin A carotenoids

To convert older RE values to micrograms RAE:

1 μg RE retinol = 1 μg RAE retinol
6 μg RE beta-carotene = 12 μg RAE beta-carotene
12 μg RE other vitamin A carotenoids = 24 μg RAE other vitamin A carotenoids

International Units (IU)
To convert IU to:

- μg vitamin D: divide by 40 or multiply by 0.025.
- 1 IU natural vitamin E = 0.67 mg alpha-tocopherol.
- 1 IU synthetic vitamin E = 0.45 mg alpha-tocopherol.
- vitamin A, see above.

Percentages

A percentage is a comparison between a number of items (perhaps your intake of energy) and a standard number (perhaps the number of calories recommended for your age and gender—your energy DRI). The standard number is the number you divide by. The answer you get after the division must be multiplied by 100 to be stated as a percentage (percent means "per 100").

Example 4
What percentage of the DRI recommendation for energy is your energy intake?

1. Find your energy DRI value in the DRI tables located at the back of this text. We'll use 2,368 calories to demonstrate.
2. Total your energy intake for a day—for example, 1,200 calories.
3. Divide your calorie intake by the DRI value:

   1,200 cal (your intake) ÷ 2,368 cal (DRI) = 0.507

4. Multiply your answer by 100 to state it as a percentage:

   0.507 × 100 = 50.7 = 51% (rounded off to the nearest whole number)

In some problems in nutrition, the percentage may be more than 100. For example, suppose your daily intake of vitamin A is 3,200 and your DRI is 900 μg. Your intake as a percentage of the DRI is more than 100 percent (i.e., you consume more than 100 percent of your recommendation for vitamin A).
The following calculations show your vitamin A intake as a percentage of the DRI value:

3,200 ÷ 900 = 3.6 (rounded)
3.6 × 100 = 360% of DRI

Example 5

Food labels express nutrients and energy contents of foods as percentages of the Daily Values. If a serving of a food contains 200 milligrams of calcium, for example, what percentage of the calcium Daily Value does the food provide?

1. Find the calcium Daily Value on the inside back cover, page Y.
2. Divide the milligrams of calcium in the food by the Daily Value standard:

$$\frac{200}{1{,}000} = 0.2$$

3. Multiply by 100:

$$0.2 \times 100 = 20\% \text{ of the Daily Value}$$

Example 6

This example demonstrates how to calculate the percentage of fat in a day's meals.

1. Recall the general formula for finding percentages of calories from a nutrient:

(one nutrient's calories ÷ total calories) × 100 = the percentage of calories from that nutrient

2. Say a day's meals provide 1,754 calories and 54 grams of fat. First, convert fat grams to fat calories:

54 g × 9 cal per g = 486 cal from fat

3. Then apply the general formula for finding percentage of calories from fat:

(fat calories ÷ total calories) × 100 = percentage of calories from fat

(486 ÷ 1,754) × 100 = 27. 7 (28%, rounded)

Weights and Measures

Length

1 inch (in.) = 2.54 centimeters (cm)
1 foot (ft) = 30.48 cm
1 meter (m) = 39.37 in

Temperature

Celsius‡		Fahrenheit	
Steam	100°C	212°F	Steam
Body temperature	37°C	98.6°F	Body temperature
Ice	0°C	32°F	Ice

- To find degrees Fahrenheit (°F) when you know degrees Celsius (°C), multiply by 9/5 and then add 32.
- To find degrees Celsius (°C) when you know degrees Fahrenheit (°F), subtract 32 and then multiply by 5/9.

Volume

Used to measure fluids or pourable dry substances such as cereal.

1 milliliter (ml) = ⅕ teaspoon or 0.034 fluid ounce or 1/1000 liter
1 deciliter (dL)= 1/10 liter
1 teaspoon (tsp or t) = 5 ml or about 5 grams (weight) salt
1 tablespoon (tbs or T) = 3 tsp or 15 ml
1 ounce, fluid (fl oz) = 2 tbs or 30 ml
1 cup (c) = 8 fl oz or 16 tbs or 237 ml
1 quart (qt) = 32 fl oz or 4 c or 0.95 liter
1 liter (L) = 1.06 qt or 1,000 ml
1 gallon (gal) = 16 c or 4 qt or 128 fl oz or 3.79 L

Weight

1 microgram (μg or mcg) = 1/1000 milligram
1 milligram (mg) = 1,000 mcg or 1/1,000 gram
1 gram (g) = 1,000 mg or 1/1,000 kilogram
1 ounce, weight (oz) = about 28 g or 1/16 pound
1 pound (lb) = 16 oz (wt) or about 454 g
1 kilogram (kg) = 1,000 g or 2.2 lb

‡Also known as centigrade.

Appendix D

Table of Intentional Food Additives

Table D.1 A Guide to Intentional Food Additives

Abbreviation	Type of Additive	Uses
AA	Anticaking agents	Keeps dry powders and crystals from clumping together (e.g., salt, powdered sugar).
C	Colors	Synthetic (laboratory-made) vegetable and fruit concentrates and other substances used to color foods (e.g., soft drinks, frosting).
E	Emulsifiers	Used to make oil and water mix (e.g., salad dressings, sauces).
EX	Extenders	"Fillers" such as fruit pulp and texturized protein (e.g., fruit drinks, hamburger).
FA	Flavoring agents	Used to add particular flavors to food (e.g., pudding, rye bread).
N	Nutrients	Used to add vitamins or minerals to foods (e.g., breakfast cereals, skim milk).
P	Preservatives	Used to keep food from spoiling (e.g., breads, breakfast cereals).
S	Sweeteners	Used to sweeten foods. Some are "artificial," such as aspartame and saccharin, and some are extracted from plants, such as sugar cane and sugar beets (e.g., soft drinks, catsup).
T	Texturizers	Used to improve the texture of food by stabilizing moisture content, dryness, volume, tenderness, or hardness (e.g., cakes, breads).
TH	Thickeners	Used to improve the consistency of foods (e.g., low-fat salad dressings, low-calorie jams).

Common Food Additives Approved for Use by the FDA and Their Primary Function	
Alginates (T)	Carrageenan (E, TH)
Alpha tocopherol (vitamin E) (N)	Cellulose gum (TH)
Alpha tocopheryl acetate (vitamin E) (N)	Chromium chloride (N)
Annatto (C)	Citric acid (FA, P)
Ascorbic acid (vitamin C) (P)	Corn syrup (S)
Aspartame (NutraSweet™) (S)	Cupric oxide (copper) (N)
Baking powder (T)	Cyanocobalamin (vitamin B_{12}) (N)
Beet juice (C)	Dextrin (TH)
Beet sugar (S)	Dextrose (S)
Beta-carotene (N)	Dibasic calcium phosphate (calcium, phosphorus) (N)
BHA (P)	Diglycerides (E)
BHT (P)	EDTA (P)
Calcium carbonate (calcium) (N)	Enzymes (FA,T)
Calcium pantothenate (pantothenic acid, a B vitamin) (N)	Extracts (FA)
Calcium propionate (P)	Ferrous fumarate (iron) (N)
Calcium silicate (AA)	Ferrous sulfate (iron) (N)
Cane sugar (S)	Fructose (S)
Carotene (C)	Fruit pulp (EX)

(*Continued*)

Table D.1 A Guide to Intentional Food Additives (Continued)

Common Food Additives Approved for Use by the FDA and Their Primary Function	
Gelatin (TH)	Sodium ascorbate (vitamin C) (N)
Glycerol (T)	Sodium benzoate (P)
Glyceryl abietate (E, T)	Sodium bicarbonate (baking soda) (T)
Guar gum (T, TH)	Sodium bisulfite (P)
High-fructose corn syrup (S)	Sodium chloride (P, FA)
Honey (S)	Sodium citrate (P)
Hydrolyzed protein (EX)	Sodium erythorbate (P)
Inulin (T)	Sodium hexametaphosphate (P, T)
Lecithin (E)	Sodium metabisulfite (P)
Magnesium oxide (magnesium) (N)	Sodium molydate (molybdenum) (N)
Maltodextrin (S, TH)	Sodium nitrate (P)
Manganese sulfate (manganese) (N)	Sodium nitrite (P)
Monoglycerides (T)	Sodium propionate (P)
Monosodium glutamate (MSG) (FA)	Sodium selenate (selenium) (N)
Monosodium monocalcium phosphate (T)	Sodium stearolyn-2-lactylate (P, E)
Natural flavorings (FA)	Sodium sulfite (P)
Niacinamide (niacin, vitamin B3) (N)	Sorbitan monostearate (E)
Nitrates (P)	Sorbitol (S, T)
Nitrites (P)	Sorghum (S, P)
Paprika (C)	Soy lecithin (E, T)
Pectin (TH)	Starches (TH)
Phosphoric acid (P)	Sucrose (S)
Phytonadione (vitamin K) (N)	Sugar (P)
Polysorbates (P)	Sulfur dioxide (P)
Potassium benzoate (P)	Sweeteners (FA)
Potassium bicarbonate (T)	TBHQ (P)
Potassium chloride (N)	Texturized protein (EX)
Potassium iodite (iodine) (N)	Thiamin hydrochloride (thiamin, B1) (N)
Potassium metabisulfite (P)	Thiamin mononitrate (thiamin, B1) (N)
Potassium sorbate (P)	Tocopherols (P, N)
Propyl gallate (P)	Turmeric (C)
Propyleneglycol (T)	Vitamin A palmitate (vitamin A) (N)
Pyridoxine hydrochloride (vitamin B6) (N)	Vitamin C (ascorbic acid) (N, P)
Reduced iron (iron) (N)	Vitamin E (N, P)
Retinol (vitamin A) (N)	Xanthan gum (T)
Riboflavin (vitamin B2) (N)	Xylitol (S)
Saccharin (S)	Yeast (T)
Salt (FA, P)	Yellow 5, 6 (C)
Silicon dioxide (AA)	Zinc oxide (zinc) (N)
Sodium aluminum phosphate (T)	

Appendix E

Cells

This appendix presents an overview of the basic structure and functions of **cells** in the human body. Cell structure and function are central to discussions of nutrition, for it is within cells that nutrients are utilized to sustain life and health. The life-sustaining processes that take place within each of the more than one hundred trillion cells in the human body are maintained by the nutrients we consume in our diet. Chemical reactions within the cells produce energy from carbohydrates, proteins, and fats; cause proteins to be broken down or built up; and in thousands of other ways keep us in a state of health.

A cell is a basic unit of life. Any substance that does not consist of one or more cells cannot be alive. Cells are the "building blocks" of tissues (such as muscles and bones), organs (such as the kidneys and liver), and systems (the respiratory and digestive systems, for example). Normal cell health and functioning are maintained when a state of nutritional and environmental utopia exists within and around the cells. A disruption in the availability of nutrients or the presence of harmful substances in the cell's environment can initiate disorders that eventually affect our health or growth. Health problems in general begin with disruptions in the normal activity of cells.

Cells The basic unit of life, of which all living things are composed. Every cell is surrounded by a membrane and contains cytoplasm, within which are organelles and a nucleus; the cell nucleus contains chromosomes.

The types and amounts of food and supplements people consume affect the cells' environment and their ability to function normally. Excessive or inadequate supplies of nutrients and other chemical substances disrupt cell functions and result in health problems. Humans remain in a state of health as long as their cells do.

Cell Structures and Functions

A generalized diagram of a human cell is shown in Illustration E.1. Although all cells have some structures and functions in common, the specific functions performed, and the structures that support those functions, can vary a good deal from cell to cell. Cells lining the esophagus, stomach, and intestines, for example, are specialized to produce and secrete mucus that helps food pass through the digestive tract. Red blood cells are specially formed to transport oxygen and carbon dioxide.

Every cell is surrounded by a cell membrane that helps move nutrients into and out of the cell. Inside the cell membrane lies the cytoplasm, a fluid material that contains many organelles, tubes, and particles. Among the organelles in the cytoplasm are ribosomes, mitochondria, and lysosomes. Each of these "little organs" is encased in a cell membrane and performs specific functions. Ribosomes assemble amino acids into proteins following the instructions of DNA and its messenger RNA. The mitochondria are made of intricately folded membranes that bear thousands of highly organized sets of enzymes on their surfaces. They contain a small amount of DNA that codes for the synthesis of proteins needed for mitochrondrial functions. Mitochondrial enzymes are actively involved in the production of energy and are found in particularly dense quantities in muscle cells. The lysosomes are like packets of enzymes. The enzymes are used to break down old cell particles that are being recycled and to destroy substances that are harmful to the body.

Cytoplasm also contains a highly organized system of membranes called the endoplasmic reticulum. When these membranes are dotted with ribosomes, they are called "rough endoplasmic reticulum." When ribosomes are absent, the endoplasmic reticulum is referred to as "smooth." Some membranes within the cytoplasm form tubes that collect certain types of cellular material and transport it out of the cell. These membranous tubes are called the "Golgi apparatus." The rough and smooth endoplasmic reticulum are continuous with the Golgi apparatus, so secretions produced throughout the cell can be collected and transported to its exterior.

Within each cell is a nucleus covered by a two-layer membrane. The nucleus contains chromosomes and the genetic material DNA. DNA encodes all of the instructions a cell needs to conduct protein synthesis and to replicate life.

All cells within the body are part of a complex communication system that uses hormones, electrical impulses, and other chemical messengers to link each cell to the others. No cell is an island that operates independently from the others.

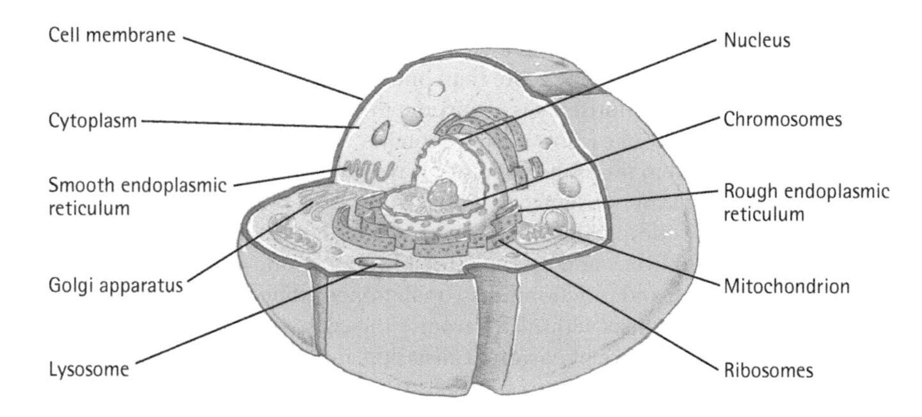

Illustration E.1 Generalized structure of a human cell.

cell membrane: the membrane that surrounds the cell and encloses its contents; made primarily of lipid and protein.
chromosomes: a set of structures within the nucleus of every cell that contain the cell's genetic material, DNA, associated with other materials (primarily proteins).
cytoplasm (SIGH-toe-plazm): the cell contents, except for the nucleus.
 cyto = cell
 plasm = a form
Golgi (GOAL-gee) **apparatus:** a set of membranes within the cell where secretory materials are packaged for export.
lysosomes: cellular organelles; membrane-enclosed sacs of degradative enzymes.
 lysis = dissolution
mitochondria (my-toe-KON-dree-uh; *singular* **mitochondrion**): the cellular organelles responsible for producing ATP aerobically; made of membranes (lipid and protein) with enzymes mounted on them. Although most DNA is located in chromosomes within the nucleus of cells, mitochondria have a small amount of their own DNA. Mitochondrial DNA contains 37 genes, all of which are essential for normal mitochondrial function.
 mitos = thread (referring to their slender shape)
 chondros = cartilage (referring to their external appearance)
nucleus: a major membrane-enclosed body within every cell, which contains the cell's genetic material, DNA, embedded in chromosomes.
 nucleus = a kernel
organelles: membrane-bound subcellular structures such as ribosomes, mitochondria, and lysosomes.
 organelle = little organ
ribosomes: protein-making organelles in cells; composed of RNA and protein.
 ribo = containing the sugar ribose
 some = body
rough endoplasmic reticulum (en-doh-PLAZ-mic reh-TIC-you-lum): intracellular membranes dotted with ribosomes, where protein synthesis takes place.
 endo = inside
 plasm = the cytoplasm
smooth endoplasmic reticulum: smooth intracellular membranes bearing no ribosomes.

Appendix F

Feedback and Answers to Nutrition Up Close and Review Questions

Unit 1

Nutrition Up Close

Nutrition Concepts Review
The nutrition concepts apply to the situations as follows:

1. 1
2. 2, 9
3. 10
4. 4
5. 3
6. 7
7. 8
8. 5
9. 6

Review Questions

1. True
2. c
3. False
4. False
5. False
6. True
7. False
8. False
9. True
10. d
11. b

12. plant chemicals/phytonutrients, antioxidants
13. Answers consist of pregnant women, breastfeeding women, infants, growing children, frail elderly, ill persons, and people recovering from illness.
14. g
15. a
16. e
17. b
18. f
19. d
20. h
21. c

Unit 2

Nutrition Up Close

Food Types for Healthful Diets

To reduce the selection of energy-dense, nutrient-poor foods, and excessive calorie intake levels, the MyPlate dietary pattern highlights foods in their basic form. That means, for example, food groups contain foods such as whole-grain breads, broccoli, baked chicken, dried beans, and oranges.

	Dietary pattern	
Food types	ChooseMyPlate	Western
Pasta		X
Cold cuts (ham, bologna, salami)		X
Fish and seafood	X	
Whole-grain breads	X	
Fruit jams and jellies		X
Potato, tortilla, and other snack chips		X
Dried beans	X	
Poultry (chicken, turkey)	X	
Fruit	X	
Vegetables	X	
Gravy		X
Ice cream		X
Soft drinks		X
Salad dressing, mayonnaise		X
Skim milk	X	
Cake		X

Review Questions

1. False
2. True
3. True
4. True
5. False

6. It is rare for modern humans to have to engage in strenuous physical activity to obtain food. Feasts are rarely followed by famines.
7. c
8. a
9. True
10. True
11. False
12. b
13. d
14. a
15. c

Unit 3

Nutrition Up Close

Checking Out a Fat-Loss Product
The product advertised earns five "yes" responses, making it highly unlikely that the product works. One or two "yes" responses provide a strong clue that advertised products may not work.

Review Questions

1. True
2. True
3. False
4. True
5. False
6. False
7. True
8. c
9. a
10. b
11. False
12. True
13. True
14. c
15. e
16. a
17. d
18. b

Unit 4

Nutrition Up Close

Comparison Shopping

Answers to the questions posed include the following:

1. The product does have less sugar than other fruit drinks but mainly consists of water, high-fructose corn syrup, preservatives, and artificial flavorings and sweeteners. Adding B vitamins and vitamin E, vitamins that are not considered nutrients of public health significance, would most likely not improve nutritional status.
2. No.
3. None.
4. To increase sale appeal/to make the product sound "good for you."
5. Positive: less sugar than other fruit drinks.

    Negative: product is called a fruit drink but it contains no fruit and has a different nutrient composition than fruit juice.

Review Questions

1. True
2. False
3. False
4. True
5. False
6. True
7. False
8. d
9. a
10. d
11. 760
12. 460
13. 6 grams, 31%
14. 3.5 grams, 17%
15. False
16. False
17. b
18. True
19. True
20. False

Unit 5

Nutrition Up Close

Improving Food Choices

There are no right or wrong answers to this unit's "Nutrition Up Close." Plans for dietary change that are specific, easy to accomplish, and highly acceptable are most likely to work in the long run.

Review Questions

1. False
2. True
3. False
4. True
5. True
6. d
7. a
8. False
9. True
10. True
11. True
12. b
13. c
14. False
15. True

Unit 6

		Calories	Fiber, g	Sodium, mg
Breakfast:				
	Jelly donuts, 2	54	2	660
Lunch:				
	Buttermilk Crispy Chicken Sandwich®	570	4	1,050
Dinner:				
	Shake Shack Double Shackburger®	855	3	1,200
	a. Totals:	1,965	10	2,910

b. Percent of Benson's daily calorie need obtained from fast foods: 89%

c. Difference in sodium intake from fast foods and her AI for sodium of 2,300 mg (add a + or a – in front of your number): −335 mg

d. By how many grams was Benson's intake of fiber less than the recommended intake amount? −18 g

Review Questions

1. False
2. True
3. d

4. c
5. False
6. True
7. True
8. False
9. True
10. c
11. a
12. True
13. d
14. b
15. b
16. a

Unit 7

Nutrition Up Close

Carbohydrate, Fat, and Protein Digestion

The primary enzymes involved in the digestion of the carbohydrate, fat, and protein in Nikki's meal, and the primary end products of carbohydrate, fat, and protein digestion that are absorbed, are shown below.

	Primary enzymes	Primary end products of digestion
Carbohydrate	1. amylase 2. sucrase 3. lactase 4. maltase	1. glucose
Fat	1. lipase	1. fatty acids 2. glycerol
Protein	1. pepsin 2. trypsin	1. amino acids

Review Questions

1. False
2. True
3. True
4. False
5. True
6. False
7. False
8. False
9. True
10. False

11. True
12. False
13. a
14. b
15. a
16. d
17. False
18. True
19. True
20. True
21. False
22. True
23. False

Unit 8

Nutrition Up Close

Food as a Source of Calories

Potato Chips
Serving size: 1 oz (about 20 chips)
Fat: 10 g __9__ = __90__ calories from fat
Carbohydrate: 15 g __4__ = __60__ calories from carbohydrate
Protein: 2 g __4__ = __8__ calories from protein
__158__ Total calories
% of calories from fat: __**90**__ = __**0.57**__ × 100 = __**57**__ %

Mini Pretzels
Serving size: 1 oz (about 17 pieces)
Fat: 0 g __9__ = __0__ calories from fat
Carbohydrate: 24 g __4__ = __96__ calories from carbohydrate
Protein: 3 g __4__ = __12__ calories from protein
__108__ Total calories
% of calories from carbohydrate: __**96**__ = __**0.89**__ × 100 = __**89**__ %

The pretzels are lower in total calories than the potato chips.

Review Questions

1. True
2. False
3. True
4. False
5. False
6. True
7. True

8. b
9. c
10. c
11. d
12. b
13. c
14. True
15. False
16. True
17. False
18. False
19. a
20. c

Unit 9

Nutrition Up Close

Are You an Apple?

If your waist circumference is over 35″ (88 cm) and you're a female, or over 40″ (102 cm) if male, you're an apple. (You are all peaches for doing this exercise.)

Review Questions

1. False
2. False
3. True
4. False
5. False
6. False
7. c
8. b
9. True
10. True
11. False
12. True
13. False
14. True
15. True
16. d

17. c
18. False
19. False
20. True
21. False
22. False
23. True

Unit 10

Nutrition Up Close

Setting Small Behavior Change Goals

The specific small steps in behavioral change that work will be different for different individuals. They can be changed over time, and may end up being a rewarding part of an enhanced lifestyle.

If you need more practice writing small behavioral change goals, take a look at a few more examples:

> I'll use a low-calorie oil and vinegar dressing on my sandwich rather than mayonnaise at lunch three times a week.
>
> I'll brighten up my plate and my diet with a serving of one of my favorite brightly colored vegetables three times a week.
>
> I'll walk up the stairs to get to my office rather than take the elevator daily.
>
> I'll lift a 10-pound weight for 15 minutes five times a week while I watch the nightly news.

Review Questions

1. False
2. False
3. False
4. True
5. False
6. False
7. c
8. b
9. True
10. True
11. True
12. True
13. b
14. a
15. d

Unit 11

Nutrition Up Close

Eating Attitudes Test
Never = 3
Rarely = 2
Sometimes = 1
Always, usually, and often = 0

A total score under 20 points may indicate abnormal eating behavior. If you think you have an eating disorder, it is best to find out for sure. Careful evaluation by a qualified health professional is necessary to exclude any possible underlying medical reasons for your symptoms. Contacting a physician, nurse practitioner, dietitian, or the student health center is an important first step. You may wish to show your Eating Attitudes Test to the health professional.

Review Questions

1. True
2. True
3. True
4. False
5. False
6. True
7. True
8. False
9. False
10. True

Unit 12

Nutrition Up Close

Does Your Fiber Intake Measure Up?
The total number of points you scored approximates the grams of total fiber you typically consume daily. Use this scale to find out if your fiber intake meets the recommended goal:

0–10 grams: You consume less than the average American. Increase your fiber intake by including more fruits, vegetables, whole grains, and legumes in your diet overall.

11–15 grams: Like other Americans, you consume too little fiber. Increase the number of servings of high-fiber foods you already enjoy, while substituting more high-fiber foods for refined food products. A quick way to add fiber to your diet is to consume more of the two fiber powerhouses: legumes and bran cereal.

15–20 grams: You currently consume more fiber than the average American. Make sure you're including 5 or more servings of fruits and vegetables. Eat 6 to 11 servings of bread, cereal, rice, and pasta daily; choose whole-grain versions of these foods often.

20–40 grams: Congratulations! Your dietary fiber intake is in the vicinity of that recommended. Keep up the good work.

Review Questions

1. True
2. False
3. True
4. True
5. False
6. True
7. False
8. True
9. False
10. True
11. a
12. True
13. False
14. c
15. False
16. True
17. d
18. b

Unit 13

Nutrition Up Close

Breakfast:	
Cornflakes, 1 cup	24
Skim milk, 1 cup	12
Orange juice, 1 cup	26
Mid-morning snack:	
Banana, 1 medium	27
Lunch:	
Dried beans, cooked, 1 cup	40
White rice, 1 cup	44
Apple, 1 medium	19
Dinner:	
Pork chop, 3 oz	0
Corn, 1 cup	32
Fruit drink, 1 cup	29
Total:	253 grams carbohydrate

Review Questions

1. True
2. False
3. True
4. True
5. True
6. True
7. False
8. False
9. False
10. c
11. d
12. d
13. d
14. False
15. False

Unit 14

Nutrition Up Close

Effects of Alcohol Intake

A. Ligia's estimated % blood alcohol = 0.03%. Mark's estimated % blood alcohol = 0.02%.

B. Three side effects of these blood alcohol levels:
1. Loss of some control over muscle movements
2. Slowed reaction time
3. Impaired thought processes

Review Questions

1. True
2. True
3. False
4. False
5. True
6. False
7. True
8. False
9. True
10. False
11. d

12. a
13. b
14. a

Unit 15

Nutrition Up Close

My Protein Intake
Compare your subtotals to find out which protein source you prefer, plant or animal. Protein from animal products is often accompanied by fat. If you are concerned about calories and fat in your diet, choose plant protein sources more often. And, if you are similar to many Americans, your intake of protein will exceed the Recommended Dietary Allowance (RDA) by quite a bit.

Review Questions

1. False
2. False
3. True
4. True
5. True
6. False
7. True
8. False
9. True
10. False
11. True
12. True
13. False
14. b
15. c
16. c

Unit 16

Nutrition Up Close

Vegetarian Main Dish Options
Vegetarians definitely had the advantage in completing this exercise! For the meat eaters, were the main dishes you identified meat-free? Here are eight vegetarian options for lunch and dinner main dishes:

- Eggplant parmesan
- Falafel

- Red beans and rice
- Pasta with broccoli
- Melted cheese, tomato, and sprout sandwiches
- Vegetable lasagna
- Veggie burgers
- Vegetable pizza

Review Questions

1. True
2. False
3. True
4. True
5. False
6. True
7. False
8. True
9. d
10. a
11. a
12. d
13. c

Unit 17

Nutrition Up Close

Gluten-Free Cuisine

Check marks should be placed in front of these gluten-free foods:

✓ fresh meats, fish
✓ tomatoes
______ pizza
______ macaroni and cheese
✓ eggs
✓ corn grits
✓ corn chips
______ sourdough rolls
✓ baked apples with sugar
✓ hamburgers
✓ grapes
✓ tossed salad with oil and vinegar dressing
______ apple pie
✓ black beans and rice
______ canned soup
______ sausage
✓ popcorn
✓ cheddar cheese
______ veggie burgers
✓ lemonade
✓ broccoli

Additional information about gluten-free foods can be found on food package labels.

Review Questions

1. True
2. True
3. False
4. False
5. True
6. False
7. True
8. c
9. d
10. b
11. d
12. False
13. True
14. False
15. d
16. a

Unit 18

Nutrition Up Close

Identify recommended dietary sources of omega-3 fatty acids
Recommended dietary sources of omega-3 fatty acids include tilelfish, catfish, DHA-fortified foods, fish oil, and perch.

Review Questions

1. True
2. True
3. True
4. False
5. True
6. True
7. True
8. True
9. False
10. False
11. c

Unit 19

Nutrition Up Close

Evaluate Your Dietary and Lifestyle Strengths for Heart Disease Prevention

How did you do? Did you find strengths in your diet and lifestyle? You are on your way.

There is no set number of right or wrong answers to this activity. The more heart-healthy choices you checked, the better for your health overall. Give yourself credit for that while becoming aware of the multiple influences of diet and lifestyle on health.

Review Questions

1. True
2. True
3. False
4. False
5. False
6. True
7. False
8. True
9. True
10. c
11. b
12. c
13. d
14. False
15. False

Unit 20

Nutrition Up Close

Antioxidant Vitamins: How Adequate Is Your Diet?

Several responses in the last two columns likely indicate adequate antioxidant vitamin consumption. If you need to boost your intake, increase the overall amount of fruits, vegetables, nuts, oils, and whole grains in your diet.

Review Questions

1. True
2. False
3. True

4. True
5. False
6. True
7. False
8. True
9. True
10. False
11. False
12. True
13. False
14. True
15. False
16. b
17. d
18. d
19. b
20. c
21. a

Unit 21

Nutrition Up Close

Have You Had Your Phytochemicals Today?
Did you eat five or more of the foods listed at least twice last week? If yes, listen carefully and you'll hear your cells say "thank you."

If you didn't eat five or more twice last week, go for the foods you didn't eat but like. Be adventurous! Try some of the phytochemical-rich foods you have never eaten before.

Review Questions

1. False
2. True
3. True
4. False
5. False
6. True
7. False
8. False
9. False

10. False
11. False
12. b
13. b
14. d
15. b
16. c
17. b

Unit 22

Nutrition Up Close

A Cancer Risk Checkup

If you answered yes to 6 out of the 8 questions you are on the right track.

If you didn't know the answer to statement 8, can you pinch an inch of fat above your ribs while you are standing? If yes, you probably have too much body fat.

Review Questions

1. True
2. False
3. True
4. True
5. False
6. True
7. False
8. False
9. True
10. False
11. d
12. b

Unit 23

Nutrition Up Close

Processed Foods in Your Diet

The purpose of this activity is to give students who often consume highly processed foods an opportunity to consider alternatives to them. For those of you who placed a check next to three or more foods listed, maybe you remembered you like nuts and would eat them instead of potato chips; or that you like and would enjoy eating apples, oranges, grapes, bananas, popcorn, raisins, or yogurt as snacks or with meals.

Review Questions

1. False
2. True
3. True
4. True
5. False
6. True
7. False
8. True
9. True
10. False
11. False
12. False
13. c
14. d
15. d
16. d
17. a

Unit 24

Nutrition Up Close

Supplement Use and Misuse

1. *Is a supplement warranted in this case?* No

   Why or why not? Martha may be getting the vitamins and minerals she needs by taking a supplement, but this cannot make up for her poor food habits. She needs to improve the overall quality of her diet. If Martha follows the ChooseMyPlate recommendations (MyPlate.gov), she should be able to get all the nutrients and beneficial phytochemicals in plant foods she needs from what she eats.

2. *Is a supplement warranted in this case?* Yes

   Why or why not? This is one of the times when a supplement is in order. Sylvia needs to follow her doctor's advice and take the prescribed iron preparation to increase her hemoglobin and replenish her iron stores. However, she is correct in consuming more iron-rich foods, too, so that after the anemia has been treated, she will not experience a relapse.

3. *Is a supplement warranted in this case?* No

   Why or why not? There is no scientific evidence that high doses of vitamins and minerals enhance physical performance. In fact, John may be setting himself up for toxicity reactions with prolonged intake of supplements with high dosages.

4. *Should Roberto try the senna, ask for a nonherbal drug, or take another action?*

   If unsure, Roberto should get more information from the pharmacist or look up senna on the National Library of Medicine's Medlineplus.gov site to get the information he needs to make a well-informed decision about senna use.

5. *Should Yuen take Hypermetabolite for weight loss?* No

Why or why not? She shouldn't take the product because the ephedra derivative may have serious side effects and because quick weight-loss strategies don't work in the longer run. Individuals who have jobs that require drug testing have been fired after ephedra derivatives appeared in their urine. Plus, there's no guarantee that the product will work or that it isn't mislabeled.

Review Questions

1. True
2. False
3. True
4. False
5. False
6. True
7. False
8. False
9. True
10. False
11. True
12. True
13. True
14. a
15. a
16. b
17. d

Unit 25

Nutrition Up Close

Foods as a Source of Water

The percentage of water is listed after each food.

1. avocado (80%); potato, boiled (77%); corn, cooked (73%)
2. egg (75%); almonds (4%); ripe (black) olives (80%)
3. watermelon (91%); celery (95%); pineapple, fresh (86%)
4. 2% milk (89%); Coke (89%); cranberry juice, low calorie (95%)
5. cheddar cheese (37%); banana (74%); refried beans (72%)
6. hot dog (53%); pork sausage, cooked (45%); ham, extra lean (74%)
7. apple (84%); mushrooms, raw (92%); orange (87%)
8. onions, cooked (88%); lettuce (96%); okra, cooked (90%)

9. peanut butter (1%); butter (16%); mayonnaise (15%)
10. cake with frosting (22%); bagel (33%); Italian bread (36%)

Review Questions

1. False
2. True
3. True
4. False
5. True
6. False
7. a
8. a
9. True
10. False
11. True
12. False
13. False
14. a

Unit 26

Nutrition Up Close

Nature and Nurture

The more *yes* responses, the better the odds that some of your family's shared dietary traits may influence disease risk. The most important responses are those you gave yourself.

Review Questions

1. True
2. False
3. True
4. False
5. False
6. True
7. True
8. True
9. False
10. True
11. False

12. True
13. b
14. d
15. c

Unit 27

Nutrition Up Close

Exercise: Your Options

For those of you who answered “Active” to the first question and “Yes” to the second and third questions, congratulations! You qualify as a mover and shaker.

Review Questions

1. True
2. False
3. True
4. True
5. True
6. True
7. False
8. True
9. False
10. False
11. True
12. b
13. d
14. c

Unit 28

Nutrition Up Close

Testing Performance Aids

1. Major limitations of the design of the study: Any of these answers would be correct: no control group, no group given a placebo, use of self-reports. You may have thought of other limitations that are warranted for this study as well.
2. Answer: No. Side effects, such as blood pressure changes, diarrhea, abnormalities in kidney function, and weight loss, were not assessed. Safety of long-term use was not addressed.

3. Answer: No. A 5% increase in strength is small and may have been due to usual differences in the number of push-ups a person can do from time to time. Any increase in strength may have been due to increased training during the week the supplement was taken. There was no control group. All participants knew they were taking phosphorus, a supplement that some may believe increases strength. This belief may have changed performance on the push-up test.

Review Questions

1. True
2. True
3. False
4. True
5. False
6. False
7. False
8. True
9. False
10. True
11. True
12. False
13. c
14. b
15. c
16. b
17. True

Unit 29

Nutrition Up Close

You Be the Judge!

Gloria's choices:

- Grains: 3 ounces (bran muffin, cereal)
- Vegetables: 3 cups (vegetable soup, carrot sticks, turnip greens, tossed salad)
- Fruits: 2½ cups (strawberries, apple, orange juice)
- Milk and milk products: 2 cups (milk, yogurt)
- Meat and beans: 3 ounces (chicken)
- Oils: 0 teaspoons

All of Gloria's choices are healthy ones, but she is missing some important foods and sufficient calories. While concentrating hard on eating enough fruits and vegetables, she has neglected to consume enough nutrient-dense foods from the remaining food groups. She needs to include in her diet more dairy products, add more meat or protein alternates, and more grains, especially whole-grain products. With a few additions to her existing choices, the quality and quantity of her diet can be greatly improved. Here is a more balanced version of Gloria's menu for a day:

Breakfast: Bran muffin, cereal with strawberries, and a glass of milk.

Lunch: Vegetable soup, carrot sticks, an apple, **burrito**, and a glass of orange juice.

Afternoon snack: Yogurt, **cheese**, and **whole-grain crackers**.

Dinner: Chicken, grits, turnip greens, tossed salad, with **2 Tbsp ranch dressing, whole grain roll**, and iced tea.

Review Questions

1. True
2. True
3. False
4. True
5. False
6. False
7. True
8. False
9. False
10. False
11. c
12. b
13. b
14. True
15. False
16. True
17. True
18. True
19. False
20. True
21. d
22. b
23. a

Unit 30

Nutrition Up Close

Obesity Prevention Close to Home

To help their children develop healthful food and activity habits, Debra and Dale can:

1. Provide lots of opportunities for fun physical activities
2. Provide a variety of healthful food choices
3. Limit meals at fast-food restaurants that promote energy-dense, low-nutrient foods to children
4. Work with other parents and school staff to increase physical activity opportunities at school

Other:

Work with other parents and school staff to increase availability of healthful food choices at school

Allow their children to eat when they are hungry and to stop eating when they are full

Never force their children to eat something or totally restrict access to their favorite foods

Offer foods in an objective, nonthreatening way

Review Questions

1. False
2. True
3. True
4. False
5. False
6. a
7. b
8. a
9. d
10. c
11. True
12. False
13. False
14. False
15. c
16. c
17. d
18. True
19. True
20. True

Unit 31

Nutrition Up Close

Does He Who Laughs, Last?
Alternate explanations:

1. The letter writer's friends may have switched to perceived health foods when they found out they were sick.
2. The letter writer may have consumed a lifelong healthy diet without choosing organic or "natural" foods.
3. The letter writer may come from a family whose members tend to have long lives.

Other possible explanations:

A sense of humor may help extend life.

Many factors in addition to diet, exercise, and genetic background influence longevity.

One individual's experience may not apply to the masses.

Review Questions

1. False
2. False
3. False
4. False
5. True
6. False
7. True
8. True
9. c
10. d
11. b

Unit 32

Nutrition Up Close

Food Safety Detective
Potential causes of the food-borne illness outbreak:

1. Disease spread from hands to foods.
2. Use of contaminated cutting board, knife, or food platter.
3. Cross-contamination between onions and lettuce or between other foods stored together.

Review Questions

1. False
2. True
3. False
4. True
5. False
6. True
7. True
8. False
9. True
10. False
11. c
12. d
13. b

Unit 33

Nutrition Up Close

Ethnic Foods Treasure Hunt

1. a. yams; b. kola nuts; c. breadfruit with corn and yams; d. cassava
2. a. chicken; b. rice; c. ripe mango; d. grapes; e. ginger; f. milk; g. many more
3. "Gallo pinto" includes the following:
   a. rice and black beans; b. eggs; c. tortillas; d. coffee; e. fruit; f. sometimes steak
4. a. Italian ices; b. calzones; c. Italian sodas; d. pizza; e. pasta; f. hot dogs
5. a. beef; b. cassava; c. milk; d. yams; e. pumpkin; f. millet; g. corn; h. dried beans;
   i. beer; j. honey; k. yogurt; l. cow's blood
6. a. reindeer; b. polar bear; c. sour milk; d. caribou; e. tea
7. Yin: a. raw kelp (seaweed); b. boiled rice; c. raw vegetables; d. fruit; e. dairy products.
   Yang: a. cooked kelp; b. cooked vegetables; c. fried rice; d. fish; e. eggs; f. chicken

Review Questions

1. True
2. False
3. True
4. True

5. True
6. True
7. True
8. True
9. True
10. True
11. c
12. d
13. a

Glossary

Note to the reader: If you have a customized book, this glossary may include some entries that will not be found in your book.

absorption
The process by which nutrients and other substances are transferred from the digestive system into body fluids for transport throughout the body.

acquired gene mutations
Defects in genes that develop over time after conception due to exposure to smoke, toxins, radiation, various dietary components, excess body fat, or other factors, or that occur randomly during cell division. They are present only in certain cells, not in every cell in the body. Acquired mutations in somatic cells (cells other than sperm and egg cells) cannot be passed on to the next generation. Acquired mutations are involved in the development of cancer.

added sugars
Sugars that are either added during the processing of foods, or are packaged as such; they include sugars (free, mono-, and disaccharides), syrups, naturally occurring sugars that are isolated from a whole food and concentrated, and other caloric sweeteners.

adequate diet
A diet consisting of foods that together supply sufficient protein, vitamins, and minerals and enough calories to meet a person's need for energy.

Adequate Intakes (AIs)
Provisional RDAs developed when there is insufficient evidence to support a specific level of intake.

adipocyte (add-dip-o-site)
A fat cell.

adiponectin (add-dip-o-neck-tin)
A hormone produced by fat cells that increases insulin sensitivity and utilization of fat stores, and decreases inflammation.

adipose tissue
A connective tissue that primarily contains adipocytes (fat cells) that store fat in the form of triglycerides. Adipose tissue also serves the purpose of cushioning and protecting organs; it also helps regulate body temperature and produces hormones and other signaling substances that affect a number of body processes.

adult
Generally defined as people 18 to 65 years old.

aerobic exercise
Physical activity in which the body's large muscles move in a rhythmic manner for a sustained period of time. Aerobic exercise involves metabolic pathways that require oxygen for energy production; it improves functioning of the cardiovascular and respiratory systems. Aerobic activities include walking, jogging, running, swimming, skiing, vacuuming, house cleaning, lawn mowing, raking, cycling, and aerobic dance.

aerobic fitness
A state of respiratory and circulatory health as measured by the ability to deliver oxygen to muscles, and the capacity of muscles to use the oxygen for physical activity.

age-related macular degeneration
Eye damage caused by oxidation of the macula, the central portion of the eye that allows a person to see details clearly. It is the leading cause of blindness in U.S. adults over the age of 65. Antioxidants provided by the carotenoids in dark green, leafy vegetables such as kale, collard greens, spinach, and Swiss chard may help prevent macular degeneration.

alcohol poisoning
A condition characterized by mental confusion, vomiting, seizures, slow or irregular breathing, and low body temperature due to the effects of excess alcohol consumption. It is life threatening and requires emergency medical help.

alcohol sugars
Simple sugars containing an alcohol group in their molecular structure. The most common are xylitol, mannitol, and sorbitol.

alcoholism
An illness characterized by dependence on alcohol and by a level of alcohol intake that interferes with health, family and social relations, and job performance.

allele
A different version of the same gene. Alleles have a different arrangement of bases than the usual version of the gene.

Alzheimer's disease
A brain disease that represents the most common form of dementia. It is characterized by memory loss for recent events that expands to more distant memories over the course of 5 to 10 years. It eventually produces profound intellectual decline characterized by dementia and personal helplessness.

amylophagia (am-e-low-phag-ah)
Laundry starch or cornstarch eating.

anaerobic exercise
Short-duration, intense activity requiring glucose for energy production. Energy formation from glucose does not require oxygen.

anaphylactic shock (an-ah-fa-lac-tic)
A serious allergic reaction that is rapid in onset and causes death. Symptoms of anaphylactic shock (or "anaphylaxis") may include abdominal cramps, chest tightness, difficulty breathing, cough, hives, flushing, swelling, and/or itching.

andropause
The time during aging when testosterone levels as well as muscle and bone mass decline.

anorexia nervosa
An eating disorder characterized by extreme weight loss, poor body image, and irrational fears of weight gain and obesity.

antibodies
Blood proteins that help the body fight particular diseases. They help the body develop an immunity, or resistance, to many diseases.

antioxidants
Chemical substances that prevent or repair damage to cells caused by oxidizing agents such as pollutants, ozone, smoke, and reactive oxygen. Oxidation reactions are a normal part of cellular processes. Vitamins C and E and certain phytochemicals function as antioxidants.

appetite
The desire to eat; a pleasant sensation that is aroused by thoughts, taste, and enjoyment of food.

association
The finding that one condition is correlated with, or related to, another condition, such as a disease or disorder. For example, diets low in vegetables are associated with breast cancer. Associations do not prove that one condition (such as a diet low in vegetables) causes an event (such as breast cancer). They indicate that a statistically significant relationship exists between a condition and an event.

atherosclerosis
"Hardening of the arteries" due to a buildup of plaque.

ATP, ADP
Adenosine triphosphate (ah-den-o-scene tri-phos-fate) and adenosine diphosphate. Molecules containing a form of phosphorus that can trap energy obtained from the macronutrients. ADP becomes ATP when it traps energy, and returns to being ADP when it releases energy for muscular and other work.

autoimmune disease
A disease related to the destruction of the body's own cells by substances produced by the immune system that mistakenly recognize certain cell components as harmful.

balanced diet
A diet that provides neither too much nor too little of nutrients and other components of food such as fat and fiber.

bariatrics
The field of medicine concerned with weight loss.

basal metabolic rate (BMR)
The rate at which energy is used by the body when it is at complete rest. BMR is expressed as calories used per unit of time, such as an hour, per unit of body weight in kilograms. Also called *resting metabolic rate* (RMR) or *resting energy expenditure* (REE).

basal metabolism
Energy used to support body processes such as growth, health, tissue repair and maintenance, and other functions. Assessed while at rest, basal metabolism includes energy the body expends for breathing, pumping of the heart, maintenance of body temperature, and other life-sustaining, ongoing functions.

bile
A yellowish-brown or green fluid produced by the liver, stored in the gallbladder, and secreted into the small intestine. It acts like a detergent, breaking down globs of fat entering the small intestine to droplets, making the fats more accessible to the action of lipase.

binge eating
The consumption of a large amount of food in a small amount of time.

binge-eating disorder
An eating disorder characterized by periodic binge eating, which normally is not followed by vomiting or the use of laxatives.

bioavailability
The amount of a nutrient consumed that is available for absorption and use by the body.

biotechnology
As applied to food products, the process of modifying the composition of foods by biologically altering their genetic makeup. Also called *genetic engineering* of foods. The food products produced are sometimes referred to as *GM* and *GMOs* (genetically modified organisms).

body mass index (BMI)
An indicator of the appropriateness of a person's weight for their height. It is calculated by dividing weight in kilograms by height in meters. It can also be calculated using the method shown in Unit 9.

bovine spongiform encephalopathy (bow-vine sponge-ah-form en-cef-a-lop-pathy)
A fatal neurodegenerative disease caused by a prion that mainly affects cattle. It can affect humans if animal tissue infected with the prior is consumed; that condition is called variant Creutzfeldt–Jakob disease.

brown fat
The type of fat primarily located above the shoulder blades, around the spinal cord, chest and neck areas, and around certain organs. Proteins found within lipid droplets in these cells activate reactions that modify energy production from glucose and fat, and produce heat upon cold exposure.

bulimia nervosa
An eating disorder characterized by recurrent episodes of rapid, uncontrolled eating of large amounts of food in a short period of time. Episodes of binge eating are followed by compensatory behaviors such as self-induced vomiting, dieting, excessive exercise, or misuse of laxatives to prevent weight gain.

calorie (calor = heat)
A unit of measure used to express the amount of energy produced by foods in the form of heat. The calorie used in nutrition is the large Calorie, or kilocalorie (kcal). It equals the amount of energy needed to raise the temperature of 1 kilogram of water (about 4 cups) from 15° to 16°C (59° to 61°F). The term *kilocalorie*, or *calorie* as used in this text, is gradually being replaced by the *kilojoule* (kJ) in the United States; 1 kcal = 4.2 kJ.

cancer
A group of diseases in which abnormal cells grow out of control and can spread throughout the body. Cancer is not contagious and has many causes.

carbohydrates
Chemical substances in foods that consist of a simple sugar molecule or multiples of them in various forms.

carbon footprint
Generally defined as the amount of the greenhouse gas carbon dioxide emitted by farming, food production, a person's activities, or a product's manufacture and transport.

cardiorespiratory fitness
The ability of the circulatory and respiratory systems to supply fuel during sustained physical activity and to eliminate fatigue products after supplying fuel. Cardiorespiratory fitness is also called *aerobic fitness* and *cardiorespiratory endurance*.

cardiovascular disease
Disorders related to plaque buildup in blood vessels in the heart, brain, legs, and other tissues and organs. Other forms of cardiovascular disease exist, including disorders related to malformation of the heart and blood vessels.

cataracts
Complete or partial clouding over of the lens of the eye.

cause and effect
A finding that demonstrates that a condition causes a particular event. For example, vitamin C deficiency causes the deficiency disease scurvy.

celiac disease
An autoimmune disease characterized by inflammation of the small intestine lining resulting from a genetically based intolerance to gluten. The inflammation produces diarrhea, fatty stools, weight loss, and vitamin and mineral deficiencies. (Also called *celiac sprue* and *gluten-sensitive enteropathy*.)

cholesterol
A fat-soluble, colorless liquid found primarily in animal products. Cholesterol is used by the body to form hormones such as testosterone and estrogen and is a component of cell membranes. Cholesterol is present in plant cell membranes, but the quantity is small and plants are not considered to be a significant dietary source of cholesterol.

chromosomes
Structures in the nuclei of cells that contain genes. Humans have 23 pairs of chromosomes; half of each pair comes from the mother and half from the father.

chronic diseases
Slow-developing, long-lasting diseases that are not contagious (e.g., heart disease, cancer, diabetes). They can be treated but not always cured.

chronic inflammation
Low-grade inflammation that lasts weeks, months, or years. Inflammation is the first response of the body's immune system to infectious agents, toxins, or irritants. It triggers the release of biologically active substances that promote oxidation and other reactions to counteract the infection, toxin, or irritant. A side effect of chronic inflammation is that it also damages lipids, cells, and tissues.

circulatory system
The heart, arteries, capillaries, and veins responsible for circulating blood throughout the body.

cirrhosis (sear-row-sis)
A disease of the liver characterized by widespread fibrous tissue buildup and disruption of normal

liver structure and function. It can be caused by a number of chronic conditions that affect the liver such as alcoholism, diabetes, and obesity.

clinical trial
A study design in which one group of randomly assigned subjects (or subjects selected by the "luck of the draw") receives an active treatment and another group receives an inactive treatment, or sugar pill, called the *placebo.*

coenzymes
Chemical substances, including many vitamins, that activate specific enzymes. Activated enzymes increase the rate at which reactions take place in the body, such as the breakdown of fats or carbohydrates in the small intestine and the conversion of glucose and fatty acids into energy within cells.

cofactors
Individual minerals required for the activity of certain proteins. For example: iron is needed for hemoglobin's function in oxygen and carbon dioxide transport; zinc is needed to activate or is a structural component of over 200 enzymes; and magnesium activates more than 300 enzymes involved in the formation of energy and proteins.

colostrum
The milk produced during the first few days after delivery. It contains more antibodies, protein, and certain minerals than the mature milk that is produced later. It is thicker than mature milk and has a yellowish color.

complementary protein sources
Plant sources of protein that together provide sufficient quantities of the nine essential amino acids.

complete proteins
Proteins that contain all of the essential amino acids in amounts needed to support growth and tissue maintenance.

complex carbohydrates
The form of carbohydrate found in starchy vegetables, grains, and dried beans and in many types of dietary fiber. The most common form of starch is made of long chains of interconnected glucose units.

control group
Subjects in a study who do not receive the active treatment or who do not have the condition under investigation. Control periods, or times when subjects are not receiving the treatment, are sometimes used instead of a control group.

C-reactive protein (CRP)
A key inflammatory factor produced in the liver in response to infection or inflammation. Elevated concentrations of CRP are associated with heart disease, obesity, diabetes, inactivity, infection, smoking, and inadequate antioxidant intake.

critical period
A specific interval of time during which cells of a tissue or organ are genetically programmed to multiply. If the supply of nutrients needed for cell multiplication is not available during the specific time interval, the growth and development of the tissue or organ are permanently impaired.

cross-contamination
The spread of bacteria, viruses, or other harmful substances from one surface to another.

cruciferous vegetables
Sulfur-containing vegetables whose outer leaves form a cross (or crucifix). Vegetables in this family include broccoli, cabbage, cauliflower, brussels sprouts, mustard and collard greens, kale, bok choy, kohlrabi, rutabaga, turnips, cauliflower, and watercress.

Daily Values (DVs)
Scientifically agreed-upon daily dietary intake standards for fat, saturated fat, cholesterol, carbohydrate, dietary fiber, and protein intake compatible with health. DVs are intended for use on nutrition labels only and are listed in the Nutrition Facts panel. "% Daily Value" on Nutrition Facts panels is calculated as the percentage of each DV supplied by a serving of the labeled food.

dehydration
A condition that occurs when the body loses more water than it takes in. Prolonged exercise accompanied by profuse sweating in high temperatures, vomiting, diarrhea, certain drugs, and decreased water intake can lead to dehydration.

delayed onset muscle soreness (DOMS)
Muscle pain, soreness, and stiffness that occurs a day or two after exercise of muscles that are not normally used to high levels of activity.

development
Processes involved in enhancing functional capabilities. For example, the brain grows, but the ability to reason develops.

diabetes
A disease characterized by abnormal utilization of glucose by the body and elevated blood glucose levels. There are three main types of diabetes: type 1, type 2, and gestational diabetes. The word *diabetes* in this text refers to type 2 diabetes, by far the most common. Diabetes is short for the term *diabetes mellitus.*

diarrhea
The presence of three or more liquid stools in a 24-hour period.

dietary fiber
Components of plants that cannot be digested by human digestive enzymes and that confer health benefits.

dietary inflammatory index
A dietary assessment tool that estimates the inflammatory potential of a person's diet.

dietary pattern
The quantity, proportions, variety, or combination of different foods, drinks, and nutrients in diets, and the frequency with which they are habitually consumed.

dietary supplements
Any products intended to supplement the diet, including vitamin and mineral supplements; proteins, enzymes, and amino acids; fish oils and fatty acids; hormones and hormone precursors; herbs and other plant extracts; and prebiotics and probiotics. Such products must be labeled "Dietary Supplement."

dietary thermogenesis
Thermogenesis means "the production of heat." Dietary thermogenesis is the energy expended during food ingestion, the digestion of food, and the absorption and utilization of nutrients. Some of the energy escapes as heat. It accounts for approximately 10% of the body's total energy need. Also called *diet-induced thermogenesis* and *thermogenic effect* of food or feeding.

dietetics
The integration, application, and communication of practice principles derived from food, nutrition, social, business, and basic sciences to achieve and maintain optimal nutrition status of individuals and groups.

digestion
The mechanical and chemical processes whereby ingested food is converted into substances that can be absorbed by the intestinal tract and utilized by the body.

diglyceride
A fat in which the glycerol molecule has two fatty acids attached to it. Also called *diacylglyceride.*

disaccharides (di = two, saccharide = sugar)
Simple sugars consisting of two molecules of monosaccharide linked together. Sucrose, maltose, and lactose are disaccharides.

DNA (deoxyribonucleic acid) (d-oxy-rye-bow-new-clay-ic)
Segments of genes that provide instructions for the synthesis, or production, of specific enzymes and other proteins needed by cells to carry out life- and health-sustaining processes. Genetic material contained in cells that initiates and directs the production of proteins in the body.

double blind
A study in which neither the subjects participating in the research nor the scientists performing the research know which subjects are receiving the treatment and which are getting the placebo. Both subjects and investigators are "blind" to the treatment administered.

double-blind, placebo-controlled food challenge
A test used to determine the presence of a food allergy or other adverse reaction to a food. In this test, neither the patient nor the care provider

knows whether a suspected offending food or a placebo is being tested.

duodenal (do-odd-en-all) and stomach ulcers
Open sores in the lining of the duodenum (the uppermost part of the small intestine) or the stomach.

edema
Swelling due to an accumulation of fluid in body tissues.

electrolytes
Minerals such as sodium and potassium that carry a charge when in solution. Many electrolytes help the body maintain an appropriate amount of fluid.

empty-calorie foods
Foods that provide an excess of calories in relation to nutrients. Soft drinks, candy, sugar, alcohol, and fats are considered empty-calorie foods.

endothelium (en-dough-thiel-e-um)
The layer of cells lining the inside of blood vessels.

energy-dense foods
Food that provide relatively high levels of calories per unit weight of the food. Fried chicken, cheeseburgers, a biscuit, egg, and sausage sandwich, and potato chips are energy-dense foods.

energy density
The number of calories per gram of food. It is calculated by dividing the number of calories in a portion of food by the food's weight in grams. May also be calculated as calories per 100 grams of food.

enrichment
The replacement of thiamin, riboflavin, niacin, and iron lost when grains are refined.

environmental trigger
An environmental factor, such as inactivity, a high-fat diet, or a high sodium intake, that causes a genetic tendency toward a disorder to be expressed.

enzymes
Protein substances that speed up chemical reactions. Enzymes are found throughout the body but are present in particularly large amounts in the digestive system.

epidemiological study
Research that seeks to identify conditions related to particular events within a population. This type of research does not identify cause-and-effect relationships. For example, much of the information known about diet and cancer is based on epidemiological studies that have found that diets low in vegetables and fruits are associated with the development of heart disease.

epigenetic
Heritable changes in gene functions that do not entail a change in DNA sequence. Epigenetic modifications play a crucial role in the silencing and expression of noncoding portions of genes.

epigenomics
Pertaining to changes in the regulation of the expression of gene activity without alteration of gene structure.

ergogenic aids (ergo = work; genic = producing)
Substances that increase the capacity for muscular work.

essential amino acids
Amino acids that cannot be synthesized in adequate amounts by humans and therefore must be obtained from the diet. They are sometimes referred to as "indispensable amino acids."

essential fatty acids
Components of fats (linoleic acid [lynn-oh-lay-ick] and alpha-linolenic acid [lynn-oh-len-ick]) required in the diet.

essential hypertension
Hypertension of no known cause; also called *primary* or *idiopathic hypertension*, it accounts for 95% of all cases of hypertension.

essential nutrients
Nutrients required for normal growth and health that the body cannot generally produce, or produce in sufficient amounts; they must be obtained in the diet.

exercise
A subcategory of physical activity that is planned, structured, and repetitive. The terms *physical activity* and *exercise* are often used interchangeably.

experimental group
Subjects in a study who receive the treatment being tested or have the condition that is being investigated.

fat
Chemical substances that are insoluble in water but soluble in fat. Triglycerides, saturated and unsaturated fats, and essential fatty acids are examples of lipids, or "fats."

fatty liver disease
A reversible condition characterized by fat infiltration of the liver (10% or more by weight). If not corrected, fatty liver disease can produce liver damage and other disorders. The condition is primarily associated with obesity, diabetes, and excess alcohol consumption. The disease is called *steatohepatitis* when accompanied by inflammation.

fermentation
The process by which carbohydrates are converted to ethanol by the action of the enzymes in yeast.

fetus
A baby in the womb from the eighth week of pregnancy until birth. (Before then, it is referred to as an embryo.)

flatulence (flat-u-lens)
Presence of excess gas in the stomach and intestines.

flexibility
Range of motion of joints.

food additives
Any substances added to food that become part of the food or affect the characteristics of the food. The term applies to substances added to food either intentionally and unintentionally.

food allergen
A substance in food (almost always a protein) that is identified as harmful by the body and elicits an allergic reaction from the immune system.

food allergy
Adverse reaction to a normally harmless substance in food that involves the body's immune system.

food-borne illness
An illness related to consumption of foods or beverages containing disease-causing bacteria, viruses, parasites, toxic chemicals, or other harmful substances.

food insecurity
Limited or uncertain availability of safe, nutritious foods or the ability to acquire them in socially acceptable ways.

food intolerance
Adverse reaction to a normally harmless substance in food that does not directly involve the body's immune system.

food security
Access at all times to a sufficient supply of safe, nutritious foods.

fortification
The addition of one or more vitamins and/or minerals to a food product.

free radicals
Chemical substances (often oxygen-based) that are missing electrons. The absence of electrons makes the chemical substance reactive and prone to oxidizing nearby molecules by stealing electrons from them. Free radicals can damage lipids, proteins, cells, DNA, and eventually tissues by altering their chemical structure and functions.

fruitarian
A form of vegetarian diet in which fruits are the major ingredient. Such diets provide inadequate amounts of a variety of nutrients.

functional foods
Generally taken to mean and foods or food ingredients that may provide a health benefit beyond the effects of traditional nutrients they contain.

genes
The basic units of heredity that occupy specific places (loci) on chromosomes. Genes consist of large DNA molecules, each of which contains the code for the manufacture of a specific enzyme or other protein.

gene variants
A different form of the same genes that direct specific protein production. Proteins produced affect every metabolic process in the body, from the digestion to glucose utilization. Although we all have the same genes, we do not all share the same gene variants. Also defined as: An alteration in the normal nucleotide sequence of a gene that affects the gene's functions. Different forms of the same genes are considered "alleles."

genetically modified organisms (GMOs)
Food that contains genetic material that has been transferred from the genetic material of another organism (usually a bacteria) to give the food or food crop specific properties.

genome
Combined term for *genes* and *chromosomes.* It represents all the genes and DNA contained in an organism. It is estimated that the human genome consists of 20,000 to 25,000 genes.

genomics
The study of the functions and interactions of all genes in the genome. Unlike genetics, it includes the study of genes related to common conditions and their interaction with environmental, or "epigenetic," factors.

genotype
The specific genetic makeup of an individual as coded by DNA.

geophagia (ge-oh-phag-ah)
Clay or dirt eating.

gestational diabetes
Diabetes first discovered during pregnancy.

glycemic index (GI)
A measure of the extent to which blood glucose levels are raised by consumption of an amount of food that contains 50 grams of carbohydrate compared to 50 grams of glucose. A portion of white bread containing 50 grams of carbohydrate is sometimes used for comparison instead of 50 grams of glucose.

glycemic load (GL)
A measure of the extent to which blood glucose level is raised by a given amount of a carbohydrate-containing food. GL is calculated by multiplying a food's GI by its carbohydrate content.

glycerol
A syrupy, colorless liquid component of fats that is soluble in water. It is similar to glucose in chemical structure.

glycogen
The body's storage form of glucose. Glycogen is stored in the liver and muscles.

growth
A process characterized by increases in cell number and size.

gut microbiota
The microbial population living in our intestine. Formerly called gut flora.

HDL cholesterol
High-density lipoprotein cholesterol. The primary function of HLD cholesterol is to transport cholesterol from body tissues to the liver for incorporation into bile and excretion from the body. HDL cholesterol is sometimes referred to as "good cholesterol" or "healthy cholesterol" because high blood levels are related to reduced risk of heart disease in most people.

health
The WHO defines health as state of complete physical, mental, and social well-being and not merely the absence of disease or infirmity.

heart disease
One of a number of disorders that result when circulation of blood to parts of the heart is inadequate. Also called *coronary heart disease.* (*Coronary* refers to the blood vessels at the top of the heart. They look somewhat like a crown.)

heartburn
A condition that results when acidic stomach contents are released into the esophagus, usually causing a burning sensation.

hemoglobin
The iron-containing protein in red blood cells.

hemoglobin A1c
A measure of the percentage of hemoglobin proteins in red blood cells that are attached to glucose. The higher the blood glucose levels, the more glucose will become attached to hemoglobin. Once glucose binds with hemoglobin, it will stay there for the lifespan of the hemoglobin (normally about 120 days). Consequently, hemoglobin A1c represents blood glucose levels over several months.

hemorrhoids (hem-or-oids)
Swelling of veins in the anus or rectum.

herbal remedy
A plant, plant part, or an extract or mixture of these believed to prevent, alleviate, or cure disease.

herbs
A plant or a part of a plant that is used as medicine or to give flavor to food. They represent a complex mixture of chemical substances and can be derived from leaves, stems, flowers, roots, and seeds of a plant. Botanists define herbs as small, seed-bearing plants with fleshy, rather than woody, parts.

high-quality protein
Proteins that contain all of the essential amino acids in amounts needed to support growth and tissue maintenance. Examples of high-quality proteins include eggs, milk, meat, and beans and rice. Also referred to as *complete proteins.*

histamine (hiss-tah-mean)
A substance released in allergic reactions. It causes dilation of blood vessels, itching, hives, and a drop in blood pressure and stimulates the release of stomach acids and other fluids. Antihistamines neutralize the effects of histamine and are used in the treatment of some cases of allergies.

homocysteine
A compound produced when the amino acid methionine is converted to another amino acid, cysteine. High blood levels of homocysteine increase the risk of hardening of the arteries, heart attack, and stroke.

hormone
A substance, usually a protein or steroid (a cholesterol-derived chemical), produced by one tissue and conveyed by the bloodstream to another. Hormones affect the body's metabolic processes such as glucose utilization and fat deposition.

hunger
Unpleasant physical and psychological sensations (weakness, stomach pains, irritability) that lead people to acquire and ingest food.

hydration status
The state of the adequacy of fluid in the body tissues.

hydrogenation
The addition of hydrogen to unsaturated fatty acids.

hypertension
High blood pressure. It is defined as blood pressure exerted inside of blood vessel walls that typically exceeds 140/90 mm Hg (millimeters of mercury) in adults aged 60 or under and 150/90 in adults over 60. Hypertension in children and adolescents is defined as systolic or diastolic blood pressure equal to or greater than the 95th blood pressure percentile of sex-, age- and height-specific blood pressure percentiles.

hypoglycemia
A disorder resulting from abnormally low blood glucose levels. Symptoms of hypoglycemia include irritability, nervousness, weakness, sweating, and hunger. These symptoms are relieved by consuming glucose or foods that provide carbohydrate.

hyponatremia
A deficiency of sodium in the blood (135 mmol/L sodium or less).

hypothesis
A statement made prior to initiating a study of the relationship sought to be tested by the research.

immune system
Body tissues that provide protection against bacteria, viruses, and other substances identified by cells as harmful.

immunoproteins
Blood proteins such as antibodies that play a role in the functioning of the immune system (the body's disease defense system). Antibodies attack foreign proteins.

incomplete proteins
Proteins that are deficient in one or more essential amino acids.

indirect calorimetry
A method of measuring energy expenditure from determination of the amount of oxygen utilized by the body during a specific unit of time.

infant mortality rate
Deaths that occur within the first year of life per 1,000 live births.

inflammation
Reactions of the body to the presence of infectious agents, toxins, or irritants. Inflammation triggers the release of biologically active substances that promote oxidation and other reactions that counteract the infectious agent, toxin, or irritant.

initiation phase
The start of the cancer process; it begins with the alteration of DNA within cells.

insulin resistance
A condition in which cell membranes have reduced sensitivity to insulin so that more insulin than normal is required to transport a given amount of glucose into cells. It is characterized by elevated levels of serum insulin, glucose, and triglycerides, and increased blood pressure.

iron deficiency
A disorder that results from a depletion of iron stores in the body. It is characterized by weakness, fatigue, short attention span, poor appetite, increased susceptibility to infection, and irritability.

iron-deficiency anemia
A condition that results when the content of hemoglobin in red blood cells is reduced due to a lack of iron. It is characterized by the signs of iron deficiency plus paleness, exhaustion, and a rapid heart rate.

irritable bowel syndrome (IBS)
A disorder of bowel function characterized by chronic or episodic gas, abdominal pain, diarrhea or constipation, or both.

kwashiorkor (kwa-she-or-kor)
A severe form of protein-energy malnutrition in young children. It is characterized by swelling, fatty liver, susceptibility to infection, profound apathy, and poor appetite.

LDL cholesterol
Low-density lipoprotein cholesterol. The function of LDL is to transport cholesterol in blood to cells, where it is used as a component of cell membranes, or for the synthesis of testosterone, estrogen, and other hormones. It is called "bad cholesterol" because high levels are associated with the development of heart disease in most people, and because it supplies the cholesterol found in plaque.

lactose intolerance
The term for gastrointestinal symptoms (flatulence, bloating, abdominal pain, diarrhea, and "rumbling in the bowel") resulting from consumption of more lactose than can be digested with available lactase.

lactose maldigestion
A disorder characterized by reduced digestion of lactose due to low availability of the enzyme lactase.

leptin
A hormone produced by fat cells involved in the regulation of energy balance and body weight. It initiates signals that suppress appetite and increase the utilization of fat stores.

life expectancy
The average length of life from birth.

lipids
Compounds that are insoluble in water and soluble in fat. Triglycerides, saturated and unsaturated fats, and essential fatty acids are examples of lipids, or "fats."

low-birthweight infants
Infants weighing less than 2,500 grams (5.5 pounds) at birth.

lymphatic system
A network of vessels that absorb some of the products of digestion and transport them to the heart, where they are mixed with the substances contained in blood.

macronutrients
The group name for carbohydrate, protein, fat, and water. They are called *macronutrients* because we need relatively large amounts of them in our daily diet.

malnutrition
A cellular imbalance between the supply of nutrients and energy and the body's demand for them to ensure growth, maintenance, and specific functions.

marasmus
A severe form of malnutrition primarily due to a chronic lack of calories and protein. Also called *protein-energy malnutrition.*

maximal oxygen consumption
The greatest amount of oxygen that can be delivered to, and utilized by, muscles for physical activity. Also called VO_2 *max* and *maximal volume of oxygen.*

menopause
Cessation of the menstrual cycle and reproductive capacity in females.

meta-analysis
An analysis of data from multiple studies. Results are based on larger samples than the individual studies and are therefore more reliable. Differences in methods and subjects among the studies may bias the results of meta-analyses.

metabolic syndrome
A constellation of metabolic abnormalities that increase the risk of heart disease, hypertension, type 2 diabetes, and other disorders. Metabolic syndrome is characterized by insulin resistance, abdominal obesity, high blood pressure and triglyceride levels, low levels of HDL cholesterol, and impaired glucose tolerance. It is also called *syndrome X* and *insulin resistance syndrome.*

metabolism
The chemical changes that take place in the body. The conversion of glucose to energy or to body fat is an example of a metabolic process.

microbes
Microscopic organisms, including bacteria and fungi. Some microbes are beneficial, some pathogenic (harmful), and some have little effect on the body. Also called *microorganisms.*

minerals
In the context of nutrition, minerals are specific, single atoms that perform particular functions in the body. There are 15 essential minerals—minerals required in the diet.

microbiome
The totality of microorganisms (bacteria, viruses, fungi) and their collective genetic material present in or on the human body.

monoglycerides
A fat in which the glycerol molecule has one fatty acid attached to it. Also called *monoacylglyceride.*

monosaccharides (mono = one, saccharide = sugar)
Simple sugars consisting of one sugar molecule. Glucose, fructose, and galactose are common monosaccharides.

monounsaturated fats
Fats that contain a fatty acid in which one carbon-to-carbon bond is not saturated with hydrogen.

mRNA (messenger RNA)
A molecule that carries copies of instructions for the assembly of amino acids from DNA to the rest of the cell.

muscular endurance
The ability of a muscle or group of muscles to sustain repeated contractions against a resistance for an extended period of time. Also called *muscular fitness.*

muscular strength
The ability of a muscle or muscle group to exert force. It is assessed by the maximal amount of resistance or force that can be sustained in a single effort.

myoglobin
The iron-containing protein in muscle cells.

neural tube defects
Malformations of the spinal cord and brain. They are among the most common and severe fetal malformations, occurring in approximately 1 in every 1,000 pregnancies. Neural tube defects include spina bifida (spinal cord fluid protrudes through a gap in the spinal cord; shown in Illustration 29.6), anencephaly (absence of the brain or spinal cord), and encephalocele (protrusion of the brain through the skull).

nitrogen balance
The difference between nitrogen intake and excretion. It is assessed as the difference between nitrogen intake and nitrogen excreted.

nonessential amino acids
Amino acids that can readily be produced by humans from components of the diet. Also referred to as "dispensable amino acids."

nonessential nutrients
Nutrients required for normal growth and health that the body can manufacture in sufficient quantities from other components of the diet. We do not require a dietary source of nonessential nutrients.

non-HDL cholesterol
The difference between total cholesterol concentration and the HDL cholesterol concentration in blood.

nucleotide
A nucleotide is one of the structural components, or building blocks, of DNA and RNA. A nucleotide consists of a base (one of four chemicals: adenine, thymine, guanine, and cytosine) plus a molecule of sugar and one of phosphoric acid.

nutrigenomics
The study of the interactions among genes and nutrients

nutrient-dense foods
Foods that contain relatively large amounts of nutrients compared to their calorie value. Broccoli, collards, bread, cantaloupe, and lean meats are examples of nutrient-dense foods.

nutrients
Chemical substances found in foods that are used by the body for growth and health. The six categories of nutrients are carbohydrates, proteins, fats, vitamins, minerals, and water.

nutrients of public significance
Nutrients, specifically calcium, potassium, vitamin D, and iron for which the U.S. population is consuming inadequate amounts and are associated with the risk of chronic disease.

nutrigenomics
The study of the effects of nutrients on gene expression, and the impact of nutrient–gene interactions on health and disease.

nutrition
The study of foods, their nutrients and other chemical constituents, and the effects of food constituents on health.

nutrition transition
A change in a population's diet from traditional foods such as grains, roots, beans, and rice to calorie-dense foods high in sugar, fat, sodium and other components that characterize "Western" diets.

obese
In adults, a BMI (body mass index) of 30 kg/m^2 or higher. In children and adolescents, a BMI at or above the 95th percentile for children and adolescents of the same age and sex.

older adult
Generally defined as individuals 65 or more years old.

osteoporosis (osteo = bones; poro = porous, osis = abnormal condition)
A condition in which bones become fragile and susceptible to fracture due to a loss of calcium and other minerals.

overweight
In adults, a BMI (body mass index) of 25–29.9 kg/m^2. In children and adolescents, overweight is defined as a BMI at or above the 85th percentile and lower than the 95th percentile for children of the same age and sex.

oxidative stress
A condition that occurs when cells are exposed to more oxidizing molecules (such as free radicals) than to antioxidant molecules that neutralize them. Over time, oxidative stress causes damage to lipids, DNA, cells, and tissues. It increases the risk of heart disease, type 2 diabetes, cancer, and other diseases.

pagophagia (pa-go-phag-ah)
Ice eating.

peer review
Evaluation of the scientific merit of research or scientific reports by experts in the area under review. Studies published in scientific journals have gone through peer review prior to being accepted for publication.

Percent Daily Value (%DV)
Scientifically agreed-upon standards of daily intake of nutrients from the diet developed for use on nutrition labels. The "% Daily Values" listed in nutrition labels represent the percentages of the standards obtained from one serving of the food product.

phenylketonuria (feen-ol-key-tone-u-reah) (PKU)
A rare genetic disorder related to lack of the enzyme phenylalanine hydroxylase. Lack of this enzyme causes the essential amino acid phenylalanine to build up in the blood.

physical activity
Body movements produced by muscles that require energy expenditure.

physical fitness
The health of the body as measured by muscular strength and endurance, cardiorespiratory fitness, and flexibility. Body composition, agility, and balance are sometimes included as components of physical fitness.

physical performance
The ability to perform a physical task or sport at a desired or particular level.

phytochemicals (phyto = plant)
Biologically active, or "bioactive," substances in plants that have positive effects on health. Also called *phytonutrients.*

pica (pike-eh)
The regular consumption of nonfood substances such as clay or laundry starch.

placebo
A "sugar pill," an imitation treatment given to subjects in research.

placebo effect
Changes in health or perceived health that result from expectations that a "treatment" will produce an effect on health.

plant stanols or **sterols**
Substances in corn, wheat, oats, rye, olives, wood, and some other plants that are similar in structure to cholesterol but that are not absorbed by the body. They decrease cholesterol absorption.

plaque (arterial)
Deposits of cholesterol, other fats, calcium, and cell materials in the lining of the inner wall of arteries.

plaque (dental)
A soft, sticky, white material on teeth; formed by bacteria.

plumbism
Lead eating (primarily from old paint flakes).

polysaccharides (poly = many, saccharide = sugar)
Carbohydrates containing many molecules of monosaccharides linked together. Starch, glycogen, and dietary fiber are the three major types of polysaccharides. Polysaccharides

consisting of 3 to 10 monosaccharides may be referred to as *oligosaccharides.*

polyunsaturated fats
Fats that contain a fatty acid in which two or more carbon-to-carbon bonds are not saturated with hydrogen.

prebiotics
Non-digestible carbohydrates (various types of dietary fiber) that serve as food for and promote the growth of beneficial microorganisms in the small intestine and colon. Also called "intestinal fertilizer."

precursor
In nutrition, a nutrient that can be converted in the body from one nutrient to another. Beta-carotene is a precursor of vitamin A.

prediabetes
A condition in which blood glucose levels are higher than normal but not high enough for a diagnosis of diabetes. It is characterized by impaired glucose tolerance, or fasting blood glucose levels between 100 to ≤125 mg/dl.

prehypertension
In adults, typical blood pressure levels of 120/80 mm Hg through 139/89 mm Hg. In children and adolescents. prehypertension is defined as systolic or diastolic blood pressure equal to or greater than the 90th percentile but less than the 95th percentile of sex-, age-, and height-specific blood pressure percentiles.

preterm
Infants born at or before 37 weeks of gestation (pregnancy).

prion
An infectious, misfolded protein that has the capability of causing normal proteins to become misfolded, thereby producing disease. The resulting diseases are called spongiform encephalopathies.

probiotics
Live microorganisms which resist digestion, and when administered in adequate amounts of appropriate strains, confer health benefits to the host. Strains of *Lactobacillus* (lac-toe-bah-sil-us) and *Bifidobacteria* (bif-id-dough bacteria) are the best-known probiotics. Also called "friendly bacteria."

processed foods
Any food other than raw agricultural commodities (such as carrots and eggs) that have been modified from their original state through processing. They are classified based on the extent of processing from minimal (frozen, canned fruits and vegetables) to ultra-processed (foods such as soft drinks, veggie "puffs," processed meats) that contain little or no whole foods.

progression
The uncontrolled growth of abnormal cells.

promotion phase
The period in cancer development when the number of cells with altered DNA increases.

prostate
A gland located above the testicles in males. The prostate secretes a fluid that surrounds sperm.

protein
Chemical substance in foods made up of chains of amino acids.

purging
The use of self-induced vomiting, laxatives, or diuretics (water pills) to rid the body of food.

Recommended Dietary Allowances (RDAs)
Intake levels of essential nutrients that meet the nutritional needs of practically all healthy people while decreasing the risk of certain chronic diseases.

refined carbohydrates
Carbohydrate-containing foods that have been processed to remove most or all of the bran coating and germ from the grain. Refined grains are lower in fiber, beneficial phytonutrients, and vitamins and minerals than are whole grains. White rice and bread, pasta, biscuits, many breakfast cereals, pizza dough, and flour tortillas are examples of refined grain products.

refined grains
Whole grains that have been processed to remove the bran covering the grain kernel and germ that stores nutrients for seedling growth. The endosperm, the portion of the grain kernel that stores complex carbohydrate and other nutrients, remains.

relative energy deficiency in sports (RED-S)
A deficiency of energy intake compared to need that impairs metabolic rate, menstrual function, bone health, immunity, protein synthesis, and cardiovascular health. It may lead to a gradual reduction in physical performance, increased risk of injury, decreased strength, and other conditions.

remodeling
The breakdown and buildup of bone tissue.

resistance exercise
Physical activities that involve the use of muscles against a weight or force. It increases muscle strength, mass, power, and endurance. Also called *resistance training, strength training,* and *weight training.*

resistant starch
Starches that do not release glucose within the small intestine but are consumed or fermented by bacteria in the colon released as fatty acids.

restrained eating
The purposeful restriction of food intake below desired amounts in order to control body weight.

ribosomal R (rRNA)
A component of ribosomes, the cell's protein factories. Strictly speaking, ribosomal RNA (rRNA) does not make proteins. It makes polypeptides (assemblies of amino acids) that go to make up proteins.

salt sensitivity
A genetically determined condition in which a person's blood pressure rises when large amounts of salt or sodium are consumed. Such individuals are sometimes identified by blood pressure increases of 5 to 10% or more when switched from a low-salt to a high-salt diet.

satiety
A feeling of fullness or of having had enough to eat.

saturated fats
Molecules of fat in which adjacent carbons within fatty acids are linked only by single bonds. The carbons are "saturated" with hydrogen; that is, they are attached to the maximum number of hydrogen atoms. Saturated fats tend to be solid at room temperature. Animal fat and palm and coconut oils are sources of saturated fats.

serotonin (sare-uh-tone-in)
A neurotransmitter, or chemical messenger, for nerve cell activities that excite or inhibit various behaviors and body functions. It plays a role in mood, appetite regulation, food intake, respiration, pain transmission, blood vessel constriction, and other body processes.

severe obesity
In children and adolescents, severe obesity is defined as BMIs for age over the 99th percentile.

simple sugars
Carbohydrates that consist of a glucose, fructose, or galactose molecule; or a combination of glucose and either fructose or galactose. High-fructose corn syrup and alcohol sugars are also considered simple sugars. Simple sugars are often referred to as *sugars.*

single-gene defects
Disorders resulting from one abnormal gene. Also called *inborn errors of metabolism.* Over 6,000 single-gene defects have been cataloged, and most are very rare.

soluble fiber
Types of dietary fiber that dissolve in water, forming a gel. Good sources of soluble fiber include oats, oatmeal, oat bran, apples, pears, dried beans, carrots, and psyllium.

starch
Complex carbohydrates made up of complex chains of glucose molecules. Starch is the primary storage form of carbohydrate in plants. The vast majority of carbohydrate in our diet consists of starch, monosaccharides, and disaccharides.

statistically significant
Research findings that likely represent a true or actual result and not one due to chance.

steatohepatitis (ste-at-oh-hepah-tie-tis)
A disease characterized by inflammation of, and fat accumulation in, the liver. It is associated with alcoholism and may occur in obesity and diabetes. Steatohepatitis may progress to cirrhosis.

stroke
The event that occurs when a blood vessel in the brain suddenly ruptures or becomes blocked, cutting off blood supply to a portion of the brain. Stroke is often associated with "hardening of the arteries" in the brain. (Also called a *cerebral vascular accident.*)

structure/function claims
A statement appearing primarily on dietary supplement labels that describes the effect a supplement may have on the structure or function of the body. Such statements cannot claim to diagnose, cure, mitigate, treat, or prevent disease.

subcutaneous fat (sub-cue-tain-e-ous)
Fat located under the skin.

sustainable agriculture
Broadly defined as the use of farming methods, environmental practices, and economic policies that meet society's present food needs without compromising the ability of future generations to meet their own needs.

sweat rat
Fluid loss per hour of exercise. It equals the sum of body weight loss plus fluid intake.

symbiosis
The interaction between two different organisms living in close physical association that typically benefits both.

symbiotics
Combinations of prebotics and probiotics that interact in ways that generally benefit both and the health of the host.

tRNA (transfer RNA)
A molecule that tranfers amino acids to ribosomes during protein synthesis.

tooth decay
The disintegration of teeth due to acids produced by bacteria in the mouth that feed on sugar. Also called *dental caries* or *cavities.*

***trans* fats**
A type of unsaturated fatty acid produced by the addition of hydrogen to liquid vegetable oils to make them more solid. Small amounts of naturally occurring *trans* fat are found in milk and meat.

triglyceride
Fats in which the glycerol molecule has three fatty acids attached to it. Triglycerides are the most common type of fat in foods and in body fat stores. Also called *triacylglycerides.*

trimester
One-third of the normal duration of pregnancy. The first trimester is 0 to 13 weeks, the second is 13 to 26 weeks, and the third is 26 to 40 weeks.

tryptophan (trip-tuh-fan)
An essential amino acid that is used to form the chemical messenger serotonin (among other functions). Tryptophan is generally present in lower amounts in food protein than most other essential amino acids. It can be produced in the body from niacin, a B vitamin.

type 1 diabetes
A disease characterized by high blood glucose levels resulting from destruction of the insulin-producing cells of the pancreas. This type of diabetes was called *juvenile-onset diabetes* and *insulin-dependent diabetes* in the past, and its official medical name is *type 1 diabetes mellitus.*

type 2 diabetes
A condition characterized by high blood glucose levels due to the body's inability to use insulin normally, or to produce enough insulin. This type of diabetes was called *adult-onset diabetes* in the past, and its official medical name is *type 2 diabetes mellitus.*

umami (u-mam-e)
A Japanese word meaning "pleasant savory taste." The taste is described as brothy or meaty and is recognized as the fifth basic taste. Foods containing glutamate, an amino acid derivative, elicit the umami taste. These include fish, meats, mushrooms, ripe tomatoes, and aged cheese.

underweight
Usually defined as a low weight for height. May also represent a deficit of body fat.

unsaturated fats
Molecules of fat in which adjacent carbons within the fatty acids are linked only by single carbon-to-carbon bonds. Saturated fats tend to be solid at room temperature, and are mainly found in animal fat and palm and coconut oil.

urea
Nitrogen released from the breakdown of proteins for energy is largely excreted in the urine in the form of urea. It can be measured in urine, or in blood as "blood urea nitrogen" (BUN).

variety
A diet consisting of many different foods from all of the food groups.

visceral fat (vis-sir-el)
Fat located under the skin and muscle of the abdomen.

vitamins
Components of food required by the body in small amounts for growth and health maintenance.

water balance
The ratio between the water consumed and that lost from the body. Individuals are in water balance when water intake equals water output.

water intoxication
A condition that results when the body accumulates more water than it can excrete due to low availability of sodium. The condition can result in headache, fatigue, nausea and vomiting, confusion, irritability, seizures, and coma. It can occur in people who consume or receive too much water without sodium and in those with impaired kidney function. Also referred to as overhydration and hyponatremia.

white fat
The primary type of fat in the body composed of triglycerides and cholesterol particles stored within large lipid droplets. White adipocytes are largely used for energy storage and production, but also produce a number of chemical messengers that influence inflammation, appetite, and that affect regulation of blood glucose levels.

whole grains
Cereal grains that consist of the intact, ground, cracked, or flaked kernel, which includes the bran, the germ, and the innermost part of the kernel (the endosperm).

zoochemicals
Chemical substances in animal foods, some of which may be biologically active in the body.

References

Unit 1

1. What is food security? www.ers.usda.gov/topics/food-nutrition-assistance/food-security-in-the-us/measurement.aspx#insecurity, accessed10/18/14.
2. Pruitt SL et al. Who is food insecure? Implications for targeted recruitment and outreach, National Health and Nutrition Examination Survey, 2005–2010. *Prev Chronic Dis* 2016; doi.org/10.5888/pcd13.160103.
3. Coleman-Jensen A et al. Household Food Security in the United States in 2010. Economic Research Report No. (ERR-125), September 2011, www.ers.usda.gov/Publications/ERR12B.
4. Stang J et al. Position of the American Dietetic Association: Child and adolescent nutrition assistance programs. *J Am Diet Assoc* 2010;110:791–99.
5. Food Security Status of U.S. Households in 2016. USDA, Economic Research Service, 2017, www.ers.usda.gov/topics/food-nutrition-assistance/food-security-in-the-us/key-statistics-graphics.aspx
6. Edge J et al. Enough for all: Household food security in Canada, Conference Board of Canada, August 2103, www.conferenceboard.ca/e-library/abstract.aspx?did=5723, accessed 10/20/14.
7. Final Review: Healthy People 2010, Nutrition and overweight. www.cdc.gov/nchs/data/hpdata2010/hp2010_final_review_focus_area_19.pdf, accessed 4/6/15.
8. Nord M et al. *Prevalence of U.S. Food Insecurity Is Related to Changes in Unemployment, Inflation, and the Price of Food.* USDA, June 2014, available at: www.ers.usda.gov/media/1489980/err167_summary.pdf.
9. *2014 Hunger in America Report.* www.eatright.org/members/eatrightweekly/article.aspx?folderid=6442452698&mycontentid=6442482198, accessed 8/28/14.
10. Bahrampour T. More college students battle hunger as education and living costs rise. April 9, 2014, www.washingtonpost.com/local/more-college-students-battle-hunger-as-education-and-living-costs-rise/2014/04/09/60208db6-bb63-11e3-9a05-c739f29ccb08_story.html, accessed 10/20/14.
11. Agriculture and the Food Supply. September 13, 2013, www.epa.gov/climatechange/impacts-adaptation/agriculture.html, accessed 10/20/14.
12. Sustainable Food Consumption and Production. www.fao.org/ag/ags/sustainable-food-consumption-and-production/en/, FAO, 2014, accessed 10/20/14.
13. Sources of Greenhouse Gas Emissions. www.epa.gov/climatechange/ghgemissions/sources.html, August 2014, accessed 10/20/14.
14. Hunter DJ et al. Preventive medicine for the planet and its peoples. *NEJM* 2017;376:105–07.
15. Feenstra G, What is sustainable agriculture? University of California at Davis, Agricultural Sustainability Institute: accessed 10/14/17 at http://asi.ucdavis.edu/programs/sarep/about/what-is-sustainable-agriculture
16. Scalbert A et al. The food metabolome: A window over dietary exposure. *Am J Clin Nutr* 2014;99:1286–308.
17. Jiang Y et al. Cruciferous vegetable intake is inversely correlated with circulating levels of proinflammatory markers in women. *J Acad Nutr Diet* 2014;114:700–08.
18. Saey TH. Your body is mostly microbes. *Sci News,* December 28, 2013, pp. 18–19.
19. Puterman E et al. Determinants of telomere attrition over 1 year in healthy older women: Stress and health behaviors matter. *Mol Psychiatry,* July 29, 2014, doi: 10.1038/mp.2014.70.
20. Fairbanks VF. Chapter 10. Iron in Medicine and Nutrition. In: Shils ME et al., eds. *Modern Nutrition in Health and Disease,* 9th ed. Baltimore: Lippincott Williams Wilkins; 1999:193–221.
21. Fairbanks VF. Chapter 10. Iron in Medicine and Nutrition. In: Shils ME, et al., eds. *Modern Nutrition in Health and Disease,* 9th ed. Baltimore: Lippincott Williams Wilkins, 1999:193–221.
22. Hayes DP. Adverse effects of nutritional inadequacy and excess: A hormetic model. *Am J Clin Nutr* 2008;88(suppl):578S–81S.
23. King JC et al. Zinc. In: Shils ME et al., eds. *Modern Nutrition in Health and Disease,* 10th ed. Philadelphia: Lippincott, Williams & Wilkins, 2006;279–83.
24. Tanumihardjo SA. Vitamin A: Biomarkers of nutrition for development. *Am J Clin Nutr* 2011;94(suppl):658S–65S.
25. Goodman, AS. Vitamin A and retinols in health and disease. *N Engl J Med* 1984;310:1023–31.
26. Mahoney DH Jr. Anemia in at-risk populations—what should be our focus? *Am J Clin Nutr* 2008;88:1457–58.
27. Agnes N. Pedersen et al. Health effects of protein intake in healthy adults: A systematic literature review. *Food Nutr Res* 2013;57, doi: 10.3402/fnr.v57i0.21245.
28. Krauss RM. Atherogenic Lipoprotein Phenotype and Diet-Gene Interactions. *J Nutr* 2001;131:340S-3S.
29. Gillman MW et al. How early should obesity prevention start? *N Engl J Med* 2013;369:2173–75.
30. Ohlhorst SD et al. Nutrition research to affect food and a healthy lifespan. *Am J Clin Nutr* 2013;143:1349–54.
31. Hiza HAB et al. Position of the Academy of Nutrition and Dietetics: Total diet approach to healthy eating. *J Am Acad Nutr Diet* 2013;113:307–17.
32. *2010 Dietary Guideline for Americans,* available at www.dietaryguidelines.gov, accessed February 2011.
33. Discretionary Calories, www.health.gov, DIETARYGUIDELINES/dga2005/report/HTML/D3_DiscCalories. htm#tabled31, accessed November 2008.

Unit 2

1. Leading causes of death. CDC, 2017, www.cdc.gov/nchs/fastats/leading-causes-of-death.htm
2. Mehta, M et al. The population health benefits of a healthy lifestyle: Life expectancy increased and onset of disability delayed. *Health Affairs,* 2017, doi: 10.1377/hlthaff.2016.1569.
3. Cody MM et al. The Academy of Nutrition and Dietetics' public policy priorities overview. *J Acad Nutr Diet* 2013:113;392–395, March 2013, doi: http://dx.doi.org/10.1016/j.jand.2013.01.008.
4. Chronic diseases: The leading causes of death and disability in the United States.

CDC 2017, www.cdc.gov/chronicdisease/overview/index.htm
5. Mortality in the United States, 2012, www.cdc.gov/nchs/data/databriefs/db168.htm. NCHS Data Brief, Number 168, October 2014.
6. Bauer UE et al. Prevention of chronic disease in the 21st century: Elimination of the leading preventable causes of premature death and disability in the USA. *The Lancet* 2014;384:45–52, doi: 10.1016/S0140-6736(14)60648-6.
7. Paez K. More Americans Getting Multiple Chronic Illnesses, available at: www.medscape.com/viewarticle/58636. Accessed January 2009.
8. Kim K et al. Dietary total antioxidant capacity is inversely associated with all-cause and cardiovascular disease death of US adults. *Euro J Nutr,* August 8 2017, https://link.springer.com/article/10.1007/s00394-017-1519-7
9. Bauer UE et al. Prevention of chronic disease in the 21st century: Elimination of the leading preventable causes of premature death and disability in the USA. *The Lancet* 2014;384(9937):45–52, doi: 10.1016/S0140-6736(14)60648-6. Published online July 2, 2014.
10. Graudal N et al. Compared with usual sodium intake, low- and excessive-sodium diets are associated with increased mortality: A meta-analysis. *Am J Hyperten* 2014;27:1129–37, doi: 10.1093/ajh/hpu028.
11. Candidio FG et al. Impact of dietary fat on gut microbiota and low-grade systemic inflammation: Mechanisms and clinical implications on obesity. *Int J Food Sci Nutr.* July 4, 2017; doi.org/10.1080/09637486.2017.1343286.
12. Wang X et al. Fruit and vegetable consumption and mortality from all causes, cardiovascular disease, and cancer: Systematic review and dose-response meta-analysis of prospective cohort studies. *BMJ,* July 29, 2014;349:g4490, doi: 10.1136/bmj.g4490.
13. Cho SS et al. Consumption of cereal fiber, mixtures of whole grains and bran, and whole grains and risk reduction in type 2 diabetes, obesity, and cardiovascular disease. *Am J Clin Nutr.* August 2013, doi:10.3945/ajcn.113.067629.
14. Foods that fight inflammation. Harvard Women's Health Watch. 2017; www.health.harvard.edu/staying-healthy/foods-that-fight-inflammation
15. Steck S. The dietary inflammatory index and risk of colorectal and breast cancers in the Women's Health Initiative. 2017, www.whi.org/researchers/sigs/nutrition/Supporting%20Docs/Steck%20CPCP%20WHI%20Nutrition%20Group%2005Sept2014.pdf
16. Grimble RF. Nutritional modulation of cytokine biology. *Nutrition* 1998;14:634–40.
17. Anderson AL et al. Dietary patterns, insulin sensitivity and inflammation in older adults. *Eur J Clin Nutr* 2012;66:18–24, doi:10.1038/ejcn.2011.162, available at: www.ncbi.nlm.nih.gov/pmc/articles/PMC3251708.
18. Cespedes EM et al. Dietary patterns: from nutritional epidemiologic analysis to national guidelines. *Am J Clin Nutr* 2015;105:899–900.
19. Mori N et al. Cruciferous vegetable intake is inversely associated with lung cancer risk among current nonsmoking men in the Japan Public Health Center (JPHC). *Study J Nutr* 2017;147: 841–49.
20. Graff M et al. Genome-wide physical activity interactions in adiposity — A meta-analysis of 200,452 adults. *PLOS Genetics,* April 27, 2017, doi.org/10.1371/journal.pgen.1006528.
21. Goni L et al. Future perspectives of personalized weight loss interventions based on nutrigenetic, epigenetic, and metagenomics data. *J Nutr* 2016;46(suppl):905S-12S.
22. Schwingshackl L et al. Food groups and risk of all-cause mortality: a systematic review and meta-analysis of prospective studies. *Am J Clin Nutr* 2017;105:1462–73.
23. Cordain L et al. Origins and evolution of the Western diet: Health implications for the 21st century. *Am J Clin Nutr* 2005:81:341–54.
24. Turner BL et al. Beyond the Paleolithic prescription: Incorporating diversity and flexibility in the study of human diet evolution. *Nutr Rev* 2013;71:501–10, doi:10.1111/nure.12039.
25. Lindeberg S, et al. A Paleolithic diet improves glucose tolerance more than a Mediterranean-like diet in individuals with ischaemic heart disease. *Diabetologia* 2007;50:1795–807.
26. Fallon S et al. Inside Japan: Surprising facts about Japanese foodways. www.westonaprice.org/health-topics/inside-japan-surprising-facts-about-japanese-foodways, January 1, 2000, accessed 10/24/14.
27. Kato H, et al. Study of the epidemiology of health and dietary habits of Japanese and Japanese Americans. *Japan J Nutr* 1989;47:121–30.
28. Kudo Y et al. Evolution of meal patterns and food choices of Japanese-American females born in the United States. *Eur J Clin Nutr* 2000;54:665–70.
29. Ma RCW et al. Type 2 diabetes in East Asians: Similarities and differences with populations in Europe and the United States. *Ann N Y Acad Sci* 2013;1281:64–91, doi:10.1111/nyas.12098.
30. Hsu WC et al. Improvement of insulin sensitivity by isoenergy high carbohydrate traditional Asian diet: A randomized controlled pilot feasibility study. *PLoS ONE,* September 16, 2014, doi:10.1371/journal.pone.0106851.
31. Kato H et al. Study of the epidemiology of health and dietary habits of Japanese and Japanese Americans. *Japan J Nutr* 1989;47:121–30.
32. Hu FB. Globalization of diabetes: The role of diet, lifestyle, and genes. *Diabetes Care* 2011;34:1249–57.
33. *World Bank. Life Expectancy at Birth, Total (Years).* data.worldbank.org/indicator/SP.DYN.LE00.IN, 2014, accessed 10/24/14.
34. Perez-Escamilla R. Acculturation, nutrition, and health disparities in Latinos. *Am J Clin Nutr* 2011;93(suppl):1163S–7S.
35. Ford ES et al. Explaining the decrease in U.S. deaths from coronary disease, 1980–2000. *N Engl J Med* 2007;356:2388–98.
36. Veronique LR, et al. Heart disease and stroke statistics—2011 update. *Circulation* 2011;123: e18–e209, December 15, 2010, doi:10.1161/CIR.0b013e3182009701.
37. *Healthy People 2020,* available at: www.healthypeople.gov, accessed 9/23/11.
38. Total Diet Study, www.fda.gov/Food/FoodScienceResearch/TotalDietStudy/default.htm, accessed 10/22/14.
39. van Woudenbergh GJ et al. Adapted dietary inflammatory index and its association with a summary score for low-grade inflammation and markers of glucose metabolism: The Cohort study on Diabetes and Atherosclerosis Maastricht (CODAM) and the Hoorn study. *Am J Clin Nutr* 2013;98:1533–42, doi:10.3945/ajcn.112.056333.

Unit 3

1. Rowe SB et al. On post-truth, fake news, and trust. *Nutr Today* 2017;52:179–82.
2. Offit PA et al. Cancer and the power of placebo. December 13, 2013, www.medscape.com/viewarticle/817524_1, accessed 7/21/14.
3. MacVean M et al. *Unscientific Beliefs about Scientific Topics in Nutrition.* Experimental Biology Annual Meeting, San Diego, April 27, 2014.
4. Who's minding the lab? Tufts University Health and Nutrition Letter May 1997:4.
5. Levine J, et al. Authors' financial relationships with the food and beverage industry and their published positions on the fat substitute Olestra. *Am J Pub Health* 2003;93:664–69.

6. Kelly JC. Randomized clinical trials: 1 in 3 not reported. *BMJ,* published online October 29, 2013, www.medscape.com/viewarticle/813447, accessed 7/23/14.
7. Fung-Berman A. Wyeth paid writers to promote hormone therapy. *PLoS Med,* September 7, 2010, dodi/10.1371/journal.pmed.1000335.
8. Katz DL. A decade of diet lies. Linked In, August 3, 2017, www.linkedin.com/pulse/decade-diet-lies-david-l-katz-md-mph-facpm-facp-faclm?trk=mp-reader-card
9. Bisphenol A (BPA): Use in food contact application. January 6, 2015, www.fda.gov/NewsEvents/PublicHealthFocus/ucm064437.htm, accessed 2/21/15.
10. Zeratsky K. What is BPA, and what are the concerns about BPA? www.mayoclinic.org/healthy-living/nutrition-and-healthy-eating/expert-answers/bpa/faq-20058331, 2103, accessed 2/1/15.
11. Harle CA et al. Quality of health-related online search results. *Decision Support Systems* 2014;57:454–462.
12. Steinbrook R. Searching for the right search—reaching the medical literature. *N Engl J Med* 2006;354:4–7.
13. Al-Ubaydli M. Using search engines to find online medical information. *PLoS Med.* October 2, 2005;2(9):e228.
14. *Academy of Nutrition and Dietetics, Definition of Terms List.* June, 2017, http://eatrightpro.org/~/media/eatrightpro%20files/practice/scope%20standards%20of%20practice/academydefinitionoftermslist.ashx
15. O'Malley R et al. Academy of Nutrition and Dietetics Debuts New Online Platform to Support Patient Care, Data Collection, Outcomes Research. August 18, 2014, www.eatright.org/Media/content.aspx?id=6442482097#.U_OkoyhWz6E.
16. Alger SA et al. Effect of a tricyclic antidepressant and opiate antagonist on binge-eating behavior in normal weight, bulimic, and obese, binge-eating subjects. *Am J Clin Nutr* 1991;53:865–71.
17. Shapiro J. Hair loss in women. *N Engl J Med* 2008;357:1620–30.
18. Wasko CA et al. The sixty-second hair count for assessing hair loss in men. *Arch Dermatol* 2008;144:759–62.

Unit 4

1. Post R et al. *A Guide to Federal Food Labeling Requirements for Meat and Poultry Products.* Food Safety and Inspection Service, USDA, August, 2007, available at: www.fsis.usda.gov/pdf/labeling_requirements_guide.pdf, accessed November I, 2011.
2. CFR—Code of Federal Regulations Title 21. www.accessdata.fda.gov/scripts/cdrh/cfdocs/cfcfr/CFRSearch.cfm?fr=101.45, 4/1/14 and www.accessdata.fda.gov/scripts/cdrh/cfdocs/cfcfr/cfrsearch.cfm?fr=101.60, 9/1/14, accessed 10/30/14.
3. McBurney MI et al. Implications of US Nutrition Facts label changes on micronutrient density of fortified foods and supplements. *J Nutr* 2017;147:1025-1030.
4. Changes to the Nutrition Facts label. 6/19/17, www.fda.gov/Food/GuidanceRegulation/GuidanceDocumentsRegulatoryInformation/LabelingNutrition/ucm385663.htm
5. FDA defines "gluten-free" for food labeling. www.fda.gov/NewsEvents/Newsroom/PressAnnouncements/ucm363474.htm, accessed 7/28/14.
6. Consumer Reports. National Research Center, Survey Research Report, Food Labels Survey. www.greenerchoices.org/pdf/ConsumerReportsFoodLabeling SurveyJune2014.pdf, accessed 10/30/14.
7. Qualified health claims. FDA Sep 2017, www.fda.gov/food/labelingnutrition/ucm2006877.htm
8. Summary of qualified health claims subject to enforcement discretion. www.fda.gov/Food/IngredientsPackagingLabeling/LabelingNutrition/ucm073992.htm, 7/1/14, accessed 10/28/14.
9. Birth defects count. www.cdc.gov/ncbddd/folicacid/global.html, December 23, 2013, accessed 10/29/14.
10. Allen AE, et al. Update of the vitamin D content of fortified foods and supplements in the UK National Diet and Nutrition Survey Nutrient Databank. *Nutr Bulletin,* doi:10.1111/nbu.12099.
11. Berner LA et al. Fortified foods are major contributors to nutrient intakes in diets of US children and adolescents. *J Acad Nutr Diet,* released online 1/27/14, doi: 10.1016/j.jand.2013.10.012.
12. Bailey RL. *Impact of Fortification in the United States.* Experimental Biology Meeting, San Diego, April 26, 2014.
13. Food Allergen Labeling and Consumer Protection Act of 2004, available at: www.cfsan.fda.gov/dms/alrgact.html, accessed 7/7/06.
14. Color additives. FDA, December 2017, www.fda.gov/ForIndustry/ColorAdditives; Organic market analysis. Organic Trade Association, 2017, www.ota.com/resources/market-analysis
15. Food irradiation: What you need to know. FDA 2016; www.fda.gov/food/resources for you.
16. Scientific opinion on the substantiation of health claims related to prune juice and maintenance of normal bowel function (ID 1166) pursuant to Article 13(1) of Regulation (EC) No. 1924/2006. *EFSA Journal* 2010;8(10):1768, available at: www.efsa.europa.eu/en/efsajournal/doc/1768.pdf, accessed 10/30/14.
17. Generally Recognized as Safe (GRAS). 2017, www.fda.gov/food/ingredientspackaginglabeling/gras, www.fda.gov/aboutfda/reportsmanualsforms/reports/economicanalyses/ucm582763.htm
18. Label claims for conventional foods and dietary supplements. 2016, www.fda.gov/Food/IngredientsPackagingLabeling/LabelingNutrition/ucm111447.htm
19. Country of origin labeling (COOL). USDA. 2017, www.ams.usda.gov/rules-regulations/cool
20. Solomon Z. Organic vs. non-organic: What's the difference? 2013, www.foodsafetynews.com/2013/08/organic-vs-non-organic-whats-the-difference/#.WfoSpzOZNBw
21. How is the term "organic" regulated? FDA, October 2017, www.fda.gov/AboutFDA/Transparency/Basics/ucm214871.htm22
23. Smith-Spangler C et al. Are organic foods safer or healthier than conventional alternatives?: A systematic review. *Ann Intern Med,* 2012 ;157:348–66.
24. Hartung F et al. Precise plant breeding using new genome editing techniques: opportunities, safety and regulation in the EU. *The Plant J* 2014:78,742–752.
25. GMO labeling. www.foodnavigator-usa.com/Trends/GMO-Labeling, 10/1/14, accessed 10/2/14.
26. GMO labeling in Canada fails. *ASN Health and Nutrition Policy Newsletter,* June2, 2017.
27. Lupton JR et al. The Smart Choices front-of-package nutrition labeling program: Rational and development criteria. *Am J Clin Nutr* 2010;91(suppl):1078S–89S.
28. Hitti M. "Smart Choices" food labels are coming. Available at: www.medscape.com/viewarticle/582810, accessed October 2008.
29. Gerrior SA. Nutrient profiling systems: Are science and the consumer connected? *Am J Clin Nutr* 2010;91(suppl):1116S–7S.
30. Nestle M. Health care reform in action—Calorie labeling goes national. *N Engl J Med* 2010;362:2343–5.
31. Sinclair SE et al. The influence of menu labeling on calories selected or consumed: A systematic review and meta-analysis. *J Acad Nutr Diet* 2014;114:1375–1388.e15, doi:10.1016/j.jand.2014.05.014.
32. Dallas ME. How many calories in your fast-food meal? Guess again. consumer.healthday.com/vitamins-and-nutritional-information-27/food-and-nutrition-news-316/how-many-calories-in-your

-fast-food-meal-guess-again-676688.html, May 24, 2013, accessed 6/23/13.
32. Breck A et al. The current limits of calorie labeling and the potential for population health impact. *J Public Policy & Marketing*, 2017 in press.
33. Kollannoor-Samuel G et al. Nutrition label use is associated with lower longer-term diabetes risk in US adults. *Am J Clin Nutr* 2017;105:1079–85.
34. Organic market analysis. Organic Trade Association, 2017, www.ota.com/resources/market-analysis

Unit 5

1. Perna S et al. Association of the bitter taste receptor gene TAS2R38 (polymorphism RS713598) with sensory responsiveness, food preferences, biochemical parameters and body-composition markers. A cross-sectional study in Italy. *Int J Food Sci Nutr,* July 24, 2017, doi.org/10.1080/09637486.2017.1353954.
2. Persoskie A et al. US Consumers' understanding of nutrition labels in 2013: The importance of health literacy. *Prev Chronic Dis* 2017;14(9), www.medscape.com/viewarticle/886592?nlid=118694_439&src=WNL_mdplsfeat_171024_mscpedit_publ&uac=61213SX&spon=42&impID=1464189&faf=1
3. Pyke M. *Man and Food.* New York: McGraw-Hill, 1972.
4. Rozin P. The acquisition of food habits and preferences. In: Matarazzo JD, Weiss SM, Herd JA, Miller NE, Weiss SM, eds., *Behavioral Health: A Handbook of Health Enhancement and Disease Prevention.* New York: John Wiley, 1984:590–607.
5. Beauchamp GK, Mennella JA. Sensitive periods in the development of human flavor perception and preference. *Ann Nestle* 1998;56:19–315
6. Hiza HAB et al. Position of the Academy of Nutrition and Dietetics: Total diet approach to healthy eating. *J Am Acad Nutr Diet* 2013;113:307–17.
7. Birch L. Development of food preferences. *Ann Rev Nutr* 1999;19:41–62.
8. Story M et al. Do young children instinctively know what to eat? The studies of Clara Davis revisited. *N Engl J Med* 1987;316:103–6.
9. Negri R et al. Taste perception and food choices. *J Pediatr Gastroenterol Nutr* 2012;54:624–29. doi: 10.1097/MPG.0b013e3182473308.
10. The Determinants of Food Choice, European Food and Nutrition Council Review. April 2005, Available at: www.eufic.org/article/en/expid/review-food-choice, accessed 11/17/11.
11. Packard V. *The Status Seekers.* New York: Random House, 1959:146.
12. Van Oudenhove L et al. Fatty acid-induced gut-brain signaling attenuates neural and behavioral effects of sad emotion in humans. *J Clin Invest* 2011;121:3094–99.
13. Parraga IM. Determinants of food consumption. *J Am Diet Assoc* 1990;90:661–63.
14. Stein K. Contemporary comfort foods: Bringing back old favorites. *J Am Diet Assoc* 2008;108:412, 414.
15. Kapsak WR et al. Putting the dietary guidelines for Americans into action: Behavior-directed messages to motivate parents—phase I and II observational and focus group findings. *J Acad Nutr Diet* 2013;113:196–204, doi:10.1016/j.jand.2012.10.019.
16. Beerman KA. Variation in nutrient intake of college students: A comparison by students' residence. *J Am Diet Assoc* 1991;91:343–44.
17. Waterlander WE et al. Price discounts significantly enhance fruit and vegetable purchases when combined with nutrition education: a randomized controlled supermarket trial. *Am J Clin Nutr* 2013;97:886–95, doi:10.3945/ajcn.112.041632.
18. Crosby G. Super-tasters and non-tasters: Is it better to be average? 2016, www.hsph.harvard.edu/nutritionsource/2016/05/31/super-tasters-non-tasters-is-it-better-to-be-average
19. Fischer ME et al. The association of taste with change in adiposity-related health measures. *J Acad Nutr Diet* 2014;114:1195–202, doi:10.1016/j.jand.2014.04.013.
20. Iwata S et al. Taste transductions in taste receptor cells: Basic tastes and moreover. *Curr Pharm Des* 2014;20:2684–92.
21. Soberg S et al. FGF21 Is a sugar-induced hormone associated with sweet intake and preference in humans. *Clin Translational Report* 2017;25:1045–53.
22. Hayes JE et al. Allelic variation in TAS2R bitter receptor gene associates with variation in sensations from and ingestive behaviors toward common bitter beverages in adults. *Chem Senses* 2011;36:311–31.
23. Quick statistics about taste and smell. www.nidcd.nih.gov/health/statistics/smelltaste/Pages/stquickstats.aspx, June 7, 2010, accessed 11/1/14.
24. Reed DR et al. Genetics of taste and smell: Poisons and pleasures. *Prog Mol Biol Transl Sci* 2010;94:213–40, doi:10.1016/B978-0-12-375003-7.00008-X.
25. Bruni F. A taste you hate? Just wait. www.nytimes.com/2014/02/26/dining/a-taste-you-hate-just-wait.html?_r=1, FEB. 24, 2014, accessed 2/27/14.
26. Bentley J et al. Food availability (per capita) data system. USDA, Economic Research Service, November 2016.
27. What's hot: 2017 culinary forecast. 2017, www.restaurant.org/Downloads/PDFs/News-Research/WhatsHot/What-s-Hot-2017-FINAL
28. Desilver D. What's on your table? How America's diet has changed over the decades. 2016, www.pewresearch.org/fact-tank/2016/12/13/whats-on-your-table-how-americas-diet-has-changed-over-the-decades
29. Lee JH et al. Popular ethnic foods in the United States: A historical and safety perspective. *Comprehen Rev Food Sci Food Safety* 2013;13:2–17. 2014. Article first published online: December 17, 2013, doi:10.1111/1541-4337.12044.
30. We are what we ate: Report details U.S. dietary trends. General Mills survey, 2014, ovens.reviewed.com/features/general-mills-details-2014-food-trends?utm_source=usat&utm_medium=referral&utm_campaign=collab, accessed 10/16/14.
31. Nutrition and you: Trends 2008. American Dietetic Association. Available at: www.eatright.org, accessed December 2008.
32. Beydoun MA. The interplay of gender, mood, and stress hormones in the association between emotional eating and dietary behavior. *J Nutr* 2014;144:1139–41, doi:10.3945/jn.114.196717.
33. Walker SP et al. Early childhood stunting is associated with poor psychological functioning in late adolescence and effects are reduced by psychological stimulation. *J Nutr* 2007;137:2464–69.
34. Stein AD et al. Nutritional supplementation in early childhood, schooling, and intellectual functioning in adulthood: A prospective study in Guatemala. *Arch Pediatr Adolesc Med 2008*;162:612–18.
36. Sandin S et al. The heritability of autism spectrum disorder. *JAMA* 2017;318:1182–84.
37. Stevens LJ et al. Mechanisms of behavioral, atopic, and other reactions to artificial food colors in children. *Nutr Rev* 2013;71:268–81, doi: 10.1111/nure.12023.
38. Wright JP et al. Association of prenatal and childhood blood lead concentrations with criminal arrests in early adulthood. *PLoS Med* 2008;5(5):e101, available at: www.medscape.com/viewarticle/576717, accessed January 2009.
39. Lozoff B et al. Iron supplementation in infancy contributes to more adaptive behavior at 10 years of age. *J Nutr* 2014;144:838–45, doi: 10.3945/jn.113.182048.

40. Walker SP et al. Early childhood stunting is associated with poor psychological functioning in late adolescence and effects are reduced by psychological stimulation. *J Nutr* 2007;137:2464–69.
41. Laus MF et al. early postnatal proteincalorie malnutrition and cognition: A review of human and animal studies. *Int J Environ Res Public Health* 2011;8:590–612.
42. Dietary Guidelines for Americans: Alcohol, 2010. Available at: health .gov/dietaryguidelines/dga20W /dietaryguidelines20W.pdf, accessed 11/17/11.
43. Kordas K. Iron, lead, and children's behavior and cognition. *Annual Review of Nutrition* 2010;30:123–48.
44. Lozoff B. Iron deficiency and child development. *Food Nutr Bull* 2007;28 (4 suppl):S560–71.
45. Jacus J et al. This Old (Pre-1978) House: Lead-Based Paint Regulation and Enforcement under TSCA. *Natural Resources Environ*;2017;32:21–25.
46. Hsueh YM et al. Association of blood heavy metals with developmental delays and health status in children. *Sci Rep* 2017 March 2;7:43608, doi: 10.1038/srep43608.
47. Kirschner K et al. Very high blood lead levels among adults—United States, 2002–2011. *MMWR Morb Mortal Wkly Rep* 2013, 29;62:967–71.
48. QuickStats: Percentage of children aged 1–5 years with elevated blood lead levels by race/ethnicity — National Health and Nutrition Examination Survey, United States, 1988–1994, 1999–2006, and 2007–2014. *Morb Mort Weekly,* October 7, 2016 / 65(39);1089.
49. Kim Y et al. Correlation between attention deficit hyperactivity disorder and sugar consumption, quality of diet, and dietary behavior in school children. *Nutr Res Pract* 2011;5:236–45.
50. Busting the sugar-hyperactivity myth. www.webmd.com/parenting/features /busting-sugar-hyperactivity-myth, accessed 11/5/14.
51. Nordenberg T. The facts about aphrodisiacs. *FDA Consumer* 1996;30(January-February).

Unit 6

1. Dietary Guidelines Advisory Committee, *Scientific Report of the 2015 Dietary Guidelines: Advisory Report to the Secretary of Health and Human Services and Secretary of Agriculture,* published February 2015, www.health.gov/dietaryguidelines/2015 -scientific-report/PDFs/Scientific-Report -of-the-2015-Dietary-Guidelines -Advisory-Committee.pdf, accessed 3/12/15.
2. Moore LV et al. Adults meeting fruit and vegetable intake recommendations— United States, 2013, *Morbidity and Mortality Weekly Report (MMWR) Weekly,* July 10, 2015; 64;709–13.
3. English LK et al. Food portion size and energy density evoke different patterns if brain activation in children. *Am J Clin Nutr* 2017;105:295–305.
4. Hiza HAB et al. Position of the Academy of Nutrition and Dietetics: Total diet approach to healthy eating. *J Am Acad Nutr Diet* 2013;113:307–17.
5. Schwingshackl L et al. Food groups and risk of all-cause mortality: A systematic review and meta-analysis of prospective studies. *Am J Clin Nutr* 2017;105:1462–73.
6. Diet/Nutrition. CDC, 2017; www.cdc.gov /nchs/fastats/diet.htm
7. Kimmons J et al. Fruit and vegetable intake among adolescents and adults in the United States: Percentage meeting individualized recommendations. Posted: January 26, 2009; *Medscape J Med* 2009;11:26.
8. Dietary guidelines from around the world. United States Department of Agriculture National Agricultural Library, www.nal.usda.gov/fnic/dietary -guidelines-around-world, accessed 11/7/17.
9. Food-based dietary guidelines Around the World. Available at: www.fao.org /ag/humannutrition/nutritioneducation /fbdg/en, accessed 11/24/11.
10. Dietary Guidelines Advisory Committee Meeting webcast, 9/16/14, 9/17/14, 11/7/14, 12/15/14, videocast.nih.gov.
11. Kimmons J et al. Fruit and vegetable intake among adolescents and adults in the United States: Percentage meeting individualized recommendations. Posted: January 26, 2009; *Medscape J Med* 2009;11:26.
12. Chen ST et al. The effect of dietary patterns on estimated coronary heart disease risk: Results from the Dietary Approaches to Stop Hypertension (DASH) trial. *Circ Cardiovasc Qual Outcomes* 2010;3:484–89.
13. *Dietary Guidelines for Americans,* Appendix A-1: The DASH Eating Plan at 1,600–3,200 calories. Available at: www .health.gov/dietaryguidelines/dga2005 /document/html/appendixA.htm.
14. *Your Guide to Lowering Your Blood Pressure with DASH.* Available at: www .nhlbi.nih.gov/health/public/heart/hbp /dash/new_dash.pdf, accessed 11/23/11.
15. Simopoulos AP. The Mediterranean diet: What is so special about the diet of Greece? The scientific evidence *J Nutr* 2001;131(suppl):S3065–73.
16. Salas-Salvado J et at. Protective effects of the Mediterranean diet on type 2 diabetes and metabolic syndrome. *J Nutr* 2016;146(Suppl):920S7S.

Appendix B. Food sources of selected nutrients

17. Maartinez-Gonzalez MA et al. Intervention trial with the Mediterranean diet on cardiovascular prevention: Understanding potential mechanisms through metabolomic profiling. *J Nutr* 2016;146(Suppl):913S-9S.
18. Young YR. How to reduce portions: What to tell patients. Available at: www .medscape.com/viewarticle/750739?src =mp&spon=2, accessed 11/23/11.
19. Portion Distortion Slide Show. National Institutes of Health, available at: http:// hp2010.nhlbihin.net/oei_ss/menu .htm#sl2, accessed 11/23/11.
20. Pierrnas C et al. Increased portion sizes from energy-dense foods affect total energy intake at eating occasions in U.S. children and adolescents. *Am J Clin Nutr* 2011;94:1324–32.
21. Rolls BJ, et al. The effect of large portion sizes on energy intake is sustained for 11 days. *Obesity* 2007;15:1535–43.
22. Wilcox S et al. Frequency of consumption at fast-food restaurants is associated with dietary intake in overweight and obese women recruited from financially disadvantaged neighborhoods. *Nutr Res* 2013;33:636–46.
23. We are what we ate: Report details U.S. dietary trends. General Mills survey, 2014, ovens.reviewed.com/features /general-mills-details-2014-food -trends?utm_source=usat&utm _medium=referral&utm_ campaign=collab, accessed 10/16/14.
24. Food-away-from-home. USDA, Sept 2017, www.ers.usda.gov/topics/food-choices -health/food-consumption-demand /food-away-from-home.aspx
25. Todd et al. The impact of food away from home on adult diet quality. ERR-90, U.S. Department of Agriculture, *Econ Res Serv,* February 2010.
26. Hammons AJ et al. Is frequency of shared family meals related to the nutritional health of children and adolescents? *Pediatrics* 2011;127:e1565–e1574.
27. Bridges A. Restaurants should shrink portions. Available at: cnn.netscape.com /news/story, accessed 6/3/06.
28. Slow Food USA. www.slowfoodusa.org, accessed 11/12/14.
29. Riley NS. The joys of cooking class. *Wall Street Journal,* March 9, 2006: D9.

30. *Dietary Guidelines for Americans, 2015–2020.* health.gov/dietaryguidelines/2015/guideline, accessed 1/9/16.
31. Moran AJ et al. Trends in nutrient content of children's menu items in U.S. chain restaurants. *Am J Preven Med* 2017;52:284–91.
32. Penney TL et al. Utilization of away-from-home food establishments, Dietary Approaches to Stop Hypertension dietary pattern, and obesity. *Am J Prev Med* 2017 November; 53(5): e155–e163.
33. Tiwari A et al. Cooking at home: A strategy to comply with U.S. dietary guidelines at no extra cost. *Am J Prev Med* 2017 February 27. pii: S0749-3797(17)30023-5. doi: 10.1016/j.amepre.2017.01.017.
34. Bahadoran Z et al. Fast food pattern and cardiometabolic disorders: A review of current studies. *Health Promot Perspect* 2016;5:231–40.
35. Willett WC et al. Mediterranean diet pyramid: A cultural model for healthy eating. *Am J Clin Nutr* 1995;61:1402S-6S.

UNIT 7

1. Hospitalization statistics, 2009. Available at: www.ct.gov/dph/cwp/view,asp?a=3131&q=397512, accessed 12/3/11.
2. Leading causes of death and hospitalization in Canada. 2008, www.phac-aspc.gc.ca/publicat/lcd-pcd97/index-eng.php, accessed 4/6/15.
3. Peptic ulcer. Available at: www.mayoclinic.com/health/peptic-ulcer/DS00242, updated January 2011, accessed 12/4/11.
4. Picco MF. Is colon cleansing a good way to eliminate toxins from your body? www.mayoclinic.org/healthy-living/consumer-health/expert-answers/colon-cleansing/faq-20058435, accessed 11/14/14.
5. Pellettieri J et al. Cell turnover and adult tissue homeostasis: From humans to planarians. *Ann Rev Genetics* 2007;41:83–105.
6. Schneeman B. Nutrition and gastrointestinal function. *Nutr Today,* January/February 1993;20–24.
7. Liao TH et al. Fat digestion by lingual lipase: Mechanism of lipolysis in the stomach and upper small intestine. *Pediatric Research* 1984;18:402–09.
8. Salles C et al. In-mouth mechanisms leading to flavor release and perception. *Crit Rev Food Sci Nutr,* January 2011;51:67–90.
9. Camilleri M et al. Human gastric emptying and colonic filling of solids characterized by a new method. *Am J Physiol Gastrointest Liver Physiol* 1989;284–89.
10. Duggan C et al. Protective nutrients and functional foods for the gastrointestinal tract. *Am J Clin Nutr* 2002;75:789–808.
11. Levitt MD et al. Use of measurements of ethanol absorption from stomach and intestine to assess human ethanol metabolism. *Am J Physiol Gastrointestinal Liver Physiol* 1997;273:G951–57.
12. Granado F, et al. Bioavailability of carotenoids and tocopherols from broccoli: In vivo and in vitro assessment. *Exp Biol Med* (Maywood) 2006;231:1733–38.
13. Perna S et al. Association of the bitter taste receptor gene TAS2R38 (polymorphism RS713598) with sensory responsiveness, food preferences, biochemical parameters and body-composition markers. A cross-sectional study in Italy. *Int J Food Sci Nutr,* July 24, 2017, doi.org/10.1080/09637486.2017.1353954.
14. Finger TE et al. Matters of taste. *The Scientist,* November 12, 2011;34–45.
15. Shanahan F et al. Feeding the microbiota: Transducer of nutrient signals for the host. *Gut* 2017;66(9):1709–17.
16. Lynch SV et al. The human microbiome in health and disease. *N Engl J Med* 2016;375:2369–80.
17. McNeil N. The contribution of the large intestine to energy supplies in man. *Am J Clin Nutr* 1984;39:338–42.
18. Conly JM et al. The production of menaquinones (vitamin K2) by intestinal bacteria and their role in maintaining coagulation homeostasis. *Prog Food Nutr Sci* 1992;16:307–43.
19. Binbuam J, et al. Biosynthesis of biotin in microorganisms v. control of vitaman production. *J Bacteriol* 1967;94:1846–53.
20. David LA et al. Diet rapidly and reproducibly alters the human gut microbiome. *Nature* 2013, doi: 10.1038/nature12820.
21. Buchta Rosean CM et al. The influence of the commensal microbiota on distal tumor-promoting inflammation. Non-celiac gluten sensitivity: All wheat attack is not celiac. *World J Gastroenterol* 2017;28; 23: 7201–10.
22. Stecher B. The roles of inflammation, nutrient availability and the commensal microbiota in enteric pathogen infection. *Microbiol Spectr* 2015 June;3(3), doi: 10.1128/microbiolspec.MBP-0008-2014.
23. Zhang Y et al. Effects of probiotic type, dose and treatment duration on irritable bowel syndrome diagnosed by Rome III criteria: A meta-analysis. *BMC Gastroenterol* 2016 June 13;16(1):62. doi: 10.1186/s12876-016-0470-z.
24. Digestive diseases. Available at: www.cdc.gov/nchs/fastats/digestiv.htm, accessed 12/6/11.
25. Lashner BA. What is recommended for chronic constipation? Available at: www.medscape.com/viewarticle/744652?src=mp&spon=17, accessed 12/6/11.
26. Health Canada. Available at: www.phac-aspc.gc.ca/publicat/lcd-pcd97/table2-eng.php, accessed 12/4/11.
27. Constipation. Available at: http://digestive.niddk.nih.gov/ddiseases/pubs/constipation, accessed 12/6/11.
28. Zelman KM et al. Dietary fiber: Insoluble vs. soluble. Available at: www.webmd.com/diet/fiber-health-benefits-11/insoluble-soluble-fiber, accessed 12/6/11.
29. Mishori R et al. The dangers of colon cleansing. *J Family Practice* 2011;60:454–57.
30. Digestive diseases statistics for the United States. June 2010, available at: http://digestive.niddk.nih.gov/statistics/statistics.aspx#specific, accessed 12/4/11.
31. Harrell L et al. Standard colonic lavage alters the natural state of mucosal-associated microbiota in the human colon. *PLoS ONE* 2012;e32545, published online February 28, 2012, doi: 10.1371/journal.pone.0032545.
32. Constipation. Available at: http://digestive.niddk.nih.gov/ddiseases/pubs/constipation, accessed 12/6/11.
33. Digestive Diseases Statistics for the United States. June 2010, available at: http://digestive.niddk.nih.gov/statistics/statistics.aspx#specific, accessed 12/4/11.
34. Suerbaum S et al. Helicobacter pylori infection. *N Engl J Med* 2002;347:1175–86.
35. Peptic Ulcer Disease: Topic Overview. Available at: www.webmd.com/digestive-disorders/tc/peptic-ulcer-disease-topic-overview, updated January 2011, accessed 12/4/11.
36. Singh N et al. Peptic Ulcer Disease: Topic Overview. Available at: www.webmd.com/digestive-disorders/tc/peptic-ulcer-disease-topic-overview, updated January 2011, accessed 12/4/11.
37. Digestive Diseases Statistics for the United States. June 2010, available at: http://digestive.niddk.nih.gov/statistics/statistics.aspx#specific, accessed 12/4/11.
38. Heartburn, gastroesophageal reflux (GER), and gastroesophageal reflux disease (GERD), May 2007. Available at: http://digestive.niddk.nih.gov/ddiseases/pubs/gerd, accessed 12/6/11.
39. Moazzez R et al. The effect of chewing sugar-free gum on gastro-esophageal reflux. *J Dent Res* 2005;84(11):1062–65.
40. Irritable bowel syndrome. www.eatright.org/Public/content.aspx?id=5547, accessed 4/14/14.
41. Irritable bowel syndrome. Available at: digestive.niddk.nih.gov/ddiseases/pubs/IBS, accessed 12/6/11.

42. Mullin GE et al. Irritable bowel syndrome: Contemporary nutrition management strategies. *J Parenter Enteral Nutr* 2014;38 :781–799, doi: 10.1177/0148607114545329.
43. Carlson MJ et al. Child and parent perceived food-induced gastrointestinal symptoms and quality of life in children with functional gastrointestinal disorders. *J Acad Nutr Diet* 2014;114: 403–13, doi: 10.1016/j.jand.2013.10.013.
44. El-Serag HB et al. Dietary intake and the risk of gastro-oesophageal reflux disease: A cross-sectional study in volunteers. *Gut* 2005;54:11–17.
45. Festi D et al. Body weight, lifestyle, dietary habits and gastroesophageal reflux disease. *World J Gastroenterol* 2009;15:1690–701.
46. National Institute of Digestive Diseases Health Information. Available at: www .niddk.nih.gov, accessed February 2009.
47. Kilgore PE et al. Trends in diarrheal disease associated mortality in U.S. children, 1968 through 1991. *JAMA* 1995;274:1143–48.
48. Meyers A. Oral rehydration therapy: What are we waiting for? *Am Earn Phys* 1993;47:740–42.
49. McRorie JW et al. Understanding the physics of functional fibers in the gastrointestinal tract. *J Am Acad Nutr Diet* 2017;117:251–64.
50. Picco MF. What makes my stomach growl when I'm hungry? Available at: mayoclinic.com/health/stomach-noise /nu00189, accessed 12/6/11.

Unit 8

1. Eiler WJ et al. Ventral frontal satiation-mediated responses to food aromas in obese and normal weight women. *Am J Clin Nutr* 2014;99:1309–18.
2. Anthanont P et al. Does basal metabolic rate predict weight gain? *Am J Clin Nutr.* 2016;104:959–63.
3. Flancbaum L et al. Comparison of indirect calorimetry, the Fick method, and prediction equations in estimating the energy requirements of critically ill patients. *Am J Clin Nutr*1999;69:461–66.
4. Ferrannini E. The theoretical bases of indirect calorimetry: A review. *Metabolism*. 1988;37:287–301.
5. Schadewaldt P et al. Indirect calorimetry in humans: A postcalorimetric evaluation procedure for correction of metabolic monitor variability. *Am J Clin Nutr* 2013;A97:763–73, doi: 10.3945/ ajcn.112.035014.
6. Frankenfield DC, et al. The Harris-Benedict studies of human basal metabolism: History and limitations. *J Am Diet Assoc* 1998;98:439–45.
7. Irwin ML et al. Estimation of energy expenditure from physical activity measures: Determinants of accuracy. *Obesity Research* 2001;9:517–25.
8. Melby CL et al. Reported diet and exercise behaviors, beliefs and knowledge among university undergraduates. *Nutr Res* 1986;6:799–808.
9. Ford ES et al. Trends in energy intake among adults in the United States: Findings from NHANES. *Am J Clin Nutr 2013*;97:848–53, doi: 10.3945/ ajcn.112.052662.
10. O'Sullivan HL et al. Effects of repeated exposure on liking for a reducedenergy-dense food. *Am J Clin Nutr* 2010;91:1584–89.
11. *Dietary Guidelines for Americans.* Available at: www.dietaryguidelines.gov, accessed 12/12/11.
12. Schroder H et al. Low energy density diets are associated with favorable nutrient intake profile and adequacy in free-living elderly men and women. *J Nutr* 2008;138:1476–81.
13. *Dietary Guidelines for Americans.* Available at: www.dietaryguidelines.gov, accessed 12/12/11.
14. Rolls BJ. The relationship between dietary energy density and energy intake. *Physiol Behav* 2009;97:609–15.
15. Schroder H et al. Low energy density diets are associated with favorable nutrient intake profile and adequacy in free-living elderly men and women. *J Nutr*2008;138:1476–81
16. Flack KD et al. Cross-validation of resting metabolic rate prediction equations. *J Acad Nutr Diet* 2016;116:1413–22.
17. Woods SC et al. Signals that regulate food intake and energy homeostasis. *Science* 1998;280:1378–83.
18. Burger KS et al. Hedonic hunger is related to increased neural and perceptional responses to cues of palatable food and motivation to consume: Evidence from 3 independent investigations. *J Nutr* 2016;146:1807–12.
19. Frank S et al. Olive oil aroma extract modulates cerebral blood flow in gustatory brain areas in humans. *Am J Clin Nutr* 2013;98:1360–66, doi: 10.3945/ ajcn.113.062679.
20. Cronise RJ et al. The "metabolic winter" hypothesis: A cause of the current epidemics of obesity and cardiometabolic disease. *Metab Syndr Relat Disord* 2014;12:355–61, doi: 10.1089/ met.2014.0027.
21. Hall KD et al. Energy balance and its components: Implications for body weight regulation. *Am J Clin Nutr* 2012;95:989–94.

Unit 9

1. Lee SW et al. Body fat distribution is more predictive of all-cause mortality than overall adiposity. *Diabetes Obes Metab* 2018, accepted Author Manuscript. doi: 10.1111/dom.13050.
2. Lyall DM et al. Association of body mass index with cardiometabolic disease in the UK Biobank: A Mendelian randomization study. *JAMA Cardiol* 2017; doi: 10.1001/jamacardio.2016.5804.
3. Flancbaum L et al. Comparison of indirect calorimetry, the Fick method, and prediction equations in estimating the energy requirements of critically ill patients. *Am J Clin Nutr* 1999;69:461–66.
4. Ferrannini E. The theoretical bases of indirect calorimetry: A review. *Metabolism* 1988;37:287–301.
5. Schadewaldt P et al. Indirect calorimetry in humans: A postcalorimetric evaluation procedure for correction of metabolic monitor variability. *Am J Clin Nutr* 2013;A97:763–73, doi: 10.3945/ ajcn.112.035014.
6. Frankenfield DC, et al. The Harris-Benedict studies of human basal metabolism: History and limitations. *J Am Diet Assoc* 1998;98:439–45.
7. Heymsfield SB et al. Mechanisms, pathophysiology, and management of obesity. *N Engl J Med* 2017;376:254–65.
8. Carbonneau E et al. A Health at Every Size intervention improves intuitive eating and diet quality in Canadian women. *Clin Nutr* 2017;36:747–54.
9. Overweight & Obesity Statistics. NIH, www.niddk.nih.gov/health-information /health-statistics/overweight-obesity, November 2017.
10. O'Sullivan HL et al. Effects of repeated exposure on liking for a reduced energy-dense food. *Am J Clin Nutr* 2010;91:1584–89.
11. Dietary Guidelines for Americans. Available at: www.dietaryguidelines.gov, accessed 12/12/11.
12. Schroder H et al. Low energy density diets are associated with favorable nutrient intake profile and adequacy in free-living elderly men and women. *J Nutr* 2008;138:1476–81.
13. Dietary Guidelines for Americans. Available at: www.dietaryguidelines.gov, Accessed 12/12/11.
14. Fryar CD et al. Prevalence of overweight and obesity among children and adolescents aged 2–19 years: United States, 1963–1965 Through 2013–2014. CDC, www.cdc.gov/nchs/data/hestat/obesity _child_13_14/obesity_child_13_14.pdf, accessed November 2017.

15. Pinho CPS et al. Effects of weight loss on adipose visceral and subcutaneous tissue in overweight adults. *Clin Nutr* May 2017, oi.org/10.1016/j.clnu.2017.05.011
16. Tschöp MH et al. Outstanding Scientific Achievement Award Lecture 2011: Defeating diabesity: The case for personalized combinatorial therapies. *Diabetes* 2012;61:1309–1314, doi: 10.2337/db12-0272.
17. Bombak A. Obesity, health at every size, and public health policy. *Am J Public Health* 2014;104:e60–7, doi: 10.2105/AJPH.2013.301486.
18. Twig G et al. Diabetes risk among overweight and obese metabolically healthy young adults. *Diabetes Care*, published online before print August 19, 2014, doi: 10.2337/dc14-0869.
19. Ahima RS et al. The health risk of obesity—better metrics imperative. *Science* 2013;341(6148):856–8, doi: 10.1126/science.1241244.
20. Lyall DM et al. Association of body mass index with cardiometabolic disease in the UK Biobank: A Mendelian randomization study. *JAMA Cardiol* 2017; doi: 10.1001/jamacardio.2016.5804.
21. Dietary Guidelines Advisory Committee Meeting webcast, 9/16/14, videocast.nih.gov.
22. Gallagher D et al. Changes in adipose tissue deposits and metabolic markers following a one-year diet and exercise intervention in overweight and obese patients with type 2 diabetes. *Diabetes Care*, October 21, 2014, doi: 10.2337/dc14-1585.
23. Nutrition, physical activity, and obesity: Trends and maps. CDC, 2017; https://nccd.cdc.gov/dnpao_dtm/rdPage.aspx?rdReport=DNPAO_DTM.
24. Defining Overweight and Obesity. Available at: http://www.cdc.gov/obesity/defining.html, accessed 12/21/11.
25. Soto L et al. Beliefs, attitudes, and phobias among medical and psychology students towards people with obesity. *Nutr Hosp* 2014;1:37–41.
26. Chrisle J et al. Fat shaming in the doctor's office can be mentally and physically harmful. *Am Psychol Assoc* Session 1051: "Weapons of Mass Distraction — Confronting Sizeism, Aug. 3, 2017, available at: http://www.apa.org/news/press/releases/2017/08/fat-shaming.aspx
27. Puhl RM. Health professional: Do you have hidden weight bias? *Medscape*, November 22, 2016.
28. Ethical issues in the care of the obese woman. Committee Opinion No. 600, *Obstet Gynecol* 2014;123:1388–93, doi: 10.1097/01.AOG.0000450756.10670.c2.
29. Rothberg AE et al. Impact of weight loss on waist circumference and the components of the metabolic syndrome. *BMJ Open Diabetes Res Care* 2017;5, www.medscape.com/viewarticle/884491.
30. Stefan N et al. Causes, characteristics, and consequences of metabolically unhealthy normal weight in humans. *Cell Metabolism* 2017;26:292–300.
31. Fogarty AW et al. A prospective study of weight change and systemic inflammation over 9 y. *Am J Clin* 2008;87:30–35.
32. Araneta AR et al. Usual BMI cut point may miss diabetes in thin Asian Americans. www.medscape.com/viewarticle/826769?nlid=59623_2521&src=wnl_edit_medp_diab&uac=61213SX&spon=22.
33. Davidson LE et al. Resistance plus aerobic exercise may be best for sedentary, abdominally obese older adults. *Arch Intern Med* 2009;169:122–31.
34. Ahmadi N, et al. Effects of intense exercise and moderate caloric restriction on cardiovascular risk factors and inflammation. *Am J Med* 2011;124:978–82.
35. Ford ES et al. Trends in mean waist circumference and abdominal obesity among US adults, 1999–2012. *JAMA* 2014;12:1151–1153, doi: 10.1001/jama.2014.8362.
36. National Institutes of Health. *The Practical Guide to the Identification, Evaluation and Treatment of Overweight and Obesity in Adults.* Bethesda, MD: National Institutes of Health, 2000, www.nhlbi.nih.gov/files/docs/guidelines/prctgd_c.pdf.
37. Tucker AM et al. Prevalence of cardiovascular disease risk factors among national football league players. *JAMA* 2009;301:2111–19.
38. Steinberg BA et al. Measuring waist circumference. *Medscape Cardiology* 2006;10(2), available at: www.medscape.com/viewarticle/542635, accessed October 2006.
39. Despres J-P et al. A focus on viscera adiposity and liver fat. 2nd International Congress on Abdominal Obesity, February 26, 2011, available at: www.medscape.com/viewarticle/738448, accessed 3/19/11.
40. Kishida K et al. Clinical significance of visceral fat reduction through health education in preventing atherosclerotic cardiovascular disease. *Nutr Metab* (Lond) 2011;8:57, doi: 10.1186/1743-7075-8-57.
41. Appendix H, Table H-1. Body measurement summary statistics. In: *Dietary Reference Intakes for Energy through Amino Acids,* Washington, DC: The National Academies Press, 1999.
42. Gibson RS. *Principles of Nutritional Assessment.* New York: Oxford University Press, 1990.
43. Guo J et al. Striatal dopamine D2-like receptor correlation patterns with human obesity and opportunistic eating behavior. *Molec Psychia,* September 9, 2014, doi: 10.1038/mp.2014.102.
44. Trumbo PR et al. Systematic review of the evidence for an association between sugar-sweetened beverage consumption and risk of obesity. *Nutr Rev,* article first published online: August 4, 2014, doi: 10.1111/nure.12128.
45. Levine JA. Poverty and obesity in the U.S. *Diabetes* 2011;60:2667–68, doi: 10.2337/db11-1118.
46. Gittelsohn J et al. A food store-based environmental intervention is associated with reduced BMI and improved psychosocial factors and food-related behaviors on the Navajo nation. *J Nutr* 2013;143:1494–500, doi: 10.3945/jn.112.165266.
47. Ross MG et al. Developmental programming of offspring obesity, adipogenesis, and appetite. *Clin Obstet Gynecol* 2013; 58: 529–36.
48. Patterson RE et al. Short sleep duration is associated with higher energy intake and expenditure among African-American and non-Hispanic white adults. *J Nutr* 2014; 144:461–66, doi: 10.3945/jn.113.186890.
49. Geha PY et al. Altered hypothalamic response to food in smokers. *Am J Clin Nutr* 2013;97:15–22, doi: 10.3945/ajcn.112.043307.
50. Aronne LJ et al. Weight gain in the treatment of mood disorders. *J Clin Pyschiatr* 2003;64(Suppl 8);22–29.
51. Kuzawa CW et al. Birth weight, postnatal weight gain, and adult body composition in five low and middle-income countries. *Am J Hum Biol* 2012;24:5–13.
52. Whitaker JA et al. Predicting obesity in young adulthood from childhood parental obesity. *N Engl J Med* 1997;337:869–73.
53. Swinburn B et al. Increased food energy supply is more than sufficient to explain the U.S. obesity epidemic. *Am J Clin Nutr* 2009;90:1453–56.
54. Ledikwe JH et al. Dietary energy density is associated with energy intake and weight status in U.S. adults. *Am J Clin Nutr* 2006;83:1362–68.
55. Duerksen SC et al. Family restaurant choices are associated with child and adult overweight status in Mexican-American families. *J Am Diet Assoc* 2007;107:849–53.

56. Rolls BJ et al. Larger portion sizes lead to sustained increases in energy intake over 2 days. *J Am Diet Assoc* 2006;106:543–49.
57. Position of the American Dietetic Association: Individual-, family-, school-, and community-based interventions for pediatric overweight. *J Am Diet Assoc* 2006;106:925–45.
58. Baidal JA et al. Protecting progress against childhood obesity—the National School Lunch Program. *N Engl J Med* 2014;372:1862–65.
59. Jeffery RW et al. Prevalence and correlates of large weight gains and losses in adults. *Int J Obes* 2002;26:969–72.
60. Mozaffarian D et al. Changes in diet and lifestyle and long-term weight gain in women and men. *N Engl J Med* 2011;364:2392–404.
61. Lombard C et al. Preventing weight gain: The baseline weight related behaviors and delivery of a randomized controlled intervention in community based women. *BMC Public Health,* January 3, 2009;9:2.
62. Rolls BJ. W. O. Atwater Memorial Lecture, presented at the Experimental Biology annual meetings, Washington, DC; May 5, 2007.
63. Benedict C et al. Acute sleep deprivation reduces energy expenditure in healthy men. *Am J Clin Nutr* 2011;93:1229–33.
64. den Hoed M et al. Postprandial responses in hunger and satiety are associated with the rs9939609 single nucleotide polymorphism in FTO. *Am J Clin Nutr* 2009;90:1426–32.
65. Leibel RL. Energy in, energy out, and the effects of obesity-related genes. *N Engl J Med* 2008;359:2603–04.
66. Kumanyika SK et al. American Heart Association's obesity statement: Population-based prevention of obesity. Circulation 2008. Available at: http://circ.ahajournals.org.
67. Levine JA. Poverty and obesity in the U.S. *Diabetes* 2011;60:2667–8.
68. Appendix H, Table H-1. Body measurement summary statistics. In: *Dietary Reference Intakes for Energy through Amino Acids,* Washington, DC: The National Academies Press, 1999.
69. Sabia S et al. BMI over the adult life course and cognition in late midlife: The Whitehall II Cohort Study. *Am J Clin Nutr* 2009;89:601–07.
70. Mattison J et al. Impact of caloric restriction on health and survival in rhesus monkeys from the NIA study. *Nature,* doi: 10.1038/nature11432, 2012.
71. Fontana L. The scientific basis of caloric restriction leading to longer life. *Curr Opin Gastroenterol* 2009;25:144–50.
72. Meyer TE et al. Long-term calorie restriction ameliorates the decline in diastolic functions in humans. *J Am Coll Cardio* 2006:47:398–402.
73. The GBD 2015 Obesity Collaborators. Health Effects of Overweight and Obesity in 195 Countries over 25 Years. *N Eng J Med,* published online, June 12, 2017, doi: 10.1056/NEJMoa1614362.
74. Boggs DA et al. Long-term diet quality is associated with lower obesity risk in young African American women with normal BMI at baseline. *J Nutr* 2013;143:1636–41, doi: 10.3945/jn.113.179002.
75. Leigh L et al. Life expectancy and BMI in old and very old women. *Brit J Nutr* 2016;116:692–96.
76. Bozorgmanesh M et al. No obesity paradox—BMI incapable of adequately capturing the relation of obesity with all-cause mortality: An inception diabetes cohort study. Int *J Endocrin* 2014, doi: org/10.1155/2014/282089.
77. Goni L et al. Future perspectives of personalized weight loss interventions based on nutrigenetic, epigenetic, and metagenomics data. *J Nutr* 2016;46(suppl):905S-12S.
78. Friedman JM. A war on obesity, not the obese. *Science* 2003;299:856-8.

Unit 10

1. Franz MJ et al. Weight-loss outcomes: A systemic review and meta-analysis of weight-loss clinical trials with a minimum of 1-year follow-up. *J Am Diet Assoc* 2007;107:1755–67.
2. Hill JO. Can a small-changes approach help address the obesity epidemic? A report of the joint task force of the ASN, IFT, and the IFIC. *Am J Clin Nutr* 2009;89:477–84.
3. Yaemsiri S et al. Perceived weight status, overweight diagnosis, and weight control among US adults: The NHANES 2003–2008 Study. *Int J Obes* (Lond) 2011;35:1063–70, doi: 10.1038/ijo.2010.229.
4. *ABC News* staff, 100 million dieters, $20 billion: The weight-loss industry by the numbers. May 2012, available at: abcnews.go.com/Health/100-million-dieters-20-billion-weight-loss-industry/story?id=16297197, accessed 11/23/14.
5. Pagoto SL et al. A call for an end to the diet debates. *JAMA* 2013;31:687–688, doi: 10.1001/jama.2013.8601.
6. Simmons L-A, et al. Taking a more personalized approach to the obesity epidemic. www.medscape.com/viewarticle/814710?nlid=41043_434&src=wnl_edit_medp_diab&uac=61213SX&spon=22, accessed 7/23/14.
7. Blackburn GL. Weight loss advertising: An analysis of current trends. 2002. Available at: www.ftc.gov, accessed 12/26/11.
8. Wadden TA et al. Very low calorie diets: Their efficacy, safety, and future. *Ann Intern Med* 1983;99:675–84.
9. Jackson SE et al. Perceived weight discrimination and changes in weight, waist circumference, and weight status. *Obesity,* article first published online: September 11, 2014, doi: 10.1002/oby.20891, accessed 9/11/14.
10. Frank A. Why is it so difficult to lose weight? *Am J Lifestyle Med,* published online before print March 31, 2014, doi: 10.1177/1559827614526353.
11. Winter JE et al. BMI and all-cause mortality in older adults: A meta-analysis. *Am J Clin Nutr,* March 2014, doi: 10.3945/ajcn.113.068122.
12. Wadden TA et al. Very low calorie diets: Their efficacy, safety, and future. *Ann Intern Med* 1983;99:675–84.
13. Obstetrics & gynecology: Committee opinion no. 600: Ethical issues in the care of the obese woman. *Obstet Gynecol* 2014;123:1388–93, doi: 10.1097/01.AOG.0000450756.10670.c2.
14. van Laack R. FTC cracks down on fad weight-loss products; Guidance to consumers and media: Don't believe ads claiming fast and easy weight loss, January 9, 2014, www.fdalawblog.net/2014/01/ftc-cracks-down-on-fad-weight-loss-products-guidance-to-consumers-and-media-dont-believe-ads-claimin, accessed 11/17.
15. Cunningham E. Are there guidelines to evaluate claims made for weight-loss products? *J Acad Nutr Diet* 2014 June;114:976, doi: 10.1016/j.jand.2014.04.008.
16. El Boghdady D. FTC continues push against fake-news web sites peddling weight-loss products. Available at: www.washingtonpost.com/business/economy/ftc-continues-push-against-fake-news-web-sites-peddling-weight-loss-products/2011/12/01/glQATgnSIO_story.html, accessed 12/26/11.
17. Taubes G. *Dietary Approaches to Weight Control. What Works?* Experimental Biology Annual Meeting, San Diego, April 14, 2003.
18. Wing RR et al. Long-term weight loss maintenance. *Am J Clin Nutr* 2005;82:222S–225S.
19. Dombrowski SU et al. Long term maintenance of weight loss with nonsurgical interventions in obese adults. BMJ, available at: www.medscape.com/viewarticle/825379_2?nlid=57826_43, accessed 6/3/14.
20. Blackburn G. Weight of the nation: Moving forward, reversing the trend using medical care. *Am J Clin Nutr* 2012;96:949–50, doi: 10.3945/ajcn.112.049643.

21. Feig EH et al. Variability in weight change early in behavioral weight loss treatment: Theoretical and clinical implications. *Obesity,* August 28, 2017 doi: 10.1002/oby.21925.
22. Page L. Getting paid for treating obesity, now that it's a "disease." July 17, 2013, www.medscape.com/viewarticle/807700, accessed 11/23/14.
23. Vega CP. Societies weigh in on guidelines for obesity treatment. CME/CE released 11/27/2013, www.medscape.org/viewarticle/813658?src=cmemp, accessed 1/20/14.
24. Thomas SL et al. "They all work . . . when you stick to them": A qualitative investigation of dieting, weight loss, and physical exercise, in obese individuals. *Nutr J* 2008;7:34 [online], November 24, 2008, doi: 10.1186/1475-2891-7-34.
25. Cunningham A et al. Is it possible to burn calories by eating grapefruit or vinegar? *J Am Diet Assoc* 2001;101:1198.
26. Weight management: U.S. consumer mindsets. Packaged Facts. www.packagedfacts.com/Weight-Management-Consumer-8351387/, August 28, 2104, accessed 9/11/14.
27. King NA et al. Dual-process action of exercise on appetite control: Increase in orexigenic drive but improvement in meal-induced satiety. *Am J Clin Nutr* 2009;90:921–27.
28. Prescription medications to treat overweight and obesity. NIHDDK, /www.niddk.nih.gov/health-information/weight-management/prescription-medications-treat-overweight-obesity, accessed 11/17.
29. The possible dangers of buying medicines over the Internet. Available at: www.fda.gov/downloads/forconsumers, updated November 2010.
30. Beware of fraudulent weight-loss "dietary supplements." March 15, 2011. Available at: http://www.fda.gov/ForConsumers/ConsumerUpdates/ucm246742.htm#2, Accessed 12/17/11.
31. Atallah R et al. Long-term effects of 4 popular diets on weight loss and cardiovascular risk factors: A systematic review of randomized controlled trials. *Circ Cardiovasc Qual Outcomes* 2014;7:815–27, doi: 10.1161/CIRCOUTCOMES.113.000723.
32. Doyle K. Diets work, but the brand doesn't matter. September 3, 2014, www.medscape.com/viewarticle/830870?nlid=64644_2521&src=wnl_edit_medp_diab&uac=61213SX&spon=22, accessed 9/5/14.
33. Huang T et al. FTO genotype, dietary protein, and change in appetite: The Preventing Overweight Using Novel Dietary Strategies Trial. *Am J Clin Nutr* 2014;99:1126–30, doi: 10.3945/ajcn.113.082164.
34. Clifton PM et al. Long-term weight maintenance after advice to consume low carbohydrate, higher protein diets—A systematic review and meta-analysis. *Nutr Metab Cardiovasc Dis* 2014;24:224–235.
35. Bazzano LA et al. The effects of low-carbohydrate versus conventional weight loss diets in severely obese adults: One-year follow-up of a randomized trial. *Ann Intern Med* 2014;161:309–318, doi: 10.7326/M14-0180.
36. Johns DJ et al. Diet or exercise interventions vs. combined behavioral weight management programs: A systematic review and meta-analysis of direct comparisons. *J Acad Nutr Diet* 2014;114:1557–68, doi: 10.1016/j.jand.2014.07.005.
37. Appel LJ et al. Comparative effectiveness of weight-loss interventions in clinical practice. *N Engl J Med* 2011;364:1959–68.
38. American College of Sports Medicine, updated guidelines for appropriate physical activity interventions for weight loss and prevention of weight regain. *MedSci Sports Exer* 2009;41:459–71.
39. May AM et al. Effect of change in physical activity on body fatness over a 10-y period in the Doetinchem Cohort Study. *Am J Clin Nutr* 2010;92:491–99.
40. American College of Sports Medicine, updated guidelines for appropriate physical activity interventions for weight loss and prevention of weight regain. *Med Sci Sports Exer* 2009;41:459–71.
41. Busko M. Contrave: Where does new obesity drug fit in armamentarium? October 27, 2014, www.medscape.com/viewarticle/833888?src=wnl_edit_medn_wir&uac=61213SX&spon=34, accessed 11/1/14.
42. Krempf M et al. Weight reduction and long-term maintenance after 18 months treatment with orlistat for obesity. *Int J Obes Relat Metab Disord* 2003;27:591–7.
43. Pray WS. New nonprescription weight loss product. *U.S. Pharmacist* 2007;32:10–5.
44. Krempf M et al. Weight reduction and long-term maintenance after 18 months treatment with orlistat for obesity. *Int J Obes Relat Metab Disord* 2003;27:591–97.
45. Yanovski SZ et al. Obesity. *N Engl J Med* 2002;346:591–602.
46. Wolfe BW et al. Long-term risks and benefits of bariatric surgery: A research challenge. *JAMA,* published online October 1, 2014, doi: 10.1001/jama.2014.12966.
47. Troke RC et al. The future role of gut hormones in the treatment of obesity. *Ther Adv Chronic Dis* 2014;5(1), available at: www.medscape.com/viewarticle/818450?nlid=50263_434&src=wnl_edit_medp_diab&uac=61213SX&spon=22, accessed 2/27/14.
48. Pontiroli AE et al. Long-term prevention of mortality in morbid obesity through bariatric surgery: A systematic review and meta-analysis of trials performed with gastric banding and gastric bypass. *Ann Surg* 2011;253:484–87.
49. Kayser BD et al. Serum lipidomics reveals early differential effects of gastric bypass compared with banding on phospholipids and sphingolipids independent of differences in weight loss. *Intl J Obesity,* 2017, doi: 10.1038/ijo.2017.63.
50. Courcoulas A et al. Safety data favors less-invasive weight-loss surgery. *Arch Surgery,* available at: http://bit.ly/sdDPNc, accessed 11/21/11.
51. Cost of gastric sleeve surgery. *Obesity Coverage,* 10/17, www.obesitycoverage.com/insurance-and-costs/how-much/average-cost-of-gastric-sleeve-surgery
52. Cardoso L et al. Short- and long-term mortality after bariatric surgery: A systematic review and meta-analysis. Short- and long-term mortality after bariatric surgery: A systematic review and meta-analysis. *Diabetes Obes Metab* 2017;0:1–10, doi.org/10.1111/dom.12922.
53. O'Riodan M et al. Lap-Band device gets wider indications: FDA lowers BMI requirements for weight-loss surgery. Available at: www.medscape.com/niewarticle/737792, accessed 2/23/11.
54. Buchwald H. Surgical intervention of the treatment of morbid obesity. *Future Lipidol* 2007;2:513–25.
55. Johnson DA. Tips for monitoring post-bariatric surgery nutritional deficiencies. Medscape 2017; www.medscape/com/viewarticle/875462.
56. Eisenberg D et al. Update on obesity surgery. *World J Gastroenterol* 2006;12:3196–3203.
57. Kenkel JM. Clarification about occurrence of complications during liposuction. *Aesthetic Surg J* 2014;34:NP75, doi: 10.1177/1090820X14539701.
58. The skinny on liposuction. Available at: http://www.fda.gov/forconsumers/consumerupdates/ucm049314.htm, accessed 12/27/11.
59. Kaoutzanis C et al. Cosmetic liposuction: Preoperative risk factors, major complication rates, and safety of combined procedures. *Aesthet Surg J* 2017; 37:680–94.

60. Swanson E et al. *Does Liposuction Offer More Than a Cosmetic Benefit?* Plastic Surgery 2011: American Society of Plastic Surgeons (ASPS) Annual Meeting. Presented September 25, 2011.
61. Benatti F et al. Liposuction induces a compensatory increase of visceral fat which is effectively counteracted by physical activity: A randomized trial. *J Clin Endocrin Metab,* published online: April 26, 2012, doi: http://dx.doi .org/10.1210/jc.2012-1012.
62. Macdonald IA. In search of the basis of successful maintenance of weight loss. *Am J Clin Nutr* 2009;90:908–09.
63. Hill JO, et al. Weight maintenance: What's missing? *J Am Diet Assoc* 2005;10B:S63–S66.
64. Macdonald IA. In search of the basis of successful maintenance of weight loss. *Am J Clin Nutr* 2009;90:908–09.
65. Field AE et al. Weight-control behaviors and subsequent weight change among adolescents and young adult females. *Am J Clin Nutr* 2010;91:147–53.
66. Thomas DM et al. Time to correctly predict the amount of weight loss with dieting. *J Acad Nutr Diet* 2014;114:857–61, doi: 10.1016/j.jand.2014.02.003.
67. Hall KD et al. Energy balance and its components: implications for body weight regulation. *Am J Clin Nutr* 2012;95:989–94.
68. Ello-Martin JA et al. Dietary energy density in the treatment of obesity: A year-long trial comparing 2 weight-loss diets. *Am J Clin Nutr* 2007;85:1465–77.
69. Cunningham E. How can I help my client who is experiencing a weight-loss plateau? *J Am Diet Assoc* 2011;101:1966–67.

Unit 11

1. Makino M et al. Prevalence of eating disorders: A comparison of Western and non-Western countries. *MedGenMed* 2004;6:49.
2. Meule A et al. Food addiction in the light of DSM-5. *Nutrients* 2014;6:3653–71.
3. Tremelling K et al. Orthorexia nervosa and eating disorder symptoms in registered dietitians nutritionists in the United States. *J Am Acad Nutr Diet* 2017;117:1612–17.
4. Agras S. Simpler therapy may successfully treat adolescents with anorexia nervosa. *Evid Based Ment Health,* 2017, doi: 10.1136/eb-2016-102535.
5. Konstantynowicz J et al. Thigh circumference as a useful predictor of body fat in adolescent girls with anorexia nervosa. *Ann Nutr Metab* 2011;58:181–87.
6. Ozier AD et al. Position of the American Dietetic Association: Nutrition intervention in the treatment of eating disorders. *J Am Diet Assoc* 2011;111:1236–41, doi: 10.1016/j.jada.2011.06.016.
7. Mangweth B et al. Body fat perception in eating-disordered men. *Int J Eat Disord* 2004;35:102–08.
8. Ulla Räisänen et al. The role of gendered constructions of eating disorders in delayed help-seeking in men: A qualitative interview study. *BMJ Open* 2014;4:e004342, doi: 10.1136/bmjopen-2013-004342.
9. Misra M et al. Percentage extremity fat, but not percentage trunk fat, is lower in adolescent boys with anorexia nervosa than in healthy adolescents. *Am J Clin Nutr* 2008;88:1478–84.
10. Marcason W. Female athlete triad or relative energy deficiency is sports (RED-S): Is there a difference? *J Am Acad Diet* 2016,106:744.
11. Mountjoy M et al. The IOC consensus statement: beyond the female athlete triad—relative energy deficiency in sport (RED-S). *Br J Sports Med,* 2014, dx.doi.org/10.1136/bjsports-2014-093502.
12. Beals KA et al. Disorders of the female athlete triad among collegiate athletes. *Int J Sports Nutr Exer Metab* 2002;12:281–93.
13. American Psychological Association. *Diagnostic and Statistical Manual of Mental Disorders: DMS-IV,* 4th ed., Text Revision. Washington, DC: APA, 2000.
14. Omizo SA et al. Anorexia nervosa: Psychological considerations for nutrition counseling. *J Am Diet Assoc* 1988;88:49–51.
15. Field AE et al. Factors related to bingeing and purging in girls and boys: The growing up today study. *Arch Pediatr Adolesc Med* 2008;162:574–79.
16. Position of the American Dietetic Association: Nutrition intervention in the treatment of anorexia nervosa, bulimia nervosa, and other eating disorders. *J Am Diet Assoc* 2006;10:2073–82.
17. Frank GKW et al. Altered structural and effective connectivity in anorexia and bulimia nervosa in circuits that regulate energy and reward homeostasis. *Translational Psychiatry* (2016) 6, e932; doi: 10.1038/tp.2016.199; published online November 1, 2016.
18. Benninghoven D et al. Body images of male patients with eating disorders. *Psychother Psychosom Med Psychol* 2007;57:120–27.
19. Agras WS et al. Comparison of 2 family therapies for adolescent anorexia nervosa: A randomized parallel trial. *JAMA Psychiatry,* published online September 24, 2014, doi: 10.1001/jamapsychiatry.2014.1025.
20. Eddy KT et al. Recovery from anorexia nervosa and bulimia nervosa at 22-year follow-up. *J Clin Psychiatry* 2016. 10.4088/JCP.15m10393.
21. Women's Health Watch May 4–5,1996; and When eating goes awry: An update on eating disorders. *Food Insight,* January/February 1997:35.
22. Naessen S et al. Women with bulimia nervosa exhibit attenuated secretion of glucagon-like peptide 1, pancreatic polypeptide, and insulin in response to a meal. *Am J Clin Nutr* 2011;94:967–72.
23. Bulimia nervosa—symptoms. 2014, www.webmd.com/mental-health /eating-disorders/bulimia-nervosa /bulimia-nervosa-symptoms, accessed 11/28/14.
24. Herzog DB et al. Bulimia nervosa—psyche and satiety (editorial). *N Engl J Med* 1988;319:716–18.
25. Hohlstein LA. *Eating Disorders Program.* American Dietetic Association annual meeting, Boston, October 27, 1997.
26. Schaefer JT et al. A review of interventions that promote eating by internal cues. *J Acad Nutr Diet* 2014;114:734–60, doi: 10.1016/j.jand.2013.12.024.
27. Wanden-Berghe RG et al. The application of mindfulness to eating disorders treatment: A systematic review. *Eat Disord* 2011;19:34–48.
28. Feldman W et al. Culture versus biology: Children's attitudes toward thinness and fatness. *Pediatrics* 1988;81:190–4.
29. Dallas ME. Eating disorders rampant on the runway. https://medlineplus.gov /news/fullstory_163358.html, 2/1/17.
30. Feldman J. Two of the biggest forces in fashion just banned ultra-thin models. *Huffpost,* September 9, 2017, www .huffingtonpost.com/entry/underweight-model-ban_us_59b15e14e4b0b5e5310444ed
31. Elgin J et al. Gender differences in disordered eating and its correlates. *Eat Weight Disord* 2006;11:e96–101.
32. Calado M, et al. The mass media exposure and disordered eating behaviours in Spanish secondary students. *Eur Eat Disord Rev* 2010;18:417–27.
33. Cooper M. Pica: *A Survey of the Historical Literature as Well as Reports from the Fields of Veterinary Medicine and Anthropology.* Springfield, IL: Charles C. Thomas, 1957.
34. Edwards CH et al. Clay- and corn starch eating women. *J Am Diet Assoc* 1959;35:810–15.
35. Woywodt A et al. Geophagia: The history of earth–eating. *J R Soc Med* 2002;95:143–46.
36. Coltman CA Jr. Pagophagia and iron lack. *JAMA* 1969;207:513–16.

37. Rainville AJ. Pica practices of pregnant women are associated with lower maternal hemoglobin level at delivery. *J Am Diet Assoc* 1998;98:293–96.
38. Blood lead levels in children aged 1–5 years—United States, 1999–2010. *Morbidity and Mortality Weekly* 2013;62(13);245–248, www.cdc.gov/mmwr/preview/mmwrhtml/mm6213a3.htm, accessed 11/28/14.
39. Ellis CR et al. Pica. September 23, 2014, available at: emedicine.medscape.com/article/914765-overview, accessed 11/28/14.
40. Wang X et al. A Western diet pattern is associated with higher concentrations of blood and bone lead among middle-aged and elderly men. *J Nutr* 2017;147:1374–83.
41. Brauser D. Is "food addiction" real? October 16, 2013, www.medscape.com/viewarticle/812650, accessed 11/28/14.
42. Hebebrand J et al. "Eating addiction," rather than "food addiction," better captures addictive-like eating behavior. *Neurosci Biobehav Reviews* 2014;47:295–306, doi: 10.1016/j.neubiorev.2014.08.016.

Unit 12

1. Hsu WC et al. Improvement of insulin sensitivity by isoenergy high carbohydrate traditional Asian diet: A randomized controlled pilot feasibility study. *PLoS ONE,* published September 16, 2014, doi: 10.1371/journal.pone.0106851.
2. Grabitske HA et al. Low-digestible carbohydrates in practice. *J Am Diet Assoc* 2008;108:1677–81.
3. Stanhope KL et al. Consuming fructose-sweetened, not glucose-sweetened, beverages increases visceral adiposity and lipids and decreases insulin sensitivity in overweight/obese humans. *J Clin Invest* 2009;119:1322–34.
4. Institute of Medicine, Food and Nutrition Board. *Dietary Reference Intakes for Energy Carbohydrates, Fiber, Fat, Fatty Acids, Cholesterol, Protein, and Amino Acids.* Washington, DC: National Academies Press, 2002.
5. Position of the American Dietetic Association: Use of nutritive and nonnutritive sweeteners. *J Am Diet Assoc* 2004;104:255–58.
6. Dietary Guidelines Advisory Committee, *Scientific Report of the 2015 Dietary Guidelines: Advisory report to the Secretary of Health and Human Services and Secretary of Agriculture.* Published February 2015, www.health.gov/dietaryguidelines/2015-scientific-report/PDFs/Scientific-Report-of-the-2015-Dietary-Guidelines-Advisory-Committee.pdf, accessed 3/12/15.
7. *Factsheet on the New Proposed Nutrition Facts Label.* www.fda.gov/Food/GuidanceRegulation/GuidanceDocumentsRegulatoryInformation/LabelingNutrition/ucm387533.htm, 2/27/14, accessed 10/27/14.
8. Labeling & nutrition. FDA, 2017, www.fda.gov/Food/IngredientsPackagingLabeling/LabelingNutrition/default.htm
9. Wittekind A et al. Worldwide trends in dietary sugars intake. *Nutr Res Rev* 2014;2:330-45.
10. Yang Q et al. Added sugar intake and cardiovascular diseases mortality among US adults. *JAMA Intern Med* 2014;174(4):516–24. doi: 10.1001/jamainternmed.2013.13563.
11. Fitch C et al. Position of the Academy of Nutrition and Dietetics: Use of nutritive and nonnutritive sweeteners. *J Am Acad Nutr Diet* 2012;112:739–58.
12. Yale J. Gain weight by "going diet"? Artificial sweeteners and the neurobiology of sugar cravings. *Biol Med* 2010;83:101–08, available at: www.ncbi.nlm.nih.gov/pmc/articles/PMC2892765, accessed 10/1/14.
13. Mattes RD et al. Nonnutritive sweetener consumption in humans: Effects on appetite and food intake and their putative mechanisms. *Am J Clin Nutr* 2009;8:1–14.
14. Allen AL et al. Bitterness of the non-nutritive sweetener acesulfame potassium varies with polymorphisms in TAS2R9 and TAS2R3. *Chem Senses* 2013;38:379–89.
15. Weihrauch MR et al. Artificial sweeteners—do they bear a carcinogenic risk? *Ann Oncol* 2004;15:1460–65, available at: annonc.oxfordjournals.org/content/15/10/1460.long, accessed 10/1/14.
16. U.S. EPA says saccharin not a threat. Available at: www.medscape.com.viewarticle/734270, accessed 12/17/10.
17. Butchko HH et al. Acceptable daily intake vs. actual intake: The aspartame example. *J Am Coll Nutr* 1991;10:258–66.
18. Spiers PA, et al. Aspartame: Neuropsychologic and neurophysiologic evaluation and chronic effects. *Am J Clin Nutr* 1998;68:531–37.
19. Magnuson B et al. Safety of aspartame. *Crit Rev Toxicol,* October 2007, available at: www.medscape.com/viewarticle/564923, accessed March 2009.
20. Broyd N. Artificial sweetener aspartame deemed safe by EU food agency. December 11, 2013, www.medscape.com/viewarticle/817687?nlid=42223_2042&src=wnl_edit_medn_obgy&uac=61213SX&spon=16, accessed 7/22/14.
21. Landberg R. Dietary fiber and mortality: Convincing observations that call for mechanistic investigations. *Am J Clin Nutr,* first published June 6, 2012, doi: 10.3945/ajcn.112.040808.
22. Park Y et al. Dietary fiber intake and mortality in the NIH-AARP diet and health study. *Arch Intern Med* 2011;171:1061–68.
23. Lupton JR. In this cohort, an apple a day could keep the doctor away. *Am J Clin Nutr* 2014;100:1409–10.
24. McRorie Jr. JW et al. Understanding the physics of functional fibers in the gastrointestinal tract. *J Am Acad Nutr Diet* 2017;107:251–64.
25. Turner ND et al. Dietary Fiber. *Adv Nutr* 2011;151–152, 2011, doi: 10.3945/an.110.000281.
26. Zelman KM et al. Dietary fiber: Insoluble vs. soluble. Available at: http://www.webmd.com/diet/fiber-health-benefits-11/insoluble-soluble-fiber, accessed 12/6/11.
27. Whitehead A et al. Cholesterol-lowering effects of oat beta-glucan: A meta-analysis of randomized controlled trials. *Am J Clin Nutr* 100:1413–21.
28. Peregrin T. Frequently asked questions of the Academy's Knowledge Center. *J Acad Nutr Diet* 1–13;103:887–91.
29. *What We Eat in America,* NHANES, 2007–2008. Available at www.ars.usda.gov/Services/docs.htm?docid=13793, accessed 2/25/11.
30. Position paper of the Academy of Nutrition and Dietetics: Health implications of dietary fiber. *J Am Acad Nutr Diet* 2015;105:1861–69.
31. Irritable bowel syndrome. www.eatright.org/Public/content.aspx?id=5547, accessed 4/14/14.
32. Scott-Thomas C. Prepare for higher calorie count for fibre, say scientists. December 19, 2008, available at: www.foodnavigator/europe.com, accessed December 2008.
33. Meule A et al. Food addiction in the light of DSM-5. *Nutrients* 2014;6:3653–71.
34. Stein K. The politics and process of revising the DSM-V and the impact of changes on dietetics. *J Acad Nutr Diet* 2014;114:350–65, doi: 10.1016/j.jand.2013.12.011.
35. Hur IY et al. Relationship between whole-grain intake, chronic disease risk indicators, and weight status among adolescents in the national health and nutrition examination survey, 1999–2004. *J Acad Nutr Diet* 2012;112:46–55.
36. Garg P. Psyllium husk should be taken at higher dose with sufficient water to maximize hydration. *J Am Acad Nutr Diet* 2017;107:681–82.

37. Liu S et al. Dietary glycemic load and type 2 diabetes: Modeling the glucose raising potential of carbohydrates for prevention. *Am J Clin Nutr* 2010;92:675–77.
38. Evans CEL et al. Glycemic index, glycemic load, and blood pressure: A systematic review of meta-analysis of randomized trials. *Am J Clin Nutr* 2017;105:1176–90.
39. Atkinson FS et al. International tables of glycemic index and glycemic load values: 2008. *Diabetes Care* 2008, doi: 10.2337/dc08-1239, available at: care.diabetesjournals.org/content/suppl/2008/09/18/dc08-1239.DC1/TableA2_1.pdf, accessed 12/2/14.
40. Evans RA et l. Chronic fructose substitution for glucose or sucrose in food or beverages has little effect on fasting blood glucose, insulin, or triglycerides: a systematic review and meta-analysis. *Am J Clin Nutr,* first published June 7, 2017, doi: 10.3945/ajcn.116.145169.
41. Juanola-Falgarona M et al. Effect of the glycemic index of the diet on weight loss, modulation of satiety, inflammation, and other metabolic risk factors: A randomized controlled trial. *Am J Clin Nutr* 2014;100:27–35.
42. Stanhope KL et al. Endocrine and metabolic effects of consuming beverages sweetened with fructose, glucose, sucrose, or high-fructose corn syrup. *Am J Clin Nutr* 2008;88(suppl):1733S–7S.
43. Glycemic index, glycemic load, and blood pressure: A systematic review and meta-analysis of randomized controlled trials. *Am J Clin Nutr* 2017;105:1176–90.
44. Cornero S et al. Diet and nutrition of prehistoric populations at the alluvial banks of the Parana River. *Medicina* 2000;60:109–14.
45. Starling S. EFSA boosts tooth-friendly developing world dental campaign. August 1, 2011, www.nutriangredients.com/content/view/print/389940.
46. Clark MB et al. Fluoride use in caries prevention in the primary care setting. *Pediatrics,* originally published online August 25, 2014, doi: 10.1542/peds.2014-1699.
47. Stanhope KL et al. Endocrine and metabolic effects of consuming beverages sweetened with fructose, glucose, sucrose, or high-fructose corn syrup. *Am J Clin Nutr* 2008;88(suppl):1733S–7S.
48. Schachtele CF et al. Will the diets of the future be less cariogenic? *J Canadian Dental Assoc* 1984;3:213–19.
49. Touger-Decker R et al. Position of the Academy of Nutrition and Dietetics: Oral health and nutrition. *J Acad Nutr Diet* 2013;113:693–701, doi: 10.1016/j.jand.2013.03.001.
50. Palmer CA et al. Position of the American Dietetic Association: The impact of fluoride on dental health. *J Am Diet Assoc* 2005;105:1620–28.
51. Bailey W et al. Populations receiving optimally fluoridated public drinking water—United States, 1992–2006. *MMWR* 2008;57:737–41.
52. National Maternal and Child Oral Health Resource Center. *Trends in Children's Oral Health.* 1999. Available at: www.ncemch.org.
53. Misselwitz B et al. Lactose malabsorption and intolerance: Pathogenesis, diagnosis and treatment. *United European Gastroenterol J* 2013 June, doi: 10.1177/2050640613484463.
54. Dykibra J et al. Lactose maldigestion revisited: Diagnosis, prevalence in ethnic minorities. *Am J Lifestyle Med* 2009;3:212–18.
55. Inman-Felton AE. Overview of lactose maldigestion (lactose nonpersistence). *J Am Diet Assoc* 1999;99:481–89.
56. Lactose intolerance: A self-fulfilling prophecy leading to osteoporosis. *Nutr Rev* 2003;61:221–23.
57. Montalto M et al. Management and treatment of lactose malabsorption. *Gastroenterol* 2006;12:187–91.
58. Inman-Felton AE. Overview of lactose maldigestion (lactose nonpersistence). *J Am Diet Assoc* 1999;99:481–89.

Unit 13

1. Dietary Guidelines Advisory Committee, *Scientific Report of the 2015 Dietary Guidelines: Advisory Report to the Secretary of Health and Human Services and Secretary of Agriculture,* published February 2015, http://www.health.gov/dietaryguidelines/2015-scientific-report/PDFs/Scientific-Report-of-the-2015-Dietary-Guidelines-Advisory-Committee.pdf, accessed 3/12/15.
2. van't Riet et al. Progression of Glucose Disturbances Prior to Onset of Diabetes. European Association for the Study of Diabetes 47th annual meeting, abstract 335, September 13, 2011.
3. Atkinson FS, Foster-Powell K, Brand-Miller JC. International tables of glycemic index and glycemic load values. *Diabetes Care* 2008;31:2281–83.
4. MacLeod JM et al. Academy of Nutrition and Dietetics Nutrition Practice Guideline for type 1 and type 2 diabetes in adults: Nutrition intervention evidence reviews and recommendations. *Am J Nutr Diet* 2017;107:1637–58.
5. *IDF Diabetes Atlas,* 6th ed. 2014 update, www.idf.org/diabetesatlas, accessed 12/12/14.
6. National Diabetes Statistics Report, 2017. Centers for Disease Control and Prevention. 7. Shulman GI. Ectopic fat in insulin resistance, dyslipidemia, and cardiometabolic disease. *N Engl J Med* 2014; 371:2236–38, doi: 10.1056/NEJMc1412427.
8. Alexander CM et al. NCEP-defined metabolic syndrome, diabetes, and prevalence of coronary heart disease among NHANES III participants age 50 years and older. *Diabetes* 2003;52:1210–14.
9. Fried SK et al. Sugars, hypertriglyceridemia, and cardiovascular disease. *Am J Clin Nutr* 2003,78(4):873S–880S.
10. Diabetes Symptoms. September 12, 2014, available at: www.diabetes.org/diabetes-basics/symptoms, accessed 12/13/14.
11. Hemmingsen B et al. Intensive glycaemic control for patients with type 2 diabetes: Systematic review with meta-analysis and trial sequential analysis of randomised clinical trials. BMJ. Available at: www.medscape.com/viewarticle/754363?src=mp&spon=2, accessed 12/16/11.
12. Bennett WL et al. Comparative effectiveness and safety of medications for type 2 diabetes: An update including new drugs and 2-drug combinations. *Ann Int Med* 2011;154:602–13.
13. Jones V. The "diabesity" epidemic: Let's rehabilitate America. *Medscape Gen Med* 2006;8:34.
14. The CDC's Database Modeling Project: Developing a new tool for chronic disease prevention and control. Available at: www.cdc.gov, accessed August 2006.
15. Diabetes in China: Mapping the road ahead. *The Lancet Diabetes Endocrinol,* September 11, 2014, doi: 10.1016/S2213-8587(14)70189-5, available at: www.thelancet.com/journals/landia/article/PIIS2213-8587(14)70189-5/fulltext.
16. Fryhofer S. The fight against type 2 diabetes: The promise of genomics. Medscape, July 11, 2014, www.medscape.com/viewarticle/828019?nlid=61363_434&src=wnl_edit_medp_diab&uac=61213SX&spon=22.
17. McEvoy CT et al. A posteriori dietary patterns are related to risk of type 2 diabetes: Findings from a systematic review and meta-analysis. *J Acad Nutr Diet* 2014;114:1759–1775, doi: 10.1016/j.jand.2014.05.001.
18. Jannasch F et al. Dietary patterns and type 2 diabetes: A systematic literature review and meta-analysis of prospective studies. *J Nutr* 2017;147:1174–82.
19. Hsueh WA et al. Prediabetes: The importance of early identification and intervention. *Postgrad Med* 2010;122:129–43.

20. Parker AR et al. The effect of medical nutrition therapy by a registered dietitian nutritionist in patients with prediabetes participating in a randomized controlled clinical research trial. *J Acad Nutr Diet* 2014;114:1739–48. doi: 10.1016/j.jand.2014.07.020.
21. Busko M. Sustained diabetes prevention achieved with lifestyle changes. November 6, 2017, www.medscape.com/viewarticle/888071?nlid=119005_3042&src=WNL_mdplsnews_171110_mscpedit_diab&uac=61213
22. Brecque DR et al. World Gastroenterology Organisation global guidelines: Nonalcoholic fatty liver disease and nonalcoholic steatohepatitis. *Clin Gastroenterol* 2014;48:467–473.
23. Harmon RC et al. Inflammation in nonalcoholic steatohepatitis. *Expert Rev Gastroenterol Hepatol* 2011;5:189–200.
24. Lazo M et al. Prevalence of nonalcoholic fatty liver disease in the United States: The Third National Health and Nutrition Examination Survey, 1988–1994. *Am J Epidemiol* 2013;178:38–45, doi: 10.1093/aje/kws448.
25. Weisenberger J. Non-alcoholic fatty liver disease. *Food Nutr.* August 29, 2017, https://foodandnutrition.org/september-october-2017/non-alcoholic-fatty-liver-disease
26. O'Connor L et al. Intakes and sources of dietary sugars and their association with metabolic and inflammatory markers. *Clin Nutr,* May 2017, doi.org/10.1016/j.clnu.2017.05.030.
27. Cook LT et al. Vegetable consumption is linked to decreased visceral and liver fat and improved insulin resistance in overweight Latino youth. *J Acad Nutr Diet* 2014;114:1776–83, doi: 10.1016/j.jand.2014.01.017.
28. Ezmaillzadeh A et al. Dietary patterns, insulin resistance, and prevalence of metabolic syndrome in women. *Am J Clin Nutr* 2007;85: 910–18.
29. Elbassuoni E. Better association of waist circumference with insulin resistance and some cardiovascular risk factors than body mass index. *Endocr Regul* 2013;47:3–14.
30. Symptoms and diagnosis of metabolic syndrome. American Heart Association, May 14, 2014, www.heart.org/HEARTORG/Conditions/More/MetabolicSyndrome/Symptoms-and-Diagnosis-of-Metabolic-Syndrome_UCM_301925_Article.jsp#, accessed 12/12/14.
31. Riediger ND et al. Prevalence of metabolic syndrome in the Canadian adult population. *CMAJ* 2011, doi: 10.1503/cmaj.110070.
32. Lerman RH et al. Enhancement of a modified Mediterranean style, low glycemic load diet with specific phytochemicals improves cardiometabolic risk factors in subjects with metabolic syndrome and hypercholesterolemia in a randomized trial. *Nutr Metab* (Lond) 2008;5:29, published online November 4, 2008, doi: 10.1186/1743-7075-5-29.
33. Steffen LM et al. A modified Mediterranean diet score is associated with a lower risk of incident metabolic syndrome over 25 years among young adults: The CARDIA (Coronary Artery Risk Development in Young Adults) study. *Brit J Nutr,* published online September 19, 2014, doi: http://dx.doi.org/10.1017/S0007114514002633.
34. MacLeod JM et al. Academy of Nutrition and Dietetics Nutrition Practice Guideline for type 1 and type 2 diabetes in adults: Nutrition intervention evidence reviews and recommendations. *Am J Nutr Diet* 2017;107:1637–58.
35. Henry RR et al. New options for the treatment of obesity and type 2 diabetes mellitus (narrative review). *J Diabetes Complications,* 2013, doi: 10.1016/j.jdiacomp.2013.04.011, available at: www.ncbi.nlm.nih.gov/pmc/articles/PMC4139280, accessed 12/15/14.
36. Understanding carbohydrates. American Diabetes Association, 2014, www.diabetes.org/food-and-fitness/food/what-can-i-eat/understanding-carbohydrates, accessed 12/15/14.
37. Slomski A. Bariatric surgery has durable effects in controlling diabetes. *JAMA* 2017;317(16):1615. doi: 10.1001/jama.2017.3787.
38. Nansel TR et al. Greater diet quality is associated with more optimal glycemic control in a longitudinal study if youth with type 1 diabetes. *Am J Clin Nutr* 2016;104:81–87.
39. Physical activity is important. American Diabetes Association, 2013, available at: www.diabetes.org/food-and-fitness/fitness/physical-activity-is-important.html, accessed 12/15/14.
40. Coen PM et al. Clinical trial demonstrates exercise following bariatric surgery improves insulin sensitivity. *J Clin Invest,* December 1, 2014, doi: 10.1172/JCI78016.
41. Hayes C et al. Role of physical activity in diabetes management. *J Am Diet Assoc* 2008;108:S19–S23.
42. Vinoy S et al. The effect of a breakfast rich in slowly digestible starch on glucose metabolism: A statistical meta-analysis of randomized controlled trials. *Nutrients* 2017;9(4);318; doi: 10.3390/nu9040318.
43. Wolever TMS et al. The Canadian Trial of Carbohydrates in Diabetes, a 1-y controlled trial of low-glycemic index dietary carbohydrate in type 2 diabetes. *Am J Clin Nutr* 2008;87:114–25.
44. Brito JP et al. Metabolic surgery in the treatment algorithm for type 2 diabetes: A joint statement by International Diabetes Organizations. *JAMA* 2017;317(6):635–36. doi: 10.1001/jama.2016.20563.
45. Finley CE et al. Glycemic index, glycemic load, and prevalence of the metabolic syndrome in the Cooper Center Longitudinal Study. *J Am Diet Assoc* 2010;110:1820–9.
46. A series of systematic reviews on the relationship between dietary patterns and health outcomes. March 2014, www.nel.gov/vault/2440/web/files/DietaryPatterns/DPRptFullFinal.pdf.
47. Busko, M. Glycemic control drives cardiac-event-free survival in diabetes. www.medscape.com/viewarticle/878984?nlid=114458_3042&src=WNL_mdplsnews_170428_mscpedit_diab&uac=61213SX&spon=22&impID=1337637&faf=1, accessed 4/24/17.
48. Busko, M. Glycemic control drives cardiac-event-free survival in diabetes. www.medscape.com/viewarticle/878984?nlid=114458_3042&src=WNL_mdplsnews_170428_mscpedit_diab&uac=61213SX&spon=22&impID=1337637&faf=1, accessed 4/24/17.
49. Diabetes Prevention Program Research Group. 10-year follow-up of diabetes incidence and weight loss in the Diabetes Prevention Program Outcomes Study. *Lancet* 2009;374:1677–86.
50. Lifestyle, Metformin, can delay diabetes, 15-Year DPP Data Show. American Diabetes Association (ADA) 74th Scientific Session, www.medscape.com/viewarticle/826824?nlid=59623_2521&src=wnl_edit_medp_diab&uac=61213SX&spon=22, June 16, 2014, San Francisco, California, accessed 6/23/14.
51. McRorie JW et al. Understanding the physics of functional fibers in the gastrointestinal tract. *J Am Acad Nutr Diet* 2017;117:251–64.
52. Mancini FR et al. Dietary antioxidant capacity and risk of type 2 diabetes in the large prospective E3N-EPIC cohort. *Diabetologia* 2017. doi.org/10.1007/s00125-017-4489.
53. Holstet C et al. Alcohol drinking patterns and risk of diabetes: A cohort study of 70,551 men and women from the general Danish population. *Diabetologia.* Published online July 27, 2017. https://link.springer.com/article/10.1007%2Fs00125-017-4359-3. Accessed 7/28/17.

54. Mitri J et al. Effects of vitamin D and calcium supplementation on pancreatic β cell function, insulin sensitivity, and glycemia in adults at high risk of diabetes: The calcium and vitamin D for diabetes mellitus randomized controlled trial. *Am J Clin Nutr* 2011;94:486–94.
55. Schienkiewitz A et al. Body mass index history and risk of type 2 diabetes: Results from the European Prospective Investigation into Cancer and Nutrition (EPIC)—Potsdam Study. *Am J Clin Nutr* 2006;84:427–33.
56. Geiss LS et al. Prevalence and incidence trends for diagnosed diabetes among adults aged 20 to 79 years, United States, 1980–2012. *JAMA* 2014;312:1218–1226, doi: 10.1001/jama.2014.11494.
57. Brown W et al. Academy policy strategies to prevent type 2 diabetes. *J Acad Nutr Diet* 2013;113:1443, 1445–6, doi: 10.1016/j.jand.2013.09.005.
58. Thomas NJM et al. Abstract from the European Association for the Study of Diabetes (EASD) 2016 Annual Meeting and reported on Medscape, Half of all type 1 diabetes develops after 30 years of age, September 20, 2016. www.medscape.com/viewarticle/869028
59. Ferrara CT et al. Type 1 diabetes and celiac disease: Causal association or true, true, unrelated? *Pediatrics,* October 2017, reviewed online at: http://pediatrics.aappublications.org/content/early/2017/10/06/peds.2017-2424?sso=1&sso_redirect_count=1&nfstatus=401&nftoken=00000000-0000-0000-0000-000000000000&nfstatusdescription=ERROR%3a+No+local+token
60. Hummel S et al. First Infant Formula Type and Risk of Islet Autoimmunity in The Environmental Determinants of Diabetes in the Young (TEDDY) Study. *Diabetes Care* 2017 January; dc161624, doi.org/10.2337/dc16-1624.
61. Lund-Blix NA et al. Infant feeding and risk of type 1 diabetes in two large Scandinavian birth cohorts. *Diabetes Care* 2017 May; dc170016. https://doi.org/10.2337/dc17-0016.
62. Harris SS. Vitamin D in type 1 diabetes prevention. *J Nutr* 2005;135:323–25.
63. Khardori R et al. Type 1 diabetes mellitus treatment & management. October 21, 2014, available at: emedicine.medscape.com/article/117739-treatment, accessed 12/15/14.
64. Chiang JL et al. Type 1 diabetes through the lifespan: A position statement of the American Diabetes Association. *Diabetes Care* 2014;37:2034–54.
65. Fowler MJ et al. Diabetes treatment, part 1: Diet and exercise. *Clin Diabetes* 2007;2:105–09, doi: 10.2337/diaclin.25.3.105, available at: clinical.diabetesjournals.org/content/25/3/105.full
66. Carbohydrate counting & diabetes. NIH, www.niddk.nih.gov/health-information/diabetes/overview/diet-eating-physical-activity/carbohydrate-counting, accessed 12/17.
67. Hall M. Understanding advanced carbohydrate counting—A useful tool for some patients to improve blood glucose control. *Today's Dietitian* 2013;15:40.
68. Haimoto H et al. Three-graded stratification of carbohydrate restriction by level of baseline hemoglobin A1c for type 2 diabetes patients with a moderate low-carbohydrate diet. *Nutr Metab* 2014, available at: www.medscape.com/viewarticle/830292_1?nlid=66725_434.
69. Wild S et al. Global prevalence of diabetes: Estimates for the year 2000 and projections for 2030. *Diabetes Care* 2004;27:1047–53.
70. Baidal DA et al. Bioengineering of an intraabdominal endocrine pancreas. Editorial, *N Engl J Med* 2017; 376:1887–89.
71. DeSisto et al. Prevalence estimates of gestational diabetes mellitus in the United States. Pregnancy Risk Assessment Monitoring System (PRAMS), 2007–2010. *Prev Chronic Dis* 2014;11:130415, available at: http://dx.doi.org/10.5888/pcd11.130415.
72. Erem C et al. Prevalence of gestational diabetes mellitus and associated risk factors in Turkish women: The Trabzon GDM Study. *Arch Med Sci* 2015;11:724–35.
73. Zhang C et al. Effect of dietary and lifestyle factors on the risk of gestational diabetes: Review of epidemiological evidence. *Am J Clin Nutr* 2011;95(suppl):1975S–9S.
74. Gestational diabetes in the United States. Available at: www.cdc.gov/diabetes/pubs/estimatesl1.htm-8, accessed 1/14/12.
75. Hartling L et al. Benefits and harms of treating gestational diabetes mellitus: A systematic review and meta-analysis for the U.S. Preventive Services Task Force and the National Institutes of Health Office of Medical Applications of Research. *Ann Intern Med* 2013;159:123–29, doi: 10.7326/0003-4819-159-2-201307160-00661.
76. Shubrook JH et al. Treat hypoglycemia fast, but how? Aug 3, 2017; www.medscape.com/viewarticle/883544?nlid=117119_1521&src=WNL_mdplsfeat_170808_mscpedit_wir&uac=61213SX&spon=17&impID=1406531&faf=1#vp_2.
77. Seaquist ER et al. Hypoglycemia and diabetes: A report of a workgroup of the American Diabetes Association and the Endocrine Society. *Diabetes Care* 2013;36:1384–95, doi: 10.2337/dc12-2480.

Unit 14

1. Heilig M. Triggering addiction. *The Scientist* 2008;December:28–2.
2. Xi B et al. Relationship of alcohol consumption to all-cause, cardiovascular, and cancer-related mortality in U.S. adults. *J Am Coll Cardiol,* August 2017; doi: 10.1016/j.jacc.2017.06.054.
3. *Dietary Guidelines for Americans.* 2015. Available at: www.dietaryguidelines.gov.
4. Alcohol use: If you drink, keep it moderate. www.mayoclinic.org/healthy-lifestyle/nutrition-and-healthy-eating/in-depth/alcohol/art-20044551, accessed 4/7/15.
5. Arevalol SP et al. Alcohol use and cardiovascular risk in a prospective cohort study of Latino Adults: The mediating effect of inflammation. *The FASEB Journal* 2016; vol. 30 no. 1 Supplement 1154.4.
6. Mostofsky E et al. Key Findings on alcohol consumption and a variety of health outcomes from the Nurses' Health Study. *American Journal of Public Health,* September 1, 2016;106:1586–91.
7. Ferreira MP et al. Alcohol consumption by aging adults in the United States: Health benefits and detriments. *J Am Diet Assoc* 2008;108:1668–76.
8. Clemente-Postigo M et al. Effect of acute and chronic red wine consumption on lipopolysaccharide concentrations. *Am J Clin Nutr* 2013;97:1053–61, doi: 10.3945/ajcn.112.051128.
9. Leifert WR et al. Cardioprotective actions of grape polyphenols. *Nutr Res* 2008;28:729–37, doi: 10.1016/j.nutres.2008.08.007.
10. Goldberg IJ. To drink or not to drink? *N Engl J Med* 2003;348:163–64.
11. WHO. Alcohol. Fact sheet, updated May 2014, www.who.int/mediacentre/factsheets/fs349/en, accessed 5/20/14.
12. Markel H. Dying for a drink: Alcohol on campus. *Medscape Public Health and Prevention,* June 7, 2006;4.
13. Cirrhosis. Updated January 2012, available at: www.nlm.nih.gov/medlineplus/cirrhosis.html, accessed 1/18/12.
14. Alcohol overdose. National Institute for Alcohol Abuse and Alcoholism, www.pubs.niaaa.nih.gov/publications/AlcoholOverdoseFactsheet/Overdisefact.htm, accessed 2017.
15. Hoyme HE et al. Updated clinical guidelines for diagnosing fetal alcohol spectrum disorders. *Pediatrics* 2016;138:e20154256
16. Cojocariu C-E et al. Alcoholic liver disease: Epidemiology and risk factors. *Rev Med Chir Soc Med Nat,* Iaşi, 2014;118:910-7

17. Guenther PM et al. Alcoholic beverage consumption by adults compared to dietary guidelines: Results of the National Health and Examination Survey, 2009–2910. *J Am Acad Nutr Diet* 2013,113;546–50.
18. Kesse E et al., Do eating habits differ according to alcohol consumption? *Am J Clin Nutr* 2001;74:322–27.
19. You YS et al. Dietary taurine and nutrient intake and dietary quality by alcohol consumption level in Korean male college students. *Adv Exp Med Biol* 2013;776:121–27, doi: 10.1007/978-1-4614-6093-0_13.
20. Breslow RA Alcoholic beverage consumption, nutrient intakes, and diet quality in the US adult population, 1999–2006. *J Am Diet Assoc* 2010;110:551–62, doi: 10.1016/j.jada.2009.12.026.
21. Latt N et al. Thiamine in the treatment of Wernicke encephalopathy in patients with alcohol use disorders. *Intern Med J* 2014;44:911–5, doi: 10.1111/imj.12522.
22. Bémeur C et al. Nutrition in the Management of Cirrhosis and its Neurological Complications. *J Clin Exp Hepatol* 2014;4:141–50.
23. Sidhu SS et al. New paradigms in management of alcoholic hepatitis: A review. *Hepatol Intl* 2017;11:255–67.
24. How alcohol affects the body. Available at: nihseniorhealth.gov/alcoholuse /howalcoholaffectsthebody/01.html, 2012, accessed 12/20/14.
25. Alcohol overdose fact sheet. National Institute for Alcohol Abuse and Alcoholism, www.pubs.niaaa.nih.gov /publications/AlcoholOverdoseFactsheet /Overdisefact.htm, accessed 12/20/14.
26. Levitt MD et al. Use of measurements of ethanol absorption from stomach and intestine to assess human ethanol metabolism. *Am J Physiol Gastrointestinal Liver Physiol* 1997;273(4):G951–G957.
27. Sayon-Orea C et al. Alcohol consumption and body weight: A systematic review. *Nutr Rev* 2011;69:419–31, doi: 10.1111/j.1753-4887.2011.00403.x.
28. Suter PM. Is alcohol consumption a risk factor for weight gain and obesity? *Crit Rev Clin Lab Sci* 2005;42:197–227.
29. Yeomans MR. Alcohol, appetite and energy balance: Is alcohol intake a risk factor for obesity. *Physiol Behav* 2010;100:82–9, doi: 10.1016/j.physbeh.2010.01.012.
30. Alcohol metabolism: An Update. July 2007, available at: http://pubs.niaaa .nih.gov/publications/AA72/AA72.htm, accessed 1/17/12.
31. Esser MB et al. Prevalence of alcohol dependence among US adult drinkers, 2009–2011. *Preven Chronic Dis* 2014, doi: 10.5888/pcd11.140329.
32. Women and alcohol. Available at: pubs. niaaa.nih.gov/publications/womensfact /womensfact.htm, February 2011, accessed 1/20/12.
33. Alcohol and your health: Weighing the pros and cons. Available at: www. mayoclinic.com/health/alcohol/SC00024, accessed August 200634.
34. Morey TE et al. Measurement of ethanol in gaseous breath using a miniature gas chromatograph. *J Anal Toxicol* 2011;35:134–42.
35. Franzen J. How do you determine the alcohol proof of a given liquid? *MadSci Network: Chemistry,* February 1998.
36. Distilled spirits labels. Available at: www.ttb.gov/pdf/brochures/p51902.pda, accessed 1/18/12.
37. Arria AM et al. The "high" risk of energy of energy drinks. *JAMA* 2011;305:600–1.
38. Fact Sheet: Caffeine and alcohol. November 2014, www.cdc.gov/alcohol /fact-sheets/caffeine-and-alcohol.htm, accessed 12/20/14.
39. Yang X et al. Common variants at 12q24 are associated with drinking behavior in Han Chinese. *Am J Clin Nutr* 2013;97:545–51, doi: 10.3945/ ajcn.112.046482.
40. Fact Sheets: Underage drinking. Available at: www.cdc.gov/alcohol/fact-sheets /underage-drinking.htm, updated July 2010, accessed 1/19/12.
41. Boyle SC et al. Facebook dethroned: Revealing the more likely social media destinations for college students' depictions of underage drinking. *Addictive Behaviors* 2017;65:63–67.

Unit 15

1. Dirks ML et al. Skeletal muscle disuse atrophy is not attenuated by dietary protein supplementation in healthy older men. *J Nutr* 2014;144:1196–203. doi: 10.3945/jn.114.194217.
2. Pasiakos SM et al. Efficacy and safety of protein supplements for U.S. Armed Forces personnel: Consensus statement. *J Nutr* 2013;143:1811S–1814S, doi: 10.3945/ jn.113.176859.
3. Nassauer S. When the box says "protein," shoppers say "I'll take it." *WSJ,* March 27, 2013: D1, D3.
4. *Dietary Intake Data: What We Eat in America,* NHANES 2007–2008. Available at: www.ars.usda.gov/ba/bhnrc/fsrg, accessed 1/19/12.
5. Van den Akker CHP et al. Human fetal albumin synthesis rates during different periods of gestation. *Am J Clin Nutr* 2008;88:997–1003.
6. *Dietary Reference Intakes: Energy, Carbohydrate, Fiber, Fat, Fatty Acids, Cholesterol, Protein, and Amino Acids.* Institute of Medicine, National Academy of Sciences. Washington, DC: National Academies Press. 2002.
7. Matthews DE. Proteins and amino acids. In: Shils ME et al., eds. *Modern Nutrition in Health and Disease,* 9th ed. Philadelphia: Lippincott, Williams & Wilkins, 1998:11–48.
8. Bihuniak JD. Dietary protein-induced increases in urinary calcium are accompanied by similar increases in urinary nitrogen and urinary urea: a controlled clinical trial. *J Acad Nutr Diet* 2013;113:447–51, doi: 10.1016/j. jand.2012.11.002.
9. Boyce B. You scream, we scream for the study of proteins. *J Acad Nutr Diet* 2014:1518–23.
10. Proud CG. Control of the translational machinery by amino acids. *Am J Clin Nutr* 2014;99:231S–236S, doi: 10.3945/ ajcn.113.066753.
11. Michelfelder AJ. Soy: A complete source of protein. *Am Earn Physician* 2009;79:43–47.
12. Dietary Guidelines Advisory Committee, *Scientific Report of the 2015 Dietary Guidelines: Advisory Report to the Secretary of Health and Human Services and Secretary of Agriculture,* published February 2015, www.health. gov/dietaryguidelines/2015-scientific-report/PDFs/Scientific-Report-of-the-2015-Dietary-Guidelines-Advisory-Committee.pdf, accessed 3/12/15.
13. Renwick AG. Establishing the upper end of the range of adequate and safe intakes for amino acids: A toxicologist's viewpoint. *J Nutr* 2004;134:1617S–24S.
14. Garlick PJ. Toxicity of methionine in humans. *J Nutr* 2006;136:1722S–25S.
15. Van de Poll MC et al. Adequate range for sulfur-containing amino acids and biomarkers for their excess: Lessons from enteral and parenteral nutrition. *J Nutr* 2006;136(6 suppl):1694S–1700S.
16. Kay LK. Melatonin. *Today's Dietitian,* November 2004:54–6.
17. Pencharz PB et al. An approach to defining the upper safe limits of amino acid intake. *J Nutr* 2009;138:1995S–2002S.
18. *2015 Dietary Guideline for Americans.* Available at: www.dietaryguidelines.gov.
19. Badaloo AV et al. Lipid kinetic differences between children with kwashiorkor and those with marasmus. *Am J Clin Nutr* 2006;83:1283–88.
20. Young VR et al. Long-term nitrogen balance studies and other criteria for protein requirement estimations. 1981, available at: http://www.fao.org /DOCREP/MEETING/004/M3021E /M3021E00.HTM, accessed 1/21/12.

21. Holwerda AM et al. Physical activity preformed in the evening increases the overnight muscle protein synthetic response to presleep protein ingestion in older men. *J Nutr* 2016;146:1307–14.
22. McLellan TM. Protein supplementation for military personnel: A review of the mechanisms and performance outcomes. *Am J Clin Nutr* 201499:86–95, doi: 10.3945/ajcn.112.055517.
23. Navarro VJ et al. Liver injury from herbals and dietary supplements in the U.S. Drug-Induced Liver Injury Network. *Hepatology.* Article first published online August 25, 2014, doi: 10.1002/hep.27317.
24. Bilsborough S et al. A review of issues of dietary protein intake in humans. *Int J Sport Nutr Exerc Metab* 2006:129-52. Review.

Unit 16

1. Jahns L et al. Position of the American Dietetic Association: Vegetarian diets. *J Am Acad Nutr Diet* 2016;106:70–78.
2. Macrobiotic diet. *CAM-Cancer,* 2017, http://cam-cancer.org/The-Summaries/Dietary-approaches/Macrobiotic-diet/What-is-it
3. Williams KA et al. Healthy plant-based diet: What does it really mean? *Am Coll Cardiol* 2017;70:423–25.
4. Macrobiotic diet. *CAM-Cancer,* 2017, http://cam-cancer.org/The-Summaries/Dietary-approaches/Macrobiotic-diet/What-is-it
5. Rizzo NC et al. Nutrient profiles of vegetarians and nonvegetarian dietary patterns. *J Acad Nutr Diet* 2013;113:1610–19.
6. *2015 Dietary Guideline for Americans.* Available at: www.dietaryguidelines.gov.
7. Cunningham E. What is a raw food diet and are there any risks or benefits associated with it? *J Am Diet Assoc* 2004;104:1623.
8. Kushi LH et al. The macrobiotic diet and cancer. *J Nutr* 2001;131:3056S–64S.
9. Rauma A-L et al. Vitamin B-12 status of long-term adherents of a strict, uncooked vegan diet (“living foods diet”) is compromised. *J Nutr* 1995;125:2511–15.
10. Dietary Guidelines Advisory Committee, *Scientific Report of the 2015 Dietary Guidelines: Advisory Report to the Secretary of Health and Human Services and Secretary of Agriculture* published February 2015, www.health.gov/dietaryguidelines/2015-scientific-report/PDFs/Scientific-Report-of-the-2015-Dietary-Guidelines-Advisory-Committee.pdf, accessed 3/12/15.
11. Tips for vegetarians. www.choosemyplate.gov/healthy-eating-tips/tips-for-vegetarian.html, accessed 1/10/15.
12. Kushi LH et al. The macrobiotic diet and cancer. *J Nutr* 2001;131:3056S–64S.
13. Messina V et al. A new food guide for North American vegetarian and vegan diets. *J Am Diet Assoc* 2003;65:771–5.
14. *Dietary Guidelines for Americans, 2015–2020.* health.gov/dietaryguidelines/2015/guideline, accessed 1/9/16.
15. Harris WS et al. Towards establishing dietary reference intakes for eicosapentaenoic and docosahexaenoic acids. *J Nutr* 2009;139:804S–19S.
16. Michelfelder AJ. Soy: A complete source of protein. *Am Fam Physician* 2009;79:43–47.
17. Sedgwick T. Meat analogs. *Food and Nutrition* (magazine), July/August 2013:22–23.

Unit 17

1. Bahna S et al. Food allergies over-estimated by public. WHO 2010 International Scientific Conference, presented December 5, 2010, available at: www.medscape.com/viewarticle/734305, accessed 124/12.
2. Jones SM et al. Food allergy. *N Engl J Med* 2017;377:1168–75.
3. Sicherer SH et al. Critical issues in food allergy: A National Academies Consensus Report. *Pediatrics* 2017;140-;1–8.
4. Jarvinen KM. Food-induced anaphylaxis. *Curr Opin Allergy Clin Immunol* 2011;11:255–61.
5. Formanek R. Food allergies: When food becomes the enemy. *FDA Consumer* (magazine), April 2004 update, available at: www.fda.gov/FDAC/features/2001/401_food.html, accessed March 2009.
6. Sicherer SH et al. Critical issues in food allergy: A National Academies Consensus Report. *Pediatrics* 2017;140-;1–8.
7. Chafen JJ et al. Diagnosing and managing common food allergies: A systematic review. *JAMA* 2010;303:1848–56, doi: 10.1001/jama.2010.582.
8. Anderson WH et al. Gut reactions—from celiac affection to autoimmune model. *N Engl J Med* 2014;371:6–7, doi: 10.1056/NEJMp1405192.
9. Neumann J. Celiac disease showing up in many forms and at all ages. December 1, 2014, www.medscape.com/viewarticle/835567?src=wnl_edit_medn_wir&uac=61213SX&spon=34
10. Catassi C et al. Celiac disease. *Curr Opin Gastroenterol* 2008;24:687–91.
11. Anderson WH et al. Gut reactions—from celiac affection to autoimmune model. *N Engl J Med* 2014;371:6–7, doi: 10.1056/NEJMp1405192.
12. Lionetti E et al. Introduction of gluten, HLA status, and the risk of celiac disease in children. *N Engl J Med* 2014; 371:1295–1303, doi: 10.1056/NEJMoa1400697.
13. Peak I et al. *Physician Prescriptions for EpiPen among Patients with a History of Anaphylaxis.* Paper presented at the American Academy of Asthma, Allergy, and Immunology Annual Meeting, March 15, 2009.
14. Lau MSY et al. Optimizing the diagnosis of celiac disease. *Curr Opin Gastroenterol* 2017;33:173–80.
15. Goebel SU et al. Celiac sprue. June 14, 2014, emedicine.medscape.com/article/171805-overview?src=wnl_ref_prac_publ&uac=61213SX, accessed 7/28/14.
16. Liu E et al. Risk of pediatric celiac disease according to HLA haplotype and country. *N Engl J Med* 2014;371:42–49, doi: 10.1056/NEJMoa131397.7.
17. Knut EA et al. Advances in celiac disease. *Curr Opin Gastroent* 2014;30:154–62.
18. Koning F et al. The million-dollar question: Is “gluten-free” food safe for patients with celiac disease? *Am J Clin Nutr* 2013, first published December 5, 2012, doi: 10.3945/ ajcn.112.053777.
19. Niewinski MM. Advances in celiac disease and gluten-free diet. *J Am Diet Assoc* 2008;108:661–72.
20. Burrowes JD. Helping adults with celiac disease to eat well. *Nutr Today* 2008;43:250–56.
21. Gluten and food labeling. FDA, 2017, www.fda.gov/Food/GuidanceRegulation/GuidanceDocumentsRegulatoryInformation/Allergens/ucm367654.htm
22. Mansueto P et al. Non-celiac gluten sensitivity: Literature review. *J Am Coll Nutr* 2014;33:39–54, doi: 10.1080/07315724.2014.869996/17/14.
23. DiGiacomo DV et al. Prevalence of gluten-free diet adherence among individuals without celiac disease in the USA: Results from the Continuous National Health and Nutrition Examination Survey 2009–2010. *Scand J Gastroenterol* 2013;48:921–25, doi: 10.3109/00365521.2013.809598.
24. Molina-Infante J et al. Suspected nonceliac gluten sensitivity confirmed in few patients after gluten challenge in double-blind, placebo-controlled trials. *Clin Gastroent Hepatol* 2017;15:339–48.
25. Skodje GI et al. Fructan, rather than gluten, induces symptoms in patients with self-reported non-celiac gluten sensitivity. *Gastroenterol* 2017; doi.org/10.1053/j.gastro.2017.10.040.
26. Position of the American Dietetic Association and Dietitians of Canada: Vegetarian diets. *J Am Diet Assoc* 2003;103:748–65.

27. What is lactose intolerance? ghr.nlm.nih.gov/condition/lactose-intolerance, May 2010, accessed 1/15/15.
28. NIH Consensus and State of the Science Statements. *NIH Consensus Development Conference Statement on Lactose Intolerance and Health.* February 22–24, 2010:27.
29. Yang WH et al. Adverse reactions to sulfites. *Can Med Assoc J* 1985;133:865–7.
30. Nyambok E et al. The role of food additives and chemicals in food allergy. *Ann Food Processing Preserv* 2016; 1(1): 1006.
31. What is food intolerance (non-IgE mediated food hypersensitivity)? www.allergyuk.org/food-intolerance/what-is-food-intolerance, October 2012, accessed 1/15/15.
32. Krymchantowski AV et al. Wine and headache. *Headache* 2014;54:967–75.
33. Definition of Chinese restaurant syndrome. Available at: www.medterms.com/script/main/art.asp?articlekey=15584, accessed September 2006.
34. Chinese restaurant syndrome. January 2015, available at: www.nlm.nih.gov/medlineplus/ency/article/001126.htm.

Unit 18

1. Rosen ED et al. What we talk about when we talk about fat. *Cell.* 2014;156(0):20–44. doi: 10.1016/j.cell.2013.12.012. Available at: www.ncbi.nlm.nih.gov/pmc/articles/PMC3934003, accessed 2/17/17.
2. Ebbeling CB et al. Dietary fat: Friend or foe? *Clin Chem* 2017, doi: 10.1373/clinchem.2017.274084.
3. Chaissaing B et al. How diet can impact gut microbiota to promote or endanger health. *Curr Opin Gastroenterol* 2017;336:417–21.
4. Clifton PM et al. A systematic review of the effects of dietary saturated and polyunsaturated fat on heart disease. *Nutr Metab Cardiovas Dis* 2017;27:1060–80.
5. Miller V et al. Fruit, vegetable, and legume intake, and cardiovascular disease and deaths in 18 countries (PURE): A prospective cohort study. *Lancet* 2017, doi: 10.1016/ S0140-6736(17)32253-5.
6. Kratz M. Dietary cholesterol, atherosclerosis and coronary heart disease. *Handb Exp Pharmacol* 2005;195–213.
7. Jones PJ. Dietary cholesterol and the risk of cardiovascular disease in patients: A review of the Harvard Egg Study and other data. *Int J Clin Pract Suppl* 2009:1–8, 28–36, doi: 10.1111/j.1742-1241.2009.02136.x.
8. Food trends and your heart. *Harvard Heart Letter,* September 2017, www.health.harvard.edu/heart-health/food-trends-and-your-heart
9. Rosen ED et al. What we talk about when we talk about fat. *Cell.* 2014;156(0):20–44. doi: 10.1016/j.cell.2013.12.012. Available at: www.ncbi.nlm.nih.gov/pmc/articles/PMC3934003, accessed 2/17/17.
10. Dehghan M et al. Relationship between healthy diet and risk of cardiovascular disease among patients on drug therapies for secondary prevention: A prospective cohort study of 31 546 high-risk individuals from 40 countries. *Circulation* 2012;126:2705–12.
11. Position of the American Academy of Nutrition and Dietetics: Dietary fatty acids for healthy adults. *J Am Acad Nutr Diet* 2014;114:136–53.
12. *Dietary Guidelines for Americans,* 2010. Part D, Section 3: Fatty Acids and Cholesterol. Available at: www.dietaryguidelines.gov.
13. Harms M et al. Brown and beige fat: development, function and therapeutic potential. *Nature Medicine* 2013;9, 1252–63.
14. Berger MM et al. Three short perioperative infusions of n-3 PUFAs reduce systemic inflammation induced by cardiopulmonary bypass surgery: A randomized controlled trial. *Am J Clin Nutr* 2013;97:246–54, doi: 10.3945/ajcn.112.046573.
15. Sowers JR. Endocrine functions of adipose tissue: focus on adiponectin. *Clin Cornerstone* 2008;9:32–8; discussion 39–40.
16. Jones PJ. Dietary cholesterol and the risk of cardiovascular disease in patients: A review of the Harvard Egg Study and other data. *Int J Clin Pract Suppl* 2009:1–8, 28–36, doi: 10.1111/j.1742-1241.2009.02136.x.
17. Appendix 11, *Dietary Guidelines for Americans,* 2010. Available at: www.dietaryguidelines.gov.
18. Appendix 11, *Dietary Guidelines for Americans,* 2010. Available at: www.dietaryguidelines.gov.
19. Oh R. Practical applications of fish oil (omega-3 fatty acids) in primary care. *J Am BoardFam Pract* 2005;18:28–36.
20. Alexander D et al. A meta-analysis of randomized trials and prospective cohort studies of eicosapentaenoic and docosahexaenoic long chain omega-3 fatty acids and coronary heart disease risk. *Mayo Clin Proc* 2017;92:15–29.
21. Oh R. Practical applications of fish oil (omega-3 fatty acids) in primary care. *J Am BoardFam Pract* 2005;18:28–36.
22. Harris WS et al. Towards establishing dietary reference intakes for eicosapentaenoic and docosahexaenoic acids. *J Nutr* 2009;139:804S–19S.
23. Dietary Guidelines Advisory Committee, *Scientific Report of the 2015 Dietary Guidelines: Advisory Report to the Secretary of Health and Human Services and Secretary of Agriculture*, published February 2015, www.health.gov/dietaryguidelines/2015-scientific-report/PDFs/Scientific-Report-of-the-2015-Dietary-Guidelines-Advisory-Committee.pdf, accessed 3/12/15.
24. FDA cuts trans fat in processed foods. 2015; www.fda.gov/forconsumers/consumerupdates/ucm372915.htm
25. Yang Q et al. Plasma trans-fatty acid concentrations continue to be associated with serum lipid and lipoprotein concentrations among US adults after reductions in trans-fatty acid intake. *J Nutr* 2017;147:896–907.
26. American Heart Association. Know your fats, July 31, 2014. Available at: www.heart.org/HEARTORG/Conditions/Cholesterol/PreventionTreatmentofHighCholesterol/Know-Your-Fats_UCM_305628_Article.jsp.
27. Sepa-Kishi D et al. Exercise-mediated effects on white and brown adipose tissue plasticity and metabolism. *Exer Sport Sci Rev* 2016:44:37–44.
28. Sterols 1. Cholesterol and cholesterol esters. January 6, 2014, available at: lipidlibrary.aocs.org/Lipids/cholest/index.htm, accessed 1/22/15.
29. Kulio IJ et al. Incorporating a genetic risk score into coronary heart disease risk estimates: Effect on LDL cholesterol levels (the MIGENES Clinical Trial). *Circulation* 2016;115.020109.
30. New label portion size. 2014, Available at: www.fda.gov/downloads/ForConsumers/ConsumerUpdates/UCM387442.pdf.
31. Diet/Nutrition. CDC, 2017; www.cdc.gov/nchs/fastats/diet.htm
32. Zaeem Z et al. Headaches: A review of the role of dietary factors. *Curr Neurol Neurosciu Repts* 2016; doi.org/10.10007/s11910-016-0720-1.

Unit 19

1. Cardiovascular disease. Available at: www.who.int/mediacentre/factsheets/fs317/en/index.html, September 2011, accessed 1/31/12.
2. Dietary Guidelines Advisory Committee, *Scientific Report of the 2015 Dietary Guidelines: Advisory Report to the Secretary of Health and Human Services and Secretary of Agriculture,* published February 2015, www.health.gov/dietaryguidelines/2015-scientific-report/PDFs/Scientific-Report-of-the-2015-Dietary-Guidelines-Advisory-Committee.pdf, accessed 3/12/15.
3. Hughes S. PURE shakes up nutritional field: Finds high fat intake beneficial. August 29, 2017, www.medscape.com/viewarticle/884937?nlid=117544_4503&src

=wnl_dne_170830_mscpedit&uac
=61213SX&impID=1421914&faf=1#vp_8.
4. Mandrola JM et al. CANTOS: Lowering inflammation, Not LDL-C, to reduce CVD. *European Society of Cardiology* (ESC) Congress 2017. Medscape September 5, 2017; www.medscape.com/viewarticle/885063?nlid=117717_1521&src=WNL_mdplsfeat_170912_mscpedit_wir&uac=61213SX&spon=17&impID=1431474&faf=1
5. Mente A et al. Association of dietary nutrients with blood lipids and blood pressure in 18 countries: A cross-sectional analysis from the PURE study. *Lancet Diabetes Endocrinol* 2017, doi: 10.1016/ S2213-8587(17)30283-8.
6. The top 10 causes of death. Fact sheet No. 310, updated May 2014, www.who.int/mediacentre/factsheets/fs310/en, accessed 1/26/15.
7. Prevalence of coronary heart disease—United States, 2006–2010. *CDC Weekly* 2011;60(40);1377–81, available at: www.cdc.gov/mmwr/preview/mmwrhtml/mm6040a1.htm, accessed 1/31/12.
8. Mosca L et al. Effectiveness-based guidelines for the prevention of cardiovascular disease in women: 2011 update. *Circulation* 2011;123:1243–62.
9. What causes heart disease? April 21, 2014. www.nhlbi.nih.gov/health/health-topics/topics/hdw/causes, accessed 1/26/15.
10. Libby P. Mechanisms of acute coronary syndromes and their implications for therapy. *N Engl J Med* 2013;368:2004–13.
11. Kopeck S et al. Think whole food, not "out of the box" for heart healthy diet. August 22, 2017, www.medscape.com/viewarticle/883879?nlid=117547_434&src=WNL_mdplsfeat_170829_mscpedit_diab&uac=61213SX&spon=22&impID=1421354&faf=1#vp_2
12. Mandrola JM et al. CANTOS: Lowering inflammation, not LDL-C, to reduce CVD. *European Society of Cardiology* (ESC) Congress 2017. Medscape September 5, 2017; www.medscape.com/viewarticle/885063?nlid=117717_1521&src=WNL_mdplsfeat_170912_mscpedit_wir&uac=61213SX&spon=17&impID=1431474&faf=1
13. Brooks M. Heavyweights weigh in on inconsistent cholesterol advice, August 14, 2017, www.medscape.com/viewarticle/884219
14. Falk E. Pathogenesis of atherosclerosis. *J Am Coll Cardiol* 2006; doi: 10.1016/j.jacc.2005.09.068.
15. Chiu S et al. Effects of a very high saturated fat diet on LDL particles in adults with atherogenic dyslipidemia: A randomized controlled trial. *PLOS One* 2017; doi.org/10.1371/journal.pone.0170664.
16. Salim S et al. Non-HDL cholesterol as a metric of good quality of care. *Tex Heart Inst J* 2011;38:160–62.
17. Kulio IJ et al. Incorporating a genetic risk score into coronary heart disease risk estimates: Effect on LDL cholesterol levels (the MIGENES Clinical Trial). *Circulation* 2016;115.020109.
18. Kim K et al. Dietary total antioxidant capacity is inversely associated with all-cause and cardiovascular disease death of US adults. *Euro J Nutr,* August 8 2017, https://link.springer.com/article/10.1007/s00394-017-1519-7
19. Women and heart disease: A different animal, August 30, 2017; www.medscape.com/viewarticle/884997?nlid=117597_3802&src=WNL_mdplsnews_170901_mscpedit_card&uac=61213SX&spon=2&impID=1423660&faf=1#vp_1
21. Kesaniemi YA et al. Intestinal cholesterol absorption efficiency in man is related to apoprotein E phenotype. *J Clin Invest* 1987;80:578–81.
22. Beynen AC et al. Hypo- and hyperresponders: Individual differences in the response of serum cholesterol concentration to changes in diet. *Adv Lipid Res* 1987;22:115–71.
23. Greene CM et al. Egg consumption in an elderly population. *Nutri Metab* 2006;3:6, doi: 10.1186/1743-7075-3-6,
24. Hayden MR. The genetic aspects of atherosclerosis and hyperlipidemia. *CMAJ* 1989;141:135.
25. Burdge GC et al. Nutrition, epigenetics, and developmental plasticity: Implications for understanding human disease. *Ann Rev Nutr* 2010;1(30):315–39.
26. Position of the Academy of Nutrition and dietetics: Dietary fatty acids for healthy adults. *J Am Acad Nutr Diet* 2014;114:136–53.
27. Fernández-Friera L et al. Normal LDL-cholesterol levels are associated with subclinical atherosclerosis in the absence of risk factors. *Am J Cardiol* 2017;104:2979–91.
28. de Oliveira Otto MC et al. Dietary fat and genotype: Toward individualized prescriptions for lifestyle changes. *Am J Clin Nutr* 2005;81:1255–56.
29. Astrup A. A changing view on saturated fatty acids and dairy: From enemy to friend. *Am J Clin Nutr* 2014;100:1407–08.
30. Fortmann SP et al. *Cardiovascular Disease and Cancer: A Systematic Evidence Review for the U.S. Preventive Services Task Force.* Rockville (MD): Agency for Healthcare Research and Quality, November 2013, Report No. 14–05199–EF–1.
31. Liu S et al. A prospective study of dietary glycemic load, carbohydrate intake, and risk of coronary heart disease in US women. *Ann Intern Med* 2014;160:398–406.
32. Khera AV, et al. Genetic risk, adherence to a healthy lifestyle, and coronary disease. *N Engl J Med* 2016; 375:2349–58.
33. Hong KN et al. How low to go with glucose, cholesterol, and blood pressure in primary prevention of CVD. *Am Coll Cardiol* 2017;70:2171–85.
34. Helgadottir S et al. Variants with large effects on blood lipids and the role of cholesterol and triglycerides in coronary disease. *Nature Genetics* 2016;48,634–39.
35. Maki AC et al. Dietary substitution for refined carbohydrate that show promise for reducing the risk of type 2 diabetes. *J Nutr* 2015;145:159S–63S.
36. Papadopoulou E et al. Questioning current recommendations on fatty acids and their role in heart health. *Nutrition Bulletin* 2014;39:253–62, doi: 10.1111/nbu.12100.
37. Mozaffarian D. The promise of lifestyle for cardiovascular health. *J Am Coll Cardiol* 2014, doi: 110.1016/j.jacc.2014.00.1191.
38. Hackethal V. New study IDs genes linking type 2 diabetes and heart disease. Medscape September 11, 2017; www.medscape.com/viewarticle/885479?nlid=117864_3042&src=WNL_mdplsnews_170915_mscpedit_diab&uac=61213SX&spon=22&impID=1434245&faf=1
39. Gurwitz JH et al. Statins for primary prevention in older adults: uncertainty and the need for more evidence. *JAMA* 2016;316(19):1971–72.

Unit 20

1. Dietary Guidelines Advisory Committee, *Scientific Report of the 2015 Dietary Guidelines: Advisory report to the Secretary of Health and Human Services and Secretary of Agriculture,* published February 2015, www.health.gov/dietaryguidelines/2015-scientific-report/PDFs/Scientific-Report-of-the-2015-Dietary-Guidelines-Advisory-Committee.pdf, accessed 3/12/15.
2. Semba RD. The discovery of the vitamins. *Int J Vitam Nutr Res* 2012;82:310–5.
3. Talegawkar SA et al. Total antioxidant performance is associated with diet and serum antioxidants in participants of the diet and physical activity substudy of the Jackson Heart Study. *J Nutr* 2009;139:1964–71.
4. Bjelakovic G et al. Mortality in randomized trials of antioxidant

supplements for primary and secondary prevention: Systematic review and meta-analysis. Available at: www.medscape.com/viewarticle/553714?src=ptalk, accessed 1/9/12.
5. Crowe FL et al. European prospective investigation into cancer and nutrition: Fruit and vegetable intake and mortality from ischaemic heart disease. *Eur Heart J* 2011, doi: 10.1093/eurheartj/ehq465, accessed 1/21/11.
6. Green R. Peripheral neuropathy risk and a transcobalamin polymorphism: Connecting the dots between excessive folate intake and disease susceptibility. *Am J Clin Nutr* 2016; 104;1495–96.
7. Quraishi SA et al. Low vitamin D status in Europe: Moving from evidence to sound public health policies. *Am J Clin Nutr* 2016;103:957–08.
8. Sawaengsri H et al. Transcobalamin 776C→G polymorphism is associated with peripheral neuropathy in elderly individuals with high folate intake. *Am J Clin Nutr* 2016;104:1665–70.
9. Odewole OA et al. Near-elimination of folate-deficiency anemia by mandatory folic acid fortification in older US adults: Reasons for Geographic and Racial Differences in Stroke study 2003–2007. *Am J Clin Nutr* 2013;98:1042–47.
10. Shelke N et al. Folic acid supplementation for women of childbearing age versus supplementation for the general population: A review of the known advantages and risks. *Int J Family Med,* May 23, 2011:173705, doi: 10.1155/2011/173705.
11. Ganji V et al. Trends in serum folate, RBC folate, and circulating total homocysteine concentrations in the US. *J Nutr* 2006;136:153–58.
12. Mayo-Wilson E et al. Vitamin A supplements for preventing mortality, illness, and blindness in children aged under 5: Systematic review and metaanalysis. *BMJ,* August 25, 2011;343:d5094, doi: 10.1136/bmj.d5094.7.
13. Whitney KM et al. Management strategies for acne vulgaris. *Clin Cosmet Investig Dermatol* 2011;4:41–53.
14. Darlenski R et al. Topical retinoids in the management of photodamaged skin: From theory to evidence-based practical approach. *Br J Dermatol* 2010;163:1157–65.
15. Panchaud A et al. Pregnancy outcome following exposure to topical retinoids: A multicenter prospective study. *J Clin Pharmacol,* December 15, 2011, doi: 10.1177/0091270011429566.
16. Amer M et al. Relation between serum 25-hydroxyvitamin d and c-reactive protein in asymptomatic adults. *Am J Cardiol* 2012;109:226–30.
17. Arnson Y et al. Vitamin D and autoimmunity: New aetiological and therapeutic considerations. *Ann Rheum Dis* 2007;66:1137–42.
18. Nikooyeh B. Daily consumption of vitamin D-or vitamin D + calcium-fortified yogurt drink improved glycemic control in patients with type 2 diabetes: A randomized clinical trial. *Am J Clin Nutr,* February 2, 2011, doi: 10.3945.
19. *Dietary Reference Intakes: Calcium and Vitamin D. Food and Nutrition Board,* National Academy of Sciences, Washington, DC: National Academies Press, 2011.
20. Holick MF. Sunlight and vitamin D for bone health and prevention of autoimmune diseases, cancers, and cardiovascular disease. *Am J Clin Nutr* 2004;80(6 suppl):1678S–88S.
21. Vitamin D. November 2014, available at: ods.od.nih.gov/factsheets/VitaminD-HealthProfessional.
22. Sawicki CM et al. Sun-exposed skin color is associated with changes in serum 25-hydroxyvitamin D in racially/ethnically diverse children. *J Nutr* 2016;146(4):751–57.
23. Godar DE et al. Solar UV doses of young Americans and vitamin D3 production. *Environ Health Perspect* 2012;120:139–43.
24. Gilchrest BA. Sun exposure and vitamin D deficiency. *Am J Clin Nutr* 2008;88(suppl):570S–7S.
25. Terushkin V1 et al. Estimated equivalency of vitamin D production from natural sun exposure versus oral vitamin D supplementation across seasons at two US latitudes. *J Am Acad Dermatol* 2010;62:929.e1-9, doi: 10.1016/j.jaad.2009.07.028.
26. Rickman JC et al. Review: Nutritional comparison of fresh, frozen and canned fruits and vegetables. Part 1. Vitamins C and B and phenolic compounds. *J Sci Food Agric* 2007;87:930–44.
27. Guallar E et al. Enough is enough: Stop wasting money on vitamin and mineral supplements. *Ann Intern Med* 2013;159:850–51.

Unit 21

1. Jacobs DR Jr. et al. Diet pattern and longevity: do simple rules suffice? A commentary. *Am J Clin Nutr* 2014; 100(suppl 1):313S–319S.
2. Harrison L. Financial gains taint debate about nutritional supplements. Medscape, November 14, 2016.
3. FooDB. Listing compounds. foodb.ca/compounds, accessed 2/4/15.
4. Yong JJ et al. Ocular nutritional supplements: Are their ingredients and manufacturers' claims evidence-based? *Opthalmology,* published online November 20, 2014, http://dx.doi.org/10.1016/j.ophtha.2014.09.039.
5. Larson AJ et al. Therapeutic potential of quercetin to decrease blood pressure: Review of efficacy and mechanisms. *Adv Nutr* 2012;3:39–46.
6. USDA, Shell eggs from farm to table. April 11, 2011, www.fsis.usda.gov/wps/wcm/connect/5235aa20-fee1-4e5b-86f5-8d6e09f351b6/Shell_Eggs_from_Farm_to_Table.pdf?MOD=AJPERES, accessed 2/4/15.
7. USDA, A series of systematic reviews on the relationship between dietary patterns and health outcomes. March 2014, available at: nel.gov/vault/2440/web/files/DietaryPatterns/DPRptFullFinal.pdf, accessed 11/8/14.
8. Ivey KL et al. Flavonoid intake and all-cause mortality. *Am J Clin Nutr* 101:1012–20.
9. Mehta AJ et al. Dietary anthocyanin intake and age-related decline in lung function: Longitudinal findings from the VA Normative Aging Study. *Am J Clin Nutr* 2016;103(2):542–50.
10. Mail KC et al. Consumption of a cranberry juice beverage lowered the number of clinical urinary tract infection episodes in women with a recent history of urinary tract infection. *Am J Clin Nutr* 2016;103;1434–42.
11. Lin T et al. Trends in cruciferous vegetable consumption and associations with breast cancer risk: A case-control study. *Curr Dev Nutr* 2017;1:e000448 doi: 10.3945/cdn.117.000448.
12. Jiang Y et al. Cruciferous vegetable intake is inversely correlated with circulating levels of proinflammatory markers in women. *J Acad Nutr Diet* 2014;114:700–08.
13. Tsumbu CN et al. Polyphenol content and modulatory activities of some tropical dietary plant extracts on the oxidant activities of neutrophils and myeloperoxidase. *Int J Mol Sci* 2012;13:628–50.
14. Ou WC et al. An eye to health: Diet and age-related macular degeneration. June 27, 2017, www.medscape.com/viewarticle/881915?src=WNL_infoc_170810_MSCPEDIT_TEMP2_MEDDEV_WETAMD&uac=61213SX&impID=1406262&faf=1
15. Harrison L. Financial gains taint debate about nutritional supplements. Medscape, November 14, 2016.
16. Dietary Guidelines Advisory Committee. *Scientific Report of the 2015 Dietary*

Guidelines Advisory Committee: Advisory report to the Secretary of Health and Human Services and Secretary of Agriculture, published February 2015, www.health.gov/dietaryguidelines/2015-scientific-report/PDFs/Scientific-Report-of-the-2015-Dietary-Guidelines-Advisory-Committee.pdf, accessed 3/12/15.
17. Bozzetto L. Diets naturally rich in polyphenols improve glucose metabolism in people at high cardiovascular risk: A controlled randomized trial. *EAS,* June 1, 2014, Madrid, Spain, Abstract M087.
18. Haytowitz DB et al. Sources of flavonoids in the U.S. diet using USDA's updated database on the flavonoid content of selected foods. www.ars.usda.gov/SP2UserFiles/Place/80400525/Articles/AICR06_flav.pdf, accessed 2/5/15.
19. Goetz ME et al. Dietary flavonoid intake and incident coronary heart disease: The Reasons for Geographic and Racial Differences in Stroke (REGARDS) study. *Am J Clin Nutr* 2016;104:1236–44.
20. Tsumbu CN et al. Polyphenol content and modulatory activities of some tropical dietary plant extracts on the oxidant activities of neutrophils and myeloperoxidase. *Int J Mol Sci* 2012;13:628–50.
21. Jiang Y et al. Cruciferous vegetable intake is inversely correlated with circulating levels of proinflammatory markers in women. *J Acad Nutr Diet* 2014;114:700–08.
22. Wang X et al. Fruit and vegetable consumption and mortality from all causes, cardiovascular disease, and cancer: Systematic review and dose-response meta-analysis of prospective cohort studies. *BMJ,* July 29, 2014;349:g4490, doi: 10.1136/bmj.g4490.
23. Mail KC et al. Consumption of a cranberry juice beverage lowered the number of clinical urinary tract infection episodes in women with a recent history of urinary tract infection. *Am J Clin Nutr* 2016;103;1434–42.
24. Wang P. The effectiveness of cranberry products to reduce urinary tract infections in females: A literature review. *Urol Nurs* 2013;33(1):38–45.
25. Salo J et al. Cranberry juice for the prevention of recurrences of urinary tract infections in children: A randomized placebo-controlled trial. *Clin Infect Dis* 2012;54:340–46.
26. Liu JJ et al. Coffee consumption is positively associated with longer leukocyte telomere length in the Nurse's Health Study. *J Nutr* 2016;146:1373–78.
27. Patil H et al. Cuppa joe: Friend or foe? Effects of chronic coffee consumption on cardiovascular and brain health. *Mo Med* 2011;108:431–38.
28. Sartorelli DS et al. Differential effects of coffee on the risk of type 2 diabetes according to meal consumption in a French cohort of women: The E3N/EPIC cohort study. *Am J Clin Nutr* 2010:91;1002–12.
29. Wedick NM et al. Effects of caffeinated and decaffeinated coffee on biological risk factors for type 2 diabetes: A randomized controlled trial. *Nutr J,* September 13, 2011, doi: 10.1186/1475-2891-10-93.
30. Crippa A. Coffee consumption and mortality from all causes, cardiovascular disease, and cancer: A dose-response meta-analysis. *Am J Epidemiol* 2014;180:763–75.
31. Hamza TH et al. Genome-wide gene-environment study identifies glutamate receptor gene grinla as a Parkinson's disease modifier gene via interaction with coffee. *PLoS Gen,* August 11, 2011, available at: www.plosgenetics.org/article/info%3Adoi%2F10.1371%2Fjournal.
32. Gorby HE et al. Do specific dietary constituents and supplements affect mental energy? Review of the evidence. *Nutr Rev* 2010;68:697–718.
33. Lucas M et al. Coffee, caffeine, and risk of depression among women. *Arch Intern Med* 2011;171:1571–78.
34. Floegel A et al. Coffee consumption and risk of chronic disease in the European Prospective Investigation into Cancer and Nutrition (EPIC)-Germany study. *Am J Clin Nutr* 2012;95:901–08, doi: 10.3945/ajcn.111.023648.
35. Harrison P. Heavy caffeine use causes severe BP spikes in MAOI patient. Medscape, May 6, 2014, www.medscape.com/viewarticle/824683?nlid=56823_2703&src=wnl_edit_dail&uac=61213SX, accessed 5/15/14.
36. Harrison L. Surge reported in energy drink emergency department visits. Available at: www.medscape.com/viewarticle/75414 6?src=mp&spon=42, accessed 12/12/11.
37. Arria AM et al. The "high" risk of energy drinks. *JAMA* 2011;305:600–1.
38. Breda JJ et al. Energy drink consumption in Europe: A review of the risks, adverse health effects, and policy options to respond. Front. *Public Health,* October 14, 2014, doi: 10.3389/fpubh.2014.00134.
39. Canada to limit caffeine in energy drinks. *Reuters,* October 7, 2011, Available at: www.medscape.com/viewarticle/751155?src=mp&spon=420TTAWA.
40. *Dietary Guidelines for Americans, 2010.* Available at: www.dietaryguidelines.gov.
41. Chang CH et al. Phytochemical characteristics, free radical scavenging activities, and neuroprotection of five medicinal plant extracts. *Evid Based Complement Alternat Med* 2012, doi: 10.1155/2012/984295.
42. Uribarri J et al. Advanced glycation end products in foods and a practical guide to their reduction in the diet. *J Am Diet Assoc* 2010;110:911–16.e12.
43. Banea JP et al. Effectiveness of wetting method for control of konzo and reduction of cyanide poisoning by removal of cyanogens from cassava flour. *Food Nutr Bull* 2014;35(1):28–32.
44. Uribarri J et al. Advanced glycation end products in foods and a practical guide to their reduction in the diet. *J Am Diet Assoc* 2010;110:911–16.e12.

Unit 22

1. Hong MW et al. Fish oil contaminated with persistent organic pollutants induces colonic aberrant crypt foci formation and reduces antioxidant enzyme gene expression in rats. *J Nutr* 2017;147:1524–30.
2. Romagnolo DF et al. Genistein prevents *BRCA1* CpG methylation and proliferation in human breast cancer cells with activated aromatic hydrocarbon receptor. *Curr Dev Nutr,* June 2017, doi.org/10.3945/cdn.117.000562.
3. Jenkins K. Most US adults unaware of two major cancer risk factors. Medscape 2017, www.medscape.com/viewarticle/887512?nlid=118727_4503&src=wnl_dne_171025_mscpedit&uac=61213SX&impID=1464872&faf=1
4. National Cancer Institute, SEER stat fact sheets: All cancer sites. seer.cancer.gov/statfacts/html/all.html, accessed 2/9/15.
5. Cancer facts and figures, 2014. www.cancer.org/acs/groups/content/@research/documents/webcontent/acspc-042151.pdf, accessed 2/9/15.
6. What causes cancer? www.cancer.org/cancer/cancercauses/index, 2015, accessed 2/9/15.
7. What is a gene mutation and how do mutations occur? February 2015. ghr.nlm.nih.gov/handbook/mutationsanddisorders/genemutation, accessed 2/9/15.
8. Zeller B et al. Familial transmission of prostate, breast and colorectal cancer in adoptees is related to cancer in biological but not in adoptive parents: A nationwide family study. *Eur J Cancer* 2014;50(13):2319–27, doi: 10.1016/j.ejca.2014.05.018.
9. Steensma DP. The beginning of the end of the beginning in cancer genomics. *N Engl J Med* 2013;368:2138–40.
10. Vogelstein B et al. The path to cancer: Three strikes and you're out. *N Engl J Med* 2015;373:1895–98.

11. Christiani DC. Combating environmental causes of cancer. *N Engl J Med* 2011;364;791–4.
12. Peto R. The fraction of cancer attributable to lifestyle and environmental factors in the UK in 2010. *British Journal of Cancer* 2011;105:S1–S1.
13. Jackson AA. Integrating the ideas of life across cellular, individual, and population levels in cancer causation. *J Nutr* 2006;135:2927S–33S.
14. Hoover RN. Cancer—nature, nurture, or both. *N Eng J Med* 2000;343:135–36.
14. Jackson AA. Integrating the ideas of life across cellular, individual, and population levels in cancer causation. *J Nutr* 2006;135:2927S–33S.
15. Jackson AA. Integrating the ideas of life across cellular, individual, and population levels in cancer causation. *J Nutr* 2006;135:2927S–33S.
16. Huang W-Q et al. A higher Dietary Inflammatory Index score is associated with a higher risk of breast cancer among Chinese women: A case-control study. *Br J Nutr*, published June 5, 2017, doi.org/10.1017/S0007114517001192.
17. Carr PR et al. Associations of red and processed meat with survival after colorectal cancer and differences according to timing of dietary assessment. *Am J Clin Nutr* 2016;192–200.
18. Lohse T et al. Adherence to the cancer prevention recommendations of the World Cancer Research Fund/American Institute for Cancer Research and mortality: A census-linked cohort. *Am J Clin Nutr* 2016;104:678–58.
19. Jemal A et al. *Annual Report to the Nation on the Status of Cancer, 1975–2014, Featuring Survival.* JNCI: Journal of the National Cancer Institute, September 2017, doi.org/10.1093/jnci/djx030.
20. Zheng J et al. American Institute for Cancer Research (AICR) 2011 *Research Conference on Food, Nutrition, Physical Activity and Cancer.* Presented November 14, 2016.
21. Romaguera D et al. Is concordance with World Cancer Research Fund/American Institute for Cancer Research guidelines for cancer prevention related to subsequent risk of cancer? Results from the EPIC study. *Am J Clin Nutr* July 2012;96:150–63, first published May 16, 2012, doi: 10.3945/ ajcn.111.031674.
22. Paur I et al. Tomato-based randomized controlled trial in prostate cancer patients: Effect on PSA. *Clin Nutr* 2017;36:672–79.
23. Chodak G. Supplement risk exposed by SELECT. March 3, 2014, www.medscape.com/viewarticle/821198?src=wnl_edit_specol&uac=61213SX.
24. Zoarsky NG t al. Men's health supplement use and outcomes in men receiving definitive intensity-modulated-radiation therapy for localized prostate cancer. *Am J Clin Nutr* 2016;104:1583–93.
25. Russo M et al. Phytochemicals in cancer prevention and therapy: Truth or dare? *Toxins* (Basel) 2010;2:517–55.
26. Chodak G. Supplement risk exposed by SELECT. March 3, 2014, www.medscape.com/viewarticle/821198?src=wnl_edit_specol&uac=61213SX.
27. Chandel NS et al. The promise and perils of antioxidants for cancer patients. *N Engl J Med* 2014;371:177–78. doi: 10.1056/NEJMcibr1405701.
28. Cruciferous vegetables and cancer prevention. June 2012, www.cancer.gov/cancertopics/factsheet/diet/cruciferous-vegetables, accessed 2/10/15.
29. Bohan Brown MM et al. Nutritional epidemiology in practice: Learning from data or promulgating beliefs? *Am J Clin Nutr* 2013;97:5–6.
30. Straus SE. Complementary and alternative medicine and cancer: What you should know. Available at: www.peoplelivingwithcancer.org, accessed October 2003.

Unit 23

1. *Lanzkowsky's Manual of Pediatric Hematology and Oncology* (6th ed.), 2016, 69–83, doi.org/10.1016/B978-0-12-801368-7.00006-5.
2. Madhur MS et al. Hypertension. Medscape, March 31, 2014, emedicine.medscape.com/article/241381-overview?src=wnl_ref_prac_publ&uac=61213SX, accessed 7/23/14.
3. Prasad AS. Discovery and importance of zinc in human nutrition. *Federation Proc* 1984;43:2829–34.
4. Thankachan P et al. Iron absorption in young Indian women: The interaction of iron status with the influence of tea and ascorbic acid. *Am J Clin Nutr* 2008;87:881–86.
5. *Bone Health and Osteoporosis: A Report of the Surgeon General.* Chapter 4: The Frequency of Bone Disease. Available at: www.surgeongeneral.gov/library/bonehealth/chapter_4.html, accessed 2/6/12.
6. Maximizing peak bone mass: Calcium supplementation increases bone mineral density in children. *Nutr Rev* 1992;50:335–37.
7. Nordin BEC. Calcium absorption revisited. *Am J Clin Nutr* 2010;92:673–74.
8. Aspray TJ et al. National Osteoporosis Society vitamin D guideline summary. *Age and Ageing* 2014;43:592–595, available at: www.medscape.com/viewarticle/831094_5?nlid=66125_905, accessed 9/23/14.
9. Cano A et al. Calcium in the prevention of postmenopausal osteoporosis: EMAS clinical guide. *Maturitas* 2018;107:7–12.
10. Orchard T et al. Dietary inflammatory index, bone mineral density and risk of fracture in postmenopausal women: Results from the Women's Health Initiative. *J Bone Miner Res* 2016, doi: 10.1002/jbmr.3070.
11. Gabel L et al. Physical activity, sedentary time, and bone strength from childhood to early adulthood: A mixed longitudinal HR-pQCT study. *J Bone Miner Res* 2017, doi: 10.1002/jbmr.3115.
12. de Jonge EAL et al. Dietary patterns explaining differences in bone mineral density and hip structure in the elderly: The Rotterdam Study. *Am J Clin Nutr* 2017;105:203–11.
13. USDA, Tables of nutrient retention in foods, Release 6 (2007). Available at: www.ars.usda.gov/Services/docs.htm?docid=9448.
14. Foods that fight inflammation. *Harvard Women's Health Watch,* 2017;www.health.harvard.edu/staying-healthy/foods-that-fight-inflammation
15. Bristow SM et al. Dietary calcium intake and rate of bone loss in men. *Br J Nutr* 2017;117:1432–38.
16. Gabel L et al. Physical activity, sedentary time, and bone strength from childhood to early adulthood: A mixed longitudinal HR-pQCT study. *J Bone Miner Res* 2017, doi: 10.1002/jbmr.3115.
17. Rehm CD et al. Dietary intake among U.S. adults, 1999–2012. *JAMA* 2016;315(23):2542-2553. doi: 10.1001/jama.2016.7491.
18. Vogel KA et al. The effect of dairy intake on bone mass and body composition in early pubertal girls and boys: A randomized controlled trial. *American Journal of Clinical Nutrition* 2017;105:1214–29,
19. Bolland M et al. Calcium intake and risk of fracture: Systematic review. *BMJ* 2015;351:h4580
20. Weaver CM et al. Choices for achieving adequate dietary calcium with a vegetarian diet. *Am J Clin Nutr* 1999;70:543S–548S.
21. Tai V et al. Calcium intake and bone mineral density: Systematic review and meta-analysis. *BMJ* 2015;351:h4183.
22. Bolland M et al. Calcium intake and risk of fracture: systematic review. *BMJ* 2015;351:h4580.
23. Collings R et al. The absorption of iron from whole diets: A systematic review.

Am J Clin Nutr, first published May 29, 2013, doi: 10.3945/ ajcn.112.050609.
24. Lopez A et al. Iron deficiency anemia. *N Engl J Med* 2016;387:907–16.
25. Stofzfus RJ. Iron interventions for women and children in low-income countries. *J Nutr* 2011;141:756S–62S.
26. Fairbanks VF. Iron in medicine and nutrition. In: Shils ME et al., eds. *Modern Nutrition in Health and Disease,* 9th ed. Philadelphia: Lippincott Williams & Wilkins, 1999:193–221.
27. Lozoff B. Early iron deficiency has brain and behavior effects consistent with dopaminergic dysfunction. *J Nutr* 2011;141:740S–6S.
28. Park J et al. Increased iron content of food due to stainless steel cookware. *J Am Diet Assoc* 1997;97:659–61.
29. Monsen ER et al. Estimation of available dietary iron. *Am J Clin Nutr* 1978;31:134–41.
30. Shannon M. Ingestion of toxic substances by children. *N Engl J Med* 2000;342:186–89.
31. Spanierman CS et al. Iron toxicity in emergency medicine. Available at: emedicine.medscape.com/article/81B213-overview, Accessed 3/7/12.
31. Calcium supplements up risk for precancerous serrated polyps. Medscape 3/18, www.medscape.com/viewarticle/893319?nlid=121049_3042&src=WNL_mdplsnews_180302_mscpedit_diab&uac=61213SX&spon=22&impID=1572267&faf=1
32. O'Donnoll M et al. Dietary sodium and cardiovascular risk. *NEJM* 2016;375:2404–06.
33. McDonough AA et al. Cardiovascular benefits associated with higher dietary K+ vs. lower dietary Na+: evidence from population and mechanistic studies. *Am J Endocrinol Metabo* 2017;312(4':E348–E356 doi: 10.1152/ajpendo.00453.2016.
34. Narula J et al. Salt intake and cardiovascular disease. *Eur Heart J* 2017;38:697–99.
35. Moore LL. "Low-sodium diet might not lower blood pressure: Findings from large, 16-year study contradict sodium limits in Dietary Guidelines for Americans." Experimental Biology Annual Meeting, Chicago, April 25, 2017.
36. Steinberg S et al. The DASH diet, 20 years later. *JAMA.* Published online March 9, 2017. doi: 10.1001/jama.2017.1628.
37. Rhee M-Y et al. Novel genetic variations associated with salt sensitivity in the Korean population. *Hyperten Res* 2011;34;606–11.
38. Steinberg S et al. The DASH diet, 20 years later *JAMA.* Published online March 9, 2017. doi: 10.1001/jama.2017.1628.
39. Soltani S et al. The effect of Dietary Approaches to Stop Hypertension (DASH) on serum inflammatory markers: A systematic review and meta-analysis of randomized trials. doi: dx.doi.org/10.1016/j.clnu.2017.02.018. Published online ahead of print: www.clinicalnutritionjournal.com/article/S0261-5614(17)30068-7/fulltext.
40. Fung TT et al. Adherence to a DASH-style diet and risk of coronary heart disease and stroke in women. *Arch Intern Med* 2008;168:713–20.
41. Schwingshackl L et al. Dietary quality as assessed by the healthy eating index, alternative healthy eating index, DASH score, and health outcomes: An updated systematic review and meta-analysis of cohort studies. *J Acad Nutr Diet* 2015;115:780–800.
42. Ostchega Y et al. Hypertension awareness, treatment, and control—continued disparity in adults United States 2005–2006. www.cdc.gov/nchs/data/databriefs/db03.pdf, accessed 2/14/15.
43. James PA et al. 2014 Evidence-based guideline for the management of high blood pressure in adults. Report from the Panel Members Appointed to the Eighth Joint National Committee. *JAMA* 2014;31:507–20, doi: 10.1001/jama.2013.284427.
44. Ingelfinger JR. Clinical practice: The child or adolescent with elevated blood pressure. *N Engl J Med* 2014;370:2316–25, doi: 10.1056/NEJMcp1001120.
45. Dugdale DC. Hypertension: Update. June 10, 2011.Available at: www.nlm.nih.gov/medlineplus/ency/article/000468.htm, accessed 3/7/12.
46. High blood pressure. October 2014, www.nlm.nih.gov/medlineplus/ency/article/000468.htm, accessed 10/23/14.
47. *What We Eat in America,* NHANES, 2007–2008. Available at: www.ars.usda.gov/Services/docs.htm?docid=13793, accessed 2/25/2011.
48. Dietary Guidelines Advisory Committee. *Scientific Report of the 2015 Dietary Guidelines Advisory Committee: Advisory report to the Secretary of Health and Human Services and Secretary of Agriculture,* published February 2015, www.health.gov/dietaryguidelines/2015-scientific-report/PDFs/Scientific-Report-of-the-2015-Dietary-Guidelines-Advisory-Committee.pdf, accessed 3/12/15.
49. Harnack LJ et al. Sources of sodium in US adults from 3 geographic regions. *Circulation* 2017;135:1775–83.
50. Bobowski N et al. Preference for salt in a food may be alterable without a low sodium diet. *Food Quality and Preference* 2015;39:40–45, doi: 10.1016/j.foodqual.2014.06.005.
51. Zhao J-G et al. Association between calcium or vitamin D supplementation and fracture incidence in community-dwelling older adults. A systematic review and meta-analysis. *JAMA* 2017;318(24):2466–82. doi: 10.1001/jama.2017.19344.

UNIT 24

1. Mole B. Vitamins, supplements effective at boosting call volume to poison centers. *ARS Technica,* July 25, 2017; https://arstechnica.com/science/2017/07/vitamins-supplements-effective-at-boosting-call-volume-to-poison-centers
2. Denham BE. Dietary supplements—regulatory issues and implications for public health. *JAMA,* July 5, 2011, doi: 10.1001/ jama.2011.982.
3. Cohen PA. The supplement paradox. *JAMA* 2016;316:1453–54.
4. Barrueto F et al. Herb poisoning. 2016; emedicine.medscape.com/article/817427-overview#a7
5. Rao N et al. An increase in dietary supplement exposures reported to US poison control centers. *J Med Toxicol,* July 24, 2017, available online at: https://link.springer.com/article/10.1007/s13181-017-0623-7
6. Duqueroy V. Dietary supplements for diabetes: What's new? Medscape, December 9, 2014, www.medscape.com/features/slideshow/dietary-supplements-for-diabetes?src=wnl_edit_specol&uac=61213SX, accessed 12/13/14.
7. Factsheet on the new proposed Nutrition Facts Label. www.fda.gov/Food/GuidanceRegulation/GuidanceDocumentsRegulatoryInformation/LabelingNutrition/ucm387533.htm, 2/27/14, accessed 10/27/14.
8. Duqueroy V. Dietary supplements for diabetes: What's new? Medscape, December 9, 2014, www.medscape.com/features/slideshow/dietary-supplements-for-diabetes?src=wnl_edit_specol&uac=61213SX, accessed 12/13/14.
9. Raynor DK, et al. Buyer beware? Does the information provided with herbal products available over the counter enable safe use? *BMC Med* 2011;9:94. doi: 10.1186/1741-7015-9-94.
10. Cassarett D. The science of choosing wisely—Overcoming the therapeutic illusion. *NEJM* 2016;374:1203–05.
11. Newmaster SG et al. DNA barcoding detects contamination and substitution in North American herbal products. *BMC Med* 2013;11:222, doi: 10.1186/1741-7015-11-222, published October 11, 2013, accessed 7/28/14.

12. Offit P. Hospital says no more dietary supplements in formulary. Medscape, October 22, 2014, www.medscape.com/viewarticle/812751?nlid=36983_1521&src=wnl_edit_medp_wir&uac=61213SX&spon=17, accessed 7/23/14.
13. Brown T. Supplement use stable, despite research questioning benefit. Medscape, October 13, 2016, www.medscape.com/viewarticle/870213?nlid=110162_439&src=WNL_mdplsfeat_161018_mscpedit_publ&uac=61213SX&spon=42&impID=1217573&faf=1.
14. Guallar E et al. Enough is enough: Stop wasting money on vitamin and mineral supplements. *Ann Intern Med* 2013;159:850–851, doi: 10.7326/0003-4819-159-12-201312170-00011.
15. Office of Disease Prevention and Health Promotion. *Scientific report of the 2015 Dietary Guidelines Advisory Committee.* Health.gov Aug 6, 2015, http://health.gov/dietaryguidelines/2015-scientific-report/10-chapter-5/d5-5.asp7.
16. Boggs W. Herbal medications risky in heart disease. Medscape, March 2, 2017, www.medscape.com/viewarticle/876576?nlid=113356_3042&src=WNL_mdplsnews_170310_mscpedit_diab&uac=61213SX&spon=22&impID=1305829&faf=1.
17. Rautiainen S et al. Multivitamin use and cardiovascular disease in a prospective study of women. *Am J Clin Nutr* 2015;101:144–53.
18. Park SY et al. Multivitamin use and the risk of mortality and cancer incidence. *Am J Epidemiol,* March 2011, doi: 10.1093/aje/kwq447.
19. Andrews KW et al. Analytical ingredient content and variability of adult multivitamin/mineral products: National estimates for the Dietary Supplement Ingredient Database. *Am J Clin Nutr* 2017105(2):526–39.
20. Hao P et al. Traditional Chinese Medicine for cardiovascular disease: Evidence and potential mechanisms. *Am Coll Cardiol* 2017, doi: 10.1016/j.jacc.2017.04.041.
21. Poisonous plants. www.cdc.gov/niosh/topics/plants, July 2014, accessed 2/19/15.
22. Brown AC. An overview of herb and dietary supplement efficacy, safety and government regulations in the United States with suggested improvements. Part 1 of 5 series. *Food and Chemical Toxicology* 2017;107:449–71.
23. Sen S. Revival, modernization and integration of Indian traditional herbal medicine in clinical practice: Importance, challenges and future. *J Trad Complem Med* 2017;7:234–44.
24. Bent S et al. Herbal medicines in the United States: Review of efficacy, safety, and regulations. *J Gen Intern Med* 2008;23:854–59, published online April 16, 2008, doi: 10.1007/s11606-008-0632-y.
25. Herbs and supplements. Medline Plus, December 2014, www.nlm.nih.gov/medlineplus/druginfo/herb_All.html, accessed 2/19/15.
26. Hulisz DT. Top herbal products: Efficacy and safety concerns. Medscape, January 2008, www.medscape.org/viewarticle/568235, accessed 2/19/15.
27. Herbs and supplements. Medline Plus. Available at: www.nlm.nih.gov/medlineplus/druginfo/herb_AII.html, accessed February 2012.
28. Newmaster SG et al. DNA barcoding detects contamination and substitution in North American herbal products. *BMC Med* 2013;11:222, doi: 10.1186/1741-7015-11-222, published October 11, 2013, accessed 7/28/14.
29. Untested stimulant found in dietary supplements: Study. Medscape, October 10, 2014, www.medscape.com/viewarticle/833056?nlid=67624_439&src=wnl_edit_medp_publ&uac=61213SX&spon=42.
30. Rao N et al. An increase in dietary supplement exposures reported to US poison control centers. *J Med Toxicol,* July 24, 2017, available online at: https://link.springer.com/article/10.1007/s13181-017-0623-7.
31. Rao N et al. An increase in dietary supplement exposures reported to US poison control centers. *J Med Toxicol,* July 24, 2017, available online at: https://link.springer.com/article/10.1007/s13181-017-0623-719.
32. Olafsdottir B et al. Dietary supplement use in the older population of Iceland and association with mortality. *Br J Nutr* 2017; doi.org/10.1017/S0007114517001313.
33. Koa RI. Adulterants in Asian patent medicines. *N Engl J Med* 1998;339:847.
34. Gilroy CM et al. Echinacea and truth in labeling. *Arch Intern Med* 2003;163:699–704.
35. Pray W. Dangers of sexual enhancement supplements. *US Pharmacist* 2007;32:10–15.
36. Pray W. Dangers of sexual enhancement supplements. *US Pharmacist* 2007;32:10–15.
37. USP dietary supplement standards. Available at: www.usp.org/dietary-supplements/overview, accessed February 2012.
38. Eslamparast T et al. Synbiotic supplementation in nonalcoholic fatty liver disease: A randomized, double-blind, placebo-controlled pilot study. *Am J Clin Nutr* 2014;99:535–42.
39. USP Verified Mark. 2015, www.usp.org/usp-verification-services/usp-verified-dietary-supplements/usp-verified-mark, accessed 2/20/15.
40. Probiotics: Topic overview. Accessed 8/20//17, www.webmd.com/digestive-disorders/tc/probiotics-topic-overv.
41. Haggans C. Office of dietary supplements. www.od.nih.gov, accessed 6/11/14.
42. Wallace TC et al. The safety of probiotics: Considerations following the 2011 U.S. Agency for Health Research and Quality Report. *J Am Coll Nutr* 2017;36(3):218–22.
43. Barengolts E et al. Gut microbiota, prebiotics, probiotics, and synbiotics in management of obesity and prediabetes: Review of randomized controlled trials. *Endocr Pract* 2016;22(10):1224–34.
44. Abutair AS et al. The effect of soluble fiber supplementation on metabolic syndrome profile among newly diagnosed type 2 diabetes patients. *Clin Nutr Res* 2018 January;7(1):31–39.
45. Raynor DK et al. Buyer beware? Does the information provided with herbal products available over the counter enable safe use? *BMC Med* 2011;9:94, doi: 10.1186/1741-7015-9-94.

UNIT 25

1. Kant AH et al. A prospective study of water intake and subsequent risk of all-cause mortality in a national cohort. *Am J Clin Nutr* 2017;105:212-220.
2. Sailer CO et al. Characteristics and outcomes of patients with profound hyponatraemia due to primary polydipsia. *Clin Endocrinol* 2017;87:492–99.
3. *Dietary Reference Intakes for Water, Potassium, Sodium Chloride, and Sulfate.* Food and Nutrition Board, National Academy of Science, http://books.nap.edu, accessed May 2004.
4. Sanders AE et al. Blood lead levels and dental caries in U.S. children who do not drink tap water. *Am J Prev Med* 2017;/dx.doi.org/10.1016/j.amepre.2017.09.0045.
5. Fluoridation of U.S. municipal water supplies. *MMWR* 2008;57:737–41.
6. Muckelbauer R et al. Association between water consumption and body weight outcomes: A systematic review. *Am J Clin Nutr,* June 26, 2013, doi: 10.3945/ajcn.112.055061.
7. Wong JM et al. Effects of advice to drink 8 cups of water per day in adolescents with overweight or obesity: A randomized clinical trial. *JAMA Pediatr* 2017;171:e170012.
8. Askew EW. Water. In: Ziegler EE, Filer LJ Jr., eds., *Present Knowledge in Nutrition.* Washington, DC: ILSI Press, 1996:98–108.

9. Kant AK et al. Intakes of plain water, moisture in foods and beverages, and total water in the adult US population—nutritional, meal pattern, and body weight correlates: National Health and Nutrition Examination Surveys 1999–2006. *Am J Clin Nutr* 2009; 90:655–63, available at: www.ncbi.nlm.nih.gov/pmc/articles/PMC2728648, accessed 2/20/15.
10. Picco M. Does drinking water during or after a meal disturb digestion? Available at: www.mayoclinic.com/health
11. Bossingham MJ et al. Water balance, hydration status, and fat-free mass hydration in younger and older adults. *Am J Clin Nutr* 2005;81:1342–50.
12. *Dietary Guidelines for Americans, 2010.* Available at: www.dietaryguidelines.gov.
13. Shmerling RH. How do you know if you're dehydrated? Available at: www.intelihealth.com/IH/ihtPrint/WSIHWOOO/24479/29730.html?hide=t&k=base, accessed 2/22/11.
14. Wingo JE. *Isolated Effects of Elevated Temperatures on Sweating in Humans.* Presented at Experimental Biology Annual Meeting, New Orleans, April 19, 2009.
15. Killer SC et al. No evidence of dehydration with moderate daily coffee intake: A counterbalanced cross-over study in a free-living population. *PLoS ONE* 2014;9:e84154, doi: 10.1371/journal.pone.0084154.
16. Goldman MB et al. Mechanisms of altered water metabolism in psychotic patients with polydipsia and hyponatremia. *N Engl J Med* 1988;318:397–403.
17. Goldman MB et al. Mechanisms of altered water metabolism in psychotic patients with polydipsia and hyponatremia. *N Engl J Med* 1988;318:397–403.
18. Cullen H. The future of water. *Discover,* December 2011;47–53.
19. A World Bank warning on water. *Washington Post,* May 13, 2016.
20. Myer SS et al. The coming health crisis. *The Scientist,* January 2011:32–37.
21. Summary of the Safe Drinking Water Act, 2015. www2.epa.gov/laws-regulations/summary-safe-drinking-water-act, accessed 2/21/15.
22. Ellis E. The Latest in Bottled Water Trends. www.foodandnutrition.org/Stone-Soup/July-2017/The-Latest-in-Bottled-Water-Trends, accessed 7/15/17.
23. Current research on the bottled water manufacturing industry. Available at: www.anythingresearch.com/industry/Bottled-Water-Manufacturing.htm, accessed 3/13/12.
24. Nassauer S. Water with some pop to it. *WSJ,* May 15, 2013: D1, D2.
25. Goldfarb S. Drinking water: What's the science? Medscape, June 18, 2014, www.medscape.com/viewarticle/826504?nlid=59723_454&src=wnl_edit_medp_peds&uac=61213SX&spon=9, accessed 6/24/14.
26. Hu Z et al. Environmental research and public health: Bottled water: United States consumers and their perceptions of water quality. *Int J Environ Res Public Health* 2011;8:565–78, doi: 10.3390/ijerph8020565.
27. Bisphenol A (BPA): Use in food contact application. January 6, 2015, www.fda.gov/NewsEvents/PublicHealthFocus/ucm064437.htm, accessed 2/21/15.
28. Quock RL et al. Fluoride content of bottled water and its implications for the genera dentist. *Gen Dent* 2009;57:29–33.
29. Tan ZY. It's in the water: The debate over fluoridation lives on. Medscape, September 26 2016; www.medscape.com/viewarticle/869269?nlid=109634_439&src=WNL_mdplsfeat_161004_mscpedit_publ&uac=61213SX&spon=42&impID=1209161&faf=1

UNIT 26

1. Church G. Compelling reasons for repairing human germlines. *NEJM* 2017;377:1909–11.
2. Stover PJ. Influence of human genetic variation on nutritional requirements. *Am J Clin Nutr* 2006;83(suppl):436S–42S.
3. Vogelstein B et al. The path to cancer—Three strikes you're out. *NEJM* 2015;373:1895–99.
4. Goni L et al. Future perspectives of personalized weight loss interventions based on nutrigenetic, epigenetic, and metagenomics data. *J Nutr* 2016;46(suppl):905S–12S.
5. Sattar N et al. Reverse causality in cardiovascular epidemiological research: More common than imagined? *Circulation* 2017;135:2369–72.
6. Genetic home reference. NIH, National Library of Medicine, 2018; https://ghr.nlm.nih.gov
7. Understanding genetics. A New York Mid-Atlantic Guide. Washington, DC: Genetic Alliance 2009.
8. Human Genome Project Information. September 19, 2011, available at: www.ornl.gov/sci/techresources/Human_Genome/faq/faqs1.shtml#genetics, accessed 3/18/12.
9. Frequently asked questions about genetic and genomic science. February 14, 2014, www.genome.gov/19016904, accessed 2/22/15.
10. Genetic home reference. ghr.nlm.nih.gov/glossary, accessed 2/24/15.
11. FDA allows marketing of first direct-to-consumer tests that provide genetic risk information for certain conditions. April 16, 2017; www.fda.gov/NewsEvents/Newsroom/PressAnnouncements/ucm551185.htm
12. Camp KM et al. Position of the Academy of Nutrition and Dietetics: Nutritional genomics. *J Acad Nutr Diet* 2014;114:299–312, doi: 10.1016/j.jand.2013.12.001.
13. Bookman EB et al. Gene-environment interplay in common complex diseases: Forging an integrative model—recommendations from an NIH workshop. *Genet Epidemiol* 2011;35:217–25, doi: 10.1002/gepi.20571, accessed 2/22/15.
14. Moore JP et al. Proteomics and systems biology: Current and future applications in the nutritional sciences. *AdvNutr* 2011;2:355–64.
15. Vaiserman AM. Early-life nutritional programming of type 2 diabetes: Experimental and quasi-experimental evidence. *Nutrients* 2017, 9(3), 236; doi: 10.3390/nu9030236.
16. Yao P et al. Effects of genetic and nongenetic factors on total and bioavailable 25(OH)D: Responses to vitamin D supplementation. *J Clin Endocrin* 2017;102:100–10.
17. Sawaengsri H et al. Transcobalamin 776C→G polymorphism is associated with peripheral neuropathy in elderly individuals with high folate intake. *Am J Clin Nutr* 2016;104:1665–70.
18. Tall AR. Increasing lipolysis and reducing atherosclerosis. *N Engl J Med* 2017;377:280–83.
19. Chen G et al. Genome-wide analysis identifies an African-specific variant in SEMA4D associated with body mass index. *Obesity,* 2017. doi: 10.1002/oby.21804.
20. Goni L et al. Future perspectives of personalized weight loss interventions based on nutrigenetic, epigenetic, and metagenomics data. *J Nutr* 2016;46(suppl):905S–12S.
21. Graff M et al. Genome-wide physical activity interactions in adiposity—A meta-analysis of 200,452 adults. *PLOS Genetics,* April 27, 2017, doi.org/10.1371/journal.pgen.1006528.
22. King JC. A summary of pathways or mechanisms linking preconception maternal nutrition with birth outcomes. *J Nutr* 2016;146(Suppl);1437S–44S.
23. Shulman GI. Ectopic fat in insulin resistance, dyslipidemia, and cardiometabolic disease. *N Engl J Med* 2014;371:2236–38, doi: 10.1056/NEJMc1412427.
23. Loos RJF et al. The bigger picture of FTO—the first GWAS-identified obesity gene. *Nat Rev Endocrinol* 2014:51–61.

24. West AA et al. Applied choline-omics: Lessons from human metabolic studies for the integration of genomics research into nutrition practice. *J Acad Nutr Diet* 2014;114(8):1242–50, doi: 10.1016/j.jand.2013.12.012.
25. Lee C et al. Structural genomic variation and personalized medicine, *N Engl J Med* 2008;358:740–42.
26. Joffe YT et al. The relationship between dietary fatty acids and inflammatory genes on the obese phenotype and serum lipids. *Nutrients* 2013;5:1672–1705, doi: 10.3390/nu5051672.
27. Brissot P et al. Molecular diagnosis of genetic iron-overload disorders. *Expert Rev Mol Diagn* 2010;10:755–63.
28. What is sitosterolemia? ghr.nlm.nih.gov/condition/sitosterolemia, accessed 2/24/15.
29. What is galactosemia? ghr.nlm.nih.gov/condition/galactosemia, accessed 2/25/15.
30. Turki A et al. The indicator amino acid oxidation methods with use of L-[113C] leucine suggests a higher than currently recommended protein requirement in children with PKU. *J Nutr* 2017;147:211–17.
31. Temelkova-Kurktschiev T et al. Lifestyle and genetics in obesity and type 2 diabetes. *Exp Clin Endocrinol Diabetes* 2012;120:1–6.
32. Drewnowski A et al. Genetic taste markers and preferences for vegetables and fruit of female breast care patients. *J Am Diet Assoc* 2000;100:191–97.
33. Feeney E et al. Genetic variation in taste perception: Does it have a role in healthy eating? *Proc Nutr Soc* 2011;70:135–43.
34. Kormaroff AL. Gene editing using CRISPR: Why the excitement? *JAMA* 2017;318:699–700.
35. Vassy JL et al. The impact of whole-genome sequencing on the primary care and outcomes of healthy adult patients: A pilot randomized trial. *Ann Intern Med*, June 27, 2017, doi: 10.7326/M17-0188.
36. Xia Li et al. Cats lack a sweet taste receptor. *J Nutr* 2006;136:1932S–34S.
37. McCall B. New genes hold clue to why many with obesity don't get diabetes. Medscape, 2018; www.medscape.com/viewarticle/894098?nlid=121473_439&src=WNL_mdplsfeat_180327_mscpedit_publ&uac=61213SX&spon=42&impID=1591747&faf=1

UNIT 27

1. *Dietary Guidelines for Americans*, 8th ed. December 2015. Available at https://health.gov/dietaryguidelines/2015/guidelines
2. Dietary Guidelines Advisory Committee, *Scientific Report of the 2015 Dietary Guidelines: Advisory Report to the Secretary of Health and Human Services and Secretary of Agriculture*, published February 2015, www.health.gov/dietaryguidelines/2015-scientific-report/PDFs/Scientific-Report-of-the-2015-Dietary-Guidelines-Advisory-Committee.pdf, accessed 3/12/15.
3. Michaelsson K et al. Leisure physical activity and the risk of fracture in men. *PLoS Med*, June 4, 2007;(6):e199, doi: 10.1371/ journal.pmed.0040199.
4. Hankinson AL et al. Maintaining a high physical activity level over 20 years and weight gain. *JAMA* 2010;304:2603–10.
5. Villareal DT et al. Weight loss, exercise, or both and physical function in obese older adults. *N Engl J Med* 2011;362:1218–29.
6. King NA et al. Dual-process action of exercise on appetite control: Increase in orexigenic drive but improvement in meat induced satiety. *Am J Clin Nutr* 2009;90:921–27.
7. Hall KD et al. Energy balance and its components: Implications for body weight regulation. *Am J Clin Nutr* 2012;95:989–94.
8. Werle COC et al. Is it fun or exercise? The framing of physical activity biases subsequent snacking. *Marketing Letters*, May 2014, doi: 10.1007/s11002-014-9301-6.
9. *Physical Activity Guidelines for Americans*, 2008. U.S. Department of Health and Human Service, available at: www.health.gov/PAGuidelines, accessed October 2008.
10. Global strategy on diet, physical activity and health. WHO, 2012, available at: www.who.int/dietphysicalactivity/pa/en/index.htm, accessed 3/21/12.
11. Garber C et al. Quantity and quality of exercise for developing and maintaining cardiorespiratory, musculoskeletal, and neuromotor fitness in apparently healthy adults: Guidance for prescribing exercise. *Med Sci Sports Exer* 2011;43:1334–59.
12. Target heart rate and estimated maximum heart rate. Available at: www.cdc.gov/physicalactivity/everyone/measuring/heartrate.html, accessed 2/27/15.
13. Target heart rates. January 8, 2015, www.heart.org/HEARTORG/GettingHealthy/PhysicalActivity/FitnessBasics/Target-Heart-Rates_UCM_434341_Article.jsp, accessed 2/27/15.
14. Maximum heart rate. January 2015, www.brianmac.co.uk/maxhr.htm, accessed 2/28/15.
15. Graff M et al. Genome-wide physical activity interactions in adiposity—A meta-analysis of 200,452 adults. *PLOS Genetics*, April 27, 2017, doi.org/10.1371/journal.pgen.1006528.
16. Shcherbina A et al. Accuracy in wrist-worn, sensor-based measurements of heart rate and energy expenditure in a diverse cohort. *Journal of Personalized Medicine* 2017;7;2:3; doi: 10.3390/jpm7020003.
17. Case MA et al. Accuracy of smartphone applications and wearable devices for tracking physical activity data. *JAMA* 2015;13:625–26, doi: 10.1001/jama.2014.17841.
18. Heart-rate monitors. 2013, www.consumerreports.org/cro/heart-rate-monitors/buying-guide.htm, accessed 2/28/15.
19. Lyons EJ et al. Behavior change techniques implemented in electronic lifestyle activity monitors: A systematic content analysis. *J Med Internet Res* 2014;16(8):e192), doi: 10.2196/jmir.3469.
20. Freeman D. The truth about heart rate and exercise. October 23, 2009, www.webmd.com/fitness-exercise/the-truth-about-heart-rate-and-exercise?page=1, accessed 2/28/15.
21. Warm up, cool down. American Heart Association. April 21, 2015. www.heart.org/HEARTORG/HealthyLiving/PhysicalActivity/FitnessBasics/Warm-Up-Cool-Down_UCM_430168_Article.jsp#.WKuNUFUrKcx, Accessed February 20, 2017.
22. Herbert RD et al. Effects of stretching before and after exercising on muscle soreness and risk of injury: Systematic review. *BMJ* 2002;325(7362):468.
23. Page P. Current concepts in muscle stretching for exercise and rehabilitation. *Int J Sports Phys Ther* 2012;7:109–19.
24. Watson S. What causes muscle soreness after exercise? Available at: www.webmd.com/fitness-exercise/features/art-sore-muscles-joint-pain, accessed 3/22/12.
25. Horton ES. Metabolic fuels, utilization, and exercise. *Am J Clin Nutr* 1989;49(suppl):931–32.
26. Jones NL et al. Exercise limitations in health and disease. *N Engl J Med* 2000;343:632–41.
27. Position of the American Dietetic Association, Dietitians of Canada, and the American College of Sports Medicine: Nutrition and physical performance. *J Am Diet Assoc* 2009;109:509–27.
28. Tanaka H et al. J Age-predicted maxima heart rate revisited. *Am Coll Cardiol* 2001;37:153–56.

29. Ho SS et al. The effect of 12 weeks of aerobic, resistance or combination exercise training on cardiovascular risk factors in the overweight and obese in a randomized trial. *BMC Public Health* 2012;12:704.
30. Morey MC et al. Medical assessment for health advocacy and practical strategies for exercise initiation. *Am J Preventive Med* 2003;29:1168–74.
31. Physical activity facts. Available at: www.cdc.gov/healthyyouth/physicalactivity/facts.htm, accessed 3/24/12.
32. Matthews CE et al. Accelerometer-measured dose-response for physical activity, sedentary time, and mortality in US adults. *Am J Clin Nutr* 2016 ;104:1424–32.
33. O'Donovan G, et al. Association of "weekend warrior" and other leisure time physical activity patterns with risks for all-cause, cardiovascular, and cancer mortality. *JAMA Intern Med* 2017;177:335–42.
34. Keating SE et al. Continuous exercise but not high intensity interval training improves fat distribution in overweight adults. *J Obes,* 2014, published online January 19, 2014, doi: 10.1155/2014/834865.
35. Ho SS et al. The effect of 12 weeks of aerobic, resistance or combination exercise training on cardiovascular risk factors in the overweight and obese in a randomized trial. *BMC Public Health* 2012;12:704.
36. Lauer EE et al. Meeting USDHHS Physical Activity Guidelines and Health Outcomes. *Int J Exerc Sci* 2017;10;1:121–27.
37. Witkowski S et al. Enhancing treatment for cardiovascular disease. *Exerc Sports Sci Rev* 2011;39:93–101.

UNIT 28

1. Position of the Academy of Nutrition and Dietetics, Dietitians of Canada and the American College of Sports Medicine: Nutrition and athletic performance. *J Am Assoc Nutr Diet* 2016;116:501–28.
2. Barnes K et al. An international comparison of nutrition education standards, occupational standards and scopes of practice for personal trainers. *Hum Kinetics J* 2017;27:507–19.
3. Beals JW et al. Anabolic sensitivity of postprandial muscle protein synthesis to the ingestion of a protein-dense food is reduced in overweight and obese young adults. *Am J Clin Nutr.* 2016;104:1014–22.
4. Pasiakos SM et al. Protein supplementation in U.S. military personnel. *J Nutr* 2013;143:1815S–19S.
5. Burke LM. Nutrition strategies for the marathon. *Sports Med* 2007;37:344–47.
6. Cermak NM et al. The use of carbohydrates during exercise as an ergogenic aid. *Sports Med* 201;43:1139–55.
7. Quinn E. What causes the "bonk" and hitting the wall during exercise? December 15, 2014, sportsmedicine.about.com/od/Exercise-Metabolism-Energy/a/Bonking-Hitting-The-Wall.htm, accessed 3/7/15.
8. *What We Eat in America,* NHANES, 2007–08. Available at: www.ars.usda.gov/Services/docs.htm?docid=13793, accessed 2/25/11.
9. Position of the American Dietetic Association, Dietitians of Canada, and the American College of Sports Medicine: Nutrition and physical performance. *J Am Diet Assoc* 2009;109:509–27.
10. Aragon AA et al. Nutrient timing revisited: Is there a post-exercise anabolic window? *J Int Soc Sports Nutr* 2013;10:5, doi: 10.1186/1550-2783-10-5.
11. Buell JL et al. National Athletic Trainers Association position statement: Evaluation of dietary supplements for performance nutrition. *J Athl Train* 2013;48:124–36.
12. Schneid S. Thermoregulation. Presented at the Experimental Biology Annual Meeting, New Orleans, April 19, 2009.
13. Dehydration. www.merckmanuals.com/home/hormonal_and_metabolic_disorders/water_balance/dehydration.html, accessed 3/7/15.
14. Helman RS et al. Heat stroke treatment & management. April 7, 2014, emedicine.medscape.com/article/166320-treatment, accessed 3/8/15.
15. Sinert RH et al. Fast Five Quiz: Are you familiar with key elements regarding heat stroke? August 2, 2017, http://reference.medscape.com/viewarticle/883580?src=wnl_fastquiz_170812_mscpref&uac=61213SX&impID=1409319
16. DiMarco NM et al. A-Z of nutritional supplements: Dietary supplements, sports nutrition foods and ergogenic aids for health and performance—part 30. *Brit J Sports Med* 2012;46:299–300.
17. Temesi J et al. Carbohydrate ingestion during endurance exercise improves performance in adults. *J Nutr* 2011;141:890–97.
18. Lloyd T et al. Interrelationships of diet, athletic activity, menstrual status, and bone density in collegiate women. *Am J Clin Nutr* 1987;46:681–84.
19. Ireland ML. *The Female Athlete.* Philadelphia: W. B. Saunders, 2003.
20. Drakh A et al. Low energy availability in female athletes. emedicine.medscape.com/article/312312-overview.1/16/14.
21. Palacio LE et al. Nutrition for the female athlete. October 31, 2008, available at: http://emedicine.medscape.com/article/108994, accessed 9/28/11.
22. Weight loss in wrestlers. ACSM Sports Md Basics 2016; www.acsm.org/docs/default-source/sports-medicine-basics/basics_wrestlers-weight-loss.pdf?sfvrsn=2
23. Case HS et al. Weight loss in wrestlers. www.acsm.org/docs/current-comments/weightlossinwrestlers.pdf, accessed 3/8/15.
24. Pasricha R et al. Iron supplementation benefits physical performance in women of reproductive age: A systematic review and meta-analysis. *J Nutr* 2014;144: 906–14.
25. Sinclair LM et al. Prevalence of iron deficiency with and without anemia in recreationally active men and women. *J Am Diet Assoc* 2005;105:975–78.
26. Case HS et al. Weight loss in wrestlers. www.acsm.org/docs/current-comments/weightlossinwrestlers.pdf, accessed 3/8/15.
27. 192 banned performance enhancing substances and methods, 2014. sportsanddrugs.procon.org/view.resource.php?resourceID=002037, accessed 3/4/15.
28. A to Z of nutritional supplements: Conclusions—What can athletes do? *Br J Sports Med* 2011;45:752–54.
29. Hartgens F et al. Effects of androgenicanabolic steroids in athletes. *Sports Med* 2004;34:513–54.
30. Spriet LL. Exercise and sport performance with low doses of caffeine. *Sports Med* 2014;44(suppl 2):S175–84.
31. Kreider RB et al. ISSN exercise and sport nutrition review: Research and recommendations. *J Int Soc Sports Nutr,* February 2, 2010;7:7. doi: 10.1186/1550-2783-7-7.
32. Morton RW et al. Systematic review. Meta-analysis and mega-regression of the effect of protein supplementation on resistance training-induced gains in muscle mass and strength on healthy adults. *Br J Sports Med* 2017; doi.org/10.1136/bjsports-2017-097608.

UNIT 29

1. MacDorman MF et al. the United States maternal mortality rate increasing? Disentangling trends from measurement issues. *Obstet Gynecol* 2016;128: 447–55.
2. Riddell CA et al. Trends in differences in US mortality rates between black and white infants. *JAMA Pediatr.* Published online July 3, 2017. doi: 10.1001/jamapediatrics.2017.1365.

3. MacDorman MF et al. Annual summary of vital statistics. *Pediatrics* 2002;110:1037–50.
4. King JC. A summary of pathways or mechanisms linking preconception maternal nutrition with birth outcomes. *J Nutr* 2016;146(Suppl);1437S-44S.
5. Good health before pregnancy: Preconception care. ACOG 2017, www.acog.org/Patients/FAQs/Good-Health-Before-Pregnancy-Preconception-Care
6. Danielewicz H et al. Diet in pregnancy—more than food. *Euro J Prediatr* 2017;176:1573-9.
7. Rosso P. *Nutrition and Metabolism in Pregnancy.* New York: Oxford University Press, 1990.
8. Zeisel SH. Is maternal diet supplementation beneficial? Optimal development of infant depends on mother's diet. *Am J Clin Nutr* 2009;89:685S–687S.
9. Ladipo OA. Nutrition in pregnancy: Mineral and vitamin supplements. *Am J Clin Nutr* 2000;72(suppl):280S–90S.
10. Miller DR. Vitamin excess and toxicity. In: Hathcock JN, ed., *Nutritional Toxicology.* New York: Academic Press, 1982:81–131.
11. Rasmussen KM et al. *Weight Gain during Pregnancy: Reexamining the Guidelines.* Institute of Medicine and National Research Council Committee to Reexamine IOM Pregnancy Weight Guidelines; Washington, DC: National Academies Press, 2009.
12. Frederick IO et al. Pre-pregnancy body mass index, gestational weight gain, and other maternal characteristics in relation to infant birth weight. *Matern Child Health* J 2008;12:557–67.
13. Ohlin A et al. Maternal body weight development after pregnancy. *Int J Obesity* 1990;14:159–73.
14. Siega-Riz AM. A systematic review of outcomes of maternal weight gain according to the Institute of Medicine recommendations: Birth weight, fetal growth, and postpartum weight retention. *Am J Obstet Gynecol* 2009;201:339.e1-14.
15. Miller DR. Vitamin excess and toxicity. In: Hathcock JN, ed. *Nutritional Toxicology.* New York: Academic Press, 1982:81–131.
16. Schwarzenberg SJ et al. Advocacy for improving nutrition in the first 1000 days to support childhood development and adult health. *Pediatrics* 2018; http://pediatrics.aappublications.org/content/pediatrics/early/2018/01/18/peds.2018;-3716.full.pdf
17. Dubois DL et al. Adequacy of nutritional intake from food and supplements in a cohort of pregnant women in Québec, Canada: the 3D Cohort Study (Design, Develop, Discover). *Am J Clin Nutr* 2017;106:541-548.
18. *Energy, Carbohydrate, Fiber, Fatty Acids, Cholesterol, Protein, and Amino Acids: Dietary Reference Intakes.* Washington, DC: National Academies Press, 2002.
19. Zeisel SH. Is maternal diet supplementation beneficial? Optimal development of infant depends on mother's diet. *Am J Clin Nutr* 2009;89:685S–687S.
20. Crider KS et al. Folic acid food fortification—its history, effect, concerns, and future directions. *Nutrients* 2011;3:370–384.
21. Chen LW et al. Maternal folate status, but not that of vitamins B-12 or B-6, is associated with gestational age and preterm birth risk in a multiethnic Asian population. *J Nutr* 2015;145:113–20.
22. Williams J et al. Updated estimates of neural tube defects prevented by mandatory folic acid fortification—United States, 1995–2011. *MMRWeekly,* January 16, 2015;64(01):1–5, available at: www.cdc.gov/mmwr/preview/mmwrhtml/mm6401a2.htm?s_cid=mm6401a2_w#fig.
23. Pfeiffer CM et al. Estimation of trends in serum RBC folate in the U.S. population from pre- to post-fortification using assay-adjusted data from the NHANES 1998–2010. *J Nutr* 2012;142:886–93.
24. Dietary reference intakes for vitamin A through zinc. Available at: www.nap.edu.
25. Brown JE et al. Predictors of red cell folate level in women attempting pregnancy. *JAMA* 1997;277:548–52.
26. Manson JE et al. Vitamin D deficiency—Is there really a pandemic? *NEJM* 2016;375:181720.
27. Abel MH et al. Suboptimal maternal iodine intake is associated with impaired child neurodevelopment at 3 years of age in the Norwegian Mother and Child Cohort Study. *J Nutr* 2017 July;147:1314-1324.
28. Pearce EN et al. Consequences of iodine deficiency and excess in pregnant women: An overview or current knowns and unknowns. *Am J Clin Nutr* 2016;104(Suppl);918S-23S.
29. Brannon PM et al. Iron supplementation during pregnancy and infancy: Uncertainties and implications for research and policy. *Nutrients* 2017, 9(12), 1327; doi: 10.3390/nu9121327.
30. Viswanathan V et al. Folic acid supplementation for the prevention of neural tube defects-An updated evidence report and systematic review for the US Preventive Services Task Force. *JAMA* 2017;317(2):190–203.
31. Gomez-Arango LF et al. Low dietary fiber intake increases *Collinsella* abundance in the gut microbiota of overweight and obese pregnant women. *Gut Microbes* 2017; doi.org/10.1080/19490976.2017.1406584.
32. Van Heertum K et al. Alcohol and fertility: How much is too much? *Fertil Res Pract* 2017; 3: 10, 2017 July 10. doi: 10.1186/s40738-017-0037-x.
33. Pyror JR et al. Pregnancy intention and maternal alcohol consumption. *Obstet Gynecol* 2017;129:727–33.
34. May PA et al. Prevalence of fetal alcohol spectrum disorders in 4 US communities. *JAMA* 2018;319:474–482, 448–449.
35. May PA et al. Prevalence of fetal alcohol spectrum disorders in 4 US communities. *JAMA* 2018;319:474–482, 448–449.
36. Hoyme HE et al. Updated clinical guidelines for diagnosing fetal alcohol spectrum disorders. *Pediatrics* 2016;138:e20154256
37. Bower C et al. New opportunities for evidence in fetal alcohol spectrum disorder. *JAMA Pediatr.* Published online June 5, 2017. doi: 10.1001/jamapediatrics.2017.1342.
38. Pearlman M et al. The association between artificial sweeteners and obesity. *Current Gastroent Rpts* 2–17; doi.org/10.1007/s11894-017-0602-1.
39. Zhu Y et al. Maternal consumption of artificially sweetened beverages during pregnancy, and offspring growth through 7 years of age: A prospective cohort study. *Am J Epidemiol* 2017;46:1499–1508.
40. Menezes FP et al. Chapter 22 – Caffeine. *Repro Developmental Toxicol* (Second Edition), 2017; pp. 399–411.
41. Basco TW Jr. Prenatal caffeine and ADHD: Is there a link? Medscape, January 19, 2018. www.medscape.com/viewarticle/890713?nlid=120149_905&src=WNL_mdplsfeat_180123_mscpedit_obgy&uac=61213SX&spon=16&impID=1540954&faf=1
42. Mandrola J. Enough with the coffee research and other distractions. Medscape, August 2, 2017; http://www.medscape.com/viewarticle/883709?nlid=117112_3381&src=WNL_mdplsnews_170804_mscpedit_fmed&uac=61213SX&spon=34&impID=1403321&faf=1#vp_2
43. Poole R et al. Coffee consumption and health: Umbrella review of meta-analyses of multiple health outcomes. *BMJ* 2017;359:j502
44. Knight CA et al. Beverage caffeine intake in US consumers and subpopulations of interest: Estimates from the Share of

Intake Panel survey. *Food Chem Toxicol* 2004;42:1923–30.
45. Asayama K et al. The impact of salt intake during and after pregnancy. *Hyperten Res* 2018;41:1–5.
46. 2017 EPA-FDA Advice about Eating Fish and Shellfish. https://www.epa.gov/fish-tech/2017-epa-fda-advice-about-eating-fish-and-shellfish, accessed 2/9/17.
47. Koutelidakis AE et al. Higher adherence to Mediterranean diet prior to pregnancy is associated with decreased risk for deviation from the maternal recommended gestational weight gain. *Int J Food Sci Nutr,* June 14, 2017, doi.org/10.1080/09637486.2017.1330403.
48. Mennella JA et al. Learning to like vegetables during breastfeeding: A randomized clinical trial of lactating mothers and infants. *Am J Clin Nutr* 2017;106:67–76.
49. Streym S et al. Vitamin D content in human breast milk. *Am J Clin Nutr* 2016;103:107–14.
50. Munblit D et al. Allergy prevention by breastfeeding: Possible mechanisms and evidence from human cohorts. *Curr Opin Allergy Clin Immunol* 2016;16(5):427–433.
51. Fewtrelll M et al. Complementary feeding: A position paper by the European Society for Paediatric Gastroenterology, Hepatology, and Nutrition (ESPGHAN) Committee on Nutrition. *JPGN,* 2017; i: doi: 10.1097/MPG.0000000000001454.
52. Netting MJ et al. An Australian consensus on infant feeding guidelines to prevent food allergy: Outcomes from the Australian Infant Feeding Summit. *J Allergy Clin Immunol* 2017;5:1617–24.
53. West AS et al. Folate-status response to a controlled folate intake in nonpregnant, pregnant, and lactating women. *Am J Clin Nutr* 2012;96:789–800.
54. Dibley MJ et al. Safety and toxicity of vitamin A supplements in pregnancy. *Food Nutr Bulletin* 2001;22:248–66.
55. Meyers DG et al. Safety of antioxidant vitamins. *Arch Intern Med* 1996;156:925–35.
56. Steinburg BJ. *Women's Oral Health Issues.* Presented at the California Dental Association Meeting, September 13, 2000, available at: www.cda.org/Library/cda_member/pubs/journal/jour0900/women.html, accessed 4/1/12.
57. Meertens LJE et al. Should women be advised to use calcium supplements during pregnancy? A decision analysis. *Matern Child Nutr* 2018; doi.org/10.1111/mcn.12479.
58. Manson JE. Vitamin and mineral supplements: Which patients will benefit? Medscape Ob/GYN, 2018, www.medscape.com/viewarticle/892051?nlid=120554_3866&src=WNL_mdplsfeat_180206_mscpedit_card&uac=61213SX&spon=2&impID=1552318&faf=1
59. Mulligan ML et al. Implications of vitamin D deficiency in pregnancy and lactation. *Am J Obstet Gynecol* 2009;202:429.e1-429.e9.
60. *Calcium and Vitamin D: Dietary Reference Intakes. Food and Nutrition Board,* National Academy of Sciences, Washington, DC: National Academies Press, 2011.
61. Mei Z et al. Assessment of iron status in U.S. pregnant women from the National Health and Nutrition Examination Survey (NHANES), 1999–2006. *Am J Clin Nutr* 2011;93:1312–20.
62. Ribot B et al. Depleted iron stores without anaemia early in pregnancy carries increased risk of lower birthweight even when supplemented daily with moderate iron. *Hum Reprod* 2012;27:1260–6.
63. Zimmermann MB. Iodine deficiency in pregnancy and the effects of maternal iodine supplementation on the offspring: A review. *Am J Clin Nutr* 2009;89:668S–72S.
64. Utiger RD. Iodine nutrition: More is better. *N Engl J Med* 2006;354:2819–21.
65. Practice paper of the Academy of Nutrition and Dietetics. Abstract: Nutrition and lifestyle for a healthy pregnancy outcome. *J Acad Nutr Diet* 2014;114:1447. Full practice paper is available to American Academy of Nutrition and Dietetics members.
66. Bartick M et al. The burden of suboptimal breastfeeding in the United States: A pediatric cost analysis. *Pediatrics* 2010;125:e1048–56.
67. Position of the Academy of Nutrition and Dietetics: Promoting and supporting breastfeeding. *J Acad Nutr Diet* 2015;115:444–449.
68. Allen LH. Adequacy of family foods for complementary feeding. *Am J Clin Nutr* 2012;95:785–6.
69. Simopoulos AP et al. Factors associated with the choice and duration of infant-feeding practices. *Pediatrics* 1984;74(suppl):603–14.
70. Vega CP et al. USPSTF endorses updated breastfeeding recommendations. 2016, www.medscape.org/viewarticle/871710?nlid=111283_2709&src=wnl_cmemp_161212_mscpedu_peds&uac=61213SX&impID=1252362&faf=1
71. Breastfeeding and the use of human milk. *Pediatrics,* published online February 27, 2012, doi: 10.1542/peds.2011-3552.
72. Breastfeeding Report Card 2014. www.cdc.gov/breastfeeding/data/reportcard.htm, accessed 3/9/15.
73. McDowell MM et al. *Breast-Feeding in the United States: Findings from NHANES, 1999–2006.* NCHS Data Brief No. 5, April 2008.
74. Committee on Nutrition, American Academy of Pediatrics. Breastfeeding and the use of human milk. *Pediatrics* 2005;115:496–506.
75. National Academy of Sciences (Institute of Medicine). *Nutrition during Lactation.* Washington, DC: National Academy Press, 1991.
76. Lovelady CA et al. The effect of weight loss in overweight, lactating women on the growth of their infants. *N Engl J Med* 2000;342:449–53.
77. Abel MH et al. Suboptimal maternal iodine intake is associated with impaired child neurodevelopment at 3 years of age in the Norwegian Mother and Child Cohort Study. *J Nutr* 2017 July;147:1314–24.
78. Mattson SN et al. Heavy prenatal alcohol exposure with or without physical features of fetal alcohol syndrome leads to IQ deficits. *J Pediatr* 1997;131:718–21.
79. Mennella J. Alcohol's effect on lactation. pubs.niaaa.nih.gov/publications/arh25-3/230-234.htm, accessed 3/17/15.
80. Grossman X et al. Neonatal weight loss at a US baby-friendly hospital. *Am Acad Nutr Diet* 2012;112:410–3.
81. The WHO child growth standards. Available at: www.who.int/childgrowth/publications/en, accessed 4/2/12
82. WHO growth standards are recommended for use in the U.S. for infants and children 0 to 2 years of age. Available at: http://www.cdc.gov/growthcharts/who_charts.htm.
83. Rappo PD et al. Pediatrician's responsibility for infant nutrition. *Pediatrics* 1997;99:749–50.
84. Lloyd-Still JD et al. Intellectual development after severe malnutrition in infancy. *Pediatrics* 1974;54:306–11.
85. Macknin ML et al. Infant sleep and bedtime cereal. *Arch Periatr Adol Med* 1989;143:1066–8.
86. Beal VA. Termination of night feeding in infancy. *J Pediatr* 1969;75:690–2.
87. Nutrition for healthy term infants: Recommendations from birth to 6 months. *Canadian J Dent Prac Res* 2012; Winter,73(4)204.
88. Markowitz DL. Fluoride supplementation: The ongoing debate. Available at: www.medscape.com/viewarticle/753193?src=mp&spon=17, accessed 2/18/12.
89. Wagner CL et al. Prevention of rickets and vitamin D deficiency in infants,

children, and adolescents. *Pediatrics* 2008;122:1142–52.
90. Davis C. Self-selection of diets by newly weaned infants: An experimental study. *Am J Dis Child* 1928;36:651–79.
91. Birch LL et al. Caloric compensation and sensory specific satiety: Evidence for self-regulation of food intake by young children. *Appetite* 1986;7:323–31.
92. Birch LL et al. Learning to eat: Birth to age 2 y. *Am J Clin Nutr* 2014;99(suppl):723S–8S.
93. Kong KL et al. Origins of food reinforcement in infants. *Am J Clin Nutr* 2015;101:515–22.
94. Pimpin L et al. Dietary intake of young twins: Nature or nurture? *Am J Clin Nutr* 2013;98:1326–34.
95. Story M et al. Do young children instinctively know what to eat? The studies of Clara Davis revisited. *N Engl J Med* 1987;316:103–06.
96. Johansen AMW et al. Maternal dietary intake of vitamin A and risk of oral-facial clefts: A population-based case-control study in Norway. *Am J Epidemiol* 2008;167:1164–70.
97. Position of the Academy of Nutrition and Dietetics: Nutrition lifestyles for a healthy pregnancy outcomes. *J Acad Nutr Diet* 2014; 114:1099–103.
98. Ostadrahimi A et al. The effect of perinatal fish oil supplementation on neurodevelopment and growth of infants: A randomized controlled trial. *Euro J Nutr,* July 21, 2017, https://link.springer.com/article/10.1007/s00394-017-1512-1
99. Lehman S. DHA Supplements during pregnancy don't raise kids' IQ. www.medscape.com/viewarticle/877872?nlid=113829_2042&src=WNL_mdplsnews_170331_mscpedit_obgy&uac=61213SX&spon=16&impID=1320249&faf=1, 3/29/17.
100. Feldman-Winter L et al. A nationwide quality improvement initiative to increase breastfeeding. *Pediatrics* 2017;140: doi: 10.1542/peds.

UNIT 30

1. Birch LL et al. Calorie compensation and sensory specific satiety: Evidence for self-regulation of food intake by young children. *Appetite* 1986;7:323–31.
2. *Nutrition and the Health of Young People.* October 2014, www.cdc.gov/healthyyouth/nutrition/facts.htm, accessed 3/21/15.
3. Dev DA et al. "Great job cleaning your plate today!" *J Am Nutr Diet* 2016;116:1803–09.
4. WHO growth standards are recommended for use in the U.S. for infants and children 0 to 2 years of age. Available at: www.cdc.gov/growthcharts/who_charts.htm, accessed 4/9/12.
5. Growth Charts. Available at: cdc.gov/growthcharts, accessed 4/9/12.
6. Skinner AC. Prevalence of obesity and severe obesity in US children, 1999–2014. *Obesity* 2018; doi.org/10.1002/oby.21497.
7. Tanner JM. *Fetus into Man: Physical Growth from Conception to Maturity.* Cambridge, MA: Harvard University Press, 1978.
8. Smith GD. *Growth and Its Disorders.* Philadelphia: W.B. Saunders, 1979.
9. Ohyama S et al. Some secular changes in body height and proportion of Japanese medical students. *Am J Phys Anthropol,* May 3, 2005, doi: 10.1002/ajpa.1330730204.
10. Bronner F. Adaptation and nutritional needs (letter to the editor). *Am J Clin Nutr* 1997;65:1570.
11. Gigante DP et al. Early life factors are determinants of female height at age 19 years in a population-based birth cohort. *J Nutr* 2006;136:473–78.
12. Growth hormone deficiency—children. March 2015, www.nlm.nih.gov/medlineplus/ency/article/001176.htm, accessed 3/24/15.
13. Beck AL et al. Full fat milk consumption protects against severe childhood obesity in Latinos. *Prev Med Rpts* 2017;8:1–5.
14. MyPlate for Kids. www.myplate.gov, 2018.
15. Beck AL et al. Full fat milk consumption protects against severe childhood obesity in Latinos. *Prev Med Rpts* 2017;8:1–5.
16. Cohen JFW et al. Dietary Approaches to Stop Hypertension diet, weight status, and blood pressure among children and adolescents: National Health and Nutrition Examination Surveys 2003–2012. *J Acad Nutr Diet* 2017;117:1437–44.
17. Nutrient intakes from food and beverages: Mean amounts consumed per individual, by gender and age. *What We Eat in America,* NHANES 2011–2012. Agricultural Research Service, 2014, www.ars.usda.gov/SP2UserFiles/Place/80400530/pdf/1112/Table_1_NIN_GEN_11.pdf, accessed 3/25/15.
18. Tan ZY. It's in the water: The debate over fluoridation lives on. Medscape, September 26 2016; www.medscape.com/viewarticle/869269?nlid=109634_439&src=WNL_mdplsfeat_161004_mscpedit_publ&uac=61213SX&spon=42&impID=1209161&faf=1
19. Office of Disease Prevention and Health Promotion. *Scientific report of the 2015 Dietary Guidelines Advisory Committee.* Health.gov Aug 6, 2015, http://health.gov/dietaryguidelines/2015-scientific-report/10-chapter-5/d5-5.asp
20. Childhood obesity facts. December 2014, www.cdc.gov/healthyyouth/obesity/facts.htm, accessed 3/21/15.
21. Hales CM et al. Prevalence of obesity among adults and youth: United States, 2015–2016. NCHS Data Brief No. 288, October 2017, www.cdc.gov/nchs/products/databriefs/db288.htm
22. Loos RJF et al. Predicting polygenic obesity using genetic information. *Cell Metab* 2017;25:535–43.
23. Seyednasrollah F et al. Prediction of adulthood obesity using genetic and childhood clinical risk factors in the Cardiovascular Risk in Young Finns Study. *Genomic and Precision Medicine* 2017;10:e001554.
24. Singer K et al. The initiation of metabolic inflammation in childhood obesity. *J Clin Invest* 2017;127(1):65–73.
25. Lee BY et al. Modeling the economic and health impact of increasing children's physical activity in the United States. *Health Affairs* 2017;36:902–08.
26. Syrad H et al. Health and happiness more important than weight: a qualitative investigation. *J Hum Nutr Diet* 2014, doi: 10.1111/jhn.12217.
27. Lou JCK et al. Association between neighbourhood walkability and metabolic risk factors influenced by physical activity: A cross-sectional study of adults in Toronto, Canada. *BMJ Open* 2017;7. doi.org/10.1136/bmjopen-2016-013889.
28. Raine LB et al. Obesity, visceral adipose tissue, and cognitive function in childhood. *J Pediatrics,* published ahead of print, accessed June 25, 2017, doi.org/10.1016/j.jpeds.2017.05.023.
29. Proposed rules. *Federal Register* 2015;80(10):2038.
30. Story M. The Third School Nutrition Dietary Assessment Study: Findings and policy implications or improving the health of U.S. children. *J Am Diet Assoc* 2009;109:S7–S13.
31. Franckle RL et al. Student obesity prevalence and behavioral outcomes for the Massachusetts Childhood Obesity Research Demonstration project. *Obes* 2017; doi.org/10.1002/oby.21867.
32. Cohen JFW et al. Dietary Approaches to Stop Hypertension diet, weight status, and blood pressure among children and adolescents: National Health and Nutrition Examination Surveys 2003–2012. *J Acad Nutr Diet* 2017;117:1437–44.
33. Goldstein RF et al. Association of gestational weight gain with maternal and infant outcomes: A systematic review and meta-analysis. *JAMA* 2017;317:2207–25.
34. Smart snacks in school. www.fns.usda.gov/sites/default/files/allfoods_flyer.pdf, accessed 9/10/14.

35. Proposed rules. *Federal Register* 2015;80(10):2038.
36. *Physical Activity Guidelines for Americans*, 2008, U.S. Department of Health and Human Services, available at: www.health.gov/PAGuidelines.
37. Behm DG et al. Effectiveness of traditional strength vs. power training on muscle strength, power and speed with youth: A systematic review and meta-analysis. *Front Physiol* 2017, doi.org/10.3389/fphys.2017.00423.
38. Small EW et al. Guidelines on strength training for children. *Pediatrics* 2008;121:835–40.
39. de Wild VWT et al. Use of different vegetable products to increase preschool-aged children's preference for and intake of a target vegetable: A randomized controlled trial. *J Acad Nutr Diet* 2017;117:859–66.
40. Podlesak AKM et la. Associations between parenting style and parent and toddler mealtime behaviors. *Curr Dev Nutr*, June 2017, doi.org/10.3945/cdn.117.000570.
41. Position of the Academy of Nutrition and Dietetics: Nutrition guidance for healthy children ages 2 to 11 years. *J Am Acad Nutr Diet* 2014;114:1257–76.
42. School-based gardening encourages healthier eating in children. *Science Daily*, May 7, 2014, www.sciencedaily.com/releases/2014/05/140507211701.htm, accessed 3/24/15.
43. Keller KL. The use of repeated exposure and associative conditioning to increase vegetable acceptance in children: Explaining the variability across studies. *J Acad Nutr Diet* 2014;114:1169–73.
44. Nutrition and the health of young people. October 2014, www.cdc.gov/healthyyouth/nutrition/facts.htm, accessed 3/21/15.45. 22x
46. Brewer JD et al. Increasing student physical activity during the school day: Opportunities for the physical educator. *Strategies: A Journal for Physical and Sport Educators*, January 1, 2009, available at: www.articlearchives.com/medicine-health/diseases-disorders-cancer-breast/2310574-1.html, accessed May 2009.
47. Story M. The Third School Nutrition Dietary Assessment Study: Findings and policy implications or improving the health of U.S. children. *J Am Diet Assoc* 2009;109:S7–S13.

UNIT 31

1. Vicinanza R et al. Aging and adherence to the Mediterranean diet: Relationship with cardiometabolic disorders and polypharmacy. *J Nutr Health Aging* 2017. doi: 10.1007/s12603-017-0922-3.
2. Zaninotto P et al. Sustained enjoyment of life and mortality at older ages: Analysis of the English Longitudinal Study of Aging. *BMJ* 2016;355:i6267.1.
3. NIH Study finds calorie restriction does not affect survival. NIA, August 29, 2012, www.nia.nih.gov/newsroom/2012/08/nih-study-finds-calorie-restriction-does-not-affect-survival, accessed 3/26/15.
4. Winter JE et al. BMI and all-cause mortality in older adults: A meta-analysis. *Am J Clin Nutr*, first published January 22, 2014, doi: 10.3945/ ajcn.113.
5. Rafii M et al. Dietary protein requirement of men >65 years old determined by the indicator amino acid oxidation technique is higher than the current estimated average requirement. *J Nutri* 2016;146:681–87.
6. Position of the American Academy of Nutrition and Dietetics: Individualized nutrition approaches for older adults" Long-term care, post-acute care, and other settings. *J Am Nutr Diet* 2018;118:724–35.
7. Reedy J et al. Higher diet quality is associated with decreased risk of all-cause, cardiovascular disease, and cancer mortality among older adults. *J Nutr* 2014;144:881–89.
8. Heidemann C et al. Dietary patterns and risk of mortality from cardiovascular disease, cancer, and all causes in a prospective cohort of women. *Circulation* 2008;118:230–37.
9. Kabat GC et al. Adherence to cancer prevention guidelines and cancer incidence, cancer mortality, and total mortality: A prospective cohort study. *Am J Clin Nutr* 2015;101:558–69.
10. Aging Statistics. www.aoa.acl.gov/Aging_Statistics/index.aspx, accessed 3/31/15.
11. Life expectancy. Available at: www.cdc.gov/nchs/fastats/lifexpec.htm, accessed 4/11/12
12. Harper S et al. Declining US life expectancy: A first look. *Epidemiol* 2017;28:e54–e56.
13. Crimmins EM et al. Gender differences in health: Results from SHARE, ELSA and HRS. *Eur J Public Health* 2011;21:81–91.
14. Vaidya D et al. Ageing, menopause, and ischaemic heart disease mortality in England, Wales, and the United States. *BMJ* 2011;343, doi: 10.1136/bmj.d5170, accessed 4/10/12.
15. U.S. Census Bureau. Decennial census data population projections, 2000.
16. Administration on Aging, USDHHS, Minority Aging. Available at: www.aoa.gov/AoRoot/Aging_Statistics/Minority_Aging/index.asp, accessed 6/3/10.
17. National Vital Statistics Report, Deaths: Preliminary data for 2010. January 11, 2012, www.cdc.gov/nchs/fastats/lifexpec.htm, accessed 4/11/12.
18. Health, United States 2008. Available at: www.cdc.gov/nchs/hus.htm, accessed May 2009.
19. Health, United States, 2013. www.cdc.gov/nchs/data/hus/hus13.pdf#017, accessed 3/28/15.
20. Fuchs VR. New priorities for future biomedical innovations. *N Engl J Med* 2010;363:704–06.
21. World Fact Book. 2014, www.cia.gov/library/publications/the-world-factbook/rankorder/2102rank.html, accessed 3/26/15.
22. Zeliadt N. Live long and proper: Genetic factors associated with increased longevity identified. *Scientific American*, July 2010, www.scientificamerican.com/article/genetic-factors-associated-with-increased-longevity-identified/, accessed 3/26/15.
23. Crandall JP et al. Genes versus healthy living to age 100. *J Am Geriatr Soc* 2011, available from: www.health.harvard.edu/blog/living-to-100-and-beyond-the-right-genes-plus-a-healthy-lifestyle-201201114092, accessed 4/12/12.
24. Gonzalez AB et al. Body mass index and mortality among 1.46 million white adults. *N Engl J Med* 2010;362:2211–19.
25. Talegawkar SA et al. A higher adherence to a Mediterranean-style diet is inversely associated with the development of frailty in community-dwelling elderly men and women. *J Nutr* 142:2161–66.
26. Olshansky SJ et al. What if humans were designed to last? *The Scientist*, March 2007:28–35.
25. Talegawkar SA et al. A higher adherence to a Mediterranean-style diet is inversely associated with the development of frailty in community-dwelling elderly men and women. *J Nutr* 142:2161–66.
26. Olshansky SJ et al. What if humans were designed to last? *The Scientist*, March 2007:28–35.
27. Lee I-M et al. Accelerometer-measured physical activity and sedentary behavior in relation to all-cause mortality: The Women's Health Study. doi.org/10.1161/CIRCULATIONAHA.117.031300, Circulation. Originally published November 6, 2017; *Circulation*, 117.031300.
28. Vitamin D. November 2014, ods.od.nih.gov/factsheets/VitaminD-HealthProfessional, accessed 3/26/15.
29. Phillips SM. Determining the protein needs of "older" persons one meal at a time. *Am J Clin Nutr* 2017;105:291–92.
30. Bossingham MJ et al. Water balance, hydration status, and fat-free mass hydration in younger and older adults. *Am J Clin Nutr* 2005;81:1342–50.

31. Bales CW. Beyond BMI: Health implications of weight and muscle mass in aging adults. *Nutr Notes, Experimental Biology,* 2017.
32. Martinez JA et al. Physical activity modifies the association between dietary protein and lean muscle mass of postmenopausal women. *J Am Acad Nutr Diet* 2017;1117:192–203.
33. Nutrient intakes from food and beverages: Mean amounts consumed per individual, by gender and age. *What We Eat in America,* NHANES 2011–2012. Agricultural Research Service, 2014, www.ars.usda.gov/SP2UserFiles/Place/80400530/pdf/1112/Table_1_NIN_GEN_11.pdf, accessed 3/25/15.
34. Manson JAE et al. Vitamin and mineral supplements: What clinicians need to know. *JAMA* 2018;330;859–60.
35. Hoeijmakers JHJ. DNA damage, aging, and cancer. *N Engl J Med* 2009;361:1475–85.
36. Phillips SM. Nutrition and the elderly: A recommendation for more (evenly distributed) protein? *Am J Clin Nutr* 2017;106:12–13.
37. Tighe P et al. Effect of increased consumption of whole grain foods on blood pressure and other cardiovascular risk markers in healthy middle-aged persons: A randomized controlled trial. *Am J Clin Nutr* 2010;92:733–40.
38. Tan AG et al. Antioxidant nutrient intake and the long-term incidence of age-related cataract: The Blue Mountain Eye Study. *Am J Clin Nutr* 2008;87:1899–905.
39. Rowe JW et al. Human aging: Usual and successful. *Clin Nutr* 1990; 9:26–33.
40. Faes MC et al. Dehydration in geriatrics. *Geriatr Aging* 2007;10:590–96, available at: www.medscape.com/viewarticle/567678, accessed 3/30/15.
41. *Dietary Reference Intakes for Water, Potassium, Sodium, Chloride, and Sulfate. Food and Nutrition Board,* 2004, available at: books.nap.edu/openbook/0309091691/gifmid/74.gif.
42. Rolls BJ. Do chemosensory changes influence food intake in the elderly? *Physiol Behav* 1999;66:193–97.
43. Schiffman SS. Taste and smell losses in normal aging and disease. *JAMA*1997;278:1357–62.
44. Dietary Guidelines Advisory Committee, *Scientific Report of the 2015 Dietary Guidelines: Advisory Report to the Secretary of Health and Human Services and Secretary of Agriculture,* published February 2015, www.health.gov/dietaryguidelines/2015-scientific-report/PDFs/Scientific-Report-of-the-2015-Dietary-Guidelines-Advisory-Committee.pdf, accessed 3/25/15.
45. Isenring EA et al. Beyond malnutrition screening: Appropriate methods to guide nutrition care for aged care residents. *J Acad Nutr Diet* 2012;112:376–81.

UNIT 32

1. Surveillance for foodborne disease outbreaks United States, 2015. CDC 2017; www.cdc.gov/foodsafety/pdfs/2015FoodBorneOutbreaks_508.pdf
2. Estimates of foodborne illness in the United States. CDC, January 2014, www.cdc.gov/foodborneburden, accessed 4/1/15.
3. Study finds widespread contamination in chicken. WebMD, December 2013, www.webmd.com/food-recipes/food-poisoning/20131219/survey-finds-widespread-contamination-in-chicken, accessed 4/3/15.
4. *Bad Bug Book,* 2nd ed. Foodborne Pathogenic Microorganisms and Natural Toxins Handbook, FDA, October 2014, accessed 4/1/15, available at: www.fda.gov/Food/FoodborneIllnessContaminants/CausesOfIllnessBadBugBook/default.htm. This resource describes the signs and symptoms of specific foodborne illnesses.
5. Foodborne illness. Available at: www.cdc.gov/ncidod/dbmd/diseaseinfo/foodborneinfections_g.htm, accessed May 2009.
6. CDC 2011 estimates: Findings. February 2012, available at: www.cdc.gov/foodborneburden/2011-foodborne-estimates.html, accessed 4/13/12.
7. Foodborne illnesses. Available at: www.cdc.gov, accessed May 2009.
8. People at risk. June 2014, www.fda.gov/Food/FoodborneIllnessContaminants/PeopleAtRisk/default.htm, accessed 4/1/15.
9. Position of the American Academy of Nutrition and Dietetics: Food and water safety. *J Am Acad Nutr Diet* 2014;114:1819–29.
10. Foodborne outbreaks. February 2015, www.cdc.gov/foodsafety/outbreaks/multistate-outbreaks/outbreaks-list.html, accessed 4/1/15.
11. CDC 2011 estimates: Findings. February 2012. Available at: www.cdc.gov/foodborneburden/2011-foodborne-estimates.html, accessed 4/13/12.
12. CDC 2011 estimates: Findings. February 2012, Available at: www.cdc.gov/foodborneburden/2011-foodborne-estimates.html, accessed 4/13/12.
13. Kent AJ et al. Antibiotic-resistant *Campylobacter:* An increasing problem. *Postgrad Med J* 2008;84:106–08.
14. Campylobacter. June 2014, www.cdc.gov/nczved/divisions/dfbmd/diseases/campylobacter/#what, accessed 4/1/15.
15. Consumer Reports: Two-thirds of chickens carry bacteria. ABC News, November 2009, available at: abcnews.go.com, accessed 4/13/12.
16. What are the most common foodborne diseases? Available at: www.cdc.gov/ncidod/dbmd/diseaseinfo/foodborneinfections_g.htm#mostcommon, accessed May 2009.
17. Welland D. E. coli . . . salmonella . . . Listeria . . . unwelcome house guests. *Envir Nutr,* October 1997;20:1,6.
18. Norovirus. CDC, December 2014, www.cdc.gov/norovirus, accessed 4/3/15.
19. Listeria (Listeriosis). CDC, January 2013, www.cdc.gov/listeria/definition.html, accessed 4/4/15.
20. Dietary Guidelines Advisory Committee, *Scientific Report of the 2015 Dietary Guidelines: Advisory Report to the Secretary of Health and Human Services and Secretary of Agriculture,* published February 2015, www.health.gov/dietaryguidelines/2015-scientific-report/PDFs/Scientific-Report-of-the-2015-Dietary-Guidelines-Advisory-Committee.pdf, accessed 3/12/15.
21. Botulism in Alaska 2005 update: A guide for physicians and health care providers. Available at: www.epi.hss.state.ak.us/pubs/botulism/Botulism.pdf, accessed 4/12/12.
22. Botulism in Alaska, 2011. www.epi.hss.state.ak.us/id/botulism/Botulism.pdf, accessed 4/1/15.
23. Botulism. April 2014, www.cdc.gov/nczved/divisions/dfbmd/diseases/botulism, accessed 4/1/15.
24. Weir E. Sushi, nematodes, and allergies. *CMAJ* 2005;1;172:329.
25. Parasites—Toxoplasmosis (Toxoplasma infection). CDC, January 2013, www.cdc.gov/parasites/toxoplasmosis, accessed 4/1/!5.
26. Braun JM et al. Impact of early-life bisphenol A exposure on behavior and executive function in children. *Pediatrics,* October 24, 2011, doi: 10.1542/peds.2011-1335.
27. Park S-R et al. Fast and simple determination and exposure assessment of bisphenol A, phenol, p-tert-butylphenol, and diphenylcarbonate transferred from polycarbonate food-contact materials to food stimulants. *Chemosphere* 2018;203:300–06.
28. Bisphenol A (BPA): Use in food contact application. FDA, January 6, 2015, www.fda.gov/NewsEvents/PublicHealthFocus/ucm064437.htm, accessed 2/21/15.

29. Health Canada. Bisphenol A. December 2014, www.hc-sc.gc.ca/fn-an/securit/packag-emball/bpa/index-eng.php, accessed 4/1/15.
30. Zeratsky K. What is BPA, and what are the concerns about BPA? www.mayoclinic.org/healthy-living/nutrition-and-healthy-eating/expert-answers/bpa/faq-20058331, 2103, accessed 2/1/15.
31. Questions and answers regarding bovine spongiform encephalopathy (BSE) and variant Creutzfeldt-Jakob disease (vCJD). CDC, October 2014, www.cdc.gov/ncidod/dvrd/vcjd/qa.htm, accessed 4/1/15.
32. Moda F et al. Prions in the urine of patients with variant Creutzfeldt-Jakob disease. *N Engl J Med* 2014;371:530–39, doi: 10.1056/NEJMoa1404401.
33. Khan ZZ et al. Kuru. Medscape, July 12, 2010, available at: emedicine.medscape.com/article/220043, accessed 2/22/11.
34. Antibiotic use in food-producing animals. CDC, September 2014, www.cdc.gov/narms/animals.html, accessed 4/1/15.
35. *Summary Report on Antimicrobials Sold or Distributed for Use in Food-Producing Animals.* FDA, 2012, www.fda.gov/downloads/ForIndustry/UserFees/AnimalDrugUserFeeActADUFA/UCM416983.pdf, published September 2014, accessed 10/3/14.
36. Perdue Farms first major poultry supplier to end routine antibiotic use. *Wall Street Journal,* October 6, 2016; www.wsj.com/articles/perdue-farms-eliminated-all-antibiotics-from-its-chicken-supply-
37. Holtz S. Antibiotics and hormone growth promoters in Canadian agriculture. April 2009, available at: www.cielap.org/pdf/AHGPs.pdf, accessed 4/13/12.
38. U.S Food and Drug Administration—Total Diet Study, 2005. Available at: www.fda.gov/downloads/Food/FoodSafety/FoodContaminantsAdulteration/TotalDietStudy/UCM291686.pdf, accessed 4/13/12.
39. *Dietary Guidelines for Americans,* 2010. Available at: www.dietaryguidelines.gov
40. Organic agriculture. USDA, January 2015, www.usda.gov/wps/portal/usda/usdahome?contentidonly=true&contentid=organic-agriculture.html, accessed 4/4/15.
41. Madigan D. After Delaney: How will the new pesticide law work? *CNI Weekly Report,* October 1996;11:4–5.
42. Food irradiation: What you need to know. FDA, 2016; www.fda.gov/food/resources for you.
43. Marcason W. Was there a recent update to the FDA food safety code? *J Acad Nutr Diet* 2014;114:336.
45. Food Facts, from the FDA: The dangers of raw milk. www.fda.gov/downloads/Food/FoodborneIllnessContaminants/UCM239493.pdf, accessed 8/28/14.
44. Food Facts, from the FDA: The dangers of raw milk. www.fda.gov/downloads/Food/FoodborneIllnessContaminants
45. Zeratsky K. Moldy cheese: Is it unsafe to eat? February 26, 2011, www.mayoclinic.com/print/food-and-nutrition/AN01024.
46. Safe minimum cooking temperatures, www.foodsafety.gov/keep/charts/mintemp.html, accessed 6/21/12.
47. Safe minimum cooking temperatures. 2015, www.foodsafety.gov/keep/charts/mintemp.html, accessed 4/3/15.
48. *Dietary Guidelines for Americans,* 2010. Available at: www.dietaryguidelines.gov.
49. Safe food handling: Myth busters. Available at: www.fightbac.org/conte Keep It Cool: Refrigerator/Freezer Storage Chart. Available at: www.homefoodsafety.org/pub/file.cfm?item_type=xm_file&id=1165, accessed 9/15/11./view/151/2, accessed October 2006.
50. Keep it cool: Refrigerator/freezer storage chart. Available at: www.homefoodsafety.org/pub/file.cfm?item_type=xm_file&id=1165, accessed 9/15/11.

UNIT 33

1. Water, sanitation and hygiene. 2013, www.unicef.org/wash, accessed 4/8/15.
2. *United Nations Human Development Report.* 2014, hdr.undp.org/sites/default/files/hdr14-report-en-1.pdf, accessed 4/8/15.
3. Obesity in the midst of unyielding food insecurity in developing countries. Available at: www.ers.usda.gov/AmberWaves/September08/Features/Obesitycountries.htm, accessed June 2009.
4. *The State of Food Insecurity in the World 2014: Strengthening the Enabling Environment for Food Security and Nutrition.* FAO, IFAD, and WFP, 2014; Rome, FAO, available at: www.fao.org/3/a-i4030e.pdf, accessed 9/19/14.
5. The Global Village. April 2015, available at: www.nationsonline.org/oneworld/global-village.htm, accessed 4/8/15.
6. May M. The challenge of changing health. *Scientific American* 2010, available at: www.sa-pathways.com/the-challenge-of-changing-health/the-challenge, accessed 4/12/12.
7. *CIA World Factbook.* Available at: www.cia.gov/library/publications/the-world-factbook/rankorder/2102rank.html, accessed 4/15/1
8. Africa tops global hunger index, driven by war and climate shocks. Medscape, October 12, 2017, www.medscape.com/viewarticle/886963?nlid=118507_454&src=WNL_mdplsfeat_171017_msedit_peds&uac=61213SX&spon=9&impID=1458753&faf=1
9. Alegre-Diaz J et al. Diabetes and cause-specific mortality Mexico City. *N Engl J Med* 2016;375:1961–71.
10. Popkin B. What is the nutrition transition? UNC Carolina Population Center, 2002, updated 2006, available at: www.cpc.unc.edu/projects/nutrans/whatis, accessed 4/6/15.
11. Global status report on noncommunicable diseases 2010. WHO, September 14, 2011, available at: www.who.int/nmh/publications/ncd_report2010/en, accessed 4/17/12.
12. Global health and aging, new disease patterns. March 28, 2012, available at: www.nia.nih.gov/research/publication/global-health-and-aging/new-disease-patterns, accessed 4/16/12.
13. Victora CG et al. Optimal child growth and the double burden of malnutrition: Research and programmatic implications. *Am J Clin Nutr* 2014;100:1611S–2S, | doi: 10.3945/ajcn.114.084475.
14. Obesity and overweight. January 2015, available at: www.who.int/mediacentre/factsheets/fs311/en, accessed 4/8/15.
15. Oddo VM et al. Predictors of maternal and child double burden of malnutrition in rural Indonesia and Bangladesh. *Am J Clin Nutr* 2012;95:951.
16. Worldwide prevalence of child malnutrition assessed as percent underweight among children under 5 years of age, 2000–2009. Available at: www.globalhealthfacts.org/data/topic/map.aspx?ind=48, accessed 4/12/12.
17. Merrill RD et al. Factors associated with inflammation in preschool children and women of reproductive age: Biomarkers Reflecting Inflammation and Nutritional Determinants of Anemia (BRINDA) project. *Am J Clin Nutr* 2017;106(Suppl 1):348S–358S.
18. Jacobs A et al. With NAFTA, Mexico receives unexpected import: Obesity. *NY Times,* 12/12/17.
19. Position of American Dietetic Association: Addressing world hunger, malnutrition, and food insecurity. *J Am Diet Assoc* 2003;103:1046–56.
20. Nordin SM et al. Position of the academy of nutrition and dietetics: Nutrition security in developing nations: Sustainable food, water, and health. *J Acad Nutr Diet* 2013;113:581–95, doi: 10.1016/j.jand.2013.01.025.
21. Popkin BM et al. Now and then: The global nutrition transition: The pandemic of obesity in developing countries. *Nutr Rev* 2012;70:3–21.
22. Narayan KMV et al. Global noncommunicable disease—where worlds meet. *N Engl J Med* 2010;363:1196–98.

23. Essential Nutrition Actions Improving Maternal-Newborn-Infant and Young Child Health and Nutrition draft. May 2011, available at: www.who.int/nutrition/EB128_18_backgroundpaper2_A_reviewofhealthinterventionswithaneffectonnutrition.pdf, accessed 4/16/12.
24. Amadi B et al. Reduced production of sulfated glycosaminoglycans occurs in Zambian children with kwashiorkor but not for marasmus. *Am J Clin Nutr* 2009;89:592–600.
25. Neumann CG. Background, symposium: Food-based approaches to combating micronutrient deficiencies in children of developing countries. *J Nutr* 2007;137:1091–92.
26. Scrimshaw NS. Historical concepts of interactions, synergism and antagonism between nutrition and infection. *J Nutr* 2003;133:316S–21S.
27. Platts-Mills JA et al. Association between enteropathogens and malnutrition in children aged 6–23 mo in Bangladesh: A case-control study. *Am J Clin Nutr* 2017;105:1132–38.
28. Merrill RD et al. Factors associated with inflammation in preschool children and women of reproductive age: Biomarkers Reflecting Inflammation and Nutritional Determinants of Anemia (BRINDA) project. *Am J Clin Nutr* 2017;106(Suppl 1): 348S–358S.
29. Colchero MA et al. The costs of inadequate breastfeeding of infants in Mexico. *Am J Clin Nutr* 2015;101:579–86.
30. Evidence shows vitamin A for children can save lives. BMJ August 25, 2011, available at: www.medscape.com/viewarticle/7486087src=mpnews&spon=9.
31. Global HIV/AIDS overview. 2014, www.aids.gov/federal-resources/around-the-world/global-aids-overview, accessed 4/5/12.
32. WHO. 90–90–90. An ambitious treatment target to help end the AIDS epidemic; 2014. www.unaids.org/sites/default/files/media_asset/90-90-90_en_0.pdf
33. Alarming Malnutrition in Sudan Conflict Zone. Available at: www.medscape.com/viewarticle/756399?scr-mp&spon=42,accessed 1/23/12.
34. Moloney GM et al. Notes from the field: Malnutrition and mortality—southern Somalia. *MMWR,* August 5, 2011;60:1026–27.
35. Alarming malnutrition in Sudan conflict zone. Available at: www.medscape.com/viewarticle/756399?scr-mp&spon=42, accessed 1/23/12.
36. Ng M et al. Global, regional, and national prevalence of overweight and obesity in children and adults during 1980–2013: A systematic analysis for the Global Burden of Disease Study 2013. *The Lancet* 2014;384:766–81, doi: 10.1016/S0140-6736(14)60460-8.
37. Decimosexta Conferencia Internaciona de Nutricion, Montreal. *Nutr View,* Fall 1997;1–36.
38. Food fortification to end micronutrient malnutrition: State of the art. *Nutr View,* Special issue 1997;1–8.
39 Food fortification to end micronutrient malnutrition: State of the art. *Nutr View,* Special issue 1997;1–8.
40. The Baby-Friendly Hospital Initiative. April 2015, www.unicef.org/programme/breastfeeding/baby.htm, accessed 4/9/15.
41. North Korea lowers height requirements for military service. April 3, 2012, available at: www.outsidethebeltway.com/north-korea-lowers-height-requirements-for-military-service, accessed 4/17/12.

Index

Page references are the chapter number followed by the individual page number.

B

C

D

E

F

G

H

M

N

O

P

Q

R

S

T

U

V

W

X

Z

Dietary Reference Intakes (DRI)

The Dietary Reference Intakes (DRI) include two sets of nutrient intake goals for individuals—the Recommended Dietary Allowance (RDA) and Adequate Intake (AI). The RDA reflects the average daily amount of a nutrient considered adequate to meet the needs of most healthy people. If there is insufficient evidence to determine an RDA, an AI is set. In addition, the Estimated Energy Requirement (EER) represents the average dietary energy intake considered adequate to maintain energy balance in healthy people.

The DRI also include the Tolerable Upper Intake Level (UL) that represents the estimated maximum daily amount of a nutrient that appears safe for most healthy people to consume on a regular basis. Turn the page for a listing of the UL for selected vitamins and minerals. Note that the absence of a UL for a nutrient does not indicate that it is safe to consume in high doses, but only that research is too limited to set a UL. Chapter 1 describes these DRI values in detail.

Estimated Energy Requirements (EER), Recommended Dietary Allowances (RDA), and Adequate Intakes (AI) for Water, Energy, and the Energy Nutrients

Age (yr)	Reference BMI (kg/m^2)	Reference Height cm (in)	Reference Weight kg (lb)	Water[a] AI (L/day)	Energy EER[b] (kcal/day)	Carbohydrate RDA (g/day)	Total Fiber AI (g/day)	Total Fat AI (g/day)	Linoleic Acid AI (g/day)	Linolenic Acid[c] AI (g/day)	Protein RDA (g/day)[d]	Protein RDA (g/kg/day)
Males												
0–0.5	—	62 (24)	6 (13)	0.7[e]	570	60	—	31	4.4	0.5	9.1	1.52
0.5–1	—	71 (28)	9 (20)	0.8[f]	743	95	—	30	4.6	0.5	11	1.20
1–3[g]	—	86 (34)	12 (27)	1.3	1046	130	19	—	7	0.7	13	1.05
4–8[g]	15.3	115 (45)	20 (44)	1.7	1742	130	25	—	10	0.9	19	0.95
9–13	17.2	144 (57)	36 (79)	2.4	2279	130	31	—	12	1.2	34	0.95
14–18	20.5	174 (68)	61 (134)	3.3	3152	130	38	—	16	1.6	52	0.85
19–30	22.5	177 (70)	70 (154)	3.7	3067[h]	130	38	—	17	1.6	56	0.80
31–50	22.5[i]	177 (70)[i]	70 (154)[i]	3.7	3067[h]	130	38	—	17	1.6	56	0.80
>50	22.5[i]	177 (70)[i]	70 (154)[i]	3.7	3067[h]	130	30	—	14	1.6	56	0.80
Females												
0–0.5	—	62 (24)	6 (13)	0.7[e]	520	60	—	31	4.4	0.5	9.1	1.52
0.5–1	—	71 (28)	9 (20)	0.8[f]	676	95	—	30	4.6	0.5	11	1.20
1–3[g]	—	86 (34)	12 (27)	1.3	992	130	19	—	7	0.7	13	1.05
4–8[g]	15.3	115 (45)	20 (44)	1.7	1642	130	25	—	10	0.9	19	0.95
9–13	17.4	144 (57)	37 (81)	2.1	2071	130	26	—	10	1.0	34	0.95
14–18	20.4	163 (64)	54 (119)	2.3	2368	130	26	—	11	1.1	46	0.85
19–30	21.5	163 (64)	57 (126)	2.7	2403[j]	130	25	—	12	1.1	46	0.80
31–50	21.5[i]	163 (64)[i]	57 (126)[i]	2.7	2403[j]	130	25	—	12	1.1	46	0.80
>50	21.5[i]	163 (64)[i]	57 (126)[i]	2.7	2403[j]	130	21	—	11	1.1	46	0.80
Pregnancy												
1st trimester				3.0	+0	175	28	—	13	1.4	46	0.80
2nd trimester				3.0	+340	175	28	—	13	1.4	71	1.10
3rd trimester				3.0	+452	175	28	—	13	1.4	71	1.10
Lactation												
1st 6 months				3.8	+330	210	29	—	13	1.3	71	1.30
2nd 6 months				3.8	+400	210	29	—	13	1.3	71	1.30

NOTE: For all nutrients, values for infants are AI. Dashes indicate that values have not been determined.

[a]The water AI includes drinking water, water in beverages, and water in foods; in general, drinking water and other beverages contribute about 70–80%, and foods, the remainder. Conversion factors: 1 L = 33.8 fluid oz; 1 L = 1.06 qt; 1 cup = 8 fluid oz.

[b]The Estimated Energy Requirement (EER) represents the average dietary energy intake that will maintain energy balance in a healthy person of a given gender, age, weight, height, and physical activity level. The values listed are based on an "active" person at the reference height and weight and at the midpoint ages for each group until age 19. Chapter 8 and Appendix F provide equations and tables to determine estimated energy requirements.

[c]The linolenic acid referred to in this table and text is the omega-3 fatty acid known as alpha-linolenic acid.

[d]The values listed are based on reference body weights.

[e]Assumed to be from human milk.

[f]Assumed to be from human milk and complementary foods and beverages. This includes approximately 0.6 L (~2½ cups) as total fluid including formula, juices, and drinking water.

[g]For energy, the age groups for young children are 1–2 years and 3–8 years.

[h]For males, subtract 10 kcalories per day for each year of age above 19.

[i]Because weight need not change as adults age if activity is maintained, reference weights for adults 19 through 30 years are applied to all adult age groups.

[j]For females, subtract 7 kcalories per day for each year of age above 19.

Recommended Dietary Allowances (RDA) and Adequate Intakes (AI) for Vitamins

Age (yr)	Thiamin RDA (mg/day)	Riboflavin RDA (mg/day)	Niacin RDA (mg/day)[a]	Biotin AI (μg/day)	Pantothenic acid AI (mg/day)	Vitamin B_6 RDA (mg/day)	Folate RDA (μg/day)[b]	Vitamin B_{12} RDA (μg/day)	Choline AI (mg/day)	Vitamin C RDA (mg/day)	Vitamin A RDA (μg/day)[c]	Vitamin D RDA (IU/day)[d]	Vitamin E RDA (mg/day)[e]	Vitamin K AI (μg/day)
Infants														
0–0.5	0.2	0.3	2	5	1.7	0.1	65	0.4	125	40	400	400 (10 μg)	4	2.0
0.5–1	0.3	0.4	4	6	1.8	0.3	80	0.5	150	50	500	400 (10 μg)	5	2.5
Children														
1–3	0.5	0.5	6	8	2	0.5	150	0.9	200	15	300	600 (15 μg)	6	30
4–8	0.6	0.6	8	12	3	0.6	200	1.2	250	25	400	600 (15 μg)	7	55
Males														
9–13	0.9	0.9	12	20	4	1.0	300	1.8	375	45	600	600 (15 μg)	11	60
14–18	1.2	1.3	16	25	5	1.3	400	2.4	550	75	900	600 (15 μg)	15	75
19–30	1.2	1.3	16	30	5	1.3	400	2.4	550	90	900	600 (15 μg)	15	120
31–50	1.2	1.3	16	30	5	1.3	400	2.4	550	90	900	600 (15 μg)	15	120
51–70	1.2	1.3	16	30	5	1.7	400	2.4	550	90	900	600 (15 μg)	15	120
>70	1.2	1.3	16	30	5	1.7	400	2.4	550	90	900	800 (20 μg)	15	120
Females														
9–13	0.9	0.9	12	20	4	1.0	300	1.8	375	45	600	600 (15 μg)	11	60
14–18	1.0	1.0	14	25	5	1.2	400	2.4	400	65	700	600 (15 μg)	15	75
19–30	1.1	1.1	14	30	5	1.3	400	2.4	425	75	700	600 (15 μg)	15	90
31–50	1.1	1.1	14	30	5	1.3	400	2.4	425	75	700	600 (15 μg)	15	90
51–70	1.1	1.1	14	30	5	1.5	400	2.4	425	75	700	600 (15 μg)	15	90
>70	1.1	1.1	14	30	5	1.5	400	2.4	425	75	700	800 (20 μg)	15	90
Pregnancy														
≤18	1.4	1.4	18	30	6	1.9	600	2.6	450	80	750	600 (15 μg)	15	75
19–30	1.4	1.4	18	30	6	1.9	600	2.6	450	85	770	600 (15 μg)	15	90
31–50	1.4	1.4	18	30	6	1.9	600	2.6	450	85	770	600 (15 μg)	15	90
Lactation														
≤18	1.4	1.6	17	35	7	2.0	500	2.8	550	115	1200	600 (15 μg)	19	75
19–30	1.4	1.6	17	35	7	2.0	500	2.8	550	120	1300	600 (15 μg)	19	90
31–50	1.4	1.6	17	35	7	2.0	500	2.8	550	120	1300	600 (15 μg)	19	90

NOTE: For all nutrients, values for infants are AI. The glossary on the inside back cover defines units of nutrient measure.
[a]Niacin recommendations are expressed as niacin equivalents (NE), except for recommendations for infants younger than 6 months, which are expressed as preformed niacin.
[b]Folate recommendations are expressed as dietary folate equivalents (DFE).
[c]Vitamin A recommendations are expressed as retinol activity equivalents (RAE).
[d]Vitamin D recommendations are expressed as cholecalciferol and assume an absence of adequate exposure to sunlight.
[e]Vitamin E recommendations are expressed as α-tocopherol.

Recommended Dietary Allowances (RDA) and Adequate Intakes (AI) for Minerals

Age (yr)	Sodium AI (mg/day)	Chloride AI (mg/day)	Potassium AI (mg/day)	Calcium RDA (mg/day)	Phosphorus RDA (mg/day)	Magnesium RDA (mg/day)	Iron RDA (mg/day)	Zinc RDA (mg/day)	Iodine RDA (μg/day)	Selenium RDA (μg/day)	Copper RDA (μg/day)	Manganese AI (mg/day)	Fluoride AI (mg/day)	Chromium AI (μg/day)	Molybdenum RDA (μg/day)
Infants															
0–0.5	120	180	400	200	100	30	0.27	2	110	15	200	0.003	0.01	0.2	2
0.5–1	370	570	700	260	275	75	11	3	130	20	220	0.6	0.5	5.5	3
Children															
1–3	1000	1500	3000	700	460	80	7	3	90	20	340	1.2	0.7	11	17
4–8	1200	1900	3800	1000	500	130	10	5	90	30	440	1.5	1.0	15	22
Males															
9–13	1500	2300	4500	1300	1250	240	8	8	120	40	700	1.9	2	25	34
14–18	1500	2300	4700	1300	1250	410	11	11	150	55	890	2.2	3	35	43
19–30	1500	2300	4700	1000	700	400	8	11	150	55	900	2.3	4	35	45
31–50	1500	2300	4700	1000	700	420	8	11	150	55	900	2.3	4	35	45
51–70	1300	2000	4700	1000	700	420	8	11	150	55	900	2.3	4	30	45
>70	1200	1800	4700	1200	700	420	8	11	150	55	900	2.3	4	30	45
Females															
9–13	1500	2300	4500	1300	1250	240	8	8	120	40	700	1.6	2	21	34
14–18	1500	2300	4700	1300	1250	360	15	9	150	55	890	1.6	3	24	43
19–30	1500	2300	4700	1000	700	310	18	8	150	55	900	1.8	3	25	45
31–50	1500	2300	4700	1000	700	320	18	8	150	55	900	1.8	3	25	45
51–70	1300	2000	4700	1200	700	320	8	8	150	55	900	1.8	3	20	45
>70	1200	1800	4700	1200	700	320	8	8	150	55	900	1.8	3	20	45
Pregnancy															
≤18	1500	2300	4700	1300	1250	400	27	12	220	60	1000	2.0	3	29	50
19–30	1500	2300	4700	1000	700	350	27	11	220	60	1000	2.0	3	30	50
31–50	1500	2300	4700	1000	700	360	27	11	220	60	1000	2.0	3	30	50
Lactation															
≤18	1500	2300	5100	1300	1250	360	10	13	290	70	1300	2.6	3	44	50
19–30	1500	2300	5100	1000	700	310	9	12	290	70	1300	2.6	3	45	50
31–50	1500	2300	5100	1000	700	320	9	12	290	70	1300	2.6	3	45	50

NOTE: For all nutrients, values for infants are AI. The glossary on the inside back cover defines units of nutrient measure.

Tolerable Upper Intake Levels (UL) for Vitamins

Age (yr)	Niacin (mg/day)[a]	Vitamin B_6 (mg/day)	Folate (μg/day)[a]	Choline (mg/day)	Vitamin C (mg/day)	Vitamin A (IU/day)[b]	Vitamin D (IU/day)	Vitamin E (mg/day)[c]
Infants								
0–0.5	—	—	—	—	—	600	1000 (25 μg)	—
0.5–1	—	—	—	—	—	600	1500 (38 μg)	—
Children								
1–3	10	30	300	1000	400	600	2500 (63 μg)	200
4–8	15	40	400	1000	650	900	3000 (75 μg)	300
9–13	20	60	600	2000	1200	1700	4000 (100 μg)	600
Adolescents								
14–18	30	80	800	3000	1800	2800	4000 (100 μg)	800
Adults								
19–70	35	100	1000	3500	2000	3000	4000 (100 μg)	1000
>70	35	100	1000	3500	2000	3000	4000 (100 μg)	1000
Pregnancy								
≤18	30	80	800	3000	1800	2800	4000 (100 μg)	800
19–50	35	100	1000	3500	2000	3000	4000 (100 μg)	1000
Lactation								
≤18	30	80	800	3000	1800	2800	4000 (100 μg)	800
19–50	35	100	1000	3500	2000	3000	4000 (100 μg)	1000

[a]The UL for niacin and folate apply to synthetic forms obtained from supplements, fortified foods, or a combination of the two.
[b]The UL for vitamin A applies to the preformed vitamin only.
[c]The UL for vitamin E applies to any form of supplemental α-tocopherol, fortified foods, or a combination of the two.

Tolerable Upper Intake Levels (UL) for Minerals

Age (yr)	Sodium (mg/day)	Chloride (mg/day)	Calcium (mg/day)	Phosphorus (mg/day)	Magnesium (mg/day)[d]	Iron (mg/day)	Zinc (mg/day)	Iodine (μg/day)	Selenium (μg/day)	Copper (μg/day)	Manganese (mg/day)	Fluoride (mg/day)	Molybdenum (μg/day)	Boron (mg/day)	Nickel (mg/day)	Vanadium (mg/day)
Infants																
0–0.5	—	—	1000	—	—	40	4	—	45	—	—	0.7	—	—	—	—
0.5–1	—	—	1500	—	—	40	5	—	60	—	—	0.9	—	—	—	—
Children																
1–3	1500	2300	2500	3000	65	40	7	200	90	1000	2	1.3	300	3	0.2	—
4–8	1900	2900	2500	3000	110	40	12	300	150	3000	3	2.2	600	6	0.3	—
9–13	2200	3400	3000	4000	350	40	23	600	280	5000	6	10	1100	11	0.6	—
Adolescents																
14–18	2300	3600	3000	4000	350	45	34	900	400	8000	9	10	1700	17	1.0	—
Adults																
19–50	2300	3600	2500	4000	350	45	40	1100	400	10,000	11	10	2000	20	1.0	1.8
51–70	2300	3600	2000	4000	350	45	40	1100	400	10,000	11	10	2000	20	1.0	1.8
>70	2300	3600	2000	3000	350	45	40	1100	400	10,000	11	10	2000	20	1.0	1.8
Pregnancy																
≤18	2300	3600	3000	3500	350	45	34	900	400	8000	9	10	1700	17	1.0	—
19–50	2300	3600	2500	3500	350	45	40	1100	400	10,000	11	10	2000	20	1.0	—
Lactation																
≤18	2300	3600	3000	4000	350	45	34	900	400	8000	9	10	1700	17	1.0	—
19–50	2300	3600	2500	4000	350	45	40	1100	400	10,000	11	10	2000	20	1.0	—

[d]The UL for magnesium applies to synthetic forms obtained from supplements or drugs only.
NOTE: An upper Limit was not established for vitamins and minerals not listed and for those age groups listed with a dash (—) because of a lack of data, not because these nutrients are safe to consume at any level of intake. All nutrients can have adverse effects when intakes are excessive.

Daily Values for Food Labels

The Daily Values are standard values developed by the Food and Drug Administration (FDA) for use on food labels. The values are based on 2000 kcalories a day for adults and children over 4 years old. Chapter 2 provides more details.

Nutrient	Amount
Protein[a]	50 g
Thiamin	1.5 mg
Riboflavin	1.7 mg
Niacin	20 mg NE
Biotin	300 μg
Pantothenic acid	10 mg
Vitamin B_6	2 mg
Folate	400 μg
Vitamin B_{12}	6 μg
Vitamin C	60 mg
Vitamin A	5000 IU[b]
Vitamin D	400 IU[b]
Vitamin E	30 IU[b]
Vitamin K	80 μg
Calcium	1000 mg
Iron	18 mg
Zinc	15 mg
Iodine	150 μg
Copper	2 mg
Chromium	120 μg
Selenium	70 μg
Molybdenum	75 μg
Manganese	2 mg
Chloride	3400 mg
Magnesium	400 mg
Phosphorus	1000 mg

[a]The Daily Values for protein vary for different groups of people: pregnant women, 60 g; nursing mothers, 65 g; infants under 1 year, 14 g; children 1–4 years, 16 g.
[b]Equivalent values for nutrients expressed as IU are: vitamin A, 1500 RAE (assumes a mixture of 40% retinol and 60% beta-carotene); vitamin D, 10 μg; vitamin E, 20 mg.

Food Component	Amount	Calculation Factors
Fat	65 g	30% of kcalories
Saturated fat	20 g	10% of kcalories
Cholesterol	300 mg	Same regardless of kcalories
Carbohydrate (total)	300 g	60% of kcalories
Fiber	25 g	11.5 g per 1000 kcalories
Protein	50 g	10% of kcalories
Sodium	2400 mg	Same regardless of kcalories
Potassium	3500 mg	Same regardless of kcalories

GLOSSARY OF NUTRIENT MEASURES

kcal: kcalories; a unit by which energy is measured (Chapter 1 provides more details).

g: grams; a unit of weight equivalent to about 0.03 ounces.

mg: milligrams; one-thousandth of a gram.

μg: micrograms; one-millionth of a gram.

IU: international units; an old measure of vitamin activity determined by biological methods (as opposed to new measures that are determined by direct chemical analyses). Many fortified foods and supplements use IU on their labels.

- For vitamin A, 1 IU = 0.3 μg retinol, 3.6 μg β-carotene, or 7.2 μg other vitamin A carotenoids
- For vitamin D, 1 IU = 0.02 μg cholecalciferol
- For vitamin E, 1 IU = 0.67 natural α-tocopherol (other conversion factors are used for different forms of vitamin E)

mg NE: milligrams niacin equivalents; a measure of niacin activity (Chapter 10 provides more details).

- 1 NE = 1 mg niacin
  = 60 mg tryptophan (an amino acid)

μg DFE: micrograms dietary folate equivalents; a measure of folate activity (Chapter 10 provides more details).

- 1 μg DFE = 1 μg food folate
  = 0.6 μg fortified food or supplement folate taken with food
  = 0.5 μg supplement folate taken on an empty stomach

μg RAE: micrograms retinol activity equivalents; a measure of vitamin A activity (Chapter 11 provides more details).

- 1 μg RAE = 1 μg retinol
  = 12 μg β-carotene
  = 24 μg other vitamin A carotenoids

mmol: millimoles; one-thousanth of a mole, the molecular weight of a substance. To convert mmol to mg, multiply by the atomic weight of the substance.

- For sodium, mmol × 23 = mg Na
- For chloride, mmol × 35.5 = mg Cl
- For sodium chloride, mmol × 58.5 = mg NaCl